Nature Conservation in Europe

Approaches and Lessons

European ecosystems and species remain under pressure from intensive agriculture and forestry, fishing, pollution, urban sprawl, invasive species and climate change. This book provides a detailed description and critical analysis of nature conservation responses, achievements and failures, motivated by the concerning state of nature and missed biodiversity targets. It summarises Europe's nature and the impact of human activities, and then gives an overview of relevant international biodiversity treaties and the EU nature conservation policy and legislative framework. The core of the book comprises chapters written by national experts, which cover the UK and 25 EU Member States, providing comparative case studies from which valuable lessons are drawn. Covering wide-ranging topics such as biodiversity pressures, legislation and governance, biodiversity strategies, species protection, protected areas, habitat management and funding, this book is of interest to a wide audience, including academics and professionals involved in nature conservation and related environmental fields.

Graham Tucker is an ecologist and a leading authority on European nature conservation policy, and has a particular interest in its achievements and failures. He has gained insights on these issues from 36 years of professional experience, including at Birdlife International and, more recently, at the Institute for European Environmental Policy, where he was Head of the Biodiversity Programme. This is his third book focusing on European nature conservation. He is now an independent consultant and proprietor of Nature Conservation Consulting.

Nature Conservation in Europe

Approaches and Lessons

Edited by
Graham Tucker

CAMBRIDGE
UNIVERSITY PRESS

CAMBRIDGE
UNIVERSITY PRESS

Shaftesbury Road, Cambridge CB2 8EA, United Kingdom

One Liberty Plaza, 20th Floor, New York, NY 10006, USA

477 Williamstown Road, Port Melbourne, VIC 3207, Australia

314–321, 3rd Floor, Plot 3, Splendor Forum, Jasola District Centre, New Delhi – 110025, India

103 Penang Road, #05-06/07, Visioncrest Commercial, Singapore 238467

Cambridge University Press is part of Cambridge University Press & Assessment, a department of the University of Cambridge.

We share the University's mission to contribute to society through the pursuit of education, learning and research at the highest international levels of excellence.

www.cambridge.org
Information on this title: www.cambridge.org/9781108496742

DOI: 10.1017/9781108654647

First published 2023

A catalogue record for this publication is available from the British Library.

Library of Congress Cataloging-in-Publication Data
Names: Tucker, G. M. (Graham M.), author.
Title: Nature conservation in Europe: approaches and lessons / edited by Graham Tucker.
Description: Cambridge, United Kingdom; New York, NY: Cambridge University Press, 2023. | Includes bibliographical references and index.
Identifiers: LCCN 2022040181 (print) | LCCN 2022040182 (ebook) | ISBN 9781108496742 (Hardback) | ISBN 9781108721912 (Paperback) | ISBN 9781108654647 (epub)
Subjects: LCSH: Nature conservation–Europe.
Classification: LCC QH77.E9 T83 2023 (print) | LCC QH77.E9 (ebook) | DDC 333.72094–dc23/eng/20220831
LC record available at https://lccn.loc.gov/2022040181
LC ebook record available at https://lccn.loc.gov/2022040182

ISBN 978-1-108-49674-2 Hardback

Additional resources for this publication at www.cambridge.org/natureconservation.

To all those striving to conserve nature in their respective ways – especially the unsung heroes.

Contents

List of Contributors *page* ix
Preface xi
Acknowledgements xiii
Glossary, Abbreviations and Country Codes xiv

1 **Introduction: Aims, Scope, Structure and Key Information Sources** 1
Graham Tucker

2 **Europe's Nature and Conservation Needs** 13
Graham Tucker

3 **The International Drivers of Nature Conservation, Their Objectives and Impacts on Nature Conservation Policies and Actions in Europe** 41
David Stroud, Euan Dunn, Wyn Jones and Graham Tucker

4 **Nature Conservation Policy, Legislation and Funding in the EU** 59
Graham Tucker, Kaley Hart, Andrew Farmer, Euan Dunn and David Baldock

5 **Conclusions, Lessons Learnt and Implications for the Future** 106
Graham Tucker

6 **Austria** 142
Wolfgang Suske and Kathrin Horvath

7 **Belgium** 160
Els Martens and Jorge Ventocilla

8 **Republic of Bulgaria** 181
Boris Barov

9 **Republic of Croatia** 200
Jasminka Radović

10 **Cyprus** 219
Clairie Papazoglou and Andreas Demetropoulos†

11 **The Czech Republic** 237
Petr Roth

12 **Denmark** 254
Anette Petersen and Jan Woollhead

13 **Estonia** 273
Aleksei Lotman and Silvia Lotman

14 **Finland** 291
Heikki Toivonen and Olli Ojala

15 **France** 311
Pierre Commenville

16 **Germany** 331
Eick von Ruschkowski

17 **Greece** 353
Ioli Christopoulou

18 **Hungary** 374
Katalin Sipos

19 **Ireland** 393
Andrew Jackson and Anja Murray

20 **Italy** 415
Elisa Cardarelli, Barbara Calaciura, Daniela Zaghi and Giuseppe Bogliani

21 **Latvia** 434
Otars Opermanis, Ilona Mendzina, Ainars Aunins and Inga Racinska

22 **Lithuania** 451
Rūta Baškytė and Žygimantas Obelevičius

23 **The Netherlands** 468
Erik Gerritsen

24 **Poland** 493
Paweł Pawlaczyk

25 **Portugal** 514
Ana Guimarães Ferreira and Domingos Leitão

26 **Romania** 534
Erika Stanciu, Ioan-Cristian Ioja, Mariana Tintarean and Mihai Pop

27 **Slovakia** 555
Viera Šefferová Stanová and Rastislav Rybanič

28 **Slovenia** 574
Peter Skoberne

29 **Spain** 593
Concha Olmeda and Juan Carlos Blanco

30 **Sweden** 612
Mats O. G. Eriksson and Mia Pantzar

31 **United Kingdom** 634
Graham Tucker, Wyn Jones, Susan Davies and Joan Edwards

Appendix: Habitats Directive Annex I Habitat Types Referred to in This Book 678
Index 682

Contributors

Ainars Aunins
University of Latvia, Riga, Latvia

David Baldock
Institute for European Environmental Policy UK

Boris Barov
Brussels, Belgium

Rūta Baškytė
State Service of Protected Areas, Vilnius, Lithuania

Juan Carlos Blanco
Consultores en Biología de la Conservación S.L., Madrid, Spain

Giuseppe Bogliani
University of Pavia, Pavia, Italy

Barbara Calaciura
Comunità Ambiente, Roma, Italy

Elisa Cardarelli
University of Pavia, Pavia, Italy

Ioli Christopoulou
The Green Tank, Athens, Greece

Pierre Commenville
Les Ecrins National Park, Gap, France

Susan Davies
Scottish Seabird Centre, North Berwick, UK

Andreas Demetropoulos†
Nicosia, Cyprus

Euan Dunn
Cambridge, UK

Joan Edwards
Royal Society of Wildlife Trusts, Plymouth, UK

Mats O. G. Eriksson
MK Natur- och Miljökonsult HB, Lindome, Sweden

Andrew Farmer
Institute for European Environmental Policy, Brussels, Belgium

Ana Guimarães Ferreira
Bragança, Portugal

Erik Gerritsen
Brussels, Belgium

Kaley Hart
Institute for European Environmental Policy UK

Kathrin Horvath
Suske Consulting, Vienna, Austria

Ioan-Cristian Ioja
University of Bucharest, Bucharest, Romania

Andrew Jackson
University College Dublin, Ireland

Wyn Jones
Nantwich, Cheshire, UK

Domingos Leitão
Sociedade Portuguesa para o Estudo das Aves, Lisbon, Portugal

Aleksei Lotman
Estonian Fund for Nature, Pärnumaa, Estonia

Silvia Lotman
Estonian Fund for Nature, Pärnumaa, Estonia

Els Martens
Vilvoorde, Belgium

Ilona Mendzina
Ministry of Environmental Protection and Regional Development, Riga, Latvia

Anja Murray
Ireland

Žygimantas Obelevičius
State Service of Protected Areas, Vilnius, Lithuania

Olli Ojala
Lahti, Finland

Concha Olmeda
ATECMA, Asesores Técnicos de Medio Ambiente S.L., Madrid, Spain

Otars Opermanis
Biota, Riga, Latvia

Mia Pantzar
Göteburg, Sweden

Clairie Papazoglou
Nicosia, Cyprus

Paweł Pawlaczyk
Naturalists Club Poland, Świebodzin, Poland

Anette Petersen
GEON, Skovlunde, Denmark

Mihai Pop
Association for the Conservation of Biological Diversity, Covasna, Romania

Inga Racinska
Latvian Fund for Nature, Riga, Latvia

Jasminka Radović
Zagreb, Croatia

Petr Roth
Praha, Czech Republic

Eick von Ruschkowski
Laatzen, Germany

Rastislav Rybanič
Cífer, Slovakia

Katalin Sipos
WWF Hungary Foundation, Budapest, Hungary

Peter Skoberne
Ljubljana, Slovenia

Erika Stanciu
Propark Foundation for Protected Areas, Brasov, Romania

Viera Šefferová Stanová
DAPHNE – Institute of Applied Ecology, Bratislava, Slovakia

David Stroud
Peterborough, UK

Wolfgang Suske
Suske Consulting, Vienna, Austria

Mariana Tintarean
Romanian Wilderness Society, Brasov, Romania

Heikki Toivonen
Kangasala, Finland

Graham Tucker
Nature Conservation Consulting, Huntingdon, Cambridgeshire, UK

Jorge Ventocilla
Belgian Biodiversity Platform – RBINS, Brussels, Belgium

Jan Woollhead
Parks'nTrails, Sorø, Denmark

Daniela Zaghi
Comunità Ambiente, Roma, Italy

Preface

The biodiversity crisis is as bad as ever over much of Europe. This is despite over 40 years of nature conservation efforts following the ground-breaking Bern Convention, the Convention on Migratory Species and the EU[1] Birds Directive, all in 1979. Since then, there has been the Convention on Biological Diversity (CBD) and the EU Habitats Directive in 1992, followed by many EU supporting measures such as the Water Framework Directive and the Marine Strategy Framework Directive, and numerous national strategies and actions.

So why is it that nature is so impoverished and still declining? It is not a lack of knowledge: we know the main things that need to be done. It is definitely not a lack of biodiversity conferences, declarations, targets and strategies. The problem is the gap between nature conservation aspirations and achievements. This book explores the reasons for this, primarily by examining nature conservation in different countries and comparing their varying problems, solutions and outcomes.

The idea for this book and the national comparisons came from various conversations I had with numerous nature conservation experts across Europe, with whom I have had the privilege to work during the last three decades. These discussions increasingly focussed on the plight of nature and the apparent ineffectiveness of many so-called nature conservation measures. Whilst there have been successes, as described in this book, they have been too few to stem the tide of biodiversity loss. At the same time, countries have taken varying approaches and some seem to be doing better than others (at least in reducing the rates of decline), whilst others (or their political leaders) do not appear to have cared much and have only done the bare minimum. Whilst there have been numerous CBD, European Commission, European Environment Agency (EEA) and other reports based on indicators, there has been a need for a deeper comparison of national actions and actual biodiversity impacts – to see what really happened, what worked, what did not and why. About 10 years ago these conversations led to a blueprint for this book, with its main part consisting of national chapters written by national experts. The original intention was to cover all EU-28 countries plus a selection of other European countries for comparison.

Plans to commence writing were delayed in 2014 when the European Commission started a Fitness Check of the Nature Directives (i.e. the Birds Directive and Habitats Directive). As part of its regulatory reform programme, it set out to establish if they were effective, efficient, relevant, coherent with other EU policies and provided added EU value. This was concerning at the time, as there were clear political signals that some wanted the exercise to simplify and reduce the strength of the legislation. However, to some extent the Fitness Check was needed, and it carried out some of the analysis that was envisaged for this book, to draw lessons and ensure the directives were achieving their full potential. As many of us involved in the book were occupied in various ways in the Fitness Check, it was not an appropriate time to start writing. When the Fitness Check was over, the Commission concluded in 2016 that the Nature Directives were fit for purpose, and their key measures including the protection of sites and species worked well when implemented properly. However, the directives were not achieving their objectives because they were not being implemented sufficiently, primarily due to a lack of political will, and underfunding.

Although the reasons for the problems are explained in detail in the Fitness Check and supporting evaluation study, it was apparent that much more could be said, especially in terms of exploring the different national approaches and outcomes. This convinced me to go ahead with this book. Although it was ambitious, I knew that there were many excellent people willing to contribute. With the essential help of an assembled advisory group of leading nature conservation experts (see Acknowledgements) a detailed publication proposal was prepared and further potential chapter authors identified and invited to take part.

Following discussions, a consistent set of key topics was agreed for each country chapter including the natural characteristics of the country, the conservation status of habitats and

[1] Although the Birds Directive was adopted by the European Economic Community (EEC), for simplicity the European Union (EU) is sometimes used as the term to also represent what was the EEC, and then the European Community (EC), before the EU was formed in 1993.

species and main pressures affecting them, governance, policies and legislation, species protection, protected areas, spatial planning, habitat management, funding, monitoring and future developments. To avoid the chapters becoming too long it was decided that they should focus on the most important developments in the country, mainly over the last 40 years and particularly over the last decade and in relation to the EU's and CBD's biodiversity targets up to 2020. It was also agreed to add some introductory chapters, covering the principal nature conservation requirements and challenges in Europe, key international conventions, and the current EU nature policy and legislative framework. With these chapters, as well as a conclusion chapter to draw lessons, it was apparent that the book could be potentially huge and expensive. To prevent this, it now focusses on the EU and UK primarily and does not include additional comparative country chapters. Nevertheless, the book has attempted to cover Europe more widely where possible.

Thankfully, it was possible to find willing authors with the required knowledge to cover 26 countries (involving 50 contributors in total), despite the wide range of issues covered in each chapter and the considerable investment in time needed. Furthermore, as there has been no funding in the project, most of the chapters have been prepared by authors in their 'spare time'. Unfortunately, although authors were initially found for Luxembourg and Malta, they were unable to prepare the chapter due to their vital nature conservation work commitments. To help partially fill this gap, key data on the countries are included in the book where they are available.

An important aim for each country chapter has been for it to represent a balanced range of views. Many of the chapters have therefore been written by multiple national experts from universities, consultancies or institutes, as well as some NGOs. Although few staff from governmental authorities have been able to take part directly, many have provided information or critical reviews. Very fortuitously, a few recently retired senior staff from nature authorities and other institutions have authored some of the chapters, providing invaluable insights based on decades of experience. To further help ensure the chapters provide balanced accounts, each has been reviewed by a range of experts, including from authorities and NGOs, and others with differing backgrounds to the authors.

The actual writing of many chapters started in 2018, once a publication agreement had been made with Cambridge University Press. Good progress was being made until the COVID-19 pandemic struck at the beginning of 2020. Fortunately, it was no worse than disruptive for any of us. But over the last two years the inevitable changes in working practices and personal situations delayed the completion of many chapters. As a result, some of the first submitted chapters became out of date and had to be updated, which the authors took on with great patience and understanding. Whilst it has not been possible to fully catch up on all issues in all chapters, the key developments in the EU and in each country up to the finalisation of the chapters in 2022 have been included as far as possible.

Now, as this book goes to the publisher in June 2022, there are more encouraging signs for nature conservation in Europe, most notably within the EU through some aspects of the European Green Deal, Biodiversity Strategy for 2030 and proposed Regulation on Nature Restoration. Whilst such grand strategies have been seen before, the proposals for legally binding restoration targets in the EU could lead to the step change in the scale of action that is needed. There are also potentially very positive developments in the UK, including legally binding targets for habitat restoration and species, the introduction of mandatory biodiversity net gain from developments, and the transition towards the use of all agricultural subsidies for public goods, including biodiversity. However, the nature conservation outcomes of all these EU and UK initiatives will very much depend on their ambitions, specific targets and actual implementation. Hopefully, some of the lessons from this book, especially the need for strong enforcement of regulation, adequate targeted funding and wider societal involvement in nature conservation will guide these coming initiatives. For the moment, it is imperative that the lessons help improve the implementation of the Nature Directives and other important existing measures – to fill the current gap between conservation aspirations and achievements.

Acknowledgements

There would be nothing to write about, or celebrate, if it were not for the countless number of people in Europe who are contributing to nature conservation in their own way. I therefore salute them all, and hope this book will go some way to showing the breadth of activities and dedication involved. In particular, I acknowledge the essential role of the often unsung heroes, including the legislators, policymakers, regulators and administrators (often rather dismissively referred to as bureaucrats), from EU to local level. Without their hard, unglamorous work, the relatively strong nature conservation legislation and comprehensive policy framework we have in Europe would remain unimplemented and unenforced; indeed, it would not exist in the first place. It is also apparent from this book that there are many outside the environmental sector who are making important contributions to nature conservation, including amongst hunters, farmers, foresters, fishers, water managers, various industries, many other sectors and wider society. All deserve appreciation and encouragement.

Obviously, I am completely indebted to the authors of the country chapters, as well as my co-authors of the other chapters. Without the depth and breadth of knowledge and insights that they have brought, together with their dedication and diligence, this book would never have been possible. They have given their time most generously, and I am especially grateful for their patience and perseverance, as this book has been more work and taken a couple of years longer than anticipated. Sadly, Andreas Demetropoulos, an author of the chapter on Cyprus, passed away before publication of this book. He was an inspiring conservationist, especially in relation to turtles and the marine environment, and it is an honour that he was able to contribute to this book. I also thank the many reviewers and other experts, as listed at the end of each chapter, who provided information and valuable comments and suggestions.

Neither would this book have materialised if it had not been for the wise guidance and positive support that I received from the following nature conservation experts who agreed to join an advisory group: David Baldock, Andreas Baumueller, Ariel Brunner, Wyn Jones, Petr Roth, Alberto Arroyo Schnell, Kerstin Sundseth and Agnes Zolyomi. The structure and contents of the book, and the guidance that was produced for the authors, benefited greatly from their wide-ranging experience and knowledge. The fact that excellent authors were found for 26 chapters is also largely due to their sharing of contacts and suggestions, and personal invitations in some cases. I especially thank Petr Roth for his support, as the idea for the book has much to do with conversations with him 10 years ago, and he has been an enthusiastic supporter and honest advisor ever since.

I am also grateful for the unwitting support for the book from my former colleagues at the Institute for European Environmental Policy, who taught me a huge amount and stimulated many ideas over the years. They have all had an important influence on the book. In particular I thank David Baldock, Andrew Farmer, Erik Gerritsen, Kaley Hart and Evelyn Underwood.

In addition, I thank Dominic Lewis of Cambridge University Press who initially took on the publication of the book, and subsequently Aleksandra Serocka for her advice and great patience and understanding over the protracted period of writing. Jenny van der Meijden at Cambridge University Press and Anne Rufina Raymond at Straive managed the book production, and Ruth Swan carried out meticulous copyediting.

Last, but not least, I thank my friends and family, and especially Claire Johnson, for putting up with my 'book obsession' over the last few years.

Glossary, Abbreviations and Country Codes

§ Section
ABNJ Area Beyond National Jurisdiction
AECM Agri-environment–climate measure
AES Agri-environment scheme
AGR Agricultural areas – CORINE land cover classes (see Table 1.1)
ALP Alpine biogeographical region
Art. Article
ART Artificial – CORINE land cover classes (see Table 1.1)
ASCI Area of Special Conservation Interest (under the Bern Convention)
ATL Atlantic biogeographical region
BAP Biodiversity Action Plan
BD Birds Directive (when referring to articles, annexes and species)
BD species Bird species covered by the Birds Directive (i.e. all wild bird species naturally occurring in the EU)
BHD Birds and Habitats Directives
BLS Black Sea biogeographical region
BOR Boreal biogeographical region
BP Before present
c. Approximately
CAP Common Agricultural Policy
CBD Convention on Biological Diversity
CDDA Common Database on Designated Areas (now the Nationally Designated Areas Inventory)
CFP Common Fisheries Policy
CITES Convention on International Trade in Endangered Species of Wild Fauna and Flora
CJEU Court of Justice of the European Union (also informally known as the European Court of Justice – ECJ)
CLC CORINE land cover
CMS Convention on the Conservation of Migratory Species of Wild Animals (Bonn Convention)
CON Continental biogeographical region
COP Conference of the Parties
CORINE Coordination of information on the environment (land cover data)
CR Critically Threatened (IUCN Red List threat status)
DG Directorate-General (of the European Commission)
EAFRD European Agricultural Fund for Rural Development
EAGF European Agricultural and Guarantee Fund
EAP Environmental Action Programme
EC European Community
ECA European Court of Auditors
EEA European Environment Agency
EEA-39 The EU-28 and Iceland, Liechtenstein, Norway, Switzerland and Turkey, and the cooperating countries: Albania, Bosnia and Herzegovina, Kosovo, Montenegro, North Macedonia and Serbia. The UK has since left the EEA.
EEC European Economic Community
EEZ Exclusive Economic Zone
EIA Environmental impact assessment
Eionet European Environment Information and Observation Network
ELD Environmental Liability Directive
EMFF European Maritime and Fisheries Fund
EN Endangered (IUCN Red List threat status)
eNGO Environmental non-governmental organisation
ENRD European Network for Rural Development
ERDF European Regional Development Fund
ESIF European Structural and Investment Funds
ETC/BD European Topic Centre on Biological Diversity
EU European Union
EU-15 The EU Member States pre-enlargement in 2004
EU-25 The EU Member States as of 1 May 2004 and the accession of Cyprus, Czech Republic, Estonia, Hungary, Latvia, Lithuania, Malta, Poland, Slovakia and Slovenia
EU-27 The EU Member States as of 1 January 2007 and the accession of Bulgaria and Romania, before the accession of Croatia. Not the EU after the departure of the UK.
EU-28 The EU Member States as of 1 January 2013, after the accession of Croatia, including the UK
EUNIS European Nature Information System
FOR Forests – CORINE land cover classes (see Table 1.1)
FTE Full-time equivalent
GAEC Good Agricultural and Environmental Condition, with respect to CAP payment conditions
GBIF Global Biodiversity Information Facility
GDP Gross Domestic Product
GEF Global Environment Facility
GIS Geographic Information System
HD Habitats Directive (when referring to articles, annexes and habitats and species)
HD habitats Habitats listed in HD Annex I
HD species Species listed in HD Annex II and/or Annex IV and V
HNV High Nature Value in relation to farmland or forestry
IAS Invasive Alien Species: animals and plants that are introduced accidentally or deliberately into a natural environment where they are not normally found, with serious negative consequences for their new environment
IBA Important Bird Area (as defined by BirdLife International)
ICES International Council for the Exploration of the Sea
IPBES Intergovernmental Science-Policy Platform on Biodiversity and Ecosystem Services
IUCN International Union for Conservation of Nature
JRC Joint Research Centre – part of the European Commission
KBA Key Biodiversity Area
LIFE Financial Instrument for the Environment
LPIS Land Parcel Identification System
MAC Macaronesian biogeographical region
MAES Mapping and Assessment of Ecosystems and their Services (EU initiative)
MARPOL International Convention for the Prevention of Pollution from Ships

MATL Marine Atlantic region
MBAL Marine Baltic region
MBLS Marine Black Sea region
MEA Multilateral Environmental Agreement
MED Mediterranean biogeographical region
MFF The Multiannual Financial Framework of the EU
MMAC Marine Macaronesian region
MMED Marine Mediterranean region
MPA Marine protected area
MS Member State
MSFD Marine Strategy Framework Directive
MSPD Maritime Strategy Planning Directive
N2k Natura 2000
Natura habitats and species Habitats and species for which a Natura 2000 site is designated.
Nature Directives The Birds and Habitats Directives
NBS/NBSAPs National Biodiversity Strategies and Action Plans
NEAFC North East Atlantic Fisheries Commission
NGO Non-governmental organisation
NSC North Sea Ministerial Conference
NT Near Threatened (IUCN Red List threat status)
OECD Organisation for Economic Cooperation and Development
OECM Other effective area-based conservation measures (see Box 1.2)
OJ Official Journal of the European Union
OSN Other seminatural and natural areas – CORINE land cover classes (see Table 1.1)
OSPAR Convention The Convention for the Protection of the Marine Environment of the North-East Atlantic
PAF Prioritised Action Framework
PAN Pannonian biogeographical region
PoM Programme of Measures
pSCI Proposed Site of Community Importance
RBMP River Basin Management Plan (under the WFD)
RDP Rural Development Programme
REFIT (European Commission) Regulatory Fitness Check and Performance
RSC Regional seas conventions
s./ss. Section/Sections
SAC Special Area of Conservation
SCI Site of Community Importance
SEA Strategic environmental assessment
SEBI Streamlining European Biodiversity Indicators (indicator set)
SPA Special Protection Area
SPA species Bird species that require the classification of SPAs (as per BD Article 4), so including species listed in BD Annex I, and other migratory species for which SPA designation is appropriate
SPAMI Specially Protected Areas of Mediterranean Importance, created under the Barcelona Convention Protocol for Specially Protected Areas and Biological Diversity in the Mediterranean
SPEC Species of European Conservation Concern (as defined by BirdLife International)
Spp. Species (plural)
STEP Steppic biogeographical region
UAA Utilised Agricultural Area
UCS Unfavourable Conservation Status, includes unfavourable–inadequate (i.e. poor) and unfavourable–bad (i.e. bad)
UNCLOS The United Nations Convention on the Law of the Sea
UNDP The United Nations Development Programme
UNEP-WCMC UN Environment Programme World Conservation Monitoring Centre
UNESCO The United Nations Educational, Scientific and Cultural Organization
USAID United States Agency for International Development
UTM Universal Transverse Mercator
VU Vulnerable (IUCN Red List threat status)
WDPA World Database on Protected Areas
WET Wetlands and water bodies – CORINE land cover classes (see Table 1.1)
WFD Water Framework Directive

Country Codes (ISO 3166)

AD Andorra
AL Albania
AT Austria
BA Bosnia and Herzegovina
BE Belgium
BG Bulgaria
BY Belarus
CH Switzerland
CY Cyprus
CZ Czech Republic
DE Germany
DK Denmark
EE Estonia
ES Spain
FI Finland
FR France
GR Greece
HR Croatia
HU Hungary
IE Ireland
IS Iceland
IT Italy
LI Liechtenstein
LT Lithuania
LU Luxembourg
LV Latvia
MD Moldova
ME Montenegro
MK North Macedonia
MT Malta
NL Netherlands
NO Norway
PL Poland
PO Portugal
RO Romania
RS Serbia
SE Sweden
SI Slovenia
SK Slovakia
TR Turkey
UA Ukraine
UK United Kingdom
XK Kosovo

Chapter 1

Introduction: Aims, Scope, Structure and Key Information Sources

Graham Tucker

1.1 The Overall Aims and Scope of This Book

The primary aim of this book is to describe, quantify and critically analyse the nature conservation approaches and practical actions that have been taken across Europe, especially in the United Kingdom (UK) and the European Union (EU), over the last 40 years or so, and to identify the key lessons that can be learned from this experience. In particular, it examines what has happened in individual EU-28 Member States (MS),[1] what has been achieved in terms of biodiversity impacts, which conservation actions were effective and why, and which were not effective or did not happen and why.

It focusses on *in situ* nature conservation, which is the protection, care, management and maintenance of ecosystems, habitats, wildlife species and populations, within their natural environments, in order to safeguard the natural conditions for their long-term permanence. It does not cover the conservation of plant and animal species that have been bred and/or domesticated for food and/or other human uses. Thus, nature conservation as considered here is broadly equivalent to the conservation of the natural elements of biodiversity, as defined by the Convention on Biological Diversity (CBD).[2]

Although this book covers most of continental Europe (as shown in Figure 1.1), it focusses on the EU and UK (as a former Member State) with 26 in-depth country-specific chapters for the UK and all EU Member States other than Luxembourg and Malta. The exclusion of these two countries is unfortunate, and occurred only because it was not possible to find authors for them who could take on the substantial task of providing complete chapters in the time available. However, key data on the status of habitats and species, protected areas, and some other information, are provided for all EU Member States. As space is limited, the country chapters focus on their principal territories, and associated habitats, species and biodiversity conservation challenges within the geographical area of Europe (not European jurisdiction). Therefore, outlying territories are only briefly referred to within the main text, and are omitted where possible from Key Statistics tables and other statistics. The Portuguese archipelagos of the Azores and Madeira, and the Spanish Canary Islands, are also only briefly covered in this book, although they are part of the EU. This is primarily because these isolated islands are very different to the rest of Europe (Madeira and the Canary Islands being part of the African continent), with many unique habitats and species and different conservation challenges. Together they form the distinct Macaronesian terrestrial and marine biogeographical regions. Therefore, given the broad scope of this book and size limits, it was felt that it would not be possible to do them justice with the amount of space that is available.

Information on land cover, ecosystems, pressures and protected areas and some other key data are also provided for other European countries where available. In particular, this includes information on the following members of the European Environment Agency (EEA): Iceland, Liechtenstein, Luxembourg, Norway and Switzerland; and the six cooperating countries: Albania, Bosnia and Herzegovina, Kosovo, Montenegro, North Macedonia and Serbia.[3] Data are also provided for Belarus, Moldova and Ukraine where these are available from other sources. The European parts of Kazakhstan, Russia and Turkey are not included for practical reasons and because they form a small proportion of their whole territory.

The main focus of this book and most country chapters is on events over the last 44 years or so, as there were two crucial developments in 1979: the Council of Europe's Bern Convention and the European Community's Birds Directive.[4] Earlier events (such as the establishment of influential organisations or legislation) are also described if they have had a long-term influence. Some chapters concentrate on more recent periods, such as those following the main periods of political change in Eastern Europe in the late 1980s/early 1990s and/or since accession to the EU and the need to implement the 1992 Habitats Directive.[5]

[1] The EU-28 refers to the EU when the UK was a Member State (until 31 January 2020).

[2] 'Biological diversity' means the variability among living organisms from all sources including, *inter alia*, terrestrial, marine and other aquatic ecosystems and the ecological complexes of which they are part; this includes diversity within species, between species and of ecosystems.

[3] Reference to the EEA-39 is to the EU, UK (until it left following Brexit) and the additional member and cooperating countries.

[4] Directive on the conservation of wild birds (79/409/EEC), which has since been codified under Directive 2009/147/EC.

[5] Directive on the conservation of natural habitats and of wild fauna and flora (92/43/EEC).

Figure 1.1 The European territory covered in this book, which excludes the European parts of Russia and Turkey.
Notes. BEL. = Belgium; KOS. = Kosovo; LIECH. = Liechtenstein; LUX. = Luxembourg; MACE. = North Macedonia; MONT. = Montenegro; NETH. = the Netherlands; SWITZ. = Switzerland.
Source. pop_jop / DigitalVision Vectors / Getty Images.

Box 1.1 The general objectives of the Birds and Habitats Directives (Nature Directives) and their two main pillars of action

The Birds Directive aims to maintain the populations of all species of naturally occurring birds in their wild state in the EU *'at a level which corresponds in particular to ecological, scientific and cultural requirements, while taking account of economic and recreational requirements, or to adapt the population of these species to that level'*.

The overall aim of the Habitats Directive is *'to contribute towards ensuring biodiversity through the conservation of natural habitats and of species of wild fauna and flora in the European territory of the Member States to which the Treaty applies'*. More specifically, Member States shall take measures *'to maintain or restore, at favourable conservation status, natural habitats and species of wild fauna and flora of Community interest'*. The measures *'shall take account of economic, social and cultural requirements and regional and local characteristics'*. 'Natural habitats and species of wild fauna and flora of Community interest' comprise the habitats listed in HD Annex I, and species listed in HD Annexes II and/or IV or V.

In simple terms, favourable conservation status can be described as *'a situation where a habitat type or species is prospering (in both quality and extent/population) and with good prospects to do so in the future as well'* (ETC/BD, 2011). Favourable conservation status is assessed across the whole national territory or across biogeographical or marine regions within the national territory if there is more than one such region within the Member State (see Figures 2.1 and 2.2, and Box 4.1 for details).

Whilst the term 'favourable conservation status' is not mentioned in the Birds Directive, it is implied; as such the aims of the Nature Directives are considered to be broadly analogous by the European Commission.

Both directives have two main approaches (pillars) by which they aim to achieve their objectives:

- The protection of sites of particular importance for BHD habitats and species, through the establishment of the Natura 2000 network (under the Habitats Directive), which comprises:
  - Sites of Community Importance (SCIs) designated as Special Areas of Conservation (SACs) under the Habitats Directive (for habitats listed in HD Annex I and species listed in HD Annex II); and
  - Special Protection Areas (SPAs) designated under the Birds Directive (for bird taxa listed in BD Annex I and for migratory species).
- General protection measures that apply to all birds (with some exceptions) and stricter protection for selected non-bird species (listed in HD Annexes IV or V) wherever they occur.

These are supported by complementary actions, including EU funding (with some requirements under the Habitats Directive), research to guide actions, and monitoring and reporting on the implementation of measures and their impacts on BHD habitats and species.

A large proportion of the book covers the Birds Directive (BD) and Habitats Directive (HD) as these form the cornerstone of nature conservation legislation and are hereafter referred to as the Nature Directives. Given their central importance, their aims and key provisions are outlined in Box 1.1 (with further detail provided in Section 4.2). The Birds Directive covers all naturally occurring wild birds, and requires special habitat conservation measures for 197 taxa (i.e. species and subspecies) listed in BD Annex I, as well as for other migratory species. The Habitats Directive concentrates on selected 'natural habitats and species of wild fauna and flora of Community Interest', which comprise the 233 habitats listed in HD Annex I, and 1 389 species listed in HD Annexes II and/or IV or V (which do not include any bird species).[6]

1.2 The Structure of the Book and Contents of Main Chapters

To provide context for the later chapters and analysis, the next chapter (2) outlines the key characteristics of Europe's environment and nature, and the effects of human actions on it. Chapter 3 then identifies and summarises the main international drivers (other than the EU) of nature conservation, their objectives and their impacts on national nature conservation policies and actions, especially within the EU-28. Chapter 4 provides a more in-depth analysis of the EU's nature conservation policy and legislative framework, especially the provisions in the Nature Directives and their interpretation. Supporting policies, legislation and funds are also described as well as the development of the EU's biodiversity strategies and action plans.

Chapter 5 sets out the book's conclusions, drawing mainly from the analysis of the conservation measures and their lessons that have been identified in each of the later 26 country chapters. The conclusions are provided before the country chapters to keep them close to the overview of conservation measures, and to avoid them being tucked away and potentially out of sight.

The country chapters (6–31) cover 25 EU Member States and the UK and describe the conservation polices and practical actions that have been taken in each of them over recent decades, and their impacts. Each country chapter covers the following key topics according to the standard general structure below (with some variation where appropriate) to help provide comparable assessments of the countries' approaches and the effectiveness of their measures:

[6] See the EEA webpage for further information on the habitats and species covered by the Nature Directives: https://bit.ly/3ohXchQ

Summary table of Key Statistics (see Table 1.1).

1. Outline of the ecological and biodiversity characteristics of the country, biogeographical regions, ecosystems, and habitats and associated species of particular importance.
2. The status of biodiversity:
    - Key biodiversity status indicators and trends, including the 2013–2018 Member States' conservation status assessments (see Table 1.2).
    - Pressures on biodiversity and causes of change.
3. Nature conservation governance, actors, legislation and policies, and related key historical developments.
4. Species protection and conservation measures, e.g.:
    - Basic protection legislation, and strict protection of selected species.
    - Regulation of hunting and fishing, and approaches to managing conflicts.
    - Species action/recovery plans.
    - Species re-introductions.
5. Protected areas and networks:
    - Types of protected area, their historical background, roles and interrelationships (e.g. integration of domestic protected areas and Natura 2000 network) and coverage.
    - Protected area coherence, ecological network approaches and other measures to increase protected area connectivity.
    - Approaches to protection and levels of protection.
    - Protected area management, such as governance arrangements, site management planning, involvement of stakeholders, and implementation of practical site management measures.
6. General biodiversity measures that do not relate solely to the protection of species and/or protected areas, but are broader and cross-cutting, including:
    - Tackling development impacts (e.g. approaches to spatial planning, impact assessments, mitigation and offsetting).
    - Habitat management and restoration.
    - Other topics of particular relevance to the country (e.g. pollution, tackling invasive alien species, and climate change adaptation).
7. Nature conservation costs and funding:
    - Estimates of costs.
    - Economic benefits.
    - Main funding sources, including EU, national and innovative funding.
8. Surveillance, monitoring and reporting, e.g.:
    - The main sources of biodiversity data (e.g. habitat maps and species occurrence records).
    - Approaches to monitoring and reporting on the conservation status of BHD habitats and species.
    - Wider biodiversity monitoring schemes (e.g. species and habitats covered, lead organisations, contributors and funding).
9. Future developments, including anticipated legislation, polices and plans.
10. Conclusions, including an overall integrated assessment of what has happened in the country, including key policy developments and actioins, their overall impacts and the main lessons learnt.

Where possible, each country chapter delves beyond the basic statistics and government and NGO statements, to establish what actually happened. In so doing, it aims to identify the approaches that have been most successful and the causes of failure when they have not. Clearly, it is not feasible to do this for all significant nature conservation actions that have taken place in each country. Therefore, within the standardised outline structure, each chapter identifies and focusses on particularly relevant habitats or issues, or measures that the country may have championed or are novel. Thus, the book aims to provide broadly comparable chapters but also some variation in their topics and lessons.

Many of the chapters have tended to focus on terrestrial nature conservation activities. This is often because practical marine conservation has lagged behind terrestrial conservation, despite early marine conventions and policy initiatives. Therefore, there is sometimes less of a story to tell at a country level. Thankfully, momentum has grown in marine conservation in recent years, such as with the establishment of marine protected areas (MPAs). It is to be hoped that in the near future another book will be able to do justice to the subject of marine conservation and have much more action to report on.

Some readers may also be surprised to see that the subject of hunting is also covered relatively briefly in most chapters. This is because, whilst excessive hunting was a major driver of conservation initiatives, such as the Birds Directive, its impact on most species populations is now much reduced. Although some significant hunting related impacts still remain, and are therefore addressed in the relevant chapters, the main conservation needs over more recent decades have been to address broader and more significant pressures from habitat loss, degradation and fragmentation, pollution and the increasing spread of invasive alien species (IAS).

The chapters vary in length to some extent, primarily reflecting the countries' size, biogeographical diversity, and complexity of nature conservation arrangements. Thus, chapters covering countries with strongly regionalised governance have needed to devote a lot of space to describing the interactions between responsibilities at the national level and lower regional levels. The chapters also reflect the length and complexity of the history of nature conservation, and the degree to which it has been documented. Consequently the chapter on the UK is particularly long, as nature conservation has been devolved, such that it is carried out very differently over its four 'nations'; the effects of Brexit have also resulted in fundamental and complex changes that need to be described.

Whilst the authors have attempted to adequately cover the most important issues, limits on the length of this book have meant that it has not been possible to include some more detailed information on some topics. Some chapters have therefore provided such information as annexes, which are available as free to access online PDFs on the Cambridge University Press website, at the following address: www.cambridge.org/natureconservation.

1.3 Definitions of Key Terms, Data Sources and Limitations

1.3.1 Species Protected under the Nature Directives

For brevity, the term 'HD habitats' is used in this book to refer to natural habitats of Community interest listed in HD Annex I, and 'HD species' are species of wild fauna and flora of Community interest listed in HD Annexes II and/or IV or V. The term 'BHD habitats and species' is used to refer to all habitats and species that are protected under the Birds and Habitats Directives (i.e. Nature Directives), which comprise all wild native birds, and natural habitats and species of wild fauna and flora of Community interest (i.e. HD habitats and HD species).

1.3.2 Definition and Data Sources Relating to the Key Statistics Table

At the start of each chapter a standard Key Statistics table, as set out in Table 1.1, provides background information of particular relevance to nature conservation in the country. As the focus of this book is on the EU 'mainland', the key information mostly excludes data from the Azores, Madeira and the Canary Islands, which comprise the Macaronesian biogeographical region and Macaronesian marine region. Also, the statistics do not take into account BHD habitats and species that only occur within the Macaronesian regions. Data are also omitted from other territories of EU Member States that are outside the European continent.

The Key Statistics table includes the country's terrestrial and marine waters area, population density, relative economic size (measured as gross domestic product (GDP) per capita) and the proportional coverage of the main types of land cover, based on CORINE land cover data (Büttner *et al.*, 2021). Unless indicated otherwise, references to land cover in the text are based on CORINE land cover data from 2018, as summarised in the chapter's Key Statistics table. The biogeographical regions that occur in the country are also listed (see Figures 2.1 and 2.2 for boundaries). To place the statistics in context, the land and marine water areas are also given as a percentage of the EU-28 total. Similarly, the population density and GDP statistics are also shown as a percentage of the EU-28 average. Although the UK left the EU a few years ago, the statistics are related to the EU-28 as it was, because this represents the main period covered by this book.

The number of HD habitats and species, and BD Annex I bird taxa, that occur in the country are included in the table, as well as the percentage that each comprises of the total that occur in the EU mainland. The number of habitats and species occurring in each Member State are those marked as 'present regularly' or 'newly arriving species' according to the reporting checklists compiled in June 2020 by the EEA and European Topic Centre on Biological Diversity (ETC/BD). Although the checklists take into account the results of the Member State reporting over 2013–2018, the numbers given in the table may differ slightly from the numbers that were actually reported on by each Member State. The number of habitats and species should in any case be treated as indicative as it can be difficult to decide whether or not some habitats definitely meet Annex I definitions, or if species occur sufficiently regularly to be considered as present. Habitats and species distributions are also constantly changing, as well as knowledge of them, which also varies across European countries. Despite the limitations, the proportions of HD habitats and species and BD Annex I birds that occur in each country give an indication of the importance of the country with respect to EU conservation objectives within the mainland (further discussed in Section 2.3).

Lastly, as an indication of their status, the proportion of HD habitats and species with an unfavourable conservation status (UCS) that have decreased over the short term is provided. The reason for the use of this metric as an indicator of conservation status is described in Section 1.3.3. As birds are assessed differently, their status is indicated by the proportion of breeding BD Annex I taxa that are declining.

1.3.3 The Conservation Status of Habitats and Species That Are the Focus of the Nature Directives

This book draws heavily on the surveillance that Member States are required to carry out on all BHD habitats and species, and the reports that must be produced in accordance with BD Article 12 and HD Article 17 (described further in Section 4.2). Although the monitoring and requirements as set out in the Nature Directives differ, they have been harmonised such that they now provide very similar information on the implementation of the directives and the status of habitats and species, over the same six-year period. Following a transition in the previous reporting period, reporting under the Birds and Habitats Directive coincided fully over the last reporting period, 2013–2018. Unless otherwise indicated, references to BD Article 12 and HD Article 17 data for 2013–2018 refer to the Member States' reports as

Table 1.1 Key Statistics presented at the start of each country chapter.

Label	Description	Source
Terrestrial area (km^2)/% of EU-28*	Terrestrial area (km^2) based on the European land area given in the Natura 2000 barometer (adjusted for Portugal and Spain mainland only)	EEA[7]
Marine waters (km^2)/% of EU-28*	For EU MS, as defined in MSFD Article 3.1, which extend to the outmost reach where the MS has and/or exercises jurisdictional rights, in accordance with UNCLOS. UK marine waters are delimited by the UK Exclusive Economic Zone (EEZ) and the UK Continental Shelf	Marine Information Systems for Europe[8]/UK JNCC[9]
Population density (p/km^2)/% of EU-28*	Average persons per km^2 over 2010–2019	Eurostat: 'Population on 1 January' (online data code: TPS00001)
GDP per capita (€)/% of EU-28	The ratio of real GDP to the average population of a specific year at market prices, averaged over 2010–2019	Eurostat: 'Real GDP per capita' (online data code: SDG_08_10)
Land cover (CORINE land cover classes 2018)*	% of area comprising CORINE land cover classes combined as: **ART** = artificial: 1.1 urban; 1.2 industry and transport; 1.3 mines, dumps and construction; 1.4 artificial vegetated areas **AGR** = agricultural areas: 2.1 arable; 2.2 permanent crops; 2.3 pastures (inc. unimproved and lightly improved); 2.4 heterogeneous agriculture (inc. areas with significant natural vegetation) **FOR** = forests: 3.1 forests, areas with native or exotic coniferous and/or broad-leaved trees (>5 m with >30% canopy closure) **OSN** = other semi-natural and natural areas: 3.2 scrub and/or herbaceous associations (inc. natural grasslands, heathlands, sclerophyllous vegetation, transitional woodland/shrub); 3.3 open spaces with little or no vegetation (e.g. rocky areas, sand dunes and beaches) **WET** = wetlands and water bodies: 4.1 inland wetlands (marshes and mires); 4.2 marine wetlands (salt marshes, salinas and intertidal flats); 5.1 inland waters (lakes and rivers, etc.)	EEA land cover and change statistics 2000–2018[10]
Biogeographical regions	Terrestrial: **ALP** = Alpine, **ATL** = Atlantic, **BLS** = Black Sea, **BOR** = Boreal, **CON** = Continental, **MAC** = Macaronesian, **MED** = Mediterranean, **PAN** = Pannonian, **STE** = Steppic. Marine: **MATL** = Atlantic, **MBAL** = Baltic Sea, **MBLS** = Black Sea, **MMAC** = Macaronesian, **MMED** = Mediterranean Sea	MS HD Article 17 reports on the EEA national summary Article 17 dashboard
Number of HD Annex I habitats/%*	HD Annex I habitats listed for the country in the checklist for Annex I habitat types (last updated: June 2020)	EEA/Eionet Reference portal for HD Article 17 reporting
Number of HD Annex II, IV and V species/%*	HD species listed for the country in the checklists for Annex II, IV and V species (last updated: June 2020)	As above
Number of BD Annex I bird taxa/%*	Breeding, winter and passage bird species and subspecies regularly occurring in the country according to the checklist for bird species (last updated: June 2020). Subpopulations distinguished in the checklist are counted together	EEA/Eionet Reference portal for BD Article 12 reporting
% HD UCS and decreasing habitats/% unknown*	% HD Annex I habitats with an unfavourable conservation status that have decreased over the short term (i.e. last two reporting periods = 2007–2018). % HD Annex I habitats with an unknown status	MS HD Article 17 reports on the EEA national summary Article 17 dashboard

7 www.eea.europa.eu/data-and-maps/dashboards/natura-2000-barometer

8 https://water.europa.eu/marine/countries-and-regional-seas/country-profiles

9 https://jncc.gov.uk/our-work/uk-marine-protected-area-network-statistics/

10 www.eea.europa.eu/data-and-maps/dashboards/land-cover-and-change-statistics

Table 1.1 *(cont.)*

Label	Description	Source
% HD UCS and decreasing species/% unknown*	% HD Annex II, IV and V species with an unfavourable conservation status that have decreased over the short term. % HD species with an unknown status	As above
% breeding BD Annex I birds decreasing/% uncertain and unknown*	% breeding BD Annex I birds that have decreased over the short term. % with uncertain or unknown trends	MS BD Article 12 reports on the EEA national summary Article 12 dashboard

Note. * Excludes areas and habitats and species within the Macaronesian terrestrial and marine biogeographical regions.

Table 1.2 Categories of EU population status of birds and EU biogeographical conservation status of HD habitats and species.

EU status	Good	Poor	Bad	Unknown
EU populations of bird species/subspecies	Secure	Near Threatened (according to IUCN Red List criteria), declining or depleted	Threatened (according to IUCN Red List criteria), i.e. Critically Endangered, Endangered, or Vulnerable	Unknown
EU biogeographical conservation status of HD habitats and species	Favourable	Unfavourable–inadequate	Unfavourable–bad	Unknown

Source. Adapted from EEA (2020).

compiled and presented by the EEA on the national summary dashboards for Article 12[11] and Article 17.[12]

Whilst there has been some harmonisation of the reporting under the Birds and Habitats Directives, BD Article 12 reports differ from HD Article 17 reports in two fundamental ways. Firstly, Member States provide assessments at a national level, rather than a biogeographical level. Secondly, the national assessments are in relation to each species', or subspecies', trends (i.e. increasing, stable, decreasing, uncertain or unknown), rather than an assessment of conservation status. An overall EU status assessment is then made by the ETC/BD using an agreed methodology (DG Environment, 2014). The first step assesses whether the bird species is threatened (i.e. Critically Endangered, Endangered, or Vulnerable) or Near Threatened according to IUCN regional Red List criteria (IUCN, 2012, 2019). In the second step, species that are neither threatened nor Near Threatened are categorised as 'declining' (i.e. EU-28 population or range declined by $\geq$20% since 1980 with continuing decline since 2007), 'depleted' (i.e. EU-28 population or range declined by $\geq$20% since 1980 but no longer declining since 2007), 'secure' or 'unknown'.

As indicated in Table 1.2, the EEA has created a common framework for summarising the results of the reporting under the Nature Directives. This framework and the categories are therefore followed in this book, although the more specific status of birds with a poor status is also referred to where appropriate. Further information on the methods used by the EEA to assess the conservation status of BHD habitats and species is provided by Röschel *et al.* (2020).

The concept of aiming to achieve the favourable conservation status of habitats and species that was introduced in the HD has been groundbreaking, and has raised nature conservation ambitions from attempting merely to prevent extinctions and declines. Now, the long-term aim is for habitats or species to be prospering (in both quality and extent/population) and with good prospects to do so in the future as well (ETC/BD, 2011). The status of each HD habitat is assessed in relation to its natural range, area, structure and functions, species, status and future prospects. Species are assessed in relation to their range, population trend, habitat sufficiency and future prospects. Importantly, this involves assessments against the habitat's and species' previous range and area in relation to favourable reference values (ETC/BD, 2011; described further in Box 4.1). Thus, to be in favourable conservation status the habitat's or species' range and area must be at least as large as it was when the directive came into force in the Member State, and sufficient for its long-term viability. Other factors affecting the habitat's or species' future prospects must also be favourable.

Whilst this is a logical and laudable aim, it does mean that the achievement of favourable conservation status in practice tends to be a long-term objective and therefore makes a rather insensitive metric of change. This is evident from the results of the last few HD Article 17 monitoring reports (EEA, 2020; Section 5.1). Furthermore, the application of favourable reference values has been inconsistently applied by Member States (Bijlsma *et al.*, 2019), which makes it more difficult to

11 www.eea.europa.eu/themes/biodiversity/state-of-nature-in-the-eu/article-12-national-summary-dashboards

12 www.eea.europa.eu/themes/biodiversity/state-of-nature-in-the-eu/article-17-national-summary-dashboards

compare the achievement of favourable conservation status across countries. Also, it is unlikely to be feasible to regain the former reference value areas and ranges of some habitats and species, the implication being that they will never achieve a favourable status, even though they may be prospering in all other respects. To reduce these problems, and to highlight habitats and species that are of highest immediate conservation concern, this book has used the following two metrics as standard indicators of progress towards the achievement of the objectives of the Nature Directives:

- Proportion of HD habitats and HD species that have an unfavourable conservation status (UCS) and have shown short-term decreases.
- Proportion of breeding birds listed in BD Annex I that have shown short-term decreases.

Under the current Nature Directives reporting framework, 'short-term' refers to the last two six-year reporting periods, which cover 2007–2018. These metrics are more comparable across the Member States because they are less influenced by historic constraints and their approaches to setting reference values and interpretation of future prospects. The short-term trends also reflect what has been happening recently, rather than long-term pressures that may no longer be very relevant. The metrics for each country are included in the summary of Key Statistics table at the start of each chapter (see Table 1.1), and are compared across countries in the conclusions chapter (Table 5.1). A more detailed breakdown of the short-term trends of HD habitats and species, and breeding and wintering BD Annex I birds is also provided in each country chapter. To provide a complete picture, each chapter also briefly summarises the overall conservation status of BHD habitats and species.

It is important to note that many changes in the status and trends of BHD habitats and species between the most recent reporting period (2013–2018) and the previous monitoring periods are likely to be due to changes in methodology or improved knowledge, requiring careful interpretation. Therefore, to avoid presenting potentially misleading data, the standard status and trend tables in this book do not include the information from previous periods.

1.3.4 Protected Areas

A key aim of this book has been to compile information on protected areas in each country, to compare their coverage, types and effectiveness. The intention has been to follow the IUCN and CBD definitions of protected areas (Box 1.2), which the IUCN and the CBD Secretariat agree have the same meaning (Lopoukhine and Dias, 2012). However, the definitions leave significant room for interpretation, which has resulted in considerable variation in what is considered to be a protected area by European countries and regions within them. The country chapter authors have therefore needed to consider carefully, using the best available information, which types of listed nationally or regionally designated areas actually meet the IUCN and CBD protected area definitions. Whilst it is clear for some protected area types, there are also many cases where it is not, and a detailed assessment would be necessary to be certain, such as was carried out in the UK (Crofts *et al.*, 2014). The protected areas listed in the country chapter tables should therefore be interpreted with caution, taking into account the available information and accompanying author's commentary.

Where possible, the chapter authors have mainly based the tables and their analysis on national official lists of designated protected areas and their categories. Unfortunately, such national data are not always available, complete or up-to-date. These include data on internationally designated sites from the Ramsar Convention, and UNESCO for World Heritage Sites Biosphere Reserves.[13] For Biosphere Reserves, only the core areas are included, as buffer zones and transition zones are generally not protected areas. Information on MPAs designated under the Regional Seas Conventions (described in Chapter 3) has also been obtained from the relevant websites for Helsinki Commission (HELCOM) MPAs in the Baltic Sea, MPAs under the Convention for the Protection of the Marine Environment of the North-East Atlantic (OSPAR) and Specially Protected Areas of Mediterranean Importance (SPAMI) under the Barcelona Convention.[14] For consistency, except where there is more recent information on significant changes, Natura 2000 data and coverage estimates are mainly taken from the June 2021 update of the EEA's Natura 2000 barometer,[15] which indicates the situation at the end of 2020.

Reference has also often been made to the listed protected areas and estimation of their coverage provided on the Protected Planet website.[16] This is based on the World Database on Protected Areas (WDPA), which is a joint project between UN Environment and IUCN, and is compiled and managed by the UN Environment Programme World Conservation Monitoring Centre (UNEP-WCMC). Amongst other things, it is used for reporting to the CBD on progress towards protected area targets. The official source of European data in the WDPA is the European Nationally Designated Areas Inventory (CDDA).[17] The CDDA is maintained by the EEA with support from the ETC/BD, with the data delivered by the Eionet partnership countries.[18] It is important to note

[13] Ramsar Sites: https://rsis.ramsar.org/ World Heritage Sites: https://whc.unesco.org/en/statesparties/ Biosphere Reserves: https://en.unesco.org/biosphere/eu-na

[14] HELCOM: http://mpas.helcom.fi/apex/f?p=103:1 OSPAR: https://mpa.ospar.org/home-ospar/key-figures Barcelona Convention – Regional Activity Centre for Specially Protected Areas: www.rac-spa.org/spami

[15] www.eea.europa.eu/data-and-maps/dashboards/natura-2000-barometer

[16] www.protectedplanet.net/en

[17] According to its former name 'Common Database on Designated Areas'.

[18] www.eea.europa.eu/data-and-maps/data/nationally-designated-areas-national-cdda-17

Box 1.2 Protected area definitions and categories

IUCN protected area definition:
'A protected area is a clearly defined geographical space, recognised, dedicated and managed, through legal or other effective means, to achieve the long term conservation of nature with associated ecosystem services and cultural values.'

(Dudley, 2008).

CBD protected area definition:
'A geographically defined area, which is designated or regulated and managed to achieve specific conservation objectives' (CBD Article 2). This definition is further expanded upon under CBD Article 8.

CBD definition of 'Other effective area-based conservation measures' (OECMs)
'A geographically defined area other than a Protected Area, which is governed and managed in ways that achieve positive and sustained long-term outcomes for the *in situ* conservation of biodiversity, with associated ecosystem functions and services and where applicable, cultural, spiritual, socio-economic, and other locally relevant values.' (14th CBD COP, Decision 14/8, 2018).

IUCN protected area categories (adapted from Dudley, 2008 with Stolton *et al.*, 2013)

Ia Strict Nature Reserve: Category Ia are strictly protected areas set aside to protect biodiversity and also possibly geological/geomorphological features, where human visitation, use and impacts are strictly controlled and limited to ensure protection of the conservation values. Such protected areas can serve as indispensable reference areas for scientific research and monitoring.

Ib Wilderness Area: Category Ib protected areas are usually large unmodified or slightly modified areas, retaining their natural character and influence without permanent or significant human habitation, which are protected and managed so as to preserve their natural condition.

II National Park: Category II protected areas are large natural or near natural areas set aside to protect large-scale ecological processes, along with the complement of species and ecosystems characteristic of the area, which also provide a foundation for environmentally and culturally compatible, spiritual, scientific, educational, recreational and visitor opportunities.

III Natural Monument or Feature: Category III protected areas are set aside to protect a specific natural monument, which can be a landform, sea mount, submarine cavern, geological feature such as a cave or even a living feature such as an ancient grove. They are generally quite small protected areas and often have high visitor value.

IV Habitat/Species Management Area: Category IV protected areas aim to protect particular species or habitats and management reflects this priority. Many Category IV protected areas will need regular, active interventions to address the requirements of particular species or to maintain habitats, but this is not a requirement of the category.

V Protected Landscape/Seascape: A protected area where the interaction of people and nature over time has produced an area of distinct character with significant, ecological, biological, cultural and scenic value, and where safeguarding the integrity of this interaction is vital to protecting and sustaining the area and its associated nature conservation and other values.

VI Protected area with sustainable use of natural resources: Category VI protected areas conserve ecosystems and habitats together with associated cultural values and traditional natural resource management systems. They are generally large, with most of the area in a natural condition, where a proportion is under sustainable natural resource management and where low-level non-industrial use of natural resources compatible with nature conservation is seen as one of the main aims of the area.

that, whilst the CDDA includes protected areas that meet IUCN and CBD definitions, it also contains other designated sites including areas closer to the 'other effective area-based conservation measures' (OECMs) definition (Box 1.2). The CDDA is also the source used by the EEA for its protected area coverage indicator in the EU Biodiversity Strategy Dashboard. This is not directly referred to in most chapters, because it was set up after the protected area analysis had been completed using WDPA data (which were in any case based on a more up-to-date dataset at the time).

According to UNEP-WCMC (2019), records should not be submitted to the WDPA if they do not meet the IUCN or CBD definition of a protected area. However, UNEP-WCMC states that 'it cannot be guaranteed that data-providers consistently follow this standard. In part, this is because countries often have national definitions of protected areas that may not fully align with the IUCN or CBD definition. It should therefore not be assumed that all records in the WDPA meet the IUCN or CBD definition.' In fact, it is clear from several of the national chapters in this book that some of the listed sites do not meet the definitions (e.g. in Denmark, Latvia and the UK). These are therefore pointed out by the authors and excluded from protected area coverage calculations where feasible.

To assess progress towards protected area targets, including whether the 2020 CBD Aichi Target 11 was met (i.e. 17% on land and 10% at sea), each chapter has estimated the combined area of protected areas on land and separately at sea, taking into account overlaps. The estimates are based on national and/or Protected Planet (WDPA) GIS analyses, depending on which are most up-to-date and in accordance with IUCN/CBD protected area definitions. It should be noted that the Protected Planet estimates do not directly include

Biosphere Reserves (UNEP-WCMC, 2019). Table 5.4 in the conclusions chapter provides a comparison of the Protected Planet estimates and national assessments, if available.

The protected area tables in each chapter also indicate the IUCN protected area management categories (Box 1.2) that apply to each type of protected area, where these are known. However, in the European context, the IUCN typology can be difficult to apply and by itself does not normally provide a clear picture of the role of a protected area in nature conservation. Discussion of the protected area types therefore tends to concentrate on them in terms of two key attributes: (1) the degree to which they focus on nature, and (2) their strength of protection. It is also important to bear in mind that the names used for types of protected areas can have very different meanings in different countries. For example, in some countries National Parks have a very high level of protection and may be dedicated to nature conservation objectives. But in others, such as the UK, National Parks are primarily designated on the basis of landscape and recreational considerations and have relatively weak protection levels primarily relating to restrictions on built developments.

Due to these issues, care needs to be taken in interpretation of the CDDA/WDPA and official country data on protected areas, especially their coverage estimates and management categories. This topic and its implications are discussed in the conclusions chapter (Section 5.4). It is also evident that the periodic updates of data sources on protected areas vary in their frequency, depending on the protected area type, country and institution involved. This book therefore aims only to provide an approximate 'snapshot' of the coverage of protected areas, and readers should refer to more up-to-date sources for definitive data.

1.3.5 Nature Conservation Costs, Benefits and Funding

Practical nature conservation is often constrained by funding limitations, for example affecting environmental authorities and the execution of their roles, but in particular practical conservation actions, such as habitat management. Therefore, another important aim of the national chapters has been to examine the costs of nature conservation in the country and to compare the costs with potentially available funds and their actual use. The sections primarily focus on funding over the last 10 years and in particular the EU Multiannual Financial Framework (MFF) for 2014–2020.

In most national chapters the information on costs has been drawn from Prioritised Action Frameworks (PAFs) that Member States prepared in accordance with the current financing arrangements for the Natura 2000 network (further discussed in Section 4.2.6). The PAFs are strategic multiannual planning tools that aim to provide a comprehensive assessment of the measures needed, and their costs, to achieve the aims of the Nature Directives through the implementation of the Natura 2000 network. The first PAFs were completed for the 2014–2020 MFF, whilst the current PAFs were developed to help access funds from the 2021–2027 MFF.

In addition to covering the measures required to meet the conservation objectives of the sites, the PAFs also include additional needs related to species protection and wider green infrastructure[19] where they contribute to the ecological coherence of the Natura 2000 network. It is therefore important to bear in mind that the costs estimated in the PAFs do not cover all nature conservation costs, although they should be a very large proportion. Comparisons of PAF cost estimates between countries also need to be carried out with caution, as Member States' interpretation of the guidance on completion of the PAFs has varied, especially in the 2014–2020 PAFs. In particular, some have ambitiously costed the full achievement of the favourable conservation status of BHD habitats and species, whilst others have estimated what they consider to be more realistic within the funding period. They also differ in their treatment of green infrastructure and other requirements outside Natura 2000 sites.

To enable a comparison of nature conservation costs and available funding, each chapter includes a standardised funding table, which sets out the allocations of the main EU and national sources of funds for nature conservation related actions. As funding is often from national funds with co-financing from the EU, these components are separated where data allow. The main focus is on funding over the 2014–2020 MFF, but longer periods are assessed to provide a more complete picture for some funds (e.g. the EU LIFE programme), or due to data limitations. As the periods covered by the various funds differ, the funding allocations have been annualised to aid comparisons. For traditional LIFE projects, the annual allocations are based on the period over which the projects were awarded. The annualised funding amounts for Integrated LIFE projects are based on the projects' duration. It should be noted that figures in the funding tables are rounded to the nearest €0.1 million, as a result of which rounded subtotals and totals may not correspond to the sum of the individual rows.

The main sources of information used for the funding tables are the European Commission database on the European Structural and Investment Funds (ESIF) 2014–2020[20] allocations as at October 2020, as well as the PAFs and websites of the individual EU funds. The ESIF database was used to ascertain the European Agriculture Fund for Rural Development (EAFRD) allocations to the Rural Development Programme measures related to Priority 4 'restoring, preserving and enhancing ecosystems related to agriculture and forestry'. Although

[19] Defined by the European Commission as 'a strategically planned network of natural and semi-natural areas with other environmental features designed and managed to deliver a wide range of ecosystem services'.

[20] https://cohesiondata.ec.europa.eu/2014-2020/2014-2020-EAFRD-allocation-by-focus-area-EU-planne/6q2m-arif downloaded 24 October 2020.

e RDP programming periods were extended to the end of 22 due to the COVID-19 pandemic, the updated funding is ot taken into account as it may not reflect typical funding locations. The information on RDP allocations was not based n the PAFs because they have a narrower focus on Natura 000 and supporting measures, whereas the table attempts to entify and include all nature conservation related funding. his is particularly important for RDP measures, because agri-nvironment schemes, and many of the other measures, often ave substantial coverage beyond protected areas. The PAFs ave generally been the information source for the other EU SIF funds, which tend to focus funding on implementation of Natura 2000, as well as national sources of financing. Unless therwise indicated, information on projects funded by the EU LIFE programme has been obtained from the European Commission's LIFE Database[21] or country factsheets.[22]

Where possible the total potentially available funds are calculated, to enable a comparison with the estimates of nature conservation costs. However, such comparisons need to be nterpreted very cautiously as in most cases the available data on costs and funds differ in their scope and completeness. Furthermore, the data in the funding tables are mainly allocations (rather than actual spending), and the potentially available funds are not always used, for example as a result of low take-up by landowners. Also, available funding may not match, or be targeted to, the most important funding needs, as further discussed in the conclusions chapter (Section 5.6). Nevertheless, in some countries the data point to obvious gaps between available funds and requirements.

Although the costs of nature conservation are considerable, there is growing evidence that these are typically outweighed by their benefits (e.g. Martín-López *et al.*, 2018; Dasgupta, 2021). Therefore a summary is provided of key studies carried out within each country of the social, health and economic benefits of ecosystems, and the added value of nature conservation and restoration measures. However, this is a complex subject, and the estimates of such benefits can be incomplete, variable and difficult to interpret depending on the available data and methods used; therefore only a brief discussion of the issue is possible within the scope of this book.

As this book covers over 40 years of nature conservation action, some of the estimated monetary costs, funds and benefits that are quoted are sufficiently old now to be significantly different from current values due to the effects of inflation. Therefore, on the one hand, the interpretation of monetary data can be aided by providing inflation adjusted current values as well as the original value. On the other hand, this would result in long and fragmented text in some sections if all monetary values were adjusted for inflation. As the cumulative effects of inflation have not been very high since 2011 (with an increase of 11% in EU prices up to 2020), monetary values after 2010 are not adjusted. Monetary values before 2011 are firstly presented according to their original source year, and then equivalent figures at 2020 prices are provided in brackets. The adjustments have been made using Eurostat annual Harmonised Indices of Consumer Prices for each country and the average annual euro exchange rate at the time.

To provide a consistent comparative reference, all monetary values that are not in euros are also presented in euros in brackets alongside, based on one currency exchange rate for 28 May 2021, as compiled by the European Commission. The relevant exchange rate for each currency is given as a footnote at the first relevant point in each national chapter.

1.4 Habitat and Species Names, Taxonomy and Global and European Threat Status

When an HD Annex I habitat is referred to, the corresponding code is provided alongside in parentheses. The addition of '*' indicates that it is a 'priority habitat', according to HD Article 1. The full names of HD Annex I habitats mentioned in this book are provided in the Appendix.

The scientific bird names used in this book follow those in the 2018 List of birds of the European Union,[23] which are based on the BirdLife International *Illustrated Checklist of the Birds of the World*, Volumes 1 and 2 (del Hoyo and Collar, 2014, 2016). The taxonomy and scientific names for other species that are used in this book follow those that are recommended for HD Article 17 reports according to the latest EEA Checklists for Annex II, IV and V species,[24] or the IUCN Red List website for other species.[25]

English names for species are generally based on those used by Birdlife International for birds and the IUCN Red List for other species, except where other names are much more frequently used in Europe. However, to shorten species lists, where there is no potentially similar regularly occurring species in Europe, the following components of names are omitted: Common, Western, Northern, Eurasian and European. English names are not used for plants or invertebrates unless they are very well known.

As the global threat status of species, and European threat status of birds, are frequently referred to in the text, the source for each is not repeated each time. In all cases the global threat status was taken between October 2021 and April 2022, from BirdLife International[26] for birds and the IUCN Red List website for other species. References to the European threat status of birds are based on BirdLife International (2021). The reference sources for the European or national status of other species are given in each chapter.

21 https://webgate.ec.europa.eu/life/publicWebsite/index.cfm
22 https://cinea.ec.europa.eu/life/life-european-countries_en

23 https://ec.europa.eu/environment/nature/conservation/wildbirds/eu_species/index_en.htm
24 https://cdr.eionet.europa.eu/help/habitats_art17/
25 www.iucnredlist.org/
26 http://datazone.birdlife.org/home

References

Bijlsma, R. J., Agrillo, E., Attorre, F. *et al.* (2019) *Defining and applying the concept of Favourable Reference Values for species and habitats under the EU Birds and Habitats Directive.* Wageningen, the Netherlands: Wageningen Environmental Research.

BirdLife International (2021) *European Red List of Birds.* Luxembourg: Publications Office of the European Union.

Büttner, G., Kosztra, B., Maucha, G. *et al.* (2021) *Copernicus Land Monitoring Service. CORINE Land Cover User Manual.* Copenhagen: EEA.

Crofts, R., Dudley, N., Mahon, C. *et al.* (2014) *Putting Nature on the Map: A Report and Recommendations on the Use of the IUCN System of Protected Area Categorisation in the UK.* IUCN National Committee UK.

Dasgupta, P. (2021) *The Economics of Biodiversity: The Dasgupta Review.* London: HM Treasury.

del Hoyo, J. & Collar, N. J. (2014) *International Illustrated Checklist of the Birds of the World.* Volume 1. Lynx Editions BirdLife International.

(2016) *International Illustrated Checklist of the Birds of the World.* Volume 2. Lynx Editions BirdLife International.

DG Environment (2014) Using the data reported by Member States under Article 12 of the Birds Directive to summarise and present species' population status at EU level and measure progress towards Target 1(ii)of the EU Biodiversity Strategy to 2020. Available at https://circabc.europa.eu/sd/a/4b101339-6e13-4379-ada5-400e5d1ec8ac/Point%203%20-%20Background-Paper-%2021%20Nov%202013%20.pdf

Dudley, N. (ed.) (2008). *Guidelines for Applying Protected Area Management Categories.* Gland, Switzerland: IUCN. WITH Stolton, S., Shadie, P. & Dudley, N. (2013) *IUCN WCPA Best Practice Guidance on Recognising Protected Areas and Assigning Management Categories and Governance Types.* Best Practice Protected Area Guidelines Series No. 21. Gland, Switzerland: IUCN.

EEA (2020) *State of nature in the EU. Results from reporting under the nature directives 2013–2018.* Copenhagen: European Environment Agency.

ETC/BD (2011) *Assessment, monitoring and reporting under Article 17 of the Habitats Directive: Explanatory Notes & Guidelines for the period 2007–2012.* Paris: European Topic Centre on Biodiversity.

IUCN (2012) *Guidelines for application of IUCN Red List Criteria at regional and national levels. Version 4.0.* Gland, Switzerland & Cambridge, UK: IUCN.

(2019) *Guidelines for using the IUCN Red List categories and criteria, Version 14.* Prepared by the Standards and Petitions Committee.

Lopoukhine, N. & Dias, B. F. de S. (2012) What does Target 11 really mean? *PARKS,* **18**, 5–8.

Martín-López, B., Church, A., Başak Dessane, E. *et al.* (2018) Nature's contributions to people and quality of life. In *The IPBES regional assessment report on biodiversity and ecosystem services for Europe and Central Asia,* eds. M. Rounsevell, M. Fischer, A. Torre-Marin Rando & A. Mader, pp. 86–266. Bonn: Secretariat of the Intergovernmental Science-Policy Platform on Biodiversity and Ecosystem Services.

Röschel, L., Noebel, R., Stein, U. *et al.* (2020) *State of Nature in the EU - Methodological paper. Methodologies under the Nature Directives reporting 2013-2018 and analysis for the State of Nature 2000.* ETC/BD report to the EEA.

UNEP-WCMC (2019) *User Manual for the World Database on Protected Areas and world database on other effective area-based conservation measures: 1.6.* Cambridge, UK: UNEP-WCMC.

Chapter 2

Europe's Nature and Conservation Needs

Graham Tucker

2.1 The Biophysical Geography and Natural History of Europe

Whilst it is only possible to provide a brief account of the natural history of Europe here, this section aims to provide some important context for the approaches that have been taken to nature conservation. It mainly summarises the biophysical influences on ecosystems, and the impacts of humans on them up to about 40 years ago, as more recent events are covered in the national chapters. This is primarily drawn from the following sources: Polunin and Walters (1985), Tucker and Evans (1997), Ellenberg (2009), Veen *et al.* (2009), Blondel *et al.* (2010) and BirdLife International (2017a).

2.1.1 Biogeographical Regions and Their Characteristics

Europe has a relatively diverse range of ecosystems and habitats, primarily as a result of its varied topography, geology and climate, as well as some historical impacts of human activities. An important influence on ecosystems in western Europe is the relatively warm oceanic waters that result from the north Atlantic current, which keep winter temperatures much higher than other areas of the world at a similar latitude. Thus the climate ranges from subtropical in parts of the Iberian Peninsula and around the Mediterranean to subpolar in the north-east of Norway and northern Russia. This underlying climatic pattern is then further modified by the continent's topography, especially its high mountain ranges but also its lowland plains. These climatic and other physical combinations give rise to 11 terrestrial biogeographical regions in Europe, as defined by the European Commission and the Council of Europe for nature conservation purposes (Roekaerts, 2002), with the 2016 version shown in Figure 2.1. It is noteworthy that the Alpine region represents areas with similar characteristics, rather than a geographical region centred on the Alps. It therefore also includes the Pyrenees, Apennines, Carpathians, Dinaric Alps, Balkans and Rhodopes, as well as the Urals and the Caucasus mountains on the border of Europe. The nine biogeographical regions that are relevant to this book are summarised in Table 2.1. A more detailed statistically derived stratification for Europe based on biogeophysical attributes has been produced by Metzger *et al.* (2012).

The marine waters covered in this book comprise the European parts of the Atlantic, Baltic Sea, Black Sea and Mediterranean Sea. In particular, they cover the marine regions and subregions as defined by the Marine Strategy Framework Directive (MSFD), as shown in Figure 2.2, with the exception of the Macaronesian subregion of the North-East Atlantic. MSFD region boundaries are, to the extent possible, harmonised with other EU legislation where maritime boundaries are of relevance, and specifically the biogeographical regions of the Nature Directives. This book mainly refers to the EU-28 waters that have been defined as the Atlantic (MATL), Baltic Sea (MBAL), Black Sea (MBLS) and Mediterranean Sea (MMED) marine regions, for the purposes of implementing the Nature Directives. In this respect, the Atlantic region comprises the following MSFD subregions: the Greater North Sea (including the Kattegat and English Channel), Celtic Seas, and the Bay of Biscay and Iberian Coast. Macaronesian waters are treated as a separate marine region under the Nature Directives.

As summarised in Table 2.2, these marine regions differ considerably in their biophysical properties such as their depth, tides, currents, exposure to ocean waves and wind, salinity and nutrient levels. These diverse properties in turn give rise to varying habitats and species communities. It should be noted that most HD Annex I marine habitats are only very broadly differentiated, and therefore most occur in all marine biogeographical regions. As listing them would be of little value, most are not referred to in Table 2.2. It is particularly noteworthy that the Mediterranean Sea is considered to be a biodiversity 'hotspot' as it has high species diversity for a temperate sea (Coll *et al.*, 2010).

2.1.2 The Legacy of the Last Ice Age

The flora and fauna of Europe have been profoundly affected by the glacial episodes that occurred over the Quaternary period, the most recent of which commenced about 110 000 years ago, with its maximum extension occurring about 20 000 years ago. The ice covered all of Scandinavia and the Baltic States and extended south to Poland and northern Germany, but no further south-west than western Denmark. In the Alps, ice extended down to the plains. The island of Ireland was mostly covered with ice, as were the northern three-quarters of Great Britain. As the sea-level was about 100 m lower than now, Great Britain and Ireland were connected by dry land to what is currently continental Europe. To the south, most of the rest of Europe was covered in steppe and tundra, with wooded tundra or forest steppe, mainly in Iberia and to the south and south-

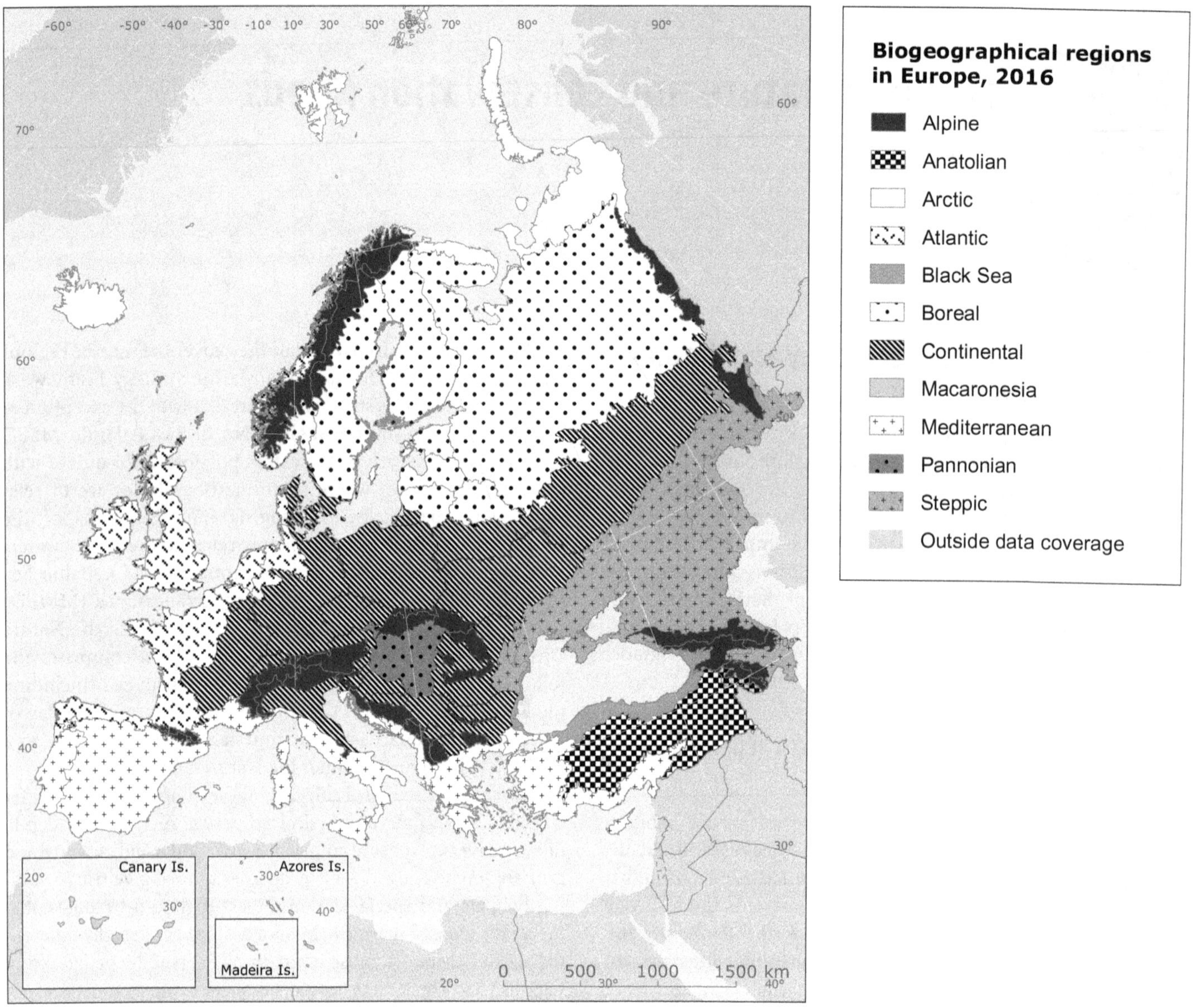

Figure 2.1 Terrestrial biogeographical regions in Europe.
Note. The Anatolian and Arctic biogeographical regions do not occur within the EU.
Source. Adapted from EEA (2017a).

east of the Alps. Mixed woodland was mainly confined to southern Italy, Greece and Mediterranean islands and the eastern Black Sea coast, but no woodland communities now present in central Europe were able to survive (Ellenberg, 2009).

With the warming of the climate and retreat of the ice about 10 000 years ago, surviving plants and animals moved north, recolonising some of their former areas. Following a succession of forest types, by 8 000–5 000 BP broad-leaved forests are thought to have become the climax vegetation across most of lowland temperate Europe. Traditionally it has been thought that this landscape was predominately closed-canopy forest, and that open habitats were limited to localised areas (e.g. too wet, dry, rocky or unstable). Open areas may also have been maintained by herbivores after fires or storms, but grazing animals did not create them by themselves.

More recently it has been suggested by Vera (2000) that in prehistoric times broad-leaved forests were more open than this, due to the activity of large herbivores, including deer, bison (*Bison* spp.), Beaver (*Castor fiber*), Wild Boar (*Sus scrofa*) and the now extinct Aurochs (*Bos primigenius*) and Tarpan/Eurasian Wild Horse (*Equus ferus ferus*). He has proposed that they created a half-open landscape with a cyclic shifting mosaic of scrub, closed-canopy stands (groves), degenerating groves and open ground with scattered trees and scrub. Some have suggested that these areas would have been similar to some pastoral woodlands in Europe today (e.g. the New Forest in the UK or *dehesa* in Spain).

Whilst the presence of half-open wood pasture is not supported by evidence (e.g. Svenning, 2002; Hodder *et al.*,

Table 2.1 The key characteristics of the terrestrial biogeographical regions in Europe.

Region	Key abiotic characteristics	Characteristic ecosystems and HD habitat types
Arctic (not present in EU)	Maritime subarctic (−3 to 11 °C) – continental subarctic (−11 to 13 °C). Extreme annual variation in sunlight gives short intensive growing seasons.	Tundra in the north and coniferous forest in the south, with numerous mires, oligotrophic (i.e. nutrient poor) lakes and rivers. Largest area of true wilderness in Europe.
Alpine (ALP)	Maximum altitudes >4 000 m in the Alps, >3 000 m in the Pyrenees, and >2 000 m elsewhere in Europe. Extreme climates, with high winds and precipitation (as rain and snow) at altitude, high seasonal, altitudinal and aspect temperature variations (e.g. −21 to +35 °C in the Carpathians), long-lying snow, ice and glaciers (although retreating), steep terrain, and extensive areas of bare rock and soil. Large areas of unbroken pristine habitat, especially in the Scandes (i.e. Scandinavian mountains).	Forests and semi-natural grasslands on the lower slopes give way at the tree line (c. 2 000 m, 1 000 m Scandes), to alpine grasslands, fells and scrub heath with increasing altitude, and rock and snow habitats with just a few specialist species. Forests are predominantly coniferous, or beech/beech-fir (e.g. in Carpathians), or have a Mediterranean character with oak or pine on south facing slopes of the southern ranges. Scandes predominantly montane scrub and tundra. A wide range of natural and semi-natural (e.g. meadows) grasslands occur, with high plant diversity and levels of endemism. Numerous rivers, but many are human altered.
Boreal (BOR)	Cool and mainly continental climate, with 500–800 mm precipitation per year, and temperatures between −15 and 20 °C. Relatively flat (mostly below 500 m). Coastline with numerous shallow inlets and islands.	Mostly managed coniferous forest (western taïga) but 16 HD forests types occur, seven priority habitats, e.g. old growth western taïga (9010*)[1], Fennoscandian deciduous swamp woods (9080*) and natural forests undergoing primary succession on coasts (9030*). Also numerous rivers, oligotrophic lakes and mires (peatland >50% of land in northern areas), including extensive Boreal aapa mires (7310*). Some important grasslands include northern Boreal alluvial meadows (6450) and, in the south, Nordic alvar (6280*), Baltic coastal meadows (1630*) and rare Fennoscandian wooded pastures (9070). Four other types of Boreal Baltic coastal habitats.
Atlantic (ATL)	Oceanic climate with mild winters, cool summers, predominantly westerly winds and moderate rainfall throughout the year. E.g. rainfall is about 3 000 mm per year in the mountains of north-west Scotland, but can be as low as 700 mm, even 550 mm in small areas in the east. Mostly lowlands, but some hills and mountains (mostly <1 000 m). Sheltered and exposed coasts with a variety of young soft sedimentary and hard rocks. Most lowland terrestrial habitats are now highly fragmented.	Forest cover is low, and endemic yew woodlands (91J0*), old sessile oak woods (91A0), and Caledonian forest (91C0*), which are confined to the region, are scarce. Large expanses of mire, mainly blanket bog (7130* if active), acid grasslands and dwarf shrubland predominate in the uplands. Lowland heathlands are characteristic of the region but are much reduced, as are lowland floodplain, acid and calcareous grasslands. Diverse coastal habitats, from muddy estuaries, salt marsh, sandy beaches and dunes, to steep tall cliffs and fjords. Machair (21A0* in Ireland) is unique to the region.
Continental (CON)	Strong contrast between cold winters and hot summers especially in the east (e.g. −3 to 19 °C in Warsaw). Precipitation highest at 1 700 mm/year in the German Black Forest, but lower on the plains, e.g. 500 mm/year in Warsaw. Mostly flat lowlands, with hills in the south that grade into some mountain ranges. Mostly fertile soils. Major rivers flow through the region (e.g. Danube, Loire, Rhine and Po), which are heavily modified.	The climax vegetation is mainly deciduous forest, typically dominated by beech (e.g. *Luzulo-Fagetum* 9110) or sometimes oak and other species, such as *Galio-Carpinetum* oak-hornbeam forests (9170). Forests are now greatly reduced. Other habitats include riverine forests, fens, marshes, and grasslands (including subalpine grasslands), but all are much reduced, especially in the west. Some large marshlands remain in Poland and further east. More local habitats include karst landscapes with caves, and inland dunes. A range of coastal habitats remain, e.g. shingle, salt meadows, lagoons, dunes and dune heaths.

[1] Numbers in brackets are the HD Annex I habitat codes (see Appendix for full names). * Indicates a priority habitat.

Table 2.1 *(cont.)*

Region	Key abiotic characteristics	Characteristic ecosystems and HD habitat types
Pannonian (PAN)	Dominated by the flat alluvial Great Hungarian Plain, which is almost completely enclosed by low-lying hills and mountains. The plain has mosaics of sand and loess and varying levels of water and salinity (due to high summer evaporation rates and shallow underground water sources). Average temperatures are −0.7 °C in January and 22 °C in July, and annual rainfall is moderate (700–800 mm), but the climate is complex and variable, as it is affected by neighbouring climate systems.	Forms a boundary between two vegetation belts: deciduous forests and forest-steppes, with a wide diversity of habitats and associated species (with high levels of endemism). The original, mostly oak, dominated forests now almost completely replaced by the extensive steppe landscape (the *Puszta*), with grasslands, including loess and sandy grasslands (e.g. 6250* and 6260*), but these are now scarce. Other localised characteristic habitats include Pannonic inland dunes (2340*), Pannonic salt steppes and salt marshes (1530*) and shallow alkaline lakes. In the hills, some woodlands occur (e.g. with *Quercus pubescens* 91H0*), as well as karst springs and calcareous grasslands.
Steppic (STE)	Mainly low-lying plains with some hills (e.g. 500 m in the Macin Mountains in Romania) and undulating high plateaus. Harsh continental climate with cold winters (−3 to −14 °C), hot summers (20–22 °C, but up to 30 °C), low precipitation (150–400 mm per year) and drying winds. The distinctive fertile black soils predominate.	Treeless steppic grasslands are the main natural vegetation type, i.e. dominated by grass species in the genera *Stipa*, accompanied by *Festuca, Agropyrum, Koeleria, Andropogon and Helictotrichon*, and other drought resistant plants. Due to the fertile soils, about 80% of the grassland has been converted to agricultural use. Woodlands occur in some damp valleys and hills.
Black Sea (BLS)	In Europe only occurs as a narrow coastal strip extending south from the Danube Delta in Romania, across low mountains in Bulgaria, to the Bosphorus outlet in Turkey. The continental climate is moderated by the sea, giving cool winters and warm summers (e.g. between −1 and 21 °C in Bulgaria). Annual rainfall is fairly low, e.g. 370 mm in Romania.	The Danube delta primarily consists of extensive reedbeds, lagoons, river channels, sandbanks and riverine woodland, but large areas have been drained and converted to agriculture. Other coastal habitats include brackish and saline lakes, and cliffs. Forests also occur outside the delta, especially on the low-lying hills in the south of Bulgaria and include a variety of rare HD habitats such as western Pontic beech forests (91S0*).
Mediterranean (MED)	The climate has a very strong influence, as it is particularly hot and dry in summer, often with long periods over 30 °C. Annual rainfall varies between 600 and 1 200 mm, but can be as low as 350 or even 100 mm. Winters are mild and humid (average 6 °C). Another characteristic is the considerable variability in weather conditions, including the influence of strong winds and rain storms. Fires are frequent, and have a strong influence on most habitats. Much of the region is hilly or mountainous, and altitude and aspect therefore have a high secondary influence on the climate. Soils lack organic matter and are prone to erosion. Most of the coastline is rocky or sandy.	A very wide range of natural and semi-natural habitats with high species diversity and exceptional levels of endemism. Sclerophyllous (evergreen hard-leaved) plants predominate, and forests are the main climax vegetation type. They normally have diverse tree species, but several HD types are dominated by evergreen oaks. Habitats with deciduous oaks and other trees and shrubs occur in less dry areas, and some distinctive conifer forests occur on mountains. Although not climax vegetation, sclerophyllous scrubland (*matorral*) is particularly characteristic and widespread. It includes tall dense shrubland (*maquis*) and vegetation with sparse dwarf shrubs (*garrigue* or *phrygana*). Semi-natural grasslands are also widespread, such as pseudo-steppe (6220*). Traditional grazing and cultivation of oak forests has created the highly distinctive *dehesa/montado* habitat (6310) of Spain and Portugal. Now much of the region is used for agricultural crops, including characteristic olive groves and vineyards. Wetlands, such as lagoons, are relatively localised.

Note. The Anatolian biogeographical region does not occur in Europe, and the Macaronesian region is not covered in this book.
Sources. Based on information in Polunin and Walters (1985), Sundseth (2009a–h), EEA (2008a) and Metzger *et al.* (2015).

Figure 2.2 Marine regions and subregions in the EU based on the Marine Strategy Framework Directive.
Note. The marine region boundaries are indicative only and do not imply any legal status. The Marine Atlantic region, as referred to in this book and used for HD Article 17 reporting, comprises three MSFD subregions: the Celtic Seas, Greater North Sea (including the Kattegat and the English Channel) and the Bay of Biscay and the Iberian Coast.
Source. EEA (2020a).

2005; Mitchell, 2005), it is now generally accepted that large herbivores probably affected forest structure more than previously thought, and that mixed landscapes occurred to some degree. However, this would have varied according to local conditions and the influence of other interacting disturbances, such as fire or floods. Vera's ideas have also triggered debate on whether conservation objectives should be based on a different, prehistoric landscape model and more naturalistic grazing as, for example, pioneered at Oostvaardersplassen in the Netherlands (see Chapter 23).

The Ice Age has had lasting impacts on the distribution and diversity of natural vegetation and associated animal communities across most of Europe. Compared to similar areas in North America and eastern Asia, Europe has fewer plants, especially trees, due to their wide extermination and the presence of the Mediterranean Sea, which has acted as a barrier to recolonisation (Ellenberg, 2009). European mountain ranges that are predominantly aligned west–east (i.e. Pyrenees, Alps and Carpathians) also had a similar effect, creating barriers that hindered species' southward retreat and then northward recolonisation. The diversity of native species in Great Britain and Ireland remains lower than on the continent, because some species were unable to return before the sea-level rose again and separated Ireland from Great Britain, and then Great Britain from the continent about 7 500 years ago. However, in some parts of Europe, plant and animal populations survived the last glacial periods in isolated refuges of suitable climate and habitat, and have evolved into new subspecies, thereby increasing biodiversity.

As the Mediterranean area escaped the worst impacts of the last Ice Age, much of its vegetation remained. Moreover, the

Table 2.2 Key characteristics of the Atlantic, Baltic Sea, Black Sea and Mediterranean Sea marine regions.

Region	Key abiotic characteristics	Characteristic ecosystems
Atlantic (i.e. Greater North Sea, Celtic Seas, Bay of Biscay and Iberian coast)	Includes deep water thousands of metres deep. Within the continental shelf depths are mostly <200 m, especially shallow in the North Sea. The north-west is influenced by the warm north Atlantic current, and so surface temperatures are typically 7–15 °C. Salinity is about 35 ‰ or above. Predominantly a high energy environment due to ocean waves and climate, high tidal ranges (up to 15 m) and strong currents. Mostly oligotrophic.	The varied abiotic conditions give rise to a wide range of habitats, including those that are characteristic of conditions of high exposure to waves and currents, as well as more sheltered bays, inlets and sea lochs. Whilst soft substrate habitats dominate most of the seabed away from the coast, there also physiographic features that form rocky reefs. The high productivity of the northern waters, especially the North Sea, supports high fish densities and large seabird populations.
Baltic Sea	Relatively shallow with almost no tide. Brackish with salinity of c. 6 ‰ in the main part but lower in the north and east. Intense seasonality in temperature (with northern areas ice covered in winter) and inflows. Stratification between low-salinity surface water and more saline water prevents the exchange of oxygen and nutrients, leading to naturally lifeless areas of seabed.	Mixed or soft substrate benthic habitats predominate, whilst rocky habitats occur closer to the coastline. Northern shores include numerous bays and inlets. Species diversity is low, but communities vary in response to the marked vertical and horizontal salinity gradients. Uniquely, there are areas where freshwater, brackish water and marine species are all present. Highest biodiversity in the south-west.
Black Sea	Tideless inland sea with only a narrow outlet to the Mediterranean. Highly stratified due to lower salinity surface layer (c. 18 ‰) overlying denser more saline water, which has created permanent anoxic conditions below 100–200 m. Surface water temperatures vary from 0 to 25 °C.	A highly specialised and sensitive marine ecosystem. The main habitats in shallow-water areas are more or less shelly, or sandy, with terrigenous muds. There are extensive biogenic reefs and Black Sea-specific 'fields' of the red alga *Phyllophora crispa*. Deep pelagic and benthic organisms are largely absent. The brackish nature of the water restricts the number of species that are present, most of which are also found in the Mediterranean Sea.
Mediterranean Sea	Average depth of 1 500 m, with 20% less than 200 m. Narrow continental shelf and littoral zone in the south, but wider in the north. Highly saline (average 38.5 ‰) and limited freshwater inflows. Mostly oligotrophic outside the coastal zone, especially in the east. Mean surface temperatures highly seasonal (16–26 °C) and no temperature boundaries at depth. Micro-tidal (mostly a few cm). Diverse environmental conditions.	Low primary production, combined with limited development of higher levels of the food chain, including low fish production, are characteristic. However, biodiversity is high (especially in the north-west) due to the variety of conditions and high endemism due to its partial isolation, estimated at 20%. Most biodiversity is concentrated in shallow coastal areas, including the characteristic seagrass beds, e.g. of *Posidonia oceanica* (1120*), and coralligenous reefs and maerl beds created by coralline red algae.

Sources. Based on information from EEA (2008b), Coll *et al*. (2010) and Gubbay *et al*. (2016).

region's natural vegetation is also highly diverse as a result of its geological history, diverse biophysical conditions and location at the intersection of three major landmasses. Approximately 25 000 species of flowering plants and ferns occur (including in non-European areas), compared with 6 000 species in non-Mediterranean Europe (Quézel, 1985). Furthermore, more than half of the plants of the Mediterranean basin are endemic, comprising 80% of the European total (Gomez-Campo, 1985). Consequently, in spite of profound human impacts (described later), the region is still considered to be a global biodiversity hotspot (Mittermeier *et al.*, 2004).

2.2 The Impacts of Human Activities on Biodiversity in Europe

2.2.1 Early Impacts and the Creation of Semi-natural Ecosystems and Cultural Landscapes

Since the arrival of humankind, European ecosystems and their species have been increasingly influenced by human activities, especially within the Mediterranean basin. This first became significant as the Neolithic agricultural transition spread from the south-east about 8 000 BP, arriving in the north-west around

5 000 BP (Isern *et al.*, 2012). Settlements were established and agriculture developed with the domestication of crops and livestock and the creation of wood pastures. More widespread clearance of forest for the creation of pastures, meadows and cropland occurred from the Iron Age. In some parts of north-west Europe, the grazing and repeated burning of forest combined with turf cutting, and resultant leaching of exposed soils in heavy rainfall areas, created and maintained extensive heathlands. Forest clearance, and a change in climate to wetter and cooler conditions around 2 500 years ago, also led to an increase in mires.

Hunting also had indirect impacts on ecosystems, through its effects on the populations of large herbivores and the extinction of some, such as the Aurochs. Domestic livestock may have replaced the ecological functions of some natural herbivores, as woodland pasturing was widespread until about the nineteenth century, but there were some gaps in grazing niches. To protect livestock, large carnivores such as the Wolf (*Canis lupus*) and Brown Bear (*Ursus arctos*) were exterminated in many countries, with knock-on impacts on their prey species and ecosystems.

Although forests and other natural habitats declined as early farming spread, the new and diverse semi-natural habitats, and their novel species communities, probably increased overall biodiversity (Pons and Quézel, 1985; Ellenberg, 2009). Diverse cultural landscapes were also created, as varied intertwined farming and cultural practices developed (Oppermann and Paracchini, 2012). As these diverse semi-natural agricultural systems evolved, so too did their vegetation and associated animal communities. New cropping systems and types of grasslands were formed, some in relatively modern times, such as litter meadows in the Alps in the nineteenth century (Poschlod *et al.*, 2009). Between the seventeenth and nineteenth centuries, major landscape changes widely occurred as a result of the enclosure of areas of common land. This gave rise to 'bocage' type landscapes with hedges and other field boundaries, as well as laws prohibiting forest grazing (Künster and Keenleyside, 2009).

2.2.2 The Industrial and Agricultural Revolutions – up to 1980

It is no exaggeration to say that the industrial revolution, which started in the late eighteenth century, and the later agricultural revolution have led to profound impacts on Europe's nature. The most significant and widespread have been changes in land use and management, particularly in agricultural systems. According to Jepsen *et al.* (2015), these were the result of three main drivers – technological, economic and institutional (e.g. land reforms) – and gave rise to similar stages of agricultural expansion, intensification and eventual industrialisation. However, the specific drivers and timing of these varied between countries, with initial intensification starting around 1850 in some (e.g. Belgium, West Germany and the Netherlands) as a result of the invention of the clay drain pipe, early machinery, the availability of some fertilisers, and railways increasing access to urban markets. By 1900, most of Europe was undergoing such initial intensification, the main exceptions being parts of Spain, Portugal and Italy.

Almost simultaneous major changes occurred across Europe from about 1945, as a result of World War II and the subsequent need to increase food production, which coincided with increased availability of mineral fertilisers, machinery and the use of irrigation. In most of Western Europe this resulted in further intensification, crop specialisation and large-scale farming (Potter, 1997), a period that is referred to as the industrialisation of farming by Jepsen *et al.* These developments were also supported by the establishment of the European Economic Community and its Common Agricultural Policy (CAP) in 1957. The CAP aimed to increase productivity, including by stabilising market prices and providing subsidies, which were initially linked to production, and grants to expand agricultural land and intensify management practices. Similar agricultural industrialisation occurred over much of the Eastern Bloc countries, driven by the political process of collectivisation (Jepsen *et al.*, 2015). However, as described in the country chapters, some mountain regions and other areas were less affected.

Much of the expansion of agriculture was at the expense of natural and semi-natural habitats (Baldock, 1990). For centuries, wetlands, especially peatlands, were targeted for drainage and conversion to agriculture, resulting in two-thirds being lost across Europe between 1900 and the mid-1980s (European Commission, 1995). Large areas of heathland and Mediterranean shrublands were converted to arable or permanent crops.

Agricultural intensification has followed similar pathways across Europe, especially in the west, but more slowly and less consistently in central and eastern Europe (e.g. Stoate *et al.*, 2001, 2009, Tryjanowski *et al.*, 2011; Kuemmerle *et al.*, 2016). Within grasslands, many of those that were wet were drained, and the use of nitrogenous fertilisers became common where it was possible to apply them using machinery, allowing increases in grazing intensity. To make best use of the fertiliser, grasslands have also been increasingly reseeded with high-yielding rye-grass cultivars (e.g. *Lolium perenne*), creating taller and denser species-poor swards. These reseeded grasslands are subject to higher grazing densities and/or conversion from hay fields to silage crops that are cut several times and earlier in the year than hay.

Industrial large-scale specialised crop production also resulted from increased usage of fertilisers, herbicides and other pesticides. This led to denser crops, reduced crop rotations, the near disappearance of fallow in the landscape and the removal of hedgerows and other field boundaries to create larger fields. Some pesticides had severe direct toxic effects on non-target species, in particular birds of prey, the most notorious being DDT[2] from the 1950s, until national bans

[2] The persistent organochlorine insecticide dichlorodiphenyltrichloroethane.

started in the late 1970s. Since then, the main impacts of herbicides and pesticides have been their disruption of food webs, such as declines in broad-leaved weeds and invertebrates.

In contrast to the dominant trends of agricultural expansion and intensification, agriculture has been abandoned in some areas. This has mainly affected areas with poor and remote agricultural land, especially in the hills and mountains of southern and eastern Europe (Pointereau *et al.*, 2008; Keenleyside and Tucker, 2010). This has had varying, but probably mostly detrimental, impacts on biodiversity so far (as well as declines in cultural landscapes, rural traditions and communities). On the one hand, abandonment has reduced human disturbance and increased scrub and forest habitat area and connectivity, to the benefit of some species such as large carnivores. On the other hand, abandonment has led to substantial declines in biodiverse semi-natural grasslands, shrublands and other open habitats, and their associated threatened species. According to a global review of studies of agricultural abandonment, in Europe the impacts were reported as negative in 65% of studies, whilst they were considered to be positive in only 6% (Queiroz *et al.*, 2014). For example, in the Alps, whilst pastoral abandonment may increase bird diversity, this is to the benefit of common species whilst it is detrimental for the more specialist and threatened grassland specialists (Laiolo *et al.*, 2004). A broader study in Sweden of fungi, vascular plants and insects (Lepidoptera, Coleoptera and Hymenoptera) suggested that abandonment of traditional management in agricultural landscapes would lead to extinction rates two or three orders of magnitude higher than global background rates (Eriksson, 2021). A further problem is that, whilst the abandonment of some areas could be beneficial in the long term, producing high biodiversity habitats, abandoned areas have often been targeted for forest plantations (and bioenergy crops more recently).

The planting of forests for timber production began at the end of the nineteenth century and was widespread across Europe in the 50 years following World War II, increasing forest cover by 30% in western Europe, 20% in central and eastern Europe, and 10% in the south (Gold, 2003). This was often concentrated in the uplands or in other areas that were unfavourable for agriculture, such as on dunes, shrubland, heathland and drained peatland. Large areas of valuable natural and semi-natural habitat were lost in the process, often replaced with even-aged forest monocultures of low biodiversity value. Furthermore, in many cases plantations consist of non-native trees, most commonly conifers (especially species of *Pinus*, as well as *Pseudotsuga*, *Picea* and *Larix*), poplars (*Populus* hybrids) and Australian eucalyptus species. These trees typically support very few native species, and so result in a very simple and species-poor ecosystem.

Within ancient forests, traditional uses such as pasturage, pollarding, charcoal production and coppicing have been largely replaced by forest management for timber. This led to a change in woodland structure, with taller mature trees and a closed canopy. Since then there has been a trend towards increasingly intensive forest management over much of Europe. Thus, forests that are commercially managed have been drained, thinned, cleared of deadwood, clear-cut over large areas and replanted. This has resulted in biodiversity losses, as biodiversity declines considerably with increasing management intensity (Sing *et al.*, 2018).

Physical changes to other ecosystems have included the creation of reservoirs for water storage or hydropower, widespread canalisation of lowland rivers (for river transportation and flood protection), reduced water levels in wetlands due to abstraction (often for agriculture), and construction of coastal flood defences (with resulting losses of intertidal habitat).

Alien species have had important influences on terrestrial and marine species, communities and ecosystems since humans settled across Europe and started introducing new species, such as for food, materials or sport. This has accelerated since the beginning of the twentieth century, due to increasing intentional imports of plants and animals (such as for horticulture and the pet trade) and accidental introductions associated with increasing international travel and trade. Many alien species have not spread widely or had noticeable detrimental effects on native species, or other aspects of the environment. However, a substantial proportion have become invasive and had significant environmental impacts; these are hereafter referred to as invasive alien species (IAS).[3] Detrimental impacts have included competing with or consuming native species, spreading disease, causing genetic changes through interbreeding, and disrupting food webs and the physical environment (Scalera *et al.*, 2012).

Pollution has affected most ecosystems since the beginning of the industrial revolution, particularly in the most densely populated and industrialised areas of western and eastern Europe. The most severe and widespread biodiversity impacts have generally resulted from nutrient enrichment, that is, eutrophication (e.g. Galloway *et al.*, 2004; Fowler *et al.*, 2013), and ecosystem acidification (e.g. Schöpp *et al.*, 2003). Apart from in upland areas, a high proportion of rivers and lakes have been affected by eutrophication, primarily from increases in phosphorus, with the main sources being silty agricultural run-off, sewage and industrial effluent. Whilst eutrophication impacts are complex, low nutrient enrichment levels generally lead to increases in submerged plants, which increases the productivity of the ecosystem, benefiting some fish and birds. At higher levels, major changes in the ecosystem and its species result as algae tend to proliferate and submerged plants die out due to the reduced light levels. At very high levels, algal blooms can lead to oxygen depletion and the death of fish and other animals. Eutrophication impacts can be long-lasting and difficult to reverse where nutrient-rich sediments have built up.

Air pollution increased substantially due to the growth of industry and cities, and their use of coal, and later the invention of the internal combustion engine and the industrialisation of agriculture. This has led to major changes in the physiochemical conditions of sensitive terrestrial and aquatic ecosystems, and

[3] As in the Regulation (EU) 1143/2014 on invasive alien species.

impacts on their species, across much of Europe (Stevens *et al.*, 2020). High concentrations of some pollutants in the air, such as sulphur dioxide, have caused the widespread destruction of lichen communities since the nineteenth century. Changes have also resulted from the deposition of air pollutants dissolved in rain or snow. This deposition, colloquially termed 'acid rain', acidifies the soil or water (if the dissolved pollutants are sulphur dioxide, nitrogen oxides or ammonia), as well as causing eutrophication through nitrogen enrichment (if the dissolved pollutants are nitrogen oxides or ammonia).

Sensitive ecosystems, such as rivers and lakes, have been significantly affected by acidification in many parts of Europe since the 1950s, especially in regions where the soils and rocks have a low buffering capacity, such as much of Fennoscandia. This resulted in wide-ranging impacts on the ecosystems and their species, including declines in species diversity and acid-sensitive groups (e.g. molluscs and amphipods) leading to massive declines in fish diversity and numbers (Muniz, 1990). Acid rain also led to the decline of coniferous forests in some mountain areas in central Europe, particularly in the Czech Republic, Germany, Poland and Slovakia (Stanners and Bourdeau, 1995). Eutrophication has mainly affected naturally low-nutrient ecosystems on acidic soils, such as mires, acid grasslands and heathlands, especially when close to areas with high livestock densities as they are a prime source of ammonia emissions (e.g. Bobbink *et al.*, 2010; Sutton *et al.*, 2011). Typically, the biodiversity value of exposed vegetation declines as the characteristic specialist species tend to be sensitive to the pollutants and outcompeted by more competitive species, such as grasses. Consequently, eutrophication is one of the most widespread and significant threats to plant species richness in natural and semi-natural ecosystems (Stevens *et al.*, 2010).

Whilst not as obvious as on land, human actions have had major impacts on the structure, properties and species communities of marine ecosystems. Most have been profoundly affected by overfishing, which has occurred historically in all EU regional seas (Jackson *et al.*, 2001). This has changed marine food webs, affecting species composition and abundance, and incidental catches of non-target species have increased the magnitude of such changes. For example, the depletion of lower trophic level species such as sand-eels, sardines and herring can result in declines in upper trophic level predators such as larger fishes, seabirds and marine mammals. Furthermore, intensive bottom-trawling and dredging for shellfish regularly disturbs large areas of seabed and damages sensitive habitats such as biogenic reefs, which can then take many years to recover, if at all (EEA, 2015). Some fish species, birds and cetaceans have also been affected by high levels of by-catch.

All seas have also been altered by pollution, especially the semi-enclosed Black and Baltic Seas, and parts of the North Sea, which have been heavily affected by eutrophication (e.g. increasing oxygen depletion, algal blooms and the death of fish and benthic fauna). In the marine environment, this mainly results from nitrogen enrichment (or phosphate in low-salinity waters), the main sources being sewage, agricultural run-off and atmospheric pollution. The other main pollutants have been synthetic organic compounds such as polychlorinated biphenyls (PCBs) (which have built up to levels that have harmed some species such as marine mammals and seabirds), oil, litter and, to a lesser extent, heavy metals (Stanners and Bourdeau, 1995).

2.2.3 The Situation since 1980

This book mainly focusses on nature conservation over the last 40 years, and the most recent decades in particular. Therefore, these periods are examined in more detail in the national chapters, and a brief overview is presented here to highlight the most widespread and severe pressures. Importantly, the main pressures on biodiversity have changed with time, and conservation strategies have therefore had to adjust accordingly. Much of the text in this section is based on the *2010 Biodiversity Baseline* (EEA, 2010) and six EEA *European Environment – State and Outlook* reports produced since 1995, especially the most recent (EEA, 2019) but also starting with the first, known as the *Dobříš Assessment of Europe's Environment* (Stanners and Bourdeau, 1995). In addition, the text draws from some chapters from the International Panel on Biodiversity and Ecosystem Services (IPBES) regional assessment report for Europe and Central Asia (Rounsevell *et al.*, 2018; Visconti *et al.*, 2018). Additional sources for marine pressures include EEA (2015), Gubbay *et al.* (2016) and Vaughan *et al.* (2019).

A particularly important change is that the overall expansion of agricultural area has largely stopped, as a result of the productivity increases resulting from intensification, socio-economic changes and increasing imports of some foods and other commodities. CORINE land cover (CLC) data indicate that net changes in land cover have been mostly very small. In 27 European countries (and Turkey), between 1990 and 2000, there were annual declines of 0.04% for both cropland and pastures, 0.05% for semi-natural vegetation and 0.07% for wetlands (EEA, 2017b). The rate of forest expansion also slowed, resulting in a net annual increase of just 0.02%, in part as a result of agricultural abandonment and natural regeneration. The main changes in land cover resulted from the expansion of housing, industry and infrastructure (i.e. artificial areas), which increased by 0.5% per year.

More complete European CLC data are available for 2000 to 2018, and these are summarised in Table 2.3 (excluding Turkey in this case). For a more detailed analysis by the EEA, see Petersen *et al.* (2021). Some CLC trends were similar to the 1990–2000 period, including the relatively small net declines in agricultural land (although less so on cropland outside the EU). The rate of wetland loss declined further, to be probably stable. According to the CLC data in Table 2.3, there was a small decline in forest area. However, EEA analysis suggests it has been stable (Petersen *et al.*, 2021), and a Forest Europe (2020) analysis has indicated an ongoing small increase in the area of forest and other wooded land of 0.3% per year for Europe over 1990–2020. These discrepancies may be due to differences in the way in which forest, transitional woodland and scrubland are categorised.

Table 2.3 CORINE land cover areas (km²) and changes in Europe between 2000 and 2018.

	EU-28 (European territories*)					Other EEA-39 countries				
	2000	**2018**	**% in 2018**	**Annual change**	**Annual % change**	**2000**	**2018**	**% in 2018**	**Annual change**	**Annual % change**
Forest	1 403 487	1 374 813	31.4	−1 593	−0.114%	194 903	190 315	28.2	−255	−0.131%
Crops	1 223 088	1 216 224	27.8	−381	−0.031%	43 308	43 243	6.4	−4	Stable?
Grass and mixed	777 275	771 824	17.6	−303	−0.039%	71 687	71 169	10.5	−29	−0.040%
Other nat/semi	545 036	572 466	13.1	1 524	0.280%	305 046	309 221	45.7	232	0.076%
Wetland	113 915	113 521	2.6	−22	−0.019%	30 163	30 108	4.5	−3	Stable?
Inland water	107 978	109 291	2.5	73	Stable?	19 751	19 780	2.9	2	Stable?
Artificial	208 884	221 612	5.1	707	0.349%	11 121	12 155	1.8	57	0.517%
Total area	4 379 662	4 379 752				675 979	675 991			

Note. * EU European territories exclude those in the Macaronesian biogeographical region. The other EEA countries covered are: Albania, Bosnia and Herzegovina, Iceland, Kosovo, Liechtenstein, Montenegro, North Macedonia, Norway, Serbia and Switzerland. 'Grass and mixed' comprises pastures and heterogeneous agricultural areas. 'Other nat/semi' are other natural and semi-natural habitats, and mostly comprise scrub and shrubland, but also sparsely vegetated areas, moorland and heathland, and natural grasslands. 'Wetland' includes inland and coastal. 'Stable?' indicates that the scale of change is insufficient to be certain of trends according to the thresholds used by Petersen *et al.* (2021).
Source. Based on EEA CORINE land cover and change statistics 2000–2018[4] (downloaded 31 December 2020).

The area of combined CLC categories for 'other natural and seminatural' increased slightly, mainly in the EU. More detailed analysis of CLC data shows that this was probably partly due to further agricultural abandonment, as there was an increase in transitional woodland-shrubland of nearly 11% over the same period in the EEA-39 (Petersen *et al.*, 2021). The same study shows that, over the same time, there were small declines in sclerophyllous vegetation (−2%), natural/semi-natural grassland (−0.7%) and moors and heathland (−0.4%). However, it is important to note that these statistics do not capture all losses of semi-natural habitats, as many will be too subtle and/or small-scale to be detected using CORINE remote sensing methods and classes.

The growth of urban areas, infrastructure and other artificial areas continued at a similar rate to the 1990–2000 period. Although the area converted is relatively small, the losses can be considered permanent and often targeted areas of low agricultural value. Thus, impacts have tended to be disproportionately high on semi-natural habitats and species. Urban sprawl and the spread of roads and other infrastructure in the countryside have also contributed to further habitat fragmentation and increased human disturbance of wildlife.

Although farmland use has not expanded in most parts of Europe, widespread major impacts have continued from previous and further **agricultural improvements and intensification**, especially in western Europe. Most semi-natural grasslands have been lost through fertilisation and other agricultural improvements, and now a high proportion of lowland grasslands are temporary sown-grass monocultures. This has led to the almost complete loss of lowland hay fields, and declines in pastures, as livestock are now often kept in stockyards for most, or even all, of the year. The total area of grasslands has also decreased as they have been converted to crops for human food, livestock feed (e.g. maize) or, in recent decades, bioenergy. Low-intensity arable crops, which had relatively high levels of biodiversity, have also almost completely disappeared, apart from in parts of Iberia and eastern Europe (Hoffmann, 2012). Within intensive cropland the rate of intensification may have declined, but new biodiversity impacts have arisen. Most notably, neonicotinoid insecticides have been found to have widespread chronic impacts upon invertebrates, especially bees and other pollinators, and aquatic insects, which appear to be especially susceptible (Hladik *et al.*, 2018). Farming has also become more specialised, such that mixed farming is much less common. Over much of lowland Europe, these changes in farming have resulted in a strong decline in landscape diversity, with scarce and fragmented patches of remaining semi-natural vegetation (Jongman, 2002).

The lasting compounded impact of all these agricultural improvements has been the widespread impoverishment of wildlife in European agricultural landscapes, especially in the West, as the biodiversity value of habitats decreases in proportion to its degree of modification. The decline in semi-natural components in the landscape has been particularly detrimental, as they contribute most to overall species richness (Hoffmann *et al.*, 2000, cited in Billeter *et al.*, 2008; Oppermann and Hoffmann, 2012).

Evidence of the biodiversity impacts of this progressive intensification of agriculture comes from long-term monitoring of birds that has revealed substantial declines in farmland bird

[4] www.eea.europa.eu/data-and-maps/dashboards/land-cover-and-change-statistics

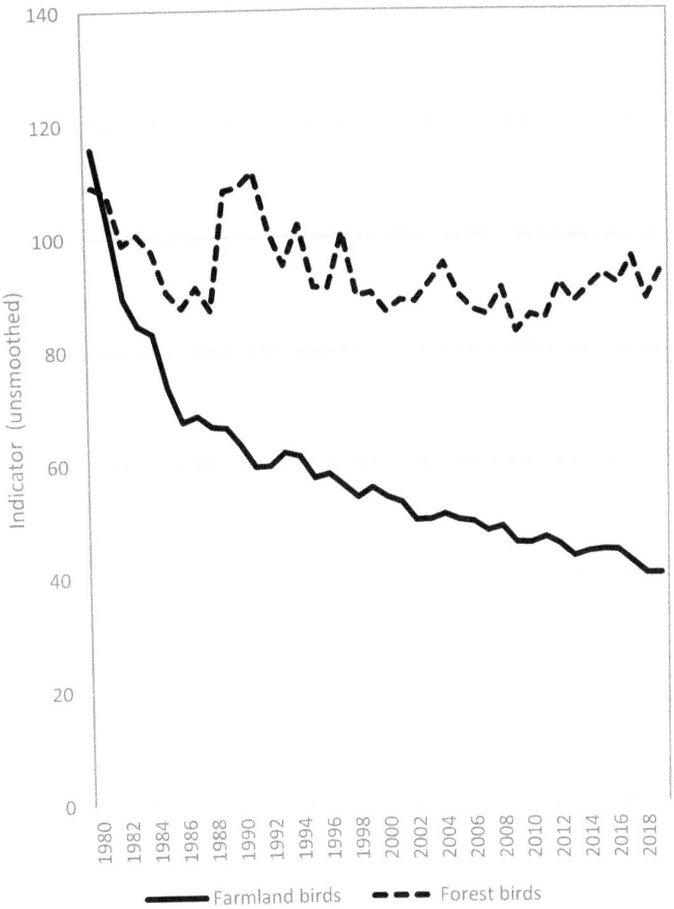

Figure 2.3 Common bird indicator values for farmland and forest species in Europe.
Notes. 1980 base year. Based on 28 countries' data. See the PanEuropean Common Bird Monitoring Scheme for methods, and included countries and indicator species (https://pecbms.info/trends-and-indicators/).
Source. EBCC/BirdLife/RSPB/CSO (2022).

populations (Figure 2.3). Between 1980 and 2000 the European farmland bird index fell by 53%. Whilst the rate of decline has since slowed, the index decreased by another 11% by 2019. Analysis of the bird trend data by Donald *et al.* (2001, 2006) revealed that the declines were closely associated with indicators of agricultural intensification. The authors also noted in 2001 that declines and intensification had been highest in the EU at the time, and therefore predicted similar agricultural trends and resulting bird declines in countries joining the EU.

For some time, there have been growing concerns in the EU over the environmental impacts of agriculture and the need to address overproduction; in the 1990s this unease stimulated a process of gradual changes towards a more sustainable CAP (Robson, 1997; Jepsen *et al.*, 2015). These changes included the replacement of most production-based subsidies with area-based subsidies, and the introduction of environmental support measures, as described further in Section 4.3 and online Annex 3.[5]

In the former Eastern Bloc, around 1990, more sudden and very different changes occurred in agriculture. These were a result of the collapse of communism, and led to state farms and collectives being dissolved, with the land distributed to private owners. This initially resulted in two diverging effects on land use. Much of the land was subject to commercialisation and agricultural intensification, especially where bought up by large agri-companies. In contrast, elsewhere de-intensification occurred, or even abandonment where the new owners were absent or uninterested in farming. For a while a large proportion of former farmland remained unused for agriculture, estimated to be 15–20% of cropland in Slovakia, Poland and Ukraine (Keenleyside and Tucker, 2010).

In more recent decades, the region has been characterised by rapid economic and social development and urbanisation, especially in countries that have joined the EU, such that they increasingly resemble those in Western Europe (Rounsevell *et al.*, 2018). Whilst semi-natural habitats and biodiversity levels had been much higher in Eastern Europe, especially in agricultural areas that had escaped collectivisation (Tryjanowski *et al.*, 2011), this difference has been gradually reduced. Despite the CAP's environmental reforms, agricultural intensification and industrialisation, with associated biodiversity declines, have continued over the EU, and especially in new Member States, as the CAP and other EU support measures have helped them 'catch up'. Evidence of this comes from bird monitoring in the Czech Republic, where EU accession was followed by increases in agricultural intensification that correlated with steep declines in bird populations (Reif and Vermouzek, 2019), as predicted by Donald *et al.* (2001).

Economic growth in the former Eastern Bloc has also led to the loss of some remaining wetlands and other semi-natural habitats, particularly along the coasts as these have been the focus of many tourist developments. This has long been the case in much of western and southern Europe, where recreational pressures have damaged coastal habitats and resulted in declines in sensitive species such as beach-nesting birds and turtles.

Within **forests** the main biodiversity impacts over the last 40 years have resulted from forestry management, as about 80% of forest area is available for wood supply. This supply has been primarily used for timber and pulp, but increasingly for fuelwood, which in 2018 accounted for 22% of roundwood use (EEA, 2019). At its most intensive, forest management has included the conversion of semi-natural forest to even-aged plantations, with profound ecosystem changes and biodiversity losses. Calculation of the proportion of European forests that is intensively managed plantations (including from afforestation) is difficult as it depends on the interpretation of definitions. The EEA (2016) estimated that plantations constitute 9% of forest area in the EEA-39 countries (based on FAO 2015 data), whilst a much lower estimate of 3.8% is given for the larger area assessed by Forest Europe (2020).[6] It should be

[5] All annexes are free to access online at CUP: www.cambridge.org/natureconservation

[6] Forest Europe data cover all of Europe, except for the Russian part, and include Georgia and Turkey.

noted that these estimates exclude planted forests that have not been subject to forestry operations for a long time. It is arguable whether this is justified, as such forests are still likely to be dense monocultures with a highly artificial even-aged structure.

About half of plantations are dominated by introduced species (as described earlier). However, the overall proportion of forest dominated by introduced tree species is much larger: estimated to be 31% according to Forest Europe (2020) for 31 European countries, and as much as 48% for the EU-28 and 41% for the EEA-39.

Most of Europe's forest is now considered to be semi-natural[7]: estimated to be 87% by the EEA (2016) and as much as 94% by Forest Europe (2020). However, these figures seem to be overestimates given the proportion of forest that is dominated by non-native species. Furthermore, 60% of forests in Europe are even-aged, and the proportion of old forests (>100 years) has decreased substantially in some northern and western EEA countries (EEA, 2016). These are all signs that the majority of so-called semi-natural forests have in fact been relatively intensively managed, with the resulting biodiversity impacts discussed earlier (Sing *et al.*, 2018). In recent years more sustainable forest management practices have been widely adopted, and there are some indications of slight improvements in the ecological condition of forests, such as small increases in deadwood, which is of particular importance for biodiversity (EEA, 2016; Forest Europe, 2020). The common forest bird indicator has also increased slightly since 2009 (see Figure 2.3). Nevertheless, most 'semi-natural' forest ecosystems are far below their potential biodiversity value, and thus need more ambitious environmentally sensitive forestry practices to increase their naturalness and undisturbed areas.

In contrast to most forests, some semi-natural forest areas have suffered from neglect following the abandonment of some traditional forms of forestry (e.g. coppicing), leading to forests with low structural diversity. High population densities of deer in some countries have resulted in little tree regeneration. Forests are also increasingly being affected by climate change, including increased temperatures, droughts, fires and extreme weather events. These effects are reducing forest ecosystem condition and resilience to pests, diseases and IAS.

Some reductions in **air pollution** have occurred across much of Europe, most notably sulphur with rapid substantial declines in deposition since the 1970s (Fowler *et al.*, 2007). Nevertheless, the deposition of airborne nitrogen across Europe as a whole only declined slightly between 1980 and 2003 (Fagerli and Aas, 2008). As a result, in 2005, 67% of European ecosystems and 78% of Natura 2000 sites were exposed to nitrogen deposition exceeding their critical loads[8] (EEA, 2019). More recent data indicate that ammonia emissions from agriculture only slightly decreased between 2000 and 2013, and then increased by about 3% by 2017 (EEA, 2019). However, ammonia emission levels vary greatly between countries, being highest in those with high-density intensive livestock production (e.g. see the Netherlands, Chapter 23).

Some progress has been made in tackling **water pollution** and other pressures on **rivers, lakes and other wetlands**, in part driven by better technology but also EU policies and legislation (described in Section 4.3). The main improvements have been in relation to point sources of pollution (e.g. sewage and industrial discharges) and nitrogen surpluses from agriculture. Despite these improvements, by 2018 only 40% of Europe's surface water bodies had achieved a good ecological status and wetlands remained widely degraded (EEA, 2018, 2019). The main pressures constraining further improvements are diffuse pollution, water abstraction (although it has declined) and hydromorphological changes, such as dams or river channel re-engineering. Over recent decades, an increasing pressure on rivers has been the growth in hydropower, and especially the cumulative effects of micro-hydro schemes on river hydrology and migratory fish (European Commission, 2018a).

Another biodiversity pressure that has particularly affected wetland species has been the poisoning of birds, and sometimes other animals, from the use of lead ammunition for hunting (Watson *et al.*, 2009). This primarily affects waterbirds that ingest some of the large amounts of lead shot that are now in the environment as a result of the use of shotguns. As a result of this it has been estimated that approximately one million wildfowl (representing 17 different species) or 8.7% of the total population may die every winter from lead poisoning caused by ingestion of lead gunshot in Europe (Mateo, 2009). Predators and scavengers, such as eagles and vultures, may also suffer secondary poisoning if they consume the flesh of animals that have been killed or wounded with lead shot or bullets.

In the **marine environment**, the main human impacts on ecosystems since 1980 have continued to be from fisheries and pollution, although the pressures have declined in some areas. Decreased fishing pressure in the north-east Atlantic Ocean and the Baltic Sea in recent years has led to signs of recovery of many stocks, but overfishing has continued in the Mediterranean and Black Seas (EEA, 2019). Furthermore, by-catch of non-target fish, seabirds and marine mammals remains of concern, and intensive fishing activities frequently continue to affect benthic habitats in many areas. This has been revealed by a spatial analysis of fishing activities, which found that, in six[9] out of nine studied areas in Europe, more than 25% of the seabed was trawled each year (Amoroso *et al.*, 2018).

[7] That is, they are influenced by forest operations but retain the characteristics of natural forest ecosystems with regard to their structures and functions.

[8] That is, a quantitative estimate of an exposure to one or more pollutants below which significant harmful effects on specified sensitive elements of the environment do not occur according to present knowledge.

[9] The Adriatic, west of Iberia, Skagerrak and Kattegat, Tyrrhenian Sea, North Sea and western Baltic.

Some marine pollutants have declined locally, particularly from point sources but less so from diffuse sources, which primarily come from agriculture. Despite the reductions in nitrogen and phosphate pollution, the Black Sea and Baltic Sea remain eutrophic. Some toxic contaminants have remained at high levels, such as certain heavy metals and PCBs (despite being banned decades ago), and marine litter levels have increased. The PCB levels are of particular concern, as they accumulate in high-level predators, and there is evidence that they are threatening the survival of the Killer Whale (*Orcinus orca*) in parts of the region (Desforges *et al.*, 2018).

Marine ecosystems have been historically affected by IAS, many of which arrived attached to ships or in their ballast water, whilst some were deliberately introduced, such as the Pacific Oyster (*Magallana gigas*). In recent decades there has been a noticeable acceleration in the occurrence of IAS in European seas (EEA, 2015), and these are increasingly affecting marine ecosystems, especially in the Black Sea and the Mediterranean Sea. Whilst the impacts of many IAS are uncertain, some have led to obvious ecosystem changes. Perhaps the most serious change arose from the predatory comb jelly *Mnemiopsis leidyi*, which became super-abundant in the Black Sea in the 1980s. In combination with over-fishing pressures, this disrupted the food web sufficiently to trigger a regime change in the ecosystem,[10] with a collapse in fish stocks (Möllmann and Diekmann, 2012). Since then, its population has declined and predation by another alien comb jelly species, *Beroe ovata*, which arrived in 1999, has meant that the Black Sea ecosystem shows signs of recovery (Vaughan *et al.*, 2019). The number and impacts of IAS are still increasing on land, as well as in the sea, despite increased legal and social responses in recent years (Rabitsch *et al.*, 2016).

The direct effects of **climate change** (e.g. increases in temperatures, changes in precipitation, and increases in extreme weather events) and indirect effects of human responses (e.g. renewable energy and infrastructure to mitigate impacts such as flood defences) have become much more significant over recent decades. The sea has also become more acidic as a result of increasing carbon dioxide concentrations in the atmosphere. In European seas, observed ecosystem impacts have included changes in species distributions and the timing of biological events. Such changes have added to the other existing pressures, including over-fishing and eutrophication, and led to regime shifts in the Baltic, Mediterranean and north-east Atlantic Seas at the end of the 1980s/early 1990s (Möllmann and Diekmann, 2012; EEA, 2015).

Terrestrial impacts of climate change have mostly not been so severe, but there is evidence of growing disruption of ecosystems in Europe by increasing temperatures, changing rainfall patterns, droughts, fires, flooding, other extreme weather events and sea-level rise (EEA, 2017c; IPCC, 2021). In freshwater lakes and streams, higher water temperatures and darkening of water colour or 'browning' by elevated concentrations of dissolved organic carbon may lead to serious ecosystem impacts (e.g. Weyhenmeyer *et al.*, 2016). These climate effects, and the knock-on complex impacts on species interactions, are already leading to changes in species distribution and declines in some vulnerable species (Hickling *et al.*, 2006; Ockendon *et al.*, 2014; Martay *et al.*, 2017). Such impacts can be confidently expected to increase due to the further climate changes that are already inevitable from greenhouse gas emissions so far. In fact, Europe is expected to be especially affected as temperatures in Europe are predicted to continue increasing faster than the global average all through the twenty-first century under all scenarios (IPCC, 2021). However, the severity and nature of the impacts will vary regionally, with the highest temperature increases in north-eastern Europe and Scandinavia during winter, and in southern Europe in summer (Jacob *et al.*, 2014).

2.3 Europe's Remaining Biodiversity and Its Priority Conservation Needs

2.3.1 What Is Left and Where Is It?

To place the following section and country chapters in context, Figure 2.4 indicates the proportion of each EEA-39 country that comprises forests, wetlands, other semi-natural habitats (e.g. natural grasslands, heathlands, shrublands and rocky habitats), agricultural land and artificial areas, based on CLC data. This has been ordered in relation to the proportion of 'other semi-natural habitats', as these are of particular importance for biodiversity in most countries. As further discussed below, a high proportion of forest habitat is also considered to be semi-natural, as are some areas of agricultural land, but it is not possible to separate them out using CLC data.

No ecosystem in Europe is pristine; all are affected by the legacy of past human actions (e.g. extinction of key species, drainage and eutrophication) and ongoing modern day pollution (nutrients, toxins and plastic, etc.). Nevertheless, despite the profound changes to ecosystems and landscapes, described in Section 2.2, Europe retains a relatively high diversity of habitats and associated species (especially in the Mediterranean region). Extensive areas of some near-natural ecosystems[11] remain, such as some rivers, lakes, tundra, mires, grasslands, karst features and caves, cliffs, screes, dunes, salt

[10] That is, a sudden, persistent reorganisation of the structure and function of the ecosystem.

[11] That is, those that would exist in the absence of human actions and show low levels of human impacts.

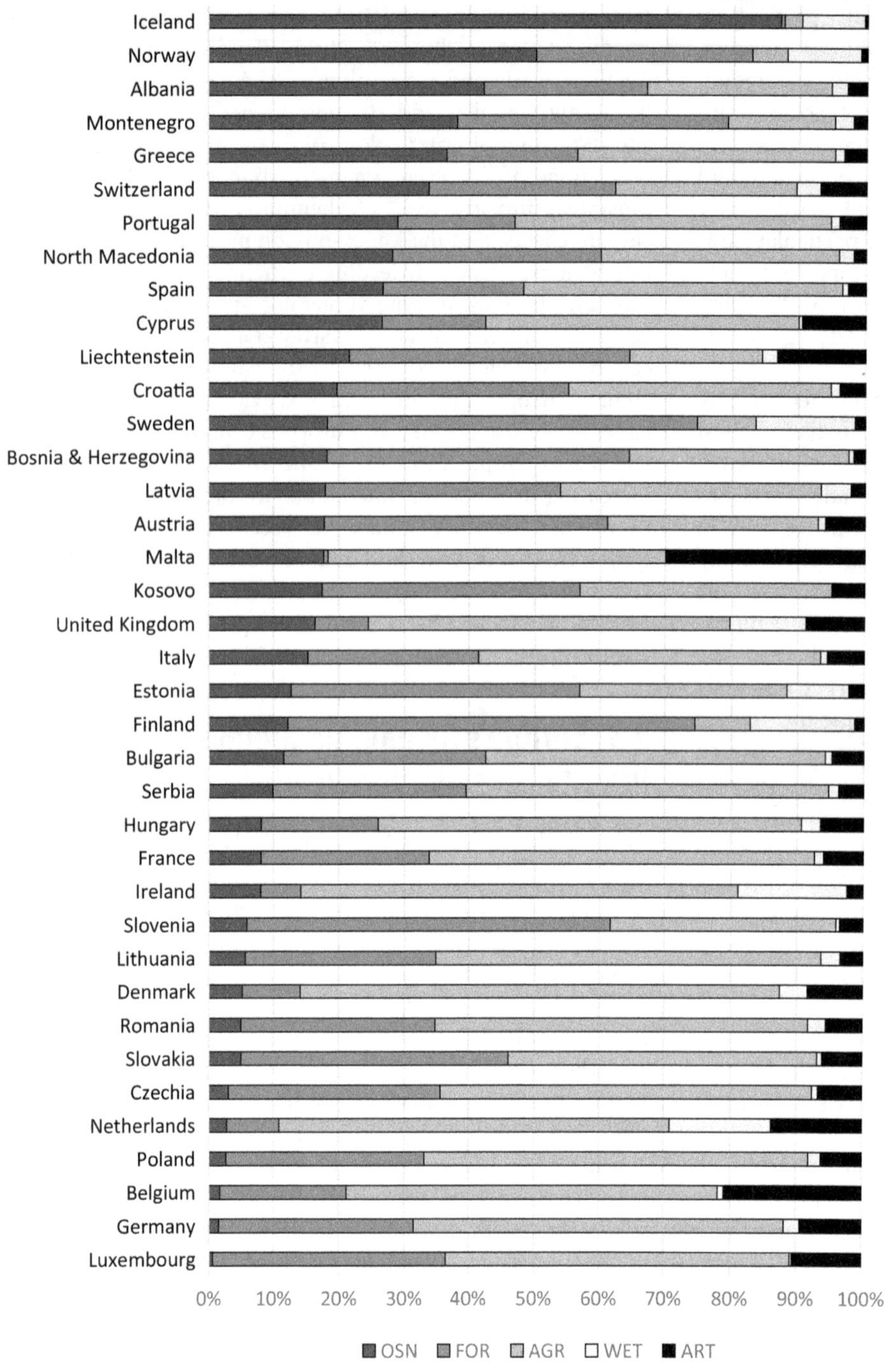

Figure 2.4 Proportional land cover in European countries in 2018 according to CORINE land cover classes.
Key. OSN = other seminatural (e.g. shrublands, rocky habitats); FOR = forests; AGR = agricultural habitats; WET = inland and coastal wetlands and inland water bodies; ART = artificial areas (see Table 1.1 for details). Covers EEA-39 countries other than Turkey, and excludes areas within the Macaronesian biogeographical region.
Source. Based on CORINE Land Cover Level 2 data for 2018, downloaded from the EEA www.eea.europa.eu/data-and-maps/dashboards/land-cover-and-change-statistics (31 December 2021).

marshes, mudflats, beaches and diverse benthic habitats – although many of these are remnants that are mostly confined to remote northern latitudes, high mountains and offshore marine areas. Some near-natural forests also occur, but they form a tiny proportion of Europe's forest area and are mainly confined to the Boreal region, as well as mountains in central and eastern Europe. According to the EEA (2016) and Forest Europe (2020), such undisturbed forests[12] have been estimated to comprise just 2% of European forests. Sabatini *et al.* (2018)

[12] That is, naturally regenerated stands of native species, with natural dynamics (which requires sufficient area and natural structures). They are always of high nature conservation value. No clearly visible indications of human activities are acceptable.

attempted to map equivalent areas of primary forest[13] across Europe (excluding Russia), but the mapped areas amounted to only 0.7% of forest area due to knowledge gaps.

Over most of Europe, terrestrial biodiversity-rich areas are now mainly associated with semi-natural habitats, the most extensive of which are forests. As discussed earlier, the EEA and Forest Europe consider that the majority of forest in Europe is 'semi-natural', although this is based on a very broad interpretation of the term. Whilst forest is the climax vegetation that would occur in most lowland areas, few such semi-natural forests closely resemble their equivalent natural communities in species composition and structure, because most are managed for forestry. Nevertheless, forest habitat types of high intrinsic nature importance remain, including 81 habitat types listed in HD Annex I (i.e. 35%). These cover 491 900 km^2, which is about 27% of the EU forest area (EEA, 2020b).[14]

HD forests support a high diversity of species, a large proportion of which are of high conservation importance, as many are habitat specialists, endemic, rare or otherwise threatened. HD Annex II species that are dependent on forest habitats include the iconic Brown Bear, Wolf, Eurasian Lynx (*Lynx lynx*) and Wolverine (*Gulo gulo*), which have recovered much of their range in recent decades (for reasons described in several chapters). Europe is alone in the world in having increasing populations of large carnivores that are not confined to protected areas, and are sharing the same land as people (Chapron *et al.*, 2014).

Most of the other remaining biodiverse semi-natural habitats are the grasslands, heathlands, Mediterranean shrublands and pastoral woodlands that are the result of traditional low-intensity farming. These are often complex labour-intensive farming systems using livestock breeds, crop types and husbandry practices that are highly adapted to local soils, vegetation and climate. Among extensive pastoral systems of high conservation, some are of particular scientific interest because to some extent they mimic natural grassland ecosystems that were formerly present and maintained by wild native herbivores. It has been suggested that species in these landscapes reflect a species pool from Pleistocene herbivore-structured environments, which, after the extinction of the Pleistocene mega-fauna, was rescued by the introduction of pre-historic agriculture (Eriksson, 2021).

Semi-natural grasslands and some other habitats also often have very high levels of small-scale species diversity. In fact, according to Wilson *et al.* (2012), pastoral woodlands in Estonia and some semi-natural grasslands in eastern central Europe have amongst the highest levels of small-scale species richness in the world (e.g. shoots of 43 species over 0.1 m^2 in an area of semi-dry basiphilous grassland in Romania). Of Europe's endemic vascular plants, 18% are bound to grassland habitats, which is nearly twice as many as in forests, despite the latter covering a much greater area (Hobohm and Bruchmann, 2009, cited in Habel *et al.*, 2013). Semi-natural grasslands are also very important for a wide range of associated animal species, for example hosting 88% of European butterflies, although the proportion of endemics is low (Wallis DeVries and van Swaay, 2009).

As a result of their high biodiversity value, and because many are now scarce and/or declining, most semi-natural agricultural habitats in the EU are listed in HD Annex I, and many associated species are listed in HD Annex II or BD Annex I. In fact 58 HD Annex I habitats (about 30%) are considered to be key farmland habitats because they are dependent on, or associated with, extensive agricultural practices (European Commission, 2018b). HD Annex II includes 197 species or subspecies that are associated with agro-ecosystems or grassland ecosystems, and 62 BD Annex I birds are considered to be key farmland species.

The nature conservation and cultural importance of the farming systems that maintain semi-natural habitats and their associated 'cultural landscapes' is now generally recognised. To highlight their biodiversity importance, the term High Nature Value (HNV) farming was coined by Baldock *et al.* (1993). This concept has been widely adopted, incorporated into EU policy and further developed, including for forests (e.g. IEEP, 2007). Three types of HNV farmland are now recognised (Paracchini *et al.*, 2008):

- Type 1: Farmland with a high proportion of semi-natural vegetation.
- Type 2: Farmland with a mosaic of low-intensity agriculture and natural and structural elements, such as field margins, hedgerows, stone walls, patches of woodland or scrub, small rivers, etc.
- Type 3: Farmland supporting rare species or a high proportion of European or world populations.

Type 1 HNV farmland comprises a wide range of habitats (many of which are HD Annex I habitats), including coastal, floodplain, steppic and alpine meadows and pastures, heathlands, sclerophyllous scrub (matorral) and wood pastures. As well as including mixed low-intensity farmland with hedges and woods, etc., Type 2 also includes extensive open dry croplands with traditional rotations and grazed stubbles, which are now scarce. Such habitats have diverse plant and invertebrate communities, and are especially important for steppe birds, such as the globally Vulnerable Great Bustard (*Otis tarda*) (Bota *et al.*, 2005). Type 2 HNV may also include traditional

[13] In accordance with the FAO and Buchwald (2005), these are 'Relatively intact forest areas that have always or at least for the past sixty to eighty years been essentially unmodified by human activity. Human impacts in such forest areas have normally been limited to low levels of hunting, fishing and harvesting of forest products, and, in some cases, to historical or pre-historical low intensity agriculture.' They include primeval, virgin, near-virgin, old-growth and long-untouched forests.

[14] This appears to be based on the EEA 2016 forest estimate which uses the FAO definition of forests. It would be 36% of EU forest cover based on the CLC 2018 data in Table 2.3.

orchards, vineyards and olive groves, when they retain large old trees and a semi-natural understory that is extensively grazed by livestock. More detailed descriptions and examples are provided in Oppermann *et al.* (2012). In reality, the HNV types are not mutually exclusive. It should also be noted that Type 3 HNV farmland often includes more intensive farmland. For example, conventionally farmed cropland on the Black Sea coast is an important feeding ground for wintering Red-Breasted Geese (*Branta ruficollis*), another globally vulnerable species.

Mapping and quantifying HNV farmland is difficult in practice, as the different types are rather subjectively and loosely defined (especially Type 2) and the necessary datasets are currently lacking. Nevertheless, mapping has been carried out biannually in Germany since 2009 in sample areas, to monitor HNV extent (Benzler *et al.*, 2015). Predictive maps of the likelihood of HNV farmland across Europe have been produced by the EEA using CLC, farming and biodiversity data (Paracchini *et al.*, 2008), the latest update being produced using CLC 2006 data (Schwaiger *et al.*, 2012). According to its estimates for each country, over the EU-28 excluding Greece (due to data gaps), the HNV area was estimated to be 787 517 km^2, which is about 34% of the agricultural area. HNV farmland in the other EEA-39 amounted to 124 925 km^2, which is about 66% of the agricultural area based on CLC data and the HNV map. This high proportion is mainly due to the predicted large HNV areas in Iceland and Norway. As the study notes, care should be taken in interpreting the estimates due to data gaps and limitations. It should also be borne in mind that the methods and data are best able to predict Type 1 HNV, and Type 3 only for some species groups. Type 2 HNV is only indirectly predicted and is possibly underestimated.

A more recent study by Keenleyside *et al.* (2014) compiled the best available national estimates, both minimum and maximum, of HNV farmland extent in each of the EU-28 Member States. The authors did not provide overall totals for the EU, as they felt the estimates were too variable and uncertain. This seems overly cautious as the national estimates are probably at least as reliable as the predictions by the EEA for most countries, and where this is not the case then the EEA data are used. Adding the EU-28 estimates (which omit Malta) gives minimum and maximum total HNV areas of 579 047 km^2 and 827 790 km^2, which equate to 25% and 36% of the agricultural area estimated by the EEA. Thus, both sets of estimates are reasonably close, which gives some reassurance that they provide useful indications of HNV area, although both have similar limitations especially regarding Type 2 HNV farming.

Building on initial concepts by IEEP (2007), the EEA (2014) has defined HNV forest as 'an area covered by forests or other wooded lands having a current ecosystem state similar to its natural state.' The EEA has also carried out exploratory studies to identify and map HNV forest and develop a related European-level forest naturalness indicator. However, HNV forest areas have yet to be mapped or quantified at an EU or wider scale.

Rivers, lakes, lagoons, mudflats, salt marshes, mires, fens and other wetlands remain relatively common and extensive over much of Europe, especially in the north and west. They make up about 5% of the EU's area and 7% of other countries in the EEA-39 (Table 2.3). Whilst most have been modified by humans and remain polluted (especially nutrient enrichment), they still provide essential habitat for a wide range of species, some of which are rare, and/or endemic to isolated water bodies. Wetlands are also very productive habitats and therefore often support huge numbers of particular species of invertebrates, fish and waterbirds. Furthermore, most of the larger sites, such as the Volga Delta, Danube Delta and Waddenzee, are also of crucial importance as migratory staging posts and wintering areas for birds (Heath and Evans, 2000).

As described earlier, most of Europe's seas show severe impacts from the effects of fishing, pollution and other pressures. Despite this, due to their variety of biophysical conditions, they still retain a wide range of marine habitats (although they are only broadly categorised in the Habitats Directive) such as rocky shores, shallow seagrass and kelp beds, dynamic sandbanks and biogenic reefs. These in turn support a high diversity of species, especially in the Mediterranean, where there is also a high proportion of endemics, and a number of rare species including turtles and the endangered Mediterranean Monk Seal (*Monachus monachus*). Part of the north-east Atlantic is also characterised by its very high productivity, which in turn supports large numbers of seabirds and sea mammals, including high proportions of the global population of certain species, such as the Manx Shearwater (*Puffinus puffinus*) and Grey Seal (*Halichoerus grypus*). Cetacean numbers remain depleted from past hunting, but they are increasing and a diverse range of species occur, including the largest animal on the planet – the globally Endangered Blue Whale (*Balaenoptera musculus*).

2.3.2 Species' Distribution Patterns and Concentrations of Species of High Conservation Concern

To provide contextual information for the later chapters, it is useful to know the relative importance of each country with respect to its species richness, especially in relation to those of high conservation concern at a European level. Ideally this should draw on comprehensive lists using consistent criteria that cover all (or at least most) species groups, and each European country. Unfortunately, such compiled information is not available for most species groups across all European countries, but it is possible to get an indication of the importance of each country with respect to bird species richness and Species of European Conservation Concern (SPEC) that have been identified by BirdLife International, most recently in 2015. SPECs include species that, according to IUCN Red List criteria, are threatened (i.e. Critically Endangered, Endangered or Vulnerable) or Near Threatened globally and/

or in Europe (BirdLife International, 2015). They also include species that are moderately declining or depleted. Table 2.4 indicates the number of species and SPECs that occur in each country. It should be noted that the list of globally threatened species has since been updated, a new Red List has been published (BirdLife International, 2021) and SPECs are to be revised. However, these updates are not likely to alter substantially the broad pattern of bird species conservation importance across Europe.

There is an obvious tendency for large countries, most notably the Ukraine, France and Spain, to have the most species and SPECs. This is not surprising as it reflects the general species richness area relationship that has been widely observed in nature (e.g. Rosenzweig, 1995). One reason for this relationship is the tendency for large areas to have greater variation in their abiotic conditions and habitats. To explore this further, and to show which countries have disproportionately high species richness, Figure 2.5 plots the number of species in relation to the terrestrial area of the country. Each country's marine area has not been included as a very small proportion of bird species spend all their time at sea other than at their nest site.

The graph confirms that there is a clear relationship between species richness and the area of the country, which is best described by the logarithmic curve. Thus, countries that are above the curve can be said to be particularly species rich, for example, as a result of their particularly high geological, topographical or climatic diversity and relatively low levels of human impact. This indicates that Ukraine, and especially France and Spain, are not just species rich because of their size. Other countries that are particularly rich in birds for their size include Bulgaria, Greece, Serbia and Romania. In contrast, the UK, Ireland and Iceland are notably species poor for their size. For the UK, and especially Ireland, this is partly due to their isolation since the Ice Age, as discussed earlier. For instance, Ireland lacks some birds with short dispersal distances, including the Tawny Owl (*Strix aluco*) and three out of four woodpeckers that occur in Great Britain. As well as being absent from Ireland, some other woodpeckers that occur close by on the continent are lacking in Great Britain, such as Black Woodpecker (*Dryocopus martius*). Similar results to those shown in Figure 2.5 are obtained if the numbers of SPECs are plotted against the area of the country.

More detailed information on species richness and the distribution of SPECs is available from the maps in the second *European Breeding Bird Atlas* (Keller *et al.*, 2020). There is clearly a high level of coincidence between areas with high overall species richness and the number of SPECs, with concentrations of both in southern Sweden, southern Finland, the Baltic States, Belarus, Poland, Bulgaria, south-east Romania and parts of Spain.

Although such detailed Europe-wide distribution information is not available for many other taxa than birds, areas of high conservation importance with respect to species richness, European endemics and threatened species have been mapped for a number of species groups as part of the preparation of European Red Lists. Summaries of such information are therefore presented in Table 2.5, primarily to highlight the varying conservation importance of the groups and the areas of particular importance for them. Whilst some Red Lists are now rather old, and some have been partly revised (Table 2, Neubert *et al.*, 2019), the distribution patterns of interest here are unlikely to have changed since they were produced. Some higher plant Red List results are not included in the table as they focussed on selected groups that would not provide a broadly representative picture as required here. Besides, plant diversity in Europe has been more broadly studied, and 24 centres of diversity identified, of which nine occur within the Iberian Peninsula and 14 are mountain ranges (e.g. the Alps, Pyrenees, Troodos Mountains and Carpathians) (WWF and IUCN 1994, cited in Bilz *et al.*, 2011).

For many species groups there is an obvious geographical gradient, with species richness increasing from north to south, such that Iberia, the Mediterranean region and the Balkans are particularly species rich (e.g. for trees, grasshoppers, etc., bees and reptiles). However, for some others, mid latitudes in central and eastern Europe tend to have the highest species richness (e.g. dragonflies, freshwater fish and amphibians). Mountains also have a marked influence on these patterns, being, for example, particularly rich in species of ferns, butterflies and mammals. Marine fish and mammals contrast in their species richness patterns: the Mediterranean is particularly rich in fish, whilst most European marine mammals are found along the continental shelf of the north-east Atlantic.

Endemics often show similar patterns to overall species richness, but tend to be more concentrated in the glacial refugia of the Mediterranean, as well as mountains and islands, where their isolation has led to speciation. Whilst it is not always apparent from the Red List maps of concentrations of endemics, many islands have their own endemic forms of some more widely distributed species. Although not covered in this book and Table 2.5, the Macaronesian islands and seas are particularly important for endemic species. Threatened species tend to have more concentrated distributions, typically where high species richness and endemism coincide with intense land use change and other human developments over recent decades (e.g. in Iberia, the Mediterranean and Black Sea coasts and the Alps).

This brief analysis reveals that, although there are obvious patterns in spatial biodiversity importance, and similarities amongst some species groups, there are also significant complex variations. Hence, it is necessary for conservation strategies and targeting to understand and take these variations into account, and not be focussed on a few selected and well-studied species groups.

For the EU-28 countries, it is possible to assess their importance more formally with respect to the species identified in the Nature Directives as requiring conservation measures, as listed in BD Annex I and HD Annexes II and/or IV or V. The proportions of the species in BD Annex I, and the combined

Table 2.4 The proportion of European* bird species and Species of European Conservation Concern (SPEC) that occur in each country.

Country	Code	All species		SPEC	
		Number	%	Number	% of SPEC
Europe*		541		218	
Albania	AL	224	41	78	36
Andorra	AD	111	21	35	16
Austria	AT	215	40	72	33
Belarus	BY	224	41	83	38
Belgium	BE	184	34	62	28
Bosnia & Herzegovina	BA	223	41	71	33
Bulgaria	BG	256	47	87	40
Croatia	HR	228	42	72	33
Cyprus	CY	95	18	39	18
Czech Republic	CZ	218	40	76	35
Denmark	DK	191	35	69	32
Estonia	EE	219	40	76	35
Finland	FI	247	46	88	40
France	FR	281	52	103	47
Germany	DE	246	45	85	39
Greece	GR	251	46	86	39
Hungary	HU	217	40	75	34
Iceland	IS	75	14	35	16
Ireland	IE	134	25	50	23
Italy	IT	250	46	85	39
Kosovo	XK	180	33	58	27
Latvia	LV	218	40	75	34
Liechtenstein	LI	134	25	36	17
Lithuania	LT	214	40	76	35
Luxembourg	LU	129	24	43	20
Malta	MT	24	4	8	4
Moldova	MD	177	33	67	31
Montenegro	ME	210	39	68	31
Netherlands	NL	186	34	67	31
North Macedonia	MK	228	42	74	34
Norway	NO	250	46	89	41
Poland	PL	234	43	82	38
Portugal (inc. Azores, etc.)	PT	213	39	90	41
Romania	RO	254	47	88	40
Serbia	RS	238	44	78	36
Slovakia	SK	221	41	75	34
Slovenia	SI	209	39	65	30

Table 2.4 *(cont.)*

Country	Code	All species		SPEC	
		Number	%	Number	% of SPEC
Spain (inc Canaries etc.)	ES	281	52	117	54
Sweden	SE	256	47	91	42
Switzerland	CH	189	35	61	28
Ukraine	UA	272	50	104	48
United Kingdom	UK	211	39	77	35

Note. * Europe here includes European Russia, all of Turkey, and the Caucasus.
Source. Based on data provided by A. Staneva, BirdLife International, from BirdLife International (2017b).

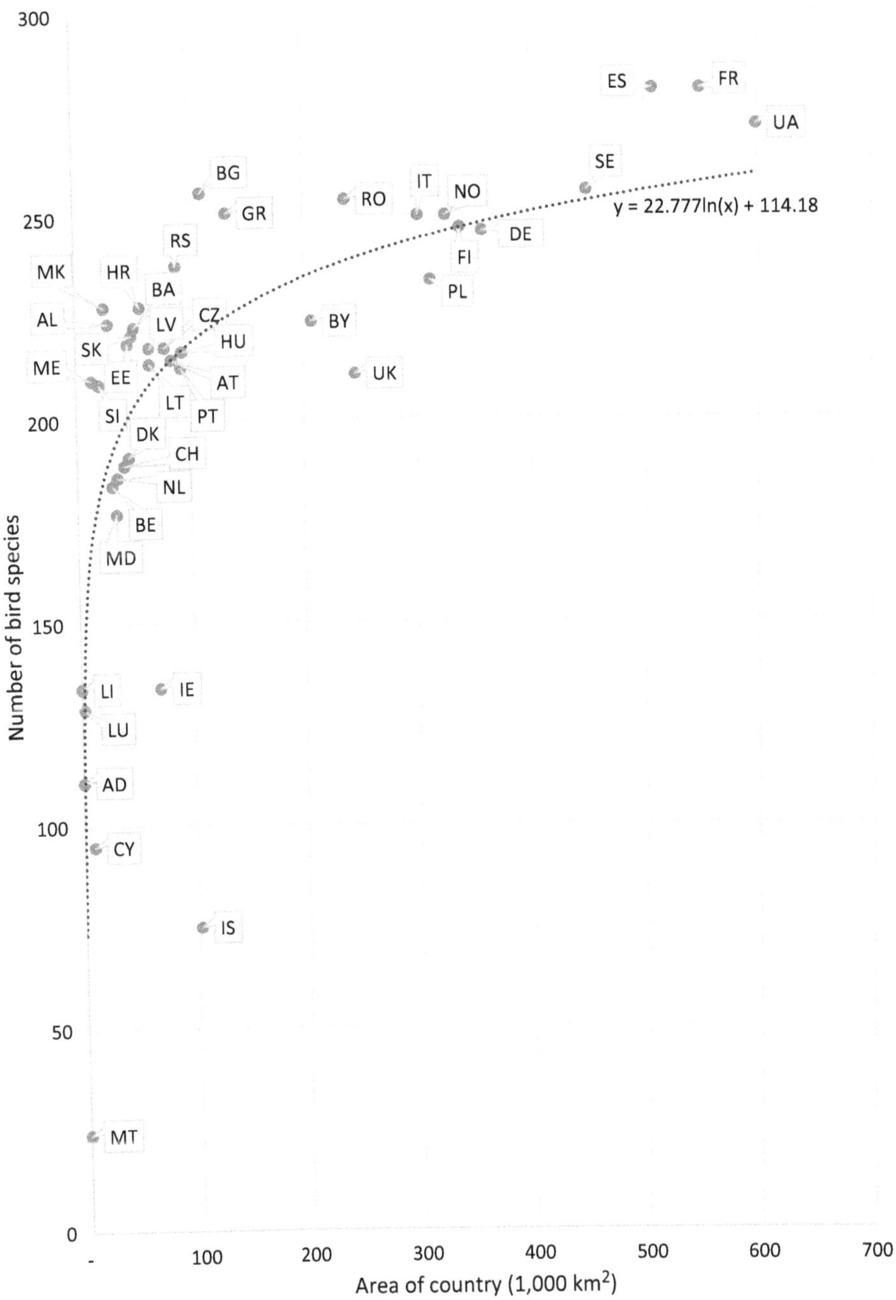

Figure 2.5 The number of bird species that occur in each European country in relation to the country's terrestrial area.
Notes. See Table 2.4 for country codes. R^2 for the logarithmic curve = 0.593.
Source. Based on data provided by A. Staneva, BirdLife International, from BirdLife International (2017b).

Table 2.5 The numbers and distribution of European species, endemics and threatened species.

Species group and source	Total species, % endemic, % threatened in Europe*	Areas with high species richness relating to		
		All species	Endemic and near endemic species	Threatened species
Mosses, liverworts and hornworts (Hodgetts *et al.*, 2019)	1 817 species, 10% endemic, 22% threatened	Central Europe, esp. Alps, also Pyrenees, Carpathians, Scandinavia and UK	Few endemics, evenly distributed with some high numbers in the Alps, UK and Ireland	Alps (esp. E), followed by the Carpathians, E Pyrenees and some Scandinavian mountains
Lycopods and ferns (García Criado *et al.*, 2017)	194 species, 27% endemic, 20% threatened	Mountainous areas, inc. Alps, Pyrenees, Massif Central and Carpathians	Alps, locally N Sardinia, mainland Italy and S Spain	Alps, esp. in Switzerland
Trees (Rivers *et al.*, 2019)	454 species, 58% endemic, 42% threatened	Mediterranean region and Balkan Peninsula	S and central Europe, inc. Alps, Pyrenees, Carpathians, Apennines and the Balkan Peninsula	Widespread, some concentrations of *Sorbus* spp., e.g. in UK, Carpathians and Hungary
Freshwater molluscs (Cuttelod *et al.*, 2011)	856 species, 87% endemic and 44% threatened	Iberia, Mediterranean, Alps, Carpathians and Balkans	As all species, esp. ancient lakes in Balkans and Mediterranean islands for narrow range species	Similar to endemics, e.g. Iberia, S France, Germany, Austria and Greece
Terrestrial molluscs (Neubert *et al.*, 2019)	2 480 species, 92% endemic and 22% threatened	Mid latitudes of central Europe, esp. Pyrenees, Alps, Carpathians and Balkans	As for all species	Greece notable hotspot
Dragonflies (Kalkman *et al.*, 2010, 2018)	138 species, 13% endemic and 15% threatened	France, central Europe, Poland, Belarus, Slovenia and parts of the Balkans	Most in S France and Iberia, some in Balkan Peninsula and islands	Iberian Peninsula, S France, Balkan Peninsula and Crete
Grasshoppers, crickets and bush-crickets (Hochkirch *et al.*, 2016)	1 082 species, 68% endemic, 26% threatened	S Europe, esp. Spain, S France, and the Balkans	As for all species, but also Italy, Alps and Carpathians	S Spain, S France, Italy and Greece
Saproxylic (deadwood) beetles (Cálix *et al.*, 2018)	c. 4 000 species and endemics unknown. Of 688 assessed, 18% threatened, but 24% data deficient	Mid latitudes, esp. mountain forests of the Pyrenees, Alps and Carpathians	Similar to all species	Central and E Europe, esp. Hungary and surrounding countries
Butterflies (van Swaay *et al.*, 2010)	482 species, 13% endemic, 8% threatened	Mountains in S Europe, esp. Pyrenees, Alps and Balkans	Esp. Alps & Pyrenees, and other mountains of Spain, Italy and the Balkans	Central and E Europe
Bees (Nieto *et al.*, 2014)	1 965 species, 20% endemic, 9% threatened	Increases towards the S, esp. the Mediterranean climatic region	High proportions in S Europe	South-central Europe, and probably more widely in the S
Freshwater fish (Freyhof and Brooks, 2011)	531 species, 80% endemic, 37% threatened	E and central Europe, Balkan Peninsula, catchments of the Elbe and S Baltic Sea basin	Central Europe (e.g. subalpine lakes in Austria, Germany, Switzerland and France), Balkan Peninsula and parts of N Europe	S central Europe, N Mediterranean coast, Balkan Peninsula and Bulgarian coastal streams

Table 2.5 *(cont.)*

Species group and source	Total species, % endemic, % threatened in Europe*	Areas with high species richness relating to		
		All species	**Endemic and near endemic species**	**Threatened species**
Amphibians (Temple and Cox, 2009)	85 species, 75% endemic, 23% threatened	Mid latitudes, inc. NW Iberia, France, Germany, Czech Republic, N Italy and Slovenia	Iberia, France and Italy	NW and SE Iberia, Italy, Slovenia and Balkan coast
Reptiles (Cox and Temple, 2009)	151 species, 48% endemic, 19% threatened	Strong increase from N to S, high in Iberia, Italy, Cyprus and esp. the Balkan Peninsula	Concentrated in Iberia, also some in S France, Balkans and Mediterranean islands	Concentrated in the Iberian Peninsula, some also in the Balkans and Cyprus
Birds (BirdLife International, 2015, 2021)	540 species, 19% endemic or near endemic and 13% threatened	Spain, S Balkans, N and W Ukraine, Belarus, Baltic States, E Poland	Spain, France, E Germany, Czech Republic, Slovakia, Poland, Lithuania, W Belarus and W Ukraine	Spain, Ukraine, Belarus and Estonia
Mammals – terrestrial (Temple and Terry, 2007)	219 species, 27% endemic, 14% threatened	Central European and Mediterranean mountains and Balkan Peninsula	Iberia, the Alps and Italy	Iberia, Italy and the Balkan Peninsula, esp. Bulgaria
Mammals – marine (Temple and Terry, 2007)	41 species, none endemic, 22% threatened	NE Atlantic, inc. to W of Ireland, Britain, France and Iberia	No European endemics occur	As all species, and some in E Mediterranean
Marine fish (Nieto *et al.*, 2015)	1 220 species, 15% endemic, 7% threatened	Mediterranean coastal waters, W coast of Portugal, shelf edge of W France and UK, S Iceland	Similar to all species, esp. high along European coast of Mediterranean, e.g. the Balearic, Ligurian, Tyrrhenian, Adriatic and Aegean Seas	Similar to all species, with highest concentration around Iberia and in the Mediterranean Sea

Notes. * Europe here includes the Canaries, Azores and Madeira, European Russia, Armenia, Azerbaijan, Georgia and all of Turkey. Distribution information on areas outside the scope of this book in Russia, Turkey and the Caucasus is not included in the table. Species totals generally refer to native or naturalised species in Europe before 1500 CE. Threat assessments exclude 'Not Applicable' species (e.g. non-native), which are normally a small proportion unless indicated. % threatened assumes that the same proportion of Data Deficient species are threatened.

HD Annexes, that occur in each EU-28 Member State are indicated in Figure 2.6. Both sets of Annex listed species show a rather similar pattern to the bird SPECs (Table 2.4), including a clear species–area relationship. A relatively large proportion also occur in areas with generally high levels of species richness and endemism: thus particularly around the Mediterranean and elsewhere in south-east Europe. For instance, it is noticeable that Greece and Italy have the highest number of Annex listed species, despite not being the largest EU countries. These results are similar to those found in a Red List assessment of plants protected by the Habitats Directive, Bern Convention and CITES, which revealed concentrations of the species within Iberia, France and the Balkan Peninsula (Bilz *et al.*, 2011).

It should be borne in mind that levels of conservation importance may vary considerably within countries. This is apparent from inspection of the smaller-scale bird atlas data, which reveals, for example, that large areas of Italy and Greece have relatively low numbers of Annex I birds (Keller *et al.*, 2020).

As EU conservation priorities for species are closely linked to the Annexes, it should be borne in mind that they are neither comprehensive nor based on clear consistent scientific criteria. As a result, they are uneven in their taxonomic and geographical representation. HD Annex II is notable for being dominated by higher plants and including very few invertebrates. Furthermore, the selection of birds for inclusion in Annex I was politically influenced, and the list is now very old. Although the species Annexes were expanded as Member States joined the EU, they have not been adjusted in response to changes in the status of species over the years.

In response to these problems, some scientists have suggested that the Annexes should be updated (e.g. Hochkirch *et al.*, 2013). This is a complex and controversial issue and cannot be adequately discussed here, but it was concluded during the Fitness Check of the Nature Directives that updating them now would be counter-productive (European Commission, 2016). This is because updating the Annexes would be a distraction when the priority is to protect and

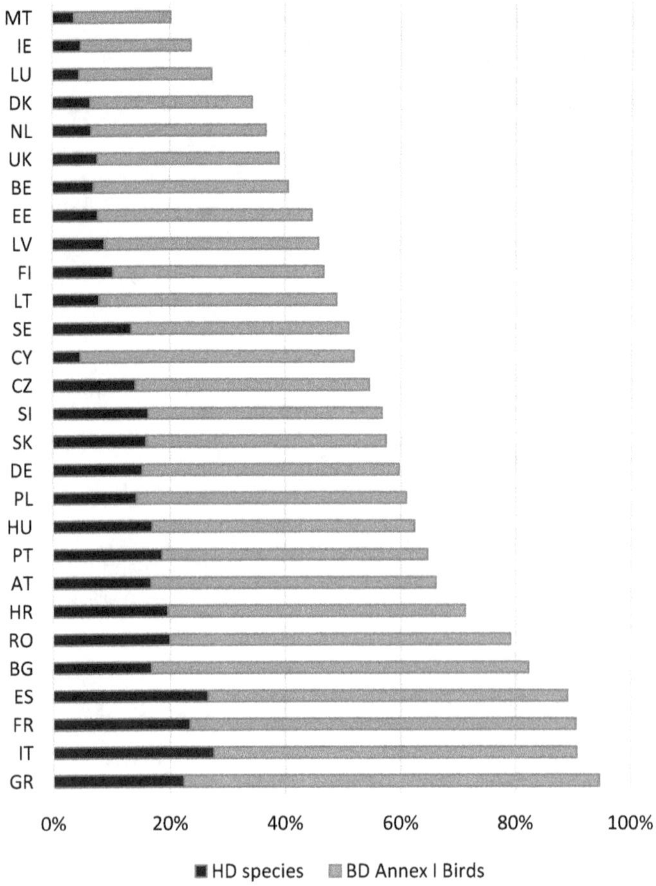

Figure 2.6 The percentages of HD species and BD Annex I species that are native and occur regularly in the European part of each EU Member State. *Notes.* HD species are 'species of Community interest' listed in HD Annexes II and/or IV or V. HD species occurring in the Macaronesian terrestrial and marine regions are excluded. UK data exclude species only occurring in Gibraltar. *Source.* BD Article 12 reporting checklist (https://cdr.eionet.europa.eu/help/birds_art12) and HD Article 17 reporting checklist (https://cdr.eionet.europa.eu/help/habitats_art17) (both June 2020 versions).

manage better the protected areas that have been designated. Furthermore, Natura 2000 sites have been found to protect a high proportion of other species besides those listed in the Annexes for which the sites were designated (Milieu *et al.*, 2016), particularly in the case of butterflies, plants and birds (van der Sluis *et al.*, 2016).

2.3.3 The Implications for Nature Conservation Approaches and Priorities

The fact that much of Europe's remaining terrestrial biodiversity is associated with semi-natural habitats that have been created and maintained by human activities has fundamental implications for the way nature conservation is carried out. In particular, it means that the conservation of many important threatened habitats usually depends on continuing the key human activities that created them. Many species are also reliant on habitat manipulation (Luoto *et al.*, 2003), including for many threatened species, a high proportion of which are dependent on open habitats and mosaics. Thus a considerable amount of effort in Europe goes into maintaining HNV farming systems for biodiversity. This contrasts with many other parts of the world where the main nature conservation aim is to minimise human influences and allow nature to look after itself. The latter is more applicable to the marine environment in Europe, where the principal priorities are to reduce pollution, the various impacts of fishing and IAS.

In recent years there have been calls from some conservationists for fewer interventions within terrestrial ecosystems, allowing instead natural processes to become predominant in greater areas, through rewilding.[15] This is often with the associated aims of re-establishing native keystone species, such as Beavers and large herbivores and carnivores. In practice, rewilding, as well as 'traditional' nature conservation, normally requires interventions to maintain biodiversity, which is highly dependent on habitat mosaics of various scales created by succession and disturbance processes. Hence, intervention and rewilding are actually complementary nature conservation approaches (Van Meerbeek *et al.*, 2019; Fuller and Gilroy, 2021). Most obviously and frequently, rewilding requires some grazing and browsing of the vegetation. Therefore, low-intensity livestock farming takes place, using appropriate ancient hardy breeds, as the former wild herbivores are normally absent. Interventions are also sometimes needed to address the legacy of past detrimental impacts, before natural processes can be allowed to become predominant. Forests represent a widespread example, as many managed forests have low structural and species diversity. Therefore, some felling and selective planting is needed to improve their naturalness and biodiversity, at least in the short term. Many highly eutrophicated wetlands require removal of the accumulated nutrients before they can return to their previous natural regime. Over the longer term, conservation interventions are usually required to address ongoing pressures, such as those from further nutrient pollution and colonisation of IAS.

Over recent decades, there has been little need to proactively promote rewilding over much of Europe as it has been occurring incidentally as a result of agricultural abandonment, especially in mountainous areas and other remote regions. This has had some conservation benefits, including contributing to the recovery of populations of large carnivores, but has probably been mostly detrimental for HD habitats and associated specialist threatened species. Whilst more rewilding of agricultural land could be beneficial if judiciously targeted, the higher priority has been to avoid large-scale agricultural abandonment of HNV farmland as it would lead to substantial losses of biodiversity as well as unique cultural landscapes. Given ongoing rural socio-economic trends, priorities in most of Europe continue to be conserve semi-natural habitats and the HNV systems that maintain them – protecting them from

15 'Letting nature take care of itself, enabling natural processes to shape land and sea, repair damaged ecosystems and restore degraded landscapes. . .' https://rewildingeurope.com/what-is-rewilding/

abandonment or conversion to more intensive farming systems, forest plantations, bioenergy crops, solar parks or other uses.

Another important consideration is the fact that remaining natural and semi-natural habitats, outside high mountains and remote northern areas, are highly fragmented in most parts of Europe, especially in densely populated western and central areas (EEA-FOEN, 2011). This fragmentation is partly a result of habitat change, as natural and semi-natural habitat patches have become surrounded by inhospitable areas in the landscape. Fragmentation also results from artificial barriers to movement, such as transport infrastructure or dams on rivers. Fragmentation impacts depend on their context and the habitats and species in question; some fragmentation can be beneficial (Fahrig, 2017; Fahrig *et al.*, 2019). More frequently, fragmentation is considered to be detrimental, as it disrupts ecosystem processes, reduces habitat quality, increases disturbance levels and external pressures, constrains species movements across the landscape and increases the risk of local extinctions of small populations of species (EEA-FOEN, 2011; Haddad *et al.*, 2015; Fletcher *et al.*, 2018). As a result, it also exacerbates the impacts of other pressures and constrains the ability of species to adapt to climate change (Opdam and Wascher, 2004)

To tackle losses and fragmentation of habitats, the first priority has been to conserve what is left, especially of the most depleted natural and semi-natural habitats, through effective protection and management. Relatively small patches of poor habitat also need to be maintained where they have important landscape-scale connectivity functions. Secondly, ecosystem/habitat restoration has been increasingly necessary over recent decades, to maintain ecological processes and the viability of small and isolated species populations. At the same time, restoration can increase connectivity, also helping to maintain the viability of meta-populations (Hanski, 1999) as well as species movements where necessary, between important patches of habitat within a wider network (Crick *et al.*, 2020).

In addition to ecosystem/habitat measures, specific additional actions are also sometimes required to conserve threatened species. In this respect, it is necessary to bear in mind that there are indications of a substantial extinction debt among European species of various taxonomic groups. As found by Dullinger *et al.* (2013), the national Red Lists for 22 countries reflected past pressures from the early and mid, rather than the late twentieth century. Thus, lags in impacts on threatened species need to be taken into account when establishing species conservation strategies and priorities. Whilst such threatened species persist, previous pressures and declines that would otherwise lead to extinction can be reversed if relevant measures are taken rather than just maintaining the status quo (Kuussaari *et al.*, 2009; Krauss *et al.*, 2010).

At present, most of Europe is dominated by highly modified or completely artificial habitats, including intensively managed forests and farmland, and cities and other human created habitats. Whilst these have a greatly impoverished biodiversity, dominated by common and generalist species, they still merit conservation and enhancement where this is possible. Indeed, their species are the most often encountered and enjoyed by people and are therefore particularly valued for aesthetic, educational, recreational and other cultural reasons. More fundamentally, it is now widely recognised that biodiversity both underpins and forms ecosystem services that provide a wide range of essential social, health and economic benefits for humankind (MEA, 2005; TEEB, 2012; IPBES, 2019; Dasgupta, 2021). Thus, to use the current terminology, all ecosystems, habitats and species are key components of natural capital that need to be valued and maintained (Barbier, 2011; Helm, 2015).

From the preceding analysis it is clear that nature conservation in Europe has had broad and evolving objectives and multiple challenges. It has been necessary to give precedence to the many species and unique habitats that are threatened, giving priority according to scale (i.e. preventing global extinctions first) and the biogeographical importance of the population/area concerned. However, nature conservation has increasingly also needed to encompass all ecosystems and native species, conserving and restoring biodiversity in the wider countryside and urban areas. Tackling these differing objectives has required a variety of approaches and interacting measures, which are described in the next two chapters.

Acknowledgements

Very helpful comments and suggestions were provided on a draft version of this chapter by Mats Eriksson, Ana Guimarães, Jan-Erik Petersen, Owen Mountford and Rainer Oppermann.

References

Amoroso, R. O., Pitcher, C. R., Rijnsdorp, A. D. *et al.* (2018) Bottom trawl fishing footprints on the world's continental shelves. *Proceedings of the National Academy of Sciences*, **115**, E10275–E10282.

Baldock, D. (1990) *Agriculture and Habitat Loss in Europe*. London: World Wide Fund for Nature International.

Baldock, D., Beaufoy, G., Bennett, G. & Clark, J. (1993) *Nature Conservation and New Directions in the EC Common Agricultural Policy*. London: Institute for European Environmental Policy.

Barbier, E. B. (2011) *Capitalizing on Nature: Ecosystems as Natural Assets*. Cambridge, UK: Cambridge University Press.

Benzler, A., Fuchs, D. & Hünig, C. (2015) Monitoring and first results of High Nature Value Farmland Monitoring in Germany: evidence of ongoing biodiversity loss within agricultural landscapes. *Nature & Landschaft*, **90** (7), 309–316.

Billeter, R., Liira, J., Bailey, D. *et al.* (2008) Indicators for biodiversity in agricultural landscapes: a pan-European study. *Journal of Applied Ecology*, **45** (1), 141–150.

Bilz, M., Kell, S. P., Maxted, N. & Lansdown, R.V. (2011) *European Red List of Vascular Plants*. Luxembourg: Publications Office of the European Union.

Birdlife International (2015) *European Red List of Birds*. Luxembourg: Office for Official Publications of the European Communities.

(2017a) *Ecosystem Profile – Mediterranean Basin Biodiversity Hotspot*. www.cepf.net/sites/default/files/mediterranean-basin-2017-ecosystem-profile-english_2.pdf

(2017b) *European Birds of Conservation Concern: Populations, Trends and National Responsibilities*. Cambridge, UK: BirdLife International.

(2021) *European Red List of Birds*. Luxembourg: Publications Office of the European Union.

Blondel, J., Aronson, J., Bodiou, J-Y. & Boeuf, G. (2010) *The Mediterranean Region. Biological Diversity in Space and Time*. 2nd ed. Wiltshire, UK: Oxford University Press.

Bobbink, R., Hicks, K., Galloway, J. *et al.* (2010) Global assessment of nitrogen deposition effects on terrestrial plant diversity: a synthesis. *Ecological Applications*, **20**, 30–59.

Bota, G., Morales, M. B., Mañosa, S. & Camprodon, J. (eds.) (2005) *Ecology and Conservation of Steppe-Land Birds*. Barcelona: Lynx Edicions & Centre Tecnològic Forestal de Catalunya.

Buchwald, E. (2005) *A hierarchical terminology for more or less natural forests in relation to sustainable management and biodiversity conservation*. Proceedings: third expert meeting on harmonizing forest-related definitions for use by various stakeholders, 17–19 January 2005. Rome: Food and Agriculture Organization of the United Nations.

Cálix, M., Alexander, K. N. A., Nieto, A. *et al.* (2018) *European Red List of Saproxylic Beetles*. Brussels: IUCN.

Chapron, G., Kaczensky, P., Linnell, J. *et al.* (2014) Recovery of large carnivores in Europe's modern human-dominated landscapes. *Science*, **346**, 1517–1519.

Coll, M., Piroddi, C., Steenbeek, J. *et al.* (2010) The biodiversity of the Mediterranean Sea: estimates, patterns, and threats. *PLoS ONE*, **5**, e11842.

Cox, N. A. & Temple, H. J. (2009) *European Red List of Reptiles*. Luxembourg: Office for Official Publications of the European Communities.

Crick, H., Crosher, I., Mainstone, C. *et al.* (2020) *Nature networks evidence handbook*. York, UK: Natural England. Research report NERR081.

Cuttelod, A., Seddon, M. & Neubert, E. (2011) *European Red List of Non-marine Molluscs*. Luxembourg: Publications Office of the European Union.

Dasgupta, P. (2021) *The Economics of Biodiversity: The Dasgupta Review*. London: HM Treasury.

Desforges, J. P., Hall, A., McConnell, B. *et al.* (2018) Predicting global killer whale population collapse from PCB pollution. *Science*, **361**, 1373–1376.

Donald, P. F., Green, R. E. & Heath, M. F. (2001) Agricultural intensification and the collapse of Europe's farmland bird populations. *Proceedings of the Royal Society B Biological Sciences*, **268**, 25–29.

Donald, P. F., Sanderson, F. J., Burfield, I. J. & van Bommel, F. P. J. (2006) Further evidence of continent-wide impacts of agricultural intensification on European farmland birds, 1990-2000. *Agriculture, Ecosystems & Environment*, **116**, 189–196.

Dullinger, S., Essi, F., Rabitsch, W. *et al.* (2013) Europe's other debt crisis caused by the long legacy of future extinctions. *PNAS*, **110**, 7342–7347.

EBCC/BirdLife/RSPB/CSO (2022) *European Indicators*. https://pecbms.info/trends-and-indicators/indicators/

EEA (2008a) *Biogeographical regions in Europe*. www.eea.europa.eu/publications/report_2002_0524_154909/biogeographical-regions-in-europe

(2008b) *Regional seas around Europe*. www.eea.europa.eu/publications/report_2002_0524_154909/regional-seas-around-europe

(2010) *EU 2010 Biodiversity Baseline*. EEA Technical Report No 12/2010. European Environment Agency.

(2014) *Developing a forest naturalness indicator for Europe. Concept and methodology for a high nature value (HNV) forest indicator*. European Environment Agency.

(2015) *State of Europe's seas*. European Environment Agency.

(2016) *European forest ecosystems – state and trends*. European Environment Agency.

(2017a) *Biogeographical regions in Europe*. www.eea.europa.eu/data-and-maps/figures/biogeographical-regions-in-europe-2

(2017b) *Landscapes in transition. An account of 25 years of land cover change in Europe*. European Environment Agency.

(2017c) *Climate change, impacts and vulnerability in Europe 2016*. European Environment Agency.

(2018) *European waters. Assessment of status and pressures*. European Environment Agency.

(2019) *The European environment – state and outlook 2020*. European Environment Agency.

(2020a) *Marine regions and subregions*. www.eea.europa.eu/data-and-maps/figures/marine-regions-and-subregions

(2020b) *State of nature in the EU. Results from reporting under the nature directives 2013–2018*. European Environment Agency.

EEA-FOEN (2011) *Landscape fragmentation in Europe*. European Environment Agency and Swiss Federal Office for the Environment.

Ellenberg, H. (2009) *Vegetation Ecology of Central Europe*. 4th ed. Cambridge, UK: Cambridge University Press.

Eriksson, O. (2021) The importance of traditional agricultural landscapes for preventing species extinctions. *Biodiversity and Conservation*, **30**, 1–17.

European Commission (1995) *Wise use and conservation of wetlands*. COM(95) 189 final.

(2016) *FITNESS CHECK of the EU Nature Legislation (Birds and Habitats Directives)*. SWD(2016) 472 final.

(2018a) *Guidance document on the requirements for hydropower in relation to EU Nature legislation*.

(2018b) *Farming for Natura 2000 Guidance on how to support Natura 2000 farming systems to achieve conservation objectives, based on Member States good practice experiences*.

Fagerli, H. & Aas, W. (2008) Trends of nitrogen in air and precipitation: model results and observations at EMEP sites in Europe, 1980–2003. *Environmental Pollution*, **154**, 448–461.

Fahrig, L. (2017) Ecological responses to habitat fragmentation per se. *Annual Review of Ecology, Evolution, and Systematics*, **48**, 1–23.

Fahrig, L., Arroyo-Rodríguez, V., Bennett, J. R. *et al.* (2019) Is habitat fragmentation bad for biodiversity? *Biological Conservation*, **230**, 179–186.

Fletcher Jr, R. J., Didham, R. K., Banks-Leite, C. *et al.* (2018) Is habitat fragmentation good for biodiversity?. *Biological Conservation*, **226**, 9–15.

Forest Europe (2020) *State of Europe's Forests 2020*. Bratislava: Ministerial Conference on the Protection of Forests in Europe – Forest Europe Liaison Unit.

Fowler, D., Coyle, M., Skiba, U. *et al.* (2013) The global nitrogen cycle in the twenty-first century. *Philosophical Transactions of the Royal Society B*, **368**, 20130164.

Fowler, D., Smith, R., Muller, J. *et al.* (2007) Long term trends in sulfur and nitrogen deposition in Europe and the cause of non-linearities. *Water Air and Soil Pollution Focus*, **7**, 41–47.

Freyhof, J. & Brooks, E. (2011) *European Red List of Freshwater Fishes*. Luxembourg: Publications Office of the European Union.

Fuller, R. F. & Gilroy, J. (2021) Rewilding and intervention: complementary philosophies for nature conservation in the UK. *British Wildlife*, **32**, 258–267.

Galloway, J., Dentener, F., Capone, D. *et al.* (2004) Nitrogen cycles: past, present, and future. *Biogeochemistry*, **70**, 153–226.

García Criado, M., Väre, H., Nieto, A. *et al.* (2017) *European Red List of Lycopods and Ferns*. Brussels: IUCN.

Gold, S. (2003) *The Development of European Forest Resources, 1950 to 2000: A Better Information Base*. Geneva: United Nations Economic Commission for Europe/Food and Agriculture Organization of the United Nations. Geneva Timber and Forest Discussion Paper 31.

Gomez-Campo, C. (ed.) (1985) *Plant Conservation in the Mediterranean Area*. Dordrecht, the Netherlands: Dr. W. Junk Publishers.

Gubbay, S., Sanders, N., Haynes, T. *et al.* (2016) *European Red List of Habitats. Part 1. Marine Habitats*. Luxembourg: Publications Office of the European Union.

Habel, J. C., Dengler, J., Janisova, M., Török, P., Wellstein, C. & Wiezik, M. (2013) European grassland ecosystems: threatened hotspots of biodiversity. *Biodiversity and Conservation*, **22**, 2131–2138.

Haddad, N. M., Brudvig, L. A., Clobert, J. *et al.* (2015) Habitat fragmentation and its lasting impact on Earth's ecosystems. *Science Advances*, **1**, e1500052.

Hanski, I. (1999) *Metapopulation Ecology*. Oxford, UK: Oxford University Press.

Heath, M. F. & Evans, M. I. (2000) *Important Bird Areas in Europe: Priority Sites for Conservation*. Two volumes. Cambridge, UK: BirdLife International.

Helm, D. (2015) *Natural Capital*. Yale: Yale University Press.

Hickling, R., Roy, D. B., Hill, J. K., Fox, R. & Thomas, C. D. (2006) The distributions of a wide range of taxonomic groups are expanding polewards. *Global Change Biology*, **12** (3), 450–455.

Hladik, M. L., Main, A. R. & Goulson, D. (2018) Environmental risks and challenges associated with neonicotinoid insecticides. *Environmental Science and Technology*, **52** (6), 3329–3335.

Hobohm, C. & Bruchmann, I. (2009) Endemische Gefäßpflanzen und ihre Habitate in Europa: Plädoyer für den Schutz der Grasland-Ökosysteme. *Berichte der Reinhold-Tüxen-Gesellschaft*, **21**, 142–161.

Hochkirch, A., Nieto, A., García Criado, M. *et al.* (2016) *European Red List of Grasshoppers, Crickets and Bush-crickets*. Luxembourg: Publications Office of the European Union.

Hochkirch, A., Schmitt, T., Beninde, J. *et al.* (2013) Europe needs a new vision for a Natura 2020 network. *Conservation Letters*, **6**, 462–467.

Hodder, K. H., Bullock, J. M., Buckland, P. C. & Kirby, K. J. (2005) *Large Herbivores in the Wildwood and Modern Naturalistic Grazing Systems*. Peterborough, UK: English Nature. Research Report 648.

Hodgetts, N., Cálix, M., Englefield, E. *et al.* (2019) *A Miniature World in Decline: European Red List of Mosses, Liverworts and Hornworts*. Brussels: IUCN.

Hoffmann, J. (2012) Species-rich arable land. In *High Nature Value Farming in Europe*, eds. R. Oppermann, G. Beaufoy & G. Jones, pp. 58–69. Ubstadt-Wieher, Germany: Verlag regionalkultur.

Hoffmann, J., Kretschmer, H. & Pfeffer, H. (2000) Effects of patterning on biodiversity in Northeast German agro-landscapes. *Ecology Studies*, **147**, 325–340.

IEEP (2007) *Final Report for the Study on HNV Indicators for Evaluation*. London: Institute for European Environmental Policy.

IPBES (2019) *Global Assessment Report on Biodiversity and Ecosystem Services of the Intergovernmental Science-Policy Platform on Biodiversity and Ecosystem Services*. E. S. Brondizio, J. Settele, S. Díaz, & H. T. Ngo (eds). Bonn: IPBES Secretariat.

IPCC (2021) *Summary for Policymakers*. In *Climate Change 2021: The Physical Science Basis*. Contribution of Working Group I to the Sixth Assessment Report of the Intergovernmental Panel on Climate Change. Cambridge, UK: Cambridge University Press.

Isern, N., Fort, J. & Vander Linden, M. (2012) Space competition and time delays in human range expansions. Application to the neolithic transition. *PLoS ONE*, **7**, e51106.

Jackson, J. B. C., Kirby, M. X., Berger, W. H. *et al.* (2001) Historical overfishing and the recent collapse of coastal ecosystems. *Science*, **293**, 629–637.

Jacob, D., Petersen, J., Eggert, B. *et al.* (2014) EURO-CORDEX: new high-resolution climate change projections for European impact research. *Regional Environmental Change*, **14** (2), 563–578.

Jepsen, M. R., Kuemmerle, T., Müller, D. *et al.* (2015) Transitions in European land-management regimes between 1800 and 2010. *Land Use Policy*, **49**, 53–64.

Jongman, R. H. G. (2002) Homogenisation and fragmentation of the European landscape: ecological consequences and solutions. *Landscape and Urban Planning*, **58**, 211–221.

Kalkman, V. J., Boudot, J.-P., Bernard, R. *et al.* (2010) *European Red List of Dragonflies*. Luxembourg: Office for Official Publications of the European Communities.

Kalkman, V. J., Boudot, J. P., Bernard, R., De Knijf, G., Suhling, F. & Termaat, T. (2018) Diversity and conservation of European dragonflies and damselflies (Odonata). *Hydrobiologia*, **811**, 269–282.

Keenleyside, C. & Tucker, G. M. (2010) *Farmland Abandonment in the EU: An Assessment of Trends and Prospects.* London: Institute for European Environmental Policy. Report prepared for WWF.

Keenleyside, C., Beaufoy, G., Tucker, G. & Jones, G. (2014) *High Nature Value Farming throughout EU-27 and Its Financial Support under the CAP.* London: Institute for European Environmental Policy, Report for DG Environment.

Keller, V., Herrando, S., Voříšek, P. *et al.* (2020) *European Breeding Bird Atlas 2: Distribution, Abundance and Change.* Barcelona: European Bird Census Council & Lynx Edicions.

Krauss, J., Bommarco, R., Guardiola, M. *et al.* (2010) Habitat fragmentation causes immediate and time-delayed biodiversity loss at different trophic levels. *Ecology Letters*, **13**, 597–605.

Kuemmerle, T., Levers, C., Erb, D. *et al.* (2016) Hotspots of land use change in Europe. *Environmental Research Letters*, **11**, 064020.

Künster, H. & Keenleyside, C. (2009) The origin and use of agricultural grasslands in Europe. In *Grasslands in Europe of High Nature Value*, eds. P. Veen, R. Jefferson, J. de Smidt & J. v. d. Straaten, pp. 9–14. Zeist, the Netherlands: KNNV Publishing.

Kuussaari, M., Bommarco, R., Heikkinen, R. K. *et al.* (2009) Extinction debt: a challenge for biodiversity conservation. *Trends in Ecology and Evolution*, **24**, 564–571.

Laiolo, P., Dondero, F., Ciliento, E. & Rolando, A. (2004) Consequences of pastoral abandonment for the structure and diversity of the alpine avifauna. *Journal of Applied Ecology*, **41** (2), 294–304.

Luoto, M., Pykälä, J. & Kuussaari, M. (2003) Decline of landscape-scale habitat and species diversity after the end of cattle grazing. *Journal for Nature Conservation*, **11**, 171–178.

Martay, B., Brewer, M., Elston, D. *et al.* (2017) Impacts of climate change on national biodiversity population trends. *Ecography*, **40** (10), 1139–1151.

Mateo, R. (2009) Lead poisoning in wild birds in Europe and the regulations adopted by different countries. In *Ingestion of Lead from Spent Ammunition: Implications for Wildlife and Humans*, eds. R. T. Watson, M. Fuller, M. Pokras & W. G. Hunt, pp. 71–98. Boise, Idaho, USA: The Peregrine Fund.

MEA (2005) *Ecosystems and Human Well-Being: Biodiversity Synthesis.* Washington, DC: World Resources Institute, Millennium Ecosystem Assessment.

Metzger, M. J., Shkaruba, A. D., Jongman, R. H. G. & Bunce, R. G. H. (2012) *Descriptions of the European Environmental Zones and Strata.* Wageningen, the Netherlands: Alterra.

Milieu, IEEP & ICF (2016) *Evaluation study to support the fitness check of the Birds and Habitats Directives.* Brussels: Milieu Ltd, Institute for European Environmental Policy and the ICF International.

Mitchell, F. G. (2005) How open were European primeval forests? Hypothesis testing using palaeoecological data. *Journal of Ecology*, **93**, 168–177.

Mittermeier, R. A., Robles-Gil, P., Hoffmann, M. *et al.* (2004) *Hotspots Revisited: Earth's Biologically Richest and Most Endangered Ecoregions.* Mexico City: CEMEX.

Möllmann, C. & Diekmann, R. (2012) Chapter 4 - Marine ecosystem regime shifts induced by climate and overfishing: a review for the northern hemisphere. In *Global Change in Multispecies Systems* Part 2, eds. G. Woodward, U. Jacob & E. J. O'Gorman. *Advances in Ecological Research*, **47**, 303–347.

Muniz, I. P. (1990) Freshwater acidification: its effects on species and communities of freshwater microbes, plants and animals. *Proceedings of the Royal Society of Edinburgh, Section B: Biological Sciences*, **97**, Acidic Deposition: Its Nature and Impacts, 227–254.

Neubert, E., Seddon, M. B., Allen, D. J. *et al.* (2019) *European Red List of Terrestrial Molluscs.* Cambridge, UK, and Brussels: IUCN.

Nieto, A., Ralph, G. M., Comeros-Raynal, M. T. *et al.* (2015) *European Red List of Marine Fishes.* Luxembourg: Publications Office of the European Union.

Nieto, A., Roberts, S. P. M., Kemp, J. *et al.* (2014) *European Red List of Bees.* Luxembourg: Publication Office of the European Union.

Ockendon, N., Baker, D. J., Carr, J. A. *et al.* (2014) Mechanisms underpinning climatic impacts on natural populations: altered species interactions are more important than direct effects. *Global Change Biology*, **20** (7), 2221–2229.

Opdam, P. & Wascher, D. (2004) Climate change meets habitat fragmentation: linking landscape and biogeographical scale levels in research and conservation. *Biological Conservation*, **117** (3), 285–297.

Oppermann, R. & Hoffmann, J. (2012) Features of HNV farmland mosaic landscapes. In *High Nature Value Farming in Europe*, eds. R. Oppermann, G. Beaufoy & G. Jones, pp. 85–96. Ubstadt-Wieher, Germany: Verlag regionalkultur.

Oppermann, R. & Paracchini, M. L. (2012) HNV farming – central to European cultural landscapes and biodiversity. In *High Nature Value Farming in Europe*, eds. R. Oppermann, G. Beaufoy & G. Jones, pp. 17–23. Ubstadt-Wieher, Germany: Verlag regionalkultur.

Oppermann, R., Beaufoy, G. & Jones, G. (eds.) (2012) *High Nature Value Farming in Europe.* Ubstadt-Wieher, Germany: Verlag regionalkultur.

Paracchini, M. L., Petersen, J-E., Hoogeveen, Y., Bamps, C., Burfield, I. & van Swaay, C. (2008) *High nature value farmland in Europe. An estimate of the distribution patterns on the basis of land cover and biodiversity data.* European Commission Joint Research Centre. Luxembourg: Office for Official Publications of the European Communities of the European Union.

Petersen, J-E., Desaulty, D., Ivits, E., Mancosu, E., Milego, R. & King, S. (2021) *Ecosystem Extent Accounts 2000–2018.* Copenhagen: EEA. A European Analysis EEA technical report produced as part of the EU project on Integrated Natural Capital Accounting (KIP INCA). Version 1.0 – February 2021.

Pointereau, P., Coulon, F., Girard, P. *et al.* (2008) *Analysis of farmland abandonment and the extent and location of agricultural areas that are actually abandoned or are in risk to be abandoned.* Ispra, Italy: JRC Scientific and Technical Reports (EUR 23411 EN).

Polunin, O. & Walters, M. (1985) *A Guide to the Vegetation of Britain and Europe.* Oxford, UK: Oxford University Press.

Pons, A. & Quézel, P. (1985) The history of the flora and vegetation and past and present human disturbance in the Mediterranean region. In *Plant Conservation in the Mediterranean Area*, ed. C.

Gómez-Campo, pp. 25–43. Dordrecht, the Netherlands: Dr. W. Junk Publishers.

Poschlod, P., Baumann, A. & Karlik, P. (2009) Origin and development of grasslands in Central Europe. In *Grasslands in Europe of High Nature Value*, eds. P. Veen, R. Jefferson, J. de Smidt & J. v. d. Straaten, pp. 15–25. Zeist, the Netherlands: KNNV Publishing.

Potter, C. (1997) Europe's changing farmed landscapes. In *Farming and Birds in Europe: the Common Agricultural Policy and Its Implications for Bird Conservation*, eds. D. J. Pain & M. W. Pienkowski, pp. 25–42. London: Academic Press.

Queiroz, C., Beilin, R., Folke, C. & Lindborg, R. (2014) Farmland abandonment: threat or opportunity for biodiversity conservation? A global review. *Frontiers in Ecology and the Environment*, **12** (5), 288–296.

Quézel, P. (1985) Definition of the Mediterranean region and the origin of its flora. In *Plant Conservation in the Mediterranean Area*, ed. C. Gómez-Campo, pp. 9–24. Dordrecht, the Netherlands: Dr. W. Junk Publishers.

Rabitsch, W., Genovesi, P., Scalera, R., Biała, K., Josefsson, M., & Essl, F. (2016) Developing and testing alien species indicators for Europe. *Journal for Nature Conservation*, **29**, 89–96.

Reif, J. & Vermouzek, Z. (2019) Collapse of farmland bird populations in an Eastern European country following its EU accession. *Conservation Letters*, **12**, e12585.

Rivers, M. C., Beech, E., Bazos, I. *et al.* (2019) *European Red List of Trees*. Cambridge, UK, and Brussels: IUCN.

Robson, C. (1997) The evolution of the Common Agricultural Policy and the incorporation of environmental considerations. In *Farming and Birds in Europe: the Common Agricultural Policy and Its Implications for Bird Conservation*, eds. D. J. Pain & M. W. Pienkowski, pp. 43–78. London: Academic Press.

Roekaerts, M. (2002) *The Biogeographical Regions Map of Europe. Basic principles of its creation and overview of its development*. Copenhagen: EEA and European Topic Centre – Nature Protection and Biodiversity.

Rosenzweig, M. L. (1995) *Species Diversity in Space and Time*. Cambridge, UK: Cambridge University Press.

Rounsevell, M., Fischer, M., Boeraeve, F. *et al.* (2018) Chapter 1: Setting the scene. In *The IPBES Regional Assessment Report on Biodiversity and Ecosystem Services for Europe and Central Asia*, eds. M. Rounsevell, M. Fischer, A. Torre-Marin Rando & A. Mader, pp. 9–85. Bonn: Secretariat of the Intergovernmental Science-Policy Platform on Biodiversity and Ecosystem services.

Sabatini, F. M., Burrascano, S., Keeton, W. S. *et al.* (2018) Where are Europe's last primary forests? *Diversity and Distributions*, **24**, 1426–1439.

Scalera, R., Genovesi, P., Essl, F. & Rabitsch, W. (2012) *The Impacts of Invasive Alien Species in Europe*. Copenhagen: European Environment Agency.

Schöpp, W., Posch, M., Mylona, S. & Johansson, M. (2003) Long-term development of acid deposition (1880-2030) in sensitive freshwater regions in Europe. *Hydrology and Earth System Sciences*, **7**, 436–446.

Schwaiger, E., Banko, G., Brodsky, L. & van Doorn, A. (2012) *Updated High Nature Value Farmland in Europe. An estimate of the distribution patterns on the basis of CORINE Land Cover 2006 and biodiversity data*. European Environment Agency.

Sing, L., Metzger, M. J., Paterson, J. S. & Duncan, R. (2018) A review of the effects of forest management intensity on ecosystem services for northern European temperate forests with a focus on the UK. *Forestry: An International Journal of Forest Research*, **91**, 151–164.

Stanners, D. & Bourdeau, P. (1995) *Europe's Environment: The Dobris Assessment*. Luxembourg: Office for Official Publications of the European Communities.

Stevens, C. J., Bell, J. N. B., Brimblecombe, P. *et al.* (2020) The impact of air pollution on terrestrial managed and natural vegetation. *Philosophical Transactions of the Royal Society A*, **378** (2183).

Stevens, C. J., Dupré, C., Dorland, E., Gaudnik, C., Gowing, D. J. G. & Bleeker, A. (2010) Nitrogen deposition threatens species richness of grasslands across Europe. *Environmental Pollution*, **158**, 2940–2945.

Stoate, C., Báldi, A., Beja, P. *et al.* (2009) Ecological impacts of early 21st century agricultural change in Europe – a review. *Environmental Management*, **91**, 22–46.

Stoate, C., Boatman, N. D., Borralho, R. J., Carvalho, C. R., de Snoo, G. R & Eden, P. (2001) Ecological impacts of arable intensification in Europe. *Journal of Environmental Management*, **63**, 337–365.

Sundseth, K. (2009a) *Natura 2000 in the Alpine region*. Luxembourg: Office for Official Publications of the European Communities.

(2009b) *Natura 2000 in the Boreal region*. Luxembourg: Office for Official Publications of the European Communities.

(2009c) *Natura 2000 in the Atlantic region*. Luxembourg: Office for Official Publications of the European Communities.

(2009d) *Natura 2000 in the Continental region*. Luxembourg: Office for Official Publications of the European Communities.

(2009e) *Natura 2000 in the Pannonic region*. Luxembourg: Office for Official Publications of the European Communities.

(2009f) *Natura 2000 in the Steppic region*. Luxembourg: Office for Official Publications of the European Communities.

(2009g) *Natura 2000 in the Black Sea region*. Luxembourg: Office for Official Publications of the European Communities.

(2009h) *Natura 2000 in the Mediterranean region*. Luxembourg: Office for Official Publications of the European Communities.

Sutton, M. A., Howard, C. M., Erisman, J. W. *et al.* (eds.) (2011) *The European Nitrogen Assessment*. Cambridge, UK: Cambridge University Press.

Svenning, J-C. (2002) A review of natural vegetation openness in north-western Europe. *Biological Conservation*, **104**, 133–148.

TEEB (2012) *The Economics of Ecosystems and Biodiversity: Ecological and Economic Foundations*, ed. P. Kumar. Abingdon, UK, and New York: Routledge.

Temple, H. J. & Cox, N. A. (2009) *European Red List of Amphibians*. Luxembourg: Office for Official Publications of the European Communities.

Temple, H. J. & Terry, A. (Compilers) (2007) *The Status and Distribution of European Mammals*. Luxembourg: Office for Official Publications of the European Communities.

Tryjanowski, P., Hartel, T., Báldi, A. *et al.* (2011) Conservation of farmland birds faces different challenges in Western and Central-Eastern Europe. *Acta Ornithoecologia*, **46**, 1–12.

Tucker, G. M. & Evans, M. E. (1997) *Habitats for Birds in Europe: A Conservation Strategy for the Wider Environment*. Cambridge, UK: Birdlife International.

van der Sluis, T., Foppen, R., Gillings, S. & Jones-Walters, L. (2016) *How Much Biodiversity Is in Natura 2000? The 'umbrella effect' of the European Natura 2000 protected area network*. Wageningen, the Netherlands: Alterra Wageningen UR.

Van Meerbeek, K., Muys, B., Schowanek, S. D. & Svenning, J. C. (2019) Reconciling conflicting paradigms of biodiversity conservation: human intervention and rewilding. *BioScience*, **69** (12), 997–1007.

van Swaay, C., Cuttelod, A., Collins, S. *et al.* (2010). *European Red List of Butterflies*. Luxembourg: Office for Official Publications of the European Communities.

Vaughan, D., Korpinen, S., Nygård, H., Andersen, J. H., Murray, C. & Kallenbach, E. (2019) *Biodiversity in Europe's seas*. European Topic Centre on Inland and Marine Waters. ETC/ICM Technical Report 3/2019.

Veen, P., Jefferson, R., de Smidt, J. & v. d. Straaten, J. (eds.) (2009) *Grasslands in Europe of High Nature Value*. Zeist, the Netherlands: KNNV Publishing.

Vera, F. W. M. (2000) *Grazing Ecology and Forest History*. Wallingford, UK: CABI.

Visconti, P., Elias, V., Sousa Pinto, I. *et al.* (2018) Chapter 3: Status, trends and future dynamics of biodiversity and ecosystems underpinning nature's contributions to people. In *The IPBES Regional Assessment Report on Biodiversity and Ecosystem Services for Europe and Central Asia*, eds. M. Rounsevell, M. Fischer, A. Torre-Marin Rando & A. Mader, pp. 267–548. Bonn: Secretariat of the Intergovernmental Science-Policy Platform on Biodiversity and Ecosystem services.

Wallis DeVries, M. F. & van Swaay, C. A. M. (2009) Grasslands as habitats for butterflies in Europe. In *Grasslands in Europe of High Nature Value*, eds. P. Veen, R. Jefferson, J. de Smidt & J. v. d. Straaten, pp. 27–34. Zeist, the Netherlands: KNNV Publishing.

Watson, R. T., Fuller, M., Pokras, M. & Hunt, W. G. (eds.) (2009) *Ingestion of Spent Lead Ammunition: Implications for Wildlife and Humans*. Boise, Idaho, USA: The Peregrine Fund.

Weyhenmeyer, G. A., Müller, R. A., Norman, M. & Tranvik, L. J. (2016) Sensitivity of freshwaters to browning in response to future climate change. *Climate Change*, **134**, 225–239.

Wilson, J. B., Peet, R. K., Dengler, J. & Pärtel, M. (2012) Plant species richness: the world records. *Journal of Vegetation Science*, **23**, 796–802.

WWF & IUCN (1994) *Centres of Plant Diversity. A Guide and Strategy for Their Conservation*. Volume 1. Europe, Africa, South West Asia and the Middle East. Cambridge, UK: IUCN Publications Unit.

Chapter 3 The International Drivers of Nature Conservation, Their Objectives and Impacts on Nature Conservation Policies and Actions in Europe

David Stroud, Euan Dunn, Wyn Jones and Graham Tucker

3.1 Outline of the Development of Nature Conservation in Response to Pressures on Biodiversity and Observed Changes

As a result of growing public and political concern relating to wide-scale environmental degradation, in Europe and elsewhere, many multilateral environmental agreements (MEAs) and other international initiatives have attempted to reduce and even reverse its impacts. As noted by Stroud *et al.* (2021), international collaborative actions have been taken by governments for two main reasons: to tackle issues that are larger than a nation state (e.g. transboundary pollution, the conservation of migratory species); and to gain collective benefits from common frameworks for action. Table 3.1 lists the MEAs and other key events that have had most influence on the development of nature conservation policies, legislation and practical actions in Europe, primarily over the last 50 years.

3.2 International Instruments and Initiatives That Have Contributed to Nature Conservation in Europe

Table 3.2 provides an overview of the objectives and obligations of the MEAs and other international initiatives that have had a significant impact on nature conservation in Europe over the last 50 years. Of these, the Convention on Wetlands of International Importance especially as Waterfowl Habitat (Ramsar Convention), Convention on the Conservation of European Wildlife and Natural Habitats (Bern Convention), Convention on the Conservation of Migratory Species of Wild Animals (CMS), Convention on Biological Diversity (CBD), and the marine Conventions and North Sea ministerial conferences, have had most influence, and are therefore described later in this section. Table 3.3 indicates which countries are Parties to the above Conventions, and the dates that the Conventions came into force.

The 'para-Scandinavian' north-west European countries (Sweden, Denmark, UK, Netherlands, Germany and Norway) were consistently first signatories of the global or European Conventions. The EU was also consistently early although it was not legally possible to adopt some MEAs such as Ramsar. Late/recent signatories comprise some of the newly independent Balkan and Baltic states (Latvia, Slovenia, North Macedonia, Lithuania, Croatia, Albania, Moldova, Serbia, and Bosnia and Herzegovina).

More detailed accounts of the Ramsar and Bern Conventions, relevant instruments under CMS, and marine Conventions and ministerial conferences are available in online Annex 1.[1] EU policies and measures are described more fully in Chapter 4 given their profound influence over the Member States (MS), and significant indirect effects on many other European countries.

3.3 The Convention on Wetlands (Ramsar Convention)

One of the principal drivers for the development of the Ramsar Convention had been concern over the impacts of wetland loss on waterfowl (i.e. waterbirds), but the wider socio-economic importance of wetlands was also recognised. With time the Convention has increasingly promoted an appreciation of these diverse wetland values and pursued holistic wetland conservation goals. Consequently, the Convention is now normally known as the 'Wetlands Convention' rather than its full legal title with the specific reference to waterfowl habitat.

The Ramsar Convention was the first of the modern environmental conventions, and it has had an important influence on nature conservation, not least in establishing a significant European network of internationally protected wetlands prior to the advent of the Natura 2000 network: 133 Ramsar Sites in 10 then, or yet to be, Member States had already been designated at the time the Birds Directive entered into force in 1979. The Convention also developed the concept of the '1% Criterion' for assessing international importance of sites by proportionate assessment of threshold waterbird numbers (Atkinson-Willes *et al.*, 1982; Stroud and Davidson, 2021). Parties accept an obligation to maintain the ecological character of Ramsar Sites (Stroud, 2018). However, this policy obligation does not necessarily result in legal protection at national level; it typically requires additional national legislative protection, which may not always occur.

[1] All annexes are free to access online at CUP: www.cambridge.org/natureconservation

Table 3.1 Key events in the evolution of nature conservation policy and legislation in Europe.

1946	International Convention for the Regulation of Whaling – established the International Whaling Commission
1956	The Treaties of Rome signed: one established the European Economic Community (EEC), comprising Belgium, France, Germany, Italy, Luxembourg and the Netherlands
1971	Ramsar Convention adopted (entered into force in 1975)
1972	UN (Stockholm) Conference on the human environment – the first major global gathering to address environmental issues
	EEC Paris Summit agreed to the development of environmental policy
	World Heritage Convention signed
	Oslo Convention for the Prevention of Marine Pollution by Dumping from Ships and Aircraft
1973	First EEC Environmental Action Programme
	Denmark, Ireland and the UK joined the EEC (nine MS)
1974	Paris Convention for the Prevention of Marine Pollution from Land-Based Sources adopted
	The Convention on the Protection of the Marine Environment in the Baltic Sea Area (Helsinki Convention), entered into force 1980
1976	Convention for Protection of the Mediterranean Sea against Pollution (Barcelona Convention), entered into force 1978
1979	Birds Directive notified and came into force 6 April 1979
	Bern Convention agreed (came into force in 1982 and ratified by the EEC and some MS)
	Convention on Migratory Species (Bonn Convention) adopted (came into force in 1985)
1980	Common Market measures can lawfully pursue environmental objectives (Court of Justice)
1981	Greece joined the EEC (ten MS)
1982	United Nations Convention on the Law of the Sea (UNCLOS), came into force in 1994
1984	First of six North Sea Ministerial Conferences, until 2006
1985	Environmental protection is 'one of the Communities' essential objectives' (Court of Justice)
	Environmental Impact Assessment (EIA) Directive
1986	Portugal and Spain joined the EEC (12 MS)
1987	Publication of the Brundtland Report *'Our Common Future'* by the UN, which popularised the term 'sustainable development'
	The Single European Act came into force, embedding environmental policy within the EEC Treaty
1988	Environmental protection became a 'mandatory requirement' of the EEC (Court of Justice)
1989	Directorate General for the Environment formed in the EEC Commission
1990	European Environment Agency (EEA) established (work began in 1994)
1991	The Nitrates Directive
	EU Urban Waste Water Treatment Directive (UWWTD)
	The Plant Protection Products Directive
1992	UN Conference on the Environment and Development (the 'Earth Summit') led to the Rio Conventions, including the Convention on Biological Diversity (CBD)
	Habitats Directive notified 21 May 1992 (came into force 10 June 1994)
	The LIFE programme established – the EU's dedicated funding instrument for the environment and climate
	Environment policy became a significant element of the Common Agricultural Policy (CAP), and CAP Regulation 2078/92 makes agri-environment schemes compulsory for MS
	The Treaty on European Union (Maastricht Treaty) signed (came into force in 1993) under which the EEC became the European Community (EC), with the precautionary principle included in the Treaty
	Convention for the Protection of the Marine Environment in the North-East Atlantic (OSPAR Convention) adopted (came into force 1998), replacing the 1972 Oslo Convention and 1974 Paris Convention
	The Helsinki Convention updated, entered into force 2000
	The Bucharest Convention on the Protection of the Black Sea against Pollution adopted, entered into force 1994
1995	Council of Europe's *Pan-European Biological and Landscape Diversity Strategy* (PEBLDS) approved at the Ministerial Conference 'Environment for Europe' in Sofia

	Austria, Finland and Sweden joined the EC (15 MS)
	The Barcelona Convention updated, to become the Convention for the Protection of the Marine Environment and the Coastal Region of the Mediterranean
1998	Aarhus Convention adopted (came into force in 2001), which establishes rights for citizens' access to information, participation and access to justice in relation to decision making on the environment
	European Commission adopted *EC Biodiversity Strategy*
	OSPAR scope widened and commitment made to establish a network of marine protected areas
1999	Amsterdam Treaty came into force: includes the promotion of sustainable development as an objective of the EC, and also requires environmental protection to be integrated across all policies
2000	Water Framework Directive (WFD)
2001	*European Community Sustainable Development Strategy* (renewed in 2006)
	Strategic Environmental Assessment (SEA) Directive
	National Emission Ceilings Directive (NECD)
	EU Heads of State adopted a target of halting the loss of biodiversity in the EU by 2010
	European Commission produced *Biodiversity Action Plans* in the areas of Conservation of Natural Resources, Agriculture, Fisheries, and Development and Economic Cooperation
2003	Fifth Environment for Europe Ministerial Conference agreed the Kyiv Resolution on Biodiversity, which extended the EU target of halting the loss of biodiversity by 2010 to the pan-European region
2004	Cyprus, the Czech Republic, Estonia, Hungary, Latvia, Lithuania, Malta, Poland, Slovakia and Slovenia joined the EU (25 MS)
	Environmental Liability Directive (ELD)
2005	The *Millennium Ecosystem Assessment* published
	CAP European Agricultural Fund for Rural Development (EAFRD) introduced with requirement for MS to spend at least 25% on environmental measures
2006	European Commission *EU Biodiversity Action Plan* to achieve the EU 2010 biodiversity target
2007	Lisbon Treaty led to the Treaty on European Union (TEU) and the Treaty on the Functioning of the European Union (TFEU), and created a more powerful European Parliament. Came into force 2009
2007	Bulgaria and Romania joined the EU (27 MS)
2008	The Integrated Pollution Prevention and Control (IPPC) Directive
	The Marine Strategy Framework Directive (MSFD)
	Directive on the Protection of the Environment through Criminal Law
2010	Industrial Emissions Directive
	Tenth CBD COP adopted a revised *Strategic Plan for Biodiversity*, including the Aichi Biodiversity Targets for 2011–2020
	EU adopted a target of halting the loss of biodiversity and the degradation of ecosystem services by 2020
2011	European Commission adopted the *EU Biodiversity Strategy to 2020*
2012	The Rio +20 Earth Summit launched development of the Sustainable Development Goals, adopted by UN General Assembly in 2015
2013	Croatia joined the EU (28 MS)
2015	The Invasive Alien Species (IAS) Regulation
2016	European Commission published the 'Fitness Check' evaluation of the EU Birds and Habitats Directives
2017	In response to Fitness Check conclusions, European Commission published the *Action Plan for Nature, People and the Economy* to improve the implementation of the Nature Directives
2019	EU European Green Deal
2020	The UK left the EU (27 MS)
	EU Biodiversity Strategy for 2030, aims to have Europe's biodiversity on the path to recovery by 2030; includes a commitment to propose legally binding targets for restoration in 2021
2022	European Commission published a proposal for a Regulation on nature restoration

Sources. Edited and expanded version of tables in Bell *et al.* (2013) and Earll (2019), information in Farmer (2012) and DG Environment website.

Table 3.2 Key objectives and obligations of the multilateral environmental agreements and initiatives that have had a significant impact on nature conservation in Europe.

Objectives	Obligations	Effects
UNESCO Man and Biosphere Programme (MAB): launched in 1970		
Objectives: An intergovernmental scientific programme that aims to establish a scientific basis for the improvement of relationships between people and their environments	**Obligations**: The nomination of Biosphere Reserves for the conservation of biodiversity, for sustainable human development locally, and for interdisciplinary research. Each reserve should qualify under one or more of the following categories: • Representative of a natural biome • Unique community or area with unusual natural features of exceptional interest • Harmonious landscape, resulting from traditional patterns of land use • Modified or degraded landscape capable of being restored to more natural conditions	**Effects**: Stimulated interdisciplinary research and novel site-based participatory approaches to environmental protection and sustainable development in Biosphere Reserves, which incorporate core, buffer and transition zones. Relatively few Reserves have been established in most European countries, only core areas provide effective protection, and achievement of Reserve goals is variable. See Bridgewater (2016) for further discussion
Convention on Wetlands of International Importance especially as Waterfowl Habitat (Ramsar Convention): adopted 1971, entered into force 1975		
Objectives: The conservation and wise use of all wetlands through local and national actions and international cooperation, as a contribution towards achieving sustainable development throughout the world	**Obligations**: In summary: • Commit to the wise use of all wetlands 'as far as possible' • Designate suitable wetlands for the list of Wetlands of International Importance (the 'Ramsar List') and ensure their effective management • Cooperate internationally on transboundary wetlands, shared wetland systems and shared species	**Effects**: Although relatively few Ramsar Sites have been designated in some countries, and designation does not provide direct legal protection, the Convention has had an important influence on nature conservation – see text for details
World Heritage Convention: adopted 1972, entered into force 1975		
Objectives: To ensure the protection of natural and cultural areas of outstanding universal value as a duty of the international community as a whole, by granting collective assistance	**Obligations**: The identification, protection, conservation, presentation and transmission to future generations of the Party's cultural and natural heritage	**Effects:** Strict obligations give a high level of protection for World Heritage Sites, and the Convention offers financial and technical assistance, but few sites are listed in Europe for their natural heritage
The Convention on International Trade in Endangered Species of Wild Fauna and Flora (CITES): adopted 1973, entered into force 1975		
Objectives: To ensure that international trade in specimens (or derivatives) of wild animals and plants does not threaten their survival	**Obligations**: Parties must authorise through a licensing system all imports, exports, re-exports and introductions of species listed in three Appendices, according to the level of protection or regulation of trade that they are assigned. Appendix I includes species threatened with extinction, and trade of these is permitted only in exceptional circumstances; Appendix II includes species not necessarily threatened with extinction, but in which trade must be controlled; Appendix III contains species that are protected in at least one country, which has asked other CITES Parties for assistance in controlling the trade Due to EU competency over trade, implementation in the EU is governed by EU regulations	**Effects**: Few listed species naturally occur in Europe, but Europe is a major trade hub for derivatives of protected species, as well as having major importation or transit trade in live species for aquaculture (live tropical fish and corals especially). EU implementing regulations include certain non-CITES species, and also contain provisions to prohibit or restrict imports of species that are considered to be a threat to native EU fauna and flora
Convention for the Prevention of Pollution from Ships, 1973, as modified by the Protocol of 1978 (MARPOL 73/78): adopted 1973, entered into force 1983		
Objectives: To prevent and control deliberate discharges and accidental spills from shipping into the marine environment	**Obligations**: Measures to prevent pollution by oil, noxious liquid substances in bulk, harmful substances carried by sea, sewage and garbage	**Effects**: Packages of maritime laws that aim to improve safety in the shipping industry and reduce environmental damage

Table 3.2 *(cont.)*

UNESCO Man and Biosphere Programme (MAB): launched in 1970

Helsinki Convention on the Protection of the Marine Environment of the Baltic Sea Area: adopted 1974, replaced with 1992 Helsinki Convention, entered into force 2000

Objectives	Obligations	Effects
Objectives: To prevent and eliminate pollution in order to promote the ecological restoration of the Baltic Sea Area and the preservation of its ecological balance	**Obligations**: • Various measures to prevent pollution and prohibit dumping and incineration at sea • Art. 15: Parties shall take measures to conserve natural habitats and biological diversity and to protect ecological processes. Such measures shall also be taken in order to ensure the sustainable use of natural resources	**Effects**: Important driver of protection, notably from shipping impacts (Baltic Sea designated a Particularly Sensitive Sea Area in 2005 under the International Maritime Organisation (IMO)). Baltic Sea also used as a pilot study for the EU MSFD, and the Convention creates a platform for that Directive and other European and national instruments. In 2007 Parties adopted an ambitious Baltic Sea Action Plan, which details measures and actions needed to achieve 'good ecological status' of the Baltic marine environment by 2021

Convention for the Protection of the Mediterranean Sea against Pollution (Barcelona Convention): adopted 1976, entered into force 1978; modified in 1995 to become the **Convention for the Protection of Marine Environment and the Coastal Region of the Mediterranean**, entered into force in 2004

Objectives	Obligations	Effects
Objectives: To counteract the introduction of certain pollutants into the Mediterranean Sea area (including internal waters), and to conserve natural species and habitats	**Obligations**: • To take measures to prevent, reduce and, as appropriate, eliminate pollution by oil and other harmful substances; dumping from ships and aircraft; from land-based sources; resulting from exploration and exploitation of the continental shelf and seabed • Address problems caused by the transboundary movement of wastes and their disposal. To address the conservation of specially protected areas and biodiversity in the marine and coastal zone	**Effects**: Important driver of protection, along with its seven Protocols, including the innovative SPAMI (Specially Protected Area of Marine Importance) Protocol (entered into force 1999) providing for collaborative establishment of MPAs between two or more Parties, and enabling designation of Areas Beyond National Jurisdiction (ABNJs) (Raftopoulos, 2011). The Convention benefits the EU MSFD and other European and national instruments by creating a platform for MS to agree and develop monitoring programmes and metrics for assessment, and draw on collective national resources

Convention on Migratory Species (CMS): adopted 1979, entered into force 1983

Objectives	Obligations	Effects
Objectives: The conservation and effective management of species over the whole of their migratory range	**Obligations**: (see text) include: • Adopt strict protection measures for migratory species that have been categorised as endangered (listed in Appendix I) • Conclude Agreements for the conservation of migratory species that have an unfavourable conservation status or would benefit significantly from international cooperation (listed in Appendix II) • Undertake research activities relating to migratory species	**Effects**: Has led to a number of Agreements, and other less formal instruments covering migratory species, including for waterbirds, cetaceans, albatrosses and petrels, and bats; giving added value in protection for EU species beyond EU territory

Convention on the Conservation of European Wildlife and Natural Habitats (Bern Convention): adopted 1979, entered into force 1982

Objectives	Obligations	Effects
Objectives: To conserve wild flora and fauna and their natural habitats, especially those species and habitats	**Obligations**: (see text) in summary: Take appropriate and necessary legislative and administrative measures to ensure the conservation of the habitats of the wild flora	**Effects**: Key driver of the EU Habitats Directive and Natura 2000 network, and the similar

Table 3.2 *(cont.)*

UNESCO Man and Biosphere Programme (MAB): launched in 1970		
whose conservation requires the cooperation of several states	and fauna, especially those specified in Appendices I and II, and the conservation of endangered natural habitats Requirement to establish Emerald Network of protected areas first introduced in 1989	complementary Emerald Network and basic species protection measures elsewhere
The Convention on Long-Range Transboundary Air Pollution (CLRTAP): open for signature in 1979 and adopted in 1983		
Objectives: To protect the human environment against air pollution and to gradually reduce and prevent air pollution, including long-range transboundary air pollution. Covers the entire United Nations Economic Commission for Europe region, which includes Russia and North America	**Obligations**: Parties are required to develop and review policies and strategies to combat the discharge of air pollutants, based on consultation, research and monitoring; and undertake consultations between those affected by air pollution or those who are the potential sources of air pollutants	**Effects**: Through scientific collaboration and policy negotiation, has contributed to declines in major air pollutants, including through eight protocols identifying specific measures to be taken by Parties to cut emissions
United Nations Convention on the Law of the Sea (UNCLOS): adopted 1982, entered into force 1994		
Objectives: A legal treaty facilitating international communication and promoting peaceful uses of seas and oceans; equitable and efficient utilisation of their resources; conservation of living resources; and study, protection and preservation of the marine environment	**Obligations**: 320 articles and nine annexes dealing with almost all aspects of ocean use. Exploitation and protection of living resources and anti-pollution measures are covered, albeit only generally	**Effects**: Main legal driver of EU cooperation on ocean governance with bilateral, regional and multilateral partners globally. The EU MSFD recognises and takes full account of UNCLOS obligations. The 1995 UN Fish Stocks Agreement builds on the UNCLOS provisions
Convention on Biological Diversity (CBD): adopted 1992, entered into force 1994		
Objectives: To ensure the conservation of biological diversity, the sustainable use of its components and the fair and equitable sharing of the benefits arising out of the utilisation of genetic resources	**Obligations**: (see text) include: • Prepare national strategies, plans or programmes for the conservation and sustainable use of the nation's biological resources, and integrate these into other sectoral or cross sectoral plans, programmes and policies • Promote the protection of ecosystems, natural habitats and the maintenance of viable populations of species in natural surroundings, and to rehabilitate and restore degraded ecosystems and promote recovery of threatened species, *inter alia*, through the development and implementation of plans or other management strategies	**Effects**: Major driver of European, EU, and national level biodiversity strategies, targets and plans, in particular in response to the Aichi Targets agreed at the Tenth COP in 2010, e.g. in relation to protected area coverage. Few substantial new measures to address most other targets
United Nations Framework Convention on Climate Change (UNFCC): adopted 1992, entered into force 1994 (elaborated by the 2015 Paris Agreement)		
Objectives: The stabilisation of greenhouse gas concentrations in the atmosphere at a level that would prevent dangerous anthropogenic interference with the climate system. Such a level should be achieved within a timeframe sufficient to allow ecosystems to adapt naturally to climate change, to ensure that food production is not threatened, and to enable economic development to proceed in a sustainable manner	**Obligations**: • Develop national inventories of greenhouse gases • Formulate policies to mitigate climate change • Cooperate in preparing for adaptation to climate change • Promote international cooperation in the transfer of technology	**Effects**: Major driver of national and international environmental policy, especially in the EU
Convention on the Protection of the Black Sea against Pollution (Bucharest Convention): adopted 1992, entered into force 1994		
Objectives: To prevent, reduce and control the pollution in the Black Sea in	**Obligations**: • Various measures to prevent pollution and prohibit dumping at sea	**Effects**: Considered a weaker governance framework than other Regional Seas Conventions, having

Table 3.2 *(cont.)*

UNESCO Man and Biosphere Programme (MAB): launched in 1970		
order to protect and preserve the marine environment	• Protect biodiversity and the marine living resources • Provide a framework for scientific and technical cooperation and monitoring activities	fewer Protocols and less effective implementation than, e.g., the Mediterranean Sea programme (Oral, 2011)
Convention for the Protection of the Marine Environment of the North East Atlantic (OSPAR Convention): adopted 1992, entered into force 1998. The result of revising and merging the Oslo and Paris Conventions		
Objectives: Prevent and eliminate pollution and take the necessary measures to protect the maritime area against the adverse effects of human activities so as to safeguard human health and to conserve marine ecosystems and, when practicable, restore marine areas that have been adversely affected	**Obligations**: • Prevent and eliminate pollution from dumping, incineration, land-based, offshore, and other sources • Undertake and publish regular assessments of the quality status of the marine environment • Carry out scientific and technical research	**Effects**: Important driver, notably of MPA network, including ABNJs. Benefits the EU MSFD and other EU instruments by creating a platform for MS to agree and develop monitoring programmes and metrics for assessment, drawing on collective national resources
Pan-European Biological and Landscape Diversity Strategy (PEBLDS): endorsed by EU environment ministers in 1995		
Objectives: Reduce threats to biodiversity and landscapes, increase resilience and coherence, and involve people, through: 1. Creation and effective management of a Pan-European Ecological Network (PEEN) 2. Sustainable management and use of Europe's biological and landscape diversity 3. Integration of biological and landscape diversity conservation and sustainable use objectives into all relevant sectors 4. Improved information and awareness of biological and landscape diversity issues, and increased public participation in actions 5. Improved understanding of the state of Europe's biological and landscape diversity 6. Assurance of adequate financial means to implement the Strategy	**Obligations**: Under existing legal obligations, such as the CBD, a wide range of actors were encouraged to implement the Strategy, through a series of five-year Action Plans, which focused on the following themes for 1996–2000: • Establishing the PEEN • Integration of biological and landscape diversity considerations into sectors • Raising awareness and support with policymakers and the public • Conservation of landscapes • Ecosystem actions: coastal and marine, rivers, wetlands, grassland, forests and mountains • Action for threatened species	**Effects**: Raised awareness of the issues, led to indicative PEEN maps, and catalysed actions on some other objectives (primarily outside the EU) but these were constrained by lack of capacity and available financing (Hajdu and Kiss, 2009; Prip, 2013) Led to Kyiv Resolution on Biodiversity at the 5th Ministerial Conference in 2003. Replaced by the Pan-European 2020 Strategy for Biodiversity – see below
Pan-European 2020 Strategy for Biodiversity, endorsed by pan-European countries in 2011		
Objectives: To improve cooperation and capacity through: • A platform for the exchange of views for the implementation of the CBD Strategic Plan, other MEAs and initiatives • Direct support for pilot activities • Regional cooperation for policy development, implementation and the identification of priority themes • Enhanced international coherence and synergies of biodiversity policies and priority actions	**Obligations**: Countries are encouraged to take the following actions: 1. Develop a resource mobilisation strategy 2. Re-establish the Biodiversity Service for National Biodiversity Strategies and Action Plans (NBSAP) delivery 3. Promote cooperation and synergies of biodiversity-related Conventions in the pan-European region 4. Develop issue-based modules under InforMEA – the UN Information Portal on MEAs project 5. Support streamlining of reporting to MEAs 6. Joint implementation of biodiversity-related Convention activities in pilot countries 7. Undertake thematic actions	**Effects**: Information on its implementation and effects does not appear to be currently available

Table 3.2 *(cont.)*

UNESCO Man and Biosphere Programme (MAB): launched in 1970		
Convention on Access to Information, Public Participation in Decision Making and Access to Justice in Environmental Matters (Aarhus Convention): adopted 1998, entered into force 2001		
Objectives: To contribute to the protection of the right of every person of present and future generations to live in an environment adequate to his or her health and well-being	**Obligations**: Parties are required to make the necessary provisions so that public authorities (at national, regional or local level) guarantee: • The right of everyone to receive environmental information that is held by public authorities ('access to environmental information') • The right to participate in environmental decision making • The right to review procedures to challenge public decisions that have been made without respecting the two aforementioned rights or environmental law in general ('access to justice')	**Effects**: In EU: 2003 EU Directive for access to information 2003 EU Directive on public participation 2017 Notice on Access to Justice in Environmental Matters
International Convention for the Control and Management of Ships' Ballast Water and Sediments (Ballast Water Management Convention or **BWM Convention)**: adopted 2004, entered into force 2017		
Objectives: To help prevent the spread of potentially harmful aquatic organisms and pathogens in ships' ballast water	**Obligations**: All international sea-going ships must: • Implement a 'Ballast Water Management Plan' to enable management of ballast water and sediment discharge to a standard set by the Convention and the International Maritime Organisation (IMO) guidelines • Comply with standard operational guidance (notably on where and how ballast water can be discharged), planning and management, sampling points and systems, etc.	**Effects**: No EU Law but the Invasive Alien Species (IAS) Regulation recognises the BWM Convention among possible management measures. The level and environmental impact of IAS are also a descriptor for assessing Good Environmental Status under the EU MSFD

Sources. Own compilation based on convention texts and websites, Farmer (2012) and Bell *et al.* (2013).

The Convention also introduced the concept of 'wise use', which has been interpreted as being synonymous with 'sustainable use' (Ramsar Convention Secretariat, 2010). Its initial definition in 1987 (COP 3) has been updated and it is now defined as 'the maintenance of [wetland's] ecological character, achieved through the implementation of ecosystem approaches, within the context of sustainable development' (Ramsar Convention Secretariat, 2010; Finlayson *et al.*, 2011). Parties commit to work towards the wise use of all the wetlands and water resources in their territory, through agreed approaches including policy measures, wetland inventories, monitoring, research, education, public awareness and integrated site management planning (Ramsar Convention Secretariat, 2013).

In the EU many or most of the Ramsar Convention's obligations are implemented by Member States through the EU's environmental instruments (e.g. the Nature Directives, MSFD and WFD). Some have argued that given this, the Convention is perhaps of greater significance in regions lacking such richness in environmental policy. However, analysis shows that the Convention has had an important role in stimulating such legislative and policy development, especially through opportunities for intergovernmental discussions at triennial COPs and other meetings.

3.4 The Convention on the Conservation of Migratory Species of Wild Animals

The Convention on the Conservation of Migratory Species of Wild Animals (CMS), also known as the Bonn Convention, derived directly from a recommendation of the 1972 Stockholm Conference on the human environment. It was agreed in 1979, entered into force in 1983, and has been ratified by all Member States and the EU. The underlying principle of the CMS is that States should take concerted action to ensure appropriate conservation and management of threatened populations of shared migratory species listed on Appendix I that occur within their borders (that is, when they are Range States). The Convention also provides (Article IV) for the conclusion, between Range States, of special international agreements for the protection of particular migratory species that have an unfavourable conservation status and are listed in Appendix II. In this respect, CMS acts as a framework Convention. The agreements may range from legally binding treaties (called Agreements) to less formal instruments, such as Memoranda of Understanding (MoU), and can be adapted to the requirements of particular regions.

Table 3.3 European Parties to the international nature conservation focussed Conventions.

Key: n/a = not applicable; NP = non-party. CMS indicates date Convention in force, and which are Parties to European Agreements: on AP = ACAP; CM = ACCOBAMS; CN = ASCOBANS; EB = EUROBATS; WA = AEWA; WS = Wadden Sea Seals.
As at 20 August 2021.

	Ramsar	Bern	CMS	CBD
	In force	In force	In force	Party since
Albania	1996	1999	2001 CM EB WA	1994
Austria	1983	1983	2019	1994
Belarus	1991	1990	2003 WA	1993
Belgium	1986	1990	1990 CN EB WA	1997
Bosnia and Herzegovina	1992	2009	2017	2002
Bulgaria	1976	1991	1999 CM EB WA	1996
Croatia	1991	2000	2000 CM EB WA	1997
Cyprus	2001	1988	2001 CM EB WA	1996
Czech Republic	1993	1998	1994 EB WA	1994
Denmark	1978	1983	1983 CN EB WA WS	1994
Estonia	1994	1992	2008 EB WA	1994
European Union	n/a	1982	1983 WA	1994
Finland	1975	1986	1989 CN EB WA	1994
France	1986	1990	1990 AP CM CN EB WA	1994
Germany	1976	1985	1984 CN EB WA WS	1994
Greece	1975	1983	1999 CM	1994
Hungary	1979	1990	1983 EB WA	1994
Iceland	1978	1993	NP WA	1994
Ireland	1985	1982	1983 EB WA	1996
Italy	1997	1982	1983 CM EB WA	1994
Latvia	1995	1997	1999 EB WA	1996
Liechtenstein	1991	1982	1997	1998
Lithuania	1993	1997	2002 CN EB WA	1996
Luxembourg	1998	1992	1983 EB WA	1994
Malta	1989	1994	2001 CM EB	2001
Moldova	2000	1994	2001 EB WA	1996
Montenegro	2006	1994	2009 CM EB WA	2006
Netherlands	1980	1982	1983 CN EB WA WS	1994
North Macedonia	1991	1999	1999 EB WA	1998
Norway	1975	1986	1985 AP EB WA	1993
Poland	1978	1996	1996 CN EB	1996
Portugal (inc. Azores, etc.)	1981	1982	1983 CM EB WA	1994
Romania	1991	1993	1998 CM EB WA	1994
Serbia	1992	2008	2008 WA	2002
Slovakia	1993	1997	1995 EB WA	1994

Table 3.3 *(cont.)*

	Ramsar In force	Bern In force	CMS In force	CBD Party since
Slovenia	1991	2000	1999 CM EB WA	1996
Spain (inc. Canaries, etc.)	1982	1986	1985 AP CM WA	1994
Sweden	1975	1983	1983 CN EB WA	1994
Switzerland	1976	1982	1995 EB WA	1995
Ukraine	1991	1999	1999 CM EB WA	1995
United Kingdom	1976	1982	1985 AP CN EB WA	1994

Sources. Based on data from the Ramsar Convention Secretariat[2]; Council of Europe[3]; CMS[4]; CBD[5].

The following CMS Agreements of European relevance have been concluded to date (see Table 3.3 for Parties):

- Wadden Sea Seals: Agreement on the Conservation of Seals in the Wadden Sea. Three Parties: Denmark, Germany and the Netherlands. Entered into force: 1 October 1991.
- EUROBATS: Agreement on the Conservation of Populations of European Bats. Thirty-two European parties. Entered into force: 16 January 1994.
- ASCOBANS: Agreement on the Conservation of Small Cetaceans in the Baltic, North East Atlantic, Irish and North Seas. Seventeen European Parties plus Russia. Entered into force: 29 March 1994.
- AEWA: Agreement on the Conservation of African-Eurasian Migratory Waterbirds. Thirty-five European Parties. Entered into force: 1 November 1999.
- ACCOBAMS: Agreement on the Conservation of Cetaceans of the Black Seas, Mediterranean and Contiguous Atlantic Area. Ten European Parties. Entered into force: 1 June 2001.
- ACAP: Agreement on the Conservation of Albatrosses and Petrels (primarily addresses southern oceanic species including those threatened by high seas EU fishing, but in European waters includes the globally Critically Endangered Balearic Shearwater (*Puffinus mauretanicus*)). Four European Parties. Entered into force: 1 February 2004.

Of these, AEWA has been of particular importance, as it has extended a comprehensive conservation framework, similar to the Birds Directive, to migratory waterbirds over 119 Range States covering roughly a third of the planet (Lewis, 2016). Further information is provided on AEWA, ACAP, ASCOBANS and ACCOBAMS in online Annex 1.

A number of less formal MoUs and Action Plans have also been developed under the auspices of the CMS; those of relevance to the EU relate to the Aquatic Warbler (*Acrocephalus paludicola*), Atlantic Turtles, middle European population of the Great Bustard (*Otis tarda*), migratory raptors of Africa-Eurasia, Mediterranean Monk Seal (*Monachus monachus*), Slender-billed Curlew (*Numenius tenuirostris*) and sharks. The Action Plan on Africa-Eurasian Landbirds adopted by COP 11 is particularly relevant, complementing – for landbirds – other bird-related instruments for the European region. However, like many of the less formal CMS instruments, it remains largely unimplemented. As a recent example, the Raptors MoU is further described in online Annex 1.

Goriup and Tucker (2005) outline the strengths and weaknesses of different types of instrument under CMS in the context of intergovernmental negotiations regarding what ultimately became the Raptors MoU. Whilst a full Agreement is legally binding, and requires financial subscriptions to fund Secretariat functions, it can be cumbersome to negotiate and enter into force: it took four years for AEWA to enter into force, and essentially the same time again for it to start to develop a coordinated body of significant conservation activities. MoUs and Action Plans have the advantage of being more flexible means of cooperation but have the potentially major weakness of being dependent on CMS (or Parties) for their funding and coordination structures – which are critical for successful implementation.

Beyond its Agreements, CMS has gained significance for the EU through its species listing process. Article III requires Parties to provide strict protection for globally threatened species listed on Appendix I of the Convention as well as conserving and restoring their habitats and removing impediments to their migration. Appendix I is typified by many species of charismatic mega-fauna. The requirement to undertake habitat-related measures has been given little attention by Parties but was stressed in Resolution 12.25 (2017) in the context of the plight of multiple Appendix I species using coastal environments.

Recent COPs have also adopted a number of useful decisions related to migratory species conservation with significant EU relevance, notably relating to developing a programme of

[2] www.ramsar.org/country-profiles
[3] www.coe.int/en/web/conventions/full-list/-/conventions/treaty/104/signatures/
[4] www.cms.int/en/cms-instruments/agreements
[5] www.cbd.int/information/parties.shtml

work on climate change and migratory species (Resolution 11.26); preventing the poisoning of migratory birds (Resolution 11.15); preventing the illegal killing, taking and trading of migratory birds (Resolution 11.16); connectivity and ecological networks with respect to migratory species (Resolution 11.25); minimising impacts of renewable energy (Resolution 11.27); and promoting the conservation of critical intertidal and other coastal habitats for migratory species (Resolution 12.25).

The discharge of lead gunshot into the environment is a long-established cause of poisoning of birds and other wildlife (see Section 2.2.3) and is a cross-cutting issue that CMS, AEWA[6] and the Raptors MoU have tried to address. Kanstrup *et al.* (2018) concluded that 'Hunting with lead ammunition cannot be justified under established principles of public/international policy and is not sustainable.' However, although CMS in 2014 urged its Parties to 'Phase-out the use of lead ammunition across all habitats (wetland and terrestrial) with non-toxic alternatives within the next three years' (Resolution 11.15), progress to that end has been very slow (Mateo and Kanstrup, 2019). Nevertheless, the CMS Resolution supported the EU's policy via REACH chemicals Regulations[7] to prohibit the use or possession of lead gunshot within and near wetlands from 15 February 2023.

3.5 The Convention on the Conservation of European Wildlife and Natural Habitats (Bern Convention)

The Convention on the Conservation of European Wildlife and Natural Habitats was drawn up under the auspices of the Council of Europe and signed in Bern on 19 September 1979, and hence is hereafter referred to as the Bern Convention.[8] The Convention became effective on 1 June 1982 with its secretariat being provided by the Council of Europe. Currently there are 51 signatories to the Convention including five non-members of the Council of Europe and the EU who ratified its signature once all the then members of the European Community had done so.

The aims of the Convention are to conserve wild flora and fauna and their natural habitats, especially those species and habitats whose conservation requires the cooperation of several states, and to promote such cooperation. Particular emphasis is given to protecting endangered and vulnerable species, including those that are migratory, that are listed in the appendices. Appendix I lists over 700 plant species, many of which are endemic. Appendix II lists over 700 species, many of which are migratory. Appendix III is a long list of animals, including many common mammals, and all birds, reptiles and amphibians not in Appendix II, with the exception of a few common bird species. Species on Appendix III are protected from unsustainable hunting or other forms of exploitation. The Convention includes provisions that allow Parties to derogate from the species protection provisions for exceptional reasons.

Parties are required to:

- take measures to maintain the population of wild flora and fauna (not just those in the appendices) at a level that corresponds in particular to ecological, scientific and cultural requirements, whilst taking account of economic and recreational requirements;
- take steps to promote national policies for nature conservation, giving particular attention to endangered and vulnerable species, especially endemic ones, and endangered habitats;
- have regard to the conservation of wild flora and fauna in its planning and development policies and in its measures against pollution; and
- promote education and dissemination of information on the need to conserve species of wild flora and fauna and their habitats.

The Convention established a Standing Committee that is responsible for the application of the Convention. It has the authority to make Recommendations to the Parties, which are an important means of giving substance to the provisions of the Convention (Jones, 2012). In turn, the Committee has established a number of expert groups to provide advice, including on the production of species action plans, codes of conduct and guidelines. Parties are required to submit biennial reports on their implementation of the Convention.

In some ways, the Convention provides a strong basis for a legislative framework for Parties – supported by the relevant species action plans, codes of conduct and guidelines – to fulfil their obligations under the Convention. Most notably, most of the provisions are mandatory and clearly worded as 'requirements', and not couched in the hortatory language used by some other wildlife treaties (Lyster, 1985; Bowman, *et al.*, 2010). However, its enforcement mechanisms are weak: there are no penalties and no requirement for restoration or compensation. The main incentive for compliance is the possible embarrassment for the relevant Party of having a 'case file' remain open so that no progress is reported at Standing Committee meetings. The case file system has further shortcomings according to Jones (2012). Therefore, the Convention primarily relies on persuasion, supported by consensus building, partnerships and guidelines, etc. (Díaz, 2010).

The Convention had a considerable impact on the drafting of the EU Birds and Habitats Directives. Both Directives mirror the provisions of the Convention but are underpinned by far more robust enforcement provisions. The Birds Directive was drafted in parallel with the Convention, whilst the Habitats Directive is based on a more considered implementation of the Convention's non-avian obligations. Importantly, the Habitats Directive replicated the Convention's style of mandatory language, and built

[6] For example, www.unep-aewa.org/en/topics/lead-poisoning

[7] https://eur-lex.europa.eu/legal-content/EN/TXT/PDF/?uri=CELEX:32021R0057&rid=1

[8] Based on the German spelling of the city, as used by the Council of Europe, rather than the French spelling of 'Berne'.

upon its aims and objectives. This has included the protection of endangered natural habitat types, not for particular species, but for their own value. As the Bern Convention had not at the time produced a list of endangered habitats, this was developed for the Habitats Directive, as Annex I, and subsequently adapted by the Convention under Resolution No. 4. The Habitats Directive also amends the Birds Directive so that robust protection measures are also afforded to bird sites (further discussed in Section 4.2).

Conversely, the development of the Habitats Directive influenced the implementation of the Bern Convention. Most importantly, under Resolution 3 in 1996, the Council of Europe's earlier network ideas (in 1989 Recommendation 16) were further developed such that the Emerald Network became modelled on Natura 2000. The intention was that it would be a pan-European network, extending the Natura 2000 concept to countries outside the EU. However, it is important to note that, unlike Natura 2000 sites, the designation of an Area of Special Conservation Interest (ASCI) does not by itself result in legal protection. Under Recommendation 16, Parties are asked to ensure, wherever possible, that ASCIs are the subject of an appropriate regime designed to conserve the species and habitats for which they were designated; but this does not necessarily mean that they become protected areas, as defined by IUCN/CBD.

3.6 The Convention on Biological Diversity

The Convention on Biological Diversity (CBD) was signed by the EC and Member States, and some other European countries, at the UN Conference on Environment and Development held in Rio de Janeiro in June 1992. The Convention came into force on 29 December 1993 (90 days after the 30th ratification). All European countries are Parties to the Convention (other than the Holy See (Vatican)), as well as the EU in its own right (see Table 3.3 for dates).

The rationale and impetus for the CBD were driven and influenced by a number of prior initiatives including the *World Conservation Strategy* (IUCN, UNEP and WWF, 1980), *UN World Charter for Nature* (UNGA, 1982), *Strategy for Sustainable Living* (IUCN, UNEP and WWF, 1991), and the Report of the World Commission on Environment and Development, *Our Common Future* (WCED, 1987).[9] These recognised the value of biodiversity, in its widest sense, to humankind, and highlighted the need for global action to conserve biodiversity by ensuring that its use is sustainable, and that all forms of development do not threaten biodiversity, and hence are also sustainable.

These considerations are reflected in the CBD's objectives, which are: (a) the conservation of biological diversity, (b) the sustainable use of its components, and (c) the equitable sharing of the benefits arising out of the use of genetic resources. The CBD is the first international treaty to address all aspects of biodiversity (Bowman *et al.*, 2010), which it defines as 'the variability among living organisms from all sources including, *inter alia*, terrestrial, marine and other aquatic ecosystems and the ecological complexes of which they are part; this includes diversity within species, between species and of ecosystems'. Sustainable use is defined as 'the use of components of biological diversity in a way and at a rate that does not lead to the long-term decline of biological diversity, thereby maintaining its potential to meet the needs and aspirations of present and future generations'. Although the rationale for the CBD is largely based upon the need to maintain the utilitarian benefits of biodiversity, Bowman *et al.*, (2010) also consider that it 'constituted an important milestone in the recognition of the intrinsic value of biodiversity and of the conservation of biodiversity as a "common concern of humankind"'.

Importantly, the CBD provides a global framework of actions that aim to complement and support existing treaties for the conservation of ecosystems and threatened species (e.g. the Ramsar Convention, CMS and World Heritage Convention). Its provisions also set out guiding principles, required approaches and broad goals rather than specific obligations. However, it is weakened by the absence of clear measurable obligations, and any requirement for uniformity of implementation. Each Party is responsible for the implementation of the CBD 'as far as possible and appropriate' (MacKenzie, 2012). Whilst the Convention's soft provisions probably helped garner political support and agreement, this has been at the expense of their potential effectiveness and added value.

The CBD places a strong emphasis on the need to carry out *in situ* conservation measures (i.e. the conservation of biodiversity in its natural surroundings), such as the establishment and management of protected areas, protection measures for threatened species, actions to control IAS, sustainable development in the wider environment, and research and monitoring (Box 3.1). Although Article 9 promotes *ex situ* conservation measures (e.g. captive breeding and seed banks), these are to be 'predominantly for the purpose of complementing *in situ* measures'.

Of particular relevance to the situation in Europe are the following Article 6 General Measures for Conservation and Sustainable Use.

'Each Contracting Party shall, in accordance with its particular conditions and capabilities:

(a) Develop national strategies, plans or programmes for the conservation and sustainable use of biological diversity or adapt for this purpose existing strategies, plans or programmes which shall reflect, *inter alia*, the measures set out in this Convention relevant to the Contracting Party concerned; and

(b) Integrate, as far as possible and as appropriate, the conservation and sustainable use of biological diversity into relevant sectoral or cross-sectoral plans, programmes and policies.'

[9] Also known as the Brundtland Report.

Box 3.1 Article 8 of the Convention on Biological Diversity: in situ conservation

Each Contracting Party shall, as far as possible and as appropriate:

(a) Establish a system of protected areas or areas where special measures need to be taken to conserve biological diversity;
(b) Develop, where necessary, guidelines for the selection, establishment and management of protected areas or areas where special measures need to be taken to conserve biological diversity;
(c) Regulate or manage biological resources important for the conservation of biological diversity whether within or outside protected areas, with a view to ensuring their conservation and sustainable use;
(d) Promote the protection of ecosystems, natural habitats and the maintenance of viable populations of species in natural surroundings;
(e) Promote environmentally sound and sustainable development in areas adjacent to protected areas with a view to furthering protection of these areas;
(f) Rehabilitate and restore degraded ecosystems and promote the recovery of threatened species, *inter alia*, through the development and implementation of plans or other management strategies;
(g) Establish or maintain means to regulate, manage or control the risks associated with the use and release of living modified organisms resulting from biotechnology which are likely to have adverse environmental impacts that could affect the conservation and sustainable use of biological diversity, taking also into account the risks to human health;
(h) Prevent the introduction of, control or eradicate those alien species which threaten ecosystems, habitats or species;
(i) Endeavour to provide the conditions needed for compatibility between present uses and the conservation of biological diversity and the sustainable use of its components;
(j) Subject to its national legislation, respect, preserve and maintain knowledge, innovations and practices of indigenous and local communities embodying traditional lifestyles relevant for the conservation and sustainable use of biological diversity and promote their wider application with the approval and involvement of the holders of such knowledge, innovations and practices and encourage the equitable sharing of the benefits arising from the utilization of such knowledge, innovations and practices;
(k) Develop or maintain necessary legislation and/or other regulatory provisions for the protection of threatened species and populations;
(l) Where a significant adverse effect on biological diversity has been determined pursuant to Article 7, regulate or manage the relevant processes and categories of activities; and
(m) Cooperate in providing financial and other support for in-situ conservation outlined in subparagraphs (a) to (l) above, particularly to developing countries.

As further discussed in Section 4.4, this requirement has led to the development of three EU biodiversity strategies (as the EU is a Party in its own right), as well as National Biodiversity Strategies and Action Plans (NBSAPs). Other CBD provisions of particular relevance to Europe include the development of incentive measures for biodiversity conservation, research and training requirements, promoting public education and awareness of the importance of biodiversity, minimising adverse effects through impact assessments, exchanging information and encouraging technical and scientific cooperation. Although there are financial support measures to help achieve the CBD's objectives, these are solely directed to Parties in developing countries and are hence not applicable to Europe. Funding contributions are supplied by Parties in developed countries, with the Global Environment Facility being the primary funding mechanism.

The CBD's governing body is the Conference of the Parties (COP), which is supported by a Secretariat, as well as a Clearinghouse Mechanism (to promote and facilitate technical and scientific cooperation) and a Subsidiary Body on Scientific, Technical and Technological Advice. Despite the absence of obligatory measures, or even clear voluntary measurable targets in the Convention text itself, a requirement that encourages action is the need for each Party to report on the measures that it has taken to implement the Convention. The timing of each report, and its specific requirements, are decided upon by the COP. So far six reports have been required, with the latest providing a review of progress up to the end of 2018 in the implementation of the *Strategic Plan for Biodiversity 2011–2020* (see Section 3.6.2). The extent to which the reports incentivise action is uncertain as they are written by the State authorities (or contractors) and therefore lack independence. Assessments undertaken by NGOs on the extent of successful national implementation sharply differ in some countries (Stroud *et al.*, 2021).

An important development at COP II in 1995 was the adoption of 'the ecosystem approach' as a consistent framework for the implementation of the Convention. Subsequently COP V defined the ecosystem approach as being 'a strategy for the integrated management of land, water and living resources that promotes conservation and sustainable use in an equitable way', and set out 12 common principles for its application.[10] Whilst the principles encourage adaptive management, through knowledge and the understanding of ecosystem processes, they promote a wider set of arguments for decentralisation and stakeholder participation in decision making. COP

[10] CBD COP Decision V/6 www.cbd.int/decision/cop/default.shtml?id=7148

VII agreed that the priority should be on facilitating implementation of the ecosystem approach.[11]

3.6.1 CBD Strategic Plan and Target for 2010

In 2002, in recognition of the ongoing threats to biodiversity and the need to promote broad-based cooperative actions amongst key actors around agreed goals, the Parties developed a *Strategic Plan* for the CBD.[12] In this, CBD Parties committed themselves 'to a more effective and coherent implementation of the three objectives of the Convention, to achieve by 2010 a significant reduction of the current rate of biodiversity loss at the global, regional and national level as a contribution to poverty alleviation and to the benefit of all life on Earth'. This target less ambitiously echoed the intention of heads of EU Member States at their 2001 meeting in Göteborg 'that biodiversity decline should be halted with the aim of reaching this objective by 2010'. The EU objective to halt rather than just reduce the rate of biodiversity loss was stronger than the CBD target, which was subsequently endorsed by the World Summit on Sustainable Development (the 'Rio + 10' summit) in Johannesburg, 2002, and by the United Nations General Assembly. It was also included as a target under one of the Millennium Development Goals – Ensure Environmental Sustainability. As a result, the 2010 CBD biodiversity target was a global commitment from all governments.

Although the CBD's 2010 target is considered to have inspired actions at many levels, these were not sufficient to address the pressures on biodiversity, and it is evident that the objective was not achieved according to the third edition of the *Global Biodiversity Outlook* (GBO-3) (CBD, 2010). In fact, none of the 21 sub-targets accompanying the overall target of significantly reducing the rate of biodiversity loss by 2010 was achieved globally, although some had been partially or locally achieved. This failure was due to a number of factors, including insufficient integration of biodiversity issues into broader policies (often described as 'mainstreaming' and as required under Article 6) and various constraints on actions, especially inadequate financial, human and technical resources.

3.6.2 CBD Strategic Plan and Target for 2020

In order to redouble biodiversity conservation efforts following the failure to meet the 2010 target, Parties at the 10th COP held in Nagoya, Japan, adopted a revised and updated '*Strategic Plan for Biodiversity 2011–2020*'.[13] The Strategic Plan consisted of a shared vision, mission, five Strategic Goals and 20 related biodiversity targets, known as the Aichi Targets (Box 3.2).

The Vision of the Strategic Plan was a world 'Living in harmony with nature' where 'By 2050, biodiversity is valued, conserved, restored and wisely used, maintaining ecosystem services, sustaining a healthy planet and delivering benefits essential for all people.'

The 2020 mission was:

> Take effective and urgent action to halt the loss of biodiversity in order to ensure that by 2020 ecosystems are resilient and continue to provide essential services, thereby securing the planet's variety of life, and contributing to human well-being, and poverty eradication. To ensure this, pressures on biodiversity are reduced, ecosystems are restored, biological resources are sustainably used and benefits arising out of utilization of genetic resources are shared in a fair and equitable manner; adequate financial resources are provided, capacities are enhanced, biodiversity issues and values mainstreamed, appropriate policies are effectively implemented, and decision-making is based on sound science and the precautionary approach.

Implementation of the *Strategic Plan* relied primarily on activities and action at the regional, national and subnational level, as coordinated through NBSAPs. Parties therefore committed to updating and revising them in order to bring them in line with the *Strategic Plan for Biodiversity 2011–2020*. In the same decision, regional organisations were also urged to consider developing or updating existing regional biodiversity strategies, including regional targets that would complement those strategies and action plans developed at the national level and would contribute to the implementation of the *Strategic Plan for Biodiversity*. As a result of this the EU updated its biodiversity strategy in 2011 (described in Section 4.4).

Following a mid-term assessment (GBO-4), GBO-5 (CBD, 2020) provided a final report on the progress made towards the targets with a 2020 deadline. This was based on a range of indicators and assessments, in particular the IPBES *Global Assessment on Biodiversity and Ecosystem Services* (IPBES, 2019), as well as the national reports. The report confirmed that the overall mission target for 2020 had evidently not been achieved, and the world was not on track to achieve its 2050 Vision for Biodiversity. Furthermore, whilst progress has been made on most targets it has been insufficient as none of the Aichi targets had been fully achieved, and only six had been partly achieved (9, 11, 16, 17, 19 and 20). The partially achieved targets included Target 11 on protected areas, which increased over 2000–2020 from about 10% to at least 15% terrestrially, and from about 3% to at least 7% in marine areas.[14] Despite the increases, their coverage falls short of the 17% and 10% targets for terrestrial and marine areas respectively. However, rather optimistically, the GBO assessment concludes that the targets would be achieved if all the existing commitments were fulfilled. Nevertheless, inadequate progress was being made with increasing protected area coverage of Key Biodiversity Areas, and improving protected area management and connectivity. Many assessments have demonstrated that

[11] CBD COP Decision VII/11 www.cbd.int/decision/cop/default.shtml?id=7748

[12] CBD COP Decision VI/26 www.cbd.int/decision/cop/default.shtml?id=7200

[13] CBD COP Decision X/2: www.cbd.int/decision/cop/?id=12268

[14] According to the Word Database on Protected Areas, which, unintentionally, includes some sites that do not meet IUCN and CBD protected area definitions (see Section 1.3.4).

Box 3.2 **A simplified list of the Aichi targets**

Note: All targets relate to 2020 unless otherwise indicated

1. Awareness is increased.
2. Biodiversity values are integrated.
3. Incentives are reformed to eliminate harmful subsidies.
4. The sustainability of consumption and production is improved.
5. Habitat loss is halved or reduced.
6. Marine living resources are sustainably managed.
7. Areas under agriculture, aquaculture and forestry are managed sustainably, ensuring conservation of biodiversity.
8. Pollution is reduced to levels that are not detrimental to ecosystem function and biodiversity.
9. Invasive alien species are prevented from becoming established, or controlled if present.
10. By 2015, pressures on vulnerable ecosystems are reduced.
11. Protected areas are increased (to at least 17% of terrestrial and inland water areas, and 10% of coastal and marine areas), especially in areas of particular importance for biodiversity, and are effectively managed, representative and well-connected.
12. The extinction of known threatened species has been prevented.
13. Genetic diversity of cultivated plants, domesticated animals and wild relatives is maintained.
14. Ecosystems and essential services are safeguarded.
15. Ecosystems are restored and resilience enhanced, including over at least 15% of degraded ecosystems.
16. By 2015, the Nagoya Protocol on Access to Genetic Resources and the Fair and Equitable Sharing of Benefits Arising from their Utilization is in force and operational.
17. By 2015, each Party has developed and adopted an updated national biodiversity strategy and action plan (NBSAP).
18. Traditional knowledge is respected.
19. Knowledge is improved, shared and applied.
20. Financial resources from all sources are increased.

Source. CBD (2013).

even following legal establishment, protected area management is widely failing to maintain the ecological character and conservation objectives of the sites.

Similar conclusions can be drawn on the achievement of the Aichi targets in Europe, based on the *IPBES Regional Assessment* (IPBES, 2018), the evaluation of the *EU Biodiversity Strategy to 2020* discussed in Section 5.2.4, and the national chapters in this book.

3.6.3 CBD Post-2020 Global Biodiversity Framework

As a result of the COVID-19 pandemic, the development of the post-2020 biodiversity framework, and COP-15 initially planned for October 2020 in Kunming, China, were delayed and carried out in two main phases. Initial drafts were prepared taking into account the IPBES (IPBES, 2019) and GBO-5 (CBD, 2020) results and online discussions amongst CBD Parties and stakeholders. COP-15 was then eventually held in Montreal, Canada in December 2022, resulting in the agreed Kunming-Montreal Global Biodiversity Framework.[15] As the agreement was reached as this book was about to go to press, only some key elements are briefly outlined in the following paragraphs.

Retaining the 2050 Vision set out in the 2011–2020 Strategic Plan (see previous section), the framework's 2030 Mission is to 'To take urgent action to halt and reverse biodiversity loss to put nature on a path to recovery for the benefit of people and planet by conserving and sustainably using biodiversity, and ensuring the fair and equitable sharing of benefits from the use of genetic resources, while providing the necessary means of implementation.'

The framework is accompanied by four long-term goals relating to the 2050 Vision. Most fundamentally, Goal A includes the commitment to maintain, enhance or restore the integrity, connectivity and resilience of all ecosystems, substantially increasing the area of natural ecosystems by 2050; halt the extinction of threatened species, reduce extinction rates and risks of all species, and increase the abundance of native wild species to healthy and resilient levels by 2050; and maintain genetic diversity. The other three goals relate to the values and benefits of nature, and the need to close the gap between financial needs and availability.

The framework sets out 23 action-oriented targets for 2030, as well as complementary supporting proposals for implementation support mechanisms, enabling conditions, responsibility and transparency, and communications, education, awareness and uptake. Of the targets, Target 3 is widely considered to be particularly important, as it is to effectively conserve and manage at least 30% of the land and sea through protected areas and other effective area-based conservation measures (OECMs). As further discussed in Section 4.4, this is a target that many European countries have supported and the EU has already committed to in its *Biodiversity Strategy for 2030* (European Commission, 2020).

[15] https://www.cbd.int/doc/decisions/cop-15/cop-15-dec-04-en.pdf

Other 2030 targets that are of particular relevance to Europe, and where more action is most needed, include using spatial planning and other measures to reduce the loss of areas of high importance to close to zero (Target 1); the restoration of at least 30% of degraded ecosystems (2); ensuring the exploitation of wild species is sustainable (5); tackling IAS (6); reducing all pollution to non-harmful levels (7); minimising the impacts of climate change (8); and ensuring agriculture, aquaculture, fisheries and forestry are sustainable (10). The achievement of all targets is likely to be dependent on other supporting targets, especially in Europe the integration of biodiversity into policies (14); increasing the contribution of business (15); further phasing out harmful subsidies (18); and increasing financial resources from all sources, to mobilise $200 million per year globally (€188 million) (19).

Initial reactions from the environmental sector to the Kunming-Montreal Global Biodiversity Framework have included relief that there is a clear 2030 mission to halt and reverse biodiversity loss. The goals and targets are also generally considered to be ambitious, but there are concerns that most are described in vague terms and the goals lack milestones. Progress will therefore be difficult to measure and monitor. A detailed plan to implement the framework is also lacking, and the intended funding is considered to be insufficient. Time will tell if better progress is made towards the vision, mission and targets than under previous CBD agreements.

3.7 Marine Conventions and the North Sea Ministerial Conferences

A number of Regional Sea Conventions (RSCs) and other instruments protect Europe's marine environment (Table 3.2) under the overarching legal regime of the United Nations Convention on the Law of the Sea (UNCLOS). UNCLOS was the basis for the introduction of the influential FAO Code of Conduct for Responsible Fisheries in 1995, and the UN Agreement on the Conservation and Management of Straddling Fish Stocks and Highly Migratory Fish Stocks ('the UN Fish Stocks Agreement' or FSA) in 2001. These have been important drivers of, among other things, the implementation of measures within the EU Common Fisheries Policy (CFP) to reduce incidental by-catch of seabirds, cetaceans and other marine wildlife.[16]

When UNCLOS was adopted, there was considerable optimism that it would 'usher in a new age of marine environmental protection' (Mossop, 2018). While acknowledging the Convention's important innovations, Mossop argues that progress has generally been impeded by a number of limitations including the lack of an institution with competence over the law of the sea. Nevertheless, recent developments offer hope of filling gaps in UNCLOS, notably ongoing negotiations for a new treaty on the protection and sustainable use of biodiversity in Areas Beyond National Jurisdiction (ABNJs) (Mossop, 2018; Tiller *et al.*, 2019).

Of the four European RSCs, the 1992 Convention for the Protection of the Marine Environment of the North East Atlantic (OSPAR Convention), as a merger of the two pioneering Oslo and Paris Conventions on marine pollution, has set an important precedent for shaping the approach to protection of other regional seas, notably the Helsinki and Barcelona Conventions with which OSPAR is geographically linked. OSPAR also has a strong functional relationship and synergy with the EU, such as the implementation and assessment of the MSFD and the international North Sea Ministerial Conferences (see below) (Skjærseth, 2006). Since its inception, signal achievements of OSPAR include a major reduction in marine pollution, whether generated on land or at sea, the development of an ecosystem-based approach to human activities, and the progressive establishment of an ecologically coherent network of Marine Protected Areas (MPAs). According to the latest available compiled data, at the end of 2020 OSPAR MPAs comprised 552 sites covering 6.5% of the OSPAR Maritime Area, with substantial gaps in coverage remaining, particularly in Arctic Waters and the Wider Atlantic (OSPAR Commission, 2020). The widespread deficit in management of sites, both within and beyond territorial waters, is acknowledged as a further major challenge (Matz-Lück and Fuchs, 2014).

The decision-making process in OSPAR and the EU was strongly influenced and hastened by the 'soft law' international North Sea Ministerial Conference (hereafter NSC) declarations which, while not legally binding, signalled strong political intent. Six NSCs were held from 1984 (Bremen) until the final conference in 2006 (Göteborg), along with two important Intermediate Ministerial Meetings (Copenhagen 1993 and Bergen 1997). In turn, OSPAR and the EU have facilitated implementation of the measures agreed at the NSCs (Skjærseth, 2006). The first NSC focussed on concern among North Sea states about the potentially irreversible damage from inputs of hazardous substances via rivers, direct discharges and dumping of waste at sea. As a measure of progress towards eliminating this threat, by the time of the 2002 NSC in Bergen most North Sea states had achieved a 70% reduction of mercury, lead and cadmium inputs to the marine environment (Skjærseth, 2006).

As a test-bed and template for other European regional seas and beyond, the NSC process has had a profound impact on marine conservation at a much wider spatial scale, notably in its pioneering recognition and application of the precautionary principle and the ecosystem-based approach, notably to fisheries management. For reviews of NSC history and process, see Ehlers (1990), Skjærseth (2006) and Earll (2019). For further and broader analysis of marine governance structures, processes and challenges in European seas, including the evolving role of Regional Seas Conventions, see Gilek and Kern (2015).

[16] For example, Regulation (EU) 2019/1241 of 20 June 2019 on the conservation of fisheries resources and the protection of marine ecosystems through technical measures.

References

Atkinson-Willes, G. L., Scott, D. A. & Prater, A. J. (1982) Criteria for selecting wetlands of international importance. In Proceedings of the Conference on the Conservation of Wetlands of International Importance especially as Waterfowl Habitat. Cagliari, Italy, 24-29 November 1980. *Supplemento alle Ricerche di Biologia della Selvaggina*, **81**, 1017–1042.

Bell, S., McGillivray, D. & Pedersen, O.W. (2013) *Environmental Law*. 8th ed. Oxford: Oxford University Press.

Bowman, M., Davies, P. & Redgwell, C. (2010) *Lyster's International Wildlife Law*. 2nd ed. Cambridge: Cambridge University Press.

Bridgewater, P. (2016) The Man and Biosphere programme of UNESCO: rambunctious child of the sixties, but was the promise fulfilled? *Current Opinion in Environmental Sustainability*, **19**, 1–6.

CBD (2010) *Global Biodiversity Outlook 3*. Montreal: Secretariat of the Convention on Biological Diversity.

CBD (2013) *Quick guides to the Aichi Biodiversity Targets*. Montreal: Secretariat of the Convention on Biological Diversity.

CBD (2020) *Global Biodiversity Outlook 5*. Montreal: Secretariat of the Convention on Biological Diversity.

Díaz, C. L. (2010) The Bern Convention: 30 years of nature conservation in Europe. *Review of European Community & International Environmental Law*, **19**, 185–196.

Earll, B. (2019) *Marine Conservation: People, Ideas and Action*. Exeter, UK: Pelagic Publishing.

Ehlers, P. (1990) The History of the International North Sea Conferences. In *The North Sea: Perspectives on Regional Environmental Cooperation*, eds. D. Freestone & T. IJlstra, pp. 3–14. Special Issue of the *International Journal of Estuarine and Coastal Law*. London: Graham & Trotman Ltd.

European Commission (2020) *EU Biodiversity Strategy for 2030 — bringing nature back into our lives*. COM(2020) 380 final.

Farmer, A. (ed.) (2012) *Manual of European Environmental Policy*. 2nd ed. London: Taylor and Francis.

Finlayson, C. M., Davidson, N. C., Pritchard, D. E., Milton, G. R & MacKay, H. (2011) The Ramsar Convention and ecosystem-based approaches to the wise use and sustainable development of wetlands. *Journal of International Wildlife Law & Policy*, **14**, 176–198.

Gilek, M. & Kern, K. (eds.) (2015) *Governing Europe's Marine Environment: Europeanization of Regional Seas or Regionalization of EU Policies?* London & New York: Routledge.

Goriup, P. & Tucker, G. (2005) *Assessment of the merits of an Instrument under the Convention on Migratory Species (CMS) covering Migratory Raptors in the African-Eurasian Region. Final Report with Draft MoU and Action Plan*. Newbury, UK: Nature Bureau report to DEFRA.

Hajdu, K. & Kiss, V. (2009) *Clear View: Regional synthesis report on the biodiversity challenges in Pan-Europe*. Budapest: CEEweb for Biodiversity.

IPBES (2018) *The IPBES regional assessment report on biodiversity and ecosystem services for Europe and Central Asia*, eds. M. Rounsevell, M. Fischer, A. Torre-Marin Rando & A. Mader. Bonn, Germany: Secretariat of the Intergovernmental Science-Policy Platform on Biodiversity and Ecosystem Services.

IPBES (2019) *Global assessment report of the Intergovernmental Science-Policy Platform on Biodiversity and Ecosystem Services*, eds. E. S. Brondízio, J. Settele, S. Díaz, & H. T. Ngo. Bonn, Germany: Secretariat of the Intergovernmental Science-Policy Platform on Biodiversity and Ecosystem Services.

IUCN, UNEP & WWF (1980) *World Conservation Strategy: Living Resource Conservation for Sustainable Development*. IUCN–UNEP–WWF.

IUCN, UNEP & WWF (1991) *Caring for the Earth: A Strategy for Sustainable Living*. IUCN (World Conservation Union), UNEP (United Nations Environment Programme) and WWF (World Wide Fund for Nature). London: Earthscan.

Jones, G. (2012) The Bern Convention and the Origins of the Habitats Directive. In *The Habitats Directive - a Developer's Obstacle Course?* ed. G. Jones, pp. 1–23. Oxford: Hart Publishing.

Kanstrup, N., Swift, J., Stroud, D. A. & Lewis, M. (2018) Hunting with lead is not sustainable: European perspectives. *Ambio*, **47**, 846–857.

Lewis, M. (2016) AEWA at Twenty: an appraisal of the African-Eurasian Waterbird Agreement and its unique place in international environmental law. *Journal of International Wildlife Law & Policy*, **19**, 22–61.

Lyster, S. (1985) *International Wildlife Law*. Cambridge, UK: Cambridge University Press.

MacKenzie, C. (2012) A comparison of the Habitats Directive with the 1992 Convention on Biological Diversity. In *The Habitats Directive - a Developer's Obstacle Course?* ed. G. Jones, pp. 25–41. Oxford: Hart Publishing.

Mateo, R. & Kanstrup, N. (2019) Regulations on lead ammunition adopted in Europe and evidence of compliance. *Ambio*, **48**, 989–998.

Matz-Lück, N. & Fuchs, J. (2014) The impact of OSPAR on protected area management beyond national jurisdiction: Effective regional cooperation or a network of paper parks? *Marine Policy*, **49**, 155–166.

Mossop, J. (2018) *Can We Make the Oceans Greener? The Successes and Failures of UNCLOS as an Environmental Treaty*. Victoria University of Wellington Legal Research Paper No. 48/2019.

Oral, N. (2011) The Black Sea: Moving Forward. In *The World Ocean in Globalisation: Climate Change, Sustainable Fisheries, Biodiversity, Shipping, Regional Issues*, eds. D. Vidas & P. J. Schei, pp. 473–494. Leiden/Boston: Martinus Nijhoff Publishers.

OSPAR Commission (2020) *Status of the OSPAR Network of Marine Protected Areas in 2020*. https://oap-cloudfront.ospar.org/media/filer_public/94/bd/94bdc35b-05f6-4efa-9e1d-f84cb5ebe0f9/assessment_sheet_mpa_status_2020.pdf

Prip, C. (2013) *Terminal evaluation of the UNEP project Supporting the Implementation of the Pan-European Biological and Landscape Diversity Strategy (PEBLDS)*. United Nations Environment Programme.

Raftopoulos, E. (2011) The Mediterranean Response to Global Challenges: Environmental Governance and the Barcelona Convention System. In *The World Ocean in Globalisation: Climate Change, Sustainable Fisheries, Biodiversity, Shipping, Regional Issues*, eds. D. Vidas & P. J. Schei, pp. 507–532. Leiden/Boston: Martinus Nijhoff Publishers.

Ramsar Convention Secretariat (2010) *Wise use of wetlands: concepts and approaches for the wise use of wetlands.* Ramsar Handbooks for the wise use of wetlands, 4th ed. Volume 1. Gland, Switzerland: Ramsar Convention Secretariat.

Ramsar Convention Secretariat (2013) *The Ramsar Convention Manual.* 6th ed. Gland, Switzerland: Ramsar Convention Secretariat.

Skjærseth, J. B. (2006) Protecting the Northeast Atlantic: One Problem, Three Institutions. In *Institutional Interaction in Global Environmental Governance: Synergy and Conflict among International and EU Policies*, eds. S. Oberthür & T. Gehring, pp. 102–125. Cambridge, Massachusetts, USA: MIT Press.

Stroud, D. A. (2018) Ramsar Convention: Ramsar Site designation process. In *The Wetland Book*, eds. C. M. Finlayson, M. Everard, K. Irvine, R. McInnes, B. Middleton, A. van Dam & N. C. Davidson, pp. 459–466. Dordrecht, The Netherlands: Springer.

Stroud, D. A., Cromie, R., Finlayson, M. *et al.* (2021) *International Treaties in Nature Conservation: A UK Perspective.* Totnes, UK: Biodiversity Press.

Stroud, D. A. & Davidson, N. C. (2021) Fifty years of criteria development for selecting wetlands of international importance. *Marine & Freshwater Research.* https://doi.org/10.1071/MF21190

Tiller, R., De Santo, E., Mendenhall, E. & Nyman, E. (2019) The once and future treaty: Towards a new regime for areas beyond national jurisdiction. *Marine Policy,* **99**, 239–242.

UNGA (1982) *World Charter for Nature.* United Nations General Assembly. (37th sess.: 1982–1983).

WCED (1987) *Our Common Future.* Report of the World Commission on Environment and Development (Brundtland Commission).

Chapter 4

Nature Conservation Policy, Legislation and Funding in the EU

Graham Tucker, Kaley Hart, Andrew Farmer, Euan Dunn and David Baldock

The EU has played an important role in advancing nature conservation in all its Member States (MS), especially through the development and implementation of the Birds Directive and the Habitats Directive (i.e. the Nature Directives). It has also had an influence in other European countries, such as through stimulating and supporting actions under other international frameworks and initiatives (e.g. the Bern Convention's Emerald Network, and the Convention on the Conservation of Migratory Species of Wild Animals (CMS)). This chapter briefly describes the evolution of EU environmental policy before focussing primarily on the two Nature Directives, in particular their objectives and key measures (Section 4.2). However, the Nature Directives depend to some extent on other EU environmental policies and legislation that complement and support them, and therefore the most important of these are also outlined, as are their key interactions with the Directives (Section 4.3). Nature conservation also requires proactive practical actions, such as those to restore and maintain habitats – and these need funding. The EU's role in funding the implementation of the Nature Directives, as well as other international and national conservation actions, is therefore also described (Section 4.2.6), providing a background to the more detailed national analysis in each country chapter.

Whilst the Nature Directives form the heart of EU policy in the conservation sphere, the development of the EU's four biodiversity strategies since 1998 (in response to the Convention on Biological Diversity (CBD) requirements) has built on these core legislative instruments to form a broader and more integrated policy framework, especially with regard to the involvement of key economic sectors. The development of the EU's biodiversity targets and related strategies and action plans is therefore also summarised (Section 4.4). However, as evidence from the national chapters shows, these action plans have had less influence on practical conservation measures than the key EU directives, and therefore their actions and their implementation are not described in detail.

Whilst the actions that are supposed to have been taken under the various EU measures are outlined in this chapter, analysis and commentary on their actual implementation, achievements and impacts are primarily to be found in the country chapters. However, some important implementation issues and policy evaluations are briefly referred to here. Finally, their overall EU impacts are further discussed in the next chapter, which draws conclusions on their combined effects on the pressures affecting biodiversity, and ultimately the status of the habitats and species that they targeted.

4.1 The Development of EU Environmental Policy: in Brief

The pivotal role that the EU has acquired in environmental policy, both within Europe and significantly beyond it, can be traced back to the 1970s. It was a significant departure from what had been envisaged in initial debates about the value of a more united Europe in the 1950s, when neither the environment generally nor nature conservation in particular was seen as a key societal challenge, in contrast to the rebuilding of a war-torn continent and establishing conditions for prosperity. There was no sign of environmental concepts or ambition in the Treaty of Rome, signed in 1957; the priority focus then was on securing peace and political stability and stimulating economic growth. The European Economic Community (EEC),[1] as it then was, decided to embark on a substantive environmental policy only in October 1972 at the Paris European Summit. This was propelled by the emergence of the environment as a significant political and international issue, brought to prominence by the ground-breaking UN Conference on the Human Environment in Stockholm,[2] held only four months before the Summit. A Declaration was signed in Paris committing the Community to an environmental policy, amongst other new directions. Signed by the Heads of both the six existing Member States and also the three countries committed to join in the following January, it proved a momentous step beyond a largely economic European entity.

In the following year, the initial manifestation of what was to become the Directorate-General (DG) for the Environment was created inside the European Commission and the first forward-looking Environmental Action Programme (EAP) was agreed, setting a pathway along which policy was expected to be developed. Both the DG and EAPs have survived and

[1] The acronyms EEC, EU and the EC are to be found in this and other chapters. The reason for this is that the original EEC was incorporated into the European Union (EU) when it was formed in 1993 and renamed the European Community (EC).

[2] www.un.org/en/conferences/environment/stockholm1972

the current EAP to 2030 is the Eighth.[3] The European Environment Agency (EEA) was established in 1990. The flow of European Community environmental legislation began to gather pace during the mid-1970s and there are now hundreds of items of environmental law. In the subsequent decades the EU has created one of the most comprehensive and arguably the most influential corpus of environmental law in the world.

The benefits of a collective approach to environmental issues provided the political foundation for this new wing of European Community policy but there has also been an economic rationale for a considerable portion of the legislation agreed. Step by step, common environmental standards have been put in place on a European scale, especially but not solely in relation to transboundary issues. The scope of legislation includes reductions in air and water pollution, improvements in natural resource and waste management, the conservation of nature, the protection of the marine environment, requirements for environmental impact assessment, standards for noise, the control of chemicals and the creation of liability for environmental damage. Standards for soil management are one of the few areas where policy drivers have remained at the national rather than EU level (although the European Commission launched a new EU Soil Strategy in late 2021, which, amongst other actions, committed the Commission to propose binding legislation in the form of an EU Soil Health Law by 2023).

Over time, global environmental challenges such as the reduction of greenhouse gas emissions and control of illegal tropical timber imports have had growing weight within the EU's environmental ambitions. There is an expanding armoury of measures concerned with creating a low-carbon economy, including through improved energy efficiency and the displacement of fossil fuels by renewables.

At the same time, economic concerns have been reflected in continued efforts to avoid major distortions to a level playing field between the Member States in the environmental domain. From the outset there was seen to be a danger that countries that chose to adopt lower environmental standards, such as laxer controls on emissions from industrial plants, potentially would benefit from lower costs and gain an economic advantage. Common EU standards, usually allowing Member States freedom to be more stringent if they wished, have been considered an important safeguard against anti-competitive behaviour. This was one of the reasons for the strong pressure on the central and eastern European countries joining the EU from 2004 onwards to adopt EU environmental legislation within a short period.

Initially the environmental legislation put in place was not underpinned by strong legal foundations in the relevant European Community (now EU) treaties. This changed progressively, starting with the Single European Act of 1986, which amended the Treaty of Rome in several respects, confirming the legitimacy of environmental policy in legal terms and underpinning it with a number of principles, such as that of the polluter pays. These principles remain fundamental to EU legislation in this sphere. The Maastricht Treaty of 1991 was a further step forward, introducing the precautionary principle as a basis for European Community law, strengthening the requirement to take account of environmental protection in other areas of policy and extending the qualified majority voting procedure to most areas of environmental policy (Wilkinson, 1992; Haigh, 2016).

The initially relatively narrow economic objectives of the Community were also widened in a succession of treaty changes. By June 1997, the formal 'task' of the Community was modified in Article 2 of the Treaty of Amsterdam to promote a number of objectives via mechanisms such as common policies and the common market. Amongst these objectives is 'a high level of protection and improvement of the quality of the environment, and the raising of the standard of living and quality of life'.

Legislation and other initiatives on nature conservation have been an important element of EU environmental policy from the outset, despite continuing tensions over the extent to which it is appropriate to act at a European as well as a more local level. The Birds Directive was agreed in 1979 with strong support from the European Parliament, creating requirements for national action well beyond the conservation of migratory species. Concerns about damage to sites of conservation importance were prominent in the case for introducing the initial Environmental Impact Assessment (EIA) Directive in 1985. The Habitats Directive followed in 1992, representing a central pivot in EU environmental law and a high point in ambition, albeit one that has proved difficult to attain in practice. The setting of strategic longer-term goals in this Directive is also a feature of the Water Framework Directive (WFD), the influential centrepiece of water law, adopted in 2000.

Political enthusiasm for major step changes of this kind in the EU then waned for a period, especially after the 2008 recession, with climate mitigation policy the major exception. Nonetheless, progress continued, including some strategic measures, such as the Marine Strategy Framework Directive (MSFD), adopted in 2008.

Obligations falling on the EU from international agreements, such as the global Convention on Biological Diversity (CBD), also have been an important driver of actions. As described in Section 4.4, these included the development of a series of EU biodiversity strategies in 1998, 2006 and 2011, and the associated targets, initially to halt biodiversity loss by 2010, and then halt the loss of biodiversity and the degradation of ecosystem services by 2020. The most recent of these strategies, adopted by the Commission in 2020, includes the target of ensuring that biodiversity will be on a path to recovery by 2030 (European Commission, 2020a). While this appears to be less ambitious and definable, it may be more realistic, especially if backed up with specific obligations on the Member States,

[3] Decision (EU) 2022/591 on a General Union Environment Action Programme to 2030.

particularly with regard to nature restoration. This current strategy is further discussed, in the light of this book's conclusions, in Section 5.2.

Implementation of environmental law within the EU is far from perfect, but the same could be said for most other jurisdictions. Unlike in individual nation states, there is external pressure on national governments from the European Commission and European Court of Justice (CJEU) to comply with the obligations that they have accepted under EU law. Persistent failure can lead to the imposition of significant financial penalties. NGOs have been active in utilising their right to launch complaints and have helped to focus considerable attention on the widespread failures to comply fully with nature legislation. The impact of this legislation on the ground has increased considerably because of the volume of infringement cases and a number of important court judgments (described later). On the other side of the coin, many governments have sought to weaken certain provisions of the Nature Directives. They have not succeeded in doing so, despite encouraging the European Commission to carry out a Fitness Check in 2014–2016 (see background in online Annex 2).[4] This major review of the Directives, described further in Section 4.2.8, upheld their value and fitness for purpose and turned the focus to better implementation (European Commission, 2016; Milieu *et al.*, 2016).

Outside the realm of environmental law itself, the environment is somewhat of a force in shaping other EU policies. Since the 1997 Amsterdam Treaty, there has been a requirement that 'Environmental protection requirements must be integrated into the definition and implementation of the Community policies…' (Article 6 of the European Community Treaty). This has underpinned the effort to 'mainstream' biodiversity concerns into other areas of policy, especially the Common Fisheries Policy (CFP) and Common Agricultural Policy (CAP), and helped to shift a significant share of agricultural policy funding into measures with greener aspirations. Whilst there is a great deal further to go in this direction, a substantial share of overall public funding for biodiversity in most Member States now derives from the EU, including the CAP and regional aid policies (further discussed in Section 4.2).

The escalation of the climate agenda and arrival of the new European Commission led by Ursula von der Leyen in 2019 opened a new chapter in EU environmental and economic policy, with significant consequences for biodiversity. Soon after it took over in December 2019, the new Commission announced the European Green Deal (European Commission, 2019a), a strategic initiative comprising a substantial suite of different policies and budget lines to be introduced over a period of years up to 2024. The overall aim is to reach net zero emissions of greenhouse gasses in the EU by 2050 and to attain a cleaner environment, affordable energy, new jobs and an overall better quality of life.

The raft of measures under this umbrella includes the new *Biodiversity Strategy for 2030* (European Commission, 2020a), the *Farm to Fork Strategy for a fair, healthy and environmentally friendly food system* (European Commission, 2020b) and the *Circular Economy Action Plan* (European Commission, 2020c). Significant climate mitigation legislation proposals were made in 2021 under the 'Fit for 55' package of measures (European Commission, 2021a) and further measures in 2022 included a Sustainable Use of Pesticides Regulation that in principle should contribute to the EU Green Deal goal of cutting the use of pesticides and nutrients by 50% by 2030. A new Regulation on Nature Restoration that sets binding targets for the restoration of nature in the Member States by 2030 was also proposed in 2022, forming a key delivery mechanism for the goals of the Biodiversity Strategy.

The Green Deal has not been universally welcomed but it has opened a new chapter after a period when new EU legislation on biodiversity seemed unlikely. It is disappointing that there is no intention in the Biodiversity Strategy to introduce the once envisaged mandatory requirement to achieve no net loss – in contrast to its recent adoption in the UK. However, there has been increased recognition of the link between the climate and biodiversity crises, and also the potential role of ecosystem restoration in addressing both. This has contributed to the inclusion of biodiversity in the European Green Deal set of policy initiatives, alongside the goals of economic growth and the recovery from COVID-19. As noted already, the proposed legally binding EU nature restoration targets could be particularly important, provided that their finally agreed scope and ambition are commensurate with the requirements. There is also an important nature dimension to many other areas where legislation continues to advance, including the promotion of the 'circular economy' and the management of bioresources.

For a history of the development of EU environmental policy, and some of the key issues that have arisen, see Haigh (2016), which this summary draws on extensively. A more focussed account of nature conservation policy and practice in Europe up to 1995 is provided by Bromley (1997), the development of the Birds and Habitats Directives is summarised in Farmer (2012) and a detailed analysis of the origins and implementation of the Directives in relation to Natura 2000 is provided by Jackson (2018).

4.2 The Birds Directive and Habitats Directive

4.2.1 Their Objectives and Key Measures

The Birds and Habitats Directives have similar objectives (see Box 1.1) and measures that are summarised in relation to their specific articles in Table 4.1. Both Directives have the general objective of contributing to the conservation of biodiversity.

[4] All annexes are free to access online at CUP: www.cambridge.org/natureconservation

Table 4.1 Summarised versions of the Articles in the EU Birds and Habitats Directives.

Birds Directive (i.e. Directive on the conservation of wild birds 2009/147/EC, codified version of Directive 79/409/EEC)
1: Scope: It applies to all naturally occurring species of wild birds in the EU, and their eggs, nests and habitats.
2: Aims: See Box 1.1 and main text.
3: MS are to 'preserve, maintain or re-establish a sufficient diversity and area of habitats' for birds. This is to be done primarily by creating protected areas, managing habitats both inside and outside protected areas, re-establishing destroyed biotopes and creating new ones.
4: Species listed in Annex I and migratory species shall be subject to special conservation measures concerning their habitat, in particular the classification of their most suitable sites as Special Protection Areas (SPAs). They must form a coherent whole that meets the requirements of their species. As a result of the Habitats Directive all SPAs are part of the Natura 2000 network. Also SPA protection measures originally contained in the first sentence of Article 4(4) of the Directive have been replaced by Articles 6(2), (3) and (4) of the Habitats Directive.
5: MS must create a general system of protection for all species referred to in Article 1, prohibiting: deliberate killing and capture; deliberate destruction or damage of nests and eggs; taking of eggs; deliberate disturbance (particularly during the breeding season) and keeping birds the hunting and capture of which is prohibited.
6: MS shall prohibit the sale of wild birds (live or dead), except for species listed in Annex III/A and, subject to consultation with the Commission, those listed in Annex III/B.
7(1–3): Species listed in Annex II may be hunted, but MS are to ensure that hunting does not jeopardise their conservation. Species listed in Annex II/A may be hunted anywhere, but the species in Annex II/B may be hunted only in certain MS.
7(4): MS are to ensure that hunting and falconry complies with the principles of wise use and ecologically balanced control of the species concerned. In particular, MS are to ensure that birds are not hunted during their breeding season and that, in addition, migratory birds are not hunted during their return to their breeding sites. MS shall send the Commission relevant information on the practical application of their hunting regulations.
8: MS shall prohibit the use of all means of large-scale or non-selective killing of birds, or methods capable of causing the local disappearance of species, especially those methods listed in Annex IV(a).
9: MS may derogate from the prohibitions under Articles 5–8, if there is no other solution, for: human safety; to prevent serious damage to crops, livestock, forests, fisheries, water; to protect flora and fauna; to enable research, breeding and re-introductions; to allow selective taking in very small numbers. Each derogation must be specified, MS must report annually on them and the Commission shall ensure they are compatible with the aims of the Directive.
10: MS are to encourage research, especially on subjects listed in Annex V.
11: MS shall ensure that introductions of non-native species of bird do not prejudice local flora and fauna, and shall consult the Commission on this issue.
12: MS must report on the implementation of the Directive every 3 years. Note: reporting is now aligned with Habitats Directive and takes place at the same time and every 6 years.
13: The measures taken in relation to the Directive may not lead to deterioration of the present situation as regards the conservation of birds.
14: MS may introduce stricter protection measures than those provided under the Directive.
15: Provisions for amending the Annexes.
16: Establishes the Committee for Adaptation to Technical and Scientific Progress, known as the ORNIS Committee, to assist the Commission.
Habitats Directive (i.e. Directive 92/43/EEC on the conservation of natural habitats and of wild fauna and flora)
1: Definitions.
2: Aims – see Box 1.1 and main text.
3: Requirement to establish a coherent network of Special Areas of Conservation (SAC) hosting habitats listed in Annex I and species listed in Annex II, sufficient to achieve their favourable conservation status across their natural range – to be titled Natura 2000. SPAs designated under the Birds Directive are included in Natura 2000.
4 & 5: Procedures, in relation to criteria in Annex III and a framework of defined biogeographical regions, for MS to provide a list of sites holding Annex I habitats and Annex II species (within 3 years), and the Commission to select Sites of Community Importance (SCIs) within 3 years. SCIs are subject to Articles 6(2–4). MS are required to designate SCIs as SACs within 6 years and establish conservation priorities. The setting of conservation objectives (as referred to in the preamble and Article 6(3)) is also a practical necessity at the same time.
6: Requirements for site management measures and protection.

Table 4.1 *(cont.)*

Habitats Directive (i.e. Directive 92/43/EEC on the conservation of natural habitats and of wild fauna and flora)
6(1): MS shall establish the 'necessary conservation measures' for SACs to ensure the ecological requirements of the Annex I habitats and Annex II species that are present.
6(2): MS shall take 'appropriate steps' to avoid the deterioration of the habitats and significant disturbance within SACs. Also applies to SCIs and SPAs.
6(3): Any plan or project likely to have a significant effect on a site (i.e. SCI, SAC or SPA) shall be subject to an 'appropriate assessment' in relation to the site's conservation objectives. It may only go ahead, following public consultation, if it is 'ascertained that it will not adversely affect the integrity of the site', subject to Article 6(4).
6(4): A plan or project having adverse effects on site integrity may exceptionally go ahead if there are no alternatives and there are 'imperative reasons of overriding public interest', with compensatory measures to protect the coherence of Natura 2000.
7: Extends Articles 6(2), (3) and (4) to all SPAs classified under the Birds Directive.
8: MS are required to submit cost estimates for establishing the necessary conservation measures on SACs holding priority habitats and species (as listed in Annexes I and II respectively) for the Commission to review and adopt a prioritised action framework of measures involving co-financing.
9: Provisions for periodic review of Natura 2000 and the declassification of SACs.
10: Where an MS considers it necessary, within its land-use planning and development policies, it should encourage the management of landscape features to improve the ecological coherence of the Natura 2000 network.
11: MS shall undertake surveillance of the conservation status of all their habitats and species of Community interest.
12–13: MS shall establish a system for ensuring the strict protection of animal and plant species listed in Annex IV, prohibiting: their capture, killing and picking, etc; deliberate disturbance; deliberate destruction; deterioration of breeding sites; and keeping and sale. MS are also required to monitor the incidental capture and killing of the protected animal species in Annex IVa, and ensure it does not have a negative impact on them.
14: MS shall ensure the taking of species listed in Annex V is in accordance with the maintenance of their favourable conservation status (in the light of surveillance under Article 11), if necessary taking measures such as: temporary bans, restrictions on the periods or methods for taking specimens, licensing systems, regulations on the sale and transport of specimens; and breeding programmes.
15: Prohibition of indiscriminate means of capture and killing species listed in Annexes IV(a) and V(a), especially in relation to the methods and modes of transport referred to in Annex VI.
16: Derogation provisions for Articles 12–15, which closely match those in the Birds Directive (BD Article 9) for Articles 5–8.
17: MS are required to provide a 6-yearly report on implementation of the Directive, particularly regarding Article 6(1) conservation measures and their impacts on the conservation status of Annex I habitats and Annex II species, as well as the results of the surveillance of all habitats and species of Community interest carried out under Article 11.
18: MS shall encourage research, and exchange information, to support the objectives of the Directive, especially regarding Articles 4 and 10, and transboundary issues.
19: Procedures for amending Annexes of the Directive, which requires a qualified majority approval from the Council for Annexes I, II, III, V and VI, but unanimous approval from the Council for Annex IV.
20: Establishes a Committee to assist the Commission (now referred to as the Habitats Committee).
22: MS shall consider re-introducing native species listed in Annex IV, and must ensure that introduction of any non-native species does not prejudice native habitats and species. The need for habitat and species conservation shall be promoted.

Specifically, the Birds Directive aims to maintain the populations of all species of naturally occurring birds in their wild state in the EU, 'at a level which corresponds in particular to ecological, scientific and cultural requirements, while taking account of economic and recreational requirements . . .' (BD Article 2). Whilst the Birds Directive indirectly contributes to broader biodiversity conservation than birds, it is complemented by the Habitats Directive, which aims to 'maintain or restore, at favourable conservation status, natural habitats and species of wild fauna and flora of Community interest' (HD Article 2(2)). Box 4.1 sets out the criteria by which favourable conservation status is assessed across the whole national territory, or across biogeographical or marine regions if there is more than one such region within the country.[5] In this context the term 'conservation status' differs from the use of the term in Red Data books or Red Lists of threatened species, in that it

[5] For detailed evaluation matrices see http://cdr.eionet.europa.eu/help/habitats_art17/

Box 4.1 Definition of favourable conservation status in the Habitats Directive

Under HD Article 1(e), the conservation status of a natural habitat will be taken as 'favourable' when:

- its natural *range* and *areas* it covers within that range are stable or increasing; and
- the specific *structure and functions* which are necessary for its long-term maintenance exist and are likely to continue to exist for the foreseeable future; and
- the conservation status of its *typical species* is favourable as defined in Article 1(i).

Under HD Article 1(i), the conservation status of a species will be taken as 'favourable' when:

- *population dynamics* data on the species concerned indicate that it is maintaining itself on a long-term basis as a viable component of its natural habitats; and
- the natural *range* of the species is neither being reduced nor is likely to be reduced in the foreseeable future; and
- there is, and will probably continue to be, a *sufficiently large habitat* to maintain its population on a long-term basis.

Assessments are made with respect to each biogeographical region within the Member State that the habitat or species occurs within, within and outside the Natura 2000 network. The European Topic Centre on Biological Diversity (ETC/BD) combines and interprets the Member State assessments to produce an overall assessment for every habitat type and every species in each biogeographical region.

For reporting under HD Article 17, three classes of conservation status are used: favourable (FV), unfavourable–inadequate (U1) and unfavourable–bad (U2). **Favourable** effectively represents a situation where the habitat or species can be expected to prosper without any further changes to existing management or policies. **Unfavourable–inadequate** (now reported as 'poor') usually describes a situation where a change in management or policy is required to return the habitat type or species to favourable status but there is no risk of extinction in the foreseeable future. It is also often used because a habitat or species is greatly depleted as a result of historic impacts, and well below its favourable reference value (see below). In practice, even if optimally managed, it will not be feasible for most of these greatly depleted habitats and species to recover sufficiently to achieve a favourable status. **Unfavourable–bad** (now reported as 'bad') is used for species or habitat types in serious danger of becoming extinct (at least regionally).

To assess the conservation status of habitats and species, the concept of favourable reference values has been developed through discussions with the Scientific Working Group (Habitats), the Habitats Committee and workshops with Member States (ETC/BD, 2011). The values are set by Member States, based on scientific grounds, for the range of habitats and species, the area of habitats and the population sizes of species. All must be at least as high as when the Directive came into force within the Member State concerned, and sufficient to ensure the long-term survival of the habitats and species. The favourable reference value for habitat and species ranges must also be sufficient to contain all significant variation in the habitat or species. Despite the importance of these reference values, they have been poorly developed and inconsistently applied, often due to data limitations (McConville and Tucker, 2015). The Commission has therefore had new guidance produced that sets out a common methodology for setting favourable reference values (Bijlsma *et al.*, 2019).

refers to the distance from a defined favourable situation of a habitat type or species rather than just an assessment of the relative risk of extinction. It is also important to note that the conservation status of species and habitats relates to their entire range within each biogeographical area that they occur in, not just within the Natura 2000 network of protected areas. Despite this, in practice, Member States tend to base their assessments on the situation within their Natura 2000 network. Table 1.2 provides more information on the current terminology used for reporting.

Whilst the term 'favourable conservation status' is not mentioned in the Birds Directive, it is implied with its aim of maintaining bird populations (Article 2). Therefore, the aims of the Nature Directives are considered to be broadly analogous according to the European Commission's legal services.

Following the precedent set in the Birds Directive, the Habitats Directive also states that measures 'shall take account of economic, social, and cultural requirements and regional and local characteristics'. Thus, both Directives implicitly recognise the need to allow, and contribute to, sustainable development. Moreover, the recitals of the Habitats Directive explicitly indicate that it aims to contribute to sustainable development.

Both Directives have a similar framework of two key pillars of measures:

1. **Site protection and conservation management** measures, which are targeted to sites of particular importance, comprising Special Protection Areas (SPAs) for selected birds, and Special Areas of Conservation (SACs) for selected habitat types and selected non-bird species – in combination creating the Natura 2000 network.
2. **Species protection** measures that apply wherever the species occurs.

These are supported by a number of complementary measures, such as to provide funding, guide actions through research, and assess implementation and impacts through monitoring and reporting. The Nature Directives' measures are described here in relation to this broad framework, following a brief discussion of the issue of transposition and the role of court cases in clarifying the legal interpretation of the Directives' provisions.

Despite the urgent need to tackle biodiversity declines, as indicated in the preamble to each Directive, neither includes any timetable for the achievement of their aims, although there are deadlines for their transposition into national law, certain site-related actions and reporting (described later).

Box 4.2 EU pilot mechanism and infringement procedures

Article 258 of the Treaty on the Functioning of the European Union (TFEU) gives the European Commission the power to take legal action against a Member State that is not respecting its obligations under EU law. The Commission identifies possible infringements on the basis of its own investigations or following complaints from stakeholders or other citizens.

Since 2008, the Commission and Member States have used the 'EU Pilot' mechanism to communicate and clarify the factual and legal background where there may be potential breaches of EU law. In appropriate cases, this has therefore been used as a first step to try to resolve problems through informal dialogue. In general, opening a Pilot case involves the following procedure:

1. The Commission sends a query to the Member State concerned, which as a rule has 10 weeks to reply.
2. The Commission then has, as a rule, 10 weeks to assess the Member State's response.
3. If the response is unsatisfactory, the Commission may start an infringement procedure.

If the Pilot mechanism does not resolve the issue, or is not used, the Commission may launch a formal infringement procedure to address the suspected violation of EU law. The procedure follows a number of steps laid out in the EU treaties, each ending with a formal decision:

1. The Commission sends a 'Letter of Formal Notice' requesting further information to the Member State, which must send a detailed reply within a specified period, usually two months.
2. If the Commission concludes that the Member State is failing to fulfil its obligations under EU law, it may send a 'Reasoned Opinion': a formal request to comply with EU law. It also explains why the Commission considers that the country is breaching EU law, and requests information on the measures being taken, within a specified period, usually two months.
3. If the Member State still does not comply, the Commission may decide to refer the matter to the CJEU. Most cases are settled before being referred to the court.
4. If the Member State fails to communicate measures that implement the provisions of a directive in time, the Commission may ask the court to impose penalties.
5. If the court finds that the Member State has breached EU law, the national authorities must take action to comply with the court judgment.

If, despite the first ruling, a Member State still fails to act, the Commission may open another infringement case under Article 260 of the TFEU, with only one written warning before referring the Member State back to the CJEU.

Source. Based on European Commission.[8]

4.2.2 Transposition and the Role of Case Law

As directives (rather than regulations, which have direct legal force) the Nature Directives need to be transposed into national law in each Member State. This means the national authorities have to adapt their laws to meet the goals of the Directives. In accordance with the subsidiarity principle, Member States have some flexibility in deciding how to meet the goals, as a directive is binding regarding its required results rather than the means. However, according to Commission guidance, case law has clarified that 'transposition must be clear and precise, faithful and with unquestionable binding force' (European Commission, 2018).

Transposition of both Directives was required within two years for the then EU Member States. Newly acceding countries are obliged to implement the Nature Directives from their date of accession. No transition periods were agreed with any Member States joining the EU, with the single exception of a strictly limited transition period (until 2008) for Malta regarding one provision of the Birds Directive concerning the traditional trapping of songbirds for captivity.

There have been widespread problems with the transposition of the Nature Directives. In fact, not a single Member State fully transposed or implemented its obligations on time (Born *et al.*, 2015), which led to several waves of numerous reported breaches of the Directives' obligations and infringements being referred to the CJEU (see Box 4.2 for an explanation of the process). From 1981 to 2016 there were 4 102 reported breaches of the Nature Directives, which led to 768 infringement proceedings (Milieu *et al.*, 2016). Many of the early CJEU rulings had important impacts on the implementation of the Birds Directives, some of which also influenced aspects of the development of the Habitats Directive, in particular regarding the provisions relating to the identification and protection of sites (see later discussion of HD Articles 6(3) and 6(4)). Subsequently, the implementation of the Habitats Directive was also driven forward and influenced by the CJEU.

Other court judgments have also had important consequences. Notably, in relation to marine sites, in 1999 a UK High Court ruling,[6] commonly referred to as the 'Greenpeace Judgment', extended the territory to which the Habitats Directive applies in the UK from the 12 nautical mile (nm) boundary of its Territorial Sea out to the limit of its 200 nm Exclusive Fishing Zone.[7] Subsequently, the decision was confirmed by the CJEU (Case C-6/04), such that all Member States

[6] R v. *Secretary of State for Trade and Industry, ex p Greenpeace Ltd* (No 2) (2000) 2 CMLR 94 (QBD).

[7] The UK used to have a 200 nm Exclusive Fishing Zone but since 2014 has had an Exclusive Economic Zone.

[8] https://commission.europa.eu/law/law-making-process/applying-eu-law/infringement-procedure_en

are likewise required to ensure the Directive is implemented throughout their Exclusive Economic Zones (EEZs).

Consequently, the development of the body of case law has made a substantial contribution to clarifying the obligations of Member States under the Nature Directives. Some of the most influential CJEU cases and their implications have therefore been compiled by the European Commission (2006a, 2008), Farmer (2012) and Sundseth and Roth (2014). Some of these cases are referred to in the sections later in this chapter, and some are further described in the respective national chapters.

4.2.3 Site Protection and Management Measures

Identification and Designation of Special Protection Areas for Birds

Site protection measures are required for birds that are listed in Annex I because they are in danger of extinction, vulnerable to habitat change, rare, localised or otherwise affected by habitat-related threats (BD Article 4(1)). For these species, Member States are required to classify the most suitable sites as SPAs in order to ensure their survival and reproduction in the area of their distribution. Whilst the term 'classify' is ambiguous, in practice it means the identification, cataloguing and designation of sites as SPAs. Similar measures are required under Article 4(2) for regularly occurring migratory species to protect their breeding, moulting and wintering sites and migration staging posts. To support the Ramsar Convention, the Directive explicitly states that particular attention is to be paid to the protection of wetlands, and particularly wetlands of international importance (i.e. Ramsar Sites).

Member States must designate at least one SPA for each Annex I species, even though this may not be ecologically justified for some common and dispersed species. This was confirmed by a CJEU ruling (C-418/04) concerning some widespread Annex I species in Ireland, such as the Common Kingfisher (*Alcedo atthis*). In practice SPAs are not designated for all migratory species as site protection is an impractical means of conserving some of them. Also, the migratory behaviour of many species varies across their range. Therefore, although the European Commission has provided a list of EU bird species[9] and indicated which are migratory according to the CMS definition,[10] Member States have discretion over which non-Annex I migratory species require SPAs, and have compiled national lists of SPA trigger species.

Although the identification of suitable sites for SPA species must be on a scientific basis, using the best available information, the process is relatively simple and the selection is made by each Member State. Unlike as described later for SCIs, there is no process or criteria for assessing whether a national SPA network is sufficient. However, case law has establishe the lists of Important Bird Areas (IBAs) in Europe that hav been produced by BirdLife International (Grimmett and Jone 1989; Heath and Evans, 2000) as key reference sources th should be taken into account in the absence of other publishe justifications for national SPA networks. The selected sites ar then to be communicated to the European Commission by th two-year formal compliance deadline (i.e. 7 April 1981) or b the Member State's accession date if that is later.

Whilst there is a requirement in BD Article 4(3) for th network to form a coherent whole for species across the are covered by the Directive, this has never been assessed b the Commission.

Identification and Designation of Natura 2000 Sites

The Habitats Directive requires the establishment of a 'coher ent European ecological network' known as Natura 2000 (HD Article 3(1)). See Evans (2012) for background history, and a description of the site selection process in practice. The Natura 2000 network comprises SPAs for birds, and SACs[11] hosting habitat types and species listed in Annexes I and II respectively. Annex I currently lists 233 'natural habitat types', which are defined in the Directive as 'terrestrial or aquatic areas distinguished by geographic, abiotic and biotic features whether entirely natural or semi-natural'. It is important to note that these are not necessarily habitats of threatened species, but are distinct entities that are considered to be worthy of conservation in their own right. In reality, many are semi-natural habitats (e.g. types of meadow or pastoral woodlands) rather than natural habitats. For brevity, these are hereafter referred to as 'habitats' rather than 'habitat types', and 'HD habitats' rather than 'HD Annex I habitats'.

A habitat is included in HD Annex I if it is in danger of disappearance within its natural range, or has a small natural range, or represents an outstanding example of one or more of nine biogeographical regions that occur in the EU: Alpine, Atlantic, Continental, Macaronesian, Mediterranean, Boreal, Pannonian, Steppic and Black Sea (Figure 2.1). There are 71 HD habitats that are given 'priority' status because the EU had a particular responsibility for them at the time they were included in Annex I. The list of HD habitats was initially based on the hierarchical classification of European habitats developed by the CORINE Biotopes project (Moss and Wyatt, 1994). However, as a result of revisions of the classification and the need for further guidance on the definitions and identification of Annex I habitat types, an interpretation manual for the EU-12 was published by the European Commission in 1995. This has subsequently been updated to take into account the habitats and biogeographical regions that have been added to the EU territory as a result of Member State accessions, with the most recent interpretation version

9 https://ec.europa.eu/environment/nature/conservation/wildbirds/eu_species/index_en.htm?msclkid=8e43c1a4ba5011ecbaee0965db5399a5

10 That is, those with 'a significant proportion of whose members cyclically and predictably cross one or more national jurisdictional boundaries'.

11 According to HD Article 3. In practice reference to Natura 2000 often includes SCIs, for example in the EEA Natura 2000 barometer.

being produced in 2013 (European Commission, 2013a). As a result of EU enlargement, some HD habitats are now better represented and no longer likely to be in danger of disappearance within the EU.

HD Annex II lists 1 036 animal and plant species that would benefit from the protection of their habitats, because within the EU they are endangered, vulnerable, rare or endemic.[12] Of the Annex II species, 262 are given 'priority' status because the EU has a particular responsibility for them, irrespective of whether they are in danger of disappearance (which is different to the criteria applied to priority habitats).

The site selection process under the Habitats Directive is based on scientific criteria listed in Annex III and relevant scientific information following a biogeographical regional approach (HD Article 4(2)). It is a two-step process whereby each Member State firstly carries out a national assessment of the relative importance of its sites, and provides a proposed list of Sites of Community Importance (SCIs), with a Natura 2000 Standard Data Form[13] for each, that should ensure the protection of a sufficient proportion of the area and populations of HD habitats and HD Annex II species within its territory. The list of sites for each habitat and species is supposed to be in proportion to its representation within its European territory (HD Article 3(2)). Proposals for SCIs for wide-ranging aquatic species are only required where suitable sites that may benefit the species can be identified. Flexibility in the interpretation of the SCI criteria is also allowed when Member States have a high proportion (>5%) of a priority habitat or priority species within their territory. Member States must submit their proposed lists of SCIs within three years of the notification of the Habitats Directive.

In the second step, the Commission, supported by the EEA and in consultation with Member States, carries out an assessment of the Community importance of each proposed site. Following any necessary modifications, an agreed list of SCIs is adopted for each biogeographical region that occurs within the Member State.

To assist with the identification of SCIs and the assessment of their sufficiency, so called 'biogeographical seminars' have been held with Member States, NGOs and other experts, and guidance has been produced by the EEA's European Topic Centre on Biological Diversity (ETC/BD). According to the guidance (ETC/BD, 1997), whilst other factors should also be considered and requirements will vary depending on their context, coverage of a HD habitat or species of 60% within SCIs in a biogeographical region of a Member State is normally considered sufficient. Higher coverage may sometimes be required, such as for particularly rare or localised species. Less than 20% coverage is normally considered insufficient, unless the habitat or species is especially widespread.

Within six years of adoption of the list of SCIs by the European Commission, Member States must designate each SCI as an SAC through a statutory administrative and/or contractual act and establish site-specific conservation measures (HD Article 4(4) and 6(1)).

Natura 2000 Coherence

It is clear that Natura 2000 is meant to be more than a collection of important sites, as HD Article 3(1) explicitly refers to it being a 'coherent network'. Similarly, the Birds Directive indicates the need for SPAs to 'form a coherent whole which meets the protection requirements of these species in the geographical sea and land area where this Directive applies' (BD Article 4 (3)). As neither of the Directives defines the meaning of 'network coherence' the practical implications for Member States have been the subject of some confusion and debate (Kettunen *et al.*, 2007). However, on the basis of a European Commission workshop and a review of relevant literature (most notably Carpenter *et al.*, 2001; Pritchard, 2007; UNEP-WCMC, 2008), a Commission study (Arcadis and IEEP, 2010) recommended that to be coherent the Natura 2000 network should meet the criteria for the following four key properties:

1. Adequacy: the individual components of the network are of sufficient size and shape, with an appropriate distribution to ensure the ecological viability and integrity (i.e. favourable conservation status) of its HD habitats and species.
2. Representativity: the components provide for all of the requirements of all HD habitats and species over their annual cycle (e.g. breeding, roosting, feeding and migrating).
3. Resilience: the network is able to undergo, absorb and respond to change and disturbance whilst maintaining its functions and controls.
4. Connectivity: the network is sufficiently connected to enable necessary movements of propagules (e.g. larvae and seeds) and individuals for breeding and dispersal, foraging, migration, climate change adaptation, and to maintain ecological processes and linkages.

The necessity to improve the coherence of the network is increased by the growing impacts of climate change on biodiversity, and consequently forms a key element of recommended adaptation strategies, including in guidance from the European Commission (2013d).[14]

The Establishment of the Natura 2000 Network

A common problem with the implementation of the Nature Directives has been the slow or inadequate identification and

[12] Based on taxonomy used in the EEA Article 17 reporting checklist (26 June 2020): https://cdr.eionet.europa.eu/help/habitats_art17 See also https://bit.ly/3ohXchQ

[13] Latest version: Commission Implementing Decision (2011/484/EU): https://eur-lex.europa.eu/legal-content/EN/TXT/PDF/?uri=CELEX:32011D0484&from=EN

[14] Updated European Commission guidance on Natura 2000 and climate change is expected in 2023.

subsequent designation of sites. This was particularly the case with the early classification of SPAs, but the rate of progress improved significantly after a number of important CJEU rulings relating to the adequacy of SPA networks. A key problem was that the Directive does not provide criteria for the identification of SPAs or their sufficiency. It has therefore been difficult to judge the adequacy of the SPA networks that have been proposed by Member States. However, the Commission assisted with the identification of SPAs by providing funding for the compilation of some ornithological information (e.g. on wetlands) and supporting the development of scientific criteria (publicly available in Appendix 3 of Stroud *et al.*, 1990).

A major advance was then made with the publication of an inventory of *Important Birds Areas* (IBAs) in Europe based on standardised criteria, by the International Council for Bird Preservation (now BirdLife International) and International Waterfowl and Wetlands Research Bureau (Grimmett and Jones, 1989). As noted earlier, this served as a key reference source to inform SPA identification. However, many IBAs remained unclassified as SPAs until a CJEU ruling in 1998 on the adequacy of the Netherlands' SPA network (Case C-3/96). This confirmed that, although not legally binding on Member States, the IBA inventory lists the most suitable territories for the conservation of wild birds in the EU and should always be considered when proposing national SPA networks. BirdLife published an updated list of IBAs in 2000 (Heath and Evans, 2000) and the Commission has continued to use the IBA inventory as a scientific reference list, including using it to issue written warnings to Member States and start infringement procedures (Milieu *et al.*, 2016).

Other court cases concerning the Santoña Marshes in Spain (Case C-355-90) and Lappel Bank in the Medway Estuary, UK (Case C-44/95), made it clear that the selection of SPAs and demarcation of their boundaries must be based solely on ornithological criteria. A similar principle applies to the identification of SCIs, as clarified by the CJEU ruling on case C-371/98 that economic considerations could not be taken into account by Member States when preparing their lists of possible SCIs.

Implementation of the Habitats Directive has also been hampered by the slow and incomplete identification of sites by Member States that should be considered for inclusion in the list of SCIs and, thereby, subsequent designation as SACs. Although the Directive includes criteria for the selection of SCIs, which avoided some of the ambiguity seen in the selection of SPAs, much less information was available on the best sites for HD habitats and species compared with birds. Therefore, extensive and time-consuming surveys and mapping were generally required. Progress was also constrained by the need to assess national proposals in the context of biogeographical needs. This led to lengthy iterative negotiations between many of the Member States and the Commission on their proposed lists of SCIs.

To speed up progress, in 2001 the Commission informed Member States that they would not receive funds for infrastructure development unless applications showed that the development would not threaten areas likely to be included in the Natura 2000 network. This is believed to be the first time that Member States had been threatened with withholding EU funding as a way to make them apply environmental legislation (Farmer, 2012). The Commission also warned Member States that failure to meet their commitments under the Nature Directives might also jeopardise their receipt of EU LIFE funds and the much more substantial funding for Rural Development Programmes (RDPs) (see Section 4.3). As progress with the identification of marine SCIs had been especially slow, a Dutch Council Presidency conference in 2004, held to celebrate 25 years of the Habitats Directive, urged the EU to complete the full extension to the marine environment by 2008. However, this turned out to be unfeasible following the CJEU's judgment on case C-6/04 requiring the Nature Directives to apply to each country's whole EEZ. The conference also called for the establishment of an effective protection regime for all SPAs, with management objectives in place and initiated for all sites by 2010. Despite these actions by the Commission, the submission of national lists of eligible SCIs was delayed for all Member States. Many received warning letters, and a high proportion led to CJEU rulings.

Spurred on by the combined effects of legal cases, the Commission's initiatives and growing national pressure from NGOs and others, the number and cumulative area of SPAs and SCIs increased more rapidly. By 2015 the terrestrial component of the Natura 2000 network covered 18.0% of the EU-28 land area, and was considered to be just about complete (EEA, 2015). According to the EEA's Natura 2000 barometer,[15] at the end of 2020 there were 26 935 Natura 2000 sites on land and at sea in the EU-27 (i.e. excluding UK sites) comprising 23 567 SCIs and 5 397 SCIs. The terrestrial components of the SCIs covered 587 172 km^2 and the SPAs 527 472 km^2, many of which overlap, giving a combined coverage of 764 222 km^2, which is 18.5% of the EU-27's land area. Therefore, even without the inclusion of nationally protected areas, the EU met the CBD's global Aichi Target 11, which aimed to conserve at least 17% of terrestrial and inland water areas by 2020.

The development of the marine Natura 2000 network has been particularly slow. This was partly due to the unexpected need to expand the coverage over a huge area, following the C-6/04 judgment, 26 years after the Birds Directive was adopted and 13 years after the Habitats Directive. The identification of sites for designation has also been constrained by limited data on the distribution of many marine habitats and species, especially in offshore areas. Although the pace of designation increased over the last decade, further sites are required. At the end of 2020 the marine area of SCIs was 354 971 km^2

[15] www.eea.europa.eu/data-and-maps/dashboards/natura-2000-barometer

and SPAs covered 294 801 km^2 (according to the Natura Barometer), giving a combined coverage of 449 497 km^2, which is 8.9% of EU-27 waters (5 038 743 km^2 as defined under the MSFD).[16] Thus, even though incomplete, the marine Natura 2000 network alone came close to meeting the CBD Aichi Target 11 of conserving at least 10% of coastal and marine areas by 2020. Taking into account an additional 3% under national protection gave a total marine protected area (MPA) coverage in 2020 of 11% (European Commission, 2020a), which was thus just above the Aichi target.

Further details of the extent of coverage of Natura 2000 sites and their sufficiency, and other protected areas, are provided in each national chapter, and overall conclusions are drawn in Section 5.4.1.

Measures to Maintain and Enhance the Coherence of the Natura 2000 Network

To help improve the coherence of the Natura 2000 network, the Habitats Directive requires Member States to 'endeavour', where they consider it necessary, to encourage the management of landscape features in their land-use planning and development policies (Article 3(3)). HD Article 10 repeats the provisions and clarifies that the landscape features in question are linear or continuous structures, such as rivers or traditional forms of field boundary (e.g. hedges), or habitat patches that act as stepping stones, such as ponds or small woods. Although the requirement for action on this is at the discretion of Member States, a DG Environment commissioned report (Kettunen *et al.*, 2007) concluded that, in principle, these measures should be taken when Member States regard them as necessary to achieve the overall objectives of the Directives (i.e. maintaining or restoring species' and habitats' favourable conservation status). The Birds Directive lacks such specific provisions on landscape features, but BD Article 3(2) indicates that the general requirement for the preservation, maintenance and re-establishment of habitats does not only apply within protected areas. Furthermore, as SPAs are incorporated into the Natura 2000 network, they are subject to the HD Article 3 (3) coherence provisions referred to above.

Some countries have sought to improve the coherence of their Natura 2000 and wider protected areas networks through initiatives to develop ecological networks (Jongman and Pungetti, 2004; Bennett and Mulongoy, 2006; Catchpole *et al.*, 2009), some of which are described in the national chapters of this book. Most pre-date the Habitats Directive and have a broad range of ecological, environmental/ecosystem service and social objectives (e.g. in the Czech Republic, Lithuania, Slovakia and the Netherlands), which do not necessarily contribute greatly to Natura 2000 coherence. More recent approaches, such as in France, have typically incorporated broader green infrastructure concepts and ideas, as taken up in the *EU Biodiversity Strategy to 2020* (see Section 4.4). As discussed in Chapter 5, in general, the added value of these initiatives in actually increasing practical protection and enhancement of ecological networks and green infrastructure is rather uncertain.

Management and Protection of Natura 2000 Sites

A key aspect of the Nature Directives is that they not only require sites to be protected from damage, but they also require Member States to ensure they are appropriately managed, and if necessary restored, to make sure each site contributes as much as possible to achieving the overall objectives of the Directives. To guide this, once listed as an SCI, site-level **conservation objectives** and priorities should be established for all the HD habitats and species for which the site was designated. Whilst the Habitats Directive does not legally require the setting of site-level conservation objectives, they are referred to in the preamble to the Directive and are a practical necessity for setting site-specific management measures and forming the basis for deciding whether a project or plan is likely to damage a site under HD Article 6(3). Therefore, conservation objectives are also required for SPAs, as they also fall under HD Article 6(3) provisions (see further discussion later in this section). Irrespective of this, they are required in practice to help achieve the requirements of the Birds Directive in relation to Articles 2 and 4(1), (2) and (4) (European Commission, 2012a).

In principle, conservation objectives are required for all habitat types and species for which the site was designated, as listed in the Natura 2000 Standard Data Form. These should include all HD habitats and species and all BD Annex I bird taxa that are significantly present, as well as regularly occurring migratory birds.[17] For brevity, the habitats and species for which a Natura 2000 site is designated are referred to hereafter as 'Natura habitats and species'.

Conservation objectives aim to define in clear and measurable terms the desired site-specific condition of the Natura habitats and species. They therefore need to be based on each habitat's and species' ecological requirements (including site-specific and wider considerations). Furthermore, to be effective in driving forward and monitoring conservation measures, they need to be clear, specific and quantified in measurable terms as far as possible.

The need to set site conservation priorities is an explicit requirement of HD Article 4(4). Their purpose is to identify the Natura habitats and species of greatest importance within the site, and the most important and urgent conservation measures that need to be taken (European Commission, 2012a). Thus, conservation objectives and conservation priorities are interrelated, and in practice are often developed together.

16 www.eea.europa.eu/themes/biodiversity/natura-2000/natura-2000-coverage-in-european-seas-3 (February 2022).

17 Habitats and species that are present but have insignificant populations are those given a category 'D' in the Natura 2000 Standard Data Form.

Commission guidance has indicated that Natura 2000 site conservation objectives should aim to contribute in the best way possible to the habitat's and species' maintenance or restoration of a favourable conservation status, as well as the coherence of the Natura 2000 network (European Commission, 2012a). Thus, as a minimum, the objectives should aim to prevent any decline in the condition[18] of the Natura habitats and species within the site concerned, and to improve those in poor condition (i.e. that are not contributing to their overall favourable conservation status). Furthermore, increasing the areas of certain Natura habitats or species numbers on the site, beyond that necessary to achieve their good condition, may also be appropriate to help achieve their overall favourable conservation status. Consequently the setting of site conservation objectives is not just about the site, and needs to consider several important factors. It is particularly important to take into account the proportion of the total national and biogeographical area of each HD habitat, and the proportion of the total national and biogeographical population of each Natura species, that occurs within the site. The condition of each Natura habitat and species should be considered, together with its wider-scale national and biogeographical level conservation status. The site's spatial relationship with others in the network, such as its role as a staging post for migratory species and/or in maintaining meta-populations, also needs to be taken into account.

Given these requirements, the setting of conservation objectives can be a difficult and complex task, especially where there are many Natura habitats and species to consider, some of which may have conflicting ecological requirements. Furthermore, gaps in knowledge on the ecology of habitats and species and their degree of conservation within sites, and wider afield, can hamper the setting of conservation objectives. This has therefore led to some Member States struggling to define conservation objectives comprehensively. Others have found that their desired beneficial objectives for some Natura habitats or species would be detrimental for others – thus potentially leading to breaches of the Habitats Directive. Others have used the process to more clearly set and monitor targets, which has led to an increase in conservation action where previously sites were deteriorating through neglect (e.g. in the UK).

To achieve the overall aims of the Nature Directives, and specific conservation objectives of each Natura 2000 site, the sites need to be protected and necessary conservation measures put in place and implemented. Specifically, under the Birds Directive, Member States are required to preserve, maintain and re-establish biotopes and habitats, and this includes their management in accordance with ecological needs, both inside and outside protected areas (Article 3(2)). In addition, appropriate steps must be taken to avoid pollution or deterioration of habitats or any disturbances affecting the birds within SPAs, where these would affect their survival and reproduction within their area of distribution (Article 4(4)). Member States are also required to strive to avoid pollution or deterioration of habitats outside SPAs.

However, a case concerning the Leybucht Dykes in Germany and the protection of an SPA had profound implications for both the Birds and Habitats Directives (Case C-57/89). The case suggested that Member States were obliged to protect sites more stringently than they were normally accustomed to under domestic protected area legislation. As this occurred during the development of the Habitats Directive, it led to new clauses in the text that introduced considerations of overriding public interest in the permitting provisions (HD Article 6(4)). As this would lead to a situation with more stringent protection of SPAs than SCIs and SACs, the situation was harmonised by the protection requirements under Article 4 (4) of the Birds Directive being replaced by those of HD Articles 6(2), (3) and (4). This reduced the impact of the Court's judgment.

A related ruling in 2000 on the Basses Corbiéres case (C-374/98) took the view that Member States should not in any way be rewarded for failure to designate a site. The Court therefore ruled that areas that have not yet been classified as SPAs, but should have been so classified, still fall under the protection regime set out in BD Article 4(4). As indicated above, this regime is stricter than the protection provisions in the Habitats Directive. This therefore effectively acts as an incentive for Member States to classify SPAs.

Article 6 of the Habitats Directive includes provisions for the management and protection of Natura 2000 sites, which comprise interlinked requirements to establish necessary conservation measures (Article 6(1)), prevent deterioration and disturbance (Article 6(2)) and establish assessment procedures and safeguards for potentially detrimental plans and projects (Articles 6(3) and (4)). As noted in the European Commission guidance on Article 6 (European Commission, 2018), in the wider context of the Treaty of the EU, 'Article 6 can be regarded as a key framework to give effect to the principle of integration since it encourages Member States to manage the Natura 2000 sites in a sustainable way and it sets the limits of activities which can impact negatively on protected areas while allowing some derogations in specific circumstances.' It is also important to note that the practical planning and implementation of Article 6 need to take into account and interact with other EU policies such as those regarding agriculture, water, fisheries and other maritime issues (see Section 4.3). Another point to bear in mind is that the provisions may require actions to be taken outside the Natura 2000 sites concerned, such as to manage ground water levels (e.g. with respect to water abstractions), prevent pollution (e.g. from upstream sources in rivers, or from airborne deposition) or disturbance (e.g. from roads).

Under **HD Article 6(1)**, as soon as a site is designated as an SAC (not SCI), Member States must establish the 'necessary conservation measures' for the site. Article 6(1) does not apply

[18] Referred to as degree of conservation in the Natura 2000 Standard Data Forms.

to SPAs as the similar provisions in the Birds Directive are retained, as described earlier. The conservation measures must provide the ecological requirements for each of the Natura habitats and species, which must be established on a case by case basis, linking to the conservation objectives for the site (although this logical link is not explicitly stated in the Directive). The required measures may be identified and set out in a management plan (which may include the prior setting of conservation objectives), but this is not mandatory. Although the European Commission has not gone as far as explicitly recommending that management plans are developed for Natura 2000 sites, it has stated that they are a useful tool for ensuring the implementation of Article 6.1 provisions in a clear and transparent way, informing stakeholders about what Natura 2000 sets out to achieve and engaging their participation in discussions. Management plans may also help to identify funding requirements and measures and help to integrate conservation measures into other plans (European Commission, 2013b). However, some types of management plans, such as integrated plans, sectoral plans and plans for other types of protected area (e.g. National Parks) may not be suitable if they do not include clear and specific targets for each of the Natura habitats and species present. Whatever means and documents are used to identify the necessary conservation measures, it is of most importance that the measures are based on sound knowledge, developed through a participatory process among all stakeholders (allowing economic activities to be carried out that respect or support the site's conservation objectives), defined in sufficient detail and allocated adequate funding to allow full implementation (European Commission, 2013b).

Member States have discretion on the most appropriate means of practically ensuring the necessary conservation measures are implemented, which according to Commission guidance (European Commission, 2018) can involve appropriate statutory, administrative or contractual measures (with no hierarchy of preference for these). A commonly used contractual approach is to use RDP measures (see discussion in Section 4.3), such as agri-environment schemes, to fund required practical actions (e.g. the continuation of extensive grazing). However, landowner participation in such schemes is voluntary, and funding may be restricted, therefore this may not guarantee that the required measures are actually implemented. The RDP Natura 2000 funding measure may also be used to compensate landowners if statutory measures restrict their activities (e.g. logging).

Article 6(2) of the Habitats Directive applies to all Natura 2000 sites, including SPAs (replacing the provisions under BD Article 4(4)), as well as sites that have been identified as SCIs but not yet designated as SACs. The provisions require Member States to take appropriate steps to avoid the deterioration of all Natura habitats and species and the significant disturbance of all Natura species within the sites. As the provisions are based on the prevention principle, case law (e.g. C-418/04 in Ireland – see Chapter 19) and Commission guidance has highlighted that 'It is not acceptable to wait until deterioration or disturbances occur before taking measures' (European Commission, 2018). Thus measures must be taken to avoid reasonably predictable events (whether man-made or not), such as floods, fires, pollutant spills or colonisation of a site by a nearby invasive species. Deterioration as a result of dynamic issues and wide-scale pressures, such as from climate change or coastal erosion, also needs to be addressed, although this can be on a case by case basis. Other important aspects of the provisions are that they apply to all activities, such as agriculture, water management and fisheries, and not just those involving projects and plans, which may require measures outside the site.

The CJEU has confirmed that the ecological characteristics of a site must not be allowed to deteriorate below the level that existed at the time that the Member State authority proposed the site for inclusion as an SCI (case C-244/05 in Germany). This has led to some Member States complaining that this requirement is too inflexible and does not take into account natural processes (e.g. succession and erosion). However, Commission guidance has indicated that the assessment of whether deterioration or significant disturbance may occur needs to be made in relation to the site's conservation objectives and its contribution to the coherence of the Natura 2000 network. This therefore provides a more dynamic basis for assessing deterioration, but also relies on the sites' conservation objectives being clearly defined in quantified terms, including with respect to network coherence.

Like other provisions in the Habitats Directive, Article 6(2) does not prescribe how Member States must avoid deterioration and significant disturbance. Therefore there is flexibility in the approach that can be taken, but Commission guidance states that the legal regime 'has to be specific, coherent and complete, capable of ensuring the effective protection of the sites concerned' (European Commission, 2018). Case law has also indicated that selected activities, such as recreational activities, hunting or commercial fishing, cannot be generally regarded as not causing disturbance in order to justify their systematic exclusion from the obligation to avoid deterioration and disturbance of species (HD Article 6(2)), or to assess their impacts on Natura 2000 sites (Article 6(3)) (C-241/08).

Habitats Directive Articles 6(3) and 6(4) address the permitting of projects and plans that may affect Natura 2000 sites, giving them strict protection from damage unless there are certain exceptional circumstances (Figure 4.1). Under **Article 6(3)**, plans or projects that, individually or in combination with others, are likely to have a significant effect on a site but are not directly connected to their management for nature conservation, are to be subject to an 'appropriate assessment' of their implications with respect to the site's conservation objectives. In this context projects and plans are defined broadly, in line with the definitions used in the EIA and Strategic Environmental Assessment (SEA) Directives respectively (European Commission, 2018). However, unlike for EIA and SEA, there are no lists of types of projects or plans that should be assessed, as any that have the potential to have an adverse effect should be

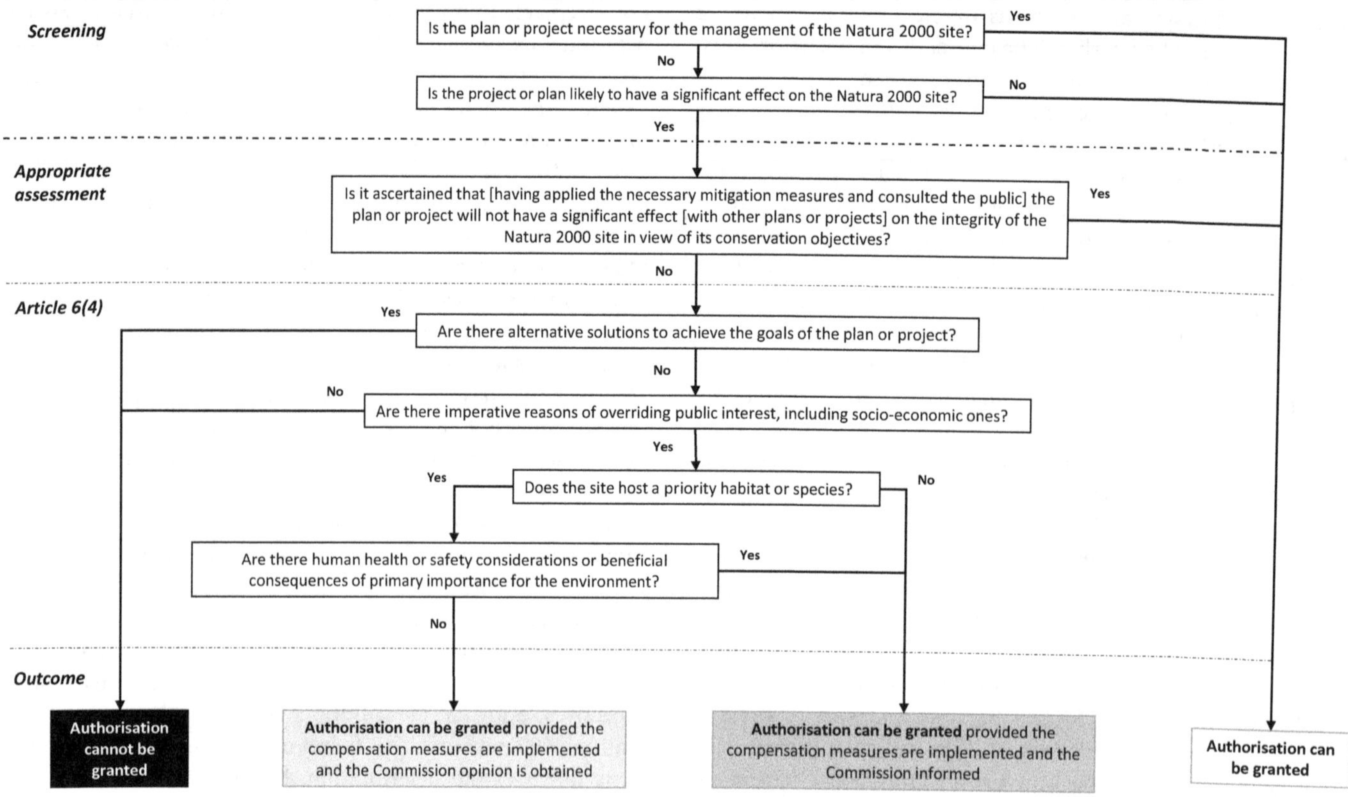

Figure 4.1 Flow chart of Article 6(3) and 6(4) procedures for assessing plans and projects in relation to Natura 2000.
Source. Redrawn with modifications from European Commission (2018, 2021c) under Creative Commons Attribution 4.0 International (CC-BY 4.0) licence (https://creativecommons.org/licenses/by/4.0/).

considered. Accordingly, in addition to projects, plans and programmes that are normally subject to an EIA or SEA (e.g. infrastructure, housing, water, energy and the extraction of minerals), other significant interventions in the environment may require an appropriate assessment, such as agricultural intensification and fisheries (the latter following the Waddenzee case C-127/02). Plans that may require an appropriate assessment include spatial plans, land-use plans and sectoral plans, such as for transport or energy. Broader policies that do not provide an indication of which Natura 2000 sites may be affected do not require an appropriate assessment.

It is important to note that appropriate assessments differ in several critical ways from impact assessments carried out in relation to the EIA and SEA Directives (see Section 4.3). Most fundamentally, the results of EIAs and SEAs are advisory, whereas those of an appropriate assessment have clear binding implications, making the procedure very robust. Nevertheless, in practice they are often interlinked and therefore carried out together to some extent. Where this is the case it is essential that the results of the appropriate assessment are presented clearly and separately in a way that distinguishes them from the assessments of broader biodiversity impacts (European Commission, 2018).

As the Article 6(3) provisions are briefly set out and rather vaguely worded, they created some initial confusion amongst Member State authorities on what is required in practice, which led to numerous transposition and implementation problems and, in turn, Commission interventions and CJEU cases. To address this, the Commission produced a number of documents that aimed to clarify the legal requirements (European Commission, 2001, 2007a,b, 2014a, 2021c), as well as compilations of key CJEU rulings on Article 6 (Sundseth and Roth, 2014). Sector-specific guidance has also been produced by the Commission, including for non-energy mineral extraction, coastal developments, inland waterway transport, aquaculture, wind energy, hydropower and energy transmission.[19]

A common subject of confusion has been over the screening of projects and plans to determine whether or not they require an appropriate assessment. For example, some projects and plans were initially excluded by Member States because they were variously already subject to periodically issued permits (e.g. for fishing), were not within a Natura 2000 site or were small or of low value; such reasons were considered by the CJEU to be irrelevant (Milieu *et al.*, 2016). Instead the key consideration is simply whether or not the project or plan is likely to have a detrimental effect, irrespective of its type, size, location, etc. The Court has also made it clear that when determining its potential effects, possible mitigation measures cannot be taken into account in the screening stage

[19] Available on the European Commission website: https://ec.europa.eu/environment/nature/natura2000/management/guidance_en.htm

but should instead be considered as part of the appropriate assessment (Case C-323/17).

A particularly important CJEU ruling, known as the Waddenzee ruling (Case C-127/02), citing Case C-157/96, clarified that Article 6(3) integrates the precautionary principle. Consequently, if there is doubt over whether or not a significant effect may occur, an appropriate assessment should be carried out. Similarly, if an appropriate assessment does not show beyond reasonable scientific doubt that there will be no adverse effects, then the project or plan cannot go ahead. Importantly, this places the onus on the proponent to prove the absence of detrimental effects.

Another area of uncertainty has been over what is meant by an 'appropriate assessment' and what should be its scope, methodology and content. The key requirement is for the assessment to enable the competent authorities[20] to ascertain whether the plan or project will adversely affect the integrity of the site concerned. In this respect the integrity of the site needs to be considered in relation to the site's conservation objectives for the Natura habitats and species present on the site (as also confirmed by C-127/02). It is therefore important to note that Natura 2000 designation does not result in the prohibition of all potential biodiversity impacts, but instead is focussed on those affecting habitats and species for which the site is designated. This enables development to proceed where it is compatible with the site's objectives. However, appropriate assessments must bear in mind that achieving conservation objectives for particular Natura habitats and species will depend on the sites' ecological integrity, which can be defined as 'the coherent sum of the site's ecological structure, function and ecological processes, across its whole area, which enables it to sustain the habitats, complex of habitats and/or populations of species for which the site is designated' (European Commission, 2018). Thus the potential for a wide range of direct and indirect effects on Natura habitats and species, as well as ecological interactions, should be considered, including long-term impacts.

As a result of the development of case law, Commission and national guidance and gradual accumulation of experience, many of the problems associated with Article 6(3) have been resolved (Sundseth and Roth, 2013). However, some concerns remain, such as in relation to screening procedures, the scope, quality and independence of assessments, and the underestimation of potential impacts, including from cumulative effects. These issues are described further in several country chapters.

Under **HD Article 6(4)**, in exceptional circumstances and under strict conditions, a plan or project that may have detrimental impacts on the Natura habitats and species within a Natura 2000 site may still be permitted, if there are no alternative solutions and there are 'imperative reasons of overriding public interest', including those of a 'social or economic nature'. Where a site affected hosts a priority habitat type or species the only considerations that may be raised are those relating to human health and public safety, or where their beneficial consequences are of primary importance for the environment or other reasons further to an opinion from the Commission. A fundamental aspect of these derogation provisions, as confirmed by the CJEU on several occasions (European Commission, 2018), is that they can only be applied after the implications of the plan or project in question have been fully studied and documented in an appropriate assessment (Figure 4.1). Furthermore, as they form an exception to the protection afforded to Natura 2000 sites, they must be strictly interpreted and only applied where all conditions are met.

The first key condition is that the potential for alternative solutions that meet the plan's or project's aims, yet maintain the integrity of the site(s) in question, is thoroughly examined. This might involve considering alternative locations, timing, types of activity and, if necessary, the no plan/project option. Should such an alternative exist, the plan or project must not be approved. The absence of alternatives must be demonstrated before considering the second condition, which is that the plan or project is required for imperative reasons of overriding public interest. Such reasons are not defined in the Habitats Directive, but case law (e.g. C-182/10) and Commission guidance (European Commission, 2007a, 2018, 2021c) have indicated that they 'refer to situations where plans or projects envisaged prove to be indispensable:

- within the framework of actions or policies aiming to protect fundamental values for the citizens' life (health, safety, the environment);
- within the framework of fundamental policies for the State and the society;
- within the framework of carrying out activities of an economic or social nature, fulfilling specific obligations of public service.'

Furthermore, the assessment should consider the longer-term public benefit rather than short-term gains.

Despite the case law and Commission guidance, two studies have concluded that some Member States' reasons for allowing damaging projects and plans to go ahead, as well as some Commission opinions on the matter, have taken a wide interpretation of the meaning of 'imperative reasons of overriding public interest' (Krämer, 2009; Clutten and Tafur, 2012).

If a competent authority decides that a project should go ahead because imperative public benefits outweigh the damage to the Natura 2000 site(s), then the third condition that must be met is for compensatory measures to be taken to protect the overall coherence of Natura 2000. The Member State must also inform the Commission of the measures that are taken. The term 'compensatory measures' is not defined in the Habitats

[20] Competent national authorities are those entitled to give authorisation or consent to a plan or project; in most cases these are not nature conservation authorities, therefore precise and clear interpretation of and understanding of the appropriate assessment are of utmost importance.

Directive, but Commission guidance has clarified that they are distinct from mitigation measures (which are part of the project, and aim to eliminate or at least reduce impacts) and should 'offset the negative effects of the plan or project so that the overall ecological coherence of the Natura 2000 Network is maintained' (European Commission, 2007a). Importantly, the Article 6(3) and 6(4) provisions, and their sequential consideration (Figure 4.1), follow the widely accepted principles of the mitigation hierarchy, with compensatory measures only being applied to residual impacts after appropriate avoidance and mitigation measures have been taken.[21]

To ensure the coherence of the Natura 2000 network, the European Commission (2007a, 2018) has stated that the compensatory measures should: 'a) address, in comparable proportions, the habitats and species negatively affected; b) provide functions comparable to those which had justified the selection criteria of the original site, particularly regarding the adequate geographical distribution.' A further fundamental requirement is that they go beyond the obligations that Member States already have under the Nature Directives in regard to protection and management of the Natura 2000 sites.

Compensatory measures are ecological measures that result in measurable outcomes for the Natura habitats and species concerned. Thus, other actions such as the creation of visitor facilities, or monetary payments to protected areas, are not acceptable in the context of these Habitats Directive requirements. According to the most recent Commission guidance (European Commission, 2021c) compensatory measures may consist of one or more of the following in proportion to the loss:

- Habitat restoration or enhancement in existing sites, provided it goes beyond the sites' conservation objectives.
- Habitat restoration or the recreation of comparable habitat on a new or enlarged site, to be incorporated into the Natura 2000 network, protected and managed.
- Designation of a new Natura 2000 site, and implementation of management measures.
- Species re-introduction, recovery and reinforcement (including reinforcement of prey species), where species have disappeared or are declining.

Other key requirements for compensatory measures concern their quantity, location and timing, as summarised here based on the latest Commission guidance (European Commission, 2021c). The amount of compensatory habitat that needs to be created or enhanced is primarily dependent on the project's residual impacts on the quantity and quality of the Natura habitats and species within the site, with respect to their conservation objectives. These residual impacts should be precisely quantified in the appropriate assessment. Therefore, whilst a compensation ratio of at least 1:1 is the normal starting point, the actual compensation requirement needs to be calculated on a case by case basis. This calculation also needs to take into account the expected effectiveness (and its time lag) of the compensatory measures. Measures that might be expected to be only partially successful will need to adjust for this, by including additional compensation as a contingency.

It is also important for the compensatory measures to be located in areas where they will be most effective in maintaining the overall coherence of the Natura 2000 network. In this regard, the Commission guidance indicates that they must take place within the same biogeographical region as the affected site, and where possible within the affected Natura 2000 site, or at least nearby. However, the ecological conditions necessary to allow this may not always be present, or other practical considerations may prevent it. In such cases compensation can occur outside Natura 2000 sites and, if necessary, in more distant areas, but always within the same EU Member State. This prioritisation of location is a risk-averse approach, which assumes that the affected habitats and species are currently in the most suitable location, but this is not necessarily the case in the highly modified, disturbed and fragmented landscapes of most of Europe. Therefore, in some cases it would be better to place compensatory measures in more suitable areas (e.g. less fragmented or disturbed). There may also be strategic advantages to locating compensation away from the impacted site, which can increase the coherence of the network, such as by increasing the size (and therefore viability and resilience of sites) or by creating new sites that act as stepping stones.

In terms of their timing, compensatory measures should be in place and providing their intended ecological functions before damage to the Natura 2000 site occurs. However, this is not always practically feasible, and therefore time lags are allowable if this does not affect the overall coherence of the Natura 2000 network and does not lead to any population losses of BD Annex I species or HD Annex II species. Where time lags are expected to occur, then additional compensation is required. Irrespective of when the compensation may be functional, all technical, legal or financial provisions needed to implement the compensatory measures must be completed before the project starts. This includes the designation of areas as Natura 2000 sites where they are outside the network but are to be the site of the compensatory measures.

Despite the good intentions of the Habitats Directive, and subsequent guidance by the Commission, to ensure that damaging projects fully and effectively compensate for their impacts, it is important to note that this is not always possible in practice. Most obviously, this is because many HD habitats, such as forests and mires, take centuries to develop and achieve the ecological quality of those that are likely to be lost (Morris *et al.*, 2006). Others may require biophysical conditions that are impossible to create (e.g. tidal estuaries). Despite this, the wording of Article 6(4) implicitly assumes that compensation is possible, and only requires Member States to inform the Commission of its compensatory measures. Therefore, this is not a robust condition for the approval of projects, and it is

[21] For example, see Business and Biodiversity Offsets Programme Principles on Biodiversity Offsets: www.forest-trends.org/wp-content/uploads/2018/10/The-BBOP-Principles_20181023.pdf

likely that some projects and plans have gone ahead that have not in reality compensated for their impacts. Unfortunately, evidence of this is lacking due to the poor reporting by Member States. It is also unknown to what extent, if at all, projects have been dropped solely on the basis that they could not fully compensate for their impacts.

4.2.4 Species Protection Measures

The Protection of Birds

The Birds Directive requires Member States to provide a basic level of protection for all species referred to in Article 1 of the Directive (i.e. all wild birds that occur naturally within the EU) whilst allowing for justified exceptions that do not undermine the overall aims of the Directive. Thus, following the typical structure of much environmental legislation (Galbraith, 2013) key provisions require certain actions to be generally prohibited, whilst other provisions allow for 'opt out' derogations from these requirements under certain conditions.

Specifically, under Article 5, Member States must generally prohibit the deliberate killing and capture of wild birds as well as the deliberate destruction of their nests or eggs and taking of eggs; and their deliberate disturbance, especially during the breeding season; as well as the keeping of birds.

Article 6 prohibits trade in wild birds, unless they are taxa listed in Annex III of the Directive. Annex IIIa lists nine taxa that can be traded in the entire EU, and Annex IIIb lists 19 taxa that Member States may allow to be traded, in all cases provided that it does not endanger them.

Article 7 allows for the hunting of 84 bird taxa listed in Annex II of the Directive. Most of these are game species (e.g. partridges, pheasant, grouse, most ducks and geese; some waders, pigeons and thrushes). Some are included in Annex II because these are considered to be pests and need to be controlled in some circumstances (e.g. most species of crow, some pigeons and some gulls). The 25 taxa listed in Annex II (A) can be hunted in all Member States, whilst the 59 taxa in Annex II(B) are only potentially huntable in particular countries.[22] Irrespective of the species, large-scale or non-selective capturing or killing of birds is prohibited under Article 8, especially through the methods or means of transport listed in Annex IV (e.g. using explosives or nets, or hunting from aircraft). Furthermore, Member States must ensure that hunting complies with the principles of wise use and ecologically balanced control of the species concerned. Member States are required to send the Commission all relevant information on the practical application of their hunting regulations. Recently, to facilitate sustainability assessments, periodic bag totals have been added to the Article 12 reporting requirements (described later).

Under Article 9, Member States may derogate from the above prohibitions on killing or capture (i.e. avoid the need to apply them) for certain reasons (described later) but only if there are no other satisfactory solutions.

Importantly, the Birds Directive's species protection provisions, particularly those concerning hunting, aim to balance two potentially conflicting objectives included in Article 2 (although it is arguable that they are not conflicting if a long-term view is taken): firstly, maintaining bird populations whilst, secondly, taking into account economic and recreational requirements (including hunting and fishing). Therefore, recognising that hunting is a legitimate activity that can also contribute to nature conservation, such as through habitat protection and appropriate management (European Commission, 2013c), the hunting provisions are based on the principles of wise use and ecologically balanced control (Article 7(4)). In this respect the Birds Directive develops the wise use concept introduced in the Ramsar Convention (Section 3.3 and online Annex 1). It also corresponds well with the more recent definition of sustainable utilisation by the CBD, which is 'the use of components of biological diversity in a way and at a rate that does not lead to the long-term decline of biological diversity, thereby maintaining its potential to meet the needs and aspirations of present and future generations'. Ecologically balanced control refers to balancing conservation and other interests. It also implies that when hunting is carried out to control species this should be based on ecologically sound principles and carried out in proportion to the problem to be solved, taking into account the conservation status of the species involved (European Commission, 2008).

Hunting is allowed for certain bird species (primarily for game species and for those that need to be controlled) with conditions that are supposed to ensure it is compatible with the aims of the Birds Directive. However, huntable species include 12 Annex I species, some of which are traditional game species but are now threatened in some countries, such as Capercaillie (*Tetrao urogallus*) and Golden Plover (*Pluvialis apricaria*). A reason for this is that the conservation of such species is normally more related to habitat factors than the off-take from hunting, or at least it was when the Annexes were drawn up. It is also notable that the hunting provisions in the Birds Directive do not include a general ban on hunting in Natura 2000 sites. Nevertheless, HD Article 6 provisions may require that hunting in Natura 2000 sites is prohibited, or restricted to some extent, to meet the site's conservation objectives, such as in relation to hunted Annex I species, or impacts from lead shot pollution or disturbance.

A fundamental aspect of wise use is that the hunting of each targeted species is sustainable (i.e. it does not lead to a decline in its population, range or breeding success). A further key requirement is that hunting is sustainable with respect to non-target species (i.e. it does not lead to their declines). Therefore, the indirect impacts of hunting on non-target species such as through disturbance or the effects of changes in habitats and species communities, such as from game management practices (e.g. habitat management, feeding the target

[22] One species, Willow Grouse (*Lagopus lagopus*), has subspecies on both Annex IIA and IIB.

species or predator control), need to be considered (ideally within the context of a site management plan) and kept within acceptable limits.

The Commission's *Guide to Sustainable Hunting under the Birds Directive* (European Commission, 2008) indicates that the general approach should be to ensure that the hunting of a species does not exceed sustainable yields, which is the capacity of the species population to replace losses through natural processes. This requires the calculation of a 'maximum sustainable yield' (resulting from density-dependent increases in survival and/or reproduction in remaining birds following reductions in their numbers from hunting) and 'optimum sustainable yields' (which take into account variations and uncertainties in yields). However, the accurate calculation of such yields is difficult, especially for migratory species, and therefore needs to be based on sound science and monitoring data, including reliable bag statistics. The Commission guidance therefore advises that 'In the absence of good information on population dynamics and hunting take of sedentary and migratory species high levels of exploitation should generally be avoided.'

A key component of wise use, and increasing sustainable yields, is the need to set appropriate start and end dates for hunting periods, to minimise the impacts of hunting on breeding population levels. In particular, it is of key biological importance to ensure that hunting does not occur during the breeding period, and to avoid the hunting of birds on their pre-nuptial migration (i.e. as they return to their breeding grounds in spring), both legal requirements of Article 7(4). Problems have arisen as a result of the discretionary authority of Member States to specify the hunting season for migratory birds under Article 7(4) and differences in their interpretation of the requirements. A number of important CJEU rulings have therefore been required to clarify the requirements, including the need to close hunting periods so that they ensure the complete protection of all birds during periods when they are most vulnerable, that is, all birds that are migrating to their breeding grounds and all birds that are in the process of breeding or have dependent young (e.g. C-157/89 and C-435/92). Thus, the hunting season in each Member State or region must not overlap with the earliest breeding and migrating birds, and latest breeders.

To help Member States interpret and comply with the requirement for complete protection, the European Commission produced ORNIS Committee approved guidance on the key concepts of Article 7(4). This was first published in 2001 and subsequently updated in 2014 (European Commission, 2014b) and then again most comprehensively in 2021 (with online species accounts, references and a database).[23] The guidance identifies the period of reproduction and of pre-nuptial migration for bird species listed in Annex II of the Directive, in each Member State, using available information on their breeding biology and migration behaviour. The start of the period of reproduction is defined for each species, but generally is the 'occupation of the breeding sites' for migratory birds, or the 'construction of the nest' for resident birds. The end of the period of reproduction is normally the 'full flight of young birds' (i.e. fledging of all broods including second or third broods for some species).

As the requirements for closing the hunting season differ amongst species, some Member States have set staggered closing dates of the hunting season with different closing dates for different species. However, in cases C-38/99 and C-435/92, the CJEU recognised that this leads to two additional difficulties: the risk of confusion between different species (which may lead to the shooting of species for which the hunting season is already closed); and potential for disturbance of a species by the hunting of other bird species. To address these risks, the Commission's *Guide to Sustainable Hunting* indicates that Member States can only allow staggered closing dates if they can show that there are no significant risks of confusion and disturbance. To address confusion risks, staggered hunting dates must be fixed for each group of 'look-alike' species that occur in the same habitat and at the same time, in a way that ensures that there are not overlaps with non-permissible hunting periods under the Birds Directive. As regards disturbance, the guidance notes that 'there is a need to demonstrate a balance between the intensity, frequency and duration of hunting, and the availability and proximity of sufficient undisturbed areas offering adequate feeding and roosting areas'. It then goes on to recommend that, in areas with staggered hunting dates, integrated planning can take full account of hunting and other potential disturbances on the birds and their use of the natural resources, with monitoring undertaken to assess impacts on the birds concerned.

Article 9 of the Birds Directive gives Member States the opportunity to derogate from the species protection provisions of Articles 5–8, where this is justifiable for the following potential reasons:

- in the interest of public health and safety;
- in the interests of air safety;
- to prevent serious damage to crops, livestock, forests, fisheries and water;
- for the protection of flora and fauna;
- for the purposes of research and teaching;
- for repopulation or re-introduction; and
- to permit, under strictly supervised conditions and on a selective basis, the capture, keeping or other judicious use of certain birds in small numbers.

However, as an overarching condition, a derogation can only be taken where there is no other satisfactory solution to the reason for its consideration. Whilst Member States do not need to consult with the European Commission before applying derogations, they are required to report on them to the Commission in annual reports. The Commission then has an obligation to ensure that the derogations are justified and

[23] https://ec.europa.eu/environment/nature/conservation/wildbirds/action_plans/guidance_en.htm

being properly applied, and act accordingly if they are not. It is important to note that, because derogation provisions allow otherwise illegal activities to be carried out, it is necessary that they are applied in strict accordance with their conditions and enforced. The requirement for strict implementation on the one hand and, on the other, the limited definition of some key terms in the legislation, initially contributed to considerable uncertainty and resulting litigation over some aspects of the provisions. Some of the key problems are briefly discussed here.

As a key requirement for all derogations, there have been important clarifications from the Court regarding when it can be considered that there is 'no other satisfactory solution'. For example, if a possible solution does exist, then any arguments that it is not satisfactory need to be strong and robust, and based on objectively verifiable information (C-10/96 and comments from the Advocate General, cited in European Commission, 2008). There has also been debate over whether the 'no other satisfactory solution' condition could ever be satisfied in relation to hunting, such as when derogations are used to extend hunting seasons for Annex II species. In this respect, whilst the Court has indicated that recreational hunting may constitute a 'judicious use', and thus be a potential reason for a derogation under BD Article 9(1)(c), it has confirmed that it is only allowable where there is no other satisfactory solution (Case C-182/02). Furthermore, it clarified that this condition would not be met if the sole purpose of the derogation is to extend the hunting period for certain species of birds where they already occur within the territory concerned during periods when the species may be hunted in accordance with Article 7 (as discussed earlier).

Another derogation condition that has led to confusion over its meaning, and practical control, concerns the requirement under BD Article 9(1)(c) for it to affect 'small numbers'. This has also led to a number of cases and guidance from the Commission that has indicated that the definition of small numbers is relative and depends on the requirements to maintain the population and the reproductive situation of the species concerned (e.g. Case C-252/85). However, in its second report on the application of the Birds Directive (European Commission, 1993), the Commission indicated that, according to the work of the ORNIS committee, 'small numbers' should be understood to mean less than 1% of the total annual mortality rate of the population in question (average value) for those species that are not hunted, and in the order of 1% for those species that may be hunted. Although not legally binding, the ORNIS Committee views have been used as a reference by the CJEU (e.g. Cases C-79/03 and C-344/03). The issue of small numbers has since been treated in more detail in the European Commission's *Guide to Sustainable Hunting*.

The *Guide to Sustainable Hunting* has also indicated that, according to case law, derogations should not be granted for species with populations that are declining, unless they are beneficial for the species concerned and are carried out as part of a conservation plan (discussed earlier). Derogations must be based on sound reasons for particular species and specific situations. Therefore, general licences that authorise landowners to control pest species (e.g. beyond the restrictions of Article 7) are not compliant with Article 9 (e.g. Case C-247/85). Furthermore, the conditions under Article 9(1)(c) relating to 'under strictly supervised conditions and on a selective basis' should be understood to imply a system of individual authorisations (or narrow-category authorisations involving a high degree of accountability) and should imply strict territorial, temporal and personal controls.

The Member States' and the Commission's composite derogation reports indicate that there are substantial numbers of derogations relating to the Birds Directive's species protection provisions each year, though there is considerable variation amongst the Member States in the numbers of requests and their reasons for such requests. There is a considerable variation in the number of derogations and the number of licences issued per derogation, and therefore the number of birds killed under the derogations is uncertain. Furthermore, the reported numbers are clearly underestimates because some national reports do not include such information. Consequently there is little value in presenting here the number reported as being killed in each Member State or for each species. However, it is likely that a large proportion of the birds killed under the derogations are the so-called 'pest species'.

In addition to the Court rulings and the Commission guidance referred to previously, there have been a number of other initiatives to improve the implementation of the bird protection provisions of the Birds Directive, especially in relation to hunting. These included the preparation of Action Plans for priority Annex I species (some jointly agreed with the Bern Convention, CMS and Agreement on the Conservation of African-Eurasian Migratory Waterbirds (AEWA)) and management plans for Annex II species with an unfavourable conservation status, which were carried out initially by the Commission between 1997 and 2000.

The Commission also launched a Sustainable Hunting Initiative in 2001. Despite these measures, a study of the status of birds in the EU by BirdLife International (2004) concluded that the status of Annex II species had worsened, with 46% having an unfavourable conservation status. As a further response to such problems, the two main organisations representing hunters and bird conservationists in Europe, the Federation of Associations for Hunting and Conservation of the EU (FACE) and BirdLife International, signed a sustainable hunting agreement in October 2004, in which they affirmed their commitment to the objectives of the Birds Directive and agreed not to seek to amend it. Amongst other things, they also agreed to support the establishment of the Natura 2000 network (recognising that hunting may be a compatible activity in a Natura 2000 site), and recognised that hunting rules should be based on the best available evidence, management plans should be developed and implemented for BD Annex II species, competent authorities should take appropriate initiatives to ensure bird conservation legislation is adhered to, and the use of lead shot in wetlands should be phased out by the year 2009 at the latest.

In 2012 the European Commission produced a *Roadmap towards eliminating illegal killing, trapping and trade of birds*[24] in response to Recommendation 155 (2011) of the Standing Committee of the Bern Convention on the illegal killing, trapping and trade of wild birds. This was subsequently updated to take into account Recommendation 164 (2013) of the Standing Committee of the Bern Convention on the implementation of the Tunis Action Plan 2013–2020 for the eradication of illegal killing, trapping and trade of wild birds. The roadmap sets out targeted actions to be carried out by Member States, stakeholders, and the Commission; the most recent version was published in 2017.

The current status of the Commission's management planning initiative for Annex II species with an unfavourable status is unclear. Some of the initial Annex II species management plans have been revised to take into account new information and the expansion of the EU-25, with plans for 13 species currently listed on the DG Environment website.[25] However, only one of these is up-to-date: the action plan for the Turtle-dove (*Streptopelia turtur*) (Fisher *et al.*, 2018). The other management plans need updating as they only cover up to 2011 at the latest.

As this chapter is finalised in April 2022, the European Commission is developing a new initiative with Member States to address the ongoing poor conservation status of most Annex II bird species (further discussed in Section 5.3). This includes prioritising species for further action and agreeing on a toolbox of measures, including the possibility of national suspensions of hunting, as well as land-use and other habitat management measures. To reverse the declining trends, the measures will clearly need to be much more effective and better implemented than those over the last three decades.

The Protection of Species Other Than Birds

Articles 12-16 of the Habitats Directive have very similar species protection provisions to those of the Birds Directive, which apply to the species wherever they occur within their natural range (i.e. within and outside Natura 2000 sites). However, unlike the Birds Directive, they only apply to selected species of animals and plants, which comprise 1 058 species listed in HD Annex IV (animals in Annex IVa and plants in Annex IVb). Some 789 species listed in Annex IV are also listed under HD Annex II, with the intention that they also benefit from the protection of core areas for the species (i.e. Natura 2000 sites). Guidance on the strict protection of Annex IV species has been provided by the European Commission (2021b, updating guidance in 2012), which draws on court judgments and the Member States' experience. The summary below is based on the Commission's guidance, unless otherwise indicated.

Under HD Articles 12(1)(a)–(c), animals listed in Annex IVa are protected from deliberate killing, disturbance, and collection of eggs. Article 12(1)d also prohibits the deterioration or destruction of their breeding and resting sites. Unlike the similar provision in the Birds Directive, this does not include the term 'deliberate' with respect to the deterioration or destruction of breeding sites or resting places. Therefore such activities must be prohibited, irrespective of whether they are deliberate or not. With respect to Articles 12(1)(a)–(c), the term 'deliberate' is interpreted by the CJEU as going beyond 'direct intention'. Thus, the provision applies not only to someone who intends to capture or kill a specimen of a protected species, but also to a person who is sufficiently informed and aware of the likely consequences of their action but goes ahead anyway. Member States are also required to monitor the incidental capture and killing of Annex IVa listed animals. Under Article 13, Annex IVb listed plants are protected from deliberate picking and cutting, etc., and keeping, transport, sale and exchange.

Importantly, Member States are required to establish systems that proactively prohibit the damaging activities listed in Articles 12(1) and 13(1) and take complementary preventative measures to address known threats, such as through some form of planning or permitting procedures and awareness raising activities. A species by species approach is also required that considers the circumstances surrounding each case. Although the measures to strictly protect species should contribute to their favourable conservation status, the CJEU has clarified that the strict protection prohibitions apply at the level of individual specimens, irrespective of the numbers,[26] and irrespective of the conservation status of the species concerned and whether or not the activities will have an adverse effect on its status.[27] Such facts should be taken into account in the derogation process instead (described later).

The use of such preventive measures is especially important because many Annex IV species have the potential to be unintentionally detrimentally affected by common ongoing activities, such as those associated with agriculture, forestry or fisheries. In practice, as is clear from many country chapters, such protection provisions have been inconsistently and rather weakly applied to land-use activities, especially in relation to agriculture. Forestry plans and operations have more widely been adapted to avoid the destruction or disturbance of some Annex IV species and their habitat. As regards fisheries, Member States should use the available mechanisms under the CFP and, more specifically, the technical measures regulation,[28] to protect species, such as from by-catch. Preventive measures can, for instance, include modifications of, or

[24] http://ec.europa.eu/environment/nature/conservation/wildbirds/docs/Roadmap%20illegal%20killing.pdf

[25] http://ec.europa.eu/environment/nature/conservation/wildbirds/hunting/managt_plans_en.htm

[26] Cases C-441/17, C-473/19 and C-474/19.

[27] Cases C 473/19 and C 474/19.

[28] Regulation (EU) 2019/1241.

restrictions on, certain types of fishing gear or spatial/temporal regulation of fishing activities.

The Habitats Directive does not contain provisions that specifically address hunting. However, under Article 14 (which has no equivalent in the Birds Directive), Member States may adopt management measures (e.g. licensing systems and closed seasons) if surveillance indicates that this is necessary to ensure that the taking of 96 species listed in Annex V (which includes some game species and plants and animals of value) is compatible with the maintenance of their favourable conservation status. As under the Birds Directive, indiscriminate means of killing are prohibited (Article 15).

In accordance with Article 16(1), derogations from the strict protection requirements are allowed for Annex IV species provided that they meet three criteria. Firstly, they are for one or more reasons listed in Article 16(1), which are very similar to those under Article 9 of the Birds Directive (as listed earlier). The main difference is that, under the Habitats Directive, derogations can be for imperative reasons of overriding public interest other than public health and safety. Secondly, as in the Birds Directive, there is no satisfactory alternative (e.g. using non-lethal means to exclude or deter species from causing damage or other problems). Thirdly, they are not detrimental to the conservation status of the populations of the species concerned. Importantly, this does not refer to the overall conservation status at a biogeographical level, but the more local impacts. No such requirement applies to derogations under the Birds Directive.

The CJEU has reiterated that the derogations should be used as a last resort, meet all three criteria, and cover precise requirements and specific situations. Consequently, for each derogation, the competent national authority must ensure that the derogation is justified, has clear objectives, is appropriate according to scientific data to meet those objectives, and specifies the terms and conditions under which the derogation is authorised (e.g. the number of specimens for which the derogation is allowed, by whom, where and when).

Member States do not require prior approval from the Commission for derogations, but under Articles 16(2) and 16(3) they must provide reports every two years that describe the derogations that have been applied. Unlike under Article 9(4) of the Birds Directive, there is no explicit obligation on the Commission to ensure that the derogations are in accordance with the aims of the Habitats Directive. Nevertheless, the Commission has undertaken numerous infringement proceedings in relation to the improper use of the derogations.

The derogations have been mostly used to allow the control of species that cause serious damage to crops, trees, livestock and fish, such as the large carnivores, Beaver (*Castor fiber*) and Otter (*Lutra lutra*). Common related problems have included the general scope of derogations, their use before other options and when they may be detrimental for the conservation status of the species. As discussed in several chapters, there have been particular problems with derogations relating to large carnivores. Other common problems have related to Annex IV species that are relatively widespread, some of which have a tendency to favour the new habitats that are created by human activities (e.g. mineral extraction or site clearance prior to building works). This, and the need to prevent the losses of all individuals no matter what the impact on the conservation status of the species, has led to significant challenges with implementing the legislation. In the UK substantial project delays and costs have occurred in relation to the Great Crested Newt (*Triturus cristatus*), which eventually led to the development of a more strategic system of licensing derogations (see Chapter 31). Problems with protected species moving into development sites have also led to the 'temporary nature' concept, as described in European Commission guidance (2021b).

The European Commission has supported multiple LIFE projects[29] and other initiatives that have developed and promoted good practice for managing conflicts involving protected species, such as the establishment of the EU Platform on Coexistence between People and Large Carnivores (Hovardas and Marsden, 2018). To help Member States more broadly conserve the species covered by the Habitats Directive, since 2008 the Commission has also supported the development of EU species action plans for several selected species or groups (e.g. for all bats and sturgeons).[30]

4.2.5 Supporting Measures

Important measures that support and complement the site-focussed and general species protection measures include the requirement in both Directives to carry out research to help achieve their objectives. The Birds Directive provides a list in Annex V of subjects that should be the focus of attention. Whilst inadequate, it includes listing endangered species, assessing the population levels of migratory species and identifying their important sites, assessing the influence of the methods of taking birds on their populations, developing methods for preventing damage by birds, and assessing the impacts of chemical pollution on birds and their potential as indicators of pollution. The Habitats Directive does not provide such a specific list of topics to focus on but indicates that research should focus on the identification of SCIs, improving the coherence of the Natura network through the management of landscape features, and cooperative investigations of transboundary issues.

Both Directives also require Member States to ensure that any introduction of a species that does not occur naturally in the wild does not harm other species of flora and fauna (and, under the Habitats Directive, HD habitats).

The Habitats Directive has two further supplementary obligations for Member States (with no equivalent in the Birds Directive). Firstly, it requires Member States to consider

[29] For example, LIFE IN QUARRIES, LIFE14 NAT/BE/000364.

[30] For full list and plans see https://ec.europa.eu/environment/nature/conservation/species/action_plans/index_en.htm

the desirability of re-introducing species listed in Annex IV that are native to their territory (if previous experience from other Member States indicates that this could help maintain or restore their favourable conservation status). There is also a requirement to promote education and general information on the need to protect species and their habitats.

4.2.6 Funding

Nature conservation is an expensive activity in Europe as it normally requires a wide range of ongoing activities to maintain the desired habitats and species, such as to provide the required biophysical conditions, maintain the required vegetation and its structure, and deal with unwanted species or disease outbreaks. Such practical activities also need to be planned, managed and monitored, from national down to local levels. It was envisaged that the implementation costs of the Nature Directives would be borne by the Member States as necessary. Consequently, the Birds Directive has no funding provisions, and the Habitats Directive has rather limited provisions for EU co-financing under Article 8. These were only added to the Habitats Directive after an intervention by Spain, stimulated by concerns that a large proportion of the country would become Natura 2000 sites (Fernández, 2003).

HD Article 8 requires Member States to identify funding requirements for the management measures that are necessary within SCIs to achieve the favourable conservation status of priority habitats and priority species. The Commission is then supposed to assess the funding requirements and subsequently adopt, in consultation with the Member State, a prioritised action framework of conservation measures involving EU co-financing, taking into account the availability of EU funds. If EU co-financing is not provided, then the required measures can be postponed by the Member State, but it must not allow affected sites to deteriorate.

Whilst the arrival of the Nature Directives stimulated an increase in some funding, there has been a substantial gap between the funding required to properly implement the Directives and the amount that has been made available. To address this, the Commission set up a working group in 2001 to address the issue, including the possibility of Community co-financing. In 2004 the Commission estimated the costs of Natura 2000 to be €6.1 billion per annum for the EU-25 (European Commission, 2004), and at the same time published draft proposals concerning funding arrangements for the 2007–2013 period. This made a case for primarily financing Natura 2000 through wider EU programmes, in particular the rural, fisheries and regional development funds under the EU's Multiannual Financial Framework (MFF). This integrated approach builds upon the concept that environment policy can deliver its best results when its objectives are mainstreamed across all relevant policy areas, including their funding instruments. It is also consistent with the legal requirement under the Treaty (Article 11 of the TFEU) that requires environmental protection requirements to be integrated into the definition and implementation of the EU's policies and activities. Moreover, it recognises that investing in the environment, including biodiversity, also contributes to wider EU goals such as in relation to sustainable economic growth, employment and health.

The integrated funding of Natura 2000 was introduced during the 2007–2013 MFF. Since then the main sources of funding have been the European Agricultural Fund for Rural Development (EAFRD), European Fisheries Fund (EFF) and its successors, the European Regional Development Fund (ERDF) and Cohesion Funds, which have complemented funding under the LIFE programme. These EU funds are summarised in Table 4.2, with indicative estimates of their past spending on biodiversity, and the 2014–2020 funding regulations and their objectives. The funding allocations for the 2014–2020 period are given in each country chapter. All of the listed funds are included, with modifications, in the MFF for 2021–2027.

The EU normally provides co-financing, and the Member States are required to meet a proportion of the costs, which varies depending on the funding instrument, the actions being taken and various other factors. It should, however, be emphasised that, with the exception of the LIFE programme, it is very difficult to estimate the amount of funding that actually led to significant biodiversity benefits because many measures and schemes have multiple objectives and the aims of spending are tracked and recorded according to broad objectives (Medarova-Bergstrom *et al.*, 2015). Information is also only available with respect to Natura 2000 funding for some funds, and not for wider nature conservation activities.

Following some earlier Community funds, the EU LIFE Programme was established in 1992 as the main EU co-financing instrument specifically dedicated to the protection of the environment, including supporting implementation of the Nature Directives and other nature conservation priorities. More information on the history of the LIFE programme and its various phases is provided on the LIFE website.[31] The Programme is managed by the European Commission, which awards projects on a competitive basis taking into account multiannual work programme priorities. EU co-financing rates for LIFE projects vary from 30% to 100%, but they typically receive 50–60% funding.

The total budget for LIFE over the 2014–2020 funding period was €3.4 billion, with around 75% allocated to the sub-programme for environment, of which at least 55% of the resources was dedicated to projects (financed by way of action grants) to support conservation of nature and biodiversity. To respond to calls for further funding, the LIFE budget for 2021–2027 increased to €5.4 billion, including an indicative allocation of €2.1 billion for the nature and biodiversity sub-programme.

[31] https://wayback.archive-it.org/12090/20210412134647/https://ec.europa.eu/easme/en/section/life/life-history-life

Table 4.2 The EU funds of most importance for supporting nature conservation up to 2021.

Legislative basis and main goals	Opportunities for biodiversity financing	Estimated EU annual funding related to nature conservation
Programme for the Environment and Climate Action (LIFE)		
Regulation (EC) No 1293/2013. Funding programme for climate and the environment with four objectives: (a) Tackle environmental challenges (b) Improve policy and legislation and promote their integration and mainstreaming (c) Support governance (d) Support the implementation of the 7th EAP	Nature and biodiversity is one of the three priority areas of the LIFE programme's 'Environment' sub-programme. At least 50% of the funds dedicated to environment in LIFE must be allocated to this target. Natura 2000 network implementation and management. Dissemination of best practices.	€107–120 million, actual and planned allocations respectively (Kettunen *et al.*, 2011)
European Agricultural Fund for Rural Development (EAFRD)		
Regulation (EU) 1305/2013. Sustainable rural development with three objectives: (a) Competitiveness of agriculture (b) Sustainable management (agriculture and forestry) and climate action (c) Balanced territorial development of rural economies and communities	Priority 4 is restoring, preserving and enhancing ecosystems. Supports biodiversity measures, most importantly agri-environment schemes, Natura 2000 payments and forest measures, as well as organic farming and payments to Areas with Natural Constraints.	Natura 2000 relevant spending: €859 million – €82 million dedicated Natura 2000 payments + €777 million from agri-environment schemes during 2007–2013 (Kettunen *et al.*, 2011)
European Fisheries Fund (EFF), for 2014–2020 European Maritime and Fisheries Fund (EMFF)		
Regulation (EU) No 508/2014. Four objectives: (a) Competitive, environmentally sustainable, economically viable and socially responsible fisheries and aquaculture (b) Implementation of the Common Fisheries Policy (CFP) (c) Balanced and inclusive territorial development (d) Integrated Maritime Policy (IMP)	Includes dedicated opportunities for biodiversity conservation: • protection and restoration of aquatic biodiversity and ecosystem and compensation regimes for sustainable fishing activities; • management, restoration and monitoring of Natura 2000 sites and marine protected areas; and • development of aquaculture providing environmental services	EFF Natura 2000 relevant spending €2.8 million (Kettunen *et al.*, 2011)
European Regional Development Fund (ERDF), European Social Fund (ESF) and Cohesion Fund		
Regulation (EU) No 1301/2013 Objectives included: • Reinforcing economic, social and territorial cohesion by redressing regional imbalances • Sustainable development and structural adjustment of regional economies • Conversion of declining industrial regions • Investments, networking and sharing facilities, infrastructure and human facilities	Include dedicated opportunities for biodiversity conservation: • protecting and restoring biodiversity and soil and promoting ecosystem services, including through Natura 2000; and • green infrastructure in the investment priority thematic point No 6	Approximately €400 million to promote biodiversity and nature protection, including Natura 2000 (ECA, 2014)
Framework Programme for Research and Innovation – Horizon 2020		
Regulation (EU) No 1291/2013. Objectives included: • Research and innovation activities • Building a society and an economy based on knowledge with three priority areas: excellent science, industrial leadership and tackling societal challenges	A strong emphasis on sustainable development and thus also biodiversity	€278 million for biodiversity related research in 2015 (DG Research, quoted in Kettunen *et al.* (2017))

Note. The goals listed in the table relate to the regulations as they were over the 2014–2020 programming period.
Source. Adapted and updated from Kettunen *et al.* (2017).

Compared to the funding requirements for nature conservation in the EU, the LIFE budget is very small. Therefore an important requirement for most LIFE projects has been to develop and/or demonstrate new approaches and techniques that have the potential for wider application (e.g. with support from larger funds, such as EAFRD funded agri-environment schemes). As indicated in the national chapters, a wide variety of projects have been funded by the LIFE Programme, but typical nature conservation projects have included habitat restoration and management (e.g. river realignment, removal of unwanted vegetation, hydrological management, and reinstatement of traditional agricultural or forestry practices), habitat creation (e.g. wetlands to compensate for coastal erosion), elimination of alien species, and captive breeding and/or translocations to re-introduce or supplement populations. Stakeholder engagement and the dissemination of project findings are also key components of most LIFE nature conservation focussed projects. Since 1997 LIFE operating grants totalling around €12 million a year have been provided to independent environmental NGOs, including to support nature conservation work. In order to achieve some large-scale impacts the LIFE Programme introduced integrated projects in 2014, which combine LIFE funding with other sources of financing to provide much higher budgets for the implementation of environmental plans or strategies, such as Natura 2000 Prioritised Action Frameworks (PAFs) – described later. In addition, since 2021 the LIFE programme has funded strategic nature projects, which implement coherent programmes of action at the national level to mainstream the EU's biodiversity objectives into other policies and financing instruments, including through coordinated implementation of the PAF.

The most important source of funding for nature conservation has been the Common Agricultural Policy's (CAP) EAFRD, particularly through its funding of agri-environment (-climate) schemes and the Natura 2000 compensation payment, although the latter has been little used by most Member States so far (see Section 4.3.8 for details). The ERDF and Cohesion Fund have also made substantial contributions to Natura 2000 and broader nature conservation goals, especially in the new Member States, under a number of budget categories directly and indirectly relating to biodiversity conservation. Funding has, for example, supported habitat mapping, preparation of management plans, practical restoration measures, monitoring and capacity building of authorities.

Although the integrated financing approach did increase the funding available for the implementation of the Nature Directives, the step change that was necessary did not occur. This was demonstrated by an analysis of Member State EU funding allocations for the 2007–2013 programming period, which found that the amount budgeted for the Natura 2000 network (€550–1 500 million per year) was only 9–19% of its estimated financing needs (Kettunen *et al.*, 2011). This finding related to a revised estimate of the costs of fully implementing the Natura 2000 network based on Member States' responses to a questionnaire (Gantioler *et al.*, 2010). The financing needs were estimated to be around €5.8 billion per annum for the EU-27 (€65/ha based on the terrestrial Natura 2000 area at the time). As in previous estimates, this was considered to be conservative, because few Member States considered the level of financial investment required to fully maintain and/or restore the favourable conservation status of habitats and species across the network. Whilst the cost of managing the Natura 2000 network is substantial, it is appropriate to consider it in relation to its economic benefits, which have been estimated to be approximately €200–300 billion per year (ten Brink *et al.*, 2011).

In 2011 the Commission published *Financing Natura 2000* (European Commission, 2011a). This recognised the shortfall in funding for Natura 2000 but maintained the Commission's view that the integrated approach is the best means of ensuring adequate financing for the network. A number of measures were therefore taken to improve the uptake of available funds. Of particular importance was the requirement for Member and States to adopt a more strategic approach to identifying financing requirements and opportunities through the development of PAFs. The purpose of these documents is to establish a national or regional strategy for achieving the favourable conservation status of BHD habitats and species, including through the protection and management of Natura 2000 and wider green infrastructure, within the context of the relevant EU financial instruments. PAFs aim to identify the required Natura 2000 conservation priorities and management measures as well as their related costs and potential financing sources, and then to match the former with the latter. The management authorities of the financial instruments are then supposed to take the PAFs into account when developing their funding priorities and programmes. They are also supposed to form the foundation for LIFE funded Integrated Projects related to Natura 2000. Whilst the Commission's initial justification for the PAF initiative is the implementation of HD Article 8 (described earlier), the aims and scope of the PAFs now clearly go beyond the legal requirements as they are not only focussed on priority habitats and priority species in SCIs/SACs.

4.2.7 Monitoring and Reporting

It is a well-known business management adage that 'You can't manage what you don't measure.' This is also highly relevant to nature conservation. Monitoring of the status and trends in habitats and species, as well as the factors affecting them and actions being taken to manage them, provides the key information that is required to establish if conservation actions are working, and to quantify progress towards defined targets. Good monitoring also facilitates adaptive management. The reporting of monitoring results is essential to ensuring the accountability of organisations affecting the environment, and local, regional and national governments. This is especially relevant in the context of the EU's environmental legislation, as it requires wide-scale multinational action to be effective.

Reporting also has a role to play in supporting the single market, by ensuring that the agreed common environmental standards are being met. Recognising such requirements, monitoring and reporting are fundamental elements of the Nature Directives, and are particularly well developed in the Habitats Directive.

Under Article 12 of the Birds Directive, Member States are required to report every three years on the practical implementation of measures, rather than on the results of monitoring their outcomes with respect to the birds' populations. This greatly limited the effectiveness of the reporting and made it difficult to quantify the impacts of the Directive. To some extent, this gap was filled by BirdLife International's assessments of the conservation status of birds in Europe (Tucker and Heath, 1994; BirdLife International, 2004). These documented the ongoing declines in many species, but also provided the data that convincingly demonstrated and quantified the positive impacts of the Birds Directive on bird populations in the EU (Donald *et al.*, 2007; Sanderson *et al.*, 2015).

To address the information gap, and stimulated by the BirdLife assessments, Member States developed new reporting procedures that provide data on the actual state and trends of bird populations, in a manner similar to the reporting under Article 17 of the Habitats Directive (described later). As a result, since 2007 the Article 12 reports have consisted of the following two parts:

- *General reporting*, which retains some reporting on progress implementing the Directive but encourages use of external links to other sources of information (e.g. work carried out for the protection of bird populations) to keep the text to a minimum.
- *Reporting on the size and trends of individual bird species' populations and distributions*, which also includes reporting of threats and pressures affecting species for which SPAs have been classified (based on the same threat and pressure categories as used for reporting under the Habitats Directive). Unlike for HD species, there is no requirement to assess the status of each species at a biogeographical or national level. Only an overall EU level status assessment is carried out by the European Commission (by contractors).

Reporting under the Birds Directive has also been changed to every six years to coincide with that of the Habitats Directive, following an initial transition period during 2008–2012. Thus, the most recent reporting period of 2013–2018 was closely synchronised with that of the Habitats Directive.

Under Article 11 of the Habitats Directive, Member States are required to undertake surveillance of the conservation status of all their HD habitats and species with particular regard to those with a Priority status. Article 17 of the Directive then requires Member States to produce a report every six years to outline the measures taken to implement the Directive, the main results of the surveillance referred to above, and an evaluation of the impact of measures taken on the conservation status of the HD habitats and species.

Reporting under the first six-year period after the implementation of the Habitats Directive, 1994–2000, focussed on progress of the legal transposition and implementation of the Directive, the establishment of the Natura 2000 network and administrative aspects. The second reporting period, 2001–2006, was therefore the first in which the conservation status of HD habitats and species was assessed. The collated Article 17 data provide an extremely valuable and accessible detailed assessment of the status of HD habitats and species over most Member States and by biogeographical region. Consequently, as well as providing the basis for measuring the achievement of the Directive's objectives, the data are used for wider biodiversity monitoring purposes, such as the Streamlining European Biodiversity Indicators (SEBI) (EEA, 2007, 2012), in the EEA Biodiversity Baseline report (EEA, 2010) and as part of the Mapping and Assessment of Ecosystems and their Services (MAES) assessments of ecosystem condition (Maes *et al.*, 2020).

However, the 2001–2006 assessment revealed that there was insufficient information to assess the status of many habitats and species, especially in south and south-east Europe. Therefore, the dataset included many gaps, and there were also some inconsistencies in assessment approaches, such as in relation to the units used to report on habitats and species. The list of pressures that was reported on was also found to cause confusion and had many overlaps, and as a result was not well used. In response, an improved and updated guidance document was produced for the third reporting period, 2007–2012 (ETC/BD, 2011). The guidance included a revised and detailed list of pressures/threats which has resulted in more useful and consistent information being provided by Member States, which is often referred to in the later country chapters.

Further refinements and additions were made to the reporting forms and guidance for the 2013–2018 period. Box 4.3 provides a summary of the main elements of the reporting requirements as they currently stand.

4.2.8 The Nature Directives Fitness Check and Action Plan

Under its commitment to Better Regulation, the European Commission launched in 2014 a comprehensive evaluation of the Nature Directives, called a 'Fitness Check'. Specifically this assessed the Nature Directives' effectiveness, efficiency, relevance, policy coherence and EU added value, based on a major review of evidence from the Member States' reporting information, the published literature and consultations with national authorities, stakeholders and the wider public (through an online survey). Online Annex 2 provides a brief account of the political background to the Fitness Check and the huge response to the public consultation (over half a million replies).

The supporting evaluation study for the Fitness Check, by Milieu *et al.* (2016), was completed in March 2016, but it and

Box 4.3 Key elements of the reporting framework under HD Article 17

General

- Main achievements under the Habitats Directive.
- Natura 2000 (including pSCIs) – site designation: number and area.
- Management plans for Natura 2000 sites (SACs): number and % of sites.
- Measures taken in relation to approval of plans and projects (Article 6.4).
- Measures taken to ensure coherence of the Natura 2000 Network (Article 10).
- Re-introduction of Annex IV species.

For each Annex II, IV and V species and biogeographical region

- National distribution map.
- National information related to Annex V species: measures taken and hunting bag if exploited and with an unfavourable status.
- Range area and population size: short-term trend (12 yr), long-term trend (24 yr), favourable reference value, reasons for any changes compared to the previous reporting period.
- Habitat for species: sufficiency, short-term and long-term trends.
- Main pressures and threats of medium and high importance.
- Main measures taken: their status, purpose, location and timeframe.
- Future prospects: range, population, habitat of the species.
- Conclusions: range, population, habitat for the species, future prospects, overall assessment of conservation status, overall trend in conservation status (improving/deteriorating/stable/unknown), reasons for any changes in status or trend.
- Natura 2000 (pSCIs, SCIs and SACs) coverage for Annex II species.

Main results for each HD Annex I habitat

As above for species where relevant plus:

- Structure and functions: area of habitat in good and unknown condition, short-term trend.

Source. Reporting under Article 17 of the Habitats Directive – format for 2013–2018.[32]

the Commission's official Communication were not released until December of that year (European Commission, 2016). The reason for this delay is uncertain, but it is widely believed that it was because the evaluation study and the Commission's draft communication were considered by some high-level officials to be too positive, and not in accordance with President Juncker's instruction to revise the Directives. However, in the end, the published Communication closely reflected the findings of the evaluation study and did not call for the Nature Directives to be opened up for legislative changes.

The evaluation study and European Commission concluded that, as part of broader EU biodiversity policy, the Nature Directives are fit for purpose, are effective where properly implemented and are making progress towards achieving their objectives, especially in relation to the general protection of most species and the establishment of the Natura 2000 network on land. Despite the slow progress and other problems with implementation, important evidence of the benefits of the Birds Directive came from a scientific study of bird population trends in Europe, which showed that the Directive has had a measurable positive impact on its targeted species (Donald *et al.*, 2007; Sanderson *et al.*, 2015) – further discussed in Chapter 5.

However, much more needed to be done, and progress had been particularly insufficient with the identification and designation of marine Natura 2000 sites, the establishment of appropriate conservation measures for Natura 2000 sites, and actions to improve site connectivity and maintain species and habitats in the wider environment.

As a follow-up to the Fitness Check, the Commission developed an *Action Plan for nature, people and the economy* (European Commission, 2017a), to improve the implementation of the Directives, their coherence with socio-economic objectives and engagement with national, regional and local authorities, stakeholders and citizens. The plan contained 15 actions under the following four priority areas:

- Priority A: Improving guidance and knowledge and ensuring better coherence with broader socio-economic objectives.
- Priority B: Building political ownership and strengthening compliance.
- Priority C: Strengthening investment in Natura 2000 and improving synergies with EU funding instruments.
- Priority D: Better communication and outreach, engaging citizens, stakeholders and communities.

The actions, further described in a separate Staff Working Document (European Commission, 2017b), included commitments for more than 100 sub-actions for Member States, the Commission, the Committee of the Regions, and other stakeholders, depending on the activity.

4.3 Other EU Policies and Legislation Contributing to Nature Conservation

Needless to say, in practice nature conservation relies on a much wider range of measures than those included in the Nature Directives. In particular, it depends on the policies and laws that affect the use of the land (especially agriculture, forestry and, increasingly, bioenergy), freshwaters (e.g. water use and hydropower) and the sea (e.g. fisheries, transport and energy). The ecological condition of the land and waters is also highly linked to the control and management of wastes, pollutants and invasive alien species. Habitats and species are also

[32] http://cdr.eionet.europa.eu/help/habitats_art17/

directly affected by built developments (e.g. housing, transport, industry, energy and infrastructure) and extractive industries, which are subject to spatial planning and impact assessments. Whilst it is beyond the scope of this book to list, let alone describe, all such relevant instruments, Table 4.3 provides a summary of some of the legislation that has had most influence on nature conservation in the EU. The table does not include legislation relating to trade (e.g. the Convention on International Trade in Endangered Species of Wild Fauna and Flora – CITES) or animal welfare.

Given their particularly high relevance, further but brief accounts are given of the aims and main provisions of the Environmental Impact Assessment Directive (EIA), Strategic Environmental Assessment Directive (SEA), Environmental Liability Directive (ELD), Water Framework Directive (WFD), and maritime policy, especially the Marine Strategy Framework Directive (MSFD), Maritime Spatial Planning Directive (MSPD) and Common Fisheries Policy (CFP), air pollution legislation and the Regulation on Invasive Alien Species (IAS). As the Common Agricultural Policy (CAP) has the widest influence on biodiversity in the EU, and is by far the largest source of funding, it is described in more detail, with further information provided in the online Annex 3.

In contrast to agriculture, the EU has limited competency over EU forest matters. Therefore, other than requirements under the SEA and EIA Directives, EU policy objectives have been promoted through non-legislation measures, primarily forest strategies and action plans (currently European Commission, 2021d) and RDP funding under the CAP. The strategies have addressed a wide range of issues, including supporting rural communities and the forest sector, protecting forests and their ecosystem services, addressing climate change, gathering information, facilitating cooperation and innovation, and the global dimension of forestry. Environmental issues, including nature conservation, have been mainly promoted through the adoption of established sustainable forestry principles and additional voluntary actions.

4.3.1 Environmental Impact Assessments

The EIA Directive (85/337/EEC) has been in force since 1985 and applies to a wide range of both public and private projects, which are defined in Annexes I and II of the Directive. All projects listed in Annex I are considered as having potential significant effects on the environment and require an EIA (e.g. long-distance railway lines, motorways and express roads, some airports, installations for the disposal of waste, large waste-water treatment plants).

For projects listed in Annex II, national authorities have to decide whether an EIA is needed. This is done by the 'screening procedure', which determines the effects of projects on the basis of thresholds/criteria or a case by case examination. Annex II projects include urban development projects and flood-relief works. They also include projects in the agriculture, silviculture and aquaculture sectors that involve: restructuring of rural land holdings; use of uncultivated land or semi-natural areas for intensive agricultural purposes; water management projects for agriculture, including irrigation and land drainage projects; afforestation and deforestation for the purposes of conversion to another type of land use; intensive livestock installations and fish farms; and reclamation of land from the sea.

Following a review of the way in which the Directive could be improved (European Commission, 2009) and a consultation process, an amended EIA Directive (Directive 2014/52/EU) entered into force in 2014. The new Directive made EIA more effective and efficient through the adoption of simplified environmental assessment procedures and increasing the accessibility of reports to the general public, as well as requiring ongoing monitoring of 'significant adverse environmental impacts'. The revision also requires mandatory assessment of 'reasonable alternatives', rather than simply an outline of the alternatives which the developer has studied, as well as description of the 'baseline scenario' (i.e. current state of the environment) and an outline of the likely evolution without the implementation of the project. Member States must ensure that developers implement the envisaged mitigation and offsetting measures foreseen in the consent decision for significant adverse effects on the environment. The revised EIA Directive also includes explicit references to biodiversity (Article 3b) including references to biodiversity in screening and the content of the environmental assessment report.

EIAs and SEAs (described in Section 4.3.2) differ significantly from the appropriate assessments carried out under HD Article 6(3). Most fundamentally, the results of an appropriate assessment are binding, such that the competent authorities must prohibit all damaging developments, unless Article 6(4) applies. In contrast, whilst EIAs and SEAs are processes that the competent authorities must undertake and take into account in their decision making, they do not lead to results that Member States are strictly bound by. The appropriate assessment also has a narrower remit, strictly considering impacts only on Natura habitats and species within Natura 2000 sites, and in relation to the site's conservation objectives, whereas EIA and SEA cover biodiversity impacts in general. On the other hand, appropriate assessments must consider all types of projects and plans (no matter how small) and all types of impacts that may be detrimental to the Natura habitats and species in a site. Despite these key differences, in practice appropriate assessments and EIA and SEA are closely linked, and according to the EIA Directive, a coordinated and/or joint procedure should be carried out if an assessment is required under both the Nature Directives and the EIA Directive.

4.3.2 Strategic Environmental Assessments

Directive 2001/42/EC on the assessment of the effects of certain plans and programmes on the environment (SEA Directive) extends EIA procedures and principles from projects to public plans and programmes (but not policies). It was adopted in 2001, with formal compliance required from July 2004.

Table 4.3 Summary of EU legislative instruments, other than the Nature Directives, that have a substantial influence on biodiversity.

Policy/Instrument	Overall objective	Biodiversity relevant measures
EIA Directive: Directive 2014/52/EU of 16 April 2014 amending Directive 2011/92/EU on the assessment of the effects of certain public and private projects on the environment	To assess and reduce the environmental impacts of projects	Explicitly requires potential biodiversity impacts to be assessed. May be carried out jointly with an HD appropriate assessment.
SEA Directive: Directive 2001/42/EC of 27 June 2001 on the assessment of the effects of certain plans and programmes on the environment	To assess and reduce the environmental impacts of plans and programmes	Requires the assessment of biodiversity, and provides the best opportunity for strategic planning and avoidance of major biodiversity impacts. May be carried out jointly with an HD appropriate assessment.
ELD Directive: Directive 2004/35/CE of 21 April 2004 on environmental liability with regard to the prevention and remedying of environmental damage (as amended)	To establish a framework of environmental liability, based on the polluter-pays principle, to prevent and remedy environmental damage	Includes the requirement to remedy damage to BHD habitats and species, and MS have the option to include broader biodiversity. It introduces the favourable conservation status concept to birds.
Water Framework Directive: Directive 2000/60/EC of 23 October 2000 establishing a framework for Community action in the field of water policy	To avoid the deterioration of water bodies, and achieve good water status (ecological and chemical) of surface waters and good quantitative and chemical status of ground water bodies	The achievement of good status contributes to wide-ranging aquatic biodiversity goals, and the Directive also adopts more ambitious objectives to meet the requirements of BHD habitats and species where necessary
Nitrates Directive: Council Directive 91/676/EEC of 12 December 1991 concerning the protection of waters against pollution caused by nitrates from agricultural sources	To prevent nitrates from agricultural sources from affecting ground and surface waters	Contributes to reductions in water eutrophication and widespread impacts on aquatic ecosystems and species
Urban Waste Water Treatment Directive: Council Directive 91/271/EEC of 21 May 1991 concerning urban waste-water treatment	To protect the environment from the adverse effects of urban waste-water discharges and discharges from certain industrial sectors. Concerns the collection, treatment and discharge of waste-water	Contributes to reductions in water eutrophication, especially from phosphates in sewage in freshwater ecosystems, and other forms of pollution in aquatic ecosystems
Floods Directive: Directive 2007/60/EC of 23 October 2007 on the assessment and management of flood risks	To reduce and manage the risks that significant floods pose to human health, the environment, cultural heritage and economic activity	Encourages the maintenance and restoration of natural floodplains, as well as other natural flood management approaches, which can provide/enhance habitats. Interacts with WFD
The Integrated Pollution Prevention and Control Directive: Directive 2008/1/EC of 15 January 2008 concerning integrated pollution prevention and control. Replaced by Industrial Emissions Directive	The prevention and reduction of pollution through the requirement for industrial and agricultural activities with a high pollution potential (e.g. energy, metals, minerals, chemicals, waste management, livestock farming) to have a permit, which can only be issued if certain environmental conditions are met	Helps to reduce the emissions of a wide range of pollutants that can have detrimental impacts on aquatic and terrestrial habitats and species
Industrial Emissions Directive: Directive 2010/75/EU of 24 November 2010 on industrial emissions (integrated pollution prevention and control)	To achieve a high level of protection of human health and the environment taken as a whole by reducing harmful industrial emissions across the EU, in particular through better application of Best Available Techniques (BAT)	Helps to reduce the emissions of a wide range of pollutants that can have detrimental impacts on aquatic and terrestrial habitats and species
The Plant Protection Products Regulation: Regulation (EC) No 1107/2009 of 21 October 2009 concerning the placing of plant protection products on the market and repealing Council Directives 79/117/EEC and 91/414/EEC	To ensure a high level of protection of both human and animal health and the environment and to improve the functioning of the internal market through the harmonisation of the rules on the placing on the market of plant protection products, while improving agricultural production	Helps to avoid and reduce the direct and indirect effects of plant protection products (i.e. forms of pesticide) on habitats and species, especially in agricultural environments

Table 4.3 *(cont.)*

Policy/Instrument	Overall objective	Biodiversity relevant measures
The Air Quality Directive: Directive 2008/50/EC of 21 May 2008 on ambient air quality and cleaner air for Europe	Sets objectives for ambient air quality to avoid, prevent or reduce harmful effects on human health and the environment; requires MS to assess their ambient air quality, share information with the public, maintain good air quality and improve it where necessary, and cooperate with other MS where necessary	Includes limits for concentration of certain pollutants in the air (ozone, nitrogen oxides and sulphur dioxide) in order to avoid environmental damage
National Emissions Ceilings Directive: Directive (EU) 2016/2284 of the European Parliament and of the Council of 14 December 2016 on the reduction of national emissions of certain atmospheric pollutants, amending Directive 2003/35/EC and repealing Directive 2001/81/EC	The reduction of air pollution by the setting of national emission reduction commitments for five important air pollutants: nitrogen oxides (NOx), non-methane volatile organic compounds (NMVOCs), sulphur dioxide (SO_2), ammonia (NH_3) and fine particulate matter (PM2.5)	Important means of reducing air pollution impacts, including acid deposition (from SO_2) and eutrophication of habitats (NH_3) – the latter being a widespread and serious pressure on many sensitive habitats
IAS Regulation: Regulation (EU) No 1143/2014 of the European Parliament and of the Council of 22 October 2014 on the prevention and management of the introduction and spread of invasive alien species	Sets out rules to prevent, minimise and mitigate the adverse impact on biodiversity of the introduction and spread within the Union, both intentional and unintentional, of invasive alien species	Provides strong legislative requirement to address IAS on the Union list, currently 66 species, including species with substantial biodiversity impacts. Fulfils Action 16 of Target 5 of the EU 2020 Biodiversity Strategy.
The Directive on Environmental Crime: Directive 2008/99/EC of 19 November 2008 on the protection of the environment through criminal law	To protect the environment more effectively, by requiring MS to criminalise conduct 'which causes or is likely to cause death or serious injury to any person or substantial damage to the quality of air, the quality of soil or the quality of water, or to animals or plants'	Supports compliance with existing environmental legislation, including the Nature Directives, and other instruments that have significant biodiversity impacts, e.g. in relation to illegal killing of species.
Marine Strategy Framework Directive: Directive 2008/56/EC of 17 June 2008 establishing a framework for community action in the field of marine environmental policy	To achieve or maintain 'good environmental status' by 2020 in the marine environment, including coastal waters, the seabed and subsoil	The achievement of good environmental status contributes to wide-ranging marine biodiversity goals, and the Directive also adopts more ambitious objectives to meet the requirements of BHD habitats and species where necessary. Its measures include the establishment of MPAs, in addition to Natura 2000 sites.
Maritime Spatial Planning Directive: Directive 2014/89/EU of 23 July 2014 establishing a framework for maritime spatial planning	Establishes a framework for maritime spatial planning aimed at promoting the sustainable growth of maritime economies, the sustainable development of marine areas and the sustainable use of marine resources	The Directive requires MS to prepare Maritime Spatial Plans, including for the preservation, protection and improvement of the environment, including resilience to climate change impacts. The plans provide a means of supporting the MSFD, Nature Directives and other EU and national environmental objectives.
Common Fisheries Policy: Regulation (EU) No 1380/2013 of the European Parliament and of the Council of 11 December 2013	To ensure that fishing and aquaculture are environmentally, economically and socially sustainable and that they provide a source of healthy food for EU citizens	Requires the implementation of an ecosystem-based approach to minimise the negative impacts of fishing activities on the marine ecosystem, and endeavours to avoid environmental degradation from fisheries and marine aquaculture. The CFP also enshrines coherence with EU environmental legislation, including the MSFD.

Table 4.3 (*cont.*)

Policy/Instrument	Overall objective	Biodiversity relevant measures
Common Agricultural Policy (multiple regulations)	Three general objectives are: viable food production; sustainable management of natural resources and climate action; and balanced territorial development	Includes incentives to avoid environmental damage, maintain and restore agricultural habitats and features of high nature value, and enhance agricultural habitats of low biodiversity value. Some payments may encourage agricultural intensification, which has detrimental impacts.

Notes. See Farmer (2012) for an overview of each directive's legislative provisions, the background for its development and its implementation, including key court judgments.

Unlike the EIA Directive, the SEA Directive does not have a list of plans/programmes, but an SEA is mandatory for plans/programmes that are prepared for agriculture, forestry, fisheries, energy, industry, transport, waste/water management, telecommunications, tourism, town and country planning or land use and which set the framework for future development consent of projects listed in the EIA Directive. The requirement for an SEA is also triggered by a requirement for an appropriate assessment of a plan or programme under Article 6(3) of the Habitats Directive. Screening is carried out for other plans and programmes (according to criteria in Annex II of the Directive) to determine whether they are likely to have significant environmental effects, in which case an SEA must be carried out. Whilst the scope of the Directive appears to be potentially very large, covering a wide range of plans and programmes, it only applies to plans and programmes that are 'required by legislative, regulatory or administrative provisions'.

The SEA contributes to the systematic and structured consideration of environmental concerns (including biodiversity) in higher-level decision making at early stages of the policy planning process. The process of identifying and developing mitigation measures, including offsets, can normally be carried out in the first instance as part of the SEA process. SEA provides the best opportunities to consider alternatives that locate developments in areas that avoid significant impacts as much as possible, and if necessary to identify sites where offsetting could provide valuable benefits. The SEA also complements other Directives and therefore provides scope for a coordinated approach. More precisely, where the obligation to carry out assessments of the effects on the environment arises simultaneously from the SEA Directive and other EU legislation, such as the Nature Directives, or the WFD, in order to avoid duplication of the assessments, the Member States may provide for coordinated or joint procedures fulfilling the requirements of the relevant Community legislation.

The SEA has the potential to overcome many of the limitations of project-based EIA by providing opportunities for conservation and sustainable use of biodiversity to be considered as a fundamental part of strategic decision making, rather than as a single specialist topic that needs considering on a more reactive basis. SEA can support and enhance EIA processes, as well as other applicable environmental assessments for a particular plan/programme, by:

- building biodiversity objectives into land-use, urban or sectoral policies, plans and programmes, at any point between international and local levels;
- identifying and managing cumulative impacts, which would be considered insignificant if assessed in isolation, but which may pose severe threats to biodiversity if assessed in combination with other similar impacts;
- identifying biodiversity-friendly alternatives and mitigation strategies that would be compatible with sustained delivery of ecosystem services;
- ensuring effective monitoring programmes are in place to provide information about biodiversity to inform baseline assessments carried out for EIA; and
- allowing biodiversity specialists and decision makers and/or planners to engage and to develop a shared understanding of biodiversity requirements.

The SEA process involves a number of steps, namely screening, scoping (including the development of the Environmental Report), public consultation and monitoring (including assessing data availability). The application and effectiveness of the SEA Directive amongst the EU Member States vary considerably, both in terms of the institutional and legal arrangements, and the actual implementation of the SEA procedures. This is partly because Member States have considerable discretion over how to implement many elements of the SEA; the SEA Directive is fairly ambiguous on many aspects and there is a lack of guidance that sufficiently clarifies some of the issues (Committee of the Regions, 2010).

Following an evaluation under the European Commission's Regulatory Fitness Check and Performance (REFIT) programme, the Commission has concluded that the SEA Directive remains fit for purpose and provides multiple environmental benefits for the EU (European Commission, 2019b). However, the Directive's effectiveness differs between sectors and the types of plans and programmes to which it is applied, and still depends significantly on how it has been transposed into national law and implemented in each Member State.

4.3.3 Environmental Liability Directive

The purpose of the Environmental Liability Directive (ELD) (2004/35/CE) is to establish a framework of environmental liability, based on the polluter pays principle, to prevent and remedy environmental damage. The strict liability standard of operators is based on a list of dangerous activities, including for example industrial activities covered by the Industrial Emissions Directive (IED); waste management activities; use, transport, and discharge of dangerous polluting substances; unauthorised discharges into surface or ground water; and transport of dangerous substances.

The ELD covers damages to land, water and biodiversity. In relation to biodiversity, it is effectively a no net loss mechanism for the specific types of damage covered under the Directive, through the following four mechanisms:

1. The incentivisation of long-term avoidance of environmental damage.
2. The requirement to take *preventative action* to *avoid* imminent threats of environmental damage.
3. The requirement to undertake *primary remediation* if damage occurs, which is nearly equivalent to *rehabilitation/ restoration* of the affected site.
4. If necessary, the requirement for *complementary and compensatory remediation*, for *residual impacts*.

The effectiveness of each of these aspects of the ELD depends on the scope and strength of implementation of the Directive. In particular, the coverage of habitats and species depends on the transposition of the Directive by a given Member State. Whilst the ELD text requires that all BHD habitats and species are included, the main added value of the ELD comes if a Member State chooses to also include biodiversity resources defined under national laws or non-protected species.

The added value of covering BHD species has been rather low in practice, because biodiversity damage is defined in the ELD as 'damage to protected species and natural habitats, which is any damage that has significant adverse effects on reaching or maintaining the favourable conservation status of such habitats or species' (Article 2). Consequently, as favourable conservation status is assessed at the national–biogeographical region level, this has allowed Member States to consider individual site level damage to normally be insignificant in relation to the habitats' or species' overall status. This is thought to be a reason for the low number of ELD cases that have arisen (Milieu and IUCN, 2014). In recognition of this weakness and related problems, detailed guidance has been provided on the assessment of damage under the ELD, including in relation to protected habitats and species (European Commission, 2021e).

4.3.4 Water Framework Directive and Related Directives

The Water Framework Directive (WFD) (2000/60/EC) has an overarching and key role to play in conserving nature in the aquatic environment as well as 'terrestrial ecosystems and wetlands directly depending on the aquatic ecosystems'. According to Article 1, the purpose of the Directive is to establish a framework for the protection and sustainable management of inland surface waters, transitional waters (estuaries) and coastal waters (up to one nautical mile from shore or the outer limit of transitional waters). Article 4 sets out the environmental objectives, which include the prevention of the deterioration of ecological quality and pollution of surface waters, and their restoration, to achieve good water status in all surface waters by 31 December 2015 (unless they are artificial or highly modified). There is also a requirement to comply with all standards and objectives relating to protected areas (e.g. Natura 2000 sites) by 31 December 2010.

The achievement of good water status is dependent on achieving at least both good chemical status and good ecological status. Chemical status relates to the concentrations of pollutants in relation to established environmental quality standards. Ecological status takes into account biological elements (including fish, macro-invertebrates, macrophytes and phytoplankton) and supporting hydromorphological and chemical aspects. The status of each water body is then judged against reference conditions for water bodies of the same class that are judged to have a 'high status'. To reach good status a water body must meet all of the necessary conditions for biological, chemical and hydromorphological elements – this is the so-called 'one-out, all-out principle' of the WFD (and is similar to that for achieving favourable conservation status under the Habitats Directive). Water bodies that are classified as artificial or heavily modified are required to achieve 'good ecological potential', which in essence means their ecological condition needs to be as good as is potentially feasible and compatible with their type of water body and its legitimate use.

It can be expected that the achievement of good ecological status of a water body should help aquatic habitats and species covered by the Nature Directives maintain or achieve their favourable status (European Commission, 2011b). This link is further strengthened by the additional requirement to meet the objectives of protected areas through planning and measures within the WFD. Furthermore, if meeting objectives for an aquatic Natura 2000 site requires measures additional to those to meet good status under the WFD, the WFD requires that they are identified and implemented.

The primary means of implementation of the WFD is through the development of river basin management plans (RBMPs) for river basin districts. These assess the status of each water body, identify objectives for it (taking into account other considerations, e.g. Natura 2000 designations) and identify pressures affecting its status. The pressures are supposed to be addressed through an integrated programme of measures, including use of existing EU environmental and other instruments (e.g. the CAP). The measures must include a number of so-called 'basic measures', incorporating those necessary to comply with the objectives of the other EU legislation, as well as any supplementary measures necessary to achieve the WFD's environmental objectives.

The WFD builds on and interacts with a number of older EU water policies, the most relevant to nature conservation being the **Nitrates Directive** (91/676/EEC), which aims to prevent nitrates from agricultural sources from affecting ground and surface waters, and the **Urban Waste Water Treatment Directive** (UWWTD) (91/271/EEC), which seeks to reduce the pollution of freshwater, estuarial and coastal waters by domestic sewage, industrial waste water and rain-water run-off – collectively, 'urban wastewater'.

4.3.5 Maritime Policy and Legislation

The **Marine Strategy Framework Directive** (MSFD) (2008/56/EC) is the environmental pillar of the EU's Integrated Maritime Policy (Farmer, 2012). It builds on and incorporates some of the principles and measures included in the Habitats Directive and especially the WFD. As set out in Article 1:

1. The MSFD 'establishes a framework within which Member States shall take the necessary measures to achieve or maintain good environmental status in the marine environment by the year 2020 at the latest.
2. For that purpose, marine strategies shall be developed and implemented in order to:
   a. protect and preserve the marine environment, prevent its deterioration or, where practicable, restore marine ecosystems in areas where they have been adversely affected;
   b. prevent and reduce inputs in the marine environment, with a view to phasing out pollution as defined in Article 3(8), so as to ensure that there are no significant impacts on or risks to marine biodiversity, marine ecosystems, human health or legitimate uses of the sea.'

The MSFD applies to all marine waters of the EU Member States, including 'coastal waters' as defined under the WFD and, where claimed, Exclusive Economic Zones (EEZ) and extended continental shelf areas. This area is divided into four main regions: Baltic Sea, Black Sea, Mediterranean Sea (with four sub-regions) and North-east Atlantic Ocean (with four sub-regions). These broadly equate to the marine biogeographical regions defined under the Habitats Directive, with the exception of the Macaronesian marine region, which is treated as a distinct region under the Habitats Directive but is a sub-region of the Atlantic region under the MSFD.

As noted earlier, the WFD applies to coastal waters up to one nautical mile from the shore. The MSFD also applies to these waters. To aid coherence between the two directives, the MSFD states that, for coastal waters, the MSFD only applies 'in so far as particular aspects of the environmental status of the marine environment are not already addressed through that Directive or other Community legislation' (Article 3.1). In other words, for coastal waters it is likely that the WFD will drive action on nutrients and chemicals. However, the MSFD includes descriptors of good environmental status (see Box 4.4) not included in the WFD, such as litter and noise.

Box 4.4 The definition of good environmental status in the MSFD (Article 3(5))

'"Good environmental status" means the environmental status of marine waters where these provide ecologically diverse and dynamic oceans and seas which are clean, healthy and productive within their intrinsic conditions, and the use of the marine environment is at a level that is sustainable, thus safeguarding the potential for uses and activities by current and future generations, i.e.:

a) the structure, functions and processes of the constituent marine ecosystems, together with the associated physiographic, geographic, geological and climatic factors, allow those ecosystems to function fully and to maintain their resilience to human-induced environmental change. Marine species and habitats are protected, human-induced decline of biodiversity is prevented and diverse biological components function in balance;
b) hydro-morphological, physical and chemical properties of the ecosystems, including those properties which result from human activities in the area concerned, support the ecosystems as described above. Anthropogenic inputs of substances and energy, including noise, into the marine environment do not cause pollution effects.'

In order to achieve good environmental status, each Member State is required to develop a marine strategy for each region, or sub-region, over which it has jurisdiction. Joint Marine Strategies for shared sea areas are not required, but coordination amongst Member States is encouraged, in particular through the work of the Regional Seas Conventions (see Section 3.7). Each Marine Strategy should take an ecosystem approach to the management of human activities, ensuring that the collective pressures are within levels compatible with the achievement of good environmental status and that the capacity of marine ecosystems to respond to human-induced changes is not compromised, while enabling sustainable use.

Specifically, each strategy requires the following sequential actions to be undertaken (with early and effective stakeholder participation):

1. An initial assessment of the current environmental status of national marine waters and the environmental impact and socio-economic analysis of human activities in these waters.
2. Determination of what good environmental status means for national marine waters.
3. Establishment of environmental targets and associated indicators to achieve good environmental status by 2020.
4. Establishment of a monitoring programme for the ongoing assessment and the regular update of targets.
5. Development of a programme of measures designed to achieve or maintain good environmental status by 2020.

Member States are required to keep their marine strategies up-to-date and renew their status assessments, targets, monitoring programmes and programme of measures every six years. This cyclical process mirrors that of the WFD. However, it is important to note that determination of good ecological status under the WFD is a tougher condition for Member States than good environmental status under the MSFD.

Reflecting the broad objectives of the MSFD, the definition of good environmental status (Box 4.4) takes into account a variety of environmental attributes including ecosystem functions and hydro-morphological, physical and chemical characteristics, as well as the status of habitats and species. Determination of environmental status is carried out with reference to a list of 11 descriptors of requirements (in Annex 1 of the Directive), which relate to, respectively, biological diversity, non-indigenous species, commercially exploited fish and shellfish populations, food webs, eutrophication, sea-floor integrity, hydrographical conditions, concentrations of contaminants in marine waters, contaminants in seafood, marine litter, and the introduction of energy into the marine environment (e.g. noise). The European Commission provided a decision in 2010 with criteria and standards for assessing good environmental status, which were subsequently updated in 2017 (Decision (EU) 2017/848). These clearly overlap with, and link to, the criteria used for the assessment of conservation status of HD habitats and species.

In developing their programmes of measures, Member States need to take into account requirements in other EU environmental legislation (e.g. the Nature Directives, WFD, CFP, Bathing Water Directive,[33] and regional conventions) and to consider sustainable development and, in particular, the social and economic impacts of the measures envisaged. Unlike the WFD, the MSFD includes a specific requirement for spatial protection measures. These aim to contribute to the establishment of coherent and representative networks of marine protected areas that will provide adequate coverage of the diversity of marine ecosystems. These need to include not only Natura 2000 sites, but also other marine protected areas designated in accordance with other international or regional agreements.

The MSFD recognises that it will not always be possible for Member States to achieve all the environmental targets, or overall good environmental status, in all sea areas (or within the intended timetable). This may be for reasons beyond their control, or for reasons of overriding public interest that outweigh the negative impact on the environment. However, Member States are still required to mitigate such impacts and prevent deterioration, and where they are for overriding public interest they must not prevent the achievement of overall good environmental status at the level of the marine region or subregion. This latter safeguard is less strict than the Habitats Directive provisions relating to projects that are allowed to damage Natura 2000 sites for similar public interests, as such projects are required to carry out compensatory measures to maintain the coherence of Natura 2000 (Article 6(4)). However, it is important to bear in mind that within Natura 2000 sites the MSFD exceptions cannot take precedence over Article 6 of the Habitats Directive, as the Treaty requires that stricter provisions take precedence when more than one applies to the same issue (European Commission, 2012b).

As with the WFD, there is considerable overlap between the objectives and measures of the MSFD and the Nature Directives, and therefore it can be reliably anticipated that the achievement of good environmental status supports the Nature Directives, and vice versa. Furthermore, with its requirement to designate marine protected areas, the MSFD can help maintain and enhance the coherence of the Natura 2000 network, as sites designated for broader biodiversity objectives for the MSFD will nevertheless probably contain habitats and species that are the focus of the Nature Directives. Also, within Natura 2000 sites, MSFD objectives will need to be consistent with the site's conservation objectives for its Natura habitats and species, and therefore measures may need to go beyond those necessary to achieve good environmental status.

Meeting the broad goals of the EU's maritime policy and specific MSFD objectives is a considerable challenge given the longstanding and increasingly intensive and widespread economic activity in the European seas, and along coastlines. There has consequently been a growing recognition that maritime spatial planning has a key role to play in supporting environmental objectives and legislation. EU initiatives to develop maritime and coastal spatial planning have included a 2002 EU Council Recommendation on Integrated Coastal Zone Management (ICZM),[34] followed by a 2007 Commission evaluation that showed the benefits of ICZM in Europe (European Commission, 2007c), and most importantly the **Maritime Spatial Planning Directive (MSPD)** (2014/89/EU). The MSPD sets a requirement for Member States to undertake maritime spatial planning and produce maritime spatial plans by 2021. The objectives of the plans, as indicated in Article 5, are to 'consider economic, social and environmental aspects to support sustainable development and growth in the maritime sector, applying an ecosystem-based approach, and to promote the coexistence of relevant activities and uses'.[35] More specifically, 'through their maritime spatial plans, Member States shall aim to contribute to the sustainable development of energy sectors at sea, of maritime transport, and of the fisheries and aquaculture sectors, and to the preservation, protection and improvement of the environment, including resilience to

[33] Directive 2006/7/EC of 15 February 2006 concerning the management of bathing water quality and repealing Directive 76/160/EEC.

[34] Recommendation of 30 May 2002 concerning the implementation of Integrated Coastal Zone Management in Europe (OJ L 148, 06/06/2002, pp. 24–27).

[35] As defined by the CBD, an 'ecosystem-based approach' is 'A strategy for the integrated management of land, water and living resources that promotes conservation and sustainable use in an equitable way.'

climate change impacts. In addition, Member States may pursue other objectives such as the promotion of sustainable tourism and the sustainable extraction of raw materials.'

The MSPD provides a set of minimum common requirements to ensure consistency and cross-border cooperation amongst Member States and ensure land–sea interactions are taken into account. It does not impose obligations on Member States on how to pursue sectoral policies, but states that maritime spatial planning should contribute to these policies through the planning process. Such planning should support the implementation of the MSFD, and Nature Directives, by managing the collective pressures of activities such that good environmental status can be achieved. For example, plans may include an element of zoning, such as in relation to Natura 2000 sites and other marine protected areas.

A further EU policy critically important to marine conservation is the **Common Fisheries Policy (CFP)**, which covers most aspects of the fish production chain, from capture/fish-farming through to landing, processing and marketing. It is important to highlight that, under the EU Treaty, competence for the management of fisheries beyond Member States' 12 nautical mile limit is the exclusive competence of the EU. This means that if a Member State were to determine that action on fisheries in its wider waters was necessary to meet obligations under EU environmental law (such as the MSFD), it does not have the competence (i.e. right) to take such action. This adds to the complexity of the delivery of marine biodiversity protection.

The CFP is reformed on a 10-year cycle. The last reform was agreed on 30 May 2013, creating a new fisheries regime, based on three main pillars:

- The CFP basic regulation.[36]
- The common organisation of the markets in fishery and aquaculture products (Regulation (EU) No 1379/2013).
- The European Maritime and Fisheries Fund (EMFF) (Regulation (EU) No 508/2014).

The 2013 CFP mandated much stronger intent than its predecessors to prevent damage from fishing activity to the wider marine environment, by setting explicit objectives to 'conserve marine biological resources and to implement an ecosystem-based approach to fisheries management, while committing to compliance with Union environmental legislation, in particular the MSFD but also the Nature Directives'. It set a goal to deliver maximum sustainable yield (MSY) of fish stocks by 2015 where possible and, at the latest, by 2020. In pursuit of this the CFP introduced measures to restrict catches of particular species in particular areas, and restrict the use of certain types of fishing gear and/or boat sizes, etc., held accountable by an enforcement regime. Under the so-called 'Landing Obligation', the discarding by vessels of catches of 'unwanted' fish was also prohibited.

Another significant change in fisheries management was the introduction in 2014 of a stronger regional management structure and more consultation with stakeholders, in efforts to decentralise governance and promote regional tailoring of fisheries management, and thereby improve compliance. Among the functions assigned to the new bodies of Member States established for each of the regional seas (e.g. North Sea, Baltic Sea) was the empowerment to propose (for subsequent adoption by the European Commission) fisheries management measures for the conservation of Natura 2000 sites and MPAs designated under the MSFD. The regional groups are required to consult with the Advisory Councils (ACs), namely stakeholder bodies comprising representatives of the fishing sector and other interest groups, notably NGOs, re-named from the Regional Advisory Councils (RACs) created in the 2002 reform of the CFP. A notable addition in the 2013 reform was the creation of an Aquaculture Advisory Council. Notwithstanding these new measures, for many years ministers have routinely set fishing quotas that exceeded the scientific advice for what was sustainable, contributing to overfishing and the depletion of fish stocks. According to Carpenter (2018), on average two-thirds of total allowable catches (TACs) in EU waters were set above scientific advice between 2001 and 2018. While the situation is improving in northern EU waters, exploitation rates of Mediterranean and Black Sea stocks remain twice as high as sustainable levels (European Commission, 2021f).

In alignment with the reform of the CFP basic regulation, two important related regulations were subsequently introduced which included new measures for addressing the ecosystem impacts of fishing. Firstly, in 2017, the new data collection regulation ((EU) 2017/1004) requires Member States to collect data on the incidental by-catch in fishing gear of birds, mammals and marine turtles, as well as fish. This was followed in 2019 by a revised regulation (EU 2019/1241) on technical conservation measures which obliges Member States to 'minimise and where possible eliminate' such by-catch. The first report by the European Commission (2021g) on the implementation of the latter regulation found that while some Member States had already started to develop additional fisheries measures to protect sensitive species and habitats, further action and more decisive steps were needed to achieve the goals of the *EU Biodiversity Strategy for 2030*. These fisheries regulations, along with, importantly, the MSFD and Birds Directive, give legal teeth to the voluntary EU Seabird Plan of Action (European Commission, 2012c) of which, with the notable exception of the UK, there has been little uptake by Member States (ICES, 2017).

Overall, the 2013 CFP was regarded as a 'big step in the right direction, albeit less ambitious than the environmental organisations had pushed and hoped for' (IEEP, 2016). Significantly, the commitment to 'implement the ecosystem-

[36] Regulation (EU) No 1380/2013 of 11 December 2013 on the Common Fisheries Policy, amending Council Regulations (EC) No 1954/2003 and (EC) No 1224/2009 and repealing Council Regulations (EC) No 2371/2002 and (EC) No 639/2004 and Council Decision 2004/585/EC.

based approach to fisheries management' was unequivocal compared with the much weaker objective of the 2002 reform, namely 'shall aim at a progressive implementation of an ecosystem-based approach. . .'. However, the heightened ambition of the 2013 reform has fallen short of effective implementation, most notably in failing to achieve sustainable exploitation rates for all stocks, with catch limits continuing to exceed scientific advice (The Pew Charitable Trusts, 2021). For example, see discussion of resulting NGO legal action in Ireland (Section 19.2). In regard to the Landing Obligation that requires fishers to retain and land fish they would previously have discarded, the evidence points to practical difficulties of implementation, generating strong industry resistance (Maynou *et al.*, 2018) such that discarding continues widely with impunity.[37] This in turn has led to calls for the introduction of innovative monitoring and control, in particular the mandatory use of remote electronic monitoring by installing cameras on fishing vessels for external scrutiny of catches and fishing practices and enhanced data collection for stock assessment purposes (WWF-UK, 2017). Assessment of the potential implications of the Landing Obligation for scavenging seabird populations suggested that, while negative impacts were foreseen on certain species, it was unlikely to present a crisis for most populations (Bicknell *et al.*, 2013).

In anticipation of the next review of the CFP, the Commission is obliged to report to the European Parliament and Council by the end of 2022 on the functioning of the CFP. In making recommendations for improvement, the NGOs considered that the 2013 CFP and related legal instruments already contain all the tools necessary to address shortfalls in implementation of the CFP (including importantly mitigating climate change) and, as such, it was premature to seek medium-term reform (Anonymous, 2021).

The fisheries funding package, EMFF (2014–2020), echoed the greater integration between the CFP 2013 reform and environmental conservation. Previous fisheries funding had been damaging to the marine environment in helping to perpetuate overfishing and the collateral damage inflicted on species and habitats. At the same time, prior to the 2014 EMFF, funding for research and data collection as well as for control and enforcement measures was relatively low with less than €50 million allocated for each funding area per year. Shifting funding away from subsidies that incentivise and support overfishing towards supporting sustainable fisheries and healthy marine ecosystems was urged by NGOs. They therefore welcomed that the 2014 EMFF expanded the scope of the fund by supporting not just the CFP but also the Integrated Maritime Policy (described earlier), in principle enabling Member States to finance activities implementing the MSFD and the Nature Directives. In practice, as indicated in many of the national chapters, very little uptake from Member States has been directed towards key activities such as financing the MSFD's Monitoring Programmes and the Programmes of Measures, or supporting management of the Natura 2000 network as set out in the PAFs.

The EMFF's 2021–2027 successor, which entered into force in July 2021, applied retrospectively from January 2021, is renamed the European Maritime, Fisheries and Aquaculture Fund (EMFAF)[38] to reflect a greater focus on aquaculture. The NGOs considered the new regulation to be retrogressive in failing to eliminate harmful subsidies – and indeed introducing new ones – that contribute to fleet overcapacity, overfishing and ocean degradation.[39] The finally agreed EMFAF rejected the European Parliament's call to ringfence 25% of the €6.1 billion fund for protecting and restoring the marine environment, albeit the Parliament's position was assessed by the NGOs as negative for nature overall.

4.3.6 Air Pollution

As described in Section 2.2 on biodiversity pressures, air pollution resulting from high concentrations in ambient air, or deposition, of pollutants (e.g. ammonia, ozone, nitrogen dioxide, sulphur dioxide) has led to major changes in the physiochemical conditions of sensitive terrestrial and aquatic ecosystems (e.g. acidification or eutrophication), and resulting impacts on their species, across much of the EU. Within EU environmental law, there are three principal approaches to addressing air quality issues, variously addressing ambient and deposited air pollution:

1. The setting of quality objectives for ambient air – some obligatory, some not (with supporting requirements on information provision to the public, monitoring, etc.).
2. The adoption of controls on specific sources of air pollutants – this is largely for industrial and transport emissions of various types.
3. The setting of overall caps (or ceilings) on emissions of specified pollutants by individual Member States.

The setting of quality objectives in ambient air in EU law goes back to the 1970s. However, this is now mostly established within the Air Quality Directive 2008/50/EC, which sets limit values (obligatory values) and target values (not obligatory). Obligatory limit values are established for pollutants such as sulphur dioxide and nitrogen dioxide (these often have local sources and therefore can be controlled by Member State action). Non-binding targets are set for pollutants such as ozone (which has much transboundary movement, and meeting an objective may require action in another Member State). The main emphasis of this legislation is on setting targets to protect human health, but it does contain some objectives for protection of vegetation. For vegetation the response to air

[37] https://publications.parliament.uk/pa/ld201719/ldselect/ldeucom/276/276.pdf

[38] Regulation (EU) 2020/11391.

[39] https://seas-at-risk.org/wp-content/uploads/2021/07/Joint-NGO-letter-to-EU-decision-makers-on-the-European-Maritime-Fisheries-and-Aquaculture-Fund-.pdf

pollution may not be simple. For sulphur dioxide a simple threshold value is possible. However, for ozone (which naturally occurs in the air), damage is based on a cumulative exposure over a growing period above a baseline value. Due to controls on emissions, impacts from sulphur dioxide are limited to a few locations in the EU (where poor quality coal is still burnt). In contrast, nitrogen oxide emissions still cause damage, as does ozone (Stevens *et al.*, 2020).

The setting of overall ceilings for a Member State is focussed on controlling deposited air pollution (although controls may lead to ambient air quality benefits also). Further, the evolution of the legal objectives has been strongly linked to the impacts they have. As the impact of acid rain became of high public concern in the 1980s, there was considerable apportionment of 'blame' within Europe. Sweden, for example, suffered serious acidification of waters and forest soils, much of which came from burning of coal in the UK, Poland and the then German Democratic Republic. In the 1980s, the principal forum for policy development in this area was the United Nations Economic Commission for Europe (UNECE), as certain key countries like Sweden were not members of the European Community. Debate and analysis focussed on two issues:

1. What level of deposition is acceptable? (It was clear that some ecosystems were more sensitive than others, but how to measure this?)
2. How to identify who needs to control what to meet levels of deposition that are acceptable?

Combined, these formed the 'effects-based approach'. The first issue led to the concept of critical loads – the quantities of, for example, acidity that a given soil or water body can absorb without changing character. Calcareous soils can absorb large amounts of acidity, but peaty soils very little. For soils the defining point was one of chemistry. For water, the critical load was also chemical, but the acid neutralising capacity objective for waters was set at a level consistent with maintaining salmonid fish populations (critical in much of northern Europe suffering acid deposition). Therefore, the objective had a direct link to biology. A similar approach is used to determine critical loads for nitrogen deposition, identifying the amount of nitrogen a habitat can absorb without significant change (e.g. nutrient-poor systems such as lowland heaths being particularly sensitive). Mapping areas where critical loads have been exceeded can provide detailed evidence of individual nature conservation sites at risk.

The second issue required detailed modelling of individual sources and where the emissions from these were eventually deposited. The models could then explore options to reduce emissions to meet the critical loads for soils and waters across Europe, so optimising emission reductions (later models added cost information). As a result, under the UNECE Convention on Long-Range Transboundary Air Pollution, the concept of different countries needing to reduce emissions to different degrees was accepted.

As the European Community expanded, the role of EU law increased. Early EU law did include overall emission reductions – under the Large Combustion Plants Directive[40] – but these caps were not based on the effects the emissions caused and some proved controversial as a result. This early legislation was replaced by the National Emissions Ceilings Directive (NECD). This integrated the approach taken under the UNECE. The 2001 NECD (2001/81/EC) set national emission ceilings to be met by 2010 and it required that national programmes be adopted in 2002 and revised in 2006. The 2016 NECD (2016/2284/EU) set revised national ceilings to be met by 2020 and 2030. It required national air pollution control programmes to be adopted by 1 April 2019.

As further discussed in the conclusions chapter (Chapter 5), the controls have led to significant decreases in some deposited air pollutants, in particular sulphur dioxide and nitrogen oxides. Ammonia emissions, especially from agriculture, have been more difficult to reduce (e.g. as described in Chapter 23 on the Netherlands).

4.3.7 Invasive Alien Species (IAS) Regulation (Regulation (EU) No 1143/2014)

The IAS Regulation entered into force on 1 January 2015, fulfilling Action 16 of Target 5 of the *EU Biodiversity Strategy to 2020*, as well as CBD Aichi Target 9. The Regulation is a vital policy tool for the prevention, surveillance, eradication and management of IAS in the EU, providing coordinated action at the EU level to prevent potentially damaging invasions, and manage or eradicate the damaging effects of IAS that are already established in the EU. Key policy mechanisms and tools for coordinated EU level action include:

- the 'list of IAS of Union concern' with the associated risk assessment process; and
- voluntary mechanisms to promote coordinated action by Member States at the regional level to tackle IAS that are considered being of regional concern, and voluntary action at the MS level.

The IAS Regulation includes three types of measures to be taken to tackle IAS included on the Union List, which follows an internationally agreed hierarchical approach to combatting IAS:

1. *Prevention*: a number of robust measures aim to prevent the intentional or unintentional introduction of IAS of Union concern into the EU (e.g. horizon scanning, pathways analysis and surveillance).
2. *Early detection and rapid eradication*: Member States must put in place a surveillance system to detect the presence of IAS of Union concern as early as possible and

[40] Directive 2001/80/EC of 23 October 2001 on the limitation of emissions of certain pollutants into the air from large combustion plants. This replaced Directive 88/609/EEC, as amended by Directive 94/66/EC.

take rapid eradication measures to prevent them from establishing.

3. *Management*: some IAS of Union concern are already established in certain Member States. Concerted actions, prioritised according to risk and cost-effectiveness, are needed to prevent them from spreading any further, to minimise the harm they cause and to assist the recovery of affected ecosystems.

The core of the IAS Regulation is the 'list of IAS of Union concern' (Union List). For those species on the Union List, the regulation establishes restrictions on intentional introduction, transport, trade, keeping, breeding and release into the environment. In addition, all necessary steps must be taken to prevent the unintentional introduction and spread of the species on the Union List. The Union List is adopted by the Commission and is reviewed and updated at least every six years.

Species are considered for inclusion on the Union List if they meet criteria defined in Article 4(3) and 4(6) of the Regulation. These include the potential to cause significant impacts and the need for concerted EU action. Proposals for species to be included on the list may be made by the Commission and Member States, with supporting peer-reviewed risk assessments. The proposals and risk assessments are firstly reviewed by the Scientific Forum of nominated experts, and those that are approved are then considered by the IAS Committee. Updates to the Union List are then made through agreement between the Commission and IAS Committee following public consultation.

The first list of IAS of Union concern included 37 species and was adopted in 2016. Species were added in 2017, 2019 and 2022, giving a current Union List of 88 species. The Union List has been criticised for being much too short. For example, a systematic study of 1 323 IAS candidates found that 900 meet the IAS Regulation criteria for inclusion on the Union List, and identified 207 that should be subject to urgent risk assessments (Carboneras *et al.*, 2018).

As described in Scalera *et al.* (2020), the protection provisions in the Nature Directives and the prevention and management provisions in the IAS Regulation complement each other and interact (likewise with the WFD, MSFD and other environmental instruments). Synergies with the Nature Directives are explicitly foreseen in the IAS Regulation, as it notes that 'In the interest of efficiency and cost-effectiveness, existing systems of customs control, surveillance and monitoring already established by Union law should be applied, in particular those set out in Directives 92/43/EEC [Habitats Directive].' Additionally, the IAS regulation recognises that 'Invasive alien species generally cause damage to ecosystems and reduce the resilience of those ecosystems. Therefore proportionate restoration measures should be undertaken to strengthen the ecosystems' resilience towards invasions, to repair the damage caused and to enhance the conservation status of species and their habitats...' and that this should be in accordance with the Habitats Directive.

The Nature Directives need to be taken into account in the implementation of the IAS Regulation, as consideration of BHD habitats and species, and of the Natura 2000 network, should be explicitly incorporated into each of the mechanisms and tools. For example, the eradication of IAS must not inadvertently result in detrimental impacts on BHD species.

The IAS Regulation also makes certain requirements of the Nature Directives' implementation, for example, integration of:

- IAS prevention/eradication/management measures as well as post-IAS restoration in Natura 2000 site conservation measures and site management planning;
- IAS surveillance with Natura 2000 condition monitoring, and with monitoring of BHD habitats and species; and
- BHD species protection measures with IAS prevention and eradication measures.

4.3.8 The Common Agricultural Policy (CAP)

Objectives and Key Reforms

The CAP is one of the oldest of all EU policies, and still based on objectives defined in the Treaty of Rome 60 years ago, which are principally to increase agricultural productivity, ensure a fair standard of living for the agricultural community, and stabilise markets to ensure the availability of supplies to consumers at reasonable prices.[41] As described in Chapter 2, and some of the national chapters, the CAP has contributed to some serious environmental problems in Europe, such as through providing financial incentives that were coupled to production (e.g. payments per head of livestock) that encouraged overgrazing of sensitive habitats, and grants for the ploughing of grasslands, drainage, irrigation and inappropriate afforestation schemes.

Since the 1980s, reforms to the CAP have responded to increasing awareness of the environmental effects of agricultural activity, such that biodiversity and other environmental concerns have gradually been taken into account. Policy changes that focussed primarily on improving the environmental sustainability of agriculture include the growing use of environmental conditionality on farm support payments, through the use of instruments such as cross-compliance, and the introduction of incentive payments to encourage the use of management practices that are environmentally beneficial, for example, the agri-environment measure. Other policy changes constrained production (quota, set-aside) or decoupled payments from production.

Since 2000 the CAP has consisted of two Pillars with differing objectives, financing, functioning and structure. Pillar 1 is financed fully from the European Agricultural Guarantee Fund (EAGF) and provides direct payments to farmers, amongst other things. Pillar 2 (known as the Rural

[41] Article 39 TFEU.

Development Regulation) is co-financed jointly by the European Agricultural Fund for Rural Development (EAFRD) and Member States, and offers a wide range of measures for the environment (such as agri-environment schemes) and other rural issues through multiannual Rural Development Programmes (RDPs).

The 2013 CAP reform introduced some fairly fundamental changes to the structure of Pillar 1 support measures for farmers. From a biodiversity perspective, the most significant change was the introduction of three main measures providing farmers with 'payments for agricultural practices beneficial for the climate and the environment', otherwise known as 'green direct payments' or 'greening'.[42] These measures were mandatory for Member States to put in place and applied to the majority of agricultural land. Permanent crops were excluded from the measures, and organic farmers were considered to be already carrying out basic environmental management and were therefore exempted.

The 2020 reform of the CAP (to be operational from January 2023) involves a fundamental change in approach, by requiring Member States to design both Pillar 1 and Pillar 2 support in a strategic way to deliver against a set of ten specific objectives in light of national and regionally identified needs. This is set out within CAP Strategic Plans (CSPs) that should focus on the delivery of results. This is the first time that Pillar 1 support has been subject to this form of programming and to approval by the Commission. From a biodiversity perspective, key developments are the strengthening of conditionality requirements for farmers receiving direct payments (to incorporate the 'greening' measures) and the introduction of a new 'eco-scheme' measure under Pillar 1 to provide payments for management practices that benefit the environment, climate and animal welfare.

The CAP reforms and measures that have had most influence on biodiversity are described here, and their impacts described in each national chapter and summarised in Chapter 5. A more comprehensive account of the development of the CAP and its environmental measures is provided in online Annex 3.

Set-Aside (Production Control Measures)

Although not introduced for environmental reasons, the set-aside measure is important to highlight, given the impacts that it had on biodiversity. Originally introduced as a voluntary supply-control measure in 1988, it became compulsory in 1992 for farmers to set aside a specified proportion of their land under cereals, oilseeds and proteins as a prerequisite for obtaining direct payments. The level of the set-aside obligation changed from year to year depending on cereal markets and ranged from 0% to 17.5%. Although the impacts of set-asi on the environment were mixed, depending greatly on how t areas were managed, studies revealed that they often provid significant biodiversity benefits (IEEP, 2008). Set-aside w abolished in 2009 as it was no longer justified once far support was decoupled from production.

Payments Linked to Production

From 2005, payments coupled to production were progre sively phased out with the exception of those for suckler bee sheep and goats. As a result, most support is now provide through a direct support payment, based on the area of lan farmed. This led to accelerated declines in the numbers o livestock, particularly within the beef sector (see for exampl Alliance Environnement, 2007, 2009; Brady, 2010; Schmi *et al.*, 2010). In some areas this reduced pressure from overgrazing, whilst in others it led to undergrazing and aban donment of farming, with detrimental impacts on semi natural grassland and shrubland habitats.

At the same time, in 2005, some flexibility was introduce into the way some Member States could target their direc payments, by allowing them to divert up to 10% of thei Pillar 1 budget to provide 'special support' to particular sectors for a variety of purposes, including protecting and enhancing the environment.[43] Greater flexibility was introduced in 2009[44] and it was extended to all Member States. However, although the majority of Member States chose to use this flexibility, only a few used it for environmental purposes, such as introducing new types of agri-environment schemes (Hart *et al.*, 2010).

The overhaul of Pillar 1 under the 2013 reforms saw these flexibilities morph into an optional 'voluntary coupled support' measure for certain sectors.[45] In practice, this provided Member States with considerable leeway, such that the measure has been widely used, including supporting half the EU beef and dairy herd. Whilst this coupled support may have helped prevent the abandonment of High Nature Value (HNV) farming, it may also have encouraged high grazing rates and intensification of cropping systems.

Environmental Conditionality (Cross-Compliance)

Over time, cross-compliance has developed as a key policy instrument for incentivising a minimum level of environmental management across the farmed countryside. It is a mechanism that ties area-based CAP payments (direct payments and some RDP measures) to compliance with environmental, animal welfare and health standards. The cross-compliance obligations include Statutory Management Requirements (SMRs), which reiterate existing farm-level obligations under EU law, and standards for Good Agricultural and Environmental

[42] Article 43 of Regulation (EU) 1307/2013 of 17 December 2013 establishing rules for direct payments to farmers under support schemes within the framework of the common agricultural policy and repealing Council Regulation (EC) No 637/2008 and Council Regulation (EC) No 73/2009.

[43] This flexibility was provided under Article 69 of Council Regulation 1782/2003.

[44] 'Special support' could be provided under specific circumstances laid out under Article 68 of Council Regulation 73/2009.

[45] Article 52(3) of Regulation (EU) 1307/2013.

ondition of land (GAECs), which are defined at the national or gional level. Farmers' payments can be reduced or withheld if ey fail to comply with these requirements.

SMR2 and SMR3 are particularly relevant to nature conser- tion because they require farmers to comply with the man- tory requirements arising from the Birds Directive and abitats Directive respectively. Over the CAP programming eriods GAEC standards have varied in their scope and ambi- on. During the 2014–2020 CAP, the only GAEC requirement f significant relevance to nature conservation was GAEC 7, hich primarily set out rules on the retention of landscape atures, such as hedges, ponds and field margins. In theory is could support the requirement to maintain landscape atures under HD Article 10 (see Section 4.2). However, ecause Member States have a lot of discretion on the details f all GAEC standards, it has relied on their levels of ambition nd enforcement, which have been highly variable.

From 2023 the new conditionality requirements will come nto play. Since these have incorporated the 'greening' meas- ures, there are more that are relevant for biodiversity (e.g. a an on the conversion and ploughing of permanent grassland lesignated as sensitive in Natura 2000 areas, and a require- nent for a minimum share of the agricultural area to be devoted to non-productive activities). In addition, a new GAEC standard has been introduced requiring Member States to ensure the protection of wetlands and peatlands. The requirement to retain certain landscape features remains as before.

The Pillar 1 'Greening' Measures

For the first time, in 2015, Member States were required to allocate 30% of their Pillar 1 budgets to three 'greening measures' supporting agricultural practices beneficial for the climate and the environment. These measures, which applied only to grassland and arable areas, comprised:

- the maintenance of permanent grassland:
    - maintaining the ratio of permanent grassland area compared to the total agricultural area within a 5% range;
    - the designation and strict protection of environmentally sensitive permanent grassland (ESPG) in Natura 2000 sites (and optionally elsewhere);
- crop diversification; and
- Ecological Focus Areas (EFAs) on at least 5% of eligible arable land (e.g. fallow fields, hedgerows, nitrogen-fixing crops) intended to provide direct or indirect benefits for biodiversity.

The two greening elements with the greatest potential to benefit biodiversity were the EFA measure and the designation of ESPG. In relation to EFAs, certain elements had the potential to be beneficial to biodiversity through increasing the area of habitats such as fallow, nectar- and pollen-rich and/or seed-rich vegetation, some multiannual fodder crops, species-rich grass margins, and landscape elements such as hedgerows. But actual effects depended significantly on Member State implementation choices (e.g. types of crop and species allowed, whether the use of fertilisers and pesticides was restricted and the area covered). In 2018, EFAs covered 9.5 million ha or 9% of the EU's arable area.

In practice, the measure mainly contributed to the expansion of nitrogen-fixing crops (alongside voluntary coupled support and the crop diversification measure), catch crops and cover crops (also required under some Nitrate Action Plans), which are generally as poor for biodiversity as other conventionally managed arable crops (Alliance Environnement and Thünen-Institut, 2017; Alliance Environnement, 2019). Pesticides were permitted on these areas until 2017 but have now been banned, probably increasing their biodiversity value. The measure does appear to have helped stabilise the negative trend in EU fallow area seen in many countries since the abolition of set-aside.

In terms of the ESPG measure, although the area designated covered 57% of permanent grassland in Natura 2000 areas, only half of this (4.6 million ha) was in receipt of CAP payments and therefore subject to the rules. A further 2% of HD Annex I grasslands outside the Natura 2000 network were designated. Although this measure banned ploughing on habitats that should already be protected under the Habitats Directive, it is still expected to have helped reduce the observed loss of semi-natural grassland of high biodiversity value habitats and associated species. However, ESPG designation did not give protection from damage from all agricultural improvements, such as fertiliser use.

Rural Development Measures

Table 4.4 provides a summary of the RDP measures with the potential to make significant contributions to nature conservation. This is based on the measures available during the 2014–2020 CAP,[46] which were mostly similar to measures included in the previous programming periods. Other RDP measures were more focussed on climate objectives, increasing the competitiveness of agriculture or the socio-economic development of rural areas more generally. These may also indirectly support nature objectives if, for example, they help maintain HNV farming systems. On the other hand, these and all other RDP measures can be detrimental for biodiversity if they lack safeguards and lead to environmentally damaging farm improvements and intensification.

Member States (and regions, in federal states) have a large degree of freedom in choosing how and where they use RDP funds, but require approval from the Commission as they must reflect EU priorities. Over the 2014–2020 CAP there were six priorities, of which Priority 4 had most direct relevance to nature conservation: restoring, preserving and enhancing ecosystems dependent on agriculture and forestry. Priority 4 included three focus areas: (a) restoring, preserving and

[46] At the time of writing the CAP programming period had been extended to the end of 2022.

Table 4.4 CAP 2014–2020 Rural Development Programme measures with the potential to significantly support nature conservation objectives.

Measures/sub-measure	CAP instrument/measure's objective	Potential support for nature conservation
M1 Knowledge transfer and information actions	Enhance the competitiveness and resource efficiency and improve environmental performance while ... contributing to the sustainability of the rural economy	Training where funding is prioritised under Focus Area 4A
M2 Advisory services, farm management and farm relief services	Improve the sustainable management and overall performance	Advice where prioritised under Focus Area 4A, e.g. compulsory training of beneficiaries of M10.1 and M15
M3 Quality schemes for agricultural products and foodstuffs	Encourage the participation of farmers in EU or national quality schemes, including farm certification schemes for agricultural products and food	The marketing of quality products from HNV farmland, thereby indirectly supporting the economic viability of HNV
M4 Investments in physical assets: M4.4 Support for non-productive investments linked to the achievement of agri-environment–climate objectives	Improve the economic and environmental performance of farms through investments	Can complement other measures (e.g. M10) to maintain or enhance habitats and species
M7 Basic services and village renewal in rural areas	Restoration and upgrading of cultural and natural heritage of villages and rural landscapes through the development of local infrastructure and services	Studies/investments in the maintenance and restoration of rural landscapes and HNV areas; and environmental awareness
M8 Investments in forest area development and improvement of the viability of forests: M8.1 Afforestation; M8.2 Agro-forestry; M8.5 Investments improving the resilience and environmental value of forests	Supports forest creation and management, which is conditional on a forest management plan (or similar) in line with sustainable forest management	Investments aimed at improving forest ecosystems
M10 Agri-environment–climate (AECM): M10.1 Payment for agri-environment-–climate commitments; M10.2 Support for conservation and sustainable use of genetic resources in agriculture	Preserve and promote the necessary changes to agricultural practices that make a positive contribution to the environment and climate Compulsory measure and must therefore be included in all RDPs	A wide range of actions for habitats and species. Genetic measures may help conserve traditional livestock breeds that are beneficial for HNV farming
M11 Organic farming: M11.1 – Payments for conversion; M11.2 – Payments to maintain organic farming	Supports organic farming, to adopt or maintain environmentally friendly farm practices and high standards for animal welfare	Organic practices, including restrictions on use of chemical pesticides and inorganic fertilisers
M12 Natura 2000 and WFD compensation payments: M12.1 for Natura 2000 agricultural areas; M12.2 for Natura 2000 forest areas; M12.3 for agricultural areas included in river basin management plans	Compensate beneficiaries for additional costs and income foregone from disadvantages in the areas concerned relative to the implementation of the BHD and WFD	Supports management measures in Natura 2000 sites and measures in river basin management plans
M13 Payments to areas facing natural or other specific constraints (ANC)	Maintaining the countryside and sustainable farming systems by continued use of agricultural land	May improve the economic viability of HNV farming systems
M15 Forest–environmental and climate services and forest conservation	Support for commitments to enhance biodiversity, preserve high-value forest ecosystems, improve their climate change mitigation and adaptation potential, and reinforce the protective value of forests	Can support a wide range of actions for habitats and species in forests including in Natura 2000 sites

Table 4.4 *(cont.)*

Measures/sub-measure	CAP instrument/measure's objective	Potential support for nature conservation
M16 Cooperation: M16.5 – Support for joint action on mitigating or adapting to climate change and for joint approaches to environmental projects and environmental practices; M16.8 – Support for drawing up of forest management plans or equivalent instruments	Encourage cooperation, to produce greater and more consistent environmental and climate benefits than those which can be produced at individual scale	Joint approaches/cooperation, e.g. for landscape-scale agri-environment–climate schemes, and preparation of forest management plans

Note. This framework of measures has been extended until the end of 2022.
Source. Adapted from Alliance Environnement (2019), based on Regulation (EU) No 1305/2013, Regulation (EU) No 1306/2013, Regulation (EU) No 1307/2013 and Regulation (EU) No 1308/2013.

enhancing biodiversity, including in Natura 2000 areas, and in areas facing natural or other specific constraints, and HNV farming, as well as the state of European landscapes; (b) improving water management, including fertiliser and pesticide management; (c) preventing soil erosion and improving soil management. The national Priority 4 budget allocation for each RDP measure is given in the funding table in each national chapter. Another requirement for Member States was to allocate at least 30% of their EAFRD funding to certain measures that support environment and climate change objectives.

The agri-environment measure, since 2014 called the agri-environment–climate measure (M10), has been of particular importance for nature conservation. It was first introduced in 1985 as a voluntary measure that allowed Member States to establish Environmentally Sensitive Area schemes. Since 1992 it has been compulsory for all Member States to use the agri-environment measure, as a means of addressing environmental priorities. Over time, the budget available for funding the measure has grown, and it is now the most important source of funding for practical biodiversity conservation and restoration in terrestrial areas, both in terms of the spatial coverage of schemes and allocated resources. The agri-environment measure used to include support for organic farming, but organic farming has been treated as a separate measure since 2014.

A very high level of subsidiarity is encouraged in the design, targeting, delivery and payment levels of schemes by Member States so that they are able to reflect their differing needs and environmental priorities. This has led to considerable diversity amongst the agri-environment schemes that currently operate in the EU and also variation in their effectiveness (ECA, 2011; Batáry *et al.*, 2015). Schemes developed under the agri-environment measure tend to fall into two main groups: those that are focussed on maintaining low-input traditional farming systems (i.e. HNV systems), often covering whole farms; and more targeted schemes that focus on the often more complex management requirements needed for the maintenance or restoration of particular habitats and species. The associated 'non-productive investments' measure can be used in conjunction with the agri-environment measure to provide support for capital expenditure, such as activities to help with habitat restoration, planting hedgerows and so on.

Although take-up of agri-environment schemes by farmers is voluntary, and some more demanding schemes have not been popular, 13.5% of the utilised agricultural area (UAA) was under some form of agri-environment agreement in the EU-27 in 2019 (amounting to 21.8 million ha).[47]

In addition to the agri-environment schemes, a potentially important RDP measure is the Natura 2000 payments measure (M12.1). This compensates landowners for restrictions on their agricultural or forestry related activities that are necessary to achieve the conservation objectives of Natura 2000 sites. The Natura measure has been little used by Member States, in part due to the need to have the necessary conservation measures for sites defined in management plans or documents. The latest available compiled data are for 2017, when this measure was being used in 18 of the EU-28 Member States but applied to only 1.3 million ha of agricultural land (c. 6.6% of the Natura 2000 agricultural area) and 0.3 million ha of forest land (c. 2% of the Annex I forest habitat within Natura 2000).[48]

Whilst the EU has limited competency over forest policy (as it is not mentioned in the Treaty on the Functioning of the EU), it has undertaken a number of initiatives to support sustainable forest management, whilst recognising the need for subsidiarity. This has included the development of the EU Forest Strategies since 1998, which have established a framework for forest-related actions that support sustainable forest management (European Commission, 1998a). The rural development part of the CAP has been the main source of funding for delivering the priorities set out in the EU Forest Strategy, including delivering environmental outcomes.

Whilst it has been possible to fund measures for afforestation and the management of woodland under the CAP since the early 1990s, they have become more prominent in the last

[47] https://agridata.ec.europa.eu/extensions/DashboardIndicators/Environment.html?select=EU27_FLAG,1

[48] Based on 157 588 km^2 of Annex I forest types in Natura 2000 (Maes *et al.*, 2020).

decade. Under the 2014–2020 rural development regulation there was a suite of measures that could apply to forests, and Priority 4 was one of the key priorities to which they are intended to contribute. Of these measures, the forest–environment–climate measure (M15) was of most relevance to achieving nature conservation and restoration objectives. In the right circumstances, support for the afforestation and creation of woodland (M8.1), establishing and maintaining agro-forestry systems (M8.2), and investments to improve the environmental value of forest ecosystems (M8.5) could also be beneficial for biodiversity. But the inappropriate use of forest measures has sometimes been damaging.

An evaluation of the CAP's forest measures concluded that, despite some good examples in individual countries, Member States had not maximised the potential of these measures, particularly to meet Member States' obligations under EU biodiversity policies (Alliance Environnement, 2017).

Since the 1970s, support has been provided to farmers in mountain areas and other areas facing natural or other specific constraints to compensate them for all, or part, of the additional costs and income foregone in relation to the constraints on agricultural production in the area concerned. Initially known as Less Favoured Areas, they were designated by each Member State according to biophysical and socio-economic criteria. A revised set of biophysical criteria only (i.e. soil, slope and climate) was subsequently developed and the areas renamed Areas facing Natural Constraints (ANC). Member States were given until 2018 to apply these new criteria.

Although payments are not linked to environmental requirements and the designations are not based on environmental criteria, indirectly the support may help maintain HNV farming systems (Keenleyside *et al.*, 2014). In 2019 these areas covered approximately 57% of the EU's agricultural area,[49] with a high proportion overlapping with Natura 2000 areas and HNV farming systems.

4.4 EU Biodiversity Strategies

4.4.1 1998 European Community Biodiversity Strategy

In order to meet its obligations as a Party to the CBD (as discussed in Section 3.6), including the requirement to produce a National Biodiversity Strategy and Action Plan (NBSAP), the European Community produced a *Communication on a European Biodiversity Strategy* in 1998 (European Commission, 1998b). In addition, the strategy sought to address the need to integrate environmental concerns into other sectoral policies (Article 130R (2) of the Treaty) and the commitment to sustainable development in accordance with the Amsterdam Treaty. It also aimed to contribute to the objectives of the Council of Europe's *Pan-European Biological and Landscape Diversity Strategy* (PEBLDS) (Council of Europe, 1996), which had been approved by the Ministerial Conference 'Environment for Europe' in Sofia, in 1995.

The Biodiversity Strategy did not have an overall goal, but instead firstly set out a number of broad objectives in relation to four themes: (1) conservation and sustainable use of biological diversity; (2) sharing of benefits arising out of the utilisation of genetic resources; (3) research, identification, monitoring and exchange of information; and (4) education, training and awareness. More specific objectives were then proposed in relation to the following key policy areas: conservation of natural resources, agriculture, fisheries, regional policies and spatial planning, forests, energy and transport, tourism, development and economic cooperation. Compared to more recent biodiversity strategies, the objectives are not well defined and not linked to measurable targets. More detailed actions were identified in four follow-up Biodiversity Action Plans (none of which appear to be available), respectively, for the Conservation of Natural Resources; Agriculture; Fisheries; and Economic and Development Cooperation.

4.4.2 2010 EU Target and Biodiversity Strategy

In response to the growing concerns over threats to biodiversity on the one hand, and the EU's relatively comprehensive environmental policy framework on the other, in June 2001 EU Heads of State and Government adopted the ambitious target of halting the loss of biodiversity in the EU by 2010. This coincided with the adoption of the EU's *Sustainable Development Strategy* (which included a headline objective 'to protect and restore habitats and natural systems and halt the loss of biodiversity by 2010') and the sixth *Environmental Action Programme*. In 2003, the fifth Environment for Europe Ministerial Conference agreed the Kyiv Resolution on Biodiversity, which extended the EU 2010 target to the pan-European region.

Not surprisingly, a less ambitious global target was agreed at the sixth CBD COP, in 2002, of 'the achievement by 2010 of a significant reduction in the current rate of loss of biological diversity'. This was subsequently also adopted at the World Summit on Sustainable Development.

To achieve the targets in the EU, it was recognised that further and better coordinated conservation efforts would be required. The European Commission carried out a consultation exercise, which firstly reviewed the effectiveness of the 1998 Biodiversity Strategy and its action plans. This was then followed up by a stakeholder conference held under the Irish presidency in Malahide in 2004, which produced a broad consensus on priority objectives for further biodiversity

[49] According to CAP Context Indicator data: https://agriculture.ec.europa.eu/cap-my-country/performance-agricultural-policy/cap-indicators/context-indicators_en

actions, as set out in the *Message from Malahide*. These were then taken into account in a European Commission Communication in May 2006, *Halting biodiversity loss by 2010 – and beyond – Sustaining ecosystem services for human well-being* (European Commission, 2006b), which was accompanied by a detailed *EU Biodiversity Action Plan*. The plan included few new measures, but instead focussed on redoubling efforts to implement intended actions, such as the designation of protected areas, appropriate management of habitats (e.g. through agri-environment schemes) and to secure adequate financing. The plan also recognised the important interlinkages between climate and biodiversity, and therefore included measures that may not only help biodiversity adapt to climate change, but may provide wider ecosystem-based adaptation and mitigation benefits.

Despite the production of the plan and significant progress with the implementation of many actions, monitoring of a range of biodiversity indicators led the Commission to conclude that the EU had failed to achieve its 2010 target (European Commission, 2010c). This was because there had been little change in the key pressures on biodiversity (in particular, changes in land and sea use, pollution and IAS). Although the *Biodiversity Action Plan* had set out to address these pressures and drivers, its effectiveness had been hampered by insufficient integration into other sectoral policies, incomplete implementation of existing legislation, policy gaps, insufficient funding, limited awareness about biodiversity, inadequacy of the policy framework and governance as well as missing administrative capacity, skills and knowledge gaps.

4.4.3 2020 EU Target and Biodiversity Strategy

Although there was some apparent political ambivalence over the failure to meet the 2010 target, there was at the same time a growing acknowledgement of the importance of conserving biodiversity. This was partly the result of a number of studies that demonstrated the fundamental role of biodiversity in underpinning ecosystem services that are of immense socio-economic value in the EU and globally, including the *Millennium Ecosystem Assessment* (2005) and The Economics of Ecosystems and Biodiversity (TEEB, 2011) initiative.[50] This led to widening concern over the situation and a desire to renew efforts to curb biodiversity losses, which helped to bolster political commitments for action.

In response to EU ambitions, and the global CBD target, in March 2010, the European Council adopted the new target to '*halt biodiversity and ecosystem service loss by 2020, to restore ecosystems in so far as is feasible, and to step up the EU contribution to averting global biodiversity loss*'. This target now explicitly recognises the importance of the services provided by biodiversity in addition to the need to protect biodiversity for its intrinsic value. A longer-term vision was also adopted: '*By 2050, EU biodiversity and the ecosystem services it provides – its natural capital – are protected, valued and appropriately restored for biodiversity's intrinsic value and for their essential contribution to human wellbeing and economic prosperity, and so that catastrophic changes caused by the loss of biodiversity are avoided.*' The vision and target were subsequently endorsed by the European Council on 26 March 2010.

To help achieve the EU 2020 headline target the Commission developed, in cooperation with Member States, *Our life insurance, our natural capital: an EU biodiversity strategy to 2020* (European Commission, 2011c), including sub-targets and related actions to achieve them (as shown in Table 4.5). The seventh EAP, developed in 2013, also underlined the importance of implementing the strategy.

4.4.4 EU Biodiversity Strategy for 2030

Although the overall 2020 biodiversity strategy target was not achieved, and there were limited achievements with respect to the sub-targets and actions (as further discussed in Section 5.2.4), political commitment to tackling the recognised 'biodiversity crisis' has remained relatively strong. In fact, as a result of the increasing recognition (e.g. in the European Green Deal) of the link between the biodiversity crisis, and the climate crisis, and other environmental, socio-economic and health issues affecting society, the Commission adopted a relatively ambitious *EU Biodiversity Strategy for 2030* (European Commission, 2020a). In support of the CBD's overall aim of ensuring that '*by 2050 all of the world's ecosystems are restored, resilient, and adequately protected*', the EU's 2030 target is to '*ensure that **Europe's biodiversity will be on the path to recovery by 2030** for the benefit of people, the planet, the climate and our economy*'. However, whilst this implies that biodiversity losses are at least halted to enable recovery, the intended outcome is not clearly defined.

The strategy has 17 key commitments for 2030, including protecting at least 30% of both the EU's land and seas (with at least 10% under strict protection), introducing legally binding targets for nature restoration, reversing the decline in pollinators, increasing organic farming to 25% coverage, other measures to reduce the impacts of pesticides and fertilisers, ensuring at least 10% of farmland is within high diversity landscape features, restoring 25 000 km of free-flowing rivers and reducing the impacts of fishing. There are also additional supporting measures, including the development of a new governance framework. As the main focus of this book is on the lessons learnt up to now, these 2030 targets and measures are not described in further detail here. Instead, the strategy and key commitments are presented in Box 5.1 and discussed in relation to some of the lessons that can be drawn from the implementation and achievements of the previous EU biodiversity strategies.

[50] Initiated at the G8+5 Potsdam meeting of Environment Ministers in March 2007. See http://teebweb.org/

Table 4.5 The EU 2020 Biodiversity Strategy targets and summarised actions for Targets 1–5.

Target 1: To halt the deterioration in the status of all species and habitats covered by EU nature legislation and achieve a significant and measurable improvement in their status so that, by 2020, compared to current assessments: (i) 100% more habitat assessments and 50% more species assessments under the Habitats Directive show an improved conservation status; and (ii) 50% more species assessments under the Birds Directive show a secure or improved status

Action 1: Complete the establishment of the Natura 2000 network and ensure good management

Action 2: Ensure adequate financing of Natura 2000 sites

Action 3: Increase stakeholder awareness and involvement and improve enforcement

Action 4: Improve and streamline monitoring and reporting

Target 2: By 2020, ecosystems and their services are maintained and enhanced by establishing green infrastructure and restoring at least 15% of degraded ecosystems

Action 5: Improve knowledge of ecosystems and their services in the EU [including the mapping and assessment of ecosystems and their services (MAES) initiative]

Action 6: Set priorities to restore and promote the use of green infrastructure

Action 7: Ensure no net loss of biodiversity and ecosystem services

Target 3

(A) Agriculture: By 2020, maximise areas under agriculture across grasslands, arable land and permanent crops that are covered by biodiversity-related measures under the CAP so as to ensure the conservation of biodiversity and to bring about a measurable improvement in the conservation status of species and habitats that depend on or are affected by agriculture and in the provision of ecosystem services as compared to the EU 2010 Baseline, thus contributing to enhance sustainable management

(B) Forests: By 2020, Forest Management Plans or equivalent instruments, in line with Sustainable Forest Management (SFM), are in place for all forests that are publicly owned and for forest holdings above a certain size (to be defined by the Member States or regions and communicated in their Rural Development Programmes) that receive funding under the EU Rural Development Policy so as to bring about a measurable improvement in the conservation status of species and habitats that depend on or are affected by forestry and in the provision of related ecosystem services as compared to the EU 2010 Baseline

Action 8: Enhance direct payments for environmental public goods in the EU Common Agricultural Policy

Action 9: Better target Rural Development to biodiversity conservation

Action 10: Conserve Europe's agricultural genetic diversity

Action 11: Encourage forest holders to protect and enhance forest biodiversity

Action 12: Integrate biodiversity measures in forest management plans

Target 4: Fisheries: Achieve Maximum Sustainable Yield (MSY) by 2015. Achieve a population age and size distribution indicative of a healthy stock, through fisheries management with no significant adverse impacts on other stocks, species and ecosystems, in support of achieving Good Environmental Status by 2020, as required under the Marine Strategy Framework Directive

Action 13: Improve the management of fished stocks

Action 14: Eliminate adverse impacts on fish stocks, species, habitats and ecosystems

Target 5: By 2020, Invasive Alien Species and their pathways are identified and prioritised, priority species are controlled or eradicated, and pathways are managed to prevent the introduction and establishment of new IAS

Action 15: Strengthen the EU Plant and Animal Health Regimes

Action 16: Establish a dedicated instrument on Invasive Alien Species

Target 6: By 2020, the EU has stepped up its contribution to averting global biodiversity loss

Note. Sub-actions are not included here. Actions for Target 6 are not included as they are for biodiversity outside Europe.

Acknowledgements

We thank the following for their very valuable feedback and suggestions on the draft chapter: Andrew Jackson, Petr Roth, David Stroud and Evelyn Underwood.

References

Alliance Environnement (2007) *Evaluation of the environmental impacts of CAP (Common Agricultural Policy) measures related to the beef and veal sector and the milk sector*. Brussels: Alliance Environnement. Report to DG Agriculture, European Commission.

(2009) *Evaluation of the environmental impact of milk quotas.* Brussels: Alliance Environnement. Report to DG Agriculture, European Commission.

(2017) *Evaluation study of the forestry measures under Rural Development.* Brussels: Alliance Environnement. Report to DG Agriculture, European Commission.

(2019) *Evaluation of the impact of the CAP on habitats, landscapes, biodiversity.* Luxembourg: Publications Office of the European Union. Alliance Environnement Report to DG Agriculture and Rural Development, European Commission.

Alliance Environnement & Thünen-Institut (2017) *Evaluation study of the payment for agricultural practices beneficial for the climate and the environment.* Brussels: Alliance Environnement. Report to DG Agriculture, European Commission.

Anonymous (2021) *Common Fisheries Policy: Mission Not Yet Accomplished.* BirdLife International, ClientEarth, The Fisheries Secretariat, Oceana, Our Fish, Seas at Risk, and WWF. https://europe.oceana.org/sites/default/files/20210615_cfp_mission_not_yet_accomplished_joint_ngo_paper.pdf

Arcadis & IEEP (2010) *Dealing with conflicts in the implementation and management of the Natura 2000 network – strategic planning (lot 2). Guidance document.* Antwerp, Belgium: Arcadis. Report to the European Commission.

Batáry, P., Dicks, L. V., Kleijn, D. & Sutherland, W. J. (2015) The role of agri-environment schemes in conservation and environmental management. *Conservation Biology,* **29** (4), 1006–1016.

Bennett, G. & Mulongoy, K. J. (2006) *Review of experience with ecological networks, corridors and buffer zones.* Montreal: Secretariat of the Convention on Biological Diversity. CBD Technical Series No 23.

Bicknell, A. W. J., Oro, D., Camphuysen, C. J. & Votier, S. C. (2013) Potential consequences of discard reform for seabird communities. *Journal of Applied Ecology,* **50** (3), 649–658.

Bijlsma R. J., Agrillo, E., Attorre, F. *et al.* (2019) *Defining and applying the concept of Favourable Reference Values for species and habitats under the EU Birds and Habitats Directive.* Wageningen, the Netherlands: Wageningen Environmental Research.

BirdLife International (2004) *Birds in the European Union: A Status Assessment.* Wageningen, Netherlands: Birdlife International.

Born, C.-H., Cliquet, A., Schoukens, H., Misonne, D. & Van Hoorick, G. (eds.) (2015) *The Habitats Directive in Its EU Environmental Law Context – European Nature's Best Hope?* London & New York: Routledge.

Brady, M. (2010) *Impact of CAP reform on the environment: some regional results.* Paper to the OECD Workshop on the Disaggregated Impacts of CAP Reform, 10–11 March 2010. Paris: OECD.

Bromley, P. (1997) *Nature Conservation in Europe.* London: Chapman and Hall.

Carboneras, C., Genovesi, P., Vilà, M. *et al.* (2018) A prioritised list of invasive alien species to assist the effective implementation of EU legislation. *Journal of Applied Ecology,* **55**, 539–547.

Carpenter, G. (2018) *Landing the Blame: Overfishing in the Atlantic 2018.* London: New Economics Foundation.

Carpenter, S. R. B., Walker, B. H., Anderies, J. M. & Abel, N. (2001) From metaphor to measurement: resilience of what to what? *Ecosystems,* **4**, 765–781.

Catchpole, R., Smithers, R., Baarda, P. & Eycott, A. (eds.) (2009) *Ecological networks: science and practice.* Proceedings of the 16th Annual IALE (UK) Conference. Edinburgh University, 1–3 September 2009. International Association for Landscape Ecology, UK Region.

Clutten, R. & Tafur, I. (2012) Are imperative reasons imperiling the Habitats Directive? An assessment of Article 6(4) and the IROPI exception. In *The Habitats Directive – A Developer's Obstacle Course?,* ed. G. Jones, pp. 167–182. Oxford: Hart Publishing.

Committee of the Regions (2010) *Opinion of the Committee of the Regions on improving the EIA and SEA Directives.* European Union Committee of the Regions 8th plenary session 14 and 15 April 2010.

Council of Europe (1996) *Pan-European Biological and Landscape Diversity Strategy.* Strasbourg: Council of Europe. Nature and Environment, No. 74.

Donald, P. F., Sanderson, F. J., Burfield, I. J., Bierman, S. M., Gregory, R. D. & Waliczky, Z. (2007) International conservation policy delivers benefits for birds in Europe. *Science,* **317** (5839), 810–813.

ECA (2011) *Is agri-environment support well designed and managed?* Luxembourg: European Court of Auditors. Special Report No 7/2011.

(2014) *Is the ERDF effective in funding projects that directly promote biodiversity under the EU biodiversity strategy to 2020?* Luxembourg: European Court of Auditors. Special Report No 12/2014.

EEA (2007) *Halting the loss of biodiversity by 2010: proposal for a first set of indicators to monitor progress in Europe.* Copenhagen: European Environment Agency.

(2010) *EU 2010 Biodiversity Baseline.* Copenhagen: European Environment Agency. EEA Technical Report No 12/2010.

(2012) *Streamlining European biodiversity indicators 2020: Building a future on lessons learnt from the SEBI 2010 process.* Copenhagen: European Environment Agency. Technical Report No 11/2012.

(2015) *State of nature in the EU: results from reporting under the nature directives 2007-2012.* Copenhagen: European Environment Agency. Technical Report No 2/2015.

ETC/BD (1997) Criteria for assessing national lists of pSCI at biogeographical level. Paris: European Topic Centre on Biodiversity. https://www.eionet.europa.eu/etcs/etc-bd/activities/hab_97_2_criter_en.pdf

(2011) *Assessment, monitoring and reporting under Article 17 of the Habitats Directive: Explanatory Notes & Guidelines for the period 2007–2012.* Paris: European Topic Centre on Biodiversity.

European Commission (1993) *Second report on the application of Directive No 79/409/EEC on the conservation of wild birds.* COM(93) 572.

(1998a) *Communication on a Forestry Strategy for the European Union.* COM(1998) 649.

(1998b) *Communication on a European Biodiversity Strategy.* COM(1998) 42 final.

(2001) *Assessment of plans and projects significantly affecting Natura 2000 sites. Methodological guidance on the provisions of Article 6(3) and (4) of the Habitats Directive 92/43/EEC.* Luxembourg: Office for Official Publications of the European Communities.

(2004) *Financing Natura 2000.* COM(2004)431.

(2006a) *Nature and Biodiversity cases. Rulings of the European Court of Justice.* Brussels: Office for Official Publications of the European Communities.

(2006b) *Halting the loss of biodiversity by 2010 – and beyond – Sustaining ecosystem services for human well-being.* COM(2006) 0216 final.

(2007a) *Guidance document on Article 6(4) of the 'Habitats Directive' 92/43/EEC: Clarification of the concepts of: alternative solutions, imperative reasons of overriding public interest, compensatory measures, overall coherence, opinion of the Commission.*

(2007b) *Assessment of plans and projects significantly affecting Natura 20000 sites. Methodological guidance on the provisions of Article 6(3) and (4) of the Habitats Directive 92/43/EEC.*

(2007c) *An evaluation of Integrated Coastal Zone Management (ICZM) in Europe.* COM(2007) 308 final.

(2008) *Guide to sustainable hunting under the Birds Directive. Council Directive 79/409/EEC on the conservation of wild birds.* Brussels: European Commission.

(2009) *On the application and effectiveness of the EIA Directive (Directive 85/337/EEC, as amended by Directives 97/11/EC and 2003/35/EC).* COM(2009) 378 final.

(2011a) *Financing Natura 2000 – Investing in Natura 2000: delivering benefits for nature and people.* SEC(2011)1573 final.

(2011b) *Links between the Water Framework Directive and Nature Directives: frequently asked questions.* Brussels: European Commission.

(2011c) *Our life insurance, our natural capital: an EU biodiversity strategy to 2020.* COM(2011) 244 final.

(2012a) *Commission note on setting conservation objectives for Natura 2000 sites*, Brussels: European Commission. Final version 23/11/2012.

(2012b) *Links between the Marine Strategy Framework Directive (MSFD 2008/56/EC) and the Nature Directives (Birds Directive 2009/147/EEC (BD) and Habitats Directive 92/43/EEC (HD)).* Brussels: European Commission.

(2012c) *Action Plan for reducing incidental catches of seabirds in fishing gears.* COM(2012) 665 final.

(2013a) *Interpretation Manual of European Union Habitats – EUR28.* Brussels: European Commission.

(2013b) *Commission note on establishing conservation measures for Natura 2000 sites.* Brussels: European Commission.

(2013c) *Sustainable hunting and Natura 2000.* Brussels: European Commission.

(2013d) *Guidelines on climate change and Natura 2000. Dealing with the impact of climate change on the management of the Natura 2000 network of areas of high biodiversity value.* Brussels: European Commission.

(2014a) *Managing Natura 2000 sites – The provisions of Article 6 of the 'Habitats' Directive 92/43/EEC.* Brussels: European Commission.

(2014b) *Key concepts of Article 7(4) of Directive 79/409/EEC period of reproduction and prenuptial migration of Annex II bird species in the 28 EU Member States.* Brussels: European Commission.

(2016) *Fitness Check of the EU Nature Legislation (Birds and Habitats Directives).* SWD(2016) 472 final.

(2017a) *An Action Plan for nature, people and the economy.* COM (2017) 198 final.

(2017b) *Factsheets providing details of actions in the Action Plan for nature, people and the economy.* SWD(2017) 139 final.

(2018) *Managing Natura 2000 sites: the provisions of Article 6 of the 'Habitats' Directive 92/43/EEC.* C(2018) 7621 final.

(2019a) *The European Green Deal.* COM(2019) 640 final.

(2019b) *EVALUATION of the Directive 2001/42/EC on the assessment of the effects of certain plans and programmes on the environment.* SWD(2019) 413 final.

(2020a) *EU Biodiversity Strategy for 2030 – Bringing back nature into our lives.* COM(2020) 380 final.

(2020b) *A Farm to Fork Strategy – for a fair, healthy and environmentally-friendly food system.* COM(2020) 381 final.

(2020c) *A new Circular Economy Action Plan – For a cleaner and more competitive Europe.* COM (2020) 98 final.

(2021a) *'Fit for 55': delivering the EU's 2030 Climate Target on the way to climate neutrality.* COM(2021) 550 final.

(2021b) *Guidance document on the strict protection of animal species of Community interest under the Habitats Directive.* COM (2021) 7301 final.

(2021c) *Commission notice – Assessment of plans and projects in relation to Natura 2000 sites – Methodological guidance on Article 6 (3) and (4) of the Habitats Directive 92/43/EEC.* C(2021) 6913 final.

(2021d) *New EU Forest Strategy for 2030.* COM(2021) 572 final.

(2021e) *Guidelines providing a common understanding of the term 'environmental damage' as defined in Article 2 of Directive 2004/35/EC of the European Parliament and of the Council on environmental liability with regard to the prevention and remedying of environmental damage.* Commission Notice C/2021/1860.

(2021f) *Towards more sustainable fishing in the EU: state of play and orientations for 2022.* COM(2021) 279 final.

(2021g) *Implementation of the Technical Measures Regulation (Article 31 of Regulation (EU) 2019/1241).* COM(2021) 583 final.

Evans, D. (2012) Building the European Union's Natura 2000 network. *Nature Conservation*, **1**, 11–26.

Farmer, A. (ed.) (2012) *Manual of European Environmental Policy.* 2nd ed. London: Taylor and Francis. Available online at https://ieep.eu/understanding-the-eu/manual-of-european-environmental-policy/

Fernández, S. A. (2003) Spanish coordination in the European Union: the case of the habitats directive. *Administration & Society*, **34**, 678–699.

Fisher, I., Ashpole, J., Scallan, D., Proud, T. & Carboneras, C. (compilers) (2018) *International Single Species Action Plan for the conservation of the European Turtle-dove Streptopelia turtur (2018 to 2028).* Brussels: European Commission.

Galbraith, J. (2013) Treaty options: towards a behavioral understanding of treaty design. *Virginia Journal of International Law*, **53** (2), 309–363.

Gantioler, S., Rayment, M., Bassi, S. *et al.* (2010) *Costs and Socio-Economic Benefits associated with the Natura 2000 Network.* Brussels: Institute for European Environmental Policy/GHK/Ecologic. Report to the European Commission.

Grimmett, R. F. A. & Jones, T. A. (1989) *Important Bird Areas in Europe.* Cambridge, UK: International Council for Bird Preservation.

Haigh, N. (2016) *EU Environmental Policy; Its Journey to Centre Stage.* Abingdon, UK: Routledge.

Hart, K., Rayment, M. & Lee, H. (2010) *Paper 1: Achieving a Transition Away from CAP Direct Payments.* London: Institute for European Environmental Policy. Paper prepared by IEEP for the Land Use Policy Group.

Heath, M. F. & Evans, M. I. (eds.) (2000) *Important Bird Areas in Europe: Priority Sites for Conservation.* Updated (2011) ed. Cambridge, UK: BirdLife International.

Hovardas, T. & Marsden, K. (2018) Good practice in large carnivore conservation and management: Insights from the EU Platform on Coexistence between People and Large Carnivores. In *Large Carnivore Conservation and Management*, ed. T. Hovardas, pp. 314–337. Routledge.

ICES (2017) *Report of the OSPAR/HELCOM/ICES Working Group on Marine Birds (JWGBIRD), 10–14 October 2016, Thetford, UK.* ICES CM 2016/ACOM:29.

IEEP (2008) *The environmental benefits of set-aside in the EU, a summary of evidence*. London: Institute for European Environmental Policy. Report for the UK Department for the Environment, Food and Rural Affairs.

(2016) *The potential policy and environmental consequences for the UK of a departure from the European Union*. London: Institute for European Environmental Policy. Report in collaboration with The Wildlife Trusts, RSPB and WWF.

Jackson, A. L. R. (2018) *Conserving Europe's Wildlife: Law and Policy of the Natura 2000 Network of Protected Areas*. 1st ed. Earthscan/Routledge.

Jongman, R. H. G. & Pungetti, G. (eds.) (2004) *Ecological Networks and Greenways. Concept, Design, Implementation*. Cambridge, UK: Cambridge University Press.

Keenleyside, C., Beaufoy, G., Tucker, G. & Jones, G. (2014) *High Nature Value farming throughout EU-27 and Its financial support under the CAP*. London: Institute for European Environmental Policy. Report for DG Environment.

Kettunen, M., Baldock, D., Gantioler, S. *et al.* (2011) *Assessment of the Natura 2000 co-financing arrangements of the EU financing instrument*. Brussels: Institute for European Environmental Policy. Report for the European Commission.

Kettunen, M., Illes, A., Rayment, M. *et al.* (2017) *Integration approach to financing of biodiversity: evaluation of results and analysis of options for the future*. Brussels/London: Institute for European Environmental Policy. Report for the European Commission.

Kettunen, M., Terry, A., Tucker, G. M. & Jones, A. (2007) *Guidance on the maintenance of landscape connectivity features of major importance for wild flora and fauna. Guidance on the implementation of Article 3 of the Birds Directive (79/409/EEC) and Article 10 of the Habitats Directive (92/43/EEC)*. Brussels/London: Institute for European Environmental Policy. Report to the European Commission.

Krämer, L. (2009) The European Commission's Opinions under Article 6(4) of the Habitats Directive. *Journal of Environmental Law*, **21** (1), 59–85.

Maes, J., Teller, A., Erhard, M. *et al.* (2020) *Mapping and Assessment of Ecosystems and their Services: An EU Ecosystem Assessment*. Ispra, Italy: Publications Office of the European Union.

Maynou, F., del Mar Gill, M, Vitale, S. *et al.* (2018) Fishers' perception of the European Union discards ban: perspective from south European fisheries. *Marine Policy*, **89**, 147–153.

McConville, A. J. & Tucker, G. M. (2015) *Review of Favourable Conservation Status and Birds Directive Article 2 interpretation within the European Union*. Peterborough, UK: Natural England. Natural England commissioned Report 176.

Medarova-Bergstrom, K., Kettunen, M., Illes, A., Baldock, D., Rayment, M. & Hart, K. (2015) *Tracking biodiversity expenditure in the EU Budget: Part I – Guidance on definition and criteria for biodiversity expenditure in the EU budget*. London/Brussels: Institute for European Environmental Policy. Report for the European Commission.

Milieu & IUCN (2014) *Experience gained in the application of ELD biodiversity damage*. Brussels: Milieu. Final report for the European Commission, DG Environment.

Milieu, IEEP & ICF (2016) *Evaluation study to support the fitness check of the Birds and Habitats Directives*. Brussels: Milieu Ltd, Institute for European Environmental Policy and the ICF International. Report for the European Commission.

Millennium Ecosystem Assessment (2005) *Ecosystems and Human Well-Being: Synthesis*. Washington, DC: Island Press.

Morris, R. K. A., Alonso, I., Jefferson, R. G. & Kirby, K. J. (2006) The creation of compensatory habitat – can it secure sustainable development? *Journal for Nature Conservation*, **14**, 106–116.

Moss, D. & Wyatt, B. K. (1994) The CORINE biotopes project: a database for conservation of nature and wildlife in the European community. *Applied Geography*, **14** (4), 327–349.

Pritchard, D. E. (2007) Towards coherence in site networks. In *Waterbirds Around the World*, eds. G. C. Boere, C. A. Galbraith, & D. A. Stroud, pp. 673–674. Edinburgh: The Stationery Office.

Sanderson, F. J., Pople, R. G., Ieronymidou, C. *et al.* (2015) Assessing the performance of EU nature legislation in protecting target bird species in an era of climate change. *Conservation Letters*, **9** (3), 172–180.

Scalera, R., Rabitsch, W. & Underwood, E. (2020) *FAQs on how the Invasive Alien Species Regulation interacts with the EU Nature Directives*. Brussels: European Commission.

Schmid, E., Sinabell, F. & Hofreither, M. F. (2010) *Phasing out of environmentally harmful subsidies: consequences of the 2003 CAP Reform in Austria*. Paper to the OECD Workshop on the disaggregated impacts of CAP reform, 10–11 March 2010. Paris, OECD.

Stevens, C. J., Bell, J. N. B., Brimblecombe, P. *et al.* (2020) The impact of air pollution on terrestrial managed and natural vegetation. *Philosophical Transactions of the Royal Society*, **378** (2183).

Stroud, D. A., Pienkowski, M. W. & Mudge, G. P. (1990) *Protecting Internationally Important Bird Sites: A Review Of the Network of EC Special Protection Areas in Great Britain*. Peterborough, UK: Nature Conservancy Council.

Sundseth, K. & Roth, P. (2013) *Study on Evaluating and Improving the Article 6.3 Permit Procedure for Natura 2000 Sites*. Brussels: Ecosystems Ltd.

(2014) *Article 6 of the Habitats Directive. Rulings of the European Court of Justice*. Brussels: European Commission.

TEEB (2011) *The Economics of Ecosystems and Biodiversity in National and International Policy Making*. London and Washington: Earthscan.

ten Brink, P., Badura, T., Bassi, S. *et al.* (2011) *Estimating the overall economic value of the benefits provided by the Natura 2000 network*. Brussels: Institute for European Environmental Policy/GHK/Ecologic. Report to the European Commission.

The Pew Charitable Trusts (2021) *Lessons From Implementation of the EU's Common Fisheries Policy: Mixed record highlights steps still needed to turn the tide towards better management*. The Pew Charitable Trusts.

Tucker, G. M. & Heath, M. F. (1994) *Birds in Europe: Their Conservation Status*. Cambridge, UK: BirdLife International.

UNEP-WCMC (2008) *National and regional networks of Marine Protected Areas: a review of progress*. Cambridge, UK: UNEP-WCMC.

Wilkinson, D. (1992) Maastricht and the environment: the implications for the EC's environmental policy on the Treaty on European Union. *Journal of Environmental Law*, **4** (2), 211–239.

WWF-UK (2017) *Remote Electronic Monitoring: why camera technology is a cost-effective and robust solution to improving UK fisheries management*. Woking, UK: WWF-UK.

Chapter 5

Conclusions, Lessons Learnt and Implications for the Future

Graham Tucker

The main purpose of this chapter is to provide an integrated assessment of what has worked in nature conservation in Europe, especially in relation to the key challenges identified in Chapter 2 and the implementation of policies and measures described in Chapters 3 and 4. It draws on wide-ranging evidence, but is primarily based on the 26 in-depth country chapters that follow this (Chapters 6–31).[1] It therefore focusses on the EU's nature conservation policies and actions and their application in all the countries covered in this book. It is difficult to draw conclusions on the experience of other European countries, because comparable information on the impacts of conservation measures is scarce, but some observations are made where evidence allows.

5.1 The Status of Nature: What Has Been Achieved

5.1.1 The Current Status of Nature and Recent Trends

To help draw conclusions, this chapter firstly provides an overview of the most recent EU-level assessment of the status of Birds and Habitats Directives (BHD) habitats and species, according to the results of the EU-28 Member States' monitoring for 2013–2018 under BD Article 12 and HD Article 17.[2] Unless otherwise stated, this information is drawn from the latest *State of Nature in the EU* report (EEA, 2020).[3] As shown in Figure 5.1, the overall picture is that few BHD habitats and species had a good status, that is, favourable for HD habitats and species, or secure for BD birds (as defined in Table 1.2). The situation is particularly serious for HD habitats, as merely 15% had a good conservation status. Although the condition of HD species is better, only just over a quarter had a good status. Furthermore, 36% of HD habitats with an unfavourable or unknown status, and 35% of HD species with an unfavourable or unknown status had deteriorating trends over the short term (i.e. over 12 years, 2007–2018). Most HD habitats cannot be easily re-created, so it is especially concerning that 30% of habitat assessments reported a decline in habitat area.

Unlike the HD Article 17 species assessments, which only cover the HD Annex II, IV and V listed species, Member States are required under BD Article 12 to report on all wild bird species, thus providing a complete picture of their status and a more representative biodiversity indicator. Across all 463 bird species that naturally occur in the EU, the EU population status of 47% was considered to be good. However, 20% had a poor status (i.e. Near Threatened, declining or depleted) and 19% a bad status (i.e. threatened). Short-term EU-level trends over 2007–2018 across all breeding species reveal that most had stable, fluctuating or increasing populations. Nevertheless, 30% decreased (with 17% uncertain/unknown trends), whilst 29% of wintering populations declined (9% uncertain/unknown).

Although the percentage of birds considered to have a bad status is based on the previous Red List of birds (BirdLife International, 2015), the more recent update also found that 19% are threatened at an EU level (BirdLife International, 2021). In comparison, across Europe as a whole, nearly 13% of all 544 assessed bird species are considered to be threatened.

Whilst the proportion of BD Annex I birds that had a good status (40%) was higher than for HD species (27%), comparisons should be made with caution as the assessments are not based on the same types of criteria. Even though BD Annex I species are subject to special conservation measures, 44% had a poor or bad status, and 27% of their breeding populations had shown short-term declines. The *State of Nature* report does not indicate the trends for wintering Annex I birds.

More detailed analysis indicates that all habitat types and species groups are a cause for concern, though there are significant variations. Comparisons of broad HD habitat types reveal that coastal habitats had the lowest proportion with a favourable status (c. 7%[4]). Similarly a very small proportion of species within the marine biogeographical regions had a favourable status, although the status of many was also unknown. The ecological status assessments of freshwater and marine habitats are further discussed in Section 5.2 in relation to the WFD and MSFD.

[1] Which cover the current EU-27 other than Luxembourg and Malta, as well as the UK.

[2] See Section 1.3 for explanation and data sources.

[3] These EU-level results include habitats and species in the Macaronesian biogeographical region.

[4] Estimated from graph as no figure given.

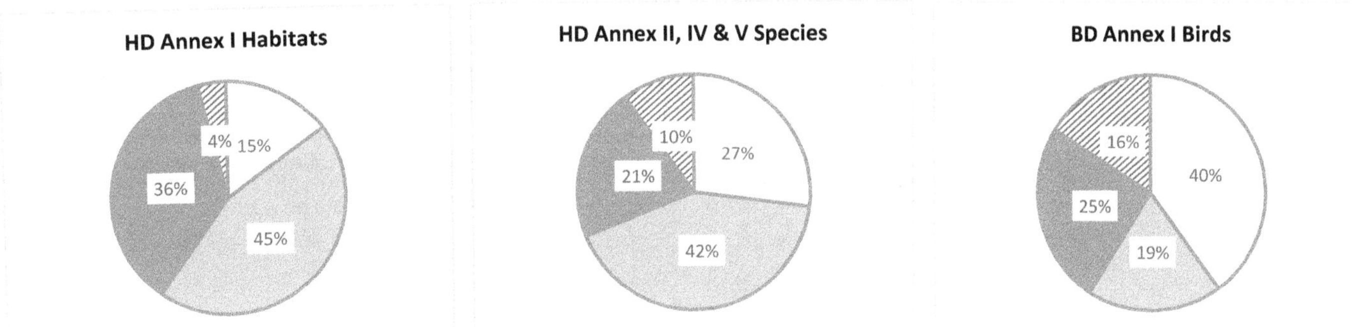

Figure 5.1 The EU-level conservation status of BD Annex I birds, HD habitats and HD species over 2013–2018.
Key. Status: White = good; light grey = poor; dark grey = bad; hatched = unknown.
Notes. Includes habitat and species assessments across the whole EU (i.e. including Canaries, Azores and Madeira).
Source. Based on data from EEA (2020).

Whilst the achievement of favourable conservation status of HD habitats and species is the ultimate aim of the Habitats Directive, the percentage of habitats and species achieving this does not provide a clear indication of current conservation performance and priorities. This is because achieving favourable status has the rather ambitious aim of restoring to some extent former ranges and depleted populations to favourable reference values (Box 4.1). The assessments of conservation status also consider whether or not the habitat or species has favourable future prospects. Therefore some habitats and species may have an unfavourable status because their range and/or population have not recovered to former levels, despite the fact that their range and populations are currently stable or increasing. Furthermore, for some it may be practically impossible to recover sufficiently to ever achieve a favourable conservation status.

Therefore, to complement the assessments of conservation status, this book has also used the percentage of HD habitats and species that have an unfavourable status and are declining as performance indicators (as discussed in Section 1.3.3). This metric provides an indication of more immediate conservation needs. Table 5.1 presents this indicator for HD habitats and species, together with the percentage decreasing for BD Annex I birds for each Member State (as included in the summary table for each chapter). The averages across the countries are rather similar for each group, such that about one-fifth of assessments of BHD habitats and species indicate a particular need for further and/or more effective conservation measures to at least halt the declines. The results in relation to each Member State are discussed in the next section.

5.1.2 The Status of BHD Habitats and Species at an EU Member State Level

An important aim of this book has been to assess the relative performance of each country in terms of its nature conservation outcomes, and the data in Table 5.1 provide one indicator of this. Figure 5.2 shows the countries ordered in relation to the average of the indicator across the three groups. Needless to say, this gives a rather crude indication of performance, and it should be borne in mind that the countries differ in the pressures they are addressing, and the practical methods used to assess the conservation status of habitats and species. Indeed, there is evidence of political interference in assessments in some countries, most obviously in Bulgaria (discussed in Chapter 8). Also, the estimates may be less reliable for those countries without well-developed monitoring systems and coverage, and those where a high proportion of trends are unknown or uncertain (as highlighted in Table 5.1). Notwithstanding these caveats, some patterns are evident and they show many similarities to the Member State level HD habitat and species conservation status and bird trends presented in the *State of Nature* report.

There seems to be a tendency for countries in eastern and south-eastern Europe to have fared better than others, and this appears most obviously the case with respect to Croatia, Cyprus, Malta, Bulgaria and Romania. However, although the results for these countries are presented as reported, they should be interpreted carefully. The low proportion of declining trends does seem unlikely, especially for Croatia and Cyprus, given the pressures described in the country chapters, and the countries have high proportions with unknown trends. In Romania the trends of many BD Annex I birds are unknown, and the data for Bulgaria are considered unreliable, as mentioned above. The results for Malta probably reflect the relatively limited number of habitats and species involved, and their localised distributions, which may make it easier to target conservation measures and prevent further declines.

Patterns are not so obvious with regards to countries with high proportions of habitats and species that are declining, with Germany and Portugal being only slightly worse than others. Table 5.1 reveals that the countries' relative performances seem to be mainly consistent across HD habitats and species, although the situation for habitats and species differs for some (most obviously Portugal). There appear to be some notable differences between HD habitats and species and BD Annex I birds, the Baltic States, Finland and Slovakia having relatively high proportions of declining bird species.

Table 5.1 The percentage of 2013–2018 assessments of HD habitats and species that revealed an unfavourable conservation status and short-term declines, a the percentage of BD Annex I taxa that had short-term declines.

EU-28 Member State		Unfavourable status and short-term declines		BD Annex I breeding birds – short-term declines (%)	Average (%)	Comments
		HD Annex I habitats (%)	HD Annex II, IV and V species (%)			
Austria	AT	14	28	17	20	59%? HD spp.
Belgium	BE	11	15	20	15	
Bulgaria	BG	1	6	20	9	Unreliable HD data – see text
Croatia	HR	10	3	11	8	84%? HD spp. 62%? birds
Cyprus	CY	14	0	10	8	62%? HD spp.
Czech Republic	CZ	32	27	9	23	
Denmark	DK	33	21	12	22	
Estonia	EE	12	9	36	19	
Finland	FI	26	14	30	23	
France	FR	32	23	20	25	
Germany	DE	41	35	20	32	
Greece	GR	4	20	18	14	
Hungary	HU	49	13	19	27	
Ireland	IE	46	15	18	26	
Italy	IT	38	33	17	29	
Latvia	LV	21	13	32	22	
Lithuania	LT	7	24	29	20	82%? HD habitats
Luxembourg	LU	32	27	21	27	
Malta	MT	3	7	14	8	
Netherlands	NL	25	25	28	26	
Poland	PL	23	14	21	19	
Portugal	PT	64	15	17	32	Exc. Macaronesia. 65%? birds
Romania	RO	7	15	6	9	72%? birds
Slovakia	SK	10	25	35	24	
Slovenia	SI	31	24	18	25	
Spain	ES	17	27	28	24	Exc. Macaronesia
Sweden	SE	35	25	22	28	
United Kingdom	UK	23	17	27	23	
Average		**24**	**19**	**21**	**21**	

Note. '?' indicates % uncertain/unknown trend if more than 50%. Austria: excludes 3 N/A for HD species.
Source. Derived from Member State BD Article 12 and HD Article 17 monitoring data for 2013–2018 as presented on the EEA dashboard.

5.1.3 Have Conservation Measures Led to Relative Improvements?

Of particular relevance to this conclusions chapter is whether or not the conservation status, or trends, of targeted habitats and species have changed in response to the nature conservation measures in place, especially those required under the Nature Directives. This is very difficult to test because it requires long-term, comparable data on population trends from many countries. However, such data have been compiled by BirdLife International for birds (Tucker and Heath, 1994; BirdLife International, 2004) and an analysis of them in 2007 provided

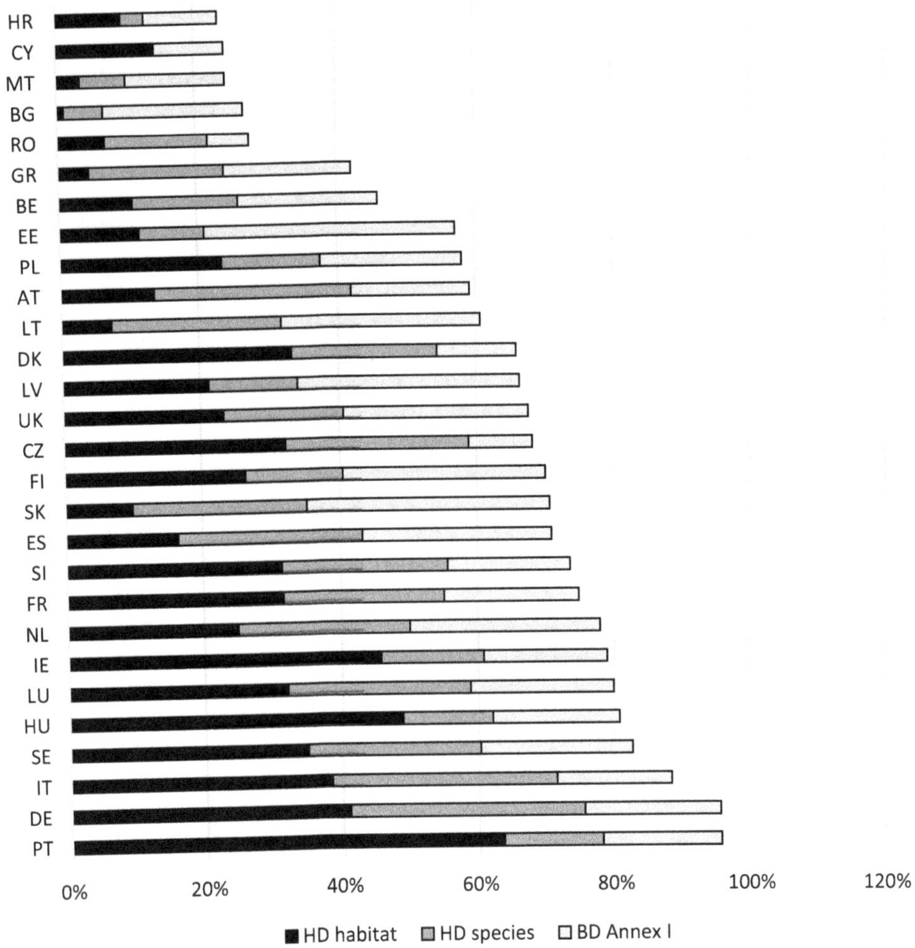

Figure 5.2 For each EU Member State and the UK, over the 2013–2018 reporting period, of the percentage HD habitats and species that had an unfavourable conservation status and declining trends, and the percentage of BD Annex I birds that had short-term declines.
Notes. HD habitat and species results for Bulgaria and HD species (0% unfavourable status) for Cyprus are considered to be unreliable.
Source. Based on Member State BD Article 12 and HD Article 17 monitoring results for 2013–2018 as presented on the EEA national summary dashboard (February 2021).

strong evidence that the Nature Directives were having significant positive effects on bird populations in the EU-15 (Donald *et al.*, 2007). This was revealed by four key findings. Firstly, BD Annex I bird species had fewer positive trends than non-Annex I species over 1970–1990, but the situation reversed over 1990–2000, revealing positive impacts once conservation measures were taken. Secondly, the improvement in Annex I species trends over 1990–2000 was greater inside the EU-15 than outside. Thirdly, the improvement in trends in Annex I species was most apparent for species that had been listed for longest (which also indicated an intervention–response time lag of at least 10 years). Lastly, for all birds there were positive associations between population trends and the proportion of land designated as SPAs, but particularly for Annex I species.

A repeat of the analysis using bird trend data up to 2012 confirmed the key findings of the 2007 study (Sanderson *et al.*, 2015). This was despite evidence of high pressures on long-distance migrant species and the increasing effects of climate change. The positive benefits of Annex I listing were found to be significantly higher in countries that had been in the EU for the longest time. A European continent-wide study of bird trends in 33 countries confirmed the benefits of being listed in Annex I when in the EU; benefits were also highest in the older Member States, but not significantly so (Koschová *et al.*, 2018). A broader study of the effects of general protection of birds in eastern Europe, including outside the EU, found that deep and narrow protection (i.e. focussed on a few species) contributed to population increases, whereas broader and shallow protection seemed to maintain population levels (Koleček *et al.*, 2014).

A key question now is whether, over a now greater period of time, the effects of the Nature Directives, and other legislation, have not just slowed or halted declines, but led to improvements in conservation status. On the face of it, one could compare the most recent status assessments and trends with those from previous reporting periods, but unfortunately this is not straightforward for two main reasons. Firstly, the conservation status of HD habitats and species is a rather insensitive indicator of short-term change, for the reasons mentioned earlier. Secondly, many of the changes are due to better information or changes in assessment methods. The *State of Nature* report provides some analysis of the reported reasons for changes, for example noting that the apparent 6% increase in the proportion of habitats showing a bad status, to 36% over 2012–2018, is mostly due to changes in methods or data quality. The same reason applies to the relatively small overall changes in HD species status.

Generally, the bird status and trend data provide a more sensitive, reliable and complete indicator of changes with time, although there are significant data gaps for some countries (Table 5.1). Of particular note in the *State of Nature* report is that, compared to the previous reporting period, the proportion of Annex I taxa with a good status decreased by 8%, while those with a poor and bad status increased by 6%, which the EEA considers to be at least partly due to genuine changes.

Overall, the scientific studies and BHD monitoring results reveal that EU nature conservation measures are having positive impacts: they are slowing or halting declines in some habitats and species, and to a lesser extent leading to recoveries. However, the Nature Directives are struggling to achieve their overall longer-term objectives. Such conclusions were drawn by the European Commission in the 2016 Fitness Check of the Nature Directives (European Commission, 2016; Milieu *et al.*, 2016). As part of broader EU biodiversity policy, the Nature Directives were considered to be effective where properly implemented, but much more needed to be done to scale up and improve their implementation. These more recent monitoring results confirm that even the short-term objectives of halting declines are not being reached for many habitats and species, let alone the longer-term aims of restoring habitats and depleted species populations.

Table 5.2 Summary of past trends and outlook for selected state and pressure indicators in the EEA-39.

	Past trends (10–15 years)	Outlook to 2030
Common species (birds and butterflies)	↘	↘
Ecosystem condition and services	↘	↕
Water ecosystems and wetlands	↕	↕
Hydromorphological pressures	↘	↕
State of marine ecosystems and biodiversity	↕	↘
Pressures and impacts on marine ecosystems	↕	↘
Urbanisation and land use by agriculture and forestry	↘	↘
Soil condition	↘	↘
Air pollution and impacts on ecosystems	↕	↕
Chemical pollution and impacts on ecosystems	↕	↘
Climate change and impacts on ecosystems	↘	↘

Key. ↕ = Trends/developments show a mixed picture. ↘ = Deteriorating trends/developments dominate (i.e. increasing pressures).
Source. Based on selected indicators from EEA (2019a).

5.1.4 Pressures on Habitats and Species

Evidence from this book's country chapters, the *State of Nature in the EU* report, and the *EU Ecosystem Assessment* (Maes *et al.*, 2020) all indicate that most of the drivers and pressures described in Chapter 2 have remained to some extent, or even intensified. This is not just the case in the EU, but over Europe more widely (IPBES, 2018; EEA, 2019a). Furthermore, some trends are expected to worsen over this current decade (Table 5.2).

Of particular concern are the ongoing impacts of agricultural and forestry production improvements: specialisation and intensification on the one hand, and abandonment of High Nature Value (HNV) systems on the other. Numerous national chapters in this book have also documented the increased ploughing, reseeding and fertilisation of semi-natural grasslands, and further industrialisation of livestock and arable farming. Semi-improved pastures in mixed farmland have been widely replaced by temporary grasslands or other forage crops for livestock that are kept in stock yards. Fallow land, which is especially important for some threatened farmland birds, has declined by 38% since 2010 (Maes *et al.*, 2020). Intensive farming also continues to be the main source of nutrient pollution affecting terrestrial habitats, freshwaters and the marine environment. Consequently, in their Article 12 and 17 reports, agriculture was the most frequently reported pressure by Member States, accounting for 21% of all reported pressures on BHD habitats and species over 2013–2018. EU land cover data in the *State of Nature* report have revealed that such pressures also apply to Natura 2000 sites, as the most frequent land cover change (18%) between 2012 and 2018 was the conversion of pastures and mosaic farmland to arable land and permanent crops. Forestry was the third most frequent pressure (affecting 11%), primarily due to intensive forest management practices, such as clear felling.

In addition to changes in land management, pressures also result from habitat loss to developments: urbanisation alone is the second most frequently listed pressure on habitats and species (13%). This, together with increasing roads, tourist developments and other built developments, is continuing to lead to habitat loss, such as of coastal wetlands, heathlands and scrublands. Whilst the overall footprint of built developments is relatively small, some habitats are greatly affected due to concentrations along coastlines and river valleys. More widely, urbanisation is resulting in further habitat fragmentation (EEA, 2011). Recreational pressures in coastal areas are increasingly high, thereby threatening sensitive habitats and species such as beach-nesting birds and turtles.

Other ongoing and significant pressures include increasing invasive alien species (IAS), over-exploitation of fish and other marine species (and the related habitat damage and by-catch) as well as hunted birds, illegal killing (e.g. predators), and freshwater and marine pollution from nutrients and toxins. Perhaps of greatest concern is the ongoing deposition of airborne nitrogen. Although emission levels have decreased over recent decades, nitrogen deposition exceeded critical loads over 65% of the area of assessed ecosystems across Europe in 2018 (Fagerli *et al.*, 2020). Over recent decades an increasing pressure on rivers has been the growth in hydropower, and especially the cumulative effects of micro-hydro schemes on river hydrology and migratory fish.

Most of these pressures are being exacerbated by the impacts of climate change. These are becoming more evident and are undoubtedly going to lead to profound changes in most ecosystems and many species over the rest of the century.

5.1.5 The Current Situation in the Context of Past Changes

The deteriorating status of nature, and ongoing pressures, might give the impression that the EU nature conservation effort has largely failed. However, the undoubtedly serious situation must be put into the context of the situation 40 or 50 years ago. At that point habitat destruction was perhaps at its most rampant with wetland drainage, agricultural expansion, and the loss of semi-natural grasslands and other habitats being at a much greater scale than now, as summarised in Chapter 2. Despite the trajectory at the time, Europe has managed to retain a relatively rich variety of habitats and associated species, including traditional 'cultural landscapes' of HNV farmland and forests. Whilst many such habitats have continued to decline, nature conservation actions appear to have at least slowed their rate of loss, although it is difficult to quantify their contribution. To some extent, reduced rates of loss of some habitats have been the result of diminishing opportunities for conversion and increasing costs as the most suitable sites were targeted some time ago. Social and economic changes have also been beneficial for some habitats and species, in particular agricultural abandonment resulting in increases in forest and scrub. Rural depopulation may have also led to reduced disturbance and hunting pressures in some areas.

As a result of favourable land-use change and some of the most effective nature conservation measures there have been successes, including for some of Europe's most iconic and treasured species of large mammals, waterbirds and raptors, as documented in Deinet *et al.* (2013). These successes have tended to be for scarce or localised species that are easier to target, including for example the Iberian Lynx (*Lynx pardinus*), which was one of the most threatened cats in the world and Critically Endangered. As described in Chapter 29 on Spain, due to concerted efforts its range and population have increased sufficiently for its threat status to be downgraded, although it remains Endangered. European and EU-wide increases in threatened species populations have been relatively rare, as to achieve such large-scale impacts clearly requires huge levels of coordinated action, funding, stakeholder engagement and societal support, as well as consistently favourable socio-economic and land-use trends. Nevertheless, notable examples have included the Brown Bear (*Ursus arctos*), Wolf (*Canis lupus*) and several species of raptor.

There have also been many smaller-scale improvements in conservation status or trends, at national or subnational biogeographical regional scales, some of which are described in a European Commission study (Tucker *et al.*, 2019) and the national chapters of this book. The 2013–2018 monitoring results revealed that improvements in the status or trends of HD habitats and species[5] were reported for 6% of assessments on average for both habitats and species. More improvements occurred in forest and freshwater habitats (10% of assessments in each), and for mammals and fish (8% of assessments in each group).

The remaining sections of this chapter consider the factors that have led to the successes in nature conservation as well as the main barriers to progress. To be consistent with the main sections in the national chapters, the discussion focusses on legislation and policies, species measures, site protection and management, general measures for the wider environment, funding, and knowledge and monitoring. Lastly, it considers the key lessons that can be drawn from the experiences documented in this book and their implications for future.

5.2 EU Nature Conservation Legislation and Policies

A number of international conventions, most notably the Bern Convention and more recently the CBD, have been important drivers of nature conservation actions in Europe, and have influenced the development of legislation and policies at the EU and country level. Conclusions on their actions and effects are not drawn again here as they are summarised in Chapter 2 and described more fully in online Annex 1. Given their fundamental role in EU nature conservation, the focus of this section is on the Nature Directives. However, these are not measures in isolation; achievement of their objectives, and the wider biodiversity aims of the EU, is highly dependent on support from other EU legal instruments and policies (see Section 4.3). It is not possible to review them all here, but the influence of the WFD and MSFD is particularly important and they are therefore considered below. Others, including the SEA and EIA Directives and Common Agricultural Policy (CAP) are discussed in later relevant sections.

5.2.1 The EU Nature Directives

Strengths and Weaknesses of the Legislation

The 1979 Birds Directive and 1992 Habitats Directive have been the principal drivers of nature conservation in the EU. Within this framework, the Natura 2000 network has provided

[5] These are habitats/species with: genuine improvements in status; or an unfavourable status and improving trends (including genuine and other than genuine changes); or an unfavourable status and a genuine stabilised trend which had been declining in 2013.

the cornerstone for achieving the favourable conservation status of HD habitats and species, and the similar conservation objectives for birds. The Directives have led to changes in approach and a substantial scaling up of action in all EU Member States, as well as significantly influencing other European countries (e.g. the Council of Europe's Emerald network, and conservation status monitoring).

Evidence from this book and a number of reviews by legal experts and conservationists (e.g. Crofts, 2014; Born *et al.*, 2015; Langhout and Brunner, 2017; Campagnaro *et al.*, 2019), and also importantly from the 2016 Fitness Check of the Nature Directives (described later), has revealed that the Nature Directives' approaches and provisions have a number of inherent strengths and weaknesses, which have greatly affected their implementation and subsequent outcomes. Perhaps their most fundamental strength is that they are outcome focussed. Whilst they set out a broad framework that includes some required actions, such as the designation of protected areas, their ultimate objectives are defined in ecological terms. They give Member States a reasonable degree of freedom in terms of how they achieve their objectives. This has helped countries to design and implement conservation measures according to their context and particular needs, although this has also turned out to be a weakness in some cases (as described later). This results-focussed framework, and the setting of objectives in ecological terms, has since been applied to the WFD and MSFD.

In defining its objectives in ecological terms, the Habitats Directive incorporated the concept of 'favourable conservation status' (Box 4.1), which had been developed under the Convention on the Conservation of Migratory Species of Wild Animals (CMS) and partly applied in the Birds Directive. Importantly, the aim of achieving favourable conservation status goes beyond that of preventing species extinctions, and instead seeks to ensure that the targeted habitats and species are thriving and have good prospects for the future. This is an ambitious aim, which has had far-reaching implications. The Habitats Directive also further developed the concept of favourable conservation status by defining it legally and applying it to habitat types as well as species. This has enabled and led to the conservation of habitats that are unique and ecologically important in their own right, as well as often being of cultural interest (e.g. hay meadows, heathlands and pastoral woodlands). Furthermore, it has indirectly contributed to the conservation of many species, including common characteristic species of the habitats, without the need for creating excessively long species lists.

The fact that the Nature Directives provide a consistent pan-EU approach for the terrestrial and marine environment, which focusses on habitats and species of EU-level conservation importance, has itself been a major strength. It has enabled the development of a common framework for reporting on their conservation status and facilitated international collaboration and sharing of resources and knowledge. However, the assessment of favourable conservation status and the identification of required Natura 2000 sites are according to defined biogeographical regions, which are not tied to national boundaries (see Figures 2.1 and 2.2). This has helped to regionalise the approach, leading to multinational cooperation, although it has also increased its complexity.

Whilst Member States have some degree of freedom over how they achieve the objectives of the Directives, two types of measures must be taken: the general protection of certain species wherever they occur, and the protection and conservation of certain habitats and species within Natura 2000 sites – to the degree necessary for them to achieve favourable conservation status. Both these complementary measures have proven to be vital and effective when properly implemented. The development of the Natura 2000 network has been particularly innovative and important. Its strengths include its aim to be a coherent network, which is identified on the basis of habitats and species of EU importance, using a biogeographical approach and purely on a scientific and ecological basis. This has resulted in a representative network of sites that has extended national protected areas (as further discussed in Section 5.4).

Most importantly, the Nature Directives provide strong legal science-based protection of species and Natura 2000 sites, together with requirements for management measures and, if necessary, restoration, to provide the necessary ecological requirements for the habitats and species concerned. This is perhaps something that is taken for granted now in the EU, although according to Campagnaro *et al.* (2019), such a robust international legal system is 'the exception rather than the rule for international conservation governance'.

Whilst the Nature Directives have undoubtedly been ground-breaking and provide a largely coherent, legally and scientifically robust framework, they have a number of weaknesses that have contributed mainly to implementation problems. Perhaps the most fundamental flaw is that they lack deadlines for the achievement of their objectives. This undoubtedly contributed to the slow progress that has occurred, especially with the implementation of site management measures and actions that are needed to actually improve the status of habitat and species. Target 1 of the *EU Biodiversity Strategy to 2020* aimed to address this issue by introducing quantitative and time-bound targets to improve the conservation status of BHD habitats and species. However, the target was not achieved, in part because it was not legally binding. To encourage swift progress, the WFD and MSFD include deadlines for the achievement of their objectives. As discussed later, this may have helped stimulate action, but it was not sufficient to ensure their objectives were met.

The other main weakness concerning the Nature Directives is their inadequate financing provisions. These are completely lacking in the Birds Directive, as it was envisaged that the Member States would be completely responsible for funding. There was also the implicit expectation that the Member States would find most of the necessary financing for the implementation of the Habitats Directive. Whilst provisions were included in Article 8 for EU co-financing of necessary management measures for SCIs, these are restricted to priority species and priority habitats. The provisions have since been broadened in practice to cover the funding requirements for all Natura 2000 sites for all BHD habitats and species. Nevertheless, as further discussed in Section 5.6, inadequate funding has led to major implementation shortfalls, which have not been solved to this day.

Another commonly held view is that the Nature Directives have an overly top-down approach, particularly concerning the selection of sites (initially as SCIs) for inclusion in the Natura 2000 network. The logic for the selection of a site is that it should be solely on its importance for nature. Therefore, it should not take into account other considerations, such as its uses, ownership or economic importance (as has been confirmed many times by the Court of Justice of the European Union (CJEU)). Instead, it was foreseen that such implications would be dealt with later in the way that the site is protected and managed. At that point consultations should be held with stakeholders, ideally through the development of agreed management plans. Whilst this approach is logical, it has been perceived negatively, as it only included input from landowners and other stakeholders into the critical site selection process if it was in the form of scientific information. This led to numerous misunderstandings and conflicts with stakeholders, which took years to resolve in some Member States and sometimes created a legacy of mistrust.

The Natura 2000 network has been criticised by some (e.g. Crofts, 2014; Verschuuren, 2015) as being too site-focussed, with inadequate requirements to connect sites through corridors and ecological networks, such as in the Netherlands. Contrary views are that it is more important to conserve large core areas of good quality habitat than to join them up with corridors (Hodgson *et al.*, 2009, 2011; Lawton *et al.*, 2010). The ecological effectiveness of corridors linking Natura 2000 sites is also uncertain, especially as the species of concern tend to have specialist habitat requirements. Furthermore, sites do not necessarily need to be structurally connected, such as by corridors, to form a network. They can be functionally connected if the sites are sufficiently close to act as 'stepping-stones', as found for vertebrates and Natura 2000 sites (Maiorano *et al.*, 2015). Some species may also be able to move between sites if the intervening landscape is not too inhospitable. If corridors are required, IUCN guidance is that they should be targeted and designed for their specific purpose (Hilty *et al.*, 2020). This is more realistic than planning large-scale generic ecological networks, which then tend not to materialise in practice, as has occurred in some European countries (e.g. Italy, Portugal and Slovakia). Difficulties in delivering such networks seem to have been due to the inadequate legal powers of the network proponents, partial support from stakeholders and their high costs (IEEP and Alterra, 2010). As a result, the National Ecological Network in the Netherlands, the 'Green and Blue Infrastructure' in France, and the Nature Frame in Lithuania, are amongst the few examples where the networks exist in a way that actually significantly influences nature conservation.

In fact the Habitats Directive includes provisions in Articles 3(3) and 10 that explicitly aim to improve the coherence of the network, where Member States consider it necessary (Table 4.1). Nevertheless, the required action is rather narrowly defined as 'maintaining, and where appropriate developing, features of the landscape...'. Implementation of these provisions has also been weak, partly because the wording in the Directive has been widely interpreted as indicating that such actions are discretionary. Instead, according to guidance produced for the European Commission (Kettunen *et al.*, 2007), Member States should consider the coherence measures to be mandatory in practice, as they are necessary to achieve the overall objectives of the Habitats Directive.

Crofts (2014) also believes that the requirement to designate Natura 2000 sites for particular habitats and species leads to static conservation objectives that do not recognise natural processes, such as vegetation succession and growing pressures such as climate change. The Fitness Check confirmed that some interpretations of the requirements have led to problems, and complaints about inflexibility from some Member State authorities. However, others have pointed out that there is in practice sufficient flexibility within the provisions of the Nature Directives to allow changes to the Natura 2000 site network, including in relation to climate change (Dodd *et al.*, 2010; Trouwborst, 2015).

As noted earlier, a strength of the Nature Directives has been that they have focussed on a common set of habitats and species that are considered to be of EU concern, as listed in the Directives' annexes. However, this strength has been undermined in some respects as a result of problems drawing up the annexes and keeping them up-to-date. These issues have included political interference with BD Annexes I and II. The conservation of habitats has suffered from the unbalanced treatment of different broad habitat types in HD Annex I. Most obviously, marine habitat types are very poorly represented and imprecisely defined. In contrast, some terrestrial habitats have been subdivided to a very high degree (e.g. coastal dunes).

The species of Community interest listed in HD Annexes II, IV and V (now 1 389 species) are largely based on the Bern Convention list, which was not compiled using objective criteria. They are dominated by higher plants, include relatively few lower plants and invertebrates and have selection biases. For example, amongst arthropods, Annexes II and IV favour well-known taxonomic groups (e.g. butterflies and dragonflies), popular species, large species, widely distributed species, and species from northern and western Europe (Cardoso, 2012). In practice, the incomplete listing of threatened species on the annexes may not be as significant a problem as it seems, as many species can be expected to be indirectly protected by the numerous BHD habitats and species captured within the scope of the legislation, especially as most Natura 2000 sites are selected for their habitats (van der Sluis *et al.*, 2016). Nevertheless, whilst there is a substantial 'umbrella effect', there appear to be some gaps in Natura 2000 coverage (e.g. Trochet and Schmeller, 2013;

Maiorano *et al.*, 2015). This has led to calls from conservation scientists for the annexes to be revised using appropriate species selection criteria (e.g. Cardoso, 2012; Hochkirch *et al.*, 2013).

The annexes have been updated a number of times, primarily in response to the accession of new Member States. Despite this, the known deficiencies and inconsistencies in the annexes have not been rectified. Nor have the lists been significantly updated in response to changes in the status of habitats and species. Consequently, a significant number of species that are now known to be threatened in Europe are not on the annexes. Conversely, some species that are on the annexes are no longer threatened and a few have become common enough to cause problems (e.g. the Beaver, *Castor fiber*). Despite this, delisting such species is risky as their favourable status may be dependent on maintaining their high level of protection. Also, whilst there are legal hurdles, the main reason the annexes have not been updated is concern that it would divert efforts away from other priorities, by triggering long and disruptive disagreements over the changes, and the eventual need for updates to Natura 2000 site designations and objectives, and possibly new sites (Maes *et al.*, 2013; Milieu *et al.*, 2016). It seems inevitable that the annexes will need to be revised or enlarged eventually; at this point it will require considerable care, objective criteria, and adequate representative data for all taxa groups.

Lessons from the Fitness Check of the Birds and Habitats Directives

In 2015–2016 the Nature Directives were subject to a Fitness Check in response to concerns that they were not performing as well as intended (see online Annex 2).[6] There were also claims from some quarters that they were costly and constraining economic development (e.g. see Chapter 31 on the UK). The Fitness Check involved a wide-ranging review of the literature, Member States' reporting data, a stakeholder questionnaire and wider public consultation. Based on this evidence, the evaluation study supporting the Fitness Check (Milieu *et al.*, 2016) and European Commission (2016) concluded that, as part of broader EU biodiversity policy, the Nature Directives are fit for their purpose: they are effective where properly implemented and are making progress towards achieving their objectives, especially in relation to the general protection of most species and the establishment of the Natura 2000 network on land. There was also no clear evidence of disproportionate costs or burdens. Instead the Nature Directives facilitate sustainable development, and their overall benefits are greater than their costs (ten Brink *et al.*, 2011). No changes to the legislation were therefore proposed. Some potential benefits from updating the annexes of habitats and species requiring protection were identified, but on balance it was considered that changes would be counter-productive for the reasons mentioned above.

However, the Nature Directives' effectiveness had been impeded by delays, especially in terms of establishing the Natura 2000 network (particularly in the marine environment) and site conservation measures. There were also deficiencies in the implementation and enforcement of protection measures for some species (primarily in relation to hunting and persecution). The flexible approach taken in the Directives, and lack of deadlines, resulted in few proactive actions being taken to achieve the favourable conservation status of habitats and species, especially outside the Natura 2000 network. It was therefore concluded that substantial improvements in the implementation of the Nature Directives are required to achieve their objectives and realise their full potential.

Reflecting the weaknesses in the Directives' implementing provisions, the main barrier to implementation was considered to be insufficient funding, especially of practical conservation measures, but also of the relevant environmental authorities and other conservation organisations. This has been due to inadequate allocations and accessibility problems (e.g. limited capacity within public authorities to set up schemes to disburse funds). Other identified key constraints were low political support and weak environmental authorities (a major reason for the inadequate funding allocations), problems with transcribing and implementing some provisions (e.g. hunting, and appropriate assessments), conflicts with other sectors (especially agriculture, forestry and fisheries) and landowners, the existence of payments that encourage damaging practices (e.g. agricultural and forestry improvements), gaps in ecological knowledge (e.g. on the distribution of EU protected species and habitats) and low levels of nature conservation expertise, capacity and standards in environmental authorities. Most of these issues are further described in the following sections.

5.2.2 The Water Framework Directive

As described in Section 4.3, one of the key specific objectives of the WFD was for all surface waters in the EU to achieve good water status (chemical and ecological) by the end of 2015, unless they are highly modified, artificial, or otherwise duly exempted. A 2018 report based on Member State monitoring data revealed that the objective was not met: only around 40% of surface water bodies had a good or high ecological status or potential, with lakes and coastal waters having better status than rivers and transitional waters (EEA, 2018). The most frequently reported surface water pressures were hydro-morphological, due to dams and other structural changes that affect the flow of rivers and hinder the movement of many species. Other pressures, in descending order of importance, were airborne deposition of pollutants, diffuse pollution (mainly from agriculture), point source pollution (mainly from urban areas) and over-abstraction of water (in particular for agriculture).

The reasons for this failure were examined in 2019 as part of a Fitness Check of water policy that covered the WFD and related

[6] Fitness Checks were introduced in COM (2010) 543 final – Smart Regulation in the European Union. They assesses whether the group of interventions is fit for purpose by assessing the performance of the relevant framework with respect to its policy objectives.

Groundwater Directive, Environmental Quality Standards Directive and Floods Directive (European Commission, 2019a). The evaluation found that the WFD and related directives had led to some successes. These included establishing the governance framework for integrated water management of more than 110 000 water bodies in the EU, increasing knowledge and monitoring of the status of water bodies and the pressures affecting them, slowing the deterioration of water status and reducing chemical pollution (mainly from point sources). Some improvements in ecological status also occurred. These have been key factors in national or biogeographical level improvements in the conservation status of some aquatic BHD habitats, such as water courses of plain to montane levels (3260),[7] and various fish, invertebrates and the Otter (*Lutra lutra*) (Tucker *et al.*, 2019).

Despite these successes, the 2015 good water status was not achieved. Furthermore, no substantial progress in water bodies' overall status was made between the first and the second river basin management cycles. This apparent limited improvement was partly due to the need for all parameters to be good (the 'one-out-all-out' principle), which is a relatively high bar to achieve (as with achieving favourable conservation status under the Habitats Directive). Ecosystems also take time to respond to restoration measures. Nevertheless, it is clear that recent progress has been insufficient.

With similar findings to the Nature Directives Fitness Check, the main WFD problems have primarily been due to slow implementation, inadequate funding in some countries (also for the Floods Directive) and insufficient integration of environmental objectives in other sectoral policies. One initial reason for slow progress was the time it took to establish the governance framework, as it had to be flexible enough to allow for the varying conditions and requirements across the EU. Funding limitations also constrained the ambition of some River Basin Management Plans, such that their programmes of measures have been often insufficiently based on their analysis of pressures and impacts. Instead, they have tended to focus on relatively easy technological fixes that have addressed point source pollution, whilst inadequately addressing diffuse sources of pollution.

The overall conclusion of the Fitness Check was that the WFD legislation remains fit for its purpose, including in relation to envisaged future challenges, and therefore no legislative changes are required. Instead, greatly improved implementation is needed, including through increased funding, the full implementation of other related EU legislation (e.g. the Nitrates Directive and the Urban Waste Water Treatment Directive), as well as better integration of water objectives in other policy areas such as agriculture, energy and transport. More specific actions for the European Commission are to share best practices on cost recovery, reduction of pollutants and the use of green infrastructure.

[7] Numbers in brackets are the HD Annex I habitat codes (see the Appendix for full names). * Indicates a priority habitat.

5.2.3 The Marine Strategy Framework Directive

As described in Section 4.3, with the MSFD (and supporting Maritime Spatial Planning Directive), the EU has established perhaps the most ambitious and comprehensive policy framework in the world for the sustainable management of the marine environment, which is also based on the ecosystem-based approach. Member States have made considerable efforts to implement the Directive and reduce pressure on their marine environment through their Marine Strategies and Programmes of Measures (European Commission, 2018). However, the measures are incomplete and have been insufficiently implemented. Consequently, despite data gaps due to only 14 Member States reporting on time, it is clear that the MSFD's 2020 legal targets were not achieved (European Commission, 2020a). Data from the MSFD and Nature Directives reporting and other assessments (OSPAR, 2017; HELCOM, 2018; UNEP-MAP, 2018) show that most marine habitats and many species do not have a good status, and some marine populations and groups of species remain under threat, including some seabirds, elasmobranchs (sharks and rays, etc.), sea turtles and sea mammals (EEA, 2019b).

Data analysis has shown that the condition of Europe's seas is directly linked to the intensity of human use, and its resulting pressures (EEA, 2019b). Fishing pressures continue to be particularly important, although reductions in exploitation of some stocks are leading to recoveries in the north-east Atlantic and Baltic Sea. In contrast, at least 87% of the commercially exploited fish and shellfish species are overfished in the Mediterranean and Black Sea. Fishing also continues to lead to extensive damage of benthic habitats and by-catch of seabirds, sharks, turtles and cetaceans (OSPAR, 2017). In terms of seabird by-catch, more than 200 000 are killed annually with at least 90 000 drowned in gillnets in the North Sea and Baltic Sea alone (Žydelis *et al.*, 2013). Thousands of cetaceans continue to die each year in European waters (Rogan *et al.*, 2021), despite this being at levels likely to be causing population level impacts in some species (ICES, 2010; Dolman *et al.*, 2016). According to the Member State MSFD reports, other remaining key pressures in the marine environment that need to be tackled are IAS, nutrient pollution leading to eutrophication (affecting nearly one-half of coastal waters), toxic pollutants, marine litter, alteration of hydrographic conditions, underwater noise and climate change.

Although the MSFD objectives have not been met, some progress has been made with the implementation of some measures, including new Marine Protected Area (MPA) designations. This helped the EU-28 achieve more than 10% coverage of its waters as MPAs, thereby meeting the 10% CBD 2020 Aichi Target 11. However, the MPAs are unevenly distributed across regions and across depth zones, and more are needed to provide the required coherent and representative network (ECA, 2020). Much more progress is also required to increase their effective management (such as through management plans), protection and enforcement. At the moment most

MPAs continue to be damaged by fishing activities, as few have strict protection or no-take zones (e.g. in Denmark and UK).

Key problems that have constrained the achievement of good environmental status have been the low ambition and lack of completeness of some Member States' measures, insufficient policy coherence and science-led collaboration with other instruments (Maritime Spatial Planning Directive, CFP and CAP) and maritime sectors (e.g. energy, transport and aquaculture), slow and inadequate implementation of measures, and time-lags between interventions and impacts (EEA, 2019b; ECA, 2020; European Commission, 2020a). Thus the MSFD is insufficiently influencing sectoral activities. Some of the causes of slow implementation have included inadequate funding, data gaps and inconsistencies, inadequate cooperation across sectors and regions, the complexity of analysing and managing the marine environment and reporting on it, inadequate enforcement (i.e. limited political support) and insufficient involvement of other economic and private sectors. As concluded by the ECA (2020), and indicated in many of this book's country chapters, most Member States have made little use of the European Maritime and Fisheries Fund (EMFF) for nature conservation purposes.

5.2.4 EU and National Biodiversity Strategies

EU and National Biodiversity Strategies and Action Plans up to 2020

It is obvious from the *State of Nature in the EU* report and the *EU Ecosystem Assessment* that the EU did not achieve its headline target of halting the loss of biodiversity and the degradation of ecosystem services by 2020, and restoring them in so far as possible by the same date. Furthermore, a study supporting the European Commission's evaluation of the *Biodiversity Strategy to 2020* (Trinomics *et al.*, 2022) and other evidence (e.g. Langhout, 2019, and the national chapters in this book) indicate that little progress was also made with respect to most of the individual targets (Table 5.3), and many of their supporting actions were not completed.

Of particular importance for the achievement of the overall headline Biodiversity Strategy target was **Target 1**, which aimed to fully implement the Nature Directives and thereby substantially improve the conservation status of habitats and species. Whilst some improvements in the implementation of the Directives occurred, particularly regarding extension of the Natura 2000 network, the habitat and species targets were not achieved. The target was almost achieved for HD species (2% gap), but there remained a 12% gap for HD habitats and a 20% gap for bird species. The *State of Nature* report noted that this was primarily due to the failure to adequately address the key pressures on biodiversity, in part due to insufficient or ineffective implementation of the Nature Directives (as discussed previously). Efforts had been made to improve implementation through the *Action Plan for Nature, People and the Economy* (European Commission, 2017), which was produced to address the needs identified in the Nature Directives Fitness Check. However, it was relatively unambitious, especially with respect to dealing with the funding gap, and clearly could not achieve much before 2020.

Although **Target 2** had the added importance of responding to CBD Aichi Target 15, the aim of restoring 15% of degraded ecosystems was clearly not achieved. This would have required the restoration of at the very least 26 700 km^2 for terrestrial HD habitats alone. This is based on estimates in the Member State Article 17 2012–2018 reports of the minimum area of HD habitats that require improvement, which was between 167 000 km^2 and 263 000 km^2, and at least another 11 000 km^2 of required habitat re-creation (EEA, 2020). Due to missing data in the reports, the actual areas requiring restoration and re-creation are likely to be much larger. Given that terrestrial HD habitats only comprise about a quarter of the EU land area, and that other ecosystems would be much more degraded, it is clear that the total restoration required would have been well beyond 100 000 km^2, or more than 10 000 km^2 per year, to achieve the 15% target.

A significant amount of restoration has been carried out in Europe in recent years, for a variety of voluntary and statutory reasons, including flood control, coastal protection, water resource management, carbon capture, post-mining rehabilitation, compensation for development impacts and nature conservation. Examples are described here in most of the country chapters, such as in relation to coastal habitats (e.g. the UK), grasslands (e.g. Estonia, Latvia), peatlands (e.g. Belgium, Estonia and Ireland), rivers (e.g. Austria, the Czech Republic, Denmark and the Netherlands) and wetlands (e.g. Hungary, Spain). Quantification of the actual restoration that took place is difficult, but between 2 850 km^2 and 5 700 km^2 of additional habitat restoration was estimated in 2017 to commence on an annual basis within the EU (eftec *et al.*, 2017). Although significant, restoration would have had to have been at least double this to have achieved the very lowest estimate of the restoration required to meet Target 2.

There is little to suggest from EU data, or the national chapters in this book, that the CBD and EU restoration targets stimulated a significant increase in restoration activities in most countries. This is indicated by the fact that only a few restoration priority frameworks were prepared in accordance with Action 6a, such as in Flanders (see Section 7.6). According to eftec *et al.* (2017), the main barriers to substantially increasing restoration have been its direct and opportunity costs, inadequate private and public funding, cultural and language issues, and regulations and policies (e.g. tax incentives and subsidies) not being designed to accommodate land-use alteration or change.

While Target 2 was probably not achieved overall, progress was made with some related actions. A ***Green Infrastructure Strategy*** was adopted by the European Commission (2013), under Action 6b. This is intended to help maintain and enhance green infrastructure, which is defined as 'a strategically planned network of natural and semi-natural areas with other environmental features designed and managed to deliver a wide range of ecosystem services'. This has much in common

Table 5.3 Summary of achievements in relation to EU Biodiversity Strategy to 2020 Targets 1–5.

Target/action	Outcome
Target 1: Fully implement the Birds and Habitats Directives; to halt deterioration and ensure:	Some improvements in implementation, but targets not achieved
• 100% more HD habitat assessments show a favourable or improved* status, i.e. **34%**	22% of assessments showed improvements
• 50% more HD species assessments show a favourable or improved* status, i.e. **35%**	33% of assessments showed improvements
• 50% more bird assessments show a secure or improved* status, i.e. **78%**	58% of assessments showed improvements
Target 2: Maintain and restore ecosystems and their services (over 15% of degraded ecosystems)	Not achieved, and very little evidence of an increase in restoration actions. Partial achievement of supporting actions
Target 3: Increase the contribution of agriculture and forestry to maintaining and enhancing biodiversity	
3A: Maximise areas under CAP biodiversity measures to improve the status of agricultural species and habitats	Not achieved: some improvements but no overall progress towards the target
3B: Use RDP funded Forest Management Plans and other instruments to improve the status of species and habitats affected by forestry	Not achieved: some improvements (e.g. in common birds) but mixed overall progress towards the target
Target 4: Ensure the sustainable use of fisheries resources and achieve good environmental status	Not achieved: mixed progress on sustainable fisheries, little action on by-catch and slow progress towards good environmental status
Target 5: By 2020, IAS and their pathways are identified and prioritised, priority species are controlled or eradicated, and pathways are managed to prevent the introduction and establishment of new IAS	Partly achieved: key IAS Regulation introduced but only currently covers about a quarter of species identified for listing by 2020

Notes. Abridged and edited targets. See Table 4.5 for full text. Target 6 and related actions concerning global biodiversity loss are not covered in this book. * The term 'improved' includes improvement in both status and trend.
Source. Based on EEA (2020), Maes *et al.* (2020), Trinomics *et al.* (2022) and chapter text and related sources.

with initiatives to promote nature-based solutions, and is increasingly being taken into account in relevant policies (such as in Denmark, France and Spain). However, practical actions have tended to be at a small scale, partly because opportunities to use EU financing instruments for green infrastructure initiatives have not been fully exploited (European Commission, 2019b).

Action 7 under Target 2 was to ensure no net loss of biodiversity, and entailed two sub-actions. Under **Action 7a**, the Commission developed **guidance on 'biodiversity proofing' of EU funds**, including the Multiannual Financial Framework (MFF) and funds under the CAP, Cohesion Policy, and the Connecting Europe Facility for energy and transport (Medarova-Bergstrom *et al.*, 2014). This guidance aims to ensure firstly that the development of the funding instruments and their implementation does not harm biodiversity, and secondly that there are biodiversity benefits from the funding, such as through promoting green infrastructure. However, it is uncertain whether the guidance is actually being taken up in practice by the relevant sectors in the Member States and leading to significant nature conservation improvements and wider benefits.

Under **Action 7b**, the Commission established a working group and conducted a number of studies (e.g. Tucker *et al.*, 2014, 2016) on options for ensuring **no net loss of biodiversity** from the unavoidable residual impacts of damaging projects, in accordance with the mitigation hierarchy (i.e. avoid–reduce–restore–compensate). This focussed on projects that do not affect Natura 2000 sites and are therefore not covered by HD Article 6(4) requirements for compensatory measures. It was concluded that biodiversity offsetting would be required to achieve no net loss, and should be applied to all sectors (including agriculture).

Progress with the no net loss initiative stalled after a public consultation revealed mixed feelings on the issue.[8] One of the main reasons for concern was that offsetting would reduce existing levels of protection outside the Natura 2000 network and could become a so-called licence to trash. Evidence from offsetting experiences in Europe indicates that this has not happened (Wende *et al.*, 2018). This may be due to the

[8] https://ec.europa.eu/environment/consultations/nnl_en.htm

fact that most offsetting has been carried out in Germany, where it is effectively regulated. The risks that offsetting could be counter-productive are greater in other countries with less well-regulated development planning and permitting. However, in such countries damaging projects are infrequently stopped on biodiversity grounds, so there might be little to lose from introducing offsetting outside protected areas.

The Commission has not advanced EU no net loss policy any further, other than to publish guidance that covers all steps under the mitigation hierarchy, which includes advice on best-practice offsetting (Tucker *et al.*, 2020). Without further policy measures, widespread substantial cumulative biodiversity and ecosystem service losses will continue, except in those countries with mandatory offsetting (i.e. primarily Germany and France in the EU). Despite this, the *EU Biodiversity Strategy for 2030* does not include a specific objective to develop further a no net loss (or net gain) policy. From November 2023, there will be a mandatory requirement for built developments to achieve a net gain in biodiversity in England (described in Section 31.6). As this will be closely monitored, it should provide further evidence on the pros and cons of biodiversity offsetting.

Target 2 was also supported by the **Mapping and Assessment of Ecosystems and their Services (MAES)** initiative, which was a collaboration between the European Commission, EEA and Member States. One of its main aims was to support the development of ecosystem accounting in national monitoring and reporting. This triggered mapping and assessment studies in many Member States and a series of EU pilots and reports that led to an EU-level analysis of ecosystems in 2020 (Maes *et al.*, 2020). However, the assessment was constrained by numerous data gaps including the lack of maps of broad ecosystems and HD habitats in most Member States. Attempts were made to use other datasets (e.g. CORINE) and indirect proxy indicators, but the results are not always easy to interpret or reliable (e.g. the reported stable HNV area is clearly incorrect). Data limitations were a particular problem for marine ecosystems and as a result many of their intended condition indicators were not assessed.

Despite the data limitations, the Commission proposes to go ahead and add a module for mandatory reporting by Member States on ecosystem accounts, by amending the existing legislation on European environmental economic accounts.[9]

Target 3 actions were intended to lead to measurable improvements, compared to 2010, in the conservation status of habitats and species that depend on, or are affected by, **agriculture and forestry**. The *State of Nature* report does not provide results compared to the 2010 baseline (unlike for Target 1), but does conclude that no real progress was made with achieving the agriculture or forest targets. This is especially obvious for agricultural habitats and species. Whilst some improvements have occurred, including as a result of CAP Rural Development Programme (RDP) measures, deteriorating trends were predominant (the reasons for which are discussed later). For example, only 8% of agricultural habitats were assessed as improving, whereas 45% were deteriorating. Furthermore, 45% of assessments revealed a bad status, compared to 31% for other HD habitats.

Less negative results were observed for forest habitats and species compared to agricultural areas. For example, 31% of the forest habitat assessments showed a bad conservation status. Nevertheless, this is still a high proportion and may even signal a deterioration in conservation status as only 26% of forest assessments were reported as bad in the previous reporting period (EEA, 2015), although the difference may be due in part to better knowledge. Trends do appear to have been more favourable: 13% of forests with an unfavourable status had improving trends – the highest for any habitat group. More forest birds also had positive trends (34%) than negative (17%). As shown in Figure 2.3, overall trends in common forest birds have been positive since 2009. Although some of the improvements in forest habitats and species populations may be due to sustainable forestry management standards and forest management plans, their biodiversity elements could be much stronger.

Target 4 aimed to **increase the sustainability of fisheries** and support the achievement of good environmental status under the MSFD. The sustainability of fisheries is not covered in detail in this book, but in summary, whilst there have been some improvements, many fish populations remain over-exploited (ECA, 2020). Under Action 14A, measures have been taken to reduce fishing by-catch, but as mentioned, it remains a major threat to some species. Although the European Commission developed an Action Plan to reduce seabird by-catch in fishing gear (European Commission, 2012), few of the identified actions have been taken and further implementation is clearly needed (Mitchell *et al.*, 2021). Similarly, little has been done to reduce the by-catch of small cetaceans (Rogan *et al.*, 2021). As described earlier, slow progress has also been made with implementing other aspects of the MSFD and the achievement of good environmental status.

The legislative actions foreseen for **IAS** under **Target 5** were completed, resulting in revisions to plant and animal health regimes and the adoption of the Regulation on IAS in 2014 (described in Section 4.3). The list of IAS of Union concern, which are subject to restrictions or other measures under the Regulation, was adopted in 2016. It has since been updated, most recently in 2022, and currently contains 88 species. However, this is a relatively small proportion of the IAS that are having significant impacts on biodiversity. This is evident from a scientific review that found that 900 species fitted the criteria for listing according to the IAS Regulation and prioritised 207 species for urgent risk assessment by 2020 (Carboneras *et al.*, 2018).

In conclusion, the *EU Biodiversity Strategy to 2020* did not achieve its main objectives and many of its actions were not

[9] Regulation (EU) No 691/2011.

completed or were ineffective. Most of the actions that were completed were the responsibility of DG Environment, and there is little evidence that the strategy greatly stimulated actions by others that would not have already been carried out in relation to existing legislation and initiatives (particularly in the case of Target 4). Almost the same thing happened with the previous *EU Biodiversity Action Plan to 2010* (European Commission, 2006), which, to a large extent, it had copied.

There are numerous reasons why the previous EU strategies and action plans had rather limited impacts, but it is widely agreed that a fundamental problem was that they were generally viewed as DG Environment strategies that lacked stronger higher-level political backing and Member State buy-in. Whilst consultations with other European Commission services, Member State authorities and other stakeholders contributed to their development, these were not deep enough to ensure wider ownership of the strategies. Consequently, many actions were deliberately ill-defined and lacked clear implementation mechanisms and responsibilities. Subsequently, the actions had limited support from key actors outside DG Environment, which led to EU policy incoherencies and inadequate implementation. Other problems that have been noted were the failure to secure and track adequate funding, mismatches between the targets' ambitions and supporting actions, and limited monitoring and reporting on national implementation (e.g. Langhout, 2019).

The country chapters in this book indicate that the EU biodiversity strategies and action plans have had varying influences on the development of national biodiversity strategies. Some followed their structure and aimed to contribute to EU goals (e.g. Greece). Others were more closely aligned to CBD requirements and, over the last 10 years, the Aichi targets. Unfortunately, many of the strategies had the same weaknesses as the EU strategies and action plans, in particular being a product of the environmental authorities without sufficient cross-government involvement. As a result, whilst they have tended to be ambitious, they have lacked wider ownership and high-level political support. There is little evidence that they made a significant difference in terms of stimulating new substantive actions that would not already have occurred. This has been particularly the case in some Eastern Bloc countries that have not had a long history of preparing such informal strategies, as noted in the Czech Republic and Poland chapters. Whilst such countries have produced biodiversity strategies, the main purpose has been to meet the obligation to do so, with little expectation that they would be substantially implemented.

The EU Biodiversity Strategy for 2030

The *EU Biodiversity Strategy for 2030* (European Commission, 2020b; Box 5.1) has high ambitions and forms a part of the European Green Deal strategy for growth and achieving climate neutrality by 2050. It seems to recognise some of the lessons from the previous strategies and action plans, including the need for some mandatory actions underpinned by legislation. However, it also appears to make some of the same mistakes. In particular, most of its targets and actions are vaguely defined and lack timelines and indications of responsibilities. It was also developed with limited external consultation and has no supporting analysis of the effectiveness of the previous strategies and the current needs and priorities. Whilst it attempts to link to a wider environmental, natural capital, climate and post–COVID-19 recovery agenda, there are few signs of wide ownership across the European Commission. This is most obvious with respect to the mismatch between the *Biodiversity Strategy* and the *Farm to Fork Strategy* (European Commission, 2020c) as the latter omits an important action: the commitment to have at least 10% of agricultural land under high-diversity landscape features. This is despite their symbolic simultaneous release.

Some of the actions are in step with the Strategy's ambitions and have the potential to make further significant contributions to nature conservation and restoration in the EU. These include a commitment to increase the EU's protected area network to 30% of both land and sea by 2030, which was in line with CBD global proposals at the time (since agreed). This is ambitious for the marine environment, where the EU-27 protected area coverage is about 11%.[10] The target should be achievable on land, as the 18% Natura 2000 coverage together with nationally designated sites gives EU-27 protected area coverage of about 26%. However, as further discussed later, some of the claimed domestic protected areas might not meet IUCN/CBD definitions. Irrespective of this, it is most important to ensure that all protected areas have objectives that correspond to the ecological requirements of the targeted habitats and species, and are effectively protected and managed. This is recognised in the Strategy, and the target of 10% coverage with strict protection would help (although it has not been defined), but there are no other clear actions to achieve this outside the Natura 2000 network.

As this chapter was being finalised, in June 2022 the European Commission published its proposals for a new Regulation on Nature Restoration, referred to by the Commission as a **Nature Restoration Law** (European Commission, 2022a). If adopted, it will oblige Member States to develop national restoration plans with specific binding targets for ecosystems and species. Hence the lesson has been learnt that such ambitions cannot be achieved by voluntary initiatives. This reflects a European Parliament resolution in January 2020[11] that asked to 'move away from voluntary commitments and to propose an ambitious and inclusive Strategy that sets legally (and, consequently, enforceable)

[10] Total marine and terrestrial protected area coverage according to the *EU Biodiversity Strategy for 2030*, based on CCDA data from 2019.

[11] Resolution on the 15th meeting of the Conference of Parties (COP15) to the Convention on Biological Diversity 2019/2824 (RSP).

binding targets for the EU and its Member States'. However, the need for a legally robust instrument brings considerable challenges, as the target(s) will need to be both meaningful and measurable.

An obvious priority, recognised in the proposed Nature Restoration Law, is to bolster the existing requirements under the Nature Directives by adding the clear measurable deadlines for their achievement that are lacking in the legislation. It also addresses the need to go beyond this and restore ecosystems and species that are not well covered by the Nature Directives and other EU environmental legislation. Overall, the restoration is proposed to cover at least 20% of the EU's land and sea by 2030, and all ecosystems in need of restoration by 2050. Importantly, it is also recognised that overall percentage-based restoration targets alone could encourage a predominance of low-cost basic restoration, with low biodiversity benefits. Therefore, complementary specific targets are proposed for all HD habitats (quantified by the percentage of the area requiring restoration to achieve favourable conservation status), all BHD species (unquantified), general restoration of agricultural, forest and urban ecosystems (according to selected ecosystem condition indicators), river connectivity and the reversal of declines in pollinators. Whilst the proposals are broadly in line with the objectives in the *Biodiversity Strategy* (Box 5.1), especially for HD habitats, they fall short for species, as there is no clear obligation on Member States to increase those with a favourable status or positive trend by 30%. Whether or not the proposed Nature Restoration Law

Box 5.1 *The EU Biodiversity Strategy for 2030* **– objectives and actions in the EU (excluding external actions supporting global biodiversity)**

Aim

To put Europe's biodiversity on the path to recovery by 2030 for the benefit of people, climate and the planet.

Nature protection: key commitments by 2030

1. Legally protect a minimum of 30% of the EU's land area and 30% of the EU's sea area and integrate ecological corridors, as part of a true Trans-European Nature Network.
2. Strictly protect at least a third of the EU's protected areas (representing 10% of the EU land and EU sea), including all remaining EU primary and old-growth forests.
3. Effectively manage all protected areas, defining clear conservation objectives and measures, and monitoring them appropriately.

EU Nature Restoration Plan: key commitments by 2030

1. Legally binding EU nature restoration targets to be proposed in 2021, subject to an impact assessment. By 2030, significant areas of degraded and carbon-rich ecosystems are restored; habitats and species show no deterioration in conservation trends and status; and at least 30% reach favourable conservation status or at least show a positive trend.
2. The decline in pollinators is reversed.
3. The risk and use of chemical pesticides is reduced by 50% and the use of more hazardous pesticides is reduced by 50%.
4. At least 10% of agricultural area is under high-diversity landscape features.
5. At least 25% of agricultural land is under organic farming management, and the uptake of agro-ecological practices is significantly increased.
6. Three billion new trees are planted in the EU, in full respect of ecological principles.
7. Significant progress has been made in the remediation of contaminated soil sites.
8. At least 25 000 km of free-flowing rivers are restored.
9. There is a 50% reduction in the number of Red List species threatened by invasive alien species.
10. The losses of nutrients from fertilisers are reduced by 50%, resulting in the reduction of the use of fertilisers by at least 20%.
11. Cities with at least 20 000 inhabitants have an ambitious Urban Greening Plan.
12. No chemical pesticides are used in sensitive areas such as EU urban green areas.
13. The negative impacts on sensitive species and habitats, including on the seabed through fishing and extraction activities, are substantially reduced to achieve good environmental status.
14. The by-catch of species is eliminated or reduced to a level that allows species recovery and conservation.

Supported by a new European biodiversity governance framework

Including:

- Stepping up implementation and enforcement of EU environmental legislation
- Building partnerships with business
- Supporting biodiversity through investments, pricing and taxation
- Measuring and integrating the value of nature into public and business decision making
- Improving knowledge, education and skills

Source. Adapted from European Commission (2020b).

lives up to its current potential, or is even further strengthened (e.g. the species targets), will depend on the outcome of negotiations under the Ordinary Legislative Procedure, which are expected to take a couple of years.

Some key actions that are necessary to support the Nature Restoration Law and achieve the Biodiversity Strategy's wider objectives appear to be weak or missing. There is no mention of clear actions to address the ongoing deposition of airborne nitrogen that remains above critical loads over 73% of the EU (Maes *et al.*, 2020). Also, more generally, much more emphasis is needed on protecting semi-natural habitats, such as from agricultural improvements, bioenergy and afforestation. As discussed in Section 5.5, this needs stronger conditionality requirements on CAP payments and a major scale-up of well-designed and targeted agri-environment–climate schemes. But such schemes and grassland protection are hardly mentioned in the Strategy or the *Farm to Fork Strategy*. Most fundamentally, whilst the Strategy recognises that a minimum €20 billion per year funding is needed for nature, there is no clear mechanism for ensuring that this is provided for the actions required to meet the Strategy's objectives. This issue is discussed later in Section 5.6 on funding.

5.3 Species Protection and Conservation Measures

The general protection of species, such as from deliberate, inappropriate and unsustainable killing, trapping and the destruction of breeding sites, forms a universal basic level of nature conservation. Such measures were therefore amongst the first to be introduced in European countries, some going back to the early twentieth century. Since then there is little doubt that the introduction of international multilateral environmental agreements, such as the Bern Convention, has driven further species protection legislation across Europe. This was initially in Western Europe, as Eastern Bloc countries were not signatories to international treaties until the 1990s. Critically, the protection components of the Bern Convention were applied and strengthened as they became legal requirements simultaneously in the Birds Directive, and later in the Habitats Directive. Whilst it is difficult to compare the current situation in EU and non-EU European countries, there is evidence from many chapters that EU accession and compliance with the Nature Directives led to further strengthening of species protection in the countries at the time.

The overall result of this species protection legislation, related awareness-raising actions and societal changes in attitude has been that the illegal killing, trapping and trade of most species has decreased considerably, particularly over the last 40 years or so. Initially, implementation of the hunting provisions was problematic, but the launch of the European Commission's sustainable hunting initiative in 2001 was a turning point. Since then there has been better understanding and collaboration for mutual benefits between hunters and conservationists in many countries. Hunting has generally declined to more sustainable levels and no longer seems to be a key factor affecting population levels in most species. This has enabled the recovery of some formerly intensively hunted and/or persecuted species, including birds of prey and large carnivores, which had become confined to fragments of their range (Deinet *et al.*, 2013). Now, these species have recovered widely and are still increasing, though it should be borne in mind that this is not the case everywhere: for example, illegal hunting and persecution have constrained the recovery of Wolves in Scandinavia (Liberg *et al.*, 2020).

Recoveries of depleted species have often been dependent on a range of complementary actions to address causes of high mortality (e.g. IAS, toxins and electrocution) and restore habitats. Such actions seem to have been most effective when concerted and coordinated through scientifically rigorous species action plans, and there is supporting evidence for this in the case of birds (EEA, 2015). However, plans for threatened birds have tended to be most successful for species that were rare and localised, and subject to well-understood acute pressures that could be relatively easily targeted with supportive stakeholders (Barov and Derhé, 2011). It has also been essential to secure adequate funds, which may need to cover ongoing long-term actions (e.g. monitoring and habitat management). In the EU, funding for recovery actions has tended to rely on LIFE projects, which has led to some uncertainties over their long-term sustainability after completion (ECA, 2009; see discussion later).

Despite the generally improved protection of most species, there have been some ongoing problems. These include inadequate application of the Nature Directives species protection measures (BD Article 9 and HD Article 12) to land- and sea-use activities, most notably forestry, agriculture and fisheries. For example, forestry management is resulting in high rates of mortality and/or low rates of breeding success in a wide range of species. As discussed earlier, fisheries by-catch remains a major problem for many BHD species.

In contrast, species protection measures have been generally more rigorously applied to the industrial sector, including in relation to building projects, mines and quarries. Some conflicts have occurred where such projects have affected HD Annex IV species that are relatively common, such as with the Great Crested Newt (*Triturus cristatus*) in the UK. Poor knowledge of their distribution and conservation status led to an inflexible approach that resulted in disproportionate costs and delays for developers, which needlessly brought the Habitats Directive into disrepute. Whilst maintaining the requirements of the Directive, a more proactive and cost-effective strategic approach to protecting the species has been introduced (before Brexit), which focusses on local populations and achieving overall favourable conservation status.

Although the illegal killing or taking of most species has declined considerably, ongoing unsustainable levels of hunting and illegal killing or taking affect some species in some parts of Europe. This is apparent from the Member States' reporting for 2013–2018, as exploitation was the most frequent pressure for wintering and passage birds, mostly related to illegal shooting or killing (27%) and hunting (19%) (EEA, 2020). Illegal killing still includes the use of indiscriminate methods of catching migrant birds, such as with lime sticks and nets in Cyprus, even though these are prohibited under BD Article 8. This is clearly unacceptable on welfare grounds irrespective of the numbers involved. Other illegal actions include the killing and trapping of protected species, and out of season hunting of legal quarry species.

Fairly recent reports of the numbers of all such illegally hunted birds have been high, especially within the Mediterranean region. Although systematic quantitative data are lacking, an expert-based survey in 2014–2015 estimated that 11–36 million individuals were being illegally killed or taken per year across the whole Mediterranean region (Brochet *et al.*, 2016). Some of the worst areas were outside Europe (Egypt, Lebanon and Syria), but 6–15 million were estimated to be illegally killed or taken per year in European Mediterranean countries, with high proportions in Cyprus and Italy. Given its small size, relatively high numbers were also reported for Malta. Whilst most of the species illegally killed and taken are relatively common and do not appear to be obviously declining as a result, some European and even globally threatened species are being killed.

Over most of the rest of Europe,[12] expert survey data from 2016–2017 suggested that the illegal killing and taking of birds was relatively low, ranging from about 217 000 to 1 million individuals per year (Brochet *et al.*, 2017). Nevertheless, as revealed in some country chapters, illegal persecution is having significant population impacts on some species, such as in the UK, where it excludes the Hen Harrier (*Circus cyaneus*) from a large proportion of what would otherwise be suitable habitat (see Section 31.4).

It is possible that the number of illegally killed and taken birds has declined more recently, perhaps as a result of initiatives by the Bern Convention, CMS, European Commission,[13] NGOs and changing public attitudes. For example, Malta has taken substantial steps to better enforce its bird protection laws in recent years (Milieu *et al.*, 2016). However, there is no clear evidence that the international initiatives have had widespread positive impacts. Therefore, it seems inevitable that the number of illegally killed and taken birds in Europe still remains too high and therefore needs to be reduced through targeted actions, including enhanced monitoring and strengthening and enforcing of protection measures. Required measures vary according to the country, but may include stricter sanctions, strengthening enforcement to increase conviction rates, training of legal professionals and others involved in enforcement, information exchange, and ensuring that NGOs have legal rights to take legal action against environmental crime (ENEC, 2016).

Whilst not illegal, there are also concerns that hunting levels of some BD Annex II bird species in some countries may be too high and thereby contributing to their declines. One indicator of this is that a higher proportion of Annex II species have shown short-term declines (46%) in their breeding populations compared to all birds (30%) (EEA, 2020). A similar, though less marked, difference occurred with wintering populations. To thoroughly investigate this, reliable consistent information on the numbers of each Annex II species hunted in Europe is required. Unfortunately this is lacking for most countries, including in the EU, even though it is a mandatory requirement to provide hunting bag information in Article 12 reports for Annex II species. In 78% of the latest reports no information on hunting bags was provided (EEA, 2020).

Notwithstanding the data gaps, Hirschfeld and Attard (2017) compiled and analysed available national hunting bag statistics for 24 EU countries,[14] Norway and Switzerland, which resulted in an estimated total annual hunting bag in the countries of at least 58 million birds per year. Comparisons with a previous analysis suggest that, for the countries with comparable data, the numbers of birds hunted had declined for most species, possibly reflecting their population declines but perhaps also declines in relative hunting rate.

Hunting levels may be high enough to contribute to the observed declines in some Annex II species, including the globally and European threatened Common Pochard (*Aythya ferina*) and Turtle-dove (*Streptopelia turtur*) and the European threatened Lapwing (*Vanellus vanellus*). A more detailed modelling study of the Turtle-dove has revealed that recent levels of legal hunting along the western flyway are more than double the sustainable fraction under all suitably conservative scenarios (Lormee *et al.*, 2020). Furthermore, such pressures are also being exacerbated by illegal hunting and trapping, including in the spring. Therefore, the international species action plan for the Turtle-dove (Fisher *et al.*, 2018) recommends a temporary hunting moratorium until an adaptive harvest management modelling framework is developed. This has not happened to date, perhaps because the action was not supported by EU hunting organisations and, more importantly, by several Member States that shoot the species. Nevertheless, infringement cases against France and Spain for not taking sufficient measures to conserve the Turtle-dove have reduced hunting of the species.

Sibille *et al.* (2020), on behalf of the European Federation for Hunting and Conservation (FACE), have pointed out that it is important to properly assess the impacts of hunting on

[12] European countries, except Russia and those bordering the Mediterranean and in the Caucasus region.

[13] European Commission 2012 Roadmap towards eliminating illegal killing, trapping and trade of birds; Bern Convention Tunis Action Plan 2013–2020 for the eradication of illegal killing, trapping and trade of wild birds; and UNEP/CMS/Resolution 11.16 on the prevention of illegal killing, taking and trade of migratory birds.

[14] Data were not available for Greece, Ireland, the UK and the Netherlands (no bird hunting occurs in the latter).

declining Annex II species, as the decline in some species may be the result of other factors, such as agricultural intensification, high levels of predation and eutrophication of water bodies. However, such alternative causes are not certain for the species concerned, and hunting may still be a significant additive contributor to population declines. Thus, in accordance with the precautionary principle, there appears to be an argument for more actions to be taken to reduce hunting pressures on declining BD Annex II species. This should include consideration of EU or national moratoriums on hunting Annex II species that are threatened and declining in Europe, and where the proportion of the declining population that is hunted is potentially significant, or unknown. In fact there is a precedent for this from a court ruling in Romania in 2020, which invalidated the hunting quotas for 36 species on the grounds that there was insufficient scientific evidence to show that the populations would not be harmed (see Section 26.4).

Resumption of hunting from a moratorium should depend upon robust evidence indicating that hunting of the population is sustainable in combination with other existing pressures. If this is not the case then new, or updated, management plans and measures could be developed to address the key pressures and allow sustainable monitored hunting when populations have returned to an acceptable level. In any case, the European Commission should also enforce the requirement to provide bag statistics for all Annex II species, as the highly incomplete reporting is undermining the effectiveness of the Birds Directive. More efforts are also needed to clarify the numbers of birds that are being killed or taken under derogations, and to halt them if there is a significant risk that they may be contributing to declines.

As described in Section 2.2, a substantial indirect impact of hunting has been the death of birds as a result of poisoning from lead ammunition (Pain *et al.*, 2019). This has been estimated to kill about one million waterbirds a year in Europe (Mateo, 2009; Andreotti *et al.*, 2018) as well as significant numbers of birds of prey. An important recent conservation measure has therefore been the January 2021 amendment of the EU REACH Regulation[15] that bans the use of lead in gunshot in, or within 100 m of, wetlands. However, it has taken over 20 years to get to this stage, even though the environmental impacts of lead ammunition have been known for some time, alternatives to lead shot have been available, parties to AEWA agreed to phase out lead ammunition (AEWA, 1999) and there has been support from key hunting groups, such as FACE in their joint agreement with BirdLife International in 2004. Although some countries have already banned the use of lead shot, this is an example of how even a relatively practical and widely accepted conservation measure can be held up by vested interests (the ammunition industry) and some stakeholders who are resistant to change. It also remains to be seen if the regulations will be enforced and effective, as a similar ban in the UK since 1999 has only resulted in a small reduction in the use of lead shotgun ammunition in wetlands in England (Stroud *et al.*, 2021).

[15] Regulation (EU) 2021/57 amending Annex XVII to Regulation (EC) No 1907/2006 concerning the Registration, Evaluation, Authorisation and Restriction of Chemicals (REACH) as regards lead in gunshot in or around wetlands.

5.4 Protected Areas and Networks

5.4.1 Protected Area Coverage and Adequacy

The designation of protected areas that are of particular importance for habitats and/or species (and related landscape or cultural values) is one of the oldest and most widespread approaches to conserving nature. Consequently, one of the most prominent CBD Aichi targets (11) was, by 2020, to designate and conserve 17% of terrestrial areas and 10% of marine areas through effectively managed networks of protected areas, and other effective area-based conservation measures (OECMs). In Europe, as elsewhere, the concept of what protected areas are, and how they should be protected and managed, has evolved in response to new ideas. Typical changes have been for 'classic' protected areas to adopt broader objectives than nature conservation, which are dynamic rather than static, fulfil local as well as national needs, involve local communities, and are part of protected area systems and networks (Phillips, 2003). Such changes have been incomplete and have also reflected the differing ideas, requirements and contexts of European countries, leading to a wide variety of protected areas and management regimes.

Protected areas now differ greatly in their focus (e.g. nature versus other values) and strength of protection, and it is therefore not always easy to establish whether designated sites that are claimed to be protected areas actually meet the IUCN or CBD criteria (Box 1.2). Unfortunately, this means that the calculation of protected area coverage is fraught with difficulties, and the Common Database on Designated Areas (CDDA) data, held by the EEA, on nationally designated areas do not provide a reliable basis for such calculations. The same can be said of the World Database on Protected Areas (WDPA), held by the UN Environment Programme World Conservation Monitoring Centre (UNEP-WCMC), as the CDDA provides the official European input to the database. Whilst UNEP-WCMC is clear that countries are not supposed to include sites in the WDPA that do not meet the IUCN or CBD criteria (UNEP-WCMC, 2019), the countries make their own decisions on this and it is obvious that some have adopted very broad interpretations. Furthermore, no verification of whether the included sites meet the protected area criteria is undertaken by the administrators of the WDPA. Several chapters provide clear evidence of cases where protected areas have been included in the lists that do not meet the IUCN or CBD criteria (e.g. Bulgaria, Denmark, Latvia, Poland and the UK). In the UK this happened despite an IUCN-UK committee review, which clearly indicated that certain designated sites (e.g. National Scenic Areas in Scotland) included in the CDDA do not meet the protected area criteria (Crofts *et al.*, 2014).

As a result, the calculations provided by UNEP-WCMC on the Protected Planet website that are based on the CDDA/WDPA databases cannot be taken as a reliable assessment of protected area coverage and whether or not the country met its Aichi targets. The same problems relate to the European Commission's protected area dashboard,[16] as it is based on the same underlying CDDA dataset. A further complication is that the CDDA data also include OECMs (which are treated separately in the WDPA) and it is not clear if they are taken into account in the dashboard statistics. At least the Protected Planet website it is possible to see the types of protected area that are included, interrogate the data and drill down to the individual site level if desired. The dashboard provides only the overall indicator statistics and no explanation of what is included.

Clearly, more consistent interpretations of the protected area criteria and reliable protected area statistics are required to monitor performance with respect to CBD targets and the *EU Biodiversity Strategy for 2030* target to legally protect at least 30% of the EU's land and sea. A European Commission Technical Note (European Commission, 2020d) recognised that some sites listed in the CDDA may not meet the protected area standards that will be necessary. Similarly the Commission's Communication on the 2030 protected area targets acknowledges that 'some protected areas reported under the CDDA are protected for reasons that are not linked to the conservation of biodiversity' (European Commission, 2022b). Consequently the Commission has also acknowledged that some form of screening of existing and additional protected areas is necessary, and is setting up a procedure to do this through an expansion of the biogeographical process. Surprisingly, the Communication includes no mention of the IUCN and CBD protected area criteria, but instead proposes new criteria for Member States to use for screening protected areas and OECMs. The criteria principally relate to the need for legal protection, a long-term commitment to protect the area (which can be through administrative or contractual means), the identification of the area's natural values and the setting of related conservation objectives and measures.

Whilst the rationale for some of the changes is sound, the experiences and lessons documented in this book highlight the need for clarity, yet the criteria and process are not described in detail (half a page). Also, the Member States will continue to make their own decisions on what are protected areas, based on what are now arguably less defined criteria. Nor is there a clear process for ensuring consistent application of the criteria and verification that protected areas included in the CDDA meet the criteria. Such verification would be complicated and a huge amount of work given the new criteria, and therefore not feasible at a European Commission level. Instead, there seems to be more than ever a need for national, transparent and independent assessments of which designated sites actually meet the protected area or OECM criteria. These could build on and adapt the verification processes used for protected areas in the UK (Crofts *et al.*, 2014) and Denmark (Woolhead *et al.*, 2020). Protected area verification should also make parallel assessments against the IUCN and CBD criteria, as their criteria will presumably continue to be used in relation to CBD protected area targets.

Without further clarification on the EU's protected area and OECM criteria, and national-level independent verification, there is a risk of confusion and increasingly diverging standards and inconsistent data on protected areas and OECMs. This would undermine the credibility of the EU and CBD protected area targets and reported progress towards them.

At this point in the book, the original intention was to provide a graph showing the protected area coverage of each country, as an indicator of its progress towards conserving nature, including meeting the Aichi targets. However, it is now clear that, by itself, this would provide a potentially misleading figure due to the variable interpretation of the protected area definition by countries. Therefore, the current percentage protected area coverage as given in the WDPA is instead presented in Table 5.4, together with the protected area coverage as estimated from national data, if available, as referred to in the relevant country chapter in this book. For EU Member States, the Natura 2000 network coverage is also included as it provides a standardised and comparable measure of protection for habitats and species of EU importance. It is important to note that the protected area coverage totals include Natura 2000 sites, which also usually overlap considerably with protected areas designated under other international and domestic legislation.

The data presented in Table 5.4 show that overall coverage of CDDA/WDPA reported protected areas in European countries varies considerably. On land, coverage ranges from less than 10% in two countries to 40% or more in six. Coverage of marine habitats is lower overall, with less than 10% in 12 countries and over 40% in two. However, it is important to bear in mind that the percentage coverage is not necessarily a good indicator of the adequacy of coverage, as that will depend on a number of other factors. On one hand, countries with high proportions of natural or semi-natural habitat might potentially require correspondingly high levels of protected area coverage. On the other hand, such habitats are often in remote or inaccessible areas that are not highly threatened and priorities for protection. As the designation of such areas contributes to coverage targets, whilst having low economic and social costs, they tend to be over-represented in protected area networks – as found globally (Venter *et al.*, 2014, 2018). In Europe, this is notably the case in Finland, where a disproportionate part of the protected area network is in remote regions in the north.

In theory it is possible to evaluate the adequacy of the extent of the Natura 2000 network, as its development within each biogeographical region has involved a systematic assessment of the sufficiency of SCI coverage for each HD habitat

[16] https://dopa.jrc.ec.europa.eu/kcbd/dashboard/

Table 5.4 The coverage of Natura 2000 sites and total coverage of protected areas (PA) in European countries (taking into account overlaps).

Country	Terrestrial % Natura 2000	Total terrestrial % PA coverage		Marine % Natura 2000	Total marine % PA coverage		Comments
		From WDPA	From national sources		From WDPA	From national sources	
Albania	–	19	–	–	3	–	
Andorra	–	27	–	–	–	–	
Austria	15	29	29	–	–	–	Note 1
Belarus	–	14	–	–	–	–	
Belgium	13	15	?	38	38	38	
Bosnia & Herzegovina	–	4	–	–	0	–	
Bulgaria	35	41	?	8	8	?	
Croatia	37	39	?	9	9	?	
Cyprus*	29	61	?	9	9	?	Note 2
Czech Republic	14	22	22	–	–	–	
Denmark	8	17	<17	15	15	5	
Estonia	18	21	19	18	19	19	
Finland	13	13	14	10	12	?	
France	13	28	27	36	46	45	Note 3
Germany	15	37	?	46	46	?	
Greece	27	35	35	5	5	5	
Hungary	21	23	23	–	–	–	
Iceland	–	20	–	–	0.4	–	
Ireland	13	14	17	2	2	?	Nat 2019
Italy	19	21	20	4	10	?	Note 4
Latvia	12	18	12	15	16	16	
Liechtenstein	–	43	–	–	–	–	
Lithuania	13	17	18	24	24	30	
Luxembourg	27	51	–	–	–	–	
Malta	13	32	–	5	5	–	
Moldova	–	11	–	–	–	–	
Montenegro	–	14	–	–	0.1	–	
Netherlands	15	21	26	26	29	24	
North Macedonia	–	12	–	–	–	–	
Norway	–	30	–	–	10	–	
Poland	20	40	?	22	22	?	
Portugal (inc. Azores and Madeira)	21	23	Nc	1	17	c. 10	
Romania	23	24	?	21	23	?	
Serbia	–	8	–	–	–	–	
Slovakia	30	37	37	–	–	–	

Table 5.4 (*cont.*)

Country	Terrestrial % Natura 2000	Total terrestrial % PA coverage		Marine % Natura 2000	Total marine % PA coverage		Comments
		From WDPA	From national sources		From WDPA	From national sources	
Slovenia	38	40	40	3	2	?	
Spain (inc. Canaries)	27	28	Nc	8	12	12	
Sweden	12	15	15	13	16	15	
Switzerland	–	12	–	–	–	–	
Ukraine	–	13	–	–	9	–	
United Kingdom	9	29	25	18	37	38	Note 5
Average EU-28	20	28	?	15	18	?	
Average EU-27	20	28	?	15	17	?	
Average non-EU-28	–	17	–	–	4	–	
Average Europe	–	25	–	–	15	–	

Notes. Kosovo not included, as no WDPA data.
* For Cyprus the data correspond to the area where the Community *acquis* applies at present, according to protocol 10 of the Accession Treaty of Cyprus. '?' = No up-to-date combined protected area coverage estimates from national data available. Nc = Not covered in chapter.
1. Austria – national estimate excludes Ramsar Sites and the natural World Heritage Site.
2. Cyprus – WDPA estimate unreliable, true area closer to Natura 2000 29%.
3. France – national estimates exclude international sites not covered by domestic protection.
4. Italy – Natura 2000 data from national source not EEA.
5. UK – Official terrestrial PA coverage indicator of 28% includes sites that are not protected areas.

Sources. Based on Natura 2000 coverage from the EEA Natura 2000 barometer, June 2021 release showing situation end 2020. % WDPA based on total terrestrial and marine protected areas from WDPA Protected Planet[17] Protected Area Profiles (UNEP-WCMC, January 2022), divided by total land area from EEA Natura 2000 barometer, or total MSFD marine waters. Protected Planet estimates do not include Biosphere Reserves. See country chapters for national data sources.

and species (Section 4.2). Summaries of the assessments for each Member State have been published, but seemingly not since 2016 in the evaluation study supporting the Nature Directives Fitness Check (Milieu *et al.*, 2016).

Similarly, the adequacy of the SPA network can to a large extent be judged by comparing it with the Important Bird Areas (IBAs) identified by Birdlife International (2013) and more recent national updates. This was most recently carried out for terrestrial IBAs through a GIS study in 2016, which found that they were generally well covered by the SPA network, with three-quarters of IBAs completely or partially covered, and 66% of the total IBA network extent covered by SPAs (Kukkala *et al.*, 2016). Nevertheless, it is clear that at the time there were low levels of coverage in some countries: below 50% in Spain, Belgium, Luxembourg and Malta. A study of marine IBAs in 2014 found that there was overall 59% coverage by SPAs, but also wide variation amongst the countries (Tarzia and Campos, 2014).

At the moment it does not appear to be possible to quantify the adequacy of either the SCI or SPA components of the Natura 2000 network using publicly available up-to-date data. However, in recent years the European Commission has stated that it generally considers the terrestrial Natura 2000 network to be virtually complete, but that gaps need to be filled within the marine environment. Up-to-date information on the coherence of the Natura 2000 network in terms of connectivity also appears to be unavailable. Nevertheless, as recognised by the European Commission (2020d), the network includes areas that are too small and too disconnected from one another.

Evidence from a number of studies (e.g. EEA, 2012; Evans, 2012), and chapters in this book, show that the requirement to designate SPAs and then SCIs/SACs to create the Natura 2000 network greatly increased protected area coverage in many Member States, including countries where it was low (e.g. Croatia, Greece, Estonia and Spain) and especially for marine sites in most countries. In addition, the Natura 2000 designation criteria (HD Annex III) and habitats and species lists also often created a more representative and coherent network of nature conservation sites than those designated under domestic legislation.

5.4.2 Protected Area Categories, Protection and Management

In the European context, it has often been difficult to classify protected areas in accordance with the IUCN management

17 www.protectedplanet.net

categories (Dudley, 2008; Box 1.2). Some countries have followed the IUCN categories, most often in relation to Category II national parks, and also Category Ia strict nature reserves and Category Ib wilderness areas. However, some protected areas do not closely match the IUCN categories, and many can fit into multiple categories. It can often be particularly difficult to categorise Natura 2000 sites, as these can vary from large areas of wilderness or highly modified cultural landscapes to very small sites for a single habitat or species. Also, whilst they allow in principle many human activities (e.g. agriculture, forestry and hunting), they arguably have amongst the strictest levels of protection (as further discussed later). Consequently, many remain uncategorised by Member States. When they are categorised, they are typically Category IV areas.

In practice, protected areas in Europe tend to vary in a gradual way along two distinct axes: the degree to which they focus on nature conservation as an explicit objective; and the strength of protection given to the site and specific features within it. A complication is that some countries have incorporated tiered zones of protection within protected areas. Despite the complexity, IUCN protected area Categories I–IV generally provide most protection for biodiversity according to many chapters in this book. This has also been recognised in a global review by Locke and Dearden (2005), who suggested that Categories V and VI should be reclassified as sustainable development areas. In reality the distinction is not clear-cut; evidence shows that all protected area types, including those designated primarily for aesthetic and cultural interests, can be expected to contribute indirectly to nature conservation to some extent, for example as was found in an influential UK study (Lawton *et al.*, 2010).

Evidence from this book's national chapters also shows that the practical degree of protection tends to decline from Categories I to VI. However, this is not always the case, as the strength of protection can be multifaceted, for example in relation to the protection of the site from destruction or damage and/or restrictions on activities that are allowed within the site. Category V and VI protected areas also normally rely more on stakeholder involvement and financial incentives than legal measures to achieve their objectives. They have therefore often been the most appropriate and practical means of including private land within protected area networks, other than within Natura 2000 sites.

Given the above, whilst Category V and IV protected areas have a significant role to play in nature conservation in Europe, it is of concern that these categories make up a high proportion of the total protected area coverage in many countries, for example in Cyprus, Germany, Lithuania, Poland and the UK. This should therefore be taken into account when assessing the adequacy of protected area coverage, whilst at the same time considering which categories are most appropriate and effective in the national circumstances.

In addition to extending the protected area network in most countries, the establishment of Natura 2000 has provided a more consistent and effective means of protecting sites than most domestic protected area legislation. Whilst Natura 2000 sites are rarely wholly dedicated to nature conservation or prohibit most human activities, they have strong legal protection from damage through HD Article 6, which also follows the precautionary principle. Member States must prevent their deterioration (Article 6(2)) and carry out appropriate assessments of potentially damaging projects and plans (Article 6(3)). Under Article 6(4), damaging activities can be permitted when of national importance, but only in the absence of alternatives, and compensation measures must be taken to fully offset residual impacts. The compensation requirement is also a further disincentive to damage sites as it can be very costly. In most countries, such compensatory actions are not normally required following damage to other types of protected area, or are rarely implemented in practice (e.g. in Sweden).

Initially there were significant problems with the transposition and practical implementation of the Habitats Directive Natura 2000 site protection measures, which led to numerous infringement procedures. This was in part due to varying interpretation of some of the provisions and legal uncertainty, as well as limited expertise, knowledge and capacity in the responsible authorities (Sundseth and Roth, 2013). As discussed in Section 4.2, jurisprudence and European Commission guidance has since dealt with many of the problematic issues and the provisions are now generally working well and giving relatively high levels of protection as envisaged (Milieu *et al.*, 2016). Nevertheless, as described in several of the national chapters in this book, there are significant ongoing problems; see also Justice and Environment *et al.* (2017) for some solutions. The problems often concern the underestimation of impacts of proposed projects and plans on Natura 2000 sites, typically due to inadequate expertise and biodiversity information. Sometimes obvious impacts have been ignored, probably due to political pressure (e.g. in Bulgaria, Cyprus, France and Ireland). A common difficulty has been the consideration of cumulative impacts, which therefore tend to be ignored in assessments. Similarly, potentially damaging agricultural and forestry improvement projects are often not subject to appropriate assessments at all.

As a result of the strict HD Article 6(4) conditions and compensation requirements, most Member States have avoided using this derogation option. On the one hand this is encouraging, because its use is supposed to be an exception. On the other hand, the desire to avoid the derogation is probably a major reason why impacts are often ignored or underestimated in appropriate assessments. In cases where damaging projects and plans have been allowed to go ahead, there have been concerns over whether they really were for justifiable imperative reasons of overriding public interest, such as the numerous cases in the UK (Clutten and Tafur, 2012). Reporting on the compensatory measures taken by Member States has also been inadequate, and it is therefore often unclear if they have achieved their objectives.

Marine Natura 2000 sites, and other MPAs, are also being inadequately protected by the Nature Directives and national legislation from the impacts of fisheries (described in Section 4.3). Whilst the CFP has measures to deal with such impacts, they require the agreement of all affected countries. This has proved to be a major barrier, and almost no fishing measures have actually been established as a result.

To achieve their objectives, protected areas need to be **effectively managed**, and this is therefore explicitly recognised in the Aichi target. In this respect, another strength of the Nature Directives is that they include comprehensive and coherent provisions that require management measures to be put in place. Furthermore, they must be focussed on achieving the site-specific conservation objectives of each habitat and species for which each site is designated. This requirement recognises the fact that, as discussed in Chapter 2, in Europe most protected areas require proactive management (e.g. to counter vegetation succession processes, eutrophication and IAS) rather than simply protection and non-intervention.

Despite this clear requirement, it appears that the management of Natura 2000 sites has mainly been inadequate so far, especially the establishment of necessary agricultural and forestry practices (discussed later). This is partly due to the length of time that it has taken to designate sites in some countries, and to subsequently develop management plans or similar management frameworks. Progress has accelerated in recent years, such that about 70% of Natura 2000 sites have management plans according to Gerritsen *et al.* (2020). Many of the plans have been prepared as part of EU funded LIFE projects, or with European Regional Development Fund (ERDF) funding (e.g. Romania and Slovakia), but the plans need to be implemented more fully to achieve their intended tangible results (ECA, 2014; Gerritsen *et al.*, 2020). In most cases the identified management measures are voluntary for landowners and other stakeholders, and only mandatory for state authorities. Implementation therefore normally depends on compensation payments for restrictions on activities and grants for more positive management actions. As discussed later, funding constraints have often limited implementation, as well as compliance monitoring and enforcement.

Problems have also occurred where the management plans have been prepared through a top-down superficial process that has not achieved sufficient ownership of the plans and support for their measures. Although the benefits of deeper participatory approaches and joint goal-setting have been known for some time, these have been the exception rather than the rule (Bouwma, 2010; Blicharska *et al.*, 2016). This has often been due to time constraints and a lack of suitably qualified staff in the nature authorities. However, where nature authorities have found the resources and time, the investment has often proved to be worthwhile in the end. This has, for example, been demonstrated in Belgium and France, where stakeholders have been closely involved in developing Natura 2000 site management plans.

Little systematic information is available on the adequacy and effectiveness of the management of the wider protected area network in most European countries. Although the Protected Planet database indicates whether a site's management effectiveness has been assessed, the result of the assessment is not provided or summarised for the country. Information from the national chapters in this book suggests that the situation is mixed, with some countries and protected area types being effectively managed (e.g. the Czech Republic and national nature reserves in France). More often, the required management is lacking and the sites' conservation objectives are not being met or are unambitious and primarily aim to maintain the status quo. A notable example is in Ireland, where the setting of conservation objectives for Natura 2000 sites, management planning and the establishment of management and restoration measures for all protected areas have been slow and incomplete. This has resulted in a reactive approach to nature conservation, including the need to use limited resources to address infringement cases.

In conclusion, it is clear that protected areas have formed a key component of nature conservation in Europe and have the potential to continue doing so, including in the face of increasing climate change. Increasing their coverage may be necessary in some countries, especially in the marine environment. However, the rationale for the EU's seemingly arbitrary 30% protected area coverage target is questionable and may have perverse consequences (Barnes *et al.*, 2018; Visconti *et al.*, 2019). Where increases in protected area coverage are required, these would be most effective if targeted to address specific requirements based on scientific evidence, which could require more than 30% protected area coverage in some areas. This might include increasing the size and connectivity of the overall network of Natura 2000 sites and other protected areas where it is fragmented and vulnerable to climate change impacts.

The highest priority seems to be to verify the currently claimed protected areas and ensure they are being effectively protected and managed and achieving their conservation objectives. Given its wide extent and focus on habitats and species of high EU conservation concern, this is especially important for the Natura 2000 network. The legal requirement for each Natura 2000 site to have appropriate conservation objectives and effective comprehensive management measures in place needs to be met, and enforced where necessary.

5.5 General Nature Conservation Measures

5.5.1 Development Planning and Impact Assessment

The protection of biodiversity from destruction from developments (e.g. buildings, infrastructure, mining) in the wider environment, and to some extent in protected areas, is highly dependent on spatial planning processes. Of particular importance is the degree to which planning processes consider nature conservation issues and the weight they give to them in making decisions. Systematic information on the extent to which biodiversity has been protected through the spatial planning system does not seem to be readily available.

ɔwever, the national chapters in this book provide some sights on the recent situation. In some countries development planning procedures and the consideration of biodiver-y continue to be very weak (e.g. Bulgaria, Cyprus and reece). The problems have often been concentrated along asts or in other areas where there is a considerable demand r holiday homes and tourist developments. This has allowed gnificant habitat loss and fragmentation, disturbance and her indirect impacts in such areas.

More generally, there have been considerable improve-ents in planning processes and the consideration of biodiver-ty in decisions. It is now common good practice for spatial lans to indicate areas of high biodiversity importance, which ıclude not only protected areas but other areas of significant iodiversity value, ecological corridors and other elements of reen infrastructure. Development zones generally avoid these reas, and project proposals that are not in accordance with the oning tend to get refused. However, further improvements re required, including in the marine environment in response o the Maritime Spatial Planning Directive, and interactions vith the MSFD (as discussed earlier). Belgium has been a ɔioneer in this regard, with marine spatial plans that have ntegrated management plans for Natura 2000, as well as the MSFD good environmental status objectives, to address all relevant pressures and (socio-)economic sectors necessary to allow the recovery of degraded habitats and species. More commonly in Europe, better integrated marine spatial planning is required, which would be helped by the development of guidance and best practices such as in relation to the ecosystem-based approach.

The treatment of biodiversity in spatial planning has also been considerably strengthened in the EU by the EIA Directive in 1985 and SEA Directive in 2001, as these have evolved since being first adopted. They have contributed to the protection of the environment by ensuring that the potential environmental impacts of projects (EIA) and plans/programmes (SEA) are analysed and publicly reported to inform decision making on their acceptability. Impact assessments were already being carried out before the Directives came into force, but they have raised standards. Revisions to the EIA Directive have also improved the treatment of biodiversity and integration with appropriate assessments under the Habitats Directive. However, there are also numerous cases where EIAs and SEAs have not dealt adequately with biodiversity issues, often for the same sort of reasons as discussed earlier in relation to appropriate assessments. Thus, biodiversity impacts, especially cumulative impacts, are sometimes overlooked or underestimated or given little weight in decision making. In some cases EIAs have been carried out after the project has been given permission (e.g. in Poland). As with appropriate assessments, EIAs have not been adequately applied to agriculture and forestry.

One of the main weaknesses of EIAs is that there is little scope for relocating the project, because the developer has usually acquired the land or a development allocation. As a result, EIAs mainly only help mitigate the impacts of the development (e.g. by changes in design) and rarely lead to a project being refused or relocated. Higher-level regional strategic planning and SEA has the potential to address this problem as it considers environmental issues early in the process. It can then examine options and keep the most damaging projects away from the most environmentally sensitive areas. Although SEA procedures have not been covered in depth in this book, there is evidence from a REFIT evaluation by the European Commission that SEAs are improving high-level decision making and contributing to biodiversity conservation (European Commission, 2019c). This could be further improved by ensuring that SEA is carried out for all relevant plans and programmes, beyond those currently within the scope of the SEA Directive, which are restricted to those that are required by legislative, regulatory or administrative provisions. SEA should also be given more weight in the planning process, start sufficiently early, adequately consider alternatives, and improve monitoring of environmental impacts.

5.5.2 Habitat Management and Restoration

As discussed in Chapter 2, the conservation of nature in Europe is highly dependent on the appropriate management of agricultural land and forests, and therefore these issues are discussed here.

High Nature Value Farmland and Other Agricultural Areas

As regards agricultural land, the most important requirement is to maintain HNV farming systems and their beneficial practices, because they are vital for the semi-natural habitats that are by far the most biodiverse and threatened. Elsewhere, it is mainly necessary to mitigate the damaging practices of intensive agricultural systems, especially by maintaining or restoring semi-natural habitat elements and heterogeneity in the landscape (Benton *et al.*, 2003; Billeter *et al.*, 2008). In the EU, meeting these requirements is critically dependent on the CAP, which has been widely blamed for encouraging environmental degradation, including the loss of HNV systems and semi-natural habitats, and increased intensification and specialisation across the farmland landscape (discussed in Section 2.2). This has been partly the result of subsidies that encouraged increases in production and/or damaging agricultural improvements. However, as described in Section 4.3 (and online Annex 3),[18] in recent decades the CAP has been reformed (albeit slowly) and most subsidies are no longer coupled to production. Moreover, it provides by far the largest potential source of funding for practical conservation actions for terrestrial habitats through RDP measures. Of these, the agri-environment (–climate) measures have been most effective, when well-designed and targeted (Evans *et al.*, 2002; ECA, 2011; Batáry *et al.*, 2015; Alliance Environnement, 2019). However, in most countries the largest proportion of RDP

[18] All annexes are free to access online at CUP: www.cambridge.org/natureconservation

funds has been allocated to payments to areas with natural constraints, without any environmental conditions. Although these payments may have helped support HNV systems, evidence of their impacts and value for money is lacking.

The agri-environment–climate measures, and some other RDP measures, have helped to conserve and restore HD Annex I habitats and other semi-natural habitats (e.g. in Austria, Estonia, Hungary, Ireland and Slovakia). Declines in populations of some species have also been reversed with well-designed and targeted schemes, such as for several declining farmland birds in the UK. However, such successes have normally been at relatively small scales, and rather few agri-environment schemes have led to national or regional improvements in the status of BHD habitats and species (Tucker *et al.*, 2019).

The limited nature conservation benefits of agri-environment schemes and other RDP measures have sometimes been due to limited cooperation between agriculture and environment authorities, poor design and a lack of tailoring and targeting of schemes to the habitat's and/or species' ecological needs. More recently, scheme designs have improved, and limitations on their beneficial impacts have been largely due to inadequate budget allocations (discussed later) and sometimes low uptake of funding opportunities by farmers, foresters and other landowners/managers (Alliance Environnement, 2019). There are a variety of reasons for the low uptake, but one of the most fundamental problems is that, under the CAP Rural Development Regulations, landowners can only be compensated for forgone profits, plus additional costs. This provides very little incentive for landowners to take on the schemes unless they are already carrying out the desired farming practices (e.g. in HNV areas) or sufficiently motivated to undertake nature conservation activities. In reality, most farmers and foresters are reluctant to substantially change their practices for nature conservation purposes, for a variety of personal, cultural and economic reasons, particularly if they require major changes that would be difficult to reverse (Lastra-Bravo *et al.*, 2015; Geitzenauer *et al.*, 2017). This reluctance to take up schemes has been exacerbated by a lack of capacity in nature authorities and RDP management authorities to promote the schemes and provide supporting guidance, which has been shown to be an important factor leading to high scheme uptake and success (Alliance Environnement, 2019).

A further disincentive to take up environmental schemes has been the competing opportunities provided to landowners to use some alternative CAP funding measures to increase their agricultural and forest profitability, such as through various agricultural or forestry improvements or modifications (e.g. drainage, replanting, irrigation and new machinery). This has not only undermined the implementation of environmental policy measures, but exacerbated the pressures that they seek to address. These issues have led to unspent RDP allocations, which have tended to be for the schemes most beneficial to biodiversity. As a result, some authorities have simplified their schemes, resulting in reduced conservation benefits, and/or switched funding to simpler, more popular measures, such as organic farming, farm modernisation grants or blanket support through payments to areas with natural constraints (Alliance Environnement, 2019). Reallocation of RDP funds away from successful and widely taken up biodiversity beneficial measures also appears to have taken place, sometimes for political reasons, as was most obviously the case recently in Hungary (Chapter 18).

The conservation of nature over wide areas in typical modern-day intensively managed farmland has proved to be particularly difficult, due to the scale of intervention that is required and its high per unit area and total costs. In an attempt to tackle this, the 2014–2020 CAP further integrated environmental considerations into the Pillar 1 direct payments through the 'greening measures': ecological focus areas (EFAs), maintenance of permanent grassland and crop diversification (described in Section 4.3). Overall these did not achieve their potential, and mainly led to small changes in management practices (Alliance Environnement and Thünen-Institut, 2017; ECA, 2017a; Alliance Environnement, 2019). In the case of EFAs, this was primarily because the options offered by Member States and chosen by farmers tended to be those that were easiest to administer, most economically beneficial and the best fit with the farming system – which happened to provide low biodiversity benefits. However, the EFA measure did help to reverse the loss of fallow land in some areas, such as Spain, with potentially significant biodiversity benefits, including for some threatened birds (e.g. bustards). The requirement to designate environmentally sensitive permanent grasslands within Natura 2000 sites may also have helped bolster their protection.

Overall, the CAP's measures and funds have not been used anywhere near the level necessary to halt farmland biodiversity declines, including in relation to supporting Natura 2000 management, despite the introduction of the Prioritised Action Frameworks (PAFs). Furthermore, there may still be unintended damaging impacts from some CAP measures, in particular the inappropriate use of some RDP measures, and voluntary coupled support payments, which are being applied to some crops and over half the EU beef and dairy herd (Alliance Environnement, 2019). The main uncoupled direct payments may also be detrimental for biodiversity, as they increase the value of land and may influence farmer attitudes, such that they are more likely to invest in agricultural improvements, some of which may be damaging for biodiversity (e.g. drainage and irrigation).

The lessons from recent years point to the urgent need for a further improvement in the protection of nature under the CAP, especially semi-natural habitats, such as through stronger and more comprehensive environmental conditions on payments and a major increase in the scale of targeted and tailored agri-environment–climate, or similar, RDP schemes.

Under the new CAP, operating from 2023, there is some strengthening and widening of conditions on payments through 'enhanced conditionality'. This incorporates revised versions of most of the 'greening' measures and previous environmental conditions (known as cross-compliance), as well as introducing new protection for peatland and wetlands. However, experience suggests that the effectiveness of these conditions will continue to depend greatly on the Member States' ambitions, interpretations and enforcement. The main new environmental measures in the CAP are eco-schemes, which Member States are required to fund using 25% of their Pillar 1 direct payment budgets. They cover a potentially wide range of climate, nature, soil and animal welfare objectives and related agricultural practices (e.g. organic farming, agro-ecological practices, agro-forestry and maintaining HNV farmland). Member States will also have considerable flexibility in terms of what they offer and how they are applied. They are voluntary for farmers and only require a one-year commitment. Given their broad scope, flexibility and potentially substantial funding, they have the potential to have wide impacts on the farmland landscape. However, lessons from similar previous schemes suggest that farmers will select the simplest options that fit best with their farming systems. Biodiversity impacts could therefore be mixed, with benefits being relatively moderate if organic farming or rather general and basic interventions predominate.

The EU seems to be relying considerably on organic farming to address environmental problems, as the target to increase coverage from the current 8% to 25% of agricultural land is ambitious and prominent in the *Biodiversity Strategy for 2030*, *Farm to Fork Strategy* and EU Green Deal. The wisdom of this decision seems questionable. Organic farmland does tend to be more biodiverse than conventional intensive farmland, on average supporting about 30% more species, with a 50% higher abundance (Bengtsson *et al.*, 2005; Tuck *et al.*, 2014). However, the benefits are very variable, tend to be overridden by landscape heterogeneity factors and are primarily for generalist species (Winqvist *et al.*, 2012; Tuck *et al.*, 2014). Furthermore, the modest positive environmental benefits are inversely correlated to crop yield, and most are not apparent on a unit of production basis (Tuomisto *et al.*, 2012).

Increasing organic farming will, therefore, do little to tackle the much more serious losses of semi-natural habitats and specialist farmland species that are occurring in Europe. In fact, increasing organic farmland on this scale might undermine other conservation actions as it will absorb a considerable amount of RDP funding that could be used for more cost-effective interventions (e.g. Hodgson *et al.*, 2010). The large increase in organic production could also lead to indirect land-use changes caused by the drop in production, as organic yields are on average 19–25% lower (Seufert and Ramankutty, 2017). This could result in commodity price increases that encourage intensification, including the conversion of semi-natural habitats to more intensive farmland in Europe and the destruction of highly biodiverse habitats elsewhere. Thus, as concluded by Gabriel *et al.* (2013), 'the relatively low yields of organic farms may result in larger areas of land being brought into agricultural production (locally or elsewhere), at a biodiversity cost much greater than the on-farm benefit of organic practice'. Such unintended negative impacts are by no means certain, but their risk should be more thoroughly assessed before the EU goes too far with implementing its organic farming action plan (European Commission, 2021a).

If eco-schemes are dominated by basic measures, then at least this could help reduce the use of Pillar 2 RDP funding for them, thereby freeing up more funds for more ambitious and targeted agri-environment–climate schemes. This could be important as the agreed MFF funding allocation for the European Agricultural Fund for Rural Development (EAFRD) of €95.5 billion (including Next Generation EU funding) is less than the €100 billion allocated for the 2014–2020 CAP (although that included the UK).[19] Whilst the minimum proportion of the RDP budget that Member States must use for environmental and climate-related objectives has increased from 30% to 35%, this does not provide the extra funding needed to substantially increase the scale of the most effective nature conservation measures, especially if a large proportion is used to meet the organic farming target.

The overall impacts of the new CAP on nature will greatly depend on how much the CAP Strategic Plans, which Member States have to prepare for Pillar 1 and 2 funds, incorporate nature conservation and restoration requirements, and especially the needs identified for Natura 2000 in the PAFs. At the time of writing, indications are that the Strategic Plans lack ambition.[20] The EU's proposed Nature Restoration Law could therefore be crucial in raising ambitions and leading to farmland targets that are practical, measurable and address biodiversity priorities.

In the longer term, more radical changes are also needed to link the CAP more closely to the wider environmental ambitions now included in the EU Green Deal. This should include broadening the scope of the CAP so that it becomes an integrated strategy for sustainable land management. The arguments for generalised income support through the Pillar 1 direct payments are now very weak; instead it is increasingly recognised that public money should be spent on public goods, including the conservation of biodiversity. This is now the aim of the new environmental land management schemes that are being established in England as a result of the new freedom from the previous CAP rules (Section 31.9). As there is a transition period it will take several years before the schemes are fully in place, but eventually it will be illuminating to compare the diverging UK and EU approaches to agricultural policy and their impacts.

Forest Management

According to the results in the *State of Nature in the EU*, the conservation status of European forests and their species

[19] https://agriculture.ec.europa.eu/common-agricultural-policy/financing-cap/cap-funds_en#eafrd

[20] As reported by BirdLife Europe, March 2022: www.birdlife.org/news/2022/03/17/cap-national-strategic-plans-intensive-agriculture/

appears to be of only slightly less concern than for agricultural and other semi-natural habitats. A low proportion of BHD forest habitats and species have a good status, and many are still deteriorating, especially specialist forest species. This is an issue of particularly high concern for the extensive forests in the Boreal biogeographical region, as for example discussed in relation to Lithuania and Sweden. Only a tiny proportion of European forests are primary forests, and most of the wider forest area is subject to fairly intensive forest management, including clear-cutting cycles and the planting of non-native species (Section 2.2). Where improvements in the biodiversity status of forests have occurred, they seem to have been partly due to a general adoption of sustainable management principles and related indicators, which have been developed through the pan-European FOREST EUROPE process. However, these are broad and rather unambitious, and have therefore only improved the situation in the most intensively managed forests.

In the EU, more ambitious environmental aims for forestry have been constrained by the EU's limited competency over EU forest policy and a powerful forestry industry lobby in several countries (e.g. Finland, Poland and Sweden). Consequently, the EU's forest strategies and action plans have mainly promoted the adoption of sustainable forestry principles and additional voluntary actions. Forestry interests have also persuaded some governments to allow increases in extraction rates, which in Latvia seem to be associated with declines in forest bird populations (Figure 21.1).

Forest management plans have been considered to be a key mechanism for integrating biodiversity objectives into forestry management. The *EU Biodiversity Strategy to 2020* therefore encouraged Member States to use them and incorporate biodiversity measures (under Actions 11 and 12 supporting Target 3b). Forest management plans are also required to access RDP funds for forests over a certain size (which varies according to Member State, from any size forest up to over a hundred hectares). According to Forest Europe (2020), 75% of Europe's forests were covered by a forest management plan at the time. However, it is uncertain whether such plans have normally addressed biodiversity requirements satisfactorily.

Academic studies (e.g. Winkel *et al.*, 2015; Sotirov, 2017) and several chapters in this book have revealed that the establishment of Natura 2000 sites in forests has been problematic. Nevertheless, the protection of important forests has been strengthened (e.g. Poland, see Annex PL.4) and the ecological condition of many has improved. Some countries have made good progress in mainstreaming forest biodiversity requirements into forest policy and forest management plans, especially within publicly owned forests (e.g. Slovenia). However, many Natura 2000 sites lack forest management plans, or measures within them that are adequate to meet the site's conservation objectives. Many problems have occurred with privately owned forests, because the establishment and management of Natura 2000 has low legitimacy and acceptance from forest owners and land-users, who feel they are not well compensated for changing their practices (Sotirov, 2017). Consequently, forest management plans, or the implementation of their actions, have tended not to be mandatory for privately owned forests.

CAP RDP measures are the main source of EU funding for forests: the Natura 2000 measure can compensate for income forgone from forestry restrictions in Natura 2000 sites, whilst other measures can also support environmental improvements. Despite the apparent need, few Member States have used the RDP measures for biodiversity objectives in forests. Where they are available, take-up by forest owners has often been low (Geitzenauer *et al.*, 2017). As a result, over 2014–2020 RDP forest measures (M8.5 and M15) were only programmed for biodiversity purposes over about 0.4% of EU forest area (Alliance Environnement, 2019). Consequently, despite some good examples in individual countries (e.g. Slovakia), Member States had not maximised the potential of these RDP measures (Alliance Environnement, 2017, 2019) and there has clearly been a substantial funding shortfall for Natura 2000 forest management (Sotirov, 2017).

It is obvious that more could be done to improve the ecological condition of Europe's forests, such as by strengthening and better integrating the nature conservation and forest policy framework, raising sustainable forestry standards, fostering better relations between forest owners and conservationists, increasing the use of well-designed and targeted RDP forest measures, and taking a larger area out of forestry management. In the EU, the latter could be supported by the 10% strictly protected areas target included in the *EU Biodiversity Strategy for 2030* and *EU Forest Strategy for 2030* (European Commission, 2021b). Other initiatives in the *EU Forest Strategy* seem unlikely to make a major difference to the ecological condition of forests, as proposed actions are relatively modest and many are voluntary. There is also concern that it proposes to boost the sustainable bio-economy, even though there is evidence that forest biomass harvesting for bioenergy is already driving forest intensification and detrimental impacts on BHD habitats and species (Bowyer *et al.*, 2020). Despite this, both the *Forest Strategy* and the Renewable Energy Directive continue to lack strong safeguards against biodiversity damage from bioenergy production.

5.6 Nature Conservation Funding

Although it was envisaged that the implementation of the Nature Directives would be primarily financed through national sources, their lack of clear provisions to ensure adequate funding has been an obvious weakness (Crofts, 2014). Whilst some EU funding instruments have since been used to provide co-financing to Member States, as described in Section 4.2.6, this has not met funding needs. The most recent EU assessment in 2011 suggested that EU funding met no more than 20% of the estimated costs of managing the Natura 2000 network, and that national funding was insufficient to fill the gap (European Commission, 2011). This underfunding has been one of the main constraints on the implementation of the Nature Directives, especially with regards to the establishment of management measures within Natura 2000 sites and other protected areas. This is evident from numerous studies (e.g. Milieu *et al.*, 2016; Geitzenauer *et al.*, 2017; Kettunen *et al.*, 2017) and almost all national chapters in this book, for instance on Croatia, Portugal, Poland, Romania and even the UK, despite its relative wealth. Funding problems have also limited the achievement of other objectives of the *EU Biodiversity Strategy to 2020*, most notably the Target 2 restoration of 15% of degraded ecosystems, as well as national nature conservation ambitions. It is also evident that this is a pan-European problem (Hajdu and Kiss, 2009). For example, in a similar way, the lack of a financial mechanism with adequate, stable and predictable funding was a major obstacle for the implementation of the *Pan-European Biological and Landscape Diversity Strategy* – PEBLDS (Prip, 2013).

Evidence from this book's national chapters, a review for the European Commission of biodiversity financing (Kettunen *et al.*, 2017) and numerous other studies indicates that the underfunding of nature conservation in Europe has resulted from two main causes: inadequate allocations due to an unwillingness to prioritise biodiversity; and problems accessing and using potentially available funds. The first and primary problem arises from the high dependency many Member States have on EU co-financed funds that are managed by sectors other than nature conservation, through the so-called integrated funding model (described in Section 4.2.6). The only dedicated biodiversity fund is the LIFE programme, which is relatively very small, meeting just 2.2% of estimated annual spending needs of Natura 2000 over 2007–2013 (European Commission, 2016). Although it has been found to be very cost-effective for its size, it has primarily functioned as a demonstration and trialling funding instrument. Often the expectation has been that larger funds, such as RDP measures, will provide the scaled-up and long-term support that is necessary to achieve the intended nature conservation ambitions. So far, this has rarely happened.

The integrated funding approach is consistent with EU policy and logical in theory, but has not worked as intended due to deep-rooted institutional factors. In practice, it has led to one-sided struggles between weak nature conservation authorities and more powerful bodies representing important economic sectors including agriculture, forestry, fisheries, energy and transport. Although there is now good evidence that biodiversity conservation and restoration provides economic benefits and high benefit ratios for investments, this remains insufficiently recognised outside the environmental sector (e.g. TEEB, 2011; Dasgupta, 2021). The economic benefits of nature conservation are also rarely perceived by local farmers, foresters and other businesses, who therefore fight to retain what they consider to be their rightful funds (Geitzenauer *et al.*, 2017). The result has been limited allocations of funds for nature conservation and restoration, especially to the most important and effective measures. There has also been relatively poor link-up with Natura 2000 management plans and follow-on from successful LIFE projects (Wätzold *et al.*, 2010).

Whilst the introduction of the PAFs helped to increase the flow of funding, some lacked ambition and quality, undermining their value (Milieu *et al.*, 2016). The funding allocations set out in this book's national chapters and studies of national/regional funding programmes (ECA, 2014; N2K Group, 2016) reveal that the integration of specific objectives and measures that are relevant to Natura 2000 and broader biodiversity conservation has been very variable. Most biodiversity funding has been allocated through EAFRD for RDP measures. The ERDF, Cohesion Fund and the European Social Fund (ESF) have been little used for biodiversity in most countries, but a few have made good use of them, such as for management planning (e.g. Greece, Lithuania and Slovakia). As mentioned earlier, the EMFF has been very little used for nature conservation purposes. In general, the total planned funding has not normally covered all the needs identified in the PAFs (ECA, 2017b), let alone the requirements for wider nature conservation measures. Incentives to allocate funds to Natura 2000 sites, and for other biodiversity purposes, have also been undermined by the lack of information on Natura 2000 expenditure (ECA, 2017b) and the crude tracking of EU biodiversity-related expenditure, which uses the simple OECD Rio markers of 0%, 40% or 100% contributions to biodiversity (Ernst & Young and Biotope, 2017).

Other responses to the ineffective integrated funding model have consisted of calls for more radical changes to nature conservation funding. These have included suggestions for an independent fund for Natura 2000 (Kati *et al.*, 2015) or a dedicated environmental fund (NABU, 2015). Another suggestion by NABU is for a more formalised legally binding approach in which some existing funds are co-managed by environmental authorities with some prior allocations 'earmarked' for biodiversity objectives. Whilst independent dedicated funds could be more targeted and cost-effective (as in the LIFE programme), there is a high risk that overall funding for biodiversity could decline (Kettunen *et al.*, 2017). This is because politicians and other decision makers would probably be reluctant to give to biodiversity, by itself, as much as is provided as part of the multiple objectives under the other funding instruments (CAP, ERDF, EMFF, etc.). For the moment, the European Commission appears to have rejected the idea as the *EU Biodiversity Strategy for 2030* lacks a dedicated biodiversity fund.

The EU Member States have, however, agreed that 7.5% of the 2021–2027 MFF will be dedicated to biodiversity objectives as of 2024, and 10% as of 2026. With a total MFF budget of €1 211 billion (in 2021 prices) this equates to about €13.0 billion per year from 2024, and €17.3 billion per year from 2026. This EU public funding has the potential to make a substantial contribution to the minimum of €20 billion per year required for Natura 2000 and green infrastructure, but additional national and private funding will be required.

Furthermore, the integrated funding framework of primarily using the CAP, regional funds and the EMFF will continue to be used. As before, this may result in inadequate national allocations and uptake of funds that provide the most significant biodiversity benefits. Much will therefore depend on the Member States' priorities, the Commission's tracking of biodiversity expenditure and, crucially, the definition of biodiversity expenditure: if it is too broad, the billions will be too thinly spread and not focussed on priorities such as Natura 2000.

The second main barrier to nature conservation funding has been capacity constraints in environmental authorities (especially at regional and local levels) and NGOs, which have hindered their ability to apply successfully for potentially available funds (ECA, 2014; Ernst & Young and Biotope, 2017). This has resulted in a rather uneven distribution of LIFE projects, because there have been no needs-based Member State allocations, and projects have been awarded purely on the basis of the quality of the application. The rather onerous application process for LIFE projects, and the requirement for significant co-funding, has also reduced the ability of some countries with limited capacities and experience to apply for, and win, projects (e.g. Croatia).

When large funds, such as ERDF Operational Programmes, have been awarded to countries, their administrative and oversight requirements have been such that some environmental authorities have struggled to absorb and disburse them. This has resulted in allocations being unspent, as for example in Romania (Ernst & Young and Biotope, 2017). Capacity constraints have also limited the ability of nature authorities, RDP fund management authorities and other actors to promote the funds and provide advice on applications and guidance on practical matters (such as habitat conservation and restoration measures). As mentioned earlier, inadequate support has been a key reason for the low take-up of some schemes by farmers, foresters and other landowners/managers.

5.7 Surveillance, Monitoring, Reporting and Applied Research

Effective and efficient nature conservation is highly dependent on comprehensive and reliable information on the ecosystems, habitats and species that need to be conserved. This includes spatial and time series data on the location of habitats and species, information on their ecological requirements and the pressures affecting them, and monitoring data on their condition and trends. The scope, detail and reliability of such biodiversity information have increased considerably across much of Europe and particularly in the EU as a result of the information requirements to implement the Nature Directives (Evans, 2012), which were supported by some EU funding. Much of this information has been gathered through successful partnerships between conservation scientists, NGOs and environmental authorities. Long-term monitoring data have also been accrued primarily based on volunteer citizen science surveys by amateur naturalists, most notably for birds and some other popular and relatively easily surveyed groups.

Whilst there have been considerable advances in biodiversity knowledge, numerous data gaps remain that are constraining some nature conservation actions, especially in the marine environment. Most countries also lack maps of HD Annex I habitats and scientifically robust assessments of functional connectivity between Natura 2000 sites. Species recording and monitoring programmes also need to cover a broader range of taxa. These are being constrained in most countries by a shortage of suitably qualified experts and volunteer amateur naturalists. Solving this problem is a long-term challenge that requires considerable efforts to increase interest in the natural world to a high level, and then support for those who wish to develop the skills necessary to become fieldworkers.

The development of the favourable conservation status concept, initially under the CMS, and as incorporated in the Habitats Directive, has also been ground-breaking and of fundamental importance as it provides the basis for ensuring meaningful objectives are identified and reached (rather than a focus on actions). Assessments therefore need to be standardised and robust, and to be comparable over time and between countries. Improvements can certainly be made, as there seems to be little transparency and consistency in the practical methods and standards (Hochkirch, 2013). Needless to say, conservation status assessments also need to be entirely science- and evidence-based and independent. Therefore, interference by politicians and lobbyists for vested interests, as has most clearly occurred in Bulgaria in relation to the 2019 assessment (Section 8.2), needs to be prevented in future or it could undermine confidence in the whole system.

A more robust, transparent and comprehensive system is required for monitoring the degree of conservation of BHD habitats and species within Natura 2000 sites, the effects of management measures on them, and their overall conservation status, including in the wider environment. The achievement of this will require an increase in funding and capacity in conservation organisations.

5.8 The Main Lessons Learnt and Their Implications

The evidence from this book, and other reviews, clearly indicates that nature conservation is dependent on a combination of two or three vital ingredients: enforced legislation, funding and motivated people. By themselves, each of these achieves little, but any two of them have been shown to be effective, whilst all three are ideal.

Motivated and informed people, initially the general public, NGOs and scientists, and then politicians and other decision makers, are firstly necessary to campaign for and establish the necessary legislation, strategies and funding. Motivated and coordinated people are then required to populate the responsible public agencies and help to enforce legislation, as is essential in this realm. NGOs have played a key supporting role in enforcement in this respect, especially in Western Europe. Motivated people are certainly necessary to make the best use of available funding for nature conservation, such as to carry out nature-friendly farming and forestry, as the financial incentives by themselves are normally weak. Motivated people are also necessary to monitor nature, as much of this is done in practice by volunteer citizen scientists. Such information then motivates and helps people to raise awareness of the plight of nature and recruit others to the cause.

Consequently, the most effective nature conservation measures have been taken in those countries where there is strong scientific, public and political support. It is therefore encouraging that there is a generally high and growing interest in nature, including wildlife-rich areas – not just green space. This depth of support for nature conservation was demonstrated during the Nature Directives Fitness Check, when about half a million people spoke out in support of the Directives in the online public consultation (a record for any Commission consultation). The role that nature plays in supporting public health and well-being has also been highlighted by the COVID-19 pandemic.

Despite the high number of nature conservation supporters, a fundamental problem is that it is still not great enough. Also, too few are sufficiently motivated to call for, support, and take the nature conservation actions that are necessary to achieve current objectives. Consequently, too many politicians see nature conservation as a low priority, and environmental authorities are weak and under-resourced. Moreover, too many people are actually antagonistic to some nature conservation actions and the use of public money on such things. This is often due to their limited involvement and resulting misconceptions and alienation. In other cases, nature conservation may lead to unavoidable and undesirable real-life changes for some, which should not be ignored but properly addressed through just transitions.

Therefore, as suggested by Rayment *et al.* (2018), there is a need for a much larger social movement to halt biodiversity loss – one that is based on a broader and deeper shared societal recognition of the value of nature to human development and well-being. This requires effective engagement with people, through more education and inclusive forms of participation, reaching all social and ethnic groups and ages, recognising values and interests and using appropriate language. There is a wealth of examples to demonstrate that this is possible, and much experience that could be better shared and promoted.

Unfortunately, it is increasingly difficult for people to personally experience inspiring nature because so much of Europe's countryside is impoverished. Many people do not realise and appreciate what they are missing, and therefore are uninterested in conserving it. To counteract this shifting baseline of expectations there is a need for a more positive and ambitious message on the restoration and revival of nature so that this can capture people's imagination in terms of what is possible. Hence, radical proposals for rewilding and the re-introduction of large iconic mammals and birds have received a lot of media interest and public support (e.g. in the Netherlands and UK). However, restoration needs to focus on priorities and not be at the expense of firstly protecting what remains. Once lost, or highly degraded, the restoration and re-creation of natural and semi-natural habitats is often very difficult and expensive, if not impossible. When it is possible, it can take decades or centuries to complete, and it never fully replaces the biodiversity and ecosystem functions that were lost. For the moment, the nature conservation imperative is to use restoration to increase the viability of the most threatened habitats, species and areas. Later, if biodiversity losses can be stabilised, there will be a place for a more radical debate on what is an appropriate balance of semi-natural habitats and cultural landscapes, such as those supported by HNV farming, and wilder more natural ecosystems.

The fundamental role that nature conservation legislation, especially the EU Nature Directives, has had in securing progress, where it has been made, and inhibiting damage that otherwise would have occurred throughout Europe, is obvious and cannot be overstated. Another very important lesson from this book has been that legislation by itself is not enough: it needs to be fully implemented and backed up with well-targeted and long-term support, including guidance, funding and reliable monitoring and reporting – with effective enforcement if necessary. In this respect, the European Commission and CJEU have played a key role in ensuring that the Nature Directives and other EU legal obligations are adhered to much more fully than would have occurred in their absence. There are numerous examples where infringement cases have advanced the implementation of measures and protected sites, habitats and species, such as for the Rospuda Valley in Poland, the Kalikra SPA in Bulgaria, birds in Ireland, Italy and Malta, and large carnivores in several countries.

In recent years there appears to have been a more softly-softly approach to enforcement by the Commission, such as through 'pilot cases'. Whilst these are sometimes appropriate and successful, many other clear infringements and failings have not been addressed. Obvious cases have included the absence of appropriate assessments for some potentially damaging projects (e.g. wind turbines and micro-hydropower installations), ploughing of grasslands in Natura sites, effects of fishing on benthic habitats, and the ongoing pollution of sensitive habitats as a result of nitrogen pollution. More infringement proceedings could also be taken to address other impacts on protected species, including from unsustainable hunting and fisheries by-catch. It is therefore encouraging that the Commission has taken action against some of these issues recently, and that the *EU Biodiversity Strategy for 2030* and Environment Commissioner, Virginijus Sinkevičius, have reiterated the important role that the Commission has in upholding the Treaty and ensuring the Nature Directives are fully implemented.

Arguably, in Europe much more could have been done through legislation rather than voluntary and incentive-based measures, particularly regarding the control of the effects of land use on biodiversity. The polluter-pays principle could certainly have been more stringently adhered to in order to regulate aspects of farming and other land-use sources of water and air pollution. This could, for example, have reduced the need to use publicly funded RDP measures to pay farmers to prevent pollution, soil loss and other damaging practices, using the funds instead for more positive actions.

Most European governments have been particularly weak in terms of legally protecting habitats and ecological connectivity in the wider environment (i.e. outside protected areas). In this respect, more could be done to protect some habitats wherever they occur, as has been demonstrated in some countries such as Denmark and Germany where all forests are protected. Similarly, blanket protection could be extended to important habitat features such as old hedgerows, ponds and trees in the landscape, especially as HD Article 10 has been largely ignored. There is also a need for governments to do more to protect forest biodiversity, especially given the increasing incentives to manage woodland and forestry for timber products, bioenergy and carbon sequestration. In the EU, this might be easier to accomplish if greater legal competence over forest policy were to be granted to the EU, giving it the opportunity to adopt a more integrated and nature-focussed approach to future land use and forestry within it. Although this is a politically sensitive topic for a number of reasons, there is a case for such a change based on the potential added value of a coherent approach and the potentially greater contribution that it could make to supporting internal market standards.

A clear policy gap in most countries is the lack of requirements to ensure no net loss (or, more positively, net gain) of biodiversity from damaging activities (e.g. built developments, mining and energy projects). The cumulative impacts of development projects therefore remain a significant cause of biodiversity loss, except in those countries with effective mandatory offsetting, most notably Germany (Wende *et al.*, 2018). However, this remains a controversial issue as poorly designed and weakly regulated offsetting could be counterproductive. A more fundamental gap is that rural land-use developments – such as those related to agriculture (e.g. new irrigation, or the conversion of grasslands to crops), forest improvements and afforestation – are not subject to no net loss requirements anywhere in Europe. Currently, the EU's former target of achieving no net loss of biodiversity is off the agenda. While this continues to be the case, the substantial cumulative biodiversity losses from developments will remain unaddressed. Meanwhile, mandatory offsetting to achieve net biodiversity gains from built developments is being introduced in England in 2023. If properly regulated, implemented and genuinely positive for biodiversity, it could point the way forward for other countries that have yet to introduce no net loss or net gain policies.

Whilst more needs to be done through wider regulation, better enforcement, and stronger public and sectoral engagement, these measures and other practical nature conservation activities will continue to depend largely on public funding. It is therefore encouraging that there has been increasing and wider recognition amongst politicians in many countries that the funding of nature conservation, and broader environmental measures, provides sound positive investments for society and the economy as a whole. This has become even more important in the face of climate change and the need to recover from the socio-economic impacts of COVID-19. In the EU, it has resulted in the requirement for 7.5% of the EU MFF to be spent on biodiversity from 2024, and 10% from 2026. It is also intended that a significant proportion of the 30% allocation of the EU budget for climate measures will contribute to nature conservation and restoration, such as through nature-based solutions.

This level of funding increase has the potential to make a real positive difference for nature conservation, and more ambitious restoration targets. But lessons to date indicate that in order to achieve such objectives funding needs to be targeted to address the most important issues, using the most effective and efficient measures that are monitored and adapted in terms of their actual biodiversity impacts. At the moment it is not clear that this will happen in the EU as the *Biodiversity Strategy for 2030* does not focus on such priorities and lacks a clear funding strategy. Furthermore, as discussed earlier, the new CAP is unlikely to tackle the main causes of biodiversity loss any better than before. It seems that it may take until at least the next reform of the CAP for it to become a comprehensive and coherent sustainable land-use policy that is fully aligned with the EU's environmental ambitions. The new environmental land management schemes in England are intended to use all of what would have been CAP direct funding solely for environment public benefits, but it is currently uncertain if the schemes will live up to expectations.

To some, this book's main conclusions that the most effective approaches to nature conservation primarily are regulation combined with the use of financial incentives supported by motivated people may seem obvious, or perhaps old-fashioned. Nevertheless, whilst respecting the value of innovation and fresh approaches, it is important to be guided by the evidence drawn from a wide range of European countries. It remains vital to be as clear as possible about what has worked and why. This is all the more important because there has been a recent tendency to move away from some core biodiversity conservation approaches, and a particular antipathy towards the use of regulation. In fact, strong legal measures for conservation are the exception internationally (Campagnaro *et al.*, 2019). There have also been increasing calls for the need for voluntary initiatives to work together with partners, such as businesses or other sectors. However, the truth is that although such approaches have a role to play and have provided some impressive results (and saved public expenditure), they have been too few and too small to make a substantial contribution to addressing Europe's biodiversity crisis.

The limited effectiveness of voluntary targets and measures for governments is demonstrated by the failure to restore the 15%

of degraded ecosystems that was part of Target 2 of the previous *EU Biodiversity Strategy*. The European Commission's decision to grasp the nettle and propose a legally binding Nature Restoration Law recognises this reality. This essential initiative has the potential to be a huge new step forward for nature conservation in Europe, provided of course that the finally agreed restoration targets are ambitious, SMART[21] and focussed on priorities that are additional and complementary to existing measures.

Whatever Europe's ambitions for nature over the coming years, they should be underpinned by measures that have the best chance of delivering urgently needed results: strong and enforced legislation, adequate well-targeted funding and motivated people. Whilst approaches should evolve, it is wise to be wary of calls for new radical approaches. Whether or not it is fashionable, firm regulation remains essential, and forms a bedrock on which innovative approaches can be built. There is potential for synergies with new climate policies, but the sometimes gruelling history of seeking to integrate conservation objectives into agricultural and other sectoral policies shows that the devil is in the detail of specific policies. Clarity of focus on better outcomes for nature is critical, however the policy panorama changes.

Acknowledgements

I am very grateful for the helpful comments received on draft versions of the chapter from Alberto Arroyo, David Baldock, Ariel Brunner, Bruna Campos, Ioli Christopoulou, Pierre Commenville, Nigel Dudley, Euan Dunn, Mats Eriksson, Andrew Farmer, Ana Guimarães, Sandra Naumann, Rebecca Noebel, Clairie Papazoglou, Petr Roth, Andrej Sovinc, David Stroud and Ágnes Zólyomi.

References

AEWA (1999) *Resolution 1.14 Phasing out of lead shot in wetlands*. First meeting of the parties to the Agreement on the Conservation of African-Eurasian Migratory Waterbirds (AEWA), 6–9 November 1999. Cape Town, South Africa.

Alliance Environnement (2017) *Evaluation study of the forestry measures under Rural Development*. Brussels: Alliance Environnement. Report to DG Agriculture, European Commission.

(2019) *Evaluation of the impact of the CAP on habitats, landscapes, biodiversity*. Brussels: Alliance Environnement. Report to DG Agriculture, European Commission.

Alliance Environnement & Thünen-Institut (2017) *Evaluation study of the payment for agricultural practices beneficial for the climate and the environment*. Brussels: Alliance Environnement. Report to DG Agriculture, European Commission.

Andreotti, A., Guberti, V., Nardelli, R. *et al.* (2018) Economic assessment of wild bird mortality induced by the use of lead gunshot in European wetlands. *Science of the Total Environment*, **610**, 1505–1513.

[21] That is, specific, measurable, achievable, realistic and time bound.

Barnes, M. D., Glew, L., Wyborn, C. & Craigie, I. D. (2018) Prevent perverse outcomes from global protected area policy. *Nature Ecology & Evolution*, **2**, 759–762.

Barov, B. & Derhé, M. (2011) *Review of the implementation of species action plans for threatened birds in the European Union 2004–2010*. Brussels: BirdLife International report for the European Commission.

Batáry, P., Dicks, L. V., Kleijn, D. & Sutherland, W. J. (2015) The role of agri-environment schemes in conservation and environmental management. *Conservation Biology*, **29**, 1006–1016.

Bengtsson, J., Ahnström, J. & Weibull, A-C. (2005) The effects of organic agriculture on biodiversity and abundance: a meta-analysis. *Journal of Applied Ecology*, **42** (2), 261–269.

Benton, T. G., Vickery, J. A. & Wilson, J. D. (2003) Farmland biodiversity: is habitat heterogeneity the key? *Trends in Ecology & Evolution*, **18**, 182–188.

Billeter, R., Liira, J., Bailey, D. *et al.* (2008) Indicators for biodiversity in agricultural landscapes: a pan-European study. *Journal of Applied Ecology*, **45**, 141–150.

BirdLife International (2004) *Birds in Europe: Population Estimates, Trends and Conservation Status*. Cambridge, UK: BirdLife International.

(2013) *Designating Special Protection Areas in the European Union*. http://datazone.birdlife.org/sowb/casestudy/designating-special-protection-areas-in-the-european-union

(2015) *European Red List of Birds*. Luxembourg: Publications Office of the European Union.

(2021) *European Red List of Birds*. Luxembourg: Publications Office of the European Union.

Blicharska, M., Orlikowska, E. H., Roberge, J. M. & Grodzinska-Jurczak, M. (2016) Contribution of social science to large scale biodiversity conservation: A review of research about the Natura 2000 network. *Biological Conservation*, **199**, 110–122.

Born, C.-H., Cliquet, A., Schoukens, H., Misonne, D. & Van Hoorick, G. (eds.) (2015) *The Habitats Directive in its EU Environmental Law Context – European Nature's Best Hope?* London & New York: Routledge.

Bouwma, I. M., van Apeldoorn, R., Çil, A. *et al.* (2010) *Natura 2000 – Addressing conflicts and promoting benefits*. Wageningen, The Netherlands: Alterra.

Bowyer, C., Tucker, G., Underwood, E. *et al.* (2020) *Potential impacts of bioenergy developments on habitats and species protected under the Birds and Habitats Directives*. Brussels: Institute for European Environmental Policy, Arcadis, BirdLife International, NIRAS, Stella Consulting and Ecosystems Ltd. Report to the European Commission.

Brochet, A. L., Van den Bossche, W., Jbour, S. *et al.* (2016) Preliminary assessment of the scope and scale of illegal killing and taking of birds in the Mediterranean. *Bird Conservation International*, **26**, 1–28.

Brochet, A. L., Van Den Bossche, W., Jones, V. R. *et al.* (2017) Illegal killing and taking of birds in Europe outside the Mediterranean: assessing the scope and scale of a complex issue. *Bird Conservation International*, **29**, 10–40.

Campagnaro, T., Sitzia, T., Bridgewater, P., Evans, D. & Ellis, E. C. (2019) Half Earth or whole Earth: what can Natura 2000 teach us? *BioScience*, **69**, 117–124.

Carboneras, C., Genovesi, P., Vilà, M. *et al.* (2018) A prioritised list of invasive alien species to assist the effective implementation of EU legislation. *Journal of Applied Ecology*, **55**, 539–547.

Cardoso, P. (2012) Habitats Directive species lists: urgent need of revision. *Insect Conservation and Diversity*, **5**, 169–174.

Clutten, R. & Tafur, I. (2012) Are imperative reasons imperilling the Habitats Directive? An assessment of Article 6(4) and the IROPI exception. In *The Habitats Directive - a Developer's Obstacle Course?*, (ed.) G. Jones, pp. 167–182. Oxford, UK: Hart Publishing.

Crofts, R. (2014) The European Natura 2000 protected area approach: a practitioner's perspective. *Parks*, **20**, 79–90.

Crofts, R., Dudley, N., Mahon, C. *et al.* (2014) *Putting Nature on the Map: A Report and Recommendations on the Use of the IUCN System of Protected Area Categorisation in the UK*. IUCN National Committee UK.

Dasgupta, P. (2021) *The Economics of Biodiversity: The Dasgupta Review*. London: HM Treasury.

Deinet, S., Ieronymidou, C., McRae, L. *et al.* (2013) *Wildlife Comeback in Europe. The Recovery of Selected Mammal and Bird Species*. London: Zoological Society of London.

Dodd, A., Hardiman, A., Jennings, K. & Williams, G. (2010) Protected areas and climate change – reflections from a practitioner's perspective. *Utrecht Law Review*, **6**, 141–150.

Dolman, S., Baulch, S., Evans, P. G., Read, F. & Ritter, F. (2016) Towards an EU action plan on cetacean bycatch. *Marine Policy*, **72**, 67–75.

Donald, P. F., Sanderson, F. J., Burfield, I. J., Bierman, S. M., Gregory, R. D. & Waliczky, Z. (2007) International conservation policy delivers benefits for birds in Europe. *Science*, **317**, 810–813.

Dudley, N. (ed.) (2008) *Guidelines for Applying Protected Area Management Categories*. IUCN.

ECA (2009) *The sustainability and the Commission's management of the LIFE-Nature projects*. Luxembourg: European Court of Auditors.

(2011) *Is Agri-Environment Support Well Designed and Managed?* Luxembourg: European Court of Auditors. Special report 7/2011.

(2014) *Is the ERDF effective in funding projects that directly promote biodiversity under the EU biodiversity strategy to 2020?* Luxembourg: European Court of Auditors.

(2017a) *Greening: a more complex income support scheme, not yet environmentally effective*. Luxembourg: European Court of Auditors. Special report 21/2017.

(2017b) *More efforts needed to implement the Natura 2000 network to its full potential*. Brussels: European Court of Auditors. Special Report No 1/2017.

(2020) *Marine environment: EU protection is wide but not deep*. Luxembourg: European Court of Auditors.

EEA (2011) *Landscape fragmentation in Europe*. Copenhagen: European Environment Agency.

(2012) *Protected areas in Europe – an overview*. Copenhagen: European Environment Agency.

(2015) *State of nature in the EU. Results from reporting under the nature directives 2007–2012*. Copenhagen: European Environment Agency.

(2018) *European waters. Assessment of status and pressures 2018*. Copenhagen: European Environment Agency.

(2019a) *The European environment – state and outlook 2020*. Copenhagen: European Environment Agency.

(2019b) *Marine messages II Navigating the course towards clean, healthy and productive seas through implementation of an ecosystem-based approach*. Copenhagen: European Environment Agency.

(2020) *State of nature in the EU. Results from reporting under the nature directives 2013–2018*. Copenhagen: European Environment Agency.

eftec, ECNC, UAntwerp & CEEWEB (2017) *Promotion of ecosystem restoration in the context of the EU biodiversity strategy to 2020*. London: Economics for the Environment Consultancy Ltd (eftec). Report to European Commission, DG Environment.

ENEC (2016) *Legal recommendations to eliminate killing and taking of birds*. Madrid: SEO/Birdlife, for the European Network Against Environmental Crime.

Ernst & Young & Biotope (2017) *Study on biodiversity financing and tracking biodiversity-related expenditures in the EU budget*. Luxembourg: Publications Office of the European Union.

European Commission (2006) *Halting the Loss of Biodiversity by 2010 – and beyond. Sustaining ecosystem services for human well-being*. COM(2006) 216 final.

(2011) *Financing Natura 2000. Investing in Natura 2000: delivering benefits for nature and people*. SEC(2011)1573 final.

(2012) *Action Plan for reducing incidental catches of seabirds in fishing gears*. COM/2012/0665 final.

(2013) *Green Infrastructure (GI) – Enhancing Europe's Natural Capital*. COM(2013) 249 final.

(2016) *Fitness Check of the EU Nature Legislation (Birds and Habitats Directives)*. SWD(2016) 472 final.

(2017) *An Action Plan for Nature, People and the Economy*. COM (2017) 198 final.

(2018) *Assessing Member States' programmes of measures under the Marine Strategy Framework Directive*. COM/2018/562 final.

(2019a) *FITNESS CHECK of the Water Framework Directive, Groundwater Directive, Environmental Quality Standards Directive and Floods Directive*. SWD (2019) 439.

(2019b) *Review of progress on implementation of the EU green infrastructure strategy*. COM (2019) 236 final.

(2019c) *Evaluation of the Directive 2001/42/EC on the assessment of the effects of certain plans and programmes on the environment*. SWD(2019) 413 final.

(2020a) *Report on the implementation of the Marine Strategy Framework Directive (Directive 2008/56/EC)*. COM(2020) 259 final.

(2020b) *EU Biodiversity Strategy for 2030. Bringing nature back into our lives*. COM(2020) 380 final.

(2020c) *A Farm to Fork Strategy for a fair, healthy and environmentally-friendly food system*. COM(2020) 381 final.

(2020d) *Draft Technical Note on Criteria and Guidance for Protected Areas Designations*. Note to NADEG, October 2020.

(2021a) *On an Action Plan for the Development of Organic Production*. SWD(2021) 65 final.

(2021b) *New EU Forest Strategy for 2030.* COM(2021) 572 final.

(2022a) *Proposal for a Regulation of the European Parliament and of the Council on nature restoration.* 2022/0195 (COD).

(2022b) *Criteria and guidance for protected areas designations.* SWD(2022) 23 final.

Evans, A. D., Armstrong-Brown, S. & Grice, P. V. (2002) Science into policy: the role of research and development in the evolution of a 'smart' agri-environment scheme. *Aspects of Applied Biology*, **67**, 253–264.

Evans, D. (2012) Building the European Union's Natura 2000 network. *Nature Conservation*, **1**, 11–26.

Fagerli, H., Tsyro, S., Jonson, J. E. *et al.* (2020) *Transboundary particulate matter, photo-oxidants, acidifying and eutrophying components.* Oslo: Norwegian Meteorological Institute, EMEP Status Report 2020.

Fisher, I., Ashpole, J., Scallan, D., Proud, T. & Carboneras, C. (compilers) (2018) *International Single Species Action Plan for the conservation of the European Turtle-dove Streptopelia turtur* (2018 to 2028). Brussels: European Commission.

Forest Europe (2020) *State of Europe's Forests 2020.* Bratislava: Ministerial Conference on the Protection of Forests in Europe – Forest Europe Liaison Unit.

Gabriel, D., Sait, S. M., Kunin, W. E. & Benton, T. G. (2013) Food production vs. biodiversity: comparing organic and conventional agriculture. *Journal of Applied Ecology*, **50**, 355–364.

Geitzenauer, M., Blondet, M., De Koning, J. *et al.* (2017) The challenge of financing the implementation of Natura 2000–Empirical evidence from six European Union Member States. *Forest Policy and Economics*, **82**, 3–13.

Gerritsen, E., Klimmek, H. & Whiteoak, K. (2020) *Management effectiveness in the EU's Natura 2000 network of protected areas.* Copenhagen: EEA. Report by IEEP, UNEP-WCMC & Trinomics.

Hajdu, K. & Kiss, V. (2009) *Clear View: Regional synthesis report on the biodiversity challenges in Pan-Europe.* Budapest: CEEweb for Biodiversity.

HELCOM (2018) *State of the Baltic Sea – second HELCOM holistic assessment 2011–2016.* Helsinki: Baltic Marine Environment Protection Commission, Baltic Sea Environment Proceedings No 155.

Hilty, J., Worboys, G. L., Keeley, A. *et al.* (2020) *Guidelines for conserving connectivity through ecological networks and corridors.* Gland, Switzerland: IUCN. Best Practice Protected Area Guidelines Series No. 30.

Hirschfeld, A. & Attard, G. (2017) Vogeljagd in Europa–Analyse von Abschusszahlen und Auswirkungen der Jagd auf den Erhalt bedrohter Arten. *Berichte zum Vogelschutz*, **53**, 15–42.

Hochkirch, A., Schmitt, T., Beninde, J. *et al.* (2013) Europe needs a new vision for a Natura 2020 network. *Conservation Letters*, **6**, 462–467.

Hodgson, J. A., Kunin, W. E., Thomas, C. D., Benton, T. G. & Gabriel, D. (2010) Comparing organic farming and land sparing: optimizing yield and butterfly populations at a landscape scale. *Ecology Letters*, **13**, 1358–1367.

Hodgson, J. A., Moilanen, A., Wintle, B. A. & Thomas, C. D. (2011) Habitat area, quality and connectivity: striking the balance for efficient conservation. *Journal of Applied Ecology*, **48**, 148–152.

Hodgson, J. A., Thomas, C. D., Wintle, B. A. & Moilanen, A. (2009) Climate change, connectivity and conservation decision making: back to basics. *Journal of Applied Ecology*, **46**, 964–969.

ICES (2010) EC Request on Cetacean Bycatch Regulation 812/2004, Item 3, October 2010. www.ascobans.org/en/document/ices-2010-general-advice-ec-request-cetacean-bycatch-regulation-8122004

IEEP & Alterra (2010) *Reflecting environmental land use needs into EU policy: preserving and enhancing the environmental benefits of "land services": soil sealing, biodiversity corridors, intensification / marginalisation of land use and permanent grassland.* London: Institute for European Environmental Policy. Report to the European Commission.

IPBES (2018) *The IPBES regional assessment report on biodiversity and ecosystem services for Europe and Central Asia*, eds. M. Rounsevell, M. Fischer, A. Torre-Marin Rando & A. Mader. Bonn, Germany: Secretariat of the Intergovernmental Science-Policy Platform on Biodiversity and Ecosystem services.

Justice and Environment, CEEweb for Biodiversity, ClientEarth & European Environmental Bureau (2017) *Making Natura 2000 Impact Assessments Truly Appropriate: NGO proposal for an Action Plan.* www.clientearth.org/latest/documents/making-natura-2000-impact-assessments-truly-appropriate-ngo-proposal-for-an-action-plan/

Kati, V., Hovardas, T., Dieterich, M., Ibisch, P. L., Mihok, B. & Selva, N. (2015) The challenge of implementing the European network of protected areas Natura 2000. *Conservation Biology*, **29**, 260–270.

Kettunen, M., Illes, A., Rayment, M. *et al.* (2017) *Integration approach to EU biodiversity financing: evaluation of results and analysis of options for the future.* Brussels/London: Institute for European Policy. Final report for the European Commission.

Kettunen, M., Terry, A., Tucker, G. M. & Jones, A. (2007) *Guidance on the maintenance of landscape connectivity features of major importance for wild flora and fauna. Guidance on the implementation of Article 3 of the Birds Directive (79/409/EEC) and Article 10 of the Habitats Directive (92/43/EEC).* Brussels/London: Institute for European Environmental Policy. Report to the European Commission.

Koleček, J., Schleuning, M., Burfield, I. J. *et al.* (2014) Birds protected by national legislation show improved population trends in Eastern Europe. *Biological Conservation*, **172**, 109–116.

Koschová, M., Rivas-Salvador, J. & Reif, J. (2018) Continent-wide test of the efficiency of the European Union's conservation legislation in delivering population benefits for bird species. *Ecological Indicators*, **85**, 563–569.

Kukkala, A. S., Santangeli, A., Butchart, S. H. M. *et al.* (2016) Coverage of vertebrate species distributions by Important Bird and Biodiversity Areas and Special Protection Areas in the European Union. *Biological Conservation*, **202**, 1–9.

Langhout, W. (2019) *The EU Biodiversity Strategy to 2020. Progress report 2011–2018.* Rotterdam, The Netherlands: Langhout Ecologisch Advies. Report to Birdlife Europe and Central Asia.

Langhout, W. & Brunner, A. L. (2017) The best idea Europe has ever had? Natura 2000 – the largest network of protected areas in the world. *The George Wright Forum*, **34**, 86–95.

Lastra-Bravo, X. B., Hubbard, C., Garrod, G. & Tolón-Becerra, A. (2015) What drives farmers' participation in EU agri-environmental schemes?: Results from a qualitative meta-analysis. *Environmental Science & Policy*, **54**, 1–9.

Lawton, J. H., Brotherton, P. N. M., Brown, V. K. *et al.* (2010) *Making Space for Nature: a Review of England's Wildlife Sites and Ecological Network.* London: DEFRA.

Liberg, O., Suutarinen, J., Åkesson, M. *et al.* (2020) Poaching-related disappearance rate of wolves in Sweden was positively related to population size and negatively to legal culling. *Biological Conservation,* **234,** 108456.

Locke, H. & Dearden, P. (2005) Rethinking protected area categories and the new paradigm. *Environmental Conservation,* **32** (1), 1–10.

Lormee, H., Barbraud, C., Peach, W. *et al.* (2020) Assessing the sustainability of harvest of the European Turtle-dove along the European western flyway. *Bird Conservation International,* **30,** 506–521.

Maes, D., Collins, S., Munguira, M. L. *et al.* (2013) Not the right time to amend the annexes of the European Habitats Directive. *Conservation Letters,* **6,** 468–469.

Maes, J., Teller, A., Erhard, M. *et al.* (2020) *Mapping and Assessment of Ecosystems and Their Services: An EU ecosystem assessment.* Ispra, Italy: Publications Office of the European Union.

Maiorano, L., Amori, G., Montemaggiori, A. *et al.* (2015) On how much biodiversity is covered in Europe by national protected areas and by the Natura 2000 network: insights from terrestrial vertebrates. *Conservation Biology,* **29,** 986–995.

Mateo, R. (2009) Lead poisoning in wild birds in Europe and the regulations adopted by different countries. In *Ingestion of Lead from Spent Ammunition: Implications for Wildlife and Humans,* eds. R. T. Watson, M. Fuller, M. Pokras & W. G. Hunt, pp. 71–98. Boise, Idaho, USA: The Peregrine Fund.

Medarova-Bergstrom, K., Kettunen, M., Rayment, M., Skinner, I. & Tucker, G. (2014) *Common Framework for Biodiversity-Proofing of the EU Budget: General guidance.* London: Institute for European Environmental Policy. Report to the European Commission.

Milieu, IEEP & ICF (2016) *Evaluation study to support the fitness check of the Birds and Habitats Directives.* Brussels: Milieu Ltd, Institute for European Environmental Policy and the ICF International.

Mitchell, D., Puymartin, A., Vulcano, A. & Campos, B. (2021) *Off the Hook? Reducing Seabird Bycatch in the EU.* BirdLife International.

NABU (2015) *The future of EU financing for nature and biodiversity.* Berlin: Naturschutzbund Deutschland. www.nabu.de/imperia/md/content/nabude/europa/150527-nabu-eu-nature-financing_discussion-paper.pdf

N2K Group (2016) *Integration of Natura 2000 and biodiversity into EU funding (EAFRD, ERDF, CF, ESF): Analysis of a selection of programmes approved for 2014–2020.* Brussels: The N2K Group.

OSPAR (2017) *OSPAR intermediate assessment 2017.* London: OSPAR Commission.

Pain, D. J., Mateo, R. & Green, R. E. (2019) Effects of lead from ammunition on birds and other wildlife: a review and update. *Ambio,* **48,** 935–953.

Phillips, A. (2003) Turning ideas on their head: the new paradigm for protected areas. *The George Wright Forum,* **20,** 8–32.

Prip, C. (2013) *Terminal Evaluation of the UNEP project Supporting the Implementation of the Pan-European Biological and Landscape Diversity Strategy (PEBLDS).* United Nations Environment Programme.

Rayment, M., Arroyo, A., Baldock, D. *et al.* (2018) *Valuing biodiversity and reversing its decline by 2030.* Brussels: Institute for European Environmental Policy. Think 2030 Policy Paper.

Rogan, E., Read, A. J. & Berggren, P. (2021) Empty promises: The European Union is failing to protect dolphins and porpoises from fisheries by-catch. *Fish and Fisheries,* **22,** 865–869.

Sanderson, F. J., Pople, R. G., Ieronymidou, C. *et al.* (2015) Assessing the performance of EU nature legislation in protecting target bird species in an era of climate change. *Conservation Letters,* **9,** 172–180.

Seufert, V. & Ramankutty, N. (2017) Many shades of gray – The context-dependent performance of organic agriculture. *Science Advances,* **3** (3), e1602638.

Sibille, S., Griffin, C. & Scallan, D. (2020) *Europe's Huntable Birds: A Review of Status and Conservation Priorities.* Brussels: European Federation for Hunting and Conservation.

Sotirov, M. (ed.) (2017) *Natura 2000 and forests: Assessing the state of implementation and effectiveness.* Joensuu, Finland: European Forest Institute.

Stroud, D. A., Pain, D. J. & Green, R. (2021) Evidence of widespread illegal hunting of waterfowl in England despite partial regulation of the use of lead shotgun ammunition. *Conservation Evidence Journal,* **18,** 18–24.

Sundseth, K. & Roth, P. (2013) *Study on Evaluating and Improving the Article 6.3 Permit Procedure for Natura 2000 Sites.* Brussels: Ecosystems Ltd.

Tarzia, M. & Campos, B. (2014) *Marine Natura 2000 progress assessment.* BirdLife International.

TEEB (2011) *The Economics of Ecosystems and Biodiversity in National and International Policy Making,* ed. P. ten Brink. London and Washington, DC: Earthscan.

ten Brink P., Badura T., Bassi S. *et al.* (2011) *Estimating the Overall Economic Value of the Benefits provided by the Natura 2000 Network.* Brussels: Institute for European Environmental Policy/GHK/Ecologic. Final Report to the European Commission.

Trinomics BV, IEEP, UNEP-WCMC, IUCN, Denkstatt & ENT. (2022) *Support to the Evaluation of the EU Biodiversity Strategy to 2020, and Follow-Up.* Final evaluation report to DG Environment. Luxembourg: Publications Office of the European Union.

Trochet, A. & Schmeller, D. S. (2013) Effectiveness of the Natura 2000 network to cover threatened species. *Nature Conservation,* **4,** 35–53.

Trouwborst, A. (2015) The Habitats Directive and climate change. Is the law climate-proof? In *The Habitats Directive in Its EU Environmental Law Context – European Nature's Best Hope?,* eds. C.-H. Born, A. Cliquet, H. Schoukens, D. Misonne & G. Van Hoorick, pp. 301–324. London & New York: Routledge.

Tuck, S. L., Winqvist, C., Mota, F., Ahnström, J., Turnbull, L. A. & Bengtsson, J. (2014) Land-use intensity and the effects of organic farming on biodiversity: a hierarchical meta-analysis. *Journal of Applied Ecology,* **51,** 746–755.

Tucker, G. M., Allen, B., Conway, M. *et al.* (2014) *Policy Options for an EU No Net Loss Initiative.* London/Brussels: Institute for European Environmental Policy. Report to the European Commission.

Tucker, G. M., Dickie, I., McNeil, D. *et al.* (2016) *Supporting the Elaboration of the Impact Assessment for a Future EU Initiative on No Net Loss of Biodiversity and Ecosystem Services.* London:

Institute for European Environmental Policy. Report to the European Commission.

Tucker, G. M. & Heath, M. F. (1994) *Birds in Europe: Their Conservation Status*. Cambridge, UK: BirdLife International.

Tucker, G. M., Quétier, F. & Wende, W. (2020) *Guidance on achieving no net loss or net gain of biodiversity and ecosystem services*. Brussels: Institute for European Environmental Policy. Report to the European Commission.

Tucker, G., Stuart, T., Naumann, S., Stein, U., Landgrebe-Trinkunaite, R. & Knol, O. (2019) *Study on identifying the drivers of successful implementation of the Birds and Habitats Directives*. Brussels: Institute for European Environmental Policy. Report to the European Commission.

Tuomisto, H. L., Hodge, I. D., Riordan, P. & Macdonald, D. W. (2012) Does organic farming reduce environmental impacts?–A meta-analysis of European research. *Journal of Environmental Management*, **112**, 309–320.

UNEP-MAP (2018) *Mediterranean 2017 quality status report*. United Nations Environment Programme.

UNEP-WCMC (2019) *User Manual for the World Database on Protected Areas and world database on other effective area-based conservation measures: 1.6*. Cambridge, UK: UNEP-WCMC.

van der Sluis, T., Foppen, R., Gillings, S. *et al.* (2016) *How much biodiversity is in Natura 2000? The "umbrella effect" of the European Natura 2000 protected area network*. Wageningen, the Netherlands: Alterra. Report for the European Commission.

Venter, O., Fuller, R. A., Segan, D. B. *et al.* (2014) Targeting global protected area expansion for imperilled biodiversity. *PLoS Biol*, **12**, e1001891.

Venter, O., Magrach, A., Outram, N. *et al.* (2018) Bias in protected-area location and its effects on long-term aspirations of biodiversity conventions. *Conservation Biology*, **32**, 127–134.

Verschuuren, J. (2015) Connectivity: is Natura 2000 only an ecological network on paper? In *The Habitats Directive in its EU Environmental Law Context – European Nature's Best Hope?*, eds. C.-H. Born, A. Cliquet, H. Schoukens, D. Misonne & G. Van Hoorick, pp. 285–302. London & New York: Routledge.

Visconti, P., Butchart, S. H., Brooks, T. M. *et al.* (2019). Protected area targets post-2020. *Science*, **364**, 239–241.

Wätzold, F., Mewes, M., van Apeldoorn, R. *et al.* (2010) Cost-effectiveness of managing Natura 2000 sites: an exploratory study for Finland, Germany, the Netherlands and Poland. *Biodiversity and Conservation*, **19**, 2053–2069.

Wende, W., Tucker, G. M., Quétier, F., Rayment, M. & Darbi, M. (eds.) (2018) *Biodiversity Offsets – European Perspectives on No Net Loss of Biodiversity and Ecosystem Services*. Cham, Switzerland: Springer.

Winkel, G., Blondet, M., Borrass, L. *et al.* (2015) The implementation of Natura 2000 in forests: A trans- and interdisciplinary assessment of challenges and choices. *Environmental Science & Policy*, **52**, 23–32.

Winqvist, C., Ahnström, J. & Bengtsson, J. (2012) Effects of organic farming on biodiversity and ecosystem services: taking landscape complexity into account. *Annals of the New York Academy of Sciences*, **1249**, 191–203.

Woollhead, J., Petersen, A. & Normander, B. (2020) *Vurdering af danske beskyttede havområder efter international standard* [*Assessment of Danish MPAs according to international standards*] (Summary in English). Copenhagen: IUCN Conservation Planning Specialist Group Europe and Zoologisk Have København.

Žydelis, R., Small, C. & French, G. (2013) The incidental catch of seabirds in gillnet fisheries: a global review. *Biological Conservation*, **162**, 76–88.

Chapter 6

Austria

Wolfgang Suske and Kathrin Horvath

Key Statistics	
Terrestrial area (km^2)/% of EU-28*	83 944/1.9%
Marine waters (km^2)/% of EU-28*	0/0%
Population density (p km^2)/% of EU-28*	102/89%
GDP per capita (€)/% of EU-28	36 592/137%
Land cover (CORINE land cover classes 2018)	FOR 43%; AGR 32%; OSN 18%; ART 6%; WET 1%
Biogeographical regions	ALP, CON
Number of HD Annex I habitats/%*	71/31%
Number of HD Annex II, IV and V species/%*	207/17%
Number of BD Annex I bird taxa/%*	87/49%
% HD UCS and decreasing habitats/% unknown	14%/3%
% HD UCS and decreasing species/% unknown	28%/2%
% BD Annex I breeding birds decreasing/% unknown	17%/9%

* Excluding MAC and MMAC biogeographical regions.
Sources and codes: See Table 1.1.

6.1 The Natural History and Characteristics of Austria

Although Austria is a landlocked country, it has a high biodiversity due to its varying climate and topography, which ranges from wide-ranging flat lowland plains to its highest peak of 3 798 m (Grossglockner). Due to its relatively wide longitudinal extent, and altitudinal variation, Austria has several climatic zones. In the east, a Pannonian-type climate prevails with low precipitation, hot summers, and cold, short winters. An alpine-type climate dominates in the mountainous regions, with high precipitation, short summers and long winters. In the westernmost parts of Austria, the climate is markedly Atlantic; in the south-east it is more continental.

Forests cover a large proportion of Austria and host about two-thirds of all known animal and plant species in the country. These include: HD species, such as Eurasian Lynx (*Lynx lynx*) and Brown Bear (*Ursus arctos*); several forest-dwelling bat species, such as Lesser Horseshoe Bat (*Rhinolophus hipposideros*); deadwood beetle species (e.g. Alpine Longhorn Beetle (*Rosalia alpina*)); butterflies, including the Scarce Fritillary (*Euphydryas maturna*) and Fenton's Wood White (*Leptidea morsei*); and forest bryophytes, such as *Dicranum viride* and *Buxbaumia viridis*. The variability of the landscape in Austria led to the development of a diverse range of farming systems. In many parts of the country, these are High Nature Value (HNV) farming systems, including semi-natural grasslands and small-scale cultural landscapes with hedges, wildflower strips or small patches of arable land that are relatively biodiverse.

The **Alpine biogeographical region** covers about 63% of the country. Most of the region comprises the central Alps, which form the eastern part of the main Alpine ridge. They consist mainly of gneiss and slate, and include the highest mountain ranges and most of Austria's glaciers. In the alpine zone, Austria has a high responsibility for the conservation of Alpine and Boreal heaths (4060).[1] On wind-exposed ridges and mountain peaks that are snow-free in winter, populations of the dwarf shrubs *Loiseleuria procumbens* and *Empetrum hermaphroditum* occur, whereas on sites with long-lasting snow cover dwarf shrub vegetation is dominated by *Rhododendron* species.

The central Alps are mainly covered by forest (up to over 2 300 m in the inner alpine areas) dominated by spruce (*Picea*). In contrast to the northern Alps, the proportion of alpine grassland and meadows in the central Alps is quite high at more than 30%. In the northern limestone Alps, spruce, fir (*Abies*) and beech (*Fagus*) forests dominate. The proportion of forest habitat types in the Alpine region (23%) is significantly higher than in the continental region (7%) (Ellmauer, 2015). These include the rare forest type Subalpine and montane *Pinus uncinata* forest (9430 *if on gypsum or limestone). The Alpine biogeographical region also includes the Klagenfurt basin, which represents the largest inner-alpine basin

[1] Numbers in brackets are the HD Annex I habitat codes (see the Appendix for full names). * Indicates a priority habitat.

Table 6.1 Short-term trends (c. 2007–2018) in Habitats Directive Annex I habitats and Annex II, IV and V species with an unfavourable conservation status in Austria.

	Poor (unfavourable–inadequate)				Bad (unfavourable–bad)			
	Imp	Sta	Dec	?	Imp	Sta	Dec	?
HD Habitat assessments								
Number	4	9	4	24	0	9	12	30
%	10%	22%	10%	59%	0%	18%	24%	59%
HD Species assessments								
Number	28	57	34	45	8	21	61	25
%	17%	35%	21%	27%	7%	18%	53%	22%

Key. Imp = improving, Sta = stable, Dec = decreasing, ? = unknown.
Source. Based on Member State HD Article 17 reported data as presented on the EEA national summary dashboard (2021).

landscape with many small hills. A characteristic of this region is its numerous lakes.

The remaining area lies within the **Continental biogeographical region.** It includes the northern alpine foothills of the Alps and the Carpathians, as well as the Austrian Gneiss and Granite Highland (which is part of the Hercynian 'Bohemian massif'). The Austrian Gneiss and Granite Highland rarely extends to more than 1 000 m in altitude. The eastern part of the country includes the Vienna basin and the western fringes of the Hungarian lowlands, the Pannonian flat lands and low mountain areas. In the Pannonian flat lands, the amount of forest is comparatively limited at around 16% (Sauberer *et al.*, 2017). Wet meadows and dry grasslands are at their most extensive, and some areas remain of high importance for globally threatened birds such as Great Bustard (*Otis tarda*), Eastern Imperial Eagle (*Aquila heliaca*) and Saker Falcon (*Falco cherrug*).

The large but shallow Lake Neusiedl, the most westerly steppe lake of Europe, is located between the easternmost foothills of the Alps in the west and the western part of the Hungarian lowlands. After the Danube delta, the reed belt of the lake is the second largest continuous reed area in Europe and is particularly rich in birdlife.

6.2 The Status of Nature

6.2.1 Habitat Status and Trends

A concerning general trend, described by Schindler *et al.* (2016), is that wide-ranging habitat types are increasing at the expense of more localised habitat types that tend to be more endangered – thus implying an overall loss of diversity.

The most recent HD Article 17 report, covering 2013–2018, shows that for Austria overall, 18% of the habitat assessments revealed a favourable status, 35% a poor status, and 44% a bad status; 3% had an unknown status.[2] Of particular concern is that a significant proportion of those habitats assessed as having an unfavourable status also had deteriorating trends, especially amongst those with a bad status (Table 6.1). It is also of some concern that the trend was unknown in a high proportion of assessments with an unfavourable status. However, as in the previous reporting period, habitats in the Alpine biogeographical region were found to be faring better than those in the Continental region. Over the 2013–2018 period, 27% of habitats had a favourable status in the former, and only 7.4% in the latter. The only habitat types that had a favourable status in the Continental region were rocky habitats.

In the case of forest habitat types, those in the Continental biogeographical region are subject to high pressures from forestry, whereas significant areas of forest in the Alps are subject to little or no forest management (hence 22% were reported as having a favourable status). Positive components of forest status in Austria include their increasing area, as well as the share of broad-leaved tree species in managed forests. The proportion of deadwood,[3] a resource particularly important for many species of animals and fungi, has increased to around 8 m^3/ha since the mid-1990s (Schindler *et al.*, 2016). The target value would be 10% deadwood in the living stock, which would be an average of 33 m^3/ha (standing and lying). Currently, however, the value is around 20 m^3/ha or 6.2%. The uneven distribution of deadwood in the Austrian forest is striking: little in commercial forests, but high levels in protected forest, with most in alpine areas (Enzenhofer and Schrank, 2019).

The HD habitat types of particularly high concern are bogs and fens, rivers and other freshwater habitats, and the priority habitats Pannonic inland dunes (2340*) and Pannonic salt steppes and salt marshes (1530*), as none were reported as having a favourable status in either the Continental or Alpine biogeographical regions. Despite being priority habitats, the latter two had a bad status and the salt steppes and marshes also showed a declining trend.

2 See Section 1.3 for background, status categories and data sources.

3 Standing deadwood >10 cm diameter at breast height.

There has been a decline in the area of biodiversity-rich habitat types within the agricultural landscape, particularly in all extensively managed grasslands. Such grasslands include 15 HD habitat types, including meadows, mountain pastures, dry calcareous grasslands and steppic grasslands. Only two of these had a favourable status in the Alpine region, none in the Continental region. Moreover, 48% of assessments of grasslands were reported as bad, including 20% that also had a declining trend. Changes in this situation for grasslands seem unlikely unless land abandonment and atmospheric nitrogen deposition are greatly reduced.

The habitat types that were found to be generally in favourable condition were rocky habitats and heath and scrub.

6.2.2 Species Status and Trends

The Article 17 reporting data on HD species over 2013–2018 provide a similar picture to that for habitats. Overall, in Austria 14% of assessments indicated that the species had a favourable conservation status, 48% a poor status, 34% a bad status and 3% an unknown status. Unlike HD habitats, there was relatively little difference between the percentage of assessments in the Alpine biogeographical region and the Continental biogeographical region, with 17% and 12% respectively having had a favourable conservation status. As for HD habitats, a particular concern is that a high proportion of species with an unfavourable status had declined over the preceding 12 years, especially amongst those with a bad status (Table 6.1).

The latest BD Article 12 reporting data indicate that BD Annex I bird populations appear to be faring better than HD species as about 74% have increasing, stable or fluctuating breeding populations (Table 6.2). Furthermore, almost the same proportion of wintering Annex I birds had increasing, stable or fluctuating populations.

The declines in some BD Annex I species are of concern, and a significant proportion of other bird species have decreased, including common farmland bird populations, as indicated by the Austrian Farmland Bird Index, which has shown a steady decline over recent decades (Teufelbauer and Seaman, 2019). In the period 1998–2018, 75% of the indicator species showed a statistically significant decline; only the Tree Sparrow (*Passer montanus*) and the Goldfinch (*Carduelis carduelis*) recovered over the period. On average, 40% of all farmland bird populations disappeared between 1998 and 2018.

Table 6.2 Short-term trends (2013–2018) in Birds Directive Annex I taxa in Austria.

	Breeding				Wintering			
	Inc	Sta/ Flu	Dec	?	Inc	Sta/ Flu	Dec	?
Number	24	23	11	6	6	2	2	1
%	38%	36%	17%	9%	55%	18%	18%	9%

Key. Inc = increasing, Sta = stable, Flu = fluctuating, Dec = decreasing, ? = uncertain and unknown.
Source. Based on Member State BD Article 12 reported data as presented on the EEA national summary dashboard (2021).

6.2.3 Drivers and Causes of Change

The Austrian Biodiversity Strategy 2020+ (BMLFUW, 2014) identified housing and infrastructure developments, which lead to habitat loss and fragmentation, climate change, the spread of invasive alien species, and the use of fertilisers and pesticides as significant threats to biodiversity. However, 'Human-induced changes in hydrological conditions' and some factors related to agriculture and forestry were found to be the most important pressure as regards HD species and habitats, according to an analysis of Austria's HD Article 17 report for 2007–2012 (Schindler *et al.*, 2016). The latest BD Article 12 and HD Article 17 reporting data on the pressures and threats affecting BHD habitats and species for 2013–2020 present a similar picture to recent decades.

These assessments are consistent with earlier evidence that showed that for many species and habitats along Alpine rivers, the Danube or the Seewinkel (with its rare saline lakes), man-made changes in hydrology were key threats (Dick *et al.*, 1994; Egger *et al.*, 2009). The proportion of the river Danube that has been used for hydropower has increased steadily since the construction of the first hydropower plants in 1955 (Schindler *et al.*, 2016); free-flowing sections are nowadays restricted to the Wachau valley and the Donau-Auen National Park. For many rivers and floodplains, natural dynamics have been lost due to power plants and other barriers, and sand and gravel banks are now scarce along most rivers. Bogs, mires and other wetlands are also sensitive to changes in their water regime. Discharges of urban waste water and nutrient inputs from agriculture are also having a negative impact on many freshwater habitats.

By the mid-nineteenth century, forest cover in Austria was at a historic low and forest habitats were in a poor state due to overexploitation and widespread pasturing. Some improvements in forest condition occurred as a result of the Austrian Forestry Act of 1975, which initiated a shift towards more sustainable forest development and management. According to the Austrian 'Forest Hemeroby Study' by Grabherr *et al.* (1998), which until today remains the only available study on forest naturalness, about one-third of the national forest area is classified as heavily modified (27%) or artificial (7%). These areas are subject to intensive forest uses, and their tree species composition is not natural. In recent years, rising temperatures and an increasing frequency of wind-storms and droughts have led to considerable bark beetle damage in artificial spruce forests – a trend that is expected to continue with accelerating climate change. For forest habitats and species, the removal of deadwood (Enzenhofer and Schrank, 2019), afforestation and clear-cutting have been reported as the most significant pressures. These pressures have intensified through the increased use of biomass in recent decades due to the promotion of renewable energies.

The drivers of ecological changes in agricultural areas have included both abandonment and intensification. Land abandonment has been the main pressure on extensively grazed semi-natural grasslands, and is particularly damaging for dry and semi-dry grassland habitat types. For example, until the 1950s, eastern Austria had vast tracts of extensively managed grazing land, most of which has either been converted to arable land or become overgrown after abandonment. On more productive grasslands and arable land, intensification has been the major driver of biodiversity loss, clearly indicated by the declines in farmland birds. For example, populations of the Corn Bunting (*Emberiza calandra*) decreased by 80% between 1998 and 2013 (Teufelbauer, 2014) due to both the abandonment of extensive grasslands and further intensification (Wrbka *et al.*, 2008; Stoate *et al.*, 2009). The application of insecticides has also increased since 1998, whereas it was declining before then.

Other more recent pressures include the increase in the number of wind turbines, which has risen to about 1 200 turbines since they first appeared in Austria in 1994 (Schindler *et al.*, 2016). While being important as a renewable source of energy, they are a pressure on local and migrating populations of birds (Wichmann and Denner, 2013) and bats.

Climate change has already led to a massive decline of glaciers, with negative impacts on a broad spectrum of alpine animal and plant species (Gottfried *et al.*, 2012; Pauli *et al.*, 2012). Its impacts on other habitats and species are also increasing, whilst at the same time their capacity to adapt is being constrained by habitat fragmentation by developments and infrastructure (Hülber *et al.*, 2020). This reduces the resilience of habitats and their connectivity, which also reduces the ability of species to move in response to changing climatic conditions, such that many habitats can no longer be recolonised (Wessely *et al.*, 2017).

6.3 Nature Conservation Governance, Actors, Legislation and Policies

Table 6.3 provides a summary list of the key nature conservation developments in Austria, most important of which are further described in this section.

6.3.1 Governance, Legislation and Policy

The state of nature conservation in Austria changed decisively as a result of the dispute over a proposal in 1982 for a hydropower plant in the Hainburger Au, a natural floodplain forest of the Danube east of Vienna. In response to the proposal, WWF Austria launched a campaign 'Save the floodplains' and, with the help of some media (in particular, the largest Austrian

Table 6.3 Key nature conservation developments in Austria.

1862	Austrian Alpine Association founded
1886	Edelweiss under protection in Austria
1872	Josef Schöffel prevents complete deforestation of the Vienna Woods
1905	First draft of a law on the protection of historical and natural monuments
1912	Foundation of the Naturschutzbund (at that time: Verein Naturschutzpark)
1924	Lower Austria legislates the first Provincial Nature Conservation Act and the first Provincial Cave Protection Act of Austria
1927	Nature conservation included in the statutes of the Alpine Club
1946	Nature conservation conference after the end of the war, including representatives of official nature conservation, the federal and state governments, as well as scientists and nature conservation associations
1953	Foundation of BirdLife Austria
1963	Foundation of WWF Austria
1973	Foundation of the Umweltdachverband (at that time: Austrian Society for Nature and Environmental Protection, Österreichische Gesellschaft für Natur- und Umweltschutz in short: ÖGNU)
1978	Referendum on the Zwentendorf nuclear power plant
1981	First National Park (Hohe Tauern National Park)
1984	Environmentalists occupy the Danube floodplains at Hainburg, to prevent their destruction by a hydropower plant
1986	The Green Party enters the National Council
1995	Entry into the EU and establishment of the National Biodiversity Commission
1998	Austrian Strategy for the Implementation of the Convention on Biological Diversity
2013	Initiation of infringement proceedings for inadequate Natura 2000 site designations
2014	Austrian Biodiversity Strategy 2020+
2016	Presidency of the Alpine Convention
2017	Green Party fails to be included in the National Council with 4% of the vote
2019	Due to new elections and an election result of 14%, the Green Party enters into government negotiations
2020	The Green Party becomes a coalition partner for the first time and provides five ministers in the new government

See Table 3.3 for signatory dates of international conventions.

tabloid), began to draw public attention to the imminent destruction of part of the Danube floodplain. The protests against the power plant and the way the debate was eventually settled became a landmark achievement for nature conservation in Austria. Soon afterwards, in 1986, the Austrian Green Party was founded. Instead of the power plant, a national park was established 10 years later at the same location (1996). These events had lasting impacts with a strong growth in awareness amongst the Austrian population of the need for the conservation of nature.

Austria's nature conservation received the next significant boost when it joined the EU in 1995. This opened up completely new financial opportunities for Austria in the field of nature conservation, which were used very quickly and professionally. EU membership has also led to significant changes in the policy and legislative framework, particularly in the procedures for assessing projects where habitats or species are likely to be harmed.

Of strategic national importance have been Austria's Biodiversity Strategies. The first Austrian strategy for implementing the CBD was adopted in 1998. In 2005 the Biodiversity Strategy was further developed, and in 2014 the government adopted the Biodiversity Strategy 2020+, which is still valid today and includes objectives of the EU Biodiversity Strategy to 2020. Twelve goals have been formulated in each of five fields of action (knowing and recognising biodiversity, using biodiversity sustainably, reducing pressures on biodiversity, conserving and developing biodiversity, securing biodiversity worldwide). In addition to these fields of action and goals, the Strategy also contains a comprehensive catalogue of measures. However, the Austrian Biodiversity Strategy Austria 2020+ does not contain a clear prioritisation or timetable for the implementation of the measures. A financing plan is also not part of the Strategy.

In 1995, a National Biodiversity Commission was established to monitor and review the implementation and achievement of the objectives of the Austrian Biodiversity Strategy. It is composed of representatives of ministries, provinces, social partners, universities, federal offices, NGOs and various interest groups (e.g. landowners) and is headed by the Federal Ministry of Climate Action, Environment, Energy, Mobility, Innovation and Technology. Its members report regularly on the measures taken in their area of responsibility to implement and achieve the Strategy.

Austria has a federal structure, with nine regional *Länder* (states). Each of these *Länder* has responsibility for nature conservation matters, and has its respective nature conservation laws, which are supplemented by additional decrees for species protection (see Section 6.4) and for each Natura 2000 site. The implementation of the EU Nature Directives and other provisions of nature conservation law (e.g. the Nature Conservation Act, species protection decrees, and the National Park Act) as well as related issues such as expert services, subsidies, contractual nature conservation and international nature conservation matters (including CITES) all fall within the remit of the nature conservation authorities of the *Länder* governments.

At a lower administrative scale, nature conservation matters are dealt with by the district officials of the district authorities (Bezirkshauptmannschaften). In Styria, for example, expert opinions in the field of nature conservation are also provided by the appointed managers of the Natura 2000 sites (Europaschutzgebiete). Natura 2000 site managers and district nature conservation officers are therefore important local and regional contacts for the population.

The national environmental protection interests of Austria are represented in each *Länder* by an environmental and nature conservation advocate (Umweltanwalt), who reports to the *Länder* administration. The Umweltanwalt's tasks include the assessment of laws and ordinances, the receipt of applications and complaints (e.g. from municipalities, but also individuals) and the monitoring of compliance with notices issued by the authorities. The Umweltanwalt is a party to official proceedings.

The Federal Environment Agency, a subsidiary body of the Federal Ministry for Sustainability and Tourism, is the leading scientific expert body in Austria's environmental sector. The collection and evaluation of environmental data as well as the development of strategies and measures represent important foundations for decision makers within Austria and throughout Europe. In this respect, it is noteworthy that the Federal Environment Agency has published a comprehensive work in Austria on the development of criteria, indicators and threshold values for the assessment of conservation status of HD habitats and species, has technically supported the Natura 2000 nomination process, and has been commissioned by the *Länder* responsible for the technical preparation of HD Article 17 reporting obligations. It is also in charge of the environmental control reports and latest versions of the Austrian Biodiversity Strategies. Despite its de facto knowledge monopoly, and being a subsidiary of a federal ministry, the Federal Environment Agency is formally a private commercial entity that is required to bid for contracts in response to calls for tender published by the *Länder*.

6.3.2 Cooperation amongst Administrations

Due to Austria's federal structure, cooperation between the federal government and the *Länder* is of particular importance. The coordination of the nature conservation authorities of the *Länder* is mainly carried out at a non-binding level by so-called *Länder* working groups, without any federal involvement. Such working groups are set up by the *Länder* on a topic-related basis (e.g. on international matters), and decisions must be taken in unanimity. If the need for cooperation extends beyond an unofficial level, a so-called Verbindungsstelle der Bundes*Länder* (liaison office of the *Länder*) is given responsibility for official decisions. The vertical cooperation between the *Länder* and the district levels works quite well in most of the *Länder*.

However, as the Site of Community Importance (SCI, i.e. Natura 2000 site) nomination process in Austria proved to be difficult and lasted for more than 20 years, the coordination structures appear to have had certain weaknesses. Furthermore, implementation of Natura 2000 in Austria is not streamlined between the *Länder*, due to a lack of stable working structures, procedures and planning processes. As a result, coordination processes between the *Länder* are complex and coordination remains rather difficult. The government programme of the ÖVP-FPÖ Federal Government (2017) therefore planned to set up its own coordination office for Natura 2000 issues to ensure efficient and targeted implementation of Natura 2000. The environmental umbrella NGO (Umweltdachverband) also suggested the establishment of such a coordinating body in its 2019 report 'Natura 2000 – Zurück in die Zukunft' (Friesenbichler and Stallegger, 2019). However, up to now such a body has not been established.

The very good cooperation that has occurred between the Federal and *Länder* governments in the area of financing should be highlighted. A permanent exchange of experience and efficient collaboration occurs through numerous events under the aegis of the Rural Development Network. Recently, a new Austrian biodiversity strategy has been developed under this framework, involving all departments and ministries concerned.

The Austrian Research Association Road–Rail–Transport, which develops state-of-the-art approaches to dealing with infrastructural projects, provides a good example of efficient cooperation between authorities and experts. It has specialised committees in each relevant subject area (e.g. protection of amphibians on roads, species protection on transport routes, and implementation of noise reduction techniques) with experts working together on guidelines, to help the practical implementation of the requirements. A unique feature of the committees is their interdisciplinary composition: ecologists and other environmental specialists cooperate with road construction experts, administrations and private engineering offices. The aim is to develop guidelines that correspond to the highest technical standards while remaining usable in practice. These guidelines contain, for example, recommendations or agreements on certain thresholds and criteria (e.g. for the permitted amount of deterioration in relation to the project area), descriptions of survey methods, or definitions of technical terms. Several guidelines have already been completed on the topic 'Animals and Plants' (FSV, 2015).

6.3.3 Environmental NGOs, Stakeholders and the Wider Public

The Austrian nature conservation movement has a long history (see Table 6.3), based on the various activities of many NGOs and other associations and citizens' initiatives. Some individuals have also had a strong influence on the development of nature conservation, such as Josef Schöffel, who in 1870 campaigned to prevent the deforestation of the Vienna Woods, today a national protected site as well as a Biosphere Reserve.

The awareness of environmental issues in Austria, and the corresponding growth of the movement, was also stimulated by two major events. The first was concern over nuclear power in the 1970s and the building of the Zwentendorf nuclear power plant. This led to a referendum on the matter, with a rejection of the nuclear power plant by a very narrow margin (50.47%[4]) and an increase in environmental concerns in Austrian society. It also led to the foundation, in 1973, of the environmental umbrella organisation (Umweltdachverband), a non-political platform that currently consists of 36 member organisations from the fields of environmental and nature protection.

The second event was the proposal in 1982 to construct a hydroelectric power plant in the Hainburger Au, a natural floodplain forest of the Danube. As described earlier, WWF's 'Save the floodplain' campaign gained remarkable support from a wide range of citizens beyond the environmental movement, including artists, scientists, politicians and journalists. As a result the plans were dropped during the winter of 1985–1986 and the wetlands have become a National Park. Whilst this success was not attributable to any political party, immediately afterwards the Green Party for the first time managed to gain seats in the National Council.

The Umweltdachverband (UWD) is the most important NGO in Austria. It acts as a 'public conscience' for decision makers and pursues non-profit goals in the field of environmental protection as well as fostering environmental policy. This includes the further development of environmental laws and standards, rights of participation for NGOs, and the implementation of the Natura 2000 network. The UWD was largely responsible for the complaint to the European Commission in 2013 concerning the insufficient nomination of SCIs for inclusion in Natura 2000. Together with further experts, the UWD drew up a shadow list of sites that should be nominated by the *Länder* in order to meet the EU's requirements. It also has expertise in EIA procedures and appropriate assessments, provides statements on EIA and HD Article 6(3) appropriate assessments and is part of the European Network for Rural Development (ENRD).

WWF Austria has more than 100 000 supporters (including regular and occasional donors). Amongst other things, WWF is an important leader of LIFE projects and also deals with the protection of large predators, being especially committed to the return of the Wolf (*Canis lupus*). WWF Austria has played a crucial role in the establishment of numerous National Parks and is at the forefront of river protection and restoration in Austria.

While the UWD monitored the sufficient nomination of the SCIs, BirdLife Austria played a decisive role in shaping the SPA network. Furthermore, BirdLife Austria plays an important role in LIFE projects and in evaluations of Rural

[4] www.zwentendorf.com/dasjahr1978.asp

Development Programmes (RDPs), and is responsible for the Austrian Farmland Bird Index and related farmland bird monitoring.

The Naturschutzbund is the oldest and largest nature conservation organisation in Austria and celebrated its successful 100-year history in 2018. It has more than 80 000 members[5] committed to the preservation of a rich flora and fauna. One of its major strengths is its particularly good regional and local level networks, with its federal structure divided into nine provincial groups with respective regional groups, the Nature Conservation Youth (Naturschutzjugend) founded in 1952 and the Institute for Nature Conservation. The Naturschutzbund implements numerous regional, but also nationwide, projects for species and habitat protection and purchases or leases land worthy of protection, which now amounts to over 1 600 ha at 2 000 sites.[6]

In addition to the nature conservation NGOs, citizens' initiatives play an increasingly important role in the nature conservation movement. Such initiatives sometimes draw on a high level of expertise and make an important contribution to impact assessments procedures. In this respect, EU nature conservation legislation is considered to be of great importance, as it provides a tool for citizens to present to the authorities their concerns regarding nature conservation. As an example, in February 2020, two citizens' initiatives prevented the construction of an expressway through a Natura 2000 site because the option chosen by the authorities would have affected one of the few remaining Austrian breeding sites of the endangered Eurasian Thick-knee/Stone Curlew (*Burhinus oedicnemus*).

6.4 Species Protection and Conservation Measures

The transposition of the Nature Directives took place in the form of the respective nature conservation laws and regulations of the *Länder* (e.g. species protection ordinance and protected area ordinances). The nature conservation laws regulate the general protection of wild species of plants, fungi and animals. Strictly protected and protected species are listed in the species protection decrees. The selection of species is based on HD Annexes II and IV, the list of species in BD Annex I and the national Red Lists. Strictly protected species are subject to a comprehensive prohibition of any impact; exceptions may be granted for research and science or injured and sick animals.

Species protection is enshrined not only in nature conservation laws and species protection ordinances but also in the hunting and fishing laws of the *Länder*. This is the reason why parallel regulations for the same species exist in different laws: in the Nature Conservation Decree on the one hand, and in the Hunting or Fisheries Decrees on the other. In Styria, for example, native HD Annex IV mammal species are protected both by nature conservation legislation and by the Hunting Decree. Fish, crayfish and mussel species listed in HD Annex IV are protected, at least in Upper Austria and Lower Austria, on the one hand by nature conservation species protection and on the other by the Fisheries Decree (Mauerhofer, 2011).

Species protection programmes in Austria support efforts to halt the decline of selected animal and plant species, including the Brown Bear, Wolf, Eurasian Lynx, European Souslik (*Spermophilus citellus*), Otter (*Lutra Lutra*) and Beaver (*Castor fiber*) and several plant species (e.g. *Gentianella bohemica, Artemisia laciniata*). The programmes are generally drawn up by experts from NGOs and scientific institutions together with the responsible authorities.

Austria is a partner, with Germany and Italy, in the LIFE+ Project 'Reason for Hope – Reintroduction of the Northern Ibis in Europe' that is attempting to re-introduce the Bald Ibis (*Geronticus eremita*) as a migratory bird in Europe. The ibis is one of the most threatened bird species in the world and disappeared in Europe in the first half of the seventeenth century, mainly as a result of persecution. The goal has been to reach a population size of at least 120 birds by 2019. By the end of the project this goal was actually surpassed, with a population consisting of 140 birds (Förderverein Waldrappteam, 2020).

One of the most successful species resettlement projects has been for the Eurasian Lynx, which was exterminated in Austria in the nineteenth century. In the mid-1990s a lynx appeared repeatedly around the Kalkalpen National Park. Proactive measures were then taken between 2011 and 2013 (as part of the national project 'Urgent Lynx Conservation Action'), when two females and one male were relocated from Switzerland to the region. They bred the year after. Since then, thanks to the re-introduction measures taken, a lynx population has become re-established. According to WWF,[7] fewer than ten adult animals are currently roaming through the Alpine region, where reproduction has so far only been proven close to the Kalkalpen National Park area. Another 15–20 individuals occur in the forests of the Bohemian Massif, as part of a larger population in neighbouring Germany and the Czech Republic

Despite the species conservation programmes and re-introduction projects, most of the targeted species continue to decline. In February 2019, WWF Austria published a species protection check called WWF-Big5 *Länder*-Barometer, which reveals deficits in the handling by certain *Länder* of their populations of Beaver, Otter, Eurasian Lynx, Wolf and White-tailed Sea-eagle (*Haliaeetus albicilla*). This so-called *Länder*-Barometer evaluated and compared the management of the species in each of the nine *Länder*. As regards the Beaver, the implementation of measures for it has been poor in all *Länder*. Upper Austria is the only *Länder* that has a management plan for the Otter that includes suitable preventative measures to avoid fish losses in fishponds. Species conservation projects

[5] https://naturschutzbund.at/mithelfen.html

[6] www.naturfreikauf.at/infos-zu-unseren-flaechen.html

[7] WWF Austria: www.wwf.at/de/luchs_verbreitung/

Table 6.4 The number and extent of international designated sites and protected areas in Austria.

	IUCN category	Number	Area (km^2)
Designated under global and non-EU international legislation			
Ramsar Sites	NR	23	1 272
Natural World Heritage Sites	NA	1	71
Biosphere Reserves (exc. buffer and transition zones)	NA	4	230
Designated under the Nature Directives			
SCIs	NR	305	9 378
SPAs	NR	100	10 334
Natura 2000 terrestrial area			*12 895*
Designated under domestic legislation			
Strict Nature Reserves and Wilderness Areas (Wildnisgebiet Dürrenstein and Wildnisgebiet Sulzbachtäler)	1a+b	2	102
National Park (Nationalpark)	II	6	2 382
Nature Reserve (Naturschutzgebiet)	Mostly IV	478	3 028
Landscape Protection Area (Landschaftsschutzgebiet)	Mostly V	249	12 825
Protected Landscape Section (Geschützte Landschaftsteile)	Mostly IV III V	332	86
Nature Park (Naturpark)		50	1 214

Sources. See Section 1.3. Based on data from: UNESCO Biosphere Reserves[8]; Environment Agency Austria[9]; Wildnisgebiet Dürrenstein[10]; EEA Natura 2000 Barometer, Natura 2000 as at end 2020; UNEP-WCMC (2021) for IUCN types. *Notes.* NR = not reported; NA = not applicable.

such as 3Lynx and ProLuchs for the Lynx are being implemented in Upper Austria. Long-term comprehensive species protection concepts for the White-tailed Sea-eagle are currently established in Burgenland and Lower Austria.

[8] https://en.unesco.org/biosphere/eu-na June 2019 update, accessed 6 November 2021.

[9] www.umweltbundesamt.at/umweltthemen/naturschutz/schutzgebiete

[10] www.wildnisgebiet.at/

6.5 Protected Areas and Networks

6.5.1 Protected Area Types and Their Coverage

According to the nature conservation legislation in Austria there are 14 different categories of protected areas with varying degrees of protection. The nature conservation departments of the federal provinces designate the protected areas by ordinance, which also prescribes their protected area categories and requirements. One exception is the National Parks, which are designated by law. Some categories of protected areas are restricted to individual federal states. For example, Plant Protection Areas are only found in Salzburg and Vorarlberg, and Ecological Development Areas only in Vienna.

Table 6.4 lists the international and national categories of protected areas and their current coverage, which are further described in the rest of this section.

National Protected Areas

National protected areas of most relevance to nature conservation are the Wilderness Areas, National Parks and Nature Reserves. All other categories have less stringent protection provisions, and some of them (e.g. the Landscape Protection Areas) are focussed on the protection of the landscape character rather than on biodiversity conservation.

Austria currently has two **Wilderness Areas** (IUCN protected areas category Ib), one in Lower Austria (Wildnsigebiet Dürrenstein, 34 km^2, established in 2001) and one in Salzburg (Wildnisgebiet Sulzbachtäler, 67 km^2, established in 2018). Dürrenstein harbours the largest remaining block of truly primeval forest in the Alps, which has been protected since 1875 on the initiative of its then owner, Albert Rothschild. In 2021 the wilderness area in Dürrenstein was expanded by another 30 km^2 into the Lassingbachtal of neighbouring Styria.[11]

There are six internationally recognised **National Parks**, which are classified according to IUCN Category II. They cover 2.8% of the federal territory, conserve ecologically valuable regions of Austria and are of great importance for development in the regions. The association Nationalparks Austria was founded in 2011 and aims to promote the dynamic development and cooperation of the parks by coordinating and implementing joint projects.

Nature Reserves are largely natural or semi-natural areas characterised by the presence of habitats worthy of protection and/or the occurrence of rare or endangered species.

Landscape Protection Areas are areas with a special character, a high aesthetic value or a high recreational value. Their primary aim is to preserve the general appearance of the landscape and the special importance of the area for the population. It is not uncommon for these areas to be characterised by near-natural forms of management and to serve as buffer zones to Nature Reserves. They are the most widespread protected areas and cover about 15% of the federal territory.

[11] www.wildnisgebiet.at/

Protected Landscape Sections are small parts of the landscape that have the same protection purpose as the large Landscape Protection Areas. Although today there are 332 of these protected areas they account for only 0.1% of the national territory.

Nature Parks are parts of protected landscapes that are maintained through environmentally friendly land use and landscape management. These exemplary cultural landscapes have special natural features, usually host a high level of biodiversity, and are designated by decree of the state governments. Nature Parks are used for recreation and leisure activities and are characterised by largely free access.

Natura 2000 Sites

Austria had problems completing the network of SPAs and SCIs, which in both cases resulted in infringement proceedings and intensive negotiations with the European Commission before they were eventually resolved. The inconsistencies and incoherence of the proposed SPAs were solved on the scientific basis of the IBAs identified according to BirdLife International criteria (Dvorak, 2009) and a methodology developed specifically for Austria that was accepted by the European Commission. BirdLife was also involved in the discussions concerning the method and designation. As a result, 52 of the 56 IBAs have been designated as SPAs (12.3% of the national territory). However, their total area of 10 334 km^2 is significantly lower than that of the IBAs, which covered 14 764 km^2. Furthermore, Uhl and Denner (2015) and Uhl and Wichmann (2013) state that some SPAs are too small to protect the habitats of the relevant bird species. For example, of 18 traditional Curlew (*Numenius arquata*) territories, 7–10 are outside the Pfeiferanger SPA.

Infringement procedures for the insufficient designation of SCIs were initiated in 1996, then extended in 2006 and 2007, and finally reopened in 2013 with a requested nomination of approximately 200 additional sites. Up to December 2018, 134 additional SCIs were proposed by Austria and the infringement case was closed in July 2019; SCI coverage is now 11.2%.

Although it has taken many years and the European Commission's intervention, the Natura 2000 network is considered to be complete for now, covering 15.4% of the country.

Overall Protected Area Coverage

Although there are some overlaps between protected area types, the Austrian Natura 2000 network is substantially supplemented by some of the protected areas designated under domestic legislation, in particular the National Parks, Biosphere Reserves, Landscape Protection Areas and Wilderness Areas. As a result, Austria has a relatively large overall protected area network, which as of July 2021 was estimated to cover 24 414 km^2, 29.9% of the country, according to World Database on Protected Areas (WDPA) data (UNEP-WCMC, 2021). However, this estimate should be treated with some caution as there are discrepancies between some WDPA data and further available evidence reviewed here (e.g. the WDPA includes 26 National Parks, whereas there are clearly only six National Parks in Austria). An assessment by the Environment Agency Austria in 2021, which carefully assessed the overlaps among categories, estimated 24 303 km^2 of overall protected area coverage (or 29.0% of the country) but excluding Ramsar Sites and Natural World Heritage Sites (EAA, unpublished data). Moreover, it should be borne in mind that a large proportion of the domestic protected area network consists of IUCN category V Protected Landscape Areas only, which in practice provide a low level of protection to habitats and species.

6.5.2 The Coherence of Protected Areas and Ecological Networks

There is a need for action to improve connectivity between some areas of the protected area network, such as through corridors for species with extensive home ranges, such as the Eurasian Lynx. Efforts to secure and establish biotope networks and corridors have progressed at different rates in the *Länder*. In Burgenland and Lower Austria, there are specifically designated corridors to facilitate dispersal and other movements in the area of the so-called Alpine Carpathian Corridor.

6.5.3 Approaches to Protection and Levels of Protection

The levels of protection for biodiversity in protected areas depend on their type. Of those of most relevance to biodiversity, Nature Reserves in principle have strict protection, but their effectiveness is often compromised by legal provisions that allow 'customary agriculture and forestry'. In contrast, Wilderness Areas provide a very strict protection regime, with all extractive forms of land use totally prohibited, and naturalness and landscape quality being of key importance. A strict non-intervention regime also applies to the core zones of National Parks. According to a ministerial decision, the label National Park in Austria is entirely reserved to protected areas that fully comply with the relevant IUCN management criteria. The strict core zones of National Parks are surrounded by buffer zones that have a less restrictive, but still clearly defined, protection regime. In contrast, Landscape Protection Areas only provide a very basic and weak amount of protection. It is also interesting to note that since the designation of Natura 2000 sites the designation of further national protected areas has become very unpopular.

Natura 2000 sites are protected through their decrees (Europaschutzgebietsverordnungen), which, in addition to listing their relevant BHD species and habitats, include conservation objectives for the area. However, the determination of site-specific conservation objectives suffers from the lack of a unified system across the Austrian *Länder*, resulting in high variation in their quality. Many of the current site-level conservation objectives therefore remain insufficiently specific, measurable and achievable.

In Austria, more than a hundred experts, and as many lawyers, operate within the framework of HD Articles 6(3) and 6(4) procedures for assessing the impacts of plans and projects on Natura 2000 sites (further discussed in online Annex AT.1).[12] There is no consistent training and no coordination or cooperation structures at their working level. Therefore, the way in which every expert and lawyer deals with difficult issues depends very much on their own level of knowledge and experience.

6.5.4 Approaches to Managing Protected Areas

In Austria, with the exception of Burgenland, there is to date no legal obligation to prepare management plans for Natura 2000 sites. Nevertheless, management plans exist for almost 80% of all Natura 2000 sites (Friesenbichler and Stalleger, 2019). Management plans are normally elaborated by the nature conservation authorities. Some management plans have also been developed by NGOs (e.g. WWF) on their own initiative. Land users and owners, municipalities and the farmers' organisation Chamber of Agriculture are the most frequently involved stakeholders. Most financial resources for planning and implementation derive from the European Agricultural Fund for Rural Development (EAFRD), as described later, and from funds of the *Länder*. Site surveillance and contractual arrangements between the authorities and land users are the main tools for ensuring the implementation of the plans.

Strengths in management planning include their contents relating to the underlying legal framework, detailed descriptions of the site (including the protected species and habitats) and its land uses (Friesenbichler and Stalleger, 2019). The greatest current deficits lie in the lack of quantification of goals and measures, missing financial plans, and unidentified resources for the implementation of measures. Additional shortcomings relate to monitoring, planning the implementation of measures and target planning. Supervision of the sites by scientific experts and public participation in the planning phase are also in need of improvement.

The fact that responsibility for management planning is split between the *Länder* authorities, with varying procedures for developing the plans, results in differences in their design. Efforts to harmonise quality standards still remain at the planning stage only. For example, in Styria, a requirements profile for practice-oriented management plans was carried out in 2021 on the basis of the Nature Conservation Strategy 2025, and existing management plans will be adapted accordingly.[13]

[12] All annexes are free to access online at CUP: www.cambridge.org/natureconservation

[13] Birgit Konecny, Amt der Steiermärkischen Landesregierung, verbal communication.

6.6 General Nature Conservation Measures

6.6.1 Development Planning and Impact Assessment

In Austria, spatial planning is organised by the federal provinces, with land-use plans and zoning maps the responsibility of the municipalities. This organisation of spatial planning makes important larger-scale conservation concepts, such as corridor networks, very difficult. At the level of the federal provinces there are different spatial planning strategies that are relevant for nature conservation. For example, Lower Austria, Burgenland and Styria have separately designated areas that are suitable or unsuitable for wind energy. For the future, the development and promotion of a comprehensive spatial planning concept across the Federal States with consideration of biodiversity and the designation of ecological functions (e.g. green infrastructure or biotope networking) is particularly important. A good example of successful spatial planning across provincial and international borders is the establishment of the Alpine–Carpathian corridor, a wildlife corridor linking the easternmost ranges of the Alps to the western Carpathians, across a 120-km stretch of lowlands in eastern Austria and western Slovakia.[14]

The implementation of EIA is very complicated in Austria. Projects of nationwide relevance fall under the competence of the federal ministries, whilst others are under the competence of the *Länder*. In all EIAs and SEAs, biodiversity and other environmental issues are generally examined according to uniform criteria, which are determined by various rules (RVS, ÖNORM) and are continuously updated. Nature conservation, which in principle falls within the competence of the *Länder*, is only examined to a certain extent in EIAs, primarily in relation to EU law including the Nature Directives. However, it is important to note that EIA and appropriate assessments carried out in accordance with HD Article 6(3) are in Austria two different assessments regulated by different laws.

6.6.2 Habitat Management and Restoration

Agricultural Habitats

In Austria, permanent grassland has played a major role in agriculture, especially in mountainous areas. However, the area of permanent grassland is declining for a variety of reasons, including abandonment and overgrowth with shrubs, afforestation, or conversion to temporary grasslands, cropland or building land. Furthermore, extensively managed grasslands in climatically suitable areas are under increasing pressure from intensification. This is being exacerbated by climate change, as the growing period begins earlier and lasts longer, which means that the grass can be mowed more often. Remaining ecologically valuable grassland areas are often not managed in a suitable way for the conservation of plant species. This is the case especially with subalpine grassland

[14] Alpen-Karpaten-Korridor – Lebensader für Europas Artenvielfalt: www.wwf.at/de/akk/

(EUNIS type E1.7 Closed non-Mediterranean dry acid and neutral grassland, E4.3 Acid alpine and subalpine grassland, silicate grass, and E2.3 Mountain hay meadows), wet meadows (E3.4 Moist or wet eutrophic and mesotrophic grassland, E3.5 Moist or wet oligotrophic grassland) and primary dry grasslands (E1.2 Perennial calcareous grassland and basic steppes, E1.D Unmanaged xeric grassland) that are used too intensively over more than one-third of their area (Suske *et al.*, 2019b). These land-use changes are leading to many problems for grassland-dependent breeding birds, numerous insect species and specialised plant communities.

To address these issues, under the EAFRD-funded RDP, the Austrian Agri-environment Programme (ÖPUL) has aimed to support and maintain extensively managed grasslands, and other HNV farmland habitats, since its introduction in 1995. This includes schemes for environmentally sound and biodiversity-promoting management (requiring the creation of biodiversity areas on 5% of the farm area and the maintenance of landscape elements), maintenance of endangered livestock breeds, avoidance of silage cutting, maintenance of mowed mountain grassland, mountain grazing and herding, and nature conservation. A specific goal of the nature conservation scheme is the maintenance and improvement of the status of agricultural habitats, in particular for BHD animals and plants.[15] The schemes target the management of sensitive agricultural areas according to specific obligations that are set by the nature conservation authority, such as late mowing, prohibition of fertilisation, or restoration of wet meadows.[16] Participating farmers receive compensation payments for additional work and costs and/or reduced yields due to the obligations.

About 20 000 farms (c. 13% of all farmers) have participated every year in the ÖPUL nature conservation scheme, covering about 75 000 ha (c. 3% of farmland), with financial support of €610 million between 1995 and 2016 (BMLFUW, 2016). The effectiveness of the ÖPUL measures has been evaluated several times; in Natura 2000 areas, they were found to be key to the achievement of the sites' conservation objectives (Suske *et al.*, 2009). One reason for the programme's success has been that, since its start, nature authorities and NGOs have been involved in discussions on its design and evaluation. Moreover, there has been excellent cooperation between the nature authorities and Austrian agricultural market operators (Agrarmarkt Austria) who are responsible for the operation of the RDP. Administrative or technical problems that might otherwise hinder the implementation of good ideas can thereby be resolved efficiently.

The additional CAP payments made to farmers within areas with natural and other constraints (formerly known as less-favoured areas – LFAs) have also helped to maintain HNV farming systems and extensive grassland. Without such payments many extensively managed grasslands would be abandoned, especially alpine meadows and pastures. Over 2007–2009, an average of 93% of farmers with alpine grassland applied for LFA payments, which corresponded to 15% of the total LFA payments in Austria (Suske *et al.*, 2012). This scheme, which is part of Pillar 2, is to be continued in the new CAP.

Austria has an Agroecology-Climate Measure (AECM) for ecological set-aside, which was equivalent to the Ecological Focus Areas (EFA) under the Pillar 1 greening measures of the 2014–2020 CAP. Under this, at least 5% of the beneficiary's arable land must be dedicated to biodiversity (e.g. with specific mixtures of seeds to be sown, a minimum period of flower cover and a ban on using fertiliser and plant protection products).

Forests

The extensive forests of Austria are of considerable importance for their habitats and species, and a means of protecting them has been the Natural Forest Reserve Programme, which has been in existence since 1995. Under this, forest areas are set aside for the natural development of their ecosystem. Neither timber extraction nor grazing, nor any other human intervention, is permitted in these areas, and they serve as study and reference areas for natural forest dynamics. Still, hunting is allowed in order to control wild ungulate impacts on the vegetation. The reserves certainly contribute to the preservation of biological diversity, especially its dynamic aspects. The selection of sites aims to fully represent Austria's natural forest communities and focusses on naturalness in species composition, structure and both the temporal and spatial dynamics of forest communities. Currently, 83 km^2 of forest is included in the programme, scattered over 192 individual forest reserves.[17]

More widely, as discussed earlier, most HD forest types have an unfavourable conservation status. The main problem is their lack of structural diversity, including insufficient amounts and dimensions of deadwood and very old trees. In spite of the poor establishment of the EAFRD-funded Forest Environment Programme from 2015 onwards, an attempt has been made to incentivise ecological measures in forests. However, the programme has only been used by a few forest managers (BMLFUW, 2019), because it was unattractive to them and insufficiently supported and promoted by forest institutions (e.g. the forestry chamber).

Rivers

Numerous rivers in Natura 2000 sites have been restored and their quality improved in recent years, mostly with the help of LIFE projects (restoring river banks, fish habitats, etc.). These include the Danube, March, Salzach, Lech, Mur, Drau and

[15] Bundesministerium für Nachhaltigkeit und Tourismus (2015): Sonderrichtlinie ÖPUL 2015. BMNT-LE.1.1.8/0032-II/3/2018. Page 64.

[16] https://info.bml.gv.at/dam/jcr:79820874-9487-49c4-b1fb-0f3a5c8d4cf3/OePUL%202015%20folder_EN_web.pdf

[17] www.naturwaldreservate.at/index.php/de/#block2

various rivers in the Alpine foothills. In addition, the construction of fish passes on numerous river sections has improved the connectivity of the rivers in recent years. There are clear synergies between the WFD and Nature Directives, which are positively used in several regions of the Austrian *Länder*. Altogether, 122 Natura 2000 sites in Austria are considered to be also within the remit of the WFD.[18]

6.7 Nature Conservation Costs, Benefits and Funding Sources

6.7.1 Nature Conservation Costs

As part of the development of the Prioritised Action Framework (PAF) for Natura 2000 for 2021–2027[19] it has been estimated that the financial requirements are approximately €68 million within Natura 2000 areas and €49 million outside Natura 2000 areas. This totals about €117 million over the period, or an average of about €17 million per year. This sum includes the EU contribution and regional and national co-financing from the *Länder* and the Republic of Austria.

6.7.2 The Economic Benefits of Nature Conservation

A question that is often discussed in connection with the designation of Natura 2000 protected areas is whether or not they result in economic disadvantages for the regions concerned. From an overarching macro-economic perspective, various studies on the overall Natura 2000 network have shown a positive net benefit from their protected status (e.g. Gantioler *et al.*, 2014). However, there are debates on whether there are regionally different distributional effects (i.e. the bearing of costs), which may well be significant at a regional or local scale. In 2016, a study of the regional economic effects of Natura 2000 areas in Austria found that communities with and without Natura 2000 have developed positively in parallel (Getzner *et al.*, 2016). The results of the analysis showed also that the decision of farmers to run their farm as a main source of income or a sideline was not influenced by Natura 2000. In the tourism sector, a positive economic effect could be observed in Natura 2000 areas. Further details of the study are provided in online Annex AT.2.

6.7.3 Funding

Public Funding

Table 6.5 provides an overview of the main funding sources used in Austria, especially over the 2014–2020 funding period. This indicates that Austria has made good use of EU funding programmes, and that a large number of key nature conservation actions would not have been possible without such sources.

The EAFRD has been by far the most important financing source for funding nature conservation in agricultural habitats in Austria, and this has provided very important synergies between Natura 2000 and agricultural policy. Table 6.5 includes the 2014–2020 allocations in relation to Priority 4, which covers biodiversity. Of key importance for nature conservation has been the funding of the ÖPUL agri-environment scheme (as described in Section 6.6.2). As discussed earlier, additional support for the maintenance of extensive HNV farming systems has also been provided through the M13 payments for areas with natural constraints, and similar previous measures.

In the 2007–2013 funding period, Austria was particularly successful in using the EAFRD measure M323a (conservation and upgrading of rural heritage). The aim of this measure was to preserve and develop high-value resources in terms of nature conservation, and to safeguard the regional character of the cultural landscape, particularly of habitats and species that are protected by the Nature Directives. Altogether, 1 026 projects, mostly on agricultural land, were implemented through the measure during the programming period, with a total funding of €75 million (Pinterits *et al.*, 2014). Out of the projects funded, 54% had a direct link to Natura 2000 areas, 29% concerned the habitat of a BD Annex I bird species, and 11% concerned HD species or habitats; 26% of the projects aimed to contribute to the development of specific biotopes and habitats, and required the purchase of land. This successful support was continued in the 2014–2020 programme period[20] using Measure 7.6.1.a 'Studies and investments for the conservation, restoration and improvement of the natural heritage (nature conservation)'.

Austria has a very large number of organic farms, and some €112 million was allocated to the EAFRD organic measure M11 over 2014–2020. Whilst organic farming has undisputed positive effects on soil life and some common farmland species, many protected species and habitats do not significantly benefit from this type of farming. In the case of grassland, this is mainly because the intensity of grassland use under organic farming is still too high to benefit species with more demanding ecological requirements (BirdLife Österreich, 2019; Holzer *et al.*, 2019; Suske *et al.*, 2019a). Therefore, it is not included in the total for RDP funding in Table 6.5 as it is not considered to be a major source of funding for nature conservation.

In contrast to agricultural areas, the use of EAFRD for forests, such as through the M15 forest environment and climate measure, has been very limited. The frequent occurrence of weather extremes, as well as bark beetle damage, has led to unattractive wood prices. As a result, small forestry enterprises

---

[18] Federal Ministry Sustainability and Tourism: www.bmnt.gv.at/wasser/wasser-oesterreich/plan_gewaesser_ngp/nationaler_gewaesserbewirtschaftungsplan-ngp/ngp2015.html

[19] Preliminary draft of the PAF 2021–2027.

[20] Under Article 20 of the EAFRD Regulation.

Table 6.5 Funding sources for nature conservation in Austria.

Source and period	Measure/category	Average annual allocation (€ m)		Comments
EU funding instruments		**EU + national**	**EU**	
EAFRD Rural Development Programme measures – allocations to Priority 4 (restoring, preserving and enhancing agriculture and forest ecosystems) 2014–2020	M10 Agri-environment–climate	292.5	148.3	See text and note 1
	M13 Areas facing constraints	252.0	124.9	
	M11 Organic farming	[112.0]	[56.9]	See text
	M7 Basic services, etc.	28.0	14.1	See text and note 2
	M8 Investments in forests	14.8	7.4	
	M1 Knowledge transfer, etc.	3.1	1.6	
	M16 Cooperation	2.3	1.2	
	M4 Investments in physical assets	2.4	1.2	
	M12 Natura and WFD	0.9	0.5	100% for Natura 2000
	M2 Advisory services, etc.	0.7	0.3	
	M15 Forest environment climate	0.1	0.1	
RDP Priority 4 subtotal		*596.9*	*299.6*	Excluding M11
LIFE (1992–2020)	Nature and biodiversity projects	7.6	3.8	
ERDF/Cohesion fund 2014–2020	Interreg V Protection and enhancement of biodiversity, green infrastructure and Natura 2000	?	?	See text
EU funding instruments subtotal		>604.5	>303.3	
National public funds 2014–2020		**Additional national***	**Comments**	
Styria	Nature conservation measures	2.0	Inc. some co-financing of EU funds	
Lower Austria	Landscape Fund, Landscape Design, Aquatic Ecology	0.77	Inc. some co-financing of EU funds	
Upper Austria	Nature conservation measures	3.6	Inc. some co-financing of EU funds	
Salzburg	Small water bodies contractual nature conservation	0.58	Inc. some co-financing of EU funds	
Carinthia	National Park Hohe Tauern	0.09	Inc. some co-financing of EU funds	
Tirol	Nature conservation	0.8	Inc. some co-financing of EU funds	
Vorarlberg	Nature conservation, landscape development	0.06	Inc. some co-financing of EU funds	
Biodiversity Fund from 2021	To support the post 2021 Biodiversity Strategy	5.0	As above. €5 m endowment plus €5 m per year	

Table 6.5 *(cont.)*

Source and period	Measure/category	Average annual allocation (€ m)		Comments
EU funding instruments		**EU + national**	**EU**	
Others				
REWE International AG (since 2014)	See text	c. 1.0		
NGOs		c. 0.5		See note 3
National subtotal	*Exc. funds co-financing EU funds*	*c. 1.5*		
	Inc. funds co-financing EU funds	*c. 14.4*		
Total	*Exc. funds co-financing EU funds*	**606.0**		
	Inc. funds co-financing EU funds	**618.9**		

Sources. Based on data from: RDP data from European Commission ESIF funds database (23 October 2020); LIFE allocations based on LIFE Austria overview (2021); BMNT 2019: Grüner Bericht, table Tab_2019_5020207_OEPUL_Flaechen_Betriebe_Leistungsabgeltungen for M10 biodiversity measures; national public funds.[21] See Section 1.3 for further details.
Notes. * Excludes co-financing of EU funds unless indicated. The subtotals and totals are only approximate indications of the average annual levels of funding as they are based on annualised data from differing periods. 1. €131 million (EU + national) actually paid per year on average (years 2016–2018) for the following measures: environmentally friendly and biodiversity-promoting management, mountain grazing and herding, maintenance of mowed mountain-grassland, and nature conservation. 2. Probably about 10–30% of M7 used for M7.6.1a, see text. 3. No figures available for NGOs' funds; approximate estimate based on known projects financed by membership fees.

in particular have greatly reduced their forest use and maintenance activities, and this has also led to a reduction in the demand for funding in the forestry sector. Previous funding measures for forest habitats do not constitute an economic alternative to the desired extent for forestry enterprises, which are partially or completely dependent on forest yield.

LIFE projects have played an essential role in nature conservation as well as regional development in Austria, as local companies have been involved in the implementation of project measures.[22] Since its accession to the EU in 1995, Austria has carried out 54 nature conservation projects under LIFE, many of which have been carried out on large rivers (as discussed earlier). Some species conservation projects (e.g. for the Great Bustard or the Brown Bear) have been carried out across borders and built on the success of their predecessors. Current LIFE projects include species conservation projects for the re-introduction of Bald Ibis,[23] the recovery of the Sterlet (*Acipenser ruthenus*),[24] the smallest Danube sturgeon species, and a cross-border project for the protection of all sturgeon species on the Danube[25] that is currently taking place involving project partners from Bulgaria, Germany, Romania, Serbia and Ukraine.

The European Regional Development Fund (ERDF) is also a possible source for funding nature conservation projects. However, it was not used by Austria in the funding period 2014–2020, other than for Interreg projects, including some transnational nature conservation projects ('Sustainable bog management', 'Balance for Nature and People', 'Wild & Cultivated').

In 2021, a new 'Biodiversity Fund' was established for Austria; it was initially (2021) endowed with €5 million and will be extended to more than €5 million per year. It is intended to fund projects that support the post 2021 biodiversity strategy (see Section 6.9) and that are not eligible for funding under EAFRD or LIFE. In 2021, more than 200 projects had already been submitted.

Private Funding

Compared to the attractive and broad public funding programmes, private funding in Austria (e.g. through associations, foundations, companies) is rather weakly developed. Nevertheless, an initiative of the REWE Group is worth mentioning. Within the Austrian market, REWE International is the leader in food retailing and one of the largest companies in the country. Against the background of the alarming threat to biodiversity, REWE International AG launched the Blossoming Austria (Blühendes Österreich)[26] initiative, together with BirdLife Austria and other partners. The initiative has committed since 2014 to preserving, protecting and enhancing valuable habitats and protecting native animals and plants. To carry out nature conservation activities competently and on a long-term basis, the private foundation Blühendes Österreich – REWE

21 https://bab.gv.at/index.php?option=com_rsfiles&layout=download&path=Gruener_Bericht%2F2019%2Fgb2019.zip&Itemid=477&lang=de
22 WWF Austria: www.wwf.at/de/eine-starke-lobby-fuer-europas-natur-25-jahre-life-programm/
23 LIFE Northern Bald Ibis – Reason for Hope – Reintroduction of Northern Bald Ibis in Europe.
24 LIFE Sterlet – Restoration of sterlet populations in the Austrian Danube.
25 LIFE for Danube Sturgeons.

26 REWE International Dienstleistungsgesellschaft m.b.H.: https://rewe-group-nachhaltigkeitsbericht.de/2021/en/index.html

International gemeinnützige Privatstiftung was established in 2015. This nature conservation foundation is advised and supported by numerous experts from various fields, and is endowed with almost €1 million per year. REWE International AG is thus the first large company in Austria to establish a charitable foundation for the protection of nature and as such it is clearly playing a pioneering role.

The foundation supports the preservation and maintenance of threatened natural areas and extensive agricultural areas as well as projects for nature and sustainability education. Farmers or other managers of valuable natural areas (e.g. nature conservation associations) receive a premium so that they can continue to carefully manage or improve existing ecologically valuable areas. The foundation is financed entirely by REWE International AG. For every sold product under the private label '*Da komm ich her*' (BILLA, MERKUR and ADEG), fruit and vegetables from '*Ich bin Österreich*' (PENNY), products from '*MERKUR Immer grün*' as well as 'bi good' (BIPA), one cent is paid into the foundation. This means that customers of the trading companies are also actively involved in supporting nature conservation projects.

Also worth mentioning are the activities of the Esterhazy Private Foundation, which regularly supports projects for the protection of Natura 2000 areas. A large part of the land within the Neusiedler See – Östliches Leithagebirge Natura 2000 site is owned by Esterhazy. The activities of the foundation range from biotope network projects, reed management and forest environmental measures to flower strips projects and species conservation activities for the protection of bats in castle grounds. Some of these projects are also funded by various EU programmes, in which case the federal province of Burgenland provides the matching funds. The activities of the Esterhazy Private Foundation are regularly summarised in 'Sustainability Reports' of Esterhazy (Esterhazy Betriebe GmbH, 2018).

Overall Funding Adequacy

Overall, comparison of the estimated funding needs with available funds indicates that Austria has made good use of EU funding programmes so far, both in Natura 2000 areas and outside protected areas. However, there is some potential for better use of funding sources, especially in the forest sector. Furthermore, it is recommended that the ÖPUL organic farming measure is modified so that it can contribute more to the conservation of protected species and habitats.

6.8 Surveillance, Monitoring, Reporting and Applied Research

Knowledge of the status of habitats and species, both inside and outside the Natura 2000 network, has improved considerably in certain areas as a result of requirements under the Nature Directives and other initiatives. These have included the development of a monitoring system for selected habitats and species in 2010–2012 (Stöhr *et al.*, 2014). Additional mapping has also occurred in response to the 2013/4077 infringement (and need to identify more SCIs) and monitoring activities for the preparation of the 2019 HD Article 17 report. Furthermore, comprehensive EUNIS level 3 biotope mapping data produced by contracted experts are available for some of the *Länder* and cover most but not all habitats. However, there are currently still no Austria-wide harmonised mapping guidelines, and as a result the quality of the data differs between the different *Länder*.

The Federal Environment Agency runs a database that compiles all available occurrence and other relevant information on all HD species and habitats in Austria. Knowledge gaps still remain in relation to the occurrence of certain species and habitat types, and the necessary extensive surveys for these cannot be carried out for budgetary reasons (Friesenbichler and Stallegger, 2019).

The bird data used for BD Article 12 reports are based on extensive research by BirdLife Austria, the results of which are provided to the *Länder*. In contrast, only about one-third of all HD species and habitats are systematically monitored (Ellmauer *et al.*, 2017). The Article 17 reporting assessments also use adapted data from publicly funded projects and specific monitoring projects on selected HD species from several *Länder*. The *Länder* are intending to enlarge their monitoring activities, to extend the share of well-monitored species and habitats, for the 2019–2024 reporting period.

The project ÖBM-Cultural Landscape is a biodiversity monitoring programme focussing on the open cultural landscape in Austria (Schindler *et al.*, 2017). All habitat types as well as vascular plants, locusts and butterflies are recorded within 625 × 625 m sample plots. The survey areas include peri-urban landscapes, but especially grassland mosaic landscapes and alpine pastures. In 2017 and 2018 the first survey was carried out on 100 ÖBM sample plots (Schindler *et al.*, 2019).

In the course of preparing the WWF-Big5 *Länder*-Barometer 2019,[27] assessments were carried out on the Beaver, Otter, Eurasian Lynx, Wolf and White-tailed Sea-eagle, including information on their current situation and distribution in the respective *Länder*. Beaver monitoring is carried out using scientifically recognised methods in a number of *Länder*, but information on the distribution of Eurasian Lynx and Wolf is rather poor.

6.9 Future Developments

Since January 2020 Austria has had a new federal government consisting of the ÖVP (conservatives) and the Greens. The new government has resulted in a completely new arrangement of the ministries, in particular regarding environmental matters.

The Biodiversity Strategy 2020+ was analysed and its achievements were evaluated in 2018.[28] In summary, many of

27 www.wwf.at/artikel/bundeslaenderbarometer2019/

28 www.umweltbundesamt.at/studien-reports/publikationsdetail?pub_id=2304&cHash=398df24b392c8481e69cf73592d594f0

its measures have been implemented in Austria, with a positive impact on biodiversity. Nevertheless, the evaluation clearly indicated an urgent need for further action. Its main findings included that biodiversity is seen in a positive light by the Austrian population, the area under biodiversity-relevant agri-environmental measures has increased slightly, the Farmland Bird Index decreased, the amount of deadwood in Austrian forests increased, the ecological status of running waters was average, the conservation status of BHD species and habitat types remains unfavourable in many cases, and land take and habitat fragmentation are still much too high.

The new biodiversity strategy has been in preparation since 2020. It has considered the lessons from the preliminary evaluation of the previous strategy, and has been developed through a broad participatory process, which should be beneficial for stakeholder engagement and implementation. At the time of writing (summer 2021), a first draft was being discussed by the National Biodiversity Commission. Compared to the Biodiversity Strategy 2020+, the strategy has an even stronger focus on the specificity, the measurability, and the relevance of the objectives, many of which are quantitative. Objectives and measures are allocated to six overarching goals (related to: status and trends of habitats and species; protected areas; restoration; land take and habitat fragmentation; transformative change and mainstreaming; increased global engagement) and four enabling conditions (legal framework; funding and green finance; education and awareness raising; scientific knowledge, research and monitoring).

Over the next few years the details of how the new CAP will be implemented in Austria will be defined, and this will also have a significant influence on the further development of biodiversity in the country.

6.10 Conclusions

To a large extent, nature conservation in Austria has been shaped by the dispute over the proposed hydropower plant on the Danube in 1982, and the subsequent successful campaign by WWF and the media to protect the wetlands from destruction. This raised the awareness of the Austrian population on the threats to nature and need for action, and boosted the green movement as a political force, which has had a long-lasting legacy.

Nature conservation received a further significant boost when Austria joined the EU in 1995, strengthening the legal framework and creating new funding opportunities that were taken up quickly and effectively. The NGOs did an excellent job in submitting successful LIFE project proposals, whilst a comparatively excellent level of cooperation between nature conservation and agriculture was established in the Ministry of Agriculture,[29] which made it possible to secure new sources of finance for nature conservation in the context of modern rural development. The Austrian agri-environment programme (ÖPUL) has been outstanding in this context, as has the crucial early establishment of support for investments, management, studies and numerous activities in educational programmes. ÖPUL has been especially important in cushioning grasslands from the strong economic pressures associated with intensification, although the changes that have occurred in grasslands with regard to biodiversity are still concerning. Various nature conservation projects were also successfully implemented in the field of transnational projects, especially Interreg. This situation improved the already quite positive attitude towards nature conservation, especially in land users, who in Austria can be said to be quite well understanding of nature.

The positive attitude towards nature conservation was dampened by the extremely lengthy designation of Natura 2000 areas in Austria, which was not finally resolved until 2019 (i.e. 25 years after accession). This caused uncertainties and fears, especially among agricultural and forestry enterprises, but also among many communities and investors. One of the main reasons for this protracted nomination of potential Natura 2000 sites was the federal structure of nature conservation in Austria. This complex structure, with its various political decision makers, prompted the federal government in 2017 to intervene more proactively in the nomination process, which was still ongoing at that time. Even today, the federal structure still hinders uniform procedures, for example in relation to assessments of the impacts of plans and projects (HD Article 6) and data collection.

Another challenge, which has yet to be resolved, has been dealing with conflicts that have arisen as a result of the ongoing recolonisation of Austria by large carnivores, especially the Wolf. While the majority of the population has a very positive attitude towards the return of the Wolf, it has caused anxiety amongst livestock farmers and hunters that has led to heated debates (as in many other European countries). There have been similar, although less intense, discussions regarding Brown Bears in Austria. To address such issues, in 2019 the Austrian Federal Government founded a new Wolf Centre to inform discussions. Various initiatives, among them a current LIFE project (Euro Large Carnivores[30]), also intend to promote the acceptance of predators and to demonstrate the feasibility of livestock protection measures.

Despite the various increases in nature conservation efforts, the status of many habitats and species continues to deteriorate, especially HD habitats in the Continental region, but also amongst bogs and fens, rivers and other freshwater habitats, and the priority habitats Pannonic inland dunes and Pannonic salt steppes and salt marshes. There has also been a decline in the area of biodiversity-rich habitats in the agricultural landscape, particularly all extensively managed grasslands. A similar situation is evident for most HD species. Whilst most BD Annex I bird populations appear to be faring better than HD species, some have decreased, and there have been widespread declines in common farmland bird populations.

29 Federal Ministry of Agriculture, Regions and Tourism.

30 www.eurolargecarnivores.eu/de/

The main future challenges for nature conservation in Austria are to better conserve extensively managed grasslands (in particular semi-natural grasslands), improve management plans with similar national standards, restore and improve connectivity in rivers, provide corridors that allow movements of animal species, and improve the ecological condition of forests, in particular increasing deadwood and old trees. Also, a substantial increase in the extent of strictly protected non-intervention areas (core zones of National Parks and Wilderness Areas) is urgently needed.

Acknowledgements

We thank Johanna Huber (Suske Consulting) for information on funding, Thomas Ellmauer for information on reporting and monitoring, and Stefan Schindler (Federal Environment Agency Wien) and Bernhard Kohler (WWF Österreich) for reviewing the draft chapter.

References

Birdlife Österreich (2019) *Evaluierung der Wirkungen von Agrarumweltmaßnahmen anhand von Vogeldaten.* Vienna: BMLFUW-LE.1.3.7/8-II/1/2017.

BMLFUW (2014) *Biodiversitäts-Strategie Österreich 2020+ – Vielfalt erhalten – Lebensqualität und Wohlstand für uns und zukünftige Generationen sichern!* Vienna: Bundesministerium für Land- und Forstwirtschaft, Umwelt und Wasserwirtschaft

(2016) *Grüner Bericht 2016.* 57th ed. Vienna: Bundesministerium für Land- und Forstwirtschaft, Umwelt und Wasserwirtschaft.

(2019) *Grüner Bericht 2019.* 60th ed. Vienna: Bundesministerium für Land- und Forstwirtschaft, Umwelt und Wasserwirtschaft.

Dick, G., Dvorak, M., Grüll, A., Kohler, B. & Rauer, G. (1994) *Vogelparadies mit Zukunft? Ramsar-Gebiet Neusiedler See-Seewinkel.* Vienna: Umweltbundesamt.

Dvorak, M. (2009) *Important Bird Areas – Die wichtigsten Gebiete für den Vogelschutz in Österreich.* Vienna: Verlag Naturhistorisches Museum Wien.

Egger, G., Michor, K., Muhar, S. & Bednar, B. (2009) *Flüsse in Österreich.* Innsbruck, Austria: Studienverlag.

Ellmauer, T. (2015) Ergebnisse des aktuellen Artikel 17-Zustandsberichts mit Fokus auf Waldökosysteme. In *Natura 2000 und Wald – Handbuch und Fachbeiträge zur Umsetzung des europäischen Schutzgebietsnetzwerks Natura 2000 in Österreichs Wäldern*, eds. J. Schrank, G. Olbrich, H. Kirchmeir & M. Huber, pp. 54–60. Vienna: Kuratorium Wald/Initiative Österreich 2025.

Ellmauer, T., Paar, M., Konrad, C., Sonderegger, G., Oberleitner, I. & Stejskal-Tiefenbach, M. (2017) 20 Jahre europäischer Naturschutz in Österreich – Erreichte Umsetzung und erzielte Wirkungen der Naturschutzrichtlinien. *Natur und Landschaft*, **92** (3), S101–S108.

Enzenhofer, K. & Schrank, J. (2019) *Alt- und Totholzverbundsysteme – eine Literaturstudie zur Schaffung von naturschutzfachlichen Grundlagen.* Vienna: WWF Österreich.

Esterhazy Betriebe GmbH (2018) *Nachhaltigkeitsbericht 2016–2018.* Eisenstadt, Austria: Esterhazy Betriebe.

Förderverein Waldrappteam (2020) *Reason for Hope – Reintroduction of the Northern Bald Ibis LIFE+12 BIO/AT/000143 LIFE Northern Bald Ibis.* Mutters, Austria: LIFE+12-BIO AT 000143.

Friesenbichler K. & Stallegger, M. (2019) *Natura 2000 – Zurück in die Zukunft.* Vienna: Umweltdachverband GmbH.

FSV Forschungsgesellschaft Straße – Schiene – Verkehr (2015) *RVS 04.03.15 Artenschutz an Verkehrswegen.* Vienna: Forschungsgesellschaft Straße – Schiene – Verkehr.

Gantioler, S., Rayment, M., ten Brink, P., McConville, A., Kettunen, M. & Bassi, S. (2014) The costs and socio-economic benefits associated with the Natura 2000 network. *International Journal of Sustainable Society*, **6** (1-2), S135–S157.

Getzner, M., Haber, G., Huber, J., Preisel, H., Suske, W. & Withalm, G. (2016) *Natura 2000 und Wirtschaft – Regionale Wirkungen von Natura 2000-Schutzgebieten in Österreich.* Vienna: Studie der Technischen Universität Wien im Auftrag des Umweltdachverband GmbH.

Gottfried, M., Pauli, H., Futschik, A. *et al.* (2012) Continent-wide response of mountain vegetation to climate change. *Nature Climate Change*, **2**, S111–S115.

Grabherr, G., Koch, G., Kirchmeir, H. & Reiter, K. (1998) *Hemerobie österreichischer Waldökosysteme.* Innsbruck, Austria: Universitätsverlag Wagner.

Holzer, T., Zuna-Kratky, T. & Bieringer, G. (2019) *Bewertung der Wirkung relevanter LE-Maßnahmen auf Heuschrecken und Tagfalter als Indikatorarten für Biodiversität.* Vienna: Bundesministerium für Nachhaltigkeit und Tourismus.

Hülber, K., Kuttner, M., Moser, D. *et al.* (2020) Habitat availability disproportionally amplifies climate change risks for lowland compared to alpine species. *Global Ecology and Conservation*, **23**, e01113.

Mauerhofer, V. (2011) *Rechtswissenschaftliche Kurzstudie zum Artenschutz nach österreichischem Naturschutz-, Jagd-und Fischereirecht anhand ausgewählter Beispiele.* Vienna: Studie im Auftrag von Naturschutzbund Österreich, BirdLife Österreich und WWF Österreich.

Pauli, H., Gottfried, M., Dullinger, S. *et al.* (2012) Recent plant diversity changes on Europe's mountain summits. *Science*, **336**, S353–S355.

Pinterits, S., Ressi, W. & Bogner, D. (2014) *Evaluierung des Programms LE 07-13 – Ganzheitliche Wirkung der Fördermaßnahme "Erhaltung und Verbesserung des ländlichen Erbes, M323a – Naturschutz".* Salzburg, Austria: Studie im Auftrag des BMLFUW.

Sauberer, N., Prinz, M. & Essl, F. (2017) Österreichs Klima, Geographie und Landbedeckung im Überblick. In *Die Heuschrecken Österreichs*, eds. T. Zuna-Kratky, A. Landmann, I. Illich, L. Zechner, F. Essl, K. Lechner, A. Ortner, W. Weißmair & G. Wöss, pp. 24-34. Linz, Austria: Biologiezentrum des Oberösterreichischen Landesmuseums.

Schindler, S., Banko, G. Moser, D. *et al.* (2017) *Österreichisches Biodiversitäts-Monitoring: Kulturlandschaft – Konzept für die*

Erfassung von Status und Trends der Biodiversität. Vienna: Federal Environment Agency, REP0635.

Schindler, S., Banko, G., Neuwirth, M. *et al.* (2019) *Österreichisches Biodiverisäts-Monitoring (ÖBM) – Kulturlandschaft. Erhebungen 2017*. Umweltbundesamt Rep-0720.

Schindler, S., Zulka, K. P., Sonderegger, G. *et al.* (2016) *Entwicklungen zur biologischen Vielfalt in Österreich – Schutz, Status, Gefährdung*. Vienna: Federal Environment Agency.

Stoate, C., Báldi, A., Beja, P. *et al.* (2009) Ecological impacts of early 21st century agricultural change in Europe – a review. *Journal of Environmental Management*, **91** (1), S22–S46.

Stöhr, O., Ragger, C., Gewolf, S. *et al.* (2014) Basiserhebung von Lebensraumtypen und Arten von gemeinschaftlicher Bedeutung in Österreich. *Natur und Landschaft*, **86/11**, S471–S477.

Suske, W., Gattermaier, S., Ellmauer, T. & Tomek, H. (2009) *Analyse der Akzeptanzen der ÖPUL-Maßnahme "WF" auf Lebensräumen der FFH- und Vogelschutzrichtlinie*. Vienna: Studie im Auftrag des Bundesministeriums für Land- und Forstwirtschaft, Umwelt und Wasserwirtschaft.

Suske, W., Gattermaier, S., Huber, J., Ellmauer, T., Tomek, H. & Steurer, B. (2012) *Evaluierung des Programms zur Ländlichen Entwicklung im Bereich der Almen*. Vienna: Studie im Auftrag des Bundesministeriums für Land- und Forstwirtschaft, Umwelt und Wasserwirtschaft.

Suske, W., Huber, J., Glaser, M. *et al.* (2019a) *Ökologische Bewertung der Bewirtschaftung von Grünlandflächen hinsichtlich Nutzungsintensivierung und Nutzungsaufgabe*. Vienna: Evaluierungsstudie im Auftrag des BMNT.

Suske, W., Huber, J., Glaser, M. & Schütz, C. (2019b) *Ecological evaluation of the management of grassland areas with regard to intensification of use and abandonment of use*. Vienna: Federal Ministry for Sustainability and Tourism.

Teufelbauer, N. (2014) *Evaluierung LE07-13: Farmland Bird Index für Österreich – Indikator 2013 und 2014 – Teilbericht 1: Farmland Bird Index 2013 für Österreich*. Vienna: Studie im Auftrag des Lebensministeriums.

Teufelbauer, N. & Seaman, B. (2019) *Farmland Bird Index für Österreich: Indikatorenermittlung 2015 bis 2020 – Teilbericht: Farmland Bird Index 2018*. Vienna: Studie im Auftrag des Bundesministeriums für Nachhaltigkeit und Tourismus.

Uhl, H. & Denner, M. (2015) *Vogelschutzgebiete nach EU-Richtlinie in Österreich –Umsetzungsstand und Handlungsbedarf, speziell für ÖPUL-Förderungen*. Vienna: Studie von BirdLife Österreich im Rahmen eines Projektes für BirdLife International und MAVA Foundation.

Uhl, H. & Wichmann, G. (2013) *Wiesen- und Kulturlandschaftsvögel in Oberösterreich 2011-2013 – Landesweite Bestandszählungen 1994 bis 2012 sowie Bildungsprojekte und Öffentlichkeitsarbeit*. Vienna: BirdLife Österreich.

UNEP-WCMC (2021) *Protected Area Profile for Austria from the World Database of Protected Areas*. Downloaded from www.protectedplanet.net on 29 July 2021.

Wessely, J., Hülber, K., Gattringer, A. *et al.* (2017) Severe constraints for the effectiveness of conservation strategies to mitigate climate change-induced range losses. *Nature Climate Change*, **7**, S823–S827.

Wichmann, G. & Denner, M (2013) *Ornithologische Grundlagen für die Windkraftzonierung in Niederösterreich*. Vienna: Studie im Auftrag der NÖ Umweltanwaltschaft.

Wrbka, T., Schindler, S., Pollheimer, M., Schmitzberger, I. & Peterseil, J. (2008) Impact of the Austrian Agri-Environmental Scheme on diversity of landscape, plants and birds. *Community Ecology*, **9** (2), S217–S227.

Chapter 7

Belgium

Els Martens and Jorge Ventocilla

Key Statistics	
Terrestrial area (km^2)/% of EU-28*	30 667/0.7%
Marine waters (km^2)/% of EU-28*	3 458/0.1%
Population density (p km^2)/% of EU-28*	365/318%
GDP per capita (€)/% of EU-28	34 313/128%
Land cover (CORINE land cover classes 2018)	AGR 57%, ART 21%, FOR 19%, OSN 2%, WET 1%
Biogeographical regions	ATL, CON, MATL
Number of HD Annex I habitats/%*	59/26%
Number of HD Annex II, IV & V species/%*	86/7%
Number of BD Annex I bird taxa/%*	59/34%
% HD UCS & decreasing habitats/% unknown	11%/1%
% HD UCS & decreasing species/% unknown	15%/8%
% breeding BD Annex I birds decreasing/ % uncertain and unknown	20%/3%

* Excluding MAC and MMAC biogeographical regions.
Sources and codes: See Table 1.1.

7.1 The Natural History and Characteristics of Belgium

Belgium is a federal constitutional monarchy, composed of communities and three highly autonomous regions: Flanders (Flemish) in the north (13 521 km^2), Wallonia (mainly French-speaking) in the south (16 800 km^2), and the Brussels-Capital Region (162 km^2). The country falls within two of the EU terrestrial biogeographical regions, with the Atlantic region covering the northern half (mainly Flanders and the Brussels-Capital Region) and a small western part (Wallonia), while the Continental region includes mainly Wallonia in the southern part and a small eastern part of Flanders. The Belgian part of the North Sea is in the Marine Atlantic biogeographical region; it has a coastal length of 67 km and an area of 3 465 km^2.

The majority of the country consists of lowlands that extend from the coast south-east to a central plateau (50–200 m) with the Ardennes hills further to the south-east (highest point 694 m). The climate is maritime temperate with average monthly temperatures ranging between 3 °C and 18 °C, and annual average rainfall of 925 mm.

The geographical and geological characteristics of Belgium, together with the longstanding human influences, have resulted in the development of a wide range of habitats, the main ones being deciduous and coniferous forests, peat bogs, wetlands, lakes and rivers, and semi-natural grasslands and heathlands. Their distribution varies greatly from region to region: about 80% of the remaining forested areas are found in the rugged and rocky Ardennes, along with other habitats associated with its caves and small gorges. Forest types with high nature importance include: European Beech (*Fagus sylvatica*), often mixed with oak (*Quercus* spp.), forming variations of habitat types from *Luzulo-Fagetum* (9110)[1] to *Asperulo-Fagetum* (9130); alluvial forest types with *Alnus glutinosa* and *Fraxinus excelsior* (91E0*); and old oak forests on sandy plains with *Quercus robur* (9190). Northern Belgium is noted for its semi-natural grasslands, wetlands, heathlands and inland dunes. The main habitats of the coast include dunes, sandy beaches and sandbanks, intertidal mudflats, salt marshes and estuaries. The subtidal marine areas consist predominantly of soft sediments, sand and some mudflats, and scarce natural hard substrates.

Currently, nearly half of the land in Belgium is farmed. Over the centuries, farming contributed to the creation and maintenance of a large variety of agricultural habitats (e.g. cropland, pastures, meadows, hedges and wood pastures), which retained a relatively high diversity of species. However, as a result of intensification and specialisation, farming in Belgium is now amongst the most productive in Europe and strongly focussed on meat and dairy production. Crop production covers about 61% of farmland and 37% is permanent grassland, much of which is used for silage rather than grazing. A significant proportion of farmland is also horticultural land, particularly in Flanders where it comprises about 8% of the land area.

[1] Numbers in brackets are the HD Annex I habitat codes (see the Appendix for full names). * Indicates a priority habitat.

As a result, most agricultural landscapes are now much simpler and less biodiverse, and remaining High Nature Value (HNV) farmland areas are scarce and mostly fragmented. According to Danckaert *et al.* (2012), a 2009 survey of Flanders revealed that semi-natural HNV farmland habitats (Type 1) were rare, with 1 350 ha containing HD habitats, and another 820 ha of regional value. In addition, there were more extensive areas of small-scale mosaic farmland (Type 2 HNV), covering 52 280 ha, and about 94 000 ha of farmland of importance for species of conservation concern (Type 3 HNV), including breeding waders, wintering geese and other characteristic farmland birds. In total, this 148 450 ha with HNV values covered about 22% of farmland in Flanders; however, only 6% of this HNV area is managed to conserve these natural values, mainly through agri-environment schemes. HNV farmland under nature-oriented management in Wallonia is not quite so scarce and was estimated in 2009 to cover about 10% of the agricultural area, of which about 20% was Type 1, 53% Type 2 and 27% Type 3.

Urban zones also now cover a significant proportion of the country, especially in the Flemish northern region, as it is one of the most densely populated areas of Europe (450 inhabitants/km^2).

7.2 The Status of Nature

7.2.1 Habitat Status and Trends

Due to the dense population, intensive agriculture and ever-growing industrialisation, the landscape became very fragmented, putting wildlife under immense pressure. As reflected in the *Living Planet Report* on Belgium (WWF, 2020), the country lost much of its natural capital in the decades before 1990, thus iconic species and vital habitats have disappeared or become fragmented. The main reasons for the decline of biodiversity are land-use change (mainly for cropland and buildings and infrastructure), destruction and fragmentation of natural habitats, pollution and eutrophication caused by intensification of agricultural and industrial practices, excessive water catchments in some areas, climate change, invasive alien species and perturbations linked to leisure and tourism (IUCN, 2019). These drivers of change are further described later in Section 7.2.3.

These past impacts and ongoing pressures (described in Section 7.2.3) have been reflected in the national reports on BHD habitats and species made under the Nature Directives.[2]

In the HD Article 17 report for 2013–2018, Belgium reported on all 59 HD habitats across the three biogeographical regions, resulting in 93 assessments. Overall, for Belgium, most habitat assessments (80%) reported a bad status, and another 15% revealed a poor status. Only 4% indicated a favourable status, with the remaining 1% having an unknown status. The status of the habitats was almost exactly the same as reported for 2007–2012. The situation was similar across the two terrestrial biogeographical regions, but habitats in the Atlantic region generally had a less favourable status and trend.

Comparing the conservation status of habitats by broad ecosystem type reveals that all had unfavourable conservation status assessments, other than one coastal habitat (mudflats, 1140), one type of dune (dunes with *Hippophaërhamnoides,* 2160) and one rocky habitat (caves, 8310). The habitats with the largest shares of assessments with a poor status were: grasslands; heaths and scrub; bogs, mires and fens; and dunes. Of these, grasslands had a particularly high percentage with a bad status (100% in the Atlantic region). All 12 EU priority habitats in Belgium were reported as having an unfavourable condition, of which 10 had a bad condition.

Of further concern is that a significant proportion of habitats with an unfavourable status showed ongoing declines, or had an unknown status (Table 7.1). These included the following habitats with a bad status and decreasing trends: shifting dunes (2120), alluvial forests (91E0*), riparian mixed forests (91F0), natural eutrophic lakes (3150), xeric sand calcareous grasslands (6120*), *Molinia* meadows (6410) and hay meadows (6510 and 6520).

7.2.2 Species Status and Trends

All Taxa and Red Lists

The *Living Planet Report* (WWF, 2020) showed a slight rising general trend for overall biodiversity over 1990–2018, which became stable over the last 10 years of the period. Further analysis reveals that the trends for specific species groups and habitats varied greatly, dropping sharply for birds that typically live in agricultural areas (−61% for 1990–2018) and to a lesser extent for the species specifically found in forests (−27%). Populations of species from open nature areas and wetlands generally increased on average by 15% and 48%, respectively, in the same period.

The general slightly upward trend is an encouraging result but should be seen against the fact that biodiversity declined considerably before 1990. Based on an evaluation of 7 725 native species in Belgium, significant proportions of most species groups are Vulnerable, Endangered or Critically Endangered at regional and/or national levels (Kestemont, 2019). These include 40% or more, at a national level, of extant amphibians, reptiles, ants, beetles, butterflies, dragonflies, mosses and macrofungi (excluding those that are rare or for whom data are deficient).

Presently, there are validated Red Lists for Flanders developed by the Research Institute for Nature and Forest (INBO) for amphibians, breeding birds, butterflies, higher plants, dragonflies, ladybirds (Coccinellidae), ground beetles, moths, reptiles, grasshoppers (Orthoptera), saproxylic beetles, freshwater fish and mammals. Of the 2 624 species on the recently validated Red Lists (INBO, 2019), 182 species, or 7%, have disappeared from Flanders in the course of the last

[2] See Section 1.3 for background on monitoring under the Nature Directives, status categories and data sources.

Table 7.1 Short-term trends (c. 2007–2018) in Habitats Directive Annex I habitats and Annex II, IV and V species with an unfavourable conservation status in Belgium.

	Poor (unfavourable– inadequate)				Bad (unfavourable–bad)			
	Imp	Sta	Dec	?	Imp	Sta	Dec	?
HD Habitat assessments								
Number	3	4	1	6	18	26	9	21
%	21%	29%	7%	43%	24%	35%	12%	28%
HD Species assessments								
Number	15	8	5	15	9	11	16	13
%	35%	19%	12%	35%	18%	22%	33%	27%

Key. Imp = improving, Sta = stable, Dec = decreasing, ? = unknown.
Source. Based on Member State HD Article 17 reported data as presented on the EEA national summary dashboard (2021).

century. Of the remaining 2 442 species, almost one-third are Critically Endangered, Endangered or Vulnerable because their populations have declined sharply and/or have reached a critical minimum. This is the case, for example, for the Hazel Dormouse (*Muscardinus avellanarius*), the Common Spadefoot Toad (*Pelobates fuscus*) and several bird species. The Red List status differs greatly between the different species groups.[3]

According to the Red Lists for species in Wallonia (SPW, 2017a), 31% of the assessed animal and plant species are threatened and almost 9% have disappeared from the regional territory. Of fish, reptiles, butterflies and dragonflies, more than half of the species are Vulnerable to Critically Endangered.

Species Listed in Annexes II, IV and V of the Habitats Directive

The overall results from the Article 17 report for 2013–2018 indicate a less concerning picture for HD species than HD habitats. Nevertheless, it is clear that Target 1 of the *EU Biodiversity Strategy to 2020* was not reached in Belgium for these species. Out of the 138 assessments of HD species across the biogeographical regions, 36% reported a bad status, 31% a poor status and 25% a favourable status. Another 8% had an unknown status. There was little change in the status of the species from the previous reporting period, and the results for the two terrestrial biogeographical regions were similar. As for HD habitats, a substantial proportion of species with an unfavourable status assessment were also considered to be declining (Table 7.1), most notably one-third of those that had a bad status, such as the Common Spadefoot Toad, the Stag Beetle (*Lucanus cervus*) and the Barbel (*Barbus barbus*). It is also of concern that many species had unknown trends.

Comparison of the broad species groups indicates that amphibians, arthropods, fish, molluscs and reptiles have fared particularly badly, as a relatively high proportion of assessments indicated a bad status, and a substantial number also showed declining trends.

[3] For details see www.vlaanderen.be/inbo/en-gb/red-lists/

Table 7.2 Short-term trends (c. 2007–2018) in Birds Directive Annex I taxa in Belgium.

	Breeding				Wintering			
	Inc	Sta/ Flu	Dec	?	Inc	Sta/ Flu	Dec	?
Number	20	11	8	1	4	1	2	15
%	50%	28%	20%	3%	18%	5%	9%	68%

Key: Inc = increasing, Sta = stable, Flu = fluctuating, Dec = decreasing, ? = uncertain and unknown.
Source. Based on Member State Article 12 reported data as presented on the EEA national summary dashboard (2021).

Birds

The results from the 2019 BD Article 12 report indicate that bird population trends have been rather mixed, both over the short term (2007–2018) and longer term (1980–2018). Over the short term, across breeding populations of all bird species, 31% increased, 28% were stable or fluctuating, 31% decreased, and the remaining 9% had uncertain or unknown trends. The longer-term trends were even more divergent, with 58% having increased and 26% having decreased, with relatively few having stable or fluctuating trends (10%) or unknown or uncertain trends (6%). The trends of BD Annex I species are of particular importance, especially their short-term trends, as these may reflect the impacts of conservation measures. The short-term trends shown in Table 7.2 are relatively encouraging as about three-quarters of the species have increased or are stable or fluctuating. However, a fifth of Annex I species have continued to decline. Similar results are apparent over the longer term, although a higher proportion increased.

As a high proportion of trends for wintering populations are unknown or uncertain, it is not possible to draw conclusions about their status.

Some of the observed increases in breeding species may reflect the results of conservation measures, such as in the case of the Black Stork (*Ciconia nigra*). Others may be the result of undesirable environmental changes such as climate change. The declining species are of clear concern, especially as some localised Annex I species, such as Grey-headed Woodpecker (*Picus canus*), are now very scarce, or on the edge of national extinction in the case of Black Grouse (*Lyrurus tetrix*) and Hazel Grouse (*Bonasa bonasia*). As regards the more common species, there is good evidence from the Common Breeding Bird Index for Flanders (INBO, 2020a) that a large proportion of the declining species are farmland birds – a pattern that has been seen over much of Europe. Between 1990 and 2015 they decreased by 47%, although there is some indication that their decline has since levelled off (WWF, 2020). Woodland birds in Flanders have shown a continued decline since 2007, whilst in Wallonia a moderate decline (−11%) may have been reversing since 2011 (SPW, 2017b).

7.2.3 Drivers and Causes of Change

Biodiversity in Belgium has been mainly affected by anthropogenic activities, often driven by economic development and influenced by sectoral policies. Land conversion, whether for urban and industrial expansion, infrastructure or tourism, and land-use change, particularly agricultural intensification, have undoubtedly been the main causes of biodiversity loss. Such activities resulted in the loss, degradation and fragmentation of habitats, and currently affect all habitat types and many of their associated species of conservation importance. Habitat loss and fragmentation from built developments is particularly high in the Brussels-Capital Region, with 80% of the surface built on, but also in Flanders which shows the highest increase in building rate in Belgium (Statistics Flanders, 2019). In 2019, more than a quarter (28% or 3 855 km^2) of the surface in Flanders was built up, of which 44% is residential areas and 30% infrastructure. This was an increase of almost 30% compared to 1990 (Departement Omgeving Vlaanderen, 2018). In the Walloon Region, the building rate was lowest at 15%. The urban environment of the Brussels-Capital Region also has other specific problems, such as very high recreation pressures in green areas and the relatively high numbers of alien invasive species (IAS). The replacement of old buildings, or their renovation, is reducing nesting sites for birds that require cavities in buildings, such as the Common Swift (*Apus apus*). This is one of the main causes of an observed decline of 76% in such cavity-nesting urban bird species in the Brussels Region over 1992–2018. Urbanisation is also leading to increasing levels of light and noise pollution.

The most widespread biodiversity pressures result from agriculture, as it is amongst the most intensive, specialised and productive in Europe, especially in Flanders. As described earlier, this has resulted in the widespread conversion of semi-natural habitats and semi-improved permanent pastures to much less diverse agricultural landscapes that have an impoverished biodiversity. For example, between 1980 and 2015 in Wallonia, the area under permanent grassland decreased by 23% (SPW, 2017b). The ongoing intensification and specialisation in agricultural practices, including the very high rates of use of fertiliser and pesticides, and the conversion of pastures to silage or other forage crops continue to be the primary causes of the observed declines in typical farmland species.

Intensive agricultural systems are also leading to high external pressures, including on freshwater and marine habitats as a result of eutrophication caused by high levels of nitrogen and phosphate in water flowing from agricultural land. Furthermore, the high densities of livestock produce high levels of ammonia emissions, which lead to eutrophication of sensitive terrestrial semi-natural habitats from the deposition of airborne nitrogen. High deposition levels of farm-based nitrogen in Natura 2000 sites remain the main issue of concern, particularly in the lowland parts of Belgium. At the end of 2018, the nitrogen critical load level was exceeded over 64% of the total surface of HD habitats in Natura 2000 sites in Flanders (INBO, 2020b). As this pressure is preventing the habitats achieving a favourable status, a strategic nitrogen abatement programme is being implemented and is expected to lead to a progressive reduction of nitrogen deposition in Natura 2000 sites over the coming decades (European Commission, 2017). Whilst this problem is probably less acute in Wallonia, eutrophication and other pollution constitute a secondary threat to biodiversity in the region.

The biodiversity of forests is threatened locally, among other things, by intensive management (especially in the Continental region), pollution, changes in groundwater levels, fragmentation, recreational activities and in some locations high population densities of large game species such as deer, or invasive species.

Some natural processes, such as in estuaries due to silt accumulation, and in forests and shrublands where vegetation growth and succession leads to losses of open spaces, form another type of important pressure and cause gradual changes in habitat structure and types. As these changes may be undesirable, measures need to be taken to counteract them, sometimes requiring high investment.

IAS are one of the most frequently identified pressures on species. Therefore growing attention is being given to them, given the rapid expansion of some introduced plants, fish, amphibians, reptiles, birds, and invertebrates (e.g. crayfish, mussels and slugs). Well-known examples include Floating Pennywort (*Hydrocotyle ranunculoides*) obstructing natural waterways, and the American Bullfrog (*Lithobates catesbeianus*) having impacts on endemic amphibian species.[4] The Belgian Forum on Invasive Species has produced the Harmonia information system, which assesses and lists IAS threatening native biodiversity in Belgium and in neighbouring areas (Belgian Forum on Invasive Species, 2020). As of 5 March 2022, this listed 102 species, with 44 on the black list, 39 on the watch list, 12 on the alert list, and 7 under further research.

[4] For more examples and information on control measures see https://ias.biodiversity.be/species/all

Marine biodiversity has been highly impacted and remains particularly threatened in the Belgian coastal zone and the shelf sea area, where direct and indirect pressures are most concentrated. Two important ongoing pressures are the overexploitation of fish stocks and the adverse effects of certain fishing methods, in particular bottom-towed fishing gear which is used by fishing vessels from Belgium and other countries in Belgian waters. The ecological quality of sandbanks and mudflats (HD codes 1110, 1140) and biogenic reefs built by diverse tube worms has been affected by the impacts of bottom-disturbing activities on their structure and function (Belgian Federal Government, 2012). There are also significant levels of accidental by-catch from fishing nets, which is a major cause of death for the Harbour Porpoise (*Phocoena phocoena*) in the North Sea. Based on evidence from stranded animals, it can be estimated that up to a few dozen porpoises die every year as a result of this in Belgium.

Climate change is a growing concern, already having a perceptible impact on biodiversity (notably on the geographical range, phenology and behaviour of organisms, such as migrating birds and insects, and on forest health). It also exacerbates other threats to biodiversity, such as habitat fragmentation and biological invasions.

7.3 Nature Conservation Governance, Actors, Legislation and Policies

7.3.1 Overview of Nature Developments in Belgium

Belgium is a federal state, composed of communities and regions. Whilst nature conservation involves the federal government, the communities and the local authorities (provinces and municipalities), the remit for it is mainly within the competence of the three regions and their respective legislation. Table 7.3 therefore provides an overview of some of the most important developments in nature conservation in Belgium for each of the regions, as well as at the federal government level. Nature conservation started in 1933 with the establishment of the first national nature organisation 'De Wielewaal' and the acquisition of nature sites for bird protection. In the 1950s a new organisation was set up at a national level, 'Belgian Nature and Bird Reserves', which split into Flemish and Walloon groups with the federalisation of Belgium in the 1980s.

The adoption of the EU Nature Directives and designation of Natura 2000 offered an important core legislative basis to enhance and regulate nature protection. The transpositions of the directives into regional and federal law in 2001 and following years also enhanced consultation and cooperation with stakeholders and private owners.

7.3.2 Governance

The **federal government** is the competent body for the management of the Belgian part of the North Sea, and for international marine environment policy issues. It also coordinates Belgian external relations with respect to biodiversity. The coherence of national responses to international environmental policy is facilitated by a coordination mechanism composed of representatives from the federal government, the regions and the communities, called the Coordinating Committee for International Environment Policy (CCIEP). This body functions under the high-level authority of the Inter-ministerial Conference for the Environment. Under the CCIEP various convention-related or thematic committees have been established, such as for biodiversity (CBD focus), nature (focus on Natura 2000 and all other biodiversity-related conventions), climate change, adaptation to climate change, and forests. The committees develop the positions on relevant thematic policy documents at EU and/or international level, support exchange of expertise and knowledge, and enhance cooperation projects.

The **regions** are in charge of terrestrial issues, which include nature conservation, forest management, agriculture, exploitation of natural resources, land use and spatial planning, hunting and inland fisheries. Responsibility for tourism has also been delegated to them by the communities. The nature conservation authorities have responsibility for the implementation of Natura 2000, species protection, and site management of nature areas including nature and forest reserves and parks. The nature authorities belong to the ministry for environment of their respective regional government: the Agency for Nature and Forests (ANB) in Flanders; the division for Nature and Green Spaces under the Department for Nature and Forests in Wallonia; Environment Brussels for the Brussels-Capital Region; and Division for Marine Environment for the federal level. It is, therefore, important to note that the main drivers of biodiversity declines that need to be tackled – agricultural practices, the building sector, industry and transport – are outside the competence of nature administrations. Each regional administration has programmes in place to implement their action plan, in relation to the national action plan (see later), the Nature Directives, and to stimulate biodiversity plans and conservation actions by local authorities/municipalities and other stakeholders.

The **communities** take care of issues linked to culture, research, education and public awareness. The regions and the federal government also conduct research and raise public awareness in their own fields of competence.

The **provinces and the municipalities** play an important role at the local level, especially concerning green infrastructure, communication and awareness activities, in accordance with regional policy. Provinces have substantial competences over the management of parks, gardens, and designing ecological corridors through the development of a green and blue network (see Section 7.5); they also have their own action plans.

7.3.3 Other Actors

Environmental NGOs contribute significantly to nature conservation in Belgium in a number of ways. Some have purchased important areas and manage them as nature reserves, for which they receive financial support from the regional administrations

Table 7.3 Key nature conservation developments in Belgium.

National	
2006	Belgium's *National Biodiversity Strategy 2006–2016*
2013	*Biodiversity 2020, Update of the National Strategy of Belgium*
2019	Protocol and cooperation agreement for the implementation of the EU IAS Regulation
Brussels-Capital Region (BCR)	
1998	Government Decree of the Brussels-Capital Region including aspects on a green network
2000	Decree on the conservation of natural habitats and wild fauna and flora
2012	Key legislation organising nature policy at the regional level
2013	Legal act on derogation procedures for tree felling and forest exploitation
2015–2016	Natura 2000 site designation act setting conservation objectives and describing protection measures (SAC I, II and III)
2015 and 2020	Designation act establishing a High Council for Nature Conservation as an advisory body for nature conservation (2015–2020), and for 2020–2025
2016	Regional plan for nature
2018	Legal act on monitoring and surveillance projects concerning nature and biodiversity of BCR
2019	Approbation of the management plan for the Brussels part of the Sonian Forest
	Designation acts for the nature reserves in Anderlecht
Wallonia	
1973	Law on nature conservation
1982	Hunting Act
2008	Decree on Forest Code (M.B. 12.09.2008)
	Decree on preventive regime applicable to Natura 2000 sites (M.B. 27.11.2008)
2011	Order of Walloon Government on general preventive measures for Natura 2000 sites (M.B. 03.05.2011)
	Decree on types of management units for Natura 2000 sites and prohibitions and preventive measures (M.B. 03.06.2011)
	Order of Walloon Government on general preventive measures for Natura 2000 sites (M.B. 03.05.2011)
2013	Circular on invasive alien plants (M.B. 11.06.2013)
2016	Decree on the conservation objectives for the Natura 2000 network (M.B. 22.12.2016)
2017	Territorial development code
Federal	
1999	Law on the protection of the marine environment and the organisation of marine spatial planning in the sea areas under the jurisdiction of Belgium
2001	Royal Decree on the species protection in the seas under jurisdiction of Belgium
2010	Royal Decree on the marine strategy in the Belgian sea areas
2016	Royal Decree on the procedure for the designation and management of marine protected areas
2018	Working protocol on border controls to prevent the introduction of IAS
2019	Royal Decree establishing the marine spatial planning for 2020–2026 in the Belgian sea areas
Flanders	
1990	Forest Decree – since adapted several times, implemented through various Executive Acts
1991	Hunting Decree – since adapted several times, implemented through various Executive Acts
1993	Dunes Decree for the protection of coastal dunes
1997	Decree for Nature Conservation – since adapted several times
1998	Cooperation Protocol between the Ministry of Defence and Nature Division of Government of Flanders for the management of nature areas and Natura 2000 on military domains
2003	Executive Act on the criteria for sustainable management of forests in Flanders
2009	Executive Act on species protection

	Executive Act on procedure for designation of HD SCIs
2010	Executive Act on established regional conservation objectives for habitats and species of EU concern
2014	Executive Act on designation of SCIs and SPAs Natura 2000 sites
2017	Natura 2000 programme for Flemish region – first cycle 2016–2020
	Executive Act on nature management plans – including new subsidies system

See Table 3.3 for signatory dates of international conventions.

and become members of the consultation fora on policy and implementation planning. All NGOs also organise awareness and educational activities, monitoring actions, guided visits and servicing public access of sites, and lobbying to enhance nature conservation. The Flemish NGO Natuurpunt is the largest in Belgium with about 114 000 members and manages about 25 000 ha of its own reserves, while the Walloon Natagora manages 4 300 ha. Other NGOs in Wallonia that manage significant areas of nature reserve are Ardenne et Gaume, with 2 500 ha, and the Ligue Royale Belge pour la Protection des Oiseaux (LRBPO) managing about 200 ha. In Flanders two smaller, local NGOs, Limburgs Landschap and vzw Durme, manage 2 300 ha and 470 ha, respectively.

In Flanders, 16 Regional Landscape groups[5] have been set up to support cooperative actions for nature, landscape and heritage protection. They have a contract with the regional Agency for Nature and Forest (ANB) to support field action for nature conservation, public access to nature sites, communication and awareness, and cooperation with various stakeholders. Eleven forest management organisations of private forest owners (Bosgroepen) that facilitate integrated management of forests are supported by a regional contact point financed by ANB.[6] In Wallonia, the NGO Natagriwal was set up to support the forestry and agricultural sectors in implementing agri-environment and climatic schemes and conservation management in Natura 2000 areas.

Hunting is a leisure activity that is carried out by about 23 000 hunters in Belgium and has generated much debate. Compromises have been reached, and attitudes have evolved over the last 20 years, such that there has been improving cooperation between hunters, foresters, farmers and conservationists. Important progress has been made in putting new wildlife management insights into practice and in recognising the ecological interactions between hunting and biodiversity.

A number of initiatives have been taken to enhance **cooperation and information sharing** within Belgium and with neighbouring countries. For Natura 2000, the LIFE Belgian Nature Integrated Project (BNIP[7]) established inter-regional cooperation between the nature administrations of Flanders, Wallonia and Federal-Marine, and the two main NGOs Natuurpunt and Natagora, to enhance implementation and harmonise approaches. A national-level platform dedicated to science–policy interfacing, the **Belgian Biodiversity Platform,**[8] hosts the national focal points for GBIF, IPBES and IUCN, as well as coordinating several communities of practice such as on IAS, Ecosystems and Society, and Biodiversity and Health. Belgium has developed inter-regional and bilateral cooperation with countries in its immediate vicinity for integrated management of transboundary ecosystems, often initiated through LIFE and Interreg projects. Some notable examples, all including Natura 2000 sites on borders, are in the Schelde estuary and the Zwin estuary (BE-NL), the transboundary parks of Hainaut (FR-BE), De Zoom-Kalmthoutse Heide (NL-BE), het Kempen-Broek (NL-BE), Rivierpark Maasvallei (NL-BE), Groot Saeftinghe (NL-BE) and a number of river valleys such as of the Semois (FR-BE) and Our (DE-LU-BE).

7.3.4 Legislation

Infringement procedures in 2000–2002 for non-compliance with requirements of the EU Nature Directives and the transposition of the directives into regional law contributed to stricter regulations for nature protection. An overview of key laws is given in Table 7.3.

In Flanders the Decree of 21 October 1997 on Nature Conservation and the Natural Environment, the so-called Nature Decree, forms the legal basis of Flemish nature policy. The Nature Decree was adapted in 2003 to include the provisions regarding Natura 2000 in a chapter on area-based policy. Executive laws to regulate implementation were adopted in the following years: in 2009 on the procedures for the designation of SCI and SPA Natura 2000 areas and on species protection; in 2010 on the regional conservation objectives; in 2014 on designation of SCIs and overlapping SPAs, including the site conservation objectives; and in 2017 on site management and related subsidies, and on the site conservation objectives of the remaining SPAs. For Natura 2000 implementation an internal action plan was developed by ANB in the Natura 2000 programme of Flanders (2016–2020) that had two overall main objectives for 2020: (1) stopping or preventing a decline of the BHD habitats and species, and (2) achieving a favourable or improved status for the 16 prioritised HD habitats.

In Wallonia the provisions implementing the Nature Directives were transposed by means of amendment decrees of the Nature Conservation Act of 1973 (Walloon Government, 1973), the last amendment of which dates from 22 December 2010. Furthermore, various executive laws followed, such as: on the reimbursements and allowances in the Natura 2000 sites; in 2011 on management units for Natura 2000 sites and on general prevention measures for Natura 2000 and related modalities; and in 2016 on the regional conservation objectives

5 Overview of the Landscape groups: www.regionalelandschappen.be/
6 For a full list of partners visit: www.privaatbeheer.be/partners/
7 For a full list of partners visit: www.life-bnip.be/en/
8 Belgian Biodiversity Platform website: www.biodiversity.be/

and on the designations. The executive Act on general prevention measures indicating which activities cannot take place at all, in all or specific Natura 2000 sites is an interesting example of setting one fixed set of regulations.

Initially, the transposition into Brussels nature legislation was achieved by the Decree of 26 October 2000 'on the conservation of natural habitats and wild fauna and flora'. This decision was amended in 2002 and 2004 in response to comments of the European Commission, but the Nature Conservation Order of 2012 is now the legal basis for the Brussels Natura 2000 project.

The protection of the marine environment in the areas under Belgian jurisdiction is the responsibility of the federal government and based on the new Belgian Marine Environment Law of 20 January 1999 (SPF, 1999). This law has provisions for the identification and designation of marine protected areas (MPAs), including in relation to the Nature Directives' requirements. Species protection provisions for marine species are found in the Royal Decree of 21 December 2001 concerning species protection in the maritime regions under the jurisdiction of Belgium (Belgian Federal Government, 2001).

To reconcile the various (economic, ecological and social) interests and to provide every activity with an appropriate place in the North Sea, the former Minister of the North Sea commissioned the first Marine Spatial Plan for the period 2014–2020 (SPF, 2014), making Belgium a pioneer in Europe and even in the world. The plan proposed measures to reduce the impact of bottom-towed gear that would contribute to the implementation of the EU Nature Directives and also to good environmental status under the MSFD, but this could not be made binding for foreign fishers. In 2016 a new Royal Decree on the designation and management of protected areas in the Belgian part of the North Sea was adopted (SPF, 2016). This decree sets out to improve the implementation of the Nature Directives in the Belgian part of the North Sea. The adoption of this decree was followed in 2017 by the adoption of the conservation objectives and management plans for the marine Natura 2000 sites (SPF, 2017a). The new six-yearly Marine Spatial Plan (which entered into force in March 2020) now contains four zones where new measures will be developed, taking into account the designated protected areas and their conservation objectives.[9]

7.3.5 Biodiversity Strategies

To comply with CBD requirements, Belgium prepared a first *National Biodiversity Strategy for 2006–2016* (Belgian National Focal Point to CBD, 2006). In November 2013, Belgium's Interministerial Conference for the Environment adopted *Biodiversity 2020 – Update of Belgium's National Biodiversity Strategy* (Belgian National Focal Point to the Convention on Biological Diversity, 2013). The national strategy is a framework document that mainly builds on the existing plans, giving political orientation in order to improve the implementation of biodiversity commitments, strengthen coherence, fill gaps and integrate biodiversity concerns at national and international levels. It aimed to guide the revision of federal and regional biodiversity action plans, and be taken into account in sectoral policy making. Based largely on the previous strategy, the update incorporated provisions aligned with the CBD global *Strategic Plan for Biodiversity (2011–2020)* and the *EU Biodiversity Strategy to 2020* (see online Annex BE.1 for further information).[10] The update also included a range of new objectives to anticipate, prevent and reduce the causes of biodiversity loss in Belgium, such as in relation to green infrastructure, the no net loss concept, and mapping ecosystems and their services and assessing their values. It was the sole national document on biodiversity that was applicable both at the federal and regional levels in order to comply with the EU and other international commitments made by Belgium.

At the time of finalisation of this chapter in March 2022, no decision had been made on a follow-up to Belgium's *Biodiversity 2020* strategy.

7.4 Species Protection and Conservation Measures

7.4.1 General Protection Measures and Action Plans

General regulations for species protection were included in the existing regional laws – as the Nature Conservation Act of 1973 in Wallonia and the Nature Decree of 1997 in Flanders – that were updated to be in line with the requirements of the EU Nature Directives.

In **Flanders**, the implementation of the requirements for the protection and management of species is laid down in the 2009 Executive Law for Species Protection (Flemish Government, 2009). This law includes the list of species to be protected in Flanders, horizontal measures and also the procedures for the development and adoption of five-year species protection programmes, with specific attention to HD species. Up to now, protection programmes have been adopted for 22 species or species groups (ANB, 2020). A priority list has been established for the development of species programmes. Species (groups) have been prioritised based on criteria such as Red List status, distribution and surface requirements, need for ecological connections between sites, and whether other species can profit from the actions. The development of the programmes is outsourced and includes consultations with all relevant stakeholders. The programmes are adopted by the Minister for Environment and are binding. Implementation is financed by the ANB, through agri-environment schemes such as for Montagu's Harrier (*Circus pygargus*) and Grey Partridge (*Perdix perdix*) (VLM, 2019), and/or LIFE projects such as LIFE Oostkustpolders[11] for meadow birds, Pink-footed Goose (*Anser brachyrhynchus*) and Marsh

[9] Information on the Marine Spatial Plan: www.health.belgium.be/en/environment/seas-oceans-and-antarctica/north-sea-and-oceans/marine-spatial-plan

[10] All annexes are free to access online at CUP: www.cambridge.org/natureconservation

[11] www.natuurpunt.be/pagina/inleiding-life-oostkustpolders

Harrier (*Circus aeruginosus*). From 2019, a process of evaluation and eventual renewal of the oldest programmes is planned.

In **Wallonia**, species protection has a legal basis in the 1973 Nature Conservation Act and is carried out through species action plans. Various species, including butterflies, reptiles and bats, have benefited and continue to do so from restoration work undertaken as part of LIFE projects, or from the implementation of conservation management measures in nature reserves. About 20 species (groups) action plans have been developed and/or implemented, notably via LIFE projects, such as LIFE in Quarries[12] for example for the Sand Martin (*Riparia riparia*), Common Midwife Toad (*Alytes obstetricans*) and Natterjack Toad (*Epidalea calamita*), or LIFE Pays Mosan[13] for various bat species.

The LIFE BNIP[14] includes the development and implementation of Belgium-level action plans for various amphibian and bird species. As for most LIFE projects and agri-environment scheme agreements, actions are focussed on restoring the habitats of the species concerned, such as meadows, which may take a long time to lead to an improvement in the status of the species' populations.

The first **national *Bee Plan*** for the preservation of pollinators, in particular bees, was produced for 2012–2014 (SPF, 2012) and assessed in 2015 (SPF, 2015). It included about 30 actions and measures on six main issues: risk assessment and management for pollinators (including pesticides risk analysis); integration of pollination objectives in other policies and measures (including the economy); orientation of markets in favour of pollinators (in the broader framework of biodiversity and ecosystem services); monitoring of honey bees and wild bees; animal health policy; and the traceability of hives (for honey bees only). This first plan established a collaborative methodology and 'bee governance' with the establishment of a multilevel coordination group between policymakers, researchers and civil society for exchange of expertise, decision making concerning planning and follow-up of actions and communication. Each of the regions also has its own pollinators' action plan based on regional specificities.

A second *Federal Bee Plan 2017–2019* (SPF, 2017b) includes four objectives: assist bee-keepers; better understand the causes of bee decline; better control the risks for bee health; and engage all stakeholders.

7.4.2 Hunting

Hunting is now under the full competence of the regions, with different regulations and organisations in Flanders, Wallonia and the Brussels-Capital Region to better fit the respective situations. In the Brussels-Capital Region, hunting has been completely prohibited since 1991. Since the 1990s, modifications of the Walloon and Flemish laws on hunting requiring structured representation and coordination of their regional organisation, along with efforts from hunters, have aimed to ensure the sustainable use of wild species and their habitats. In Flanders and Wallonia, hunting plans in general are drawn up every year, for the most part by game management units, and approved by the regions in order to guarantee coordinated management of the game species. These approaches have improved monitoring of game population densities, control of hunting plans and bags, and consultations on nature policy planning and reporting. For some species the control of hunting could still be improved, such as for the Grey Partridge (now a Red List species), or for Wild Boar (*Sus scrofa*), which has some dense populations that are causing detrimental impacts on nature and agriculture.

7.4.3 Species–Human Conflicts

To deal with various conflicts that occur between species protection needs and human interests (e.g. relating to farming, livestock, fisheries and other land uses), compensation is provided for economic losses arising from certain species. In Flanders, damage from all species protected under the Species Act, including all HD species, can in principle be compensated for. In Wallonia, HD species for which compensation can be paid are the Beaver (*Castor fiber*), Otter (*Lutra lutra*), Wolf (*Canis lupus*), Grey Heron (*Ardea cinerea*) and Great Cormorant (*Phalacrocorax carbo*).

Numerous conflicts have arisen between the conservation of the Beaver and landowner interests and concerns over flooding. This has occurred because, since the late 1990s, the species has quickly recolonised the majority of Belgium, in particular in the east. To address this, Flanders adopted in 2015 a Beaver action plan, which includes action on evaluating the system of compensation and a protocol for possible future population control (ANB, 2016a). As for other protected species, compensation is given for the loss of crops, or damage to forests or other property of any private owner, if prevention and control structures have been implemented. Wallonia does not have a management plan for the Beaver, but published in 2015 a brochure with detailed management guidance (SPW, 2015). More widespread communication and training on preventive measures could improve conflict management.

It is also anticipated that conflicts will arise with the Wolf: the species has been sighted in Wallonia and has bred in eastern Flanders. Both regions have therefore added the Wolf to their regulations on compensation for damage caused by protected animal species.[15,16] A Wolf management plan has also been developed in Flanders (INBO, 2018) in close consultation with the relevant stakeholders.

12 www.lifeinquarries.eu

13 www.lifepaysmosan.eu

14 www.life-bnip.be/en

15 SPW webpage on the wolf: http://biodiversite.wallonie.be/fr/le-reseau-loup.html?IDC=6418

16 Wolf education and awareness building website by Agentschap Natuur & Bos (in Dutch): www.natuurenbos.be/wolf and www.vlaanderen.be/publicaties/wolvenplan-vlaanderen

7.5 Protected Areas and Networks

7.5.1 Protected Area Types and Their Coverage

A diverse range of protected area types exist in Belgium, with those of most relevance to biodiversity conservation and consistent with the IUCN definitions, although not always linked to a specific category, listed in Table 7.4. All these protected areas are established for the conservation of biodiversity but may differ in their protection level. Multiple protection types and levels are sometimes attributed to the same site, for example a Nature Reserve can also be a Natura 2000 site.

In the three regions, all types of reserves (i.e. nature and forest) have to have the highest protection level, be recognised by the Minister after approval of the management plan (including measures for conservation and restoration of habitats and species listed in the related inventory), and have monitoring and public access information. This approval is given by a Higher Council for Nature Conservation (in Wallonia) or by the regional nature administration. The State Nature Reserves in Wallonia and the Regional Nature Reserves in Flanders and the Brussels-Capital Region are areas owned by the government, while Certified Nature Reserves are privately owned, mainly by NGOs, in the respective regions. Within these sites, human activities are prohibited unless aimed at environmental management or public access as specified in the designation order.

Scientific Interest Sites in Wallonia, such as wetlands and underground caves, are designated for specific ecological characteristics; they have management measures to maintain and monitor them, but do not have the status of a reserve. In the forest areas, specific zones can be designated as Integral Forest Reserves in Wallonia, or Protected Zones in the Brussels region; in these, interventions are prohibited, to allow forests to age and express their natural dynamics.

Some other types of protection exist in Flanders. The Flemish Ecological Network (VEN) has a protection status that prohibits changes in protected vegetation types such as dunes, heaths, swamps, fens, permanent grassland and landscape elements. Any activity that may have an impact has to follow the EIA assessment procedures, requiring an opinion advice from ANB on its acceptability. The same protection level covers sites with regional or community natural values on the Military Zones included under the Protocol of 1999 and revised in 2009 with the Ministry of Defence, and also the Protected Dunes under the Dunes Decree of 1993. Extended Forest Management Plans for privately owned forests have to be based on the Sustainable Forest Management criteria, taking into account the maintenance of biodiversity, and be approved by ANB.

In 2017 a new legal system was adopted for site protection and management, with one common format for the management plan for Natural Reserves, forests, parks and other areas with natural values (e.g. HNV farmland, wetlands and rural private estates). This management plan can have one of four increasing levels of protection ambition, going from Level 1 to Level 4 (ANB, 2018). Level 1 aims only to maintain the existing nature levels of the site. Levels 2 and 3 aim to sustainably manage the site, and achieve 25% and 90% of the site's nature objectives respectively. Level 4 aims to achieve all objectives and attain the status of a 'Nature Reserve' for the protection of its natural values. This system is based on the 'triple P approach', seeking a balance between measures for People (e.g. access to recreation), the Planet (e.g. conservation and protection measures), and Profit (e.g. the use of resources).

The system provides an opportunity for any landowner, nature conservation organisation or local authority to develop a 24-year management plan, in which they choose one of the four levels of ambition to contribute to the site's nature objectives developed by ANB on the basis of scientific information on BHD habitats and species and Regional Important Habitats. The plans are approved by the area and/or regional consultation forum.

Together with this new regulation a 'result-based' subsidy system[17] has been introduced that is linked to the level of ambition in the agreed management plan: the higher the ambition, the higher the subsidy. The calculation of the subsidy is based on yearly standard costs per habitat type as published in the Executive Act of 2017 on subsidies for nature management. Sites with Level 4 aims may also have reduced property taxes and costs of inheritance. ANB also provides technical and financial support for revising existing or developing new site management plans (the process is described further in Section 7.5.4) as well as training and online guidance through Ecopedia.[18] As a large part of Natura 2000 is located on private land with varied types of owners, this new approach aims to enhance their conservation efforts. There is certainly interest from private owners but the development of the plan, which has to include detailed maps and area visions with targeted objectives, is quite cumbersome and may take longer than expected. In the longer term, all existing management plans will be revised to comply with this new system.

The National Park Hoge Kempen[19] in Flanders covers 120 km^2 of forest, heath, inland dunes, water systems and specific landscapes. Despite the implication of its name, it is not in itself considered a protected area. However, it includes reserves, Natura 2000 areas and privately owned protected areas. It also receives financial support from the government for its management, public access infrastructure, and guidance and cooperation with local actors to enhance sustainable maintenance of the Park's natural values. It is now a government objective to establish more National Parks over the next years.

Taking into account overlaps, the areas with a legal nature protection status regulation, as listed in Table 7.4, including Forest Stewardship Council (FSC) certified forests integrating biodiversity concerns and other effective area-based conservation measures (OECMs), in 2019 covered 17% of the land

17 Overview of subsidies for nature management (in Dutch): www.natuurenbos.be/subsidie-voor-opmaak-en-wijziging-natuurbeheerplan

18 Online information (in Dutch): www.ecopedia.be

19 www.nationaalparkhogekempen.be/en

Table 7.4 The number and extent of international designated sites and protected areas in Belgium.

	IUCN category	Number	Area (km^2)
Designated under global and non-EU international legislation			
Ramsar Sites	IV, V, NR	9	469
Natural and Mixed World Heritage Sites	NA	1	920
OSPAR Marine Protected Areas	NR	2	1 232
Designated under the Nature Directives			
SCIs (terrestrial + marine)	NR	281	3 284 + 1 178
SPAs (terrestrial + marine)	NR	255	3 189 + 316
Natura 2000 terrestrial area			*3 894*
Natura 2000 marine area			*1 317*
Designated under domestic legislation*			
Certified Forest Reserve FL (Erkend bosreservaat)	IV	16	3
Certified Nature Reserve BCR (Erkend natuurreservaat)	NR	3	0
Certified Nature Reserve FL (Erkend natuurreservaat)	IV	379	205
Certified Nature Reserve WA (Réserve naturelle agréée)	IV, except 1 VI	178	37
Extended Forest Management Plan FL (Uitgebreid bosbeheerplan)	VI	5	1 136
Flemish Ecological Network FL (Vlaams Ecologisch Netwerk)	NR	231	937
Forest Reserve BCR (Bosreservaat van regio BXL)	NR	2	2
Forest Reserve FL (Bosreservaat van regio VL)	IV	62	29
Forest Reserve WA (Réserve forestière)	VI	20	11
Military Zones FL (Militaire domeinen)	NR	23	117
Nature Management Plan Type 2 FL (Natuurbeheerplan Type 2)	VI	63	41
Nature Management Plan Type 3 FL (Natuurbeheerplan Type 3)	IV	23	6
Nature Management Plan Type 4 FL (Natuurbeheerplan Type 4)	IV	21	34
Protected Dunes FL (Beschermde duinen)	NR	104	11
Protected Zone in the Forest BCR (Beschermde zone in het bos)	NR	4	6
Regional Nature Reserve BCR (Regionaal Natuurreservaat)	NR	66	56
Regional Nature Reserve FL (Regionaal Natuurreservaat)	IV	70	66
Scientific Interest Site WA (Site d'intérêt scientifique)	IV	153	14
State Nature Reserve WA (Réserve naturelle domaniale)	IV	239	111

Note. NR = not reported.
* A site can have several designation types.
Sources. See Section 1.3. Based on data from: Ramsar Convention and OSPAR Convention websites (December 2021); EEA Natura 2000 Barometer, Natura 2000 as at end 2020; IUCN types and other protected area types from WDPA (December 2021).

surface of Flanders (Belgian National Focal Point to the CBD, 2019). For Wallonia, including areas with another or no protection status but effectively managed with nature conservation as an important objective (e.g. Prosylva forest with close to nature management), the overall surface was also 17% of the territory (Belgian National Focal Point to the CBD, 2019). As much as 14% of the Brussels-Capital Region is designated as a Natura 2000 area (Belgian National Focal Point to the CBD, 2019), which includes the Brussels Sonian Forest (1 654 ha). This part of the Sonian forest, together with the part in Flanders and Wallonia, is part of a multinational UNESCO World Heritage Site (Ancient and Primeval Beech Forests of the Carpathians and Other Regions of Europe).

As regards the Natura 2000 network, at the end of 2020 it covered 5 211 km^2 on land (Table 7.4), which is 12.7% of Belgian terrestrial territory (see Key Statistics table). Taking into account all the types of protected areas listed in Table 7.4, overall national protected area coverage was estimated to be 15.5% at the end of 2021 (UNEP-WCMC, 2021). This coverage is based only on protected areas that are supposed to meet

IUCN/CBD standards, and is therefore lower than that for the areas mentioned previously in Flanders, Wallonia and the Brussels-Capital Region, because they include OECMs. Territorial or land-based aspects, such as nature and forest conservation and management, are fully regional competences for which public data on an integrated national coverage of combined protected areas together with OECMs do not appear to be available.

In the marine environment, the Natura 2000 network, as designated by the Marine Spatial Plan (SPF, 2019) covered 1 317 km^2 at the end of 2020, which is 38% of the Belgian part of the North Sea. There are no additional domestic sites and the Belgian OSPAR sites totally overlap with the four existing Natura 2000 sites. It is assumed that the management and conservation of these sites and the habitat types present will support the Natura 2000 network as well as the OSPAR MPA network. The most recent fifth marine Natura 2000 site (Vlakte van de Raan) is not designated as an OSPAR MPA.

7.5.2 The Coherence of Protected Areas and Ecological Networks

As discussed earlier, habitat fragmentation is a major concern, especially in Flanders. Whilst the spatial cohesion is still good for 20 HD habitat types (mainly coastal, dune, heathland and freshwater habitats), it is poor for six and bad for 20 (forest, grassland and marsh habitats) (EEA, 2018). To address such issues, in the Natura 2000 network and the wider rural and urban environment, the **Flemish Ecological Network** (*Vlaams Ecologisch Netwerk*, VEN) was introduced in 1997 in the Nature Decree. The implementation is based on the Spatial Structural Plan Flanders (*Ruimtelijk Structuurplan Vlaanderen*) as the sites are designated on the formal spatial plans. The VEN comprises large natural units (sites with important natural values) and large natural units in development (sites where natural values will be further developed and/or extended), and will cover about 9% of Flanders (125 000 ha). Natura 2000 sites and Nature Reserves form core areas of the ecological network. Up to now, about 74% has been designated. An Integral Connecting and Supporting Network (*Integraal Verwevend en Ondersteunden Netwerk*, IVON) – including river systems, parks, verges of roads and crop fields and grasslands with nature-oriented management – buffers, supports and connects these core natural areas, of which only 5% has been designated. The government of Flanders uses positive incentives such as agri-environment agreements to promote ecological quality of these supporting areas.

Another important strand of work that aims to reduce fragmentation is the programme 'Roads and Nature', run by the Department of Roads and Transport. The programme has implemented a wide range of measures such as ecoducts, tunnels, fencing and ecological roadside management.[20] Surveillance reports indicate that the infrastructure is effectively being used by animals and contributes to restoration of populations and also to road traffic safety.

In **Wallonia**, a definition was developed in 2004 for the main ecological structure (*Structure Écologique Principale*, SEP[21]) (Dufrêne, 2004), which includes the Natura 2000 network as well as other sites of high biological interest, of which a central inventory is being further developed(SPW, 2010).[22] The Walloon Nature Network (*Réseau Wallonie Nature*, RWN) is the principal policy instrument for encouraging wider landscape connectivity and green infrastructure in the region, and creating the SEP in practice. This is an ongoing process based on an updatable catalogue of voluntary actions to be implemented, many of them building on existing initiatives, such as agri-environment schemes, forest certification and river contracts, as described later. Others include:

- *Plans Communaux de Développement de la Nature*: municipal development plans including delayed mowing and river contracts.
- Agreements with the private sector (e.g. quarries, electricity companies and railway companies).
- Green spaces and forests on spatial plans under the *Code wallon de l'Aménagement du Territoire, de l'Urbanisme et du Patrimoine* and related catalogue of actions of the *réseau Wallonie nature.*
- *Landscape Plateform paysage* DGO4,[23] which also deals with development of a landscape charter in Natural Parks.
- River basis plans under the Water Code and some resulting plans and programmes.

Connectivity in the urban context has been a recognised concern in the **Brussels-Capital Region** since 1998 when a government decree set out objectives for a Green Network Programme. In 2002, further developed Green and Blue Network programmes were established (Bruxelles Environnement, 2015). The initial Green Network Programme intends to gradually build a network of green spaces (parks, woods, forests and gardens) linked together by green corridors (e.g. green avenues, road and railway embankments). The programme emphasises the cohesion and continuity of green spaces and semi-natural areas in the urban environment. Its purpose is to integrate the scenic, aesthetic, social, recreational and ecological functions of green spaces and develop their interconnectivity. One of the prime objectives of the programme is to increase biodiversity through the creation and adapted management of green areas. The initial Blue Network Programme aims to have an integrated, durable and

[20] For more details, visit the Flemish Agency for Infrastructure and Mobility webpage Roads and Nature (in Dutch): www.wegenennatuur.be/

[21] For more details, visit the SPW Wallonnie webpage on the Walloon Structure Écologique Principale (SEP) (in French): http://biodiversite.wallonie.be/fr/structure-ecologique-principale.html?IDC=2997

[22] For more details, visit the SPW Wallonnie webpage on the Walloon Sites de Grand Intérêt Biologique (SGIB) (in French): http://biodiversite.wallonie.be/fr/sgib-sites-de-grand-interet-biologique.html?IDC=824

[23] For further information (in French): www.plateformepaysagedewallonie.be

ecologically justified management of open waterways in Brussels, mainly tackling pollution aspects. The network is made up of small rivers, ponds and marshes.

Amongst its 27 measures, the *2016 Brussels Nature Plan* (Leefmilieu Brussel, 2016) promotes the multifunctionality of areas while preserving biodiversity at the same time. Building on the already existing blue and green networks, the plan promotes the development of the Brussels ecological network, with the aim of reconnecting areas of value. This Brussels Ecological Network is made of central, development and linking zones. If the central areas have an active protection status (Reserve or Natura 2000 status), the surrounding areas are under urban pressure. Some of those areas already have the status of green areas under the regional plan soil assignment (e.g. regional or municipal parks), which gives them a certain level of protection, but most of them are private gardens/ property (building areas) with no strong protection status. The Nature Plan will target those areas that are priorities for consolidating the network for sustainable management of its natural values. The plan has already developed tools, including a standard for sustainable management of public parks, and an eco-potential tool that targets real estate and building promoters to encourage them to maximise the integration of biodiversity into their project.

7.5.3 Approaches to Protection and Levels of Protection

It is important to note that a large part of the Belgian Natura 2000 network is on private land or land owned by municipalities, and under economic use. This made the acceptance of the protection status and regulations much more difficult and necessitated considerable time for consultations (further discussed in Section 7.5.4).

Environmental assessments for development projects or plans with possible impacts on protected areas need advice from the nature administrations. Together with the transposition of HD Article 6 into regional and federal law, the detailed guidance documents of the European Commission were used to develop assessment procedures and specific information and training sessions for public authorities and EIA experts.

In Flanders, a derogation for overriding interest has to be approved by the government. This was the case for developments in the port of Antwerp and for the port of Zeebrugge; for both projects extensive compensation measures were also implemented. More recently, an online scoping system[24] has been developed for preliminary screening to assess possible impacts of a project or plan on Natura 2000 and to identify whether further impact studies may be required.

The North Sea is a sensitive ecosystem, and the Belgian part is one of the most densely used marine areas in the world with important pressures from sea- and land-based activities, as discussed earlier. Addressing these pressures within a complex state structure is an important overarching management issue. The established marine Natura 2000 areas are incorporated in the Marine Spatial Plan adopted in 2014. This Royal Decree forbids a number of human activities in the Natura 2000 areas (e.g. industrial activities). The management plans for the marine Natura 2000 sites adopted in 2018, as well as the MSFD good environmental status objectives (to be reached by 2020) reviewed in 2018, and related measures, address all relevant pressures and (socio-) economic sectors necessary to allow the recovery of degraded habitats and species, and the achievement of the Natura 2000 and MSFD objectives. For example, in the Vlaamse Banken Natura 2000 site, which covers about one-third of the Belgian part of the North Sea, all recreational fisheries with bottom-disturbing gear are prohibited (zonation has been proposed for commercial fisheries, but not yet approved).

7.5.4 Approaches to Managing Protected Areas

All Natura 2000 sites, SACs and SPAs (except for one SPA in Flanders: IJzer- en Handzamevallei) have been formally designated with the publication of the regional and, for marine sites, federal designation acts (see Table 7.4). In each of the regions, regional objectives have been developed for each HD habitat and species, and BD species for overlapping SPAs, and translated into site-specific objectives and priority measures that are included in the designation acts. Site management plans (which can cover Natura 2000 sites and their surrounding area) as well as habitat action plans and species protection programmes are being developed. Several action plans and programmes are outsourced under the LIFE BNIP project. In addition to the Natura 2000 network, some species and habitats of species in other areas are conserved through other conservation measures, such as agri-environment schemes and sustainable forest management measures (see Section 7.6).

To involve stakeholders in protected area management, formal **consultation groups** have been set up at both regional and site level. These have enabled transparent discussion and adoption of Natura 2000 objectives and measures, and the species conservation programmes. This process requires a considerable amount of human and financial resources, and takes time, but is essential for effective implementation as the nature authorities cannot achieve the objectives on their own, given their limited areas of competence. In Flanders, the basis for the groups has been established through a partnership agreement with the related administrations, provincial authorities and most relevant stakeholders.[25] The management of Natura 2000 sites by NGOs and private landowners is financially and technically supported by the public authority through management agreements. Wallonia also uses **River**

[24] Available (in Dutch) at: https://natura2000.vlaanderen.be/voortoets-passende-beoordeling and https://voortoets.omgeving.vlaanderen.be

[25] Overview of stakeholders (in Dutch): www.natura2000.vlaanderen.be/partners

Contracts,[26] which are participative management structures that bring together all actors within a river valley with the aim of reaching a consensus on an action programme for the restoration of the water course, the river banks and surroundings, and the water resources. In early 2017, 14 River Contracts were in place, with 236 Walloon municipalities (90%) being a partner.

Inter-regional cooperation has been established for the Sonien Forest (Zoniënwoud, 2008), the principles for which were set out within the structure vision and new management plan for the site. As well as being recognised as a UNESCO World Heritage Site it is part of the European Rewilding Network, which strives for more wild nature and wild animals in Europe. The LIFE+ project OZON for Sonien enabled large-scale restoration, construction of the Groenendaal ecoduct, and installation of ecofences and public access infrastructure.

7.6 General Nature Conservation Measures

7.6.1 Development Planning and Impact Assessment

The assessment of biodiversity impacts from projects and plans is already provided for by the EU legislative framework (i.e. the EIA Directive and HD Article 6), which has been transposed into the Belgian Federal and Regional legal framework. SEA procedures include biodiversity criteria and refer to relevant national policy documents such as the *Belgian Biodiversity Strategy*, the CBD and biodiversity-related conventions and agreements.

For Natura 2000 sites, the biodiversity criteria and objectives to be realised are set out in the designation acts. Stricter use of impact assessments and environmental permitting procedures of new developments has encouraged the introduction of more general mitigation measures into project plans, reducing their pressures on the environment and biodiversity. Stricter enforcement of existing environmental law and restrictions on emissions, land-use changes and water use has also contributed to reducing pressures by requiring revision of projects and plans before granting permits or even refusing permits. Consultations are ongoing to integrate the value of ecosystems and ecosystem services in the environmental impact assessment procedure and regulation.

7.6.2 Management of HNV Farmland and Use of Agri-environment Measures

In addition to the mandatory CAP direct payment conditions (i.e. cross-compliance, and the greening measures over 2014–2020), agri-environment–climate measures have also been used to provide incentives to maintain or increase biodiversity-beneficial farming practices. Within the framework of the Flemish Rural Development Programme (RDP) for 2014–2020 (VLM, 2014), farmers had the opportunity to receive subsidies and technical assistance for organic agriculture, orchards, preservation of local livestock breeds and fruit tree varieties, mechanical weed control instead of using herbicides, agroforestry, botanical management to restore semi-natural grasslands or meadows, farmland species, nature management to improve the ecological structure of the farmland landscape (field margins and small landscape elements), and actions to reduce erosion. However, the schemes for the conservation and restoration of specific farmland species and communities showed varying success. The scheme for botanical management (294 ha) was, together with the scheme for nature management, amongst the least popular with farmers and achieved only 9% of its 6 000 ha target. On the other hand, schemes for farmland birds covered 964 ha, and about 960 ha for the Common Hamster (*Cricetus cricetus*). The latter scheme has been stopped and replaced by a longer-term species protection programme funded by ANB to ensure the continuity of measures. The area under organic farming has increased since 2010 up to 7 300 ha, about 1.2% of the total farmland area in Flanders.

Some of the measures in Wallonia were modified in the 2014–2020 RDP in order to further develop extensive production patterns and to include more measures supporting nature conservation and Natura 2000. Under the RDP, actions were included to improve surface water and groundwater, soil, landscape, biodiversity and climate (Walloon Government, 2014). By 2019 the RDP objectives had been achieved for three – winter soil cover (49 800 ha), grasslands of high biological value (6 690 ha, 900 breeding birds) and managed field strips (1 275 km) – and almost achieved for hedgerows and forested strips (about 12 000 km), trees, bushes and isolated thicket (124 000 elements), natural grasslands (13 000 ha), extensive cereal crops (2 865 ha), local breeds (7 900 animals) and keeping low levels of livestock (34 100 ha) (Walloon Government, 2019). More mixed results were observed for ponds (4 040, 67% of target achieved), grassy strips (2 555 km, 64% of target achieved) and strips of extensive grasslands (737 km, 37% of target achieved). The number of farms converted to organic farming has continued to increase, such that it now covers 11% of farmland in Wallonia.

7.6.3 Forest Management

Up to 2017 the Flemish Forest Decree allowed two types of **forest management plan**: limited and extended. Limited management plans had to comply with only minimum standards, whereas extended plans had to meet sustainable forest management criteria including biodiversity concerns, as described in the Executive Act of 2003. Forest management plans are voluntary for smaller private forests but by decree compulsory for all forests larger than 5 ha and are supported for private owners through grants and forest owner groups. In 2019, of the c. 140 000 ha of forest (of which about 60% is privately owned), about 16 022 ha (29%) of public forest areas and 29 401 ha (35%) private forests had an extended forest

[26] For more details visit the SPW website (in French): http://environnement.wallonie.be/contrat_riviere/

management plan.[27] At the same time, 26 471 ha of forest (19%) were certified at the end of 2019 according to FSC standards.

A new Forestry Code entered into force in Wallonia at the end of 2008. It had the aim of safeguarding the regeneration and sustainability of forests, as well as achieving an optimum dynamic balance between its economic, ecological and social roles. One of the objectives of the Code is to combat climate change and to preserve biodiversity through the application of sustainable forest management principles. This Code applies to the total forest cover in Wallonia (c. 540 000 ha). As a result, certain restrictions are imposed on both public and private landowners, such as regarding the choice of species appropriate to local conditions, diversification, measures favouring biodiversity, restricting clear-cutting, drainage or water inputs, and public access. The code also requires that 3% of broadleaved forests, greater than 100 ha, are delimited as Integral Reserves requiring nature-oriented management over the whole surface. Monitoring of forest biodiversity and management is carried out by the Walloon Observatory for forest health. The surface area covered by forest management plans amounted to 270 582 ha in 2017, but the plans covering the largest part have to be revised to comply with the current Forest Code. Research on, and promotion of, close-to-nature forestry based on natural processes is carried out by the Pro Silva association[28] (including owners, managers and professionals), which manages 5% of the Walloon forest. Forests under the Programme for the Endorsement of Forest Certification (PEFC), which have less stringent biodiversity criteria, cover nearly 298 000 ha (54%) of the region's forest area. Forest habitat quality is improving at a local level but needs to be upscaled (SPW, 2017b).

Important efforts are being made in the Brussels-Capital Region to integrate biodiversity considerations into forestry policy, in particular in relation to the Sonian Forest. The management of the forest has been in accordance with FSC standards, with the first accreditation given in 2003. In addition, as a Natura 2000 site, it benefits from protection under the Habitats Directive and has a specific management plan (Leefmilieu Brussels, 2004). In this plan, buffer zones around domestic protected sites, areas with fragile or threatened vegetation, and wildlife refuges were designated in 2007 to keep visitors on paths and reduce their impacts (Zoniënwoud, 2008). In 2019 the plan was revised into a multifunctional plan that promotes sustainable multifunctional use of the forest with the aim of providing ecosystem services while reconnecting people to nature. The plan promotes the concept of 'equipped and connected entry doors', 'contact zones' and 'diffusion zones' for channelling public access. Public use and climate change are still the most important pressures.

7.6.4 Habitat Restoration

Recognising that protected areas are necessary but not sufficient to restore biodiversity to a favorable conservation status across the country, and to maintain the provision of ecosystem services, Objective 3.3 of the Belgian *National Biodiversity Strategy* to 2020 is to ensure 'Ecosystems, their resilience and their services are maintained and enhanced by establishing, *inter alia*, a green infrastructure and restoring at least 15% of degraded ecosystems.' The regions have also had ambitious restoration objectives and plans, especially concerning Natura 2000. In particular, the Natura 2000 programme of Flanders (2016–2020) had two overall main objectives for 2020: (1) stopping or preventing a decline in the habitats and species of EU concern; and (2) achieving a favourable or improved status for the 16 prioritised HD habitats.

Furthermore, in 2016 Flanders submitted a *Restoration Prioritisation Framework* (ANB, 2016b) to the European Commission, as required under Target 2 (Action 6a) of the *EU Biodiversity Strategy to 2020*, to which very few other Member States responded. The framework focussed in the first place on Natura 2000 (with maps of the areas and action sites for species protection), water bodies (maps of priority action zones) and larger-scale projects, such as the Sigma plan and Grensmaas. The overlapping areas of all these maps helped to identify zones for priority action to contribute to the different policy objectives and realise win-wins. An overall assessment of the state of habitat types was made for the NARA 2020, the two-yearly Nature Report. The cover of estuarine habitats increased due to the nature developments, but salt marshes decreased in cover, being replaced by brackish water vegetation such as reeds (INBO, 2020c).

Ecosystem restoration actions are also included in the *Flemish Regional Climate Policy Plan* (2013–2020).[29] The same is the case for Wallonia, which updated its climate policy through the new *Air-Climate-Energy Plan* (PACE) in 2016.

Many large-scale projects have been carried out to attempt to achieve the various restoration objectives, mainly with a focus on Natura 2000, and mainly funded by the LIFE Programme (see Section 7.7.3), but also Interreg or projects financed by the regional governments. These have, for example, addressed peat bogs and wetlands in the Ardennes,[30] wetlands and estuarine systems along the Schelde river[31] and the Zwin estuary,[32] river banks along the

[27] Further information on forest cover and management (in Dutch): www.natuurenbos.be/vlaamse-bosinventaris/bosareaal.html and www.inbo.be/nl/natuurindicator/oppervlakte-met-effectief-natuurbeheer

[28] Further information: www.prosilva.org/members/full-members/pro-silva-wallonie/

[29] Plan and information documents (in Dutch): www.vlaanderen.be/publicaties/vlaams-klimaatbeleidsplan-2013-2020-overkoepelend-luik

[30] Vlaams klimaatbeleidsplan 2013-2020. Overkoepelend luik | Vlaanderen.be and 25613 (vlaanderen.be)

[31] https://life-sparc.eu/

[32] www.zwinstreek.eu/info/zeno

Maas,[33] and heath and grassland within military areas in Wallonia and in Flanders.[34] RDP measures have also supported restoration actions and investments in Natura 2000.

However, despite the political commitments and objectives, and resulting enormous efforts and investments in Natura 2000 implementation and restoration projects over the last decade, the Natura 2000 objectives and wider restoration aims have not been reached. Whilst the initiatives led to local improvements in the status of some habitats and species, wider-scale action appears to be needed.

7.7 Nature Conservation Costs, Benefits and Funding Sources

7.7.1 Nature Conservation Costs

Nature conservation incurs costs at various governance levels – federal and regional and local – and in various governmental departments and sectors. Whilst there has never been systematic tracking of nature conservation financing and needs in Belgium, the Prioritised Action Framework (PAF) for Natura 2000 required by Member States includes an estimate of the costs of the Natura 2000 network and related green infrastructure. PAFs for 2021–2027 have been submitted by Flanders (ANB, 2019) and Wallonia (SPW, 2019). The PAF for the Brussels-Capital Region remains in development.

The preparation of the PAFs required extensive consultations of the main actors (public administrations) and stakeholders. It used scenario planning to identify the restoration and conservation measures needed up to 2027 and then estimated their costs. In Flanders, the PAF covers all BHD habitats and species, restoration and development of connectivity between sites, research and monitoring, communication and capacity building, and is estimated at about €250 million per year (with 61% for one-off actions). In Wallonia, the PAF is less inclusive, not taking into account all habitats and species, and gives an estimate of about €100 million per year (with 71% on recurrent actions). The two PAFs therefore contrast in the proportion of their actions and costs that are recurrent and one-off. They also differ substantially in some of the allocations to the main broad types of measures: the main costs in Flanders are on additional actions for species (26%), whilst in Wallonia the main costs are on supporting green infrastructure for Natura 2000 (48%) and additional costs for species comprise only 1% of the total.

7.7.2 The Economic Benefits of Nature Conservation

Various studies have been conducted in Belgium on the importance and value of biodiversity and ecosystem services, including through an informal network of Belgian ecosystem services experts (the Belgian Ecosystems and Society community).[35] For example, a study assessed the value of the 1 663 km^2 Natura 2000 network in Flanders (VITO, 2013) and found that it provided, amongst others, the following benefits: more than 34 million tons of CO_2 stored each year, 4 000–8 000 tons of fine dust eliminated from the air each year, 16 million m^3 of water purified each year, a gain of 2 100 healthy life years (for about 1.8 million people), and 26–43 million visitors annually. The monetary value of the Natura 2000 network's ecosystem services was estimated to be €780 million to €1.4 billion per year, with the achievement of its conservation objectives expected to improve the ecosystems' quality and quantity and therefore add another €15–94 million per year. However, this evaluation was incomplete as 25 of the 36 known ecosystem services were not evaluated, including flood protection. A case study on the monetary value of the forest ecosystem services in the Wallonian region found that three of them (wood, big game and carbon sequestration) together represented a stock value of more than €6.5 billion, and related benefits such as water purification and wild forest fruits were worth a minimum of €1.45 ha/year (SPW, 2013).

Information from the studies on ecosystem service values has been used in awareness activities and to enhance integration in planning processes and sectoral policies. The Nature Value Explorer website[36] serves as a calculation tool to value ecosystem services and assists in the mapping of the socio-economic importance of ecosystems.

7.7.3 Funding Sources for Nature Conservation

Table 7.5 provides a summary of the main funding sources that have been used for nature conservation in Belgium. This clearly indicates that, as in other EU countries, by far the most important source of funding for practical nature conservation measures has been EU funding instruments, and in particular the EAFRD RDP measures. These have been particularly important for supporting the HNV farmland management and Natura 2000 implementation actions described in Section 7.6, especially in Wallonia.

The European Regional Development Fund (ERDF) Operational Programmes are decentralised to the regions, which prioritise to varying degrees broad-based economic and social considerations. Environmental considerations were given a lower priority, especially over the 2014–2020 financial period. Nevertheless, several Interreg projects in which Belgium participated focussed on biodiversity, such as enhancing nature by businesses,[37] tackling the spread of invasive species,[38] action for pollinators,[39] peat restoration[40] and the Schelde estuary restoration.[41]

33 www.lifepaysmosan.eu

34 LIFE Nature project on military domains Wallonia After LIFE 2MIL | LIFE Natura2MIL 2006-2010 | La biodiversité en Wallonie and Flanders www.natura2000.vlaanderen.be/project/life-project-danah

35 www.biodiversity.be/3949

36 Online nature value estimation: www.natuurwaardeverkenner.be/#/

37 https://grensregio.eu/projecten/2b-connect

The European Maritime and Fisheries Fund (EMFF) programme in Belgium falls under the management of the fisheries department of Flanders, and did not allocate funding to biodiversity conservation measures over 2014–2020.

The Department of Agriculture and Fisheries in Flanders published a report '*What can the farmer do for European habitats and species?*' (Zwaenepoel *et al.*, 2016), which provides 26 practical measures that could support Natura 2000 conservation objectives. The report strongly emphasises the benefits of more collective actions and integrated agricultural and nature management approaches to contribute to habitats and species occurring in the agricultural environment. In a recent inter-regional project the agricultural sector also financed some demonstration projects on agro-biodiversity and agroforestry, taking into account nature-oriented practices that contribute to improving environmental quality, where information is gathered and shared to stimulate other farmers.

Belgium has benefited from EU funding under LIFE-Nature, with 80 projects in Belgium since 1992 with a total investment of c. €301 million (c. €165 million EU contribution).[42] This is thanks to a strong network of nature NGOs and public authorities promoting a strong habitat restoration agenda. Conservation strategies in Belgium are substantially based on land purchase and habitat restoration.

Grants are also provided by the public authorities for activities that benefit biodiversity, such as private sustainable management of protected areas and sustainable forestry (as described earlier). Other financial incentives for biodiversity include exemptions from succession rights for private forests, including a levy for land in the Flemish Ecological Network, exemptions from death duties and land in Natura 2000 Walloon sites, and exemption from succession rights for non-profit associations that make nature areas accessible to the public, etc. These types of levies and tax exemptions are listed in the new Executive Law on Natura 2000 of 2016 in Wallonia, and the Executive Law of 2017 on nature management plans and subsidies in Flanders (see Table 7.3). Financial compensation for the installation and use of wind farms at sea is re-invested in measures for improving marine biodiversity.

NGOs, private landowners and local authorities can obtain financial support from the regional public authorities for development and implementation of nature management plans.

Despite the availability of substantial EU funds, insufficient resources, both human and financial, have been a significant barrier to full implementation of the Nature Directives and broader commitments for biodiversity conservation in Belgium. Based on the funding sources listed in Table 7.5, available annual funding amounts to about €292 million, which is substantially below the €350 million per year estimated in the 2021–2027 PAFs as being needed to meet the Natura 2000 objectives alone. This shortfall is largely because uptake of EU financing has been limited by a lack of administrative capacity in the authorities to support both project development and implementation, as well as inadequate national/regional project budgets. Nevertheless, it should be emphasised that the availability of EU co-financing has increased regional investments in nature conservation.

To improve the uptake of EU funds in Flanders, ANB is prioritising objectives for projects, developing multi-year planning for project topics and increasing cooperation opportunities. To enhance project initiatives by NGOs and private landowners, the new subsidy system as described in the Executive Laws of 2017 for nature management also includes financial support for LIFE projects covering up to 80% of the regional co-budget for LIFE Nature projects of NGOs and private owners.

The payments for fishing and hunting permits in Flanders flow to dedicated financial funds that re-invest in fisheries (*Visserijfonds*) and game management (*Jachtfonds*). Financial compensation for deforestation is re-invested in acquisition of lands for afforestation.

Where appropriate, payments for ecosystem services are considered a useful policy tool, notably when they promote measures that go beyond the sustainable management of natural resources or contribute to restoration. Although regulations for this new approach are not yet developed, the scientific arguments are important for raising awareness and as a basis for environmental assessments.

7.8 Surveillance, Monitoring, Reporting and Applied Research

The monitoring of biodiversity and remedies to its threats is inherent in all the objectives of Belgium's *National Biodiversity Strategy*, especially Objectives 1 and 2. Objective 1 is to 'Identify and monitor priority components of biodiversity in Belgium.' This aimed to enable the evaluation and communication of progress made by Belgium towards its 2020 target, and help fulfil reporting obligations. It supported adaptive management (in particular with regard to climate change) and strengthening of policies addressing threats to biodiversity. The strategy focussed efforts where they are most needed: on components of biodiversity that are most at risk or could be subject to high risks in the near future. Priority components of biodiversity requiring the most urgent protective measures were identified and their status monitored, including: (1) ecosystems and habitats that are unique, rare, in danger of disappearance, or play a crucial role for priority species; (2) species that are rare, endangered, vulnerable, or are endemic or live in specific habitats; (3) genomes and genes of particular social, scientific or economic importance; and (4) functional

38 https://grensregio.eu/projecten/invasieve-exoten
39 www.natuurpunt.be/pagina/interreg-project-sapoll-0
40 www.natuurpunt.be/pagina/interreg-project-canape
41 https://grensregio.eu/projecten/grenspark-groot-saeftinghe
42 European Commission – LIFE overview (May 2020 update) https://ec.europa.eu/easme/sites/easme-site/files/belgium_update_en_final_may20.pdf

Table 7.5 Funding sources for nature conservation in Belgium.

Source and period	Measure/category	Average annual allocation (€ m)		Comments
EU funding instruments		**EU + national**	**EU**	
EAFRD Rural Development Programme measures – allocations to Priority 4 (restoring, preserving and enhancing agriculture and forest ecosystems) 2014–2020	M10 Agri-environment–climate	34.6	15.9	Important for N2k, etc.
	M11 Organic farming	16.2	6.7	
	M13 Areas facing constraints (WA)	8.3	3.3	
	M12 Natura and WFD (WA)	5.3	2.1	All for N2k, etc.
	M7 Basic services, etc.	4.1	1.9	Important for N2k, etc.
	M1 Knowledge transfer, etc.	0.9	0.6	
	M2 Advisory services, etc. (FL)	0.6	0.4	
	M4 Investments in physical assets (FL)	0.5	0.3	
	M8 Investments in forests	0.1	0.06	All for N2k, etc.
	M16 Cooperation (FL)	0.1	0.09	
RDP Priority 4 subtotal		*70.6*	*31.2*	
LIFE Programme	Nature and biodiversity (1992–2020)	10.4	5.7	
	Integrated projects (2015–2026)	3.1	1.8	Two biodiversity projects
ERDF/Cohesion fund 2014–2020	Interreg V Protection and enhancement of biodiversity, green infrastructure and Natura 2000	c. 1.2	0.6	For N2k, etc.
	Other	c. 0.8	0.4	For N2k, etc.
	M85: Protection and enhancement of biodiversity, nature protection and green infrastructure	c. 0.6	0.3	
EU funding instrument subtotal		*86.7*	*40.0*	
National funds within 2014–2020		**Additional national***	**Comments**	
Public funds				
Brussels City Region	Mean yearly expenditure of Nature Dept only and exc. personnel	1.0		
Flanders authorities	Expenditure on nature policy, Natura 2000 and green infrastructure (overall budget)	126.7	For N2k, etc.	
Walloon authorities	Expenditure on nature policy, Natura 2000 and green infrastructure (overall budget)	60		
FED marine authorities	Anti-pollution measures and restoration (from off-shore wind farm compensation)	0.3		

Table 7.5 *(cont.)*

Source and period	Measure/category	Average annual allocation (€ m)		Comments
EU funding instruments		**EU + national**	**EU**	
Others				
Main NGO Natuurpunt- Flanders	Overall expenditure	8.2		These may include some co-funding of EU instruments already included above, e.g. LIFE projects
Main NGO Natagora - Wallonia	Overall expenditure	8.9		
Additional national subtotal		*c. 205.0*		
National total inc. co-financing of EU instruments		*c. 251.7*		
Total (EU + all national financing)		**c. 291.7**		

Notes. 'For N2k, etc.' indicates funding relevant to Natura 2000, related green infrastructure and BHD species (based on the PAFs). ERDF co-financing proportion estimate based on EU average of 50%.

* Excludes co-financing of EU funds unless indicated. The subtotals and totals are only approximate indications of the average annual levels of funding as they are based on annualised data from differing periods.

Sources. Based on data from: RDP data from European Commission ESIF funds database (23 October 2020); PAF FL (ANB, 2019), PAF WL for relevance to Natura 2000 and FL national public funds; LIFE Belgium factsheet, European Commission (updated April 2021) and database; national public funds based on communications from relevant institutions. See Section 1.3 for further details.

components of biodiversity that are essential for the provision of ecosystem services.

Flanders and Wallonia already use biodiversity indicators closely linked to the EU's Streamlined European Biodiversity Indicators (SEBI) and CBD indicators to evaluate and report on the status of the priority components of biodiversity, as listed earlier, in their region on a yearly basis.[43] The City-biodiversity index (CBI indicators) is tested in the Brussels-Capital Region. The Statistical Office (Statistics Belgium[44]) publishes and sends to the OECD a national biodiversity indicator (% existing and threatened species).

In each of the regions an extensive monitoring programme has been developed by the nature authorities and their research institute to monitor HD habitats and species and birds, and the effects of management measures. This uses monitoring network systems in Natura 2000 areas, nature and forest reserves, and private and government domains. Cooperation in monitoring of species has been established with NGOs through the system of Waarnemingen.be in Flanders and Observations.be in Wallonia, which also promotes citizen science. To streamline and support coordination of monitoring of species by volunteer groups, monitoring blueprints and methodologies and common databank systems have been developed.[45] Harmonised approaches and tools are now being developed by the LIFE Integrated project (BNIP) involving all the nature administrations and the two main NGOs Natuurpunt and Natagora.

In Flanders, information on the characteristics of habitats and species, distribution, main threats, management and restoration measures, is published through the online knowledge platform Ecopedia.[46] Annual reports on specific indicators can be consulted on the INBO website.[47]

The results of monitoring in Flanders are published in two-yearly reports on the status of, and pressures affecting nature and, separately, the environment (NARA[48] and MIRA[49] reports), which together describe the state of the environment. The *State of Environment Report – Wallonia 2017* (SPW, 2017b) presents an assessment of the environmental situation and performance of Wallonia through a compilation of environmental, socio-economic, administrative and health indicators.

7.9 Future Developments

Over recent decades a lot of effort has been invested into gaining public and political support for nature conservation actions and developing ambitious objectives for Natura 2000 and wider biodiversity. More recently, reports on the

[43] For Flanders Nature Report Flanders 2020 | Instituut voor Natuur- en Bosonderzoek (vlaanderen.be); and for Wallonia http://etat.environnement.wallonie.be/home/fr/a-propos.html

[44] Website of the Belgian statistical office: https://statbel.fgov.be/fr/themes/environnement/biodiversite#panel-12

[45] www.natuurpunt.be/pagina/waarnemingenbe and https://observations.be/

[46] Available on (in Dutch): www.ecopedia.be/

[47] www.inbo.be/en

[48] www.vlaanderen.be/inbo/en-gb/inbo-nature-reports-flanders/

[49] www.milieurapport.be/publicaties/overzicht-mira-rapporten

very poor status of HD habitats and species, and other more common species, and the impacts of deteriorating nature and environment quality on public health, have led to outcries from society for more effective actions. Greater attention is therefore being paid to the implementation of previously agreed strategies, such as through strengthening integrated approaches in relation to climate, health and nature.

Policy objectives are being periodically reviewed and updated under the five-year political cycle of legislation of the federal and regional governments, and the level of ambition may therefore depend on public demands and political will. For this legislation period, 2019–2024, enhanced actions are foreseen, including greening urban and peri-urban areas and infrastructure (e.g. extending city park areas and nature re-development), afforestation (e.g. 4 000 ha additional forest in Flanders), improving Natura 2000 site management and habitat restoration, stimulating nature-based solution projects, enhancing ecological connectivity with development of green–blue networks, strengthening cooperation with other administrations and stakeholders, including the private sector, strengthening mainstreaming of biodiversity into other sectors, extending nature-friendly farming practices and stimulating bio-farming (i.e. organic farming) and agro-forestry. An overall priority objective will be to improve and/or maintain the favourable conservation status of all BHD habitats and species through wider and more effective practical implementation of conservation measures: increasing the area of habitats and restoring their structure and functions, extending green infrastructure to support their sustainable management, and improving enforcement. Other actions will depend greatly on which habitats, species and landscapes are prioritised in each region and on the threats affecting them and their drivers.

A review of policy objectives and biodiversity strategy and actions will also be needed to align with the *EU Biodiversity Strategy for 2030* and global-level ambitions, including the required further integration of biodiversity objectives in other strategies and policies, especially those concerning climate change, agriculture and land use, and health.

7.10 Conclusions

With the creation of the Nature Directives more attention was given to nature conservation, the expansion of protected areas, and stricter laws for protection of habitats and species. During the past decade Belgium has built up a comprehensive knowledge of its biodiversity and especially BHD habitats and species, on ecosystems and their services, and on nature–climate and nature–health relationships. Extensive and time-consuming consultations with stakeholders and relevant administrations have improved understanding of the objectives and expectations of other actors, and increased expertise in finding ways to build consensus, establish partnerships and strengthen the mainstreaming of biodiversity into other sectors and policies. A wide range of approaches, tools and practical guidelines for restoration and management of natural values and sites have been developed at the regional and federal levels and are still being refined. Acquisition of sites has proven to be the most effective approach for conservation but requires high investments, mainly from the public authorities, for land purchase and management. Therefore, a high priority is being given to creating partnerships with private landowners, including through technical and financial support, to help them contribute to nature objectives.

Improved monitoring practices and methodologies have allowed better evaluation of the status of habitats, species and protected areas and environmental quality, which is helping to identify gaps in implementation. This knowledge, expertise and experience should now be used to enhance public and political support for better implementation of nature conservation actions in the field, and to upscale them urgently to reverse ongoing declining trends.

Acknowledgements

Myriam Dumortier (INBO) for reviewing the first draft and Jean-Paul Herremans (NEMO) for reviewing aspects of EU financial support projects.

References

ANB (2016a) *Soortenbeschermingsprogramma voor de Europese bever (Castor fiber) in Vlaanderen.* Brussels: Agentschap voor Natuur en Bos.

(2016b) *Prioriteitenkader voor ecosysteemherstel in Vlaanderen.* Brussels: Agentschap voor Natuur en Bos.

(2018) *Nature Management Plan (Flanders).* Brussels: Agentschap voor Natuur en Bos.

(2019) *PRIORITAIR ACTIEKADER (PAF) VOOR NATURA 2000 IN VLAANDEREN, BELGIË – voor het meerjarig financieel kader voor de periode 2021 – 2027.* Brussels: Agentschap voor Natuur en Bos.

(2020) *Soortenbeschermingsprogramma's.* Brussels: Agentschap voor Natuur en Bos.

Belgian Federal Government (2001) Chapter 3 of the *Royal Decree of 21 December 2001 Arrêté royal visant la protection des espèces dans les espaces marins sous juridiction de la Belgique.* [Royal Decree for the protection of species in marine areas under the jurisdiction of Belgium.] *Belgian State Journal.*

(2012) Belgische Noordzee - Levend water! Biodiversiteit en Natura 2000 in het Belgische deel van de Noordzee. Brussels: Federale Overheidsdienst Volksgezondheid, Veiligheid van de Voedselketen en Leefmilieu. 40 pp.

Belgian Forum on Invasive Species (2020) Harmonia Database. https://ias.biodiversity.be/species/all

Belgian National Focal Point to CBD (ed.) (2006) *Belgium's National Biodiversity Strategy 2006-20.* Brussels: Royal Belgian Institute of Natural Sciences.

Belgian National Focal Point to the Convention on Biological Diversity (ed.) (2013) *Biodiversity 2020 – Update of Belgium's National Biodiversity Strategy.* Brussels: Royal Belgian Institute of Natural Sciences.

Belgian National Focal Point to the CBD (ed.) (2019) *Sixth National Report of Belgium to the Convention on Biological Diversity*. Brussels: Royal Belgian Institute of Natural Sciences. www.cbd.int/doc/nr/nr-06/be-nr-06-en.pdf

Bruxelles Environnement (2015) *State of the Environment Report 2011-2014*. Brussels: Environment Brussels.

Danckaert, S., de Rijck, K., Mulders, C. & Peeters, A. (2012) Belgium. In *High Nature Farming in Europe*, eds. R. Oppermann, G. Beaufoy & G. Jones, pp. 129–137. Ubstadt-Wieher, Germany: Verlag regionalkultur.

Departement Omgeving Vlaanderen (2018) *Milieurapport 2018 Vlaanderen*. Brussels: Departement Omgeving Vlaanderen. https://researchportal.be/nl/publicatie/milieuverkenning-2018-0

Dufrêne, M. (2004) *Réseau écologique et structure écologique principale: concepts - structure - stratégie d'élaboration*. DGARNE/CRNFB/OFFH. Jambes, Belgium: Direction Générale Opérationnelle Agriculture, Ressources naturelles et Environnement, Service Public Wallonie.

EEA (2018) Reports on land use and land use change. www.eea.europa.eu/data-and-maps/indicators/mobility-and-urbanisation-pressure-on-ecosystems/assessment

European Commission (2017) *The EU Environmental Implementation Review: Common Challenges and How to Combine Efforts to Deliver Better Results*. COM(2017) 63 final.

Flemish Government (1997) Natura 2000 Partners (Flanders). www.natura2000.vlaanderen.be/partners

(2009) Executive Law for Species Protection of 15th May 2009. *Belgian State Journal*.

INBO (2018) *Wolvenplan Vlaanderen*. Brussels: Instituut voor Natuur- en Bosonderzoek Vlaanderen.

(2019) *Rode-Lijststatus per groep*. Brussels: Instituut voor Natuur- en Bosonderzoek Vlaanderen.

(2020a) *Algemene broedvogelindex*. Brussels: Instituut voor Natuur- en Bosonderzoek Vlaanderen.

(2020b) *Overschrijding van de kritische stikstofdepositie in het Natura 2000- areaal*. Brussels: Instituut voor Natuur- en Bosonderzoek Vlaanderen.

(2020c) *Natuurrapport 2020*. Brussels: Instituut voor Natuur- en Bosonderzoek Vlaanderen.

IUCN (2019) *Country-focus: Belgium*. www.iucn.org

Kestemont, B. (2019) The bottom-up assessment of threatened species. *Nature Conservation Research*, **4** (3), 93–106.

Leefmilieu Brussel (2016) *The 2016 Brussels Nature Plan*. Brussels: Leefmilieu Brussel.

(2004) *Beheerplan voor het Zoniënwoud*. Brussels: Leefmilieu Brussel.

SPF (1999) *20 JANUARI 1999. - Wet ter bescherming van het mariene milieu in de zeegebieden onder de rechtsbe*. EMIS (vito.be)

(2012) *Le Plan Abeilles 2012–2014*. www.health.belgium.be/fr/le-plan-abeilles

(2014) 2014 *Marine Spatial Plan for the Belgian part of the North Sea*. Brussels: Service Public Federal.

SPF (2015) *Assessment of the Federal Bee Plan 2012-2014*, ed. D. Cuypers, D/2015/2196/7. Brussels: Service Public Federal.

(2016) *Royal Decree of 27 October 2016*. Brussels: Service Public Federal.

(2017a) *Ministerieel besluit betreffende de aanname van instandhoudingsdoelstellingen voor de mariene beschermde gebieden*. Brussels: Service Public Federal.

(2017b) *Le Plan fédéral Abeilles 2017-2019*. Brussels: Service Public Federal.

(2019) Royal Decree on marine spatial plan dd 19/5/2019. *Belgian State Journal*.

SPW (2010) Présentation de l'inventaire des SGIB.

(2013) *Dossier scientifique sur les services rendus par les écosystèmes en Wallonie*. Jambes, Belgium: Service Public Wallonie.

(2015) *Cohabiter avec le castor en Wallonie*. Jambes, Belgium: Service Public Wallonie.

(2017a) *Red Lists for species in Wallonia*. Jambes, Belgium: Service Public Wallonie.

(2017b) *State of Environment Report - Wallonia 2017*. Jambes, Belgium: Service Public Wallonie.

(2019) *PRIORITISED ACTION FRAMEWORK (PAF) FOR NATURA 2000 in WALLONIA/BELGIUM – for the Multiannual Financial Framework period 2021 – 2027*. Jambes, Belgium: Service Public Wallonie.

Statistics Flanders (2019) Built-up area. www.vlaanderen.be/en/statistics-flanders/land-use/built-up-area

UNEP-WCMC (2021) *Protected Area Profile for Belgium from the World Database of Protected Areas, December 2021*. www.protectedplanet.net

VITO (2013) *Raming van de baten geleverd door het Vlaamse NATURA 2000-netwerk*. Vlaamse Instelling Technologisch Onderzoek.

VLM (2014) *Flemish Rural Development Programme (RDP) for 2014–2020*. Brussels: Vlaamse Landmaatschappij.

(2019) *Samenwerken met landbouwers*. Brussels: Vlaamse Landmaatschappij. www.vlaanderen.be/publicaties/samenwerken-met-landbouwers-aan-biodiversiteit-milieu-en-landschap-ervaringen-met-beheerovereenkomsten

Walloon Government (1973) Nature Conservation Act (Wallonia). *Belgian State Journal*.

(2014) *Walloon Rural Development Programme (RDP) for 2014–2020*. Jambes, Belgium: Direction Générale Opérationnelle Agriculture, Ressources naturelles et Environnement, Service Public Wallonie.

(2019) *Rapport d'évaluation du programme wallon de développement rural 2014-2020*. Jambes, Belgium: Service Public Wallonie, Agriculture. https://agriculture.wallonie.be/evaluation-du-pwdr

WWF (2020) *Living Planet Report - Natuur in België*. O. Szczodry, H. Eggermont, J-Y. Paquet, M. Herremans, S. Luyten. Brussels: WWF.

Zoniënwoud (2008) *Inter-regional management plans*. Zoniënwoud, Groenendaal. zonienwoud.be

Zwaenepoel, E., Danckaert, S. & Van Gijseghem, D. (2016) *Wat kan de landbouwer doen voor Europese habitats en soorten?* Brussels: Departement Landbouw en Visserij.

Chapter 8

Republic of Bulgaria

Boris Barov

Key Statistics

Terrestrial area (km^2)/% of EU-28*	110 995/2.5%
Marine waters (km^2)/% of EU-28*	35 139/0.8%
Population density (p km^2)/% of EU-28*	65/57%
GDP per capita (€)/% of EU-28	5 817/22%
Land cover (CORINE land cover classes 2018)	AGR 52%, FOR 31%, OSN 12%, ART 5%, WET 1%
Biogeographical regions	ALP, BLS, CON, MBS
Number of HD Annex I habitats/%*	92/41%
Number of HD Annex II, IV and V species/%*	206/17%
Number of BD Annex I bird taxa/%*	115/65%
% HD UCS and short-term decreasing habitats/% unknown	1%/3%
% HD UCS and short-term decreasing species/% unknown	?%/?% (see text)
% breeding BD Annex I birds decreasing/% uncertain and unknown	20%/17%

* Excluding MAC and MMAC biogeographical regions.
Sources and codes: See Table 1.1.

8.1 The Natural History and Characteristics of Bulgaria

Bulgaria is a medium-sized south-eastern European country located in the centre and east of the Balkan Peninsula. Approximately 28% of the territory is mountainous and plains take up about another third, while the rest is made up of plateaux and hills. The northern border with Romania is almost entirely traced by the river Danube, and northern Bulgaria is dominated by the hilly Danuban plain, which gradually transitions into the Dobrudzha plateau. To the east is the 354 km Black Sea coastline of alternating sandy beaches and rocky promontories. The Balkan range, a distinct biogeographical zone itself, crosses Bulgaria from west to east and divides the country into northern and southern parts with quite distinctive climatic and biogeographical characteristics. South of the Balkan range lies the fertile Thracian lowland. Further south are the mountains of the Rila-Rhodope massif, with the horsts of the Rila (Mount Mousala, at 2 925 m, is the highest point on the Balkan Peninsula), the Pirin, the Belasitsa and Slavyanka Mountains being the most prominent. The Rhodope Mountains are an expansive system of high, rounded mountain ridges, and occupy most of the southern border with Greece. The western Rhodopes are higher, carved by deep valleys and gorges, many of karst origin. The eastern Rhodopes slope towards the Thrace lowlands, which makes them open to Mediterranean climatic and biogeographical influences. Importantly, these mountains include the oldest land masses in the Balkans and remained unaffected by Quaternary glaciation, which is the reason for the presence of numerous relict species.

Despite its relatively small size, Bulgaria has a rich flora and fauna and relatively well preserved natural and semi-natural ecosystems (Fet and Popov, 2007). The high biodiversity is due to the country's geographical position, between the temperate and subtropical zones, and its underlying topographical and geological variety. From a phytogeographical point of view, the vegetation is a complex of Boreal (in the high mountains), central European (most widely distributed), Steppic (second in frequency), Arctic, Alpine, Balkan (including Mediterranean) and local vegetation (Roussakova, 2011). The complex nature of the Bulgarian flora, its high species diversity and the relatively high percentage of endemism determine its high conservation value (Peev and Delcheva, in Fet and Popov, 2007, pp. 571–579; Petrova and Vladimirov, 2009, 2010). Limestone areas with calcareous soils and karst relief cover approximately 23% of the country, contributing to their own specific vegetation and fauna. For some animal groups, such as the saproxylic beetles, dragonflies and butterflies, the proportion of European species that occur in Bulgaria is particularly high. Two major distinct zoogeographical complexes shape the faunal diversity: the northern (Eurosiberian) complex of cold-tolerant species, and the southern (Mediterranean) complex of thermophilic species (Josifov, 1988).

Although the Bulgarian lands were home to ancient civilisations, farming and other activities did not transform the landscape and dense forest prevailed during medieval times. With the advancement of the Ottomans from the fifteenth century onwards, forests were increasingly felled for fuel, mining and military purposes until most of the remaining forest cover was in mountainous areas. Mountain landscapes, despite being heavily deforested over past centuries, have preserved a more natural character due to the difficulty of access and centralised ownership. Their predominant land use has been extensive livestock grazing, and the practice of transhumance was common until the early twentieth century. Livestock grazing and seasonal burning by shepherds led to a significant artificial expansion of non-forested montane pasture.

Mining, road and rail construction, as well as the lumber trade, remained an important economic driver of deforestation. By the beginning of the twentieth century state control over forestry and logging concessions was established. Since the establishment of the forestry service in the late nineteenth century, forests have been managed for economic, anti-erosion and water supply protection goals with organised planting and watershed restoration measures as early as the 1900s. The majority of the productive forests were managed through felling rotations, conversions and seedling regeneration and planting. At the same time the exploitation of natural forests by the local population for timber and forage continued. With the development of nature conservation, significant areas of natural beech, spruce and pine forest have been protected as Nature Reserves since the 1930s.

Throughout the twentieth century, and especially after World War II, large areas of previously eroded or degraded forest land were planted with coniferous monocultures to curb erosion. The natural lowland riverine forests were largely cleared, degraded or replaced by hybrid poplar plantations, but further areas of natural forests were gradually protected. With the economic crisis of the last years of the communist regime, and especially after the democratic changes between 1990 and 2013, the total forested area increased by 11% to about 31–38% (CLC 2018 data; FISE 2020[1]). This increase in forest cover was largely due to natural succession of abandoned agricultural land in unproductive or depopulated areas, and not to further large-scale afforestation.

The dominant land use in the country is agriculture. Farming is mostly arable in the plains and lowlands, while grasslands and mosaics occur at higher elevations. Due to their particular importance for biodiversity and their threatened status, it is worth describing the grasslands further. They have been traditionally used for communal grazing in hilly and drier parts of the country (pastures) and for hay production in mountain areas and river valleys (meadows). Only a small proportion originates from remnant steppe, mostly in northeast Bulgaria, or limestone vegetation (Meshinev *et al.*, 2005). The area of permanent grassland has decreased significantly in recent decades: from about 16% of the country at the beginning of the twentieth century (Ganchev *et al.*, 1964), to less than 8% (National Statistical Institute, 2019), of which only about a half is covered by Natura 2000 (Apostolova *et al.*, 2018).

Wetlands only occupy about 1% of Bulgaria's territory, but their conservation value is very high as they contain diverse habitats that are rare in the country, as well as protected plants and animals. Their importance for migratory birds is substantial, especially along the Black Sea coast and the Danube where several species congregate in internationally important numbers, such as Great White Pelican (*Pelecanus onocrotalus*), Dalmatian Pelican (*Pelecanus crispus*) and Pygmy Cormorant (*Microcarbo pygmaeus*). Especially noteworthy are the Black Sea coastal lakes that host globally significant numbers of wintering waterbirds, such as White-headed Duck (*Oxyura leucocephala*) and Red-breasted Goose (*Branta ruficollis*), with up to 90% of the global population of the latter present in cold winters. During the twentieth century, industrialisation, urbanisation, farming and other land uses, pollution and changes in water body morphology have destroyed or damaged most of the inland wetlands. Rough estimates suggest that the area occupied by wetlands has decreased by a factor of 20 since the beginning of the twentieth century (Vassilev *et al.*, 2013).

8.2 The Status of Nature

8.2.1 Habitat Status and Trends

For the first time in 2011 the *Red Data Book* by the Bulgarian Academy of Sciences (RDB, 2011) assessed 166 EUNIS habitats of 'conservation value'. Based on the IUCN criteria and categories, 146 were considered to be threatened (CR 28, EN 71, VU 47) and 20 were Near Threatened. However, as discussed more fully in Section 8.8, systematic monitoring and assessment of the status of HD habitats and species is relatively new to the country. Furthermore, the Executive Environmental Agency (ExEA) responsible for the National Biodiversity Monitoring Scheme (NBMS) has limited capacity to organise monitoring and analyse the results. These factors have constrained the ability to provide accurate and comprehensive HD Article 17 assessments.[2]

In 2013 Bulgaria submitted its first HD Article 17 report, which covered 2007–2012. The assessments were made by consultants based on information used for the 2011 *Red Data Book*, supplemented by additional surveys and models by scientists. Out of 184 conservation status assessments, 11% were favourable and 88% poor. Only two assessments were categorised as unknown, and none were bad. This latter result raises some doubt over the credibility of the assessments, given the contrast with the *Red Data Book* assessment.

[1] https://forest.eea.europa.eu/countries/bulgaria/bulgaria-basic-data

[2] See Section 1.3 for background on monitoring under the Nature Directives, status categories and data sources.

Table 8.1 Short-term trends (c. 2007–2018) in Habitats Directive Annex I habitats with an unfavourable conservation status in Bulgaria.

	Poor (unfavourable– inadequate)				Bad (unfavourable–bad)			
	Imp	Sta	Dec	?	Imp	Sta	Dec	?
HD Habitat assessments								
Number	37	49	1	62	0	0	1	9
%	25%	33%	1%	42%	0%	0%	10%	90%

Key. Imp = improving, Sta = stable, Dec = decreasing, ? = unknown.
Source. Based on Member State HD Article 17 reported data as presented on the EEA national summary dashboard (2021).

The organisation of the HD Article 17 report covering 2013–2018 was very late, despite the availability of a sufficient budget from EU funds; the contract with an external contractor was only signed at the beginning of 2019. As the HD report was due in April 2019, no new data collection or meaningful involvement of the scientific community was feasible in the short time remaining. The consultants had no choice but to rely on existing data and a handful of recent studies of habitats and species.

The Ministry of Environment and Water (MoEW) submitted the first draft of the report on the conservation status of HD habitats and species at the end of August 2019. The assessment methodology reflected the new reporting guidelines and considered the critical comments given by the European Commission on the 2014 report (particularly in relation to the use of species distribution models). In contrast to the previous reporting cycle, many of the habitats and species previously considered favourable were assessed as unfavourable, mostly due to the changes in methodology. Consequently, over the 2013–2018 reporting period, 12% of assessments were favourable, 79% were poor, 5% were bad, and the status of 3% was assessed as unknown. As indicated in Table 8.1, only two habitat assessments with an unfavourable status were reported as declining, but for a large proportion their trend was unknown. Whilst the overall picture was probably more realistic than the previous reporting period, these results should be treated with considerable caution as they still appear to be inconsistent with the findings of the relatively recent 2011 *Red Data Book* assessment and are not based on sufficient field data.

8.2.2 Species Status and Trends

The difficulties around HD habitat reporting described previously apply equally to species. As with habitats, the latest comprehensive scientific assessment of threatened species was the 2011 *Red Data Book*, which included 800 plants and fungi (CR 273, EN 442, VU 79) and 442 species of animals (87 CR, 70 EN, 100 VU). As the NBMS had only been partly operational during the reporting period, sufficient data were not available.

Similarly to the HD habitats assessment, the initial draft report for 2013–2018 submitted to the European Commission in August 2019 showed a significantly worse situation for most species in relation to the previous reporting period, also due to methodological changes. As the report had been submitted without consultation with Bulgarian scientists or conservation stakeholders it was heavily criticised for its lack of transparency and inconsistent results. The Institute of Biodiversity and Ecosystem Research and the National Natural History Museum of the Bulgarian Academy of Sciences openly criticised the MoEW for failing to organise the monitoring and reporting process effectively, and formally refused to participate in further consultations. Overall, the exclusion of Bulgarian experts was a strategic mistake that derailed the reporting process and poisoned the atmosphere for collaboration.

At the same time, the MoEW was put under growing pressure by business organisations and media to report more favourable results. For example, in an open letter[3] to the Prime Minister and in numerous public statements, business representatives and politicians urged the government to 'correct the highly negative report it submitted to the EC' because otherwise the consequences would lead to sanctions and require further increases in the Natura 2000 network and would thus 'block the development of the country'.

In January 2020, in a step that could only be explained as succumbing to pressure, and without adequate explanation to the public, the MoEW submitted a second version of the report. The status of habitats was unchanged, but the species assessments were very different in comparison with the first draft. Favourable assessments increased from 84 to 169, poor assessments decreased from 308 to 138, and bad assessments decreased from 36 to 14. The number of assessments reported as unknown increased tenfold – from 12 to 123. Given their rejection by official scientific institutions, the HD Article 17 species trends are not presented here as they are in other chapters. To do so could provide the opportunity for the data to be quoted out of context.

As discussed in Section 8.8, the status of birds is better known because several monitoring schemes cover common breeding and wintering birds, and a number of threatened species. Nevertheless, the compilation of the BD Article 12 report suffered from the same delays and deficiencies of approach. The reported short-term trends for BD Annex I species indicate mostly stable or increasing populations

[3] https://bica-bg.org/писмо-на-аобр-до-министър-председател-4/

Table 8.2 Short-term trends (c. 2007–2018) in Birds Directive Annex I taxa in Bulgaria.

	Breeding				Wintering			
	Inc	Sta/ Flu	Dec	?	Inc	Sta/ Flu	Dec	?
Number	14	39	17	14	6	17	4	7
%	17%	46%	20%	17%	18%	50%	12%	21%

Key. Inc = increasing, Sta = stable, Flu = fluctuating, Dec = decreasing, ? = uncertain and unknown.
Source. Based on Member State Article 12 reported data as presented on the EEA national summary dashboard (2021).

(Table 8.2), but the results should be treated with caution as monitoring of a large number of species was not done and many of the references provided pre-date the reporting period.

8.2.3 Drivers and Causes of Change

Bulgaria's political and economic transition was slow and painful, resulting in the collapse of certain primary and secondary economic sectors. One of the positive aspects of this process during the 1990s was the large-scale recovery of some ecosystems, as well as a significant reduction in industrial pollution. The 1990s land reform severely impeded farm modernisation and led to a pronounced decline of the agriculture sector, resulting in large areas of arable land reverting to grassland and scrub. The extent to which land abandonment resulted in biodiversity loss, as widely debated by many authors (Queiroz *et al.*, 2014), remains unclear as studies from this period are scarce and inconclusive.

The main pressures affecting biodiversity at the beginning of this century have been identified in Bulgaria's 2005 *National Biodiversity Strategy and Action Plan*[4] (NBSAP), and confirmed in more recent references, such as the *Red Data Book* (2011) and the HD Article 17 reports of 2014 and 2020. These pressures have included: human-induced habitat degradation, fragmentation and loss; industrial, agricultural and household waste pollution; direct exploitation of biological resources; genetic ingression and invasive alien species; and global climate change effects. Such pressures have been enabled by environmental policy and implementation weaknesses related to poor enforcement of regulations, low institutional and expert capacity, and local government complacency with opportunistic and illegal real estate development. The economic transition was harsh on rural and disadvantaged communities who reverted to contingent natural resource use for survival (e.g. firewood and non-timber forest products), which seemed to be tolerated.

The situation changed dramatically with EU accession, which quickly swung the agricultural model to the other extreme (Rangelova and Vladimirova, 2017). CAP subsidies were targeted preferentially to the cereal sector (area-based payments), causing extreme concentration of arable farming among a few large owners, rapid increases in the size of arable fields and farms, and the near complete loss of natural vegetation and landscape elements in the most productive lowland farmlands. Agricultural intensification has had the worst impact, particularly on farmland biodiversity and semi-natural grasslands, which have been in decline ever since. Two studies of an SPA located within a typical agricultural region documented the loss of important habitat features for biodiversity caused by newly introduced agricultural practices inconsistent with traditional grassland management (Dobrev *et al.*, 2014; Popgeorgiev *et al.*, 2014). According to the authors, the Single Area Payment System (SAPS) had incentivised the conversion of grassland in this SPA to arable land, vineyards and orchards since 2006. This widespread problem was the consequence of an inappropriate conditionality introduced by the national authorities on the SAP payments that excluded pastures with landscape features (e.g. more than 100 trees and/or shrubs per ha) from eligibility for SAPS payments. Thus, more than one-third of the grasslands in Bulgaria became ineligible for Pillar 1 payments (Stefanova and Kazakova, 2015) which, in parallel with the decline in livestock numbers, led to their economic undervaluation and stimulated conversion to other uses and loss of natural vegetation.

A growing problem with unsustainable grassland management in the National Parks has been reported since the adoption of particular CAP instruments, such as the submeasure 'Pastoralism' under the agri-environment–climate measure, which has resulted in a significant increase in grazing in most National and Nature Parks. Unfortunately, due to inadequate assessment of the grazing potential and grassland ecology, and lack of specialised plans for grazing, in many cases this measure has had mostly negative effects (Gussev *et al.*, 2016) such as overgrazing, deterioration of grassland structure and altered species composition, soil pollution and eutrophication of wetlands.

As far as acute loss of rare and sensitive habitats and species is concerned, the most important drivers have been the construction of buildings and tourist developments. This is despite Bulgaria's very low population density and low land take rate (growth of artificial areas) of 0.14% in Bulgaria over 2006–2012 as indicated by CLC data – well below the EU average of 0.41% (European Commission, 2019b). Nevertheless, a disproportionate amount of construction has affected habitats of restricted distribution and small coverage, such as coastal dunes, steppe grasslands and wetlands. Over 2004–2008, almost all remaining undeveloped coastal areas were included in building plans at the expense of unique habitats for the country, which have been irreversibly lost.

Another group of drivers for the loss of biodiversity is related to the abandonment of traditional land and wetland management practices. A particular example is the large number of fishponds that have disappeared along the Black Sea coast, the Danube and inland rivers. Although many had

[4] Национална стратегия за опазване на биологичното разнообразие (Протокол 15.3 на МС/1998).

been converted from natural wetlands in the second half of the twentieth century, their low-intensity management provided suitable conditions for wetland vegetation and birdlife. Over the last 20 years, this type of fish farming has become less profitable, resulting in the almost complete demise of these valuable wetlands and affecting the populations of many vulnerable species, such as the Ferruginous Duck (*Aythya nyroca*). Further effects of this abandonment have included habitat loss and the proliferation of invasive vegetation such as False Indigo-bush (*Amorpha fruticosa*), Tree of Heaven (*Ailanthus altissima*) and Black Locust (*Robinia pseudoacacia*).

All forests outside natural reserves and National Parks are subject to silvicultural management under the Forestry Act (2011), which defines three classes of forest management functions. Firstly, protective forest areas must be managed to protect soil, water and infrastructure from erosion. Secondly, special forest areas (including protected areas and Natura 2000 sites) must be managed to protect biodiversity. These two types cover 68% of the forest area. Thirdly, economic areas are managed for timber production. The Forestry Act grants the right to take timber for personal use (for a fee) and a free of charge use of non-timber forest products (herbs, berries and mushrooms) from state forests.

Illegal logging is a widespread issue of social significance and public concern due to its large volume and subversive character. According to WWF analysis,[5] over 2013–2017 the volume of illegal logging was 2.4–2.7 million m^3 per annum, which was close to 25% of the official legal harvest. To make things worse, alleged illegal logging was behind hundreds of deliberate forest fires that had devastating effects on local fauna and flora. Floodplain forests were widespread in Bulgaria but currently only a small fraction remains on the Danube islands, along the lower Kamchia, Ropotamo and Tundzha rivers, and on the islands of the Maritza river. These forests are almost completely destroyed along the reaches of the Iskar, Vit, Osam, Yantra and Struma rivers (WWF-Bulgaria, 2019). These trends coincided with a continuous reduction in the annual budget for protection, restoration and afforestation activities. Pest outbreaks and forest fires have recently affected conifer plantations throughout the country.

Bulgaria experienced an investment boom in renewable energy between 2008 and 2015, particularly in wind and mini-hydropower. As a result of poor environmental impact assessments, many projects caused widespread damage to natural habitats such as rivers and streams, riparian forests, grasslands and steppe vegetation, all with an unfavourable conservation status. Once constructed, there is little control on the operations of hydropower facilities, and it is not uncommon for entire sections of the affected river to dry out completely.[6] Another problem is occasional discharges of anoxic sediments accumulated behind the dam, which cause de-oxygenation of the ecosystem downstream. Finally, the still common illegal extraction of aggregates from riverbeds is another cause of habitat degradation.

Development of wind power plants in Bulgaria was accelerated through economic incentives between 2008 and 2012, and the cumulative installed wind power capacity reached 699 MW.[7] Several wind farm projects were contested on nature conservation grounds, due to poor impact assessment and insufficient monitoring data presented to the public. An emblematic case is the Cape Kaliakra wind farm, which, alongside other cumulative habitat loss impacts, caused the European Commission to open a protracted infringement procedure. This finally resulted in a CJEU ruling[8] that Bulgaria had failed to comply with EU nature conservation legislation, in particular the Nature Directives and EIA Directive. The case is also subject to an ongoing Bern Convention case file and Recommendation 130 (2007).[9]

With the recovery and growth of industry in recent years there have been increasing incidents of pollution (e.g. water pollution from point sources such as individual industrial facilities or mine tailings) that have caused local die-outs of aquatic fauna. Problems with waste management and control, including illegal landfilling or illegal dumping of municipal solid waste, remain unresolved and the effects are visible throughout the country.

Finally, many plants and animals are subject to unsustainable use. Examples include the economically important harvesting of wild medicinal plants and mushrooms, for which Bulgaria is a leading exporter globally. Hunting and hunting tourism were traditionally well developed, with significant areas of state forest land allocated to game management in the past, including dedicated state game-breeding stations and game management concessions in often well-protected forest areas. In recent years the organisation of hunting has become less effective, which has led to an increase in poaching and decline in herbivores, and, sadly, to increasing use of poison by herders to control predators. Several cases of wildlife poisoning have been reported in recent years, affecting vultures and other birds of prey, mostly driven by livestock–predator conflicts. Another illegal practice has been recently uncovered among pigeon breeders who use poison-smeared pigeons as live decoys to kill Peregrine Falcons (*Falco peregrinus*) and other raptors.

Among the species of high conservation importance affected by various specific pressures other than those described above are raptors, storks and pelicans being killed

5 WWF-Bulgaria (2018) https://wwfeu.awsassets.panda.org/downloads/______2013_2017.pdf

6 Balkanka Fishing Club and WWF: https://dams.reki.bg/Dams/Dams

7 Renewable Market Watch (2020): https://renewablemarketwatch.com/country-reports/europe/bulgaria/bulgaria-wind-power-market-outlook-2020-2030-single-user-detail

8 ECJ ruling C- 141/14 from 2016.

9 T-PVS/Files(2020)67 Wind farms in Balchik and Kaliakra – Via Pontica (Bulgaria).

by powerline collisions and electrocution, and the disturbance of raptor breeding sites (Spasov *et al.*, 2012).

8.3 Nature Conservation Governance, Actors, Legislation and Policies

8.3.1 Overview of Nature Developments in Bulgaria

Table 8.3 presents the timeline of the most important developments in Bulgaria's nature conservation sector since its origins in the late nineteenth century. The first legislative acts in the principal directions of nature conservation concerned forestry (1890), the protection of certain species (1890s) and hunting (1897). Early civil society movements marked the establishment of the Union of Nature Protection (1928), which soon after initiated the designation of several Nature Reserves (1933) and the first National Park (1934). The first specific nature protection regulation was the Ordinance for the Protection of Nature (1936). In 1993, a *National Action Plan* for the most significant wetlands was elaborated by the Institute of Ecology of the Bulgarian Academy of Sciences in cooperation with the MoEW. For the first time, priority actions and a management budget were identified with a variety of stakeholders for each Ramsar Site.

Bulgaria was among the first to prepare a *National Strategy for Biodiversity Protection* (1993, adopted in 1998) following the Rio process, and since then a *National Plan for Biological Diversity Protection* has been adopted for 1999–2003, 2005–2010, and 2011–2020, in which the threats, objectives, and measures for their achievement have been revised.

8.3.2 Legislation

The Environmental Protection Act[10] (SG No. 91/2002) is the key framework law for environmental protection. It defines the national environmental authorities and their areas of responsibility; guarantees the right of access to information on the environment in line with the Aarhus Convention; sets the regulations for the financing and organisation of environmental protection activities; provides the mandate for key environmental strategies and programmes; defines the framework for EIAs and SEAs; introduces the precautionary and the polluter pays principles (especially in relation to the prevention and limitation of industrial pollution); and establishes the National Environmental Monitoring System, part of which is the NBMS, further discussed in Section 8.8.

The Protected Areas Act[11] (SG No. 133/1998) defines the categories of protected areas based on, but not fully identical to, the six IUCN protected areas management categories (see Box 1.2). The development of this legislation, and the ensuing protected area administration and first management plans, were among the main outcomes of the US–Bulgaria Partnership on Biodiversity Conservation (see online Annex

[10] Закон за опазване на околната среда (ДВ 91/2002).
[11] Закон за защитените територии (ДВ 133/1998).

Table 8.3 Key nature conservation developments in Bulgaria.

1890s	Adoption of the first forestry and hunting laws
1928	Establishment of the Union of Nature Protection, the first such NGO in the country
1933	Designation of the first Nature Reserves
1934	Designation of the first National Park, Vitosha Mountain
1936	First Ordinance for the Protection of Nature
1948	Suspension of NGOs and establishment of total state control on all spheres of social life, including on nature protection
1967	The Nature Protection Law introduced the framework for site and species protection
1971	Establishment of the Ministry of Forests and Protection of the Environment
1975	Establishment of the Committee for the Environment at the Council of Ministers
1984	Publication of the first national *Red Data Book*
1988	Foundation of the Bulgarian Society for the Protection of Birds, the first nature conservation NGO in the recent history of the country
1989	Political changes – important role of environmental protection and NGOs
1990	Establishment of the Ministry of Environment, later becoming the Ministry of Environment and Water (MoEW)
1991	Adoption of the new Constitution of Bulgaria, with Art. 18 proclaiming exclusive state ownership of 'parks of national significance, nature and archaeological reserves designated by law'
	Establishment of the National Nature Protection Service at the MoEW
	Start of the USA–Bulgarian partnership for the protection of nature, with the technical and financial support of the Government of the USA
1992–1995	Development of *National Biological Diversity Conservation Strategy* under CBD with the financial support of USAID
1993	First *National Action Plan for the Protection of Wetlands of International Significance*
1994	Launch of the Bulgarian–Swiss Biodiversity Conservation Programme, financed by the Swiss Agency for Development Cooperation
1997	Publication of *Important Bird Areas* inventory of 50 IBAs of global and regional importance
1998	Adoption of the Protected Areas Law introduced the IUCN categories of protected areas leading to pre-categorisation of many of the national protected areas

	Adoption of the Medicinal Plants legislation
1999	Bulgaria joined the EEA and Eionet and established the National Executive Environmental Agency as its reference centre
2001	Biodiversity Law replaced older legislation and introduced the ecological network concept
2003–2005	First large restoration project on the Danubian island of Persin funded by GEF/World Bank. Several other GEF-funded projects underway in this pre-EU accession period
2004	Development of the *National Biodiversity Action Plan*
	National Ecotourism Strategy and Action Plan
2007	Accession to the EU
2008	Identification of 114 IBAs as candidate SPAs under the Birds Directive
	Designation of the first Natura 2000 sites
2013	*Second National Action Plan for the Protection of Wetlands of International Significance*
2021	*National Action Plan Against the Illegal Use of Poisons in Nature*

See Table 3.3 for signatory dates of international conventions.

BG.1[12]). The Act regulates the ownership of protected areas: National Parks and Reserves are exclusive state property in accordance with the Constitution and any use of territory within them can only be made through an economic concession. The Act regulates the general regimes of designation as protection, management and use and ascribes the responsibilities to the respective authorities. The MoEW has the overall responsibility, with the National Nature Protection Service (NNPS) in charge of protected areas (see Section 8.3.3 for further details).

The Biological Diversity Act[13] (SG No. 77/2002) introduced novel concepts in relation to the protection, management and sustainable use of biodiversity. For the first time, a legal definition was given to important terms such as biodiversity, habitat and conservation status. The Act envisages the establishment of a National Ecological Network to integrate the existing protected areas with the protected zones to be designated under the Nature Directives. Building on EU case law, the Act specifically pointed to Important Bird Areas (IBAs), Ramsar Sites and CORINE Biotopes[14] sites as priority sites for integration into the network. Several secondary legislative acts further detailed the transposition of relevant EU law and practice, including the 2009 Ordinance on terms and procedure for elaboration and adoption of management plans of protected zones (SPAs and SACs) in Natura 2000, and the 2007 Ordinance on conditions and procedures for assessing the compatibility of plans, programmes, projects and development proposals with the protection purposes of protected zones in Natura 2000 (with reference to HD Article 6).

Bulgaria's accession to the EU in 2007 was the main driving force behind the modernisation and strengthening of the legal framework for environmental and nature protection. However, since accession there has been a visible setback in terms of implementation of this legislation, which has resulted in a large number of infringement procedures. Policy-related weaknesses still relate to poor enforcement of conservation laws and environmental regulations, including absence of conservation objectives and measures; ineffective management and administration of protected areas; and insufficient financing or inefficient spending of available financing (European Commission, 2019b).

8.3.3 Governance

The institution ultimately responsible for all aspects of environmental policy in the government is the **Ministry of Environment and Water** (MoEW). It was established for the first time as a dedicated ministry in 1990. Previously, the functions concerning nature protection belonged to various state committees and/or were subordinated to the forestry service (depending on the period). The following description of governance, actors and policies reflects those at the time of finalisation of the chapter (early 2022), which have been in place since 2010. The MoEW elaborates and implements the national nature conservation policy and has the primary responsibility for the implementation of the Nature Directives. Using its regional and specialised structures, the ministry monitors and manages most protected area categories, enforces the legally established permitting regimes in those protected areas, and plays a leading role in the definition of nature conservation priorities and their relevant financing from the national budget and EU funds. The **National Nature Protection Service** (NNPS) within the MoEW is responsible for the protection and management of biodiversity, protected areas and the Natura 2000 network. The service also oversees the protection of rare and threatened species and issues permits and derogations. However, as described later, the responsibility and decision-making power in relation to spatial and land-use and natural resources use and control are scattered among many institutions beyond the MoEW and the NNPS, even in relation to protected areas and Natura 2000 sites, which leads to challenges in coordination and lack of clarity in responsibility (BNAO, 2019).

The **Executive Environment Agency** (ExEA) under the MoEW performs management, coordination, monitoring and information activities for environmental protection and control. The agency designs and manages the National Environmental

[12] All annexes are free to access online at CUP: www.cambridge.org/natureconservation.

[13] Закон за биологичното разнообразие (ДВ 77/2002).

[14] www.eea.europa.eu/publications/COR0-biotopes/file.

Monitoring System, part of which deals with biodiversity (NBMS). ExEA manages all aspects of information collection in relation to the status of the environment, including biodiversity. Since 2001 ExEA has been the national reference centre for the EEA.

Sixteen **Regional Inspectorates of Environment and Water** (RIEWs) implement the environment protection policy at the regional level. Their functions include monitoring, control, regulation and provision of information. The RIEWs are the competent authorities in charge of the management of domestic protected areas not assigned to a dedicated administrative unit (i.e. not part of a National or Nature Park). This arrangement also applies to Natura 2000 sites. The three National Parks are managed by National Park Directorates, specialised structures of the MoEW, while Nature Parks' Directorates are subordinate to the Ministry of Agriculture Food and Forests.

The (Water) Basin Directorates are administrative structures in the MoEW system tasked with management of water at the basin and territorial levels. The basin management principle follows the natural catchment basins and covers the four main river systems in the country: Danuban, Black Sea, Western Aegean and Eastern Aegean.

The **Ministry of Agriculture Food and Forests** (MAFF) is the institution responsible for the national agricultural policy and for managing the CAP. Under the umbrella of MAFF is the **Executive Forest Agency**, whose regional structures are responsible for the management and protection of forests, regulation of the use of forest resources, forest restoration, management of Nature Parks, and hunting and game management (including its control). The management of state-owned forests is carried out by six state forest management enterprises under the governance of the MAFF and the Executive Forest Agency. Forests that are municipal property are governed by the relevant municipal council.

Also under MAFF, the **Executive Agency on Fisheries and Aquacultures** deals with regulation and control of coastal and freshwater fisheries.

The **Ministry of Regional Development and Public Works** coordinates territorial and urban development, construction, and regional planning and development. It oversees the use of land, energy and other resources and sustainable development at regional and municipal levels, including the use of some basic natural resources such as drinking water and green space. The ministry is chiefly responsible for land-use planning and regulation of urbanisation and major infrastructures.

8.3.4 Non-governmental Actors and Public Support

Environmental protection is a sensitive issue among Bulgarians. Overall, they tend to attach patriotic value to the natural riches of the country and demonstrated high levels of awareness about biodiversity and Natura 2000 (74%) according to Eurobarometer[15] (2019). Eurobarometer figures from 2017 show quite a high level of agreement (73% of respondents) that an individual can play a role in protecting the environment. This has fallen since the 2014 Eurobarometer results and is below the average for the EU (87%).

Bulgarians are far more sceptical about NGOs nowadays, including those engaged in nature protection, a paradox attributed to the tainted negative presentation of civil society by governments and media in recent years. NGOs in the country face the double challenge of operating with limited public support in a state that offers very few financial incentives for their activities. As a result, the green NGO movement is financially weaker in comparison to Western European countries, due not only to cultural, economic and political reasons, but also as a result of the confrontational stance adopted by recent governments. This has been brought about through political speeches and media pressure, but mainly by a deliberate policy of closing communication and funding channels. Practically all nature conservation NGOs are fully dependent on grant funding since EU accession, which coincided with the phasing out of many donor programmes. At the same time, access to EU funds is managed by the government through cumbersome and bureaucratic procedures that often lack transparency.

The environmental civil society movement in the country has gone through three stages of evolution: the 'dissidence' of the late 1980s, the 'NGO period' of the 1990s, and the 'new green wave' of the first decade of the twenty-first century (Krastanova, 2012). At critical moments for nature conservation it managed to mobilise public support and has more or less successfully advocated for the public interest. Among the key organisations in the nature conservation civil society sector are the Bulgarian Society for the Protection of Birds/BirdLife Bulgaria (BSPB) and the Green Balkans Federation. Both were established as early as 1988 and have been active ever since, mainly in the conservation of birds and wildlife and the development of the protected areas network. WWF, which established its Bulgarian office in 2001 as part of the Danube-Carpathian Programme, has had a major contribution to the conservation of the Lower Danube ecological corridor (including its tributaries and threatened fish), as well as in forest protection, sustainable forest management and Forest Stewardship Council (FSC) certification.

Bilateral cooperation between governments has played an important positive role for nature conservation in Bulgaria (e.g. USAID). Cooperation programmes with some donor countries were established in the 1990s, part of which also involved NGOs in project implementation (e.g. the Bulgarian–Swiss Biodiversity Conservation Programme). A short summary of the most important bilateral programmes with their specific contributions is provided in online Annex BG.2.

8.4 Species Protection and Conservation Measures

The earliest reference to species protection dates back to the 1930s Ordinance for the Protection of Nature, which

[15] Eurobarometer Special report 481 doi:10.2779/456395.

mentioned specifically endangered and rare animal species but did not include specific conservation measures for them. The Nature Protection Act (1967) introduced the term 'protected species'. Based on this law, in the 1980s the Committee for the Environment at the Council of Ministers approved lists of protected plants and animals prepared by the Bulgarian Academy of Sciences (BAS), based on scientific studies and the *Red Data Book*. At that time, by order of the chairman of the committee, the use of commercially important species (e.g. frogs, snails, and later sand vipers due to the use of their venom for serum) was regulated by secondary legal acts.

At present, the basic law for the conservation of species is the Biological Diversity Act, which lists 113 habitat types for protection and 756 wild animal and plant species for conservation in their habitats. All bird species, except those listed as strictly protected, or under a regime of regulated use (e.g. hunting; Annex 4), are subject to 'conservation and control in view of the favourable status of their populations' (Art. 45). There are 426 strictly protected species listed in Annex 3. In relation to birds and strictly protected species a suite of general restrictions apply: all forms of taking, damage and disturbance of specimens and their nests and breeding and hibernating sites are prohibited. As regards 10 species listed in Appendix 2 that are marked with '*', Article 6 of the Act requires priority protection of their habitats, including their designation as protected areas in the National Ecological Network (and thus Natura 2000). Under the provisions of the Penal Code[16] (Art. 278c, para. 4) violators may face up to five years imprisonment and a fine of €2 500 to €10 000.

Under special conditions, derogations from strict protection can be given by the MoEW. For example, concerning the Brown Bear (*Ursus arctos*), a protected species since 1992, exceptions can be given in the absence of other alternative solutions where there is a need to deal with problematic animals to protect people and property, etc. Approximately seven to 10 bears are killed annually under such derogations, and perhaps many more by poachers. Despite the protection, the bear population declined considerably between 2011 and 2018.[17] The Wolf (*Canis lupus*) and Golden Jackal (*Canis aureus*) are listed as huntable species with shooting authorised year-round as their populations are not considered threatened and despite the EU protection status of the wolf.

In addition to strict protection, the Biological Diversity Act introduces mechanisms for preventive protection and sustainable use through individual restrictions, rules and regulations, permits and licensing. The MoEW issues annual quotas for the harvesting of medicinal plants, snails and mushrooms (by species, regions and in certain quantities and periods). Species Action Plans can introduce specific measures for threatened species (to date such plans have been adopted for 10 bird and 40 plant species).

Organised bird conservation measures in Bulgaria date back to the establishment of NGOs in the 1990s. For a long time, these conservation efforts relied mainly on international donors and voluntary action by NGOs and local groups. Nevertheless, some significant results in the protection and recovery of many rare species have been achieved, examples of which include recovery of threatened species, re-introductions of extinct raptors and restoration of formerly lost breeding colonies. In recent years, funding opportunities have increased mainly through EU programmes (see Section 8.7). Over 2013–2018, a total of 57 projects were carried out, wholly or partially related to the conservation of threatened bird species and their habitats, both inside and outside the Natura 2000 network. Of these, 50 focussed entirely on the conservation of specific bird species (or groups of species) and their habitats, to varying degrees containing components of field and scientific studies, *in situ* and *ex situ* conservation, stakeholder engagement and communication activities. Most of these projects were co-funded by the EU LIFE programme and international donors, but also by private companies, such as the electricity utilities. Some successful pilots using the CAP and other EU funds to support conservation objectives have also been carried out. For example, a successful link between the implementation of the EU Nature Directives, the LIFE funding programme and the CAP agri-environment measures has been a scheme to help conserve the internationally important wintering population of the Red-breasted Goose on arable land.

The Bulgarian nature conservation community has played a leading role in the re-introduction or recovery of conservation-dependent species in eastern Europe, with more than 20 recent programmes involving birds, mammals, fishes, reptiles, various invertebrates and plants. These have, for example, included pioneering projects for the Griffon Vulture (*Gyps fulvus*), Lesser Kestrel (*Falco naumanni*) and Saker Falcon (*Falco cherrug*) (Stoynov, 2019), and a Cinereous Vulture (*Aegypius monachus*) re-introduction programme started recently.

8.5 Protected Areas and Networks

8.5.1 Protected Area Types (Domestic)

The current framework for protected areas and networks is based on two pillars: the domestic protected areas (Protected Areas Act, 1998), and the special protected zones (Biological Diversity Act, 2002) that correspond to SPAs and SCIs/SACs. Both types of protected areas are included in the National Ecological Network, part of which has also been designated within the Natura 2000 network since 2007.

The Protected Areas Act (1998) introduced six categories of domestic protected areas, superseding the earlier system. Between 1998 and 2003 the then existing protected areas went through a process of recategorisation into one of the six categories that to a great extent match the IUCN management categories:

[16] Наказателен кодекс на Република България/1968 (Penal Code of the Republic of Bulgaria/1968).

[17] http://eea.government.bg/bg/bio/nsmbr

- **Reserves** (IUCN category I) are representative areas of natural ecosystems, including characteristic and/or remarkable wild plant and animal species, as well as their habitats. They are exclusively state owned and have rather restrictive protection regimes that typically only allow access to visitors on designated paths, and fire prevention activities.
- **National Parks** (IUCN Category II) preserve complexes of self-regulating ecosystems and their inherent species diversity, habitats of rare and endangered species and communities, characteristic and remarkable landscapes and inanimate objects that are of global importance for science and culture. They must be larger than 1 000 ha and have no human settlements; they are exclusively state property. In accordance with their goal of preserving natural processes, forest and water management is limited and tightly regulated.
- **Natural Phenomena** (IUCN category III) include characteristic or remarkable objects of inanimate and living nature that are small and/or occupy a single site (e.g. caves, canyons and waterfalls are typically designated under this category).
- **Managed (maintained) Reserves** (IUCN category IV) preserve ecosystems including rare and/or endangered wild plant and animal species, as well as their habitats, that require some form of regular maintenance, management or restoration activities. Typically, such management activities cover maintenance of the water regime, traditional salt production or forest restoration activities.
- **Nature Parks** (IUCN Category IV or V) are large areas with diverse ecosystems, plants and animals and their habitats, characteristic remarkable landscapes, and natural and semi-natural areas shaped by traditional human activities. Nature Parks have rather relaxed management regimes, seeking a balance between the protection of natural resources and economic development.
- **Protected Localities** (IUCN category IV and/or V) contain characteristic or remarkable landscapes and habitats of endangered, rare or vulnerable plant and animal species and communities. They are small, and their management regimes are focussed on the specific needs of the target species or habitats.

A special category of forests are designated under the forestry and hunting laws as State Game Husbandries; although commercial forestry is excluded, they do not have the status of protected areas.

Marine and coastal waters have not been included in the domestic protected area system. The first protected marine areas were those designated as SPAs and SCIs in the Natura 2000 network.

The first nature reserves in Bulgaria were established in 1933, followed in 1934 by the first National Park in the Balkans, Vitosha. Both types of protected areas were designated under forestry legislation, followed by the Ordinance for the Protection of Nature in 1936. The further expansion of the protected areas network in the second half of the twentieth century included different types of designations: People's Parks, Nature Reserves, Protected Landscapes, Nature Monuments and Historic Sites. In 1967, a new wide-ranging Nature Protection Act[18] was adopted. With the participation of Bulgaria in the UNESCO Man and Biosphere Programme in the 1970s, 16 Biosphere Reserves were established as Reserves with reference ecosystems. Among the greatest achievements of the 1990s was the designation of the Rila and Central Balkan National Parks, incorporating the already existing Strict Nature Reserves and doubling the total area under protection.

At the time of the adoption of the *National Biodiversity Conservation Strategy* in 1998 the total area of domestic protected areas in the country did not exceed 4.5%, with a target of 7.5–8%, and has only increased slightly since then. By May 2020,[19] there were 1 020 domestic protected areas covering 5 676 km^2 (as listed in Table 8.4), which is 5.1% of the territory of the country. Since EU accession and the designation of SPAs and SCIs, very few protected areas have been designated under national law.

8.5.2 Natura 2000 (Special Protected Zones)

According to the Biological Diversity Act, the National Ecological Network encompasses SACs and SPAs (Special Protection Zones) and domestic protected areas. Whether domestic protected areas become designated as Natura 2000 sites depends on them meeting the SCI or SPA criteria. If they do, they retain their legal protection status and management regime as domestic protected areas (e.g. the status of Reserves and National Parks remains unchanged). The National Ecological Network incorporates as a priority the CORINE Biotope Sites,[20] Ramsar Sites, Important Plant Areas and Important Bird Areas (IBAs), the latter now fully designated as SPAs. As indicated in Table 8.4, in 2021 the Natura 2000 network covered 38 728 km^2 of land or 34.9% of the territory; of this area 56% is forest and 32% is agricultural land.

The designation of the Natura 2000 network has increased the area of land under legal protection, at least on paper, seven times. According to the latest available national data, from August 2021, domestic protected areas and Natura 2000 sites overlap almost completely, with only a small percentage of national protected areas not included. However, an official figure for the total area of coverage of protected areas was not available at the time of writing. This differs from the World Database on Protected Areas (WDPA) coverage of 41% (UNEP-WCMC, 2022),[21] as the latter figure includes the State Game Husbandries, which are not considered to be protected areas according to the law, although they have regulated use.

[18] Закон за защита на природата (ДВ 49/1967г.).

[19] http://eea.government.bg/zpo/bg/ Protected areas register, accessed on 14 May 2020.

[20] http://natura2000.moew.government.bg/Home/CmsDocument/3114

[21] http://eea.government.bg

Table 8.4 The number and extent of international designated sites and protected areas in Bulgaria.

	IUCN category	Number	Area (km²)
Designated under global and non-EU legislation			
Ramsar Sites	IV	11	499
Natural and Mixed World Heritage Sites	III	3	517
Biosphere Reserves (Parks) (exc. buffer and transition zones)	VI	6	4 414
Designated under the Nature Directives			
SCIs (terrestrial + marine)	NA	234	38 737 + 2 821
SPAs (terrestrial + marine)	NA	120	25 609 + 550
Natura 2000 terrestrial area			38 728
Natura 2000 marine area			2 827
Designated under domestic legislation			
Reserves (Резервати)	I	55	774
National Parks (Национални паркове)	II	3	1 934
Protected Sites (Защитени местности)	IV	596	757
Managed Reserves (Поддържани резервати)	IV	35	45
Nature Parks (Природни паркове)	V	11	2 759

Note. NA = Not assigned.
Sources. See Section 1.3. Based on data from: MoEW Annual report for the Implementation of PAF (2021)[22]; Executive Environmental Agency (November 2021).[23]

The requirement to designate Natura 2000 sites also led to the first, and currently only, protected areas in coastal and marine waters. At the end of 2020, predominantly marine Natura 2000 sites covered 2 827 km^2, which is 8.0% of Bulgaria's marine waters.

8.5.3 Protected Area Types: Global Legislation

Bulgaria has designated some of its domestic protected areas under international agreements. All internationally designated protected areas coincide with domestic protected areas and their boundaries almost fully overlap, with the exception of the Dragoman Marsh Ramsar Site that remains unprotected under national law. Three of the Ramsar Sites along the Lower Danube were designated as Transboundary Protected Wetlands in 2013, with shared management between Bulgaria and Romania.

A significant development in recent years has been the re-designation of 14 former Biosphere Reserves from the Reserve category as six so-called 'post-Sevilla' Biosphere Parks, which have now incorporated larger areas around the existing Reserves (i.e. strictly protected) and Managed Reserves and the Central Balkan National Park. There is no separate legal designation category for Biosphere Parks in the national legislation.

8.5.4 Approaches to Protection and Levels of Protection

Domestic Protected Areas

Despite 90 years of experience with protected areas, the approach to their protection and management is mostly passive, that is, through legally established restrictions and without time-bound management objectives. The National Park Directorates, under the MoEW, have gained considerable experience in managing large areas, dealing with stakeholders and applying different types of conservation measures. However, their rights and executive powers are limited, and many controlling functions are out of their hands (e.g. the traffic police regulate access roads, and the forestry services control forestry operations and hunting in Nature Parks). The mixed ownership Nature Parks are managed by the Ministry of Agriculture and Forests' Executive Forest Agency, although they are mostly responsible for forestry. All other protected areas are managed by their owners, which can be state, municipal or private. Several protected areas are managed by NGOs.

Since 1998, management plans have been obligatory for National Parks, Nature Parks, Managed Reserves and Reserves, and optional for Protected Sites and Natural Monuments. The management plans are valid for 10 years, while short-term annual activity plans and budgets are approved each year. The process of evaluation, review and renewal of expired management plans is not streamlined due to slow approval procedures and, in several cases, pressure from resource users to moderate the protection regimes.

Public–private partnerships in protected areas management are an exception. A controversial private concession over a ski station has been in place since 2001 in the Pirin National Park. It has been the subject of numerous legal challenges, due to a lack of clarity over its borders, permitted size and planning regulations. This has resulted in numerous complaints and monitoring by national and international institutions, including the UNESCO World Heritage Committee.

22 www.eea.europa.eu/data-and-maps/data/corine-biotopes

23 For the same reason, the European Commission Biodiversity Strategy Dashboard also gives 41% protected area coverage (April 2022), based on EEA CDDA data (from 2021), which is the same underlying data sent to the WDPA.

Ineffective Management of Natura 2000

Bulgaria has not yet met the Habitats Directive requirements to designate all SCIs as SACs within six years, and to define site-specific conservation objectives and establish conservation measures for them (European Commission, 2019b). As of August 2021,[24] 186 out of 234 SCIs had been formally designated as an SAC, a subject being given close attention by the European Commission, and this process has accelerated since then.

Although all 120 SPAs have been formally designated, the approach to their protection by the authorities remains passive. Furthermore, the only restrictions on activities within them are those written in the designation order. In HD Article 6(3) appropriate assessments and EIAs, the conservation status of habitats and species, and site conservation objectives, are not given due regard. Instead, the general principle followed seems to be that if an activity is not explicitly prohibited in the designation order then it is authorised and, in most cases, no significant impact on the habitats and species subject to conservation is assumed by the authorities.

The MoEW and its 16 Regional Inspectorates for Environment and Waters (RIEW) are responsible for supervising the management of Natura 2000 and for the assessment of plans and projects that can affect the sites. Until 2016, the Natura 2000 Department of NNPS was the responsible unit for Natura 2000 management. Since 2016, the MoEW has changed policy and merged the Natura 2000 and the domestic protected areas departments, causing staff redundancies and a loss of valuable expertise.

As a result of the above problems, the poor application and enforcement of environmental legislation are major causes of infringement proceedings, with appropriate assessment procedures remaining among the most problematic issues (European Commission, 2019a). So far, two infringement cases for nature protection have gone through the entire process and resulted in the condemnation of Bulgaria in the CJEU[25] covering: (1) failure to designate appropriate sites in full; (2) failure to implement appropriate assessments, and frequently issuing permits with inadequate or no EIA; (3) failure to measure the cumulative impact of investments on Natura 2000 sites; and (4) failure to protect globally threatened species, including the Red-breasted Goose, from these impacts. The ensuing rectification process is still ongoing.

In November 2021, the Commission referred Bulgaria to the EJC of the European Union for failing to protect and manage its Natura 2000 sites and for not respecting key obligations under the Habitats Directive, because of failures to designate SACs within the legally required period, and because Bulgaria 'generally and persistently failed to set site-specific detailed conservation objectives and measures for any of them' (European Commission 2021).[26] At the end of 2021, management planning had been completed on 10.4% of the area covered by Natura 2000 sites (MoEW, 2021).

Another problem has been failure to apply the legal protection to Natura 2000 sites in the face of business investment projects, which may be due to insufficient capacity and understanding of the law, to political pressure on the responsible administration (local, regional or central) or to corruption. According to a targeted study by NGOs (BSPB, 2019, pers. comm.), only 0.7% of project applications in Natura 2000 sites were subject to a complete appropriate assessment procedure. Such procedures were screened out for 87% of all project applications, with a subjective and non-transparent decision by the competent authority that 'an appropriate assessment was not necessary due to an absence of expected significant impacts'. Over 2007–2017, at least 74 925 approvals for investment proposals were granted as part of an appropriate assessment procedure and not a single project was rejected due to expected significant impacts (BSPB, 2019, pers. comm.).

An audit performed in 2019 by the National Audit Office (BNAO, 2019) concluded that governance of Natura 2000 was ineffective, for the following reasons: (1) unclearly assigned responsibilities for the establishment and management of the network; (2) poor communication and coordination between the participating bodies at national, regional and municipal level and the non-governmental sector; (3) failure to implement many of the management measures foreseen in the national Priority Action Framework (PAF); (4) a non-functional system for monitoring and evaluation of the implementation of the PAF; and (5) lack of reliable financial information and accountability on the financing of the management measures and related public expenditure and their outcomes.

8.6 General Nature Conservation Measures

8.6.1 General Provisions of the Environmental Legislation

As described in Section 8.3.2, the Environmental Protection Act[27] (2002) and Biological Diversity Act are the principal laws relating to nature conservation, including general measures such as EIA and SEA. The *National Development Programme Bulgaria 2020*[28] is the leading strategic and programming document detailing the objectives of the development policies of the country to 2020. It acknowledges that 'the preserved nature in Bulgaria and the unique biodiversity are prerequisite not only for the provision of a supportive and healthy environment, but also for the development of perspective environmental industries – sustainable forms of tourism (ecological,

[24] www.moew.government.bg/bg/priroda/natura-2000/registri-za-zastiteni-zoni/obnarodvani-zapovedi-po-reda-na-chl-12-ot-zbr/ (6 May 2022)

[25] Cases C-141/14 (Kaliakra SPA) and C-97/17 (Rila SPA).

[26] https://ec.europa.eu/commission/presscorner/detail/en/ip_21_5351

[27] Закон за опазване на околната среда, ДВ бр. 91 (2002).

[28] www.strategy.bg/StrategicDocuments/View.aspx?Id=765

health, spa, etc.), organic farming, protection of lands with high natural value and protected areas, traditional farming, herbs-gathering, etc'.

Spatial planning is regulated by the Spatial Planning Act[29] (2001), a complex piece of legislation with numerous amendments. It is the responsibility of municipalities to develop local spatial plans; the Ministry of Regional Development and Public Works approves them and oversees their implementation. The complex procedures have shown many loopholes as far as the conservation of domestic protected areas and Natura 2000 sites is concerned, especially in the absence of site management plans and conservation objectives.

8.6.2 Habitat Management and Restoration

The concept of habitat management and restoration was first introduced in legislation by the Biological Diversity Act (2002) in relation to the designation of Natura 2000 sites and achievement of the favourable conservation status of habitats and species.

Outside of the Natura 2000 network, and to ensure connectivity between protected areas, the Biological Diversity Act stipulates the need for 'measures and activities for protection of the elements of the landscape, which on the basis of their linear and continuous structure or connecting function are significant for the migration, geographical distribution and genetic exchange in the plant and animal populations and species'. The elements of the landscape that are considered are wide-ranging, comprising:

1. Rivers and their banks and flooded old riverbeds.
2. Natural swamps, lakes, wet meadows and other wetlands.
3. Caves, rock ridges, cliffs and dunes.
4. Mountain passes and other natural territories, connecting separate mountain massifs.
5. Field borders, field protection strips, meadows and pastures.
6. Floodplain river terraces and riparian vegetation.
7. Forests located less than 500 m above sea level.

Unfortunately, the application of these well-meaning provisions in practice is very limited, due to the general problem with the application of nature protection legislation, which is not objective driven, but formalistic and reactive. This low priority is also confirmed by the absence of funds for management of green infrastructure in the PAF (see Section 8.7). Therefore, actual habitat restoration initiatives have, so far, mainly originated from non-legal instruments such as strategies, action plans and donor-funded conservation projects.

For example, in the forest sector, a definition of old-growth forests was agreed in the context of FSC certification and, in 2015, enshrined into the forest legislation with the goal of ensuring their protection. Natural, undisturbed forests cover an area of no more than 200 000 ha, or about 4.7% of all forests in the country (WWF 2014).[30] At the end of 2016, a conservation campaign for the protection of these forests was supported by MAFF with a special ordinance by which 109 300 ha of old-growth forests in Natura 2000 sites were designated for strict protection and excluded from harvesting plans. These forests constitute a 10% increase in the forest area under strict protection in the Natura 2000 network, representing 25 out of the 27 HD forest habitats that occur in Bulgaria. This is a significant increase in strictly protected forests compared to the 79 000 ha of forests protected in Reserves in the country.

Forest landscape restoration along the Lower Danube was initiated in Bulgaria in 1999. In 2000, a joint declaration was signed by the Environment Ministers of Bulgaria, Romania, Ukraine and Moldova to establish a Lower Danube Green Corridor. This agreement commits the four countries to preserve a total of 935 000 ha of floodplain forests, including enhanced protection for 775 000 ha of existing protected areas and new protection of another 160 000 ha; to restore 223 000 ha of former wetland areas; and to promote sustainable development along the Lower Danube. The only large-scale floodplain restoration project implemented so far was the restoration of the marshes on the largest Danuban island in Bulgaria, Belene. Plans and small-scale projects are further described in online Annex BG.2.

A *National Action Plan for the Conservation of Wetlands of International Significance* (Vassilev *et al.*, 2013) was developed by the Bulgarian Biodiversity Foundation and adopted by the government with goals for protection, maintenance and restoration, as well as horizontal measures for the conservation and sustainable use of wetlands. The plan includes measures for spatial and functional re-connection of wetland habitats. It identifies 11 Ramsar Sites as priority areas for conservation and 25 additional wetlands that meet one or more of the Ramsar nomination criteria or have substantial potential for protection and restoration. Unfortunately, so far, this plan has not received financial backing and only one wetland restoration project has been implemented.

The most significant wetland restoration initiative took place in 2002–2008 prior to the national wetland plan, in the framework of a Global Environment Facility (GEF)/World Bank wetland restoration and pollution reduction project. It included the re-flooding of the largest marsh complex (385 ha in total) on the River Danube island of Persin (a Ramsar Site and IBA). As a result of the restored connections to the river, the marsh significantly improved in condition, resulting in the re-establishment of a breeding colony of Dalmatian Pelicans and other waterbirds.

8.7 Nature Conservation Costs, Benefits and Funding Sources

8.7.1 Nature Conservation Costs

Few data exist on the actual or estimated costs of nature conservation prior to the 2000s. In 2000, the World Bank

[29] Закон за устройство на територията, ДВ бр. 1 (2001).

[30] https://wwfeu.awsassets.panda.org/downloads/analysis_on_old_growth_forests_wwf_2012_2.pdf

estimated the total costs of Bulgaria's compliance with EU environmental standards to be in the range of €5.5–8 billion (World Bank, 2000) (€10.8–15.7 billion at 2020 prices compared to 2000). In the study, nature protection was lumped under the category 'others' and the average annual cost of nature conservation was estimated to be €16 million (€31 million at 2020 prices). At about the same time, 1999–2003, the *National Biodiversity Conservation Plan* made the first assessment of the necessary funding requirements for the entire nature protection system, which were estimated to be €35 million per year (€69 million at 2020 prices compared to 2000). These low figures reflect the very high inflation rate and depreciation of the national currency towards the end of the 1990s. Other factors at the time were the underdevelopment of the protected areas system, the low number of management bodies and almost complete lack of management plans and conservation objectives for the protected areas.

The assessment of the financial costs of Natura 2000 provided in the 2014–2020 PAF (MoEW, 2014) averaged €216 million per year, including €0.93 million for completing the designation of SACs, €13 million for management planning, €146 million for ongoing habitat management and monitoring and €56 million for investments. These costs reflected the priorities and objectives of the PAF, which excluded more expensive operations such as land purchase, large compensatory payments to private landowners and/or expropriation costs and large-scale restoration projects.

The PAF for 2021–2027 (MoEW, 2021) has a lower annual cost estimate of €102.3 million, of which €15.0 million is for horizontal measures and administration, €86.0 million for habitat management and restoration and €1.3 million for additional species-specific measures. No budget is specifically identified for supporting green infrastructure measures outside of Natura 2000.

8.7.2 The Economic Benefits of Nature Conservation

Direct Revenues

There is no overall national assessment or accounting of the direct revenues from various natural resources in Bulgaria, but sectoral assessments for forests, water and energy do exist. It is, however, noteworthy that non-timber forest products, including medicinal plants, are of significant economic importance, and an annual fee is gathered for their collection (unless it is for personal use). The collection of medicinal and aromatic plants is governed by special provisions based on the Medicinal Plants Act.[31] There are around 770 plant species that are considered to be medicinal in Bulgaria; some 250 of these are used in large quantities for trade and processing and represent a traditional export product that is well placed on the international markets. The collection of herbs is carried out by so-called 'herb processors', based on a corresponding permit issued for a fee. The revenues from permit fees are earmarked for the Enterprise for Management of Environmental Protection Activities (EMEPA) when the medicinal plants are gathered within National Parks, or for the respective forest enterprises, municipalities or the state budget according to the ownership of the land from which the medicinal plants are harvested.

Ecosystem Services

The role of ecosystems in service provision has been recognised in recent years in legislation and strategic plans. For example, the *National Strategy for the Development of the Forest Sector 2013–2020*[32] includes 'Measure 4.4: Establishment of conditions for sustainable and compensational use of the ecosystem services provided by the forest territories'. An attempt to evaluate in monetary terms the ecosystem services provided by floodplain forests was made for the Lower Danube Corridor.

Several coordinated projects were implemented between 2009 and 2018 to define and map green infrastructure and estimate the provision of services by the ecosystems not covered by protected areas and Natura 2000 designations. Although these studies are given as positive examples (Stanchev *et al.*, 2015), they fall short of providing an applicable monetisation, transaction and accounting approach with potential for practical implementation by national policymakers and land managers.

8.7.3 Funding Sources for Nature Conservation

The main financial sources for nature conservation in Bulgaria have been the state budget, government funds established under the Environmental Protection Act and the Forestry Act, several international and bilateral donor programmes (1990–2000), pre-accession aid in the early 2000s, and EU funds since accession in 2007 – now by far the biggest contributor (Table 8.5). The past trends and recent funding priorities are presented in the following paragraphs. As the structure of the available budgeting and financial reporting information does not reflect exactly the classification and types of environmental activities, it is not straightforward to compare periods and funding programmes. Hence, the presented information is approximate and should be interpreted with care, mainly as an indication of trends rather than precise figures.

State Budget

In 2020, the state budget included €1.26 billion national expenditure for the protection and restoration of the environment, representing 1.9% of GDP (NSI 2021[33]). Most was allocated to the management of waste (59%), wastewater

[31] Закон за лечебните растения, ДВ бр. 29 (2000).

[32] www.mzh.government.bg/bg/politiki-i-programi/politiki-i-strategii/nacionalna-strategiya-za-razvitie-na-gorskiya-sektor-v-republika/?msclkid=f928ab1eb8ec11ec95efc143dfda1c0a

[33] www.nsi.bg/en/content/2584/expenditure-protection-and-restoration-environment-use

(19%) and air pollution (14%), while the share for biodiversity/nature protection (0.3%) and forest management (0.3%) remained low. Between 2014 and 2020, state funding for biodiversity, through the budget of the MoEW, ranged between €2.8 and 4.5 million, of which personnel costs represented the bulk (70%). The programme budgets of the MoEW and MAFF present no clear data on the funds earmarked for the management of Natura 2000, which were anyway very small. This left few options for domestic protected area and Natura 2000 management due to dependency on the success of project applications and/or short-term (political) decisions of the Minister, or inadequate funds. This situation has hindered effective planning and implementation of management activities, and risks the accountability and cost-effectiveness of public spending on the Natura 2000 network (BNAO, 2019).

Enterprise for Management of Environmental Protection Activities (EMEPA)

EMEPA is a state-owned non-commercial enterprise, chaired by the MoEW, that funds (e.g. through grants or interest-free and low-interest loans) implementation of environmental projects on water management, waste management, biodiversity conservation and protection of natural ecosystems. The beneficiaries are natural and legal entities (i.e. municipalities, enterprises and companies, etc.). EMEPA's funding sources are the revenues from environmental fees, penal collections and fines, and earmarked funds from the state budget. During 2008–2015 the total revenue from all these sources amounted to €225 million, with annual expenditure on environmental projects totalling c. €32 million. Over the same period, EMEPA's spending on biodiversity-related measures varied between €21 and 410 thousand (0.3–1.5%).

EU Funds

During the pre-accession period, three main pre-accession funding instruments[34] played a significant role in improving the state of the environment in the country, including the integration of Bulgarian protected areas into the Natura 2000 network. Over 2001–2011, international projects and EU pre-accession programmes, and the EU LIFE programme, invested slightly over €50 million in the nature protection sector.

Since 2007 the reliance on EU funds for environment-related expenditure has increased significantly. Operational Programme (OP) Environment has been a particularly important instrument through which funding from ERDF and Cohesion Funds has been allocated for nature conservation purposes. During 2007–2013, under OP Environment, €92.6 million was allocated to biodiversity (UNECE, 2017). Over 2014–2020, the total OP Environment budget increased to €1.8 billion (85% EU financing), of which €103 million was allocated to Priority Axis 3 Natura 2000 and Biodiversity. However, by the end of 2018 only one project had been completed successfully under this priority and €35 million had been spent. One call funded 13 projects, with a total value of approximately €20 million, which aimed to improve the conservation status of habitats and species within Natura 2000 sites covered by National Parks, Nature Parks and Managed Reserves. Effective absorption of EU funds remains a major challenge due to weak mobilisation of co-financiers, burdensome grant application, management and monitoring procedures, and the quality of the projects.

A further major source of funding for the practical management and restoration of habitats has been the CAP RDP measures, with some €123 million per year allocated in relation to Priority 4 which covers, among others, restoring, preserving and enhancing biodiversity, including in Natura 2000 sites (Table 8.5). Of these, the agri-environment–climate measure has been particularly important for nature conservation. A successful use of the measure to help implement the Nature Directives, whilst also linking to the LIFE Programme, was a scheme focussed on conserving the Red-breasted Goose population that winters on arable land. The scheme extended conservation measures that had been developed and trialled under a LIFE project.[35] In the first year, 240 farmers applied to join the scheme, covering 18 000 ha, with funding of €1.8 million (European Commission, 2018).

The use of the RDP Natura 2000 and WFD Measure (M12) increased in the 2014–2020 programming period, providing payments to land managers operating within designated Natura 2000 sites as compensation for restrictions on the use of agricultural land and forests (e.g. retention of field borders and hedges, late mowing, bans on pesticide and mineral fertiliser use on meadows). However, so far these have only been used for farms in SPAs, because the payments need to be linked to specific restrictions arising from the site's conservation objectives and related management measures, which have yet to be devised for SACs. The RDP measure that received the largest proportion of funding (33%) was payments for areas with natural and other constraints (M13). Although the payments are not linked to any specific obligations on the farmer, they could be helping to reduce the abandonment of farmland; however, their contribution to nature conservation is uncertain. It is notable that, despite the importance of Bulgaria's forests, no budget allocation was given under Priority 4 for the forest environment climate measure (M15).

The use of the European Maritime and Fisheries Fund (EMFF) for Natura 2000 objectives was rather limited over 2014–2020 because no marine SCIs have been fully designated yet. Funding from the EMFF has mostly been used to support aquaculture, port reconstruction and decommissioning of small vessels.

[34] Poland and Hungary Assistance for Restructuring their Economies (PHARE), Instrument for Structural Policies for Pre-Accession (ISPA), Special Accession Programme for Agriculture and Rural Development (SAPARD).

[35] LIFE09 NAT/BG/000230 Safe Ground Redbreasts.

Table 8.5 Funding sources for nature conservation in Bulgaria.

Source and period	Measure/category	Average annual allocation (BGN m)		Average annual allocation (€ m)		Comments
EU funding instruments		**EU + national**	**EU**	**EU + national**	**EU**	
EAFRD Rural Development Programme measures – allocations to Priority 4 (restoring, preserving and enhancing agriculture and forest ecosystems) 2014–2020	M13 Areas facing constraints	79.4	59.6	40.6	30.5	
	M10 Agri-environment–climate	57.4	43.0	29.3	22.0	
	M11 Organic farming	53.5	40.1	27.4	20.5	
	M12 Natura and WFD	39.0	29.3	20.0	15.0	
	M8 Investments in forests	4.8	4.1	2.4	2.1	
	M4 Investments in physical assets	4.3	3.2	2.2	1.6	See 1
	M16 Cooperation	1.4	1.3	0.7	0.7	
	M1 Knowledge transfer, etc.	1.0	0.9	0.5	0.5	
	M2 Advisory services, etc.	0.4	0.3	0.2	0.2	
RDP subtotal		*241.2*	*181.7*	*123.3*	*92.9*	
ERDF/Cohesion Fund 2014–2020	85 Protection and enhancement of biodiversity, nature protection and green infrastructure	?	?	?	?	For N2k, etc.
	86 Protection, restoration and sustainable use of Natura 2000	11.7	10.0	6.0	5.1	
Interreg		?	?	?	?	
EMFF 2014–2020	Priorities 1–7	?	?	?	?	
LIFE 2007–2000	Nature and biodiversity	10.0	6.8	5.1	3.5	37 projects
EU funding instrument subtotal		>262.9	>198.5	>134.4	>101.5	
National funds within 2014–2020		**Additional non-EU* (BGN)**		**Additional non-EU* (€ m)**		**Comments**
Public						
State budget of the MoEW 2014–2020	Personnel (70%) and running costs	7.0		3.6		
Enterprise Environmental Management 2014–2020		0.5		0.25		Range over €0.2–0.3
Others						
Bulgarian–Swiss Cooperation Programme 2011–2019		15.6		8.0		
EEA Financial Mechanism 2014–2018		2.2		1.1		
Additional national and others subtotal		*25.3*		*13.0*		
National total including co-financing of EU instruments		>71.8		>36.7		
Total		**>288.2**		**>147.3**		

Notes. * Excludes co-financing of EU funds. 'For N2k, etc.' indicates funding relevant to Natura 2000, related green infrastructure and BHD species (based on the PAF). The subtotals and totals are only approximate indications of the average annual levels of funding as they are based on annualised data from differing periods. Currency conversion rate (29 May 2021): €1 = BGN (Bulgarian lev) 1.95. 1. The PAF 2021–2027 includes a much higher allocation for M4, but this does not appear to be under Priority 4 for ecosystems, etc., and is for broader agricultural and environmental issues.
Sources. Based on RDP data from: European Commission ESIF funds database (23 October 2020); Bulgaria LIFE factsheet (European Commission, March, 2022); PAF 2021–2027 (MoEW, 2021), National Statistical Institute. See Section 1.3 for further details.

Other Important Funding Sources

Over 2004–2009, the European Economic Area (EEA) Financial Mechanism (Norway, Iceland and Liechtenstein) provided €16 million in grants for energy, biodiversity and environment protection. Out of that, €1 million was used for biodiversity (Tisma *et al.*, 2012). In 2011, a new agreement was concluded providing €8 million for biodiversity projects. National co-financing amounted to €1.4 million. The implementation period ended in April 2017. For 2016–2024 the Programme 'Environment, Climate Change Adaptation and Ecosystems'[36] has a budget of €20 million, but it is not known what part of this will be used for financing biodiversity projects.

Overall Funding Adequacy

Given currently available information, it is not possible to estimate reliably all the funding allocations in Table 8.5 and calculate an overall level of recent funding for nature conservation in Bulgaria. This is primarily because the most up-to-date information available for the 2014–2020 funding period, from the PAF for 2021–2027, gives implausibly high allocations for Natura 2000 under ERDF/Cohesion funding Priority 85. Although the PAF is only supposed to include ERDF/Cohesion funding of relevance to Natura 2000, it seems to include funding for other objectives, such as regional tourism and transport infrastructure, competitiveness, energy efficiency and education. Similarly, the PAF has a high allocation for Natura 2000 related measures under the EMFF, but includes mainly activities that would not benefit nature conservation, such as (re)construction of fishing ports and landing facilities, development of aquaculture and decommissioning of vessels.

Consequently, the overall adequacy of potentially available funding for nature conservation is uncertain, including in relation to the requirements estimated in the current PAF for Natura 2000. Whilst the minimum annual total of €147 million appears to be significantly more than the PAF estimated requirement of €102 million per year, it is important to bear in mind that this covers wider nature conservation funding than just Natura 2000, and the PAF does not include funding needs for wider green infrastructure. Furthermore, a high proportion of the funding is from RDP payments to areas with natural constraints and organic farmers, with uncertain biodiversity benefits. An additional funding gap is also likely due to the need for further funding for national nature conservation requirements beyond those of Natura 2000 and wider measures under the Nature Directives.

It is also important to bear in mind that, as discussed earlier, a significant proportion of funds that have been allocated to nature conservation have not been fully utilised. The main impediment to this has come from the incomplete management planning process and the inadequate management capacity of the responsible institutions.

8.8 Surveillance, Monitoring, Reporting and Applied Research

Natural history research in Bulgaria dates back to the nineteenth century and was initially carried out by visiting European authors. In 1889, the Royal (now National) Museum of Natural History was founded and continues to be the oldest and the richest museum of this kind in the Balkans (Fet and Popov, 2007). Over time, several research institutes were established under the umbrella of the BAS, nowadays the Institute for Biodiversity and Ecosystem Research, the Institute of Oceanology and the Institute of Forests, among others. Their role in the study and evaluation of biodiversity status and trends and in management planning is potentially very important, but has not been effectively utilised (as illustrated in the case of the Nature Directives reporting). Applied research is also carried out at Sofia University 'St. Kliment Ochridski' and the Forestry Technical University, as well as in the smaller universities of Plovdiv and Shumen.

The rather complex NBMS[37] was established in the 2010s whereby ExEA is responsible for the organisation, collection, processing and storing of data. The species currently covered are: 14 fungi, 13 mosses (6 spp. and 1 genus HD taxa), 199 species of higher plants (26 spp. and 1 genus HD taxa), 156 invertebrates (52 HD species), 176 non-bird vertebrates (135 HD species) and 414 birds. In addition, 93 HD habitat types and 22 others of national importance (identified by EUNIS classification codes) are included. However, it is important to note that, despite significant experience with monitoring some species, many of the official methodologies included in the NBMS are new, insufficiently mature and require significant expert capacity and resources, which are not readily available. Hence, the implementation of the NBMS depends on the cooperative participation of many experts from the scientific institutes of the BAS, universities and NGOs. Although funding for such collective effort is available through EU funds, its full implementation is limited as a collaborative framework and effective leadership and coordination by the state authorities are lacking.

Bird monitoring schemes are implemented mainly by NGOs. For example, BSPB/BirdLife Bulgaria has run a common breeding bird monitoring scheme since 2004, which produces the Common Bird Index (Spasov *et al.*, 2017). Single species monitoring schemes for a small number of species subject to long-term conservation programmes, such as the Eastern Imperial Eagle (*Aquila heliaca*), vultures, Saker Falcon and Red-breasted Goose, gather information on their

[36] https://eeagrants.org/sites/default/files/resources/Programme%20agreement%20RO-ENVIRONMENT.pdf

[37] Национална система за мониторинг на биологичното разнообразие (National Biodiversity Monitoring System).

distribution, numbers, breeding success, productivity, diet and movements, etc., mainly as part of LIFE projects.

8.9 Future Developments

Since 2018, the MoEW has promoted a new approach to Natura 2000 management with the project 'Management Approach for Natura 2000' funded by EU Operational Programme – Environment 2014–2020. It aims to establish a new management system, to increase institutional capacity, to decentralise management responsibilities and to increase the involvement of local communities (Consortium Communication about Natura 2000, 2019) through a two-tier governance model, including National and Regional Network Management Authorities. Stakeholder involvement is to be promoted through a National Natura 2000 Advisory Board and Stakeholder Committees at the regional level. Management planning for Natura 2000 is to become mandatory and a four-level approach to setting conservation objectives developed: at biogeographical (for HD habitats and species) or national level (birds), at the network level, at site level and at specific localities. A national document incorporating the Natura 2000 conservation objectives and 28 regional management plans (corresponding to the 28 administrative regions) is also to be developed.

However, the introduction of the new Natura 2000 management approach requires changes to the Biological Diversity Act, which have been vigorously opposed by the main conservation NGOs due to a lack of trust and cooperation with the Ministry and the attempts to isolate the scientific organisations on substantial issues. At the time of writing, in early 2022, this process had been revived by the new government, elected in 2021, although it has focussed on more immediate biodiversity-related priorities concerning the ongoing infringement procedures.

8.10 Conclusions

Despite the modern legal framework adopted during the EU accession period and the large coverage of the Natura 2000 network, Bulgaria has faced serious challenges in the governance and effectiveness of its institutions responsible for nature conservation and biodiversity. The momentum of the 1990s and early 2000s seems to have been lost. This is despite, or perhaps a result of, the significant financial resources that were attached to environmental policies by the EU, which led to heavy political and economic pressure on the responsible institutions and their failure to address public interests and follow up on declared priorities on numerous occasions. The contradictory policies of the MoEW under several ministers also undermined the Natura 2000 network (demonstrated by numerous public statements over the years by the governing political parties and the ministers). This has led to a low level of trust, a lack of cooperation, and bitterness that dominate the relationships with key stakeholders from the scientific and conservation community, whose cooperation the Ministry needs to support its work.

The undesired, but unavoidable, consequences of these unproductive relationships have resulted in numerous conflicts and infringement procedures, as well as the complete block on the development of an effective Natura 2000 governance and management model based on a vision, strategy and clear objectives supported by the stakeholders. Consequently, much of the funding for Natura 2000 remains underused or misused. The funds earmarked in the state budget for biodiversity conservation cover basic running costs and are not sufficient for management planning and implementation, or for investments such as the acquisition of especially valuable or threatened sites and/or restoration of habitats. The PAF for 2014–2020 focussed almost entirely on the allocation of EU funds, but this was ineffective as few of the planned management, conservation and restoration measures were actually implemented. In the absence of objective-driven management plans and implementation capacity, even if potentially sufficient funds became available, they would remain unspent or misspent. It is, therefore, important to reformulate the national policy in this direction, so that the Natura 2000 network and Bulgaria's rich natural capital receive the attention and consideration they deserve, including as profitable investments in social well-being and a sustainable economy.

Acknowledgements

Mira Mileva and Dimitar Stoev for providing historical and recent data and valuable insights, and to Petar Iankov and Pavel Stoev for their review of the draft text.

References

Apostolova, I., Sopotlieva, D., Velev, N. & Vassilev, V. (2018) Ecosystem condition assessment of semi-natural grasslands outside the Natura 2000 network in Bulgaria, using vegetation data. *Tuexenia*, **38**, 385–404.

BNAO (2019) *Audit Report No 0300101818 auditing the implementation of "Management effectiveness of the Natura 2000 network aiming to protect the environment and the local population in the zones covered by the network" for the period 01.01.2016–31.12.2018*. Sofia: Bulgarian National Audit Office.

Consortium Communication about Natura 2000 (2019) Conservation objectives for Natura 2000. Version for consultation. Project report BG16M1OP002-3.006-0001-C02 Knowledge about Natura 2000. Sofia.

Dobrev, V., Popgeorgiev, G. & Plachiyski, D. (2014) Effects of the Common Agricultural Policy on the Coverage of Grassland Habitats in Besaparski Ridove Special Protection Area (Natura 2000), Southern Bulgaria. *Acta Zoologica Bulgarica*, **2014**, 147–155.

European Commission (2018) 25 years of LIFE and still going strong... *Natura 2000 Newsletter*, **43**, 10–13.

(2019a) *Environmental Implementation Review 2019: A Europe that protects its citizens and enhances their quality of life*. COM (2019) 149 final.

(2019b) *EU Environmental Implementation Review 2019 Country Report – BULGARIA*. SWD(2019) 113 final.

Fet, V. & Popov, A. (eds.) (2007) *Biogeography and Ecology of Bulgaria*. Dordrecht, The Netherlands: Springer.

Ganchev, I., Bondev, I. & Ganchev, S. (1964) [Vegetation of meadows and pastures in Bulgaria.] (In Bulgarian). Sofia: Bulgarian Academy of Sciences Press.

Gussev, C., Tzonev, R. & Dimitrov, M. (2016) The future of submeasure "Pastoralism" of Measure 214 "Agroecological payments" in the Rural Development Programme of Bulgaria: advantages, disadvantages and challenges. *Phytologia Balcanica: International Journal of Balkan Flora and Vegetation*, **22**, 167–177.

Josifov, M. (1988) Über den zoogeographischen Charakter der südeuropäischen Insektenfauna unter besonderer Berücksichtigung der Heteropteren. *Berichte des naturwissenschaftlich-medizinischen Vereins*, **75**, 177–184.

Krastanova, R. (2012) *Green Movement and Green Parties in Bulgaria: Between System Integration and System Change*. Sofia: Friedrich Ebert Foundation.

Meshinev, T., Apostolova, I., Georgiev, G., Dimitrov, V., Petrova, A. & Veen P. (2005) *Grasslands of Bulgaria. Final report on the National Grasslands Inventory Project – Bulgaria, 2001–2004 (PINMATRA/2001/020)*. Sofia: Dragon 2003 Publishers.

Ministry of Environment and Water, MoEW (2014) Министерство на околната среда и водите, Национална приоритетна рамка за действие (НПРД) за Натура 2000. България. За многогодишен финансов период на ЕС 2014-2020 (Актуализирана версия, август 2019 г). [National Priority Action Framework for Natura 2000. Bulgaria. For the EU financial period 2014-2020. Updated version, August 2019.]

(2021) Министерство на околната среда и водите, Рамка за приоритетни действия (РПД) за Натура 2000 в България съгласно член 8 от Директива 92/43/ЕИО на Съвета за опазване на естествените местообитания и на дивата флора и фауна (Директива за местообитанията) за многогодишната финансова рамка за периода 2021–2027 г. [National Priority Action Framework for Natura 2000. Bulgaria. For the EU financial period 2021–2027.]

National Statistical Institute. (2019) Statistical Yearbook. Republic of Bulgaria. www.nsi.bg/sites/default/files/files/publications/StatBook2019.pdf

Petrova, A. & Vladimirov, V. (2009) Red List of Bulgarian Vascular Plants. *Phytologia Balcanica*, **15** (1), 63–94.

(2010) Balkan endemics in the Bulgarian flora. *Phytologia Balcanica*, **16** (2), 293–311.

Popgeorgiev, G., Tzankov, N., Kornilev, Y., Plachiyski, D., Naumov, B. & Stoyanov, A. (2014) Changes in agri-environmental practices pose a threat to the herpetofauna: a case study from Besaparski Ridove Special Protection Area (Natura 2000), Southern Bulgaria. *Acta Zoologica Bulgarica Suppl.*, **5**, 157–169.

Queiroz, C., Beilin, R., Folke, C. & Lindborg, R. (2014) Farmland abandonment: threat or opportunity for biodiversity conservation? A global review. *Frontiers in Ecology and the Environment*, **12**, 288–296.

Rangelova, R. & Vladimirova, K. (2017) Agricultural sector in Bulgaria during the transition to market economy and the integration into the European Union. *Agricultural and Resource Economics: International Scientific E-Journal*, **3**, 30–43.

RDB (2011) *Red Data Book of the Republic of Bulgaria*. Three volumes. Sofia: Bulgarian Academy of Sciences & Ministry of Environment and Water. Digital edition: http://e-ecodb.bas.bg/rdb/en/

Roussakova, V. (2011) Natural habitats as a part of the biological diversity of Bulgaria. In *Red Data Book of the Republic of Bulgaria*. Volume 3. Sofia: Bulgarian Academy of Sciences & Ministry of Environment and Water. Digital edition: http://e-ecodb.bas.bg/rdb/en/

Spasov, S., Arkumarev, V., Dobrev, D. & Dobrev, V. (2012) An overview of monitoring for raptors in Bulgaria. *Acrocephalus*, **33** (154–155), 181–189.

Spasov, S., Hristov, I., Eaton, M. & Nikolov, S. (2017) Population trends of common birds in Bulgaria: is their status improving after the EU Accession? *Acta Zoologica Bulgarica*, **69**, 95–104.

Stanchev, R., Vergiev, S., Gocheva, K., Nedkov, S. & Bratanova-Doncheva, S. (2015) *Country Fact Sheet: Bulgaria (BG)*. ESMERALDA – Enhancing ecosystem services mapping for policy and decision making. https://catalogue.biodiversity.europa.eu/uploads/document/file/1302/Esmeralda_country_fact_sheet_Bulgaria.pdf

Stefanova, V. & Kazakova, Y. (2015) *Country report on the implementation of the new CAP and its possible effects on permanent pastures: Bulgaria*. Lampeter, UK: Report for the European Forum on Nature Conservation and Pastoralism.

Stoynov, E. (2019) Reintroduction of the Griffon Vulture (*Gyps fulvus*) in model territories in Bulgaria (in Bulgarian). PhD thesis, Sofia University. doi: 10.13140/RG.2.2.18269.67043.

Tisma, S., Boromisa, A-M. & Kaselj, A. P. (2012) *Environment Finance and Development*. London and New York: Routledge.

UNECE (2017) *Environmental Performance Reviews. Bulgaria*. Third Review. Environmental Performance Reviews Series No. 46. New York and Geneva: United Nations Economic Commission for Europe.

UNEP-WCMC (2022) *Protected Area Profile for Bulgaria from the World Database of Protected Areas, April 2022*. www.protectedplanet.net

Vassilev, V., Vassilev, R., Yankov, P. *et al.* (2013) *NATIONAL ACTION PLAN for Conservation of Wetlands of High Significance in Bulgaria 2013-2022*. [Bulgarian/English]. Sofia: Bulgarian Biodiversity Foundation.

World Bank (2000) *Bulgaria: The Challenges of Complying with EU Environmental Directives*. Washington, DC: World Bank.

WWF-Bulgaria (2018) *Анализ на незаконния дърводобив в България и ефективността на мерките по контрол в горските територии*. [Analysis of the illegal logging in Bulgaria and the effectiveness of control measures in forests.] Sofia: WWF-Bulgaria.

(2019) *After-LIFE Conservation Plan for LIFE13 NAT/BG/000801 Project: Conservation and restoration of riparian forests (habitat type *91E0) in Natura 2000 sites and model areas in Bulgaria*. Sofia: WWF-Bulgaria.

Chapter 9

Republic of Croatia

Jasminka Radović

Key Statistics

Terrestrial area (km²)/% of EU-28*	56 434/1.3%
Marine waters (km²)/% of EU-28*	55 492/1.2%
Population density (p km²)/% of EU-28*	74/65%
GDP per capita (€)/% of EU-28	10 979/41%
Land cover (CORINE land cover classes 2018)	AGR 40%, FOR 35%, OSN 20%, ART 4%, WET 1%
Biogeographical regions	ALP, CON, MED, MMED
Number of HD Annex I habitats/%*	74/33%
Number of HD Annex II, IV and V species/%*	240/20%
Number of BD Annex I bird taxa/%*	91/52%
% HD UCS and decreasing habitats/ % unknown	10%/14%
% HD UCS and decreasing species/ % unknown	3%/47%
% breeding BD Annex I birds decreasing % uncertain and unknown	11%/62%

* Excluding MAC and MMAC biogeographical regions.
Sources and codes: See Table 1.1.

9.1 The Natural History and Characteristics of Croatia

Croatia has a very diverse range of natural features in its relatively small territory. With its continental and Mediterranean climatic influences and biogeographical regions, and Alpine belt in-between, it encompasses not only the characteristic vegetation of these three zones, but also many kinds of rather specific transitional habitats with their accompanying species. It also has a high proportion of endemic species (2.7%) (ZZOP, 2019), most of which are freshwater fishes of the Adriatic basin rivers, vascular plants of calcareous rocky habitats and, especially, subterranean invertebrates – with 70% being endemic taxa with extremely restricted distributions (DZZP, 2014).

The Continental, Alpine and Mediterranean biogeographical regions stretch through Croatia as three belts, reflecting their three main natural, geomorphological and socio-geographical characteristics.

The **Continental** biogeographical region covers almost 55% of Croatia and includes the lowland areas between the rivers Kupa, Sava, Drava and Danube, interspersed with oak–hornbeam and beech forest–covered hills and mountains up to about 1 000 m. At one time all of the region's lowlands were covered with forests, except for river valley wetlands, but now urban and agricultural landscapes prevail, including the capital of Zagreb. Nevertheless, some natural areas remain, such as the large floodplains along the River Sava with riparian mixed forests of European Oak (*Quercus robur*) and Narrow-leaved Ash (*Fraxinus angustifolia*) (91F0),[1] the willow–poplar galleries (91E0*) along the Drava River (which is still rich in hydromorphic features, such as gravel islands and old oxbows) and the confluence of the Drava and River Danube at Kopački Rit, which is a key stop-over site for migratory waterbirds. The White-tailed Sea-eagle (*Haliaeetus albicilla*), Black Stork (*Ciconia nigra*), Spoonbill (*Platalea leucorodia*) and Ferruginous Duck (*Aythya nyroca*) are a few of the species that have important populations in this lowland area. Such species have also benefited from the creation of extensive complexes of carp fishponds, which often become overgrown with aquatic and marsh vegetation.

The **Alpine** biogeographical region, situated in the heart of Croatia, covers about 15% of the country but contains only 2% of the population. It remains the most natural region, being almost completely covered with dense complexes of beech and beech–fir forests (91K0). This practically unfragmented zone connects the Alpine and Dinaric mountain chains as well as their populations of large carnivores: the Brown Bear (*Ursus arctos*), Wolf (*Canis lupus*) and Eurasian Lynx (*Lynx lynx*). As a result of its porous calcareous bedrock, it is very rich in underground habitats like caves and sinkholes, with lakes and water courses, inhabited by numerous endemic cave troglobionts (i.e. species restricted to caves). Large karst *poljes*[2] are

[1] Numbers in brackets are the HD Annex I habitat codes (see Appendix for full names). * Indicates a priority habitat.

[2] Karst polje is a large plain in the karst region with a specific water regime, often connected to intermittent rivers (ponornice) which

covered with wet *Molinia* grasslands (6410), while the flora of mountain rocky habitats abounds with tertiary and glacial endemics and relics. The unique biogenic tufa cascades (i.e. carbonate precipitates with algae, bryophytes and ferns, etc.) of the region were added to HD Annex I as the new habitat type 'Tufa cascades of karst rivers of the Dinaric Alps' (32A0) when Croatia joined the EU.

The **Mediterranean** biogeographical region occurs within the coastal belt, including the coastal flanks of the Velebit, Dinara and Biokovo mountains. Centuries of forest clearance and livestock grazing have created a landscape dominated by eastern sub-Mediterranean dry grasslands (62A0) and eu-Mediterranean dry grasslands[3] (6220*). However, in recent times livestock have become rare due to agricultural abandonment. As a result, large areas of these grasslands are becoming overgrown with maquis and other shrubs. In the sub-Mediterranean part of this region karst *poljes* are also present, with the distinctive vegetation of 'sub-Mediterranean grasslands of the *Molinio-Hordeion secalini*' (6540, added to Annex I by Croatia). Although small, the karstic rivers of the Adriatic basin contain 15 endemic fishes. More than 90% of the eastern Adriatic coast is rocky and highly indented, with over 1 000 islands and many sea cliffs, and is important for its endemic flora, reptiles and invertebrates. Although the coast is under ever-growing pressure from urbanisation and tourism, valuable remnants of diverse small vulnerable habitats remain, such as Mediterranean wetlands, sand and shingle beaches, sandflats, mudflats and complex estuarine habitats.

The **marine** ecosystems of Croatia also add to its rich biodiversity. These include extensive *Posidonia* beds (1120*) and well-preserved reefs (1170), as well as several types of submerged karst that are exclusive to the east Adriatic coast. The offshore zone is important habitat for, amongst others, the Bottlenose Dolphin (*Tursiops truncatus*), Loggerhead Turtle (*Caretta caretta*) and seabirds that breed on remote islands.

9.2 The Status of Nature

Croatia is still developing its national monitoring framework and there is a significant lack of data for a number of habitats and species. In 2019, the first national reports under the Habitats and Birds Directives were prepared. Due to insufficient information, the conservation status and trends of a very high proportion of species were assessed as unknown. The situation with habitat types is somewhat better, thanks to two recently completed habitat mapping projects from 2004[4] and 2016.[5]

include underground flows linked with the surface through sinkholes (ponori). In springtime these areas are mostly wet, while during the dry season water levels fall below ground level.

3 As known in Croatia, but called 'Pseudo-steppe with grasses and annuals of the *Thero-Brachypodietea*' in HD Annex I.

4 www.haop.hr/hr/baze-i-portali/karta-stanista-rh-2004

5 www.haop.hr/hr/baze-i-portali/karta-kopnenih-nesumskih-stanista-republike-hrvatske-2016

9.2.1 Habitat Status and Trends

Based on the HD Article 17 national report for 2013–2018,[6] 39% of the 125 assessments of HD habitats across the four biogeographical regions revealed a favourable conservation status, 26% a poor status and 21% a bad status; 14% had an unknown status. Although only a small proportion of those with a poor status were reported as having a declining trend, 42% of those with a bad status and known trend were declining (Table 9.1). Furthermore, nearly a third of habitat assessments amongst those with a poor or bad status had unknown trends.

Among the seven marine habitats, four had a poor status, including sea caves (8330) as well as *Posidonia* beds and reefs, which are locally threatened. The others, as well as coastal lagoons (1150*), had an unknown status, although they are exposed to numerous anthropogenic pressures related to the development and intensive use of the coastal area. The coastal wetland habitats (salt marshes) are mainly within small sites that are highly exposed to anthropogenic impacts and most had a poor conservation status. There is a similar situation with coastal shifting dunes (2110) and inland dunes (2340*), which were reported as having a bad status.

Croatia has a high diversity of grassland habitats (dry, wet and mesophilic), most of which were reported as having an unfavourable status, especially in the Continental region. Moreover, many had a bad status in all regions. These include steppic habitats represented within only a few small localities that are on the edge of their natural distribution (1530*, 6240*, 6250*, 6260*), and some other grasslands that have significantly decreased (e.g. 6230*). On the other hand, two dry grasslands had a favourable status (6540, 6220*), as did eastern sub-Mediterranean dry grasslands (62A0) in the Alpine Region, and Alpine and subalpine grasslands (6170).

Bogs and fens are rare in Croatia and occur within very small localities with special microclimate conditions. They were all considered to have an unfavourable status as they are threatened with vegetation succession. River habitats dependent on natural dynamics (3230, 3260, 3270, 32A0) as well as caves (8310) also had an unfavourable status in all regions, except for 3260 in the Alpine (favourable status) and Continental (unknown status) regions.

Habitats that generally had a favourable status were rock and scree habitats, as well as heathland and shrubland, except for declining and threatened dry heaths (4030). Most of Croatia's forest types also had a favourable status in most biogeographical regions. However, the European Oak forests (91F0), oak or oak–hornbeam forests (9160) and mixed forests of screes and ravines (9180*) had a bad status in the Continental region. Montane spruce forests (9410) also had a bad conservation status.

6 See Section 1.3 for background on monitoring under the Nature Directives, status categories and data sources.

Table 9.1 Short-term trends (c. 2007–2018) in Habitats Directive Annex I habitats and Annex II, IV and V species with an unfavourable conservation status in Croatia.

	Poor (unfavourable– inadequate)				Bad (unfavourable–bad)			
	Imp	Sta	Dec	?	Imp	Sta	Dec	?
HD Habitat assessments								
Number	0	21	1	10	0	7	11	8
%	0%	66%	3%	31%	0%	27%	42%	31%
HD Species assessments								
Number	2	0	14	84	0	14	0	78
%	2%	0%	14%	84%	0%	15%	0%	85%

Key. Imp = improving, Sta = stable, Dec = decreasing, ? = unknown.
Source. Based on Member State HD Article 17 reported data as presented on the EEA national summary dashboard (2021).

9.2.2 Species Status and Trends

The HD Article 17 report for 2013–2018 indicates that only 7% of assessments of HD species across all biogeographical regions showed a favourable conservation status, and most (47%) resulted in an unknown status. Despite the limited knowledge, 24% were considered to have a poor status, and 22% a bad status. Of those with an unfavourable assessment, the vast majority had unknown short-term trends (Table 9.1). Nevertheless, 14% of those with a poor status were considered to have decreased.

According to the BD Article 12 assessment of bird population in Croatia, the short-term trends of most Annex I species are also uncertain or unknown (Table 9.2). Whilst most with known trends had increasing, stable or fluctuating populations, the breeding populations of eight species had decreased, including Ferruginous Duck, Little Tern (*Sternula albifrons*), Corncrake (*Crex crex*) and Capercaillie (*Tetrao urogallus*).

The national threat status of 8% of recorded species has been assessed in relation to IUCN criteria and 18 Red Lists prepared (ZZOP, 2019), including all groups of vertebrates and several groups of invertebrates. Results show that 40% of assessed species belong to one of the three threatened categories (CR, EN and VU), while 21% have been identified as Data Deficient. The most threatened groups are corals, freshwater fishes, decapodes (crabs and crayfish, etc.) and cave fauna. However, it should be borne in mind that not all known species were evaluated in these groups, but mainly those considered to be the most threatened. All species of amphibians, reptiles and birds have been assessed, amongst which 23% of breeding birds and 17% of amphibians and reptiles are considered to be nationally threatened.

Table 9.2 Short-term trends (c. 2007–2018) in Birds Directive Annex I taxa in Croatia.

	Breeding				Wintering			
	Inc	Sta/ Flu	Dec	?	Inc	Sta/ Flu	Dec	?
Number	5	16	8	47	2	2	0	17
%	7%	21%	11%	62%	10%	10%	0%	81%

Key. Inc = increasing, Sta = stable, Flu = fluctuating, Dec = decreasing, ? = uncertain and unknown.
Source. Based on Member State Article 12 reported data as presented on the EEA national summary dashboard (2021).

9.2.3 Drivers and Causes of Change

According to the Report on the State of Nature in the Republic of Croatia for 2013–2017 (ZZOP, 2019), nature in Croatia has been, and continues to be, most threatened by human activities, especially through the alteration of natural ecosystems, the use of biological resources, and pollution. Particular problems have arisen from water management and use (e.g. the building of dams), agricultural and municipal wastewater, as well as from construction of residential and urban areas. The regulation of water courses, which causes changes in water regimes, is highly detrimental to all water-dependent habitat types, especially river habitats and riparian forests, particularly those of the European Oak (91F0).

The process of vegetation succession, and resulting changes in species composition, following the abandonment of traditional agricultural systems and practices, has been the dominant pressure affecting open habitats, like grasslands (ZZOP, 2019). This longstanding pressure has resulted in the overgrowth and disappearance of many pastures, meadows and small-scale mosaics of arable land. However, since Croatia's accession to the EU, the situation has changed, with fallow land and often also grasslands being turned into arable land or orchards. On the other hand, some EU CAP measures have helped to maintain grasslands, as further discussed in Section 9.6.

Building and tourism-related activities have significant negative impacts on a number of small, threatened coastal habitats, including beaches, mudflats and sandflats, coastal lagoons, Mediterranean salt meadows and halophilous (salt-tolerant) scrub. Marine habitats of *Posidonia* beds and reefs are locally threatened by fishing, nautical tourism and mariculture.

Certain forest habitats have been affected by changes in their structure and composition, at the expense of old forest stands. Fragmentation and isolation of forest stands is another continuous pressure (ZZOP, 2019). In recent years, stands of the Narrow-leaved Ash have declined rapidly due to infection with the pathogenic fungus *Chalara fraxinea*.

The most substantial pressures affecting wild species in Croatia have resulted from the degradation of their habitats (as described previously), while ever-growing threats are arising from the spread of new invasive alien species (IAS) and increasing climate change issues. In 2017, 907 alien species were recorded in Croatia, the majority of them being vascular plants, followed by insects, fishes, crustaceans and mammals. Of these alien species, 14% are considered to be IAS. Recently, 17 IAS from the list of IAS of EU concern have been recorded in Croatia (ZZOP, 2019). Although climate change issues are poorly understood, they have been identified as a significant threat to many Red List species, especially butterflies, dragonflies, decapodes and amphibians.

A number of animal species, especially birds of prey and bats, are also affected by numerous wind power projects (Anonymous, 2017). Furthermore, the cumulative effects of these projects are becoming increasingly significant, as future plans are not always being adequately considered in strategic impact assessments. Another pressure on birds has been from illegal killing, including by hunters, which has affected more Red List bird species in Croatia than any other identified threat (Tutiš *et al.*, 2013).

9.3 Nature Conservation Governance, Actors, Legislation and Policies

9.3.1 Overview of Nature Conservation in Croatia

The beginnings of nature conservation–related activities can be traced to the end of the nineteenth century when a number of initiatives from naturalists through civil society tried to influence authorities and legislation (Kamenarović, 1961) (see Table 9.3). During the first decades of the twentieth century there were significant efforts from the scientific community, which was active through the Croatian Society of Natural Sciences and the Croatian Mountaineering Society.

After World War II, Croatia became a republic of the former Socialist Federal Republic of Yugoslavia. Official systematic nature conservation activities then started in 1946 through the establishment of the Institute for Protection of Natural Rarities of Croatia in Zagreb. Natural and cultural heritage were dealt with together, but nature conservation became a separate issue through the 1960 Nature Protection Act (NPA).[7] The Institute for Nature Protection was founded in 1961 and the role and strength of the nature conservation sector started to increase. Although the Institute's activities were concentrated on the designation of protected areas and species, it also had significant involvement in spatial planning at all levels. The first public institutions for management of protected areas were also established. These institutions and their responsibilities continued following the creation of the independent **Republic of Croatia** in 1991.

Table 9.3 Key nature conservation developments in Croatia.

1893	The first nature conservation–related legislation – Hunting Act (regulating protection of songbirds); initiative for the first protected area
1900	The Act on Protection of Caves
1910	The Order on Conservation of Masterpieces of Nature
1922	Foundation of the Committee for the Protection of Natural Monuments of the Croatian Society for Natural Sciences
1924	Initiative of the Croatian Mountaineering Society to draft the Nature Protection Act and to protect mountain flora
1929	Proclamation of four National Parks (Plitvice Lakes, Paklenica, Štirovača and Bijele stijene) by the Financial Act for 1928/29 (valid only for this period)
1946	Foundation of the Institute for Protection of Natural Rarities of Croatia
1949	Act on Protection of Cultural Monuments and Natural Values of Croatia; Designation of National Parks Plitvice Lakes and Paklenica
1950	Establishment of the network of honorary commissioners (volunteers) for nature protection in Croatia
1960	New Nature Protection Act (NPA)
1961	Foundation of the Institute for Nature Protection
1999	The first *National Biodiversity and Landscape Strategy and Action Plan*
2003	New NPA, a turning point in the approach to nature protection; creation of the State Institute for Nature Protection
2006	Regulation on the National Ecological Network (NEN)
2008	Revision of the *National Biodiversity and Landscape Strategy and Action Plan*
2009	Ordinance on Assessment of Plans, Programmes and Projects on Ecological Network
2013	1 July: EU accession. New NPA and the Regulation on Ecological Network (Natura 2000)
2017	*Strategy and Action Plan for Nature Protection for 2017–2025*

See Table 3.3 for signatory dates of international conventions.

[7] In Croatia, the officially used term in the nature conservation legislative and administrative framework is 'nature protection'; 'nature conservation' is used in a broader sense.

Following ratification of the CBD in 1996, the first National Report and the first *National Biodiversity and Landscape Strategy and Action Plan*[8] were prepared and adopted by the Croatian Parliament in 1999. It was a turning point in the approach to nature conservation, widening the focus from protected areas and species to the whole of biodiversity. Consequently, the **2003 NPA** adopted this new approach and the requirements of relevant international conventions, including the main elements of the Nature Directives, as Croatia was starting to prepare for EU accession. This NPA established the concept of the National Ecological Network (NEN), together with the mechanism for carrying out nature-related impact assessments that would be adequate for appropriate assessments under Article 6(3) of the Habitats Directive. This was a form of transition mechanism towards Natura 2000 and the appropriate assessment procedure in the pre-accession period. The NEN was designated in 2006, and comprised key sites for species and habitat types threatened at European and national levels, as listed in the annexes of the Bern Convention and Nature Directives, as well as the former Red List of Croatia.

The State Institute for Nature Protection (SINP), established in 2003, had a key role in this work, which involved intensive preparations to meet the nature conservation obligations resulting from the coming accession of Croatia to the EU. This included significant inventory work, through engagement with the scientific community and a number of NGOs, as well as wider participatory processes through LIFE, PHARE (Poland and Hungary: Assistance for Restructuring their Economies[9]) and IPA (Instrument for Pre-accession Assistance) projects. All this resulted in the NEN being replaced by the Natura 2000 network[10] (legally referred to as the Ecological Network Natura 2000) and the harmonisation of the NPA with the EU directives immediately upon accession in 2013. In 2015 the SINP was merged with the Croatian Environment Agency and continued to work as the Croatian Agency for Environment and Nature (CAEN) until government reorganisations resulted in it being annexed to the Ministry of Environment and Energy in 2019 as the Institute for Environment and Nature Protection.

9.3.2 The Current Situation

Currently, nature conservation **legislation** in Croatia is based on the 2013 NPA (amended in 2018 and 2019)[11] and its several by-laws, some of the most important being the Regulation on the Ecological Network (Natura 2000)[12] and the Ordinance on Strictly Protected Species.[13] In addition, there is the Act on Transboundary Movement and Trade in Wild Species and the Act on Prevention and Management of the Introduction and Spread of Alien and Invasive Alien Species. Beside these, there is a wide range of environmental and sectoral legislation that is highly relevant for the implementation of the NPA, including: the Act on Environment (covering EIA/SEA procedures, which are consolidated with the Habitats Directive appropriate assessment procedure), the Act on Forests, the Hunting Act, the Freshwater Fishery Act, the Marine Fishery Act and the Agriculture Land Act.

The main strategic document on nature protection is the ***Strategy and Action Plan for Nature Protection of the Republic of Croatia for 2017 to 2025*** **(NPSAP)** (Anonymous, 2017), prepared on the basis of the five-year Report on the State of Nature and adopted by the Croatian Parliament. This updates previous similar strategies from 1999 and 2008, and was adopted in 2017. It defines the vision, strategic goals and priorities for the period, in relation to the Aichi Targets of the CBD's *Strategic Plan for Biodiversity 2011–2020* as well as the targets of the *EU Biodiversity Strategy to 2020*.

Governance is divided between national and regional levels. The key national institution is the Ministry of Economy and Sustainable Development (MESD),[14] with its Directorate for Nature Protection and the recently reorganised expert unit for environment and nature (Institute for Environment and Nature Protection – formerly SINP/CAEN). Regional-level governance is through County Offices, which are responsible for appropriate assessment/EIA/SEA procedures and spatial planning at regional and local levels. There are 19 public institutions (PIs) for the management of National Parks and Nature Parks established by the government. County administrations form PIs for the management of other categories of protected areas. PIs are also responsible for the management of Natura 2000 sites, including the preparation of management plans, which are obligatory. While the state-level PIs focus on specific sites (National and Nature Parks) and mostly have significant resources, including earnings from visitor-related activities, the county-level PIs are responsible for all the other protected areas and Natura 2000 sites within their territories but suffer from a significant lack of human and financial resources. Several local authorities have also established PIs for the management of certain protected areas that have significant touristic potential.

Regarding the responsibilities for other related **sectors**, the MESD also covers water management and energy. The Ministry of Physical Planning, Construction and State Assets

[8] Strategija i akcijski plan zaštite biološke i krajobrazne raznolikosti Republike Hrvatske ('Narodne novine' br. 81/1999).

[9] Which was extended to Croatia and some other Member States.

[10] The NEN and Natura 2000 network are very similar, but not identical as the NEN also included some major corridors (for birds and for sea turtles) and some habitat types and species of national importance.

[11] Zakon o zaštiti prirode ('Narodne novine' br. 80/2013, 15/2018, 14/2019, 127/2019).

[12] Uredba o ekološkoj mreži i nadležnostima javnih ustanova za upravljanje područjima ekološke mreže ('Narodne novine' br. 80/2019).

[13] Pravilnik o strogo zaštićenim vrstama ('Narodne novine' br. 144/2013, 73/2016).

[14] Until 2020 it was the Ministry of Environment and Energy.

is responsible for coordinating the implementation of the Building Act and the Spatial Planning Act, including preparation of the *Spatial Development Strategy of Croatia*. Other highly relevant sectors – agriculture, forestry, hunting and fisheries – are under the responsibility of the Ministry of Agriculture (MOA). Some of the important institutions under the supervision of the respective ministries are: Croatian Waters (responsible for the management of water bodies, which are public property); Croatian Forests (responsible for management of state-owned forests); and the Paying Agency for Agriculture, Fisheries and Rural Development.

Upon Croatia's accession to the EU, nature conservation issues gained in importance and cooperation with other sectors has continued to grow. In particular, the Habitats Directive appropriate assessment procedure has become a powerful mechanism that influences sectoral strategic planning as well as development projects, not only in the Natura 2000 network but also more widely. Systematic intersectoral cooperation has been established for a number of other issues, such as: the incorporation and implementation of nature protection requirements into the annual maintenance plans of water bodies; defining the obligatory content for management plans for forest Natura sites (being prepared by Croatian Forests); management of populations of large carnivores, the Beaver (*Castor fiber*) and the Wildcat (*Felis silvestris*); and the drafting and implementation of agri-environment–climate measures.[15] This cooperation has been effective in terms of planning and preparing the legislative framework, but stronger implementation is needed. In addition, there are many issues that need further cooperation, including strategic planning of locations for energy projects (e.g. hydro and wind power plants) and the assessment of their cumulative effects, the conservation of carp fishponds in Natura 2000 (and addressing losses from fish-eating birds), and implementation of the Mediterranean Regulation[16] and other fishing regulations in marine protected areas.

There are a very large number of **environmental NGOs** in Croatia – 944 on the register in 2014 (Anonymous, 2017) – including many that primarily focus on nature conservation (even though they only comprise 5% of NGOs). Some of the nature NGOs, especially those that joined relevant international organisations, have significantly strengthened their capacities, employ staff and implement large projects (e.g. LIFE or Interreg). Some of the most active are, for example, BIOM (BirdLife partner), Institut Plavi svijet (Blue World Institute), Zelena akcija (Friends of the Earth Croatia), WWF Adria, Zagreb (who also act in the wider region). Some NGOs are research- and science-oriented, providing a great contribution to national biodiversity inventory and monitoring programmes. As well as BIOM and Institute Plavi svijet, these include Hrvatsko botaničko društvo (Croatian Botanical Society), Hrvatsko biospeleološko društvo (Croatian Biospeleological Society), Hrvatsko društvo za zaštitu ptica i prirode (Croatian Society for Bird and Nature Protection), Hrvatsko ihtiološko društvo (Croatian Ichthyological Society), Udruga Sunce (Association Sunce), Udruga Hyla (Association Hyla) and others. NGOs participate in the preparation of Natura 2000 management plans and often hold workshops and training for public institutions involved in nature protection, helping to raise the capacities of their employees (Anonymous, 2017). Some of them also actively scrutinise the enactment of nature protection legislation and appropriate assessment/EIA/SEA procedures.

Stakeholders and the wider public have some influence over official decisions, such as through regular internet-based public consultations during appropriate assessment, EIA and SEA processes, the adoption of nature protection legislation and planning processes. According to the methodology prescribed by the MESD (Ministarstvo zaštite okoliša i energetike and Hrvatska agencija za okoliš i prirodu, 2018), and already established practice, stakeholders participate in the preparation of management plans for protected areas/Natura 2000 and action plans for strictly protected species through individual meetings and facilitated workshops. Although this practice is relatively new, it has shown good results because the process is open to all interested stakeholders. However, a problem is that stakeholders are often unfamiliar with nature conservation issues, so it will probably take a while until they are more interested and able to have significant inputs into the planning and decision-making process.

9.4 Species Protection and Conservation Measures

9.4.1 Legislation

The NPA defines the general protection of all wild indigenous species so that no species should be exterminated or their wild populations influenced in such a way that they become endangered. Threatened species included in the Red List of Croatia,[17] locally distributed endemics, and species in need of strict protection according to the Habitats Directive (i.e. Annex IV species) or international agreements (i.e. Bern Convention, Barcelona Convention and Bonn Convention) have a strictly protected status. Adequate measures are prescribed in accordance with the Birds Directive (Articles 5 and 6) and Habitats Directive (Article 12), which also apply to all naturally occurring bird species, even if they are not mentioned in the list of strictly protected species, which is produced under the Ordinance on Strictly Protected Species. This Ordinance also defines the content, methodology and procedure for adoption of 'management plans for strictly protected species with action plans' (MPAPs), which are required for selected priority species.

15 https://ruralnirazvoj.hr/program/

16 Council Regulation (EC) No 1967/2006 of 21 December 2006 concerning management measures for the sustainable exploitation of fishery resources in the Mediterranean Sea.

17 www.haop.hr/hr/tematska-podrucja/prirodne-vrijednosti-stanje-i-ocuvanje/ugrozenost-vrsta-i-stanista/crveni-1

Three strictly protected species are also included on the list of game species under the Hunting Act,[18] but with a permanently closed hunting season: the Beaver, Wildcat and Brown Bear. Their populations are managed in cooperation between the MESD and the Ministry of Agriculture (MOA), based on management plans enacted by the MOA. Currently, a management plan exists for the Brown Bear (Ministarstvo poljoprivrede and Ministarstvo zaštite okoliša i energetike, 2019), while plans for the other two are in preparation.

An important legal measure for large carnivores, and some other species that are threatened by the development of transport infrastructure, is the requirement under the Ordinance on Wildlife Crossings (based in the NPA). This requires the construction of special crossings over highways ('green bridges') that connect fragmented habitats and reduce the number of animals killed by traffic. The locations in which they are required are based on EIAs and their maintenance needs are defined in the Ordinance. Regular monitoring indicates that the bridges are effective in enabling animals to cross the roads.[19]

Species that can be exploited are mostly covered by the Hunting Act, the Freshwater Fishery Act and the Marine Fishery Act, which regulate their management. The rest, such as medicinal plants, edible fungi and snails, fall under the Ordinance on the Collection of Native Wild Species.[20] This is enacted by the MESD and, based on expert studies, prescribes the permits that must be obtained to collect the species, the specific means of their collection and any conditions.

Conservation objectives and measures for Natura species in SPAs and SCIs must be defined according to two respective ordinances that also define responsibilities for implementation, which are elaborated in detail in Natura 2000 site management plans.

The appropriate assessment procedure for strategies, plans or programmes and the issuing of permits for activities in protected areas are also used as a mechanism for prescribing detailed measures for the conservation of strictly protected species in the form of nature protection requirements. These are an integral part of the permit, issued by the competent authority (MESD or regional authorities, in accordance with the NPA) and often based on the opinion of the expert nature protection unit of the MESD.

9.4.2 Species-Specific Conservation Measures

Management plans for strictly protected species with action plans (MPAPs) are to be adopted by the MESD for priority species, which are chosen according to defined criteria.[21] MPAPs for the first set of 21 priority species have been elaborated through an EU-funded Operational Programme Competitiveness and Cohesion (OPCC) project[22] (further described in Section 9.7.3), which includes revision of existing action plans for the Wolf and the Eurasian Lynx. The expert background documents for the MPAPs have been prepared by consultants, with the participation of stakeholders.

Species re-introductions require the approval of the MESD, which may be dependent on an ecological risk assessment.

The first official re-introduction was for the Beaver in the late 1990s in several locations in the northern part of Croatia, with animals originating from Bavaria in Germany. The re-introduction was successful due to the existence of suitable habitat along the Drava, Sava and Kupa rivers, and the species has since spread rapidly. Today the Beaver is a strictly protected species, falling under the management of the Hunting Act. However, the MESD issues derogations where it is necessary to remove beaver dams that are causing problems. As mentioned, a management plan for the Beaver is in preparation, due to its spread and the resulting conflicts.

Although the Eurasian Lynx had become extinct as a result of hunting by the end of the nineteenth century, it is once again present in Croatia. This is the consequence of the re-introduction of six animals in neighbouring Slovenia during the 1970s, which soon resulted in a population that spread into Croatia.

Croatia has been active on the issue of the **management of large carnivore populations** for a long time. In the past this was largely to deal with conflicts, but today there is more of a focus on their conservation. The most problematic species has been the Wolf, as it can result in significant losses of livestock (Ministarstvo kulture, 2010). Hunters have also traditionally seen the Wolf as a pest because they consider that it reduces game populations. Consequently, Wolves were systematically persecuted as pests and their population drastically declined in the second half of the twentieth century. As noted above, the Eurasian Lynx was also persecuted, resulting in its extinction in Croatia. The Brown Bear has been in a better position as it was a valuable hunting trophy, for which high prices have been paid. Consequently, there has been a tradition of careful management of the bear population in Croatia.

Conservation of the Wolf started in the 1990s, with strict protection afforded in 1995 and a temporary management plan prepared in 1999 by the National Committee for Large Carnivores. This Committee also initiated systematic research and monitoring of all three large carnivore species (e.g. through radio-tracking). Hunters were also involved, collecting data on legal and illegal killing and other causes of mortality, as well as monitoring (e.g. recording tracks in the snow and collecting faeces for DNA analysis). At the same time, an administrative framework of authorised experts was developed to deal with damage. Transboundary cooperation with

[18] Zakon o lovstvu ('Narodne novine' br. 99/2018, 32/2019, 32/2020).

[19] https://oikon.hr/our-departments/laboratory-for-research-and-monitoring-of-large-carnivores/

[20] Pravilnik o sakupljanju zavičajnih divljih vrsta ('Narodne novine' br. 114/2017).

[21] www.haop.hr/hr/tematska-podrucja/prirodne-vrijednosti-stanje-i-ocuvanje/planovi-upravljanja-i-mjere-ocuvanja-6

[22] Operational Programme Competitiveness and Cohesion 2014–2020, project 'Preparation of draft management plans for strictly protected species with action plans (MPAPs)'.

Slovenia was also established to collect data on the distribution and abundance of large carnivores. The LIFE project 'Conservation and Management of Wolves in Croatia' (CROWOLFCON), implemented by the SINP over 2002–2005, significantly contributed to development of good management practices.

Management plans for all three species were prepared in 2005 by the National Committee for Large Carnivores through a participatory process and adopted by the Ministry of Culture, which was then responsible for nature protection.[23] Culling quotas for the Wolf and the Eurasian Lynx were still issued annually for a certain period, until monitoring results indicated that the lynx is Critically Endangered in Croatia, due to its very small population, while the Wolf population has declined (Jeremić and Štrbenac, 2017). The Brown Bear remained a game species until EU accession in 2013. Currently its population is considered to be secure (approximately 1 000 individuals) and is being managed in accordance with the management plan from 2019 (Ministarstvo poljoprivrede and Ministarstvo zaštite okoliša i energetike, 2019). This provides the basis for the preparation of annual management plans with culling quotas, which fall under derogations under the HD Article 16(1)(e), that is, taking in limited numbers (see Section 4.2). According to the latest publicly available information on reported derogations for Croatia, from the annual culling quota of 130 bears (plus 20 expected from other reasons of mortality), 122 animals were actually culled in 2015 and 123 in 2016.[24]

Dealing with conflicts caused by strictly protected species has been an important issue in Croatia, and therefore measures have been established to tackle conflict, including compensation schemes, although they have significant costs. There is a defined procedure for reporting damage to the MESD or to authorised experts. Compensation is then paid according to documented investigations from authorised experts, prescribed criteria and standard price lists. However, field data clearly show that if livestock are guarded, losses from Wolves are small or non-existent (MENP, 2014). Therefore, to reduce damage, a donation programme for livestock breeders was organised; this provided sheep dogs from 1997 and electric fences from 2003, until it had to stop in 2012 due to a lack of funding (MENP, 2014). Over the period, 147 Tornjak sheep dogs (an indigenous breed) and 80 electric fences were donated to farmers. To reinstate and extend these activities, the CAP Rural Development Programme (RDP) for 2014–2020 included under Measure 4 'Investments in physical assets', a submeasure 4.4.1. 'non-productive investments related to environmental protection' which included financial support for purchasing electric fences and/or Tornjak sheep dogs.

The conservation of large carnivores has been the main initiative for threatened species in Croatia. Over the last three decades of systematic work, fruitful cooperation developed between authorities, scientists and representatives of the various key stakeholder groups (hunters, forest managers, farmers). Whilst some problems have not been solved, an effective framework for dealing with them has been established, including the compensation payments to farmers for livestock lost to Wolves.

9.5 Protected Areas and Networks

9.5.1 Protected Area Types and Their Coverage

Protected Area Types

The designation of protected areas in Croatia goes back almost a century, but it was not until the 1960 NPA that the term 'protected objects of nature' (sites and species) was introduced and the first categories of protected areas were defined. These were revised by the 1976 NPA and subsequently have been little changed (DZZP, 2014). The large majority of domestic protected areas that exist today were designated in the second half of the twentieth century and reflect the typology at the time. Subsequently, in 2003, the new category of Regional Park was introduced as well as the concept of the NEN, which was replaced by Natura 2000 by the 2013 NPA upon Croatia's accession to the EU. While Natura 2000 is defined by the current NPA[25] in accordance with the Habitats Directive, the sites are considered to differ from protected areas, which Croatia defines as 'a geographically clearly defined area intended for nature protection and managed for the long-term conservation of nature and associated ecosystem services' (differing slightly from the IUCN definition). Nevertheless, Natura 2000 sets out to cover the conservation needs of its target habitats and species, which also contributes to wider biodiversity protection. Consequently, not many new domestic protected areas are expected to be designated in future, with the exception of marine protected areas (MPAs) as discussed later.

There are **nine types of domestic protected areas**, most of them aimed at biodiversity conservation (among other objectives). Apart from Monuments of Park Architecture and Forest Parks, the other seven listed in Table 9.4 are considered to meet IUCN protected area criteria and broadly correspond to IUCN categories (DZZP, 2014). However, the system has not been harmonised with IUCN criteria, following the former tradition in Croatia of combining the protection of cultural and natural heritage, and putting the accent on landscape values. As a result, the Croatian categories are defined according to the values of protected areas (i.e. their characteristics), with no clear reference to management objectives. Therefore, the proper categorisation of the protected areas according to the IUCN system requires individual interpretation of each site, bearing in mind its prime management objectives.

23 www.life-vuk.hr

24 https://rod.eionet.europa.eu/obligations/268/deliveries

25 Zakon o zaštiti prirode, 'Narodne novine' br. 80/2013; 15/2018; 14/2019.

Table 9.4 The number and extent of international designated sites and protected areas in Croatia.

	IUCN category	Number	Area (km^2)
Designated under global and non-EU international legislation			
Ramsar Sites	NC	5	943
Natural and Mixed World Heritage Sites	NC	2	437
UNESCO Global Geoparks	NC	2	7 175
Biosphere Reserves (exc. buffer and transition zones)	NC	2	914
Designated under the Nature Directives			
SCIs (terrestrial + marine)	NC	745	16 036 + 4 919
SPAs (terrestrial + marine)	NC	38	17 050 + 1 112
Natura 2000 terrestrial area			20 716
Natura 2000 marine area			5 238
Designated under domestic legislation			
Strict Reserve (strogi rezervat)	Ia	2	24
National Park (nacionalni park)	II	8	980
Nature Monument (spomenik prirode)	III	79	2
Special Reserve (posebni rezervat)	IV	79	408
Nature Park (park prirode)	V	12	4 950
Regional Park (regionalni park)	V	2	1 026
Significant Landscape (značajni krajobraz)	V	81	1 379

Note. NC = Not categorised.
Sources. See Section 1.3. Based on data from Ramsar Convention and UNESCO World Heritage Site websites (accessed December 2021); EEA Natura 2000 Barometer, as at end 2020; other sites from www.bioportal.hr

The NPA divides categories of domestic protected areas into two classes: sites of state-level importance and those of regional/local importance. The highest level of protection is given to the categories of National Park, Strict Reserve and Special Reserve (as further described in Section 9.5.3). Nature Parks are, like National Parks, also designated by special laws as these are always large areas, but the regime is not so strict (in National Parks the economic use of natural resources is forbidden, while in Nature Parks it is allowed if it does not threaten the key characteristics and roles of the site). The definitions and levels of designation and management of each type of domestic protected area category are presented in Table 9.5. Most of them, with the exception of the two Strict Reserves and a set of relatively small Special Reserves, focus on aesthetic, educational and recreational values, while their management often concentrates on visitor management and required infrastructure.

Protected Area Designation Criteria

Although the criteria for protected areas differ, depending on their category, they have not been defined in detail beyond the NPA definitions (Table 9.5). Instead, from the outset, the designation of protected areas has been based on technical work and expert judgement by the nature conservation expert service, supported by available scientific information. The most intensive period of establishing protected areas was the second half of the twentieth century. Due to the authority of the service, and systematic cooperation with local authorities, valuable sites identified as potentially requiring protection were included in spatial plans (so-called recorded sites), thereby giving them a form of preliminary protection. Formal designation was preceded by a technical description of the key values of the area. In these former times, the designation of protected areas did not include a participatory process, as it does now, and private property could be more easily subject to protected area legislation. Due to this, the protected area network was developed relatively quickly with very low financial input.

The situation has changed since the 2003 NPA,[26] as it requires that the background document for designation has to list the consequences of protection, especially with respect to property rights and current economic activities, as well as providing an assessment of costs and sources of finance. Public access to the designation documentation also became obligatory, enabling the participation of stakeholders in the process. As a result, protected area designation has become more complex and onerous, but benefits from increased stakeholder and wider public input.

Protected Area Coverage

According to the MESD,[27] **domestic national protected areas** in 2021 comprised 410 sites with a total area of 8 213 km^2; although there are overlaps, this represents about 13.4% of the terrestrial territory and 1.9% of the coastal sea (i.e. the Territorial Sea and Internal Waters).[28] Nature Parks are the category that covers the largest area (60% of the total protected area). Whilst the number of sites is rather high, most of them are small: 28% of them, mostly in the categories of Nature Monuments and Monuments of Park Architecture, are smaller than 1 ha, while almost 90% are less than 1 000 ha (DZZP, 2014). The largest are Nature Parks and Regional Parks.

26 Zakon o zaštiti prirode, 'Narodne novine' br. 162/2003.
27 www.haop.hr/hr/tematska-podrucja/zasticena-podrucja/zasticena-podrucja/sto-je-zasticeno-podrucje
28 https://mingor.gov.hr/UserDocsImages/slike/Vijesti/2021/INFOGRAFIKE.pdf

Table 9.5 Definitions of the different types of domestic protected area in Croatia.

Category	Definition	Level of management	Proclamation
Strict Reserve	A mainland and/or a marine area with an unmodified or slightly modified overall nature, intended exclusively for the conservation of its original nature	State and county	Government of RC
National Park	A large, predominantly unmodified mainland and/or marine area of outstanding and multiple natural values, which includes one or more preserved or slightly modified ecosystems, primarily intended for the conservation of original natural and landscape values	State	Croatian Parliament
Special Reserve	A mainland and/or a marine area of a special importance due to its unique, rare or representative natural values, or a threatened habitat, or a habitat of a threatened wild species, primarily intended for the conservation of these values	State, county, also municipality and town in individual cases	Government of RC
Nature Park	A large natural or partly cultivated mainland and/or marine area of great biodiversity and/or geodiversity, with valuable ecological features, emphasised landscape and cultural and historical values	State	Croatian Parliament
Regional Park	A large natural or partly cultivated mainland and/or marine area of great biodiversity and/or geodiversity, with valuable ecological features and landscape values characteristic of the area in which it is located	County	Representative body of the competent unit of regional self-government
Nature Monument	An individual unmodified part of nature that has an ecological, scientific, aesthetic or educational value	County, also municipality in individual cases	Representative body of the competent unit of regional self-government
Significant Landscape	A natural or cultivated area of great landscape value and biodiversity and/or geodiversity, or a landscape of conserved unique features characteristic of a particular area	County, also municipality in individual cases	Representative body of the competent unit of regional self-government
Forest Park	A natural or a planted forest of greater biodiversity and/or landscape value, intended for rest and recreation	County, also municipality and town in individual cases	Representative body of the competent unit of regional self-government
Monument of Park Architecture	An artificially formed space (a public garden, a botanical garden, an arboretum, a city garden) that has aesthetic, stylistic, artistic, cultural, historical and educational value	County	Representative body of the competent unit of regional self-government

Source. Based on the Nature Protection Act.

Geographically, protected areas are not distributed evenly in Croatia. As might be expected, the highest coverage is in the Alpine biogeographical region, which is the most natural and the least populated. This area, together with the Mediterranean region in Croatia, includes most of the sites with strict protection. These are rare in the more densely inhabited Continental region, where 96% of the protected area comprises Nature Parks, Regional Parks and Significant Landscapes (DZZP, 2014).

Significant protected areas have also been designated as **international protected areas** under respective international conventions. Although the NPA does not give them direct protection, all overlap with some form of domestic protected area. UNESCO World Heritage Sites include the Plitvice Lakes National Park and primeval beech forests on Velebit Mountain (within a multinational site). Velebit mountain, which is protected as a Nature Park, and the Mura-Drava Regional Park are included in the UNESCO MaB programme, while Papuk Mountain and the Vis Archipelago have been designated as UNESCO Global Geoparks (UGGp).

Natura 2000 in Croatia consists of 783 sites and covers 36.7% of the land and 9.4% of Croatia's entire marine waters (16.3% of the Territorial Sea and Internal Waters[29]). Around 26% of Natura 2000 is protected under one of the nine categories of domestic protected areas, while 87% of all domestic protected areas are included in Natura 2000 (those which satisfied scientific criteria according to the Nature Directives) (Anonymous, 2017). SCIs have been defined for 73 HD habitat

[29] https://mingor.gov.hr/UserDocsImages/slike/Vijesti/2021/INFOGRAFIKE.pdf

types and for 141 species (126 HD Annex II species and 15 species on the Croatian Red list). Their selection was based on extensive scientific research from 2002, when the LIFE CRO-NEN project started, until EU accession in 2013. It was also informed by the Biogeographical Seminar for Croatia, which was held in 2014. The new Regulation on the Ecological Network Natura 2000 was also adopted in 2015[30] and again in 2019[31] in order to incorporate conclusions from the seminar and new scientific information. While the terrestrial part of the SCI network can be considered complete (European Commission, 2017), at the time of writing this account at the end of 2021, important scientific reserves are required for some marine habitats and species, and research is underway to address this (ZZOP, 2019).

Only five SCIs have been designated as SACs[32] so far upon adoption of adequate management plans.

There are 38 SPAs covering 30.2% of the land and 2.0% of Croatia's marine waters (3.2% of Territorial Sea and Internal Waters). These have been defined for 126 bird species, including 84 BD Annex I taxa, 28 regularly occurring migratory birds and 14 nationally threatened species. The proposal for SPAs was prepared by the Institute for Ornithology in cooperation with the SINP in accordance with BirdLife International's European IBA criteria, with the input of the wider ornithological community. Comparing the SPA network with the IBA network (BirdLife International, 2020), it can be concluded that the terrestrial and coastal SPA network is sufficient and fully compatible with IBAs. However, there may be a potential need to extend the network to offshore waters, depending on the results of currently ongoing research on marine birds.

Recent estimates of the combined coverage of protected areas do not appear to be publicly available from national sources. According to the World Database on Protected Areas (WDPA), taking into account overlaps, the combined terrestrial coverage of international sites (other than Biosphere Reserves), all nine categories of domestic protected areas and Natura 2000 sites is 21 811 km^2, which is 38.6% of the land (UNEP-WCMC, 2022). Although the estimate includes Forest Parks and Horticultural Monuments, which do not meet IUCN Protected Area criteria, their total area is only 38 km^2, so their influence on the overall coverage is insignificant. The WDPA data indicate that total marine protected area coverage is 4 985 km^2, which is 15.8% of the Territorial Sea and Internal Waters, or 9.0% of Croatia's total marine waters.

Detailed spatial information on the sites is publicly available via the Bioportal service of the MESD.[33]

9.5.2 The Coherence of Protected Areas and Ecological Networks

As mentioned earlier, it was not envisaged that protected areas in Croatia would create a network of sites for biodiversity conservation. In fact, many have been designated for their landscape values and/or recreation. Nevertheless, they all contribute to biodiversity conservation as they preserve space for nature, protecting the environment to some degree from damage, such as built developments and the intensive use of natural resources.

In combination with Natura 2000, a number of domestic protected areas have gained a new dimension, and their importance is heightened as their network both helps to support the Natura network ecologically and is also supported by it. The surveys and research that were carried out to identify Natura 2000 sites also increased the understanding of the importance of most domestic protected areas for biodiversity protection, even though many were designated for rather general landscape and aesthetic reasons. Consequently, all National Parks and Nature Parks, which are considered to contain the most valuable parts of nature in Croatia, are included in Natura 2000 (comprising 23%). In addition, those domestic protected areas that have become part of Natura 2000 now have an added focus on biodiversity conservation. Furthermore, as the majority of domestic protected areas are situated inside Natura 2000, these two systems of site protection in fact largely share the same space, the same management authorities and the same management planning mechanism. Thus the two systems complement each other, and the infrastructure of protected areas provides a good opportunity to present the key features of Natura 2000 sites and to educate visitors. In some cases, domestic protected areas have stronger and more concrete legal provisions that directly or indirectly contribute to biodiversity conservation, while Natura 2000 currently relies mostly on the appropriate assessment mechanism, as management measures and plans are still under preparation. Both systems have an important influence on spatial planning as they generally prevent damaging activities to some extent over a large part of the Croatian territory.

Overall, national protected areas and Natura 2000 as a whole comprise a functionally connected network because Natura sites are large and close to each other, or even border each other, especially in the Alpine and Mediterranean biogeographical regions. In the Continental region, which contains more densely populated and agricultural areas, the network could benefit from additional sites or corridors created through various types of green infrastructure.

9.5.3 Approaches to Protection and Levels of Protection

The NPA defines the general framework and a wide spectrum of legislative mechanisms for the protection of domestic protected areas, while regimes for individual sites vary, depending

[30] Uredba o ekološkoj mreži, 'Narodne novine' br. 105/2015.

[31] Uredba o ekološkoj mreži i nadležnostima javnih ustanova za upravljanje područjima ekološke mreže, 'Narodne novine' br. 80/2019.

[32] Proclaimed by the Regulation on Ecological Network from 2019.

[33] www.bioportal.hr/gis/

on their category. National Parks, Strict Reserves and Special Reserves, which are of state-level importance and aimed at biodiversity conservation, have a more restrictive regime compared to others. All nine categories of protected areas are supposed to have management plans adopted, while for National Parks and Nature Parks spatial plans are also obligatory, adopted by the Croatian Parliament. The plans are based on expert studies by the MESD and define zones according to conservation objectives and levels of use.

A number of activities are prohibited in all domestic protected areas (e.g. release of genetically modified organisms; camping, swimming or disposing of waste outside designated locations; visiting without a permit if required) or certain categories (e.g. straying from marked hiking trails and outside the zone for visitors in a Strict Reserve, Special Reserve and National Park). Many of the areas refer to the rules of conduct for visitors (e.g. rules for using vehicles or recreational activities), for which penalties may be incurred if they are breached. Detailed rules for the protection, conservation, improvement and use of National Parks, Nature Parks, Strict Reserves and Special Reserves are prescribed by ordinances for individual sites, enacted by the competent minister. For other categories, similar rules can be set in ordinances by county authorities. Unlike a management plan, which is defined as a planning document, an ordinance has legal force.

One of the most effective protection mechanisms is issuing Nature Protection Requirements in relation to a proposed built development (under building legislation) or sectoral management plans and spatial plans (in which case these requirements are a part of the final permit following an appropriate assessment or SEA procedure). Nature Protection Requirements may be included in permits that are required for other potentially damaging activities and projects as well as for research activities in protected areas.

Equal access to visit protected areas is provided for, according to the prescribed regime for each site. Private owners and holders of rights are obliged to allow visitors, but have a right to compensation if this causes restrictions for them.

The State has a right to expropriate, or restrict the use of, private real estate (typically land or sometimes buildings) if it is necessary for the conservation of protected parts of nature within domestic protected areas. If new restrictions prevent the landowner from continuing their activities then they are entitled within two years to offer it for sale to the State/county (depending on the protected area category) so that it can be bought or exchanged for equally valuable real estate. If an owner of real estate in a National Park, Strict Reserve or Special Reserve wishes to sell it, it must be offered to the State in the first instance, then to the respective county and finally to the local community. If none of these wish to buy it, it can be sold to another person but not for a lower price than was offered to the State/county.

Where protected area restrictions result in reduced revenues (which cannot be compensated through another allowed activity), then financial compensation may be paid by the State/county (except to the institutions that are majority owned by the State, like Croatian Forests and Croatian Waters), although this is rare.

Regarding the Natura 2000 network, the NPA states that its conservation is ensured by the Habitats Directive appropriate assessment requirements, implementation of conservation measures prescribed by ordinances in relation to SPA and SAC conservation objectives and their required measures, implementation of site management plans and other adequate legal, administrative and contractual measures. To date, whilst some proposed projects were found to be potentially detrimental to Natura 2000, none have been permitted to go ahead for imperative reasons of overriding public interest (under HD Article 6(4)).

9.5.4 Approaches to Managing Protected Areas

Management plans for Natura sites and all domestic protected areas are obligatory documents that are to be implemented by the competent public institutions responsible for their management (except for marine sites outside of territorial waters, which are the responsibility of the MESD). Management plans are planning documents that in the case of Natura 2000 sites set out in detail the implementation of conservation measures prescribed in two respective ordinances.[34] If a Natura 2000 site is also a domestic protected area, one management plan is prepared, covering overall issues. Plans can be prepared for multiple Natura 2000 and/or other protected sites. Guidelines for the preparation of management plans for domestic protected areas and/or Natura 2000 sites have been produced by the expert group of the competent ministry and published on the website (Ministarstvo zaštite okoliša i energetike and Hrvatska agencija za okoliš i prirodu, 2018). They emphasise the need for a participatory approach during their preparation and adoption. Zonation approaches are also promoted, which for Natura 2000 sites are based on the distribution of the target habitats and species.

Currently, only National Parks and Nature Parks have adopted management plans,[35] out of which only the few newest have also covered Natura 2000 issues. However, a large EU-funded OPCC project on Natura 2000 management[36] aims to develop management plans for at least 40% of Natura 2000 sites by 2023. Generally, management plans tend to define extensive sets of required activities and relevant sources of financing, but their implementation is/will be (for future ones) significantly restricted by existing resources, especially at

34 According to the NPA, the site-level conservation objectives and conservation measures that correspond to the ecological requirements of target species and habitat types are to be enacted by two ordinances (for SPAs and SCIs/SACs).

35 www.haop.hr/hr/tematska-podrucja/zasticena-podrucja/upravljanje-zasticenim-podrucjima/dokumenti-upravljanja

36 https://mingor.gov.hr/o-ministarstvu-1065/djelokrug/uprava-za-zastitu-prirode-1180/ekoloska-mreza-natura-2000/projekt-razvoj-okvira-za-upravljanje-ekoloskom-mrezom-natura-2000/5990

a county level. Despite their obligatory status, there is no strong commitment for financing their implementation. Hopefully this will change with the recently prepared Prioritised Action Framework (PAF) for implementation of the Natura 2000 network for 2021–2027.

The care of protected areas, including Natura 2000 sites, or parts thereof, can be conferred to the landowner or rights holder, or even to another person or organisation (e.g. NGO), by signing a contract of care, following a public tender. However, this mechanism has not been used, despite the low capacities of county PIs.

9.6 General Nature Conservation Measures

9.6.1 Development Planning and Impact Assessment

The umbrella national development policy document is the *Spatial Development Strategy of the Republic of Croatia* from 2017.[37] Based on the Law on Physical Planning,[38] there is a set of county spatial plans and local municipality/town plans that have to be harmonised with regional-level documents and the overarching national strategy. They have all incorporated nature protection requirements, identified through SEA procedures that include distinct appropriate assessments of potential impacts on Natura 2000 sites. However, a problem has been the lack of a methodology for the assessment of the cumulative effects of projects, which may negatively impact on Natura 2000.

The spatial planning system is currently undergoing substantial changes as the State Spatial Development Plan of Croatia is being prepared. After its adoption, new county plans will be created, followed by municipalities and cities ('new generation of spatial plans', as stated in the *Spatial Development Strategy*). For all of these new plans new SEAs will be mandatory, and, if relevant, appropriate assessments.

The HD Article 6(3) **appropriate assessment** procedure is the key mechanism for the protection of Natura 2000 sites, especially outside domestic protected areas. It is streamlined with the EIA/SEA procedures but, to meet the requirements of the Habitats Directive, it is a clearly recognisable part of the EIA/SEA documentation. It is implemented by the MESD or the county-level administrative body, depending on criteria defined by the NPA. In all cases it is based on the professional opinion of the nature protection expert service (currently a unit of the MESD). In 2008, upon proclamation of the NEN, and as preparation for EU accession, a procedure for appropriate assessment was established that corresponded to the requirements of the Habitats Directive. This was very helpful, as it enabled the framework to be in place at the time of accession. However, initially implementation of the appropriate assessment procedure was hampered by very limited administrative capacities and a weak legal framework. Upon accession, legislation was additionally harmonised with the Nature Directives, while administrative capacities were strengthened, including through training. Despite this, limited administrative resources remain a problem. The MESD has prepared guidance documents on the implementation of appropriate assessment procedures,[39] but, as for other impact assessments, there remains a lack of guidance on the assessment of cumulative effects.

9.6.2 Habitat Management and Restoration

Forests, grasslands and freshwater habitats are mostly managed through the forestry, agriculture and water management sectors. Hunting and freshwater and marine fisheries also have a large impact on habitats. In Croatia, nature protection requirements concerning these sectors are incorporated into sectoral planning documents through appropriate assessment procedures. In accordance with the NPA, these requirements refer to protected areas, strictly protected species and threatened/rare habitat types in the whole area covered by the sector management plan, even if these are not part of Natura 2000. Requirements are issued based on technical assessments carried out by the Institute for Environment and Nature Protection of the MESD, taking into account all significant natural values present in the area covered by the planning document. However, influence on the marine fishery sector through this mechanism is weak, as there is a lack of planning documents covering this sector and relatively limited domestic protected area coverage, while the marine Natura 2000 network has not yet been finalised.

The majority of **forests** in Croatia can be considered as semi-natural, and forest management policy and legislation are broadly compatible with nature conservation requirements, being based on 'principles of sustainable use, economic and ecological acceptability and social responsibility' (Forestry Act[40]). Sustainable forest management is also aided by the fact that 81% of forests are state-owned, managed by Croatian Forests Ltd[41] based on management plans that include nature protection requirements.

Agriculture has a much larger influence on habitats, which can be beneficial or detrimental. Semi-natural grasslands and small-scale cropland mosaics, rich with hedges and tree lines, are of high nature value. However, they are especially threatened due to the disappearance of traditional agricultural practices in areas being abandoned, or agricultural improvements and intensification in others. One of the steps taken to

37 Strategija prostornog razvoja Republike Hrvatske ('Narodne novine' 106/2017).

38 Zakon o prostornom uređenju ('Narodne novine' 153/2013, 65/2017, 114/2018, 39/2019, 98/2019).

39 www.haop.hr/hr/tematska-podrucja/odrzivo-koristenje-prirodnih-dobara-i-ekoloska-mreza/ekoloska-mreza/ocjena

40 Zakon o šumama ('Narodne novine' br. 68/2018, 115/2018, 98/2019).

41 Hrvatske šume d.o.o.

deal with this problem during the 2014–2020 CAP was the designation of all grasslands in Natura 2000 sites as Environmentally Sensitive Permanent Grasslands (ESPG) under the CAP greening measures (see Section 4.3), which was important due to the fact that one-third of all grassland in Croatia is inside the Natura 2000 network. All permanent grasslands in the Natura 2000 network were designated as ESPG in 2016, but none were designated outside as most biodiverse grasslands are in the network. Farmers who declared these ESPG as being eligible for direct payments were banned from ploughing them and changing them into another type of agricultural land use (e.g. arable land, orchards or vineyards) (Act on Agriculture[42]).

CAP RDP measures have also been a key tool for the conservation management of agricultural habitats in Croatia, with the agri-environment–climate measure (AECM), which is voluntary for farmers, being most important (see Section 9.7 for budgets). The non-productive investments measure has also been used to finance restoration of overgrown grasslands. RDP Natura 2000 payments to farmers have not yet been used, due to the fact that restrictive conservation measures that have required compensation have not been prescribed in Natura 2000 sites. Some of the most important AECM submeasures for biodiversity included in the 2014–2020 RDP were: preservation of HNV grasslands; preservation of hedges; protection of the Corncrake; and protection of butterflies. Unfortunately, as mentioned in the latest PAF (Ministarstvo gospodarstva i održivog razvoja, 2022), the uptake of AECM was very low, possibly due to, among other reasons, the very small land parcels that prevail in Croatia. In addition, AECM payments have been unattractive and the administrative procedures for obtaining them rather complicated and off-putting, especially for older farmers, who still predominate in Croatia. More tailor-made measures with result-based payments may be more appropriate, at least in some key Natura 2000 sites designated for agricultural habitats and species.

Water management is the responsibility of Croatian Waters,[43] which therefore has a key role to play in the conservation management of freshwater habitats. Long-term cooperation with the nature conservation sector has been established through the Nature Protection Requirements that are issued in relation to annual maintenance work programmes. There have also been several oxbow restoration projects, although these were primarily for recreational purposes rather than biodiversity. Other restoration activities have been carried out through a 2015–2020 LIFE project on Integrated River Management,[44] coordinated by Croatian Waters in partnership with several county public institutions for nature conservation.

[42] Zakon o poljoprivredi ('Narodne novine' br. 118/2018, 42/2020, 127/2020, 52/2021).

[43] Hrvatske vode.

[44] www.drava-life.hr/en/project/

9.7 Nature Conservation Costs, Benefits and Funding Sources

9.7.1 Nature Conservation Costs

As there has been no systematic tracking, information on the past costs of nature conservation in Croatia is lacking. Generally, costs rose during the pre-accession period, when extensive biodiversity research was carried out to support the NEN and coming Natura 2000 network. After accession, the implementation of the Nature Directives created additional costs, but, at the same time, new EU funding opportunities arose, as described further in Section 9.7.3.

Current regular costs relate to the activities of the MESD and other public institutions involved in the management of protected areas, management costs and compensation for damage resulting from Wolves and other strictly protected species. The total costs of nature conservation are unknown, but those of implementing the Natura 2000 network have been estimated as part of the development of the PAF for 2021–2027. According to the PAF (Ministarstvo gospodarstva i održivog razvoja, 2022), the total estimated annual costs averaged over 2021–2027 for Natura 2000 amount to about €212 million (€126 million annual costs, and €86 million one-off costs). These annual costs include €51 million general cross-cutting measures and administrative costs, 66% of which are one-off costs (mostly management planning, communication activities and visitor access, etc.). Averaged annual maintenance and restoration measures for the network are estimated to be €80 million (73% running costs). The annual costs of green infrastructure to support the Natura 2000 network are estimated to be €70 million (65% recurring actions). The vast majority of costs for both categories of the network and the green infrastructure are intended for agricultural incentive payments (for grasslands and other agricultural ecosystems) and for implementation of conservation measures in forests. The estimated annual costs of additional species-specific nature conservation measures are €6.6 million (80% one-off), and those for prevention, mitigation or compensation of damage caused by protected species are €3.7 million (98% recurring costs). The PAF seems to be quite ambitious but at the same time realistic, mostly based on the follow-up activities and projects from the previous MFF.

9.7.2 The Economic Benefits of Nature Conservation

Although several studies have evaluated ecosystem services in a variety of sites and habitats in Croatia,[45] no comprehensive calculation of the economic benefits of nature conservation has been produced. Nevertheless, it is clear that the large ecological network with numerous protected areas is an important basis for the development of green jobs and revenues from visitors and recreation activities, especially in National Parks and Nature Parks.

[45] www.haop.hr/hr/tematska-podrucja/prirodne-vrijednosti-stanje-i-ocuvanje/stanista-i-ekosustavi/usluge-ekosustava

9.7.3 Funding Sources for Nature Conservation

EU financial instruments, with state co-financing, are the main source of funds for nature conservation and related environmental measures (Table 9.6). Croatia has used **EU Cohesion Policy** funding for priority activities primarily requiring capital investments, including preparation of management plans for Natura 2000 sites, preparation of action plans for priority species, developing the monitoring and reporting framework, mapping marine habitats, and establishing a national monitoring system for IAS. During the 2014–2020 budget period, the OPCC was used to fund such activities, with a total investment of €354 million (EU contribution €266 million). Interreg funding has also been significant for transboundary and terrestrial cooperation projects. A number of projects were financed in which institutions and NGOs from Croatia were the fund holders or partners: cross-border projects (with Italy, Hungary, Slovenia and Serbia) and transnational projects (Danube, Adriatic-Ionian, Mediterranean and Central Europe). Most of the projects involved public institutions for the management of protected areas as partners.

The European Agricultural Fund for Rural Development (EAFRD) has also been a major source of funding for nature conservation in Croatia, especially the management of agricultural areas. The total EAFRD allocation for Croatia over 2014–2020 was €2.4 billion (with €2 billion from the EU budget). Out of this, about €626 million was allocated to measures under Priority 4 that contribute to ecosystem restoration and conservation. However, about 80% of this was allocated to payments to areas with constraints, and organic farming, probably with limited biodiversity benefits. Regarding the payments to areas with constraints, the submeasure that relates to extensive karst pastures is important for their conservation as it requires recipients to maintain grazing with at least 0.3 livestock units/ha. Only around 10% was allocated to AECM schemes, which, as described earlier, are the most effective for nature conservation, although their uptake has been low.

The OP for Maritime Affairs and Fisheries over 2014–2020 had €344 million available, including €252 million of EU funding through the **European Maritime and Fisheries Fund (EMFF)** for 36 defined measures. The main expected results for nature protection from the EMFF investments relate to reduced impacts of fishing on the environment and more sustainable use of marine biological resources, with the most relevant measures listed in Table 9.6. According to the PAF, whilst the uptake of these measures has been low, this is expected to change with the adoption of management plans for Natura 2000 marine sites, and the establishment of national monitoring for marine species and habitat types.

Compared to most other EU Member States, the uptake of **LIFE Programme** funding for nature conservation projects has been relatively low; one of the main reasons has been obstacles with national co-financing. Since 2013, Croatia has participated in five LIFE projects for nature and biodiversity, with total funding of €15 million (EU contribution €10 million). These included three for which Croatia was the main beneficiary: on sea turtles (EUROTURTLES), integrated river management (DRAVA LIFE) and a seabird conservation network in the Adriatic (LIFE Artina).

National and regional financing of nature conservation in Croatia is still highly dependent on the **State Budget** (Anonymous, 2017), which primarily covers regular ongoing administrative expenses. The NPA from 2018 introduced a new financial mechanism requesting each National Park and Nature Park to contribute 3% of its entrance fee revenue, as well as 10% of its excess income, to the State Budget. These funds help to finance the priority needs of the whole system of public institutions that support the National and Nature Parks. **Counties** also contribute to the funding of regional public institutions and administration. The remaining national public funding resources are now focussed on providing necessary co-financing of EU-funded projects, mainly through the Environmental Protection and Energy Efficiency Fund (EPEEF). **Private funding** is practically insignificant.

Over 2011–2017, two internationally funded nature protection projects were of particular importance in Croatia (Ministarstvo gospodarstva i održivog razvoja, 2022). The largest was the **'EU Natura 2000 Integration Project – NIP'**, financed by a World Bank loan of €18.9 million. The project strengthened the capacity of public institutions to manage protected areas; provided substantial investments in visitor infrastructure; collated and integrated data on Croatia's biodiversity; developed the Nature Protection Information System; produced a map of non-forest terrestrial habitats of Croatia; drafted the Agriculture, Environment and Climate Change RDP measure; and prepared strategic projects that were financed from the OPCC for 2014–2020. The other significant project was **'Strengthening the Institutional and Financial Sustainability of the National Protected Area System – PARCS'**, which received a €4.4 million UNDP Global Environment Facility grant.

From this short overview of the financing for nature conservation in Croatia, it can be concluded that significant financial resources have potentially been available, especially from EU funds, but their uptake has hardly matched their availability for the various reasons mentioned earlier. Strengthening relevant administrative capacities, as well as information and education activities aimed at those stakeholders that have the potential to implement conservation measures, seems to be the main requirement to change this situation.

9.8 Surveillance, Monitoring, Reporting and Applied Research

Although there is a long history of biodiversity research in Croatia, the first attempt to collect all relevant information and to prepare an overview of biodiversity was not until 1999, during preparation of the first *National Biodiversity and Landscape Strategy and Action Plan* and the first National

Table 9.6 Funding sources for nature conservation in Croatia.

Source and period	Measure/category	Average annual allocation (€ m)		Comments
EU funding instruments		**EU + national**	**EU**	
EAFRD Rural Development Programme measures – allocations to Priority 4 (restoring, preserving and enhancing agriculture and forest ecosystems) 2014–2020	M13 Areas facing constraints	45.4	38.6	May help maintain important karst pastures
	M11 Organic farming	26.5	22.5	Large uptake, popular measure
	M10 Agri-environment–climate	9.3	7.9	Most important for biodiversity but low uptake
	M8 Investments in forests	6.7	5.7	For N2k, etc. Mostly conversion of forest monocultures
	M4 Investments in physical assets	0.9	0.8	According to PAF, not yet realised
	M1 Knowledge transfer, etc.	0.3	0.3	
	M2 Advisory services, etc.	0.3	0.3	
RDP Priority 4 subtotal		*89.4*	*76.0*	
ERDF, European Social Fund and Cohesion Fund 2014–2020	85 Protection and enhancement of biodiversity, etc.	4.8	3.7	For N2k, etc. Mostly the projects of MESD and Croatian Forests
	86 Protection, restoration, etc. of N2k.	19.9	16.0	For N2k, etc. Mostly the projects of MESD and Croatian Forests
	91 Development of tourism potential of natural areas	25.9	18.3	For N2k, etc. Including for PAs mostly for visitor infrastructure
ERDF – Interreg 2014–2020	Projects related to biodiversity, etc.	5.3	4.5	For N2k, etc.
EMFF 2014–2020	M II.10 – see notes	0.4	0.3	For N2k, etc.
	M I.15 – see notes	0.3	0.2	
	M I.17/I.18 – see notes	0.4	0.3	
	M VIII.2/VIII.3 – see notes	0.2	0.1	
LIFE Programme	Nature projects (2013–2020)	1.9	1.3	Five projects included; others have also contributed
EU funding instruments subtotal		*148.5*	*120.7*	
Within 2014–2020		**Additional non-EU***	**Comments**	
National funds				
State budget	Administration, inc. salaries and running cots	9.4	Based on costs over 2014–2018	
State budget, EPEEF and regional and local government	Protected area management	5.3	Based on costs over 2014–2018	
MEE (now MESD)	Compensation for strictly protected species	1.5		
EPEEF	Several SINP/CAEN projects	0.1	For N2k, etc.	
Others				
World Bank loan	EU N2k Integration Project (NIP)	2.1	See comment 1	

Table 9.6 *(cont.)*

Source and period	Measure/category	Average annual allocation (€ m)		Comments
EU funding instruments		**EU + national**	**EU**	
GEF grant	One project	0.6		See comment 2
Other international sources	Several long-term projects	0.1		See comment 3
NGOs		0.6		Based on costs over 2014–2018. Mostly from Swiss MAVA foundation
Additional national and others subtotal		*19.8*		
National total including co-financing of EU instruments		*47.6*		
Total		**168.3**		

Notes. 'For N2k, etc.' indicates funding relevant to Natura 2000, related green infrastructure and BHD species (based on the PAF). EMFF: M II.10 = Aquaculture providing environmental services; M I.15 = Restriction of influence of fisheries on marine environment, and protection of species; M I.17/I.18 = Conservation of marine biodiversity and ecosystems, and sustainable fishing activities; M VIII.2/VIII.3 = Marine environment conservation, sustainable use and improving knowledge including for protected areas and N2k. * Excludes co-financing of EU funds. The subtotals and totals are only approximate indications of the average annual levels of funding as they are based on annualised data from differing periods. *Comments.* 1. Full project duration 2011–2017, with €18.9 million funding, of which €14.7 million during 2014–2020 funding period. 2. Project on Strengthening the Institutional and Financial Sustainability of the National Protected Area System. 3. Projects inc. 'Towards strengthened conservation planning in South-Eastern Europe'; ESENIAS; MedMPAnet; WWF MedPAN South project–phase II – SEA-Med and WWF Dinaric Arc Parks.

Sources. Based on data from: RDP data from European Commission ESIF funds database (23 October 2020); PAF 2021–2027 (Ministarstvo gospodarstva i održivog razvoja, 2022); Croatia LIFE factsheet, European Commission (March 2021). See Section 1.3 for further details.

Report to the CBD.[46] In 2003, the SINP was founded as the central nature conservation expert service, with one of its tasks being to establish a biodiversity inventory and monitoring. This initiated a period of intensive work, engaging the scientific community and competent NGOs. It was financed from the State Budget as well as from various projects, including the EU Natura 2000 Integration Project – World Bank loan; PHARE project on Natura 2000; Pre-Accession Assistance (IPA) Programme NATURA 2000: Management and Monitoring (MAN-MON); IPA Preparation of the marine part of Natura 2000 and others. A national monitoring framework has also been gradually developed, supported by a number of manuals and training.[47]

The former CAEN, in cooperation with experts for individual taxonomic groups, has prepared 58 national monitoring programmes for 81 species and habitat types based on a methodological framework set up through the project MAN-MON in 2012.[48] The best covered groups are birds (31) and freshwater fishes (20). However, some of the adopted monitoring programmes have not started yet due to a lack of financing. The aim is to prepare monitoring programmes for at least 400 species and habitat types by 2023, primarily through an OPCC project.[49] Other OPCC projects are collecting data on marine habitats and establishing a national system for monitoring IAS. All these OPCC projects started in 2018.

The first Habitat Map of Croatia, which is based on a national classification,[50] was prepared in 2004 (scale 1:100 000), followed by the new Map of Terrestrial Non-forest Habitats (scale 1:25 000) in 2016. A methodology for field mapping of marine habitats has been developed and the first individual mapping projects have started. A Nature Protection Information System has also been established, which includes publicly available data through a web GIS-based Bioportal.[51] It is linked with a previously developed Flora Croatica Database on vascular plants,[52] led by the Department of Botany of the Faculty of Natural Sciences.

9.9 Future Developments

According to the already mentioned NPSAP, which is the main strategic document of nature protection in Croatia, the priorities of nature protection in the forthcoming period are strongly connected with the establishment of implementation mechanisms that aim to ensure the favourable conservation status of species and habitat types, primarily through setting up the management framework for the Natura 2000 network. Its **strategic goals** are to: increase the effectiveness of key nature protection mechanisms; reduce direct pressures on

46 www.cbd.int/reports/
47 www.haop.hr/hr/taxonomy/term/153
48 www.haop.hr/hr/tematska-podrucja/prirodne-vrijednosti-stanje-i-ocuvanje/pracenje-stanja-prirode/provedba-pracenja
49 OPCC Project 'Development of national system for monitoring conservation status of species and habitats of EU interest' (2018–2023).
50 www.haop.hr/sites/default/files/uploads/dokumenti/03_prirodne/stanista/NKS_2018_opisi_ver5.pdf
51 www.bioportal.hr/gis/
52 https://hirc.botanic.hr/fcd/

nature and promote sustainable use of natural resources; strengthen the capacities of the nature protection system; increase the knowledge and availability of data on nature; and raise the level of knowledge, understanding and support for nature conservation among the general public.

The implementation of priority activities has started primarily through the **OPCC projects** that concentrate on the management of Natura 2000, preparation of MPAPs for priority species, development of the national monitoring framework, the mapping of marine habitat types, and dealing with IAS and other issues.

However, the NPSAP's strategic goals are concentrated on core needs for which the nature conservation sector is directly responsible according to the NPA and Nature Directives. This leaves a number of other important broader and longer-term issues that need to be addressed, such as how to conserve biodiversity in the face of agricultural development, how to restore habitats in protected areas and how to adapt habitats and species management to ever-increasing climate changes. It is therefore important that these issues are also addressed by the most relevant sectors in such a way that they become an integral part of their policy, legislation and strategic planning, especially where they relate to agriculture, forestry, fisheries and hunting.

9.10 Conclusions

Over its long tradition of nature conservation, Croatia has been relatively successful in preserving much of its outstanding natural assets and their values. One important reason for this has been the systematic, devoted work of its nature conservation expert service, established in 1961. Since then the service has carried out key technical work, including the designation of all protected areas, including Natura 2000, numerous conservation projects, organisation of systematic biodiversity inventories and monitoring, and regularly advising relevant national and regional authorities and public institutions.

A large network of domestic protected areas has been established; this has been further extended by the Natura 2000 network and now covers 38.6% of Croatia's land, including 36.7% Natura 2000 coverage. As a result, the terrestrial Natura 2000 network is the second highest amongst EU Member States (Table 5.4). However, the marine components of the Natura 2000 network and the set of national MPAs have still to be completed – one of the key challenges for the coming years.

Besides its large network of designated sites, some good examples of other nature conservation actions have included increasing large carnivore populations whilst also addressing conflicts with farmers. Although the Eurasian Lynx remains Critically Endangered in the country, and the Wolf still suffers from illegal killing, these issues are being dealt with. Public awareness regarding Wolf protection is rising while the planned lynx repopulation will hopefully improve its conservation status. All these efforts demonstrate how long-lasting, comprehensive and demanding the work on conservation of such species can be.

Currently, nature protection in Croatia, although struggling with inadequate administrative and financial capacities compared to other sectors, is managing to meet its principal national, EU and international commitments. Building on the results of a number of important EU-supported pre-accession projects, today Croatia continues to systematically address its priority activities (e.g. preparation of Natura 2000 site management plans and MPAPs for priority species, and development of the national monitoring framework) through a number of further EU-funded projects, which will hopefully be followed up. One of the greatest challenges will be to fully establish the necessary management measures in Natura 2000 sites and to ensure regular adequate biodiversity monitoring. Conservation efforts outside the networks of protected areas will hopefully also be intensified in coming years (Ministarstvo gospodarstva i održivog razvoja, 2022), especially bearing in mind the significant number of habitat types and species reported to have an unfavourable conservation status.

Acknowledgements

I would like to warmly thank Professor Dr. sc. Jasenka Topić (retired professor from the Department of Botany of the University of Zagreb) and Dr. sc. Vesna Tutiš from the Institute for Ornithology of the Croatian Academy of Sciences and Arts, who reviewed and supported this chapter, and provided a number of valuable comments.

References

Anonymous (2017) *Strategija i akcijski plan zaštite prirode Republike Hrvatske za razdoblje od 2017. do 2025. godine.* Narodne novine br. 72/2017.

BirdLife International (2020) *Country profile: Croatia.* www.birdlife.org/datazone/country/croatia

DZZP (2014) *Analiza stanja prirode u Republici Hrvatskoj za razdoblje 2008. - 2012.* Zagreb: Državni zavod za zaštitu prirode.

European Commission (2017) *The EU Environmental Implementation Review. Country Report – Croatia.* European Commission.

Jeremić, J. & Štrbenac, A. (2017) *Izvješće o stanju populacije vuka u Hrvatskoj u 2016. godini.* Zagreb: Hrvatska agencija za okoliš i prirodu.

Kamenarović, M. (1961) *Zaštita prirode u Hrvatskoj.* Zagreb: Zavod za zaštitu prirode.

Ministarstvo gospodarstva i održivog razvoja (2022) *Prijedlog Odluke o donošenju Programa financiranja ekološke mreže Natura 2000 Republike Hrvatske i Prioritetnog akcijskog okvira (PAO) za mrežu Natura 2000 u Republici Hrvatskoj za razdoblje 2021.-2027.*

Ministarstvo kulture (2010) *Plan upravljanja vukom u Republici Hrvatskoj za razdoblje 2010.-2015.godine.* Zagreb: Ministarstvo kulture.

Ministarstvo poljoprivrede & Ministarstvo zaštite okoliša i energetike (2019) *Plan gospodarenja smeđim medvjedom u Republici Hrvatskoj.* Zagreb: Ministarstvo poljoprivrede & Ministarstvo zaštite okoliša i energetike.

Ministarstvo zaštite okoliša i energetike & Hrvatska agencija za okoliš i prirodu (2018) *Smjernice za planiranje upravljanja zaštićenim područjima i/ili područjima ekološke mreže*. Zagreb: UNDP, Hrvatska.

MENP (2014) *Fifth National Report of the Republic of Croatia to the Convention on Biological Diversity*. Zagreb: Ministry of Environmental and Nature Protection.

Tutiš, V., Kralj, J., Radović, D., Ćiković, D. & Barišić, S. (eds.) (2013) *Crvena knjiga ptica Hrvatske [Red Data Book of Birds of Croatia]*. Zagreb: Ministarstvo zaštite okoliša i prirode, Državni zavod za zaštitu prirode.

UNEP-WCMC (2022) *Protected Area Profile for Croatia from the World Database of Protected Areas, February 2022*. www.protectedplanet.net

ZZOP (2019) *Izvješće o stanju prirode u Republici Hrvatskoj za razdoblje od 2013. do 2017. godine*. Zagreb: Ministarstvo zaštite okoliša i energetike, Zavod za zaštitu okoliša i prirode.

Chapter

10

Cyprus

Clairie Papazoglou and Andreas Demetropoulos†

Key Statistics	
Terrestrial area (km²)/% of EU-28*	9 249/0.2%
Marine waters (km²)/% of EU-28*	98 040/2.1%
Population density (p km²)/% of EU-28*	92/80%
GDP per capita (€)/% of EU-28	22 387/84%
Land cover (CORINE land cover classes 2018)	AGR 48%, OSN 27%, FOR 16%, ART 9%, WET 0.5%
Biogeographical regions	MED, MMED
Number of HD Annex I habitats/%*	42/19%
Number of HD Annex II, IV and V species/%*	58/5%
Number of BD Annex I bird taxa/%*	83/47%
% HD UCS and decreasing habitats/ % unknown	14%/0%
% HD UCS and decreasing species/ % unknown	0%/13%
% breeding BD Annex I birds decreasing/% uncertain and unknown	10%/7%

* Excluding MAC and MMAC biogeographical regions.
Sources and codes: See Table 1.1.

This chapter includes a description of the biodiversity of the whole island of Cyprus, but since the focus of this book is on nature conservation policy, and, especially, the effect of EU policy, it focuses on the situation in the part of the island where the *acquis communautaire* applies. The Republic of Cyprus has been part of the EU since 1 May 2004. Whereas the whole island acceded to the EU, according to Protocol 10 of the Treaty of the Accession of Cyprus to the EU, the application of the *acquis* has been suspended in those areas where the Government of the Republic does not exercise effective control, that is to say the northern part of the island. Furthermore, another Protocol (no. 3) has been signed in relation to the two Sovereign Base Areas (SBAs) on the island, according to which the Accession Treaty shall not apply to the SBAs in Cyprus, except to the extent necessary to ensure the implementation of the arrangements set out in the Protocol. In practice, the SBAs in Cyprus enforce their own legislation on nature protection, which largely mirrors the Republic's transposition of the Nature Directives.

10.1 The Natural History and Characteristics of Cyprus

The wealth of the natural environment of Cyprus is a consequence of its proximity to three continents (Africa, Europe and Asia), geology, morphology, climatic conditions and partial isolation as an island. Moreover, it lies within the Mediterranean Basin, which is one of the 36 global biodiversity hotspots (Cuttelod *et al.*, 2008).[1] As a result of these combined influences Cyprus has a high degree of endemism, especially amongst plants and insects. Its habitats and associated species are now greatly influenced by human activities, with evidence suggesting that Cyprus has been permanently inhabited since about 8 500 BCE.

Historically, Cyprus was covered in dense conifer forests and maquis, but from around the fourth millennium BCE, when copper was discovered, a lot of forests were cut and burnt. Forest clearance continued with the expansion of agriculture and more recently because of tourism development, especially near the coast. Today around 18% of the island is covered in tall forests, 13% in maquis, 9% in phrygana (garrigue) and around 0.5% in other vegetation categories (Hadjikyriakou, 2017). In 2012 the Utilised Agricultural Area (UAA) of Cyprus was around 21% of the area under the control of the Government of the Republic. However, the UAA has greatly diminished since 1975, especially as regards extensive crops (e.g. vines, cereals, olive and carob trees), compared to more intensive crops (i.e. potatoes, other fruit and vegetables), which comprise around 10% of the UAA (Zomeni *et al.*, 2018). The natural environment has also been changed substantially by the damming of rivers (with Cyprus having more dams than any other European country for its size), which has caused chain effects on coastal estuaries and freshwater ecosystems as well as erosion of coastal areas.[2]

1 www.cepf.net/our-work/biodiversity-hotspots/hotspots-defined
2 Fourth National Report to the CBD (Cyprus) Nicosia, September 2010.

In addition, a relatively high proportion of land is now taken up for housing and tourism, with artificial areas covering 9% in 2018 according to CORINE data.

Despite the major changes to its natural environment, Cyprus still has an exceptionally diverse flora. According to the latest update, Cyprus has 1 640 species of indigenous plant taxa, of which 110 species and 32 subspecies are endemic (Christodoulou *et al.*, 2018). The main centres of endemism for plants are the Troodos and Pentadaktylos mountains, as well as the Akamas area in the Pafos (Paphos) District in the west of the island. Cyprus, upon accession to the EU, added 18 species of plants to HD Annex II, including 16 endemic species (MANRE, 2013).

The floristic diversity is also reflected in the island's habitat diversity: there are 52 HD habitat types recorded on the island of Cyprus, of which 15 are priority habitats.[3] The Republic reported on 42 HD habitats in its HD Article 17 report for 2014–2018. The most characteristic habitat types on the island are forests of pine, olive and carob and Oriental Planes (*Platanus orientalis*), and Mediterranean mattoral with Juniper, and areas with Mediterranean phrygana vegetation with species such as *Sarcopoterium spinosum*. As a result of the accession of Cyprus, five new habitat types were added to HD Annex I and the definitions of four others were amended in order to better represent Cyprus habitats. The new habitat additions were: serpentinophilous grassland of Cyprus (62B0*),[4] peat grasslands of Troodos (6460), scrub and low forest vegetation with *Quercus alnifolia* (9390*), *Cedrus brevifolia* forests (*Cedrosetum brevifoliae*) (9590*) and woodland with *Quercus infectoria* (93A0) (MANRE, 2013).

In terms of animals, the Cyprus Mouflon *Ovis orientalis ophion* (*Ovis gmelini ophion*) is of particular interest as it is a subspecies of wild sheep that was probably introduced by the first people that inhabited the island and has since developed into an endemic subspecies. Other endemic animal species include the Cypriot Mouse (*Mus cypriacus*), one snake (*Coluber cypriensis*) and one lizard (*Phoenicolacerta troodica*). As for birds, out of 90 regular breeders (60 residents and 30 migrants), three are endemic. These include the Cyprus Warbler (*Sylvia melanothorax*) and the Cyprus Wheatear (*Oenanthe cypriaca*), which were added to BD Annex I upon accession. A third recently approved endemic is the Cyprus Scops-owl (*Otus cyprius*).[5]

Cyprus hosts very important populations of Loggerhead Turtle (*Caretta caretta*) and Green Turtle (*Chelonia mydas*), the two species of sea turtle nesting in the EU, the latter nesting only in Cyprus among the EU countries (Demetropoulos & Hadjichristophorou, 2017). The marine environment around Cyprus is home to cetaceans (mainly dolphins) and the extremely rare Mediterranean Monk Seal (*Monachus monachus*), of which only 7–15 individuals were recorded in the HD 17 report for 2014–2018. The shallow seas around Cyprus are rich in *Posidonia* beds (1120*).

10.2 The Status of Nature

10.2.1 Habitat Status and Trends

Cyprus submitted its first report in accordance with HD Article 17[6] for the period 2001–2006, but as it joined the EU in May 2004 this covered only a short period. In the first reporting period of 2001–2006, 36% of terrestrial habitats and 75% of marine habitats were reported as unfavourable, and a large percentage of habitats and species were reported as 'unknown'. That changed with the 2007–2012 reporting period, when Cyprus reported 97% of habitats as favourable. Habitats were evaluated using field sampling and aerial photographs for that period, not using all parameters required in the national report. The 2013–2018 report was more complete, using mapping of all habitat types (including some outside Natura 2000), monitoring protocols and covering all parameters required. In that report only 45% of habitats were reported as favourable, with 19% having a poor status, and 36% a bad status. This 52% reduction in favourable status is justified as the result of using other methods and/or improved knowledge. There were 23 habitats reported as unfavourable, with six of them having a bad status and a decreasing trend (Table 10.1). These six habitats belong in the coastal (especially salt marsh habitats), dunes and freshwater categories (especially intermittently flowing rivers).

10.2.2 Species Status and Trends

Similarly, Cyprus reported on the HD species for the first time in the period 2001–2006. At that time only 50 species were assessed, whereas in the two subsequent assessments (2007–2012 and 2013–2018) 55 species were assessed. In the 2007–2012 report 69% of species were reported as having a favourable status, while in the 2013–2018 report this proportion dropped to 64% – the highest proportion in the EU (EEA, 2020). Remarkably, none was considered to have a declining trend in the most recent report (Table 10.1) or its predecessor. These species trends do not seem to be credible given the status of habitats and known pressures on species discussed below. The high proportion with unknown trends and significant changes from one reporting period to the other, especially for habitats, clearly points to a lack of robust biodiversity data. This problem is particularly notable for snakes, lizards and some mammals (mainly bats). Monitoring of most plants and habitats is much better, and it is exceptional for sea turtles. As

[3] Based on the check list for HD Article 17 reporting https://cdr.eionet.europa.eu/Converters/run_conversion?file=cy/eu/art17/envxw_f9q/CY_habitats_checklist.xml&conv=591&source=remote.

[4] Numbers in brackets are the HD Annex I habitat codes (see the Appendix for full names). * Indicates a priority habitat.

[5] BirdLife International (2020) Species factsheet: *Otus cyprius*. www.birdlife.org (2 February 2020).

[6] See Section 1.3 for background on monitoring under the Nature Directives, status categories and data sources.

Table 10.1 Short-term trends (c. 2013–2018) in Habitats Directive Annex I habitats and Annex II, IV and V species with an unfavourable conservation status in Cyprus

	Poor (unfavourable–inadequate)				Bad (unfavourable–bad)			
	Imp	Sta	Dec	?	Imp	Sta	Dec	?
HD Habitat assessments								
Number	0	7	0	1	0	6	6	3
%	0%	88%	0%	12%	0%	40%	40%	20%
HD Species assessments								
Number	2	3	0	6	0	0	0	2
%	18%	27%	0%	55%	0%	0%	0%	100%

Key. Imp = improving, Sta = stable, Dec = decreasing, ? = unknown.
Source. Based on Member State HD Article 17 reported data as presented on the EEA national summary dashboard (2021).

covered later, the most comprehensive and reliable data are generally available for birds.

The species most recently reported as unfavourable include a number of species of bats, the Mediterranean Monk Seal, the Green Turtle, the Striped-neck Terrapin (*Mauremys caspica*/*Mauremys rivulata*) and three species of plants. Species listed as having a bad status included the Cyprus Grass Snake (*Natrix natrix cypriaca*) and the Egyptian Fruit Bat (*Rousettus aegyptiacus*), while the trend for both was reported as unknown. These species clearly constitute conservation priorities. Regarding the plants, it is worth noting that of the 19 species of plants assessed for the last Article 17 report, 14 have part or all of their population in the part of Cyprus under the effective control of the Republic of Cyprus. Of the 14, a total of 11 are listed as Vulnerable, Endangered, Critical or Near Threatened in the *Cyprus Red Data Book* (Tsintides *et al.*, 2007, in Greek) and/or at global, European and Mediterranean level (The IUCN Red List of Threatened Species online), but were nevertheless reported as Favourable in the Article 17 report.

In Cyprus, the Game and Fauna Service (GFS) regularly monitors birds, especially raptors, some game species and waterbirds, while BirdLife Cyprus contributes with several medium-term monitoring programmes (see Section 10.8) that provide reliable and comparable data for birds, as well as a wealth of birdwatching data, which can provide useful information.

The BD Article 12 report (2013–2018) indicates that, whilst most birds had stable or increasing populations, short-term declines were reported for 13% of species, and trends were unknown or uncertain for 7%. As illustrated by Table 10.2, which shows trends for Annex I species, the situation for Annex I breeding taxa was rather similar to that shown by all breeding species. No declines were reported for wintering Annex I taxa, but a high proportion had unknown or uncertain trends.

Among the Annex I species showing declines, the most worrying perhaps is in the European Roller (*Coracias garrulus*), a species conservation priority for Cyprus as it holds an important population at EU level, and, of course, the endemic Cyprus Warbler, which is also showing declines, although the cause is ultimately unknown. There is also high concern over the Annex I listed Griffon Vulture (*Gyps fulvus*), as well as two declining BD Annex II species, Quail (*Coturnix coturnix*) and Turtle-dove (*Streptopelia turtur*), especially the latter, as it is threatened in Europe and globally.

Table 10.2 Short-term trends (c. 2007–2018) in Birds Directive Annex I taxa in Cyprus.

	Breeding				Wintering			
	Inc	Sta/ Flu	Dec	?	Inc	Sta/ Flu	Dec	?
Number	10	15	3	2	5	3	0	7
%	33%	50%	10%	7%	33%	20%	0%	47%

Key. Inc = increasing, Sta = stable, Flu = fluctuating, Dec = decreasing, ? = uncertain and unknown.
Source. Based on Member State Article 12 reported data as presented on the EEA national summary dashboard (2021).

10.2.3 Drivers and Causes of Change

Panayides (2005) identified six main aspects of land-use change that affect biodiversity in Cyprus. These include urbanisation/housing and tourism development, damming of rivers, agricultural intensification and abandonment, the very dense forest road/track network and transport infrastructure. Similar major pressures and threats to BHD habitats and species are highlighted in the HD Article 17 and BD Article 12 reports for 2014–2018: construction and land-use change, including urbanisation, tourism and recreation developments, and construction and use of transport infrastructure. Another main threat is the intensification of agriculture. Other important threats mentioned include modification/alteration of natural processes, climate change and invasive alien species (IAS).

Leading up to EU accession (i.e. the late 1990s to 2004) and until today, there have been substantial **infrastructure projects and tourism developments**, especially in coastal areas, leading to high land-use pressures. Unfortunately, mitigation of such pressures has been hampered by a lack of sustainable spatial planning and, in certain cases a lenient application of EU law, which has allowed some incompatible developments in sensitive or protected areas. Examples include villas built above the Mediterranean Monk Seal breeding caves at Thalassines Spilies, a marina development within and adjacent to Ayia Thekla-Liopetri SPA, and the authorisation of the Limni Bay golf resort behind an important sea turtle beach. Badly placed renewables, such as wind farms and lately photovoltaic parks and thermo-solar plants in the countryside are also problematic. It is worth noting that the European Commission also identified the problem of incompatible activities or developments in its 2019 Environmental Implementation Review for Cyprus (European Commission, 2019).

A study analysing the effects of roads on bird species richness in Natura 2000 sites in Cyprus found that roads have detrimental effects on overall species richness for BD Annex I birds, farmland birds and wetland species (Mammides *et al.*, 2015).

Water scarcity and high demand is another important driver of biodiversity pressures in Cyprus, as indicated by its water stress index being the highest in the EU (Sofroniou & Bishop, 2014). Water scarcity has two effects on nature. Firstly, all main natural water courses are dammed and used primarily for water supply and irrigation so only minute quantities of water reach the sea, resulting in increased coastal erosion (due to reduced sediment supply) and impacts on estuarine habitats and species. Secondly, the resulting low flows directly affect aquatic habitats and species. To mitigate this, according to the EU WFD, water should be released from dammed water courses as necessary to maintain the 'minimum ecological flow' that is required to achieve the Directive's objectives. However, despite the Directive's hierarchy for supply that mandates ensuring drinking water demands are met first, then environmental requirements and finally irrigation, the actual practice prioritises agriculture over environment. Although the requirement to ensure minimum ecological flows is included in the measures in the second River Basin Management Plan (RBMP) for 2016–2021 for Cyprus, these flows are not defined clearly in terms of quantities, months or frequency. As a result, compliance cannot be verified, but the impression is that the low flow measures have not been implemented. The third RBMP (2022–2027) is expected to address this issue in detail and will also fully review implementation of the second RBMP. Similarly, the restructuring of irrigation in agriculture has not been implemented.

Climate change is already being felt in Cyprus and is a major pressure and growing threat for its habitats and species. Weather stations have recorded substantial increases in temperature (e.g. an increase of 2.3 °C from 1903 to 2016 in Limassol) and declines in rainfall, from a mean rainfall of around 559 mm annually between 1901 and 1930 to only 463 mm over 1970–2000 (DoE, 2017). So, desertification is a real threat.

The most vulnerable habitats are wetlands, which already suffer from much-reduced water availability and intense competition from other uses, and high-altitude sites. In addition, some priority habitats such as coastal lagoons (1150*), (sub-) Mediterranean pine forests with endemic black pines (9530*) and the peat grasslands of Troodos (6460) are expected to suffer. Recognising the threat of climate change, the Cyprus government adopted in 2017 a National Strategy for Adaptation to Climate Change (DoE, 2017). Although the strategy contains useful actions and ideas, these have not yet been implemented.

Finally, **invasive alien species** are a concern in Cyprus, especially as it is an island country. Project Ris-Ky has tried to compile all sources on IAS in Cyprus and has produced a number of papers on IAS in the SBAs.[7] Apart from the well-known invasive plant species like *Acacia saligna* and *Atriplex semibaccata*, Cyprus is also affected by marine IAS, as it is practically opposite the Suez Canal and therefore amongst the first countries to be affected by Lessepsian species (i.e. Red Sea species) migrating into the Mediterranean Sea.

10.3 Nature Conservation Governance, Actors, Legislation and Policies

10.3.1 Governance

Four government agencies share the responsibilities related to nature conservation: the Department of Environment (DoE), the Department of Forests, the Department of Fisheries and Marine Research (DFMR) and the Game and Fauna Service (GFS). The first three are departments of the Ministry of Agriculture, Rural Development and Environment (MARDE), while the GFS is an agency under the Ministry of Interior. The reasons why the GFS is under the Ministry of Interior are historical and linked to the fact that all enforcement agencies, including the police, were originally under that Ministry and the GFS (Game Fund, as it was then called) was primarily an enforcement agency with game wardens.

The **DoE** has overall responsibility for the implementation of environmental legislation and represents Cyprus in international fora (e.g. EU Committees, CBD, CMS, Bern Convention and RAMSAR). It is also solely responsible for implementing the law on EIA and SEA in accordance with the EU directives as well as the assessment of ecological evaluations and HD Article 6(3) appropriate assessments. The Department is also responsible for overseeing the designation of proposed SCIs (pSCIs) and SACs, according to relevant national Law 153(I)/2003 and with the advice of a Scientific Committee, which includes members from relevant government departments, the union of Municipalities and

7 https://ris-ky.info/resources

Communities, two NGOs, and two independent experts. The Committee is also responsible for approving site management plans, Management Decrees for conservation measures, derogations and other potentially important conservation-related decisions. Finally, the DoE is responsible for the conservation of the Cyprus Grass Snake and the Striped-neck Terrapin, as well as all other species that do not fall under the responsibility of other departments mentioned here.

The **Department of Forests** is one of the oldest government departments, having been founded in 1879, one year after Cyprus was annexed to the British Empire. Today, the Department has prime responsibility for the protection and management of all state forest land (including enforcement of EU legislation), focussing on issues such as monitoring, protection and management of flora and habitats, forest fire prevention and control (together with the Fire Service), and environmental education and leisure activities in nature[8] (including running three environmental education centres). The Department is also involved in biodiversity projects and surveys, including some on bats and marine mammals, although those responsibilities are not clearly ascribed in the forest law.

The **DFMR** has responsibility for the monitoring and protection of marine habitats and species (mainly the two sea turtles, the Mediterranean Monk Seal and cetaceans). The DFMR is the department that issues restrictions, regulations and decrees for activities in the marine and coastal environment, and is also responsible for the management of all freshwater fish.

The **Water Development Department (WDD)** issues restrictions and regulations for dams and is responsible jointly with the DoE for wetland monitoring.

Bird protection has followed another path, mainly due to historical reasons for the way hunting and poaching were regulated. For this reason, the **GFS**, which is under the Ministry of Interior, is responsible for implementing the Birds Directive (i.e. the conservation of wild birds and their habitats, classifying SPAs and controlling hunting and illegal killing). It is also responsible for the Cyprus Mouflon and monitoring and management of all game birds and the European Hare (*Lepus europaeus*). The GFS wardens primarily patrol for hunting and poaching violations, but they increasingly also look for site protection violations. In fact, following a modification of Law 152(I)/2003 in 2017, out-of-court penalties were introduced for unauthorised interventions or development in SPAs.

This governance structure has led to fragmentation of responsibilities and overlapping, unclear roles between the different departments. This fragmentation has been discussed in a number of studies: firstly by Jean-Francois Verstrynge, Honorary Director-General, DG Environment, European Commission (Verstrynge, 2004); secondly in a study by the World Bank focussed the Ministry of Agriculture and Rural Development, proposing restructuring (The World Bank, 2014); and finally in 2018 the Scottish Environment Protection Agency (SEPA) provided support to the Cypriot Government to develop a national environmental permitting and inspection agency. Following these studies, in January 2022 it was announced that a new Directorate General will be created in the MARDE that will include all major departments and agencies and will be responsible for all environmental issues, including nature protection.

10.3.2 Environmental NGOs

Cyprus in general has quite a weak base of NGOs and civil society, and the operating environment for NGOs is not empowering (Expert Council on NGO Law of the Council of Europe's Conference of INGOs *et al.*, 2015). This is mainly due to the very strong influence of political parties or other special interest groups, the negative public perception of NGOs, the lack of domestic funding opportunities and lack of professionalism in the NGO sector itself. Nevertheless, there are a number of environmental NGOs in Cyprus, most of which are very small and entirely based on volunteers. Among the first were those dealing with birds – the Cyprus Ornithological Society of 1957 and another with the same name of 1970, which in 2003 merged to found BirdLife Cyprus. The main issue they dealt with, apart from observing and recording birds, was the fight against bird trapping. Today BirdLife Cyprus has professional staff and is one of the biggest environmental NGOs in Cyprus. Another environmental NGO with staff is Terra Cypria, founded in 1992. Other environmental NGOs include Friends of the Earth Cyprus (founded in 1980), Friends of Akamas (founded in 1986) and Cyprus Ecological Movement (founded in 1989), but their impact is limited by their weak membership base and/or few or no staff.

There are also a few NGOs specifically focussed on the marine environment: Cyprus Wildlife Society (founded in 1984), AKTI Project and Research Centre (founded in 2000), Cyprus Marine Environment Protection Association and Enalia Physis Environmental Research Centre (founded in 2009). Cyprus Wildlife Society plays an important role in monitoring and protecting sea turtles, and in the activities of the Federation of Environmental Organisations (see later).

Cyprus Natural Coastline is a citizens' voluntary group initiative that has been very active in the last few years, campaigning for the protection of coastal areas of significant ecological value and other Natura 2000 sites, fighting against poor appropriate assessments, raising public awareness, and organising public events that attract a lot of publicity.

In 1988 the Federation of Environmental Organisations of Cyprus was founded to act as an umbrella organisation for environmental NGOs. Currently it has around 20 member organisations, mostly small volunteer-based organisations, but neither of the bigger NGOs with permanent staff

[8] www.moa.gov.cy/moa/fd/fd.nsf/fd03_gr/fd03_gr?OpenDocument (Department of Forests, Our History) (22 February 2022).

(i.e. BirdLife Cyprus or Terra Cypria) is a member, and a lack of permanent staff hinders its potential impact.

Despite the challenges, NGOs have made quite an impact, especially during Cyprus's accession negotiations.

The Cypriot hunting community is numerous and powerful politically. Indeed, data from around 2010 suggest that hunting density is among the highest in Europe,[9] with c. 45 000 licensed hunters in the part of the island under the effective control of the Republic of Cyprus and around 11 500 in the northern part (Betz-Heinemann, 2018). Hunters are organised in a federation named Cyprus Hunting Federation and Wildlife Conservation (established in 1952), which enhances their political influence. It is worth noting that the GFS, which is an agency set up under public law, has a management committee that comprises 50% government employees and 50% representatives of the hunting federation, even though the Ministry of Interior alone is responsible for the implementation of the hunting law. The hunters' federation relations with the conservation NGOs are indifferent. Historically, the hunting federation has mainly focussed on hunting and game issues.

10.3.3 Legislation and Policies

The first laws for the protection of nature in Cyprus were adopted shortly after the creation of the Forest Department in 1879 (Table 10.3). Forests started being delineated from the late nineteenth century, the Forest Law entered into force in 1939, and in 1940 the Main State Forests (Κύρια Κρατικά Δάση) were designated. In 1967, more modern legislation for forest protection and management was adopted. Today forest management is regulated by Law 25(I)/2012 (the 2012 law about forests).

The protection of birds goes back to the first legislation introduced in 1974, but this was substantially strengthened after Cyprus ratified the Bern Convention in 1988 (see Section 10.4.2).

The first laws for the protection of sea turtles were adopted around 1979, and more specific laws and regulations were adopted in 1984 and 1989, specifically to protect the Lara/Toxeftra beaches in Akamas for nesting turtles.

Subsequently, preparations for accession to the EU and the transposition of EU legislation into Cypriot laws undoubtedly strengthened the existing legislation and brought about significant changes in governmental practice and understanding around the importance of nature and the assessment of threats to it.

The two main laws transposing the EU Directives are Law 153(I)/2003[10] (Law about the Protection and Management of Nature and Wildlife), which transposed the Habitats Directive, and Law 152(I)/2003 (Law about the Protection and Management of Wild Birds and Game),[11] which transposed the Birds Directive but also covers the hare as a huntable species and the Cyprus Mouflon as an HD species. The first main impact of the Nature Directives in Cyprus was the designation of SCIs and SPAs (see Section 10.5).

According to the two national nature laws, protection of a Natura 2000 site becomes official through the issuance of a Ministerial Decree, signed by either the Minister of Agriculture, Rural Development and Environment (HD law) or the Minister of Interior (BD law), depending on the national law. Additionally, decrees are also issued for declaring sites as SACs and for setting conservation objectives and establishing management measures in Natura 2000 sites. The first conservation objective decree was not adopted until 2011 (for three SACs); more then followed, and in 2021 there were 37 SACs with conservation objectives. Currently there is only one decree setting conservation objectives for one SPA.

In 2017, a catch-all decree was published (238/2017) that provides for general management measures for the SACs. The main reason for publishing this decree was to have a legally binding document with management obligations for Natura 2000 sites, in order to implement Natura 2000 payments (Measure 12) under the Rural Development Programme (RDP) for the 2014–2020 funding period. However, the uptake was very low (by end of 2019 only around 25% had been taken up, and the budget of the measure was revised to 25% of its original)[12] and the decree was revised in 2019 and made more flexible. Whereas numerous Natura 2000 sites have been designated, the actual, *in situ*, protection of these sites has not yet been adequately integrated in government policies and practice. This was partly because the initial laws transposing the Habitats Directive in 2003 contained errors in the transposition of Article 6. The laws included general exceptions from HD Article 6 for any projects/works done under the provisions of the laws for Forests, Land Planning, Fisheries, the Wild Bird and Game Law, and the Nature and Wildlife Law. These exceptions were removed in 2015 through two amendments,[13] after a European Commission letter of formal notice in 2013.

As a consequence, approval of damaging or incompatible projects continued for a number of years following Cyprus's accession. In fact, HD Article 6(3) appropriate assessments were hardly used until about 2011, some seven years after EU accession. NGOs played an important role in raising concerns about this, and the European Commission started

[9] www.face.eu/sites/default/files/attachments/data_hunters-region_sept_2010.pdf

[10] Ο Περί Προστασίας και Διαχείρισης της Φύσης καιτης Άγριας Ζωής Νόμος (153(I)/2003).

[11] Ο Περί Προστασίας και Διαχείρισης Αγρίων Πτηνών και Θηραμάτων Νόμος (152(I)/2003).

[12] RDP Implementation Report for 2019. www.paa.gov.cy/moa/paa/paa.nsf/All/ED19D257267F30EEC225866C002566D4/$file/Implementation%20report_2014CY06RDNP001_2019_0_en.pdf

[13] Amendment 67(I)/2015 to the Law about Protection and Management of Nature and Wildlife Law 153(I)/2003 to 2012, and amendment 68(I)/2015 to the Law about Management of Birds 152(I)/2003 to 2014.

Table 10.3 Key nature conservation developments in Cyprus.

1879	First laws for the protection of forests, under British rule
1971	Fisheries regulations adopted, protecting sea turtles, dolphins and the Mediterranean Monk Seal
1974	Catching of songbirds using limesticks or nets made illegal
1978	Fisheries Department launched a pioneering project to conserve the two sea turtle species in Cyprus (the first sea turtle project in the Mediterranean)
1988	Ratification of the Bern Convention, with reservation for some birds (larks, bee-eater) and Levantine Viper (*Macrovipera lebetina*). Songbird killing made illegal
1989	Fisheries Law Chapter 135, and Regulations, protecting sea turtles
1991	Spring shooting banned for the first time
2001	Bern Convention Recommendation No. 90 (2001) on the catching, killing or trading of protected birds in Cyprus
2002	Royal Society for the Protection of Birds (UK) started a monitoring programme for illegal bird killing on Cyprus, later taken on by BirdLife Cyprus
2004	Accession to the EU, EU Nature Directives and other environmental legislation came into force. Three SCIs designated
2011	Larnaca Declaration, Decl (2011) – European Conference on Illegal Killing of Birds
2012	Judgment of the CJEU, C 340/10, *Commission* v. *Republic of Cyprus*
	30 SPAs classified, 3 SACs and 37 SCIs designated
2019	First LIFE Integrated Project for nature approved
2020	*Cyprus Biodiversity Strategy* adopted at ministerial level

See Table 3.3 for signatory dates of international conventions.

infringements on nature-related issues from around 2006. The infringement procedures concerned the lack of appropriate assessments, as well as damage to, or deterioration of, specific sites (e.g. Paralimni and Oroklini Lakes). Such actions, and the only nature-related CJEU case concerning Cyprus,[14] led to the adoption of new procedures (see Section 10.6) and played an important role in ensuring better implementation of the Nature Directives.

Following the recovery from the economic crisis in Cyprus in 2015, the approval of damaging and incompatible projects continued and coincided with a new approach by the European Commission that focusses more on bilateral discussions (such as through the pilot process – see Section 4.3) and less on legal interventions. Since 2015, a number of cases have been closed and then reopened, at an earlier stage of the infringement procedure. Arguably the most striking example was a case on HD Article 6 that was opened in 2011 (letter of formal notice), with an additional letter of formal notice in 2013, and then closed in 2015 after the five projects that it included were halted or addressed. However, as the problems with HD Article 6 implementation continued, the Commission sent a new Pilot about HD Article 6 in 2015, and in November 2019[15] sent a letter of formal notice on the same issue, claiming that Cyprus has been systematically implementing HD Article 6 wrongly for numerous cases since accession. The extremely slow processes of the European Commission have meanwhile allowed more sites to be affected.

Cyprus adopted a ***Biodiversity Strategy and Action Plan*** in June 2020 (DoE, 2020). The Frederick University of Cyprus produced a draft in 2012 that went through stakeholder consultation and prioritisation, leading to the final strategy. The strategy includes a thorough description of the biodiversity of Cyprus and the threats to it, proposes 13 strategic goals and 86 different measures, and sets targets for 2020–2030 in accordance with the EU and CBD biodiversity obligations. The budget for its implementation is nearly €27 million, excluding government salaries and those measures funded under the RDP.

10.4 Species Protection and Conservation Measures

The main species protection measures have historically included Ministerial Decrees or Regulations. Those have been

[14] Case C-340/10, *Commission* v. *Republic of Cyprus*, about the failure to include Paralimni Lake as an SCI; and protection of the Cyprus Grass Snake. Judgment 15 March 2012.

[15] INFR (219) 2303 https://ec.europa.eu/atwork/applying-eu-law/infringements-proceedings/infringement_decisions/index.cfm?lang_code=EN&typeOfSearch=true&active_only=1&noncom=0&r_dossier=INFR%282019%292303&decision_date_from=&decision_date_to=&DG=ENV&title=nature&submit=Search

adopted under the Fisheries Law for the protection of sea turtles, dolphins and the Mediterranean Monk Seal, or under the hunting law for the regulation of hunting activities and prohibition of illegal killing of birds.

10.4.1 Hunting

The regulation of hunting of game species and its enforcement has been historically effective in Cyprus, given the very high number of hunters and the demand for game. The agency responsible, the GFS, is probably the only agency with such a high number of wardens (around 80 at Republic level), who patrol day and night, seven days a week. Nevertheless, there are regularly cases of poaching, especially of European Hare, a favoured game species, as well as birds. The bird species favoured as game are Chukar (*Alectoris chukar*), Black Francolin (*Francolinus francolinus*), Turtle-dove (*Streptopelia turtur*), Woodpigeon (*Columba palumbus*) and thrushes, mainly in the winter. Hunting is generally allowed, both on state and private land, except in designated game reserve areas, and provided that certain minimum distances from built areas (e.g. 500 m from villages) are respected. Game reserves are either permanent (i.e. Pafos forest) or temporary, and in total take up about 30% of the area under the effective control of the government. Temporary game reserves might change from year to year. Hunters regularly blame Red Foxes (*Vulpes vulpes*) for prey fluctuations and persecute them even if fox shooting or poisoning is not allowed.

The hunting period starts in August, mainly for Woodpigeon and Turtle-dove, but the 'main' hunting period is from November to December, when hunting of Chukar, Black Francolin and hare is usually allowed for 18 hunting days (Wednesdays and Sundays). Hunting continues until the end of February, mainly for thrushes, Skylark (*Alauda arvensis*) and Woodpigeon.

Spring hunting of Turtle-dove was allowed in Cyprus until it was banned in 1991, following the ratification of the Bern Convention in 1988. Spring hunting was not therefore an issue during accession to the EU. Nevertheless, some hunters did demand it after accession, and it was allowed for one year in 2007, on the premise of BD Articles 9.1.b and 9.1.c derogations (see Section 4.3). However, the Commission immediately opened a case against this, and spring hunting of the species has not been repeated, nor have there been further requests for it. Incidents of illegal shooting of Turtle-doves in spring are generally isolated.

10.4.2 Killing of Songbirds

Despite the relatively good management of regular hunting activities, the illegal killing of songbirds remains a long-term, persistent and serious problem and one for which Cyprus has attracted a lot of negative international publicity. Trapping of songbirds for food is a practice recorded in Cyprus since ancient times and re-emerged in the Middle Ages. It involves trapping birds using a natural glue (made from the fruit of the Syrian Plum) spread on twigs placed in possible perching places in trees and bushes. Historically, trapped birds were a food supplement and form of trade for the poor island inhabitants living off the land, with some sent to nobles and rulers in Venice and Rome during the Frankish Era (1191–1489) and Venetian Era (1489–1571).[16] The main species targeted are songbirds during their autumn migration, mostly Blackcaps (*Sylvia atricapilla*), when they arrive on the island full of fat, ready for the rest of their migration leg. It is this fat that makes them 'delicacies' for some in Cyprus.

Trapping with limesticks and mist nets was forbidden for the first time in 1974 under the Protection and Development of Game and Wild Birds Law of 1974 (39/1974). According to that law, the method (limesticks) and the trade were prohibited but not killing of the actual species. After Cyprus ratified the Bern Convention in 1988, whereby a long list of songbirds became 'protected' species, the law was amended and prohibited for the first time the killing of the actual bird species (i.e. the Blackcap and other songbirds). Ahead of Cyprus joining the EU in May 2004, the law was changed again in order to transpose the provisions of the Birds Directive into Law 152(I)/2003, which prohibited the use of methods such as limesticks, mist nets and calling devices as well as the possession of all the banned equipment, trapped birds and devices. Anyone caught breaking the law went to court, but the courts usually imposed very lenient penalties of a few hundred euros.

Although illegal, since the mid-1980s the songbird 'business' has become very lucrative, attracting huge profits for those operating 'professionally', because the birds are sold in restaurants where they are served as expensive delicacies, costing as much as €80 for 12 birds. The GFS estimated the annual illegal turnover from the sale of the birds to be around €15 million in a paper submitted to the Cyprus Parliament in 2010 (GFS, 2010).

According to BirdLife Cyprus, which has monitored this illegal activity on the island since 2002, the total number of species affected is as many as 155 (BirdLife Cyprus, 2019). This includes all songbirds, such as warblers, but also shrikes, swallows, wagtails, Robins (*Erithacus rubecula*) (in the winter), and many species that could be characterised as 'by-catch', such as small raptors, owls and nightjars, which are discarded when caught. According to the winter 2018–2019 report of BirdLife Cyprus, during 2018 as many as 365 000 birds could have been trapped across the island (BirdLife Cyprus, 2019). Although this is a very high number, it represents a significant reduction of as much as 85% in the number of birds trapped compared to the previous years. The NGO estimated that in 2016 as many as 2.5 million birds could have been trapped across the island (BirdLife Cyprus, 2017b).

This reduction is attributed to three main factors: systematic monitoring of illegal bird trapping by NGOs, better

[16] http://foodmuseum.cs.ucy.ac.cy/web/guest/parsintages/civitem/1195, downloaded 16 February 2020.

enforcement on the ground, and higher financial penalties for offenders. BirdLife Cyprus, in collaboration with the RSPB, started monitoring bird trapping in 2002, mostly by mist nets, and the Committee Against Bird Slaughter (CABS) started monitoring bird trapping in 2009, mostly by limesticks. These systematic monitoring programmes provided sufficient documentation of this wildlife crime activity and put pressure on law enforcement agencies to take the appropriate action. Historically, the few hundreds of euro penalties imposed by the Courts were insignificant compared to the huge profits from selling the birds. This all changed in 2017 when the law N.152 (I)/2003 was amended and all offences stipulated under the law became high 'out of court' fines. The motivation for this change was to impose high penalties that would act as a deterrent, especially for those engaging in illegal killing of birds and other game (e.g. hares). The fines generally are quite high (e.g. minimum €2 000 for the use of nets) except for the use of limesticks (e.g. €200), which is problematic, although limesticks are not generally used by the 'professional' trappers. However, in autumn 2020 the Parliament amended that law again and reduced the fine for 14 species, including the Blackcap, when killed in autumn, and that was seen as an encouragement for those trapping with limesticks or shooting these species; consequently trapping seems to be rising again (BirdLife Cyprus, 2022).

At the same time, the police in the SBAs in Dekeleia in the east of the island have increased enforcement during the last three years, working with BirdLife Cyprus and the RSPB from the UK, as well as NGO CABS. The autumn 2018 report of NGO BirdLife Cyprus notes that the trapping index in the SBAs in Dekeleia was 91% lower than the all-time high of 2016, when trapping levels in the SBAs used to be around seven times more intense than in the Republic (BirdLife Cyprus, 2017a), Cape Pyla in the SBAs being a particular hotspot for this illegal activity.

The issue of illegal killing of birds over the years has attracted a lot of international attention, resulting in recommendations from the Bern Convention (Rec. No. 90, from 2001, following Rec. No.5 of 1986), with the file remaining open to today, and special reference to the trapping of birds in Cyprus in the Bern Convention Larnaca Declaration in 2011 (Decl. 2011, 7 July 2011). Interestingly, the European Commission, despite the serious trapping problem that was especially pronounced during 2009–2016, has never opened an infringement against Cyprus, while at the same time it has taken Malta to court twice. It seems this is because the Commission considers the problem is not related to legal transposition of the Birds Directive, but one of enforcement, which is the responsibility of Member States under the principle of subsidiarity and much more difficult to argue in court.

10.4.3 Conservation of Sea Turtles and the Mediterranean Monk Seal

It is worth noting that the DFMR started a pioneering marine turtle conservation project in 1978, which is the longest wildlife monitoring scheme on the island and the earliest such programme in Europe. The programme primarily monitors the nests and nesting success of the two species of sea turtle along the beaches that are under the effective control of the Republic of Cyprus. In 1989, specific habitat protection legislation was introduced for the Lara/Toxeftra area, where the main nesting beaches of these two species are located, in Akamas, in the Pafos District. Further pioneering measures included the use of protective ring cages around the nests, starting in around 1995, to protect against Red Fox predation, which was quite a significant problem.

There is also a *National Action Plan for the Conservation of the Monk Seal* (Demetropoulos, 2011), and Ministerial Decrees regulate use of marine areas around their breeding caves.

10.5 Protected Areas and Networks

10.5.1 Protected Area Types and Their Coverage

There are three types of protected areas in the part of Cyprus under the effective control of the Republic of Cyprus, in the SBAs and the Buffer Zone, which often overlap with each other. They are areas protected under national legislation, non-EU internationally protected areas (like Ramsar Sites and the UNESCO Global Geopark), and EU Natura 2000 sites (SPAs and SCIs), as listed in Table 10.4. The areas protected under national legislation fall into two main categories: those characterised as nature areas or landscapes under the land planning policy, and those protected for their importance for nature conservation under the Law for Forests and the hunting law.

Sites Protected under National Legislation

Nationally protected sites designated under the Law on Forests are all forest land but have some special features to make them worthy of special protection. There are four types of designation, all protected under the Law on Forests: National Forest Parks, Nature Reserves, Forest Monuments, and Natural Micro-reserves. They are all proposed by the Department of Forests according to a set of criteria listed in the Law on Forests, and are discussed and approved by a Forest Consultative Body that is formed under the Law, with representatives of different government agencies, professional bodies, NGOs and representatives of local authorities. Nature Reserves are areas hosting important habitats or species (and probably closest to IUCN Category Ia). National Forest Parks host important biodiversity elements and provide leisure opportunities to visitors. Protective Forests surround important water reserves; they probably do not meet IUCN protected area criteria (and are not therefore included in Table 10.4).

There are also three types of protected areas under the Land Planning Policy for the countryside[17]: Coasts and Areas

[17] bit.ly/3Fso2h1

Table 10.4 The number and extent of international designated sites and protected areas in Cyprus.

	IUCN category	Number	Area (km^2)
Designated under global and non-EU international legislation			
Ramsar Sites	Ia	2	33
Specially Protected Area of Mediterranean Importance (SPAMI)	NAs	1	6
Troodos UNESCO Global Geopark	VI	1	1 150
Designated under the Nature Directives			
SCIs (terrestrial and marine)	NR	39	959 + 8 464
SPAs (terrestrial and marine)	NR	30	1 544 + 110
Natura 2000 terrestrial area			*1 680*
Natura 2000 marine area			*8 464*
Designated under domestic legislation			
National Forest Park (Εθνικό Δασικό Πάρκο)	II	11	234
Nature Reserve (Φυσικό Απόθεμα)	Ia (6), III (1)	7	47
Forest Monument (Δασικό Μνημείο)	III	73	0.04
Natural Micro-reserve (Φυσικό Μικροαπόθεμα)	IV	3	0.56
Protected Landscape (Προστατευόμενο Τοπίο)	III (5), V (2)	7	6
Areas of Outstanding Natural Beauty (Περιοχές εξαιρετικής Φυσικής Καλλονής)	IV (1), VI (3)	4	2 530
Coasts and Areas for Nature Protection (Ακτές και Περιοχές Προστασίας της Φύσης)	IV (10), VI (1)	11	377
Permanent Game Reserves (Μόνιμα απαγορευμένη περιοχή κυνηγίου)	IV	9	724

Note. NR = Not reported; NAs = not assigned. The Ramsar Sites, SPAMI and part of the UNESCO Global Geopark are all within Natura 2000 sites.
Sources. See Section 1.3. Based on data from: EEA Natura 2000 Barometer, Natura 2000 as at end 2020; Law on Forests Law 25(I)/2012; Land Planning Policy for the Countryside (Δήλωση Πολιτικής για την Ύπαιθρο); Nature Reserves, Forest Monuments and Micro-reserves from correspondence with Department of Forests, February 2020; all other protected areas from WDPA (UNEP-WCMC, 2022).

for Nature Protection, Protected Landscapes, and Areas of Outstanding Natural Beauty (or Areas of Special Aesthetic Value). These types of protected areas recognise the importance of nature and landscape protection, but confer only land planning restrictions at the site and around it. The three types can overlap.

There are seven Coasts and Areas of Nature protection listed in the Policy Statement on the Countryside. There are also further sites under the Local Development Plans (Τοπικά Σχέδια Ανάπτυξης), but the most important are listed in the Policy Statement on the Countryside.

There are dozens of Protected Landscapes, which are areas selected for the beauty of the landscape, protecting also traditional agriculture. They are often areas around dams or other geological features, mountaintops, cliffs, gorges and other geomorphological features. Although a lot are villages designated for their traditional architecture, seven Protected Landscapes are important as nature sites and included in the WDPA.

The Areas of Outstanding Natural Beauty include large areas of the island with high aesthetic and ecological value. These are lands selected for their natural beauty and importance, where the objective is to maintain their beauty and promote sustainable local activities for local populations. Although these areas have been given an IUCN category status of IV or VI, given their objectives it is debatable whether they actually meet IUCN/CBD protected area criteria.

Game Reserves are areas delineated under the Law for the Protection of Wild Birds and Game Species (152(I)/2003) and are areas where hunting is not allowed. They can be permanent or temporary (i.e. change from year to year). There are no conservation objectives linked to them other than prohibition of hunting. Given their limited, and sometimes temporary, protection measures and lack of objectives, it is also debatable whether these are recognisable as protected areas with respect to IUCN/CBD criteria.

As regards the marine environment, several areas are protected through decrees issued under the national fisheries legislation and managed by the DFMR. These decrees primarily prohibit fishing activities and sometimes also boat traffic. One of the sites is the 'Eratosthenes Seamount', which covers 10 298 km^2 and is within the EEZ declared by the Republic of Cyprus since 2004. The southern and western parts of this EEZ have been divided into exploration plots and tendered by the Republic to companies for hydrocarbon exploration.

Natura 2000 Sites

During the designation of Natura 2000 areas, National Forest Parks, Nature Reserves and several Forest Monuments and Nature Micro-Reserves were incorporated into the Natura 2000 network, but there is no direct linkage in the law between pre-existing domestic protected areas and Natura 2000 sites. The main approach used for the selection of Natura 2000 areas was consultation with the Scientific Committee (see Section 10.3) – a completely new and independent procedure. This was informed by an early LIFE Third Countries project,[18] which produced a first draft list of sites for designation, especially the SCIs. The draft list for SPAs was the list of Important Bird Areas (IBAs) identified by BirdLife International (Heath and Evans, 2000). The proposed SCIs generally avoided the inclusion of private land, whereas 36% of SPAs are on private land.

As indicated in Table 10.4, as of the end of 2020 Cyprus had a total of 63 Natura 2000 sites, 30 SPAs and 39 SCIs/SACs. As a result, Cyprus then had 29.3% of its land area[19] designated as Natura 2000 sites – one of the largest proportions in the EU (Table 5.4). Importantly, Natura 2000 sites include substantial areas that were not considered as worth protecting before EU accession, such as low hills and valleys, and agricultural areas.

In addition to some existing SPAs and SCIs with coastal marine components, a large SCI called 'Oceanis' with an area of 8 326 km^2 has been designated in the EEZ. As a result, at the end of 2020, the total Natura 2000 marine coverage was 8 464 km^2, or 8.6% of Cyprus's marine waters.

Since the most recent Natura 2000 data were compiled by the EEA in 2020, a few more SCIs and SPAs have been designated or enlarged, but they are not expected to significantly increase the overall Natura 2000 area when the statistics are updated.

Total Protected Area Coverage

Estimates of the overall coverage of all different types of protected area in Cyprus do not appear to be publicly available. According to WDPA data and analysis by UNEP-WCMC (2022), the total terrestrial protected area coverage,[20] taking into account overlaps, is 3 509 km^2, which is 61.2% of the land area. This is a very high proportion, and appears to be the highest in Europe (Table 5.4). However, a very high proportion of this comprises Areas of Outstanding Natural Beauty and Game Reserves, whose protected area status is debatable. Whilst it is not possible to estimate here the coverage of other protected areas, due to the unknown extent of overlaps, it is clear that their total protected area coverage would be much lower, and probably close to the 29.3% Natura 2000 coverage.

In the marine environment, according to the WDPA, the combined coverage of domestic protected areas (not including the large fishing protected areas), the one small Specially Protected Area of Mediterranean Importance (Lara–Toxeftra Turtle Reserve) and Natura 2000 sites amounts to 8 472 km^2 according to UNEP-WCMC (2022). This is virtually the same as the Natura 2000 area, presumably due to almost complete overlaps, giving the same 8.6% marine coverage.

10.5.2 The Coherence of Protected Areas and Ecological Networks

The concept of coherence of protected area networks is not really part of the national strategy for protected areas. However, an assessment found that the SCI network is adequate for Red Data Book plants, as only two hotspots for endemic species were insufficiently covered (Christodoulou *et al.*, 2018). An analysis by BirdLife Cyprus found that terrestrial SPAs cover 75% of the network of 34 IBAs identified in Hellicar *et al.* (2014).

Finally, it is worth noting that in October 2020 the European Commission sent a reasoned opinion to Cyprus for insufficient designation of SCIs (infringement number 2018/2058), which highlighted a number of insufficiencies regarding particular taxa as well as a lack of marine Natura 2000 sites. Following this, Cyprus proposed several more SCIs (one marine and one terrestrial) and unified and extended some other sites. Further additions to SCIs are expected, and in addition to a large marine SCI also classified as an SPA in 2022, ongoing research is expected to identify further marine SPAs.

10.5.3 Approaches to Protection and Levels of Protection

Before EU accession, the main protection afforded to areas designated under national legislation was through planning policy or the Law for Forests. The sites designated under the land planning law, such as Protected Landscapes, Areas of Outstanding Natural Beauty, and Coasts and Areas for Nature Protection receive some protection through land planning restrictions. For example, in Coasts and Areas of Nature Protection no development is allowed, except small facilities (up to 30 m^2), hiking trails and Department of Forests management infrastructure.

In Protected Landscapes only very small-scale and specific developments are allowed; these must be appropriate for the landscape and not pose a risk to the water quality of reservoirs. A minimum distance of 300 m from the protected feature is used for other developments.

In Areas of Outstanding Natural Beauty, quarrying, mining, industrial zones, urban zones and similar activities are forbidden, whilst small-scale developments are permitted. More extensive developments are possible but under certain terms and conditions. Of the three types of protected area, the Coasts and Areas of Nature Protection are afforded the highest

---

[18] LIFE98 TCY/CY/172.

[19] Compared to the area of island under the effective control of the Republic of Cyprus, plus the Buffer Zone.

[20] The WDPA includes four Permanent Game Reserves in the north, outside the area under effective control of the Republic of Cyprus, because the reserves existed before 1974.

protection, the Protected Landscapes are second and finally the Areas of Outstanding Natural Beauty, which mainly exclude intensive land uses such as mining, quarrying and urbanisation.

The decrees declaring the marine protected areas outline a list of restrictions including fishing, fishing gear or fishing times, as well as other recreational activities that are forbidden or controlled.

The nationally protected sites designated under the Law for Forests are protected through development restrictions and administrative fines. Forests are patrolled by forest wardens; anyone breaking the law, if caught, is fined or prosecuted in court. Examples of prohibited activities include illegal logging of trees (i.e. without a permit from the Forest Department), setting a forest fire (intentionally or accidentally) and raking the ground for mushrooms. According to the Law of Forests the only exception for development in those areas is public roads.

Protection in permanent and temporary Game Reserves is through patrols by game wardens to ensure that no hunting is taking place in them.

The land planning policy also recognises Natura 2000 areas and makes reference to HD Article 6, which inevitably makes the correct implementation of HD Article 6 the most important safeguard for nature sites in Cyprus.

10.5.4 Approaches to Managing Protected Areas

All 30 SPAs have management plans, which were compiled in 2016. These built on the lessons learnt from a LIFE project for a site, called Oroklini, which included pioneering work on the development of favourable reference values (see Section 4.2) for its qualifying species. The LIFE project report[21] evaluated four different methods for setting the favourable reference values, proposing the most appropriate for each SPA qualifying species. Similar methods were then evaluated and used accordingly for all other qualifying species for all SPAs in Cyprus,[22] which acted as the basis for the conservation objectives and management plans for all remaining SPA sites. However, the Ministerial Decrees that are required to enact the management plans have not yet been adopted.

Management plans also exist for most SACs but they are outdated and do not offer clear management specifications or conservation targets. The Ministerial Decrees that were issued for the sites recently (in 2019) are clearer than the management plans themselves in terms of conservation objectives and management prescriptions or prohibitions, but they do not set quantitative measurable targets. Furthermore, they are not based on favourable reference values as these have not yet been defined for any taxa other than birds.

In practice, the main areas that are under existing management are state forests and some coastal areas, mainly sea turtle nesting beaches (managed by the DFMR). The state forests are managed by the Department of Forests and it is the Department's responsibility to monitor the state of the forests and habitats within these areas and take action for their conservation. The Department's main concern over the years has been fire prevention and a lot of its budget and management actions relate to that. This involves activities that are not always positive for nature, such as creating road access to the most remote areas of forests and opening fire breaks, which in turn cause disturbance, increase erosion and access, and fragment the habitat; the scientific justification for this kind of management is lacking. It is worth noting that public access is unrestricted in the countryside.

The Department of Forests has participated in a number of LIFE projects that have carried out practical conservation actions in forest areas, such as a project focussing on Cedar of Lebanon (*Cedrus brevifolia*) forests, flora micro-reserves, and *Juniperus* forests, as well as drier habitats such as arborescent matorral with *Zyziphus* (5220*) and Iberian gypsum vegetation (*Gypsophiletalia*) (1520*).

The DFMR has historically managed coasts and marine areas through the 1989 Regulations of the Fisheries Law and Ministerial Decrees, in order to protect the main sea turtle nesting beaches in Cyprus, which are now all included in the Natura 2000 network. All nests, on practically all beaches, protected or not, are protected from predation by special non-magnetic cages.

All other habitats are currently without a clear management regime and/or management body and consequently management is exercised in an ad hoc manner, for example through LIFE projects such as the Oroklini LIFE project mentioned earlier, which restored a small wetland.

Land purchase and other similar measures (transfer of development rights or exchange of land parcels), though discussed, have not been used as a conservation measure as yet in Cyprus.

10.6 General Nature Conservation Measures

10.6.1 Development Planning and Impact Assessment

This is one of the biggest challenges for the conservation of nature in Cyprus. Land development and real estate have been, and remain, amongst the most important economic sectors in Cyprus. Consequently, the Town Planning and Housing Department is one of the most powerful departments and existing land planning zones have ultimately influenced the shape of Natura 2000 sites and whether or not developments are approved within them or around them. Land planning policy recognises four types of protected area (Protected

[21] Setting Favourable Reference Values for Annex I bird species at Oroklini marsh as part of the LIFE project: 'Restoration and Management of Oroklini Lake SPA in Larnaca, Cyprus' (2014) LIFE10 NAT CY 000716 OROKLINI.

[22] Game and Fauna Service website, SPA Management Plans (30): www.moi.gov.cy/moi/Wildlife/wildlife_new.nsf/All/6D33B5DB087AA3C6C22580E60031B10C?OpenDocument. Done under contract with GFS, by IACO Environment & Water Consultants and BirdLife Cyprus (2016).

Landscapes, Areas of Outstanding Natural Beauty, Coasts and Areas for Nature Protection, and Natura 2000 areas). The land development restrictions corresponding to those types of planning zones are set out in relevant chapters of the Policy for the Countryside.

Problems relating to land development permissions and the Nature Directives have resulted, among other reasons, from the erroneous general exceptions given to other Cypriot laws, including the Land Planning Law, from the provisions of HD Article 6, which were not corrected until 2015 (as discussed in Section 10.3.3). There have also been problems with **appropriate assessment** procedures. The process was hardly used at all until about 2011 when an 'Ad Hoc Committee on Appropriate Assessment' was created. This informal committee was formed by the DoE in order to assess appropriate assessment studies submitted by consultants and help the department draft its opinions for the EIA Committee (which provides the DoE's official view to the Department of Town Planning and Housing). The Ad Hoc Committee is made up of relevant government agencies and the three main NGOs (Federation of Environmental NGOs, BirdLife Cyprus and Terra Cypria).

Implementation of the appropriate assessment procedure still presents some of the most serious challenges for the protection of Natura 2000 sites in Cyprus, especially given the rate of land development and land-use changes. In November 2019, the European Commission sent a letter of Formal Notice to Cyprus, arguing in the accompanying Press Release that Cyprus has 'generally and persistently, since the date of accession to the EU, failed to ensure that its authorities subject plans or projects to appropriate assessment of their implications on the Natura 2000 sites and/or that it has agreed to plans or projects without having ascertained that they will not adversely affect the integrity of the Natura 2000 site concerned'. The problems that have been observed range from not conducting an appropriate assessment at all (i.e. incorrect/inappropriate screening), to conducting one and taking a very lax view of what is a significant impact, to not reflecting negative opinions of the Ad Hoc Committee within the final opinion of the DoE.

Another serious problem is that Natura 2000 sites and the countryside in general suffer greatly from the cumulative impacts of many smaller projects resulting from the policy of '**isolated housing**' (Μεμονωμένη κατοικία). Although this was introduced in the early 1990s for the benefit of less privileged families, it has ended up being the standard practice for gaining a land development right for everyone, anywhere in the countryside, provided that there is access to a road. Given the extremely dense road network in Cyprus, this means that there are thousands of 'isolated houses' all over the countryside, causing fragmentation, disturbance and a consistent threat of forest fires. A recent much-needed attempt to restrict and regulate this indiscriminate housing policy to a maximum radius of 500 m from existing villages and towns was unfortunately overturned in March 2017 by the Council of Ministers after an outcry by some Members of Parliament, so the problematic policy remains in place.[23]

Equally important are the cumulative impacts of large-scale and specialised types of developments, such as golf courses or marinas with related residential, commercial and tourist developments, based on Decisions by the Council of Ministers, as well as various developments and tourist resorts in the countryside, based on specific provisions of the 'Policy Statement of the Countryside'.

Given these ongoing problems with managing land development impacts, capacity building in the administration to better implement appropriate assessment procedures has been highlighted among the recommendations for Cyprus in the Environmental Implementation Review (European Commission, 2019). However, arguably the most fundamental problem is a lack of environmental education in society and a lack of knowledge and appreciation of the benefits of a healthy natural environment. This lack of interest in the environment weakens the DoE, and other environmental actors, in the face of political leaders who often exercise pressure for approval of projects.

10.6.2 Invasive Alien Species

As briefly described earlier, a range of IAS are affecting biodiversity in Cyprus. Efforts to tackle the problems for the moment, however, are limited. Some clearance of *Acacia saligna* has taken place as part of LIFE projects, and locally by the Department of Forests. The LIFE ICOSTACY project and the DFMR removed some Yellow-bellied Slider turtle (*Trachemys scripta*) individuals from several sites. A LIFE project[24] attempted to prevent the invasion of the Lionfish (*Pterois miles*), and there have been some efforts by DFMR to combat the spread of the Silver-cheeked Toad-fish (*Lagocephalus sceleratus*), a puffer fish, by paying compensation to fishers for fishing for them. There is currently no concerted effort to tackle other invasive species in Cyprus. The European Commission sent a reasoned opinion regarding the absence of a national Action Plan on IAS in February 2022.[25]

10.6.3 Habitat Restoration

There are few examples of habitat restoration to date. Two wetland restoration projects have brought significant restoration improvements for sites. A LIFE project (Oroklini – see earlier), has restored a small natural wetland SPA (improving water availability, protection from disturbance and providing nesting islets for birds), and a UK Darwin Plus funded project

[23] Press release by Commissioner for Environment 2 April 2018, Press releases by Technical Chamber of Cyprus 2018 and 2017 opposing policy of isolated housing and Position of the Environment Department approving the restricted policy in December 2015.

[24] RELIONMED-LIFE (LIFE16 NAT/CY/000832).

[25] https://ec.europa.eu/commission/presscorner/detail/en/ip_22_748

in Akrotiri wetland targeted the hydrology and overgrowth of reeds, as well as promoting animal grazing, better awareness and monitoring. There have also been some LIFE projects dealing with specific habitat types, such as the endemic forests with *Juniperus* species (9560*)[26] and habitat (LIFE JuniperCY), Troodos peat grasslands (Life COMANACY) and Cedar of Lebanon (*Cedrus brevifolia*) forests (9590*),[27] the last focussing on the main threats to the restricted habitat type (including a lack of regeneration, competition from other trees, and fires) and expanding the habitat to new areas.

10.7 Nature Conservation Costs, Benefits and Funding Sources

10.7.1 Nature Conservation Costs

The only available estimates for nature conservation costs in Cyprus are from the Priority Action Framework (PAF) documents in relation to the implementation of the Natura 2000 and related costs for green infrastructure and BHD species. According to the PAF for 2014–2020, the cost of fully implementing Natura 2000 in Cyprus over the period was estimated to be around €125 million per year (MANRE, 2013). This assessment was extremely high, given that the annual cost for the EU-27 was estimated at around €5.8 billion (Gantioler *et al.*, 2010), and Cyprus only has 0.1% of the total EU Natura 2000 network (0.2% of land sites). The main reason for this proportionately very high figure was because Cyprus originally envisaged that all, or most, private Natura 2000 land would/could be purchased, at an assumed high price, and that this alone would cost around €750 million. In fact, to date, no land has been purchased.

The current PAF for 2021–2027 estimates the total Natura 2000 related costs to be just €7.4 million per year (DoE, 2021). Of this total, 38% of the estimated costs are for administration and other horizontal activities, 32% for habitats and species within Natura 2000 sites, 13% for supporting green infrastructure (e.g. to improve connectivity) and 18% for additional species-specific actions not connected to specific habitats or ecosystems.

10.7.2 The Economic Benefits of Nature Conservation

Very little work has been done in assessing ecosystem services or other benefits of nature conservation in Cyprus so far. However, a LIFE project[28] is underway that is seeking to increase public awareness of the natural values of Troodos National Forest Park and the ecosystem services it provides, in particular in relation to nature tourism. There have also been some efforts by the Cyprus Tourism Organisation to stimulate nature tourism in Cyprus.

10.7.3 Funding Sources for Nature Conservation

As indicated in Table 10.5, funding for conservation is extremely limited. As a result, conservation management has been mainly carried out through traditional EU LIFE projects, even though the LIFE programme normally aims to support innovative and demonstration actions rather than ongoing management. Up to May 2021, 12 LIFE Nature and Biodiversity projects were underway or completed in Cyprus since EU accession in 2004, with a total budget of around €15.1 million and an EU contribution of €9.6 million. Some other international sources have been used, for example Territorial Cooperation Greece-Cyprus 2007–2013 funding, although this seems to be largely the exception for such projects.

A much larger LIFE Integrated Project, 'Managing the Natura 2000 Network in Cyprus and shaping a sustainable future' (Pandoteira), started in November 2019 and will run for 10 years. The project has a total budget of around €17 million (60% from the EU). If properly implemented, this will play a critical role in solving some of the challenges affecting biodiversity in Cyprus, as identified here. The main actions focus on achieving a better understanding of the distribution of key species and habitats, monitoring, clarifying the required methods for managing sites that are not in forest land, developing action plans for threatened BHD species and habitats, studying the requirements of certain BD Annex I bird species, actions on IAS, and management and restoration actions in specific Natura 2000 sites.

The Integrated Project also leveraged a further €8 million from EU Regional Funds during 2014–2020 for the management of Natura 2000 areas. These are mainly ERDF, RDP and EMFF funds. Of these, about €2 million was for certain actions in the Akamas area and the remaining €6 million for other nature-related actions such as monitoring, restoration, and habitat and species management.

The EU CAP European Agricultural Fund for Rural Development (EAFRD) has been the largest potential source of funds for nature conservation management and restoration on agricultural and forest land. Over the 2014–2020 funding period, Cyprus was allocated €251 million for its RDP. Most of the funding was not used for measures with nature conservation objectives, but out of the funding allocated in relation to Priority 4 on ecosystems, around €52 million was allocated for agri-environment and climate measures, €3 million for investments in forest area development and improvement of the viability of forests, and €1 million for Natura 2000/WFD payments. Additionally, around €3 million were spent from the M7 'basic services' measures (not included in Priority 4 funding), specifically for works in Akamas.

In practice, the agri-environment schemes have not been targeted to nature conservation requirements and have not necessarily been implemented in important biodiversity areas.

[26] JUNIPERCY (LIFE10 NAT/CY/000717).
[27] LIFE KEDROS (LIFE15 NAT/CY/000850).
[28] iLIFE-TROODOS (LIFE16 GIE/CY/000709).

Table 10.5 Funding sources for nature conservation in Cyprus.

Source and period	Measure/category	Average annual allocation (€ m)		Comments
EU funding instruments		**EU + national**	**EU**	
EAFRD – Rural Development Programme measures – allocations to Priority 4 (restoring, preserving and enhancing agriculture and forest ecosystems) 2014–2020	M10 Agri-environment–climate	7.5	4.0	See text
	M13 Areas facing constraints	5.9	3.1	
	M11 Organic farming	2.0	1.1	
	M8 Investments in forests	0.4	0.2	
	M12 Natura and WFD	0.1	0.1	
	M16 Cooperation	0.1	0.0	
	M1 Knowledge transfer, etc.	0.02	0.01	
RDP Priority 4 subtotal		*16.0*	*8.5*	
EMFF (Thalassa fund) 2014–2020	Actions to improve the knowledge of marine habitats and species, including deep sea research, management actions	0.4	0.3	
Additional European Structural and Investment Funding 2014–2020		1.1	0.9	€8 million over 2019–2029
LIFE Programme	Nature and biodiversity projects (2004–2020)	0.9	0.6	
	Integrated Project 2019–2029	1.7	1.0	LIFE IP Pandoteira
EU funding instruments subtotal		*20.1*	*11.3*	
National funds within 2014–2020		**Additional non-EU***	**Comments**	
DFMR	Sea turtle monitoring programme	c. 0.02		
Research and Innovation Foundation		?	Not regular	
MAVA and BirdLife International projects	Illegal bird killing, by-catch, flora studies, capacity building, wetland inventory, etc.	c. 0.03	To NGOs and Government	
Leventis Conservation Foundation	Running costs and project co-funding for conservation NGOs	?		
Others				
Darwin Plus projects	Restoration Akrotiri marsh, IAS, Addressing Ecological Change	c. 0.13	£117 386. Only in the SBAs	
Additional national and other subtotal		*0.2*		
National total including co-financing of EU instruments		*8.9*		
Total		**20.3**		

Notes. * Excludes co-financing of EU funds unless indicated. The subtotals and totals are only approximate indications of the average annual levels of funding as they are based on annualised data from differing periods.
Sources. Based on data from: RDP data from European Commission ESIF funds database (23 October 2020); European Commission LIFE database (February 2022); PAF 2021–2027 (DoE, 2022); other EU funds, DFMR Annual Report 2020 (16 February 2022); Darwin Plus Database (16 February 2022). See Section 1.3 for further details.

Furthermore, according to the last available annual report,[29] uptake of RDP measures, especially Natura 2000 payments, was extremely poor, with only €0.2 million committed, out of €1 million, at the end of 2019. The two submeasures available were: (1) leaving 10% of the crop unpicked on the trees, or if it is an annual crop (e.g. wheat) 10% not collected until July; and (2) prohibition of grazing between February and March in the Akamas area and Diarizos Valley.

During the consultation period, the NGOs and nature government agencies, including the GFS, had proposed much more ambitious Natura 2000 payment measures and ideas for agri-environment schemes. These included leaving large strips of uncultivated land, especially in cereal fields (like the old set-aside scheme), delaying harvesting until after May (harvests often start in February/March in Cyprus, which is extremely early for ground-nesting birds), supporting the use of less pesticide in fields and using manual methods, and using Barn Owls (*Tyto alba*) for rodent control with parallel ending of widespread use of rodenticides in the countryside. The GFS also proposed some special measures for the Cyprus Mouflon, which included leaving some of the crop unpicked, subsidising better fences and other avoidance measures. However, these proposals were largely rejected by the RDP Management Authority as they considered them to be too complicated to implement and difficult to monitor objectively. This reflects a general problem of inadequate integration of biodiversity requirements into other policies, because other departments have their own priorities that do not usually match those of the Nature Directives.

Finally, the European Fisheries and Maritime Fund was also used during 2014–2020, primarily for cetacean and other marine surveys in Cyprus's EEZ. Furthermore, DFMR annually allocates national funds for the implementation of the sea turtle programme.

In general, although there has been substantial potentially available funding, especially through EU instruments, its actual use has been quite low and untargeted. Furthermore, the integration of biodiversity concerns into other funds could be much more targeted and effective.

10.8 Surveillance, Monitoring, Reporting and Applied Research

As highlighted in Section 10.2 outlining the results from Article 17 and Article 12, monitoring of and research on biodiversity in Cyprus is very limited with some exceptions. The only species regularly monitored are the birds, some plants, some habitats, the two species of sea turtles, the Cyprus Mouflon and to some extent the Mediterranean Monk Seal.

Additionally, there is very limited research taking place in Cypriot universities on Cypriot biodiversity, and very little applied research related to nature. Some notable exceptions include the work done by Blosat (2008) on the endemic Grass Snake, and by Jones (2006) on the ecological relationship between the Cyprus Warbler and Sardinian Warbler (*Sylvia melanocephala*), and some work on the Cyprus Mouflon.

The DoE, which does not have field personnel and wardens, occasionally contracts external assistants as wardens and surveyors of sites, who then provide the Department with regular reports about possible threats to sites. However, this is done in a very ad hoc manner, and as a consequence there is no real coherent monitoring of the state of sites or the threats they are facing, especially those sites that are not in state forest areas.

The state forest areas, which are under the responsibility of the Department of Forests, are surveyed by the Department for interventions, threats or illegal activities, and the SPAs are surveyed by the GFS for poaching and illegal trapping as well as unauthorised activities/developments in protected areas or other countryside areas. However, it is unclear whether this field monitoring is sufficient as there is no substantive information on its real impact. Field monitoring is also conducted by DFMR in relation to issued decrees, but similarly it is unclear whether this monitoring is sufficient.

Improving biodiversity monitoring protocols and methods, and scientific understanding of the ecology of species, and ensuring a permanent system of monitoring in the long term are important areas to develop in the future. Such requirements are identified as a priority in the LIFE Integrated Project.

10.9 Future Developments

The main challenge for the future remains the management of habitats and species and Natura 2000 areas, as well as their monitoring, surveillance and wardening. The 10-year LIFE IP 'Pandoteira' therefore has a crucial role to play in meeting these requirements, as it aims to put in place methodologies and strategies to manage sites and species, monitor sites and species and generally implement the Nature Directives fully. Importantly, it also aims to complement existing measures by clarifying management structures and strategies for Natura 2000 areas that are not in state forest areas.

10.10 Conclusions

The fact that Cyprus has designated 29% of the territory that is under the effective control of the Government of the Republic of Cyprus as Natura 2000 areas is of great importance for nature conservation. Furthermore, the fact that Cyprus has the oldest sea turtle conservation project in the Mediterranean and a population of sea turtles that is showing good signs of recovery due to conservation measures taken decades ago is also very significant. Additionally, the state forests of Cyprus, having been protected since the

[29] Annual Implementation Report Rural Development Plan, Year 2019 (Ετήσια Έκθεση Υλοποίησης Πρόγραμμα Αγροτικής Ανάπτυξης Έτος 2019): www.paa.gov.cy/moa/paa/paa.nsf/All/ED19D257267F30EEC225866C002566D4/$file/Implementation%20report_2014CY06RDNP001_2019_0_en.pdf

mid-twentieth century, are prominent in the national psyche so that the public are supportive of their protection. This in turn means that they are generally better managed and less threatened than other habitats, especially the coastal ones.

On the other hand, the experience in Cyprus so far has shown that, whereas EU nature legislation has brought important benefits for nature conservation, the learning required to fully and effectively implement the Nature Directives by national authorities should not be underestimated – especially when nature protection was not high on the agenda in the first place. The early euphoria about joining the EU and keenness to comply with the new legislation encouraged the inclusion of a significant percentage of land in Natura 2000. This, however, does not guarantee adequate protection and management for those sites. If other conflicting drivers are operating, such as pressures for incompatible developments, robust implementation and enforcement is needed if habitat loss, degradation and species declines are to be avoided. Good procedures for appropriate assessment are among the most important to put in place early on, with adequate penalties for illegal activities backed up by a system of wardening and surveillance. These are all essential ingredients of nature protection. Also, strategic and environmentally conscious spatial planning and effective integration of biodiversity goals and targets in other horizontal policies is crucial.

The story of the implementation of the Nature Directives in Cyprus illustrates the European Commission's very important role as Guardian of the Treaty, supporting and checking compliance with the legislation. Timely monitoring and action by the Commission continues to be essential to ensure that sites are not damaged.

Finally, the legislation cannot perform miracles on its own: nature conservation also needs political will and public support, strong environment departments, sufficient and better targeted funding, commitment to implementation and enforcement at national level, and strong and vocal NGOs.

Acknowledgements

CP: I am especially grateful to the invaluable help from Panicos Panayides, my husband, who works for the GFS, who provided me with key information and data about nature in Cyprus. He was particularly helpful in explaining some of the key issues related to planning zones. I also thank Klitos Papastylianou for useful explanations, data and references.

CP and AD: We thank Despo Zavrou and Elena Stylianopoulou from the Environment Department for key information provided in a timely manner and Gerald Dorflinger from the Water Development Department for information and updates regarding the RBMP. We also thank the Forest Department for clear information on nationally protected areas and classifications.

Shortly before the publication of the book, Andreas Demetropoulos sadly passed away. Andreas was a pioneer for marine conservation, especially sea turtle conservation in the Mediterranean and the world. In recognition of his exceptional, pioneering work with sea turtles, in 1988 he was included in the United Nations 'Global 500 Roll of Honour'. He has left a very important legacy for the protection of the marine environment and is sorely missed.

References

Betz-Heinemann, K. A. V. (2018) How to count. In Hunting, a technology of belonging in Northern Cyprus, pp. 155–162. Doctor of Philosophy thesis, University of Kent, UK. https://kar.kent.ac.uk/id/eprint/72786 and https://khalilavi.org/2021/04/16/a-lesson-in-how-to-count/

BirdLife Cyprus (2017a) *Update on Illegal Bird Trapping Activity in Cyprus: Covering the autumn 2016 Findings of BirdLife Cyprus' Continuing Monitoring Programme for Illegal Bird Trapping in Cyprus and Providing an Overview of the Latest Developments Regarding the Problem.* Nicosia: BirdLife Cyprus. BirdLife report downloaded from: www.paa.gov.cy/moa/paa/paa.nsf/All/ED19D257267F30EEC225866C002566D4/$file/Implementation%20report_2014CY06RDNP001_2019_0_en.pdf

(2017b) *Update on Illegal Bird Trapping Activity in Cyprus: Covering the winter 2016-17 Findings of BirdLife Cyprus' Continuing Monitoring Programme for Illegal Bird Trapping in Cyprus and Providing an Overview of the Latest Developments Regarding the Problem.* Nicosia: BirdLife Cyprus. BirdLife report downloaded from: https://birdlifecyprus.org/wp-content/uploads/sites/3/2022/02/BirdLife-Cyprus_Autumn_2021_Trapping_report-FINAL_for-publication.pdf

(2019) *Update on Illegal Bird Trapping Activity in Cyprus: Covering the Winter 2018–2019 Findings of BirdLife Cyprus' Continuing Monitoring Programme for Illegal Bird Trapping in Cyprus and Providing an Overview of the Latest Developments Regarding the Problem* Nicosia: BirdLife Cyprus. https://birdlifecyprus.org/wpcontent/uploads/sites/3/2020/05/BirdLife-Cyprus_Winter-2018-19_Trapping-report_For_Publication.pdf

BirdLife Cyprus (2022) *Update on Illegal Bird Trapping Activity in Cyprus: Covering the Autumn 2021 Findings of BirdLife Cyprus' Continuing Monitoring Programme for Illegal Bird Trapping in Cyprus and Providing an Overview of the Latest Developments Regarding the Problem.* Nicosia: BirdLife Cyprus. https://birdlifecyprus.org/wp-content/uploads/sites/3/2022/02/BirdLife-Cyprus_Autumn_2021_Trapping_report-FINAL_for-publication.pdf

Blosat, B. (2008) Population status, threats and protection of the grass snake, *Natrix natrix cypriaca* (Hecht, 1930) on Cyprus. *Mertensiella*, **17**, 246–271.

Christodoulou, C. S., Griffiths, G. H. & Vogiatzakis, I. N. (2018) Using threatened plant species to identify conservation gaps and opportunities on the island of Cyprus. *Biodiversity and Conservation*, **27**, 2837–2858.

Cuttelod, A., García, N., Abdul Malak, D., Temple, H. & Katariya, V. (2008) The Mediterranean: a biodiversity hotspot under threat. In *The 2008 Review of the IUCN Red List of Threatened Species*, eds. J.-C. Vić, C. Hilton-Taylor & S. N. Stuart. Gland, Switzerland: IUCN.

Demetropoulos, A. (2011) *National Action Plan for the Conservation of the Mediterranean Monk Seal in Cyprus.* Contract RAC/SPA: N° 20/RAC/SPA_2011. Nicosia: United Nations Environment

Programme/Mediterranean Action Plan, and Department of Fisheries and Marine Research.

Demetropoulos, A. & Hadjichristophorou, M. (2017) Η θαλάσσια ζωή της Κύπρου-Κατάσταση και Προοπτικές [Marine life of Cyprus – status and prospects]. In *Geography of Cyprus*, eds. K. Demetriades, A. Sophocleous & P. Argyrides, pp. 431–450. Nicosia: Cyprus Geographical Association.

DoE (2017) *Εθνική Στρατηγική για την Προσαρμογή στην Κλιματική Αλλαγή.* [National Strategy for Adaptation to Climate Change.] Nicosia: Department of Environment, Ministry of Agriculture, Rural Development and Environment, Cyprus.

(2020) *Στρατηγική και Σχέδιο Δράσης για την Βιοποικιλότητα στην Κύπρο.*[Biodiversity Strategy and Action Plan in Cyprus.] Nicosia: Department of Environment, Ministry of Agriculture, Rural Development and Environment, Cyprus.

(2021) *Πλαίσιο Δράσης Προτεραιότητας (PAF) για το Δίκτυο Natura 2000 στην Κυπριακή Δημοκρατία (Τελικό).* [Prioritised Action Framework (PAF) for Natura 2000 in the Republic of Cyprus (Final).] Nicosia: Department of Environment, Ministry of Agriculture, Rural Development and Environment, Cyprus.

EEA (2020) *State of Nature in the EU. Results from Reporting under the Nature Directives 2013–2018.* Copenhagen: European Environment Agency.

European Commission (2019) *The EU Environmental Implementation Review Country Report – CYPRUS.* SWD (2019) 115 final.

Expert Council on NGO Law of the Council of Europe's Conference of INGOs, Cyprus NGO Initiative on Law Reform & Office of the Commissioner of Volunteerism and NGOs (2015) Civil society in Cyprus: building for the future. Nicosia: Unpublished report.

Gantioler, S., Rayment, M., Bassi, S. *et al.* (2010) *Costs and Socio-Economic Benefits Associated with the Natura 2000 Network.* Final report to the European Commission, DG Environment on Contract ENV.B.2/SER/2008/0038. Brussels: Institute for European Environmental Policy/GHK/Ecologic.

GFS (2010) Τροποποίηση του Περί Προστασίας και Διαχερίσης Αγρίων Πτηνών και Θηραμάτων Νόμου. 17 Μαρτίου 2010. Game and Fauna Service.

Hadjikyriakou, G. (2017) Η ιθαγενής χλωρίδα της Κύπρου από την Πλειόκαινη Εποχή (5,3-2,6 εκατομμύρια χρόνια) μέχρι σήμερα. [The indigenous flora of Cyprus from the Pliocene (5.3-2.6 million years) till today.] In *Geography of Cyprus*, eds. K. Demetriades, A. Sophocleous & P. Argyrides, pp. 189–256. Nicosia: Cyprus Geographical Association.

Heath, M. F. & Evans, M. I. (eds.) (2000) *Important Bird Areas in Europe: Priority Sites for Conservation. 2: Southern Europe.* Cambridge, UK: BirdLife International.

Hellicar, M. A., Anastasi, V., Beton, D. & Snape, R. (2014) *Importa Bird Areas of Cyprus.* Nicosia: BirdLife Cyprus.

Jones, V. R. (2006) Comparative ecology of the endemic Cyprus warbler: implications of recent coexistence. PhD thesis. Universi of Cambridge, UK.

Mammides, C., Kadis, C. & Coulson, T. (2015) The effects of road networks and habitat heterogeneity on the species richness of birds in Natura 2000 sites in Cyprus. *Landscape Ecology*, **30**, 67–75.

MANRE (2013) *A Prioritised Action Framework (PAF) for Natura 2000: For the EU Multiannual Financing Period 2014–2020.* Nicosia: Ministry of Agriculture, Natural Resources and Environment.

Panayides, P. (2005) Six aspects of land use and development activi that result in adverse effects to Cypriot wildlife resources. In *Proceedings of the XXVth International Congress of the International Union of Game Biologists – IUGB and the IXth International Symposium Perdix, Limassol 2001*, ed. E. Hadjisterkotis, pp. 182–196. Nicosia: Ministry of Interior.

Sofroniou, A. & Bishop, S. (2014) Water scarcity in Cyprus: a review and call for integrated policy. *Water*, **6**, 2898–2928.

The World Bank (2014) *Analysis of the Function and Structure of th Ministry of Agriculture, Natural Resources and Environment (MANRE) in the Republic of Cyprus.* Poverty Reduction and Economic Development Unit, Southern Europe Programme. www .crcs.gov.cy/crcs/crcs.nsf/0/ 92ED3646DD22C2C2C2257E340030CBB8/$file/Final%20Report %20MANRE%20and%20SOEs.pdf

Tsintides, T., Christodoulou C. S., Delipetrou, P. & Georgiou, K. (2007) *Το κόκκινο βιβλίο της χλωρίδας της Κύπρου. Φιλοδασικός Σύνδεσμος Κύπρου, Λευκωσία, Κύπρος.* [*The Red Data Book of the Flora of Cyprus.*] Nicosia: Cyprus Forestry Association.

UNEP-WCMC (2022) *Protected Area Profile for Cyprus from the World Database of Protected Areas, April 2022.* www .protectedplanet.net

Verstrynge, J.-F. (2004) The institutional organisation of the environmental policy of Cyprus in the light of the accession of the Republic of Cyprus to the European Union. Unpublished report.

Zomeni, M., Martinou, A. F., Stavrinides, M. C. & Vogiatzakis, I. N. (2018) High nature value farmlands: challenges in identification and interpretation using Cyprus as a case study. *Nature Conservation*, **31**, 53–70.

Chapter 11

The Czech Republic

Petr Roth

Key Statistics

Terrestrial area (km^2)/% of EU-28*	78 874/1.8%
Marine waters (km^2)/% of EU-28*	0
Population density (p km^2)/% of EU-28*	134/116%
GDP per capita (€)/% of EU-28	16 705/61%
Land cover (CORINE land cover classes 2018)	AGR 57%, FOR 33%, ART 7%, OSN 3%, WET 0.9%
Biogeographical regions	CON, PAN
Number of HD Annex I habitats/%*	60/27%
Number of HD Annex II, IV and V species/%*	173/14%
Number of HD Annex I bird taxa/%*	71/42%
% HD UCS & decreasing habitats/% unknown	32%/1%
% HD UCS & decreasing species/% unknown	27%/6%
% breeding BD Annex I birds decreasing/% uncertain and unknown	9%/42%

Note: * Excluding MAC and MMAC biogeographical regions.
Sources and codes. See Table 1.1.

11.1 The Natural History and Characteristics of the Czech Republic

The Czech Republic is a landlocked medium-sized country located in the temperate geographical centre of Europe. Two different major geological units, the Bohemian Massif and Western Carpathians, cover the territory. Lowlands below 300 m only cover 20%, whilst 39% of the country comprises hills up to 600 m altitude, 30% is up to 900 m and 12% is mountains (the highest being 1 602 m). The climate is mild, with warm summers and no dry season, being transitional between temperate oceanic (west) and continental (east). Two biogeographical regions are recognised: the Continental (96%), and the Pannonian in the warm south-east.

Since the end of the last ice age the whole territory has been heavily modified by humans. Population growth resulted in vast deforestation with its peak around 1750 when only 14% of territory remained afforested. The nineteenth century brought plantation forests, mainly of monocultures of spruce and pine, since when forests have covered 34% of the country. From the middle of the century, major changes resulted from industrial developments, including the canalisation of all major rivers. A quarter of small water courses have been straightened and their banks partly paved (Vašků, 2011). Over the last three decades small hydropower plants have been constructed on most upland water courses, changing their aquatic habitats significantly.

The start of Communist rule in 1948 changed the country's landscapes forever. Countrywide nationalisation of all private land and the so-called process of collectivisation led to consolidated farm units, with contiguous blocks of up to 200 ha that lacked baulks,[1] rain-fed farm ponds and scattered seminatural vegetation. The intensification and industrialisation of agriculture was a pillar of the official Communist policy, as were massive residential settlements and industrial development from the 1960s.

When the political regime changed in 1989, the artificially supported agriculture collapsed, and land and property restitution started. The majority of the land was returned to individuals who were not interested in farming, so it was soon leased to large holdings, the average size of which is 112 ha – the highest in the EU, and more than seven times the EU average.[2] Consequently, although the use of pesticides declined rapidly from the 1990s, most of the other negative effects of intensive agriculture remained (e.g. the predominance of arable land, extremely large fields, eutrophication and compacted soils). As a result, the agricultural landscape has a highly impoverished biodiversity.

As well as being subject to intensive management, forests have been affected by air pollution. Since 2010, serious droughts have weakened the spruce-dominated forests (spruce represents 50% of all Czech forests), leading to massive bark beetle outbreaks and the death of most spruce forests below

[1] That is, 2–5 m strips of non-arable land with grass and fruit trees that separated fields. They were used as a source of fruit as well as fresh grass for poultry and rabbits.

[2] https://ec.europa.eu/eurostat/statistics-explained/index.php/Farm_structure_statistics

Table 11.1 Short-term trends (c. 2007–2018) in Habitats Directive Annex I habitats and Annex II, IV and V species with an unfavourable conservation status in the Czech Republic.

	Poor (unfavourable–inadequate)				Bad (unfavourable–bad)			
	Imp	Sta	Dec	?	Imp	Sta	Dec	?
HD Habitat assessments								
Number	0	32	9	5	0	6	21	1
%	0%	70%	20%	11%	0%	21%	75%	4%
HD Species assessments								
Number	18	43	37	4	8	24	34	3
%	18%	42%	36%	4%	12%	35%	49%	4%

Key. Imp = improving, Sta = stable, Dec = decreasing, ? = unknown.
Source. Based on Member State HD Article 17 reported data as presented on the EEA national summary dashboard (2021).

300 m in 2015–2019. It is estimated that within the next 10–15 years, all spruce forests below 500 m will disappear.

Despite all the profound changes, the Czech Republic retains many seminatural habitats and species of high conservation importance (especially in numerous protected areas but also in the wider countryside), including floodplain habitats, extensive hay meadows, beech-dominated forests, raised bogs, steppes and alpine habitats.

11.2 The Status of Nature

Biodiversity monitoring in the Czech Republic has been little developed (as further discussed in Section 11.8), and currently the main information available relates to birds and HD habitats and species. Other important sources of information include an overview of the status of botanical diversity by Härtel (2017) and a new Red List of habitats of the Czech Republic (Chytrý *et al.*, 2019).

11.2.1 Habitat Status and Trends

According to the national HD Article 17 monitoring report for 2013–2018,[3] only 19% of assessments of HD habitats revealed a favourable status (with 1% unknown). However, as discussed in Chapter 1, the achievement of a favourable status is a relatively high and long-term ambition. Given this, of most concern are those habitats that have a declining status, especially those with a bad status (i.e. threatened). Overall, 44% of all habitat assessments were bad, and in 75% of these the habitat's condition has declined over the short term (Table 11.1). The most threatened habitats are bogs (50% with a bad status and declining), forests (29%) and grasslands (24%).

11.2.2 Species Status and Trends

The Article 17 2013–2018 monitoring data also show that a small proportion of species assessments revealed a favourable conservation status (14%, with 2% unknown). Furthermore, 34% of assessments were bad, and in 49% of these the species were considered to have decreased (Table 11.1). The most threatened groups are molluscs (50% bad status and declined), fish (23%), arthropods (16%), vascular plants (14%) and amphibians (14%). Despite an improvement in 25 species with bad status, another 71 declined – three times as many. However, the situation was better for some species, including large carnivores, which attract public attention and are therefore monitored by the nature conservancy. The Wolf (*Canis lupus*) and Eurasian Lynx (*Lynx lynx*) were exterminated by the nineteenth century but have slowly been returning recently.

Birds have been subject to a long tradition of systematic surveillance (Bejček and Šťastný, 2015). Data from such surveys have been used by the Czech Society for Ornithology to calculate a Farmland Bird Indicator for 20 common species since 1982. With the exception of a brief increase in the 1990s, the Indicator has declined since its inception, with its 2017 level being 28% lower than in 2000 (Vermouzek and Zámečník, 2017). This reflects important changes in Czech agriculture: initially the collapse of the intensive collectivised agriculture in the early 1990s resulting in extensification of production and the abandonment of vast farmland areas. This was followed by land consolidation and large-scale agricultural recovery, which have in part been assisted by EU CAP policies and funds.

Recent population trends in most BD Annex I birds have been relatively satisfactory: only 10% of breeding populations and no wintering populations have decreased over the short term (Table 11.2). However, the population trends of a substantial number of species are unknown, despite the endeavours of hundreds of volunteers participating in censuses. There is also particular concern over the survival prospects of two iconic bird species: Capercaillie (*Tetrao urogallus*) and Black Grouse (*Lyrurus tetrix*). While the former's current population of 470 individuals is believed to be above its critical threshold (Bečka and Rösner, 2019), the latter has declined to several hundred individuals in three isolated areas, which are subject to increased disturbance, habitat fragmentation and predation by a high population of Wild Boar (*Sus scrofa*) (Flousek, 2019).

[3] See Section 1.3 for background on monitoring under the Nature Directives, status categories and data sources.

Table 11.2 Short-term trends (c. 2007–2018) in Birds Directive Annex I taxa in the Czech Republic.

	Breeding				Wintering			
	Inc	Sta/ Flu	Dec	?	Inc	Sta/ Flu	Dec	?
Number	18	13	6	27	3	2	0	8
%	28%	20%	9%	42%	23%	15%	0%	62%

Key. Inc = increasing, Sta = stable, Flu = fluctuating, Dec = decreasing, ? = uncertain and unknown.
Source. Based on Member State Article 12 reported data as presented on the EEA national summary dashboard (2021).

Recently, some evidence has suggested that even relatively common species, such as Common Toad (*Bufo bufo*) and Smooth Newt (*Lissotriton vulgaris*), have rapidly declined in abundance due to general deterioration of habitats in the countryside (Fischer *et al.*, 2015).

11.2.3 Drivers and Causes of Change

The key driver of biodiversity change has been the conversion of the relatively low-intensity and small-scale farmland landscape that existed before the 1950s to the current intensive agricultural landscapes with large-scale farmland blocks and an absence of boundary habitats and wetlands. A further recent cause of agricultural change has been an increase in the growing of oilseed rape due to biofuel subsidies (planted on 24% of arable land in 2019[4]), with biofuel crops also allowed higher levels of pesticide use than food crops. Elsewhere, non-farmed habitats have been subject to overgrowth and afforestation.

Most forests are spruce-dominated even-aged plantations with unnatural species composition, managed predominantly by a clear-felling system persisting from the Austrian Empire times; small remnants of semi-natural forests survive mostly in protected areas. In the 1970s–1980s, very high levels of sulphur dioxide and nitrogen oxides air pollution from coal power plants affected hundreds of thousands of hectares of mountainous spruce forest in the country, accompanied by bark beetle (*Ips typographus*) outbreaks and subsequent clearcuts. To counter this, the government at the time had a policy of designating hundreds of protected areas. However, this was unable to compensate for the scale of habitat loss, which has been detrimental for biodiversity and especially for invertebrates (Kuras *et al.*, 2017). Induced changes in soil chemistry have also persisted until today and have made forestry restoration attempts unsuccessful for decades.

It is sometimes uncertain if the observed return of a species is a consequence of conservation measures or occurs for other reasons. It is therefore encouraging that in the case of the Peregrine Falcon (*Falco peregrinus*), which disappeared in the 1950s following the mass use of organochlorine pesticides, a study has concluded that its return was in part due to conservation actions (Hlaváč and Beran, 2011). Whilst the recovery was spontaneous, following the DDT ban in 1980s, a viable breeding population would not have become established without proactive measures, including reducing electrocutions, creating and guarding artificial nests and reducing disturbance from rock climbing.

11.3 Nature Conservation Governance, Actors, Legislation and Policies

What is currently the Czech Republic used to be part of the Austrian Empire between 1526 and 1918, during which nature conservation did not occur, other than with respect to game species. This resulted in the overhunting of some favoured game, introduction of some alien species (subsequently becoming a threat to some native fauna and flora) and intentional extermination of large carnivores. However, in 1838, a private initiative by the enlightened Prince George Buquoy resulted in the establishment of two primeval forests as the first protected areas. Some of the most important developments affecting nature conservation since then are listed in Table 11.3.

11.3.1 Governance

The initial conservation activities of the independent Czechoslovak Republic focussed on the designation of small-scale nature monuments, and it was not until 1956 that the Act on State Nature Protection[5] introduced full responsibility of the State for nature conservation and its funding through seven regional and 76 county authorities, supervised by the Ministry of Culture, and set the framework for a systematic approach to both site and species protection. In 1992, it was replaced by the current Act on Nature and Landscape Protection (NPA),[6] which has embraced both new national developments and later EU law requirements and established a new system of public administration authorities and expert bodies.

The current administrative system was established in 2000 when the previous system of 76 counties was abolished and 14 regions established instead. As the regional and municipal authorities are a mixture of self-government and state administration, there are unavoidable differences in nature conservation implementation among the regions due to slightly different interpretation of the NPA and regional political pressures. After EU accession in 2004, the EU requirements were combined with those for national nature conservation and are dealt with by the same competent authorities (as summarised in Table 11.4). Of particular importance is the Nature Conservation Agency of the Czech Republic (NCACR), which

[4] Czech Statistical Office, 24 January 2019: www.czso.cz/csu/czso/osevni-plochy-ozimych-plodin-pro-sklizen-v-roce-2019-

[5] Zákon č. 40/1956 Sb., o státní ochraně přírody.

[6] Zákon č. 114/1992 Sb., o ochraně přírody a krajiny, v platném znění.

Table 11.3 Key nature conservation developments in the Czech Republic.

1838	The first two protected areas designated
1919	Ministry of National Education and Awareness of the Czechoslovak Republic (independent since 1918) established an office of 'Conservator General of the Czechoslovak Republic' with four staff dedicated to nature conservation
1918–1938	Over 100 protected areas declared as 'natural monuments' on private land with agreement of landowners
1945–1946	German-speaking population expelled from one-third of the country's territory, changing land use from intensely managed farmland to extensive meadows and pastures, with vast areas abandoned and overgrown by forest
1948	Communist rule starts. All industry, forests and farmland nationalised, centrally governed large-scale agriculture and forestry introduced, freedom of assembly and association abolished; official policy promotes heavy industry supported by coal mining and coal power stations
1955	The first large-scale Protected Landscape Area (PLA) Český ráj established, followed by 25 PLAs covering about 15% of the country by 2016
1956	Act on State Nature Protection No. 40/1956 Coll. adopted
1958	State Institute for Historic Heritage and Nature Conservation (SÚPPOP), central methodological body for nature conservation established, and, subsequently, seven Centres for Historic Heritage and Nature Conservation (KSPPOP) in all regions, with SÚPPOP coordinating their work. KSSPPOPs established their branches in each of 19 PLAs as well as one National Park (NP) designated pursuant to the first Act
1963	Krkonoše NP established, protecting the favourite high-mountain ski resort area with relic alpine habitats
1969	Yew – Union for Nature and Landscape Conservation (TIS) established as the only NGO-like conservation organisation. Officially terminated in 1979 and replaced by the Czech Union of Nature Conservationists (ČSOP) established by the Ministry of Culture in all 76 counties and many bigger towns
1989	The Communist regime collapsed, being peacefully replaced by a democracy
1990	Ministry of Environment (MoE) established; nature conservation left the sector of culture
1992	Act on Nature and Landscape Protection No. 114/1992 Coll. (NPA) adopted, establishing the current system of nature protection and conservation
1990s	Several national funding schemes for nature conservation independent of the state budget established, some of them persisting until today to complement EU-funding mechanisms
1998–2004	Mapping and identification of 41 SPAs and 863 pSCIs representing the bulk of the Natura 2000 network; establishment of nature conservation information system
2000	New regional arrangement of the country introduced, which changed the nature conservation administrative system completely
2004–2009	EU accession, completion of Natura 2000 and its integration into national system of protected areas including regular management
2004	Fundamental change in financing nature conservation, from national public budgets to EU co-financed projects
2017	NPA amendment enforced position of National Parks but reduced public participation rights

See Table 3.3 for signatory dates of international conventions.

has responsibility for many Natura 2000 sites, 24 out of 26 Protected Landscape Areas (PLAs), other protected areas outside National Parks, and species protection within PLAs. In addition to its headquarters it has 14 regional branches, with almost 600 staff. Its combination of administrators and technical staff is rather unusual in the EU and enables its high level of decision making to be supported by field expert data.

Although exact figures are not available, it can be estimated that the state and regional nature conservation authorities and bodies employ over 1 000 full-time staff members. In addition, tens of experts serve as temporary project managers under various EU-funded projects at the Ministry or regional levels.

11.3.2 Public Involvement and NGOs

The functioning of civil society was broken by the Nazi regime and then by Communism, with freedom of association abolished in 1949 and NGOs prohibited until 1989. However, there were some opportunities for citizen activities within official structures, such as county voluntary conservationists or members of PLA advisory boards. In 1989, the only existing 'interest conservation organisation', the Czech Union of Nature Conservationists (Český svaz ochránců přírody) transformed itself into a federation of about 380 small NGOs (only 6 000 members altogether) under identical names; some new

Table 11.4 Overview of nature conservation authorities and their competencies.

Administrative body	Role, structure and staffing
Central level	
Ministry of Environment (MoE)[7]	Ultimate nature conservation authority (although not directly supervising regional or municipal authorities); guides the NCACR and four National Park Administrations
Nature Conservation Agency of the Czech Republic (NCACR)[8]	Authorisation of any development inside PLAs; governance and management of PLAs and small-scale protected areas and all Natura 2000 sites overlapping their territory; state administration and conservation management of National Nature Reserves and National Nature Monuments across the entire country outside National Parks, and expert activities (opinions, support of state administration); in charge of all Natura 2000 sites overlapping the above-mentioned national protected areas; in charge of strict species protection and issuance of derogations from that regime inside PLAs. Both administrative and expert staff.
Four National Park Administrations[9]	The same competencies over the National Parks as NCACR; in addition, state administration and management of forests in National Parks
Ministry of Defence – four Military District Authorities	Nature conservation administration within four military training areas (MTA) (82 333 ha, 1% of the country); concluding and implementing contracts for conservation of Natura 2000 sites in MTA
Administration of Caves of the Czech Republic[10]	Management of 10 publicly accessible cave systems
Czech Inspectorate for Environment (ČIŽP)[11]	Enforcement body: investigation of cases based on denunciations; regular inspections of the work of authorities of all levels. Headquarters and 10 regional inspectorates, with about 70 staff in charge of nature protection.
Regional level	
14 Regional Authorities	State administration and supervision of management of Nature Reserves, Nature Monuments and Natura 2000 sites as well as administration of strict species protection and issuing derogations from the strict protection regime outside PLAs, National Parks, and MTAs; designation of nature parks (part of general nature protection). Only administrative staff.
Municipal level	
Three types of municipal authorities (about 6 000 altogether)	206 municipal 'authorities with expanded scope of authority' carry out specialised activities within general nature protection (e.g. general protection of species and birds, protection of significant landscape elements, creation of 'territorial system of ecological stability' – TSES). Two lowest municipal levels manage Monumental Trees[12] and authorise felling of non-forest trees. Only administrative staff.

ones were founded too. For political reasons, however, they never became strong and have never received major political support.

The only representative of international NGOs in the country is the Czech Society for Ornithology[13] (founded 1926; 3 900 members), which is the BirdLife International partner and has an important role to play in practical conservation including policy making and advocating bird protection. A number of scientific societies (which are NGOs according to the law), under the Czech Academy of Sciences, also contribute to nature conservation. Of these, the Czech Botanical Society and Czech Entomological Society are the most important.

Despite their low membership and recognition, conservation NGOs were given very strong powers by the 1992 NPA. Under the Act, any conservation NGO was entitled to apply to become a party to any administrative proceedings affecting nature conservation concerns, and was then automatically approved. This concerned proceedings pursuant to the NPA and the Building Act, as well as many other pieces of law governing important infrastructure developments. As a result of this, conservation NGOs significantly improved the state administrative procedures and prevented some potentially damaging projects. For these reasons, however, NGOs were not welcomed by the administration and politicians, who alleged that they had been misusing their strong powers. Although this was never proven, the NPA was changed: since 1 January 2018, NGOs have only had the right to participate in proceedings related to the NPA.[14]

7 www.mzp.cz
8 www.nature.cz
9 www.mzp.cz/cz/narodni_parky
10 www.caves.cz/
11 www.cizp.cz/
12 5 487 designated as of 31 December 2019.
13 www.birdlife.cz/

14 In any other cases, they can only become parties to EIA and SEA proceedings through the general public consultation procedures.

11.3.3 Policy and Legislation

Nature conservation policy is based on the NPA and its implementing (secondary) legislation. All domestic protected areas, Natura 2000 sites and strictly protected species are designated by particular pieces of legislation at various levels (e.g. National Parks by the NPA, PLAs and Natura 2000 sites by governmental decrees, other protected areas and species by ministerial decrees and ordinances of regional authorities, NCACR and National Park administrations).

Secondary legislation designating small protected areas has always been accepted relatively smoothly. In contrast, there have been longstanding substantial political pressures related to the NPA. Every occasion on which the NPA has had to be amended (e.g. due to implementation of EU requirements) has been used by politicians and interest groups, including other sectors, as an opportunity to weaken the powers and competencies of nature conservation. Consequently, to avoid the risk of major weakening of nature conservation, the MoE has chosen not to attempt to replace the 1992 NPA (even though it is severely affected by its 42 amendments) with a new, consistent and logically arranged nature conservation law.

Attempts to draw up usable strategies for nature conservation have not been very successful. In 1998, the State Programme of Nature and Landscape Conservation was elaborated by the MoE and adopted by the government, and an updated version valid to 2021 was adopted by the government in 2009.[15] However, both documents proved to be difficult to implement as their authors did not take into account existing policy tools, and many proposed measures went beyond the existing legislative framework. The update was not preceded by any assessment of the previous document, and it proposed equally unrealistic measures and tasks as the original document. Besides, since 2000, most regions have drawn up their own regional nature conservation policies, which are coordinated neither with the State Programme nor with policies of adjacent regions. The 2016 *National Biodiversity Strategy* (Mach *et al.*, 2016) is very general and does not provide guidance for implementation of the proposed measures, as its main intention was to formally meet obligations stemming from both EU and CBD biodiversity strategies. Thus, in practice, nature conservation policy is mostly directed by ad hoc decisions by the MoE, NCACR and individual regional authorities, but sometimes also by political pronouncements at higher levels.

11.4 Species Protection and Conservation Measures

11.4.1 Protection Measures

The 1956 Act introduced the first species protection measures for selected flora and fauna. These were further developed in the 1992 NPA through the setting up of measures for 'gene species protection', including protecting all species fro destruction and damage, restricting damaging activities, a requiring authorisation of the intentional distribution of no native species (e.g. Larch (*Larix decidua*) for forestry), gene bird protection pursuant to the Birds Directive (since 2004) a 'special species protection'. The last is analogous to that of H Articles 12 and 13: mostly reactive and focussing on protecti (rather than conservation) of individuals and populations selected species rather than of their habitats, with the possibili for official derogations to be granted in case of other overridi public interest. The NPA contains the 'basic conditions f protection' of strictly protected species listed in the irregular amended implementation decree (currently 846 plant a animal taxa, including those from HD Annex IV).[16]

The NPA recognises three categories of protected specie critically endangered, strongly endangered and endangere They neither correspond to global or European IUCN Re Lists nor are they based on the national Red Lists, even thoug several high-quality national Red Lists have been produced fo all major groups of fauna and flora since 2001 (Chobot, 2012) However, the Red List methodologies differ (some do not follo the IUCN categorisation), and they have always been viewed a 'scientific papers' without direct links to practical nature conservation (Chobot, 2012). The NPA protected species categorie serve to place particular species into groups requiring identica protection conditions, but the names are misleading. In particular, the 'critically endangered' category includes man species, such as the Great Crested Newt (*Triturus cristatus*), tha are ecologically valuable but not really critically endangered; the only reason for their inclusion in that category was that conditions for granting derogations are stricter than for the other two categories. The system is considered unnecessarily complex and confusing; the species lists are outdated and do not correspond to current scientific knowledge. It has therefore been subject to wide criticism[17] and the MoE has promised to change the system (Šíma and Zmeškalová, 2018).

Another reason for change is increased evidence of the low effectiveness of this type of species protection. Both the 'national' and the HD Article 12 and 13 approaches put emphasis on the protection of individuals rather than their habitats, when in many cases merely reactive protection cannot ensure their survival if there is no obligation to carry out additional proactive complementary conservation measures that are required to maintain the whole population (especially for species depending on meta-populations, such as some butterflies).

Derogations from the protection provisions have existed in the NPA since 1992 (the legal construction was similar to that of the Nature Directives and was only slightly amended after EU accession). They can be issued by the authorities, if

[15] www.ochranaprirody.cz/o-aopk-cr/strategicke-dokumenty/statni-program-ochrany-prirody/

[16] Vyhláška č. 395/1992 Sb., v platném znění.

[17] For example, in the dedicated issue of the internet journal *Fórum ochrany přírody* (Nature Conservation Forum) 1/2018: www.casopis.forumochranyprirody.cz/magazin/archiv/15/

requested by the applicant and if the applicant can prove there are no alternative solutions to the proposed activity and it is of overriding public interest. Such protection measures have become one of the strongest nature protection tools, as the derogations have been issued under very strict conditions. As a result, many developments have been stopped (including some major projects, even though they had low impacts on nature) as they could not comply with the strict derogation administrative requirements. NGOs have played an important role in derogation proceedings (see Section 11.3), advocating the conservation of nature and providing field data on protected species to the competent authorities.

11.4.2 Species-Specific Conservation Measures

The Czech Republic has a number of birds of high conservation importance that have been subject to high rates of mortality from electrocution, such as the White Stork (*Ciconia ciconia*) and some birds of prey (Hlaváč *et al.*, 2017). To address this, since EU accession in 2004, there has been a legal obligation for all new or reconstructed powerlines and pylons to be designed in a way that prevents the electrocution of birds. In 2016, the MoE, in agreement with the Ministry of Industry and Trade, the main electricity providers, NCACR and the Czech Society for Ornithology, issued instructions to the powerline operators on how to prevent birds injuring themselves on powerlines (Ministerstvo životního prostředí, 2016), accompanied by a regularly updated list of recommended equipment on the NCACR webpage.[18] However, the major electricity providers had already been developing and installing such equipment even before the methodology was issued, as they considered it as an opportunity to improve their image in the eyes of the public.

Some active conservation activities are being undertaken for species of conservation interest, but these are rather limited as knowledge of most species is still insufficient to develop effective 'rescue schemes'[19] for them, irrespective of potential funding and capacity challenges. To date, rescue schemes have been approved (1999–2011) and implemented for the following HD species: *Angelica palustris, Dianthus arenarius* ssp. *bohemicus, Gentianella bohemica*, Freshwater Pearl Mussel (*Margaritifera margaritifera*), Scarce Fritillary (*Euphydryas maturna*), Aesculapian Ratsnake (*Zamenis longissimus*) and European Souslik (*Spermophilus citellus*).

A strategic framework was adopted by the MoE and NCACR in 2014,[20] identifying many species in potential need of rescue. The strategy also introduced 'management programmes', which are non-binding schemes for protected species that are currently not endangered but require specific attention (e.g. to address conflicts or regulate populations where they are high in certain areas). Currently, two species are subject to such programmes: Otter (*Lutra lutra*) and Beaver (*Castor fiber*). A programme for Eurasian Lynx is under preparation.

Several other species re-introduction activities are also underway outside mainstream nature conservation. These include re-introduction of the Atlantic Salmon (*Salmo salar*) by the Czech Union of Anglers since 1997 (which so far has been unsuccessful[21]). Other projects attracting the attention of the public are repatriation activities (rewilding) of Bison (*Bison bonasus*), 'aurochs', and wild horses in a former MTA implemented by a private company.[22]

Several rather theoretical projects aimed at the identification and demarcation of 'corridors' to maintain or improve landscape connectivity for large carnivores have been carried out or are under preparation by the state nature conservancy.[23]

11.5 Protected Areas and Networks

11.5.1 Protected Area Types and Their Coverage

The first protected areas were Nature Monuments, which were mainly established for educational purposes and under the concept that nature conservation was 'preservation' and required no interventions. Over 100 such protected areas were established between 1918 and 1938 on an ad hoc basis, all of which were on private land with designation requiring the agreement of the landowners. The 1956 Act on State Nature Protection established a protected area system, introducing official categories of protected areas: two large-scale (National Parks and Protected Landscape Areas – PLAs[24]), and several small-scale (reserves, monuments, protected parks and gardens, etc.). To conceal its anti-environmental development policy, the state supported designation of hundreds of protected areas even in regions with heavy industry and mining (Kender, 2004). The 1992 NPA modified the 1956 system and established the domestic protected area legislation that is currently in place. Since 2004, the Natura 2000 network has been integrated into the national system of protected areas. Table 11.5 lists the international designated sites and the current types of protected areas, together with the most recent information available on their number and extent.

18 https://bit.ly/3HQgpUo
19 Czech NPA analogy of the term 'action plan'.
20 https://bit.ly/3HQgpUo
21 www.rybsvaz.cz/?page=rada/hosp_odd/losos_navrat
22 www.ceska-krajina.cz/projekty/
23 www.nature.cz
24 The Czech name of this category, which corresponds by its contents to IUCN category V, 'chráněná krajinná oblast', should correctly be translated into English as 'Landscape Protected Area'. However, the term 'Protected Landscape Area' has been used officially since 1990s, so this settled translation is used here.

Table 11.5 The number and extent of international designated sites and protected areas in the Czech Republic.

	IUCN category	Number	Area (km²)
Designated under global and non-EU international legislation			
Ramsar Site		14	635
Mixed World Heritage Site		1	283
Biosphere Reserve (exc. buffer and transition areas)	II, IV, V	6	4 616
Designated under the Nature Directives			
SCIs	I–V	1 113	7 952
SPAs	I–V	41	7 034
Natura 2000 area		*1 153*	*11 148*
Designated under domestic legislation			
National Park (národní park)	II, V	4	1 190
Protected Landscape Area (chráněná krajinná oblast)	V	26	11 381
National Nature Reserve (národní přírodní rezervace)	Ia, Ib, III, IV, V	110	305
National Nature Monument (národní přírodní památka)	III, IV, V	126	81
Nature Reserve (přírodní rezervace)	Ib, III, IV	816	432
Nature Monument (přírodní památka)	Ib, III, IV, V	1 580	337
Contractually Protected Site (smluvně chráněné území)	IV	49	467

Sources. See Section 1.3. Based on data from: Ramsar Sites, Nature Conservation Agency of the Czech Republic (NCACR)[25]; World Heritage Sites, National Heritage Institute[26] and Biosphere Reserves, Regional Development Agency Šumava[27]; EEA Natura 2000 Barometer, Natura 2000 as at end 2020; national summary of areas under conservation.[28]

Since 1992, there have been six categories of '**specially protected areas**'[29] **(ZCHÚ)**: the former large-scale categories, and four new categories of small-scale sites. The reasons for the designation and basic rules for the protection of each category are prescribed by the NPA and are strictly followed in practice. Each individual ZCHÚ must be designated by a specific law or bylaw, which may include, in addition to the general NPA protection rules, site-specific conditions of protection.

National Parks are extensive areas with prevailing occurrence of natural or only slightly altered ecosystems that are unique and important at the national or international level from the ecological, scientific or educational point of view. Any exploitation within them must be dependent on the maintenance of their natural ecosystems and achievement of the natural level of biodiversity, and must be in line with their official conservation objectives. Four conservation management zones are identified in each park from the natural zone to cultural landscape.

Protected Landscape Areas (PLAs) are extensive areas with a harmoniously shaped landscape, a characteristic relief, a significant share of natural forest and grassland ecosystems, with abundant forests, alternatively with preserved monuments of historical settlements. Economic use of their territory is differentiated through zoning (into four zones), which serves both maintenance and improvement of biodiversity and ecological functions and recreational use that does not impair their natural values.

National Nature Reserves protect smaller areas with extraordinary ecosystems of national or international importance. Any economic use is forbidden, and access is only allowed on marked paths.

Natural Reserves are smaller areas with concentrated natural values and typical ecosystems. Any intense economic use, development, or gathering of fauna and flora is prohibited.

National Nature Monuments are small areas of national or international ecological, scientific or aesthetic significance, with concentrated valuable phenomena of natural or anthropogenic origin (e.g. remnants of hay meadows, occurrences of protected plants or animals). Economic use is possible if it does not damage the sites.

[25] http://mokrady.ochranaprirody.cz/o-mokradech-mokrady-mezinarodniho-vyznamu-v-ceske-republice/

[26] www.npu.cz/

[27] www.br-sumava.cz/br-v-ceske-republice/

[28] https://drusop.nature.cz/portal/

[29] The NPA differentiates between 'specially protected areas' that belong under the section of 'special protection and conservation' and other 'protected areas' that can be either SPAs and SACs that do not overlap with specially protected areas, areas protected under 'general nature protection' (such as Nature Parks), international 'labels' not recognised by law (e.g. Ramsar or Biosphere Reserves) or areas providing some level of protection to nature but designated for different purposes pursuant to other pieces of law (e.g. Act on Waters, mining law).

Nature Monuments are sites with the same characteristics and limits as National Nature Monuments but with regional or local importance.

No charge is imposed in any protected area. In those of 'national' categories the state has the pre-emptive right to have the first opportunity to purchase any privately owned land that is offered for sale.

The Czech system of ZCHÚ was never built with the aim of becoming compatible with the IUCN management category system. Instead of being based on management needs, it relates to the level of importance of particular categories ('national' categories being of national or supra-national importance, 'nature' ones of regional to local importance). Thus, almost every category of ZCHÚ includes sites belonging to several IUCN categories. Long-lasting discussions have been held concerning Czech National Parks: all of them have claimed to be IUCN category II as a matter of prestige, although only one meets the criteria in full, two are transitional between II and V, and one corresponds solely to category V.

As of 1 July 2021, there were 2 672 ZCHÚ[30] covering 13 260 km^2 (17% of the country).

Although not officially recognised in law, or by the administration, there are also almost 600 small sites managed for nature conservation by NGOs (Pešout, 2014). Most of them lie within the current protected areas.

The **Natura 2000 network** now represents an intrinsic part of Czech nature conservation, although its integration into the system has not been easy. Initially some felt it was unnecessary, given the established domestic protected area system, but it was realised that it was a major opportunity to increase the effectiveness of nature conservation. Therefore, preparatory works (1999–2005) were taken extremely seriously, the Czech Republic being the only EU Member State where site identification was preceded by detailed habitat mapping of the entire territory as well as field surveys for most HD species. As a result, all site proposals were underpinned by accurate and up-to-date scientific data. In addition, a decision was taken in 1999 to propose Natura 2000 sites irrespective of existing protected areas and on the sole basis of criteria developed for the Nature Directives. Since 2018, negotiations with ETC/BD on several insufficiencies have been underway (MoE, 2020).

The original idea was that SPAs ('bird areas' in Czech) would become a new, seventh, category of ZCHÚ, while 'sites of European importance' (SCIs as well as SACs – Czech law uses the same term for these) would be designated within any of the six national categories of ZCHÚ. When adopting the NPA amendment transposing the Nature Directives in 2004, the Parliament excluded SPAs from the section of ZCHÚ in the Act and established them as an additional category of 'protected areas per se' with specific (albeit very weak) protection rules. Between 2004 and 2009, 41 SPAs – all proposed strictly according to the BirdLife IBA criteria for SPAs – were designated, covering 8.9% of the country (which is still the case as of July 2021).

In addition, the law enabled an additional means of safeguarding SCIs – Contractually Protected Sites – instead of designation. It suits SCIs consisting of individual buildings (e.g. those hosting bat roosts) and MTAs where the military does not allow designation of protected areas but agrees to carry out necessary conservation management. Since 2004, this tool has been applied in 49 SCIs covering 0.59% of the country.[31]

Soon after the European Commission's adoption of the SCI national lists (2005 for the Pannonian region, 2007 for the Continental) it became obvious that the idea of designating the majority of SCIs as ZCHÚ was not feasible. Many SCIs have never fulfilled the characteristics of 'regular' protected areas: the protection conditions would be too restrictive, and their designation would not have been feasible for administrative reasons. Therefore, the NPA was amended in 2009 to introduce a new form of protection for those SCIs that do not require specific conservation management due to their character: so-called basic protection. Thus, an SCI or its part can be protected by: (1) basic protection without further designation, (2) designation within one of the national categories of ZCHÚ, or (3) contractual protection.

As of 1 August 2021, out of 1 113 SCIs covering 10% of the country, 1 034 have been designated as SACs and 1 102 have specific conservation objectives and measures.[32]

Taking into account overlaps of both SPAs and SCIs/SACs, altogether the Natura 2000 network covers 14.1% of the country. The most recent national published data (2019) indicate that protected areas, when defined as ZCHÚ and Natura 2000 sites, covered some 17 259 km^2, which is about 21.9% of the country.[33] Current data in the WDPA for all the protected area types included in Table 11.5 (except Biosphere Reserves) give almost the same total area, due to overlaps, which rounds up to the same 21.9% coverage (UNEP-WCMC, 2022).

Biosphere Reserves have never been legally anchored in Czech law, but instead represent only a 'label' that is used mostly for their recreational promotion by regional agencies as well as transboundary cooperation. As a result, they do not officially count as 'protected areas'. Nevertheless, most of their territory overlaps with 'real' protected areas and Natura 2000 sites, which provide them with the necessary level of protection.

30 https://drusop.nature.cz/ost/chrobjekty/sumarizace/

31 https://drusop.nature.cz/ost/chrobjekty/schru/index.php?pageposschru=0&SELECT_ID_USE_FILTERschru=&SELECT_ID_CHECK_ALLschru=&EXPORT_ALL_step=&ORDER_BYschru=v1.nazev

32 https://drusop.nature.cz/ost/chrobjekty/sumarizace/; data from MoE.

33 www.cenia.cz/publikace/zpravy-o-zp/

11.5.2 The Coherence of Protected Areas and Ecological Networks

The system of large-scale PLAs has been developed since 1955 to protect valuable landscapes, habitat complexes and unique ecosystems. The principal idea was that the PLA system should be 'representative', embracing all landscape types in the country. After the majority of planned PLAs were designated by the mid-1990s, the system was assessed in relation to bioregions and the phytogeographical divisions of the country and was found to be representative (Pelc, 2000). It was completed in 2016 and consists of 26 PLAs covering 14% of the country. It is complemented by the four National Parks.

The ZCHÚ have never been intended to create a coherent network, given the character of the Czech landscape with countless scattered habitats of high natural value that have been separated from each other, often for centuries. Furthermore, as large carnivores had been absent for over a century, it did not seem necessary to connect distant habitats for their sake. Instead, a rather different and specific Czech concept of the 'territorial system of ecological stability' (TSES)[34] – analogous to ecological networks – was developed during the 1980s under the influence of H. T. Odum's ecological ideas from the 1960s. The concept was based on the (now outdated) idea of ecosystems being bound to ultimately reach a climax stage at which point they become most stable and resilient, forest ecosystems being most stable whilst grasslands and arable land are least stable. According to this concept, it was necessary to identify, protect and artificially amend the 'skeleton of ecological stability' in landscapes consisting of 'biocenters' linked by 'biocorridors'. Once established, this 'skeleton' was supposed to support the spread of 'ecological stability' into the surrounding 'unstable' landscape, thus improving its ecological functions (Roth, 2021). This concept was included in the 1992 NPA and started to be implemented in terms of identification and planning of biocentres and biocorridors within spatial plans. Whilst this protected these areas from development, there was negligible real implementation in the field. The TSES never had protected areas in its focus and developed separately.

Originally, the TSES did not take into account landscape connectivity requirements for animals. This issue was considered later, but often focussed on the needs of game or several 'flagship animal species' rather than threatened and/or ecologically valuable species. Whilst the potential for the TSES to provide some nature conservation benefits has been generally high, and it has become a recognised and respected tool in spatial planning, its actual achievements have not been assessed as effectiveness indicators are lacking.

[34] www.nature.cz/uses

11.5.3 Approaches to Protection and Levels of Protection

The NPA defines a wide range of bans and restrictions defining an obligatory protection regime specific to each ZCHÚ category (as well as to particular zones of large-scale ZCHÚ). Individual designation decrees prescribe additional activities needing individual authorisation. Restrictions mostly affect economic uses, waste deposits, camping and collecting of animals and plants other than forest fruits. Access is only restricted in the 'rest zones' of National Parks and in the National Nature Reserves. Derogations from the bans can be issued by the nature conservation authorities in case of overriding public interest or in the interest of nature conservation.

HD Article 6(3) appropriate assessments (see Section 11.6) have consistently been applied in and around all Natura 2000 sites, thus positively protecting also the ZCHÚ that overlap them.

11.5.4 Approaches to Managing Protected Areas

All specially protected areas must have management plans (called 'management principles' in National Parks) covering 10 years and commissioned and adopted by the respective authorities following a procedure involving landowners. For newly proposed sites, draft management plans must be in place simultaneously with the site proposal, therefore no protected area can be designated without being underpinned by conservation management rules based on good-quality field data. Due to the long-lasting tradition of natural history research, the quality of the plans is generally high – a view shared by both conservation and scientific communities. Since the 1990s, the MoE has also issued guidelines for management planning,[35] which are obligatory for NCACR, National Park and PLA administrations and are also generally respected by regional authorities.

As the current system of ZCHÚ was founded during the previous regime when property rights were not taken into account, the then protected areas were designated without any communication with the landowners. Even today, after reprivatisation of most of the land in the country, designation of ZCHÚ is still viewed as an exclusively expert exercise that does not necessarily need to involve landowners or the public. The NPA still does not require any consent of landowners for site designation, although in practice such consents are sought by the authorities. According to the law, landowners and stakeholders are provided with the draft management plan prepared by the authority and only then can any dialogue begin, should the stakeholders be interested. Therefore, participatory approaches in management planning have only rarely been used in some PLAs and National Parks. On the other hand, due to a custom that still survives – that there is an

[35] For recent versions valid since 2019, see mzp.cz/cz/osnova_planu_pece

obligation on the state to secure conservation management and its funding – many landowners are not interested in entering consultations as they are often unconcerned if their property is to become part of a protected area.

As Natura 2000 sites are not ZCHÚ, the NPA prescribes for them different management documents called 'summaries of recommended measures' (SDO). SDOs for SCIs are more concise than management plans and focus on particular conservation measures that are necessary to meet the site's conservation objectives. SDOs for SPAs are optional, depending on the ecological character of target species and local circumstances. Up to 2020, the MoE issued SDOs for 11 SPAs; in some others their role has been overtaken by 'ordinary' management plans of overlapping National Parks or PLAs.[36] For SCIs/SACs, SDOs are obligatory even where they overlap with a ZCHÚ. Up to 2020, the MoE had adopted SDOs for 97% of SCIs.[37]

Conservation management prescribed by management planning documents is implemented in practice by both nature conservation authorities themselves and contracted companies and individuals.

11.6 General Nature Conservation Measures

The 1992, the NPA introduced a completely new idea of 'general nature protection'; before this, the landscape outside protected areas was completely unprotected. According to the NPA, the following should be protected: the TSES (Section 11.3), significant landscape elements (by law: all forests, bogs, mires, water courses, fishponds, lakes, river floodplains; optionally: other valuable features), caves, areas with characteristic landscape scenery (through Nature Parks), temporarily protected areas, fossils, and all flora and fauna. Consent is required from the nature conservation authorities for any use of land or interventions likely to affect its ecological functions. The general protection of birds and the bird protection derogations were amended after EU accession.

According to the NPA, an implementation decree(s) should have been adopted that would have further specified the necessary approaches to dealing with these provisions, but it has not happened. Unfortunately, most of these NPA provisions (except for bird protection and Nature Parks,[38] which mostly serve public recreation and help promote particular regions) have been implemented in a way that meets administrative requirements but does not necessarily provide their intended biodiversity conservation outcomes. The conservation community, therefore, considers that this has been a wasted opportunity to strengthen the protection of nature.

36 www.mzp.cz/cz/souhrn_doporucenych_opatreni

37 CZ PAF 2021–27.

38 Over 100 of them designated by regional authorities in all regions by 2019.

11.6.1 Development Planning and Impact Assessment

The supreme governmental spatial planning strategy is the *Spatial Development Policy of the Czech Republic 2008*, as amended in 2018 and 2019,[39] which coordinates planning activities of regions and municipalities, sectoral and intersectoral concepts, policies and strategies and other documents of ministries and other central administration authorities. Each of the 14 regions has its own 'Principles of spatial development' that must not contradict the principles of the Policy, and most municipalities have adopted spatial plans, which must be in line with regional principles. All spatial planning documentation is subject to SEA and must be in line with the conservation policy set by the NPA, including Natura 2000.

EIA and SEA procedures have been in force since 1992. In 2004, appropriate assessment was incorporated into the NPA. Procedurally, it has been merged with SEA/EIA but is carried out by a separate pool of individual experts licensed and supervised by the MoE (who must pass a complex exam and have an education in ecology). This licensing and regular quality checks by the MoE seem to be the main reason for the generally very high quality of appropriate assessment studies. Within the project-level appropriate assessment there is a two-step procedure: while the appropriate assessment study carried out by an expert is commissioned by the developer, the competent authority pays for an independent peer review by another licensed expert, who can request amendments to the study or reject it if there are serious mistakes or omissions. Licensed appropriate assessment experts also advise developers before the start of such procedures on how to amend, or even drop, their projects in order to avoid conflicts with Natura 2000. Up to 2019, only two derogation procedures pursuant to HD Article 6(4) have occurred, as Natura 2000 is generally considered 'untouchable' and is avoided by developers as much as possible.

11.6.2 Invasive Alien Species

Of the overall number of non-native plant (1 454) and animal (595) species, 61 plants and 113 animals are considered invasive in the Czech Republic (of which 20 plant and seven animal species are considered most dangerous under Czech conditions) (for the Black List, see Pergl *et al.*, 2016). This list only partially overlaps with the list of EU IAS, with the latter containing species that are not of greatest concern in the Czech Republic, and omits some that are. Many invasive plants only pose a risk to certain habitat types that mostly remain in protected areas (e.g. remnants of steppes), whilst the impacts of some aquatic animal IAS are much more detrimental (especially two North American crayfish species that have transmitted crayfish plague).

39 www.mmr.cz/cs/Ministerstvo/Stavebni-pravo/Koncepce-Strategie/Politika-uzemniho-rozvoje-Ceske-republiky

To tackle IAS issues, several eradication projects (including some funded by the EU LIFE Programme – see Section 11.7.3) have been implemented, and guidelines for eradication exist for many species.[40] Although most of the projects have met their goals, their overall impacts on IAS are very low due to their local scale.

11.6.3 Habitat Restoration, Non-intervention and Rewilding

Habitat restoration is an essential part of conservation management in ZCHÚ areas where this is appropriate, especially in those with aquatic habitats and wetlands (e.g. raised bogs and fishponds). Outside ZCHÚ, the River System Restoration Programme (1992–2011) has facilitated restoration of dozens of artificially regulated small water courses and their floodplains.

As the majority of protected areas host anthropogenic ecosystems that depend on agricultural and forestry management, only a few areas would benefit from a lack of human intervention. A non-intervention approach is generally accepted for remnants of natural steppes, which are mostly in small-scale protected areas (and where this approach might be incorrect as steppes need regular disturbance to prevent overgrowth). In contrast, in forests it is hardly imaginable, even by many conservationists, because traditional plantation forestry has created a conviction that Czech forests cannot be left to develop naturally. Consequently, for example, it took more than 25 years to reach a consensus on non-intervention management in clearly defined areas representing a small part of the forests in the largest National Park, Šumava, even though the law requires National Parks to be left to natural development. In contrast, in Podyjí National Park the predominant xerothermic forests have a high natural potential for non-intervention, which is being enforced by the National Park. However, this has led to disagreements with some scientists who consider that this policy is detrimental for certain invertebrates that require active forest management such as coppicing to survive.

The conservation community has long been aware of the benefits of non-traditional methods of conservation management of some ZCHÚ, such as allowing motorbikes, off-road vehicles, or even military vehicles to disturb the ground. Experiments have proven their high efficiency in terms of the restoration of important vegetation communities and habitats threatened by succession,[41] but this is not yet something that the public is ready to accept. Several forms of more established conservation management intervention also remain 'taboo', such as intentional forest burning (which is a crime according to fire regulations) and forest grazing (prohibited by law since the 1700s despite the absence of any freely moving stock for a century). Nevertheless, the number of positive examples is increasing.

One NGO (Hnutí Duha), and some scientists, have been promoting the 'return of wilderness' into the landscape outside of protected areas. However, a unified definition of wilderness is lacking; therefore, the concepts differ and range from scientifically underpinned proposals to rewild areas that would historically have been non-intervention areas to halting all human interventions in heavily modified anthropogenic areas.

11.7 Nature Conservation Costs, Benefits and Funding Sources

11.7.1 Nature Conservation Costs

An overall assessment of the costs of nature conservation has never been made, for four principal reasons. Firstly, nature conservation is executed by three different levels of public administration, the budgets of which are not linked; thus, for example, management costs of ZCHÚ are set independently by regional authorities based on management plans and the real situation in the field, without any reporting to the centre. Secondly, the wages of nature conservation staff of all levels have not been included in cost calculations, partly because some staff deal with wider issues than just nature conservation. Thirdly, it is not always clear what are the 'real' costs needed for nature conservation and what are the costs that result from it (e.g. compensation for livestock losses to increasing Wolves) or from political pressures (remuneration of potential opportunity costs, which are often hypothetical). Fourthly, EU accession enabled new payments to be made, such as CAP Rural Development Programme (RDP) Natura 2000 payments that serve to motivate landowners, but these are not necessarily fully required to achieve the desired conservation outcomes.

As regards Natura 2000 costs, these cannot be easily separated from the other costs as they are born by the same authorities in charge of 'national' nature protection and conservation. The first Prioritised Action Framework (PAF) (2013) estimated the costs of fully implementing the Natura 2000 network as €79 million per year. However, this estimate is potentially misleading for two reasons. Primarily, its purpose was misunderstood by the Czech authorities, so that it became a 'wish list' of activities that did not consider constraints on funding sources and overall capacities (e.g. available experts) in the country. Secondly, the PAF included costs linked directly to Natura 2000 as well as broader costs relating to species protection, surveillance, research and species rescue programmes. The PAF 2021–2027 (MoE, 2020) estimates the annual costs as €149 million, the increase being due to greater management costs for the completed Natura 2000 network.

11.7.2 The Economic Benefits of Nature Conservation

Nature conservation is part of the Czech culture and there is an understanding that the current population is obliged to conserve the remaining natural assets for future generations.

[40] https://invaznidruhy.nature.cz/metodiky-prirucky

[41] www.beleco.cz/militarylife/managementy/regulacni-management/pojezdy-vojenskou-technikou.html

Therefore, whilst nature conservation is of recognised cultural value, its full economic benefits have not been considered until recently and no study has attempted to quantify them.

In 2014, a 'Methodological framework for integrated assessment of ecosystem services in the Czech Republic'[42] was developed to enable a national assessment of ecosystem services. According to the *National Biodiversity Strategy* (Mach *et al.*, 2016), this national assessment was due in 2020 in order to contribute to the EU Biodiversity Strategy 2020, but it had not been completed as of July 2021. The ecosystem services concept is not widely understood in the conservation community and there is some doubt as to whether it will persuade politicians and the public of the necessity to protect and conserve nature. Although a study estimated that the value of ecosystem services in the Czech Republic is one and a half times GDP (Frélichová *et al.*, 2014), a problem with this is that the services are provided both inside and outside of areas managed for nature conservation. It is therefore not normally possible to quantify the added effects of nature conservation activities.

11.7.3 Funding Sources for Nature Conservation

Table 11.6 summarises the main sources of funding for nature conservation, mostly over the 2014–2020 EU funding period. Over 2014–2020 the main source of funding for the most effective and targeted nature conservation measures was the ERDF and Cohesion Fund Operational Programme Environment (OPE) Axis 4, which had three specific objectives: ensuring favourable status of protected areas (management planning, one-off interventions, monitoring); strengthening biodiversity (species conservation); and strengthening of natural functions of the landscape (habitat restoration). Over the funding period, some CZK 15 596 million (€604 million) was allocated, with a 58% EU contribution. The previous OP Environment, over 2007–2013, was also of considerable importance as it focussed on Natura 2000 sites and spent CZK 15 025 million (€582 million) over the period.[43] However, the OPEs could not fund recurrent activities, such as regular site management, which reduced their potential use.

While the regular management of protected areas is mostly covered by national management schemes of the MoE and regional authorities, management of the wider environment is greatly dependent on CAP (EAFRD) RDP measures, especially the agri-environment–climate measures, with 44% of the RDP funding over 2014–2020. Natura 2000 payments were relatively low as they only apply to farmland in core zones of large-scale protected areas, the representation of which is negligible. However, the major part of the other RDP funds was used for measures with low or uncertain benefits for nature conservation. Quite often, conservation benefits of these measures have been claimed by the agriculture sector as a justification for traditional (and sometimes explicitly harmful) agricultural approaches maintained from the past. This includes the payments for areas facing constraints (M13), which used 36% of the budget. Although it is expected that they reduce the abandonment of High Nature Value farming systems, the extent and actual biodiversity benefits are uncertain. Organic farming also received a substantial proportion of the RDP budget (17%); whilst it may have some benefits (e.g. food quality and public health), those for nature are less certain and it could even be detrimental overall for biodiversity, as the field experience of nature conservation has shown. As discussed in Chapter 5 (Section 5.5), organic farming's modest per unit area biodiversity benefits could be outweighed by the need for agricultural expansion and/or intensification elsewhere in response to their lower yields. This is also a reason why some of these measures were not listed among the nature conservation funding sources in the PAF 2021–2027 (MoE, 2020).

The main RDP allocations to forests (under M8) were for replanting, often in response to bark beetle outbreaks. Even if applied in some protected areas, replanting was primarily to restore tree cover in accordance with the obligations of the Forest Act, and not directly for nature conservation purposes.

The main national nature conservation funding schemes have been in place since 1992 and were modified or replaced upon EU accession. They complement EU sources as they can support state bodies in charge of protected areas and caves and enable financing of recurrent management activities and land purchase in protected areas including Natura 2000. In addition, state and public budgets cover the salary and operational costs of nature conservation authorities and organisations, which altogether represent a very high cost item.

Unlike in many EU countries, the EU LIFE programme has not been of major importance due to the low number of projects and their limited scope. It is hoped that this will change with a current large Integrated Project, which aims to establish, *inter alia*, a more effective management system for the Natura 2000 network.[44] Interreg is locally important in boundary regions, focussing on smaller transboundary projects. EEA grants are an important funding source for conservation NGOs.

Each of the 14 regions has its own nature conservation funding scheme (several regions also run grant schemes for NGOs) funded from their regional budgets.

The overall sum of resources available for nature conservation is high, and could probably potentially meet the justified needs, but much of the largest source of funds (EAFRD) is used for measures with low or unproven nature conservation benefits. The challenge with the current funding arrangements is that project-related financing (which has become the main source

[42] www.ecosystemservices.cz/cs/metodologicky-ramec-integrovaneho-hodnoceni-ekosystemovych-sluzeb-v-ceske-republice/

[43] www.opzp.cz/o-programu/cerpani-a-schvalene-projekty/

[45] https://webgate.ec.europa.eu/life/publicWebsite/index.cfm?fuseaction=search.dspPage&n_proj_id=7002

Table 11.6 Funding sources for nature conservation in the Czech Republic.

Source and period	Measure/category	Averaged annual allocation (CZK m)		Averaged annual allocation (€ m)		Comments
EU funding instrument		**EU + national**	**EU**	**EU + national**	**EU**	
EAFRD – Rural Development Programme – allocations to Priority 4 (restoring, preserving and enhancing agriculture and forest ecosystems) 2014–2020	M10 Agri-environment–climate	3 768.1	2 502.7	146.0	96.9	Actual use depends on uptake
	M13 Areas facing constraints	3 106.0	2 173.1	120.3	84.2	Uncertain benefit for conservation
	M11 Organic farming	1 437.5	942.4	55.7	36.5	Probably little benefit for conservation
	M8 Investments in forests	174.9	86.6	6.8	3.4	Not used for conservation
	M1 Knowledge transfer, etc.	32.7	16.2	1.3	0.6	General training support
	M15 Forest environment climate	29.1	21.6	1.1	0.8	Comment 1.
	M12 Natura and WFD	17.0	12.2	0.7	0.5	
RDP Priority 4 subtotal		*8 565.2*	*5 754.8*	*331.8*	*222.9*	
ERDF/Cohesion fund 2014–2020	OP Environment. Axis No. 4 Conservation and management of nature and landscape	2 228.0	1 296.0	86.3	50.2	
ERDF Interreg Central Europe, Danube, AT, PL, SK 2017–2021	11 conservation projects	30.2	?	1.1	?	Comment 1.
LIFE Programme	Nature and biodiversity projects 2004–2019	94.4	59.7	3.7	2.3	14 projects
	Integrated project 2019–2026	65.7	39.4	2.5	1.5	LIFE-IP: N2k Revisited
EU funding instruments subtotal		10 981.8	>7 150.0	425.4	>276.9	
		Additional non-EU* (CZK m)		**Additional non-EU* (€ m)**		**Comments**
National funds						
MoE						
Landscape Management Programme (non-capital scheme) 2004–2018	Implementation of management plans in protected areas and N2k sites and landscape improvements outside protected areas	160.1		6.2		
Management of Inalienable State Property within Protected Areas (both non-investment and capital scheme) 2003–2017	Conservation and management measures demanded both by law and site management plans and forest management plans	64.0		2.5		Established due to many protected areas being owned by the state and managed by NCACR
Restoration of Natural Functions of the Landscape 2009–2017 (capital scheme)	Climate and conservation measures in protected areas managed by NCACR and National Parks	39.2		1.5		Used for state bodies not eligible for EU funding

Table 11.6 *(cont.)*

Source and period	Measure/category	Averaged annual allocation (CZK m)	Averaged annual allocation (€ m)	Comments
MoE budget annual	Small NGO grant scheme	c. 5.0	0.2	
Regional				
14 regional public budgets	Nature conservation schemes	Millions per region		Content and scope vary among regions
Other funds				
EEA and Norway Grants 2015–2017*	NCACR projects (e.g. surveillance, species rescue programmes)	29.1	1.1	
	NGO-implemented projects	?	?	Comment 2.
Additional national and other subtotal		>297.3	>11.5	
National total including co-financing of EU instruments		>4 100.0	>158.8	
Total		>11 279.1	>436.9	

Notes. Currency conversion rate (29 May 2021): €1 = CZK 25.817. * Excludes co-financing of EU funds unless indicated. The subtotals and totals are only approximate indications of the average annual levels of funding as they are based on annualised data from differing periods. *Comments.* 1. Only a certain part is used for conservation; however, official data do not enable an estimate of the proportion. 2. Allocations variable, no summary data available.
Sources. Based on data from: RDP data from European Commission ESIF funds database (23 October 2020); LIFE Czechia factsheet (March 2022); LIFE Public database; Kešner *et al.* (2017); MoE (2020); Operational Programme Environment;[45] public web pages of the individual funds. See Section 1.3 for further details.

since EU accession) does not guarantee long-term sustainability, is very difficult to obtain and involves onerous application processes. Also, the administration of EU funds places a substantial burden on authorities, which have a limited capacity to deal with them and meet their other objectives. Therefore, whilst the independent national funding schemes that complement the EU funds are relatively small, they are essential for ensuring nature conservation remains operational.

11.8 Surveillance, Monitoring, Reporting and Applied Research

Before EU accession, only birds and a few other species were routinely monitored. Some of these were used as indicators of the status of nature in MoE Environmental Reports, which have been published annually since the 1990s. As these indicators are not sufficiently representative, the 2016 *National Biodiversity Strategy* included a goal 'to create a comprehensive set of indicators of status, changes and trends of biological diversity'. These will be used to evaluate the Strategy and better inform the public about biodiversity conservation issues in future.

Requirements under the Nature Directives have led to the mapping and systematic surveillance of BHD habitats and species, as well as some others of national importance. Detailed mapping of HD habitats was carried out by 550 field workers from 2000 to 2005 (covering 3.5 million ha) to obtain data for the identification of Natura 2000 sites and provide reference habitat data for the entire country.[46] It also led to the launch of a publicly accessible, expert verified database of HD species records to help fill data gaps.[47]

The development of the habitat and species surveillance methodologies and data gathering has been organised in a systematic manner and funded by the Department for Biodiversity Monitoring of the NCACR[48] (which is responsible for reporting under the Nature Directives). This enabled the Czech Republic to contribute to the first round of HD Article 17 reporting in 2007 and, due to its expertise, data and systematic approach, the NCACR has been an ETC/BD member institution since 2006. Data processed by the NCACR are gathered both by staff of the nature conservation authorities (hundreds of people) and outsourced experts, mainly from the academic sphere. Bird data have traditionally been provided by the Czech Society for Ornithology. On the other hand, greatly needed long-term data on diversity and abundance of insects are still missing.[49]

The NCACR has coordinated some nature conservation related R&D projects (funded by the Technological Agency of the Czech Republic or EEA grants), such as on surveillance methodologies and an inventory of 9 605 fish migration barriers in the Czech river network.[50]

45 www.opzp.cz/o-programu/cerpani-a-schvalene-projekty/

46 https://portal.nature.cz/publik_syst/ctihtmlpage.php?what=1035
47 https://portal.nature.cz/nd/
48 www.biomonitoring.cz/
49 https://butterfly-monitoring.net/bms-schemes
50 http://vodnitoky.ochranaprirody.cz/migracni-bariery-a-mve-migracni-bariery/

Although the expertise of the staff of the NCACR and National Parks is high, there is often a need for focussed conservation research that goes beyond their capabilities. Although the MoE supervises several research institutes, applied conservation research has never been systematically supported (Härtel *et al.*, 2014). Thus, where necessary, research projects are contracted from universities and the Academy of Sciences when they are within their capacities for applied research.

11.9 Future Developments

While 1998–2009 was characterised by enthusiasm for the establishment of the Natura 2000 network and surveillance system, which contributed hugely to Czech nature conservation, current ambitions and activity are at a low ebb. This is in part because biodiversity conservation has ceased to be a political priority; instead, over the long term, political pressures against nature conservation have grown as it allegedly 'prevents developments'. Furthermore, staff in the nature authorities are demoralised by low wages and requirements to carry out tasks that merely formally fulfil actions set out in strategic documents, which has led to staff resignations.

Meanwhile, there is the need to confront major challenges, such as the poor condition of forests caused by bark beetle outbreaks and the persistent adverse effects of large-scale intensive agriculture. On the positive side, the level of governance and management remains very high, NCACR and the National Park administrations still have highly qualified expert staff, and nature conservation has its unchallenged place in the system of public administration. Furthermore, the current EU budget period brings new ways of funding Natura 2000, such as through the LIFE Integrated Project, although its impacts are difficult to predict.

11.10 Conclusions

The tradition of Czech nature conservation is relatively longstanding and has been characterised by its systematic approach. Dedicated institutions have been in place since 1919. Since 1950, protected areas have had a system of clearly defined categories, which were further developed in the 1990s and amended by establishment and integration of the Natura 2000 network in the 2000s. Czech nature conservation has also been based on good governance, a high number of qualified and dedicated staff, and relatively adequate funding. The nationalisation of land in the 1950s led to an expectation that the state should be responsible for the funding and implementation of conservation management. This view has persisted to this day, and although much of the nationalised land is now back in the hands of private landowners, their role in nature conservation remains limited.

A key component of nature conservation has been the designation of large-scale protected areas that cover over 15% of the country and are administered by authorities with sufficient expertise and effective powers. Overall, protected areas cover 22% of the country, and all domestic sites, and most Natura 2000 sites, have management plans that are implemented and result in real conservation management measures in the field. Species protection is at a crossroads; new approaches are being discussed to make the efforts effective, as the traditional reactive approach has proved to be ineffective. Proactive measures in the open landscapes addressing habitat fragmentation, lack of connectivity and the decline of traditional land use need to be developed at a much greater scale, with the participation of other sectors, landowners and the public.

Another strength of the systematic approach in the Czech Republic has been the reliance on gathering and managing scientific information, in particular to identify Natura 2000 sites. The establishment of surveillance and monitoring activities is considered as important as site and species protection. A weakness has been inadequate communication of the importance of nature conservation to the public, which has contributed to low public support for nature conservation, especially in the countryside. This has recently been recognised by the nature conservation authorities, and communication skills and abilities have started to be appreciated in equal measure to technical expertise.

Acknowledgements

This chapter would never have originated without the very existence and everyday endeavours of thousands of professional and voluntary conservationists and their predecessors who established contemporary nature conservation in the Czech Republic. My sincere thanks go to the reviewers of this chapter, Handrij Härtel, Michael Hošek and Ondřej Sedláček.

References

Bečka, P. & Rösner, S. (2019) Tetřev bez hranic. *Ochrana přírody*, **74** (1), 16–20.

Bejček, V. & Šťastný, K. (2015) Čtvrté mapování hnízdního rozšíření ptáků v České republice. *Ochrana přírody*, **70** (6), 30–36.

Chobot, K. (2012) Červené seznamy: zpráva o stavu. *Ochrana přírody*, **67** (4), 17–19.

Chytrý, M., Hájek, M., Kočí, M. *et al.* (2019) Red List of Habitats of the Czech Republic. *Ecological Indicators*, **106**, 105446.

Fischer, D., Jeřábková, L. & Vlach, P. (2015) Jsou čolek obecný a ropucha obecná stále ještě obecní? *Ochrana přírody*, **70** (2), 32–36.

Flousek, J. (2019) Přežijí krkonošští tetřívci rok 2040? *Ochrana přírody*, **74** (2), 6–9.

Frélichová, J., Vačkář, D., Pártl, A., Loučková, B., Harmáčková, Z. V. & Lorencová, E. (2014) Integrated assessment of ecosystem services in the Czech Republic. *Ecosystem Services*, **8**, 110–117.

Härtel, H. (2017) Conservation of Botanical Diversity in the Czech Republic. In *Flora and Vegetation of the Czech Republic*, eds. M. Chytrý, J. Danihelka, Z. Kaplan & P. Pyšek, pp. 401–444. Cham, Switzerland: Springer International Publishing.

Härtel, H., Vrška, T. & Hruška, J. (2014) Jak je na tom ochranářský výzkum? *Fórum ochrany přírody*, **1/2014**, 11–13.

Hlaváč, V. & Beran, V. (2011) Návrat sokola stěhovavého - úspěch ochrany přírody, nebo důsledek samovolného vývoje? *Ochrana přírody*, **66** (1), 8–11.

Hlaváč, V., Škorpíková, V. & Janoška, Z. (2017) Na sloupech elektrického vedení stále hynou desetitisíce dravců. *Ochrana přírody*, **72** (2), 7–9.

Kender, J. (ed.) (2004) *Péče o krajinu*. Prague: Ministry of the Environment.

Kešner, M., Fryčová, L., Hošek, M., Kundrata, M. & Dušek, J. (2017) *Analytické podklady pro přípravu strategie mobilizace zdrojů pro ochranu biologické rozmanitosti v ČR*. An unpublished study for MoE, Annex 2 – nature conservation costs overview. Prague: DHP Conservation.

Kuras, T., Šarapatka, B., Mazalová, M., Tuf, I. H. & Bednář, M. (2017) Krajinná struktura. Klíč k ochraně biologické rozmanitosti, půdy a vody. Část I – Ochrana biodiverzity. *Ochrana přírody*, **72** (6), 18–23.

Mach, J., Pojer, F., Plesník, J., Hošek, M. & Dušek, J. (eds.) (2016) *National Biodiversity Strategy of the Czech Republic 2016–2025*. Prague: Ministry of the Environment.

Ministerstvo životního prostředí (2016) Zajištění ochrany ptáků před úrazy na elektrických vedeních podle zákona č. 114/1992 Sb., o ochraně přírody a krajiny, ve znění pozdějších předpisů. *Věstník Ministerstva životního prostředí*, **26** (10), 1–17.

MoE (2020) *Prioritní akční rámec (PAF) pro soustavu Natura 2000 v České republice podle článku 8 směrnice Rady 92/43/EHS o ochraně přírodních stanovišť, volně žijících živočichů a planě rostoucích rostlin (směrnice o stanovištích) ve víceletém finančním rámci na období 2021–2027*. Prague.

Pelc, P. (ed.) (2000) *Program rozvoje chráněných krajinných oblastí (analýza, cíle, opatření)*. Prague: SCHKO ČR.

Pergl, J., Sádlo, J., Petrusek, A. *et al.* (2016) Black, Grey and Watch Lists of alien species in the Czech Republic based on environmental impacts and management strategy. *NeoBiota*, **28**, 1–37.

Pešout, P. (2014) "Soukromé rezervace" v České republice. *Ochrana přírody*, **69** (2), 19–22.

Roth, P. (2021) ÚSES aneb Císařovy nové šaty po česku. *Fórum ochrany přírody*, **01/2021**, 6–10.

Šíma, J. & Zmeškalová, J. (2018) Aktualizace seznamu zvláště chráněných druhů a změny v druhové ochraně. *Fórum ochrany přírody*, **01/2018**, 11–13.

UNEP-WCMC (2022) *Protected Area Profile for Czech Republic from the World Database of Protected Areas*, January 2022. www.protectedplanet.net

Vašků, Z. (2011) Zlo zvané meliorace. *Vesmír*, **90**, 440–444.

Vermouzek, Z. & Zámečník, V. (2017) *Indikátor ptáků zemědělské krajiny za rok 2017*. Studie pro Ministerstvo zemědělství ČR. Česká společnost ornitologická.

Chapter 12 Denmark

Anette Petersen and Jan Woollhead

Key Statistics	
Terrestrial area (km^2)/% of EU-28*	43 167/1.0%
Marine waters (km^2)/% of EU-28*	122 815/2.7%
Population density (p km^2)/% of EU-28*	130/113%
GDP per capita (€)/% of EU-28	45 989/172%
Land cover (CORINE land cover classes 2018)	AGR 73%, FOR 9%, ART 8%, OSN 5%, WET4%
Biogeographical regions	ATL, CON, MATL, MBAL
Number of HD Annex I habitats/%*	60/27%
Number of HD Annex II, IV and V species/%*	79/6%
Number of BD Annex I bird taxa/%*	49/28%
% HD UCS and decreasing habitats/% unknown	33%/0%
% HD UCS and decreasing species/% unknown	21%/20%
% breeding BD Annex I birds decreasing/% uncertain and unknown	12%/12%

* Excluding MAC and MMAC biogeographical regions.
Sources and codes. See Table 1.1.

12.1 The Natural History and Characteristics of Denmark

Denmark is situated in northern Europe and is the southernmost country of Scandinavia, which by local convention comprises the three kingdoms of Denmark, Norway and Sweden. The Kingdom of Denmark includes the autonomous territories of Greenland and the Faroe Islands, which are not covered in this book.

Denmark is a relatively small country, but has a very long coastline of 8 750 km (Danmarks Statistik, 2019) as it consists of the peninsula Jylland (Jutland) and around 1 419 islands of over 100 m^2 (Folke, 2020). Of these islands, the largest are Sjælland (Zealand) and Fyn (Funen). It has a temperate climate and moderate rainfall. The country is divided into two terrestrial biogeographical zones: the Atlantic to the west and the Continental to the east, with the boundary broadly down the centre of Jutland.

Bedrock forms the foundation of the Danish landscape, but since the Precambrian the land has been largely sea-covered. Therefore, the majority of the country is covered by thick, mainly marine deposits. Only on the island of Bornholm, which forms the eastern-most part of the Danish territory, can the bedrock be seen at the surface. The landscapes encountered on the surface today are thus very young, being predominantly formed during the Quaternary period, in particular the last stage of the Weichsel Ice Age. It also means that the land is fairly flat with little variation in height (average of 31 m, and a maximum of 171 m). At the end of the ice age Denmark was at first connected by land to Sweden and England, and the tundra was gradually replaced by forest, but then a period of alternating transgressions of the sea and post-glacial rebound followed until the present distribution of land and sea was attained (Sand-Jensen and Larsen, 2017).

Since then, with its temperate climate, Denmark's natural climax vegetation has been deciduous forest. However, open areas would also have occurred such as where affected by wind and wave erosion and wet or saline soils. Wild herbivores also probably kept open areas in the forests. Later, grazing livestock had the same effect until woodland grazing was prohibited in 1805 (but is now being reinstated in some areas) (Sand-Jensen and Vestergaard, 2017).

Today most of Denmark is open country, mainly due to human activities. Deforestation started during man's transition from being hunters to farmers in the Neolithic Period, about 6 000 years ago. This created a number of new extensive semi-natural habitats. Grazed areas on the moraine soil became open areas with grasslands. In damp areas, within and outside forests, marshy meadows were created by grazing and haymaking. On sandy soils forests were cleared and the land was then cultivated. Later, where cultivation ceased and grazing took over, these areas became heathlands. As grazing and deforestation continued, these semi-natural habitats increased to cover about half the country. Forest cover declined to about 4%, its lowest extent, in 1820 (Sand-Jensen and Møller, 2017).

In the 1850s, further expansion of agricultural land began, including the cultivation of the heathland, which at that time constituted about 15–20% of Denmark (Buttenschøn and Schmidt, 2015). At the same time, extensive drainage of large wetlands occurred (although conservation activities are currently restoring some areas). Over the last 150 years, Denmark's open country has been further changed by afforestation and urban development (as its population increased from 1.4 to 5.8 million inhabitants (2019)).

Different sources give slightly different numbers for land use, depending on how the numbers are calculated. According to CORINE data for 2018, agriculture covered 73% of the land, 61% being cultivated cropland, which is the highest proportion in Europe. According to Statistics Denmark,[1] arable land with agricultural crops excluding renewable grassland constitutes 54% of the area. Forest cover has increased since the 1990s and covers nearly 9% of the country (CLC data for 2018) or 13% (Danmarks Statistik, 2017). However, it is important to point out that the majority of forest areas are under productive forestry management to varying degrees. Consequently, there is very little room on the land for nature: according to CLC data for 2018, semi-natural areas now cover only about 5.2% of the country (amongst the lowest in the EU).

Considering the size of the country, Denmark has a very large marine area. The country is bordered by the North Sea, the Skagerrak and the Kattegat (in the Atlantic biogeographical region) and the Baltic Sea (in the Continental biogeographical region); its Territorial Sea and EEZ sea area is about 2.5 times that of its land area. The marine habitats in Danish waters are relatively natural and probably do not differ radically from those that would occur in the absence of humans (Fenchel, 2017). They are characterised by their shallow water, and the inundated coastline gives rise to numerous bays and peninsulas. However, marine habitats and their associated species are affected by pollution, raw material extraction and especially fisheries.

The extensive coastal habitats of Denmark are relatively unspoiled and undoubtedly of highest international biodiversity importance (Ejrnæs *et al.*, 2011). Coastal dynamics occur that are not found in many places in Europe, and there are extensive areas of sand dune and salt meadows. These include HD habitats of which Denmark has a relatively high proportion of the total area in its biogeographical region, including decalcified fixed dunes (2140*),[2] dry sand heaths (2320) and Atlantic salt meadows (1330). The shallow seas and bays hold high densities of many invertebrates, fish and internationally important concentrations of wintering waterbirds (e.g. divers, grebes and sea-ducks), especially Common Scoter (*Melanitta nigra*) and Common Eider (*Somateria mollissima*) (Rasmussen *et al.*, 2000). In addition, some of these sites are particularly important for birds in harsh winters as they remain ice-free longer than other areas. The intertidal mudflats of the Wadden Sea (which is a World Heritage Site shared with the Netherlands and Germany) are of very high importance for geese, ducks and waders, especially on passage and in winter.

12.2 The Status of Nature

12.2.1 Habitat Status and Trends

Over the past 50 years, the loss of natural and semi-natural areas has continued. This has been mainly due to the decline in the remaining small patches of habitat in the farmland landscape, conversion of meadows to cropland, or increasing use of land for building. The overall diversity of habitat types has declined since the early 1970s, both in number and distribution (Asbirk, 1988).

The detrimental impacts of the changes in land use, and other human activities described above, are reflected in the HD Article 17 reports[3] that have been provided on the conservation status of HD habitats. The most recent report, for 2013–2018, shows that the unfavourable status of most habitats has continued and perhaps even worsened. Of 111 biogeographical level habitat assessments (of 60 HD habitats), only 5% indicated a favourable conservation status. Of particular concern is that most assessments revealed a bad status (77%) whilst the remainder were poor (18%). Comparison of the different biogeographical regions indicates that they all showed similar results. Furthermore, a high proportion of the assessments that revealed an unfavourable status also indicated declines over the short term (Table 12.1), although this was not the case in the marine biogeographical regions.

Comparison of the different HD habitat types reveals that the situation is particularly concerning for heathlands and forests, as all 12 types had a bad status in each biogeographical region where they occurred. Furthermore, both of the heathland habitats showed a deteriorating trend, as did 69% of the forest assessments. Despite the fact that semi-natural grasslands, and bog, mire and fen habitats now only occur in small remnants (and presumably have a high proportion protected), all their assessments also showed an unfavourable status, with 89% for grasslands and 85% of those for bogs, etc., revealing a bad status. All lake and river habitat types also had an unfavourable status, with 64% having a bad status.

Despite coastal habitats being the least spoilt in Denmark, the conservation status of these particularly important HD habitats is little better than others. Two types had a favourable assessment: *Spartina* swards (1320) in both the Atlantic and Continental regions, and embryonic shifting dunes (2110), but only in the Atlantic region. Most other assessments revealed a bad conservation status: 67% for dunes (including inland dunes) and 73% for other coastal habitats. For example, Atlantic salt meadows, the most widespread coastal habitat, had a bad conservation status in the Continental region and a poor status in the Atlantic region.

1 www.statistikbanken.dk/statbank5a/default.asp?w=2048

2 Numbers in brackets are HD Annex I habitat codes (see Appendix for full names). * Indicates a priority habitat.

3 See Section 1.3 for background on monitoring under the Nature Directives, status categories and data sources.

12.2.2 Species Status and Trends

Species Listed in Annexes II, IV and V of the Habitats Directive

In the 2013–2018 HD Article 17 report, Denmark reported on 122 conservation status assessments covering 84 HD species. The results suggest that the situation for these species is not quite as concerning as it is for HD habitats. Across all species and regions, 20% of species assessments revealed a favourable conservation status, whilst 27% had a poor status and 34% a bad status. The species' conservation status was recorded as unknown for 20% of assessments. As for habitats, the two terrestrial biogeographical regions had similar results, as did the marine regions. Whilst most species with a poor status had improving or stable short-term trends, about a third had decreasing trends, including those with a bad status (Table 12.1). Of further concern is that nearly a third of species with a bad status had an unknown trend.

Of the different species groups, it is clear that mammals are mostly faring better than all the others, with 48% of assessments revealing a favourable conservation status, and most others having stable or increasing trends. However, one notable exception is the Hazel Dormouse (*Muscardinus avellanarius*), which occurs in the Continental region where it has a bad conservation status and declining trend. It lacks adequate scrub and hedgerow habitat, and its populations are highly fragmented. On the positive side, the Wolf (*Canis lupus*) has returned to Denmark, after 200 years of absence (Naturstyrelsen, 2014a).

Most of Denmark's marine mammal populations have a favourable status in the Atlantic marine region. The exception to this is the Grey Seal (*Halichoerus grypus*), which has a bad status in the region and in the Baltic, as its occurrence and breeding are far below previous levels. However, the situation is improving in both marine regions.

In the Baltic biogeographical marine region, there is concern over the Harbour Porpoise (*Phocoena phocoena*). There are two populations in the region: the Belt Sea one has a favourable status, whilst the Baltic Sea population is considered to be Critically Endangered by IUCN (Hammond *et al.*, 2008). To address this, ICES have published a recommendation on the combination of spatial–temporal closures and the use of pingers in static net fishing (ICES, 2020).

Amongst the other groups, no other species has a favourable conservation status apart from one mollusc, Desmoulin's Whorl Snail (*Vertigo moulinsiana*), which occurs in the Continental region, and one fish, the Brook Lamprey (*Lampetra planeri*), in both terrestrial biogeographical regions. In contrast, most other fish have a bad conservation status, including the Sea Lamprey (*Petromyzon marinus*), River Lamprey (*Lampetra fluviatilis*) and Houting (*Coregonus oxyrinchus*). Although a major LIFE project from 2005 to 2013 focussed on the Houting, its total population is still very low, and further restoration is required in hinterland waters (Fredshavn *et al.*, 2019a).

Amongst the other species groups, it is clear that non-vascular plants and vascular plants are under particularly high pressure, as 67% and 64% respectively have a bad status. Furthermore, amongst vascular plants, 45% have a bad status and are decreasing. The plants are affected by the same influencing factors as the habitats they grow in. Altered hydrology and eutrophication are a problem for water-related species, such as Slender Green Feather Moss (*Hamatocaulis vernicosus*), and two aquatic plants, the Floating Water-plantain (*Luronium natans*) and Slender Naiad (*Najas flexilis*). For species with very small, scattered populations, genetic isolation is a problem, as in the case of the Lady's Slipper Orchid (*Cypripedium calceolus*).

Birds

The latest BD Article 12 assessment indicated that, in general, bird populations are doing relatively well compared to other species. Over 2007–2018, 61% of 196 breeding bird populations increased, were stable or fluctuated, while 30% decreased (8% had unknown or uncertain trends). Amongst 46 key wintering species' populations, only 9% declined, 14% had unknown trends and the rest had increasing, stable or fluctuating trends. Table 12.2 shows the situation for BD Annex I taxa and indicates that declines were observed in only 12% of breeding populations, and no wintering populations, although the trends were unknown or uncertain for some.

Table 12.1 Short-term trends (c. 2007–2018) in Habitats Directive Annex I habitats and Annex II, IV and V species with an unfavourable conservation status in Denmark.

	Poor (unfavourable–inadequate)				Bad (unfavourable–bad)			
	Imp	Sta	Dec	?	Imp	Sta	Dec	?
HD Habitat assessments								
Number	0	1	5	14	5	13	32	35
%	0%	5%	25%	70%	6%	15%	38%	41%
HD Species assessments								
Number	14	4	11	3	6	7	14	12
%	44%	13%	34%	9%	15%	18%	36%	31%

Key. Imp = improving, Sta = stable, Dec = decreasing, ? = unknown.
Source. Based on Member State HD Article 17 reported data as presented on the EEA national summary dashboard (2021).

Table 12.2 Short-term trends (c. 2007–2018) in Birds Directive Annex I taxa in Denmark.

	Breeding				Wintering			
	Inc	Sta/ Flu	Dec	?	Inc	Sta/ Flu	Dec	?
Number	11	22	5	5	3	3	0	3
%	26%	51%	12%	12%	33%	33%	0%	33%

Key. Inc = increasing, Sta = stable, Flu = fluctuating, Dec = decreasing, ? = unknown or uncertain.
Source. Based on Member State Article 12 reported data as presented on the EEA national summary dashboard (2021).

Despite the predominance of positive or stable trends, 59 breeding bird species are declining; these include characteristic farmland species, for example the Lapwing (*Vanellus vanellus*), Skylark (*Alauda arvensis*) and buntings, such as the Corn Bunting (*Emberiza calandra*). In fact, specialist farmland birds have been found to have exhibited some of the most rapid declines, and may indicate broader biodiversity losses in agricultural habitats (Heldbjerg *et al.*, 2018). As a result of ongoing declines in suitable grazed wet coastal grasslands, the globally Endangered BD Annex I Baltic subspecies of the Dunlin (*Calidris alpina schinzii*) is now Endangered in Denmark (Moeslund *et al.*, 2019). Examples of Annex I species that declined over the short term, but were formerly stable or fluctuating, are the Sandwich Tern (*Thalasseus sandvicensis*) and Little Tern (*Sternula albifrons*), the latter now considered to be Vulnerable in Denmark.

The Danish Red List

The latest update of the Danish Red List in 2019 evaluated the risk of national extinction in 10 662 species of fungi, animals and plants, using IUCN regional criteria and categories (Moeslund *et al.*, 2019). The results, which are similar to those of the 2010 assessment, show that a large proportion of Denmark's species are at risk of extinction. Over all evaluated taxa, 17% were considered to be threatened (4% Critically Endangered, 6% Endangered and 7% Vulnerable) and another 5% Near Threatened. In addition, 4% were judged to be regionally extinct. It is also possible that in reality more species are threatened, as available data were insufficient to assess the risk of extinction of 15% of the species.

12.2.3 Drivers and Causes of Change

Most of the changes in habitats and associated impacts on species, as described previously, are the result of pressures from human activities, which are driven by high levels of economic development and competition for land. Overall, the fact that most of the land is used for food production has an overwhelming influence on nature. Very high levels of agricultural intensification and specialisation have impoverished farmland biodiversity and had wider-ranging external impacts, in particular from pollution. High levels of atmospheric nitrogen deposition, due to the very high densities of livestock, have led to the eutrophication of sensitive habitats, such as semi-natural grasslands, dunes and bogs, etc., and are a major cause of their poor condition (Fredshavn *et al.*, 2019a). Pollution from farmland is also a major cause of the unfavourable status of freshwater habitats in addition to hydrological changes. Although the discharge of nutrients (phosphorus and nitrogen) has decreased since the 1970s, it remains a problem because it was from a high starting point (Hansen and Søndergaard, 2019).

The hydrological changes affecting bog, fen and other wetland habitats are due to drainage for agricultural purposes, resulting in local declines in the water table. The straightening and deepening of streams and rivers, and their maintenance, also impact freshwater fauna.

The majority of forests in Denmark are commercially used and therefore also subject to intensive management and harvesting. This results in the felling of large trees, thinning of vegetation, removal of dead and dying wood, and reduced water tables through drainage. Although most HD forest types are stable and large enough for long-term maintenance, this is not the case with two rare forest types: medio-European limestone beech forests of the *Cephalanthero-Fagion* (9150) and *Galio-Carpinetum* oak-hornbeam forests (9170) (Fredshavn *et al.*, 2019a).

Whilst most coastal areas are much less affected by human activities, some pressures occur along the shoreline as a result of flood prevention measures. These include measures that aim to stop sand movements and erosion, because they inhibit the natural dynamic coastal geomorphological processes, which in turn affect the building of sand ridges and dunes and eventually lagoons. In contrast, along the unprotected coasts, which are particularly exposed to the influence of the sea and the wind, there is extensive transport of sand and other materials that build coastal landforms. Coastal areas are also subject to very high levels of recreational disturbance, especially as there are about 223 000 holiday homes in Denmark. Disturbance is thought to be one of the major causes of the decline in the Little Tern (*Sternula albifrons*) (Therkildsen *et al.*, 2013).

In contrast to the pressures described above, some terrestrial habitats and their associated species are detrimentally affected by a lack of human activities, or changes in traditional practices. As mentioned earlier, many of the open habitats, such as grasslands and heathlands, are the result of human activities. In particular, these habitats are dependent on low-intensity grazing and other traditional management practices to prevent their natural succession to scrub and eventually forest. However, there has been a decline in such traditional grazing, and this is leading to overgrowth of the vegetation (a cause of the decline in breeding Dunlin) and exacerbating the spread of invasive alien species (IAS), such as *Rosa rugosa* and the Dwarf Mountain Pine (*Pinus mugo*).

As a result of the pressures and resulting impacts over the years, many areas of semi-natural habitat have become small and isolated patches, especially away from the coastal areas.

This fragmentation has ongoing impacts, as it degrades the habitats and constrains the movements of species (measures to counteract this are described in Section 12.4).

The marine environment is also subject to pressure from human activities that are impacting on a wide range of habitats and species. Fisheries exert the most influence on the marine environment, along with the supply of the nutrients nitrogen and phosphorus from agricultural areas (Riemann *et al.*, 2015; Petersen, 2018). Bottom trawling and the use of scrapers for shellfish is particularly damaging to habitats and removes important key species. Furthermore, Danish waters are subject to intensive fishing activities, with disturbance rates of about 85% of the seabed in the North Sea and 67% in the Baltic Sea in 2013 (Havstrategi II, 2019). Similarly, some sandbanks are affected by marine raw material extraction.

The full effects of climate change in relation to nature protection have yet to be seen, and it can be expected that further indirect impacts will arise. For example, Denmark has many interests on the coasts that are at risk from increasing coastal erosion and flooding. As well as the cities, many cottages that are built on the foreshore will require protection, and there are increasing examples of this happening illegally. This conflicts with the desire for natural processes to be unhindered along many parts of the coasts.

12.3 Nature Conservation Governance, Actors, Legislation and Policies

Table 12.3 lists the key events that have influenced nature conservation in Denmark, the most important of which are further described in this section.

12.3.1 Governance and Actors

Denmark has a three-level political system, comprising the state, regions and municipalities. In 2007, the 14 counties were merged into five regions, and the number of municipalities was reduced to 98. Overall policy development and legislation takes place at the national government level. As part of the reform, the main responsibilities for implementing nature conservation were further devolved from the regional level to the municipalities. The only nature conservation actions for the regions now consist of supporting some projects, such as applying for UNESCO recognition of a World Heritage Site or Geopark.

As the counties previously did, the municipalities deal with nature protection, management of protected nature, and spatial planning for it on private and municipality owned land (e.g. with respect to green corridors). The reform has meant that nature conservation issues are dealt with by many more authorities than before, and as a result some now lack the required expertise. The municipalities also differ in how much effort they put into nature conservation. Often the sector lacks personnel and finances. An advantage of the change is that the municipalities now have full competence for planning within both the towns and the countryside, which provides some possibilities for synergies, such as planning for green corridors along their full length.

The Ministry of Environment (Miljøministeriet), through The Danish Nature Agency (Naturstyrelsen), manages state-owned forests and nature sites (e.g. some protected areas as well as other important areas for nature) and all marine protected areas (MPAs). The Danish Environmental Protection Agency deals with the legislation and enforcement of nature protection. During the 2007 reform, the Ministry of Environment was merged with the Ministry of Agriculture and Fisheries. National planning, including for green infrastructure, moved to the Ministry of Business, Industry and Financial Affairs. In 2021, national planning moved to the Ministry of the Interior and Housing. At the end of 2020, the Environment Ministry was restored – although it still did not include physical planning – indicating a strengthening of nature and environmental interests within the Government. In 2021, national planning moved to the Ministry of the Interior and Housing. The idea behind the national-level changes was to improve cooperation between nature protection and agriculture industry, housing and planning.

In Denmark, there is one dominant NGO, Danmarks Naturfredningsforening (The Danish Society for Nature Conservation). It has about 160 000 members, 82 employees and local branches in all municipalities. Its main activities are nature conservation, environmental protection and sustainability, and it has the authority to make proposals for new Conservation Areas (as discussed in Section 12.4). It acts on a wide variety of issues, as well as running focussed campaigns, such as on saving the sea, regaining biodiversity, and climate mitigation in municipalities. Other important players are Dansk Ornitologisk Forening (BirdLife Denmark), which has had a major influence on nature conservation in Denmark, for example through its lobbying of Parliament, despite being quite small. Furthermore, Verdensnaturfonden (WWF Denmark) and Danmarks Sportsfiskerforbund (Danish Anglers' Association) are important players. The Outdoor Council, an umbrella organisation for outdoor recreation, also contributes to nature protection and runs the national scheme of Nature Parks.

The Aage V. Jensen Nature Foundation, the largest private landowner in Denmark, plays a key role in nature conservation as it has some of the most important reserves in the country. In addition, the foundation finances a lot of nature monitoring projects, information campaigns, authors and conferences, which provides it with considerable influence. There are also other foundations that own reserves, including the Danish Nature Foundation (which can also support nature projects), Denmark's Nature Foundation and the Bird Protection Foundation.

It is difficult to judge the general views of the public on nature conservation in Denmark. Most people enjoy walking on paths through farmland and forest areas, and regard them as natural areas, although they are highly influenced by

Table 12.3 Key nature conservation developments in Denmark.

1917	Nature Conservation Act agreed in the Parliament making it possible to establish conservation areas
1937	Protection zones of 100 m from the shoreline and 300 m for forest established
1963	Conservation planning committees were set up in the counties
1970	Regional planning by the counties began, including planning for nature protection, and the counties became responsible for nature protection on land owned privately and by municipalities and counties
1983	The first Special Protection Areas were designated according to the EU Birds Directive
1992	The Nature Protection Act replaced the Nature Conservation Act, including general protection of most wetlands, grazed lands and heathlands
	First national strategy on ancient forest and other forests of conservation
1995	First national biodiversity strategy presented
	The White-tailed Sea-eagle returned as a breeding bird after being extinct since 1917
1998	First SCIs were designated according to the EU Habitats Directive
1999	Beavers released into Klosterheden Plantation after being extinct in Denmark for several thousand years
2001	The Wilhjelm Committee report on how to improve protection of biodiversity and nature in Denmark
2007	The power of the counties within nature conservation was handed over to the municipalities as part of a national reform of governance
2008	The first National Park in Denmark was established: Thy National Park
2012	The Wolf returned to Denmark after being extinct for 200 years
2013	The Danish Outdoor Council started its national certification system for Nature Parks (at the regional level)
2014	First natural World Heritage Sites designated: Stevns Klint and the Wadden Sea
	First UNESCO Global Geopark designated: Geopark Odsherred
	The National Commission on Nature and Agriculture delivered its conclusions
	Naturplan Danmark, an updated national biodiversity strategy, was published. It includes a proposal on a new green infrastructure called Green Map Denmark
2016	The annual Nature meeting in Hirtshals began and created a new momentum for nature protection in Denmark
2017	The first UNESCO Biosphere Reserve was designated: Biosfære Møn

See Table 3.3 for signatory dates of international conventions.

humans. Nevertheless, there seems to be general support for widespread conservation areas. Furthermore, nature gained prominence on the political agenda at the beginning of 2016 due to the public 'Nature Meeting' in Hirtshals' harbour.[4] The Nature Meetings have been repeated since 2016, hosted by the municipality of Hjørring and partly financed by the Aage V. Jensen Nature Foundation. About 20 000 people attended the meeting in 2019, and they participated in debates and lectures, exhibitions, workshops and visiting stalls. Recently, political party leaders have taken part in the meeting, and the environment minister has devoted much time to it. It seems to have become an event that is going to stay, although it had to go online in 2020 and 2021 due to the COVID-19 pandemic.

[4] The People's Festival of Nature: https://naturmoedet.dk/

12.3.2 Legislation

Nature conservation in Denmark is primarily organised via the Nature Protection Act[5] of 1992. Its predecessor, the Nature Conservation Act,[6] came into force in 1917, making it possible to establish conservation areas. The legislation has gradually evolved to include protection of the coast, forests, rivers and lakes, prohibiting changes within a certain distance of them. Later, the general protection of natural habitats and the requirements of the EU Nature Directives were incorporated into Danish law.

Shortly after the 1972 UN Environment Conference in Stockholm, Denmark introduced its first Environmental Protection Act,[7] which restricted the use of pesticides and

[5] Naturbeskyttelsesloven LBK nr. 240 af 13/03/2019.
[6] Naturfredningsloven, LBK nr. 530 af 10/10/1984.
[7] Miljøbeskyttelsesloven, LBK nr. 1218 af 25/11/2019.

established a permitting regime for air, water and land pollution by industries and farmers. This Act was later used to restrict farming practices with the aim of reducing nutrient pollution, such as from ammonia, nitrates and phosphorus.

Other relevant environmental legislation includes the Planning Act,[8] which contains the regulation that has established the Green Map Denmark (i.e. green infrastructure of protected areas and corridors, see Section 12.5), and the Wildlife Act,[9] which enables the establishment of Wildlife Reserves.

12.3.3 Strategies and Plans

As a follow-up to the Rio Conference in 1992, Denmark published in the same year a national strategy on ancient forest and other forests of conservation value (Skov- og Naturstyrelsen, 1992). It was decided that high-value state-owned forests should be managed to secure biodiversity, and the owners of private forests should be paid to give up forestry in some of their most valuable forests. This approach has in recent years been extended and gradually more state-owned forest has changed to 'untouched forest', most recently with the Nature Package in 2020 agreed by a majority in the parliament.[10]

To meets the obligations of the CBD, which it joined in 1993/1994, Denmark produced its first biodiversity strategy in 1995 (Prip and Wind, 1995). The strategy raised the status of nature protection in Denmark, but did not really set a new agenda for improving nature. The most recent biodiversity strategy from 2014, called *Naturplan Danmark*, has also not set new standards except with regard to a Green Map Denmark (Miljøministeriet, 2014). In general, there has been a lack of nature planning at the national level. This is the case with the Green Map, which is primarily planned and realised by the municipalities, such that connections are made with neighbouring municipalities. A new national biodiversity strategy is being developed, to be published in 2023.

In 2001, the Wilhjelm Committee, named after its chair Niels Wilhjelm, carried out an in-depth analysis on the state of nature protection in Denmark. It was highly critical of the efforts that had been made up to then (Wilhjelmudvalget, 2001). It pointed out that the natural sites lacked management and were too small, fragmented and polluted with fertilisers from nearby agriculture. One important proposal from the Committee's work was to establish National Parks in Denmark. In 2012, the National Commission on Nature and Agriculture was established with the purpose of connecting environment, nature and economic issues related to the agricultural sector (Natur- og Landbrugskommissionen, 2014). One of their conclusions was that peatland drained for farming could be changed to protected areas with meadows or marshes to halt the release of climate gasses and nutrient pollution of rivers, and to improve the network of natural corridors. Furthermore, it emphasised that each farm should restrict its nutrient loading on the surrounding environment because of negative impacts on especially nutrient-poor habitat types.

12.4 Species Protection and Conservation Measures

Denmark protects species of plants and animals (other than birds) according to the National Nature Protection Act, connected to which are lists of nationally protected species and HD species. An executive order from the Ministry of Environment and Food further added protected species and set rules for their protection.[11] The list of protected species is a mixture of species on the HD list and species under national protection, which is not linked to the national Red List of threatened species. Species on the HD list have stronger protection with regard to ensuring their habitats are not destroyed, as set out in the Environment Damage Act.[12]

Species protection is also regulated by the Hunting and Wildlife Act,[13] which restricts or prohibits the hunting of specific species of mammals and birds, at a national level or at specific sites (Wildlife Reserves), and the Fishery Act, which enables fishing for specific species at specific sites to be restricted. In general, most bird species are protected, including the nests of breeding birds, but some species such as shorebirds and birds of prey are often affected by disturbance. A recent study found that the establishment of a mountain bike trail in a state forest caused 19 out of 24 pairs of birds of prey to stop breeding (Skriver, 2017).

With regard to hunting, a special advisory board for the Environment Ministry was set up with representatives from nature, outdoor and hunting organisations as well as wildlife researchers.[14] The board advises on which species can be hunted and when, and also deals with the regulation of species that can give rise to conflicts with humans. In general, the minister has followed the board's advice, if they have reached agreement, which creates a strong incentive to do so. In Denmark, most people accept hunting, partly due to the updating of hunting restrictions and the strict non-acceptance of any illegal activities by the Danish Hunters' Association. Despite this, and considerable improvements since the 1980s, hunting of ducks and illegal killing of birds of prey is having some negative impacts on populations (Fredshavn *et al.*, 2019b). Since the winter of 2020–2021 hunting has been banned of the national Red Listed wintering diving ducks: Long-tailed

[8] Planloven, LBK nr.1157 af 01/07/2020.

[9] Lov om jagt og vildtforvaltning, LBK nr. 265 af 21/03/2019.

[10] https://mim.dk/nyheder/2020/dec/ny-natur-og-biodiversitetspakke-klar-historisk-loeft-til-dansk-natur/

[11] Bekendtgørelse om fredning af visse dyre- og plantearter og pleje af tilskadekommet vildt. BEK nr. 1466 af 06/12/2018.

[12] Lov om undersøgelse, forebyggelse og afhjælpning af miljøskader (miljøskadeloven). LBK nr. 277 af 27/03/2017.

[13] Lov om Jagt og Vildforvaltning. LBK nr. 265 af 21/03/2019.

[14] Vildtforvaltningsrådet.

uck (*Clangula hyemalis*), Velvet Scoter (*Melanitta fusca*) and ommon Pochard (*Aythya ferina*).

The release of new species in Denmark requires the approval f the Ministry of Environment. Nevertheless, it has also appened by accident, as was the case with Red Deer (*Cervus aphus*) in Western Zealand, and no action has been taken to emove the escaped animals. Strong measures are, however, eing taken to prevent Wild Boar (*Sus scrofa*) spreading in the ountry, as they can be carriers of African swine fever and herefore pose a risk to Denmark's very important pig industry nd its exports. To counter this, Wild Boar are now normally hot when they occur in Denmark. In 2019, a fence was erected long the German border to hinder the free movement of Wild Boar into Denmark. This has been the subject of a lot of criticism, because it may not be effective (the animals can swim), and it may interfere with the movement of other species.

A number of species re-introduction and recovery programmes have been carried out in Denmark. Beavers (*Castor fiber*) were released in Jutland in 1999 in a huge state-owned coniferous plantation (Klosterheden) with the idea that they would improve the habitats by creating dams and open areas. The population later spread outside the plantation, causing a lot of conflicts with private landowners. Other re-introductions have included the Stag Beetle (*Lucanus cervus*), at five locations in 2013, and the Elk (*Alces alces*) in Lille Vildmosen in 2015. To bolster their populations, toads and frogs have been introduced into newly created ponds. This has especially included the HD species Fire-bellied Toad (*Bombina bombina*) and European Tree Frog (*Hyla arborea*) and they are more widespread as a result of this today.

In 2012, wolves were observed in the wild in Denmark for the first time for more than nearly 200 years. It has been proven that the animals came from Germany, and they have bred since 2017. As of September 2020, the population in Denmark was about 10 individuals (Olsen *et al.*, 2020). As the wolves have attacked some farm animals, their return has caused a lot of debate. A national management plan was adopted for the Wolf (Naturstyrelsen, 2014a), and the government introduced compensation for farmers whose animals are killed by wolves. There is one documented case from 2018 of a wolf being illegally killed leading to prosecution and punishment of the hunter.

Table 12.4 The number and extent of international designated sites and protected areas in Denmark.

	IUCN category	Number	Area (km^2)
Designated under global and non-EU international legislation			
Ramsar Sites	NR	28	7 482
Biosphere Reserve (core area)	NR	1	6 165
Natural and Mixed World Heritage Sites	NAp	2	1 157
HELCOM Marine Protected Areas	NR	66	11 224
OSPAR Marine Protected Areas[1]	NR	34	12 470
Designated under the EU Nature Directives			
SCIs (terrestrial + marine)	NR	261	3 178 + 16 492
SPAs (terrestrial + marine)	NR	113	2 605 + 12 184
Natura 2000 terrestrial area			*3 594*
Natura 2000 marine area			*19 053*
Designated under domestic legislation			
Conservation Areas (Fredninger)	Mostly IV and V	381	1 782
Protected Habitat Areas (§3-beskyttede områder)[2]	IV and NAp	≫194	4 458
Marine Strategy Areas (Havstrategiområder)	NR	9	590

Note. NR = Not reported; NAp = not applicable. 1. Some of the Danish MPAs on the OSPAR website have incorrect boundary polygons. Numbers are therefore based on a project from 2020 assessing marine MPAs carried out by Woollhead *et al.* (2020). 2. The areas are not reported individually but for each municipality in two groups according to whether they are private or public. The number is therefore much higher (304 194 polygons on the Environment Portal, downloaded 12 April 2020). Many are unlikely to meet the IUCN Protected Area criteria.

Sources. See Section 1.3. Based on data from: UNESCO website (January 2022); HELCOM database (downloaded 27 January 2022); EEA Natura 2000 Barometer, Natura 2000 as at end 2020; others and IUCN categories from WDPA (January 2022).

12.5 Protected Areas and Networks

12.5.1 Protected Area Types and Coverage

Sites Designated under Domestic Legislation

The national terrestrial protected area types that contribute most to nature conservation are the designated Conservation Areas (*Fredninger*) and Protected Habitat Areas (*§3-beskyttede naturtyper*) that can be established under the Danish Nature Act. Both protection systems are unique to Denmark and may provide ideas and lessons for the international conservation community. These and other protected areas are listed in Table 12.4.

It has been possible to designate **Conservation Areas** since the 1917 Nature Conservation Act. Their objective is to protect specific natural or cultural sites or features, or to make natural sites accessible for recreation as well as protection of landscapes or viewpoints. As of 2013, there were more than 5 000 Conservation Areas, which covered approximately 5% of the land area. A review in 2018 of 1 720 Conservation Areas identified by the Danish Environment Protection Agency as the most important revealed that 375 fulfilled the IUCN definition of protected areas (Woollhead and Petersen, 2018; Garn *et al.*, 2019). Most did not, because their main purpose was not nature conservation. Furthermore, many areas lacked proper management. Now 381 Conservation Areas are listed as protected areas in the WDPA.

There are three different types of Conservation Areas in terms of the way they are designated. The Environment Ministry can declare them on government-owned land or at sea. Others are based on proposals that are decided on by a three-person Conservation Board, which is led by a judge and also has one person appointed by the Environment Ministry and one by the municipality concerned. Proposals can be put forward by the government, municipalities and the Danish Society for Nature Conservation (which has special recognition in law for this). Today, most Conservation Areas are the result of a proposal by the Society, which often is prepared together with the Danish Agency of Environment Protection and Local Government Denmark.[15] The main criteria for new proposals for Conservation Areas are that the site is threatened and not protected by other efficient means. It should also contribute to green infrastructure, biodiversity protection and the conservation of landscapes.

Until 1970, private landowners could also directly ask the Conservation Board to declare a Conservation Area on their land. Now, they can only put forward a proposal indirectly, via a body that can make such proposals, which seems more difficult than before. The change was made in combination with a stricter planning law that makes it possible to set restricted zones in regional plans. Some landowners used the conservation of their land to restrict, or hinder, the possibility for new developments of roads, etc.

A critical review after a hundred years of designation of Conservation Areas pointed out that, although most of the areas would not have existed today without their designation, many have not been managed and protected properly (Hansen, 2017). The problem has often been a lack of management by the municipalities. It is important that local conservationists continue to put pressure on their authorities to keep track of their responsibilities and encourage them to take action. The great advantage of Conservation Areas is that their regulations and purpose can be adapted to the specific site and its conditions, taking into account the views of local stakeholders.

Protected Habitat Areas, designated under Section 3 (§3) of the Nature Protection Act, were introduced in 1978 and are all areas of a protected habitat that are above a minimum size. They comprise ponds and lakes larger than 100 m^2, and heaths, bogs, salt marshes, meadows and dry grassland when the individual habitat or group of habitats covers an area larger than 2.5 ha, as well as designated water courses. The main weakness with this type of protection is that there are no demands for management. Therefore, although these areas are estimated to cover 4 458 km^2 according to the WDPA, a large proportion would probably not fulfil the IUCN protected area criteria.

In addition to the above two protected area types, Wildlife Reserves, National Parks and Nature Parks also add to the mix of designated areas. Many of them are not protected areas according to the IUCN definition. Nevertheless, they are described briefly here as they contribute to the protection of the environment and nature conservation.

Denmark has almost 100 **Wildlife Reserves** (equivalent to game reserves in other countries), which cover 3 300 km^2, mostly at sea. These are designated under the Hunting and Wildlife Act with the objective of protecting breeding or roosting birds or mammals. At sea, hunting and access are allowed unless restricted by a Wildlife Reserve or, in some cases, a Conservation Area. As hunting and outdoor recreation can cause serious disturbance of birds and seals, the reserves have rules restricting these activities (prohibiting hunting, sailing and windsurfing, etc.). Many are set up within Natura 2000 areas and act as a form of core zone.

Until relatively recently, Denmark chose not to establish any **National Parks**, mainly because most of the country is farmland. However, inspired by the UK, around the year 2000 it was concluded that cultural landscapes could be designated as National Parks. In 2007 the National Park Act[16] was passed, and the first National Park was opened in 2008 in Northwest Jutland (Thy National Park), almost 100 years after the first was established in Europe. There are now five National Parks covering 2 316 km^2. However, there has been much criticism of the system, because the National Parks are set up as foundations that have very few legal powers concerning the protection of the areas and they often include much cultivated and inhabited land. Whilst they mainly function as planning and interpretation units, they receive funding (mainly from the state) and try to encourage landowners to improve management. The National Park Act is to be reviewed within the next few years, hopefully to improve some of its weaknesses. In 2021, to counteract some of the criticisms of the present National Parks, the Parliament agreed to establish 15 so-called **Nature National Parks** in state-owned forests or natural areas, building on the principles of rewilding.

In 2013, the Danish Outdoor Council set up a national scheme for the recognition and regional certification of **Nature Parks.**[17] One or more municipalities can apply for the recognition of a Nature Park, which is then decided upon by a board consisting of representatives from a wide set of sectors, including landowners, the Danish Nature Agency and the Danish Society for Nature Conservation. To be certified, the parks have to fulfil certain criteria, including that at least 50% of their area should be encompassed by other types of protection. Certification also provides a platform for government, landowners and environmental organisations to work together, to make common management plans and disseminate them, and to attract visitors. In general, the Parks' focus is mostly on

[15] Kommunernes Landsforening (Local Government Denmark).

[16] Lov om Nationalparker. LBK 749 af 21/06/2017.

[17] Danske Naturparker: https://friluftsraadet.dk/danskenaturparker

recreation, but they also tend to increase the level of management and protection as they have to be included in municipal plans. At the end of 2021, there were 14 Nature Parks, covering 1 526 km^2, one of which is defined as a marine Nature Park. The latter makes sense because Denmark has no specific system for designation of marine parks.

It is also important to mention that several **private nature foundations** in Denmark have purchased and now manage land for nature conservation. Of these, the Aage V. Jensen Nature Foundation has the largest area, which, in 2021, comprised 193 km^2 over 31 reserves,[18] including some of the most important natural sites in Denmark (e.g. Vejlerne, Lille Vildmose, Fiilsø and Maribosøerne). In addition to this, the Danish Bird Foundation owns 23 sites, Denmark's Nature Fund (established by the Danish Society for Nature Conservation) has 15 sites, and the Danish Nature Fund (established by the state, the Villum Foundation and the Aage V. Jensen Foundation) has 13 sites and supports nature management on many more sites.

International and EU sites

Denmark has designated a number of sites under international conventions (Table 12.4), including many Ramsar Sites in recognition of their importance as wetlands and for their waterbird populations (Koester, 1989). In 2018, Denmark started a process to reduce many of the Ramsar Sites by area, in conjunction with a similar process within Natura 2000 areas. This is mainly to exclude intensive farmland used by waders, geese and swans. At the beginning of 2022 there were still no conclusions from the process. Denmark has also designated UNESCO sites, although this was rather late compared to other countries, and the sites are not underpinned by national legislation. Two World Heritage Sites were designated in 2014 – Stevns Klint and the Wadden Sea (a transnational site with Germany and the Netherlands) – and three more natural sites are on the tentative list. One Biosphere Reserve was established on the island of Moen in 2017 and two Global Geoparks in 2014 and 2021 respectively, in Odsherred, in the north-western corner of Zealand, and Western Jutland.

When Denmark started to implement the Birds Directive in the 1980s and designate SPAs, all Ramsar Sites were included, thereby improving their protection. Additional SPAs were also designated, and these now cover 6.0% of the land area and 12.1% of the marine area. However, in 2018 BirdLife Denmark considered that the designation of terrestrial SPAs, as well as some marine SPAs, was unsatisfactory (BirdLife International *et al.*, 2018).

In the 1990s, SCIs were identified in accordance with the Habitats Directive. This led to some protests by landowners who were concerned that the protection regime could restrict their options for managing their land. Now, the SCIs cover 7.4% of the land area. Together with the SPAs, which overlap with SCIs considerably, this gives a total Natura 2000 terrestrial coverage of 8.3%. This is the lowest Natura 2000 terrestrial coverage in the EU (Table 5.4) and reflects the high proportion of Denmark's landscape that consists of intensive farmland and forestry. SCI coverage at sea is proportionally much higher, amounting to 13.4% of Denmark's marine waters, and, combined with SPAs, gives rise to 15.5% combined coverage. However, damaging use of bottom trawling gear for fishing continues to be allowed in many of the areas and as a consequence the IUCN definition of protected areas is not fulfilled for many of the marine protected areas (Woollhead *et al.*, 2020).

In addition to its marine Natura 2000 sites, Denmark has designated marine protected areas (MPAs) in accordance with the requirements of the HELCOM (Baltic Sea) and OSPAR (North Sea) regional conventions. All these MPAs are also Natura 2000 sites. As part of Denmark's programme of measures to implement the MSFD, six additional MPAs, covering 590 km^2, have been set up to protect soft seabed habitats not included in the list of HD habitats. They are mainly protected through international or national fishery regulations, which prohibit physical disturbance of the seabed, such as from bottom trawling, as well as dredging and clapping (i.e. dumping) of dredged material on the sea floor. More areas are to be established in the Baltic Sea, around the island of Bornholm.

Protected Area Coverage

In all, a diverse system of protected areas is gradually developing in Denmark. The idea of National Parks and regional Nature Parks, together with UNESCO designations, is expected to develop further in the years to come, with more protected sites added to the respective lists. National estimates of the current combined area of the sites listed in Table 12.4 do not appear to be available. According to WDPA data, as of January 2022, terrestrial sites covered 7 691 km^2, amounting to 17% of the land (UNEP-WCMC, 2022). This does not include land owned and managed by nature conservation organisations, which can be expected to increase the coverage of protected areas to some extent despite overlaps with other designations. More importantly, the WDPA includes all of the Protected Habitat Areas, many of which most likely do not meet IUCN protected area criteria. It is therefore likely that the WDPA's 17% land coverage is a significant overestimate.

In the marine environment, overall protected area coverage as of January 2022 was estimated to be 18 408 km^2 according to the WDPA, equalling 15.0% of Denmark's marine waters. As noted earlier, according to the European Commission there are 19 053 km^2 of marine Natura 2000 sites, indicating effectively complete overlap with the other MPAs (and some differences in data and analysis). However, whilst a desktop analysis in 2020 estimated 19 758 km^2 (16.1% according to the MSFD waters) coverage of all MPA types (including national, Natura 2000 and other international designations), only about one-quarter fulfilled the IUCN definition for protected areas, giving

[18] According to www.avjf.dk/avjnf/naturomraader/ 3 February 2022.

4.8% (4.1% according to the MSFD waters) protected area coverage (Woollhead *et al.*, 2020; Normander *et al.*, 2021). This low proportion meeting the IUCN criteria was because damaging activities continue to be allowed in most of the areas, including the Natura 2000 sites.

12.5.2 The Coherence of Protected Areas and Ecological Networks

The problems that arise from the high level of habitat fragmentation that is typical of much of Denmark have been recognised for some time, and a number of measures have been taken to address them. Until the governance changes in 2006, 'ecological corridors' for wildlife were included in the regional plans prepared by counties and renewed every fourth year. These were mainly rivers and riverbanks, but could also include areas of forest, dry grassland and coastal habitats. Although the corridors were protected from building developments, there was insufficient funding to manage and enhance them to make them fully functional. Since the end of 2006, the now bigger municipalities have the responsibility for the corridors.

In 2014, as part of its biodiversity strategy (*Naturplan Danmark*), the government introduced a new initiative for drawing up a 'Green Map of Denmark'. This aimed to provide a strategic framework for ensuring that conservation actions take place where they have the potential to be most beneficial. The initiative also aimed to get wide inputs from farmers, foresters and environmental organisations, so about 25 boards were set up with the municipalities acting as secretariats. In most cases the exercise improved understanding between landowners and environmental NGOs, but the outputs were very diverse. Some boards only agreed on key principles, others provided concrete proposals and some failed to achieve any agreement. The municipalities were expected to reflect the outcomes in their municipality plans, but in practice their thoughts on ecological corridors did not change much. However, there is clearer differentiation between potential and present corridors and larger natural areas.

In 2017, the Danish parliament passed a change in the Planning Act so that the municipalities must make a comprehensive plan for nature protection interests as well as completing the Green Map. They are therefore required to describe the protection of existing nature conservation interests (e.g. Natura 2000 sites, other protected natural areas, and ecological connections), potential natural areas and potential ecological links, and to prioritise the municipal council's nature efforts within the Green Map. These areas are supposed to be taken into account in the municipalities' spatial plans (discussed later in Section 12.6.1).

12.5.3 Approaches to Protection and Levels of Protection

Every Conservation Area has its own regulations, the advantage of this being that its protection regime can be designed according to its specific needs. The regulations often primarily concern the protection of cultural values or recreational opportunities so may not contribute greatly to nature conservation (hence many are not nature protected areas). Many of the Conservation Areas are managed by municipalities, but their implementation or supervision of management has sometimes lapsed for the older sites. Conservation Areas on private land are often established against the will of landowners, as the regulations restrict their activities. This has led to hostile attitudes to nature conservation, although the owners are compensated. Nevertheless, landowners often become more positive with time as their knowledge about conservation values increases, and they receive public support for management.

As described earlier, if a Protected Habitat Type is above the required size, the area concerned is automatically protected from alteration. In most parts of the country, these areas have been mapped (and added to a public app) so that landowners, conservationists and others can see which are protected. The main weakness of this protection is that there is no requirement for management, and the habitat can therefore disappear through neglect (e.g. if hay cutting or grazing stops on a meadow). Furthermore, there may be a disincentive for landowners to allow areas of protected habitats to develop, as it could constrain their use of the land in future. Investigations have shown that some areas of these habitats have disappeared without anybody noticing. For this reason, a national survey was carried out in 2011–2013.

As regards the protection of Natura 2000 sites, to begin with SPAs did not get much attention, but this gradually changed, mainly because of complaints by NGOs to the European Commission. Today, the protection of Natura 2000 sites is given highest priority because of the legal obligations to take care of them. In general, Denmark is quite strict in not allowing new development projects in Natura 2000 sites (Danmarks Naturfredningforening, 2018). Appropriate assessments of the potential impacts of projects and plans are normally adequate. National legislation prohibits any activity that could conflict with the protection of the BHD habitats and species for which the Natura 2000 sites were designated. By law, private landowners have to notify municipalities beforehand if they want to make any change to the management of a Natura 2000 site. If the municipality disagrees, they cannot go ahead.

12.5.4 Approaches to Managing Protected Areas

In the early days of nature conservation in Denmark it was common practice for protected areas to be left alone, but this led to problems as many areas needed some form of traditional management, such as grazing, to maintain their habitats and nature conservation value. When this was realised, there were sometimes problems deciding on the appropriate management requirements as the sites lacked clear conservation goals. To address such problems, the conservation law was changed in the 1980s so that all new Conservation Areas would have a clear conservation goal and a possible management plan.

Subsequently, the requirement to manage Natura 2000 sites has further improved the planning of protected areas in Denmark. All Natura 2000 sites in Denmark now have a 'Natura 2000 Plan', which is reviewed every sixth year. Although it is not a comprehensive management plan, it sets out the conservation objectives for the site, starting with a Baseline Analysis (which maps the species and habitats for which the site was designated and assesses their status and potential threats). Each plan is subject to public consultation and, once agreed, the Danish authorities must ensure it is followed. A weakness is that management focusses on the requirements of BHD habitats and species and does not take into account the potentially different needs of other important habitats or Red Listed species.

Despite the efforts that are being made to protect and manage Natura 2000 sites, the achievement of their conservation objectives is often constrained by external factors. In particular, nearly all terrestrial Natura 2000 sites are subject to high levels of ammonia deposition, due to the predominance of high-intensity livestock production. Requests to expand or create new livestock farms can be refused by the authorities if they may harm nearby Natura 2000 sites, but it is difficult to reduce impacts from existing farms. At sea, nearly all Natura 2000 sites are also affected by high levels of nutrients, most of which originate from farmland. The nutrient pollution issue is being addressed through measures under the WFD, with a whole catalogue of actions such as control of storage of cattle and pig dung, keeping most fields covered by vegetation during the winter, and regulating the amount of manure allowed per hectare of farmland. Although this has reduced pollution, there is still a long way to go.

12.6 General Nature Conservation Measures

12.6.1 Development Planning and Impact Assessment

Protected areas, ecological corridors and other areas of nature conservation importance must be taken into account in planning processes at national, regional and municipal level. The potential impacts of individual plans and projects on these areas must also be considered in SEAs and EIAs.

Since the structural reform in 2007, the main responsibility for spatial planning in Denmark lies with the municipalities (i.e. in both the cities and the rural landscape), as set out in the Planning Act.[19] They must ensure that the areas in the municipality are used in the best way, based on an overall balance of the relevant considerations. The overall part of the municipality's planning is determined in the municipal plan, which forms the basis for local plans. These plans must not be in conflict with national planning directives and state interests, or Natura 2000 plans.

[19] Planloven LBK nr. 1157 af 01/07/2020.

In order to ensure good coherence between the municipal plans and to avoid conflict with national interests, the Minister for the Interior and Housing has an overall national planning role. The minister publishes an overview of the state's interests in municipal planning, and announces so-called 'national planning directives', which are binding rules for municipal planning. National planning directives are executive orders or circulars by which the minister can establish a binding framework for the content of municipal planning according to national areas of interests like nature, environmental and landscape protection defined by the Ministry of Environment. Some national planning directives target a specific geographical area, while others apply throughout the country. Current national planning directives deal with, for example, planning in the capital area, designation of new areas for holiday cottages in the coastal zone and locating national test centres for wind turbines.

The state interests in municipal planning are set out by the Housing and Planning Agency every four years, prior to the preparation of the municipal plans. Thus, the state has set out national interests that cover nature conservation and restoration including lowland areas, coastal protection and climate adaptation, afforestation and ground water.

It is likewise of national interest that the planning of the rural landscape helps to ensure that valuable landscapes are preserved, including that the open coasts and geological values continue to constitute a significant natural and landscape value. The state therefore objects to municipal plans if they conflict with significant national interests within the announced areas.

12.6.2 Habitat Management and Restoration

As mentioned in the introduction, Denmark has lost a great deal of natural and semi-natural habitat as result of its long history of drainage and land reclamation for agriculture. Now some of these losses are being reversed, such as through restoration of raised bogs and marshes,[20] re-meandering of streams, and removal of dykes to re-flood reclaimed farmland to recreate lakes, intertidal habitats, marshes and lagoons (such as at Gyldensteen Strand[21]). The restoration projects have predominantly been carried out by the Danish Nature Agency, some with substantial LIFE funding, but also with contributions from the municipalities, citizens and private foundations. More than DKK 3.4 billion (€0.46 billion)[22] was spent on nature restoration over 1989–2007 (see Section 12.7.3 for the main sources). In 2020, the Danish government allocated DKK 888 million (€119 million) for 2021–2024 to

[20] https://naturstyrelsen.dk/nyheder/2015/sep/sjaelden-natur-faar-millionindsproejtning/

[21] www.avjf.dk/wp-content/uploads/2018/07/Gyldensteen-Folder-UK_Web.pdf

[22] 29 May 2021 exchange rate: €1 = DKK 7.44.

expand the area of untouched forest from about 45 000 ha to 75 000 ha, and establish 15 Nature National Parks.

The largest and most expensive nature restoration project, financially supported by the EU, has been that of the river Skjern Å in West Jutland, which cost DKK 283 million (€38 million) and covered an area of 2 200 ha. It partly reversed the canalisation and embankment of the lowest 20 km of the river, which occurred in the 1960s and resulted in the conversion of 4 000 ha of meadows to arable land. As well as incurring loss of biodiversity, the river straightening increased nutrient flows into Ringkøbing Fjord as the river's ability to trap nutrients was reduced. In addition, the lowered ground water level enabled oxygenation of iron compounds, which led to harmful iron oxide pollution.

The overall aim of the restoration was to re-establish a large, coherent nature conservation area (Dubgaard *et al.*, 2005). To achieve this, the natural meandering of the river has been re-created to form a delta, a lake and an extensively grazed floodplain grassland. With the restoration, many good fish spawning grounds were re-created and the conditions for the original Skjern Å Atlantic Salmon (*Salmo salar*), which had been threatened, were significantly improved. The Otter (*Lutra lutra*) has also recolonised and now has a strong population in the area. Few streams have as many special aquatic plants as Skjern Å, including the rare River Water-dropwort (*Oenanthe fluviatilis*) at its only site in Denmark. Today the river valley is designated as a Natura 2000 area due to its great importance for a number of rare and protection-dependent plant and animal species.

Further information about the costs and benefits of the restoration is provided in the next section and online Annex DK.1.[23]

Further restoration is planned in other areas in order to fulfil WFD objectives, with DKK 112 million (€15 million) allocated for 2018–2022 to cover two projects: Søborg Sø and Ekkodalen.[24] Søborg Sø was a large lake until it was drained for agricultural purposes. This will be restored as a shallow lake of approximately 500 ha. The project in Ekkodalen will re-create a lake formed in connection with the last ice age. Part of the lake has naturally developed into a bog, whilst other parts have been drained to create agricultural land. Approximately 170 ha will be restored.

12.7 Nature Conservation Costs, Benefits and Funding Sources

12.7.1 Nature Conservation Costs

Up-to-date and comprehensive estimates of the costs of achieving Denmark's nature conservation objectives do not appear to be publicly available. The only estimates focus on the financial needs of managing Natura 2000 and come fro the Prioritised Action Framework (PAF) that was produced f 2014–2020 (Anonymous, 2013). Over 2010–2015, it was est mated that total annual Natura 2000 costs would be €40. million (DKK 301 million), of which 56% comprised one-o costs averaged over the six years. These estimates were base on the government decision within the framework of th Green Growth Agreement, but it is unclear from the infor mation provided as to what degree the objectives of Natur 2000 would be met with this level of funding.

In 2011–2012, a combined biological and economic analy sis was carried out to help guide what can be done to preserv terrestrial biodiversity in Denmark, and to estimate its costs The recommendations were included in a so-called wise-ma report prepared by researchers at the Secretariat of th Economic Council and the Center for Macroecology Evolution and Climate at the University of Copenhagen (D Økonomiske Råds Sekretariat, 2012). An important require ment was to identify initiatives that provide more space fo nature and create more habitats, by focussing on the areas where species are found today. Against this background, areas were identified that will be most cost-effective in conserving and restoring biodiversity.

The results of the analysis identified a number of forests and open habitats that should be targeted for biodiversity conservation. Within the forest areas, the most effective measures are the creation of untouched deciduous forest, which means cessation of forestry management for wood production. The measures in the open habitats should include grazing or the cutting of hay to prevent overgrowth with trees and shrubs, and reducing the impact of nitrogen pollution. In addition, the designated natural areas should be expanded by converting the surrounding cultivated fields to the desired habitats. The efforts in the forests are both most important and cheapest in relation to biodiversity according to the report.

The study estimated that it would cost approximately DKK 850 million (€114 million) annually to carry out the described measures in the selected areas covering a total area of 1 700 km^2. The cost corresponds to less than 0.1% of Denmark's total public expenditure and includes lost earnings from the halting of timber production in the selected forests. In the open habitats, the costs relate to lost income from the cessation of farming in the designated areas and loss of capital from livestock buildings in the nitrogen reduction zones. Although it was not considered feasible to reliably estimate the socio-economic benefits of the biodiversity measures, it was thought that they could potentially be very high.

12.7.2 The Economic Benefits of Nature Conservation

Of particular note are the results of a cost-benefit analysis that was carried out of the Skjern Å restoration project described earlier (Dubgaard *et al.*, 2005; Dubgaard, 2018; see Annex DK.1 for further details). Although it was a very expensive

[23] All annexes are free to access online at CUP: www.cambridge.org/natureconservation

[24] Finanslov for finansåret 2020, Finansministeriet.

and controversial project, especially in agricultural circles, the analysis turned upside down the previous perception of what could be considered advantageous from a socio-economic point of view. It was found that the project's costs (mainly from the extensive construction work and the loss of agricultural land rent) were outweighed by the benefits at present values, even on rather conservative estimates. The main benefits come from the improvement in outdoor recreation opportunities, followed by the reduction in pollution (i.e. from nitrogen, phosphorus and ocher), improved angling and the existence value of increased biodiversity. Whilst it was found that the net present values for the scheme were sensitive to the applied discount rates,[25] they were positive up to high discount rates of 7%, ranging from DKK 238 million (€32 million) at 3% to DKK 5.5 million (€0.74 million) at 7%.

Another large nature restoration project, covering up to 1 800 ha on Zealand, includes the restoration of the wetland Åmosen, where benefits in addition to nature, species and biodiversity also include the conservation of important cultural–historical values. A study has shown that there is great willingness of the population to pay for nature restoration in Åmosen, amounting to DKK 500 (€67.2) per year in extra income tax (Lundhede *et al.*, 2005). The project has yet to be completed.

Another way to illustrate the cost benefit of nature conservation is exemplified by the Danish Nature Fund, a private, non-profit fund that works to improve the state of nature in Denmark. The fund values the purchase of land for nature purposes to be DKK 12 (€1.6) per m^2.[26]

12.7.3 Funding Sources for Nature Conservation

Table 12.5 provides an overview of the main funds that have been used to support nature conservation actions in Denmark, especially over the 2014–2020 EU funding period, including in relation to Natura 2000 according to the PAF for 2021–2027 (MoEF, 2019), as well as wider biodiversity objectives.

Table 12.5 clearly shows that, over 2014–2020, the principal source of funds for nature conservation measures in Denmark was the European Agricultural Fund for Rural Development (EAFRD). This was also the case in the previous funding period, as indicated by the 2013 PAF. Over 2014–2020, Denmark allocated about DKK 883 million (€119 million) of EU money to Rural Development Programme (RDP) measures related to Priority 4 on restoring, preserving and enhancing ecosystems. Most of this funding has been used to support organic farming (doubling its area) and physical investments in agriculture (e.g. to reduce pollution or create low-intensity pastures in carbon-rich areas), which have rather general biodiversity benefits. Of most importance for nature conservation was the use of the agri-environment–climate measure, which supported schemes targeted at Natura 2000 sites, such as to support hydrological and grassland management.

Although much smaller than the CAP funds, the EU LIFE programme has also played an important role in supporting targeted nature conservation actions. From 1992 to 2020, 39 Nature and Biodiversity projects were awarded in Denmark, with a total investment of DKK 997 million (€134 million), 54% being from the EU. Most projects, which have been mainly led by the Danish Nature Agency, as well as by regional and local authorities, have restored habitats, such as dunes, coastal lagoons, raised bogs, dry grasslands, heathlands and river systems. Others have focussed on targeted actions for endangered species, including the Marsh Fritillary (*Euphydryas aurinia*), meadow birds, Fire-bellied Toad and Houting.

According to the PAF for 2021–2027 (MoEF, 2019) national public funding, excluding co-funding of EU instruments, has been about DKK 141 million per year (€19 million) over recent years, much of which has focused on Natura 2000 and related requirements under the Nature Directives. Although not mentioned in the PAF, private funding of nature conservation has also been important in Denmark for some time. To further encourage this, in 2014 the Danish Government, together with the Villum Foundation and the Aage V. Jensen Foundation, set up a new national nature foundation, the Danish Nature Fund, to buy up land and manage it for the benefit of nature. The Government allocated DKK 500 million (€67 million) to the fund as initial capital spread over four years, and DKK 375 million (€50 million) from private funds.[27]

Based on the data in Table 12.5, total annual funding available for nature conservation appears to be about €174 million, which compares favourably with the estimated requirements for Natura 2000, etc., of about €40 million per year. However, as noted previously, much of the available funding is for broader conservation objectives, the costs of which have not been quantified.

Information on expenditure on all nature protection activities is available from a fairly comprehensive estimate of the resources that have been used for nature conservation by the Danish Society for Nature Conservation (Danmarks Naturfredningsforening, 2016), drawing on the Danish Nature Agency's fifth report to the CBD (Naturstyrelsen, 2014b). In 2014, spending for nature and biodiversity purposes in Denmark amounted to DKK 2.5 billion (€0.33 billion) from all sources: state, municipality, EU and private funds. The calculation included expenditure on direct biodiversity conservation and nature restoration actions, as well as indirect biodiversity-related activities such as research, information campaigns and education. Such broad calculations need to be

25 That is, the annual percentage rate at which the present value of future monetary values is estimated to decrease over time.

26 Den Danske Naturfond (The Danish Nature Fund): https://naturfonden.dk/

27 Cf. articles of association at the homepage of The Danish Nature Fund. www.naturfonden.dk/

Table 12.5 Funding sources for nature conservation in Denmark.

Source and period	Measure/category	Average annual allocation (DKK m)		Average annual allocation (€ m)		Comments
EU funding instruments		**EU + national**	**EU**	**EU + national**	**EU**	
EAFRD Rural Development Programme measures allocations to Priority 4 (restoring, preserving and enhancing agriculture and forest ecosystems) 2014–2020	M11 Organic farming	361.5	271.1	48.6	36.4	
	M4 Investments in physical assets	218.7	180.8	29.4	24.3	E.g. N2k hydrology
	M10 Agri-environment–climate	214.4	160.8	28.8	21.6	Targeted to N2k, etc.
	M8 Investments in forests	51.4	44.3	6.9	6.0	
	M12 Natura and WFD	24.4	24.4	3.3	3.3	For WFD as not listed in PAF.
	M1 Knowledge transfer, etc.	8.1	4.3	1.1	0.6	
	M15 Forest environment climate	4.0	3.0	0.5	0.4	
	M7 Basic services, etc.	0.9	0.9	0.13	0.12	
RDP Priority 4 subtotal		*883.3*	*689.5*	*118.7*	*92.7*	
ERDF 2014–2020	Interreg: CANAPE (Creating a New Approach to Peatland Ecosystems) 2017–2021	1.4	0.7	0.2	0.1	
EMFF 2014–2020	MOI.18	8.9	?	1.2	?	For N2k, etc.
	OVIII.3 Improving knowledge on marine env.	1.2	?	0.2	?	For N2k, etc.
LIFE	Nature and biodiversity 1992–2020	34.4	15.6	4.6	2.1	NATUREMAN
	Integrated project 2018–2026	15.2	9.1	2.0	1.2	IP Natureman for N2k, etc.
EU funding instruments subtotal		*944.4*	*>715.0*	*126.9*	*>96.1*	
National funds		**Additional national*** **(DKK m)**		**(€ m)**		**Comments**
Ministry of Environment and Food (2014–2020)	N2k planning, monitoring and management	30.5		4.1		€3 m for management
2019–2026	For biodiversity in private forests	9.7		1.3		
2016–2021	Lake restoration	6.0		0.8		Note 1
2016–2020	Freshwater aquaculture	7.1		1.0		Note 1
2016–2021	Municipalities nature management	50.0		6.7		In N2k
2018	Compensation for farming extensification	9.7		1.3		In N2k
2016–2019	Compensation for halting forestry	9.7		1.3		Mostly in N2k

Table 12.5 *(cont.)*

Source and period	Measure/category	Average annual allocation (DKK m)	Average annual allocation (€ m)	Comments
2018–2020	Nature restoration	18.2	2.5	Søborg Sø and Ekkodalen
Danish Nature Fund (2014–2017)	Land purchase	217.6	29.3	See text
Other NGOs, etc.		?	?	See text
Additional national subtotal		*358.4*	*48.2*	
National total inc. co-financing of EU instruments		*587.8*	*79.0*	
Total		**1 302.8**	**175.1**	Incomplete

Notes. Currency conversion rate: €1 = DKK 7.44. 'For Nk2, etc.' indicates funding relevant to Natura 2000, related green infrastructure and BHD species (based on PAFs). * Excludes co-financing of EU funds unless indicated. The subtotals and totals are only approximate indications of the average annual levels of funding as they are based on annualised data from differing periods. 1. RBMP supplementary measures.
Sources. RDP data from European Commission ESIF funds database (23 October 2020); ERDF, EMFF and national funding from PAF (MoEF, 2019); Denmark LIFE factsheet, European Commission (updated January 2021); Danish Nature Fund, see text. See Section 1.3 for further details.

treated with caution as there is no standard agreement on what expenses can be attributed to nature protection, and therefore estimates vary in terms of what they include. Whilst the available funding seems to be substantial, it is uncertain if it is sufficient given the ongoing poor state of habitats and species.

12.8 Surveillance, Monitoring, Reporting and Applied Research

Monitoring of nature in Denmark is mainly carried out at the national level to make the most efficient use of resources. Since 2004, a National Survey of Water, Environment and Nature, known as NOVANA, has been carried out, with the main aim of meeting EU and other international monitoring and reporting requirements (Miljøstyrelsen, 2017). The programme is coordinated by the Danish Agency of Environment and involves the Danish Centre for Environment and Energy at Aarhus University and the Geological Survey of Denmark and Greenland. Fieldwork is carried out by the three partners or external consultants and covers all Natura 2000 sites as well as most BHD species and habitats. This enables a standardised survey, but may mean that the management authorities are not so directly aware of the situation in their area.

Habitat types in Natura 2000 sites have been mapped for almost all sites, the exception being some marine sites. For most of the terrestrial sites, monitoring of habitat types has been carried out twice, making it possible to compare changes.

Several atlas projects have been carried out on vascular plants, fungi with fruiting bodies, fish, amphibians, mammals and birds. These have been conducted by NGOs, sometimes in connection with universities, with the main fieldwork done by volunteers. The work has been supported by the Aage V. Jensen Nature Foundation. BirdLife Denmark coordinates an annual bird census to provide indicators of the status of wintering and breeding bird populations.

The Agency of Environment runs a web page called Miljøportalen (the Environment Portal).[28] Besides giving a map of Danish protected areas, it also provides monitoring data for a large number of sites, consisting mainly of botanical surveys. A national database, used both by amateurs and professionals, called arter.dk[29] is hosted by Miljøportalen and organises many observations on plants, invertebrates, amphibians, mammals, etc. It is especially useful for rare species that are difficult to find in standardised surveys restricted to specific Natura 2000 sites. BirdLife Denmark has a database with bird observations, called DOF-basen.[30]

The Danish Society for Nature Conservation, in cooperation with the Centre of Macro Ecology (Copenhagen University), has developed an app called 'Naturtjek'. It encourages people interested in nature to send in information using the app and, if possible, to document the observation with a photo. More than 20 000 people contributed with 530 000 observations over 2015–2020.[31]

The National Centre of Environment and Energy and the Nature Museum, both at Aarhus University, mainly conduct applied research supporting nature conservation. In addition, Copenhagen University is a key contributor with its Centre of Macro Ecology, the National Museum of Natural History and the Institute of Geoscience and Nature Management. The

[28] https://arealinformation.miljoeportal.dk/
[29] https://arter.dk/landing-page. Previously Naturbasen.dk was the main database. Arter.dk is managed by Miljøstyrelsen, Copenhagen and Aarhus Universities and DanBIF.
[30] https://dofbasen.dk/
[31] https://biodiversitet.nu/natur-over-tid

Department of People and Technology in the University of Roskilde also has an important role, running a Centre for Nature Park and National Park Research. The Technical University and Southern Danish University carry out nature conservation related research on marine fish and marine mammals, respectively.

12.9 Future Developments

The attention given to nature conservation in Denmark is currently increasing. During the last parliamentary election, nature, together with climate change, was high on the agenda for the first time in many years. Political parties, particularly the Social Democrats, who won the 2019 election and formed a minority government, have made many promises about improving nature, for example by establishing so-called Nature National Parks where nature has top priority. These would be established on state-owned land with high densities of 'wild' grazing animals (but fenced off from the farmland). In all, the goal is to create 15 such Nature National Parks. The idea originated from the Danish Nature Conservation Society, which found that the existing five National Parks were overly dominated by cultural landscapes, and as a result were insufficiently rich in nature. This, together with reports of decreasing biodiversity and the 'Nature Meetings' held since 2016, has put the political parties under pressure to act.

Another area-focussed initiative is the development of a high density of regional Nature Parks, as has been seen in Germany. It is considered to be a way of recognising that one's local area is something special and worth visiting. Furthermore, the certification system of Nature Parks can also increase local involvement, dissemination and setting long-term goals through a common planning system. Many local stakeholders have also understood that if they have nature of international value, they can apply for UNESCO recognition. For example, several municipalities or local groups are now working to become UNESCO Global Geoparks. Denmark could have more natural World Heritage sites, given that several sites are on the Danish tentative list for nomination. Besides this, there is pressure from municipalities and NGOs to have more National Parks, apart from the Nature National Parks.

It is increasingly recognised that standardised measures of protected area coverage are essential, and that the use of the IUCN definition could lift their general management and protection level in Denmark. Accordingly, the Danish Environmental Protection Agency has updated the WDPA using the results of an assessment of Conservation Areas (Woollhead and Petersen, 2018). Hopefully, the WDPA will also be updated following a similar assessment of marine protected areas in Danish waters (Woollhead *et al.*, 2020).

Suggestions that drained and farmed peatlands should be rewetted and converted to nature seem to be gaining ground because this can help reverse the high CO_2 emissions from peat oxygenation. Furthermore, many river valleys have drained peatland, and these could become important green infrastructure, binding large natural areas together. Government programmes intend to buy large areas of farmland on non-peat soils and then exchange them with privately owned fields on peatlands.

Since the autumn of 2020, the Social Democrat government has been developing a new national biodiversity strategy (delayed due to the COVID-19 crisis). The idea is to set targets for natural areas in Denmark, probably reflecting the targets in the *EU Biodiversity Strategy for 2030*. Meeting such targets for terrestrial nature may be very difficult, as it would be very expensive and could lead to conflicts with farmers. Furthermore, there is pressure on the government to have a biodiversity law to set legally binding targets for nature conservation, to counteract the biodiversity crisis.

12.10 Conclusions

Over the past 50 years, natural and semi-natural habitats have continued to decline as areas under intensive use have increased. Denmark is now largely an agricultural country and its original biodiverse forest has almost disappeared, together with its iconic large mammals. Recent efforts have led to an increase in forest cover, but a large proportion of the forest area is managed for wood production. Other terrestrial semi-natural habitats, such as grasslands, heathlands and bogs, are now very scarce and rely largely on intensive nature conservation management to prevent them turning into forest because there are no longer the traditional low-intensity farming systems that created and maintained them. Many of these habitats, and also aquatic ecosystems, have also been affected by high levels of nutrient pollution, especially from the intensive livestock industry. Whilst some measures have reduced pollutant levels, further action is required. Coastal habitats, including Denmark's internationally important dunes and intertidal habitats and shallow seas, are less impacted by human activities. Nevertheless, they remain under pressure in some areas, such as from built developments, recreation, intensive fishing and nutrient pollution.

Denmark has used some innovative approaches to tackle its conservation challenges; these approaches include its Conservation Areas, certification scheme for Nature Parks and giving general protection to certain habitats. Whilst it has been slow to establish some types of protected areas, it is now developing a more diverse system. According to the WDPA, Denmark meets the CBD's goals in terms of nature protection both at sea and on land. However, recent analyses have shown that this is not the case when the designated protected areas are assessed according to the IUCN's standard for nature protection, which is the basis of the CBD goals. These approaches, and some other measures, such as large-scale habitat restoration projects and targeted species re-introductions, have resulted in some successes, such as slowing habitat losses and increasing some species (e.g. White-tailed Sea-eagle). However, biodiversity has continued to decline overall. This decline has especially affected common and widespread species, such as on farmland, but it also includes most of the habitats and species that are the focus of the Nature Directives, as shown by the recent HD Article 17 and BD Article 12 reports.

In recent years some additional efforts have been taken to address biodiversity declines, including an attempt to build a network that includes existing and potential natural areas and their ecological connections (the Green Map) and improve nature management across the country. Many other measures have been launched, such as increasing the share of untouched forest and making it illegal to spray and fertilise protected nature, but it remains to be seen what their effects will be. The biodiversity crisis has risen up the political agenda in recent years, and more funds have been allocated to 'investments' in nature. Time will tell if these new trends can reverse some of the declines over the last 100 years.

Acknowledgements

We want to give a big thanks to Johannes Birk Schjelde (The Danish Society for Nature Conservation) for giving our draft a thorough review. Also, thanks to Alex Dubgaard (University of Copenhagen), for helping us to get the figures right on the Skjern Å cost-benefit analysis, and for checking the related text. We thank Lisbeth Bjørndal Andersen (Ministry of Environment of Denmark) who assisted with finding the most up-to-date PAF. Kevin Vejrup (Statistics Denmark) helped in clarifying data from Statistics Denmark, and Laura Beck advised on the right English words for some types of nature. Geir Tveit (Copenhagen University) and Per Larsen (The Danish Society for Nature Conservation) both provided requested publications.

References

Anonymous (2013) *Prioritised Action Framework (PAF) for Natura 2000 Denmark. For the EU Multiannual Financing Period 2014–2020.*

Asbirk, S. (1988) *Naturen i Danmark – Status og udviklingstendenser. Naturovervågning 1988.* Randbøl, Denmark: Naturstyrelsen.

BirdLife International, European Environmental Bureau, Friends of the Earth Europe & World Wildlife Fund (2018) *The State of Implementation of the Birds and Habitats Directives in the EU. An analysis by national environmental NGOs in 18 Member States.*

Buttenschøn, R. M. & Schmidt, I. K. (2015) *De danske hedetyper; deres udbredelse og tilstand.* Flora og Fauna 121 (3+4), pp. 75–77. Aarhus, Denmark: Jydsk Naturhistorisk Forening.

Danmarks Naturfredningsforening (2016) *Sådan ligger landet - Tal om naturen 2016.* Copenhagen: Danmarks Naturfredningsforening. www.dn.dk/om-os/publikationer/sadan-ligger-landet/

(2018) *Nature Score Card Denmark.* In *The State of Implementation of the Birds and Habitats Directives in the EU,* eds. BirdLife International, European Environmental Bureau, Friends of the Earth Europe and WWF. wwfeu.awsassets.panda.org/downloads/nature_scorecard_denmark_march2018.pdf

Danmarks Statistik (2017) *Nyt fra Danmarks Statistik - Afgrøder i dansk landbrug 2017 - Erhvervslivets sektorer.* No. 302, 2017. Copenhagen: Danmarks Statistik.

(2019) *Danmark i Tal 2019.* https://bit.ly/3BPigFe

De Økonomiske Råds Sekretariat (2012) *Økonomi og Miljø 2012. [Economics and environment 2012]* (Summary in English). Copenhagen: De Økonomiske Råds Sekretariat.

Dubgaard, A. (2018) *Cost-benefit analysen af Skjern Å-projektet.* In *Sæt pris på naturen,* eds. J. S. Schou, J. Jensen & B. J. Thorsen, pp. 142–151. Copenhagen University, Department of Food and Resource Economics.

Dubgaard, A., Kallesøe, M. F., Ladenburg, J. & Petersen, M. L. (2005) Cost–benefit analysis of river restoration in Denmark. In *Cost-Benefit Analysis and Water Resources Management,* eds. R. Brouwer & D. Pearce, pp. 124–150. Cheltenham, UK: Edward Elgar Publishing.

Ejrnæs, R., Wiberg-Larsen, P., Holm, T. E. *et al.* (2011) *Danmarks biodiversitet 2010 – status, udvikling og trusler.* Danmarks Miljøundersøgelser, Aarhus Universitet, 152 pp – Faglig rapport fra DMU nr. 815.

Fenchel, T. (2017) Naturen i Havet. In *Naturen i Danmark - Havet,* ed. K. Sand-Jensen, pp. 17–34. Copenhagen: Gyldendal.

Folke, H. (2020) *ø - i Den Store Danske på lex.dk.* (Retrieved 20 December 2022 from https://denstoredanske.lex.dk/%C3%B8).

Fredshavn, J., Nygaard, B., Ejrnæs, R. *et al.* (2019a) *Bevaringsstatus for naturtyper og arter – 2019. Habitatdirektivets Artikel 17-rapportering.* Aarhus University, DCE – Nationalt Center for Miljø og Energi. Videnskabelig rapport fra DCE – Nationalt Center for Miljø og Energi nr. 340.

Fredshavn, J. R., Holm, T. E., Sterup, J. *et al.* (2019b) *Størrelse og udvikling af fuglebestande i Danmark – 2019.* Artikel 12-rapportering til Fuglebeskyttelsesdirektivet. Aarhus University, DCE – Nationalt Center for Miljø og Energi, 46 pp. – Videnskabelig rapport nr. 363.

Garn, A.-K., Woollhead, J. & Petersen, A. (2019) Lessons learned from a desktop review of conservation areas in Denmark: Applying IUCN management categories for protected areas. *PARKS,* **25,** 93.

Hammond, P. S., Bearzi, G., Bjørge, A., Forney, K. A., Karczmarski, L. & Kasuya, T. (2008) *Phocoena phocoena* (Baltic Sea subpopulation) (errata version published in 2016). The IUCN Red List of Threatened Species 2008. www.iucnredlist.org (30 October 2020).

Hansen, K. (2017) *Det store svigt. Beretningen om 100 års naturfredning i Danmark.* Copenhagen: Gads Forlag.

Hansen, K. S. & Søndergaard, M. (2019) *Danske søer ca. 1970–2018,* Miljøets fodspor nr. 6.

Havstrategi II – Første del (2019) [*The Danish Marine Strategy II*]. Copenhagen: Miljø- og Fødevareministeriet.

Heldbjerg, H., Sunde, P. & Fox, A. D. (2018) Continuous population declines for specialist farmland birds 1987–2014 in Denmark indicates no halt in biodiversity loss in agricultural habitats. *Bird Conservation International,* **28,** 278–292.

ICES (2020) EU request on emergency measures to prevent bycatch of common dolphin (*Delphinus delphis*) and Baltic Proper harbour porpoise (*Phocoena phocoena*) in the Northeast Atlantic. In Report of the ICES Advisory Committee, 2020. ICES Advice 2020, sr.2020.04. https://ices-library.figshare.com/articles/report/EU_request_on_emergency_measures_to_prevent_bycatch_of_ common_dolphin_Delphinus_delphis_and_Baltic_Proper_harbour_porpoise_Phocoena_phocoena_in_the_Northeast_Atlantic/18636260

Koester, V. (1989) *The Ramsar Convention on the Conservation of Wetlands. A Legal Analysis of the Adoption and Implementation of the Convention in Denmark.* Gland, Switzerland: Ramsar Convention Bureau/IUCN.

Lundhede, T., Hasler, B. & Bille, T. (2005) *Værdisætning af naturgenopretning og bevarelse af fortidsminder i Store Åmose i Vestsjælland.* Copenhagen: Nature Agency.

Miljøministeriet (2014) *Naturplan Danmark. Vores Fælles Natur.* Copenhagen: Miljøministeriet

Miljøstyrelsen (2017) *NOVANA. Det nationale overvågningsprogram for vandmiljø og natur 2017-21. Programbeskrivelse.* DCE – Nationalt Center for Miljø og Energi ved Aarhus Universitet and GEUS – De Nationale Geologiske Undersøgelser for Danmark og Grønland.

MoEF (2019) *Prioritised Action Framework (PAF) for Natura 2000 in Denmark – for the Multiannual Financial Framework period 2021–2027.* Copenhagen: Danish Ministry of Environment and Food.

Moeslund, J. E., Nygaard, B., Ejrnæs, R. *et al.* (2019) *Den danske Rødliste* [*The Danish Red List*]. Aarhus University, DCE – Nationalt Center for Miljø og Energi. Aarhus. www.redlist.au.dk

Natur- og Landbrugskommissionen (2014) *Natur og Landbrug – en ny start.* Copenhagen: Miljø- og Fødevareministeriet.

Naturstyrelsen (2014a) *Forvaltningsplan for ulv i Danmark.* Copenhagen: Ministry of Environment of Denmark.

(2014b) *The 5th Danish Country Report to The Convention on Biological Diversity.* Copenhagen: Ministry of Environment of Denmark. https://naturstyrelsen.dk/media/nst/8399384/31marts2014_landerapport.pdf

Normander, B., Woollhead, J., Petersen, A. & Garn, A.-K. (2021) Short Communication: Denmark's marine protected areas assessed according to IUCN's international definition. *Parks*, **27** (2) 69–73.

Olsen, K., Sunde, P., Vedel-Smith, C., Hansen, M. M. & Thomsen, P. F. (2020) *Statusrapport fra den nationale overvågning af ulv (Canis lupus) i Danmark. 3. kvartal 2020.* Aarhus, Denmark: Notat fra DCE – Nationalt Center for Miljø og Energi og Naturhistorisk Museum.

Petersen, J. K. (ed.) (2018) *Menneskeskabte påvirkninger af havet:– Andre presfaktorer end næringsstoffer og FFklimaforandringer.* Lyngby: Institut for Akvatiske Ressourcer, Danmarks Tekniske Universitet. DTU Aqua-rapport nr. 336-2018

Prip, C. & Wind, P. (1995) *Biologisk mangfoldighed i Danmark – status og strategi.* Randbøl, Denmark: Naturstyrelsen. www2.skovognatur.dk/udgivelser/2003/mangfoldighed/indholddk.htm

Rasmussen, J. F., Nielsen, M. & Flensted, K. N. (2000) In *Important Bird Areas in Europe: Priority sites for conservation. 1: Northern Europe*, eds. M. F. Heath & M. I. Evans, pp. 137–177. Cambridge, UK: BirdLife International.

Riemann, B. (ed.), Carstensen J., Markager, S. *et al.* (2015) *Status og muligheder for det danske hav.* Aarhus University, DCE – Nationalt Center for Miljø og Energi.

Sand-Jensen, K. & Larsen, G. (eds.) (2017) *Naturen i Danmark – Geologien.* 2nd ed. Copenhagen: Gyldendal.

Sand-Jensen, K. & Møller, P. F. (eds.) (2017) *Naturen i Danmark – Skovene.* 2nd ed. Copenhagen: Gyldendal.

Sand-Jensen, K. & Vestergaard, P. (eds.) (2017) *Naturen i Danmark – Det åbne land.* 2nd ed. Copenhagen: Gyldendal.

Skov- og Naturstyrelsen (1992) *Strategi for bevarelse af de danske naturskove og andre bevaringsværdige Skove.* Copenhagen: Miljøministeriet.

Skriver, J. (2017) *Rovfugle forlader rederne, når mountainbikere kommer for tæt på.* DOF-Nyheder (Bird Life Denmark), 17 February 2017. www.dof.dk/om-dof/nyheder?nyhed_id=1533

Therkildsen, O. R., Andersen, S. M., Clausen, P., Bregnballe, T., Laursen, K. & Teilmann, J. (2013) *Vurdering af forstyrrelsestrusler i NATURA 2000-områderne.* Aarhus University, DCE – Nationalt Center for Miljø og Energi. Videnskabelig rapport fra DCE - Nationalt Center for Miljø og Energi nr. 52.

UNEP-WCMC (2022) *Protected Area Profile for Denmark from the World Database of Protected Areas, January 2022.* www.protectedplanet.net

Wilhjelmudvalget (2001) *En rig natur i et rigt samfund.* Copenhagen: Skov- og Naturstyrelsen. www2.skovognatur.dk/udgivelser/2002/87-7279-378-3/pdf/helepubl.pdf

Woollhead, J. & Petersen, A. (2018) *Rapport om rubricering af danske fredninger i IUCN-kategorier.* Copenhagen Zoologisk Have København. Udarbejdet til Miljøstyrelsen på vegne af IUCN-Nationalkomiteen.

Woollhead, J., Petersen, A. & Normander, B. (2020) *Vurdering af danske beskyttede havområder efter international standard* [*Assessment of Danish MPA's according to international standards*] (Summary in English). Copenhagen: IUCN Conservation Planning Specialist Group Europe and Zoologisk Have København.

Chapter 13

Estonia

Aleksei Lotman and Silvia Lotman

Key Statistics	
Terrestrial area (km²)/% of EU-28*	45 366/1.0%
Marine waters (km²)/% of EU-28*	36 609/0.8%
Population density (p km²)/% of EU-28*	29/25%
GDP per capita (€)/% of EU-28	13 364/50%
Land cover (CORINE land cover classes 2018)	FOR 44%, AGR 31%, OSN 13%, WET 9%, ART 2%
Biogeographical regions	BOR, MBAL
Number of HD Annex I habitats/%*	60/27%
Number of HD Annex II, IV and V species/%*	95/8%
Number of BD Annex I bird taxa/%*	65/37%
% HD UCS and decreasing habitats/ % unknown	12%/0%
% HD UCS and decreasing species/ % unknown	9%/7%
% breeding BD Annex I birds decreasing/% uncertain and unknown	36%/2%

* Excluding MAC and MMAC biogeographical regions.
Sources and codes. See Table 1.1.

13.1 The Natural History and Characteristics of Estonia

Estonia is located on the eastern European platform, with crystalline bedrock covered by Paleozoic sedimentary rocks. The latter are limited to Ordovician limestone in the north, being covered by Silurian limestone in the west and middle, and, in turn, by Devonian sandstone in the south. These bedrocks are covered by glacial and post-glacial deposits. The soils vary significantly due to the country's geology (Tomingas, 2018). Estonia is topographically flat and low, the highest point being 315 m (Suur Munamägi Hill). Estonia is part of the Boreal and Marine Baltic biogeographical regions.

The Estonian coastline is bounded by the Baltic Sea, which is brackish (with salinity mostly below 0.7%), shallow (max. 100 m in the Gulf of Finland) and little affected by tides. The northern coast towards the Gulf of Finland is relatively high (max. 230 m), whilst the western coasts are mostly low-lying. The retreat of the glaciers from present-day Estonia, 12 to 10 thousand years ago, resulted in land uplift that has changed the coastline during the course of history and is still taking place (Tomingas, 2018).

The current climate is intermediate between maritime and continental. Sea ice lasts 30–40 days on the western coasts of the islands of Saaremaa and Hiiumaa, and 120–130 days on Väinameri (Tomingas, 2018). An increasing trend in temperature and a slight increase in precipitation have been observed over the last half century (Klasen *et al.*, 2017).

Precipitation exceeds evaporation and hence Estonia is rich in inland waters and wetlands, with lakes covering about 5% of the country. Peat started to accumulate soon after the retreat of the glaciers, such that the proportion of the country covered by mires peaked at about 20%, or even 25% by some estimates (Valk, 1988), before the start of drainage and peat extraction. The best mire habitats, including fens (7210*, 7230),[1] transitional mires (7140) and active raised bogs (7110*), are now mostly protected.

Humans probably entered the area of present-day Estonia on the heels of the receding glaciers and influenced the developing ecosystems to some extent. However, significant human impacts originate from the time when agriculture became the main source of food, leading to a gradual clearance of the woods and an increase in open areas, driven by slash-and-burn agriculture and pastoralism (Kriiska, 2004; Tomson *et al.*, 2015). Forest area declined as a result. However, pastoralism and its low-intensity grazing and haymaking, which we now call High Nature Value (HNV) farming, gave rise to new biodiverse habitats (Kull *et al.*, 2003; Lotman, 2003, 2004). Some of these are of EU importance and HD Annex I listed, including boreal Baltic coastal meadows (1630*), seminatural dry grasslands on calcareous substrate (6210 (*important orchid sites)), Nordic alvar[2] and precambrian calcareous flatrocks (6280*), northern boreal alluvial meadows (6450),

[1] Numbers in brackets are HD Annex I habitat codes (see Appendix for full names). * Indicates a priority habitat.

[2] Sparse vegetation on thin calcareous soils formed over calciferous bedrock, in a fairly dry climate.

Fennoscandian wooded meadows (6530*) and Fennoscandian wooded pastures (9070).

Relatively recent human impacts on ecosystems resulted from the gradual abolition of the feudal land-use system during the nineteenth century (Troska, 2004) and its final dismantling by the land reform of the newly independent Republic of Estonia in the 1920s, leading to the replacement of the land tenure based on large estates with that based on small farms. This was then replaced with statecontrolled large agricultural production units following the Soviet occupation during the World War II. This, and rural depopulation (in part due to deportations), led to significant abandonment of traditional HNV farmland, with the loss of most wooded meadows, although a significant proportion of large coastal pastures and alluvial meadows remained in use. Forest cover increased again, due to natural succession and planting, to about 50% of the Estonian territory by the turn of the last century. The most significant wetland losses and straightening of rivers also occurred during Soviet times: fens and mires were drained for agriculture and forestry, reducing their area from approximately 650 000 ha to 58 000 ha, while the decline of bogs for peat extraction was less severe, from over 380 000 ha to about 250 000 ha (Paal and Leibak, 2011).

The return of Estonian independence in 1991 restored farm-based land ownership and enabled a radical shift to a market economy (Lotman, 2003, 2004). This resulted in the abandonment of most of the remaining seminatural meadows and pastures. EU accession and the introduction of the CAP also had further profound environmental impacts, including helping to further modernise and intensify agriculture. Thus, whilst pollution of inland waters decreased after the end of the Soviet period (Tomingas, 2018), diffuse nitrogen pollution increased again following the introduction of CAP funds (Klasen *et al.*, 2017). On the other hand, agri-environment schemes and other CAP funds have helped reverse the trend of abandonment, thereby saving the last remaining semi-natural pastures and meadows (Lotman, 2003, 2004).

Over the same period, industrialisation of forestry led to a shift towards younger, more productive, but less biodiverse forests (Etverk, 2003). Further intensification started about 10 years ago (Valgepea *et al.*, 2018) and the increase in forest area has levelled off (Raudsaar *et al.*, 2018). Even intensively used forest areas are still dominated by native tree species (Raudsaar *et al.*, 2018).

Some major changes in fauna occurred as a result of excessive hunting, as well as declines in forest habitat. The Beaver (*Castor fiber*) went locally extinct in the nineteenth century, and the Wild Boar (*Sus scrofa*) probably even earlier. Elk (*Alces alces*), Wolf (*Canis lupus*), Eurasian Lynx (*Lynx lynx*) and Brown Bear (*Ursus arctos*) all became extremely rare by the beginning of the twentieth century. All these species have now fully recovered, mostly through recolonisation and natural population growth that has been facilitated by appropriate management (including protection of some of the species at the initial stages of recovery).

13.2 The Status of Nature

13.2.1 Habitat Status and Trends

According to the HD Article 17 report[3] on the assessments of the conservation status of HD habitats over 2013–2018, 57% had a favourable status, 37% a poor status and 7% a bad status. Forests, bogs and grasslands had the highest proportions of habitat types with an unfavourable status. The status of 9 out of 10 HD forest habitats was assessed as unfavourable, being poor for 7, and bad for 2 priority types: western taïga (9010*) and deciduous swamp woods (9080*).

Boreal Baltic coastal meadows and four out of eight semi-natural HD grasslands (6210*, 6270, 6280*, 6450) had a poor status, whilst Fennoscandian wooded meadows (6530*) had a bad status. Thus, the current trends for grasslands are mixed; whilst there has been overall deterioration, improvements have occurred in some sites (Klasen *et al.*, 2017; Linder, 2017; Roasto and Tampere, 2020). Although important localities are included in the Natura 2000 network, they remain threatened by a lack of appropriate management (see Section 13.5.4).

Similarly, five of the eight mire habitats had a poor status, whilst alkaline fens (7230) had a bad status, as well as declining trends.

The terrestrial habitat types with highest proportions in favourable condition were rocky slopes, heath, scrub and dunes. However, three out of eight dune habitat types had a poor status. The status of freshwater habitats was also mixed: four out of seven had a favourable status, but three had a poor status.

In contrast to the situation on land, all the marine coastal habitats were considered to have a favourable status. Furthermore, the most important known localities for marine habitats and seabirds are protected within the Natura 2000 network. However, the extent of Annex I habitats is not sufficiently known over around 60% of the Estonian marine area.

Of particular concern is that a substantial proportion of HD habitats that had an unfavourable status also showed declining trends, as shown in Table 13.1.

13.2.2 Species Status and Trends

The situation in forests is indicated by the forest bird index (a geometric mean of population indices of 53 species), which showed a decline of 26% over 1983–2018 (Nellis and Volke, 2019). Out of the 53 species, trends were stable in 26, increasing in 6, decreasing in 14 and uncertain (fluctuating) in 7. Many old forest species (including ground-nesting birds of wet forests) have suffered both long-term and short-term declines in spite of strict protection (Roasto and Tampere, 2020). For example, the abundance of Capercaillie (*Tetrao urogallus*) has decreased by an average of 31% over three generations (19 years). The Siberian Flying Squirrel (*Pteromys volans*) and Black Stork (*Ciconia nigra*) have fared similarly. Forest cutting

[3] See Section 1.3 for background on monitoring under the Nature Directives, status categories and data sources.

and drainage are probably the main cause, as even though overall forest cover has been increasing, old stands have become rarer and fragmented, and forest streams and wetlands have been replaced with ditches. However, several iconic raptors nesting in old forests, like the White-tailed Sea-eagle (*Haliaeetus albicilla*) and Golden Eagle (*Aquila chrysaetos*), are increasing (Roasto and Tampere, 2020) as their known nesting sites are effectively protected, whilst the DDT ban has also contributed to their recovery.

Waders typical of meadows and pastures, especially the Baltic Dunlin (*Calidris alpina schinzii*), Black-tailed Godwit (*Limosa limosa*) and Ruff (*Philomachus pugnax*), are declining (Klasen *et al.*, 2017; Roasto and Tampere, 2020), as are many other species of traditional agricultural landscapes. The overall trend of farmland birds is negative, in spite of some species having improving trends (Marja and Nellis, 2018). Bumble bees seem to be faring better: as indicators in agricultural lands over 2009–2019 were stable, or in some cases increasing, the latter trend being evident in the more mosaic agricultural landscapes of south-east Estonia (Viik, 2019).

The populations of most bird species nesting on coastal islands have increased, especially amongst fish-eating species. Whilst the decline of some bottom-feeding waterbirds has stopped and stabilised at a low level, several species like Velvet Scoter (*Melanitta fusca*), Common Eider (*Somateria mollissima*) and Goosander (*Mergus merganser*) are still declining (Roasto and Tampere, 2020). Therefore the overall conservation status of such birds remains of concern.

The Freshwater Pearl Mussel (*Margaritifera margaritifera*) is Critically Endangered in Estonia (Roasto and Tampere, 2020) and has remained in only one north Estonian river. There have been multiple attempts to save the species but the habitat quality is very low with many pressures still present (draining of surrounding wetlands and forests, agricultural run-off as well as Beaver dams) and there is no clear recovery plan for the species.

Five of the 10 Estonian native amphibian species have deteriorating populations (Leivits, 2019). By now, the Green Toad (*Bufotes viridis*) is possibly regionally extinct and the status of the Natterjack Toad (*Epidalea calamita*) is assessed as bad (Rannap, 2019). However, there have been some significant successes in restoring Natterjack populations through habitat restoration, captive breeding and re-introduction supported by Danish experts.

The status of mammals is generally more favourable than that of most other species groups, and some threatened species are stable or increasing. Most noteworthy is the native European Mink (*Mustela lutreola*), which had gone extinct in the wild by the end of twentieth century due to competition from the alien American Mink (*Neovison vison*). Following an international captive breeding programme for this species led by Tallinn Zoo (Maran *et al.*, 2017), it was successfully re-introduced to the island of Hiiumaa after the latter species had been eradicated there. The American Mink, however, is still common throughout the rest of the country, preventing the return of the native species, which is therefore still considered to be Critically Endangered. The Otter (*Lutra lutra*) has made a strong comeback in this century after years of low numbers. Seals declined in the twentieth century, initially because of excessive hunting and later due to contamination of the Baltic Sea by DDT, PCBs and possibly other contaminants. The Grey Seal (*Halichoerus grypus*) has rebounded from extremely low numbers during the 1970s but has not yet fully recovered; the Ringed Seal (*Pusa hispida*) subpopulations in Estonia still have a bad status (Jüssi and Jüssi, 2017; Lotman *et al.*, 2018).

The trends in species and their habitats, as described here, are reflected in the most recent HD Article 17 report, which covered 2013–2018. Out of 96 assessed HD species, 56% were considered to have a favourable status, 26% had a poor status and 10% had a bad status; the status of 7% was unknown. As for habitats, it is also of concern that a significant proportion of species with an unfavourable status had also declined over the short term (Table 13.1). The declining HD species included four with a bad status: Green Toad and Siberian Flying Squirrel (as mentioned earlier), as well as the Sand Lizard (*Lacerta agilis*) and Large Blue butterfly (*Maculinea arion*).

Table 13.1 Short-term trends (c. 2007–2018) in Habitats Directive Annex I habitats and Annex II, IV and V species with an unfavourable conservation status in Estonia.

	Poor (unfavourable– inadequate)				Bad (unfavourable–bad)			
	Imp	Sta	Dec	?	Imp	Sta	Dec	?
HD Habitat assessments								
Number	4	11	6	1	0	1	1	2
%	18%	50%	27%	5%	0%	25%	25%	50%
HD Species assessments								
Number	10	9	5	1	1	4	4	1
%	40%	36%	20%	4%	10%	40%	40%	10%

Key. Imp = improving, Sta = stable, Dec = decreasing, ? = unknown.
Source. Based on Member State HD Article 17 reported data as presented on the EEA national summary dashboard (2021).

Table 13.2 Short-term trends (c. 2007–2018) in Birds Directive Annex I taxa in Estonia.

	Breeding				Wintering			
	Inc	Sta/ Flu	Dec	?	Inc	Sta/ Flu	Dec	?
Number	12	26	22	1	2	1	1	0
%	20%	43%	36%	2%	50%	25%	25%	0%

Key. Inc = increasing, Sta = stable, Flu = fluctuating, Dec = decreasing, ? = uncertain and unknown.
Source. Based on Member State Article 12 reported data as presented on the EEA national summary dashboard (2021).

For birds, the situation appears to be more favourable, as most Annex I species have stable/fluctuating, or even increasing populations (Table 13.2). Nevertheless, just over a third of Annex I taxa decreased over 2007–2018. Furthermore, as described earlier, some commoner species, such as of agricultural habitats, have been declining.

13.2.3 Drivers and Causes of Change

One clear habitat trend during the second half of the twentieth century, and a continuing pressure today, is the loss of meadows and pasture dependent on HNV farming systems (Kull *et al.*, 2003; Lotman, 2003, 2004). The main cause has been vegetation overgrowth (i.e. natural succession) due to a lack of mowing and grazing, but in some cases conversion, afforestation and building have played a part. To counter this, nature conservation actions in the form of support for appropriate grazing or mowing have been in place since the mid-1990s, and since EU accession have been incorporated into agri-environment schemes. Despite this, pressure to abandon meadows and pastures is still present (Koorberg *et al.*, 2012). Furthermore, as these habitats have experienced a major long-term decline, there is an extinction debt. Hence, in order to secure their long-term biodiversity, restoration of at least some of the overgrown sites is necessary (Linder, 2017).

In contrast, some HNV agricultural land and related habitats have been affected by agricultural intensification. In fact, whilst abandonment and intensification are locally opposite, they are two sides of one process at a larger scale, as intensification in one place usually leads to abandonment in another. Therefore, HNV agricultural land and related habitats remain under twin pressures. EU CAP support has also encouraged some intensification, as funding can support modernisation, such as renovating drainage, which in turn can lead to the use of more mineral fertilisers and pesticides. Animal husbandry has also become more concentrated and more animals are kept indoors year round.

Widespread intensification of forestry has also occurred, which has led to ever younger stands. This has put pressure on many forest species in spite of an overall increase in forest cover. Current cutting intensity is so high that the increase in forest cover has basically stopped whilst the trend for younger stands has remained. Renovation of ditches also puts pressure on forest biota (Rosenvald *et al.*, 2014, Remm *et al.*, 2018). Natural old-growth forests nowadays cover no more than 2% of forest area, with most known old stands now found in protected areas (Sims *et al.*, 2018). Thus the old-growth forest specialist species are under pressure, since protected area coverage is not sufficient to provide habitats on the necessary scale. Therefore, a need to increase protected area coverage is present, especially for meso-eutrophic and nemoral forests (Linder, 2017).

Wetlands have declined considerably over the twentieth century due to drainage and peat extraction (Paal and Leibak, 2011). By the end of 2019, 13 721 km^2 had been drained according to the Estonian Register of Land Improvement Systems. The loss of peat due to old but still functioning drainage is also ongoing. Hence there is an urgent need to restore wetlands (Linder, 2017).

Coastal habitats are somewhat threatened by building development, but the designation of protected areas and general protection of the coastal strip has reduced pressures significantly, hence most terrestrial coastal habitats have a favourable status (Linder, 2017). One of the remaining threats to some coastal biodiversity is the overgrowth of habitats with reed and other tall vegetation (Rannap *et al.*, 2007; Klasen *et al.*, 2017). This is partly a result of anthropogenic eutrophication from nutrient-enriched water, including from farmland. However, the main driver appears to be the cessation of grazing, as coastal habitats adjacent to well-managed pastures remain more open.

Although the HD habitats have been recently assessed as having a favourable status, they remain affected by some pressures. Eutrophication is currently the main pressure affecting marine habitats (Linder, 2017; Lotman *et al.*, 2018). Pollution by hazardous substances is also still a problem in spite of some progress. Similarly, whilst oil spills have decreased, they are still a cause of concern. Pollution by plastic litter is also present (Lotman *et al.*, 2018). The direct impacts from dredging and deposition of dredged material, as well as coastal developments, are somewhat less important.

Inland lakes are also impacted by eutrophication, while rivers are mostly impacted by dredging and dams, both largely legacies from the twentieth century, although some dredging continues (Klasen *et al.*, 2017). Dam removal, the introduction of fish-passes and other riverine habitat restoration activities have mitigated the pressures to some extent (Roasto and Tampere, 2020).

Of about a thousand alien species currently known in Estonia, 63 are considered invasive and 72 potentially invasive (Linder, 2017). Such invasive alien species (IAS) cause concern on land, within inland waters and in the sea. As described earlier, the American Mink is the main threat to the European Mink. In addition, it, and the Raccoon Dog (*Nyctereutes procyonoides*), predate the eggs and young of ground-nesting birds. Other examples include alien hogweeds (*Heracleum* spp.) taking over grasslands, invasive species of crayfish threatening the populations of European crayfish, and the Round Goby (*Neogobius melanostomus*) and invasive crabs depleting the bottom fauna of coastal seas.

The exploitation of some species also has some impacts, on the exploited species and on some others indirectly. While hunting does not seem to be on the scale that leads to population declines, some concern has been raised that bird hunting, especially by tourists, can cause excessive disturbance (Nellis *et al.*, 2018). Overfishing and eutrophication are a concern for some of the commercially exploited stocks, while the by-catch of seals is clearly under-reported and needs more attention (Vanhatalo *et al.*, 2014; Lotman *et al.*, 2018; Saks *et al.*, 2018).

13.3 Nature Conservation Governance, Actors, Legislation and Policies

13.3.1 Legislation and Policies

Table 13.3 provides an overview of the main developments in nature conservation in Estonia. While some consider that medieval laws limiting forest cutting were the start of nature conservation in Estonia (Kongo, 1998; Ratas, 1998), a more established view is that it began at the beginning of the twentieth century with the protection of bird-rich Vaika islets (Kumari, 1973). Estonian independence resulted in further protected areas, and the first nature conservation law was enacted in 1935. The Soviet occupation initially destroyed most aspects of nature conservation. However, with some loosening of the regime after Stalin's death, a Soviet nature conservation law for Estonia was adopted in 1957, which created several protected areas (Randla and Sillaots, 1998).

Restoration of Estonian independence enabled the development of modern environmental legislation. It also enabled the country to become a party to several important conventions, including the CBD, Ramsar, Bern and CITES (Randla and Sillaots, 1998; see Table 3.3 for dates). Of the legislative developments, the most important for nature conservation was initially the 1994 Protected Natural Objects Act[4] that established the legislative framework for protection of species and sites. This included categories of protected species and protected areas, types of zones present in the protected areas and main principles of management. The 1995 Law on the Protection of Marine and Freshwater Coasts, Shores and Banks was also important for nature conservation as it established strict building restrictions on the coastal zone. The Nature Conservation Act[5] of 2004 incorporated these into a single legal text. The Act includes a general legislation framework for protected areas, with individual protected areas having their own detailed protection rules (see Section 13.5). The Nature Conservation Act has since been amended many times, with the latest version in force since 1 June 2021.

In 1997, the first *Estonian Environmental Strategy* was approved by the Parliament. Following that the *Environmental Action Plan* (1998–2006) was developed, which had five biodiversity and landscapes related goals. In 1999, the Ministry of Environment produced the *Estonian Biodiversity Strategy and Action Plan*,[6] which set long-term aims and goals for protection of landscapes, habitats and species to meet the requirements of the CBD. A new *Environmental Action Plan* was then compiled for 2007–2013. In 2012, the government approved the *Nature Conservation Development* plan until 2020. The plan set out the main directions for awareness building and the protection of species and habitats, and attempts to integrate biodiversity conservation into the sustainable use of nature. The plan included 10 targets and indicators for habitats and species protection, of which half were met. For example, the goal to restore 10 000 ha of wetlands was more than met, with 11 000 ha, but only 37 500 ha of grassland were restored, compared to the goal of 45 000 ha. The goal to increase the favourable status of BD species from 65% to 80% was not met; instead it actually declined to 47%. The intention was that, in addition to the *Nature Conservation Development Plan*, action plans for different habitat types would be periodically developed and approved; so far, only semi-natural grassland habitats and mire habitats have such plans. Currently a new *Environmental Field Framework Development Plan* is being compiled and a new biodiversity related action plan will be developed under it.

Estonia joined the EU in 2004, transposing into national legislation and implementing the *aquis*, including the Nature Directives. Preparations for this had started well in advance and enabled an initial proposal for the Natura 2000 network to be presented to the European Commission prior to accession. The Nature Conservation Act of 2004 incorporated the most important provisions of the Nature Directives from the start, but minor adjustments have been made since then. For example, the Commission notified the Estonian government that HD Annex IV species general protection had to be secured, and these provisions were therefore added to the law.

Another law of more general relevance to nature conservation is the 1995 Sustainable Development Act,[7] which sets an overall goal of protecting biological diversity and using natural resources in a sustainable way. It also provides the legal basis for Estonia's Biosphere Reserve (see Section 13.5). While this law is quite generic and does not include strict rules and penalties, it signifies an overall direction that is favourable for nature conservation. More recently, important overall principles of environmental protection and nature conservation were set out in the 2011 General Part of the Environmental Code Act.[8] Important elements for nature conservation are the protection of biodiversity, the precautionary principle, public rights to participate in environmental decision making, and

4 Kaitstavate loodusobjektide seadus (RT I 1994, 46, 773).
5 Looduskaitseseadus (RT I 2004, 38, 258).

6 National Nature Conservation Plan: www.envir.ee/sites/default/files/lak_lop_0.pdf
7 Säästva arengu seadus (RT I 1995, 31, 384).
8 Keskkonnaseadustiku üldosa seadus (RT I, 28.02.2011, 1).

Table 13.3 Key nature conservation developments in Estonia.

1297	Danish King Erik Menved, being the authority of the region, prohibited forest cutting on three small islands near Tallinn
1853	Establishment of Estonian Naturalists' Society that promoted research and protection of nature, starting activities for sustainable fisheries
1910	Establishment of first Estonian nature conservation area, Vaika bird sanctuary. Local activist Artur Toom, local church and Riga Naturalists' Society signed a rental agreement and prohibited hunting on small islets. This event is considered to be the official start of Estonian nature conservation
1924	Establishment of Järvselja Primeval Forest Reserve that has since been untouched by forestry and is considered the best documented primeval forest site in Estonia
1933	First publication of Estonian nature magazine *Eesti Loodus* that has covered nature research and conservation topics regularly since then
1935	First Estonian Nature Protection Act compiled and enforced. Established State Nature Protection Council and governing body for state conservation areas
1936	First Environmental Inspector (Gustav Vilbaste) recruited. He compiled a comprehensive list of all state protected objects, implemented a sign system for protected objects and developed a network of volunteer inspectors
1941	By the time of World War II, there were 47 state restriction areas, 80 parks under protection, 202 protected ancient trees and 210 boulders under protection. There was an established list of rare plant and animal species and a conservation network had the cooperation of more than 500 volunteer conservation inspectors
1957	First Nature Conservation Act enforced under the Soviet regime. Establishment of four state nature protection areas (Matsalu, Nigula, Vaika and Viidumäe), 28 restriction areas and a list of conservation objects. Establishment of State Nature Conservation governing body
1970	Start of official international nature conservation networking under IUCN. Start of the 'war for mires'
1971	Establishment of the first National Park (Lahemaa)
1976	Establishment of first Ramsar Site (Matsalu)
1979	Compilation of the first Estonian Red Data Book
1981	As a result of the 'war of mires' 28 new mire and bog conservation areas were established
1987	So-called Phosphorite War created a movement that achieved its immediate goals, and also encouraged and strengthened the nationalist movement, which led to the restoration of Estonian independence in 1991
1989	Establishment of the Ministry of the Environment
1992	Independent Estonia took part in UNCED in Rio de Janeiro
1994	Protected Natural Objects Act
1995	Sustainable Development Act
1999	First *Estonian Biodiversity Strategy and Action Plan*
2000	Start of site inventories for the establishment of the Natura 2000 network
2004	EU accession, establishment of the Natura 2000 network, and new Nature Protection Act
2009	Establishment of the Environmental Board and State Forest Service under the Ministry of the Environment
2011	Environmental Code Act includes important principles relevant to nature conservation
	Finalised nationwide Estonian Mire Inventory
2012	*Nature Conservation Development* plan 2012–2020
2013	First *Estonian Seminatural Communities Action Plan* for 2014–2020 approved
2016	First *Estonian Mires Action Plan* for 2016–2023 approved
2019	The 2019 Water Act transposes into Estonian legislation the WFD and MSFD
2021	Second *Estonian Seminatural Communities Action Plan* for 2021–2027 approved

See Table 3.3 for signatory dates of international conventions.
Source. Chapter authors based on Ader and Tartes (2010).

regulations relating to rights of way. The 2019 Water Act[9] transposes into Estonian legislation the WFD and MSFD, thus affecting all aquatic biodiversity. Other pieces of environmental legislation that are important for nature conservation include the laws on hunting, fishing, forestry and environmental impact assessment, and these are further discussed in Section 13.4.

13.3.2 Governance

In the Parliament, nature conservation issues are overseen by the Environmental Committee, which leads on environmental legislation. In the Cabinet, the Minister of the Environment is responsible for environmental protection, including nature conservation, and heads the Ministry of the Environment, which includes the Nature Conservation Department as one of its subdivisions. Under the Ministry, the Environmental Board is the key institution in charge of environmental permits and overall protected area management (including management planning) and relevant law enforcement; the Environmental Agency manages monitoring data; and the State Forest Management Centre is tasked with many practical nature conservation related works as well as managing state-owned forests. Nature conservation is thus divided between several institutions.

13.3.3 Public Involvement and NGOs

Scientists, environmental NGOs and the wider public have played an important role in the development of nature conservation in Estonia over a long period. This started during the Soviet era, despite the political oppression, when some important non-governmental naturalists' movements were formed. Tartu Students' Nature Conservation Circle was founded in 1958 and the Estonian Society for Nature Conservation in 1966.

In response to massive drainage programmes, several leading scientists and nature conservationists published more than 20 articles in the leading nature journal *Eesti loodus* (Estonian Nature) in the early 1970s. These were unusually straightforward in their message for the time and initially provoked strong reactions from the authorities, but later the official position softened. The conflict was nicknamed '*soodesõda*', meaning 'war of mires'. As a result of this, many mires were saved from drainage and 28 mire conservation areas were officially designated.

In 1987, the Estonian public became aware of the Soviet leadership's plan to start mining phosphorite in Viru county. Public opposition to the plan led to the development of the Estonian Green Movement, which rapidly became a massive environmental NGO and soon a member of the international Friends of the Earth. In 1989, the wildlife protection cooperative Kotkas ('Eagle') was founded; it later evolved into a successful NGO devoted to the protection of eagles and the Black Stork and is now called the Eagle Club.

During the 1990s, many other new environmental NGOs were founded, including the Estonian Fund for Nature in 1991. Although it has cooperated with the WWF, it has remained an independent foundation. Its main initial activity was mapping natural habitats and proposing new conservation areas, the most important achievements from which have been the creation of Soomaa National Park and Alam-Pedja Nature Reserve.

In the beginning of the 2000s, the Estonian government supported the development of a new cruise harbour in Saaremaa close to an important bird sanctuary. Led by the Estonian Ornithological Society and Estonian Fund for Nature, NGOs formed a coalition that managed to oppose the plan. As a result of this success, the Estonian Council of Environmental NGOs was formed in 2002 to coordinate the conservation efforts of its 10 member organisations. In 2004, NGOs put together a shadow list of possible Natura 2000 sites and managed to get most of the valuable sites into the official proposal. Environmental NGOs have also been involved in designing conservation policies, keeping an eye on possible harmful developments and keeping the wider public interested and involved in environmental decision making.

The last five years have seen a rise in public awareness about forest and climate issues. While existing environmental NGOs have always been active on these issues, the broadening of public interest has also led to the birth of new environmental movements. These include both national NGOs striving to protect forests, and the involvement of the Estonian youth in international movements, such as Fridays for Future and Extinction Rebellion.

13.4 Species Protection and Conservation Measures

13.4.1 Species Protection Legislation

There are three categories of protected species according to the nature conservation law. The first category has the highest level of protection, and some species are listed in the Nature Conservation Act, including Siberian Flying Squirrel, White-tailed Sea-eagle, Short-toed Snake-eagle (*Circaetus gallicus*), Osprey (*Pandion haliaetus*), Golden Eagle, Greater Spotted Eagle (*Clanga clanga*), Lesser Spotted Eagle (*Clanga pomarina*) and Black Stork. The Act also delegates the listing of the protected species in the first and second categories to the government, and the list of protected species in the third category to the Minister of the Environment.

Although the Nature Directives have been transposed into the law, the protection categories do not directly correspond to BD Annex I for birds or HD Annexes II and IV for other species. Neither do the protected species' categories correspond directly to any other international lists, such as the appendices of the Bern Convention or any of the Red Data

[9] Veeseadus (RT I, 22.02.2019, 1).

lists, either national or international. However, whilst they were not directly copied, all such lists were taken into account when the species were assigned to the national protected categories. The national authorities thus consider that the protection requirements of the Nature Directives are adequately transposed into the domestic law. Neither the European Commission nor any other international body has seriously challenged this. However, minor adjustments to the legislation have been made in response to the requests by the Commission, for example relating to the protection of habitats of the animals listed in HD Annex IV.

There is an overall prohibition on selling and buying protected species. In addition, all the known habitats of the first category species have to be protected; for the second category the share of protected habitats must be at least 50%; and for the third category at least 10%. In practice, 'all known habitats' of the first category species means that their known breeding sites are protected. The protection of the habitats is secured either by the main categories of protected areas, or so-called Species Protection Sites, which are in effect small reserves designated by the Minister of the Environment for one species. There are currently 1 811 Species Protection Sites. The Nature Conservation Act also stipulates the minimum size for the protection of the above-listed species' habitats measured as the radius from the nesting tree (ranging from 25 m for the Siberian Flying Squirrel to 500 m for the Golden Eagle). The law also protects the spawning sites of salmonids and the hibernation sites of bears. Deliberate destruction of bird nests or eggs, as well as deliberate disturbance of nesting birds, is prohibited.

The law also regulates possible derogations from the overall prohibition on killing protected species, which for the first category is limited to when there is a direct threat to human lives. For the second and third categories, derogations may be considered for significant agricultural damage and scientific needs. The permits for derogations are issued by the Environmental Board. For the third category of protection, the Minister of the Environment has the option to list species that can be hunted outside the sites designated for their protection without a special permit (but with a valid hunting permit). Currently the list includes the Barnacle Goose (*Branta leucopsis*), the Hazel Grouse (*Bonasa bonasia*) and the Grey Seal. Damage by protected species to crops or farm animals (including fish-farms) is compensated according to the law. So far, both the European Commission and the Bern Convention secretariat have considered the Estonian system of derogations to be satisfactory.

The protection and use of wild species are also influenced by hunting, fisheries and forestry legislation. The Hunting Act[10] is particularly important as it not only sets the rules for hunting species (e.g. preventing their hunting during sensitive periods), but also because large carnivores are not listed as protected species in Estonia due to their overall favourable conservation status. Therefore, their protection is largely dependent on quotas and permits allocated in accordance with the hunting legislation. This system seems to largely work, as the Wolf and Brown Bear continue to have healthy populations. The same has been the case with the Eurasian Lynx, but due to a temporary decrease in their main prey, Roe Deer (*Capreolus capreolus*), their status over 2013–2018 was assessed as bad. Quotas are set based on the species action plans (see next paragraph) and yearly ecological monitoring. The quotas are reduced for threatened populations, sometimes to zero, as was the case with the Lynx when its prey numbers dropped. While there are calls for bigger quotas and those who ask for stricter protection, there have been no legal challenges to the system.

13.4.2 Species Action Plans and Management Plans

Species conservation action plans are created for all species of the first protection category and for the other protected species if there is a need for active measures. The plans are compiled by Environmental Boards, mainly with species experts; usually stakeholders are not very well consulted. There are over 60 plans at present, most of them for a single species, including a hermit beetle (*Osmoderma eremita*), Large Blue butterfly, Sand Lizard, Natterjack Toad, Osprey, Golden Eagle, European Mink, Siberian Flying Squirrel and Lady's Slipper Orchid (*Cypripedium calceolus*). Some plans cover groups of species with similar protection needs, such as large carnivores, bats and plants of open dry habitats. Many of the plans have been compiled with funding by the EU (see Section 13.7). Measures for improving the conditions for the protected species must be based on the relevant conservation plan or site management plan for the protected area where the species is present. Similarly structured plans are also adopted for the management of IAS, such as hogweeds, and native species causing economic losses, such as the Great Cormorant (*Phalacrocorax carbo*).

Plans are usually time-limited and subject to renewal. So far, over 15 species already have a second or even third plan. Writing a new plan for a species or species group implies a review of the old plan, and some plans have also been subject to a mid-term review. Reviews include taking stock of both the implementation of the planned conservation measures and their overall success in securing favourable conservation status. However, since the format of the plans has changed from time to time, it is hard to provide summary assessments, and indeed an overall assessment has so far not been carried out by the authorities. Their results vary, for example from the largely successful protection of large carnivores, to the failure to reverse the decline in the Greater Spotted Eagle.

13.5 Protected Areas and Networks

13.5.1 Protected Area Types and Their Coverage

Table 13.4 summarises the different types of protected areas designated under international conventions, the Nature

[10] Jahiseadus (RT I, 16.05.2013, 2).

Directives and Estonian domestic legislation, as well as their number and coverage. Those of most importance for nature conservation are described in the following paragraphs.

The international sites include the Biosphere Reserve designated under the UNESCO Man and Biosphere programme, wetlands of international importance designated under the Ramsar Convention, and HELCOM Marine Protected Areas (MPAs) – Baltic Sea coastal and marine protected areas registered under the Baltic Marine Environment Protection Commission (Helsinki Commission, HELCOM). The Biosphere Reserve, which has a total area of 15 183 km^2, is located in western Estonia, covering Saaremaa, Hiiumaa and Vormsi islands and numerous smaller islands. Its main purpose is to develop the region as a model of sustainable development. This includes monitoring, research and education focussing on protection and sustainable use of natural resources. There is a significant overlap between different designations. For example, Matsalu is also a Ramsar Site and part of the HELCOM MPA network. Also, several Ramsar Sites and HELCOM MPAs are wholly or partly included in the Biosphere Reserve. It is also noteworthy that Matsalu National Park holds a European Diploma for Protected Areas, granted by the Council of Europe, which recognises natural and semi-natural areas and landscapes of European importance that are managed in an exemplary way.

While there is a broad consistency between Estonian protected area categories and those defined by IUCN, some reflect domestic approaches that are rather different. Estonian protected areas are zoned, and this also affects their protection level and IUCN type (as further discussed later). National Parks are in general considered to match the IUCN definition but (as in several other European countries) they do contain, in addition to public lands and 'wild' areas, private land and seminatural habitats in need of active management, and even agricultural fields and built-up areas. In Estonian National Parks there is a legal obligation to protect, in addition to natural features, their cultural heritage. All six National Parks are named in the Nature Conservation Act and are relatively large (the oldest and largest being Lahemaa National Park at 74 784 ha).

Nature Reserves share with National Parks a diversity of land ownership and variety of nature values, which include both natural and semi-natural habitats. They thus belong to IUCN categories I or IV (often a combination of both). The size of the reserves varies, some like Alam-Pedja Nature Reserve (34 396 ha) or Peipsiveere Nature Reserve (34 610 ha) being as large as the National Parks, but the smallest, Sopimetsa, just 3.5 ha (Roasto and Tampere, 2020).

Protected Landscapes may also contain both private and public lands. Some sites are quite typical in the sense that their picturesque landscape is the prime reason for their protection, although they are usually also of biodiversity value. In some cases, however, the protection of habitats and species is the main reason for their designation, similarly to Nature Reserves.

The Special Conservation Area category was specifically introduced in order to secure protection of Natura 2000 sites (both SCIs/SACs and SPAs) not covered by any other existing national protection category. In the Nature Conservation Act these are not formally regarded as protected areas, as their management is goal-based and mainly based on HD Article 6, rather than the detailed rules-based protection and zoning typical of Estonian protected sites. In effect, however, they are protected areas, because they give a high level of protection to the targeted habitats and species as required for Natura 2000 sites by the Nature Directives. Their appropriate IUCN category depends on the habitats they are meant to protect, which in the case of semi-natural habitats means they clearly belong to category IV. Their land ownership is mixed.

Species Protection Sites are also not formally considered to be protected areas, but are seen as part of species conservation in the law. As discussed earlier, they are usually relatively small and dedicated to important habitats (for example nesting sites) of one species or a group of closely related species with similar ecology, like haul-out sites of seals or bat roosts.

The Nature Conservation Act also permits municipalities to gazette their own municipal protected areas. There are 23 objects or sites protected by a local municipality.

About 60 protected sites from Soviet times still have not been allocated to any of the current legal categories.

According to the Estonian Environmental Agency,[11] in 2020 the combined area of the sites listed in Table 13.4 was 8 452 km^2 on land (18.6% coverage) and 6 845 km^2 at sea (18.7% of the marine waters). At the end of 2020, the Natura 2000 component of the protected area network covered 17.9% of the Estonian land area and 27% of the Territorial Sea. However, with only small parts of the sites beyond the Territorial Sea, the overall Natura 2000 coverage of Estonia's marine waters was about 18.4%. Natura 2000 overlaps with National Parks, Nature Reserves, Protected Landscapes, Special Conservation Areas and Species Protection Sites, and covers 92% of their overall area (Roasto and Tampere, 2020).

13.5.2 The Coherence of Protected Areas and Ecological Networks

The protected area coverage is about 20% of all water courses (rivers and creeks), 47% of lake area, 75% of mires, 63% of meadows, and 20% of forests, with 13% considered strictly protected (Roasto and Tampere, 2020). The current area of strictly protected forest is in response to an analysis that showed that there is a need to strictly protect at least 10% of all Estonian forest types (Lõhmus *et al.*, 2004). However, the typological representation and territorial cover of the protected forest network is still insufficient to balance the impacts of current intensive forestry (Lõhmus, 2016). Productive forests on fertile land are under-represented, with most of the protected forests being on very wet or unproductive sites.

[11] https://kaur.maps.arcgis.com/apps/MapJournal/index.html?appid=2c6a3fc7ed4641f4b69d20c670732077

Table 13.4 The number and extent of international designated sites and protected areas in Estonia.

	IUCN category	Number	Area (km²)
Designated under global and non-EU international legislation			
Ramsar Sites	I, II, IV, V	17	3 064
Biosphere Reserves (exc. buffer and transition zones)	VI	1	360
HELCOM Marine Protected Areas	I, II, IV, V, VI	7	7 237
Designated under the Nature Directives			
SCIs (terrestrial + marine)	I, II, IV, V, VI	541	7 806 + 3 883
SPAs (terrestrial + marine)	I, II, IV, V; VI	66	6 203 + 6 480
Natura 2000 terrestrial area (incl. inland waters)	I, II, IV, V, VI		8 106
Natura 2000 marine area	I, II, IV, V, VI		6 754
Designated under domestic legislation			
National Parks (rahvuspark)	II	6	2 440
Nature Reserves (looduskaitseala)	I and IV	231	3 954
Protected Landscapes (maastikukaitseala)	V	156	1 871
Special Conservation Areas (hoiuala)	IV and VI	319	7 284
Local municipality protected area (kohalik kaitseala)		22	32
Protected parks and arboretums (kaitstavad pargid ja arboreetumid)	V	470	41
Protected single objects (kaitstav üksikobjekt)	III	1 076	11
Protected areas with old legislative regime (uuendamata eeskirjaga kaitseala)		38	30

Sources. See Section 1.3. Based on data from: Ramsar Convention and UNESCO websites (October 2021); EEA Natura 2000 Barometer, as at end 2020; Environmental Agency website, as at end 2020, Environmental Board website,[12] as at end 2020.

Several important marine sites for migrating birds are still not protected; there are also no protected areas in the EEZ, even though there are indications of sites with important benthic habitats. The programme of measures under the MSFD therefore foresees the designation of additional protected sites.

In addition to the nature conservation legislation, planning law has a role in the conservation of ecological networks. Among the themes covered in the spatial plans are the so-called green networks. The networks consist of core areas and corridors that are included in General Plans at a municipal level and shown on relevant maps, and should therefore influence planning decisions. There is some mismatch in the system because nature conservation is mostly administered by the central government while green networks are dealt with by municipalities as part of their planning responsibilities; this results in uneven levels of implementation. Sepp *et al.* (2018) have demonstrated that the green network serves forest species well, but has little added value for the meadow biota.

[12] www.kaitsealad.ee

13.5.3 Approaches to Protection and Levels of Protection

As discussed in Section 13.5.1, the level of protection for habitats and species varies according to the type of protected area. However, of greater practical importance is the level of protection that is given in relation to the categories of protected area zones, which can also be assigned to IUCN protected area categories.

The strict nature reserve zone is meant to protect nature so that it is only affected by natural processes (matching IUCN category Ia), and all human activities, including visitor access, are prohibited, with the exception being made for emergency services, law enforcement and limited scientific work. The land is state-owned.

Special management zones can be also dedicated to 'leaving it to nature' but at least some visitor access is allowed, even though there might be temporary restrictions. The zone can be viewed as similar to IUCN category Ib. However, this kind of zone can also be used for protection of semi-natural habitats, and then active protection measures are mandatory, in which case this corresponds to IUCN category IV. The land may be state-owned or private.

The limited use zone is the part of a protected area where people can actually live; it may include buildings and roads, for example, but also agricultural land, including in some cases semi-natural habitats, and woods with less strict protection. It can be seen as similar to IUCN categories IV, V or VI. The land might be state-owned or private.

National Parks and Nature Reserves have all three types of zones, while Protected Landscapes and Species Protection Sites lack the strict nature reserve zone, and municipal protected areas have only limited use zones. There is no zoning for Special Conservation Areas.

The zoning for each protected area is defined by its protection rules, a legal document adopted by the government based on the Nature Conservation Act.

13.5.4 Approaches to Managing Protected Areas

Protected areas have protection rules that define the zoning and apply restrictions to specific areas from the menu set out in the Nature Conservation Act. Special Conservation Areas do not have such rules and their protection stems directly from the Act. Both can, however, have a management plan, which is recommended but not strictly mandatory. Features protected in the areas, and the relevant conservation goals, are listed in both the rules and the plans; in the case of Natura 2000 sites, these include the habitats and species present from the relevant BHD annexes. The management plan is not a legal act but a planning document that foresees the active measures needed to secure protection. These include monitoring, marking the boundaries, building trails, controlling invasive species, and restoration and management of habitats. Preliminary multi-annual budgets are also included in the management plans. The management plans are supposed to be taken into account when visitor management or forestry planning is carried out, but there is no systematic check for consistency. Although they are not mandatory, about 200 protected areas have an official management plan.[13]

Management responsibilities are divided between different bodies under the Ministry of the Environment. The Environmental Board is the authority in charge of management planning and permits, the Environmental Agency is responsible for monitoring, and the Environmental Inspectorate carries out enforcement. The State Forest Management Centre has many practical responsibilities, especially those related to state-owned property, including visitor management, marking boundaries, trail management and habitat restoration activities. Due to the fragmented nature of protected area management, no consistent review of the management plans' implementation has been carried out.

CAP agri-environment payments are used to support grazing and mowing of semi-natural habitats and have contributed to doubling the managed area within protected areas from 152 km^2 to 322 km^2 between 2007 and 2018. Payments are made by the paying agency under the Ministry of Rural Affairs, whilst most of the practical work with the farmers, including on-the-spot checks, is performed by the Environmental Board, but so far guidance for the farmers has been lacking. Restoration of such habitats (mostly clearing vegetation) is paid for by the Environmental Board from the state budget, or by the Environmental Investment Centre from the EU Cohesion Fund and in some cases LIFE projects, since restoration has not been included in the Rural Development Programme (RDP). There has been a broad consistency between the Prioritised Action Frameworks (PAFs) and RDPs, and the site management plans, although details might differ because of differences in the plans' structures and insufficient coordination between the authorities.

13.6 General Nature Conservation Measures

13.6.1 Development Planning and Impact Assessment

Spatial planning is regulated by the Planning Act[14] and must take into account protected areas, requirements for the protection of the coastal strip, and 'green networks' (i.e. ecological networks). It is also influenced by more than 40 national strategic documents of varying legal status, some being very general, like the Estonian sustainable development strategy, whilst others are more specific, such as the national tourism development plan. Such strategies are based on relevant laws, which in the examples above relate to the Sustainable Development Act and Tourism Act. According to the Environmental Impact Assessment and Environmental Management System Act,[15] which regulates both EIA and SEA, as a rule such strategies and plans are subject to SEA if significant environmental impacts are expected. However, the SEAs that have been carried out in response to this obligation vary in quality, and integration of environmental concerns into various sectoral strategies has not been consistent.

13.6.2 Habitat Management and Restoration

Forestry and Biodiversity Conservation

The Forest Act[16] regulates forestry and impacts on biodiversity, for example by setting the minimum tree age and maximum size that can be felled. Both of these regulations have unfortunately been loosened many times over recent years.

The Forest Act also gives a framework for voluntary protection of Woodland Key Habitats, which are relatively small sites (average 2.4 ha) important for specialist, threatened or rare species. At the beginning of 2020, 10 829 Woodland Key Habitats covering 265 km^2 were protected, 80% of which were

[13] A list of current management plans is published on the Environmental Board website: https://keskkonnaamet.ee/keskkonnateadlikkus-avalikustamised/kaitsekorralduskavade-kinnitamise-teated

[14] Planeerimisseadus (RT I, 26.02.2015, 3).

[15] Keskkonnamõju hindamise ja keskkonnajuhtimissüsteemi seadus (RT I 2005, 15, 87).

[16] Metsaseadus (RT I 2006, 30, 232).

in state-owned forest. In private forests compensation is paid to the owners who voluntarily decide not to cut Woodland Key Habitats (Roasto and Tampere, 2020). Assessments of the effectiveness of the key habitats measure, for example as stepping stones between the protected areas, still need to be carried out.

The preparation of a National Forestry Development Plan is required every 10 years under the Forest Act. Currently the plan for 2020–2030 is still under preparation as significant controversies between industry interests, which tend to enjoy political and administrative support, and sustainable forestry proponents, who feel more and more alienated by the official process, are still unresolved after years of discussions in various working groups (Lõhmus, 2020).

In June 2021 the European Commission published in its infringement package a decision related to Estonian forest habitats in Natura 2000 areas. In this, the Commission called for Estonia to bring its national legislation in line with the Habitats Directive and SEA Directive, in particular with regard to logging activities in Natura 2000 sites. It remains to be seen if this will change logging practices in the sites and improve forest habitat protection.

Another initiative to improve the condition of forests is the current preparation of a Wet Forest Habitat Action Plan and a Dry Forest Habitat Action Plan, under the LIFE IP project ForEst&FarmLand. The aim is to restore 3 500 ha of wet forest habitats and 500 ha of dry forest habitats over the coming years.

Grassland Restoration and Management

Estonian nature conservation has put a lot of effort into the protection, restoration and management of semi-natural habitats. As the vast majority have been affected over the last century by agricultural intensification or abandonment, this has mainly involved grazing and mowing, and clearing excessive bushes and reed, as well as protecting them from ploughing, fertilisation or building developments. After a nationally funded pilot of semi-natural grassland management in Matsalu National Park, which started in 1996, all semi-natural grassland in protected areas became eligible for state-supported conservation management in 2001. Since 2007, the scheme has become part of the RDP and is funded by the European Agricultural Fund for Rural Development (EAFRD), as further discussed in Section 13.7.

In 2006, scientists from the University of Tartu published an analysis of extinction debt (i.e. species expected to go extinct due to past habitat loss) amongst alvar-dependent plant species (Helm *et al.*, 2006). This found that around 40% of current plant species of Estonian alvars would go extinct in Estonia unless required restoration and management took place. In response, an action plan for Estonian semi-natural grasslands was approved by the Minister of the Environment in 2013 (Keskkonnaamet, 2013). The plan stated that there is a need to manage 60 000 ha of these grasslands, and also set an interim target to manage 45 000 ha by 2020, and sub-targets for all the grassland habitat types (Keskkonnaamet, 2013). By 2020, 37 770 ha of these grasslands were restored and managed, with 880 farmers engaged in the management of the areas (Keskkonnaamet, 2021). Most of the habitat management and restoration is carried out inside protected areas. As part of the plan, restoration of alvar habitats in the 2014–2019 LIFE project (LIFE to Alvars) resulted in 2 500 ha of restored and managed alvars, mainly by sheep grazing.

Nowadays, in most of the managed semi-natural grasslands, biodiversity decline has slowed down, and even stopped or reversed in the best cases. An encouraging development is that the economic viability of some HNV farming in larger areas has improved. This has, for example, been seen with beef cattle grazing on coastal meadows, sheep grazing on alvars and, in one case, the use of hay from alluvial meadows as biomass for district heating. Nevertheless, the management of most semi-natural grasslands is not sufficiently good for biodiversity, and their long-term viability is not assured. In 2020, an analysis of the ecological coherence of semi-natural meadows in Estonia was conducted (Helm and Toussaint, 2020). The calculations in the analysis showed that all meadow types need significant restoration and improved connectivity to ensure the survival of all their threatened species. These challenges have been confronted as part of the 2021–2027 action plan compiled by the Environmental Board, and the CAP Strategic Plan.

River Restoration

There are about a thousand dams in Estonia, and over 300 have been identified as needing action in order to better enable fish migration (Viirma, 2013). Opening up of migration routes by either dam removal or building fish passes is required under the Water Act, but this has not yet been completed. Dams are still considered to be the main reason why rivers fail to reach good environmental status in accordance with the WFD, but pollution, drainage and dredging also play a role (Normak, 2019).

Two LIFE projects (HAPPYFISH and HAPPYRIVER) by NGO Wildlife Estonia, which carried out river restoration in the Alam-Pedja Natura 2000 site, gained best LIFE project of the year awards and national recognition. The projects used excavators to restore connectivity of 10 oxbow lakes and the Laeva stream with the Emajõgi river, thus enhancing their quality as fish habitats, especially for the Asp (*Aspius aspius*) as this was the focal species (Roasto and Tampere, 2020).

Currently two large river restoration projects are being carried out. Restoration of Pärnu river habitats is financed by the EU Cohesion Fund. The most important part of the project, removal of the 151 m wide and 4.5 m high Sindi dam, has already been completed. Other activities include restoration of spawning sites and further removal of obstacles to migration. The LIFE IP CleanEST project is opening up migration routes, restoring habitats and mitigating historic chemical pollution in north-eastern Estonian rivers.

Mire Restoration

Before major drainage for agriculture, forestry and peat mining, mires covered over a million hectares, which is slightly more than one-fifth of the national territory (Paal and Leibak, 2011). Many of the peat extraction sites have been abandoned without restoration and the use of some drained peatlands has not proved to be economically viable. In north-eastern Estonia, some 500 ha of mires have been destroyed directly and another 1 500 ha indirectly by excavation of oil shale in open-cast mines (Ilomets, 2015). This has impacts on biodiversity, and these peatlands emit greenhouse gasses as long as they remain drained. Wetland conservation and restoration in Estonia has been substantially led by academic and NGO initiatives, backed by governmental bodies, since the 1990s. In 1997, the Estonian Wetlands Conservation and Management Strategy (Paal *et al.*, 1998) was compiled with financial support from the World Bank's Norwegian Trust Fund. This provided the basis for an Estonian Mires Inventory, which covered the whole country in 2009–2012 and found that 240 000 ha of mire habitats remained, covering 5.3% of the land (Paal and Leibak, 2013). The Estonian Fund for Nature and Estonian Wetland Association were the first to develop detailed plans for restoration activities in a number of pilot areas since 2007. In 2010, the first pilot mire restoration was carried out in Soomaa National Park.

To tackle mire degradation issues, the Estonian Nature Conservation Action Plan (Keskkonnaministeerium, 2012) set a target to restore 10 000 ha of mire habitats by 2020, and included a list of 89 sites in need of restoration. In 2015, the Action Plan for Protected Bogs (Keskkonnaministeerium, 2015) was officially approved. Since then the State Forest Management Centre has used the EU Cohesion Fund, as well as the Estonian government budget, to reach a target of restoring 3 800 ha of mire habitats and 2 000 ha of abandoned peat extraction sites by 2023. In addition, project LIFE Mires Estonia, led by the Estonian Fund for Nature, helped to restore a natural water regime across 7 900 ha of mire habitats in 2015–2021 (Estonian Fund for Nature, 2021).

13.7 Nature Conservation Costs, Benefits and Funding Sources

13.7.1 Nature Conservation Costs

Estimates of the overall costs of nature conservation in Estonia have not been made. However, the official estimate of the costs of Natura 2000 for 2020–2027 is €304 million, or €43.5 million per year, of which 60% is for one-off actions averaged over the period, according to the PAF (Keskkonnaministeerium, 2019). The main costs (c. 78%) are for maintenance and restoration actions within Natura 2000 sites, especially for various degraded habitats: grasslands, woodlands and forests, wetlands, rivers and lakes. The estimate can be considered to be moderately ambitious as it takes into account possible sources as well as the current economic situation. To reach all biodiversity targets, the actual costs would probably be considerably higher. It is also noteworthy that the PAF only includes €0.9 million per year (c. 2%) for supporting green infrastructure outside Natura 2000 sites, and €2.7 million per year (c. 6%) for additional species-specific measures.

13.7.2 The Economic Benefits of Nature Conservation

Several projects have been carried out to map ecosystem services, but few have included quantitative monetary assessments. The current economic benefits of nature conservation in Estonia come from, amongst others, commercially exploited fish and other wildlife, and the provision of their habitats, protecting the resource base for nature-based tourism and recreation, and supporting the pollination of many crops. The Estonian State Forest Service has found that there are more than 2.6 million visits per year to official state-owned nature sites (RMK, 2019). It has also been estimated that these visitors generate annual income to other service providers of €10 million, and the total annual economic value of Estonian nature tourism as a market service is €20 million (Ehrlich, 2013).

There have been few studies that have attempted to estimate the non-market economic value of conserved nature. However, the Jägala waterfall was estimated to be worth €10 million annually, which is 35 times more than the potential value of electricity production if a plant was constructed at the site (Ehrlich and Reimann, 2011). The potential prices of timber production versus carbon sequestration in Järvselja nature protection area were estimated by Adermann *et al.* (2015). They found that the theoretical long-term average value of timber was €325/ha and the value of carbon sequestration was €1 473/ha of forest land, based on a carbon price of €4.62/ton of tradable carbon equivalent in autumn 2013 (Adermann *et al.*, 2015).

The Estonian Environment Agency has recently finalised countrywide mapping and assessment of four ecosystems (forests, grasslands, mires and agro-ecosystems) and 75 of their services. The assessment was funded by the EU Cohesion Fund and Estonian Environmental Investments Centre. The result is a working tool for planners consisting of a base map and layers of ecosystem services for the whole of Estonia that can be used to calculate the economic value of ecosystems, and to look for hot and cold spots, and valuable landscapes, etc.[17] (Helm *et al.*, 2021).

There is significant potential for economic benefits from increasing the carbon sequestration and storage of Estonian peatlands, forests and grasslands, which the Estonian government is aiming to realise through carbon trading.

[17] https://loodusveeb.ee/et/themes/uleriigiline-hindamine-ja-kaardistamine/okosusteemide-seisundi-ja-huvede-pakkumise

13.7.3 Funding Sources for Nature Conservation

As nature conservation is within the remit of the Ministry of the Environment, national funds and EU structural and cohesion funds are allocated by it, or through the Environmental Investment Centre. The Ministry of Rural Affairs allocates CAP funding, including the RDP special agri-environment–climate payments for the management of semi-natural meadows and pastures, and also Natura 2000 payments.

As indicated in Table 13.5, the recent principal sources of funds for practical conservation measures in Estonia have been from EU funds (with national co-financing). In particular, EAFRD RDP measures, and especially the agri-environment climate measure, have provided support for management of semi-natural grasslands in Natura 2000 sites. The Natura 2000 measure has also provided compensation for income foregone for conservation-related restrictions on private forest and agricultural land. The ERDF has supported the restoration of protected habitats and investments in semi-natural grassland management, protected areas and visitor infrastructure management, *ex situ* species conservation, and restoring landscape values, as well as inventories and the estimation of ecosystem services.

Although at a much smaller scale, many important habitat conservation and restoration actions, and some species-focussed actions, have been supported through the EU LIFE Programme, some of which have been mentioned previously. In total, since 2001, when the first projects helped the country prepare for EU accession, 20 nature and biodiversity LIFE projects have been carried out in Estonia, with a total value of €20.5 million (EU plus national co-financing). More recently, Estonia has been awarded a large Integrated LIFE Project (LIFE-IP ForEst&FarmLand), the main objective of which is to implement the most critical part of the Estonian PAF for Natura 2000 relating to forests and agricultural land.

Compared to most other countries covered in this book, Estonia has recently allocated a relatively high proportion of funding to nature conservation, mostly through EU instruments and especially the RDP agri-environment–climate measure. On the face of it, the total averaged annual funding in Table 13.5 of €78 million per year appears to be more than sufficient to meet the €43 million per year required for Natura 2000 for 2020–2027 as identified in the PAF. However, the PAF indicates that only about €5.3 million per year of the agri-environment–climate funding is relevant to Natura 2000, and none of the organic farming allocation. Adjusting for this gives a total annual funding level of €39 million, which does leave a funding gap for achieving the Natura 2000 objectives, if the same level of recent funding is taken forward to 2027.

As just over 80% of Estonia's land and sea is outside the Natura 2000 network, available funding seems to be far from sufficient to meet Estonia's wider overall nature conservation and restoration needs. Additional funding requirements include those for forest biodiversity conservation and the restoration of degraded wetlands and streams. Furthermore, whilst the agri-environment–climate funding for the management of semi-natural grasslands, within and outside the Natura 2000 network, is relatively high, it could perform better if it was not competing with other CAP payments that are supporting the intensification of agriculture.

13.8 Surveillance, Monitoring, Reporting and Applied Research

Environmental monitoring and reporting (e.g. to the EEA), including that for nature conservation purposes, is managed by the Estonian Environmental Agency, with most of the fieldwork carried out by subcontracts to universities, specialist NGOs or independent experts. The Agency collects and manages information regarding nature protection in the Estonian Nature Information System (EELIS).[18] EELIS is used for keeping track of existing as well as planned natural features, alien species, hunting areas and the related legal acts and documents. In cooperation with the Estonian Naturalists' Society, the Estonian Society of Lepidopterologists and the University of Tartu, the Agency also hosts the Nature Observations Database (LVA), which is a database for volunteer and freelance records of species sightings. Every fourth year the agency prepares a regular overview of nature conservation in Estonia on the basis of the information collected and managed. The latest of these (Roasto and Tampere, 2020) includes an overview of protected areas and their management, status of species and habitats, ecosystem services and green networks, etc.

Monitoring related to nature conservation is included in the environmental monitoring programme. Most relevant is the sub-programme on 'living nature', which includes monitoring of species, habitats and landscapes. Species monitoring includes thematic areas relating to plants, fungi and lichens, invertebrates, fish, amphibians, reptiles, birds and mammals. The focus is on protected species, but some other species are covered, such as small mammals that are important prey for protected species, and game species. Habitat monitoring includes forest, grassland, mire, coastal and marine habitats. Monitoring of habitats and species is carried out both inside and outside of protected areas and forms the basis of HD Article 17 reporting. Landscape monitoring is meant to cover both man-made and more natural landscapes, and is partly now under review with the role of remote sensing increasing. Other sub-programmes relevant to nature conservation are forest monitoring, soil monitoring, marine monitoring and inland water monitoring, the latter two including biodiversity and forming the basis for MSFD and WFD assessments. The Estonian

[18] www.eelis.ee

Table 13.5 Funding sources for nature conservation in Estonia.

Source and period	Measure/category	Average annual allocation (€ m)		Comments
		EU + national	EU	
EU funding instruments				
EAFRD Rural Development Programme measures – allocations to Priority 4 (restoring, preserving and enhancing agriculture and forest ecosystems) 2014–2020	M10 Agri-environment–climate	31.5	25.1	Inc. for semi-natural grassland in N2k
	M11 Organic farming	13.5	10.3	Not for N2k etc.
	M12 Natura and WFD	4.8	3.6	For private forest and agriculture land in N2k
	M1 Knowledge transfer, etc.	0.9	0.8	Not for N2k, etc.
	M16 Cooperation	0.4	0.4	Not for N2k, etc.
	M4 Investments in physical assets	0.4	0.3	Not for N2k, etc.
	M8 Investments in forests	0.2	0.1	Not for N2k, etc.
	M2 Advisory services, etc.	0.2	0.1	Not for N2k, etc.
RDP Priority 4 subtotal		*51.9*	*40.8*	
ERDF 2014–2020	85 Protection and enhancement of biodiversity, etc.	8.9	7.6	For N2k, etc. See text
EMFF 2014–2020	Improvement of fishing gear	0.7	0.5	For seal-proof fishing gear and seal pingers
	Marine biodiversity and ecosystem services	0.4	0.3	For improvement of fish spawning sites and preceding studies
LIFE Programme	Nature and biodiversity (2001–2020)	1.6	1.0	
	Integrated project (2020–2029)	2.0	1.2	LIFE-IP ForEst&FarmLand
Interreg		0.3	?	
EU funding instruments subtotal		*65.7*	*51.4*	
National funds within 2014–2020		**Additional national***		
State budget	Conservation-related costs	7.4		E.g. salaries of officials
Environmental Investment Centre	Nature conservation sub-programme – small projects	4.7		E.g. monitoring and habitat restoration
State Forest Management Centre	Habitat restoration on state lands, information signs, restocking fish, care for veteran trees	0.7		
Additional national subtotal		*12.8*		
National total including co-financing of EU instruments		*27.1*		
Total		**78.5**		

Notes. * Excludes co-financing of EU funds unless indicated. 'For Nk2, etc.' indicates funding that is relevant to Natura 2000, related green infrastructure and BHD species (based on the PAF). Average annualised figures for Interreg and national funding based on allocations over 2014–2017. The subtotals and totals are only approximate indications of the average annual levels of funding as they are based on annualised data from differing periods.
Sources. Based on data from: RDP data from European Commission ESIF funds database (23 October 2020); RDP allocations to Natura 2000, ERDF, EMFF and national funding from 2021 to 2027 PAF (Keskkonnaministeerium, 2019); European Commission LIFE Estonia factsheet (updated 5 February 2021). See Section 1.3 for further details.

Environmental Agency is also in charge of collecting various forestry-related data, for example on felling rates.

The main constraints in interpreting monitoring data are the very low number of monitoring stations and the effects of frequent changes in monitoring methodologies. Thus, there is uncertainty over the relevance of monitoring data to the assessment of the success of protected area management.

13.9 Future Developments

The compilation of an updated *Environmental Field Framework Development Plan* has been initiated by the Ministry of Environment, the horizontal aims of which relate to climate policy, and the biodiversity strategy and circular economy. The framework is planned to consist of different action plans, and a *Nature Conservation Action Plan* is under development. The framework and the action plan are intended to be approved in 2023.

Some important pressures on biodiversity are expected to increase due to major road and railway building plans, new permits for mining, and a clear trend towards the intensification of agriculture and forestry. This will put a significant strain on nature conservation management and will make the achievement of the favourable conservation status of habitats and species covered by the Nature Directives a major challenge.

13.10 Conclusions

Estonian nature conservation has had some notable success stories: several species of European conservation concern are doing remarkably well in the country. In the case of the globally threatened European Mink, the success has followed years of research, captive breeding and intensive actions to reintroduce the species on a suitable island, after the eradication of the American Mink at the site. In contrast, the successful conservation of the White-tailed Sea-eagle and some other raptors, as well as the Grey Seal, has been achieved through a combination of strict protection, including that of breeding sites, and environmental measures (a ban on DDT and PCBs). With large carnivores, whilst hunting was not totally banned, its strong regulation has been enough to restore their populations.

However, a mixed picture can be seen with the conservation of habitats. This includes semi-natural habitats, and their associated species, where major efforts to increase proactive conservation management and integrate it into agricultural practices have managed to slow down negative trends and improve the situation at the best sites. Forest habitats are relatively well protected in designated sites but are under severe pressure in the wider landscape. Whilst many wetlands are still impacted by the legacy of drainage and excavation, some are now being restored to some extent. Consequently, the fate of habitat-specialist species mirrors that of their respective habitats. Several species dependent either on ancient woodland (e.g. the Flying Squirrel and Black Stork) or semi-natural habitats (e.g. Ruff) or a combination of both (e.g. Greater Spotted Eagle) are still declining, in spite of legal protection and partly successful habitat management. More encouragingly, some of the meadow specialists (e.g. Black-tailed Godwit, Baltic Dunlin and Natterjack Toad) that also had long-term declines have shown recent modest short-term recoveries. Whether this can be interpreted as success remains to be seen over the coming years.

Most successful conservation projects and policies have depended on EU financial support, either from the LIFE Programme (e.g. for restoration of large alvar and bog areas as well as rivers and wetlands), Cohesion Fund (e.g. for restoration of wetlands and removal of dams) or EAFRD for RDP measures (e.g. management of semi-natural meadows). The key elements for success have been sound scientific knowledge of the targeted habitats as well as good communications with stakeholders.

Much remains, however, to be done in order to integrate biodiversity and sustainability concerns into economic sectors and activities at a landscape scale. The twin problems of agricultural abandonment and intensification, ever increasing felling volumes and drainage in forests, mining, large infrastructure, and pollution from energy and industry are still threatening biodiversity.

Better conservation of all major habitat types is needed, including enlarging the protected area network, implementing green infrastructure projects, and restoring and appropriately managing habitats. Although nature conservation considerations are starting to be integrated into economic decisions and spatial planning, this needs to be significantly boosted. The capacity of both public administrations and NGOs must also be strengthened if successes are to be maintained and ongoing biodiversity declines halted and reversed.

Acknowledgements

We thank reviewers Riinu Rannap and Lauri Klein for their insights and comments that helped to improve the chapter. We also thank Kadri Möller and Eleri Pulk from the Ministry of Environment and Kaire Sirel from the Environmental Agency for providing the needed data for the chapter.

References

Ader, A. & Tartes, U. (2010) *Eesti looduskaitse.* Tallinn: Keskkonnaamet.

Adermann, V., Padari, A., Sirgmets, R., Kosk, A. & Kaimre, P. (2015) Valuation of timber production and carbon sequestration on Järvselja nature protection area. *Forestry Studies*, **63**, 29–43.

Ehrlich, Ü. (2013) *Eesti loodusturism kui majandusharu.* Tallinn: Tallinna Tehnikaülikool.

Ehrlich, Ü. & Reimann, M. (2011) Hydropower or non-market values of nature: a contingent Valuation Study of Jägala Waterfalls, *Estonia. International Conference on Energy, Environment, Devices, Systems, Communications, Computers*, **11**, 232–237.

Estonian Fund for Nature (2021) Layman report of the project 'Conservation and restoration of mire habitats – LIFE Mires Estonia'. Tartu, Estonia: Estonian Fund for Nature (Bilingual publication).

Etverk, I. (2003) *20. sajand Eesti metsades. [20th Century in Estonian Forests.]* Teine trükk. [Second impression.] Eesti Metsaselts. Tallinn: Estonian Society of Foresters (Bilingual publication).

Helm, A., Hanski, I. & Pärtel, M. (2006) Slow response of plant species richness to habitat loss and fragmentation *Ecology Letters*, **9**, 72–77.

Helm, A., Kull, A., Veromann, E. *et al.* (2021) *Metsa-, soo-, niidu- ja põllumajanduslike ökosüsteemide seisundi ning ökosüsteemiteenuste baastasemete üleriigilise hindamise ja kaardistamise lõpparuanne.* ELME projekt. Tallinn: Keskkonnaagentuur

Helm, A. & Toussaint, A. (2020) *Poollooduslike koosluste ökoloogilise toimimise hinnang.* Tartu, Estonia: Tartu Ülikool, Ökoloogia ja Maateaduste Instituut.

Ilomets, M. (2015) Estonia. In *Mires and Peatlands of Europe: Status, Distribution, and Nature Conservation*, eds. H. Joosten, F. Tanneberger & A. Moen. Stuttgart, Germany: Schweizerbart Science Publishers.

Jüssi, I. & Jüssi, M. (2017) *EL merestrateegia raamdirektiivi (2008/56/EÜ) kohane merekeskkonna seisundihinnang teemal mereimetajad (D1).* Tallinn: Pro Mare.

Keskkonnaamet (2013) *Poollooduslike koosluste tegevuskava aastateks 2014–2020.* Kinnitatud keskkonnaministri käskkirjaga nr 897. Tallinn: Keskkonnaamet.

(2021) *Pärandniitude tegevuskava aastateks 2020–2027.* Kinnitats keskkonnaministri käskkirjaga. Tallinn: Keskkonnaamet.

Keskkonnaministeerium (2012) *Looduskaitse arengukava aastani 2020.* Tallinn: Keskkonnaministeerium.

(2015) *Kaitstavate soode tegevuskava.* Tallinn: Keskkonnaministeerium.

(2019) *A Prioritised Action Framework (PAF) for Natura 2000 in Estonia pursuant to Article 8 of Council Directive 92/43/EEC on the conservation of natural habitats and of wild fauna and flora (the Habitats Directive) for the Multiannual Financial Framework period 2021 2027.* Tallinn: Keskkonnaministeerium.

Klasen, H., Linder, M. & Kaukver, K. (eds.) (2017) *Eesti keskkonnaseire 2011–2015.* Tallinn: Keskkonnaagentuur.

Kongo, L. (1998) Early history of nature conservation in the Baltic region. *Estonia Maritima*, **3**, 75–86.

Koorberg, P., Raa, I. & Kikas, T. (2012) Estonia. In *High Nature Value Farming in Europe*, eds. R. Oppermann, G. Beaufoy & G. Jones, pp. 190–199. Ubstadt-Weiher, Germany: Verlag regionalcultur.

Kriiska, A. (2004) Inimene ja loodus esiajal. In *Pärandkooslused*, ed. T. Kukk, pp. 13–47. Tartu, Estonia: Pärandkoosluste Kaitse Ühing.

Kull, K., Kukk, T. & Lotman, A. (2003) When culture supports biodiversity: the case of the wooded meadow. In *Imagining Nature: Practices of Cosmology and Identity*, eds. A. Reopstorff, N. Bubandt & K. Kull, pp. 76–96. Aarhus, Denmark: Aarhus University Press.

Kumari, E. (ed.) (1973) *Looduskaitse.* Tallinn: Valgus.

Leivits, A. (2019) Liikide ohustatuse hindamine Eestis: praegune seis ja edasised vajadused. In *Liigikaitse Eesti ajateljel*, ed. T. Kull, pp. 8–12. Tartu, Estonia: Eesti Loodusfoto.

Linder, M. (ed.) (2017) *Estonian Nature Conservation in 2015.* Tallinn: Estonian Environmental Agency.

Lõhmus, A., Kohv, K., Palo, A. & Viilma, K. (2004) Loss of old-growth, and the minimum need for strictly protected forests in Estonia. *Ecological Bulletins*, **51**, 401–411.

Lõhmus, A. (2016) *Eesti rangelt kaitstavate metsade tüpoloogiline analüüs.* Tallinn: Keskkonnaagentuur.

(2020) *Metsapoleemika neljas aasta: aeg rääkida riigist. Sirp*, 04.09.2020. Tallinn: Kultuurileht

Lotman, A. (2004) *Management of coastal habitats and grasslands.* Background paper for the LIFE-Nature Co-op project 'Experience exchange on habitat management among Baltic LIFE-Nature projects'. Matsalu, Estonia: Baltic Environmental Forum.

Lotman, A., Martin, G., Viik, K. & Lips, U. (eds.) (2018) *Eesti mereala keskkonnaseisund 2018.* Tallinn: Keskkonnaministeerium.

Lotman, K. (2003) Wood pasture, abandonment and EU enlargement issues in Estonia. *La Cañada*, **17**, 19.

Maran, T., Põdra, M., Harrington, L. A. & Macdonald, D. W. (2017) European mink: restoration attempts for a species on the brink of extinction. In *Biology and Conservation of Musteloids*, eds. D. W. Macdonald, C. Newman, & L. A. Harrington, pp. 370–389. Oxford, UK: Oxford University Press.

Marja, R. & Nellis, R. (2018). Farmland bird population changes in Estonia in the period 1984–2017 and their relationship with agriculture management and predators. *Hirundo*, **31** (1), 49–68.

Nellis, R. & Volke, V. (2019) Metsalindude arvukuse muutused perioodil 1983–2018. *Hirundo*, **32** (1), 63–80.

Nellis, R., Volke, V., Kalamees, A. *et al.* (2018) *Kaitstavate alade linnujahi analüüs.* Tartu, Estonia: Eesti Ornitoloogiaühing & Keskkonnaõiguse Keskus. www.eoy.ee/pics/old/Kaitstavate%20alade%20linnujahi%20anal%C3%BC%C3%BCs.pdf

Normak, K. (2019) *Vesikonna pinnavett mõjutava koormuse ülevaade.* Tallinn: Maves.

Paal, J., Leibak, E. & Lutsar, L. (1998) *Estonian wetland inventory 1997: publication of the project 'Estonian Wetlands Conservation and Management Strategy'.* Tartu, Estonia: Eesti Loodusfoto.

Paal, J. & Leibak, E. (2011) *Estonian mires: inventory of habitats.* Tartu, Estonia: Estonian Fund for Nature.

(2013) *Eesti soode seisund ja kaitstus.* [Status and protection of Estonian mires.] Tartu, Estonia: Estonian Fund for Nature.

Randla, T. & Sillaots, T. (1998) On the development of nature conservation in the 20th century Estonia. *Estonia Maritima*, **3**, 87–92.

Rannap, R., Lõhmus, A. & Jakonson, K. (2007) Consequences of coastal meadow degradation: The case of the natterjack toad (Bufo Calamita) in Estonia. *Wetlands*, **27** (2), 390–398.

Rannap, R. (2019) Decline and conservation of amphibians in Estonia. In *Amphibian Biology. Volume 11: Status of Conservation and Decline of Amphibians: Eastern Hemisphere, Part 5, Northern*

Europe, eds H. Heatwole & J. W. Wilkinson, pp. 115–122. Exeter, UK: Pelagic Publishing

Ratas, R. (1998) Importance and role of nature conservation in Estonia. *Estonia Maritima*, **3**, 33–39.

Raudsaar, M., Sims, A., Timmusk, T., Pärt, E. & Nikopensius, M. (2018) *Mets. 1. Metsavarud. Forest resources.* Aastaraamat Mets 2017. Yearbook Forest 2017. Tallinn: Keskkonnaagentuur (Bilingual publication).

Remm, L., Vaikre, M., Rannap, R. & Kohv, M. (2018) Amphibians in drained forest landscapes: conservation opportunities for commercial forests and protected sites. *Forest Ecology and Management*, **428**, 87–92.

RMK (2019) *RMK aastaraamat 2019.* Sagadi, Estonia: RMK/

Roasto, R. & Tampere, U. (eds.) (2020) *Eesti looduse kaitse aastal 2020.* Tallinn: Keskkonnaagentuur.

Rosenvald, R., Järveküilg, R. & Lõhmus, A. (2014) Fish assemblages in forest drainage ditches: degraded small streams or novel habitats? *Limnologica*, **46**, 37–44.

Saks, L., Hommik, K. & Svirgsden, R. (2018) *EL merestrateegia raamdirektiivi (2008/56/EÜ) kohane merekeskkonna seisundihinnang teemal kalastik ja kaubanduslikel eesmärkidel kasutatavad kalad (D1, D3, D4).* Tartu, Estonia: University of Tartu.

Sepp, K., Eschbaum, K., Kutsar, R. *et al.* (2018) *Rohevõrgustiku analüüs ja planeerimisjuhendi koostamine. Rohevõrgustiku analüüsi vahearuanne.* Tallinn-Tartu: Keskkonnaagentuur.

Sims, A., Raudsaar, M., Tamm, U. & Timmusk, T. (2018) *Mets. 9. Keskkond. Environment.* Aastaraamat Mets 2017. Yearbook Forest 2017. Tallinn: Keskkonnaagentuur (Bilingual publication).

Tomingas, A. (ed.) (2018) *Republic of Estonia 100. People. History. Culture.* 1st ed. Tallinn: TEA Publishers.

Tomson, P., Bunce, R. G. H. & Sepp, K. (2015) The role of slash and burn cultivation in the formation of southern Estonian landscapes and implications for nature conservation. *Landscape and Urban Planning*, **137**, 54–63.

Troska, G. (2004) Külaelu ja pärandkoosluste majandamine Teise maailmasõjani. In *Pärandkooslused*, ed. T. Kukk, pp. 48–64. Tartu, Estonia: Pärandkoosluste Kaitse Ühing.

Valgepea, M., Sims, A., Raudsaar, M. & Timmusk, T. (2018) *Mets. 3. Raied. Fellings.* Aastaraamat Mets 2017. Yearbook Forest 2017. Tallinn: Keskkonnaagentuur (Bilingual publication).

Valk, U. (ed.) (1988) *Eesti sood.* Tallinn: Valgus.

Vanhatalo, J., Vetemaa, M., Herrero, A., Aho, T. & Tiilikainen, R. (2014) By-catch of Grey Seals (*Halichoerus grypus*) in Baltic Fisheries – A Bayesian Analysis of Interview Survey. *PLoS ONE*, **9**, e113836.

Viik, E. (2019) *Kimalaste mitmekesisuse ja arvukuse uuring. Eesti maaelu arengukava 2014–2020 4. ja 5. prioriteedi hindamine* Saku, Estonia: Põllumajandusuuringute Keskus.

Viirma, M. (2013) *Tõkestusrajatiste inventariseerimine vooluveekogudel kalade rändetingimuste parandamiseks. Koondaruanne.* Tallinn: Keskkonnaagentuur.

Chapter 14

Finland

Heikki Toivonen and Olli Ojala

Key Statistics

Terrestrial area (km^2)/% of EU-28*	337 547/7.7%
Marine waters (km^2)/% of EU-28*	82 466/1.8%
Population density (p km^2)/% of EU-28*	16/14%
GDP per capita (€)/% of EU-28	35 528/133%
Land cover (CORINE land cover classes 2018)	FOR 62%, WET 16%, OSN 12%, AGR 8%, ART 1%
Biogeographical regions	ALP, BOR, MBAL
Number of HD Annex I habitats/%*	68/30%
Number of HD Annex II, IV and V species/%*	127/10%
Number of BD Annex I bird taxa/%*	64/36%
% HD UCS and decreasing habitats/ % unknown	26%/1%
% HD UCS and decreasing species/ % unknown	14%/8%
% breeding BD Annex I birds decreasing/% uncertain and unknown	30%/0%

* Excluding MAC and MMAC biogeographical regions.
Sources and codes. See Table 1.1.

14.1 The Natural History and Characteristics of Finland

Finland is a northern country, situated between 60° and 70° latitude. Due to the Gulf Stream, the climate is more favourable than in areas of corresponding latitudes in Siberia and North America. The country is mostly covered by boreal forests (taïga), dominated by conifers. Only the Åland islands, the Archipelago Sea and the south-western coast belong to the hemiboreal zone between the boreal coniferous and nemoral deciduous vegetation zones. The very northernmost part of Finland is within the hemiarctic (subarctic) zone, being dominated by alpine heaths, scrubland and rocky areas.

About 75% of Finland's land area is covered by forests and other wooded land, 90% of which is dominated by Scots Pine (*Pinus sylvestris*) and Norway Spruce (*Picea abies*), and 10% by broad-leaved deciduous trees, mostly birches (*Betula pendula*, *B. pubescens*). Also alders (*Alnus incana*, *A. glutinosa*), Aspen (*Populus tremula*) and Rowan (*Sorbus aucuparia*) are common. Southern broad-leaved deciduous tree species are rare and restricted to sites with favourable soils and climate in the south. In northernmost Finland the Mountain Birch (*B. pubescens* subsp. *czerepanovii*) forms the forest line in fjeld areas (i.e. gently sloping Scandinavian mountains and barren uplands).

The vast boreal forests in the country hold large populations of Elk (*Alces alces*), Mountain Hare (*Lepus timidus*), and some grouse species: Capercaillie (*Tetrao urogallus*), Black Grouse (*Lyrurus tetrix*) and Hazel Grouse (*Bonasia bonasia*). Red Squirrel (*Sciurus vulgaris*) and Siberian Flying Squirrel (*Pteromys volans*) are characteristic species. Four species of large carnivore occur: the Brown Bear (*Ursus arctos*) and Eurasian Lynx (*Lynx lynx*), which are rare over most of the country, and the very rare Wolf (*Canis lupus*) and Wolverine (*Gulo gulo*). The Finnish Forest Reindeer (*Rangifer tarandus fennicus*), which was close to global extinction, has partly recovered, but the present population is quite low.

The microrelief of the country is largely the result of glaciation processes during the Ice Age. Due to Finland's relatively flat lowlands and suboceanic climate, wide areas formed peatlands (paludified) over the post-glacial period, resulting in some 10 million ha of mires (30% of the land area). Raised bogs prevail in the southern part of the country, whilst more open Aapa mires with wet hollows and dry hummocks occur in the middle and northern part of the country. A wide diversity of mire types occur. Typical are various pine bogs with ericoid dwarf-scrubs, sedge fens, spruce–birch mires, *Alnus* swamps, etc. There are some Palsa mires (Palsas are large peat mounds having permanent ice cores) in the northernmost part of the country. Mires provide breeding habitats for many waders and other waterbirds, with populations of European importance. However, owing to large-scale drainage for forestry and land clearance to agriculture, the proportion of natural peatlands has decreased considerably to 13% (Kontula and Raunio, 2019).

Finland is a country of thousands of lakes, creating very long shorelines, along with the Baltic Sea coast. About 10% of the area of the country consists of freshwater bodies, mostly

Table 14.1 Short-term trends (c. 2007–2018) in Habitats Directive Annex I habitats and Annex II, IV and V species with an unfavourable conservation status in Finland.

	Poor (unfavourable– inadequate)				Bad (unfavourable–bad)			
	Imp	Sta	Dec	?	Imp	Sta	Dec	?
HD Habitat assessments								
Number	3	15	14	0	0	16	10	3
%	9%	47%	44%	0%	0%	55%	34%	10%
HD Species assessments								
Number	4	27	10	13	0	6	12	1
%	7%	50%	19%	24%	0%	32%	63%	5%

Key. Imp = improving, Sta = stable, Dec = decreasing, ? = unknown.
Source. Based on Member State HD Article 17 reported data as presented on the EEA national summary dashboard (2021).

with low nutrient levels (i.e. oligotrophic). Large lakes are numerous, about 50 being over 100 km^2. The lakes have maintained their typical biota of waterbirds, fish and invertebrates. They also retain some isolated relict populations from glacial times, such as some salmonids and crustaceans, and the globally endangered Saimaa Seal (*Pusa hispida saimensis*), which only occurs in Lake Saimaa. Beside inland waters, Finland has a good representation of various brackish water habitats, which have been created by the gradual uplifting of the land along the Baltic coastline and archipelago area.

About 8% of the land area consists of farmland, a relatively low percentage compared to most European countries. Various semi-natural meadows, including those that occurred along many open lake shores, and alluvial meadows along rivers, were previously common, but their area has declined by more than 90% over the last hundred years. Consequently, many habitats and species bound to traditional rural livelihoods have become threatened during the twentieth century.

Most urban areas in Finland have a good proportion of semi-natural forests, as well as parks and other green areas.

14.2 The Status of Nature

14.2.1 Habitat Status and Trends

Habitats Listed in HD Annex I

Of all Annex I habitat types assessed in the Finnish Boreal region, only 15% had a favourable conservation status, whilst 44% had a poor status and 39% a bad status (i.e. threatened).[1] In the Alpine region, the situation is markedly better: 80% of habitats had a favourable conservation status, 12% a poor status and 8% a bad status. In the Alpine region, 12% of habitats showed short-term declines (over 2007–2018), whereas more than one-third of the Boreal habitats declined. All HD marine habitats had an unfavourable conservation status (40% poor and 60% bad). In general, the conservation status of HD habitats has changed little since the previous reporting round.

Table 14.1 presents the trends of HD habitats and species with an unfavourable status. Whilst most habitats had stable short-term trends, a sizeable proportion showed declines, including about one-third of those with a bad status.

In general, habitat types with a bad conservation status are those that are threatened by vegetation growth and succession processes. These include open semi-natural grassland, although some genuine improvements in the status of Boreal Baltic coastal meadows (1630*)[2] have occurred as a result of effective agri-environment schemes (discussed later in Section 14.6). Habitats with a poor conservation status are typically those with extensive areas, such as Western taïga and Aapa mires, which are subject to multiple pressures including the ongoing effects of historic changes. Their large areas also make it difficult to improve their conservation status, especially as measures in the protected areas alone are not sufficient.

Habitat types with the lowest pressures and most favourable and stable conservation status are northern fjeld and rocky habitats.

Red List Habitats

Finland is one of the few countries in the world that has conducted a thorough assessment of the threat status of its habitat types, firstly in 2008 and secondly in 2018 (Kontula and Raunio, 2019).[3] The second assessment used the IUCN Red List categories and criteria for ecosystems, and covered all natural habitats as well as semi-natural grasslands and grazed woodland. Whilst the change in methods makes it difficult to compare the two assessments directly, it is clear that negative trends have continued for most habitats.

1 See Section 1.3 for background on monitoring under the Nature Directives, status categories and data sources.

2 Numbers in brackets are the HD Annex I habitat codes (see Appendix for full names). * Indicates a priority habitat.

3 English summary information: www.ymparisto.fi/en-US/Nature/Natural_habitats/Assessment_of_threatened_habitat_types_in_Finland

The 2018 assessment found that 186 habitats (48% of all habitat types) are threatened in the whole country. These were grouped into eight main ecosystems: the most threatened are semi-natural grasslands and grazed woodlands (100% threatened), then forests (76%), coastal habitats (58%), mires (57%), fjeld habitats (38%), the Baltic Sea (24%), rocky habitats (25%), and inland waters and shores (20%). As found with the HD habitats assessments, a higher proportion of habitat types are threatened in southern Finland (59%) than in northern Finland (32%).

14.2.2 Species Status and Trends

Species Listed in HD Annexes II, IV and V

As with HD habitats, the overall conservation status of most HD species over 2013–2018 was unchanged compared to the previous reporting period. The conservation status of species was favourable for 39% of the species in the Boreal region, 40% in the Baltic Sea and 69% in the Alpine region. In the Boreal region, 40% of species had a poor status, and 13% a bad status. Similarly, in the Baltic Sea, 40% of species had a poor status and 20% a bad status. In comparison, species in the Alpine region were faring better, as only 11% had a poor and 9% a bad status. The conservation status was unknown for 8% of species in both the Boreal and Alpine regions, and none in the Baltic Sea.

Of particular concern is that 63% of species with a bad status also declined (Table 14.1).

Birds

Table 14.2 indicates that the short-term trends amongst Annex I birds in Finland have been mixed. Although most have had stable or fluctuating breeding populations, 27% increased and 30% decreased. Out of the four wintering Annex I species populations, three increased and one decreased.

Red List Species

Finland has published five Red Lists of nationally threatened species since the mid-1980s. The most recent assessment (Hyvärinen *et al.*, 2019) covers approximately 22 400 species (i.e. 47% of known species in Finland). In total, 2 667 species (12%) are classified as threatened and 312 species as regionally extinct. The number of threatened species is highest in the hemiboreal and southern boreal zones in southern Finland. The proportion of species that are considered to be threatened has increased.

Amongst Red Listed bird species, between 2010 and 2019, positive changes were recorded for 15 species, but the situation deteriorated for 64. Those with the most negative trends were especially dependent on aquatic habitats, but also forests and mires.[4]

[4] https://punainenkirja.laji.fi/en/results

Table 14.2 Short-term trends (c. 2007–2018) in Birds Directive Annex I taxa in Finland.

	Breeding				Wintering			
	Inc	Sta/Flu	Dec	?	Inc	Sta/Flu	Dec	?
Number	17	28	19	0	3	0	1	0
%	27%	43%	30%	0%	75%	0%	25%	0%

Key. Inc = increasing, Sta = stable, Flu = fluctuating, Dec = decreasing, ? = uncertain and unknown.
Source. Based on Member State Art 12 reported data as presented on the EEA national summary dashboard (2021).

14.2.3 Drivers and Causes of Change

The Red List assessments (Hyvärinen *et al.*, 2019) have revealed that the majority of threatened species occur in forests and in various farmland habitats, including those resulting from traditional agricultural practices. Other important habitats for threatened species are shorelines, rock cliffs and outcrops, and aquatic habitats.

Changes in forest habitats, mostly due to extensive and intensive forestry, constitute the major threat to forest biota (see Section 14.6). Forestry is the primary cause of the threatened status of more than 700 species, according to the latest assessment (Hyvärinen *et al.*, 2019). The main forestry pressures result from declines in dead and decaying wood, changes in tree species composition and age structure, declining numbers of old and large trees, and a reduction in burnt forest areas (fire being a natural process in boreal forests).

In grasslands, along shorelines and other open habitats, the main pressure on species is from vegetation growth and succession, which is often the result of the cessation of grazing and/or eutrophication. Further pressures arise from construction activities, human-induced changes in water courses (regulation of water tables, dredging and water abstraction) and the drainage of peatland for forestry. Future threats are generally expected to be similar to past and current pressures, except for an expected increase in climate change–related threats.

Similarly, according to Kontula and Raunio (2019), the most important pressures on threatened habitats are from forestry, drainage for forestry, the clearance of forests and peatlands for agriculture, construction activities, eutrophication of waters and shores, regulation of waters and construction activities in water courses, and vegetation growth in open habitats. Future threats are also expected to be similar to current pressures, except for the increasing effects of invasive alien species (IAS), and particularly climate change. Also, mining activities may have greater impacts in the future, particularly in the northern part of the country, due to the liberal Mining Act.[5]

Whilst there has been little change in many environmental pressures, Finland has been able to reduce pollution from

[5] 621/2011 Kaivoslaki.

many sources over recent decades to levels that are no longer acutely harmful to biodiversity. Successes include curbing emissions of sulphur dioxide, DDT, PCB and lead. Phosphorus and nitrogen loads from point sources (municipal and industrial wastewater treatment plants) have also markedly reduced in three decades.[6] At present, the most challenging pollution issue is diffuse nutrient loading from agriculture and forestry, including from drained peatlands. Eutrophication and increases in suspended matter are the most important threats to many aquatic habitats, both in the Baltic Sea and in inland waters.

Climate change was considered to be a threat for more than 150 habitat types in the 2018 Red List assessment (Kontula and Raunio, 2019) – twice the number in the 2008 assessment. Several climate change–induced changes in environmental conditions are anticipated in coastal and underwater habitats of the Baltic Sea. These include lower salinity due to increasing rainfall in the Baltic catchment, rising water levels in littoral habitats (which counteract the effects of land uplift) and effects of increasing eutrophication. Other ecosystems that are the most vulnerable to climate change include the Subarctic and Alpine habitats in the fjeld areas in northern Lapland, including mountain birch forests.

14.3 Nature Conservation Governance, Actors, Legislation and Policies

Table 14.3 summarises some of the key developments in nature conservation in Finland, the earliest of which date back to the nineteenth century. Later advances included the first Nature Conservation Act (NCA) in the early 1920s, and the establishment of the first National Parks before World War II.

14.3.1 Governance

Since its establishment in 1983, the Ministry of the Environment (MoE) has managed the Finnish environmental administration. The administration consists of the Finnish Environment Institute (SYKE), Parks and Wildlife Finland (PWF; previously Metsähallitus Natural Heritage Services) and a regional administration. Of the national organisations, SYKE is a research institute that conducts long-term monitoring and provides expert services, while PWF is the practical management body for most of the network of protected areas.

The regional administration comprises 15 regional bodies called Centres for Economic Development, Transport and Environment (ELY-Centres). Previously, between 1995 and 2010, there were 13 Regional Environmental Centres that were under direct supervision of the MoE. Currently, environmental issues in the ELY-Centres are steered by the MoE, but they receive their budget from the Ministry of Employment and the Economy of Finland.

Almost all of the protected area network is exclusively managed by PWF – a public service unit in charge of nature conservation, recreational services and cultural heritage within the state-owned enterprise Metsähallitus (see also Sections 14.5 and 14.6). Having one organisation controlling the management of all protected areas has created a unique situation where management can be considered at the level of the whole network. This facilitates effective management and the use of the most advanced methods and scientific knowledge, including prioritisation of conservation measures within the network (Gilligan *et al.*, 2005; Heinonen, 2007).

As well as forestry and agriculture, the Ministry of Agriculture and Forestry (MoAF) has responsibility for economically important fish, game species and IAS, whilst the Finnish Wildlife Agency has a role in the management of large carnivores. Close cooperation between the MoE and MoAF is very important given the high proportion of threatened species and habitats in semi-natural farmland biotopes and forests. However, the trends in these ecosystems indicate that the measures taken so far have been insufficient.

In addition, the autonomous Åland islands have their own legislation and government (Ålands Landskapsregering).

14.3.2 Public Involvement and NGOs

Environmental NGOs are active and have had significant impacts on environmental policy in Finland. Three NGOs are especially noteworthy: WWF Finland, Greenpeace, and the Finnish Association for Nature Conservation (FANC)[7], which has 30 000 individual members belonging to 180 local associations and an independent youth organisation (The Nature League[8]). These, and other NGOs, have had a major role – for instance in the partly continuing debate on the protection of old-growth forests in northern Finland. Besides active lobbying, media, public participation and litigation work, they have conducted field inventories to support their proposals, and searched for areas for conservation. The Finnish Natural Heritage Foundation, founded in 1995 by eco-philosopher Pentti Linkola, has promoted protection of old-growth forests. The foundation purchases representative areas of old-growth forests with donated funds.

The forestry and landowners' organisations, particularly the Central Union of Agricultural Producers and Forest Owners (MTK), have major influence on environmental policies, including nature conservation, and on natural resource use policy. Moreover, the Finnish constitution gives strong protection to owners' rights. Also, the status of Finnish forestry and its institutions in the Finnish economy and society are strong.

Forest policy in Finland has traditionally been primarily targeted at promoting timber production over ecological

[6] www.environment.fi: Indicators on Nitrogen and Phosphorus discharges into surface waters from point sources (updated February 2021).

[7] www.sll.fi/en/

[8] www.luontoliitto.fi/en

Table 14.3 Key nature conservation developments in Finland.

1842	Foundation of Crown Parks of the Imatra Rapids and the Esker Punkaharju
1870	Foundation of 'Spring Society' for conservation of songbirds (initiative by Professor Zacharias Topelius)
1880	Proposal for establishment of nature parks in Nordic countries, made by Professor A. E. Nordenskiöld (explorer of the Northern Sea Route)
1886	The first Forest Act
1903	National Board of Forestry established first forest reserves
1916	Establishment of first official conservation area
1923	The Nature Conservation Act
1938	Establishment of first National Parks and Strict Nature Reserves (SNRs)
1954	Nature conservation studies introduced at the University of Helsinki
1956	Establishment of several new National Parks and SNRs
1962	The Water Act
1973	The Outdoor Recreation Act
1977	Report of the National Committee on establishment of network of National Parks and SNRs
1979	Demonstration by young nature conservation activists for conservation of the Koijärvi Bird wetland (start of the green movement in Finland)
	Programme on the development of new National Parks and SNRs
	The Mire Conservation Programme
1981	Land Extraction Act
1982	Offices for environmental protection at the Provincial Governments founded
	The Waterfowl Habitats Conservation Programme
1983	Foundation of the Ministry of the Environment
1984	The Esker Conservation Programme
1987	The Act on the Protection of Rapids
1989	The Herb-rich Forest Conservation Programme
1990	The Shore Habitats Conservation Programme
1991	The Wilderness Act
1995	Foundation of Regional Environmental Centres
	Foundation of the Finnish Environment Institute (SYKE)
	Accession to the European Union
	The Programme for the Protection of Old-growth Forests
1996	New Nature Conservation Act (NCA) and new Forest Act
	First Finnish NBSAP (1997–2005) approved, initiated by the National Biodiversity Committee
1998	The government drew up the first proposal for the EU Natura 2000 network in Finland
2002	Programme for the protection of the Baltic Sea
	The METSO Programme for forest protection in Southern Finland
2006	Second Finnish NBSAP (2006–2016)
2012	The national strategy for invasive alien species (IAS)
	Third Finnish NBSAP (2012–2020)
2014	Reform of the Forest Act
2020	The Helmi habitats programme

See Table 3.3 for signatory dates of international conventions.

concerns. This tradition is changing, but only slowly, despite pressure from environmental NGOs and the rising ecological concerns of the general public (Kröger and Raitio, 2017).

The polarisation between environmental NGOs and landowners, and also industry, has caused severe conflict in forest conservation. For example, thousands of complaints were filed for all kinds of sites when the Natura 2000 network was being established in the late 1990s.

In addition to lobbying and campaigning, NGOs and the wider public (through citizen science) have a significant role in carrying out monitoring and surveys (see Section 14.8).

14.3.3 Policy and Legislation

From the late 1970s to the 1990s, the option to establish nationwide ecosystem-based Nature Conservation Programmes, as stipulated in the Nature Conservation Act (NCA), was a key tool for nature conservation. Important programmes were established for National Parks and Strict Nature Reserves (SNRs), mires, herb-rich forests, waterbird habitats, shoreline zones conservation areas and old-growth forests. The programmes included the identification and mapping of a network of sites, based on scientific analysis and criteria, that would best serve the conservation aims of the particular programme and complement the network of existing protected areas (see protected areas in Section 14.5.1). The proposed networks of sites were subject to a public hearing, after which they were approved by the government, usually mostly as initially proposed. Areas included in the approved programmes are immediately protected against damaging actions, although relevant derogation provisions apply. In most cases the rules are strict, banning the economic use of the site.

Sites protected under the programmes are mostly on state-owned land, but many sites are privately owned. Privately owned sites are purchased by the State, or they are established as Privately-owned Protected Areas, with the government compensating through a lump sum, if required, for economic losses suffered by the landowners. Expropriation can be carried out where necessary to achieve conservation objectives.

In the 1970s and 1980s, implementation of conservation programmes was slow due to insufficient funding, but it accelerated after accession to the EU as part of the formation of the Natura 2000 network. Implementation of conservation programmes was mostly completed by 2007, but some work still continues (e.g. legal designation of protected areas in state-owned areas).

The intensified implementation caused conflicts with private landowners, and in spite of its merits in systematic conservation planning, no new Nature Conservation Programmes have been started since the late 1990s. Such conflicts may also have occurred in the 2010s, as a complementary conservation programme for mires was prepared. However, a change in government led in 2014 to a decision to protect only sites on state-owned land. Sites on privately owned land were recommended for voluntary protection, such as through actions in the Forest Biodiversity Programme METSO[9] and Habitats Programme HELMI.[10]

14.3.4 Nature Conservation Act

Nature conservation in Finland, including planning, establishment of protected areas, protection of nationally important habitats, landscape protection, species protection, preservation orders, coercive measures and sanctions, is primarily regulated under the 1996 NCA.[11] The NCA covers marine, freshwater and terrestrial nature.

Finland transposed the Nature Directives to the Finnish NCA at the end of 1996. This change added the provisions related to the Natura 2000 network as an independent section to the law. New elements for Finland included the protection of breeding sites and resting places of strictly protected species. However, the existing NCA provisions concerning the nature conservation programmes and establishment of protected areas could be used as the main implementation measures for the Natura 2000 network. Consequently, most of the Natura 2000 network has been established by designating the sites as protected areas under NCA legislation. In 2014, some additional important changes were made to the NCA, such as the introduction of a general prohibition on deterioration of Natura 2000 sites, and in relation to the process of establishing SACs.

The government started updating the NCA in 2019 to revise it and better meet new challenges, such as climate change.

14.3.5 Other Legislation

The Forest Act regulates, in addition to the use of forests, the protection of specific valuable habitat types (so-called key biotopes) throughout the area where forestry is practised. This is discussed in more detail in Section 14.6 on general conservation measures.

The Water Act[12] is also important for nature conservation because it controls pollution and modification of waterways, protects a set of aquatic habitat types, and partly implements the WFD by regulating projects impacting waters. The Act on the Organisation of River Basin Management and the Marine Strategy[13] is the main tool implementing the WFD and MSFD and their planning mechanisms. The Acts on the Protection of Rapids[14] protect a list of valuable waterways by prohibiting the construction of power stations on them.

Species protection in general is regulated under the NCA. However, game species are covered by the Hunting Act, and the quotas of huntable species are determined by ministerial

9 https://mmm.fi/en/forests/biodiversity-and-protection/metso-programme

10 www.ym.fi/en/helmi-habitats-programme

11 1096/1996 Luonnonsuojelulaki.

12 587/2011 Vesilaki.

13 1299/2004 Laki vesienhoidon ja merenhoidon järjestämisestä.

14 35/1987 Koskiensuojelulaki; 703/1983 Laki Ounasjoen erityissuojelusta; 1139/1991 Laki Kyrönjoen erityissuojelusta.

decrees under the Hunting Act. The Hunting Act covers, for example, large carnivores and Wild Forest Reindeer. Hence, derogations from their protection are stipulated under this law.

In addition, the Land Use and Building Act, Land Extraction Act, and Act on Environmental Impact Assessment Procedure are important for nature conservation. They are therefore described later in the sections on protected areas (Section 14.5) and general nature conservation measures (Section 14.6).

14.3.6 National Biodiversity Strategy and Action Plans

Finland's nature conservation policy has been included and partly determined in its National Biodiversity Strategies and Action Plans (NBSAPs), particularly over the two last decades. These have been prepared by the MoE in collaboration with other ministries. In the first Finnish NBSAP, prepared for 1997–2005, cooperation was based on the principle of sectoral responsibility. That means that each ministry was responsible for conservation and sustainable use of biodiversity within its field of jurisdiction. Since the second NBSAP (2006–2016), a Biodiversity Working Group has been established to work as a collaborative body and help implement the plan. In response to the CBD biodiversity target to halt biodiversity loss by 2020, the third Finnish NBSAP[15] was approved in 2012 for 2012–2020. Its Action Plan included over 100 actions that aimed to mainstream biodiversity conservation across different policy sectors to achieve the 2010 CBD Aichi biodiversity targets.

According to an impact assessment of the third NBSAP (Auvinen *et al.*, 2020), mainstreaming of biodiversity progressed well in different administrative branches, and the private and third sector supported the policy. However, although a knowledge base and structural framework to safeguard biodiversity have been established, the action taken has not been effective enough to halt the loss of biodiversity. The research group also presented recommendations for the next Finnish biodiversity action plan, stating, *inter alia*, that its actions should be focussed, concrete and measurable so that their implementation and impacts can be monitored and assessed.

The Biodiversity Working Group presently includes almost 40 representatives from the MoE and 11 other ministries, interest groups, NGOs, research institutes, and the Sámi Parliament (a representative of the indigenous Sámi people). According to Sarkki *et al.* (2015), the Finnish NBSAP process has been able to build responsibility in different policy sectors by providing new knowledge and initiatives, and developing institutional linkages. Despite this, the responsibilities have not diffused from the environmental administration to other policy sectors to a sufficient extent.

Due to shared substantial responsibilities in the realm of nature conservation, important policies are set out also in other strategies, such as the *National Forest Strategy* and agricultural and rural policy. However, so far, the different strategies have had conflicting impacts. For example, the connectivity between protected areas is far from optimal due to the forestry management practices that affect most of the landscape.

[15] https://ym.fi/en/national-biodiversity-policy.

14.4 Species Protection and Conservation Measures

There are two main tools in the NCA that aim to directly protect species. Firstly, all mammal and bird species naturally occurring in Finland are protected, except those that are huntable, non-protected or economically exploited. Thus, most species are protected under the NCA. Also, other species can be protected if necessary. Their killing, catching, taking at any stage of development, or deliberate disturbance (especially during the breeding season and at sites that are important for their life cycles) is prohibited. Also protected are specifically marked nesting trees of protected birds and all clearly visible and actively used nesting trees of Golden Eagle (*Aquila chrysaetos*), White-tailed Sea-eagle (*Haliaeetus albicilla*), Greater Spotted Eagle (*Clanga clanga*), Lesser Spotted Eagle (*Clanga pomarina*) and Osprey (*Pandion haliaetus*).

Secondly, threatened species can be prescribed as specially protected species. For specially protected species a management plan can be drawn up and the locations at which they occur protected by specific delineations.

In practice, the general species protection regime is rather feeble as the law allows impacts caused by agriculture, forestry and construction, if they cannot be avoided without significant additional costs. The protection of specially protected species is more effective at ensuring the physical integrity of the delineated sites where they occur. However, these sites are generally tightly delineated, such that most sites are small and are on their own insufficient to maintain the species in the long term. An action programme for threatened species (MoE, 2017) identifies important measures in different habitats, and horizontal measures such as providing guidance and communication. The importance of ensuring the availability of species occurrence data is highlighted. The programme is also central to the coordinated prioritisation of regional species conservation actions.

Generally, HD Annex IV species, and naturally occurring birds, have the same protection scheme as other protected species in Finland, with the exception that a derogation from their protection can only be granted by the ELY-Centres, if the prerequisites of the respective Directives are fulfilled. Additionally, HD Annex IV species have a specific paragraph for protection of their breeding sites and resting places.

Of the HD Annex IV species, some, such as the Otter (*Lutra lutra*), Brown Bear and Eurasian Lynx, are game species under the national legislation. As for all game species, including birds, hunting and provisions for derogations from protection are stipulated in the Hunting Act[16] under the competence of MoAF. The Commission has taken Finland to the CJEU concerning hunting of Wolf (2007), spring hunting of five huntable duck species (2005), and spring hunting of Common Eider (*Somateria mollissima*) on Åland (2020). All these cases relate to the use of derogations under the Nature Directives.

The endemic Saimaa Seal has been a flagship species for Finnish nature protection for decades. Around 2010, further conservation measures were taken in response to intensified demands for action, including in relation to the increasingly serious threat posed by climate change to breeding conditions. Most importantly, gillnet bans were substantially expanded, and a national conservation strategy and action plan was drawn up in 2011. Between 2013 and 2018, a €5 million LIFE project (Safeguarding the Saimaa Ringed Seal)[17] helped to implement the strategy and action plan and to reduce threats caused by fishing, disturbance and climate change. As a result, the population has increased from 270 in 2010 to 420–430 in 2020.

14.4.1 Large Carnivores

Four species of large carnivore occur in Finland, three of which are protected across the whole territory: Brown Bear, Eurasian Lynx and Wolverine. The Wolf also occurs, but in line with the exemption in the Habitats Directive, it is protected in Finland only to the south of the Reindeer herding area. The Natura 2000 network has had a limited role in protecting these species as they require such large territories. The bear and lynx populations are relatively stable (being Near Threatened and Least Concern in the last Red List of threatened species).

The Wolf and Wolverine populations are low and both species are endangered (Hyvärinen *et al.*, 2019). The two species are subject to sometimes strong local opposition and are illegally killed in some areas. This is largely because the granting of legal derogations to kill the animals has in many areas been considered insufficient by local communities (Suutarinen and Kojola, 2018). Moreover, the Finnish application of the use of derogations was lax and considered illegal by the CJEU (C-674/17). The LIFE project 'Towards better human coexistence with wolves'[18] aims to tackle these difficult issues.

14.5 Protected Areas and Networks

14.5.1 Protected Area Types and Their Coverage

Sites Designated as Protected Areas under Domestic Legislation

National Parks and SNRs are important elements of the Finnish protected area system (Table 14.4). National Parks are large, representative areas with outstanding values in terms of landscape, wilderness or ecosystems. National Parks are freely accessible to visitors (with more than three million visitors in 2018). Unlike in many other countries, inhabited areas are excluded from Finnish National Parks. SNRs are representative examples of undisturbed ecosystems with their natural succession processes, and are not accessible to the public. The first National Parks and SNRs were established in 1938, but most were designated in the latter half of the twentieth century. There are currently 40 National Parks and 19 SNRs, all on state-owned land.

Other protected areas on state-owned land vary greatly in size, habitats and protection. Many of them belong to nationwide Nature Conservation Programmes (described earlier) and are treated as protected areas. There are Mire Reserves, Old-growth Forest Reserves, Herb-rich Forest Reserves and Waterfowl habitats, for example. The establishment of all such designated sites takes at least two years, as the administrative procedure includes many steps. However, the natural values of these sites are protected during their designation process.

Under the Wilderness Act (1991), 12 Wilderness Areas have been established on state-owned land in northern fjeld and forest areas. Besides maintaining nature conservation values, Wilderness Areas also aim to help to preserve Sámi culture and sustain traditional livelihoods, particularly reindeer herding.

There are more than 13 000 protected areas on private land, mostly small in area. The older Privately-owned Protected Areas were created for their special or rare species or habitats, or remarkable natural formations or outstanding natural beauty. Since the 1980s, many protected areas on private land have been established to implement the Nature Conservation Programmes. In more recent years many have been established through the METSO Forest Biodiversity Programme. Protected areas on private land also include 1 455 statutory habitat or species protection areas, both types established under the NCA.

At the beginning of 2020, the total number of sites in Finland's protected area network designated under domestic legislation (i.e. the NCA), was approximately 17 600, covering 37 911 km^2 of the land (c. 11%) and 9 529 km^2 of water, out of marine waters (c. 82 466 km^2) and freshwaters (c. 34 539 km^2). However, these statistics are only approximate; it is not possible to calculate an accurate marine protected area coverage percentage, due to Finland's extremely long, crenulated and dynamic shoreline.

Although they are not protected areas in accordance with the IUCN definition, some other laws give protection to the environment that indirectly benefits wildlife. Hiking Areas, established under the Outdoor Recreation Act, are in many respects comparable to protected areas, although not as strictly protected. The Acts on the protection of rapids protect listed waterways by prohibiting the construction of hydropower on

[16] 615/1993 Metsästyslaki.

[17] LIFE12 NAT/FI/000367.

[18] LIFE BOREALWOLF, LIFE18 NAT/FI/000394.

Table 14.4 The number and extent of international designated sites and protected areas in Finland.

	IUCN category	Number	Area (km^2)
Designated under global and non-EU legislation			
Ramsar Sites		49	7 995
Natural and Mixed World Heritage Sites		1	1 944
Biosphere Reserves (exc. buffer and transition zones)		2	4 907
HELCOM Marine Protected Areas		34	7 913
Designated under the Nature Directives			
SCIs (terrestrial + marine)	Ia, Ib, II, IV, V	1 730	42 200 + 7 700
SPAs (terrestrial + marine)	Ib, II, IV, V	468	24 545 + 7 142
Natura 2000 total terrestrial area			42 498
Natura 2000 total marine area			8 142
Designated under domestic legislation			
Nature reserves and other protected areas on state land			
National Parks (Kansallispuistot)	II, Ib	40	10 033
Strict Nature Reserves (Luonnonpuistot)	Ia, IV	19	1 542
Mire Conservation Reserves (Soidensuojelualueet)	Ib, IV	166	4 747
Old-growth Forest Reserves (Vanhojen metsien suojelualueet)	Ib, IV	77	86
Herb-rich Forest Reserves (Lehtojen suojelualueet)	Ib (even Ia), IV	46	11
Other State Nature Reserves (Muu luonnonsuojelualue valtion maalla)	Ib, IV	630	3 742
Sites reserved for nature conservation by government decision	Ib (even Ia), IV	2 579	5 528
Protected state forests	Ib, IV	426	614
Other protected sites on state land	IV	171	2 649
Wilderness Areas in Lapland (Erämaa-alueet)	Ib	12	14 891
Protected areas on private land			
Privately-owned Protected Areas (Yksityiset suojelualueet)	Ib (even Ia), IV	11 759	3 557
Fixed-term nature reserves (contract of 10–30 years)	IV	181	12
Statutory habitat protection areas	IV	1 188	22
Statutory species protection areas	IV	267	6

Sources. Based on data from: Ramsar Convention, UNESCO World Heritage Site and Biosphere Reserve, and HELCOM websites (February 2022); EEA Natura 2000 Barometer, Natura 2000 as at end 2020; website of the Finnish MoE[19] (April 2020); Heinonen, M. (2020) Metsähallitus Parks & Wildlife Finland.

them. The Land Extraction Act[20] enables, for example, the protection of nationally representative glaciofluvial esker formations and rocky areas, as well as landscapes of outstanding natural beauty.

A special tool for promoting biodiversity in urban areas is the Finnish National Urban Park (NUP) concept. The NUPs are established to protect and maintain the beauty of the cultural or natural landscape, biodiversity, historical characteristics or related values concerning the townscaping, social, recreational or other special values of an area in an urban environment. NUPs are established through provisions of the Land Use and Building Act. Today the network of NUPs includes areas from nine cities.

Natura 2000 Sites

Finland's original proposal in 1998 for the Natura 2000 network (i.e. proposed SCIs) was mostly based on the existing protected areas and areas in the nature conservation programmes. Most proposed Natura 2000 sites were on state-owned land. Several complementary areas were added in

[19] www.ymparisto.fi/en-US/Nature/Protected_areas/Natura_2000_areas_in_Finland

[20] 555/1981 Maa-aineslaki.

1999, 2002 and 2004, in particular representing inland waters and coastal and marine habitats in the Baltic Sea. The latest amendments were made in December 2018. Complementary sites include areas where many activities, such as forestry, building of summer residences or water construction works, are regulated by the Land Use and Building Act, the Forest Act and the Water Act, and harmful impacts are assessed where necessary. At the end of 2020, 1 875 sites had been designated as Natura 2000 sites, covering c. 13% of the terrestrial area and c. 10% of Finland's marine waters. The network is considered complete, although there are minor scientific reservations remaining for two insect species.

Sites Designated under Non-EU International Legislation

The list of UNESCO World Heritage Sites includes the Quark Archipelago with the High Coast (on the Swedish coast of the Bothnian Bay). The Quark is located in the centre of the area representing considerable land uplift (about 8 mm per year) and is one of the best places in the world to see post-glacial land uplift processes in action. There are 34 Baltic Sea Protected Areas (BSPA) designated under the HELCOM Convention.

Additional designated sites that are not protected areas by themselves include 49 Ramsar Sites, all of which are included in the Natura 2000 network, as well as the majority of the BSPA areas. There are also two UNESCO MaB Programme Biosphere Reserves (Archipelago Sea Area and North Karelian Biosphere Reserve), parts of which are within some important NCA protected areas.

According to an assessment of the impacts of implementation of the EU Biodiversity Strategy in Finland in 2021 (Kärkkäinen and Koljonen, 2021), **overall protected area coverage** – taking into account all overlaps and sites pending national designation, and excluding sites designated under global legislation (unless they have protected area status under domestic legislation) – amounted to 55 825 km^2, which is about 13.4% of the total area of Finland. These areas cover about 13.8% of Finland's land area, including inland waters. As noted previously, distinguishing between marine and some freshwater areas is very difficult in Finland, but UNEP-WCMC (2022) estimate that marine protected areas total 9 526 km^2 based on WDPA data, which is about 12% of Finland's marine waters.

14.5.2 The Coherence of Protected Areas and Ecological Networks

The Finnish protected area network (PAN), which includes all domestic protected areas and the Natura 2000 sites, is mainly representative of the northern half of the country, and most large protected areas (representing particularly mires, forests and fells) are in the north. A comparison between northern and southern Finland, made in the late 1990s, revealed poor protection coverage of many habitat types in southern Finland, especially old-growth forests with high levels of decaying wood (Virkkala and Toivonen, 1999; Virkkala *et al.*, 2000).

The PAN has also been studied by ecologists and conservation biologists at universities. In particular, the late Professor Ilkka Hanski and his colleagues highlighted the problems of small, isolated reserves, and advocated the need for larger areas with sufficient connectivity (Hanski, 2000; Kotiaho, 2017). In the early 2000s, specialised conservation prioritisation software (Zonation) was developed to study the coherence of protected area networks and select and prioritise complementary areas.[21] This is now regularly used in the conservation planning of protected areas.

Connectivity in the network is also likely to be insufficient in the light of climate change and population dynamics of threatened species (Aapala *et al.*, 2017, and references therein). The greatest challenges lie in southern Finland, where habitats are fragmented and land-use pressures are highest. Evidence for this comes also from the regional distribution of threatened species and habitat types (Hyvärinen *et al.*, 2019; Kontula and Raunio, 2019).

Finland has cooperated on nature conservation with Russia and the Nordic countries for more than 40 years, and now also with the Baltic States.[22] The cooperation has been successful, resulting in a couple of representative transboundary conservation areas along the 1 300 km long Fennoscandian Green Belt area on the both sides of the Finnish–Russian border. Furthermore, the cooperation has also produced a wide knowledge base for planning and development of the PAN in northwest Russia and the Barents area. Further information on this is provided in online Annex FI.1.[23]

14.5.3 Approaches to Protection and Levels of Protection

The NCA contains provisions on the use of protected areas, as well as on protection orders and derogations. If the protected area is located on private land, the decisions are made by the regional ELY-Centre.

As regards specific provisions for land use, the majority of Finnish protected areas are located on uninhabited land. Generally prohibited land-use activities include forestry, extraction of peat, soil or minerals, and construction of roads or buildings (except for visitors). Removal of animals or plants is also generally prohibited, with the exception of harmful species. A number of activities that are part of so-called Everyman's right are generally permitted, including light fishing and berry and mushroom picking. In large state-owned Wilderness Areas, and some other protected areas in sparsely populated northern Finland, hunting of game by locals is permitted.

21 www.helsinki.fi/en/researchgroups/digital-geography-lab/software-developed-in-cbig#section-52992

22 www.bpan.fi/en/

23 All annexes are free to access online at CUP: www.cambridge.org/natureconservation

The main part of the Finnish Natura 2000 network is established and governed according to the domestic protected area designations. For other areas, their land-use and management actions are based on the relevant sectoral legislation (e.g. for forestry, water protection, environmental protection). Most commonly, actions limit possible land-use practices in the area, which are governed through permit procedures managed by ELY-Centres. If necessary, active management is also carried out, such as maintaining traditional grazing of farmland habitats.

Since the late 1990s, nature conservation has shifted towards voluntary means. The most important instrument for supporting this has been the MoE and MoAF funded METSO Forest Biodiversity Programme for Southern Finland, which began in the early 2000s. Under this, private landowners can offer their forest areas of high biodiversity value for protection. The proposed area is assessed according to scientific criteria and, if representative and important enough, can be established as a Privately-owned Protected Area. Landowners receive a lump sum for compensation based on the economic value of the forest. Less often, the landowner receives compensation for protecting the forest through a fixed-term agreement (usually for 10–20 years). Often the forest areas are purchased by the state for protection as well.

Municipalities and other public agencies can offer their forests for conservation as well, but payments for their forests are lower. State-owned forests are protected as a complementary part of the programme. The aim of the project is to secure 1 000 km^2 of new protected forest areas. METSO has a wide acceptance in Finnish society due to its voluntary character, and many valuable forest sites have been protected during the programme. However, from the viewpoint of nature conservation planning, many protected sites are small and are not situated optimally in terms of connectivity to larger protected areas.

Privately-owned Protected Areas have also been established without compensation, as happened in a special campaign in 2017 when Finland had its 100 years' anniversary.

14.5.4 Approaches to Managing Protected Areas

Over 90% of the PAN is owned by the state and managed by PWF, which works as a nationwide organisation divided into four regional units. The 13 regional ELY-Centres, together with PWF regional units, draw up regional general management plans for Natura 2000 sites in mainland Finland. According to the plans, in 2016, around 58% of the area of the Natura 2000 network had up-to-date site-specific management plans, 7% had no planning needs as maintaining the existing natural condition was sufficient, 25% required new plans, and 10% required updated plans.

ELY-Centres and PWF also monitor and evaluate the status of all Natura 2000 sites through the so-called NATA process. The aim is to define the necessary actions and type of plans that are needed to achieve the conservation objectives of the site. Municipalities are often involved in operational management work. The autonomous province of Åland is responsible for the Natura 2000 sites within the province.

Protected sites are managed by unified principles, as most recently set out by PWF (Metsähallitus, 2016). Most habitat restoration and management activities focus on forests, mires and semi-natural grasslands. During 2010–2018, PWF carried out restoration measures on 13 800 ha of forests and mires. The management of protected areas on private land (owned by private landowners, municipalities, NGOs, enterprises, etc.) is coordinated by national and regional conservation authorities (ELY-Centres), in collaboration with PWF.

The contribution of voluntary workers to habitat management is mostly organised by NGOs (WWF, FANC). This contributes to management of semi-natural grasslands and some wooded habitats, restoration of mires and small water bodies such as brooks, or eradication of invasive alien plants.

An extensive international evaluation of the effectiveness of protected area management in Finland was carried out in 2004 (Gilligan *et al.*, 2005). Its recommendations led to regular reporting on the state of protected areas to analyse management effectiveness. A report on Finnish protected areas and their management from 2000 to 2005 was also published by PWF (Heinonen, 2007). Unfortunately, no such comprehensive reports have been published since, although management in many protected areas has been assessed.

14.6 General Nature Conservation Measures

14.6.1 Development Planning and Impact Assessment

Finland has a hierarchical land-use planning system that is steered at the highest level by the National Land Use Guidance given by the government. This guidance ensures that authorities in charge of land-use planning at all lower levels take into account issues of national importance, such as pollutant emissions and biodiversity. Thus, the 18 regional land-use plans implement the national goals by setting regional goals and the framework for local land-use planning. At municipal levels, Master Plans steer and coordinate Detailed Plans that consider individual buildings and areas.

Municipalities have the sole right to draw up General and Detailed Plans for their land use while regional public authorities ensure that the National Land Use Guidance is followed and common goods, including biodiversity, and obligations arising from the NCA, are appropriately taken into account. In 2017, the role of regional public authorities in participating in the process and filing complaints was restricted by a change in the Land Use and Building Act. The change aimed to streamline the land-use planning process and eliminate unnecessary regulations. The change was controversial as, whilst it may have saved some time and resources, environmental NGOs felt that it reduced the quality of the planning process.

National Land Use Guidance and Regional Land Use Plans aim for ecological sustainability of land use, including through planning ecological corridors and other green infrastructure across municipal-level planning. They also enable many other environmental issues to be addressed that are indirectly linked to biodiversity, such as sufficient suitable areas for recreation, sustainable use of water and extractable land resources, and environmentally sustainable arrangements for transport and technical services.

Forestry is the prevailing use of most of the Finnish landscape and is not limited in practice through the planning system, unless another land use is designated in the plans. However, important areas for biodiversity, such as mires in a natural state, are indicated in land-use plans. The planning authority must give consideration to the presence of these areas in its more detailed planning. These types of reservations can affect forestry use in the area, and in principle constitute an obligation for the public authority to compensate economic loss to the landowner. In practice, this may mean establishing a protected area with appropriate economic compensation for the landowner.

In spite of the relatively effective land-use planning system, the Red List of habitats found that construction activities are contributing to the threatened status of many habitat types (Kontula and Raunio, 2019). However, such threats are mostly of local or regional importance and are seldom the primary cause of the threatened status of habitats.

SEA and EIA procedures cover biodiversity. Avoidance and mitigation are pursued but the residual impacts are normally not compensated for by offsetting. However, there are plans to develop biodiversity offsetting in Finland: studies have been made[24] and pilots for large public infrastructure projects are planned.

14.6.2 Habitat Management and Restoration: Agriculture

Farmland covers only 8% of Finland overall, but in the southwest it accounts for more than a quarter of the land area. Farmland hosts up to 18% of the country's threatened species, most of which are dependent on traditional low-intensity farming practices that have created diverse habitats, which partly resemble natural open habitats. Small and mid-scale mosaics of arable land and forest patches with many ecotones are also essential to maintain a rich biota. Traditional agriculture in Finland practically ceased during the twentieth century (Pykälä, 2001). Changes associated with the intensification of agriculture – including the loss of meadows, open ditches, field margins, natural pastures, small farms and grazing livestock – are the most important factors threatening farmland biodiversity.

As an EU Member State, Finland implements the EU's CAP. During various CAP periods both positive and negative advances have taken place.

The CAP for 2014–2020 introduced the new requirement for farmers to undertake certain 'greening measures': minimum crop diversity, maintaining permanent grassland and having at least 5% of arable land as ecological focus areas (see Section 4.3 for details.) However, as found over much of the EU, the measures only had a modest positive impact on biodiversity in Finland (Hyvönen *et al.*, 2020).

The RDP for mainland Finland 2014–2020 included more effective biodiversity measures. The most important was the continuation of effective and well-established agri-environment schemes that support the management of semi-natural farmland habitats (i.e. various semi-natural grasslands and wooded pastures), with the highest support given to the most valuable sites. Other important schemes were for the management of wetlands and buffer zones by water bodies. Compensation payments are given to farmers whose fields are used by migrating Common Cranes (*Grus grus*), geese and swans (Hyvönen *et al.*, 2020). Other important actions include completion of the nationwide inventory of traditional farmland biotopes, with several management projects that have followed.

The RDP also provided support for organic farming. Organic livestock farming has been found to increase bird abundance in boreal regions, and thus contributes to maintaining some threatened species (Santangeli *et al.*, 2019).

According to an evaluation (Hyvönen *et al.*, 2020), the 2014–2020 RDP biodiversity targets were mostly achieved, but the areas covered remained low for many measures and need to be increased in the future. The declining trend of the High Nature Value (HNV) farmland area indicator also highlights the need to continue promoting farmland biodiversity management.

14.6.3 Habitat Management and Restoration: Forestry

Biodiversity Changes in Forests

Forests are the country's most significant renewable natural resource. They are also the focus of much economic and social discussion, not least because almost one million Finns own some forests. Most of the country, 200 000 km^2, is available for wood production, 61% of which is privately owned. The Finnish state owns about one-third of forest land, mostly in less productive areas in eastern and northern Finland. The forestry sector has a significant coordination and advice system, consisting of the Finnish Forest Centre[25] with regional offices, and local forest management associations.

Historic uses of forests, such as subsistence felling of trees, slash-and-burn agriculture, cutting of pine forests to produce tar, and cattle grazing, have shifted to the production of sawn timber and cellulose for paper. Since the middle of the twentieth century the old practice of subsistence felling, mostly cutting old, large trees, and leaving the smaller trees to grow, has been replaced by new, shorter rotation systems. These

[24] See for example: https://blogs.helsinki.fi/habitaattipankki/habitat-bank-research-consortium/habitat-bank/?lang=en

[25] www.metsakeskus.fi/en/

produce fast-growing even-aged homogenous tree stands that are thinned two or three times and then clear-felled. After World War II, large areas of the northern forests in Lapland, mostly state owned and in a more natural condition, were clearcut and the forest soils ploughed to enhance regeneration of pine trees (which was done by natural regeneration, sowing of seeds or planting saplings).

Since 1950, many measures have been adopted to intensify forest management over the whole country (Virkkala and Toivonen, 1999). Forest roads have been built systematically to improve the efficiency of harvesting, which has greatly homogenised Finnish forests and reduced and fragmented old-growth areas. Forestry has also been improved by the drainage of mires, most intensively in the 1960s and 1970s. As a result, about 60 000 km^2 of peatlands have now been drained, particularly in the south, even though 5 000–10 000 km^2 of this is not economically profitable (Kontula and Raunio, 2019). The drainage has led to the decline of many types of mire, many of which are threatened habitats (Kontula and Raunio, 2019). The drainage of peatlands has also led to harmful humus and nutrient levels in many inland waters.

As a consequence, the proportion of old-growth forests has diminished considerably (especially in the north), such that only 14% of forests are over 140 years old.[26] At the same time, forest fires have been effectively suppressed; as fire is an important ecological process in boreal forests, this has greatly changed their natural disturbance and regeneration regimes. Along with these changes, the amount of dead and decaying wood has decreased substantially. The amount of decaying wood in ordinary commercial forests is currently some 3–6 m^3 per ha,[27] which is only a fraction of the 60–120 m^3 per ha that occurs in natural old-growth forests (Virkkala and Toivonen, 1999).

Forestry activities were previously carried out mostly during the winter when the ground was frozen and covered by snow. Due to climate change, as well as the demand for continuous supplies of wood to industries, tree-felling is now also common in the summertime and has many ecologically harmful consequences. For example, the large-scale destruction of nests of breeding birds is raising public concern.

Biodiversity Considerations in Commercial Forests

In order to maintain biodiversity in commercial forests, the Forest Act was renewed in 1996. The Act recognised 10 valuable habitats of high natural value. These 'key biotopes' include natural springs, brooks and small ponds with their surroundings, patches of herb-rich woodland, nutrient-rich mires, rock outcrops and cliffs. Representative sites with these habitats should be managed with special care or set aside. HD Annex I habitat types bog woodland (91D0*), Fennoscandian herb-rich forests with *Picea abies* (9050) and Fennoscandian deciduous swamp woods (9080*) are examples of habitat types that are subject to the Forest Act.

Whilst the concept is good, the protection of habitats in this way (altogether 1 070 km^2 in 2017) in the wider landscape and actual impact on their conservation status has been mostly minor. The impact has also been reduced by changes introduced in 2014 to the Forest Act, which redefined the sites to be of limited size and of minor economic significance.[28]

Amendments to the Forest Act made it easier for forest owners to decide how to manage their forests. This has led to a slight increase in continuous-cover forestry instead of the dominant periodic-cover forestry that results from interval growing and clearcutting. Continuous-cover forestry maintains the forest canopy structure, soil carbon and forest biota better than interval growing (Peura *et al.*, 2018), and some forest owners, particularly those preferring multi-use of forests, are interested in this form of forestry. However, companies and forest organisations prefer interval growing, reflecting their desire to maximise harvest rates.

Besides the Forest Act, since the 1990s best practice guidelines have been produced by Tapio (a Finnish government-owned company that is a leading provider of forest management advice) that intend to introduce more environment-friendly forestry practices in managed forests. Recent best practice guidelines (Äijälä *et al.*, 2019) are widely distributed.

Forestry in State-Owned Forests

Metsähallitus Forestry Ltd is responsible for harvesting and selling timber as well as maintaining state-owned forests in accordance with the principles of multiple use. More than €100 million of its profits are transferred to the state annually. The forestry business also has statutory social obligations,[29] such as preserving biodiversity, ensuring recreational access to forests, promoting employment, and securing favourable conditions for the practice of reindeer husbandry and Sámi traditions.

Natural resource planning by Metsähallitus refers to the long-term sustainable use of state-owned land and water areas. These plans have been drawn up since 1997, each covering a 10-year period. Since 2002, the planning system has included a landscape–ecological assessment. When making decisions about multiple use of forests, protected areas are complemented on state-owned land with sites of high natural value. These sites are treated with special care or set aside. Guidance on forest planning and management has been provided by Metsähallitus (Päivinen *et al.*, 2011).

These efforts in state-owned forests have resulted in some biodiversity benefits compared to ordinary commercial forests on private land, such as a higher proportion of decaying wood and the better maintenance of threatened species. However, this success has been overshadowed by the higher levels of extraction that have occurred in recent years to increase profits

[26] Indicator FO8: www.biodiversity.fi
[27] indicator FO6: www.biodiversity.fi
[28] Assessment of reform of Forest Act: http://urn.fi/URN:ISBN:978-952-326-897-5
[29] Act on Metsähallitus (234/2016).

and meet budgetary demands. This has led to severe criticism by NGOs (FANC, Greenpeace, WWF Finland), as the high levels of wood extraction damage biodiversity.

14.6.4 Invasive Species

Invasions of alien species have increased during recent decades, but their influence has remained modest compared to many other countries in Europe. However, some have been damaging, such as the American Mink (*Neovison vison*) and Raccoon Dog (*Nyctereutes procyonoides*), which escaped from fur farms and have caused substantial local declines in some species, particularly amongst waterbirds. Roe Deer (*Capreolus capreolus*) and White-tailed Deer (*Odocoileus virginianus*), both introduced for game hunting, are rapidly increasing in the southern part of the country, partly because of game feeding by hunters. Their increases have, paradoxically, led to calls from hunters to increase derogations from the protection of their predators, particularly the Eurasian Lynx.

Finland's National Strategy on Invasive Alien Species,[30] which was published in 2012, includes a list of the most harmful IAS. There is now national legislation on IAS listed by EU and national authorities. Management plans have since been drafted for some of those. The national IAS portal[31] includes information on these species, maps of their ranges and instructions for combating them, and enables citizens to identify IAS and report their observations. There has been a clear trend towards better awareness of the threats and risks related to IAS in recent years, and municipalities, local communities and NGOs have organised activities to control them, including participation in EU-funded projects.

14.6.5 Climate Change Adaptation

Finland's *National Strategy for Adaptation to Climate Change* (2005) and the *National Climate Change Adaptation Plan 2022* of Finland were prepared as a joint effort by several ministries (mainly MoAF, MoE and the Ministry of Economic Development).[32] The strategy and plan cover most aspects of society. The role of renewable natural resources, particularly wood as an energy source and substitute for products made from fossil fuels, has been seen as a very important component of Finnish climate change mitigation and adaptation.

The role of forests in climate change mitigation has recently led to increasing debate on forest harvesting rates. The government policy introduced in 2015, called a '*Bioeconomy Boom*', stated that timber harvesting could be sustainably increased from the recent 68 million m^3 to 85 million m^3 per year in the near future. However, from a carbon balance viewpoint, recent scenario analyses suggest a lower harvesting intensity (60 million m^3), which also means harvesting tree stands at an older age (Heinonen *et al.*, 2017). These analyses have been ignored by the forestry lobby but welcomed by environmental NGOs (FANC, WWF Finland), stating that an increase in harvesting rate overlooks the importance of maintaining and enhancing carbon stocks in tree stands and in forest soils.

There have been many studies on the effects of climate change on butterflies, birds and other species (Aapala *et al.*, 2017). More recently, a three-year research project entitled the 'Finnish protected area network in a changing climate' (SUMI project) started in 2018 at the Finnish Environment Institute. The project aims to produce information on the capability of the protected area network to maintain biodiversity under the simultaneous threats of climate change and land-use change. Preliminary research suggests that the Finnish protected area network requires an improvement in south-to-north connectivity (Aapala *et al.*, 2017). A similar conclusion has been made concerning the marine protected areas of the Baltic Sea (HELCOM, 2016).

14.7 Nature Conservation Costs, Benefits and Funding Sources

14.7.1 Nature Conservation Costs

A number of estimates have been made of the costs of nature conservation and restoration in Finland. These include an estimate by Kotiaho *et al.* (2015) of financing needs to achieve the 15% restoration target (Aichi target 15) set under CBD biodiversity targets for 2020 and the EU 2020 Biodiversity Strategy (Target 2). The study examined restoration prioritisation scenarios and estimated the costs of each (based on the costs per hectare of different restoration measures). The scenario that would reach the set target of 15% restoration of degraded ecosystems alone was estimated to require €445 million annual investments between 2016 and 2050.

The costs of implementing Natura 2000 have been most recently estimated in the finalised Prioritised Action Framework (PAF) for Natura 2000 for 2021–2027 (MoE, 2021). It includes an analysis and prioritisation of habitat restoration and management measures within and outside the Natura 2000 network that would result in the highest increase of ecological values in the Natura 2000 network. Preparation of the PAF involved around 100 experts, from the Finnish Board on Ecological Restoration (FBER)[33] and other restoration experts.

The budget needs outside Natura 2000 sites are several times greater than those within. In particular, supporting green infrastructure measures form by far the largest running costs, amounting to €605 million annually, with one-off costs being around €18 million annually averaged over the seven years. Natura 2000 site-related maintenance and restoration

[30] https://vieraslajit.fi

[31] www.vieraslajit.fi

[32] www.mmm.fi/en/nature-and-climate/climate-change-adaptation

[33] www.metsa.fi/en/nature-and-heritage/habitats/finnish-board-on-ecological-restoration-fber/

costs for species and habitats are estimated at €19 million running costs and €7.5 million for one-off costs annually.

Horizontal measures and administration related to Natura 2000 also lead to major annual running costs of €126 million. The one-off costs of developing the whole network total €64 million annually. Additional species-specific measures, not related to specific ecosystems or habitats, are estimated to cost €14 million annually in running costs and require average annual one-off investments of €9 million.

The total cost of the Natura 2000 priority financing needs was estimated to be €862 million per year on average, amounting to €6.0 billion over the financing period 2021–2027. These measures would mainly safeguard the current status of the HD habitats, induce positive trends and achieve the favourable conservation status of some habitats. The structure and function of many habitats would substantially improve and this would be reflected in the status of the habitats' typical species as well.

Whilst the measures included in the PAF for Natura 2000, including those outside the network and species-specific measures, are expected to provide very wide-ranging biodiversity benefits beyond BHD habitats and species, it is important to bear in mind that complementary uncosted measures are required to conserve overall nature in Finland.

14.7.2 The Economic Benefits of Nature Conservation

The National Parks had an estimated total income of €219 million in 2019, and supported about 1 726 full-time jobs.[34] Additionally, it was estimated that five hiking areas (with less stringent conservation) contributed €13 million and about 112 full-time jobs. However, there was considerable variation between the parks.

Broader assessments of the socio-economic benefits of ecosystem services have been conducted in Finland and in the Nordic countries (e.g. Kettunen *et al.*, 2012; Jäppinen and Heliölä, 2015; Saarikoski *et al.*, 2015). These show that the ecosystem services that depend heavily on the condition of the underlying ecosystems are socio-economically important for Finland; they also highlight some particularly important services in the region, such as reindeer herding (especially in the north), wood-based bioenergy, non-timber forest products (e.g. berries, mushrooms and game), recreation and tourism. Some of these ecosystem services are expected to become more important in the future, including carbon storage and sequestration, water purification (e.g. in the context of eutrophication of the Baltic Sea), pollination, and biodiversity increasing resilience against pest outbreaks and extreme weather events.

The relationship between biodiversity and people's health is also becoming more apparent and important, such as in relation to their immune systems (Roslund *et al.*, 2020) and reducing the occurrence of allergies (Hanski *et al.*, 2012).

14.7.3 Funding Sources for Nature Conservation

Table 14.5 provides an overview of the main funding sources that have been used for nature conservation in Finland, clearly indicating that EU funding instruments, with national co-financing, have provided most funds. Whilst the LIFE programme has played a key role through its dedicated projects, the largest source of funding for nature conservation in Finland has been the RDP measures, supported by the EU European Agricultural Fund for Rural Development (EAFRD). Of these, by far the most effective for biodiversity in agroecosystems has been the agri-environment–climate measure, which has supported five-year environmental management contracts that aim to maintain semi-natural grasslands. Over 30% of the area under these contracts is located in Natura 2000 areas and around 60% of semi-natural grasslands in the Natura 2000 areas are covered by these contracts. In contrast, the RDP measures available for forests have not been used.

Whilst the EU European Maritime and Fisheries Fund (EMFF) can support nature conservation measures (e.g. opening fish migration routes), its potential has also not been fully used, mostly due to domestic administrative obstacles. Similarly, although the European Regional Development Fund (ERDF) has financed many projects that have some biodiversity elements, the Operational Programme 'preserving and protecting the environment and promoting resource efficiency' was not selected as one of the investment priorities in Finland (as was also the case with the European Social Fund). Therefore, most of their funding has not been directed towards nature conservation.

Other EU funding instruments have included the European Neighbourhood Initiative (ENI), which has contributed somewhat to nature conservation in Finland by facilitating projects, most importantly between Finland and Russia. Although ENI projects are not normally focussed on nature conservation, it can be one of their objectives.

Compared to many other EU countries, a significant proportion of funds for nature conservation in Finland has come from national sources, especially since a significant increase was made to the budget line in 2019 (totalling €186 million in 2020, including co-financing of EU funding instruments). The government channels the funds to nature conservation through several sectoral policies. Most of the budget allocated to the environmental administration is used for conservation measures, such as land acquisition, management of protected areas and compensating landowners for economic losses. It also covers most of the co-funding of EU LIFE projects. Municipalities also provide a significant amount of co-financing for EU projects, but information on this is lacking.

There are also substantial independent sources of funds, such as private foundations (e.g. the Baltic Sea Action

[34] www.metsa.fi/en/economic-benefits-of-national-parks

Table 14.5 Funding sources for nature conservation in Finland.

Source and period	Measure/category	Average annual allocation (€ m)		Comments
EU funding instruments		**EU + national**	**EU**	
EARDF Rural Development Programme measures – allocations to Priority 4 (restoring, preserving and enhancing agriculture and forest ecosystems) 2014–2020	M10 Agri-environment–climate	251.0	105.3	Covers >90% UAA, similar proportion for N2k
	M13 Areas facing constraints	241.9	101.4	May help maintain management by grazing/ mowing
	M11 Organic farming	51.3	21.5	Not included in PAF for N2k
	M2 Advisory services, etc.	1.7	0.7	Not included in PAF for N2k
	M16 Cooperation	1.7	0.7	Not included in PAF for N2k
	M1 Knowledge transfer, etc.	1.0	0.4	Not included in PAF for N2k
	M4 Investments in physical assets	0.9	0.4	High biodiversity benefits from restoration of semi-natural grasslands and small wetlands
RDP Priority 4 subtotal		*549.6*	*230.5*	
LIFE Programme	Nature and biodiversity projects (1995–2020)	6.0	3.3	64 projects
	Integrated project 2016–2022	3.5	1.7	FRESHABIT (LIFE14 IPE/FI/000023)
EMFF 2014–2020	Article 40.1 and Art. 80.1 measures for marine biodiversity, ecosystems and N2k	1.1	0.5	
ERDF 2014–2020		1.1	0.5	Mainly projects related to nature tourism in N2k areas
Others	Interreg, ENI CBC and ESF programmes	5.2	4.1	See text
EU funding instruments subtotal		*566.4*	*240.7*	
National funds within 2014–2020		**Additional national***		**Comments**
Metso programme		26.5		
Government budget	Financing for N2k	7.6		
HELMI programme		6.0		Operated 2019–2020. See text.
Government budget	BHD species, inventories and N2000	3.0		
VELMU programme		1.5		See Section 14.8
Water Protection Programme	Mitigation and restoration measures	1.4		Operated 2019–2020
Municipalities		?		Not known
Finnish Natural Heritage Fund		1.0		Annualised between 2015 and 2019
Additional national subtotal		*>47.0*		
National total including co-financing of EU instruments		*>372.8*		
Total		**>613.5**		

Notes. * Excludes co-financing of EU funds unless indicated. The subtotals and totals are only approximate indications of the average annual levels of funding as they are based on annualised data from differing periods.
Sources. Based on data from: RDP data from European Commission ESIF funds database (23 October 2020); RDP allocations to Natura 2000, ERDF, EMFF and national funding from 2021 to 2027 PAF; Finland LIFE factsheet (January 2021) and LIFE Project database; Ministry of Finance website – state budget.[35] See Section 1.3 for further details.

[35] https://budjetti.vm.fi/indox/sisalto.jsp?year=2021&lang=fi&maindoc=/2021/tae/hallituksenEsitys/hallituksenEsitys.xml&id=/2021/tae/hallituksenEsitys/YksityiskohtaisetPerustelut/35/35.html

roup)[36] and NGOs that run nature conservation projects. verall funding levels are not documented and are likely to variable, but a recent example is the purchase of an area for rotection by the Kone Foundation in 2021 that cost .2 million.

The recent budget increase has helped to reduce the nding gap identified in the recent PAF and earlier by otiaho *et al.* (2015). However, at the same time, the in-depth nalysis for the Natura 2000 PAF identified new major funding eeds for green infrastructure (i.e. measures conducted in the ider landscape) and horizontal measures for the management f the protected area network. Consequently, the PAF's average annual funding needs of €862 million are considerably igher than the recent financial allocations. Furthermore, the llocation for the implementation of the NBSAP in the state udget needs to be even higher, considering the wider scope of he NBSAP compared to the PAF. However, there is currently o thorough analysis of the funding needs of the whole NBSAP.

14.8 Surveillance, Monitoring, Reporting and Applied Research

14.8.1 Research Institutes and Programmes

Finnish universities provide well-organised teaching and research that train nature conservation professionals as well increasing knowledge on biodiversity issues. Governmental research institutes also have a major role, especially SYKE (under the MoE), which is the R&D centre covering nature conservation, marine and freshwater environments, global change and environmental policy. In the 2010s, three formerly separate research institutes for forestry, agriculture, and game and fisheries were combined to form the Natural Resources Institute Finland (LUKE), under the MoAF.

PWF has wide expertise on habitat restoration and management, has coordinated many EU LIFE projects, and has a central role in Finnish international cooperation on nature conservation. Together with SYKE it conducts the large Finnish Inventory Programme for Marine Underwater Environment (VELMU).[37] VELMU collected data on species, communities and habitats in the Baltic Sea in 2004–2016 and is now scheduled to continue until the end of 2025. The development of the network of marine protected areas has been greatly enhanced by the programme.

There have been some major research programmes in the field of biodiversity science. The National Research Council (The Finnish Academy) launched an interdisciplinary research effort, the Finnish Biodiversity Research Programme (FIBRE), which took place over 1997–2002. It was followed by a more applied research-oriented programme, MOSSE, funded by the MoAF and the MoE. A part of the METSO Forest Biodiversity Programme for Southern Finland (see Section 14.5) was the large-scale Research Programme on Deficiently Known and Threatened Forest species (PUTTE), conducted over 2003–2016.[38]

14.8.2 Biodiversity Monitoring and Reporting

Monitoring of biodiversity in Finland is a shared task. The MoE is responsible for reporting on the EU Nature Directives, with SYKE having a coordinating role. Various organisations run separate monitoring schemes that produce information for multiple uses, which can contribute to reporting on the state and trends of biodiversity and use of natural renewable resources. The main actors are LUOMUS (The Finnish Museum of Natural History, part of the University of Helsinki), LUKE, SYKE and PWF. Key inventories of natural resources such as the National Forest Inventory, fish stock inventories and the wildlife censuses, as well as evaluations of the agri-environment support schemes, produce important biodiversity data. Many monitoring activities relate to species, including threatened ones (Toivonen and Liukko, 2005).

In many species monitoring schemes, most of the field observations are made by volunteers or as a joint effort between authorities, NGOs and amateur naturalists. The project coordination, data analysis and reporting are usually carried out by LUOMUS, LUKE and SYKE. For example, BirdLife Finland has a crucial role in collecting bird observation data and annually publishes, in cooperation with LUOMUS and SYKE, a comprehensive research-oriented publication on Finnish bird monitoring (The Yearbook of the Linnut Magazine).[39]

Biodiversity indicators have been used successfully to evaluate the first Finnish NBSAP for 1997–2005 (Auvinen *et al.*, 2007) and subsequent NBSAPs, and to report to the CBD (e.g. see the Fifth National Report[40] and the Sixth National Report[41]). The online national biodiversity indicator collection[42] presently includes more than 110 indicators reflecting the state and development of biological diversity as well as factors driving changes in Finnish nature. In addition to covering the main broad habitats (e.g. forests, mires and the Baltic Sea) it includes indicators relating to climate change and alien species. The same portal also includes the Finnish ecosystem service indicators and reporting on the implementation of actions in the Finnish NBSAP.

LUOMUS has an increasingly important role in collecting and sharing species data. It coordinates a national virtual service infrastructure, Lajitietokeskus, the Finnish Biodiversity Information Facility (FinBIF), established in 2012. This

[36] www.bsag.fi/en/home/
[37] www.ymparisto.fi/en-US/VELMU

[38] www.ymparisto.fi/en-US/Nature/Species/Threatened_species/Research_Programme_PUTTE_20092016
[39] www.birdlife.fi/jarjesto/julkaisut/vuosikirja/
[40] www.cbd.int/doc/world/fi/fi-nr-05-en.pdf
[41] https://chm.cbd.int/database/record?documentID=243215
[42] www.biodiversity.fi

distributes species data in an open access format through a portal[43] that also provides any registered user the opportunity to record and share their own observations.

In spite of these advances, the running and coordination of monitoring work and reporting of results have suffered during the previous decade, due to governmental budget savings. A promise of change may be the establishment of the Finnish Ecosystem Observatory (FEO)[44] in SYKE in 2020. FEO aims to manage information on ecosystem condition and to develop modern ways to monitor it (e.g. monitoring increasing ecosystem restoration efforts).

14.9 Future Developments

Efforts to conserve and restore nature have to tackle the increasing impacts of climate change, especially in northern countries such as Finland. This is evident from some marked ecological consequences of rising temperatures and changes in rainfall, including changes in the ranges of species, both in southern Finland and, in particular, in fjell areas in the north. In the European context, Finland will become more important for species from the continental biogeographical region, but gaps in the ecological network in the south may hamper species redistribution.

At the same time, climate change will lengthen the growing season and increase forest growth and agricultural production. These changes, together with increasing demand for the production of timber and forest biomass, may further encourage intensive forest management with shorter rotation times. However, there is also a growing need to maintain or increase carbon stocks and sinks in forests. These partially conflicting aims may lead to greater differences between ordinary commercial and protected forests. To counter this, there is a need to develop forest management practices, such as increasing continuous-cover forestry with mixed trees, and increasing the carbon stock in the forest ecosystem. There is also an urgent need to develop the network of protected areas, in terms of area, regional balance and south–north connectivity.

Urbanisation in Finland will continue, causing more crowding and construction activities around the few largest cities, with decreasing populations in rural areas. This may lead to further declines of semi-natural habitats in urban, and also in rural areas. However, the consequences of urbanisation could be partly compensated for by new tools for habitat management and restoration, such as habitat banking, ecological compensation, assisted immigration of species, and *ex situ* conservation. Although studies on biodiversity offsetting have been carried out in Finland, such as on the suitability of habitat types (Raunio *et al.*, 2019), concrete large-scale pilots are required before the introduction of an ambitious offsetting policy mechanism.

In addition to these pressures, there are also positive signals that may modify future nature conservation policy, such as the increasing pressure to tackle climate change through nature-based solutions (such as increasing natural carbon stocks) and the growing understanding of the links between biodiversity and health. There is also an increasing acceptance that those who damage biodiversity should bear the costs, in accordance with the polluter-pays principle.

Two developments to enhance the knowledge base for biodiversity measures are promising, if given adequate funding. The Finnish Ecosystem Observatory (FEO) described earlier provides a good framework for further developing biodiversity monitoring and bridging monitoring with political decision making. The SUMI[45] project is assessing the impacts of climate change on the network of protected areas in Finland and will help guide climate-smart nature conservation.

The HELMI Habitats Programme aims to strengthen Finnish biodiversity and safeguard vital ecosystem services. The programme's actions particularly cover protecting and restoring mires, wetlands (including aquatic bird habitats) and forest habitats, as well as managing semi-natural grasslands. Restoration will remain high on the biodiversity agenda in the coming years.

However, experience has shown that the outcome of nature conservation initiatives will be highly influenced by funding levels, and hence the political and economic situation. In particular, the consequences of the COVID-19 pandemic remain to be seen. It could lead to new financial cuts and increased pressure to exploit natural resources. On the other hand, there is an urgent need to build coherence between different sectoral policies to meet the targets set for biodiversity and climate. This opens new possibilities for nature conservation, habitat restoration, and climate change mitigation and adaptation. Much depends on how funds for recovery and resilience measures for the economic losses caused by the pandemic will be used.

14.10 Conclusions

Although Finland has a low population density and is dominated by vast areas of boreal forests, mires and inland waters, their widespread use has considerably degraded their ecosystems and reduced their biodiversity. This is evident from the Red List assessments of species and habitats, as well as reporting under the Nature Directives and WFD. This is also despite significant efforts to conserve biodiversity, particularly within protected areas.

Since the 1970s, Finland has built up a comprehensive protected area system (now covering 13.8% of the land area), through systematic protected area planning and ecosystem-based conservation programmes. This was sufficient to form

[43] www.laji.fi
[44] https://feosuomi.fi/en/
[45] www.syke.fi/fi-FI/Tutkimus__kehittaminen/Tutkimus_ja_kehittamishankkeet/Hankkeet/Suojelualueverkosto_muuttuvassa_ilmastossa_SUMI

the basis for the Natura 2000 network on land, which was mainly established over a relatively short time following EU accession in 1995.

The Nature Directives led to further important nature conservation advances, including new legislation, CJEU decisions that strengthened conservation efforts, and EU funding for LIFE projects. Besides the Nature Directives, HELCOM cooperation has stimulated nature conservation in the marine environment. Additionally, the three consecutive NBSAPs have mainstreamed the efforts to halt biodiversity loss across sectors.

Nature conservation has also benefited from a well-established system of governance and a good knowledge base on biodiversity and ongoing monitoring (with field work largely carried out by volunteers), supporting conservation planning. The management of most protected areas is also carried out by one single organisation (PWF), which facilitates both their management and the prioritisation and efficient use of resources. PWF has also conducted large-scale inventories, particularly the VELMU programme studying underwater habitats in the Baltic Sea. Finland has developed wide experience in restoration of boreal forests and mires as a result of projects since the 1980s.

So far, these efforts have not been sufficient to improve the conservation status of many habitats and species, which remains particularly challenging outside the Natura 2000 network. In particular, ecological sustainability of forestry has not been achieved, and more coherent forest, biodiversity and climate policies are required. In the marine environment, the eutrophication of the Baltic Sea needs to be halted before wide-scale profound changes in the status of the ecosystem can be seen. This requires true cross-sectoral policy coordination and implementation, which remains a challenge.

Over the last two decades, implementation of nature conservation policy and action plans seems to have advanced in waves depending on the governments' attitudes to nature conservation. After the dry years of the 2010s, resources for nature substantially increased in 2020. However, there needs to be a permanent increase in the financing of biodiversity. In addition to adequate funding, the key policy and practical measures that are needed to achieve Finland's nature conservation objectives are better coherence of the most important policy sectors, including forestry, agriculture, biodiversity and climate, with a fair transition to sustainable practices. Development of the protected area network also remains a priority, particularly in the south of the country.

Acknowledgements

We are very grateful to biologist Marita Arvela (formerly Senior Officer AT ETC/BD), Professor Eeva Primmer and Senior Advisor Tapani Veistola for many valuable and insightful comments on the manuscript. Senior Advisor Mervi Heinonen from the Wildlife and Park Service Finland gave us updated statistics on the Finnish network of protected areas. Dr Johanna Suutarinen introduced us to the article on wolf poaching. They are warmly acknowledged.

References

Aapala, K., Akujärvi, A., Heikkinen, R. *et al.* (2017) *Suojelualueverkosto muuttuvassa ilmastossa – esiselvitys*. Suomen ympäristökeskuksen raportteja 23/2017. Helsinki: Suomen ympäristökeskus.

Äijälä, O., Koistinen, A., Sved, J., Vanhatalo, K. & Väisänen, P. (eds.) (2019) *Metsänhoidon suositukset*. Tapion Julkaisuja. Helsinki: Tapio Oy.

Auvinen, A.-P., Hildén, M., Toivonen, H. *et al.* (2007) Evaluation of the Finnish National Biodiversity Action Plan 1997–2005. *Monographs of the Boreal Environment Research*, **29**, 1–54.

Auvinen, A.-P., Kemppainen, E., Jäppinen, J.-P. *et al.* (2020) *Suomen biodiversiteettistrategian ja toimintaohjelman 2012–2020 toteutuksen ja vaikutusten arviointi.* [*Executive Summary: Impact assessment of the implementation of National Strategy and Action Plan for the Conservation and Sustainable use of Biodiversity in Finland (2012–2020)*.] Publications of the Government's analysis, assessment and research activities 2020: 36.

Gilligan, B., Dudley, N., Fernandez de Tejada, A. & Toivonen, H. (2005) *Management Effectiveness Evaluation of Finland's Protected Areas*. Nature Protection Publications of Metsähallitus. Series A 147. Helsinki: Metsähallitus.

Hanski, I. (2000) Extinction debt and species credit in boreal forests: modelling consequences of different approaches to biodiversity conservation. *Annales Zoologici Fennici*, **37**, 271–280.

Hanski, I., von Hertzen, L., Fyhrquist, N. *et al.* (2012) Environmental biodiversity, human microbiota, and allergy are interrelated. *Proceedings of the National Academy of Sciences of the United States of America*, **109**, 8334–8339.

Heinonen, M. (ed.) (2007) *State of the Parks in Finland. Finnish Protected Areas and Their Management from 2000 to 2005*. Nature Protection Publications of Metsähallitus. Series A 170. Helsinki: Metsähallitus.

Heinonen, T., Pukkala, T., Mehtätalo, L., Asikainen, A., Kangas, J. & Peltola, H. (2017) Scenario analyses on the effects of harvesting intensity on development of forest resources, timber supply, carbon balance and biodiversity of Finnish Forestry. *Forest Policy and Economics*, **80**, 80–98.

HELCOM (2016) *Ecological coherence assessment of the Marine Protected Area network in the Baltic Sea*. Baltic Sea Environment Proceedings No. 148.

Hyvärinen, E., Juslén, A., Kemppainen, E., Uddström, A. & Liukko, U.-M. (eds.) (2019) *The 2019 Red List of Finnish Species*. Helsinki: Ministry of the Environment & Finnish Environment Institute.

Hyvönen, T., Heliölä, J., Koikkalainen, K. *et al.* (2020) *Maatalouden ympäristötoimenpiteiden ympäristö- ja kustannustehokkuus (MYTTEHO): loppuraportti*. Luonnonvara- ja biotalouden tutkimus 12/2020. Helsinki: Luonnonvarakeskus.

Jäppinen, J.-P. & Heliölä, J. (eds.) (2015) *Towards a Sustainable and Genuinely Green Economy. The value and social significance of ecosystem services in Finland (TEEB for Finland) – Synthesis and roadmap*. The Finnish Environment 1en/2015. Helsinki.

Kärkkäinen, L. & Koljonen, S. (eds.) (2021) *Arvio EU:n biodiversiteettistrategian 2030 vaikutuksista Suomessa*.

Luonnonvara- ja biotalouden tutkimus 75/2021. Helsinki: Luonnonvarakeskus.

Kettunen, M., Vihervaara, P., Kinnunen, S. *et al.* (2012) *Socio-economic importance of ecosystem services in the Nordic Countries. Synthesis in the context of The Economics of Ecosystems and Biodiversity (TEEB).* TemaNord 2012: 559.

Kontula, T. & Raunio, A. (eds.) (2019) *Threatened Habitat Types in Finland 2018. Red List of Habitats: Results and Basis for Assessment.* The Finnish Environment 2/2019. Helsinki: Finnish Environment Institute & Ministry of the Environment.

Kotiaho, J. S. (2017) On effective biodiversity conservation, sustainability of bioeconomy, and honesty of the Finnish forest policy. *Annales Zoologici Fennici,* **54**, 13–25.

Kotiaho, J. S., Kuusela, S., Nieminen, N. & Päivinen, J. (eds.) (2015) *Elinympäristöjen tilan edistäminen Suomessa. ELITE-työryhmän mietintö elinympäristöjen tilan edistämisen priorisointisuunnitelmaksi ja arvio suunnitelman kokonaiskustannuksista.* Suomen ympäristö 8/2015. Helsinki: Ympäristöministeriö.

Kröger, M. & Raitio, K. (2017) Finnish forest policy in the era of bioeconomy: a pathway to sustainability? *Forest Policy and Economics,* **77**, 6–15.

Metsähallitus (2016) *Principles of Protected Area Management in Finland.* Nature Protection Publications of Metsähallitus. Series B 217. Helsinki: Metsähallitus.

MoE (2017) *Uhanalaisten lajien suojelun toimintaohjelma.* Suomen ympäristö 2/2017. Helsinki: Ministry of the Environment.

(2021) *Prioritised Action Framework (PAF) for Natura 2000 in Finland including the Province of Åland – for the Multiannual Financial Framework period 2021–2027.* https://ym.fi/documents/1410903/39422803/PAF+2021-2027.pdf/64cdd518-e03a-77d9-b66d-6f47eda8c833/PAF+2021-2027.pdf?t=1644406670703

Päivinen, J., Björkvist, N., Karunen, L. *et al.* (eds.) (2011) *Metsähallituksen metsätalouden ympäristöopas.* Metsähalituksen metsätalouden julkaisuja 67. Helsinki: Metsähallitus.

Peura, M., Burgas, D., Eyvindson, K., Repo, A. & Mönkkönen, M. (2018) Continuous cover forestry is a cost-efficient tool to increase multifunctionality of boreal production forests in Fennoscandia. *Biological Conservation,* **217**, 104–112.

Pykälä, J. (2001) *Perinteinen karjatalous luonnon monimuotoisuuden ylläpitäjänä. [Executive Summary: Maintaining biodiversity through traditional animal husbandry]* (Summary in English). The Finnish Environment 495. Helsinki: Suomen ympäristökeskus.

Raunio, A., Anttila, S., Pekkonen, M. & Ojala, O. (2019) *Suitability of habitat types for biodiversity offsetting in Finland.* Helsinki: Publications of the Ministry of Environment. http://urn.fi/URN:ISBN:978-952-361-012-5

Roslund, M., Puhakka, R., Grönroos, M. *et al.* (2020) Biodiversity intervention enhances immune regulation and health-associated commensal microbiota among daycare children. *Science Advance,* **6** (42), 2578.

Saarikoski, H., Jax, K., Harrison, P. A. *et al.* (2015) Exploring operational ecosystem service definitions. The case of boreal forests. *Ecosystem Services,* **14**, 144–157.

Santangeli, A., Lehikoinen, A., Lindholm, T. & Herzon, I. (2019) Organic animal farms increase bird abundance in the boreal region. *PLoS ONE* **14**, e0216009.

Sarkki, S., Niemelä, J., Tinch, R. *et al.* (2015) Are national biodiversity *strategies and action plans appropriate for building* responsibilities for mainstreaming biodiversity across policy sectors? The case of Finland. *Journal of Environmental Planning and Management,* **59**, 1377–1396.

Suutarinen, J. & Kojola, I. (2018) One way or another: predictors of wolf poaching in a legally harvested wolf population. *Animal Conservation,* **21**, 414–422.

Toivonen, H. & Liukko, U.-M. (eds.) (2005) *Ehdotus biodiversiteetin tilan valtakunnallisen seurannan järjestämisestä.* Suomen ympäristö 759. Helsinki: Ympäristöministeriö.

UNEP-WCMC (2022) *Protected Area Profile for Finland from the World Database of Protected Areas, April 2022.* www.protectedplanet.net

Virkkala, R., Korhonen, K. T., Haapanen, R. & Aapala, K. (2000) *Metsien ja soiden suojelutilanne metsä- ja suokasvillisuusvyöhykkeittäin valtakunnan metsien 8. inventoinnin perusteella.* Suomen ympäristö 395. Helsinki: Suomen ympäristökeskus.

Virkkala, R. & Toivonen, H. (1999) *Maintaining biological diversity in Finnish forests.* The Finnish Environment 278. Helsinki: Finnish Environment Institute.

Chapter 15

France

Pierre Commenville

Key Statistics	
Terrestrial area (km^2)/% of EU-28*	548 936 (inc. Corsica 8 685)/12.6%
Marine waters (km^2)/% of EU-28*	370 746/8.1%
Population density (p km^2)/% of EU-28*	115/100%
GDP per capita (€)/% of EU-28	31 753/119%
Land cover (CORINE land cover classes 2018)	AGR 59%, FOR 26%, OSN 8%, ART 6%, WET 1%
Biogeographical regions	ALP, ATL, CON, MED, MATL, MMED
Number of HD Annex I habitats/%*	130/58%
Number of HD Annex II, IV and V species/%*	287/23%
Number of BD Annex I bird taxa/%*	118/67%
% HD UCS and decreasing habitats/ % unknown	32%/4%
% HD UCS and decreasing species/ % unknown	23%/13%
% breeding BD Annex I birds decreasing/% uncertain and unknown	20%/26%

* Excluding MAC and MMAC biogeographical regions.
Sources and codes. See Table 1.1.

15.1 The Natural History of France

This chapter considers nature conservation in the European part of France, which comprises continental France and the Mediterranean island of Corsica, as well as other islands in the Mediterranean, Atlantic Ocean and English Channel (*La Manche*). This is often referred to as 'Metropolitan France', but for brevity it is referred to here simply as 'France'.

France has a very high biodiversity, mainly due to its large size and location at a crossroads of four biogeographical zones: the Atlantic, Continental, Alpine and Mediterranean. Its flora and fauna have been considerably affected by glaciation, with relict populations of ice-age species remaining in the coldest locations, notably the high mountains. The sequence of past glaciations also led to speciation in glacial refuges (*nunataks*),[1] creating endemic species. Out of the 82 536 identified animals and plants, 2 697 (3.3%) are endemic to France – mostly in Corsica, closely folowed by the Pyrenees and Alps, with some in other Mediterranean islands and the Massif Central.[2]

The mountains are varied in their topography and geology, with the warm valleys of the maritime Alps and volcanic cones and craters of the Auvergne, the acidic rock of the rounded Vosges, the limestone peaks and valleys of the Jura, and the vast high peaks of the northern Alps. The last include the highest peak in western Europe, Mont Blanc, on the border with Italy, at 4 808 m altitude. These distinct mountain ecosystems are thriving and present comparatively large areas of unfragmented semi-natural habitats, such as Alpine and subalpine calcareous grasslands (6170)[3] or Alpine pioneer formations of the *Caricion bicoloris-atrofuscae* (7240*). These habitats support important characteristic species, including 45% of protected plant species[4] in France, 29% of which are strictly mountain species. Out of 53 protected mammal species, 46 occur in the mountains.

A relatively large proportion of the country is forested (see Key Statistics table), and forest area has increased in the twentieth century following land abandonment, reforestation to combat land erosion and subsidised afforestation to increase timber production. This has benefited forest habitats but has also reduced other important habitats, such as natural grasslands and moors. Atlantic acidophilous beech forests with *Ilex* and sometimes also *Taxus* in the shrublayer (9120) dominate and represent more than 60% of the forest cover. Also well-represented are Galicio-Portuguese oak woods (9230), constituting more than 20% of the forest cover. Also worthy of note are the *Tilio-Acerion* forests of slopes, screes and ravines (9180*) that are rare and localised, for which France has a major responsibility for conservation (Maciejewski, 2016).

Five major rivers (Seine, Loire, Garonne, Rhône and Rhine) flow across the country in a dense and evenly distributed

1 Glacial islands of land that rise above surrounding glaciers.
2 https://inpn.mnhn.fr/espece/indicateur see France Métropolitaine.
3 Numbers in brackets are HD Annex I habitat codes (see Appendix for full names). * Indicates a priority habitat.
4 In reference to legal protection at both national and regional levels.

network, and inland lakes include some of the largest in Europe (e.g. Leman, Bourget, Annecy and Serre-Ponçon). The Atlantic side of the country still contains very large marshes, important at continental scale (e.g. Marais Poitevin, Brenne, Brière). Together, water bodies and other wetland habitats cover about 1.3% of the country (see Key Statistics table). Despite their relatively small area they hold large and diverse populations of many species. About 20% of plants and about 50% of birds depend on these areas.

The coastline stretches for at least 5 860 km, bordering the North Sea (60 km), English Channel (1 700 km), Atlantic (2 400 km) and Mediterranean (1 700 km). Half of the length of the littoral zone is rocks and cliffs, a third is sandy, and the remaining part comprises marshes and mudflats. Within the 500 m strip bordering the coast, about 45% of the land still retains relatively natural aquatic and terrestrial habitats. These coastal habitats support on average 3.5 million waterbirds each winter, and more than 10% of the Atlantic flyway population of the following species (in decreasing order of importance): Avocet (*Recurvirostra avosetta*), Dunlin (*Calidris alpina*), Common Ringed Plover (*Charadrius hiaticula*), Turnstone (*Arenaria interpres*) and Grey Plover (*Pluvialis squatarola*) (Rocamora and Yeatman-Berthelot, 1999).

Because human presence dates back several millennia, the anthropic use of resources has shaped the original post-glacial habitats. Extensive pastoral farming has influenced natural habitats, even in the most remote mountains and coasts, and has led to new semi-natural habitats including grasslands. High Nature Value (HNV) farming systems have supported characteristic and species-rich habitats that are of high importance, which in France include hay meadows (6510 and 6520) and tall humid grasslands (6410 and 6430), hosting important populations of Whinchat (*Saxicola rubetra*) and Corncrake (*Crex crex*). France now has landscapes that are mostly massively modified by intensive agriculture, whilst the number of farmers has plummeted. For example, hedges have disappeared or strongly regressed in most places. In addition, most of the expansion of urban areas and other housing and infrastructure has occurred on agricultural land. As a result, the remaining areas of semi-natural habitat have declined considerably, to about 8% according to CORINE 2018 data (although a small proportion of agricultural grasslands are also semi-natural). Nevertheless, France still retains some important areas of semi-natural habitats, including dry grasslands in the mountains and in the rocky areas of other regions, as well as wet grasslands and their breeding waders in the Central West (Marais Poitevin). The country also hosts a significant fraction of the European populations of Hen Harrier (*Circus cyaneus*), Montagu's Harrier (*Circus pygargus*) and Grey Partridge (*Perdix perdix*).

15.2 The Status of Nature

15.2.1 Habitat Status and Trends

France faces a considerable challenge in conserving its large number and extent of semi-natural habitats, and evidence from the most recent Member State HD Article 17 monitoring results for HD Annex I habitats indicates that it is having limited success.[5] Overall, 76% of habitat assessments over 2013–2018 were unfavourable, and 32% had declining short-term trends (i.e. over the last 12 years). This is worse than in many other countries in the EU (see Table 5.1). Of particular concern is that a large number of habitats had a bad status, and of these 61% had a declining trend (Table 15.1).

Nevertheless, there are some variations within the country. Habitats in the Alpine biogeographical region were doing best (37% with a favourable status). Clearly it is the region with the best opportunities for maintaining semi-natural habitats as a result of strong nature-related tourism. In contrast, only 18% of habitats in the Mediterranean region had a favourable conservation status. Furthermore, 45% had a declining trend, including 36% with a bad status. Habitats in the Atlantic region had the lowest proportion with a favourable status (11%), but only 22% had a declining trend.

Comparison of the habitat types reveals that the status of mires and bogs, dunes, and other coastal and freshwater habitats was especially poor (less than 10% with a favourable status). Whilst 20% of grasslands had a favourable status, a particularly high proportion had declining trends (52%). Rocky habitats appear to be the least threatened, as about half had a favourable status.

15.2.2 Species Status and Trends

The 2013–2018 monitoring results for HD species also indicate a very poor situation, although slightly less so compared to habitats. Overall, 28% of HD species had a favourable conservation status and 23% had a declining trend. About 25% had a bad status, which is slightly less than for HD habitats, but a slightly higher proportion were declining (Table 15.1). As was the case with habitats, those in the Alpine biogeographical region are faring a little better than others (37% with a favourable status). However, in contrast to habitats, a similar proportion of Mediterranean species had a favourable status to those in the Alpine region. Of particular concern is that a high proportion of species in the Atlantic (36%) and Continental (28%) biogeographical regions were declining. Information on marine species is lacking so the status of most has not been assessed.

In France, 14% of mammals, 24% of reptiles, 23% of amphibians, 32% of breeding birds, 8% of vascular flora, 6% of diurnal butterflies, 12% of damselflies and dragonflies, 19% of freshwater fishes and 28% of freshwater crustaceans are considered threatened according to the IUCN Red List national assessment.[6]

As elsewhere in Europe, bird species have been best monitored (see Section 15.7) and provide the most documented

[5] See Section 1.3 for background on monitoring under the Nature Directives, status categories and data sources.

[6] https://uicn.fr/liste-rouge-france/ (20 September 2021).

Table 15.1 Short-term trends (c. 2007–2018) in Habitats Directive Annex I habitats and Annex II, IV and V species with an unfavourable conservation status in France.

	Poor (unfavourable–inadequate)				Bad (unfavourable–bad)			
	Imp	Sta	Dec	?	Imp	Sta	Dec	?
HD Habitat assessments								
Number	4	61	33	27	0	19	61	20
%	3%	49%	26%	22%	0%	19%	61%	20%
HD Species assessments								
Number	22	78	50	69	6	14	101	40
%	10%	36%	23%	32%	4%	9%	63%	25%

Key. Imp = improving, Sta = stable, Dec = declining, ? = unknown.
Source. Based on Member State HD Article 17 reported data as presented on the EEA national summary dashboard (2021).

indicators of species trends. According to Article 12 monitoring results, 18% of all breeding bird populations have shown long-term declines (c. 1980–2018) and 26% decreased over the short term. Whilst this might indicate a deterioration in the situation for birds, this is not certain as a significant proportion of long-term trends are unknown. It is also necessary to bear in mind that an approximately equal number of species increased over both periods. Therefore, it is particularly important to consider the trends in BD Annex I species as these are of most concern (Table 15.2). This indicates that slightly fewer BD Annex I species declined than other species. Furthermore a larger proportion of species increased. This is particularly the case with wintering birds, none of which declined.

Table 15.2 Short-term trends (c. 2007–2018) in Birds Directive Annex I species in France.

	Breeding				Wintering			
	Inc	Sta/ Flu	Dec	?	Inc	Sta/ Flu	Dec	?
Number	35	17	19	25	12	5	0	8
%	36%	18%	20%	26%	48%	20%	0%	32%

Key: Inc = increasing, Sta = stable, Flu = fluctuating, Dec = decreasing, ? = uncertain and unknown.
Source. Based on Member State Article 12 reported data as presented on EEA national summary dashboard (2021).

Whilst bird population trends have been mixed, a concerning pattern has been a very high rate of decline of farmland species. Results[7] from citizen surveys, described in Fontaine and Jiguet (2016), have revealed a decline in birds of agricultural environments since the 1970s that has accelerated, as shown by a 38% decline in the indicator over 1989–2019. Specialist species of agricultural habitats, such as the Skylark (*Alauda arvensis*), Common Whitethroat (*Sylvia communis*) and Ortolan Bunting (*Emberiza hortulana*), have lost on average one-third of their populations over the past 15 years. However, common bird species are faring better in Natura 2000 sites than elsewhere (Pelissier *et al.*, 2013).

15.2.3 Drivers and Causes of Change

The most severe and widespread impacts on biodiversity in France over the last 40 years or so have resulted from agricultural changes in particular. After World War II, conservation has been confronted by the gradual disappearance of HNV farming systems (Lefeuvre, 1992). In many areas this has been the result of uncontrolled intensification, such as in the Atlantic biogeographical region, where there has been drainage, cutting of hedges and ploughing of grasslands and meadows – trends that are hard to reverse. There is evidence that the observed declines in farmland birds coincide with the intensification of agricultural practices over the last 25 years, especially since 2008 (Chiron *et al.*, 2013). Over this period, amongst other things, the CAP requirement for set-aside (i.e. fallow) ended, and there was a surge in cereal prices, a return of the overuse of fertilisers and the widespread introduction of neonicotinoids, highly persistent neurotoxic insecticides. In more recent years, there has also been a surge in organic farming, which now represents 7.5% of the national UAA.

In some parts of France, the abandonment of HNV farming has been a major driver of habitat change. It has led to declines in species that are dependent on semi-natural grasslands and other open habitats, and increases in those associated with scrubland and early-stage forest. Such abandonment has primarily occurred in areas where intensification was confronted by natural handicaps, such as in the mountains and in the Mediterranean region, despite a high proportion of CAP Rural Development payments being allocated to farmers in areas with natural constraints (see Section 15.6). Habitats in the Mediterranean region have also been affected by forest fires that have had severe negative impacts due to their high frequency and the landscape homogenisation that has resulted (Cheylan and Jacquet, 2008). The country has

[6] https://uicn.fr/liste-rouge-france/ (20 September 2021).
[7] https://uicn.fr/liste-rouge-france/ (20 September 2021).

made a considerable effort to decrease by half the surface burnt during the decade 2010–2019 compared to the preceding decade (Vogel, 2019).

Member State reports for 2013–2018 indicate that such drivers persist, as agriculture was by far the most frequently reported high pressure across all HD habitats, and especially for grasslands (reported as a pressure in 60% of assessments). Agriculture was also the most frequently reported high pressure affecting HD species (about 29% of assessments), and the second most frequent for BD Annex I birds (about 10%), especially through drainage, as well as the ploughing of grasslands. Furthermore, the reports indicate that these agricultural pressures are not expected to decline in the near future.

The other main cause of habitat loss and fragmentation, and the most important pressure and threat for birds, is urban sprawl and, to a lesser extent, related infrastructure (e.g. roads, railways and power lines). The rate of this artificialisation in France has been higher than the average in Europe and has increased more rapidly than the human population between 1985 and 2018 (Fosse, 2019). Furthermore, two-thirds of this is on agricultural land, and semi-natural habitats with low productivity have probably been most affected. Indeed, 60% of the artificial areas in cities or their suburbs are on recently abandoned agricultural land (Béchet *et al.*, 2017). Meanwhile, some artificial habitats, such as quarries or reservoirs, remain the only suitable habitats for some species and now play a major role in their conservation.

Intensive forestry practices, such as planting softwoods on abandoned land, introduction of intensive silviculture with exotic species like the Douglas Fir (*Pseudotsuga menziesii*) and intensification of timber harvesting as a source of renewable energy, are causing direct pressure on forest habitats and species, even in the few remaining areas of old-growth forest. Other important ongoing threats are the damming and canalisation of rivers (although France is a recognised leader in removing dams), invasive alien species (IAS) and climate change, which is a major contributor to coastal erosion.

15.3 Nature Conservation Governance, Actors, Legislation and Policies

15.3.1 Overview of Nature Developments in France

The key steps in the development of nature conservation policy and legislation in France are summarised in Table 15.3. These were rather few until the 1960s, as France conspicuously put little effort into nature conservation: no National Parks, no statutory reserves, no adequate system of identifying the areas and species in need of protection and little official enthusiasm for nature conservation. Whilst these early steps were isolated initiatives, they were nevertheless useful for conservation at their scale.

The decades between 1960 and 1980 saw an important expansion of conservation measures. They concerned the establishment of protected areas (National Parks in 1960, Nature Reserves in 1961, Regional Nature Parks in 1967) and sectoral measures (water-related habitats in 1964, protection of coasts in 1975, protection of species in 1976).

Building on these foundations, the legal frameworks went through a real transformation between 1980 and 2020 (Prieur, 2014). An important step was taken in 2007 with the *Grenelle de l'environnement*, a state-led consultation that debated and set broad policy objectives and measures, including cross-sectoral reforms (Cans, 2010). This was replicated in 2009 by a *Grenelle de la Mer*, which focussed on marine environments and triggered an exponential increase in the creation of marine parks and reserves (Féral, 2012). In 2016, the Biodiversity Act added more tools and measures; currently, 50 different policies and tools with a direct link to nature conservation have been implemented (Collectif, 2017a). The next three subsections of this chapter analyse the networks of power and responsibilities concerning nature conservation, the roles of non-governmental actors and, finally, some of the principal policies to illustrate their broad concepts.

15.3.2 Governance

When consideration of environmental issues emerged in the public sector in the 1960s, state authorities played a major role. The first ministry dedicated to the environment was created in 1971, with immense ambitions but very few means. As a relatively young ministry, it has lacked stability in its competences and political leadership, changing its name 21 times over the past 50 years. At the time of writing this chapter, it is the Ministry for Ecological Transition (MTE) and has a broad remit, including water and biodiversity, housing, infrastructure, risk prevention, sustainable development, energy and climate.

Although MTE has the principal responsibility for nature conservation, 30 state agencies are actively involved, and the administration is largely devolved to the 13 administrative regions and 96 departments. In 2020, a single state agency, the French Agency of Biodiversity (OFB), was formed by merging several specialised state operators, including the former National Hunting and Wildlife Agency (ONCFS) and the National Water and Aquatic Habitats Agency (ONEMA). It is obviously too early to evaluate its efficiency. The National Forest Office (ONF) and the Coastline and Lakeshore Protection Agency (CELRL) are also important actors of nature conservation at local level. MTE operates with elaborate consultation mechanisms that influence the decision-making process even though this remains centralised. The mechanisms include a National Committee of Biodiversity and 13 Regional Committees and national and regional scientific consultative bodies. The National Council of Nature Protection Council, established in 1946, has a consultative role that has been made mandatory in many case-by-case decisions (e.g. concerning the destruction of protected species and their habitats, establishment of protected areas).

Table 15.3 Key nature conservation developments in France.

1861	*Série Artistique de Fontainebleau* established – possibly the first strict nature reserve in the world
1906	First Law on Landscape Protection
1930	Second Law on Landscape Protection
1946	Establishment of the National Council for Nature Protection, a consultative body on public policies and decisions affecting nature
1960	Law on National Parks
1964	First Law on Water, which established the six Committees of Watersheds and initiated the conservation of wetlands and river ecosystems
1967	Decree established the Regional Nature Parks
1971	First independent ministry in charge of environment
1975	Establishment of Coastal and Lakeshores Protection Agency (CELRL)
1976	First law on species protection, impact assessments and Nature Reserves
2001	Transposition of the Nature Directives into national legislation after years of disputes
2005	A constitutional act introduced several principles (e.g. precautionary and polluter-pays principles)
2006	Law on National Parks and Marine Parks, creation of the Agency for Protected Marine Areas Law on Water, establishing the National Water and Aquatic Habitats Agency
2007–2009	*Grenelle de l'environnement*, large consultation process that produced several laws and commitment of state, followed by a similar process dedicated to the sea environment
2014	A law of decentralisation gave a leadership role to regional authorities in the matter of biodiversity
2016	Biodiversity Act reformed most national approaches to nature conservation and created a single state operator for biodiversity, the French Agency of Biodiversity (OFB)
2018	Issue of a special national biodiversity plan with short-term state commitments for 2018–2020

See Table 3.3 for signatory dates of international conventions.

Before the first laws of decentralisation in 1982, nearly all public policies related to nature conservation were initiated, funded and implemented by the state. There were a few exceptions, such as the Committees of Watersheds, which have made decisions on the conservation of rivers and wetlands since 1964, and the Regional Nature Parks established by regions in 1967. Regions have since gained the ability to create nature reserves and are now responsible for economic and spatial planning, and thus the capacity to integrate nature conservation in sectoral policies. Departments have been given a role in purchasing land for nature (Natural Sensitive Areas) and in organising recreation in natural areas.

15.3.3 Non-governmental Actors

Indicators of public attitudes towards biodiversity in France are in line with the EU average.[8] Biodiversity knowledge is considered to be poor and people tend to ascribe aesthetic and cultural values to nature rather than economic ones, thus the willingness to allocate financial resources to conservation is rather low (CREDOC, 2015).

More than 8 000 nature and environment protection NGOs are recognised in France, but only a few play a major role in nature protection. Many have been established to defend very local environmental issues and act as a counter-power to local authorities. Due to this focus on local issues, their average membership is 150 people (Cornu and Thierry, 2010). Only 60 of them have a legal standing, about half of which are genuinely active in the courts in cases related to nature protection. The birds protection league, Ligue pour la Protection des Oiseaux (LPO), which is the national BirdLife partner, has 57 000 members and is by far the biggest nature NGO in the country. This amounts to 0.07% of the population, which is a relatively small proportion compared to some other prosperous EU countries. There is an important federation, France Nature Environnement, that strengthens and coordinates the actions and campaigns of 3 500 local or specialised NGOs.

French NGOs are rather active in the field of direct management, for instance the Société Nationale de Protection de la Nature and LPO, which own and manage Nature Reserves. There are also Natural Areas Conservation Societies, a network of 29 non-profit associations that own 3 440 sites representing

[8] https://europa.eu/eurobarometer/api/deliverable/download/file?deliverableId=69077

178 000 ha and sustain their own federation[9] (FCEN 2019). They also act as site managers on the request of private or public landowners. This network employs 1 000 staff and is the most important private actor managing Natura 2000 sites. The individual associations have national approval of their activities and are therefore partly funded by the state and the regions.

In contrast to the situation with nature NGOs, hunting associations have many members, because hunters have to be members of a hunting association to be able to hunt. In 2015, there were 1.17 million active registered hunters in France (BIPE, 2015), which still could be the highest number in the EU. Furthermore, it is estimated that 5 million people in the country have at some point held a hunting licence. Thus, hunting is an important political force in the country and hunters are very active in local governance and habitat management, especially in the c. 30 000 municipalities with less than 2 000 inhabitants; they are also very influential at the national level.

15.3.4 Legislation and Policies

It is estimated that 70% of the French environmental legal framework is of EU origin (Lacroix and Zaccai, 2010). Indeed, of the 50 strategies, legal acts and policies that the government of France reported to the CBD in 2019, two-thirds are directly inspired by international frameworks or obligations. France officially claims to give strong support to the policies and targets set for the past 2010–2020 decade at the international level.[10] However, the transposition of the Nature Directives into national legislation was a very late and slow process, which generated much confusion in terms of its practical application. Hesitation amongst the central powers, and mistakes in public information and consultation, fuelled a widespread rejection of Natura 2000, which still has lasting impacts at the local level (Pinton *et al.*, 2006). Scientists, especially from the National Museum of Natural History (MNHN), played a central role in the selection of SCIs, but the local processes of defining the appropriate measures to ensure their conservation by local authorities did not work well. It was as late as 2005 before the whole process of site designation, objective setting and management was reformed, with the strong involvement of local stakeholders, particularly local authorities – a process that is working appropriately today (see Section 15.4 for further details on Natura 2000 sites management).

Although inspired by international initiatives, most nature conservation policies are formulated and implemented by the central government, with MTE having the primary, and sometimes sole, responsibility for them. The lack of inter-ministerial coordination and actions is regularly cited as a problem for this ministry, which is still especially weak compared to the Ministry of Finance and the Ministry of Agriculture (Laville, 2010). Several attempts have been made to counteract this weakness of MTE, such as through environmental round tables. However, the resulting plans are susceptible to political changes as they are not clearly linked to legal frameworks. Only when a permanent inter-ministerial committee is established is decision-making effective, which is rare and with a limited scope.

The Biodiversity Act in 2016 added important concepts to the legal framework that might help to ensure terrestrial, aquatic and marine natural habitats are taken into account in the decisions of other ministries: 'ecological inclusiveness', 'no net loss', 'non-regression of environmental law' and 'reparation of environmental damage in the Civil Code'.[11] Finally, the last attempt to build a more inter-ministerial approach to nature conservation was the national biodiversity plan encompassing several short-term state commitments, intended to be more operational than the *National Biodiversity Strategy 2011–2020* (Ministère de l'Écologie, 2012) that was intended to meet CBD requirements. However, once again, the actions that make most progress are mostly those that are implemented by MTE itself.

Regulatory protection tools are numerous and include standard protected areas (regional or national Nature Reserves, National Parks), hunting reserves and the designation of Natura 2000 sites. Following the long-lasting transposition process of the Nature Directives, some elements that applied to Natura 2000 were considered to be inadequate by the European Commission. This led to two important rulings by the CJEU that strengthened the protection and management of Natura 2000 sites in France, and indirectly elsewhere across the EU (see Box 15.1).

Planning tools are also very developed in four areas: water resources, territorial and urban matters, flood risks and marine waters. It is also noteworthy that France's ecological network (*Trame verte et bleue*), described in Section 15.5, has been one of the inspirations for the EU's green infrastructure initiative.

In addition to regulatory tools, France uses a number of non-binding tools, including to implement nature-related international agreements (e.g. Eurobats, AEWA, ASCOBANS and ACCOBAMS under the CMS; see Chapter 3) and action plans for species. As further discussed in Section 15.6, fiscal tools include several incentives for nature conservation. France also relies largely on contractual and financing tools, such as Regional Nature Parks or agri-environment measures and other operational instruments, to implement the Nature Directives.

Two other laws have had a strong effect on land protection: the Coastline Law[12] and the Mountain Law.[13] However, their role in land management has been regularly reduced.

[9] https://reseau-cen.org/

[10] See national report to CBD 2019.

[11] Loi n° 2016-1087 du 8 août 2016 pour la reconquête de la biodiversité, de la nature et des paysages.

[12] Loi n°86-2 du 3 janvier 1986 relative à l'aménagement, la protection et la mise en valeur du littoral.

[13] Loi n° 85-30 du 9 janvier 1985 relative au développement et à la protection de la montagne.

Box 15.1 French cases that helped strengthen the implementation of HD Article 6

Case C-96/98, Commission v. France – 'Poitevin Marsh SPA', helped strengthen the link between the protection of species and ecological functions. In this case, France was found to have an insufficient legal framework on water management as it had instead only put forward contractual measures (such as the use of an agri-environment scheme) for the conservation of a vast wetland SPA. This triggered the establishment of a specific, differentiated regulatory mechanism for the management for the whole area.

Case C-241/08, Commission v. France, made a major contribution to determining the scope of activities that should be subject to an impact assessment. At that time, France considered in its national legislation that fishing, aquaculture and hunting did not constitute activities that cause disturbance, in relation to HD Article 6(2), and that related projects and plans did not require an appropriate assessment in accordance with HD Article 6(3). The Court found that with this general approach France was failing to fulfil its obligations under Article 6(2) and Article 6(3).

15.3.5 Species Protection and Conservation Measures

The national legal and regulatory framework on species protection, established in 1976,[14] still forms a solid pillar of French nature conservation that was reinforced in 2007 (Untermaier, 2016). This established which species are protected and what actions are forbidden (e.g. their destruction, capture and sale). It also enables the protection of the habitat of each protected species, the extent of which can be determined by the local state representative, depending on what is necessary (Biotope Protection Orders). More than 9 500 species benefit from this protection status in France, with some geographical distinctions, from the national to the 96-department level; these are not limited to those listed in the Nature Directives' annexes.

The direct enforcement of these regulatory and legal provisions has always been very difficult. Despite a dedicated force, with about 2 000 personnel working for OFB, and the police being given powers to identify offences, the number of prosecuted cases has stayed very low (and mainly targeted towards international wildlife trafficking). This has also been made difficult because of the high number of species that can be hunted and the complex hunting regulations (see below). The destruction of protected species is hardly controlled, and the enforcement focusses on hunting practices that are easier to control and overlooks the trafficking of specimens. Therefore, the importance of the legal framework is clearly mainly in the prevention, mitigation or offsetting of damage to populations, individuals or habitats of protected species by permit-seeking development projects (see Section 15.5).

Hunting regulations in France are complex, with considerable rules that have been directly affected by seven laws over the past 20 years. In total, France lists 89 species that may be hunted, of which 64 are birds, which is much higher than the average of 24 bird species across the EU (Hirschfeld *et al.*, 2019). Moreover, one-third of these 64 bird species are listed as Vulnerable or more threatened in France according to IUCN Red Lists, including some iconic species such as the Capercaillie (*Tetrao urogallus*). Hunting has therefore been the focus of several disagreements between France and the European Commission over the application of the Birds Directive, including about the length and dates of hunting seasons, and the perpetuation of traditional practices concerning threatened species. Several cases are still open at the time of writing this chapter, concerning the Greylag Goose (*Anser anser*)[15] and the Turtle Dove (*Streptopelia turtur*).[16] Some others are in preparation, for example for the Curlew (*Numenius arquata*) and the Black-tailed Godwit (*Limosa limosa*).

In 2019, the government adopted a reformed approach to hunting practices and regulations with the aim of better reconciling hunting with biodiversity conservation. A national consultative committee for adaptive management was formed and now gives recommendations to MTE on hunting dates and quotas. The hunting associations in each of the 96 departments are responsible for deciding quotas and local hunting dates, which should respect the state-controlled five-year regional management strategies for all hunted species. The impact of this new system on the conservation status of the relevant species has not been assessed yet, but concern has been raised about its reliability and the state's effective capacity to both influence and control it. With a less centralised system it is rather unlikely that the divergence of views between the state and the Commission will be rapidly reconciled.

Another important effect of the legal protection of species has been the establishment of non-binding national species recovery plans. Over the last decade there has been a strong push to increase their use, with the objective of covering 131 species. The plans have also been made more robust through an effective objective setting process and state funding for their implementation. An assessment of the plans found that they generally had positive impacts, such as improving forestry and quarry industry practices in relation to amphibians and pioneer plant species, addressing the risks of powerlines for large birds, and providing carrion for vultures (Challéat and Lavarde, 2014). However, the assessment also pointed out that their funding is insufficient and that the species only improved under certain circumstances: when their distribution was very limited; when there was a clearly identified major pressure; when the targeted populations responded directly to the state

[14] Loi n° 76-629 du 10 juillet 1976 relative à la protection de la nature.

[15] Hunting is still permitted after they have begun migration to their breeding grounds.

[16] Hunting has been suspended in 2020 and 2021.

of their habitat (e.g. reduced water pollution); and where there has been a long history of recovery planning or well-established networks of stakeholders (e.g. for bats and the Otter (*Lutra lutra*)).

As in much of Europe, France has also experienced an increase in the range and populations of the Wolf (*Canis lupus*), which was eradicated in the country in 1937. However, the conservation of this emblematic species is becoming problematic due to its cost (€40 million in 2019, a great part coming from the CAP Rural Development Programme). This is about ten times the amount dedicated to all other recovery plans and includes the direct payments made to compensate for killed domestic livestock. Herding practices are adapting slowly (Boisseaux *et al.*, 2020). The establishment of a culling quota in 2015, up to 19% annually in 2019, has also been criticised and its effectiveness in reducing livestock losses is not clearly established (Duchamp *et al.*, 2017). The conservation of the Brown Bear (*Ursus arctos*) in the Pyrenees, with intensely negative local reactions to re-introductions, is also illustrative of the difficulties of adapting livestock farming to the renewed presence of predators (Benhammou and Coquet, 2008).

15.4 Protected Areas and Networks

15.4.1 Protected Area Types and Their Coverage

Historically, the protected areas system developed in a rather spontaneous way and as a result of the influences of various stakeholders, including foresters, naturalists and scientists, land planners and curators of cultural heritage. France claims to have established the first strict nature reserve in the world in 1861 (*Série artistique de Fontainebleau*). The oldest Nature Reserve, established by LPO, dates back to 1912, and the first National Park was designated in 1963. Classified and registered sites, which protect landscapes from built developments, date back to 1906.

However, the development of terrestrial protected areas for nature really only increased because of the emphasis given to the fauna and flora that should be preserved according to large-scale inventories prepared by MNHN and numerous volunteers, and launched in 1982, as well as maps of ecological importance (*Zones d'intérêt écologique, floristique et faunistique*). This investment in knowledge triggered the creation of all kinds of protected areas. Now France has a rather complex and very large system of protected area types, all of which have explicit roles for biodiversity conservation, as set out in Table 15.4. The country has 15 different types of protected areas under domestic legislation, encompassing the full range of IUCN categories. Adding these together, as well as the Natura 2000 sites, and taking into account overlaps, as of 2020 their total terrestrial area network covered approximately 148 950 km^2 (Léonard *et al.*, 2020), which equates to 27% of the land area (based on the country's area given in the Key Statistics table). The total coverage is 176 540 km^2 if sites designated under international conventions are included (Léonard *et al.*, 2020), which equates to 32% of the land area.

The terrestrial network of Natura 2000 now covers some 13% of the country's land area. Although this is significant, it did not drive further substantial development of the overall protected area network in France, as there is significant overlap with existing domestic protected areas, calculated as 75% (EEA, 2012), which may not have changed much since because terrestrial Natura 2000 reached its current extent at that time.

Protected areas that are managed with a regulatory approach are generally designated by the state and include the core areas of National Parks, Nature Reserves and Biotope Protection Orders. Together these only cover about 1.7% of France. Land that has been purchased for explicit and permanent conservation goals represents only 0.4% of the country.

An outcome of the 2007–2009 national consultation process (*Grenelle de l'Environnement*) was a state commitment to conserve 2% of the terrestrial area under a regulatory protected area regime. The completeness of the regulatory protected area network was therefore assessed in 2009, based on a sample of 500 animal and plant species (Coste *et al.*, 2010). This showed that 96% of the species did not occur within the existing regulatory protected area system. To address this, a national plan for creating regulatory protected areas was established. Its success has since been assessed, based on 2 000 species (Léonard *et al.*, 2020). This found that some progress has been made in increasing the coverage of species in regulatory protected areas, but only from 1.2% to 1.7% of the terrestrial area.

France was also very slow in designating SPAs and SACs in its maritime waters, especially beyond its territorial waters. However, Natura 2000 now covers 36% of France's marine waters, and there has been a more recent expansion of MPAs, including several large Marine Parks. This has increased the area of MPAs from 37 200 km^2 in 2009 to 168 628 km^2 in 2020 (Léonard *et al.*, 2020), equalling an increase in coverage of marine waters from 10% to 45%.[17] These designations have been mainly carried out by the state and aimed to meet international commitments, translated into France's 2010–2020 NBSAP. National strategies for the designation and management of MPAs were adopted in 2007, and then in 2015, and should have been the framework for these developments. However, the lack of state funding, and the relative simplicity of creating contractual protected areas versus regulatory ones (only 0.3% of the marine territory is covered by a marine protected area managed through regulation), has made these plans relatively ineffective and concerns have been raised about the country's real capacity to effectively protect and manage this immense network (Gomei *et al.*, 2019).

[17] Based on the MSFD waters area given in the key statistics table, or 9% to 45% according to Léonard *et al.* (2020).

Table 15.4 The number and extent of international designated sites and protected areas in France.

	IUCN category	Number	Area (km^2)
Designated under global and non-EU legislation			
Biosphere Reserves – core areas	IV	12	1 736
OSPAR Marine Protected Areas	V–VI	39	21 439
Ramsar Sites	IV	37	8 555
SPAMI Marine Protected Areas (Barcelona Convention)	V–VI	5	1 441
Natural and Mixed World Heritage Sites	II–III	3	666
Designated under the Nature Directives			
SCIs (terrestrial + marine)	IV	1 352	48 865 + 105 785
SPAs (terrestrial + marine)	IV	403	44 276 + 115 976
Natura 2000 terrestrial area	IV		*71 039*
Natura 2000 marine area	IV		*132 144*
Designated under domestic legislation			
Regional Nature Park (Parc naturel régional)	V	54	91 203
Classified and registered sites (Sites classés et inscrits)	III	2 700	11 337
National Park cooperation area (Parc national aire d'adhésion)	V	8	9 338
National Park core area (Parc national coeur)	II–V	8	4 854
National Nature Reserve (Réserve naturelle nationale)	III–V	148	1 531
Biotope, Natural Habitat or Geological Interest Protection Order (Arrêté préfectoral de protection du biotope, des habitats naturels ou des géotopes)	IV	960	1 785
Corsican Nature Reserve (Réserve naturelle de Corse)	IV	7	869
National Hunting and Wildlife Reserve (Réserve nationale de chasse et de faune sauvage)	IV	10	383
Regional Nature Reserve (Réserve naturelle régionale)	III–IV	177	395
Managed Forest Biological Reserve (Réserve forestière biologique dirigée)	IV	173	245
Wilderness Forest Biological Reserve (Réserve forestière biologique intégrale)	I	81	229
Strict Nature Reserve within the core of a National Park (Réserve intégrale de parc national)	I	2	8
Marine Nature Park (Parc naturel marin)	V–VI	6	23 586
Other			
Land owned by the departments as Sensitive Natural Areas (Espaces naturels sensibles)	V–VI	4 000	2 000
Land owned by the Coastline and Lakeshore Protection Agency CELRL (Conservatoire du littoral et des rivages lacustres)	IV–V	590	1 301
Land owned by the Natural Areas Conservation Society (Conservatoires d'espaces naturels)	IV–V–VI	1 448	688

Sources. See Section 1.3. Data from: Biosphere Reserves: UNESCO website (11 November 2021); others from National Natural Heritage Inventory (INPN) 2021. It should be noted that the INPN does not consider Natura 2000 sites as protected areas. Two pages should be visited as a consequence: https://inpn.mnhn.fr/site/natura2000/stats for Natura 2000 and https://inpn.mnhn.fr/programme/espaces-proteges/presentation for all other protected areas. Data for marine protected areas are available in the National Biodiversity Inventory as a stand-alone indicator.

15.4.2 The Coherence of Protected Areas and Ecological Networks

Before the 2007–2009 consultation process (*Grenelle de l'Environnement*), there was no coherent strategy for the development of protected areas, nor for the establishment of ecological networks. This process brought both: a national strategy and an ambitious national policy to conserve and restore ecological connectivity. This state-led programme, called the green and blue infrastructure (*Trame verte et bleue*) consists of recognising biodiversity reservoirs and ecological corridors. Biodiversity reservoirs should be areas where the biodiversity is the richest or the best represented. These include all, or part of, protected areas and other important natural areas for conservation. Ecological corridors should provide connections between the biodiversity reservoirs, to facilitate the movement of species and the achievement of their life cycle. The corridors can be linear (e.g. riverside vegetation) or discontinuous patches, and may include natural and semi-natural areas (Bonnin, 2006). Together, the terrestrial biodiversity reservoirs and corridors form a network referred to as the green infrastructure. Water courses that are designated as protected from any obstruction, and others that may be important for biodiversity, constitute both biodiversity reservoirs and ecological corridors and are referred to as the blue infrastructure.

At the national level the network must be avoided by large-scale built infrastructure developments, and at regional scales it must be taken into account in land planning schemes. Whilst state authorities have claimed that progress has been made in recognising ecological connectivity in effective land planning processes, the artificialisation of natural areas is still considered a major threat, as emphasised in the 2018 national biodiversity plan. Moreover, few assessments of the efficacy of this policy have been undertaken (Amsallem *et al.*, 2018), and indicators are still under development, such as the use of genetic tools to evaluate functional connectivity (Sordello *et al.*, 2012). Therefore, although the increase in terrestrial protected area coverage will undoubtedly have contributed to this policy's goal, further investigations are needed to measure its impact, including the possible counter-effects of recent land planning reforms.

15.4.3 Approaches to Protection and Levels of Protection

Nature Reserves have regulations that restrict most human activities but also enable the manager to identify and carry out restoration works where necessary. Some reserves also have a buffer zone that has few or no restrictions and is managed and protected through contractual approaches. In some cases the buffer zone may consist of a Regional Park. Under the same model, Nature Reserves that are marine habitats or islands may occur within Marine Nature Parks that form a de facto buffer zone.

The eight National Parks also have strict legal protection within their core areas where human activities are restricted and priority is given to nature conservation. They were complemented with a partnership system, introduced by the law on protected areas in 2006[18] and based on a cooperation area dedicated to sustainable use and functioning like a Regional Nature Park. Some concerns have been raised about the level of protection of the two newly established National Parks, especially because hunting is far less restricted in the core areas than it is in the older parks, which resulted in a greater acceptance of traditional practices.

Most protected areas in France are established and managed through a contractual approach, based on the establishment of a governance for the sustainable use of resources and implementation of regulations that are freely consented by local communities. This is described further in online Annex FR.1[19] in relation to Regional Nature Parks, which currently cover about 16% of the country. 'Cooperation Areas' of National Parks have adopted approximately the same approach as Regional Nature Parks. Marine Nature Parks have also a central governing body for inclusiveness and participation of local communities that adopts the management plan.

Protection of biodiversity within the protected areas managed with a contractual approach relies mostly on incentives, especially agri-environment measures, and on spatial planning requirements. The key to their success is the development of a supportive culture for nature conservation among local stakeholders and land-users. In the longest established sites, such efforts have been rewarded with the creation of Nature Reserves, or the adoption of strict rules contributing to conservation. For example, in the oldest Marine Nature Park in Brittany (*Mer d'Iroise*), created in 2007, stakeholders have progressively and freely consented to increasing regulations and overall control on fishing (Van Tilbeurgh, 2006).

Although there is no official assessment of the extent to which protected areas are damaged or destroyed (and de-gazetted), there is now a strong incentive for this, as a result of an increase in public concern and scrutiny in recent years. However, it is important to mention France's application of the HD Article 6(4) derogation that allows damage to a Natura 2000 site where it is of overriding public importance and unavoidable. Only seven cases were reported to the Commission as of 2019 (Barthod *et al.*, 2015). Whilst some potentially damaging projects may have been modified or abandoned after their appropriate assessment, this is a low figure for a country that has numerous built developments.

Appropriate assessments, carried out under HD Article 6 (3), have been judged to be satisfactory, but they fail generally to take into account the conservation state of populations or habitats at a larger scale than the site impacted by the project

[18] Loi n°2006-436 du 14 avril 2006 relative aux parcs nationaux, aux parcs naturels marins et aux parcs naturels régionaux.

[19] All annexes are free to access online at CUP: www.cambridge.org/natureconservation

(Barthod *et al.*, 2015). The most badly assessed threats are caused by spatial planning programmes, farming improvement projects and urbanisation, and the impacts that are most difficult to assess concern open habitats, the coastal environment, birds and bats (Espinosa, 2019).

Finally, it is important to mention that there is no nationwide overall assessment process in place on the efficiency of the protected areas system in the country. Perhaps because of the diversity and high number of stakeholders involved, France has not evaluated the system, even though it absorbs substantial resources and is regularly promoted as an asset in international fora.

15.4.4 Approaches to Managing Protected Areas

In France, planning is widely developed for all kinds of protected areas. Because the system for managing protected areas relies very much on contractual approaches, there are a large number of contracts, agreements and other kinds of partnerships that contribute to the achievement of the sites' objectives. These have a major influence on the allocation of funding structure to protected areas, as it very much depends on the projects that need to be implemented. For Regional Nature Parks, their funding requirements are taken into account in the negotiations between the state and regions in the programming of EU funds, which are then decisive for the parks' capacities for delivering action on the ground (Baron and Lajarge, 2015).

The preparation of the management plans was initially standardised in the 1990s, and since then there has been continuous progress in making them increasingly relevant to nature conservation and requiring better participation of stakeholders (Collectif, 2017b). This standard includes a well-designed prioritisation process, with high importance given to BHD habitats and species. The plans also consider funding requirements.

All of the 350 Nature Reserves have elaborated and adopted management plans; this is a legal requirement for the receipt of state funding, which provides the core budget for every reserve. It is important to note that Nature Reserves can be equally designated by state authorities or by regional authorities that have similar funding obligations (but in Corsica only the regional authority has this ability). Practical conservation management of Nature Reserves can be carried out by a wide variety of organisations, among which 60% are NGOs, others being state institutions and local authorities. All reserves must also have a consultation committee.

Standardised management plans are also prepared for National Parks, which require high standards of objective-setting that clarify the conservation purposes in their core area and provide the basis for partnership contracts in their buffer area. Every park has its own management organisation (which receives about 80% of its funding from the state) and its own governing bodies, including a large number of local stakeholders and authorities.

The only protected areas that do not have management plans are classified and registered sites, as their rules are set by a special state service with expertise in landscape and architecture.

Some protected areas comprise land that is owned by organisations with a nature conservation–related aim (see Table 15.4). Since 1975, CELRL has acquired vulnerable and threatened coasts and lake shores to ensure their permanent protection. The land is acquired amicably, by pre-emption, and only rarely by expropriation. Once incorporated into the public domain, it becomes virtually inalienable. This agency sets up management partnerships, primarily with local communities. In 2015, CELRL approved an ambitious strategy of purchasing 1 100 km^2 from private owners by 2050 to reach the threshold of placing one-third of France's coastline under protection. This has been judged unrealistic because it has purchased 900 km^2 in 45 years of existence[20] and has no real prospects of an increase in its resources. Departments also own and protect 2 000 km^2 of land under the denomination 'Sensitive Natural Areas'.

There are very few restrictions on public access in the protected area regulations, which are restricted to Integral Reserves, currently only located in core areas of National Parks. However, it is important to note that many areas are private and thus access to them can be restricted by the landowner. Whatever the type of protected area, land-owning rights remain supreme and de facto public access is not guaranteed unless they are the properties of the institutions that aim to conserve them, such as the Natural Area Conservation Societies. Nevertheless, the public cares deeply about its ability to access natural areas, and there is a very dense network of marked pathways in France – therefore few restrictions to access on private land actually occur. In fact, the improvement of public access, and in certain cases the granting of right-of-way privileges, is often a condition of local acceptance of protected areas, and some management authorities spend considerable energy attracting and managing visitors.

France has developed a particular approach to the management of Natura 2000 sites that differs from most other European countries in several ways. Most fundamentally, there is no specific regulatory foundation for the management of Natura 2000 sites, and therefore contractual measures are the only way of managing the sites unless they overlap with a type of regulatory protected area. A contractual approach has therefore been developed and applied to all sites, following successful testing in 37 pilot sites, financed by a LIFE project.[21] Contracts are prepared based on each site's conservation objectives and required measures, as set out in a management plan, which is required for every site. These arrangements and the governance system are described further in online Annex FR.2.

[20] In 2015, 700 km^2 of public domain was also managed by CELRL, and still are.

[21] LIFE95 NAT/F/000533.

The implementation of the French management approach to Natura 2000 has been judged positively by the European Commission, but is highly dependent on the level of state funding and on the involvement of regional authorities that manage EU funds (see Section 15.6).

Apart from the Natura 2000 condition assessment and reporting process, there is no national official system for evaluating the management of protected areas in the country, despite the recent development of the IUCN green list standards with 30 sites in France. However, many management plans are evaluated for their effectiveness and efficiency, and sometimes also their utility. The culture of evaluation is also increasing, as in all public policies in the country, but overall there is still insufficient reporting and publication of these results (Laslaz *et al.*, 2012).

Finally, it is also important to mention here the human network involved in the management of protected areas that helps close the gaps in protection between the designated areas and the wider environment. This has been aided by the establishment of several federations of protected area managers that exchange practical knowledge, teach, experiment and lobby both central and local powers. A network of these managers' associations also exists, supported by a dedicated NGO that pioneered the professional training of some 10 000 protected area management staff (*Atelier technique des espaces naturels*), now merged with OFB.

15.5 General Nature Conservation Measures

With such a plentiful toolbox for nature conservation, France is well equipped to cope with cross-cutting challenges. By way of examples, the next section focusses on lessons from policies or projects dealing with development planning and impact assessment, and habitat management and restoration.

15.5.1 Development Planning and Impact Assessment

Spatial planning plays a major role in nature conservation in France as habitat loss due to urbanisation and other built developments is a major driver of biodiversity loss. Planning of built developments and other changes in land use, such as quarries or renewable energy projects, is carried out at various scales and requires the consideration of nature protection requirements. The protection of wooded areas within urban landscapes has been promoted since the 1980s and is now a key component of urban planning. The consideration of wider aspects of nature conservation in spatial plans has also grown, especially since the inception of the green and blue infrastructure policy.

SEA and EIA are also well developed in the country and based on robust and regularly updated legislation. In line with the EU directives, independent bodies at the national and regional scales evaluate the quality of these assessments, although not all are scrutinised. It has been established that 5 000–10 000 environmental assessments are produced every year in France. As this quantity exceeds the administrative capacity of the responsible services, their independent control has to be prioritised. Recent improvements in the response time and coverage of issues have been observed (Autorité Environnementale, 2021).

Requirements to avoid, minimise and, if necessary, offset impacts on biodiversity, in accordance with the mitigation hierarchy, were introduced into legislation in the 1970s. The field of application of these requirements widened progressively, covering impacts on habitats inscribed within the Natura 2000 network, wetlands, forests, HD Annex 4 species, BD Annex 1 species and nationally or regionally protected species. Since 2006, the requirements have also applied to impacts on the habitats of HD protected species. However, until recently, private and public actors have not had the capacity to comply fully with the requirements (Jacob *et al.*, 2014). The administration also lacks technical skills to evaluate the relevance of assessments and to further control their effective implementation. Several court cases have also influenced this system. A consequence has been a profusion of local guidelines and advice that has failed to effectively avoid and reduce impacts before accepting offsetting (Bigard *et al.*, 2017). In particular, all kinds of offsetting measures may be accepted, such as land purchase, restoration works, green bridges, financing and research (CGDD, 2013).

One of the aims of the 2016 Biodiversity Act was to improve the application of the mitigation hierarchy. It therefore introduced into French legislation the 'no net loss principle'. It also allows developers to select from three mechanisms to implement their biodiversity offsetting requirements: a permittee responsible system and two third-party systems, namely through an 'offset operator' that can deliver the developer's offset needs or a mitigation bank that has secured gains independently of the developer's needs (Vaissière *et al.*, 2018). Thus, the law now accepts the possibility to buy offsets from accredited habitat banks. To date, a single bank has been accredited by the state and five others were allowed as experiments. This mechanism has been already criticised for enabling developers to circumvent impact avoidance (Vaissière *et al.*, 2017). To clarify what qualifies as avoidance, minimisation and offsetting, MTE issued comprehensive guidance with a useful typology of measures and actions (Alligand *et al.*, 2018).

15.5.2 Habitat Management and Restoration

Wetlands have been subject to special consideration in habitat management implemented in France. Three national action plans were dedicated to their conservation over 2005–2018[22] that aimed to develop a territorial approach to wetland management and mobilise interested stakeholders in their conservation, including through reinforced implementation of the

[22] http://pnmh.espaces-naturels.fr/accueil

Ramsar Convention in the designated Ramsar Sites (c. 9 000 km²). The last assessment found that this had provided knowledge and guidance for management (Lavoux *et al.*, 2013). However, there had been no real improvements in the declining status of habitats and species, nor in water quality, based on the EU Nitrates Directive requirements (OMH, 2014). The growing of crops has tended to supplant traditional outdoor livestock production on pastures with detrimental effects on biodiversity and water quality. Regulations introduced in 2006 have enhanced protection of wetlands against built developments, with the principle that their functionality should be preserved or offset. Tax exemptions encouraging conservative management have also been put in place, but they lack effectiveness because they do not include all wetlands (e.g. they exclude peat-bogs) and because they relate to an ancient tax regime when grasslands were more valuable than arable or buildable land (Sainteny, 2012). Thus the conservation of wetlands is still far from adequate, especially regarding their importance for providing essential ecosystem services.

Another important issue for habitat management is the regulation of the use of pesticides. France adopted in 2008 an Ecophyto plan that aimed 'if possible' to reduce pesticide use by 50% within 10 years. The target has clearly been missed since the number of dose-units of pesticides for agriculture increased by 12% over 2014–2016, compared to the 2009–2011 reference period; recent figures show no change in direction (Guichard *et al.*, 2017). The key elements of the plan were a plant health monitoring 'bulletin', a database allowing real-time assessments of pest risks, and a set of 'DEPHY farms', which was an experimental network that was supposed to disseminate good practice. The failure to achieve the target was mainly due to the plan's main focus on farmers' and advisors' practices, regardless of the broader effects of the 'socio-technical lock-in' (Guichard *et al.*, 2017). Nevertheless it is thought that the Ecophyto plan has had a symbolic effect that could be beneficial in the future, because public authorities have sent a signal that the massive use of pesticide in agriculture will end.

Direct restoration in France used to be focussed on emblematic species, such as Beaver (*Castor fiber*) or Griffon Vulture (*Gyps fulvus*), and is only recently being applied to more common habitats and species. The six Water Agencies (one for each water basin in the country) have dedicated major investments to restore aquatic habitats, mainly through the removal of artificial barriers to the rivers' free flow. The main restoration works have consisted of the removal of dams, the installation of fish passes and the restoration of quarries following exploitation. At the national scale, various incentive devices have also led to the development of ecological restoration operations. Agri-environment schemes also include actions that contribute to the restoration of natural habitats, such as the creation of grass strips along water courses. CELRL, National Parks and the departments responsible for Sensitive Natural Areas have also developed their own programmes of restoration. EU LIFE projects have helped to develop several other restoration programmes, including for important species.[23]

More recently, the green and blue infrastructure policy has increased the level of attention given to restoration. Consequently, every regional ecological connectivity scheme identifies restoration priorities in its operational plan. Their implementation then remains the responsibility of private actors (e.g. those building motorways or powerlines) or the public administrations (e.g. overcoming obstacles in rivers). The Biodiversity Act put official emphasis on biodiversity restoration, and OFB now has a clear mandate for such actions. These laws, regulations and instruments have widely stimulated restoration practices and resulted in the formation of a strong professional network, supported by scientific research (Gallet *et al.*, 2017). Investment facilities are also developing and large-scale programmes are emerging, such as Nature 2050, operated by the public banking group *Caisse des Dépôts*, which aims to link restoration needs and investors.

It is also important to mention that climate change adaptation is a growing subject of public debate, with concerns raised about the tendency to encourage climate adaptation actions, such as planting more climate-resilient tree species, in protected areas where simple protection and no intervention had been the accepted standard (Maris, 2018). Several LIFE projects have been launched to develop and integrate climate change adaptation in site and habitat management.[24]

15.6 Nature Conservation Costs, Benefits and Funding Sources

15.6.1 Nature Conservation Costs

France has carried out few studies of the real costs of effective nature conservation, perhaps because there is no strategic funding framework that has been shared by the public and private sectors. The French *National Biodiversity Strategy 2011–2020* also had no estimate of costs and no strategy for mobilising resources, although it is a recommendation of the CBD.

Only recently, on the occasion of the passing of the Biodiversity Act in 2016, did the public sector attempt an overall assessment of the financing needed for the full implementation of public policies (Lavarde *et al.*, 2016). The approach combined policies relating to water resources, terrestrial biodiversity and the marine environment. It concluded that 70% of the needs for the terrestrial biodiversity, and 74% of the needs for marine biodiversity, are currently covered. The policy that is most in need of financing is green and blue infrastructure, with an estimated requirement of €90 million per year over a period of 15 years for terrestrial ecosystems alone. Restoration is considered to be very expensive and, until

[23] Important examples include LIFE11 NAT/FR/734, LIFE 13 NAT/FR/762, and LIFE 16 NAT/FR/593.

[24] See especially LIFE17 CCA/FR/89 and LIFE18 IPC/FR/7.

now, public bodies have not seen the benefits of investing at such a high level.

Protected areas also lack funding, with some €75 million needed annually to fill the estimated gap (Lavarde *et al.*, 2016). Interestingly, Nature Reserves are pioneering the independent evaluation of their costings. Since they have a common methodology for establishing their management plans, it is straightforward to add their costs and compare them to available funding. The most recent study established that 85% of their needs are covered, but the gap is increasing because new reserves are being created whereas public funding is stagnating (VertigoLab, 2018). National Parks have also seen their individual resources decline due to the need to allocate some of the budget to new Parks.

It is also considered that the overall nature conservation funding needs can be broken down as follows: 25% for actions to reduce pressures and 75% for the management of habitats and species. This can be explained by the high proportion of semi-natural habitats that require continuation of traditional human activities, and by the importance given to the contractual approach to protected areas.

Implementation of the 2016 Biodiversity Plan was estimated by the ministry to cost €600 million over four years, including €150 million for payments for ecosystem services, €250 million for restoration actions, and €200 million for the conversion to organic farming.

The cost estimates of full implementation of Natura 2000 claimed by state authorities have changed considerably over time, with annual cost estimates of €350 million for 2000–2006, then €587 million for 2007–2013 and €200 million for 2014–2020. This can be explained by the fact that the two different ministries, ecology and agriculture, that co-finance different EAFRD Rural Development Programme (RDP) measures have had notable difficulties cooperating and working in a common direction (Barthod *et al.*, 2015).

15.6.2 The Economic Benefits of Nature Conservation

As the French contribution to EU and CBD actions, MTE has been carrying out an assessment of ecosystems and ecosystem services (EFESE) since 2012. It has produced evaluations of the main habitats in France and in 2019 delivered its messages to policymakers (CGDD, 2020). Among the findings were a detailed assessment of the use of nitrogen and pesticides in French agriculture, a survey of jobs derived from the wood industry, identified knowledge gaps relating to urban and high-mountain ecosystems, a detailed evaluation of the commercial value of wetlands and an assessment of the share of tourism depending on coastal and marine ecosystems. Nevertheless, this initiative has had little influence on public policies and private decisions (OCDE, 2019).

Few other initiatives have emerged at the national scale, but work undertaken on the socio-economic benefits of the National Parks provided results that helped their adaptation to the 2006 legislation, and this work is still used to cal more financial resources (CREDOC, 2008).

15.6.3 Funding

Based on the European Classification of Environme Protection Activities, the National Biodiversity Observato estimated that national expenditure on biodiversity €2 233 million in 2018, comprising €115 million co-finan of biodiversity-related EU funds, and an additional €2 118 lion of national funding (Table 15.5). Between 2000 and 2(this expenditure doubled, mostly due to the reinforcemen public funding from 50% to 75% of the total, including the of EU funds (OCDE, 2016). Local authorities have more t doubled their support over this period. Half of this spend was dedicated to restoration and management, mainly of aq tic ecosystems funded by Water Agencies and local authorit Less than a quarter was devoted to protection measures, s as for protected areas. This should be compared to the fact t the amount dedicated to terrestrial ecosystems tended decrease slightly over 2006–2018, which may have beer consequence of the lack of an efficient national priority-sett mechanism (Lavarde *et al.*, 2016).

Beyond the core funding of public institutions by the st budget, most environmental protection activities are funded allocated taxes. These include a series of taxes on the use water, based on the polluter-pays principle; these have be found to inadequately reflect the pressures on water resour and insufficiently fund the restoration of ecological connect ity of water courses (Levraut *et al.*, 2013). Another importa resource is from taxes on built developments that can be rais by the departments. Before 2011, a specific tax was allocated purchase natural areas for nature conservation and pub recreation (Sensitive Natural Areas, see Table 15.4). At th time it provided 13% of national funding for biodiversi (€173 million annually), which was assigned entirely to t management of the natural areas (Lavarde *et al.*, 2016). Th has changed with the merging of the tax with a broader tax o built developments, which could potentially provide up €350 million annually, but there is a great lack of accountab ity, each department remaining free to determine the amou going to biodiversity (Lavarde *et al.*, 2016). There is also a ta on the registration of boats that provides resources for CELR and a tax on the transportation of passengers to and fro MPAs that partly finances the management of the MPA concerned. Finally, the hunting and fishing licences main fund the operations of OFB, but these also tend to decreas under the pressure of hunters and fishers. Although all thes taxation resources are relatively small and leverage littl European funding, their mechanisms are of particular importance in terms of their high visibility to the taxpayers.

[25] https://naturefrance.fr/indicateurs/effort-financier-national-pour-la-biodiversite

Table 15.5 Funding sources for nature conservation in France.

Source and period	Measure/category	Average annual allocation (€ m)		Comments
EU funding instruments		**EU + national**	**EU**	
EAFRD – Rural Development Programme measures – allocations to Priority 4 (restoring, preserving and enhancing agriculture and forest ecosystems) 2014–2020	M13 Areas facing constraints	907.5	680.9	Indirect links with biodiversity, no ear-marking for N2k
	M10 Agri-environment–climate	232.8	174.3	c. 30% for N2k, etc.
	M11 Organic farming	146.3	109.9	
	M7 Basic services, etc.	61.3	34.1	
	M4 Investments in physical assets	10.9	6.7	
	M8 Investments in forests	8.73	5.59	
	M16 Cooperation	5.55	4.21	
	M1 Knowledge transfer, etc.	1.8	1.4	
	M2 Advisory services, etc.	0.5	0.4	
	M12 Natura and WFD	0.2	0.2	
RDP Priority 4 subtotal		*1 375.7*	*1 017.6*	
ERDF 2014–2020	Objective 6 environmental	160.0	80.0	For N2k, etc.
	Objective 5 climate	47.4	19.0	
LIFE 1992–2020	Nature and biodiversity	10.3	5.6	
	Integrated project – LIFE IP Marine Habitats	2.8	1.6	LIFE16 IPE/FR/000001LIFE IP
EMFF 2014–2020	Mainly Measure 40 (risk analysis) and secondly Measure 80	2.2	1.6	Affecting N2k sites
EU funding instrument subtotal		1 598.4	1 125.4	
National funds within 2014–2020		**Additional national***	**Comments**	
State public administration (2018–2020)		325	In addition c. €25 m for co-financing EU funds come from MTE and ministry of agriculture	
Water agencies (2013–2019)		526	In addition €30 m for co-financing EU funds	
Departments (2018)		247		
Regions (2017–2021)		63	In addition €60 m for co-financing EU funds	
Other public (including municipalities)		427	May include landscape protection, with no direct link to biodiversity	
Private sector (annual)		450		
Households (annual)		80		
Additional national subtotal		*2 118*		
National total including co-financing of EU instruments		*2 591*		
Total		***3 716***		

Notes. * Excludes co financing of EU funds unless indicated. 'For N2k, etc.' indicates funding relevant to Natura 2000, related green infrastructure and BHD species (based on PAF). The subtotals and totals are only approximate indications of the average annual levels of funding as they are based on annualised data from differing periods.

Sources. Based on data from: RDP data from European Commission ESIF funds database (23 October 2020); RDP allocations to Natura 2000 based on Barthod *et al.* (2015); European Commission LIFE France factsheet (updated 15 January 2021); ERDF information from Ministry for local authorities 2019; national expenditure from an assessment in 2018, issued in 2021 and based on the European Classification on Environmental Protection Activities (CEPA), amended to exclude funds co-financing EU funds according to Lavarde *et al.* (2016). Ponton *et al.* (2016) provided an evaluation of these estimates. See Section 1.3 for further details.

It is important to note that the state is the funder of protected areas with strong regulations, such as Nature Reserves and National Parks, and that local authorities mostly fund Nature Regional Parks and other local measures. The state provides the only funds for actions for marine ecosystems – up to €125 million annually (including EMFF co-funding) – of which one-third is allocated to various actions supporting fishing activities with questionable links with biodiversity (Lavarde *et al.*, 2016).

Co-financing provided by the state and its agencies, regions and other local authorities has leveraged substantial EU funding for biodiversity, as indicated in Table 15.5 for the 2014–2020 funding period. There has also been a trend of increasing the EAFRD RDP budget allocations to biodiversity-related actions, despite the complexity introduced by the decentralisation of management authorities from the state to the regions (Botti and Boutot, 2019). However, France has consistently chosen to dedicate the majority of RDP funds to supporting agriculture in areas under natural constraints and there is no evidence of actual benefits for HNV farming and nature conservation. The agri-environment schemes have provided a substantial source of funding, which has been prioritised to groundwater protection areas, receiving 70%, with Natura 2000 receiving 30%. However, the schemes are considered to have lacked relevance, adaptation to local contexts and an integrated approach with farming systems. Moreover, the schemes are hard to combine within a single farm and are insufficiently results-oriented (Panarin *et al.*, 2017). The Natura measure has hardly been used, because management plans result in few restrictions that need compensation. They are restricted to restoration of non-agricultural habitats and have been strongly discouraged by the state authorities in order to control their budgets.

Some 40% of the ERDF project funds directly supported protection measures, such as implementation of a Nature Reserve management plan or restoration to improve terrestrial ecological connectivity, while 30% has gone to ecotourism development (Botti and Boutot, 2019). It is also interesting to point out that France relies less on EU LIFE Programme funding than its neighbouring countries, especially for nature-related projects. NGOs mainly carry out LIFE projects in France (rather than local authorities), but often have problems due to their relatively small budget.

The 2016 Biodiversity Act took stock of the growing difficulty that the state and regional authorities have had in raising more money for biodiversity and addressing the decline in funding for terrestrial biodiversity. This resulted in a decision to use some of the resources of the Water Agencies to fund actions on terrestrial biodiversity. Concerns have been raised that this will only re-orient funding without adding more money, as was found for the funding of National Parks and of OFB, which now rely mostly on funding from the Water Agencies.

There is also a plan to allow 16% of green bonds issued by the state in 2017 to be for biodiversity action. This could represent an important financial instrument in some ye since €7 000 million have been issued with a maturity o years. Banks are also starting to develop specific investn facilities in biodiversity. The public banking group *Caisse Dépôts* pioneered this approach by becoming a biodiver offset operator.

As indicated in Table 15.5, the main source of biodiver financing, other than public EU and national funds, is from private sector. Companies are mostly acting through phil thropic reasons. However, biodiversity is rarely a stand-al objective, and is therefore included in more gen environment-related funding portfolios. Nevertheless, recent strengthening of offsetting requirements has led to m tractable expenditure by the private sector on habitat rest ation and conservation.

The estimated annual budget of households related biodiversity amounts to €80 million, which represents than €3 per household. This includes an undefined part financing NGOs. There are no major sources of biodivers funding from citizen-based foundations in France. The con bution of citizens to biodiversity conservation is undere mated in this accounting, because it does not include value of volunteering, which is substantial and represents important lever of action for the NGOs (Ponton *et al.*, 201

15.7 Surveillance, Monitoring, Reporting ar Applied Research

France has been preparing biodiversity inventories for seve centuries, with a prominent role played by the Natior Museum of Natural History (MNHN). This over 200-ye old institution still provides scientific support on biodiversi data collection and use, working with many state institutior regional authorities, private actors and NGOs that colle biodiversity data.

To assist with the coordination and dissemination of bi diversity data, the Nature and Landscape Information Syste (SINP)[26] was launched in 2006 as the Global Biodiversi Information Facility equivalent for France, covering all da components for species, landscapes and genetic diversity. has provided a framework with published standards for co lecting, structuring and sharing data, which has massivel changed the way people gather information as well as enablin various kinds of data to feed into common datasets. It has als helped major NGOs to develop digitalisation and make th best use of their observers. Furthermore, it has invested con siderably in developing a mass citizen science programme an given great care to strengthening the growing network c observers.[27] It is now mandatory for data collected durin baseline surveys conducted for development projects to b shared on the SINP.

[26] http://naturefrance.fr/

[27] http://vigienature.fr/

The MNHN is the main stakeholder of the SINP for species data. It is supporting the National Inventory of Natural Heritage (INPN),[28] which was initiated in 2005 to develop a national databank on French biodiversity. INPN has the aim of ensuring a standardised reporting of data that could be needed for expertise, conservation strategies, public information, and reporting in relation to national and international commitments. It distributes all data from national programmes related to protected areas, habitats, species occurrence and species status.

France has also been very active in developing IUCN Red Lists, with the publication of 130 national and regional lists between 2011 and 2018 by IUCN France and MNHN, including on some fungi groups. It is also worth mentioning the national method for assessing the ecosystem services of wetlands that has been successfully used in several good practical cases by landowners and local authorities (Gayet *et al.*, 2016).

National reporting in relation to HD Article 17 and BD Article 12 is carried out by a mixed unit, from the MNHN, OFB and the National Committee of Scientific Research (CNRS), following EU standards and using INPN data. Important work on methodologies has been facilitated by the fact that the MNHN is the lead institution of the ETC/BD.

The transposition of HD Article 6 in France also requires that the conservation status of habitats and species should be monitored at the scale of every Natura 2000 site. To assist with this, the MNHN has developed standardised methodologies for monitoring habitats at the site level that allow site-by-site comparisons. However, it is only recently that the overall efficiency of the French Natura 2000 network has been examined (Rouveyrol, 2016). This found that national reporting is appropriate but the application of the common methodology is incomplete and there are serious gaps in monitoring habitats at the site level. It also showed that site data are not combined and used to provide a coherent monitoring system at the national scale, including for Natura 2000 sites and outside. Work is still ongoing to build this national monitoring system. There is also a national programme producing a large-scale vegetation map, but this is only expected to be completed by 2025.

Beyond the reporting framework, there is also a National Biodiversity Observatory[29] that was created in 2009 as a consequence of the *Grenelle de l'environnement*. It comprises 149 indicators organised on themes, and is now supported by OFB. It has considerably increased transparency on biodiversity data since 2017 at the national level.

There is a strong scientific community that is actively involved in biodiversity in the country, with several internationally recognised applied research centres, working often at the European level. Research efforts on biodiversity have been coordinated by two important organisations: the Foundation for Research on Biodiversity (founded by 10 national research centres and the firm LVMH as donor in 2008), and the National Alliance of Research for the Environment (currently 27 members).

15.8 Future Developments

With IPBES having delivered its first results, the French government seems to be giving increasing importance to the question of biodiversity conservation, perhaps because the country has pushed for the establishment of this science–policy platform on the international stage for many years. The special plan on biodiversity for 2018–2020 gave a new impetus that was followed by the production of the *2030 National Strategy for Protected Areas* (MTE and Ministère de la mer, 2021), which aims to have 30% of the surface of the country covered by protected areas, with 10% under strict protection. This aligns with the protection target of the EU. There is also another three-year operational plan with some funding commitments.

Furthermore, the current government has stated that the conservation of biodiversity should be as important as climate change, and indeed national expenditure on biodiversity (€2.5 billion, of which 19% is from the private sector) is comparable to the national expenditure on greenhouse gasses and air pollution reduction (€3.7 billion in 2019, of which 57% is from the private sector).[30]

Because nature conservation in France is very dependent on public funding, its funding base is fairly stable but difficult to increase. Developing a more charity-oriented funding system would need significant cultural shifts. More likely is a rise in the use of payments for ecosystem services, which could have positive effects if they are effectively dedicated to changes in agriculture.

The country is also experiencing a rise in private nature conservation initiatives, such as the creation of reserves or rewilding. Although limited in area, they could influence public opinion by demonstrating that conservation can work outside the public sphere.

Other recent institutional changes may also be important. The establishment of OFB is promising, although its financial resources, based on water-related taxes, are inadequate. The continuous decentralisation process could also lead to more effective conservation action. Regions are creating their own biodiversity agencies and adopting strategies, and they may take a more important role in Natura 2000. Although the involvement of local authorities has previously limited the efficiency of conservation policies (OCDE, 2016), they are increasingly active and supportive of conservation actors.

28 http://inpn.fr/

29 https://naturefrance.fr/observatoire-national-de-la-biodiversite

30 France's 2021 Environmental Performance Review, see www.statistiques.developpement-durable.gouv.fr/depenses-de-protection-de-lenvironnement-synthese-des-connaissances-en-2021

At the time of writing in 2021, the government is about to publish a 2030 national biodiversity strategy. Some important issues that have been presented at the 2021 IUCN World Conservation Congress, hosted by France, will be pushed forward, such as stopping land artificialisation and reducing the use of pesticides. On their side, France's major NGOs involved in nature conservation published their concerns in a single manifesto (LPO, 2020).

15.9 Conclusions

France has obviously a great role to play in the conservation of European biodiversity. Whilst it is therefore encouraging that the overall trend of some habitats and species is very slowly improving, it is worsening for others, such as wetlands and common birds of the countryside.

Problems have arisen as biodiversity is only one of the many issues that the ministry of environment has to deal with, and implementation and enforcement of its plentiful legal framework have been insufficient. Furthermore, the legislation and subsequent regulations are not well known or appear too complex for many French citizens. Although there are numerous conservation programmes and actions, they are rather dispersed, have insufficient direct funding or strategies for mobilising financial resources, lack evaluation and continuity to be effective, and are overlooked by the general public.

On the other hand, there have also been successes among the plentiful production of regulations, plans and programmes, some lessons from which may be beneficial to other countries. First, the approach that has been used to implement Natura 2000 seems to have been appropriate and efficient. As the country has such a high proportion of semi-natural habitats, it has been important to give priority to contract-setting and open governance. However, it is important to note that the development of this approach took some time, and was in part due to important CJEU rulings, which also benefited the whole EU network.

Protected areas are also very developed and there is evidence that they have achieved real successes in terms of conservation. They have been genuinely adapted to local socio-economic contexts, and have implemented relevant governance. They cover an appropriate proportion of both area and conservation issues, although a higher coverage of strongly regulated protected areas is required. Most protected areas rely on contractual approaches. These allow a wide range of local stakeholders to be involved in the conservation and the management of the area. All these tools, including rules, agreements and practices, are a great source of experience for others wanting to involve stakeholders. However, the system of protected areas, and underlying issues, still lack understanding by the general public. The diversity of approaches also has its weaknesses, including failing to reduce major pressures (e.g. land take from developments and the use of pesticides) and significant variability in conservation results.

Finally, good plans and intentions have been made to address artificialisation of farmland and habitat fragmentation through the policy to identify, protect and enhance green and blue infrastructure. Although nature conservation and building resilience against climate change have been considered as major issues for more than a decade, political changes, decentralisation and a lack of national priority-setting have hampered action on the ground. For example, land restoration is still insufficiently funded, the control of artificialisation is only just starting and the reform of harmful subsidies is still waiting.

The level of consideration given to biodiversity conservation by the general public will be a key factor affecting the future. A major lesson for conservationists in the country is therefore the need to communicate more on their actions, to popularise concepts and initiatives, and to demonstrate the costs and benefits of conservation.

Acknowledgements

I would like to thank warmly the reviewers of this chapter: Dominique Aribert (LPO), Lionel Laslaz (University Savoie Mont-Blanc), Fabien Quétier (Biotope), Jean-Philippe Siblet (MNHN) and Marc Thauront (Ecosphère). A significant amount of documentation has been collected and synthesised by Gérard Collin and I am grateful to him for this initial work.

References

Alligand, G., Hubert, S. & Legendre, T. (2018) *Évaluation environnementale. Guide d'aide à la définition des mesures ERC.* Rapport Service de l'économie, de l'évaluation et de l'intégration du développement durable & CEREMA. Paris: MTE.

Amsallem, J., Tonneau, J.-P. & Chauraud, J. (2018) Méthode d'élaboration d'un dispositif de suivi et d'évaluation de la politique Trame verte et bleue: application aux Schémas régionaux de cohérence écologique. *Science Eaux & Territoires,* **25**, 20–25.

Autorité Environnementale. (2021) *Rapport annuel 2021 de l'Autorité environnementale.* Paris: MTE.

Baron, N. & Lajarge, R. (2015) *Les parcs naturels régionaux: des territoires en expériences.* Versailles: Eds Quae.

Barthod, C., Allag-Dhuisme, F., Velluet, R., Domallain, D. Reichert, P. & Jourdier, G. (2015) *Analyse du dispositif Natura 2000 en France.* Rapport CGEDD n° 009538-01, CGAAER n° 15029. Paris: Ministère de l'environnement.

Béchet, B., Le Bissonnais, Y. & Ruas, A. (2017) *Sols artificialisés et processus d'artificialisation des sols, Déterminants, impacts et leviers d'action.* Paris: Institut National de la Recherche Agronomique.

Benhammou, F. & Coquet, M. (2008) La restauration de l'ours brun (*Ursus arctos*) dans les Pyrénées françaises: entre politique environnementale et crise-mutation du monde agricole. *Norois,* **208**, 75–90.

Bigard C., Pioch S. & Thompson J. D. (2017) The inclusion of biodiversity in environmental impact assessment: policy-related progress limited by gaps and semantic confusion. *Journal of Environmental Management,* **200**, 35–45.

BIPE (2015) *Evaluation de l'impact économique, social et environnemental de la chasse française*. Paris: Fédération Nationale des Chasseurs.

Boisseaux, T., Stefanini-Meyrignac, O., Démolis, C. & Vallance, M. (2019) *Le loup et les activités d'élevage: comparaison européenne dans le cadre du plan national d'actions 2018-2023*. Rapport CGEDD n°012414601 et CGAAER n°18097. Paris: MTES.

Bonnin M. (2006) Les corridors, vecteurs d'un aménagement durable de l'espace favorable à la protection des espèces. *Natures Sciences Sociétés*, **14**, 67–69.

Botti, R. & Boutot, L. (2019) *Mise en œuvre des PO FEDER 2014-2020 et 2021-2027: Approche thématique: Changement climatique et environnement*. Brussels: European Commission.

Cans, C. (2010) La protection de la nature dans la loi portant engagement national pour l'environnement ou comment picorer dans une loi pour reconstruire une génétique des normes. *Revue juridique de l'environnement*, **5**, 95–113

CGDD (2013) *Lignes directrices nationales sur la séquence éviter, réduire et compenser les impacts sur les milieux naturels*. Rapport Collection 'Références' SEEIDD. Paris: MTE.

(2020) *Rapport de première phase de l'évaluation française des écosystèmes et des services écosystémiques*. Paris: La Documentation française.

Challeat, M. & Lavarde, P. (2014) *Les plans d'action en faveur des espèces menacées. Une politique à refonder*. Rapport CGEDD n° 00929001. Paris: Ministère de l'environnement.

Cheylan, G. & Jacquet, K. (2008) *Synthèse des connaissances sur l'impact du feu en région méditerranéenne*. Rapport. Paris: Ministère de l'environnement.

Chiron, F., Princé, K., Paracchini, M. L., Bulgheroni, C. & Jiguet, F. (2013) Forecasting potential impacts of CAP associated land-use changes on farmland birds at the country level. *Agriculture, Ecosystems and Environment*, **176**, 17–23.

Collectif (2017a) *Des outils pour la mise en œuvre de la TVB*. Coll. Cahiers techniques n°91. Montpellier, France: Agence française pour la biodiversité.

(2017b) *Guide d'élaboration des plans de gestion des espaces naturels*. Coll. Cahiers techniques n°88. Montpellier, France: Agence française pour la biodiversité.

Cornu, S. & Thierry, D. (2010) *Les associations environnementales, approches du secteur*. Paris: France Bénévolat

Coste, S., Comolet-Tirman, J., Grech, G., Poncet, L. & Siblet, J. P. (2010) *Stratégie Nationale de Création d'Aires Protégées: Première phase d'étude – Volet Biodiversité*. Rapport SPN 2010/7. Paris: MNHN & MTE.

CREDOC (2008) *Les retombées écoomiques et les aménités des espaces naturels protégés*. Rapport n°055. Paris: Credoc.

(2015) *L'opinion des français sur la participation des citoyens à une agence pour la biodiversité*. Factsheet. Paris: Credoc.

Duchamp, C., Chapron, G, Gimenez, O. *et al.* (2017) *Expertise collective scientifique sur la viabilité et le devenir de la population de loups en France à long terme sous la coordination ONCFS-MNHN*. Paris: MTE.

EEA (2012) *Protected areas in Europe – an overview*. Copenhagen: European Environment Agency.

Espinosa, E. (2019) *Evaluation de l'efficacité de l'évaluation des incidences Natura 2000*. Paris: UMS Patrimoine Naturel, MNHN/AFB/CNRS.

Féral, F. (2012) L'évolution de l'administration française des aires marines protégées. *Revue juridique de l'environnement*, **5**, 123–135.

Fontaine, B. & Jiguet, F. (2016) Le STOC, support de travaux scientifiques: des points d'écoute à la Politique Agricole Commune. *Ornithos*, **23-1**, 16–27.

Fosse, J. (2019) *Objectif "Zéro artificialisation nette". Quels leviers pour protéger les sols*. Rapport France Stratégie. Paris: La documentation française.

Gallet, S., Jaunatre, R., Regnery, B. *et al.* (2017) *L'écologie de la restauration en France. Dynamique actuelle et role d'un réseau multi-acteurs*. Paris: Eds. Naturae, MNHN.

Gayet, G., Touroult, J., Barnaud, G. *et al.* (2016) *Méthode nationale d'évaluation des fonctions des zones humides. Fondements théoriques, scientifiques et techniques*. Rapport SPN n°2016–91. Paris: Onema, MNHN.

Gomei, M., Abdulla, A., Schröder, C. *et al.* (2019) *Towards 2020: how Mediterranean countries are performing to protect their sea*. Paris: WWF.

Guichard, L., Dedieu, F., Jeuffroy, M. H., Meynard, J. M., Reau, R. & Savini, I. (2017) Ecophyto, the French action plan to reduce pesticide use: a failure analyses and reasons for hoping. *Cahiers Agricultures*, **26**, 14002.

Hirschfeld, A., Attard, G. & Scott, L. (2019) Bird hunting in Europe: an analysis of bag figures and the potential impact on the conservation of threatened species. *British Birds*, **112**, 153–166.

Jacob, C., Quétier, F., Aronson, J., Pioch, S. & Levrel, H. (2014) Vers une politique française de compensation des impacts sur la biodiversité plus efficace: défis et perspectives. *VertigO-la revue électronique en sciences de l'environnement*, **14**(3).

Lacroix, V. & Zaccai, E. (2010) Quarante ans de politique environnementale en France: évolutions, avancées, constante. *Revue française d'administration publique*, **134**(2).

Laslaz, L., Guyot, S., Depraz, S. & Héritier, S. (2012) *Atlas mondial des espaces protégés. Les sociétés face à la nature*. Paris: Autrement, coll. Atlas/Monde.

Lavarde, P., Bellec, P. & Correze-Lenne, P. (2016) *Propositions de scénarios de financement des politiques publiques de préservation des ressources en eau, de la biodiversité et des milieux marins*. Paris: CGEDD.

Laville, B. (2010) Du ministère de l'impossible au ministère d'Etat. *Revue française d'administration publique*, 2010/2 n° 134: 277–311.

Lavoux, T., Perret, B., Rathouis, P. & Barrey, G. (2013) *Évaluation du Plan national d'action pour les zones humides 2010-2013 (PNZH)*. Rapport CGEDD n° 008343-01. Paris: MTE.

Lefeuvre, J.-C. (1992) Biodiversité et territoires agricoles. *Économie rurale*, **208–209**, 79–84.

Léonard, L., Witté, I., Rouveyrol, P. & Hérard, K. (2020) *Représentativité et lacunes du réseau d'aires protégées métropolitain terrestre au regard des enjeux de biodiversité*. Rapport UMS Patrinat. Paris: MNHN.

Levraut, A., Payen, D., Coppinger, N. *et al.* (2013) *Évaluation de la politique de l'eau*. Paris: Ministère de l'Écologie, du Développement durable et de la Mer.

LPO (2020) *Livre blanc, pour que vive la nature. Biodiversité en danger, les ONG mobilisées*. Paris: Ligue pour la Protection des Oiseaux.

Maciejewski, L. (2016) *État de conservation des habitats forestiers d'intérêt communautaire, Evaluation à l'échelle du site Natura 2000, Version 2. Tome 2: Guide d'application*. Mars 2016. Rapport SPN 2016-75, Paris: MNHN.

Maris, V. (2018) Les aires protégées dans l'Anthropocène. Quelques pistes pour penser l'adaptation aux changements climatiques. In *Penser l'Anthropocène*, eds R. Beau, & C. Larrère, pp. 265–282. Paris: Presses de Sciences Po.

Ministère de l'Écologie (2012) *National Biodiversity Strategy 2011–2020*. Paris: Ministère de l'Écologie.

MTE and Ministère de la mer (2021) *2030 National Strategy for Protected Area*. Paris: Ministère de la Transition and Ministère de la mer.

OCDE (2016) *Examens environnementaux de l'OCDE: France 2016*. Paris: OCDE.

(2019) *Financer la biodiversité, agir pour l'économie et les entreprises*. Rapport préparé pour la réunion des ministres de l'Environnement du G7, les 5 et 6 mai 2019. Paris: OCDE.

OMH (2014) *Agriculture, aquaculture et milieux humides*. Chiffres clés. Paris: Observatoire national de la biodiversité.

Panarin, M., Contou, C., Le Borgne, G., Penouilh-Suzette, J. & Robert, M. (2017) *Etat des lieux de la mobilisation des Programmes de Développement Rural Régionaux en faveur de la politique agro-écologique*. Paris: Ministère de l'agriculture et de l'alimentation.

Pelissier, V., Touroult, J., Julliard, R., Siblet, J. P. & Jiguet, F. (2013) Assessing the Natura 2000 network with a common breeding bird survey. *Animal Conservation*, **16**, 566–574.

Pinton F., Alphandéry, P., Billaud, J. P. *et al.* (2006) *La construction du réseau Natura 2000 en France*. Paris: La Documentation Française.

Ponton, D., Terreaux, J.-P. & Aubertie, S. (2016) Evaluation scientifique de l'indicateur 'Effort financier national pour la biodiversité'. In *Evaluation scientifique de 55 indicateurs de la Stratégie Nationale pour la Biodiversité*, eds. B. Livoreil & S. Aubertie, pp. 60–62. Paris: Fondation pour la recherche sur la biodiversité.

Prieur, M. (2014) Vers un droit de l'environnement renouvelé. *Cahiers du conseil constitutionnel* n°15. Paris: Conseil constitutionnel.

Rocamora, G. & Yeatman-Berthelot, D. (1999) *Oiseaux menacés et à surveiller en France*. Paris: SEOF, LPO, MNHN.

Rouveyrol, P. (2016) *Evaluer l'efficacité de la mise en oeuvre des directives Nature en France: synthèse bibliographique et perspectives de travail*. Paris: MNHN-SPN.

Sainteny, G. (2012) *Les aides publiques dommageables à la biodiversité*. Rapport n°43. Paris: Centre d'analyse stratégique.

Sordello, R., Amsallem, J. & Dubus, V. (2012) *Trame verte et bleue, suivi et évaluation, faisabilité d'utiliser l'outil génétique*. Paris: MNHN-SPN, IRSTEA, UMR TETIS.

Untermaier, J. (2016) Une innovation durable: la protection de la faune et de la flore dans la loi du 10 juillet 1976, *Revue juridique de l'environnement*, **41** (4), 647–663.

Vaissière, A.-C., Quétier, F. & Levrel, H. (2017) Le nouveau dispositif des sites naturels de compensation: est-ce trop tôt? In *La loi biodiversité. Ce qui change en pratique. Avec les témoignages de juristes, chercheurs, avocats, associations*, eds. C. Cans, & O. Cizel, pp. 126–135. Éditions Législatives.

Vaissière, A.-C., Quétier, F., Bas, A. *et al.* (2018) France. In *Biodiversity Offsets: European Perspectives on No Net Loss of Biodiversity and Ecosystem Services*, eds. W. Wende, G. Tucker, F. Quétier, M. Rayment & M. Darbi, pp. 99–121. Cham, Switerzerland: Springer.

Van Tilbeurgh, V. (2006) Quand la gestion intégrée redessine les contours d'une aire protégée: le cas du parc marin en mer d'Iroise. *VertigO-la revue électronique en sciences de l'environnement*, **7** (3).

VertigoLab (2018) *Evaluation des besoins en financement des réserves naturelles et identification de scénarios de financement*. Dijon, France: Réserves Naturelles de France.

Vogel, J.-P. (2019) *Rapport d'information fait au nom de la commission des finances sur la lutte contre les feux de forêts*. Paris: Sénat.

Chapter 16

Germany

Eick von Ruschkowski

Key Statistics	
Terrestrial area (km²)/% of EU-28*	358 265/8.2%
Marine waters (km²)/% of EU-28*	55 970/1.2%
Population density (p km²)/% of EU-28*	226/197%
GDP per capita (€)/% of EU-28	34 135/128%
Land cover (CORINE land cover classes 2018)	AGR 57%, FOR 30%, ART 9%, WET 2%, OSN 2%
Biogeographical regions	ALP, ATL, CON, MATL, MBAL
Number of HD Annex I habitats/%*	93/41%
Number of HD Annex II, IV and V species/%*	187/15%
Number of BD Annex I bird taxa/%*	78/44%
% HD UCS and decreasing habitats/ % unknown	41%/1%
% HD UCS and decreasing species/ % unknown	35%/11%
% breeding BD Annex I birds decreasing/% uncertain and unknown	20%/17%

* Excluding areas, and habitats and species only occurring in the EU-28 within the MAC and MMA.
Sources and codes. See Table 1.1.

16.1 The Natural History and Characteristics of Germany

The Federal Republic of Germany consists of 16 federal states (*Länder*) and is the result of the reunification with the former German Democratic Republic (East Germany) on 3 October 1990. It is a large country that extends almost 1 000 km from its North Sea and Baltic coasts in the north to the Alps in the south (with its highest point of 2 962 m). Thus, it has a range of temperate climatic conditions and three biogeographical regions and two marine regions.

Like much of Europe, Germany's landscape characteristics and natural history have been shaped by humankind since the end of the last glacial period approximately 11 700 years ago. Human settlements have encroached on formerly remote areas, such as mountains, coastal wetlands and bogs, since medieval times at the latest, converting large areas to agriculture and forestry (Küster, 2010). As a result, no more than 0.7% of the terrestrial area of Germany remains somewhat 'untamed' as wilderness today (BfN, 2013).

Today, 24 different landscape types have been classified based on human use, including various natural, seminatural and artificial habitats (Gharadjedaghi *et al.*, 2004). In relation to the implementation of Natura 2000, the refence is a standard system of 73 natural landscape regions with a biogeographical focus (Ssymank, 1994; BfN, 2009a). The predominant land cover is now agricultural land (Oppermann, 2012; BfN, 2016). Historically, it is also important to note that Germany's political separation between 1945 and 1990 led to very different agricultural structures as a result of the socialisation of land in East Germany after 1953. This is illustrated by the huge variation in average farm sizes. The overall average for Germany is 61 ha per farm (but as low as 35 ha in the southern state of Bavaria). In contrast, large-scale farming dominates the former East German states, such as Mecklenburg-West Pomerania (275 ha) or Saxony-Anhalt (270 ha) (BMEL, 2017). Although smaller farm sizes usually result in a higher diversity of landscape elements, they are not necessarily richer in biodiversity because the structure within them is also important (Bredemeier *et al.*, 2015; Sybertz *et al.*, 2017).

Grasslands, which currently comprise about 28% of agricultural land, have shown a long-term decline, for example by 11.5% between 1991 and 2018 (Oppermann, 2012; BfN, 2016; BMEL, 2017; UBA, 2019a). As a result of this, and intensification of grassland management, seminatural grasslands and other types of High Nature Value (HNV) farmland have been much depleted. Nevertheless, Germany has retained large areas of some HD habitats. According to the last Article 17 report from 2019, Germany retains approximately 272 202 ha of grassland HD habitats (BfN, 2022a), which was about 1.6% of UAA and about 5.8% of all permanent grassland (UBA, 2019a). Of the grassland area, 59% consists of lowland

and mountain hay meadows (6510 and 6520),[1] 23% of calcareous grasslands (6210 and 6170), and 4% of tall herb communities (6430). In addition to the grasslands, HD heathland (4010, 4030 and 4060) adds another 51 056 ha of habitats (BfN, 2022a). There are also other HNV elements that do not qualify as HD habitats, which tend to be scattered through the landscape, including patches of relatively species-rich grasslands, arable land, vineyards and traditional orchards. A survey in 2009–2010 estimated that all types of HNV farmland accounted for 13.1% of German farmland (Fuchs *et al.*, 2011). However, between 2009 and 2017, HNV farmland decreased to 11.4% of the agricultural area, which was a loss of more than 325 000 ha or 12.8% (BfN, 2018a; EEA, 2019).

A substantial proportion of Europe's remaining temperate forest remains in Germany, although cover ranges from 11% in Schleswig-Holstein to 42% in central German states as a result of low mountain ranges in these regions (BMEL, 2020b). Overall, conifers – mainly Norway Spruce (*Picea abies*) and Scots Pine (*Pinus sylvestris*) – account for 57% of the tree cover while the remaining 43% are deciduous trees, dominated by European Beech (*Fagus sylvatica*), European Oak (*Quercus robur*) and Sessile Oak (*Quercus petraea*). However, the regional variation in forest characteristics is high. According to federal forest inventories, over 2002–2012, deciduous forest area increased by 7% whilst coniferous forest decreased by 4%, mainly due to forest conversion measures (BMEL, 2020b).

Historically, one of the most dramatic losses is that of bog habitats. About 14 190 km^2 (c. 4% of the total land area) in Germany was bogs, mostly in the northern German lowlands (78%) with another concentration in the alpine foothills (20%). Today, more than 90% of these bogs have been drained and are used for farming crops, pasture or forestry, while the few remaining areas are impacted by high atmospheric nitrogen depositions, leaving only approximately 140 km^2 of intact, peat-accumulating bogs (Succow and Joosten, 2001; Joosten, 2012).

Germany also has coastal habitats along the North Sea and the Baltic Sea, including several thousand hectares of primary, white and grey dune habitats. Germany, Denmark and the Netherlands share the world's largest intertidal mudflat system, the Wadden Sea, which covers more than 11 500 km^2 and has 500 km of coastline. The area is a World Heritage site, as it is one of the most important for bird migration along the east Atlantic flyway (Wadden Sea World Heritage, 2020a). In total, 43 bird species occur in internationally important numbers (i.e. more than 1% of their flyway population). Amongst these, the Wadden Sea is vital for migrating and wintering Dunlin (*Calidris alpina*) and Knot (*Calidris canutus*) and, in combination with coastal salt marshes, geese such as the dark-bellied form of the Brent Goose (*Branta bernicla*), Barnacle Goose (*Branta leucopsis*) and Greater White-fronted Goose (*Anser albifrons albifrons*). The Wadden Sea region is also an important breeding site for birds, including a quarter of the European breeding populations of Spoonbill (*Platalea leucorodia*), Avocet (*Recurvirostra avosetta*), Gull-billed Tern (*Gelochelidon nilotica*) and Sandwich Tern (*Thalasseus sandvicensis*) (Wadden Sea World Heritage, 2020b). Both the Baltic and North Sea also provide important habitats for Harbour Porpoise (*Phocoena phocoena*), Harbour Seal (*Phoca vitulina*) and Grey Seal (*Halichoerus grypus*), the latter with a steadily increasing population.

16.2 The Status of Nature

16.2.1 Habitat Status and Trends

The most recent national report under HD Article 17,[2] which covers 2013–2018, reported on 93 HD habitats in each of the Atlantic, Continental and Alpine biogeographical regions, giving a total of 195 assessments (BfN, 2022c). Across Germany as a whole, 30% of the assessments indicated that the HD habitats had a favourable status over the reporting period, 32% had a poor status, and 37% had a bad status. The status was recorded as unknown in only 1% of assessments. Of further concern is that the majority of habitats with an unfavourable status showed declining short-term trends (Table 16.1).

Comparison of the biogeographical regions indicates that habitats in the Alpine biogeographical region fared better than others, as 67% had a favourable conservation status. The Atlantic, Marine Atlantic and Continental regions each had similar proportions of habitats with a favourable status (19–20%). All of the five habitats in the Marine Baltic region had an unfavourable status.

The overall results show that some broad habitats have fared relatively badly, including bogs etc., coastal habitats, dunes and freshwater habitats: only about 20% of assessments of each group indicated a favourable status. However, the status of grassland habitats is of particular concern, as only 10% of assessments showed a favourable status. Most alarmingly, 47% of the assessments revealed a bad status. For such reasons, and inadequate completion of the Natura 2000 network (see Section 16.5), the EU Commission has ongoing infringement proceedings against Germany (European Commission, 2019a, b). The reasons include the failure to set sufficiently detailed and quantified conservation objectives, and preventing the deterioration of two particular HD habitats: lowland hay meadows and mountain hay meadows.

As well as the HD habitats, many others are of conservation importance in Germany. These have been monitored systematically since the 1980s, with differing *Länder*-based biotope mapping schemes but somewhat streamlined by the

[1] Numbers in brackets are HD Annex I habitat codes (see Appendix for full names). * Indicates a priority habitat.

[2] See Section 1.3 for background on monitoring under the Nature Directives, status categories and data sources.

Table 16.1 Short-term trends (c. 2007–2018) in Habitats Directive Annex I habitats and Annex II, IV and V species with an unfavourable conservation status in Germany.

	Poor (unfavourable–inadequate)				Bad (unfavourable–bad)			
	Imp	**Sta**	**Dec**	**?**	**Imp**	**Sta**	**Dec**	**?**
HD Habitat assessments								
Number	7	22	32	1	4	14	48	7
%	11%	35%	52%	2%	5%	19%	66%	10%
HD Species assessments								
Number	15	32	55	10	12	22	71	13
%	13%	29%	49%	9%	10%	19%	60%	11%

Key. Imp = improving, Sta = stable, Dec = decreasing, ? = unknown.
Source. Based on Member State HD Article 17 reported data as presented on the EEA national summary dashboard (2021).

national biotope classification system. In the last Red List assessment, almost two-thirds of the total 863 habitat types were listed as either Endangered or Critically Endangered, while only 25% were of Least Concern (BfN, 2016; Heinze *et al.*, 2019, 2020). This reflects the fact that many threatened habitats have their roots in historical, rather extensive HNV farming systems that have declined and degraded.

16.2.2 Species Status and Trends

Notable historic declines in species include the three large carnivores: Brown Bear (*Ursus arctos*), Wolf (*Canis lupus*) and Eurasian Lynx (*Lynx lynx*), which were extirpated between 1830 and 1850 (Ott, 2004). Many species also declined during the second half of the twentieth century.

Over recent decades, trends in flora and fauna in Germany have been more mixed and differ to some extent to those observed for the habitats. For example, of those species that were on the brink of extinction in the 1960s, some (flagship) species such as the Black Stork (*Ciconia nigra*), White-tailed Sea-eagle (*Haliaeetus albicilla*) and Peregrine Falcon (*Falco peregrinus*) have made remarkable comebacks, often due to individually tailored conservation measures, but also due to the banning of pollutants and toxic substances such as DDT. In addition, other species such as the Beaver (*Castor fiber*) and the Eurasian Lynx have increased since combined re-introduction and habitat protection measures were introduced (BfN, 2016). One rare, but notable exception has been the return of the Wolf to Germany, as this has occurred without any specific human assistance, aside from its legal protection in many European countries as a result of the Bern Convention and the Habitats Directive.

However, despite these positive cases, the conservation status of many HD species remains unfavourable. Overall, only 26% of assessments indicated that the species had a favourable status in the biogeographical region concerned, 30% had a poor status, and 33% had a bad status. The status was recorded as unknown for 11% of assessments. Furthermore, a high proportion of HD species with an unfavourable status declined over the short term, including 60% of those with a bad status (Table 16.1). As was the case with habitats, substantially more species in the Alpine biogeographical region had a favourable status than in the other terrestrial regions.

Table 16.2 Short-term trends (c. 2007–2018) in Birds Directive Annex I taxa in Germany.

	Breeding				Wintering			
	Inc	**Sta/ Flu**	**Dec**	**?**	**Inc**	**Sta/ Flu**	**Dec**	**?**
Number	26	21	15	13	8	1	2	3
%	35%	28%	20%	17%	57%	7%	14%	21%

Key. Inc = increasing, Sta = stable, Flu = fluctuating, Dec = decreasing, ? = uncertain and unknown.
Source. Based on Member State Article 12 reported data as presented on the EEA national summary dashboard (2021).

The most recent BD Article 12 report also indicates that BD Annex I birds appear to be doing better than the HD species, as only a small proportion of breeding (20%) and wintering (14%) birds showed declining short-term trends, whereas the majority are either increasing or stable/fluctuating (Table 16.2).

There have also been widespread declines in many species that were formerly considered abundant and common, which have not been the focus of special measures under the Nature Directives. This has been especially evident for farmland birds, which also probably indicate wider biodiversity losses in agricultural habitats. Monitoring over 1992–2016 has shown dramatic declines in key species of grasslands and other farmland habitats, which for some have accelerated in recent years. These include population declines of 89% for the Grey Partridge (*Perdix perdix*), 88% for the Lapwing (*Vanellus vanellus*), and 45% for the Skylark (*Alauda arvensis*), which clearly illustrate the overall negative impacts of intensified land use (Fachgruppe Vögel in der Agrarlandschaft, 2019; Gerlach *et al.*, 2019). Similar trends have been observed in many wetland bird species (Hötker *et al.*, 2007; Wahl *et al.*, 2015). In response to the Grey Partridge population collapse, the conservation NGO NABU filed an additional complaint to the European Commission in October 2020 (NABU, 2020a).

Within the *German Sustainable Development Strategy*, the federal government has included 'species diversity and landscape quality' as a success indicator. It focusses on the populations of representative bird species in different landscape and habitat types, and has a target value based on the (modelled) population in the 1970–1975 period; the aim is to achieve the target by 2030 (Bundesregierung, 2016). This indicator value has remained static at 71% between 2000 and 2016, the most current data available for the 2021 progress report (UBA, 2018a; Statistisches Bundesamt, 2021).

Wider, less visible, dramatic changes in biodiversity are also currently underway. This was recently revealed by a long-term study that showed that insect biodiversity and biomass have significantly declined, even in protected areas (Hallmann *et al.*, 2017; Homburg *et al.*, 2019).

16.2.3 Drivers and Causes of Change

For most habitats and species, the observed declines are the result of a multitude of interacting drivers and causes (BfN, 2015). Although land consumption and land-use change remain the two most important drivers (SRU, 2016a; Heinze *et al.*, 2019), other contributors have to be considered, such as airborne nitrogen deposition, as this exceeds the critical load for 68% of all vulnerable ecosystem areas in Germany (UBA, 2018b).

In farmland habitats, negative impacts from land-use change have mostly occurred through the intensification of agriculture. Aside from increased pesticide and fertiliser use, this includes structural changes, as landscape diversity decreases with increasing field size, as documented for northern and central Germany over the last four decades (Leuschner *et al.*, 2014). Declines in landscape diversity and biodiversity have also resulted from large increases in maize (which supports very few species), shorter crop rotations and more homogenous crops, and increases in drainage and fertiliser use (BfN, 2015). Moreover, landscape elements such as hedgerows, tree lines and ponds have steadily declined. In the remaining permanent pastures, mowing has often replaced grazing, with the addition of fertilisers to allow more frequent cuts. As observed in many EU countries, the permanent abolition of compulsory set-aside in 2008 increased the pressure on farmland biodiversity even further (Hötker and Leuschner, 2014; Busch *et al.*, 2020).

Rivers and floodplains remain some of the most dramatically altered habitats in Germany. The Upper Rhine was especially heavily engineered, widened and straightened by Johann Gottfried Tulla at the beginning of the nineteenth century, making it the predecessor of what happened to all other major rivers in Germany later on (Bernhardt, 2000). A BfN (2009b) inventory based on 79 German rivers with a total length of about 10 000 km indicates that two-thirds of the alluvial floodplains have been lost to dike/levee construction and other flood protection measures. Along major rivers, such as the Rhine, Danube or Elbe, the remaining floodplains rarely exceed 20% of their original extent. Ecologically intact floodplains account for less than 10% of the remaining extent. Most of the area is under agricultural use (BfN, 2009b). In addition, according to the 2017 WFD monitoring data report, only 9% of the water bodies in Germany were classified as either in very good or good ecological condition. The main causes were a lack of connectivity (poor hydromorphology and barriers for migrating water species), their mostly engineered state and the risk of eutrophication (UBA, 2017). One notable river restoration project has been implemented at the Havel River in Brandenburg and Saxony-Anhalt since 2005. Over more than 20 years it aims to reconnect 15 oxbows and remove bank protections and levees over a total length of 90 km (BfN, 2019d; NABU, 2020b).

The coastal and marine environment has also always been subject to intensified use, with detrimental impacts on habitats and species. Recent monitoring results along the coastal flyway show a mix of increasing, stable and declining species in the Wadden Sea, indicating that major, long-term environmental changes are underway (van Roomen *et al.*, 2017). The main pressures on marine habitats include nitrogen- and phosphorus-related eutrophication, chemical pollution (PCBs) and increasing amounts of ocean waste (Common Wadden Sea Secretariat, 2016). The increase in offshore wind farms in the North and the Baltic Seas has added negative pressures especially on Harbour Porpoises (*Phocoena phocoena*) and wintering divers (*Gavia* spp.) (BfN, 2015; Brandt *et al.*, 2018; Mendel *et al.*, 2019).

Most natural areas in German suffer from a certain degree of general recreational use, which is often significant due to the relatively high population density. In addition, some outdoor recreation activities can create more intense pressures in certain locations, such as mountain biking in forests, and surfing and kite surfing along the coasts (Job-Hoben *et al.*, 2009; Frohn *et al.*, 2009; Pröbstl, 2009). Apart from the traditional approach of managing recreation and visitors through trail planning, German large protected areas place a rather low emphasis on adaptive and integrated management approaches to dealing with social and ecological conflicts caused by recreation, in comparison to other countries. As a result, there is a lack of sufficient strategies to reduce recreational impacts on habitats and sometimes the significant disturbance of sensitive species (von Ruschkowski *et al.*, 2013).

Several studies have already established that climate-induced changes are contributing to habitat degradation and loss across all habitat types from marine to alpine (Pompe *et al.*, 2011; Heinze *et al.*, 2019). As climate change has a different effect on individual habitat types in relation to their geographical location, in-depth analysis will take place on regional or local levels.

3 Nature Conservation Governance, ors, Legislation and Policies

3.1 Overview of Nature Developments in many

le 16.3 provides an overview of some of the most import- nature conservation developments in Germany. The main nents of the nature conservation legislative framework and liversity strategy are further described in the following agraphs.

3.2 Nature Conservation Governance

early as 1906, with the establishment of the 'Governmental ice for Natural Monument Preservation' (Staatliche Stelle Naturdenkmalpflege) in Prussia, conservation was recog- ed as a governmental affair. Today's governance is best cribed as multifaceted, based on increasingly cooperative proaches between different levels of government and the GO sector. On the governmental side, the key players are federal, state (*Länder*) and regional/local administrations h defined responsibilities derived from the primary source legislation, the Federal Nature Conservation and Landscape anagement Act (BNatSchG).[3] At a federal level, the Ministry the Environment, Nature Conservation and Nuclear Safety undesministerium für Umwelt, Naturschutz und Nukleare cherheit – BMU) is complemented by the Federal Agency for ature Conservation (Bundesamt für Naturschutz – BfN), hich provides professional and scientific assistance in all ture conservation and landscape management issues and in ternational cooperation activities. BfN serves as the executive thority for CITES and also for the Nature Reserves within e German EEZ.

Following the principles of federalism, the legislation dele- tes the majority of the implementation and enforcement to e federal states. Most federal states have a similar structure to state ministry, with a supporting scientific and/or advisory gency. Depending on the size and the administrative structure f the state, its actual competences and responsibilities may be hared by up to three different governance levels (state, district, ounty/municipal). Studies have shown that, during a period 1995–2010) when several administrative reforms took place, hose states that dissolved their middle level of district govern- nent (*Bezirksregierung*) and adopted a higher degree of dele- ation of nature conservation issues to county/local authorities enerally had less effective nature conservation (SRU, 2007). The main reasons for this were a lack of financial and uman resources (as county and local governments received insufficient resources to deal with their additional roles), the absence of specific expertise (which was formerly concentrated at the higher administrative levels), and an increase in conflicts due to the balancing of national (or EU) interests such as the implementation of Natura 2000 with the interests of local stakeholders involved in land-use matters (SRU, 2007; Niedersächsischer Landtag, 2014).

16.3.3 Public Involvement and NGOs

Germany's NGO sector has traditionally been strongly inter- twined with the key governmental actors. Civic engagement for conservation purposes has occurred since the advent of industrialisation in the nineteenth century. Local activities then often centred around the preservation of small areas from resource extraction or for the protection of birds. By the turn of the twentieth century, the League for Bird Protection had been founded in 1899 and now serves as an example of the long history of conservation NGOs. While these organisations today certainly fulfil their role as watchdogs over governmental activities, cooperative approaches have always been fundamen- tal to practical conservation management in Germany. As an example, the vast majority of biodiversity monitoring data has been collected by volunteers and NGO members, to be passed on to the relevant administrative bodies to inform conserva- tion measures and management plans. The value of this public contribution to nature conservation cannot be underestimated (Bonn *et al.*, 2016).

The combined membership of 87 conservation NGOs, who are represented by the umbrella organisation Deutscher Naturschutzring (DNR), is around 10 million, although mul- tiple memberships occur (DNR, 2022). The DNR has a diverse range of member organisations, as it includes environmental foundations with no members at all and, in contrast, the German Alpine Club (Deutscher Alpenverein) with close to 1.3 million members. In general, there are four core NGOs that can be considered as the opinion leaders on conservation matters and who have the biggest share of members, activities and media presence: Bund für Umwelt und Naturschutz in Deutschland (BUND – Friends of the Earth Germany) with 650 000 members; Naturschutzbund Deutschland (NABU – BirdLife Germany), with 820 000 members; Greenpeace with 610 000 members; and WWF Germany with 700 000 members.[4] All four organisations each have an annual budget of €30–80 million. While the latter two organisations mainly work in a global and national context, BUND's and NABU's local and regional presence of a combined 3 000 or more chapters is often aligned with the administrative governance structures.

Bundesnaturschutzgesetz in der Fassung vom 29. Juli 2009 (BGBl. I S. 2542). As no official translation of the BNatSchG exists, this chapter mostly uses the terminology from the 'unofficial translation' provided by the Ministry (BMU, 2020a).

4 All membership and financial data derived from the organisations' respective annual reports for 2018–2019. Members might include donors/supporters.

Table 16.3 Key nature conservation developments in Germany.

1899	Bund für Vogelschutz (League for Bird Protection) is founded in Stuttgart
1906	Staatliche Stelle für Naturdenkmalpflege (governmental office for natural monument preservation) is established in Prussia
1909	Verein Naturschutzpark (VNP) is founded with objective of establishing large protected areas
1920	Prussian Feld- und Forstpolizeigesetz (field and forestry police law) establish Nature Reserves (Naturschutzgebiet) as protected area category
1921	First Nature Reserves are designated (Neandertal and Lüneburg Heath)
1935	*Reichsnaturschutzgesetz*, the first comprehensive law on nature conservation, with references to Nazi- and blood-and-soil ideology in its preamble, is passed during the Third Reich
1953	The Federal Institution for Nature Conservation and Landscape Preservation (Bundesanstalt für Naturschutz und Landschaftspflege) in the Federal Republic of Germany is founded
1954	Conservation law (*Naturschutzgesetz*) replaces *Reichsnaturschutzgesetz* in the German Democratic Republic
1956	VNP declares strategy to establish 25 Nature Parks to protect 5% of the German surface area
1970	Designation of the first National Park in West Germany, Bavarian Forest National Park
1971	Establishment of the Advisory Council on the Environment (Sachverständigenrat für Umweltfragen, SRU)
1975	Founding of Bund für Umwelt- und Naturschutz in Deutschland (Friends of the Earth Germany)
1976	Federal Nature Conservation and Landscape Management Act (*Bundesnaturschutzgesetz*, BNatSchG) is passed in the Federal Republic of Germany and replaces *Reichsnaturschutzgesetz*
1985–1986	Schleswig-Holstein (1985) and Lower Saxony and Hamburg (1986) declare the Wadden Sea National Park
1986	Six weeks after the Chernobyl disaster, the West German Federal Ministry of the Environment, Nature Conservation and Nuclear Safety (BMU) is founded
1990	With the German reunification treaty, 14 new large protected areas are established
1993	The German Federal Agency for Nature Conservation (BfN) is founded
1998	Major amendment of BNatSchG introduces Biosphere Reserves as new federal large protected area category, sets regulations for the establishment of Natura 2000
2000	The federal government assigns 50 000 ha of federal land as the first tranche of the so-called National Natural Heritage (Nationales Naturerbe) to conservation foundations and associations to manage
2001	First CJEU ruling against Germany in regards to Natura 2000 site information (C-71/99)
2002	Major BNatSchG amendment adds habitat network articles (Natura 2000 implementation, especially marine sites) and the representative action instrument
2006	Environmental Appeals Act (*Umwelt-Rechtsbehelfsgesetz*) enters into force
2006	CJEU declares Natura 2000 implementation in Germany as deficient (C-98/03)
2007	German cabinet passes the *National Strategy for Biological Diversity*
2007	Further amendments of BNatSchG following CJEU ruling to implement Natura 2000 conformity
2009	Major amendment of BNatSchG provides a change from framework law to direct law, taking effect in March of 2010
2014	First progress report on the *National Strategy for Biological Diversity*
2014	Black Forest National Park and 2015 Hunsrück-Hochwald National Park are founded; the number of National Parks has grown to 16
2015	The Federal Ministry of the Environment launches the *Nature Conservation Action Plan* (*Naturschutzoffensive*) 2020
2015	The National Natural Heritage grew to 156 000 ha
2018–2020	Successful petitions for referendums on conservation policy improvements lead to amended nature conservation laws, cooperation agreements and additional funding initiatives in Bavaria, Baden-Württemberg and Lower Saxony

See Table 3.3 for signatory dates of international conventions.

16.3.4 Policy and Legislation

Federal Nature Conservation and Landscape Management Act as a Legal Framework

The origins of the Federal Nature Conservation and Landscape Management Act (BNatSchG) are unfortunately rooted in the *Reichsnaturschutzgesetz* of 1935 during the Third Reich. Although today's modern legislation bears no relation to Nazi ideology, the fact that the original law was not replaced in West Germany until 1976, and former academic regime collaborators remained in their professions after World War II, has left a historic burden upon landscape planning and nature conservation in Germany (Wolschke-Bulmahn and Gröning, 2002; Marschall, 2018b). Under the provisions of the Basic Law of the Federal Republic of Germany, the BNatSchG was a so-called framework law. This means that the federal government defined general obligations and principles, which the federal states enacted through further legislation that was consistent with the framework.

A constitutional reform in 2006 abolished the principle of framework federal legislation in favour of directly applicable federal legislation that nevertheless still follows the concept of federalism. Under this, federal law defines the detailed, generally binding, rules while their implementation and enforcement are carried out by the federal state (except for the EEZ, where the BfN has jurisdiction). The BNatSchG, in most instances, ensures a high level of minimum standards for conservation and compliance with EU regulations. In a few defined matters, the states have a right to deviate from the legislation (e.g. extending the list of habitats under special protection). Also, in some cases the federal standards are not legally binding, but instead promote good practice approaches. All 16 federal states have used their right to produce supplemental regulations; hence the correct reference for conservation law is always the combination of the BNatSchG and the relevant state legislation.

The full transposition of the Natura 2000 legislation into BNatSchG took place incrementally between 1998 and 2010, more often than not only after the German government had received censure from the European Commission, resulting in infringement proceedings with subsequent rulings of the CJEU (e.g. Gellermann, 2007; Möckel, 2008). No sites had been identified during the first phase as foreseen by the European legislation by 1995, so in the subsequent years the dispute related to both the insufficient selection of Natura 2000 sites and the lacklustre transposition of Natura 2000 requirements into the BNatSchG. The site selection was finally completed in 2006 (BfN, 2022f) and legislation has been complete since the last BNatSchG amendments in 2010. Implementation of Natura 2000 is again under the jurisdiction of the *Länder*. By the spring of 2020, though, 88 Natura 2000 sites (all located in Lower Saxony) were still lacking legal protection, with total completion projected for the end of 2022 (Deutscher Bundestag, 2020), which would be 12 years behind the original deadline of 2010. In February 2020, the European Commission issued a Reasoned Opinion, urging Germany to properly implement the Habitats Directive and threatening to refer the case to the CJEU (European Commission, 2020).

The current BNatSchG version has been in effect since March 2010 and provides the core guidelines and instruments for nature conservation in Germany. These cover landscape planning and the impact mitigation regulations (described here in Section 16.6), the protection of certain parts of nature and landscape (Section 16.5), the protection of wild fauna and flora species (Section 16.4), of their living sites and of their biotopes (Section 16.5), marine nature conservation (under federal jurisdiction), and the participation of recognised nature conservation associations.

Participation of Recognised Nature Conservation Associations: Environmental Appeals Act

Under the BNatSchG, in combination with the Environmental Appeals Act,[5] recognised nature conservation associations can appeal against violations of environmental law (e.g. request an internal review or take legal action in the form of representative action). To be eligible, environmental associations require specific recognition from the federal government (UBA, 2017). The Act is based on the European Public Participation Directive (2003/35/EC), which goes back to the Aarhus Convention (ratified by Germany in 2007). It was amended in 2017 after Germany had received a corresponding ruling from the CJEU and from the Aarhus Convention Compliance Committee (Milieu, 2019). Recognised associations may take legal action against decisions of public authorities, such as in relation to the approval of plans (e.g. for infrastructure projects such as roads) or building permits.

As a result of the legal rights associated with this law, its use by environmental NGOs has been under close surveillance since its inception. A study (UBA, 2014) carried out for the German Environment Agency analysed the extent to which recognised NGOs made use of the opportunity to appeal and their proportion of successful appeals over 2006–2012. It concluded that the relative (and absolute) number of appeals was quite low: on average 12 appeals versus approximately 750 permits per year. In addition, out of those 58 cases where legal action was taken, 49% ruled in favour of the NGOs. This exceeds the typical success rate in administrative court cases by a factor of four, indicating that the opportunity to use legal action has not been abused and used to delay projects, but rather stands on substantive claims (UBA, 2014). More recently, similar conclusions have been drawn by a meta study (SRU, 2016b) and Schmidt and Zschiesche (2018).

[5] *Umwelt-Rechtsbehelfsgesetz*, also described as the 'Law on Supplementary Provisions Governing Actions in Environmental Matters under Directive 2003/35/EC'.

National Strategy on Biological Diversity

Aside from the formal legislation, more action-oriented approaches have been taken in political strategies with the aim of integrating biodiversity into other policy fields. Most prominently, the *National Strategy on Biological Diversity* (NBS) (BMU, 2007) was adopted by the German cabinet as a cross-sectoral strategy in 2007. It contains many quantified objectives and targets for 2010 or 2020. In the public eye, two of the 2020 objectives came under scrutiny:

1. The objective to achieve wilderness on 2% of Germany's land (currently, all 16 National Parks without marine areas account for 0.6% of the land; BfN, 2020b).
2. The objective to raise the share of forests left to natural dynamics without intervention ('Naturwälder') to 5% of the forested area, of which currently about 2.8% have been accounted for in an inventorying project (Engel *et al.*, 2016).

Achieving both objectives requires cooperation with landowners, for example because about half of the forest in Germany is privately owned and federal and state reserves are not sufficient to reach either objective.

The NBS also set additional objectives for species and habitats of significance that were more or less achievable, some of which are further discussed in the next section. After the passing of the 2010 intermediate milestone with very little accomplished, the federal government launched the *Conservation Action Programme for Biological Diversity 2020* (*Naturschutzoffensive 2020*; BfN, 2015). Based on the NBS, 10 priority fields and 40 related urgent actions were described. At the time of writing, the exact achievements of the action plan remain unclear and a reporting deadline has not been announced. Hence it remains currently quite difficult to evaluate the impact of both the NBS and the *Conservation Action Programme* as the amount of accountable achievements is rather negligible. After some initial momentum, the programme has mostly disappeared from public awareness. Of the originally identified six action fields, all remained in unsatisfactory states in 2021: the implementation of Natura 2000 had not been completed, the recent reports on the state of nature indicated concerning states, neither the wilderness nor the forest objectives were achieved, and most measures that have been accomplished are of symbolic value.

16.4 Species Protection and Conservation Measures

The protection of species is addressed in Chapter 5 of the BNatSchG and aims, in combination with the Federal Species Protection Regulations (*Bundesartenschutzverordnung*), to implement CITES (in conjunction with EU regulations) and elements of the Nature Directives. As these regulations also apply to all project approvals that fall under nature conservation laws, species protection has become one of the most disputed topics between developers, planning authorities and conservation NGOs. To some they have evolved into a necessarily strict layer of protection, whilst others argue that they result in excessive restrictions that choke financial investments, such as in renewable energy (mainly wind turbines) or infrastructure projects. The focus lies specifically on the legal interpretation of Section 44 (1) BNatSchG that prohibits both the killing and the significant disturbance of specially protected species, which include all HD Annex IV species and all European bird species. Written with the intention to adhere to the EU provisions, whilst offering some flexibility for project developers, Section 44 (5) defines criteria that need to be fulfilled once an intervention impacts nature or an authority wants to grant a permit for a project, without necessarily undergoing the derogation procedure. For that, Section 44 (5) states that an intervention does not breach the strict species protection provisions if its adverse effects do not *significantly* increase the risk of killing or injuring individuals of the affected species and when such adverse effects cannot be lowered even when all appropriate and scientifically accepted measures prove to be unsuccessful. The legal text has enough loopholes to cause disputes over whether the adverse effects should be measured at an individual or local population level, and especially what is regarded as 'significant'.

In practice, most cases concern the installation of wind turbines and appropriate measures to reduce bird and bat collisions, such as through species-specific minimum distances between the turbines and the areas used by the species of concern. This approach is necessary as flight and hunting behaviours vary between species, which makes them more or less prone to collisions. For example, the Red Kite (*Milvus milvus*) has a high risk of colliding with wind turbines. For bird species, distance recommendations have been produced by the Working Group of German State Bird Conservancies (Länderarbeitsgemeinschaft der Staatlichen Vogelschutzwarten, 2014). Despite these recommendations, wind farm impacts on biodiversity are often highly disputed at a local level. At times this is due to genuinely significant collision risks, but sometimes also to others who oppose the planned development for other reasons and use the conservation argument for their own aims (Hoeft *et al.*, 2017).

16.5 Protected Area Types and Their Coverage

Table 16.4 lists the types of protected areas that are included in the WDPA, as well as Nature Parks.

The listed protected areas include four types that were included in the original German system for protected areas laid out in the original 1976 law. Of most relevance to nature conservation are Nature Reserves and National Parks, as these have a strict protection status and focus on species and habitats. Most of the Nature Reserves in Germany are rather small; only 15% of the reserves are larger than 200 ha, although 208 exceed 1 000 ha (BfN, 2019a). They are designated at *Länder* or regional level, varying by state in accordance with BNatSchG implementation. In general, public use is prohibited other than recreation along existing trails. Land uses such as

Table 16.4 The number and extent of international designated sites and protected areas in Germany.

	IUCN category	Number	Area (km^2)
Designated under global and non-EU international legislation			
Ramsar Sites	NR	35	8 693
Natural and Mixed World Heritage Sites	NA	3	See Note 1
Biosphere Reserves (UNESCO only, exc. buffer and transition zones)	NAs	16	343
HELCOM	NR	12	4 853
OSPAR	NR	6	17 030
Designated under the Nature Directives			
SCIs (terrestrial + marine)	NR	4 544	33 550 + 20 938
SPAs (terrestrial + marine)	NR	742	40 263 + 19 738
Natura 2000 terrestrial area			*55 228*
Natura 2000 marine area			*25 603*
Designated under domestic legislation			
National Park (Nationalpark)	II	16	10 479
Nature Park (Naturpark) (not included in the WDPA)	V	104	101 000
Nature Reserve (Naturschutzgebiet)	IV/V/NAs	8 878	26 697
Landscape Protection Area (Landschaftsschutzgebiet)	V/NAs	8 903	102 674
National Natural Monument (Nationales Naturmonument)	III	6	116

Notes. NR = Not reported; NAs = not assigned. 1. Messel Pit WHS 0.42 km^2, German part of ancient beech forests WHS 44 km^2, no published data on Wadden Sea WHS part in Germany.
Sources. See Section 1.3. Based on data from: Ramsar Convention, UNESCO World Heritage Site and Biosphere Reserve, OSPAR and HELCOM websites; IUCN categories, Nature Reserve and Landscape Protection Areas, WDPA (February 2022); others BfN, 2018b; 2019a, b; 2020a; 2022a, b, d, e; Deutsche UNESCO-Kommission, 2019.

agriculture, fishing and hunting can also be prohibited or limited (e.g. date limits on fishing or hunting, no mowing of grasslands before a certain date). Landscape Protection Areas are also included in the WDPA, although these have a very low protection status and a primary focus on outstanding beauty and recreational value. Hence, they are on average four times larger than the Nature Reserves. They often serve as buffers for the latter. The 1976 BNatSchG also established Nature Parks as a large protected area category, the vast majority of which are set aside for their scenic beauty and recreation purposes. Although they are not all included in the WDPA, they are classified under IUCN category V and should be protected as Nature Reserves or Landscape Protection Areas over the majority of their extent.

Over the course of the years, new types of protected area have been added for practical or political reasons. These include Biosphere Reserves, which were added in 1998 as this category was known to the former German Democratic Republic and several such areas had been established before reunification. In today's legislation, the designation of a Biosphere Reserve under BNatSchG has to occur before the Biosphere Reserve can be recognised by the UNESCO Man and Biosphere (MAB) Programme. Of the 18 existing Biosphere Reserves in Germany, 16 have been recognised by UNESCO so far. In 2009, National Natural Monuments were added, with the objective of protecting scientific, naturally or culturally historic areas that are rare and special in accordance with IUCN category III. The most commonly known of the current six monuments is the Thuringian portion of the Green Belt, the former death strip of the Iron Curtain.

The designation of SPAs and SCIs sites was very slow in Germany and was only completed after the European Commission carried out several infringement procedures. The current set of SPAs was only substantially completed in 2012, and after slight further expansion now covers 60 002 km^2. By comparison the IBA network covers 77 206 km^2,[6] and about half of the SPA/IBA area overlaps (Underwood *et al.*, 2014). The designation of SCIs is now also considered to be complete for marine sites and nearly complete for terrestrial sites (BfN, 2022a, e). Based on the total Natura 2000 areas in Table 16.4, the terrestrial sites cover 15.4% of the land area and the marine sites cover 45.7% of Germany's marine waters.

[6] www.birdlife.org/partners/germany-nature-and-biodiversity-conservation-union-nabu/ (accessed 18 August 2020).

Germany has also been slow in the designation of its SCIs as SACs, which is the responsibility of the federal states per the Federal Nature Conservation Act. Although this should have been completed 10 years ago, 88 out of the 4 544 SCIs remained to be designated as SACs as of September 2020 (Deutscher Bundestag, 2020). Consequently, this remains the subject of a European Commission infringement procedure (2014/2262).

National estimates of the combined coverage of all protected areas based on the data in Table 16.4 do not appear to be publicly available. The estimates here are therefore based on the WDPA data from February 2022. The data indicate that total marine protected area coverage was 25 621 km^2, which approximately equals the Natura 2000 coverage (which also overlaps with OSPAR and HELCOM sites), and is 45.8% of Germany's marine waters. According to the WDPA, terrestrial protected areas cover 133 920 km^2, which is 37.4% of the land area. However, it is important to note that the WDPA does not include all Nature Parks, which cover about 28% of Germany's land area. Therefore, whilst they are not strictly protected and are mainly designated for landscape and recreational reason, they also help contribute to the wider protection of biodiversity. However, this cannot be quantified as the areas overlap to some extent with other protected area types.

16.5.1 The Coherence of Protected Areas and Ecological Networks

The adequacy and coherence of Germany's protected area network has often been the focus of criticism due to its mostly circumstantial genesis. Habitat connectivity, or rather the lack thereof, and the overwhelmingly smaller-scale approach to protected areas have been commonly cited as the biggest obstacles to successful conservation. A number of steps have therefore been taken in recent years to try to address these deficits.

An early initiative to reduce habitat fragmentation by roads has been to increase the proportion of the country with low traffic areas, each of which cover at least 100 km^2 of open space and fulfil specific criteria (BfN, 2022e). The federal government's sustainability strategy has included the percentage coverage of these areas as a progress indicator, with the aim being to return to the 2005 level of 25.4% from the current level of 23.5% (UBA, 2019b).

The establishment of a nationwide habitat network was first included in the BNatSchG in 2002, with a quantified target to cover at least 10% of the total land area, mainly comprising National Parks, National Monuments, Nature Reserves, Natura 2000 sites, Biosphere Reserves, the Green Belt, and National Natural Heritage sites (Landscape Protection Areas and Nature Parks are not considered as core contributors to the 10% objective). The *National Strategy for Biological Diversity* re-emphasised this target and called for its achievement by 2010 (BMU, 2007). By 2010, concepts for the functional design of the habitat network had been developed (e.g. Fuchs *et al.*, 2010). However, an analysis revealed that the implementation strategies of the federal states varied significantly in their ambitions and approaches (Bannas *et al.*, 2017). Currently, the degree of completion of the habitat network remains unclear.

Another approach being taken to achieve coherence and connectivity is through the enhancement of 'green infrastructure'. This initiative has been taken up by BfN to establish the needs for habitat connectivity and integrity in order to be able to deliver ecosystem services. The concept developed by the agency (BfN, 2017) not only identifies specific protected area categories and ecological networks as the backbone of green infrastructure, it also adds the measures necessary to re-establish linkages through the 2012 Federal Defragmentation Programme, which includes the construction of wildlife crossings in order to reduce barrier effects. The initial idea was promoted by NABU 14 years ago, when 910 conflict areas were identified that led to the demand to build 125 priority crossings in wildlife hotspots by 2020 (NABU, 2007). The Federal Defragmentation Programme identified about 90 priority locations for wildlife crossings over federal roads to be built by 2020, of which by 2015 44 locations were 'under examination', 33 with no activity, 12 projects under planning, 3 under construction and 1 being completed (Deutscher Bundestag, 2015). By 2018, the Federal Defragmentation Programme listed 21 crossings under planning, 11 completed and 2 under construction (Deutscher Bundestag, 2018); a year later, 8 additional crossings had been completed and 9 were (still) under construction (Deutscher Bundestag, 2019). While the overall tendency indicated progress, the programme is far from completion. An evaluation report was announced as early as 2017, but has not been released yet. Hence, no solid data are available on the efficiency of these crossings, especially since the monitoring is by law another *Länder*-level task (Deutscher Bundestag, 2015), leading to another gap in reporting.

The cost of the wildlife crossings varies, such as in relation to location, width and materials used, ranging from a little over €1 million up to almost €10 million per site (Deutscher Bundestag, 2019). Conservation standards and requirements for proper crossings (green bridges, wildlife tunnels and fauna culverts) have been published in a recent manual (Reck *et al.*, 2019) and hence are available to planners, developers and conservation agencies.

16.5.2 Approaches to Protection and Levels of Protection

Historically, nature protection was rooted in either a single-object–focussed (Nature Monument) or a tiered approach with the two categories of Nature Reserves and Landscape Protected Areas. They still serve as the basis for the designation of large protected areas (e.g. more than 50% of the area of a National Park has to meet the requirements of a Nature Reserve). Both categories serve as the main tools to protect Natura 2000 sites. Hence, it is important to look at the objectives of these two designations, which are described in the legally binding Federal Nature Conservation Act (BNatSchG), but are designated, managed and monitored at a state or district level.

Nature Reserves, the strictest protection BNatSchG category, serve the purpose of the protection and restoration of habitats or communities of wild animal and plant species, for scientific, natural heritage or landscape reasons, or because of their rarity, uniqueness or distinctive beauty (s. 23 BNatSchG). Landscape Protected Areas are designated to protect the natural diversity, uniqueness and beauty of the landscape, and also serve to protect or restore ecosystem functions and natural capital, especially for recreational purposes (s. 26 BNatSchG). While Nature Reserves cover approximately 4.4% of the terrestrial area, Landscape Protection Areas comprise roughly 28% of the surface area.

Out of the two categories, Nature Reserves are able to impose stricter land-use restrictions to avoid habitat degradation, but these restrictions need to be clearly stated in the reserves' individual statutes to be legally binding. Even Nature Reserves often still contain rights to agriculture, hunting, fishing and recreation. Given their mostly small size with no buffer zones, external effects may damage the reserves, thus compromising their conservation objectives. The protection of dry and nutrient-poor habitats in Nature Reserves is often cited as a major challenge because of the increasing pressure of eutrophication from airborne deposition of nitrogen from surrounding intensive agricultural practices (Underwood *et al.*, 2014). In many cases, Landscape Protection Areas serve as buffer zones around the smaller Nature Reserves to help reduce negative external environmental impacts, although most LPA regulations contain few to no restrictions, such as for agriculture (BfN, 2019j).

Some of these internal and external pressures can also be addressed by management plans, although these were not legally binding in the past. It is likely that the requirement for Natura 2000 management plans in those areas where Nature Reserves are Natura 2000 sites will help to improve their management.

Zoning occurs in large protected areas, mainly National Parks, which strive to achieve the IUCN category II target of 75% core area. As of 2020, only five out of the 16 National Parks had achieved the IUCN's target under a strict interpretation, but German law allows a transition process under which most National Parks should reach the IUCN target by 2030, using a three-stage zoning approach with a core, a 'natural dynamics/development' and a recreational zone (BfN, 2019h). Although similar zoning exists for Biosphere Reserves, the UNESCO requirement for a minimum core zone of 3% is currently failed by four and barely reached (3–4% core) by six out of 16 UNESCO-recognised reserves.

16.5.3 Approaches to Managing Protected Areas

The majority of protected areas in Germany are managed by public authorities, with their governance systems commonly related to their size and their required conservation measures. As they are normally smaller in size, Nature Reserves and Landscape Protection Areas are almost exclusively managed by one local or regional authority. Very often, the financial and personnel resources allocated to these authorities are insufficient (SRU, 2007). Thus, strategies are required that focus on close cooperation with local conservation groups on issues such as monitoring and interpretive/educational programmes. While empirical data are lacking, there is anecdotal evidence of the use of a broad range of informal partnerships and strictly formal contractual agreements with targets and objectives.

Of the large protected areas, National Parks have the most advanced level of organisational structure and complete staffing. In contrast, even though they may be at least a similar size, Nature Parks rarely have the same staffing level as most have very limited budgets. Consequently, the Nature Parks' umbrella organisation (VDN) has called for increased funding so that every park can employ the minimum required level of staff by 2030. Currently, about half of the 104 Nature Parks are run as a non-profit organisation where the local communities and businesses may pay a membership fee. The others are administered by district or *Länder* agencies (Verband Deutscher Naturparke, 2018).

Although practised from the early twentieth century, privately managed large protected areas were an exception. However, this situation has changed as a result of the creation of the 'Green Belt', which is a green corridor along the former highly secured border between East and West Germany. Under the auspices of the National Natural Heritage (NNH) Programme, the federal government has assigned large parcels of this land for free to private conservation foundations and NGOs. Since 2005, 156 000 ha have been transferred, with most (70 000 ha) now administered by the German Federal Environmental Foundation (DBU Naturerbe, 2020), whose aim for the areas is to allow natural processes to prevail and to transition to non-use Nature Reserves. The federal government in the form of the Institute for Federal Real Estate (BImA) and BfN holds roughly 33 000 ha while nature conservation NGOs have responsibility for 26 000 ha (BMUV, 2022a).

Management plans are supposed to be set up for all Natura 2000 sites and other protected areas, but in practice there have often been long delays between the designation of sites and the preparation of their plan. On the other hand, large protected areas usually have individual planning and management authorities with their own master plans and a related planning process. These plans have to be updated every 10 years, and depending on the individual site's features, may contain general habitat management provisions or specific species protection measures.

The delay in the designation of SCIs as SACs has had knock-on impacts on the management of the sites. Consequently, specific conservation objectives have not been set for many SCIs/SACs, and the European Commission has found that where they have, they are still insufficiently detailed and quantified. This is therefore also a subject that is currently being addressed through the SAC infringement case described earlier, with a Reasoned Opinion sent by the Commission on 12 February 2020.[7]

16.6 General Nature Conservation Measures

16.6.1 Development Planning and Impact Assessment

The public authorities charged with development and other planning tasks in Germany are also structured according to the principle of federalism, meaning that in principle three administrative levels (federal, state and local) with specific mandates on spatial and development planning exist (Turowski, 2002). Naturally, nation- or state-wide spatial planning provides less detailed goals and regulations than the regional and local planning mandates such as urban land-use planning. There are several laws that regulate spatial, land-use and building planning, many of them containing references to nature conservation matters, but two key issues are addressed directly in the BNatSchG itself: landscape planning, and the requirement to compensate for/offset unavoidable biodiversity impacts. These are further described in subsequent paragraphs. These and the other planning laws and processes are supported by impact assessments, carried out in accordance with the Federal Environmental Impact Assessment Act (based on the EU EIA Directive) and the Federal Strategic Environmental Assessment Act (based on the SEA Directive).

Landscape Planning

Landscape planning as a strategic conservation instrument was introduced with the BNatSchG in 1976 (Herberg, 2018). Sections 8–12 BNatSchG provide the legal basis for up to four levels of landscape planning in either a supra-local or local context. While landscape programmes cover nature conservation and landscape management at a federal state level, these are elaborated in greater detail in county landscape master plans. At the local (community) level, two additional instruments exist: landscape plans and open space structure plans, the latter addressing conservation on a neighbourhood- or project-related level. Each plan has to be in compliance with its respective higher-level plan (von Haaren, 2004).

The task of landscape planning as defined by law is to specify 'the purposes of nature conservation and landscape management' and to identify 'applicable requirements and measures for achieving such purposes' within the respective planning area (BMUV, 2022b). Thus, landscape planning is meant to transfer the overarching BNatSchG conservation objectives onto a detailed specified spatial plan. The resulting plans regularly contain an assessment of the existing status of nature and the required measures to maintain or achieve conservation objectives (e.g. including the favourable conservation status of habitats and species).

Most studies and evaluations of the landscape planning processes (e.g. von Haaren, 2004; Jessel, 2016; Walz *et al.*, 2018) have indicated that the intermediate planning level (county landscape master plans) seems to best implement conservation objectives and measures. These plans facilitate coordination between broader goals (e.g. habitat connectivity) and more specific goals (e.g. local conservation areas) and better balance potential land-use conflicts, such as priority areas for wind farms versus those for conservation. While the overall ability of landscape planning as a tool to reduce land consumption and the loss of habitats and open space remains questionable (see Jedicke *et al.*, 2016), other studies indicate that communities that have landscape plans (about 72% of all German municipalities) are better able to conserve habitats (Walz *et al.*, 2018). If regularly updated, local landscape plans also provide an information source on existing biodiversity, which can inform the planning of habitat banks or offsetting schemes.

Biodiversity Compensation/Offsetting

Together with the introduction of landscape planning, the two-tier principle of firstly avoiding impacts and then, if necessary, offsetting any residual adverse impacts on nature from interventions, such as built developments, has been anchored in the BNatSchG since 1976 as the so-called Impact Mitigation Regulation (*Eingriffsregelung*).[8] This has evolved over the decades, and whilst it has not turned into a broader no-net-loss instrument, it has been implemented in practice and has effectively offset many development impacts (Bruns, 2007; Wende *et al.*, 2018).

An impact on nature or the landscape is defined as '. . . any changes affecting the shape or use of areas, or changes in the groundwater level associated with the active soil layer, which may significantly impair the performance and functioning of the natural balance or landscape appearance'. Thus the regulation has a broad scope, covering nature, the integrity of the landscape and some key ecosystem functions and services. Importantly, it relates to all unprotected, as well as protected, areas. However, impacts from agriculture, forestry and fisheries are exempt.

Where impacts cannot be avoided, according to the mitigation hierarchy the project proponent must compensate for the residual impacts (normally measured through some form of metric). One way is to restore the impaired functions in an equivalent way, by on-site restoration of the same habitat type/features that are impacted (i.e. in-kind restoration). Alternatively, off-site replacement compensation (substitution) can be carried out to restore the impaired functions of the natural balance, in the relevant natural area, to an equivalent

[7] https://ec.europa.eu/commission/presscorner/detail/EN/INF_20_202

[8] The BMU translation of BNatSchG refers to the *Eingriffsregelung* as the 'intervention provisions', but this term is not commonly used in practice.

value. This can create or restore the same type, or another type, of habitat/feature (of equivalent or higher value). Off-site replacement compensation is also known as offsetting, but according to international definitions, restoration compensation is also a form of offsetting. Initially there was a clear legal requirement to prioritise on-site restoration compensation over off-site replacement compensation, but this caused practical difficulties. Changes to the BNatSchG in 2009 therefore allowed more flexibility in the choice of compensation type, so that the best and most proportionate environmental outcome can be achieved. The amendment to the legislation also introduced the possibility of monetary compensation when practical compensation is not feasible. This mainly relates to aesthetic impacts such as from wind farms.

Another important requirement is for the compensation/offsetting to be provided within an appropriate period and ideally at the same time as the impacts take place. To help achieve this, since 2000 it has been possible to establish habitat banks (i.e. pre-impact stocks of compensation), commonly referred to in Germany as 'eco-accounts' or 'land pools'. These are, for example, set up by public authorities on allocated land (identified as part of landscape planning) and run by compensation agencies. The developers can then compensate for their impacts by purchasing sufficient credits from the habitat bank (according to the metric used to measure the project's impacts). From a planning and conservation management perspective, this approach allows compensation to be combined in larger areas, which have higher ecological value than smaller and fragmented individual compensation measures. Also, this approach helps to ensure the principle of continuous ecological functionality (CEF) is followed, in alignment with Articles 12 and 13 of the Habitats Directive, as the usual time lag between the actual intervention and the completion of the compensation measure is reduced or completely avoided (Müller-Pfannenstiel, 2009).

Some aspects of the effectiveness of compensation under the Impact Mitigation Regulation have been under scrutiny for more than 20 years now. This has provided evidence that the initial implementation of compensation measures has quite frequently been incomplete compared to the planning/permit requirements, and that long-term maintenance and performance monitoring have been unsatisfactory (Mayer, 2006). However, it should be borne in mind that the study is now based on rather old evidence, and that such documented problems may have resulted in improved regulatory follow-up.

16.6.2 Habitat Management and Restoration

Maintenance of HNV Farmland and Agri-environment Schemes

As in all EU Member States, organic farming and agri-environment schemes (AES) help to conserve biodiversity on farmland. In total, about 9% of the utilised agricultural area (UAA) was being farmed organically in early 2020, with high regional variations – between 15% in Hesse and less than 4% in Lower Saxony (BMEL, 2020a).

AES can be of particularly high importance for biodiversity, but this varies according to the way they are programmed at a *Länder* level. Evaluations of two selected examples from the states of North Rhine-Westphalia and Lower Saxony/Bremen indicate why, over the 2007–2013 programming period, HNV farmland remained mostly in unfavourable condition (Reiter *et al.*, 2016a, b). First of all, AES that benefit biodiversity often covered an insignificant portion of agricultural land, such as only 5% of UAA in Lower Saxony and Bremen. In contrast, North Rhine-Westphalia was able to achieve 13% UAA coverage for AES containing biodiversity objectives. The reasons for the differing popularity of AES were the regional economic pressures, their competitiveness in relation to crops or dairy production, and the administrative burden associated with the design of individual measures.

Secondly, the biodiversity efficacy of each AES plays an important role in its overall success. The majority of AES were evaluated as being of low or medium efficacy (70% in Lower Saxony/Bremen and 87% in North Rhine-Westphalia). Measures with high efficacy – mostly contractual agreements for nature conservation (*Vertragsnaturschutz*), and very often on HNV farmland and/or areas in Natura 2000 sites – accounted for the minority of measures (10–15%). A key lesson from the evaluation was that, whilst more complex site-specific, targeted, result-based schemes usually cost more, they had higher efficacy with positive effects on biodiversity and should be used more in future AES programmes.

An evaluation of the Lower Saxony AES programme for the 2014–2020 funding period found that all indicator trends (HNV, farmland birds, etc.) were still negative, and the AES still only reached 5% of the UAA (Sander and Bathke, 2020). However, compared to the previous period, a shift towards measures with a medium and high biodiversity efficacy had occurred, and there was greater targeting of Natura 2000 sites (29% of UAA in Natura 2000 sites). In contrast, the AES contribution to other areas of HNV farmland remained low. The authors provided several recommendations for further improvements, such as retaining the focus on high-efficacy measures, improving advice to farmers, additional land purchases by local or state administrations to increase the AES-targeted UAA, and the permanent safeguarding of a sufficient AES budget to demonstrate reliable funding to the farming community (Sander and Bathke, 2020). Although these conclusions were made for the Lower Saxony and Bremen context, they are also applicable to other *Länder*.

Forest Management

In principle, three categories of forest ownership exist in Germany: state (30%)/federal (4%) government, legal entity ownership (19%, e.g. local government, churches and universities), and private ownership (44%) (BMEL, 2017). The degree of naturalness of forests varies greatly depending on their ownership, with only about 30% of private forests being in a semi- or near-natural state, compared to 40% and 42% of legal entity and state/federal forests, respectively (BMEL, 2017). As described in Section 16.3, the 2007 NBS focusses on the objective to raise the share of natural forests without intervention to 5% of the forested area. This has led to probably one of the most controversial and disputed conservation objectives over the last decade. Although the overall achievement cannot be achieved without the participation of private forest owners, recent emphasis has been given to the secondary target of a 10% share of natural forests on federal and state forest lands (BMUV, 2022c). In addition, the federal and state governments are trying to increase the amount of contractual conservation agreements to 10% of the private forest land area (BMUV, 2022c).

At a state level, all *Länder* have adopted strategies to increase or maintain their 10% share of natural forests. The mission of the respective state agencies, or state corporate entities, may vary in detail but generally includes four strategic objectives for forest management: providing an economic basis through sustainable resource use, maintaining or improving ecological functions, providing recreational opportunities to the public, and maintaining their protective function (especially in montane regions). As these entities are all governed by public law, this approach to the three dimensions of sustainable development is easier to anchor in the mission statement than on private lands that are governed by private law. The ecological measures in the forest often focus on conserving particularly important trees as habitat, increasing the amount of deadwood and the conversion of conifer-dominated forests into mixed forests (BMEL, 2020b). One additional challenge is forest adaptation to climate change, such as by the selection of tree species that have a higher resilience to increasing drought periods which currently specifically affect the Norway spruce (*Picea abies*), the former principal productive tree in German forests (Nordwestdeutsche Forstliche Versuchsanstalt, 2017). Priority should be given to native species, but the mix may include neophytes, whilst trying to avoid the introduction of potential invasive species (Bayerisches Staatsministerium für Ernährung, Landwirtschaft und Forsten, 2020).

The desired increase in contractual agreements with private forest owners is facing several challenges. Although no empirical data exist on this specific issue, it is relatively safe to assume that the previously mentioned dispute has generally not increased trust in the conservation measures offered by the respective authorities. However, a survey found that once forest owners were convinced and took part in such measures, 75% described their overall experience as positive, even though the administrative burden was seen as quite heavy (Kownatzki *et al.*, 2017). The same study concluded that income from conservation measures rarely exceeded 5% of the specific forest owner's income, hence stating that whilst the measures seemed to be currently economically unattractive, there were opportunities for significant growth in the future.

Overall, the research on contractual conservation measures in forests demonstrates that there are several parameters that could be adjusted to increase their popularity with forest owners. Of course, in Natura 2000 sites, while legislative action could always be taken to enforce conservation objectives, voluntary agreements seem to be the better accepted tool to secure forest owners' cooperation.

16.7 Nature Conservation Costs, Benefits and Funding Sources

16.7.1 Nature Conservation Costs

In 2010, as part of an EU-wide study (Gantioler *et al.*, 2010), the annual costs of the Natura 2000 network in Germany were estimated at €620 million. In 2016, an expert working group from federal and *Länder* agencies prepared an updated cost estimate for the implementation of Natura 2000 as well as meeting CBD requirements through the favourable conservation status for species, resulting in an amount of €1.416 billion (LANA, 2016). The estimate included one-off investments and ongoing costs covering conservation measures within and outside Natura 2000 sites, species protection, monitoring and public outreach. The current Prioritised Action Framework (PAF) for 2021–2027 (BMU, 2020b) picked up from there and now estimates the costs at between €1.329 and €1.547 billion per year. This annual total can be broadly split into three categories: horizontal measures and administrative costs amount to €262 million, Natura 2000 site management €377 million for one-off investments (averaged over the period) plus €494–636 million for ongoing costs, and species-focussed programmes between €198 and 272 million.

As nature conservation includes more than the implementation of Natura 2000, a previous study by Wüstemann *et al.* (2014) assessed the cost of implementing the NBS. For the modelling, the authors developed a conservation programme based on the strategy, which consisted of land-use changes for six ecosystem types and the provision of two ecosystem services (water purification and climate mitigation). This scenario resulted in estimated costs of about €3.26 billion per annum over 2010–2020.

Taking the current funding (see Section 16.7.3) for nature conservation in Germany, either the estimated cost figure from the 2021–2027 PAF or the Wüstemann *et al.* study supports several conclusions that nature conservation remains 'blatantly underfunded' (SRU & The Scientific Advisory Board on Forest Policy, 2017; Kreiser, 2018). Although the costs of effective nature conservation might seem high, they are outweighed by the benefits that nature conservation provides, as discussed in the following paragraphs.

16.7.2 The Economic Benefits of Nature Conservation

Various assessments have been carried out in Germany, covering all the provisioning, regulating, supporting and cultural categories of ecosystem services (TEEB DE, 2010). These have revealed the difficulties that arise with the valuation and the monetisation of ecosystem services, especially cultural services (e.g. see Matzdorf *et al.* (2010) with respect to HNV farmland). Nevertheless, they also show that many synergies exist between biodiversity conservation measures and ecosystem services. More recently, TEEB Germany has provided a number of comprehensive studies that have resulted in four major reports on ecosystem services in Germany: Natural Capital and Climate Policy; Ecosystem Services in Rural Areas; Ecosystem Services in the City; and a Synthesis Report (TEEB DE, 2020). In terms of the economic benefits, the TEEB Germany studies include an effort to provide at least partially quantifiable data on the economics of measures.

For example, with respect to flood protection, examined case studies demonstrated that the benefits from the relocation of dikes to increase the floodplain along the Elbe River between Dresden and Lauenburg outweighed the costs by a factor of 3:1. The benefits, including avoided flood damage, added up to €1.2 billion compared to an investment of €407 million. Retrofitting existing dikes at the same cost would only lead to economic benefits of €177 million (TEEB DE, 2017).

16.7.3 Funding

Table 16.5 provides an overview of the main funding sources available for nature conservation in Germany, in particular over the EU 2014–2020 MFF. This should be interpreted with caution; the calculation of the overall funding available for nature conservation poses quite a challenge, as there is a variety of public EU, federal, state and private funding sources. In particular, the collation and analysis of federal state data would be required to provide a complete and robust estimate of overall expenditure, but comprehensive and comparable data are not readily available to enable this. In addition, nature conservation is not only funded through the BMU, but also by other ministries such as the Ministry of Food and Agriculture (BMEL) and the Ministry of Education and Research (BMBF), sometimes in interagency programmes or as a part of EU co-financing. Hence, a summation of funds could lead to overestimates due to double-counting.

Despite the uncertainties, the data clearly show the disproportionate dependence on EU funds, which was estimated as 70% of all public funding by Kreiser (2018). Rural Development Programme (RDP) measures, under the European Agricultural Fund for Rural Development (EAFRD), have been especially important, with over a billion euros allocated each year to Priority 4 objectives, which include biodiversity. As described earlier, of particular importance has been the agri-environment–climate measure funding. However, it is important to bear in mind that over one-half of the RDP Priority 4 funding is for organic farming and supporting areas facing natural and other constraints, which have rather broad environmental objectives.

A more focussed study by Horlitz *et al.* (2018) has provided the most recent comprehensive estimate of funding sources for Natura 2000 conservation in Germany, which was also taken up in the PAF for 2021–2027 (BMU, 2020b). According to their estimates, 13% of the total 2014–2020 German EARDF funding could be categorised as principal biodiversity measures (with organic farming excluded from this category), resulting in €324 million annually. Overall, all available sources amounted to about €536 million annually. This is only about 38% of the estimated annual requirement of €1.4 billion for Natura 2000 alone (see earlier), leaving an annual gap of approximately €900 million (SRU and The Scientific Advisory Board on Forest Policy, 2017). Data from the more recent 2021–2027 PAF (BMU, 2020b) provide a very similar estimate of €515 million total EU, national and private funds annually available over the 2014–2020 funding period.

It should be also be borne in mind that the preparation of RDPs is the responsibility of the federal states, and their use of the various measures for nature conservation or other purposes varies greatly depending on their priorities. For example, allocations to the agri-environment–climate measure range between 4% and 31% of the budget (BMEL, 2019). For such reasons, comparative studies on nature conservation expenditure on agri-environment measures should ideally be at a federal state level where possible (von Ruschkowski and von Haaren, 2008). In addition, a BMU report (2017) questioned the ‘Biodiversity Tracking’ method used by the European Commission, which allocated €350 million of EU funding alone for direct and indirect measures for Germany 2014–2020, leading to a huge overestimate of available funds for conservation.

Particularly efficient sources of EU co-financing have been from the EU LIFE Programmes. Over 1992–2020, 143 LIFE Nature and biodiversity-themed projects have been implemented in Germany; although the overall funding has been relatively modest (€408 million, 57% EU contribution), the projects have contributed significantly to local or regional-level conservation actions.

Exclusively national funding sources, without NGO contributions, have amounted to approximately €152 million annually (excluding co-funding of EU funding instruments), with sources almost evenly distributed between federal and *Länder* programmes. It is very difficult to estimate the contribution of NGO funding to nature conservation in Germany, because no compiled relevant data are available and some NGO income is from national public sources. The minimum €30 million per year estimated here in Table 16.5 should therefore be treated as indicative. Furthermore, some of the funds are used to co-finance EU or federal projects, so it is safe to assume that a significant portion of NGO spending is already accounted for in the funding table. Nevertheless, it is clear that their contribution to nature conservation financing is significant as the annual budget of the major conservation NGOs in Germany varies from €42 million (BUND) to €62 million (NABU) to €92 million (WWF),[9] although not all funds are used for conservation projects.

[9] All based on data compiled from annual reports 2020.

Table 16.5 Funding sources for nature conservation in Germany.

Source and period	Measure/category	Average annual allocation (€ million)		Comments
EU funds		**EU + national**	**EU**	
EAFRD Rural Development Programme measures – allocations to Priority 4 (restoring, preserving and enhancing agriculture and forest ecosystems) 2014–2020	M10 Agri-environment–climate	369.4	259.7	
	M11 Organic farming	291.3	190.5	
	M13 Areas facing constraints	224.6	137.5	
	M7 Basic services, etc.	77.2	51.5	
	M8 Investments in forests	27.9	18.3	
	M4 Investments in physical assets	16.5	11.8	
	M12 Natura and WFD	14.1	9.8	
	M1 Knowledge transfer, etc.	8.5	6.9	
	M16 Cooperation	4.7	3.6	
	M2 Advisory services, etc.	3.1	1.8	
	M15 Forest environment climate	1.6	1.3	
RDP Priority 4 subtotal		*1 039.0*	*692.8*	
ERDF 2014–2020	085 Protection and enhancement of biodiversity, nature protection and green infrastructure	11.7	8.0	For N2k, etc.
	086 Protection, restoration and sustainable use of Natura 2000 sites	3.0	2.1	
EMFF 2014–2020	Nature conservation measures	3.0	2.2	
LIFE Programme	Nature and biodiversity projects (1992–2020)	14.1	8.0	
	Integrated Project – Environment (2016–2026)	1.7	1.0	Atlantic region DE
EU funding instrument subtotal		*1 072.5*	*714.1*	
National Funds within 2014–2020		**Additional national***	**Comments**	
Federal Conservation Programme 'chance.natur'	Securing sites of national importance for conservation	18.7	For N2k, etc.	
Federal Programme 'Biological Diversity'		29.5		
Joint Task for Agricultural Structure and Coastal Protection (GAK)	Nature conservation–related measures	14.3		
Wilderness fund		2.8	For N2k, etc. €10 m each in 2019 and 2020. Annualised over 2014–2020.	
Länder Programmes	Conservation programmes exclusively funded by the *Länder*	67	For N2k, etc.	
German Federal Environmental Foundation (DBU)		6.7		
Länder Foundations		13.2		
NGOs	Conservation programmes in Germany	>30.0	Exc. public funds from federal or state programmes	

able 16.5 *(cont.)*

ource and period	Measure/category	Average annual allocation (€ million)		Comments
U funds		**EU + national**	**EU**	
dditional national subtotal		*>182.2*		
lational total including co-financing of EU instruments		*540.6*		
otal		>1 254.7		

lotes: * Excludes co-financing of EU funds unless indicated. '% N2k, etc.' indicates actions for Natura 2000, protection of species and green infrastructure (based on he PAF). The subtotals and totals are only approximate indications of the average annual levels of funding as they are based on annualised data from differing eriods. Basis for the NGO estimate: selected NGOs' 2020 annual reports BUND/NABU/WWF, including Naturstiftung David and NABU-Stiftung Nationales laturerbe. As income is often a combination of donations mixed with public funds, the latter have been excluded.
ources. Based on data from: RDP data from European Commission ESIF funds database (23 October 2020); allocations to ERDF, EMFF and national funding from 021–2027 PAF (BMU, 2020b); Germany LIFE factsheet European Commission (December 2020); BMU (2020a, b); BMEL (2019); Horlitz *et al.* (2018). See Section 1.3 or further details.

.8 Surveillance, Monitoring, Reporting and plied Research

e monitoring of nature and its assets is regulated in Section of the Federal Nature Conservation and Landscape anagement Act as a joint duty for the federal and state encies. In practice, most of the monitoring currently relates the reporting that is required under BD Article 12 and HD ticle 17. During the early 2000s, many states drastically cut ck their staff resources in conservation agencies, which par- ularly reduced long-term monitoring of protected areas and ecies (SRU, 2007). In response to this recognised deficit, veral combined efforts between governmental and non- vernmental organisations have been launched to support proved citizen science engagement (e.g. Bonn *et al.*, 2016). his builds on a long history of biodiversity monitoring rough data generated by volunteers (Roy *et al.*, 2012).

Such citizen science initiatives are aided by increasing gitalisation but hampered by limited knowledge transfer om older to younger generation volunteers, often resulting om a lack of resources in the state conservation agencies and its in taxonomic training at German universities and other ducational institutions. This trend has been described as an rosion of taxonomic experts' by Frobel and Schlumprecht 2016), which has served as a wake-up call to rethink biodiver- ity monitoring and volunteer engagement. Although the trend as not been reversed, the topic remains high on the conser- ation agenda, and multiple projects have been launched over he past five years to train a new generation of volunteers.

Several well-established monitoring programmes exist as a tarting base. The national bird monitoring programme was onsolidated in 2008 and has since been carried out under the uspices of the Federation of German Avifaunists, jointly unded by the BfN and the *Länder* conservation agencies BfN, 2012). A web portal[10] also plays a vital role in collecting bird data, with more than five million data entries per year. Another example is the 'Butterfly Monitoring Germany' project that was launched in 2005 and is coordinated by the Helmholtz-Centre for Environmental Research (UFZ), with about 500 volunteers contributing nationwide (Kühn *et al.*, 2014). In 2019, BfN also launched an initiative for a multiple-purpose national insect monitoring programme, which is currently under development (BfN, 2019e). This programme aims to be intertwined with the bird monitoring described above, and the BfN-led HNV farmland monitoring programme, which has taken place on about 1 375 representative 1 km^2 sample areas since 2009 (Hünig and Benzler, 2017).

Applied research is partially facilitated by BfN-funded testing and development projects, which are mandated in the federal government's environmental research plan (Umweltforschungsplan) and also, since 2011, through the federal programme 'Biological Diversity' (Bundesprogramm Biologische Vielfalt).

16.9 Future Developments

As one of the most important remaining challenges – the legal designation of Natura 2000 sites – finally nears completion, the focus will shift towards the compilation and implementation of the required management plans. The status of management planning completion is not available at a national level, but for many sites this task has not been completed and is currently underway. BfN has developed management concepts for selected species and habitats (Ackermann *et al.*, 2016), which need to be tailored and implemented in relation to site-level needs.

The *National Strategy on Biological Diversity* requires an evaluation and an update to reflect UN and EU developments of biodiversity strategies and the Agenda 2030. As of October 2021, this requirement has been announced by the federal ministry, but no details have been revealed other than the continuation of stakeholder dialogues. The evaluation needs to be based on the 'Conservation Action Programme for

[10] www.ornitho.de

Biological Diversity 2020', which was launched in 2015 and is currently still awaiting an assessment.

Some notable efforts to incorporate conservation in agricultural practices have been launched in several German states. This started with a petition for a referendum on biological diversity called 'Save the Bees!', which was launched in Bavaria in 2018. No referendum was required in the end, as the Bavarian Parliament accepted the petition and took action (e.g. comprehensive changes to the Bavarian state conservation laws were adopted in 2019), resulting in about 80 detailed measures. However, progress after the first year was considered slow by the initiators (Volksbegehren Artenvielfalt, 2020). Similar efforts to impose stricter conservation measures within agricultural areas have been successful in Baden-Württemberg, North Rhine-Westphalia and Lower Saxony. Efforts are directed to more cooperative and collaborative, site-specific approaches between agriculture, forestry and conservation management.

16.10 Conclusions

Considerable efforts have been made to advance nature conservation in Germany over the last 40 years, such as through the accelerated designation of large protected areas (especially in the marine environment). The two major drivers were the German reunification in 1990, with the sudden creation of five National Parks and additional large protected areas, and the long-winded implementation of Natura 2000. Despite this, and some successes with the protection of several flagship species, the overall impacts of nature conservation have been disappointing and the current situation is rather bleak. In particular, the excessive loss of farmland birds and semi-natural grasslands over the last 25 years has indicated concerning developments.

One reason for this situation is that whilst Germany has probably had amongst the most sophisticated conservation laws in Europe since the 1970s, their implementation on the ground has remained a challenge. This has often been as a result of limited staffing and financial resources. There is therefore an obvious undisputed need for a major increase in financial investment in nature to address the documented annual funding gap described earlier. This should be achievable, as the required annual amount of €1.4 billion for the proper management of Natura 2000 is only 0.4% of the annual federal budget. Furthermore, such investments bring wider socio-economic benefits, but these are often overlooked during budget negotiations.

Aside from identifying and addressing the main drivers of biodiversity loss in the wider environment (e.g. habitat fragmentation, pesticide use, eutrophication and intensive agriculture), it is important to emphasise the need for a less fragmented approach to the conservation sites. As well as increasing their overall extent, larger individual sites are required so that landscape-scale approaches to conservation play a bigger role. This does not only apply to wilderness areas – the focus needs to shift to the 'ordinary' landscapes that often historically had rich biodiversity because of low-intensity practices such as grazing and hay making.

In future, the efficiency of conservation must also be improved through more cooperative approaches with land-users. In this respect, one of the important lessons from past experience is that successful nature conservation requires a much higher investment in the social science aspects of conservation. The complex interdependences between human land use and ecosystem services need to receive more attention when communicating conservation objectives. The degradation of ecosystem services that society directly benefits from should serve as a storyline that people can relate to and thereby better understand the need for conservation. Positive communication alone will not solve the implementation deficit, but it will be a powerful tool if used to infuse and underline evidence-based conservation rationalism with emotional stories that are based on evidence from the social sciences. The key lies in the integration of the latter into the former (Bennett *et al.*, 2017).

As the world's fourth-largest economy, Germany has the opportunity to be a role model for other EU Member States and at a global scale, by demonstrating that its economy and political system can contribute more to the conservation of its own nature, as well as wider international environmental goals of the EU, CBD and UN's SDGs.

Acknowledgements

I would like to thank Hermann Hötker, Till Hopf, Uwe Riecken, Volker Scherfose, Axel Ssymank and Raphael Weyland, who all provided additional information, reviewed parts of the chapter and helped to make sure that the information provided is up to date. This chapter is dedicated to Hermann Hötker, who passed away way too early in July of 2019.

References

Ackermann, W., Streitberger, M. & Lehrke, S. (2016) *Maßnahmenkonzepte für ausgewählte Arten und Lebensraumtypen der FFH-Richtlinie zur Verbesserung des Erhaltungszustands von Natura 2000-Schutzgütern in der atlantischen biogeografischen Region.* Bonn-Bad Godesberg: Bundesamt für Naturschutz. BfN-Skripten 449.

Bannas, L., Löffler, J. & Riecken, U. (2017) *Die Umsetzung des länderübergreifenden Biotopverbunds – rechtliche, strategische, planerische und programmatische Aspekte.* Bonn-Bad Godesberg: Bundesamt für Naturschutz. BfN-Skripten 475.

Bayerisches Staatsministerium für Ernährung, Landwirtschaft und Forsten (2020) *Baumarten für den Klimawand.* Leitlinien der Bayerischen Forstverwaltung. Munich.

Bennett, N., Roth, R., Klain, S. *et al.* (2017) Conservation social science: understanding and integrating human dimensions to improve conservation. *Biological Conservation*, **205**, 93–108.

Bernhardt, C. (2000) *Die Rheinkorrektion.* Der Bürger im Staat 2/2000, 76-81. Stuttgart: Landeszentrale für politische Bildung Baden-Württemberg.

BfN (2009a) *Biogeografische Regionen und naturräumliche Haupteinheiten Deutschland*. www.bfn.de/daten-und-fakten/biogeografische-regionen-und-naturraeumliche-haupteinheiten-deutschlands

(2009b) *Auenzustandsbericht. Flussauen in Deutschland*. Bonn-Bad Godesberg: BfN.

(2012) *Organisation des Vogelmonitorings*. www.bfn.de/vogelmonitoring

(2013) *Weitere Nationalparke für Deutschland?! Argumente und Hintergründe mit Blick auf die aktuelle Diskussion um die Ausweisung von Nationalparken in Deutschland*. Bonn-Bad Godesberg: BfN.

(2015) *Artenschutz-Report 2015. Tiere und Pflanzen in Deutschland*. Bonn-Bad Godesberg: BfN.

(2016) *Daten zur Natur*. Bonn-Bad Godesberg: BfN.

(2017) *Federal Green Infrastructure Concept*. Bonn-Bad Godesberg: BfN.

(2018a) *10 Jahre HNV-Farmland-Monitoring*. Dokumentation der Fachtagung vom 13. September 2018 in Erfurt. www.bfn.de/sites/default/files/2021-10/Zusammenfassung_HNVJub_Tagung_fuer_HP-1.pdf

(2018b) *Naturparke in Deutschland*. www.bfn.de/sites/default/files/2022-03/Naturparke%20in%20Deutschland%20Stand%2001_01_2022_pac.pdf

(2019a) Naturschutzgebiete. www.bfn.de/naturschutzgebiete

(2019b) Landschaftsschutzgebiete. www.bfn.de/landschaftsschutzgebiete

(2019c) *Unzerschnittene verkehrsarme Räume in Deutschland*. www.bfn.de/sites/default/files/2021-08/BKGI_Karte_UZVR.pdf

(2019d) *Large-scale conservation projects*. www.bfn.de/en/topic/large-scale-nature-conservation-projects

(2019e) *Bundesweites Insektenmonitoring*. www.bfn.de/insektenmonitoring

(2020a) *Nationale Naturmonumente*. www.bfn.de/nationale-naturmonumente

(2020b) *Wildnisgebiete*. www.bfn.de/wildnisgebiete

(2022a) *FFH-Gebiete in Deutschland*. www.bfn.de/sites/default/files/BfN/natura2000/Dokumente/meldeueb_ffh_20191213_bf.pdf

(2022b) *Vogelschutzgebiete in Deutschland*. www.bfn.de/sites/default/files/BfN/natura2000/Dokumente/meldeueb_spa_20191213_bf.pdf

(2022c) *Ergebnisübersicht – Nationaler Bericht 2019*. www.bfn.de/ffh-bericht-2019

(2022d) *Nationalparke*. www.bfn.de/nationalparke

(2022e) Nationale Meeresschutzgebiete: Übersicht, Kurzfakten. www.bfn.de/nationale-meeresschutzgebiete

(2022f) *Zum Stand der Umsetzung von Natura 2000 in Deutschland*. www.bfn.de/natura-2000-gebiete

BMEL (2017) *Daten und Fakten. Land-, Forst und Ernährungswirtschaft mit Fischerei und Wein- und Gartenbau*. Berlin: Bundesministerium für Ernährung und Landwirtschaft.

(2019) *Umsetzung der ELER-Förderperiode 2014 bis 2020 für ländliche Räume in Deutschland*. www.bmel.de/DE/themen/laendliche-regionen/foerderung-des-laendlichen-raumes/eu-foerderung/eler-2014-2020-umsetzung.html

(2020a) Organic Farming in Germany. February 2020. Bonn: BMEL. www.bmel.de/SharedDocs/Downloads/EN/Publications/Organic-Farming-in-Germany.pdf?__blob=publicationFile&v=4

(2020b) *Ausgewählte Ergebnisse der dritten Bundeswaldinventur*. Bundeswaldinventur 2012. www.bundeswaldinventur.de/dritte-bundeswaldinventur-2012

BMU (2007) *Nationale Strategie zur biologischen Vielfalt*. Berlin: Bundesministerium für Umwelt, Naturschutz und nukleare Sicherheit.

(2017) Schriftlicher Bericht für die 59. Amtschefkonferenz und die 88. Umweltministerkonferenz vom 3.-5. Mai 2017 in Bad Saarow, TOP 21: Naturschutzfinanzierung. www.umweltministerkonferenz.de/documents/88umk_top21_finaler-bmub-bericht_eu-naturschutzfinanzierung_1543308677.pdf

(2020a) BMU-Haushalt 2020. www.bmu.de/download/bmu-haushalt-2020/, accessed 24 July 2020.

(2020b) *Prioritärer Aktionsrahmen (PAF) für NATURA 2000 in der Bundesrepublik Deutschland gemäß Artikel 8 der Richtlinie 92/43/EWG des Rates zur Erhaltung der natürlichen Lebensräume sowie der wildlebenden Tiere und Pflanzen (Habitat-Richtlinie) für den Zeitraum 2021-2027*. Bonn: BMU.

BMUV (2022a) Nationales Naturerbe. www.bmuv.de/themen/naturschutz-artenvielfalt/naturschutz-biologische-vielfalt/gebietsschutz-und-vernetzung/nationales-naturerbe

(2022b) Federal Nature Conservation Act – BNatSchG. www.bmuv.de/en/law/federal-nature-conservation-act-bnatschg/

(2022c) Nationale Waldschutzpolitik. www.bmuv.de/themen/naturschutz-artenvielfalt/naturschutz-biologische-vielfalt/waelder/nationale-waldschutzpolitik

Bonn, A., Settele, J. & Löw, M. *et al.* (2016) *Lebendiger Atlas – Natur Deutschland: Machbarkeitsstudie*. Leipzig: Helmholtz-Zentrum für Umweltforschung – UFZ und Deutsches Zentrum für integrative Biodiversitätsforschung (iDiv).

Brandt, M., Dragon, A. C., Diederichs, A. *et al.* (2018) Disturbance of harbour porpoises during construction of the first seven offshore wind farms in Germany. *Marine Ecology Progress Series*, **596**, 213–232.

Bredemeier, B., von Haaren, C., Rüter, S. *et al.* (2015) Evaluating the nature conservation value of field habitats: a model approach for targeting agri-environmental measures and projecting their effects. *Ecological Modelling*, **295**, 113–122.

Bruns, E. (2007) *Bewertungs- und Bilanzierungsmethoden in der Eingriffsregelung. Analyse und Systematisierung von Verfahren und Vorgehensweisen des Bundes und der Länder*. Dissertation. TU Berlin.

Bundesregierung (2016) *Deutsche Nachhaltigkeitsstrategie – Neuauflage 2016*. Berlin: Bundesregierung.

Busch, M., Katzenberger, J., Trautmann, S., Gerlach B., Dröschmester, R. & Sudfeldt, C. (2020) Drivers of population change in common farmland birds in Germany. *Bird Conservation International*, **30**, 335–354.

Common Wadden Sea Secretariat (2016) *Report on the State of Conservation of the World Heritage property 'The Wadden Sea (N1314)'*. Wilhelmshaven: Common Wadden Sea Secretariat.

DBU Naturerbe (2020) DBU-Naturerbe – 70.000 Hektar für den Naturschutz. www.dbu.de/naturerbe

Deutsche UNESCO-Kommission (2019) *UNESCO-Welterbestätten in Deutschland*. Bonn: Deutsche UNESCO-Kommission.

Deutscher Bundestag (2015) *Wiedervernetzung durch Wildbrücken. Antwort der Bundesregierung*. Bundestags-Drucksache 18/6075.

(2018a) Schriftliche Fragen mit den in der Woche vom 28. Mai 2018 eingegangenen Antworten der Bundesregierung. Bundestags-Drucksache 19/2419.

(2019) Grünbrücken in Deutschland. Antwort der Bundesregierung vom 12.04.2019. Bundestags-Drucksache 19/9409.

(2020) Drohendes Vertragsverletzungsverfahren der EU-Kommission gegen Deutschland wegen mangelhafter Natura-2000-Schutzgebiete. Antwort der Bundesregierung. Drucksache 19/22032.

DNR (2022) Über den DNR. www.dnr.de/ueber-den-dnr

Engel, F., Bauhaus, J., Gärtner, S. *et al.* (2016) *Wälder mit natürlicher Entwicklung in Deutschland: Bilanzierung und Bewertung*. Naturschutz und Biologische Vielfalt 145. Bonn-Bad Godesberg: BfN.

European Commission (2019a) January infringements package: key decisions, infringement decisions, 25 January 2019. https://ec.europa.eu/commission/presscorner/detail/en/MEMO_19_462

(2019b) July infringements package: key decisions, infringement decisions, 25 July 2019. https://ec.europa.eu/commission/presscorner/detail/en/INF_19_4251

(2020) February infringements package: key decisions, infringement decisions, 12 February 2020. https://ec.europa.eu/commission/presscorner/detail/en/inf_20_202

EEA (2019) Loss of HNV farmland due to agricultural intensification per NUTS3 www.eea.europa.eu/data-and-maps/figures/loss-of-hnv-farmland-due

Fachgruppe Vögel in der Agrarlandschaft (2019) Weiterentwicklung der Gemeinsamen Agrarpolitik ab 2021: Erfordernisse zum Erhalt unserer Agrarvögel. *Berichte zum Vogelschutz*, **56**, 93–107.

Frobel, K. & Schlumprecht, H. (2016) Erosion der Artenkenner. *Naturschutz und Landschaftsplanung*, **48** (4), 105–113.

Frohn, H. W., Rosenbrock, J. & Schmoll, F. (2009) *"Wenn sich alle in der Natur erholen, wo erholt sich dann die Natur?" Naturschutz, Freizeitnutzung, Erholungsvorsorge und Sport - gestern, heute, morgen*. Naturschutz und Biologische Vielfalt 75. Bonn-Bad Godesberg: BfN.

Fuchs, D., Hänel, K., Lipski, A. *et al.* (2010) *Länderübergreifender Biotopverbund in Deutschland – Grundlagen und Fachkonzept*. Naturschutz und Biologische Vielfalt (96). Münster: Landwirtschaftsverlag.

Fuchs, D., Oppermann, R. & Krismann, A. (2011) *Umsetzung des High Nature Value Farmland-Indikators in Deutschland*. Report UFOPLAN FKZ 3508 89 0400.

Gantioler, S., Rayment, M., Bassi, S. *et al.* (2010) *Costs and Socio-Economic Benefits associated with the Natura 2000 Network*. Final report to the European Commission, DG Environment on Contract ENV.B.2/SER/2008/0038. Brussels: Institute for European Environmental Policy/GHK/Ecologic.

Gellermann, M. (2007) Die "Kleine Novelle" des Bundesnaturschutzgesetzes. *Natur und Recht*, **29**, 783–789.

Gerlach, B., Dröschmeister, R., Langgemach, T. *et al.* (2019) *Vögel in Deutschland – Übersichten zur Bestandssituation*. Münster: DDA, BfN, LAG VSW.

Gharadjedaghi, B., Heimann, R., Lenz, K. *et al.* (2004) Verbreitung und Gefährdung schutzwürdiger Landschaften in Deutschland. *Natur und Landschaft*, **79** (2), 71–81.

Hallmann, C., Sorg, M., Jongejans, E. *et al.* (2017) More than 75 percent decline over 27 years in total flying insect biomass in protected areas. *PLoS ONE*, **12** (10), e0185809.

Heinze, S., Finck, P., Raths, U. *et al.* (2019) Analyse der Gefährdungsursachen von Biotoptypen in Deutschland. *Natur und Landschaft*, **94** (11), 453–464.

et al. (2020) Revised criteria system for a national assessment of threatened habitats in Germany. *Nature Conservation*, **40**, 39–64.

Herberg, A. (2018) Naturschutz und Landschaftsplanung – zwischen Ambition und Wirklichkeit. In *Landschaftsplanung im Prozess und Dialog*, ed. I. Marschall, pp. 8–21. Bonn-Bad Godesberg: BfN. BfN-Skripten 498.

Hoeft, C., Messinger-Zimmer, S. & Zilles, J. (eds.) (2017) *Bürgerproteste in Zeiten der Energiewende. Lokale Konflikte um Windkraft, Stromtrassen und Fracking*. Bielefeld: Transcript-Verlag.

Hötker, H., Jeromin, H. & Melter, J. (2007) Entwicklung der Brutbestände der Wiesen-Limikolen in Deutschland - Ergebnisse eines neuen Ansatzes im Monitoring mittelhäufiger Brutvogelarten. *Die Vogelwelt*, **128** (2), 49–65.

Hötker, H. & Leuschner, C. (2014) *Naturschutz in der Agrarlandschaft am Scheideweg – Misserfolge, Erfolge, neue Wege*. Hamburg: Studie im Auftrag der Michael Otto Stiftung für Umweltschutz.

Homburg, K., Drees, C., Boutaud, E. *et al.* (2019) Where have all the beetles gone? Long-term study reveals carabid species decline in a nature reserve in Northern Germany. *Insect Conservation and Diversity*, **12** (4), 268–277.

Horlitz, T., Achtermann, B., Pabst, H. & Schramek, J. (2018) *Ermittlung des geplanten finanziellen Umfangs von Naturschutzmaßnahmen im Rahmen der ELER-Programme zur Entwicklung des ländlichen Raums 2014 – 2020 – Herausforderungen, Methode und Ergebnisse*. Ad hoc-Arbeitspapier im Rahmen des Forschungs- und Entwicklungsvorhabens "Biodiversitätsförderung im ELER" (ELERBiodiv) (FKZ 3515 880 300). Hannover.

Hünig, C. & Benzler, A. (2017) *Das Monitoring der Landwirtschaftsflächen mit hohem Naturwert in Deutschland*. Bonn-Bad Godesberg: Bundesamt für Naturschutz. BfN-Skripten 476.

Jedicke, E., Reinke, M. & Riedel, W. (2016) Landschaftsplanung – ein Ausblick. In *Landschaftsplanung*, eds. W. Riedel, H. Lange, E. Jedicke & M. Reinke, pp. 519–525. Berlin/Heidelberg: Springer Spektrum.

Jessel, B. (2016) Ökologie, Naturschutz und ökologisch orientierte Landschaftsplanung. *Natur und Landschaft*, **31** (9/10), 456–463.

Job-Hoben, B., Pütsch, M. & Erdmann, K. H. (2009) Naturschutz und Freizeitnutzung – Trends und Perspektiven. *Naturschutz und Biologisch Vielfalt*, **75**, 305–316.

Joosten, H. (2012) Zustand und Perspektiven der Moore weltweit. *Natur und Landschaft*, **87** (2), 50–55.

Kownatzki, D., von Blomberg, M., Demant, L. *et al.* (2017) *Status quo der Umsetzung von Naturschutz im Wald gegen Entgelt in Deutschland - Ergebnisse einer Befragung von Forstbetrieben.* Braunschweig: Johann Heinrich von Thünen-Institut. Thünen Working Paper 81.

Kreiser, K. (2018) EU-Naturschutzfinanzierung: Geschätzter Bedarf, aktuelle Defizite und Forderungen der Umweltverbände. *Naturschutz und Landschaftsplanung*, **50** (10), 360–367.

Kühn, E., Musche, M., Harpke, A. *et al.* (2014) Tagfalter-Monitoring Deutschland. *Oedippus*, **27**, 1–47.

Küster, H. J. (2010) *Geschichte der Landschaft Mitteleuropas.* Von der Eiszeit bis zur Gegenwart. 4th ed. Stuttgart: C.H. Beck.

LANA (2016) *Wirksamkeit der derzeitigen EU-Naturschutzfinanzierung in Deutschland und Anforderungen für die nächste Förderperiode ab 2020.* Positionspapier der LANA-Expertengruppe "EU-Naturschutzfinanzierung/GAP 2020", Stand: 01.09.2016.

Länderarbeitsgemeinschaft der Staatlichen Vogelschutzwarten (LAG VSW) (2014) Recommendations for distances of wind turbines to important areas for birds as well as breeding sites of selected bird species. *Berichte zum Vogelschutz*, **51**, 15–42.

Leuschner, C., Krause, B., Meyer, S. & Bartels, M. (2014) Strukturwandel im Acker- und Grünland Niedersachsens und Schleswig-Holsteins seit 1950. *Natur und Landschaft*, **89** (9/10), 386–391.

Marschall, I. (2018b) 40 Jahre Landschaftsplanung im BNatSchG – Ideen und Entwicklungen. In *Landschaftsplanung im Prozess und Dialog*, ed. I. Marschall, pp. 22–37. Bonn-Bad Godesberg: Bundesamt für Naturschutz. BfN-Skripten 498.

Matzdorf, B., Reutter, M. & Hübner, C. (2010) *Bewertung der Ökosystemdienstleistungen von HNV-Grünland (High Nature Value Grassland).* Münchberg, Germany: ZALF.

Mayer, F. (2006) *Qualitätssicherung in der Eingriffsregelung – Nachkontrolle von Ausgleichs- und Ersatzmaßnahmen.* Bonn-Bad Godesberg: BfN. BfN-Skripten 182.

Mendel, B., Schwemmer, P., Peschko, V. *et al.* (2019) Operational offshore wind farms and associated ship traffic cause profound changes in distribution patterns of Loons (Gavia spp.). *Journal of Environmental Management*, **231**, 429–438.

Milieu (2019) *Study on EU implementation of the Aarhus Convention in the area of access to justice in environmental matters.* Final report, September 2019. 07.0203/2018/786407/SER/ENV.E.4. Brussels: Milieu Consulting Sprl.

Möckel, S. (2008) Die Novelle des Bundesnaturschutzgesetzes zum europäischen Gebiets- und Artenschutz – Darstellung und Bewertung. *Zeitschrift für Umweltrecht*, **19** (2), 57–64.

Müller-Pfannenstiel, K. (2009) *Anforderungen an vorgezogene Ausgleichsmaßnahmen.* Laufener Spezialbeiträge 1/09, 55-61. Laufen: Bayerische Akademie für Naturschutz und Landschaftspflege (ANL).

NABU (2007) *Der NABU-Bundeswildwegeplan.* Bonn/Berlin: Naturschutzbund Deutschland.

(2020a) *NABU reicht EU-Beschwerde ein – Rebhuhnbestände um über 90 Prozent eingebrochen.* www.nabu.de/news/2020/10/28749.html

(2020b) *Ein Fluss wird wieder lebendig: Das NABU-Projekt zur Renaturierung der Unteren Havel.* www.nabu.de/natur-und-landschaft/fluesse/untere-havel

Niedersächsischer Landtag (2014) *Was tun die Landkreise zur Ausweisung der Natura-2000-Gebiete als Schutzgebiete? Kleine Anfrage zur schriftlichen Beantwortung mit Antwort.* Landtags-Drucksache 17/1867.

Nordwestdeutsche Forstliche Versuchsanstalt (ed.) (2017) *Hochwertiges Forstvermehrungsgut im Klimawandel.* Symposium des Verbundprojektes FitForClim vom 14. bis 15. Juni 2016 in Chorin. Beiträge aus der NW-FVA 16, Göttingen.

Oppermann, R. (2012) Germany. In *High Nature Value Farming in Europe*, eds. R. Oppermann, G. Beaufoy & G. Jones, pp. 221–233. Ubstadt-Wieher, Germany: Verlag regionalkultur.

Ott, W. (2004) *Die besiegte Wildnis. Wie Bär, Wolf, Luchs und Steinadler aus unserer Heimat verschwanden.* Echterdingen, Germany: DRW-Verlag.

Pompe, S., Berger, S., Bergmann, J. *et al.* (2011) *Modellierung der Auswirkungen des Klimawandels auf die Flora und Vegetation in Deutschland.* Bonn-Bad Godesberg: Bundesamt für Naturschutz. BfN-Skripten 304.

Pröbstl, U. (2009) Naturschutz und Natursport: Faszination, Konflikte und Lösungen. In *"Wenn sich alle in der Natur erholen, wo erholt sich dann die Natur?" Naturschutz, Freizeitnutzung, Erholungsvorsorge und Sport - gestern, heute, morgen*, eds. H. W. Frohn, J. Rosenbrock, & F. Schmoll, pp. 331–344. Naturschutz und Biologische Vielfalt 75. Bonn-Bad Godesberg: BfN.

Reck, H., Hänel, K., Strein, M. *et al.* (2019) *Green Bridges, Wildlife Tunnels and Fauna Culverts The Biodiversity Approach.* Bonn-Bad Godesberg: BfN. BfN-Skripten 522.

Reiter, K., Roggendorf, W., Sander, A., Liebersbach, H., Schmelmer, K., Techen, A.-K. (2016a) *Ex-post-Bewertung NRW-Programm Ländlicher Raum 2007 bis 2013 Agrarumweltmaßnahmen (ELER-Code 214).* Braunschweig: Thünen-Institut für Ländliche Räume.

Reiter, K., Roggendorf, W., Sander, A., Liebersbach, H. & Techen, A.-K. (2016b) *Ex-post-Bewertung PROFIL–Programm zur Förderung im ländlichen Raum Niedersachsen und Bremen 2007 bis 2013 Agrarumweltmaßnahmen (ELER-Code 214).* Braunschweig: Thünen-Institut für Ländliche Räume.

Roy, H. E., Pocock, M. J. O. Preston, C. D. *et al.* (2012) *Understanding citizen science and environmental monitoring.* Final report on behalf of UK-EOF. Wallingford, Oxfordshire, UK: NERC Centre for Ecology and Hydrology.

Sander, A. & Bathke, M. (2020) *PFEIL –Programm zur Förderung im ländlichen Raum 2014 bis 2020 in Niedersachsen und Bremen. Beiträge zur Evaluation des Schwerpunktbereichs 4A Biologische Vielfalt. 5-Länder-Evaluation.* Thünen-Institut für Ländliche Räume. https://literatur.thuenen.de/digbib_extern/dn062128.pdf

Schmidt, A. & Zschiesche, M. (2018) *Die Klagetätigkeit der Umweltschutzverbände im Zeitraum von 2013 bis 2016. Empirische Untersuchung zu Anzahl und Erfolgsquoten von Verbandsklagen im Umweltrecht.* Berlin: Studie im Auftrag des SRU.

SRU (2007) *Umweltverwaltungen unter Reformdruck. Herausforderungen, Strategien, Perspektiven.* Berlin: Sondergutachten des Sachverständigenrates für Umweltfragen.

(2016a) *Environmental Report 2016. An integrated approach to environmental policy: the way forward. Summary, May 2016.* Berlin: German Advisory Council on the Environment (SRU).

(2016b) *Verbandsklage wirksam und rechtskonform ausgestalten: Stellungnahme zur Novelle des Umwelt-Rechtsbehelfsgesetzes.* Berlin: German Advisory Council on the Environment (SRU).

SRU & The Scientific Advisory Board on Forest Policy (2017) *Für eine bessere Finanzierung des Naturschutzes in Europa nach 2020.* Berlin: Stellungnahme.

Ssymank, A. (1994) Neue Anforderungen im europäischen Naturschutz: Das Schutzgebietssystem Natura 2000 und die FFH-Richtlinie der EU. *Natur und Landschaft*, **69** (9), 395–406.

Statistisches Bundesamt (2021) *Nachhaltige Entwicklung in Deutschland. Daten zum Indikatorenbericht 2021.* Wiesbaden: Statistisches Bundesamt.

Succow, M. & Joosten, H. (2001) *Landschaftsökologische Moorkunde.* Stuttgart: E. Schweizerbart'sche Verlagsbuchhandlung.

Sybertz, J., Matthies, S., Schaarschmidt, F., Reich, M. & von Haaren, C. (2017) Assessing the value of field margins for butterflies and plants: how to document and enhance biodiversity at the farm scale. *Agriculture, Ecosystems and Environment*, **249**, 165–176.

TEEB DE (2010) *Die Ökonomie von Ökosystemen und Biodiversität: Die ökonomische Bedeutung der Natur in Entscheidungsprozesse integrieren. Ansatz, Schlussfolgerungen und Empfehlungen von TEEB – eine Synthese.* www.teebweb.org/wp-content/uploads/Study%20and%20Reports/Reports/Synthesis%20report/Synthesis_German.pdf

(2017) *Neun Fallbeispiele zur ökonomischen Bedeutung der Natur. Aus: Naturkapital Deutschland – TEEB DE: Neue Handlungsoptionen ergreifen – Eine Synthese.* Leipzig: Helmholtz-Zentrum für Umweltforschung – UFZ.

(2020) Project reports. www.ufz.de/teebde/index.php?en=43782

Turowski, G. (2002) *Spatial planning in Germany: Structures and Concepts.* Hannover: Verlag der ARL -Akademie für Raumforschung und Landesplanung. Studies in Spatial Development, No. 1.

UBA (2014) *Evaluation von Gebrauch und Wirkung der Verbandsklagemöglichkeiten nach dem Umwelt-Rechtsbehelfsgesetz (UmwRG).* Dessau-Roßlau, Germany: Umweltbundesamt.

(2017) *Gewässer in Deutschland: Zustand und Bewertung.* Dessau-Roßlau, Germany: Umweltbundesamt.

(2018a) Indicator: Species diversity and landscape quality. www.umweltbundesamt.de/en/indicator-species-diversity-landscape-quality

(2018b) Indicator: Nitrogen eutrophication. www.umweltbundesamt.de/en/indicator-nitrogen-eutrophication

(2019a) Indicator: Grasslands. www.umweltbundesamt.de/en/indicator-grasslands

(2019b) Indicator: Landscape dissection. www.umweltbundesamt.de/en/indicator-landscape-dissection

Underwood, E., Ashcroft, R., Kettunen, M., McConville, A. J. & Tucker, G. M. (2014) *Protected Area Approaches in the EU.* London/Brussels: Institute for European Environmental Poli

van Roomen, M., van Turnhout, C., Blew, J. *et al.* (2017) East At Flyway. In *Wadden Sea Quality Status Report 2017,* eds. S. Kloepper, J. M. Baptist, A. Bostelmann *et al.* Wilhelmshaven, Germany: Common WaddenSea Secretariat. http://qsr.wadde worldheritage.org/reports/east-atlantic-flyway

Verband Deutscher Naturparke (2018) *Naturparke 2030 – das Wartburger Programm der Naturparke in Deutschland.* Bonn: VDN.

Volksbegehren Artenvielfalt (2020) Holprige Umsetzung des Volksbegehrens Artenvielfalt. https://volksbegehren-artenvielf .de/2020/07/16/meint-es-die-regierungskoalition-ernst-mit-de artenschutz-2-2

von Haaren, C. (ed.) (2004) *Landschaftsplanung.* Stuttgart: Verla Eugen Ulmer.

von Ruschkowski, E., Burns, R. & Arnberger, A. (2013) Recreati management in parks and protected areas: a comparative stud resource managers' perceptions in Austria, Germany, and the United States. *Journal of Park and Recreation Administration,* (2), 95–114.

von Ruschkowski, E. & von Haaren, C. (2008) Agrarumweltmaßnahmen in Deutschland im europäischen Vergleich. *Naturschutz und Landschaftsplanung,* **40** (10), 329–

Wadden Sea World Heritage (2020a) One Wadden Sea. One Glo Heritage. www.waddensea-worldheritage.org/one-wadden-sea-one-global-heritage

(2020b) Richly diverse. www.waddensea-worldheritage.org/rich diverse

Wahl, J., Dröschmeister, R., Gerlach, B. *et al.* (2015) *Vögel in Deutschland – 2014.* Münster: DDA, BfN, LAG VSW.

Walz, U., Wende, W. & Stein, C. (2018) Landschaftsplanung eine Erfolgsstory? Ergebnisse aus dem DFG-Projekt: Analyse Landschaftswandel und Landschaftsplanung (LANALYS). In *Landschaftsplanung im Prozess und Dialog,* ed. I. Marschall, pp. 48–54. Bonn-Bad Godesberg: Bundesamt für Naturschutz. BfN-Skripten 498.

Wende, W., Albrecht, J., Darbi, M. *et al.* (2018) Germany. In *Biodiversity Offsets - European Perspectives on No Net Loss of Biodiversity and Ecosystem Services,* eds. W. Wende, G. Tucker, Quétier, M. Rayment & M. Darbi. Cham, Switzerland: Springer

Wolschke-Bulmahn, J. & Gröning, G. (2002) The National Socialis Garden and Landscape Ideal. Bodenständigkeit (Rootedness in t Soil). In *Art, Culture, and Media under the Third Reich,* ed. R. Etling, pp. 73–97. Chicago, London: The University of Chicago Press.

Wüstemann, H., Meyerhoff, J., Rühs, M., Schäfer, A. & Hartje, V. (2014) Financial costs and benefits of a program of measures to implement a National Strategy on Biological Diversity in Germany. *Land Use Policy,* **36**, 307–318.

Chapter

17 Greece

Ioli Christopoulou

Key Statistics

Terrestrial area (km²)/% of EU-28*	132 026/3.0%
Marine waters (km²)/% of EU-28*	488 520/10.6%
Population density (p km²)/% of EU-28*	83/72%
GDP per capita (€)/% of EU-28	17 461/65%
Land cover (CORINE land cover classes 2018)	AGR 39%, OSN 36%, FOR 20%, ART 3%, WET 1%
Biogeographical regions	MED, MMED
Number of HD Annex I habitats/%*	89/39%
Number of HD Annex II, IV and V species/%*	274/22%
Number of BD Annex I bird taxa/%*	127/72%
% HD UCS and decreasing habitats/ % unknown	4%/4%
% HD UCS and decreasing species/ % unknown	20%/10%
% breeding BD Annex I birds decreasing/% uncertain and unknown	18%/13%

* Excluding MAC and MMAC biogeographical regions.
Sources and codes. See Table 1.1.

17.1 The Natural History and Characteristics of Greece

Greece is a country of profound natural richness and diversity, and, in proportion to its area, the most diverse Mediterranean country in all levels of biodiversity (Legakis *et al.*, 2018). This is due to a number of factors (Catsadorakis, 2003; Blondel *et al.*, 2010). First is the compound impact of its geography, topography and geological history. Greece's location in Europe is very close to Africa and Asia, within the Mediterranean biogeographical region, one of the world's global biodiversity hotspots (CEPF, 2020). Its dynamic geological history has formed a mosaic of landscapes, including mountains, river gorges, cliffs, caves, a long 16 300 km coastline with rocky and sandy beaches, and thousands of islands and islets, with the Aegean archipelago being the fascination of many biogeographers (Pafilis *et al.*, 2018). Surrounded as it is by sea, it is not often realised that about 80% of Greece is mountainous. Along the Hellenic arc, the Pindus mountain range forms the country's backbone in parallel with the Hellenic trench in the Ionian Sea, which extends to the island of Crete as it stretches south. In fact, the country has only a few extensive areas of lowland plains. Moreover, a large part of Greece escaped being covered by ice during the last ice age, allowing many species to remain in the area (Georghiou and Delipetrou, 2010).

The second important factor is its climate. Whilst Greece generally exhibits a typical Mediterranean climate (long, hot, dry summers and mild, cool, wet winters), due to the variety in its landscapes, more diverse conditions are found across the country. As a result, temperate and even alpine climates occur inland, while semi-arid and arid conditions are found in the southernmost parts of the country.

Third, and perhaps paradoxically, Greece's natural diversity is deeply interlinked with the country's rich cultural heritage. The human presence in the region over the millennia has led to close interrelationships with nature, and the creation of delicate and biodiverse semi-natural habitats, which many species now rely on.

Greece's natural richness is evident in its variety of terrestrial, freshwater, coastal and marine ecosystems. Vegetation zones range from the thermo-Mediterranean zone at lower altitudes to the coniferous forests and even alpine zones in the country's mountains. This high habitat diversity gives rise to a high diversity of flora and fauna, and a large proportion of European species occur in Greece (e.g. 80% of all Mediterranean fish species). Furthermore, due to Greece's location and history, a large proportion of species are endemic (22% of plants and 16% of animals) (NCESD, 2018), or range-restricted, as they may be also found in neighbouring countries (Dimopoulos *et al.*, 2016). Greece is also located on a major international bird migration flyway.

A substantial amount of natural and semi-natural habitat has been lost or degraded, especially over the last 70 years or so. Nevertheless, relatively large areas of biodiverse habitat remain, and consequently Greece still has an important role in safeguarding Europe's natural heritage. This is indicated by the fact that 39% of HD habitats (excluding those in Macaronesia) occur in the country, even though its area is

Table 17.1 Short-term trends (c. 2007–2018) in Habitats Directive Annex I habitats and Annex II, IV and V species with an unfavourable conservation status in Greece.

	Poor (unfavourable–inadequate)				Bad (unfavourable–bad)			
	Imp	Sta	Dec	?	Imp	Sta	Dec	?
HD Habitat assessments								
Number	11	23	3	0	2	2	1	0
%	30%	62%	8%	0%	40%	40%	20%	0%
HD Species assessments								
Number	5	19	38	64	1	1	20	15
%	4%	15%	30%	51%	3%	3%	54%	41%

Key. Imp = improving, Sta = stable, Dec = decreasing, ? = unknown.
Source. Based on Member State HD Article 17 reported data as presented on the EEA national summary dashboard (2021).

only 3% of the EU. The variety is significant. It includes the typical, for the Mediterranean, priority marine habitat type, *Posidonia* beds (1120*).[1] Many of the Mediterranean forest habitat types also occur, notably the rare riparian forests with the endemic palm tree, *Phoenix theophrasti* (9370*), as well as several types of temperate forest. It also supports a relatively large proportion of HD species and BD birds, and is a stronghold for some. For example, it has the most important nesting sites of the globally Vulnerable Loggerhead Turtle (*Caretta caretta*) in the Mediterranean, more than half of the population of the globally Endangered Mediterranean Monk Seal (*Monachus monachus*) and the largest breeding colony of the Eleanora's Falcon (*Falco eleonorae*).

17.2 The Status of Nature

Greece is no exception to global biodiversity loss trends. Although nature degradation has been noted for many years (Gerakis, 1992; Daffis *et al.*, 1996), insufficient data have been available to quantify its impacts and assess the state of nature in Greece until recently. Largely as a result of the need to implement the EU Nature Directives, as well as growing academic interest, knowledge of the status of nature has improved considerably, initially through the studies needed to identify and delineate Natura 2000 sites and, subsequently, through the monitoring undertaken to fulfil the country's reporting obligations (Daffis *et al.*, 1996; Dimopoulos *et al.*, 2018). The 2019 HD Article 17 and BD Article 12 reports for 2013–2018 provide the most up-to-date indication of the status of most BHD habitats and species, and form the basis of the summary of the situation Sections 17.2.1 and 17.2.2.[2] Despite the improvements, important knowledge gaps remain.

17.2.1 Habitat Status and Trends

According to the summary data presented by the EEA, 48% of habitats in Greece had a favourable conservation status, 42% a poor status, and 6% a bad status; the remaining 4% had an unknown status. However, it is important to note that there is considerable variation in the status of habitats. Rocky and sclerophyllous scrub (e.g. matorral and phrygana) habitats were in best condition, with close to 85% of each category having a favourable status. Forest and freshwater habitats were also faring reasonably well, as each had close to 65% with a favourable status. The status of grasslands is of greater concern, as only 40% had a favourable status, and the status of 20% was unknown. However, of most concern is the status of dunes and coastal habitats as these two habitat categories had the lowest percentages with a favourable status: 12% and 7%, respectively. Furthermore, whilst most habitats with an unfavourable status had an improving trend (Table 17.1), the four that were reported as declining are all coastal habitats. These include coastal lagoons, which had a bad status. As discussed later, the situation with coastal habitats can be largely explained by the concentration of urban and tourist activities in coastal areas.

17.2.2 Species Status and Trends

Although information on species is less complete than on habitats, it indicates a particularly worrisome situation in Greece: only 35% of the country's HD species had a favourable status, whilst 42% had a poor status, and 12% had a bad status. Furthermore, a high proportion of species with an unfavourable status were reported as declining, and the trend of many others was unknown (Table 17.1). None of the 18 marine species had a favourable conservation status: seven had a poor status and eight had a bad status, with the remaining three listed as unknown.

Among the other species listed as having an unfavourable conservation status are most of Greece's freshwater species. This is not very consistent with the reported favourable status

[1] Numbers in brackets are HD Annex I habitat codes (see Appendix for full names). * Indicates a priority habitat.

[2] See Section 1.3 for background on monitoring under the Nature Directives, status categories and data sources.

Table 17.2 Short-term trends (c. 2007–2018) in Birds Directive Annex I taxa in Greece.

	Breeding				Wintering			
	Inc	Sta/Flu	Dec	?	Inc	Sta/Flu	Dec	?
Number	25	40	17	12	12	4	1	14
%	27%	43%	18%	13%	39%	13%	3%	45%

Key. Inc = increasing, Sta = stable, Flu = fluctuating, Dec = decreasing, ? = uncertain and unknown.
Source. Based on Member State Article 12 reported data as presented on the EEA national summary dashboard (2021).

of most freshwater habitats, but it corresponds to findings in the 2009 *Greek Red Data Book* (Legakis and Maragou, 2009), which concluded that 38% of the country's freshwater fish were under threat, the highest proportion of any animal group.

The 2019 BD Article 12 report on the status of birds indicates that across all breeding species only 11% were reported as having declined over both the short term (i.e. 2007–2018) and longer term (i.e. since c. 1980). Trends were unknown for 11% of species over the short term and 27% over the long term. Whilst this seems to show a fairly favourable situation, with stable or increasing status for birds with known trends, the situation for breeding Annex I species is more concerning as 18% have shown short-term declines (Table 17.2). This is confirmed by recent studies that indicate that one-fifth of breeding birds show declining populations, with raptors being the most threatened (Vavylis *et al.*, 2021). Wintering populations of Annex I species may be doing better. Greek international waterbird census data indicate that across all wintering species with known trends most have been stable over the long term, despite significant annual variations (Handrinos *et al.*, 2015). However, trends are unknown for too many – 41% over the short term and 75% over the long term – to draw overall conclusions on them.

17.2.3 Drivers and Causes of Change

The Article 12 and 17 reports for 2013–2018 indicate the main pressures and threats, at least in terms of those affecting BHD habitats and species. In the terrestrial environment, land-use change is the most significant threat for Greek nature as attested by the fact that agriculture and construction, development and use of residential, commercial and recreational infrastructure and areas are listed as the most frequent threats and pressures. In the marine and coastal environment, overfishing, construction, development and use of residential, commercial and recreational infrastructure, as well as pollution, are the main threats. For marine species and habitats, climate change is also listed among the top threats.

The conclusions of these reports are linked to underlying drivers and causes of change that are connected with Greece's development. Until the early twentieth century, the human impact on Greece's nature remained relatively mild and not dissimilar to that of the previous centuries. Wetland drainage has been a typical pressure that has intensified since the mid-1920s (Papayannis, 1992). From then on, and especially during the so-called Greek economic miracle of 1950–1980, population increase, movements of people (in and out of the country, and within), economic development, policy and political frameworks and social perceptions led to a changed relationship with Greece's landscape and nature. Such changes were also enabled by a generally weak environmental governance and culture. Once Greece joined the EU in 1981, they were also supported by EU funds that became available.

Changes in land use, resulting from various drivers, have had the main impacts on Greece's biodiversity. Agricultural intensification has been particularly evident in the few large plains. This shift away from traditional agricultural practices has upset the longstanding balance in the rural landscape and has led to the degradation of semi-natural habitats and the loss of local crop varieties and livestock breeds. Industrial activities, although not as extensive as in other European countries, have concentrated in specific suitable areas of the country, thus leading to significant localised environmental degradation and instances of heavy pollution.

More extensive, in terms of the area influenced, has been the impact of urban, tourism and commercial developments (Papayannis and Howard, 2012). For many decades, people have sought a better future far from the remote areas of the country, leading to rural depopulation. Today, two urban centres, Athens and Thessaloniki, account for more than half of Greece's population. The resulting urban expansion has been a particular pressure on natural and semi-natural forests and agricultural lands. As the economy grew, commercial activities expanded and Greece became a popular tourist destination. Tourism investments remained, until recently, relative small and mostly for small-scale units, homes to rent, and holiday homes. As a result of this, and the absence of a national conservation strategy (Apostolopoulou and Pantis, 2009) and planning regulations and tools (discussed in Section 17.6), the impact has been spread around the country. An extensive road network (Kati *et al.*, 2020) and supporting infrastructure was created, while an additional pressure has been from the need for quarries to supply the construction sector.

In contrast, remote islands and mountainous areas of Greece have suffered from the abandonment of rural areas and of traditional pastoral and farming activities. Whilst this may have had benefits for some species, it has been detrimental for those that depend on semi-natural grasslands and open habitat mosaics. For example, as forestry and animal herding have been gradually abandoned, forest coverage has increased, but without management some forests have a higher fire risk and lack open areas that are important for some plants and foraging areas for birds of prey (Poirazidis, 2017; Poirazidis *et al.*, 2019).

In the 2010s an unprecedented financial crisis struck the country for about a decade, as a consequence of the global 2008 financial crisis. As a result, some of the pressures on the environment eased and some people returned to the countryside, seeking to rejuvenate past practices. The Greek debt crisis set in motion a joint EU and international mechanism to avoid the country's bankruptcy. Boosting the economy became the top priority, leading to concerns for an environmental rollback, especially in terms of easing environmental permit procedures, as well as a commercialisation of environmental assets (Paliogiannis *et al.*, 2019). So far, impacts from such a rollback have not been clearly evident, as they are expected to occur mainly once the economy has picked up again, which was further stalled by the COVID-19 pandemic. As the economy is recovering and funds are becoming available it remains to be seen whether the rollback trends will persist or whether they will be reversed in line with the EU Green Deal commitments and the introduction of more environmental conditions.

17.3 Nature Conservation Governance, Actors, Legislation and Policies

17.3.1 Origins of Nature Conservation

The first modern type of nature conservation provisions in Greece can be traced back to the early decades of the twentieth century (Table 17.3).

Following global and European trends of the period, a forest law was passed that enabled the designation of National Forests in the late 1930s,[3] with Mount Olympus being designated as the first in 1938.[4] For many subsequent decades nature conservation continued to be dependent on forest legislation and was under the authority of the forest service.

Two wider developments that shaped the evolution of Greece's recent history can also be identified as landmarks in the evolution of the country's nature policy.

The first milestone was Greece's 1975 Constitution, which was concluded upon the re-instatement of democracy in Greece following the seven-year dictatorship (1967–1974). Article 24 provides that: 'the protection of the natural and cultural environment constitutes a duty of the State. The State is bound to adopt special preventive or repressive measures for the preservation of the environment'. This provision reflected the emergence of global environmental politics and was one of the reasons the Constitution was considered progressive for its time (Weale *et al.*, 2003). Article 24, which comprises the legal basis for regulatory measures for nature conservation, includes specific provisions on forest protection as well as outlining various land planning tools.

Article 24 was amended in 2001 with the second revision of the Constitution, so as to provide that the protection of the environment also constitutes 'a right of every person' while measures taken are 'in the context of the principle of sustainable development'. Following the 2001 revision, sustainable development became an overarching context for later specific environmental measures.

The second milestone was Greece's entry into the EEC in 1981 as its tenth Member State. While this was still the early period of European environmental law and prior to the recognition by the Single European Act of the environment as a shared European competence (see Section 4.1), the EEC had already issued its first two Environment Action Programmes and several environmental directives, including the Birds Directive. Since then Greece has responded to the evolution of EU environment law, which to a great extent has provided the policy priorities, legal instruments and financial resources needed to enact its own domestic constitutional environmental provisions.

17.3.2 Legislation and Policies

The adoption of the Law on the Protection of the Environment[5] in 1986 translated the provisions of Article 24 of the Constitution into a regulatory act and constituted an attempt to organise the requirements of the transposed EU environmental legal framework. In effect, the law established Greece's environmental governance system. The section on nature and landscape protection established for the first time various categories of protected areas (beyond those foreseen in the forest legislation or established under international law) and provided for the protection of indigenous species of flora and fauna. Other sections of the law relating to impact assessment, pollution and penalties have broader environmental relevance and complement the specific nature-related provisions. Since 1986, this fundamental environmental law has been amended on several occasions, yet remains the backbone of national environmental legislation.

Implementation of the nature conservation and protected areas provisions was delayed for more than a decade, as discussed later. Over this period, forest provisions remained in place and continued to make a significant contribution to nature conservation for several years.

Additionally, other provisions were introduced, for example, relating to marine pollution in line with either international (e.g. MARPOL) or Mediterranean-wide obligations (e.g. the Barcelona Convention) and those aiming to protect the Greek coastline.

Indeed, the evolving body of international environmental law has been a notable influence on Greece's nature conservation legislation and policies. Greece has ratified almost all multilateral and regional nature-related conventions (Table 3.3). Implementation of, and compliance with, these agreements has mostly been through the implementation of the relevant EU legislation and particularly the Nature Directives.

Even so, the transposition of the Nature Directives was marked by delays and required significant interventions by

[3] N. 856/1937 Περί εθνικών δρυμών (On national forests) (A΄ 368).

[4] Β.Δ. Περί ιδρύσεως Εθνικού Δρυμού Ολύμπου (Designation of Mount Olympus National Forest) (A΄ 248/1938).

[5] N. 1650/1986 Για την προστασία του περιβάλλοντος (A΄ 160).

Table 17.3 Key nature conservation developments in Greece.

1937	Law on national forests
1938	Designation of Mount Olympus as Greece's first National Forest
1950	Law 1469 on the protection of monuments and the establishment of Landscapes of Particular Natural Beauty
1969	Law 86 establishing a Forest Code
1971	Law 996 on the replacement and supplementation of Law 86/1969 of the Forest Code adding new protected area categories
1975	Constitution of Greece: Article 24 provides for the protection of the natural and cultural environment as a duty of the State and, following a 2001 revision, the right of every person
1979	Law 998 on forests
1981	Accession to the EU
	Presidential Decree 67 on the protection of Greece's indigenous flora and fauna species
1985	First elaboration of management measures in compliance with the Birds Directive
1986	Law 1650 on the Protection of the Environment
1987	Bern Convention Recommendation No 9 on Zakynthos
1998	Transposition of the EU Habitats Directive and establishment of Nature 2000 Committee
1999	Zakynthos designated as the first National Park with Greece's first Management Body for a National Park
	Law 2742 on Land Planning and Sustainable Development laying out the governance system of protected areas
2001	Law 2971 on the Seashore (Coast), Beaches and other provisions
2002	CJEU C-103/00 judgment against Greece on failure to adopt an effective system of protection for the Loggerhead Turtle on Zakynthos
	Law 3044 establishing 24 Management Bodies of Protected Areas
2003	Law on the Protection of waters
2006	EC Decision on the first approval of the Mediterranean SCI list
2010	Signing of the Agreement on the Protection and Sustainable Development of the Prespa Park Area between Albania, Greece, North Macedonia (FYROM at the time) and the EU
	Ministerial decision on the transposition and designation of measures in compliance with the Birds Directive
2011	Law 3937 on Biodiversity Conservation
2012	Legal recognition and protection of island wetlands via a Presidential Decree
	Correction and completion of the transposition and designation of measures in compliance with the Birds Directive
2014	Adoption of Greece's first *National Biodiversity Strategy* (2014–2029) and first *Action Plan* (2014–2020)
2016	Adoption of Greece's *National Climate Adaptation Strategy* via Law 4414
2017	Expansion of Natura 2000 sites, especially in the marine area
2018	Law 4519 on the Management Bodies of Protected Areas and other provisions
	Approval of Greece's *National Forest Strategy*
2020	Law 4685 on the Modernisation of Environmental Law establishing a new Natural Environment and Climate Change Agency and revamping Greece's protected areas system

See Section 3.3 for signatory dates of international conventions.

the European Commission (as in many other EU Member States). In the case of the Habitats Directive, Greece transposed the Directive in 1998, following a decision of the CJEU against Greece (C-329/96). In the case of the Birds Directive, while the initial decision to designate management measures in line with the Directive dates back to 1985,[6] the transposition was not fully completed until 2010[7] and, in fact, only following a complementary decision in 2012.[8]

About 25 years after the adoption of the Law on the Protection of the Environment, the 2011 Law on Biodiversity Conservation[9] revamped the whole system of nature conservation. The law was drafted following a WWF Greece-led NGO proposal on a new integrated legal framework for biodiversity conservation (WWF Greece *et al.*, 2009). The 2011 Biodiversity Law concluded the transposition of the Nature Directives, updating some of the provisions relating to Natura 2000 and species protection (Frederiksen *et al.*, 2017), and included additional nature conservation provisions. This law provided for a coherent system of protected areas, and outlines overarching provisions for nature conservation, species conservation, and invasive alien species, amongst others. The provisions relating to protected areas have since been revised, while many others have yet to be fulfilled. Among its implemented provisions was the adoption in 2012 of a Presidential Decree[10] on the protection of small island wetlands, which had not until then been recognised in the country's environmental regulatory system.

Greece adopted a *National Biodiversity Strategy and Action Plan* in 2014 (MEECC, 2014a), concluding a long drafting process initiated in 1999. This was the first time that a comprehensive approach for nature conservation was attempted. The strategy, which has a 15-year outlook, aims 'to halt the loss of biodiversity and the degradation of ecosystem services in Greece by 2029, and restore them, as far as it is feasible, while communicating the value of biodiversity as our national capital, and stepping up the Greek contribution towards averting global biodiversity loss'. It includes a long-term vision to 2050 and a set of 13 general targets that aimed to contribute to the global CBD 2020 biodiversity targets, as well as the *EU Biodiversity Strategy to 2020*. To better coordinate Greece's nature policy with global and EU biodiversity developments, the first *Action Plan* implementation period extended to 2020, aiming primarily for the country to comply with its global and EU policy obligations. However, the implementation period concluded without demonstrating significant progress on any of the general or specific targets (The Green Tank, 2020). A new five-year biodiversity action plan was expected in 2021, but its preparation has been delayed, awaiting the adoption of the CBD's post-2020 global biodiversity framework.

The EU requirements under the Nature Directives have been the main driver for establishing the current national framework for nature conservation. In fact, it would not be an exaggeration to say that the transposed Nature Directives, as well as the legal framework established to implement them, constitute the legal foundations of nature conservation in Greece today (Christopoulou, 2015; Tryfon, 2015). This has happened to a large extent with consistent EU financial support, as well as following various infringement procedures and CJEU rulings through which the European Commission has sought to improve implementation of, and compliance with, the EU provisions.

17.3.3 Nature Conservation Governance

Greece included the environment portfolio in a ministry for the first time in 1980. In 2009 the first autonomous environment ministry was established (then the Ministry of Environment, Energy and Climate Change), which since September 2015 has been called the Ministry of Environment and Energy (MEEN).[11] While numerous actors are involved, Greece's environmental governance system is largely centralised and has been controlled by this ministry.

The Ministry oversees the country's protected areas' management bodies and the Green Fund, the main state environmental funding entity. The new Natural Environment and Climate Change Agency (NECCA) was established in 2020 as an evolution of the country's National Centre for the Environment and Sustainable Development, taking on some of the tasks of the relevant environment authorities. NECCA aims 'to implement the policy drawn by the MEEN on protected areas' management, biodiversity conservation and promotion of sustainable development actions and tackling climate change'.[12] The establishment of such a body comes

[6] ΚΥΑ 414985/1985 Μέτρα διαχείρισης της άγριας πτηνοπανίδας (Β΄ 757).

[7] ΚΥΑ Η.Π. 37338/1807/Ε.103/2010 Καθορισμός μέτρων και διαδικασιών για τη διατήρηση της άγριας ορνιθοπανίδας και των οικοτόπων/ενδιαιτημάτων της, σε συμμόρφωση με τις διατάξεις της Οδηγίας 79/409/ΕΟΚ, «Περί διατηρήσεως των άγριων πτηνών», του Ευρωπαϊκού Συμβουλίου της 2ας Απριλίου 1979, όπως κωδικοποιήθηκε με την Οδηγία 2009/147/ΕΚ (Β΄ 1495).

[8] ΚΥΑ Η.Π. 8353/276/Ε103/2012 Τροποποίηση και συμπλήρωση της υπ' αριθ. 37338/1807/2010 κοινής υπουργικής απόφασης «Καθορισμός μέτρων και διαδικασιών για τη διατήρηση της άγριας ορνιθοπανίδας και των οικοτόπων/ενδιαιτημάτων της, σε συμμόρφωση με την Οδηγία 79/409/ΕΟΚ. . .» (Β΄ 1495), σε συμμόρφωση με τις διατάξεις του πρώτου εδαφίου της παραγράφου 1 του άρθρου 4 της Οδηγίας 79/409/ΕΟΚ «Για τη διατήρηση των άγριων πτηνών» του Ευρωπαϊκού Συμβουλίου της 2ας Απριλίου 1979, όπως κωδικοποιήθηκε με την Οδηγία 2009/147/ΕΚ (Β΄ 415).

[9] Ν. 3937/2011 Για τη διατήρηση της βιοποικιλότητας (Α΄ 60).

[10] ΠΔ 2012 Έγκριση καταλόγου μικρών νησιωτικών υγροτόπων και καθορισμός όρων και περιορισμών για την προστασία και ανάδειξη των μικρών παράκτιων υγροτόπων που περιλαμβάνονται σε αυτόν (Adoption of a list of small island wetlands and definition of conditions and restrictions for the protection and enhancement of the small coastal wetlands included in the list) (ΑΑΠ 229).

[11] The acronym MEEN will be used in the following sections to cover the different formations of the same ministry.

[12] Article 27 par. 4. N. 4685/2020 Εκσυγχρονισμός περιβαλλοντικής νομοθεσίας, ενσωμάτωση στην ελληνική νομοθεσία των Οδηγιών

years of calls for the need for strengthened centralised ort for the country's national protected areas' governance m (Maragou and Christopoulou, 2012).

The MEEN and NECCA are also responsible for the oper- n of the Nature 2000 Committee. Initially established to port the implementation of the Nature Directives, it has e evolved to become the national committee of protected s and the State's main national advisory body on liversity conservation.

A consolidation of most of the environmental authorities occurred gradually over the past decade. Until then, nature servation was fragmented across several ministries, includ- particularly those relating to the environment, agriculture marine affairs. In 2009, the longstanding Forestry Service relocated from the Ministry of Agriculture to the newly nded environment ministry, before being merged into one eral secretariat in 2014. As a result, for the first time, the central authorities mainly responsible for nature conser- ion were placed under the same ministry, even though tially under separate general secretaries. Water management nained separate from nature conservation within the envir- ment ministry until 2019, when nature, water and forests re included in the same general secretariat. Following gov- nment structure changes, in late 2021 forests were again ced under a separate Forestry General Secretariat within e MEEN. The secretariat integrates the decentralised forest vices, which had been under the authority of the Ministry of e Interior, into the central MEEN structure. Marine protec- n has remained split between the MEEN and the Ministry of aritime Affairs and Insular Policy. The former, for example, responsible for marine protected areas and the implementa- n of the MSFD; the latter deals with the implementation of ARPOL and other marine protection provisions and has der its supervision the Coast Guard, the relevant enforce- ent authority.

While the MEEN has had the main political responsibility r the environmental portfolio, other ministries are also volved in the country's nature conservation governance stems. The Ministry of Rural Development and Food is sponsible for issues regarding the sustainable use of sources and particularly agriculture, fisheries and genetic sources. The relationship between the two ministries has een tense, with a turf war between them delaying the trans- osition of the Habitats Directive (Frederiksen *et al.*, 2017). Iore recently, the newly founded Ministry for Climate Crisis nd Civil Protection will be involved in climate change adap- ation planning.

Other ministries that are involved in nature conservation overnance include the Ministry of Finance, which is responsible for the country's public property, including management of the country's coastline; the Ministry of Development and Investments as it has oversight of the design, allocation, and monitoring of the EU programming periods and the relevant EU funding priorities; and the Ministry of Interior, with respect to communication between the Ministry and the decentralised, regional and local authorities. Such authorities share the responsibility on nature conservation, fisheries and marine conservation, etc. Coordination among those authorities, and with the central government, remains a challenge.

Other actors that contribute to the country's nature conservation include environmental NGOs (see Section 17.3.4), the universities and research institutes, as well as more broadly the scientific community, whether organised in the country's several scientific and thematic associations or not. As further described in Section 17.8, the academic and scientific community provides important knowledge and expertise.

17.3.4 Environmental NGOs

NGOs have an important role in Greece's nature conservation. Some have a long history, having been established in the 1970s or even the 1950s, whereas others are much younger. The NGOs include national representatives of global conservation organisations or networks, while others are national or local initiatives. Major NGOs have close contact with Mediterranean, European and international partners and associations. The thematic focus of NGOs varies from those with a broad scope to those that have a more focussed concentration, for example on specific species. The NGOs play a multitude of roles, often substituting in practice some of those of national authorities due to reduced capacity of the latter. They undertake extensive fieldwork and monitor species and habitats, raise public awareness, follow and inform policy developments, advocate for improved implementation and take on legal action in national courts, contributing to the development of Greece's environment case law. They also submit complaints to regional fora, such the Bern Convention and the European Commission, which have often opened cases against Greece.

17.4 Species Protection and Conservation Measures

In 1981, with the adoption of a Presidential Decree, as amended, Greece acquired a first list of protected species. The Nature Directives complemented Greece's legal provisions for species, even though many indigenous species are not included in their annexes. With the emphasis given to implementing the Directives, other non-EU protected species have not been given focussed conservation attention, although they may be implicitly protected due to the extent of the coverage of the Natura 2000 network (van der Sluis *et al.*, 2016).

However, even for the BHD species, conservation actions have only been partial. Failure to ensure an effective system of

2018/844 και 2019/692 του Ευρωπαϊκού Κοινοβουλίου και του Συμβουλίου και λοιπές διατάξεις (Modernisation of environmental legislation, transposition of Directives 2018/844 and 2019/692 of the European Parliament and the Council and other provisions). (Α΄ 92).

strict protection for the country's protected species under HD Article 12 has led to several CJEU rulings against Greece. Of particular note are two cases, 15 years apart, relating to the failure to ensure the conservation of one of the country's most emblematic species – the Loggerhead Turtle – at the two most important nesting sites in the Mediterranean, namely Zakynthos and Kyparissiakos (C-103/00 and C-504/14, respectively). As a result of the rulings, Greece designated national protected areas and issued measures to protect both the species and the Natura 2000 sites. Even so, compliance has yet to be ensured. The Bern Convention, which has also followed and opened case files against Greece for the same two areas, re-opened the Zakynthos case file in 2020 and has retained the Kyparissiakos case file under consideration (Bern Convention, 2020, 2021).

In 2006, the CJEU (C-518/04) ruled against Greece for having failed to comply with the Habitats Directive because it had not taken the necessary measures to ensure an effective system of strict protection for the endemic viper (*Vipera schweizeri*) on the island of Milos. Since then, measures to protect the viper have been instituted and implemented.

Eventually, it was recognised that site protection is not adequate for species protection and additional species-specific measures are needed. Greece adopted its first species action plans in 2017, as a deliverable of LIFE projects related to the conservation of BD Annex I birds, which were carried out by the Hellenic Ornithological Society. Two of the action plans were national and covered the Egyptian Vulture (*Neophron percnopterus*) and the Lesser White-fronted Goose (*Anser erythropus*). A regional action plan was also adopted the same year for the Lesser Kestrel (*Falco naumanni*) in Thessaly.

One of the main actions of the LIFE Integrated Project (LIFE-IP 4 NATURA) project that is currently being carried out by the MEEN has been the preparation and adoption of additional action plans. As a result, in 2021, eight action plans, some of which are for multiple species, were adopted. The action plans cover: one HD habitat type (2250* Coastal dunes with *Juniperus* spp.); one globally Endangered endemic plant species (*Silene holzmannii*); the globally Critically Endangered endemic Karpathos Frog (*Pelophylax cerigensis*); the Balkan Chamois (*Rupicapra rupicapra balcanica*); the Apollo butterfly (*Parnassius apollo*); Bearded Vulture (*Gypaetus barbatus*), Griffon Vulture (*Gyps fulvus*) and Cinereous Vulture (*Aegypius monachus*); and five salmon species (*Salmo farioides, Salmo lourosensis, Salmo macedonicus, Salmo pelagonicus* and *Salmo peristericus*). Additionally, an action plan for the Loggerhead Turtle was prepared via the LIFE Euroturtles project.

The adoption of these actions plans demonstrates the contributions of LIFE projects to the conservation of Greece's nature. Indeed, in the past, several other action plans have been prepared and implemented under LIFE or other projects, mainly by NGOs and to a lesser extent by universities and academic or research centres. Although they have not been officially adopted, they have helped to identify priorities and conservation actions, and to develop new project proposals. They have also been taken into account occasionally in I and HD Article 6(3) appropriate assessments. Aiming update past proposals, three additional action plans have b prepared and are to be approved in 2023. They will cover Brown Bear (*Ursus arctos*), the Mediterranean Monk Seal the globally Vulnerable Cyprinid fish *Ladigesocypris gh* which is endemic to Rhodes.

17.4.1 Illegal Poisoned Baits

Illegal poisoning is a significant threat to Greece's wildlife many protected species have been killed, both intention and accidentally. Poisoned baits are used for many reas but among the most common is the targeting of large ca vores that lead to conflicts as a result of crop damage, livest losses or the killing of hunting dogs. Following a severe ir dent of poisoning in the region of the Nestos Straits in nor ern Greece in 2012, which resulted in the accidental loss of entire colony of Griffon Vultures, the European Commissi initiated an infringement case against Greece. Despite Gre taking some actions, such as the adoption of local ar poisoning plans, leading the Commission to close the ca Greece still lacks adequate measures to prevent the killing protected birds, especially vultures, and other wild spec from poisoned baits. In order to raise awareness and to galva ise national action in response to this threat, the MEEN collaboration with the Ministry of Rural Development a Food recognised February 22 as a National Action Day agair Illegal Poisoned Baits in 2022 (MEEN, 2022b). The idea f such a day originated from an NGO-led working group agair Illegal Poisoned Baits[13] in order to commemorate the trag Nestos incident.

17.4.2 Hunting

Hunting is a popular yet currently declining activity Greece. Indicatively, the number of members that the Gre Hunters Confederation represents has declined from abo 300 000 hunters in the early 2000s to 180 000 today.

The species that can be hunted are listed in an annu regulatory decision. Hunted species are mostly birds, and fiv mammals, including European Hare (*Lepus europaeus*), Wil Boar (*Sus scrofa*) and Red Fox (*Vulpes vulpes*). Hunting primarily regulated by Law 86/1969,[14] as an integral part the forest legislation, but has been updated over time t comply mostly with the Birds Directive, as well as othe European or international requirements. The legislation se the framework for the annual hunting regulatory decision tha lists the species hunted, the hunting period, the hunting day and maximum yields per species restrictions on specie

[13] Members of the working group include: ANIMA, Arcturos, Callisto, The Hellenic Society for the Protection of Nature, the Hellenic Ornithological Society, the Natural History Museum of Crete and WWF Greece.

[14] Ν.Δ. 86/1969 Δασικός Κώδιξ (Forest Code) (Α΄ 7).

Table 17.4 The number and extent of international designated sites and protected areas in Greece.

	IUCN category	Number	Area (km^2)
Designated under global and non-EU international legislation			
Ramsar Sites	NR	10	1 635
Mixed World Heritage Sites	NAs	2	352
Biosphere Reserves (exc. buffer and transition zones)	NAs	3	141
UNESCO Global Geopark	NAs	6	7 645
Designated under the Nature Directives			
SCIs (terrestrial + marine)	NR	265	21 912 + 17 528
SPAs (terrestrial + marine)	NR	207	27 761 + 10 764
Natura 2000 terrestrial area			35 982
Natura 2000 marine area			22 796
Designated under domestic legislation			
National Park (Εθνικό Πάρκο, including National Marine Parks)	IV/V/VI	17	15 298
National Forest (Εθνικός Δρυμός, both core and peripheral zones)	II/VI	10	768
Aesthetic Forest (Αισθητικό Δάσος)	III	19	319
Protected Forest (Προστατευόμενο Δάσος)	VI	3	417
Absolute Nature Reserves (Περιοχές Απολύτου προστασίας, beyond National Park zones)	Ia/II/IV	4	30
Nature Reserves (Περιοχές Προστασίας της Φύσης, beyond National Park zones)	II/IV/V/VI	9	?
Wildlife Refuge (Καταφύγια Άγριας Ζωής)	IV	603	10 630
Game Breeding Station (Εκτροφείο Θηραμάτων)	NAs	21	31
Controlled Hunting Areas (Ελεγχόμενη Κυνηγετική Περιοχή)	NAs	7	1 115
Natural Monuments and Landmarks (Protected as Strict Nature Reserve) (Διατηρητέο Μνημείο της Φύσης)	Ia/III	51	160
Protected Significant Natural Formation, Protected Landscape and Landscape Elements (Προστατευόμενος Φυσικός Σχηματισμός)	III/IV	3	37

Note. NR = Not reported; NAs = not assigned.
Sources. See Section 1.3. Based on data from: Ramsar Convention and UNESCO websites (February 2022); EEA Natura 2000 Barometer, Natura 2000 as at end 2020; WDPA (November 2021); NCESD (2018, 2020).

hunted, amongst others. In reality, only slight, if any, changes are made each year. Whilst hunting is permitted in the entirety of the Greek countryside, restrictions apply relating to safe distances and additional ones can be introduced following the designation of an area as protected. A few game breeding areas and controlled hunting areas have also been established (Table 17.4). The Hunters Confederation has financed the self-regulating, yet certified by MEEN, Game Keeping Body, which is active in the enforcement of hunting regulations.

Greece is one of the countries for which data on hunting bags and harvest are either not available or not reliable. Several severe instances of illegal hunting have occurred both outside and within nationally protected areas, Ramsar and Natura 2000 sites and affecting protected species (Kazantzidis *et al.*, 2020). It is estimated that 32% of the country's bird species are illegally killed or taken (Brochet *et al.*, 2016).

In 2016, the MEEN organised a national dialogue on hunting with the active participation of all relevant stakeholders. The national dialogue concluded with a report including key findings and a set of proposals (MEECC, 2016). Amongst other things, the legislation and relevant provisions were considered out of date, fragmented and unable to ensure the conservation and sustainable management of Greece's wildlife. It was also noted that there is a need for better training of hunters and a revision of the permit granting system. Despite this, there has been limited implementation of the report's proposals.

Of particular ongoing concern is the hunting of migrating Turtle-doves (*Streptopelia turtur*). Hunting during the spring is not permitted under the Birds Directive, and an international action plan for the species (Fisher *et al.*, 2018) has called for a temporary hunting moratorium due to the species being considered Vulnerable in Europe and globally. However, it is estimated that 69 000 birds are illegally killed each year in the Ionian Islands, although there is dispute over the numbers (Fisher *et al.*, 2018). Efforts have been made to curtail illegal hunting, and the maximum number of birds that can be taken legally per outing was restricted in the 2021 annual hunting regulatory decision to six compared to 12 in earlier years. Moreover, attempts have been made to conclude local management plans that aim to coordinate involved stakeholders in collaboration with the Game Keeping Body in a unified response, especially on the Ionian Islands, where the problem is most acute.

17.5 Protected Areas and Networks

17.5.1 Protected Area Types and Coverage

Domestic Protected Areas

The notion of protected areas in Greece originated, as already mentioned, in the 1930s forest legislation that enabled the designation of National Forests or National Forest Parks. Following the designation of Mount Olympus as the first such protected area, another nine National Forests were designated in subsequent decades. National Forests fall under modern IUCN category II (National Parks), with peripheral zones surrounding them falling under IUCN category IV (Habitat/Species Management Areas). Over time, the forest legislation was amended to add new categories of protected areas, including Aesthetic Forests, Monuments of Nature in 1971, and Wildlife Refuges in 1998, as an institutional evolution of sites designated originally as Game Reserves.

The second type of protection established was that of Landscapes of Particular Natural Beauty, based on provisions for protection of monuments under the archaeological legislation of the 1950s.[15] It is estimated that more than 400 such areas have been designated. A requirement to integrate these sites in the more modern categories of protected areas was included in the 2011 Biodiversity Law, but such a project has not been undertaken.

At the end of the seven-year military dictatorship, in 1974, Greece was eager to demonstrate that it was a modern, liberal democracy and active member of the global community. Thus, only four months later, it ratified the Ramsar Convention[16] and almost immediately designated 11 of its wetlands as Ramsar Sites.[17]

From this overview, it is apparent that for a long time protected areas in Greece were either forests or large wetlands. It was not until 1986 that the Law on the Protection of the Environment introduced and established five new categories of protected areas. It also established a clear, yet cumbersome, process for the designation of protected areas[18] and contained basic provisions for their management. In order for Natura 2000 sites, both SACs and SPAs, to be granted domestic recognition, they had to be designated also as national protected areas, in line with these provisions. The procedure has been marked by delays, such that only a few Natura 2000 sites have been designated fully as nationally protected areas. For others, only interim measures were concluded, which did not meet the legal requirements; as a result, the inappropriate legal acts used (a Joint Ministerial Decision rather than a Presidential Decree) were overruled by the Greek Council of State.

In 2011 the Law on Biodiversity Conservation was adopted, which revamped Greece's nature conservation provisions, including those related to the country's protected areas. The law established a unified and coherent national system of protected areas by consolidating the earlier categories created under the forest legislation and the Law on the Protection of the Environment with Natura 2000 sites into one legal framework that had biodiversity conservation as a primary objective. The timing was crucial, as the law was adopted during the fifth year of the six-year period since the European Commission had approved the first Mediterranean SCI list in 2006. It had become evident that with the long process of national designation, Greece would not meet its six-year deadline of designating the first SCIs as SACs. The new law outlined the criteria and processes for designating sites under the new protected areas categories and required a management plan for all protected areas. Importantly, the Natura 2000 sites were integrated into the country's national protected system and all SCIs were granted the status of SACs pending the definition of conservation objectives and, importantly, conservation measures for each one.

In 2020, Law 4685 on the Modernization of Environmental Law[19] was adopted. Once more the system of protected areas

[15] N. 1469/1950 περί προστασίας ειδικής κατηγορίας οικοδομημάτων και έργων τέχνης μεταγενέστερων του 1830 (On the protection of special categories of buildings and works of art dating after 1830) (A΄ 169).

[16] N.Δ. 191/1974 Περί κυρώσεως της εν Ραμσάϊρ του Ιράν κατά την 2αν Φεβρουαρίου 1971 υπογραφείσης Διεθνούς Συμφωνίας περί προστασίας των διεθνούς ενδιαφέροντος υγροτόπων ίδια ως υδροβιοτόπων. (A΄ 350).

[17] Today, there are 10 because two have been merged into one Ramsar Site.

[18] The process includes the conclusion of a Special Environment Study documenting the protection needs and measures, the adoption of a Presidential Decree designating the area as protected, and a Management Plan.

[19] N 4685/2020 Εκσυγχρονισμός περιβαλλοντικής νομοθεσίας, ενσωμάτωση στην ελληνική νομοθεσία των Οδηγιών 2018/844 και 2019/692 του Ευρωπαϊκού Κοινοβουλίου και του Συμβουλίου και λοιπές διατάξεις (Modernisation of environmental legislation, transposition of Directives 2018/844 and 2019/692 of the European Parliament and the Council and other provisions) (A΄ 92).

was reformed and the categories of protected areas were revised. The new categories of protected areas are:

- Biodiversity Conservation Areas: natural or near-natural areas with recorded presence of natural habitats and species of international, EU or national importance that require protection and conservation. All Natura 2000 sites are included *de jure* in this category.
- National Parks: natural or near-natural areas where large-scale ecological processes take place, and species and habitats of EU or national importance that require protection and conservation are found.
- Wildlife Refuges: areas suitable for wild flora and fauna conservation, acting as reproduction, feeding or wintering habitats. They may also act as ecological corridors between protected areas. Hunting and fishing is prohibited by law.
- Protected Landscapes and Protected Natural Formations: areas or unique elements of nature with significant ecological, geological or geomorphological value or areas that contribute to the conservation of natural processes and the protection of natural resources.

The new law also provides that for the first two categories (i.e. Biodiversity Conservation Areas and National Parks), one or more of the following protection and management zones can be delineated: absolute nature protection zone, nature protection zone, species and habitats conservation zone, and natural resources management zones. In these zones progressively more land uses, from a given list, and other activities can take place, following the completion of a special environmental study and a management plan for the area.

Table 17.4 presents existing protected areas designations. It is based on the previous legislation as the new framework has yet to result in new designations. At the time of writing, the MEEN has commissioned an EU co-financed €17 million project, which will complete the designation of Natura 2000 sites as nationally protected areas. Following the recent changes, the designations will be completed on the basis of the new categories.

Late in 2021, a new government initiative was announced to designate with a special protection status selected segments of mountain areas and designate them as roadless (MEEN, 2022a). Two-year ministerial decisions were issued for six mountains. They include provisions to not open new roads or allow any artificial surfaces or infrastructure in those areas in order to retain their roadless character.

Natura 2000 Sites

Natura 2000 sites cover a significant part of the country's territory and form the heart of Greece's current nature conservation policy.

The identification of Natura 2000 sites was undertaken through the preparation, with LIFE support, of a national list of proposed sites (Daffis *et al.*, 1996). Most of the older protected areas, such as National Forests or Ramsar Sites, were also designated as Natura 2000 sites. Thereafter, Greece's Natura 2000 network was a story of gradual expansion. The first list of officially proposed sites has since expanded often in response to infringement procedures by the European Commission, especially in relation to SPAs. In 2007, the CJEU ruled against Greece for having designated an insufficient number of SPAs (C-334/04). As a result, Greece increased the number of SPAs. The most recent expansion took place in 2017, when several sites, mostly coastal and marine, were added to the list. The Mediterranean biogeographical list of SCIs, which includes all Greece's sites, was formally adopted first in 2006 by the European Commission and has since been updated in line with the expansion of the network by Greece.

The Natura 2000 network in Greece, with its 446 terrestrial and marine sites, is considered almost complete (European Commission, 2019). Based on the data in Table 17.4, it covers 27.3% of its land territory. This coverage is higher than the EU average, and reflects the richness and diversity of Greece's nature. The expansion of the Natura 2000 network in marine areas was delayed due to a lack of knowledge, as well as concerns in the relevant fisheries authorities that reflected the positions of the fisheries sector. Whilst the current marine Natura 2000 sites cover 19.4% of Greece's Territorial Sea (NCESD, 2020), this is only 4.7% of its wider MSFD marine waters.

While the designation of Natura 2000 sites is considered almost complete, Greece has not defined conservation objectives within the six-year deadline since the first approval of Greece's SCIs list, nor has it set out the administrative and management measures for each Natura 2000 site (see Section 17.5.3). As a result, following an infringement procedure launched by the European Commission, the CJEU ruled against Greece in December 2020 (C-849/19). A first set of national conservation objectives was adopted in 2021,[20] but site-specific conservation objectives have yet to be defined.

International Protected Areas and Other Important Biodiversity Areas

Over time, Greece has also designated a few additional areas under other international agreements, including three Biosphere Reserves (the most recent designation in 2020), and two mixed cultural/natural World Heritage Sites. More recently, Geoparks have also been designated. These categories of protected areas are not integrated per se in the country's national protected areas system, although most have also been designated as nationally protected areas or as Natura 2000 sites.

[20] ΥΑ ΥΠΕΝ/ΔΔΦΠΒ/30339/982/2021 - Εθνικοί στόχοι διατήρησης φυσικών τύπων οικοτόπων και ειδών ενωσιακού ενδιαφέροντος (National conservation objectives of natural habitat types and species of union interest) (Β΄ 1375) και Διόρθωση σφάλματος στην ΥΠΕΝ/ΔΔΦΠΒ/30339/982/31.3.2021 απόφαση του Υφυπουργού Περιβάλλοντος και Ενέργειας (Correction of the decision) (Β΄ 1915).

Overall Protected Area Coverage

Overall, the combined coverage of protected areas is about 35% of Greece's land area according to UNEP-WCMC (2022) and NCESD (2020). There is significant overlap between domestic protected areas, Natura 2000 sites and the international protected areas. National protected areas, excluding Natura 2000 sites, account for over 17% of the country's territory (NCESD, 2020). The increase in land designated as protected in recent years is largely due to the designation process for Natura 2000 sites, which must also be designated as nationally protected.

Natura 2000 has also greatly contributed to the protection of marine areas, with the current total protected marine areas being the same as the total marine Natura 2000 network, that is, 19.4% of Greece's Territorial Sea or 4.7% of its total MSFD marine waters.

17.5.2 The Coherence of Protected Areas and Ecological Networks

The 2014 *National Biodiversity Strategy* and the 2016 *National Climate Change Adaptation Strategy* identified a need to improve the coherence of the country's Natura 2000 network and national protected areas network. However, no specific policy initiative has been undertaken to address this. Arguably, the wildlife refuges designated under earlier forest provisions, designated sites under other legal frameworks and the protection of coastal island wetlands since 2012 complement the country's landscape protection and may provide some connectivity between protected areas.

Greece shares terrestrial borders with Albania, North Macedonia, Bulgaria and Turkey. The most notable transboundary protected area is the Prespa Park, shared between Greece, Albania and North Macedonia, to which the European Commission is also a party, since signing a binding agreement in 2010. The agreement entered into force in 2019.

17.5.3 Approaches to Protection and Management of Protected Areas

With the adoption of the Law on the Protection of the Environment in 1986, Greece updated and streamlined its legal framework relating to protected areas to match that of the most advanced in the world at the time. However, the change was too significant to be absorbed and implemented in a country that did not have a strong conservation history or culture of environmental protection. It took almost 15 years to operationalise the Framework Law and create the foundations for a modern system of governance for Greece's protected areas in 1999.[21] The notable novelty of this new system of governance was the establishment of protected area management bodies with a long list of responsibilities relating to the administration and management of the areas. Nature c servation for the first time was delegated to decentralised i tutions beyond the scope of the authority of the Forest Serv These management bodies were governed by a participa board whose members included representatives from cen and local authorities, the scientific community, local st holders and NGOs. As such they introduced the concep multi-stakeholder engagement in the management of p tected areas (Vokou *et al.*, 2014).

The first management body was established in 19 together with the designation of Laganas Bay on the islanc Zakynthos as a National Marine Park, in response to press by the Bern Convention and the pending ruling of the CJ against Greece. The timing of the establishment of the r management bodies also followed the CJEU ruling agai Greece for a delayed transposition of the Habitats Direct and increasing pressure by the European Commission to up a system to manage the country's Natura 2000 sites.

By 2018, 28 management bodies had been established, b these only covered about 25% of the Natura 2000 netw (Vokou *et al.*, 2014). With the delays in the designation proc of protected areas noted previously, the remaining Natu 2000 sites lacked management schemes. As a result, with t exception of some general provisions, targeted managem measures were neither established nor implemented for a lar proportion of Natura 2000 sites.

Whilst the 2011 Biodiversity Law broke new ground relation to the coherence of the protected areas system, it d not provide for its governance. Indeed, for a long peri protected areas were divided into two categories: those th fell under the authority of a management body or the Fore Service (e.g. Wildlife Refuges), and those, especially Natu 2000 sites, that did not fall under any management authori or scheme. For this reason, as early as 2011, several propos were made to restructure Greece's national system of protect areas (e.g. WWF Greece, 2011; Nature 2000 Committee, 201 2012; GNPAMB, 2014). In the midst of the financial crisis, provision was adopted to merge and abolish the existing ma agement bodies as part of the country's administrativ reforms.[22] Following various periods of public consultatio and national dialogues, this provision was never enacted an instead a new law on the Management Bodies of Protecte Areas, adopted in 2018, required that all protected areas fa under the authority of a management body.[23] The area covere

[21] N. 2742/1999 Χωροταξικός Σχεδιασμός και αειφόρος ανάπτυξη και άλλες διατάξεις (Spatial Planning and Sustainable Development and other provisions) (A΄207).

[22] N. 4109/2013 Κατάργηση και συγχώνευση νομικών προσώπων το Δημοσίου και του ευρύτερου δημόσιου τομέα - Σύσταση Γενικής Γραμματείας για το συντονισμό του κυβερνητικού έργου και άλλε διατάξεις (Abolition and merging of legal persons of the public sector and the wider public sector – Establishment of the Genera Secretariat for the Coordination of Government Work and other provisions.). (A΄ 16).

[23] N 4519/2018 Φορείς Διαχείρισης Προστατευόμενων Περιοχών κα άλλες διατάξεις (Management Bodies of Protected Areas and othe provisions) (A΄ 25).

by the existing 28 management bodies was expanded to include neighbouring Natura 2000 sites, while eight new management bodies were created. Moreover, provisions were included for the Greek state to finance the national protected areas system, for the first time, from the national budget and for permanent staff to be employed in the management bodies. In addition, EU structural funds were earmarked for financing the start-up of the eight new management bodies. However, the operationalisation of this new system was not completed.

Following a change of government, a new approach was proposed. Law 4685/2020 reformed completely the governance system of protected areas as it had evolved over the last 20 years, without evaluating the potential effectiveness of the one in place. The MEEN set up NECCA, granting it the main responsibility for managing the country's protected areas system. By merging the previous management bodies, the agency has 24 decentralised units, covering all Natura 2000 sites, supported by mandatory participatory committees aimed at the engagement of relevant stakeholders. It attempts to implement principles of multilevel governance in the protected areas' management (OECD, 2017). The transition to the new system was completed in 2022.

Management plans constitute the main tool for organising conservation actions within a protected area. Over the last two decades, only a couple of management plans of protected areas have been adopted officially, despite the fact that several draft plans have been elaborated through LIFE- or ERDF-funded projects. These plans are not uniform, however, as specifications and guidelines for elaborating management plans have not been formally established. Nonetheless, they offer guidance on practical management measures.

The MEEN commissioned project to complete the designation of Natura 2000 sites as nationally protected areas will also prepare management plans for all of them as this is a prerequisite of the 2020 Law. Management plans are to be approved by the MEEN and implemented locally through NECCA's management units. Moreover, the LIFE-IP 4 NATURA project that is being implemented will finance the pilot implementation of some of the plans, while other national and EU funds will support the implementation of all other management plans in the coming years.

Protected areas have been the focus of Greece's nature conservation policy for at least the past 35 years. Whilst there has been steady progress on the designation of more than a third of the country's land territory and a significant part of its national waters, especially in line with the Nature Directives' requirements, the challenges of governance and effective management persist. Inadequate protection, which results from both the complex national legal provisions as well as weak enforcement of applicable EU provisions, means that various damaging activities, ill-planned or uncontrolled, have taken place within Natura 2000 sites, often with a cumulative effect.

17.6 General Nature Conservation Measures

17.6.1 Development Planning and Impact Assessment

Since the 1975 Constitution, efforts have been made to create the necessary regulatory framework and develop the tools necessary to reduce the environmental impacts incurred during Greece's development path. Spatial and sectoral planning is crucial in this respect. However, progress has been slow. For decades, important planning tools (e.g. spatial plans, a national cadastre, proper delineation of forests, streams and wetlands) have been missing. Without the necessary rules to guide land use and the siting of various activities, as well as vague property rights, development has been haphazard, unauthorised or based on private interests or convenience, leading to a steady degradation of natural areas.

This lack of planning has led to widespread and often illegal construction, especially in forested and coastal areas, in both protected and non-protected areas, which remains an important environmental problem (OECD, 2020). This, in turn, creates particular social pressure to which Greece's clientelism often succumbs, allowing for retroactive permits or legalisation, with lasting impact on the natural environment.

Gradual progress has been made. In 2008 a spatial plan was adopted. The national spatial plan was the result of a long-standing request from all involved stakeholders, including the business community which sought a clear framework on which to base their investments, and the environmental community, which sought clear rules to conserve nature. Adoption of several sectoral spatial plans – on tourism, renewable energy, industry and aquaculture – followed. Most of these plans are currently under revision, one of the reasons being the need for proper integration of nature conservation. The case of the renewable energy spatial plan is indicative, as an HD Article 6(3) appropriate assessment was not properly carried out at the stage of the SEA. This resulted in the Commission launching an infringement case in 2014, which remains open to date, awaiting a revision of the renewable energy spatial plan, which is expected to be completed by 2023.

One environmental silver lining of the financial crisis was that the completion of land-use–related tools such as forest maps and the national cadastre became a requirement under Greece's economic adjustment programme as a necessary precondition of economic growth. Hence, recent governments have sped up the process of completing these tools; however, they are not complete to date. Moreover, local spatial and urban plans have yet to be developed across the country and are now expected to be financed by the EU Recovery and Resilience Facility.

With weak overall planning tools in place, prevention of impact lies primarily with individual project impact assessments. These have been largely perceived as an additional bureaucratic burden that delays or hinders development, rather than as an instrument to improve environmental performance of projects and enhance public participation. In recent years, especially under the country's adjustment programme, the focus has been on reducing administrative burdens. There has been a constant concern that such reforms will lead to weakening of environmental safeguards. This concern is not without basis, as often exemptions to the process have been introduced by law and processes fast-tracked without the necessary personnel and resources to always meet tighter deadlines, which exacerbates the overall weakness of nature conservation policy (Paliogiannis *et al.*, 2019).

HD Article 6(3) appropriate assessments have been integrated as a distinct, yet obligatory, part of EIA procedures. However, the specifications for appropriate assessments (e.g. minimum days of field monitoring, the area to be examined for impacts) were not set out until two distinct ministerial decisions were adopted in 2013 and 2014. A further problem is that plans or projects that do not fall under the EIA requirements do not undergo appropriate assessments – a practice that goes against the requirements of the Habitats Directive and raises concerns on the transposition of Article 6(3) into national law. In 2020, revisions to the legislation clarified that an appropriate assessment is required for all activities that may affect a Natura 2000 site; however, specifications on the implementation of these provisions are still missing. It is foreseen that NECCA will take on the responsibility for reviewing the assessments with the final decision lying with the MEEN.

Due to weak assessments and procedures, disputes have arisen over the years, and several rulings, especially by the Council of State, have cancelled environmental permits of various projects that had not assessed appropriately the impacts on Natura 2000 sites. While an improvement in the quality of EIA has been noted in recent years, the need to strengthen the appropriate assessment process remains (European Commission, 2019). This is due to longlasting problems that are linked to a lack of capacity, expertise and resources, difficulties accessing data, and the somewhat complex and unclear provisions regarding appropriate assessments.

17.6.2 Habitat Management and Restoration

The nature conservation provisions, and particularly those relating to protected areas, constitute the cornerstone of habitat management in Greece. Additional and broader measures derive from other policy fields, such as forestry, agriculture, water management and fisheries.

Protection and management of **forests** is particularly important, as many of the country's protected areas and Natura 2000 sites overlap with forests. As noted earlier, Greece's nature conservation originates in the country's forest policy and legislation. Forest protection provisions have been revised on numerous occasions, especially in recent years, in response to the requirements of the country's economic adjustment programme and a longlasting perception that forest provisions, enshrined in the 1974 Constitution, constitute an obstacle to development. In 2003, an explicit provision was added in the forest legislation that biodiversity requirements must be taken into account in forestry management. However, this was difficult to put into practice as the forestry management plan specifications date back to the 1950s and 1960s. In 2018, intermediate specifications, pending the conclusion of their full revision, were issued integrating specific biodiversity considerations. In 2018, a national Forestry Strategy was adopted, which includes a recognition that biodiversity conservation is a characteristic of Mediterranean forestry practices.

In 2020, the Ministry of the Environment announced a major reforestation programme, which has since been included in Greece's Recovery and Resilience Plan. This is very much a response to the continuous and growing impact of forest fires, which became particularly grave in the summer of 2021. Although forest fires form an integral component of healthy Mediterranean forest ecosystems, they have become more frequent and intense due to socio-economic changes, land-use changes, lack of forest management, and inappropriate fire management policies, measures, equipment and capacity, as well as intensifying climate change (Goldammer *et al.*, 2019). Repeated forest fires mean that the natural regeneration of Mediterranean forests is not possible. Instead well-planned restoration measures are needed, including soil erosion prevention and planting of native trees.

Another policy area that influences nature conservation is **water management**. Implementation of the WFD in Greece has been weak and marked by delays according to the European Commission (2019) and OECD (2020). More significantly, integration in the implementation and monitoring of the WFD and the Nature Directives is limited, as is steady and systematic communication among relevant authorities (The Green Tank, 2020).

Greece has one of the highest percentages (96%) among EU Member States of beaches with an excellent quality, while 98% of its bathing waters meet the sufficiency requirement of the Bathing Water Directive (EEA (2020). While this is true of its beach water, coastal zone management has not been effective. Development pressures have concentrated along the coastlines, leading to increased pressures on coastal ecosystems. It is indicative that Greece's coastline is under the jurisdiction of the Ministry of Finance, rather than the MEEN, which only has an advisory role. Moreover, Greece has not ratified the ICZM Protocol of the Barcelona Convention despite having to comply with its provisions because the EU has ratified the Protocol.

Similarly, Greece has transposed the MSFD and other relevant instruments relating to **fisheries**, mainly drawing from the FAO General Fisheries Commission for the Mediterranean. Among the additional measures that Greece has taken is a horizontal ban on trawling over *Posidonia* meadows. Although a requirement under the Law on Biodiversity Conservation, this was enacted only after mapping and adoption of relevant ministerial decisions by the Ministry of Rural Development and Food. Overall, the integration of fisheries and nature conservation has been amongst the weakest and most difficult.

The 2014 *Biodiversity Strategy* included specific **measures to mainstream biodiversity** into other sectors. However, the least progress has been made in achieving the strategy's relevant mainstreaming targets (The Green Tank, 2020). An OECD 10-year Environment Performance Review in Greece completed in 2020 identifies the need to further integrate biodiversity in the sectoral policies, particularly tourism and agriculture (OECD, 2020). Where the mainstreaming-related measures are governed by EU instruments, such as the CAP and CFP, Greece is seeking to comply with the requirements, but limited additional effort is being noted. It is indicative that resources available for Natura 2000 and other nature conservation priorities from the Rural Development Programme (RDP) and the Marine and Fisheries Programme are often those that remain inactive or are absorbed the least. Nonetheless, some agri-environment–climate measures have financed schemes that are aimed at species conservation (see Table 17.5). These include schemes that support human–wildlife co-existence, for example providing funding to help prevent livestock attacks from large carnivores, or compensation if it occurs.

Few **restoration projects** have been implemented. One celebrated case is the restoration of Lake Karla in Thessaly, which was one of the largest lakes that were drained in Greece in the 1960s. Through the application of EU structural funds, the lake, which had been designated as a Natura 2000 site, has been restored. Although labelled as an ecosystem restoration project, initially it was treated mostly as a reservoir to meet neighbouring needs. Nonetheless, fish and other wildlife, especially waterbirds, have returned and dedicated efforts are being made to manage the lake as a natural ecosystem. Other projects have been of a smaller scale, demonstrating the academic and scientific interest on the topic, as well as the potential for Greece to implement the 2022 proposed EU Nature Restoration Law, once adopted (Aubert *et al.*, 2022).

17.6.3 Invasive Alien Species

The impact of invasive alien species (IAS) on Greece's biodiversity had not been adequately examined and the species had not been well documented. Even though the 2011 Law on Biodiversity Conservation had already included provisions relevant to IAS, both in recognition of the threat that such species pose to indigenous biodiversity as well as in anticipation of the EU Regulation, implementation has been weak and marked by delays. A national list of IAS has been now developed, awaiting adoption, as is the identification of the main arrival pathways of such species. Moreover, systems for IAS surveillance, early detection and response are yet to be established, leading the European Commission to launch an infringement case against Greece. A decision adopted in late 2021,[24] laying down the necessary regulatory procedures to implement the Regulation and penalties in cases when the IAS Regulation is not complied with, may partially address the implementation gaps, without fulfilling all requirements.

Several universities and research institutes are working on the topic, identifying numerous species as invasive, including in protected areas. Management bodies have also been noting the presence of IAS in protected areas. Nonetheless, knowledge of IAS is fragmented. An Interreg Europe project, INVALIS, led by NECCA (initiated by its predecessor the National Centre for the Environment and Sustainable Development) brought together in 2019 for the first time relevant authorities, researchers and civil society, engaging them since in a valuable exchange of insights and best practices (INVALIS, 2022).

17.7 Nature Conservation Costs, Benefits and Funding Sources

17.7.1 Nature Conservation Costs

Greece attempted the first complete estimate of the funding needs of the Natura 2000 network with its Prioritised Action Framework (PAF) for 2014–2020, which it submitted in 2014 (MEECC, 2014b). The PAF estimated the annual needs of the Natura 2000 network at €84 million and one-off actions at €247 million. A revised PAF (MEEN, 2021) has been completed for the 2021–2027 programming period and submitted to the European Commission. The revised PAF estimates the average annual needs of the Natura 2000 network at €148 million (€124 million for recurring costs and €24 million for one-off actions).

17.7.2 The Economic Benefits of Nature Conservation

There is growing recognition of the value of nature and ecosystem services (Dimopoulos *et al.*, 2017). A Hellenic Ecosystem Services Partnership has been established and several projects on mapping ecosystem services are making progress. The mapping of ecosystem services of Natura 2000 sites in Greece is being carried out as part of the LIFE-IP 4 NATURA project (Kokkoris *et al.*, 2020). It is believed that these assessments will form the basis for an economic valuation of the services.

17.7.3 Funding

Nature conservation in Greece has consistently been dependent on EU funds, as national funds have been limited. Some changes have been noted recently. National resources committed to the implementation of the Nature Directives have

24 ΚΥΑ ΥΠΕΝ/ΔΔΕΔ/115162/3055/2.12.2021 Μέτρα εφαρμογής του Κανονισμού (ΕΕ) αριθ. 1143/2014 του Ευρωπαϊκού Κοινοβουλίου και του Συμβουλίου της 22ας Οκτωβρίου 2014 «για την πρόληψη και διαχείριση της εισαγωγής και εξάπλωσης χωροκατακτητικών ξένων ειδών» (Measures implementing Regulation (EU) No 1143/2014 of the European Parliament and of the Council of 22 October 2014 "on the prevention and management of the introduction and spread of invasive alien species") (Β΄ 5697).

Table 17.5 Funding sources for nature conservation in Greece.

Source and period	Measure/category	Average annual allocation (€ million)		Comments
EU funding instruments		**EU + national**	**EU**	
EAFRD Rural Development Programme measures – allocations to Priority 4 (restoring, preserving and enhancing agriculture and forest ecosystems) 2014–2020	M13 Areas facing constraints	142.6	135.7	
	M11 Organic farming	83.6	76.1	
	M10 Agri-environment–climate	55.9	51.4	
	M8 Investments in forests	20.0	18.0	All for N2k, etc.
	M2 Advisory services, etc.	16.3	12.2	
	M1 Knowledge transfer, etc.	2.1	2.0	
	M4 Investments in physical assets	1.6	1.5	
	M12 Natura and WFD	1.3	1.1	All for N2k, etc.
	M16 Cooperation	0.9	0.8	
	M7 Basic services, etc.	0.1	0.1	
RDP Priority 4 subtotal		*324.4*	*298.8*	
ERDF/Cohesion Fund 2014–2020	Priority 85	11.3	9.3	Most for N2k, etc.
	Priority 86	3.5	2.7	
EMFF 2014–2020	Nature conservation measures (mainly under Art. 40.1, also 38 and 39)	5.3	4.0	Most for N2k, etc.
LIFE Programme	Nature and biodiversity projects (1992–2020)	3.9	2.6	
	Integrated project (2017–2025)	2.1	1.3	LIFE-IP 4 NATURA
Other EU (including Interreg)		1.2	1.0	
EU funding instrument subtotal		*351.7*	*319.7*	
National funds within 2014–2020		**Additional national***	**Comments**	
Green Fund 2018–2021	For salaries and other expenses of protected areas management bodies.	7.0	All for N2k, etc.	
National Budget via MEEN since 2018	For operational costs of protected area management bodies	1.7	All for N2k, etc.	
Other public funds		>6		
NGOs, private/foundations, businesses, etc.		?	See text	
National subtotal		>14.7		
National total including co-financing of EU instruments		>46.7		
Total		**>366.4**		

Notes. * Excludes co-financing of EU funds unless indicated. 'For N2k, etc.' indicates funding relevant to Natura 2000, related green infrastructure and BHD species (based on PAFs). The subtotals and totals are only approximate indications of the average annual levels of funding as they are based on annualised data from differing periods.
Sources. Based on data from: RDP data from European Commission ESIF funds database (23 October 2020); RDP allocations to Natura 2000, ERDF, EMFF and national funding from 2021 to 2027 PAF (MEEN, 2021); LIFE factsheet, European Commission (April, 2021); EMFF funding based on 2020 Annual report.[25] See Section 1.3 for further details.

[25] Special Management Authority OP Fisheries and Maritime (2021) Ετήσια έκθεση Υλοποίησης του ΕΤΘΑ 2020 (National Implementation Report of EMFF 2020). https://alieia.gr/wp-content/uploads/2021/07/%CE%95%CF%84%CE%AE%CF%83%CE%B9%CE%B1-%CE%88%CE%BA%CE%B8%CE%B5%CF%83%CE%B7.pdf

tly covered the personnel costs in the relevant public orities, whether in the responsible ministries or in the entralised departments. The practical costs of implementing Nature Directives, and contributing to wider biodiversity ls, have so far been met almost exclusively by EU co-nced projects, as indicated in Table 17.5. Specifically, the ironmental Operational Programmes under the European ional Development Fund (ERDF) have been especially ortant for Natura 2000, with smaller contributions from Regional Operational Programmes. RDP measures co-nced by the European Agriculture and Rural Development nd (EARDF) have also contributed to Natura 2000 as well as ader biodiversity objectives, primarily through the agri-ironment–climate measure. Most of the funding has gone supporting farms and farmers in areas with natural nstraints, which include the mountainous or remote areas the country, while a significant amount has also been directed organic farming. Other measures with biodiversity benefits ve not been given priority. The same is true also for funds ailable via the European Marine and Fisheries Fund (EMFF), hough this has recently financed projects relating to IAS and ew other biodiversity-related actions.

Of special importance has been the LIFE financial instru-ent, as 75 LIFE projects with a focus on nature and biodiver-y conservation have been implemented in Greece since 1992 able 17.5). Beneficiaries have included NGOs, academic stitutions, national and local authorities and occasionally e private sector. In particular, conservation measures for ablematic and globally threatened species such as the Monk al, Loggerhead Turtle and Egyptian Vulture have been ghly dependent on LIFE funding. A LIFE Integrated project, FE-IP 4 NATURA, involving a consortium of 10 partners ordinated by the MEEN is currently underway. Over its ght years (2017–2025) and with its €17 million budget, it ill help implement the Nature Directives, including by sup-orting the development and implementation of species action lans and management plans for selected sites, mapping the ountry's ecosystem services and offering an extended capacity uilding programme.

EU funds have been particularly relevant in financing the peration of Greece's protected areas system. Until 2015, the ersonnel salaries and core operational and management costs f the country's protected area management bodies were also argely met by the EU co-financed Operational Programmes. A few management bodies also secured additional funding rom other projects, such as Interreg or LIFE, or supplemented heir income through other means.

A major problem arose when core costs became ineligible for EU funding in the 2014–2020 programming period. With the strict restrictions in place on Greece's public finances, and an absence of relevant institutional and legal provisions, there was a risk that no resources would be available for protected areas. An interim solution was found through the use of the Green Fund, the country's main national environment fund. Since it was established in 2010, the Green Fund had supported only a few conservation actions by the protected area management bodies, through grants of less than €1 million per year. Over 2015–2018, the Green Fund covered the salaries and core expenses, at a cost of about €8 million per year.

In 2018, a provision was included in the Management Bodies Law to finance the core needs of protected areas for the first time through the country's national budget. Hence, the national budget covered the operational needs of the established management bodies (€1.7 million per year), while the Green Fund continued to cover salaries (€7 million per year). New management bodies established in 2018 continued receiving EU funding as start-up assistance. In 2021, the responsibility to cover salaries and the allocation of the relevant funds was transferred from the Green Fund to NECCA. In addition, NECCA has been eligible since 2020 for funding from Greece's EU Emissions Trading System revenues. The percentage of funds allocated has varied: 1% in 2020, 0.6% in 2021 and 2022. These funds are used mostly to cover the core costs of NECCA.

An additional €1 million has been committed by the Green Fund to finance the implementation of the first three species action plans. Other Green Fund resources have been dedicated to co-finance LIFE projects implemented in Greece. The Green Fund also manages both the national forest and marine protection funds, and has been issuing annual calls for small budget projects (<€50 000) to be implemented by NGOs, some of which have a nature conservation focus.

In addition to EU and other public funding, biodiversity-related actions are being implemented with support from other sources, including private domestic or international foundations, individual donations and corporate social responsibility funds. Amounts can be estimated at a few million annually (some of which probably contributes to national co-financing of EU-funded projects). An increase in both the interest in allocating available funding for nature conservation, as well as the sources and amounts available, has been noted in recent years. Such actions have made substantial contributions to nature conservation in Greece.

While more than €366 million seems to have been spent annually on nature conservation in Greece in recent years, the gap between financing and needs has nevertheless remained significant. In fact, if the amounts spent on the RDP measures supporting farmers in areas with natural constraints and organic farming are excluded, the level of funding is more clearly inadequate, does not transcend sectors and is far lower than the needs estimated in the 2021–2027 PAF. Furthermore, the PAF estimates only the needs for Natura 2000 and the implementation of the Birds and Habitats Directives; it does not reflect the wider costs of nature conservation and particularly the effort needed to shift practices and apply integrated approaches across sectors. The trends from the regulations of the 2021–2027 programming period, the Recovery and Resilience Facility and the priorities of private donors set the tone for the coming years to secure additional funding that must be directed primarily to cover core needs and to apply new approaches.

17.8 Surveillance, Monitoring, Reporting and Applied Research

Monitoring of protected areas, Natura 2000 sites and the conservation status of EU protected species and habitat types is undertaken through projects implemented or commissioned by MEEN, the management bodies of protected areas, and several NGOs that are active in specific areas or focus on specific species. Monitoring, therefore, is largely project-based. The most recently completed national monitoring project by the MEEN provided up-to-date information on the status of protected species and habitat types for the 2007–2012 and the 2013–2018 reporting periods. The project was co-financed by the EU with the involvement of many scientific groups and the collaboration of the protected areas' management bodies. Initially aimed to be completed on time for the 2013 reporting deadline, it was substantially delayed, resulting in an infringement case. The missing data were finally submitted in 2015 and updated in 2019. A new monitoring project will provide data for the 2019–2024 period, while NECCA aims to establish a more permanent monitoring system with funds from Greece's Recovery and Resilience Plan.

Greece's Red Data Books on plants and animals were published in 2009 (Legakis and Maragou, 2009; Phitos *et al.*, 2009), approximately 15 years after their first edition. They have not been updated since. A new partnership between NECCA, the Hellenic Zoological and Botanical Societies and IUCN has been announced in 2021, aiming to update and expand Greece's Red Data Lists of plants and animals and also to add a Red Data List for fungi.

Numerous universities and research institutes, scientific associations and NGOs are significantly active in applied research in nature conservation. For example, bird monitoring is undertaken by the Hellenic Ornithological Society's Caretakers programme, in coordination with the Greek Bird Ringing Centre, which includes a number of volunteers who participate in a mid-winter monitoring programme (Handrinos *et al.*, 2015). However, a particular concern for Greece is that data on its nature are not easily available. In fact, central databases offering easy access to data barely exist. Such tools were foreseen in the 2014 *National Biodiversity Strategy* but have not been developed. It is expected that the LIFE-IP project, with support also from a separate dedicate LIFE project (LIFE EL-BIOS) overseen by NECCA, will provide such a database.

17.9 Future Developments

Nature conservation in Greece is in flux. While the drivers that have led to biodiversity loss remain to date, a gradual shift in policy is noted, despite the recent financial crisis and the COVID-19 pandemic. This is consistent with the pressing need to comply with the EU environmental provisions, the aspiration to be part of the transformative agenda launched by the EU Green Deal and the *EU Biodiversity Strategy for 2030*, and a growing level of awareness on environmental issues. Greeks have heard and know about the Natura 2000 network, and, according to recent Eurobarometer studies, have a higher preference than the EU average for nature conservation over developments that may damage the environment (European Commission, 2018).

In the near future it is expected that many of the processes initiated in recent years will bear fruit, assuming that Greece overcomes its longstanding issues of delayed implementation. Protected areas, including Natura 2000, are expected to be fully designated as such, and to have management plans and a new governance structure to implement them. Similarly, new species action plans are to be completed and implemented. This is necessary, among others, in order to comply with the CJEU ruling against Greece and to be in line with the new EU Biodiversity Strategy and the post-2020 global biodiversity framework.

In fact, turning the tide for the country, from becoming a laggard to a champion of EU nature policy, was the commitment of Greece's Prime Minister at the 2021 IUCN World Conservation Congress. He specifically stated that all Natura 2000 sites will have been fully designated by the end of 2022, and declared that Greece has endorsed the target to designate 30% of its waters as protected and committed to designate 10% of its seas as no-take zones to address the threat of overfishing (Mitsotakis, 2021).

17.10 Conclusions

Greece is very rich in its biodiversity, with a wide variety of habitats and species to protect. While pressures and threats in Greece remain, as everywhere, and many threatened species can be found, Greece also benefits from a relatively large area of remaining semi-natural habitat. This can allow the country to focus on maintaining, effectively managing and restoring its remaining nature – demonstrating the potential of nature recovery.

Nature conservation has not been part of the country's policy heritage, and although many national initiatives have been good in theory, a recurring problem has been inadequate implementation (Papageorgiou and Vogiatzakis, 2006). Nature conservation has therefore been largely driven by EU policy, particularly the Nature Directives. With implementation remaining weak, legal enforcement by the European Commission and EU funding have been essential. Nonetheless, even during difficult periods, such as the financial crisis of the 2010s, important national initiatives and improvements have occurred. Greater focus is needed to further improve implementation and to shift priorities so that nature conservation becomes an integral component of the country's sustainable development path. The challenge of mainstreaming biodiversity across sectors remains. There is an urgent need to ensure that sustainable, green and fair development forms the core priority of the country. The tone set by the EU Green Deal and the *EU Biodiversity Strategy for 2030* signals the direction. However, to achieve this shift, there is a need for a repositioning in strategic planning to give priority to implementation and enforcement of environmental and nature conservation policy, as well as capacity building, awareness-raising and communication.

Greece has already expressed its support for the *EU Biodiversity Strategy for 2030* as well as the main components of the post-2020 global biodiversity framework. It is in a good place to focus on effective protection. Greece must also seek to create a national nature brand. By highlighting the socio-economic and health benefits of nature and identifying synergies with Greece's cultural heritage, the country could turn its natural environment to a comparative development advantage, creating more green jobs and competitive opportunities in all sectors of the economy, from farming and fishing to tourism and services.

Acknowledgements

I thank the reviewers of the draft chapter for their valuable comments: Dr Savas Kazantzidis (Research Director at the Forest Research Institute of the Hellenic Agricultural Organization 'DEMETER'), Fotios Papoulias (former senior official at DG Environment, European Commission), Dr Eleni Tryfon (National Expert at the European Environment Agency) and Dr Ioannis Mitsopoulos (Director General of the Natural Environment and Climate Change Agency).

References

Apostolopoulou, E. & Pantis, J. D. (2009) Conceptual gaps in the national strategy for the implementation of the European Natura 2000 conservation policy in Greece. *Biodiversity Conservation*, **142**, 221–237.

Aubert, G., Costa Domingo, G., Christopoulou, I., Underwood, E. & Baroni, L. (2022) The Socio-Economic Benefits of Nature Restoration in Greece: Showcasing the potential benefits of upscaling nature restoration in Greece to meet the targets of the proposed EU Nature Restoration Law. Policy brief by the Institute for European Environmental Policy. https://thegreentank.gr/wp-content/uploads/2022/07/202206_IEEP_PolicyBrief_Socio-economicBenefitsNatureRestorationGreece.pdf

Bern Convention (2020) List of Decisions and Adopted Texts. Standing Committee 40th meeting (virtual) Strasbourg, 30 November-4 December 2020 (T-PVS(2020)Misc). https://rm.coe.int/draft-list-of-decisions-of-the-40th-standing-committee/1680a09902 (15 December 2020).

(2021) List of Decisions and Adopted Texts. Standing Committee 41st meeting (online) Strasbourg, 29 November-3 December 2021 (T-PVS(2021)Misc). http://rm.coe.int/misc-e-2021-master-draft-02-12/1680a4bea2 (2 January 2021).

Blondel, J., Aronson, J., Bodiou, J.-Y. & Boeuf, G. (2010) *The Mediterranean Region: Biological Diversity in Space and Time*. 2nd ed. New York: Oxford University Press.

Brochet, A. L., Van Den Bossce, W., Jbour, S. *et al.* (2016) Preliminary assessment of the scope and scale of illegal killing and taking of birds in the Mediterranean. *Bird Conservation International*, **26** (1), 1–28.

Catsadorakis, G. (2003) *Greece's Heritage from Nature*. Athens: WWF Greece.

CEPF (2020) *Explore the Biodiversity Hotspots*. The Critical Ecosystem Partnership Fund, www.cepf.net/our-work/biodiversity-hotspots

Christopoulou, I. (2015) *Evaluation study to support the Fitness Check of the Birds and Habitats Directives*. Greece: WWF Greece. http://ec.europa.eu/environment/nature/legislation/fitness check/index en.htm (last accessed 1 November 2020).

Daffis, S., Papastergiadou, E., Georghiou, K. *et al.* (1996) *Directive 92/43/EEC. The Greek Habitat Project NATURA 2000: an overview*. Life Contract B4-3200/94/756. Commission of the European Communities DGXI. The Goulandris Natural History Museum – Greek Biotope/Wetland Centre.

Dimopoulos, P., Drakou, E., Kokkoris, I. P. *et al.* (2017) The need for the implementation of an Ecosystem Services assessment in Greece: Drafting the national agenda. *One Ecosystem*, **2**, e13714.

Dimopoulos, P., Raus, T., Bergmeier, E. *et al.* (2016) Vascular plants of Greece: An annotated checklist. Supplement. *Willdenowia*, **46**, 301–347.

Dimopoulos, P., Tsiripids, I., Xystrakis, F., Kallimanis, A. & Panitsa, M. (2018) *Methodology for Monitoring and Conservation Status Assessment of the Habitat Types in Greece*. Athens: National Centre of Environment and Sustainable Development.

European Commission (2018) EU attitudes towards biodiversity. Flash Eurobarometer – Greece (418). https://ec.europa.eu/commfrontoffice/publicopinion/index.cfm/survey/getsurveydetail/instruments/special/surveyky/2194 (30 November 2020).

(2019) *The EU Environmental Implementation Review 2019 Country Report – Greece*. SWD (2019) 138 final.

EEA (2020) *Country report – Greek bathing quality in 2019*. European Environment Agency.

Fisher I., Ashpole, J., Scallan, D., Proud, T. & Carboneras, C. (compilers) (2018) *International Single Species Action Plan for the conservation of the European Turtle-dove Streptopelia turtur (2018 to 2028)*. Luxembourg: European Commission.

Frederiksen, P., van der Sluis, T., Vadineanu, A. *et al.* (2017) Misfits and compliance patterns in the transposition and implementation of the Habitats Directive – Four cases. *Land Use Policy*, **62**, 337–350.

Georghiou, K. & Delipetrou, P. (2010) Patterns and traits of the endemic plants of Greece. *Botanical Journal of the Linnaean Society*, **162**, 130–422.

Gerakis, P.A. (1992) *Conservation and Management of Greek Wetlands: Proceedings of a Greek Wetlands Workshop, held in Thessaloniki, Greece, 17–21 April 1989*. Gland, Switzerland: IUCN.

GNPAMB (2014) *Plan - Recommendations for a functional and effective national system of administration and management of protected areas*. Athens: Green Network of Protected Areas Management Bodies.

Goldammer, J. G., Xanthopoulos, G., Eftychidis, G., Mallinis, G., Mitsopoulos, I. & Dimitrakopoulos, A. (2019) *Report of the Independent Committee tasked to Analyse the Underlying Causes and Explore the Perspectives for the Future Management of Landscape Fires in Greece*. Athens: Global Fire Monitoring Centre.

Handrinos, G., Kazantzidis, S., Alivizatos, Ch., Akriotis, T. & Portolou, D. (2015) *International waterbird census in Greece*

(1968-2006) – Analysis of the populations of wildfowl (Anseriformes) and the Coot (Fulica atra). Athens: Hellenic Ornithological Society – Hellenic Bird Ringing Centre.

Kati, V., Kassara, C., Psaralexi M. *et al.* (2020) Conservation policy under a roadless perspective: Minimizing fragmentation in Greece. *Biological Conservation*, **252**, 108828.

Kazantzidis, S., Astaras, C., Vasiliadis, I., Makrygianni, E., Ilias, V. & Spyroglou, G. (2020) Factors affecting waterfowl hunting and harvest at the Evros delta, Greece. *Belgian Journal of Zoology*, **150**, 135–152.

Kokkoris, I. P., Mallinis, G., Bekri, E. S. *et al.* (2020) National Set of MAES Indicators in Greece: Ecosystem Services and Management Implications. *Forests*, **11** (5), 595.

INVALIS (2022) Interreg Europe – INVALIS: Protecting European Biodiversity from Invasive Alien Species. https://projects2014-2020.interregeurope.eu/invalis/

Legakis, A. & Maragou, P. (2009) *Το Κόκκινο Βιβλίο των Απειλούμενων Ζώων της Ελλάδας [The Red Book of Endangered Animals of Greece]*. Athens: Greek Zoological Society.

Legakis, A., Constantinidis T. & Petrakis, P.V. (2018) Biodiversity of Greece. In *Global Biodiversity – Volume 2: Selected Countries in Europe*, ed. T. Pullaiah, pp. 71–113. Oakville, ON, Canada: Apple Academic Press – CRC Press.

Maragou, P. & Christopoulou, I. (2012) Προστατευόμενες περιοχές: Βασικές έννοιες και η αποτελεσματικότητα τους στη διατήρηση της βιοποικιλότητας στην Ελλάδα [Protected areas: Main concepts and their effectiveness in conserving biodiversity in Greece]. In *Το Δάσος: Μία ολοκληρωμένη προσέγγιση [The Forest: A Holistic Approach]*, eds. A. C. Papageorgiou, G. Karetsos & G. Catsadorakis, pp. 155–171. Athens: WWF Greece.

MEECC (2014a) *National Biodiversity Strategy and Action Plan – NBSAP*. Athens: Ministry of Environment, Energy and Climate Change.

(2014b) *Prioritised Action Framework (PAF) for Natura 2000 for the EU Multiannual Financing Period 2014-2020*. https://ypen.gov.gr/wp-content/uploads/legacy/Files/Perivallon/Diaxeirisi%20Fysikoy%20Perivallontos/Ektheseis%20-%20Meletes/20200326_action_framework_natura2000.pdf

(2016) *Πόρισμα ομάδας εργασίας αξιολόγησης απόψεων του Εθνικού Διαλόγου για το Κυνήγι [Findings of the working group evaluating the views of the National Dialogue on Hunting]*. Athens: Ministry of Environment, Energy and Climate Change.

MEEN (2021) ΠΛΑΙΣΙΟ ΔΡΑΣΕΩΝ ΠΡΟΤΕΡΑΙΟΤΗΤΑΣ ΓΙΑ ΤΟ ΔΙΚΤΥΟ NATURA 2000 στην Ελλάδα σύμφωνα με το άρθρο 8 της Οδηγίας 92/43/ΕΟΚ του Συμβουλίου για τη διατήρηση των φυσικών οικοτόπων καθώς και της άγριας πανίδας και χλωρίδας («Οδηγία για τους οικοτόπους») για την περίοδο του πολυετούς δημοσιονομικού πλαισίου 2021–2027 [Priority Action Framework for the Natura 2000 network in Greece according to article 8 of Directive 92/43/EEC of the Council on the conservation of natural habitats and of wild fauna and flora (Habitats Directive) for the period of the multiannual framework 2021-2027]. https://ypen.gov.gr/plaisio-draseon-proteraiotitas-gia-to-diktyo-natura-2000-stin-ellada-2021-2027/

MEEN (2022a) Press Conference of the Minister of Environment and Energy, Kostas Skrekas and the Deputy Minister of Environment and Energy, George Amyras, on the presentation of the initiative 'Roadless Mountains'. Press Release 18 January 2022. https://ypen.gov.gr/synentefxi-typou-tou-ypourgou-perivallontos-kai-energeias-kosta-skreka-kai-tou-yfypourgou-perivallontos-kai-energeias-giorgou-amyra-gia-tin-parousiasi-tis-protovoulias-apatita-vouna/

(2022b) 'MEEN and MRDPF declare the 22nd of February as National Action Day Against Illegal Poisoned Baits'. Press Release 22 February 2022. https://ypen.gov.gr/ypen-kai-ypaat-kiryssoun-tin-22i-fevrouariou-os-ethniki-imera-drasis-kata-ton-dilitiriasmenon-dolomaton/

Mitsotakis, K. (2021) Prime Minister Kyriakos Mitsotakis' speech at the IUCN World Conservation Congress in Marseille, https://primeminister.gr/en/2021/09/03/27343 (20 October 2020).

Nature 2000 Committee (Επιτροπή Φύση 2000) (2011) Λειτουργική αναδιάρθρωση των σχημάτων διαχείρισης για βιώσιμη και αποτελεσματική προστασία: Προϋποθέσεις και κριτήρια [Operational restructuring of management schemes for sustainable and effective protection: conditions and criteria]. https://ypen.gov.gr/wp-content/uploads/legacy/Files/Perivallon/Diaxeirisi%20Fysikoy%20Perivallontos/Epitropi%20Fysi%202000/2011_Nov_EpitrFysi-anadiarthrosi_Final.pdf

(2012) Διαχείριση προστατευόμενων περιοχών μη υπαγόμενων σε φορείς διαχείρισης [Management of protected areas not included in management bodies]. http://ypen.gov.gr/wp-content/uploads/legacy/Files/Perivallon/Diaxeirisi%20Fysikoy%20Perivallontos/Epitropi%20Fysi%202000/2012_Feb_EpitFysi-ektosFDperiohes_Final.pdf

NCESD (2018) *State of the Environment 2018*. Athens: National Centre for the Environment and Sustainable Development.

(2020) *State of the Environment 2019*. Athens: National Centre for the Environment and Sustainable Development.

OECD (2017) *Multi-level Governance Reforms: Overview of OECD Country Experiences*. OECD Multi-level Governance Studies, Paris: OECD Publishing.

(2020) *OECD Environmental Performance Reviews: Greece 2020*. Paris: OECD Publishing.

Pafilis, P., Parmakelis, A., Poulakakis, N., Sfenthourakis, S. & Triantis, K. (eds.) (2018) *Biogeography and Biodiversity of the Aegean*. Nicosia: Broken Hill Publishers Ltd.

Paliogiannis, C., Koedam, N. & Cliquet, A. (2019) The impact of the economic crisis on the implementation of the EU Nature Directives in Greece: An expert-based view. *Journal of Nature Conservation*, **48**, 36–46.

Papageorgiou, K. & Vogiatzakis, I. N. (2006) Nature protection in Greece: An appraisal of the factors shaping integrative conservation and policy effectiveness. *Environmental Science & Policy*, **9**, 476–486.

Papayannis, T. (1992) Greek wetlands: factor of change. In *Conservation and Management of Greek Wetlands: Proceedings of a Greek Wetlands Workshop, held in Thessaloniki, Greece, 17-21 April 1989*, ed. P. A. Gerrakis, pp. 207–227. Gland, Switzerland: IUCN.

Papayannis, T. & Howard, P. (eds.) (2012) *Reclaiming the Greek Landscape*. Athens: Med-INA.

Phitos, D., Constantinidis, T. & Kamari, G. (eds.) (2009) *Βιβλίο Ερυθρών Δεδομένων των Σπάνιων και Απειλουμένων Ειδών της* λληνικής Χλωρίδας [Red Data Book on Rare and Threatened lants of Greece]. Volumes I & II. Patra, Greece: Hellenic Botanical ociety.

azidis, K. (2017) Systematic raptor monitoring as conservation ɔol: 12 year results in the light of landscape changes in Dadia-efkimi-Soufli National Park. *Nature Conservation*, **22**, 17–50.

azidis, K., Bontzorlos, V., Schindler, S. & Vasilakis, D. 2019) Lesser Spotted Eagle *Clanga pomarina* Brehm, 1831 (Aves: ccipitridae) in Dadia-Lefkimi-Soufli National Park, Greece: opulation Trends and Spatial Use in Respect to Continuous andscape Changes during the Last 35 Years. *Acta Zoologica Bulgarica*, Suppl. **14**, 7–14.

Green Tank (2020) *Priority to Nature: An assessment of the mplementation of Greece's National Biodiversity Strategy.* https://thegreentank.gr/en/2020/12/09/greece-biodiversity-strategy-assessment/

fon, E. (2015) Evaluation study to support the Fitness Check of the Birds and Habitats Directives. Greece: Ministry of Reconstruction of Production, Environment and Energy. http://ec.europa.eu/environment/nature/legislation/fitness check/index en.htm (last accessed 1 November 2020).

IEP-WCMC (2022) *Protected Area Profile for Greece from the World Database of Protected Areas, May 2022.* www.protectedplanet.net

van der Sluis, T., Foppen, R., Gillings, S. *et al.* (2016) *How much Biodiversity is in Natura 2000? The "Umbrella Effect" of the European Natura 2000 protected area network.* Wageningen, The Netherlands: Altera Technical Report.

Vavylis, D., Bounas, A., Karris, G. & Triantis, K. (2021) The state of breeding birds in Greece: trends, threats, and implications for conservation. *Bird Conservation International*, **31**, 495–509.

Vokou, D., Dimitrakopoulos, P. G., Jones, N. *et al.* (2014) Ten years of co-management in Greek protected areas. An evaluation. *Biodiversity and Conservation*, **23** (11), 2833–2855.

Weale, A., Pridham, G., Cini, M., Konstadakopulos, D., Porter, M. & Flynn, B. (2003) *Environmental Governance in Europe.* Oxford, UK: Oxford University Press.

WWF Greece (2011) *Ενίσχυση της προστασίας της φύσης: Πρόταση του WWF Ελλάς για ένα συνεκτικό, ισχυρό και αποτελεσματικό σύστημα προστασίας του φυσικού περιβάλλοντος* [Strengthening nature protection: WWF Greece's proposal for a coherent, strong and effective system for the protection of the natural environment]. Athens.

WWF Greece, Greenpeace, Arcturos *et al.* (2009) *Πρόταση για ένα ολοκληρωμένο νομικό πλαίσιο για την προστασία της βιοποικιλότητας[Proposal for a comprehensive legal framework for the protection of biodiversity].* Athens.

Chapter 18

Hungary

Katalin Sipos

Key Statistics	
Terrestrial area (km^2)/% of EU-28*	93 013/2.1%
Marine waters (km^2)/% of EU-28*	0
Population density (p km^2)/% of EU-28*	106/92%
GDP per capita (€)/% of EU-28	11 196/42%
Land cover (CORINE land cover classes 2018)	AGR 65%, FOR 18%, OSN 8%, ART 6%, WET 3%
Biogeographical regions	PAN
Number of HD Annex I habitats/%*	45/20%
Number of HD Annex II, IV and V species/%*	207/17%
Number of BD Annex I bird taxa/%*	80/45%
% HD UCS and decreasing habitats/ % unknown	49%/0%
% HD UCS and decreasing species/ % unknown	13%/0%
% breeding BD Annex I birds decreasing/% uncertain and unknown	19%/11%

* Excluding MAC and MMAC biogeographical regions.
Sources and codes. See Table 1.1.

18.1 The Natural History and Characteristics of Hungary

Hungary, a country of central Europe, is located in the Carpathian basin, a lowland surrounded by the high mountain ranges of the Alps (to the west), the Carpathians (north to south-east) and the Dinaric range (south to south-west). It was a land of water: before the river regulation of the nineteenth century it had over 21 000 km^2 of wetlands with permanent or temporary surface water. The largest river crossing Hungary is the Danube (from 1 850 to 1 433 rkm), the whole country forming part of its watershed.

The entire country is within the Pannonian biogeographical region, approximately 70% of which is in Hungary. The diversity of landforms, microclimate and soils gives rise to a series of habitat types, several of which are only found within the EU in the Pannonian biogeographical region: semi-desert habitats in the sand dunes of the Duna–Tisza Interfluve; expansive salty grasslands and marshes on the Great Plain; the largest freshwater lake in Central Europe (Balaton); riverine forests, meadows and oxbows along the large rivers; and oak and beech forests up to the highest peak of Hungary, Kékestető (1 014 m). The Carpathian basin is a biogeographical melting pot, blending Alpine, Boreal, Carpathian, Continental and Illyric (i.e. Submediterranean and Balkan) species, creating a particularly rich flora and fauna. However, the landscape and vegetation have been substantially transformed by human land use over the centuries. According to a recent habitat mapping survey (MÉTA, see Section 18.8), the remnants of natural vegetation cover only 18 000 km^2, which is 19% of the area of Hungary. O this area, only 3% is described as having the highest naturalness, 33% is close to natural, 47% is moderately degraded and 17% is highly degraded (Bölöni *et al.*, 2008).

The main drivers of this habitat loss and degradation have generally been agriculture, water regulation, forest management and infrastructure development. The most profound changes were:

- the transformation of sand steppes, loess steppes, and Euro-Siberian steppic oak habitats to arable lands that currently cover over half the area of the country;
- expansive river regulation and drainage that led to the dramatic loss of over 90% of wetlands;
- afforestation with invasive alien species (IAS), especially Black Locust *(Robinia pseudoacacia)*, which covers a quarter of all Hungarian forest areas, transforming mainly sand steppes and Euro-Siberian steppic oak habitats; and
- a high degree of habitat fragmentation caused by infrastructure development and the land transformations mentioned above.

However, land use was not always a driver of degradation. It also created valuable habitat types, such as the impressive salty grasslands and marshes of Hortobágy (the '*Puszta*'), which is one of the oldest and most recognised man-made habitat complexes in Europe, formed and maintained by grazing.

The impacts of climate change are already present in Hungary. The average annual temperature has increased over

able 18.1 Short-term trends (c. 2007–2018) in Habitats Directive Annex I habitats and Annex II, IV and V species with an unfavourable conservation status in ungary.

	Poor (unfavourable–inadequate)				Bad (unfavourable–bad)			
	Imp	Sta	Dec	?	Imp	Sta	Dec	?
D Habitat assessments								
umber	0	13	4	0	0	4	18	0
6	0%	76%	24%	0%	0%	18%	82%	0%
D Species assessments								
umber	5	45	20	43	1	2	8	14
6	4%	40%	18%	38%	4%	8%	32%	56%

ey. Imp = improving, Sta = stable, Dec = decreasing, ? = unknown.
ource. Based on Member State HD Article 17 reported data as presented on the EEA national summary dashboard (2021).

past century, accompanied by an increasingly extreme tribution of precipitation and an increasing frequency of reme floods, droughts and heat waves. These changes will nplify the harmful impacts of current land uses, especially ter management, which disproportionately focusses on frastructure-based flood mitigation instead of integrated anagement of floods and droughts with nature-based water tention measures and partial restoration of former wetlands.

8.2 The Status of Nature

ased on the data from Natura 2000 state assessment surveys nd the National Biodiversity Monitoring System, the latest D Article 17 report[1] indicates a worrying situation regarding e conservation status of HD habitats and species.

8.2.1 Habitat Status and Trends

ungary has 45 HD Annex I habitat types, of which only six 3%) had a favourable conservation status over 2013–2018: our types of rocky slope vegetation, calcareous fens and *uzulo-Fagetum*) beech forests. Of the remaining habitats, 17 38%) had a poor status and 22 (49%) a bad status. Furthermore, s shown in Table 18.1, nearly a quarter of those with a poor tatus showed short-term declines, as did 82% of those with a ad status, the latter including such endemic habitat types as annonic sand steppes (6260*)[2] and Pannonic loess steppic rasslands (6250*). Concerning woodlands, the lowland flood-lain forest types, Euro-Siberian steppic woods (91I0*) and annonic inland sand dune thickets (91N0*) are the most hreatened. All of the rare wetland types with small and frag-nented areas (e.g. bogs, transition mires and alkaline fens) also ad a bad conservation status. Unfortunately, even more common widely distributed HD Annex I grassland types had a bad status, including lowland hay meadows (6510), alluvial meadows (6440) and *Molinia* meadows (6410).

18.2.2 Species Status and Trends

According to the Article 17 2013–2018 report, 35% of the 212 HD species had a favourable conservation status, while 53% had a poor status and 12% a bad status. Although the situation is not as serious as for habitats, a substantial proportion of species with an unfavourable status also showed declines (Table 18.1). The remarkably high proportion of species with an unfavourable status and unknown trends is also of concern, and may hide as many deteriorating trends as for habitats.

The latest BD Article 12 bird report reveals that, over 2007–2018, the majority of breeding and wintering populations of Annex I bird taxa were stable, fluctuated or increased. According to a recent analysis by Birdlife Hungary,[3] based on systematic monitoring data (see Section 18.8), the populations of several rare and threatened species have stabilised as a result of species-specific conservation efforts over the last 20 years, for example the Great White Egret (*Ardea alba*) – a nature conservation flagship species. However, it is of concern that about one-fifth of breeding and wintering populations of BD Annex I species have declined.

Common bird species of agricultural landscapes and habitats have also declined dramatically. For example, since 1999 the Garden Warbler (*Sylvia borin*), one of the most impacted species, has decreased by over 95%, and the Grey Partridge (*Perdix perdix*) has declined by 52%.

18.2.3 Drivers and Causes of Change

As part of the national MÉTA survey (see Section 18.8), the threats to natural and semi-natural vegetation were thoroughly analysed (Molnár *et al.*, 2008). The study identified 28 pressures, of which the spread of IAS, damage caused by game

[1] See Section 1.3 for background on monitoring under the Nature Directives, status categories and data sources.
[2] Numbers in brackets are HD Annex I habitat codes (see Appendix for full names). * Indicates a priority habitat.

[3] https://mmm.mme.hu/charts/trends

Table 18.2 Short-term trends (c. 2007–2018) in Birds Directive Annex I taxa in Hungary.

	Breeding				Wintering			
	Inc	Sta/Flu	Dec	?	Inc	Sta/Flu	Dec	?
Number	10	39	13	8	3	5	2	0
%	14%	56%	19%	11%	30%	50%	20%	0%

Key. Inc = increasing, Sta = stable, Flu = fluctuating, Dec = decreasing, ? = uncertain and unknown.
Source. Based on Member State Article 12 reported data as presented on the EEA national summary dashboard (2021).

species, desiccation of landscapes, improper land use (e.g. overgrazing or abandonment) and intensive forest management have been the most detrimental. Timber production based on clearcutting and shelterwood cutting has degraded the naturalness of 59% of forested areas, whilst improper management threatens 19% of all grasslands, and IAS have impacted 28% of wetlands and 33% of grasslands. Unfavourable land transformations, such as ploughing, construction, mining or inappropriate afforestation, could destroy 13% of wetlands, 23% of grasslands and 6% of forests, according to the experts' opinion. The survey found that only 20% of natural habitats have not yet been affected by any of these threats.

The most recent Article 17 country report describes a rather similar situation. The most frequent human pressures with high impacts on HD habitats have resulted from agriculture (affecting over 35% of habitats), followed by the extraction and cultivation of biological resources, and IAS (both affecting 29% of habitats) and forestry (24% of habitats). Although natural processes were recorded as the most frequent pressures (with high impacts on 42% of HD habitats), these consisted mainly of habitat succession following the decline of traditional land uses (described later in this section) and the desiccation of the Carpathian basin. With the exception of climate change, the report expects that the proportion of habitats highly impacted by each pressure will remain the same in the future (i.e. as threats). In contrast, the percentage of habitats with high impacts from observed climate change pressures is expected to increase from 31% to 55%.

The HD Article 17 and BD Article 12 reports also show that the most frequent pressures and threats with high impacts on species are rather similar to those affecting habitats. Although certain highly destructive land uses, such as infrastructure development, mining and transport, play a role in the loss of habitats and mortality of protected species, the most widespread negative impacts are mainly related to agriculture, forestry, and game and water management.

The detrimental impacts of agriculture are the most complex and least analysed in Hungary. Nevertheless, diffuse pollution from chemicals, landscape degradation caused by the expansion of monocultures, growing irrigation needs and soil erosion from intensive practices have evidently played a ro the loss of biodiversity. The low proportion of organic farm in Hungary (4% of the total Utilised Agricultural Area (U in 2018[4]) is also an indicator of the dominance of inten conventional farming over much of the landscape. Elsewh in less suitable areas for farming, the decline of traditional l uses (e.g. extensive grazing) and the abandonment biodiversity-rich grasslands have been the drivers behind n ural succession, which has had unfavourable biodiver impacts, such as declines in mountain hay meadows (6520

Intensive timber production, particularly the excess dominance of clearcutting and shelterwood cutting in fo management,[5] is the root cause of the degradation of fo ecosystems in Hungary. While the extent of the forested ar is rather well preserved, and further afforestation is helping tackle climate change, forest biodiversity is continually dete orating. Forests cover about 18% of Hungary (CLC 2018 da of which over 90% were subject to intensive forest mana ment according to data from 2020. As 40% of the forest are within the Natura 2000 network, it is evident that harm forest management has been authorised even within the m important and vulnerable natural areas.

The excessively high populations of some game species another conflicting and disputed driver of the degradation habitats, especially forests and the grasslands of rocky slop According to the factsheet of the Hungarian Central Statisti Office,[6] in 2019 Hungary held altogether 648 600 Red De (*Cervus elaphus*), Fallow Deer (*Dama dama*), Europe Mouflon (*Ovis aries musimon*), European Roe De (*Capreolus capreolus*) and Wild Boar (*Sus scrofa*), with gro ing populations. Although debates are ongoing on the caus of this high population, the scattered presence of large carn vores, abundant food sources in the landscape and intention winter feeding may all be contributory factors. The high gan population is a critical pressure on forest regeneration, and major obstacle to the required transition towards selecti cutting and close-to-nature forest management.

The detrimental impacts of water management appear be partly hidden in the Article 17 reporting; the Carpathia basin is gradually drying out, leading to the degradation wetlands. The *National Water Strategy*[7] describes a deficit i the multiannual national water balance of Hungary, which mainly caused by the dominance of artificial built (grey) floo risk mitigation infrastructure, rather than multipurpos nature-based water retention investments and integrated wate management (e.g. extending the active floodplain). The hig rate of climate change is adding to the problem due to th prolongation of droughts and the more extreme distribution o

[4] https://ec.europa.eu/eurostat/statistics-explained/index.php/Organic_farming_statistics
[5] NFK Erdészeti Főosztály: Magyarország erdeinek összefoglaló adatai 2020.
[6] www.ksh.hu/docs/hun/xstadat/xstadat_eves/i_ux001b.html
[7] Nemzeti Vízstratégia 2017.

precipitation, which are the main changes in the climate that are predicted for Hungary. According to data since 1990, the frequency of serious droughts has significantly increased, resulting in nine extraordinarily dry years until today, with 2022 being the driest since 1901.[8]

18.3 Nature Conservation Governance, Actors, Legislation and Policies

The history of nature conservation in Hungary dates back to the fifteenth century when the first royal measures came into force to prevent deforestation and overhunting. However, a clear and conscious need for conservation did not emerge until the early twentieth century. The first comprehensive concept was created and published in 1931 by Károly Kaán, former state secretary and head of the forest administration, who is accordingly seen as the founder of conservation in Hungary.

The history of the sector up to 2002 is thoroughly described by Rakonczay (2009), who was a prominently influential figure between 1972 and 1990 as the head of the National Conservation Office of the government (Table 18.3).

18.3.1 Governance

As a legacy of socialism, the centralised governance system is dominant over local and civil governance in Hungary. This also applies to conservation, where the capacity and power of the responsible ministry and its regional executive bodies calibrate the efficiency of the sector. The first national conservation authority was established in 1962 (National Conservation Office), which transformed into a ministry in 1987. The importance of the environment within the government is reflected in its ministerial status, which changed over the following periods:

- Ministry of Environment and Water Management (1987–1990)
- Ministry of Environment and Regional Development (1990–1998)
- Ministry of Environment (1998–2002)
- Ministry of Environment and Water (2002–2010)
- Ministry of Rural Development (2010–2014)
- Ministry of Agriculture (since 2014).

The ministry responsible for the environment has an expansive list of conservation duties: preparation and implementation of national strategies and action plans; coordination and reporting of all relevant international conventions, including the EU conservation *acquis*; designation of protected areas and other conservation areas; co-sponsoring and supervision of nature projects (e.g. LIFE and relevant Operative Programmes); coordination of the National Biodiversity Monitoring Programme; coordination and supervision of state property management within protected areas; and many more.

In the implementation of these duties, the most important executive instrument is the regional system of National Park Directorates operating under the direct supervision of the environment ministry. Unlike most other countries, the National Park Directorates in Hungary are responsible for all nationally protected areas (not just National Parks) and Natura 2000 sites within their regional working area, as well as all protected species. This responsibility is declared by law and is independent of the ownership of the land. Since 2002, 10 directorates have made up a uniform institutional system covering the whole of Hungary that has not been fundamentally restructured over the past 20 years. The National Park Directorates maintain the Ranger Service, which had a total of 250 rangers in 2018, each tasked with patrolling an average of 80 km^2 of protected area/Natura 2000.[9]

The development track of the National Park Directorates changed significantly in 2005, when the duties of the conservation administration (permissions, legal restrictions, authorisation on protected areas and species, IAS procedures, etc.) were transferred to the Environmental Inspectorates. The former regulatory, restrictive character of the Directorates started to turn into a nature manager profile, where EU-funded projects, active habitat management and reconstruction, environmental education and awareness raising, research and ecotourism offered opportunities for repositioning. However, the resolution of local conflicts with municipalities and farmers over infrastructure development and unsustainable land use remained partly the duty of the National Park Directorates as they are responsible for the Ranger Service, which creates a challenge in the allocation of limited resources and the refreshment of the institutional brand.

Nature conservation administration is an important, but the most criticised, component of the sectoral tasks due to its restrictive character. In the early phase of conservation, passive protection was dominant over active management, partly due to the sector's rather low capacity. After several organisational and functional changes, the environmental and conservation administration merged into the regional government offices in 2015 and its supervision became the duty of the Prime Minister's Office, replacing the ministry responsible for the environment, which led to a loss of environmental competence and capacity.

Established by the 53/1995 Act on Environment Protection, the National Environmental Council works as a consultant body to the government.[10] Comprising altogether 22 environmental experts from academia, NGOs and trade associations, the Council is a platform of diverse socio-economic views and approaches. The Council has the right and duty to evaluate governmental regulations and strategies before approval, identify the environmental impacts and

[8] www.ksh.hu/docs/eng/xstadat/xstadat_annual/i_uw002.html

[9] https://termeszetvedelem.hu/a-termeszetvedelmi-orszolgalat-bemutatasa-es-feladatai/

[10] 1995. évi LIII. törvény a környezet védelmének általános szabályairól.

Table 18.3 Key nature conservation developments in Hungary.

1893	Hungarian Ornithology Centre established
1901	First protected species declared
1935	Act on Forest and Nature Conservation
1939	National Conservation Council (Országos Természetvédelmi Tanács) established
	First protected area designated (Debreceni Nagyerdő Nature Reserve)
1961	Act on Nature Protection (legislation of forest and nature conservation split)
1962	National Conservation Office (Országos Természetvédelmi Hivatal) established
	Legislation on protected area designation procedure and nature management
1971	Legislation on protected area types and introduction of National Parks
1972	First National Park established (Hortobágy National Park)
1974	Birdlife Hungary (Magyar Madártani Egyesület) established
1976	First long-term conservation strategy (1976–1990)
1979	Regional Environment and Nature Conservation Inspectorates established
1987	Ministry of Environment and Water Management established
1988	Regional Environmental and Water Management Directorates established
1989	Friends of the Earth Hungary (Magyar Természetvédők Szövetsége) established
	Constitution of Hungary declared citizens' fundamental right to a healthy environment
1990	Ministry of Environment and Regional Development established
1991	First annual conference of green NGOs
	WWF Hungary established
1995	Act on the Restoration of the Level of Protection of Protected Natural Areas came into force
1996	Act on Nature conservation (in force) and uniform institutional system of National Park Directorates established
1997	First *National Conservation Plan* (1997–2002)
	National Biodiversity Monitoring System launched
1998	Ministry of Environment established
2002	Ministry of Environment and Water established
	First agri-environment scheme (ÉTT) introduced
	Greenpeace Hungary established
2003	Second *National Conservation Strategy* (2003–2008)
2004	Accession to the EU: Natura 2000 network declared in national legislation
2005	Conservation authority duties transferred from National Park Directorates to Environmental Inspectorates
2008	Green Ombudsman's office established
2009	Third *National Conservation Strategy* (2009–2014)
2010	Independent Ministry of Environment merged into Ministry of Rural Development (currently Ministry for Agriculture)
2011	European Commission approved the Hungarian Natura 2000 network
2012	Green Ombudsman's Office lost its independence and merged into Ombudsman's Office for Fundamental Rights as deputy
2015	Fourth *National Conservation Strategy* and the *National Biodiversity Strategy* (2015–2020)
2017	Environment and Conservation Inspectorates lost their independence and merged into County Government Offices
	National Landscape Strategy (2017–2026)

See Table 3.3 for signatory dates of international conventions.

propose modifications. The Council also has the right to submit proposals proactively, although the government has no obligation to apply these.

The institution of a Green Ombudsman (introduced in 2008) addresses cases, regulations and complaints related to each citizen's fundamental right to a healthy environment, and the protection of natural heritage. It also has the authority to submit complaints directly to the Constitutional Court, which also plays an important role in safeguarding conservation by preventing the erosion of nature legislation (e.g. harmful modification of the Forest Code in 2017[11]). Despite being a relatively new institution, its independence was ended in 2012 and it was placed under the General Ombudsman of Fundamental Rights, restricting the Green Ombudsman's autonomy in launching new cases and investigations or releasing environmental complaints.

18.3.2 Legislation and Policies

The Constitution of Hungary (in force in this form since 2011) declares that the natural resources including biodiversity are part of the national heritage and the State and everyone has the obligation to protect them. It also recognises and endorses the right of everyone to a healthy environment.[12]

In implementing this high-level legal commitment, the Constitutional Court of Hungary plays a significant role in safeguarding natural heritage and, accordingly, has issued several progressive decisions such as the 28/1994 decree prohibiting a deterioration in conservation status.[13]

Hungary is a party to all the major international conventions on biodiversity and conservation, including (with official ratification dates) the Ramsar Convention (1979), CITES (1985), CMS (1986), the Bern Convention (1990), and the CBD (1993). All conventions were transposed into the national legislation, and the relevant implementation and reporting are part of the duties of the ministry responsible for the environment. Compared to the domestic protection and the Natura 2000 designation, these conventions play subordinate roles in the practice of habitat management, stakeholder involvement and monitoring, but are key instruments for international collaboration and steering conservation approaches and procedures.

A key instrument for strategic conservation planning is the *National Conservation Plan*, defined by the Nature Conservation Act (see later in this section) and the Act on Environmental Protection.[14] It is an integrated part of the *National Environmental Programme*, which has had a series of six-year planning periods since 1997, and shapes comprehensively the mid-term goals and key developments for the conservation sector, especially regarding the planning and management of protected areas and Natura 2000 sites, monitoring and research, biological data management, international collaborations, sectoral funding, organisational and legislative developments, and communication and education. The planning and implementation are one of the duties of the ministry responsible for the environment, and the Prioritised Action Framework (PAF) for the implementation of Natura 2000 is an annex of the National Conservation Plan.

Fulfilling the recommendations of the CBD, Hungary has also compiled and implemented a *National Biodiversity Strategy* since 2009.[15] This strategy focussed on the themes directly relevant to the maintenance of biodiversity and was an independent strategy, although it was harmonised with the *National Conservation Plan*. The second *National Biodiversity Strategy* (2015–2020)[16] defined six major fields of intervention, including the implementation of the Nature Directives, green infrastructure development and the maintenance of ecosystem services, sustainable agriculture, forest and game management, addressing IAS, the protection of genetic resources and the prevention of GMOs.[17]

The central piece of nature conservation legislation is the 53/1996 Nature Conservation Act (NCA), which determines the key tools and processes for conservation, including: the types of protected areas, designation procedures, preventive measures and land-use rules, authorisation processes within protected areas, measures for protected species and caves, major roles and responsibilities of state and local bodies, tools of strategic planning and public participation.[18] The Act considers that maintenance of biodiversity and natural heritage requires measures outside the protected areas, and it therefore introduces legal provisions for the general protection of habitats including protection from IAS and landscape protection. It creates the legally sound opportunity to use subsidies and compensation as tools for conservation, and sets out the procedures and rules for eligibility. As the Act has only had minor modifications since 1996, this advanced legal framework for conservation has been stable for over two decades.

The legal requirements for the Natura 2000 network, and other obligations under the Nature Directives, were transposed by the 275/2004 Governmental Decree, which is a complementary regulation to the NCA in terms of spatial relevance.[19] The rules and restrictions on domestic protected areas defined

11 www.alkotmanybirosag.hu/kozlemeny/alaptorveny-ellenes-az-erdotorveny-2017-es-modositasa

12 Magyarország Alaptörvénye (2011. április 25.).

13 28/1994. (V. 20.) AB határozat.

14 27/2015. (VI. 17.) OGY határozat a 2015–2020 közötti időszakra szóló Nemzeti Környezetvédelmi Programról*.

15 96/2009. (XII. 9.) OGY határozat a 2009-2014 közötti időszakra szóló Nemzeti Környezetvédelmi Programról.

16 28/2015. (VI. 17.) OGY határozat a biológiai sokféleség megőrzésének 2015-2020 közötti időszakra szóló nemzeti stratégiájáról.

17 Az Országgyűlés 53/2006. (XI. 29.) OGY határozata a géntechnológiai tevékenységgel, annak mezőgazdasági és élelmiszer-előállítási alkalmazásával kapcsolatos egyes kérdésekről és az ezeket érintő magyar stratégiáról.

18 1996. évi LIII. törvény a természet védelméről.

19 275/2004. (X. 8.) Korm. rendelet az európai közösségi jelentőségű természetvédelmi rendeltetésű területekről.

by the NCA are rather strict and explicit, and have not changed as a result of the Nature Directives. The Natura 2000 decree introduces a 'lighter-touch' approach that focusses on parts of the Natura 2000 sites that are not protected under the domestic legislation. Its precept is to avoid the most harmful land uses with the introduction of the Art. 6(3) appropriate assessments for Natura 2000 sites and a couple of new authorisation procedures (e.g. permission for land transformation such as the ploughing of grassland). It also declares that activities that do not have significant negative impacts on BHD species and habitats for which the site was designated should be permitted. This modest appetite for the restrictive regulation of the Natura 2000 network is aligned with a general conceptual shift to a more subsidy-based and supportive conservation approach. It is also a tactic to decrease the social and economic conflicts that resulted from the designation and substantial extension of areas labelled for conservation.

The regulation of conservation has expanded significantly since the 1990s, such that there are currently over 250 pieces of legislation that designate the specific protected areas, enforce the management plans, or regulate the trade, uses and restrictions on protected species. However, certain sectoral duties, delegated by the NCA, have still not been enforced (e.g. zonation within national parks, with one recent exception, or the buffer zones around protected areas). Although the legal framework for nature conservation is well-developed, implementation and enforcement of the legislation has been only partially successful, especially regarding land-use change and achieving sustainable farming – as indicated by the unfavourable status of many habitats and species discussed in Section 18.2.

18.3.3 Public Involvement and Environmental NGOs

As in other central-eastern European countries, the size and importance of the green civil movement, public involvement and NGOs have varied hugely, depending on the dominant political regime. During the socialist period (1949–1989), public movements were strictly controlled, organisations such as the Scouts were oppressed, and grassroots initiatives were generally stigmatised or inhibited. However, some environmental education activities as well as the establishment of Birdlife Hungary date back to the 1970s; civil protests against the Gabčíkovo hydropower plant (the Danube Movement), which started in 1980, were precursors of the fundamental political changes that took place in 1989.

As Hungary turned into a democracy, the number of NGOs significantly increased and major international NGOs (e.g. Friends of the Earth, Greenpeace and WWF) established offices in the country. During the 1990s the number of environmental foundations and associations increased by almost 70%, to 973 in 1998, almost half of which focussed on nature and conservation (Mészáros, 2001). According to data from the Hungarian Central Statistical Office, there were 61 491 NGOs in 2018, of which 3.4% were established for environmental goals, showing a moderate fall from a peak of 65 561 NGOs in 2011.[20]

An exceptional level of structured collaboration amongst environmental NGOs has occurred in Hungary as a result of the development of the Green Movement. Since its inception in 1991, the Green Annual Conference has been a national summit for environmental organisations where elections take place (of green NGO representatives to different multi-stakeholder platforms, committees and councils), as well as workshops, plenary sessions, and community and field programmes. The continuity of collaboration is ensured by the Coordination Council of the Green Movement (22 elected members from different NGOs), which decides on joint actions, statements and green delegations between the Annual Conferences, operates the website and the mailing lists, and coordinates the preparation of the next summit.

In alignment with the Aarhus Convention, which Hungary ratified in 2001, green NGOs have the right to participate in environmental authorisation procedures, submit conservation positions, participate in public hearings and appeal against harmful decisions.[21] However, the duration of the public consultations, especially in the case of laws and national strategies, is often unreasonably short and insufficiently transparent; therefore, their limited capacity and legal experience can be a major obstacle for NGOs. This is partly connected with the recent phenomenon of the shrinking civic space and recurring governmental communications discrediting international NGOs. The most controversial governmental measure was the introduction of the law in 2017 on the transparency of organisations receiving foreign funds, which was formally connected to the legislation against terrorism and money laundering. This obliged those NGOs, including the major environmental organisations, that receive over c. €26 000 per year from foreign sources to register as an 'organisation receiving foreign funds' and to display this stigmatising label on all of their publications.[22]

The outlook for conservation will depend on the strength of public support and environmental activism amongst citizens. According to the latest report of the European Commission (2019), awareness of biodiversity shows a promising increase in Hungary, similar to other Member States.[23] Growing awareness of nature and climate issues is also confirmed by the increasing inclusion of environmental news as

[20] KSH Statisztikai Tükör: A nonprofit szektor legfontosabb jellemzői 2018-ban.

[21] 2001. évi LXXXI. törvény a környezeti ügyekben az információhoz való hozzáférésről, a nyilvánosságnak a döntéshozatalban történő részvételéről és az igazságszolgáltatáshoz való jog biztosításáról szóló, Aarhusban, 1998. június 25-én elfogadott Egyezmény kihirdetéséről.

[22] 2017. évi LXXVI. törvény a külföldről támogatott szervezetek átláthatóságáról.

[23] www.eea.europa.eu/data-and-maps/data/external/special-eurobarometer-481

mainstream media content. The impact of public pressure has led to successes in specific conservation causes, such as the Csarna-valley case, where the planned core zone of the Duna-Ipoly National Park was selected to be a tourism rail development area. The petition against it in 2018 (launched by WWF Hungary with the strong support of the Green Movement) was signed by over 66 000 citizens within seven days – an exceptionally strong social reaction that helped lead to the withdrawal of the plan.

18.4 Species Protection and Conservation Measures

18.4.1 Protected Species in Hungary

Species protection was one of the first conservation measures in Hungary, dating back to 1901 when 132 bird and 30 mammal species were listed as animals of importance for nature and agriculture. This was followed by the protection of five bird species for conservation purposes in 1912, including the Great White Egret, the flagship species of Hungarian state nature conservation. The legal protection of species prohibits the deterioration of their important habitats, wherever they occur, and enables fines for the poaching, collection or killing of rare plant and animal species. Whilst the trade and exploitation of domestic protected species is not generally prohibited, a series of ministerial and governmental decrees set out detailed conditions. There are also special restrictions that can be applied by the relevant authorities to the habitat of strictly protected species, such as temporarily or permanently prohibiting harmful land management measures such as mowing. The killing or endangerment of these species is also categorised as a wildlife crime.

The list of protected species has been extended several times and currently comprises 1 986 species as well as protecting the nests of six ant species based on a ministerial decree in 2001.[24] Within these species, 272 are strictly protected: the list includes 58 fungi and 17 lichens, besides plant and animal species. The law on protected species was also modified in 2008, when the species listed in the Natura Nature Directives were added.

18.4.2 Species Action Plans

Although the conservation of protected species is principally based on the management of their habitats in Hungary, species action plans determine specific measures for critically threatened species. The protection of 26 animal and 20 plant species is currently based on such nationwide action plans,[25] including species such as the Hungarian Meadow Viper (*Vipera ursinii rakosiensis*), Great Bustard (*Otis tarda*), Damon Blue butterfly (*Polyommatus damon*), Lesser Mole-rat (*Spalax leucodon*), Lady's Slipper Orchid (*Cypripedium calceolus*) and Fen Orchid (*Liparis loeselii*). In addition to having a specific permanent budget, the implementation of the plans is funded by EU projects or as part of the regular protected area management carried out by the National Park Directorates.

As in other countries, a group of protected species commonly give rise to human–wildlife conflicts. In Hungary, these include the HD Annex II strictly protected species Wolf (*Canis lupus*) and Otter (*Lutra lutra*), and some birds, including the Great Cormorant (*Phalacrocorax carbo*); activities of these species have led to numerous complaints from farmers, hunters and fishpond owners. Importantly, the NCA and the Civil Code determine the responsibility of individuals in tolerating and preventing the impacts of wildlife, and prescribe compensation only if the authority failed to take measures (regardless of their success) despite the request of the farmer. Although the lack of compensation is often criticised by stakeholders, there have been very few cases where damage has led to compensation payments, because the farmers have often failed to meet the criteria for demonstrating the damage or the causal link between the damage and nature conservation measures.

Online Annex HU.1[26] gives some examples of successful conservation actions for specific species, including the Hungarian Meadow Viper, and birds of prey such as the Eastern Imperial Eagle (*Aquila heliaca*) that suffer high mortality from electricity infrastructure and illegal poisoning.

18.5 Protected Areas and Networks

18.5.1 Protected Area Types and Their Coverage

The designation of protected areas has been ongoing since 1939; however, the speed and scale of growth in their coverage has fluctuated considerably.[27] The first prolific phase lasted until World War II and resulted in approximately 200 new protected areas. In total, these only covered 200 km^2, as the majority were small and protected specific or well-known places of value such as the city forest of Debrecen, the lime alley of Nagycenk or the intact sandy habitats of Csévharaszt. After a break, 135 additional protected areas, covering over 100 km^2, were announced between 1950 and 1965. Designations gained new momentum in the 1970s, starting with the establishment of Hortobágy National Park in 1972. After a few years of stagnation around 1980, the extent and number of domestic protected areas grew continuously and significantly until 1998, reaching over 8 000 km^2 by 2000.[28]

24 13/2001. (V. 9.) KöM rendelet a védett és a fokozottan védett növény- és állatfajokról, a fokozottan védett barlangok köréről, valamint az Európai Közösségben természetvédelmi szempontból jelentős növény- és állatfajok közzétételéről.

25 https://termeszetvedelem.hu/fajmegorzesi-tervek-keszitese-es-megvalositasa/

26 All annexes are free to access online at CUP: www.cambridge.org/natureconservation

27 www.termeszetvedelem.hu/_user/downloads/eves-adatok/1939-2006.pdf

28 https://termeszetvedelem.hu/orszagos-jelentosegu-egyedi-jogszaballyal-vedett-termeszeti-teruletek/

Table 18.4 The number and extent of international designated sites and protected areas in Hungary.

	IUCN category	Number	Area (km²)
Designated under global and non-EU legislation			
Ramsar Sites	NR	29	2 606
Natural and Mixed World Heritage Sites	NAs	1	?
Biosphere Reserves (exc. buffer and transition zones)	NAs	6	248
Designated under the Nature Directives			
SCIs	NR	479	14 442
SPAs	NR	56	13 747
Natura 2000 area	NR		19 949
Designated under domestic legislation			
National Parks (nemzeti parkok)	II, V	10	4 807
Landscape Protection Areas (tájvédelmi körzetek)	V	39	3 369
Nature Conservation Areas (természetvédelmi területek)	IV	172	314
Natural Monuments (természeti emlékek)	III	90	1
Ex lege protected natural values qualified as nature conservation areas (only mires and alkaline lakes *not overlapping* with other domestic protected areas)	NR	1 590	?
Ex lege protected natural values qualified as natural monuments (kurgans, earth fortifications, springs and sinkholes)	NR	9 294	?
Ex lege protected natural assets (caves)	NR	4 160	?
Protected natural areas of local interest (*N.B. inaccurate national database*)	NR	1 806	421

Note. NR = Not reported; NAs = not assigned; ? = not defined or applicable.
Source. See Section 1.3. Based on data from: EEA Natura 2000 Barometer for end 2020; online conservation database of the Ministry for Agriculture in Hungary for the domestic protected areas (Védett Természeti Területek Törzskönyve)[29] and for the international designations.[30]

Table 18.4 summarises the international protected areas, Natura 2000 sites and domestic protected areas in Hungary. The current types of domestic protected areas and the procedures for their designation are defined in the NCA. The following three major types of territorial protection are recognised:

- **Protected areas of national importance** are designated by the ministry responsible for the environment; their conservation administration is under a specifically appointed regional agency (currently the regional government offices), and their nature management is the responsibility of the relevant National Park Directorate. Based on their extent, conservation goal and importance, four subtypes are defined: National Park, Landscape Protection Area, Nature Conservation Area and Natural Monument.
- **Protected areas of local importance** are designated by local decree and supervised by the relevant municipality, who also designate the responsibility for their management. Areas of local importance can only be designated as Nature Conservation Areas or Natural Monuments.
- ***Ex lege* protection** applies to the following precisely defined features of particular importance and uniqueness: mires, alkaline lakes, caves, springs, sinkholes, kurgans (burial mounds) and earth fortifications. Wherever these are discovered and meet the definition in the Act, they are given automatic and immediate protection. However, to register the *ex lege* protected features, the relevant conservation authority (currently the regional government offices) identifies and designates the affected land parcels. The administration and management of these features are the same as for the protected areas of national importance.

The map of protected areas was substantially reshaped by EU accession in 2004 and the need to establish the Natura 2000 network. The process was carried out by the National Park Directorates under the coordination and supervision of the Ministry for Environment and Water, and involved a series

[29] https://termeszetvedelem.hu/kereso/vedett-termeszeti-teruletek/
[30] https://termeszetvedelem.hu/category/nemzetkozi-egyezmenyek-es-egyuttmukodesek/

of specialists and a scientific advisory board.[31] As in many other countries, the designation of SPAs was based on BirdLife International's IBA inventory, and Birdlife Hungary had a major role in the work. To provide scientific support for the identification of pSCIs, a series of biotic state assessment projects were carried out, mainly funded by the PHARE Fund. This included the establishment of systematic national habitat mapping (Intensive Botanical Data Collection; IBOA), which gathered and analysed vegetation data provided by over 171 researchers and subsequently became the foundation of the MÉTA programme (see Section 18.8).[32] After a round of amendments as part of the Pannonian Biogeographic Seminar process, the pSCIs were officially approved by the European Commission in 2011, without the need to identify any additional sites to complete the system. The current 19 949 km^2 of Natura 2000 sites (Table 18.4) amount to 21% of the country's territory, which doubled the area protected for nature conservation.

Although a detailed scientific evaluation of the adequacy of protected areas is lacking, it is possible to get an indication from their known coverage. The national protection, excluding the *ex lege* areas, currently covers 8 489 km^2 (9.1% of the country).[33] The overlap between Natura 2000 sites and domestic protected areas is 39%, while 91% of the nationally protected areas form part of the Natura 2000 network.[34] Taking into account these overlaps, in combination the domestic protected areas and Natura 2000 sites in Hungary cover 21 048 km^2, which is 22.6% of the land area.[35] According to recent habitat surveys, the extent of natural and semi-natural vegetation is 19%, with only 3.3% in the highest naturalness category (see Section 18.1). This suggests that, by itself, the set of domestic protected areas is insufficient to preserve the country's natural heritage. However, the Natura 2000 network is much more extensive, and, as described above, the European Commission considers it to be adequate.

18.5.2 The Coherence of Protected Areas and Ecological Networks

As Hungary joined the Pan European Ecological Network initiative in 1995, the importance of connectivity amongst protected areas and the need for a National Ecological Network was legally stated in the NCA. A more detailed concept and design for the network at a scale of 1:50 000 was shaped by 2000, covering 36% of the country and comprising 55% core zones, 25% corridors and 20% buffer zones.[36] The network was officially designated first in 2003 in the National Land Use Framework Plan (see Section 18.6), which is renewed every six years. The national, regional and local development plans are obliged to take into account and refine the core, buffer and ecological corridor zones. The network serves as an important means of safeguarding connectivity against infrastructure development and highly detrimental land uses (e.g. mining).

18.5.3 Approaches to Protection and Levels of Protection

The most detailed and restrictive regulation applies to the domestic protected area types, established by the NCA. The legislation explicitly prohibits certain detrimental land uses such as the planting of invasive species, the clearcutting of forests in excess of three hectares, and harmful infrastructure development. It also specifies authorisation procedures for potentially harmful land uses such as land transformations, the use of chemicals, collection of biomass, and research activities. This regulation equally applies to all types of domestic protection, such that there is no substantial difference among National Parks, Nature Reserves or Locally Protected Areas.

The Act also defines the strict protection status that can be applied within protected areas and conceptually functions as a core zone where only measures and land uses that serve conservation purposes are allowed. Although this was intended to be the strongest safeguard for biodiversity, the law has been weakly enforced in practice; therefore certain detrimental economic activities such as shelterwood cutting are officially permitted. Another problem is that some of the early strictly protected designations are no longer justified and have not changed in response to the evolution of more modern conservation approaches or natural changes. However, whilst there is a reasonable case for revising the strictly protected areas, any major modification of the valid conservation laws would carry a high risk due to the currently limited political support for nature conservation.

The NCA prescribes the establishment of zonation within National Parks, including the designation of core zones, which are supposed to be areas for future rewilding and sheltering the most sensitive species. After more than a decade of preparation, the first such law came into force in 2020, establishing the zonation of Hortobágy National Park.

As described earlier, the concept behind Natura 2000 status differs to some extent from domestic protected area protection methods. Instead of numerous explicit restrictions, Natura 2000 focusses more on a broad and flexible use of impact assessments, occasional prohibitions and the use of subsidies and compensation. This is partly to mitigate conflicts related to the past designation procedure, but is also a shift towards a more supportive conservation approach.

[31] https://termeszetvedelem.hu/natura-2000-altalanos-tudnivalok/

[32] The Preparation for Implementing the Habitats Directive in Hungary (HU9807-01-02-02).

[33] https://termeszetvedelem.hu/orszagos-jelentosegu-egyedi-jogszaballyal-vedett-termeszeti-teruletek/

[34] Prioritised Action Framework of Hungary 2021–2027.

[35] www.protectedplanet.net/country/HUN

[36] 2003. évi XXVI. törvény az Országos Területrendezési Tervről 3/1 számú melléklet.

Concerning the maintenance of the necessary level of protection and regulation, the Constitutional Court has fairly consistently applied the precautionary principle and revised harmful legislative changes related to conservation. Some decisions, such as the rejection of the privatisation of state-owned Natura 2000 parcels in 2017,[37] or the weakening of the Natura 2000 regulation by the modification of the Forest Code in 2020,[38] provide examples of this important safeguarding role.

18.5.4 Approaches to Managing Protected Areas

Given Hungary's profoundly transformed landscape and environment, the presence of biodiversity-rich man-made habitats and the fragmentation of natural areas, active nature management is the dominant approach to domestic protected areas and Natura 2000 sites. Among others, extensive grazing and mosaic mowing of grasslands, elimination of IAS, restoration of natural water regimes and beneficial land transformations are typical maintenance and restoration measures in protected areas.

The ownership structure of Natura 2000 sites is diverse and differs markedly among the specific sites, but the National Park Directorates as legally appointed nature managers are regularly minority owners. Therefore there is a high degree of dependence on management agreements with stakeholders and subsidies, and payments are essential to incentivise required land management activities (e.g. agri-environment schemes to maintain meadows).

As in other European countries, management plans have a major role in identifying nature conservation goals and designing the required measures to achieve them, in consultation with stakeholders. However, the approaches, procedures and levels of enforcement significantly differ between domestic protected areas and Natura 2000 sites.

Regarding domestic protected areas, management plans are mandatory under the NCA and the requirements for their content and preparation are defined by a specific ministerial decree. Plans must be prepared and consulted on as part of the designation of new protected areas (at all levels), and they are revised on a 10-year cycle. The plans are legally binding documents that enter into force by a minister's decree for areas of national importance and by a municipality decree for sites of local importance. Up to the end of 2019, management plans for 228 domestic protected areas had come into force.[39]

Natura 2000 management plans are advisory, and the proposed measures are partly developed in a format compatible with the agri-environment measures; however, there is no clear mechanism or legal obligation for their inclusion in the national RDP. Up to September 2020, 349 out of the 535 Natura 2000 sites had approved management plans, which are accessible online in full (except those overlapping with military areas).[40]

The involvement of stakeholders and public consultation are obligatory requirements for all types of management plans, and inputs from key stakeholders have to be considered (e.g. relevant municipalities and state organisations, development and other sectoral plans, land users). In the case of overlapping domestic protected areas and Natura 2000 sites, the harmonisation of the management plans is the responsibility of the relevant National Park Directorate.

With the exception of a few very specific strictly protected parts, Natura 2000 and domestic protected areas are open to the general public with no charge. Access is even allowed in the planned core zones of the National Parks along marked hiking trails.

18.6 General Nature Conservation Measures

As mentioned earlier, a systematic approach to conservation is taken in Hungary; therefore the NCA determines a series of responsibilities beyond the Natura 2000 network and domestic protected areas. In this respect, important horizontal conservation tools include landscape protection measures, the designation and preservation of the National Ecological Network, principles for nature-friendly forest and game management, preventing the introduction of IAS, and the general protection of geological values. However, as the majority of the requirements have only been partially incorporated into the relevant sectoral policies and legislation, the effectiveness of the law remains limited.

18.6.1 Development Planning and Impact Assessment

Since 2003, Hungary has had a National Land Use Framework Plan that defines the long-term spatial structure and land-use system of the country and aims to protect the natural and cultural heritage of national importance.[41] It includes a series of thematic maps that indicate the National Ecological Network, the zone of forests and afforestation, the zone of landscape scenery protection, World Heritage Sites and candidates, and water protection areas, besides Natura 2000 sites

---

37 28/2017. (X. 25.) AB határozat a védett természeti területnek nem minősülő Natura 2000 földrészletek értékesítése és hasznosítása természetvédelmi szempontjai érvényesítésével kapcsolatos mulasztásban megnyilvánuló alaptörvény-ellenesség megállapításáról, valamint a Nemzeti Földalapba tartozó földrészletek hasznosításának részletes szabályairól szóló 262/2010. (XI. 17.) Korm. rendelet 31. § (3) bekezdés 9. pontja megsemmisítésére irányuló indítvány elutasításáról.

38 Az Alkotmánybíróság 14/2020. (VII. 6.) AB határozata a természet védelméről szóló 1996. évi LIII. törvény, valamint az erdőről, az erdő védelméről és az erdőgazdálkodásról szóló 2009. évi XXXVII. törvény egyes rendelkezései alaptörvény-ellenességének megállapításáról és megsemmisítéséről.

39 http://termeszetvedelem.hu/_user/browser/File/Tenyek&Adatok/Termeszetvedelmi_adatok-20191231.pdf

40 www.termeszetvedelem.hu/elfogadott-fenntartasi-tervek

41 2018. évi CXXXIX. törvény Magyarország és egyes kiemelt térségeinek területrendezési tervéről.

and nationally protected areas. As regards the National Ecological Network, it determines general restrictions on building, new mining areas and power plants, and obligations for the maintenance of connectivity in response to the construction of linear infrastructure. The National Land Use Framework Plan predominates over the regional and local development plans, which are obliged to transpose the designated national spatial zones.

According to the Construction Act of Hungary,[42] local development plans have to give regard to the environment, nature and landscape heritage, and harmonise future land uses with the preservation of water, soil, air, climate and biodiversity. In accordance with the NCA, the local plan has to identify unique landscape values, maintain the traditional characteristics of the landscape, and steer land-use changes in a way that avoids the loss of remaining fragments of natural and semi-natural vegetation. Although the law defines a clear conceptual approach, remaining challenges are controlling urban sprawl, addressing predominant landscape changes, restoring degraded ecosystems and development of the ecological network, especially around major cities and popular tourist destinations.

Since the Act on Environment Protection came into force in 1995,[43] Hungary has regularly applied EIA and SEA procedures, which were revised and aligned with the EU directives in 2005. Experts authorised to carry out EIA/SEA surveys are licensed by the Ministry for Agriculture and the Hungarian Chamber of Engineers, depending on their field of expertise. Although the assessment is commissioned by the developer, and expected damage has been underestimated, recent experiences have shown a limited number of problematic assessments. The EIA and SEA procedures provide opportunities for civil participation and public hearings, and the environmental authority is obliged to publish online the ongoing cases and the major actions. However, the introduction of a special procedure for high priority state investments and the restructuring of the environmental authority have reduced the effectiveness of the regulatory protection. The Parliament approved a law in 2006 for speeding up authorisation procedures for state investments that are labelled by governmental decree as high priority.[44] Amongst other effects, it restricted EIA procedures to a maximum of 60 days, which is substantially shorter than the regular duration. Furthermore, the Inspectorates for Environment and Nature lost their organisational independence in 2015 and now have a subordinate unit in the complex regional state administration system led by government commissioners; this has made the environmental authority more exposed to the influence of sectoral interests and politics.[45]

18.6.2 Agri-environment Programmes

The relationship between agriculture and biodiversity is particularly important in Hungary, given that 65% of the country consists of agricultural crops and grasslands (see Key Statistics table). The first attempts to develop a systematic agri-environment programme started in 1999, prior to EU accession. The programme focussed on vulnerable natural sites of special importance (ÉTT), such as the key landscapes for the Great Bustard, or the largest areas of intact saline steppes and marshes. The payments were based on the farmers' voluntary commitment to the application of spatially determined management zones and measures, including beneficial land transformations (e.g. arable land to grassland). Payments were based on five-year contracts, which allowed sufficient time for transformative land management but slowed uptake, since it needed significant commitment and trust from the farmer at a time of rapidly changing circumstances due to EU accession. The ÉTT programme was incorporated into the CAP national strategy and Rural Development Programme (RDP) in 2004, and was continued, extended and simplified as the programme for High Conservation Value Sites (MTÉT). This agri-environment scheme defined land-use packages for higher incentives in 28 selected Natura 2000 sites for the protection of unique values such as Great Bustard habitat or Pannonian salty grasslands.[46] Owing to the communication and promotion efforts of the environment ministry and Birdlife Hungary, the impacted areas gradually increased to 200 000 ha in 2009, although they took up a very small proportion (max. 1.4%) of the RDP budget (Tóth, 2016). This progress was reversed in 2016, when the government reallocated a substantial portion of the RDP budget to less environmentally beneficial horizontal schemes and halved the subsidised area of MTÉT, ignoring the objection of major green NGOs.[47]

Following the concept of MTÉT, the approach and structure of the Natura 2000 management plans were designed to be the basis for future agri-environmental zonal sub-programmes, offering opportunities for more sophisticated and multilevel commitments. Although the compilation of Natura 2000 management plans is well advanced, and has been funded partly by RDP measures, the programmes have yet to start, and the current limited political support for MTÉT makes their future application unpredictable.

During the 2014–2020 CAP programming period, RDP payments were available for forest management. Amongst

42 1997. évi LXXVIII. törvény az épített környezet alakításáról és védelméről.

43 1995. évi LIII. törvény a környezet védelmének általános szabályairól.

44 2006. évi LIII. törvény a nemzetgazdasági szempontból kiemelt jelentőségű beruházások megvalósításának gyorsításáról és egyszerűsítéséről.

45 71/2015. (III. 30.) Korm. rendelet a környezetvédelmi és természetvédelmi hatósági és igazgatási feladatokat ellátó szervek kijelöléséről.

46 https://termeszetvedelem.hu/magas-termeszeti-erteku-teruletek-mtet/

47 www.mme.hu/egy-ujabb-eszkozzel-gyengul-termeszet-vedelme

other things, the payments supported afforestation, gene preservation, Natura 2000 compensation and improvement of forest ecosystems. However, uptake was insufficient to use the allocated budgets under all subcategories. Also, whilst harmful subsidies were avoided, the payments would have had greater future conservation benefits if there had been a stronger differentiation between native and non-native plantations and an increased shift towards selective cutting.

18.6.3 Invasive Alien Species

The spread of IAS is one of the most difficult ecological problems impacting HD habitats. According to the latest Article 17 country report, IAS seriously degrade about 30% of HD habitat types, with floodplain vegetation, sandy grasslands and lowland forest types being particularly threatened. A forum of botanists in 1998 identified 33 terrestrial and eight aquatic plant species as the most damaging, of which 15 are particularly harmful and widespread. This was followed by a book series on invasive plant species (Botta-Dukát and Mihály, 2004, 2006). Floodplains and wetlands are strongly exposed to Box Elder (*Acer negundo*), False Indigo-bush (*Amorpha fruticosa*), American daisy species (*Aster* sp.), Indian Balsam (*Impatiens glandulifera*) and goldenrod species (*Solidago gigantea* and *S. canadensis*), while dry habitat types are invaded by Tree of Heaven (*Ailanthus altissima*), Black Cherry (*Padus serotina*) and Common Milkweed (*Asclepias syrica*).

The Black Locust tree is a particular challenge. As well as being one of the most harmful species in dry habitat types, it is a preferred wood species for lowland forestry and honey production, and 25% of Hungarian forests are now Black Locust plantations. Debates around the species are continuous, and are spiced with such controversial actions as the decision by the Ministry for Agriculture in 2014 to label Black Locust as a 'Hungaricum' (i.e. a unique value of Hungary) in response to lobbying by a coalition of foresters, beekeepers and farmers.[48]

Eradication efforts against IAS plants are permanent within protected areas, especially in dry habitat types, where pinpoint treatments with chemicals are authorised; however their impacts on the IAS remain local.

The IAS Advisory Committee of the Ministry for Agriculture identified 33 particularly dangerous invasive animal species, mainly fishes, arthropods and molluscs, including such well-known and widespread species as the Harlequin (*Harmonia axyridis*), the European Mouflon (*Ovis aries musimon*) and the Goldfish (*Carassius auratus*).

Hungary transposed the EU IAS Regulation in 2016,[49] incorporating the necessary changes into the sectoral legislation concerning forest, game and fish management and agriculture, and introduced preventative measures and penalties.

18.6.4 Nature-Based Solutions and Climate Change Adaptation

Owing to mainly EU funds (see Table 18.5), there have been several successful habitat restoration projects that have improved the conservation status of habitats and species within protected areas over the last 20 years. Important examples include the restoration of the natural water regime of saline steppes or degraded wetlands, eradication of IAS and larger-scale transformations of non-native forest plantations to natural woodlands, reconnection of side-arms along large rivers, and habitat management and restoration for particular species, such as the Hungarian Meadow Viper (see Annex HU.1) and the Great Bustard. However, the majority of these were designed and executed by the National Park Directorates and located within the Natura 2000 network. Consequently, the participation and ownership of other potential actors (e.g. water directorates, forest managers, farmers, municipalities or businesses) has remained limited, and the connections among domestic protected areas and Natura 2000 sites have not significantly improved. With the impacts of climate change and the general decline of biodiversity, including common species, broader nature conservation measures are required, such as through the application of nature-based solutions for water retention, prevention of soil erosion and flash floods, local climate cooling and enhancing carbon sinks. This will require broader engagement of relevant stakeholders and sectors.

The acceptance and application of nature-based solutions are at an early stage in Hungary, but interesting pilot programmes have started. The LIFE Climate project 'Municipalities as integrators and coordinators in adaptation to climate change'[50] aims to improve local water retention with small-scale wetland recreation as a preventative measure against flash floods and droughts.[51] The project is being carried out as a partnership between the Ministry for Internal Affairs, municipalities, NGOs and the National Water Directorate. The applied measures such as the creation of semi-natural ponds for retaining treated wastewater or rainfall from residential areas, and the installation of wooden leaky dams, are innovative in Hungary, therefore dissemination and multiplication are a priority for the partnership.

The Hungarian Parliament approved the second *National Climate Change Strategy* in 2018, which provides the roadmap for decarbonisation, the adaptation strategy and a *Partnership for Climate*[52] dissemination plan. The importance of natural resources has been recognised at a strategic level and their conservation has been assigned as a key tool for climate

[48] www.hungarikum.hu/en/content/hungarian-acacia

[49] 2016. évi CXXXVII. törvény egyes törvényeknek az idegenhonos inváziós fajok betelepítésének vagy behurcolásának és terjedésének megelőzésével és kezelésével összefüggésben történő módosításáról.

[50] LIFE16 CCA/HU/000115

[51] https://vizmegtartomegoldasok.bm.hu/en

[52] Melléklet a 23/2018. (X. 31.) OGY határozathoz.

Table 18.5 Funding sources for nature conservation in Hungary.

Source and period	Measure/category	Average annual allocation (HUF m)		Average annual allocation (€ m)		Comments
EU funding instruments		**EU + national**	**EU**	**EU + national**	**EU**	
EAFRD – Rural Development Programme measures – allocations to Priority 4 (restoring, preserving and enhancing agriculture and forest ecosystems) 2014–2020	M10 Agri-environment–climate	38 983.9	31 107.0	108.2	86.3	Important for N2k, etc.
	M11 Organic farming	10 688.1	8 516.1	29.7	23.6	Not included in PAF
	M12 Natura and WFD	9 443.5	7 082.6	26.2	19.7	All for N2k, etc.
	M13 Areas facing constraints	2 131.57	1 598.7	5.9	4.4	No relevance to N2k, etc.
	M15 Forest environment climate	1 967.3	1 475.5	5.5	4.1	All for N2k, etc.
	M4 Investments in physical assets	978.6	734.0	2.7	2.0	No relevance to N2k, etc.
	M1 Knowledge transfer, etc.	692.9	619.7	1.9	1.7	Not included in PAF
	M2 Advisory services, etc.	516.9	426.5	1.4	1.2	Not included in PAF
	M16 Cooperation	498.1	445.5	1.4	1.2	Not included in PAF
	M8 Investments in forests	333.2	274.9	0.9	0.8	
RDP Priority 4 subtotal		*66 234.0*	*52 280.4*	*183.8*	*145.1*	
ERDF/Cohesion fund 2014–2020	OP. 4. Conservation and biodiversity	5 753.7	4 696.5	16.0	13.0	For N2k
LIFE Programme	Nature and biodiversity (2001–2021)	1 549.7	1 099.2	4.3	3.1	
	Integrated project (2019–2026)	777.5	466.5	2.2	1.3	LIFE-IP GRASSLAND-HU
ERDF Interreg 2014–2020	85 Protection of biodiversity, etc.	422.7	367.6	1.2	1.0	Estimated spending in HU
	86 Protection of N2k, etc.	352.3	306.3	1.0	0.8	Estimated spending in HU. See notes
EMFF 2014–2020	Art. 40. E.g. for fish passes, aquaculture, fish spawning grounds	302.7	227.0	0.8	0.6	
EU funding instruments subtotal		*75 392.6*	*59 443.5*	*209.2*	*164.9*	
		Additional non-EU*				**Comments**
National Funds within 2014–2020		**(HUF m)**	**(€ m)**			
Ministry of Agriculture 2014–2020	Compensatory fund for conservation	648.7	1.8			
	Grant for green NGOs ('Zöld Forrás')	72.1	0.2			For environmental NGOs

Table 18.5 *(cont.)*

Source and period	Measure/category	Average annual allocation (HUF m)		Average annual allocation (€ m)		Comments
EU funding instruments		**EU + national**	**EU**	**EU + national**	**EU**	
Others						
Swiss Contribution	Protecting the environment			252.3	0.7	Six biodiversity projects over 2011–2017 – see notes
Additional national and others subtotal				973.1	2.7	
National total including co-financing of EU instruments				16 669.9	46.3	
Total (EU + all national financing)				76 365.9	211.9	

Notes. Currency conversion rate (29 May 2021): €1 = HUF 360.39. 'For N2k, etc.' indicates allocations relevant to Natura 2000, green infrastructure and protection of BHD species (based on the PAF). Interreg includes HU, AT, SK, HR, RS, RO bilateral cooperations. * Excludes co-financing of EU funds unless indicated. Swiss funding annualised over 2014–2020 based on €0.7 million from a total of €1.3 million over 2011–2017. The subtotals and totals are only approximate indications of the average annual levels of funding as they are based on annualised data from differing periods.
Sources. Based on data from: RDP data from European Commission ESIF funds database (23 October 2020); RDP allocations to Natura 2000, ERDF, EMFF and national funding from 2021 to 2027 PAF (Ministry for Agriculture, 2020); Hungary LIFE factsheet, European Commission (2021); online public sources (termeszetvedelem.hu; kormany.hu). See Section 1.3 for further details.

adaptation. However, the governmental communication and recent public action plans focus mainly on increasing forest cover to 27% from the current 21%[53] (18% according to CLC 2018 data).

18.7 Nature Conservation Costs, Benefits and Funding Sources

18.7.1 Nature Conservation Costs

According to the 2014–2020 PAF for Hungary, the estimated annual cost of implementing Natura 2000 was about €180 million (Ministry for Rural Development, 2013). The recently published PAF for 2021–2027 (Ministry for Agriculture, 2020) has significantly increased the estimated costs based on a more detailed analysis and concluded that the average annual funding needed is €361 million, including €194 million annual recurring costs. The main recurring annual costs include the administrative management of the Natura 2000 network, capacity building and maintenance of the Ranger Service, the National Biodiversity Monitoring System, regular habitat management and IAS elimination measures. Since these tasks can only be partly financed by LIFE projects and CAP RDP measures, national funding is also clearly needed. However, supportive governmental decisions on the state budget allocation towards these responsibilities are a recurring challenge every year.

18.7.2 The Economic Benefits of Nature Conservation

The policy recognition of ecosystem services dates back to 2009, when the National Environmental Programme (2009–2014) incorporated this concept into the national conservation strategy.[54] Since then, individual research and pilot surveys have slowly strengthened, and the concept has started to penetrate into broader conservation thinking, although debates around the appropriateness of monetising ecosystem services and its methodological challenges have not been settled. Researchers at the Szent István and Corvinus Universities have played a key role in converging nature and social sciences, as well as methodologies, and communicating this to conservation professionals (Kovács *et al.*, 2011).

Starting in 2017 under the coordination of the Ministry for Agriculture, a strategic operative programme has funded the mapping and evaluation of ecosystem services at a national level, involving major ecological, land-use and GIS research institutions and the National Park Directorates.[55] The project involves complex and nationwide assessment and mapping of natural and semi-natural ecosystems, and development of a four-level ecosystem service modelling and evaluation tool with an accentuated inclusion of economic and well-being dimensions. The selected 12 target ecosystem service types cover key natural functions such as food and energy production, pollination, flood risk mitigation, water supply and

[53] https://docplayer.hu/182810204-Klima-es-termeszetvedelmi-akcioterv.html

[54] 96/2009. (XII. 9.) OGY határozat a 2009-2014 közötti időszakra szóló Nemzeti Környezetvédelmi Programról.

[55] www.termeszetem.hu/en/

microclimate control. The involvement of potential key stakeholders and users (e.g. regional development, agriculture, forestry and civil society) in the advisory board of the project is important, as is the open access to its results, maps and data.[56] The completion of the programme is expected in 2021 and may significantly increase awareness of ecosystem services and their integration in various sectoral strategies.

18.7.3 Funding Sources for Nature Conservation

As in most EU countries, the majority of conservation investments and habitat restoration is funded by various EU sources (Table 18.5). Of these, the Operative Programme for Environment and Energy Efficiency Axis 4, supported by the European Regional Development Fund (ERDF) and Cohesion Fund, has played a central role in governmental conservation work. Besides financing major capacity building and strategic planning activities, it has supported medium and large-scale habitat restoration, infrastructure developments for livestock, grazing and water retention, equipment and machinery for habitat management, and infrastructure for environmental education. The predetermined priority conservation projects aim to support the preparation and implementation of larger-scale investments, carried out mostly by the National Park Directorates.

Although relatively small, another important funding source has been the EU LIFE programme, due to its targeting and effectiveness. However, the required high proportion of co-financing was a serious challenge for potential applicants until 2018, when a state co-financing mechanism was introduced. Hungary has participated in the programme since 2001, with 38 Nature and Biodiversity projects awarded up until May 2021.[57] In addition, an Integrated Project that is implementing the PAF measures for the long-term conservation of Pannonian grasslands and related habitats is underway.[58]

The national budget allocation for environmental funds is very limited and mainly covers the LIFE co-financing scheme available for all Hungarian applicants with an annual budget of about €2.9 million. A small grant scheme is accessible for green NGOs with an annual budget up to €200 000, which is the diminished remnant of the former National Environmental Fund. The Ministry for Agriculture also manages a compensatory fund for conservation with an annual budget of about €1.8 million, mostly for wildlife damage compensation and research into human–wildlife conflicts, prevention measures and management planning.

Although the RDP provides the largest source of funding for wide-scale biodiversity conservation, the proportion that is environmentally sound and significantly beneficial is rather limited. The funding allocation for agri-environment schemes remains by far of greatest importance for biodiversity overall and for Natura 2000, although, as mentioned earlier, its budget and coverage were cut in 2016. Funding for forest measures is noticeably very low mainly because of the limited uptake. According to the PAF for 2021–2027, out of the total €3.4 billion national EAFRD allocation for 2014–2020, only about €0.89 billion (26%) was allocated for schemes with expected positive impacts on the Natura 2000 network (Ministry for Agriculture, 2020).

The PAF also indicates that the average total annual budget allocation for Natura 2000 was about €159 million over 2014–2020, slightly below the estimated need at the time and substantially less than the current cost projection (€362 million). The significant gap between the foreseen costs and the currently available financial sources shows clearly that funding is insufficient in the conservation sector, especially for the regular costs of maintenance and administration, which are not eligible for the main EU funding sources. In addition, this cost analysis partially covers environmental investment needs in other sectors (e.g. the stronger implementation of the WFD) and the broader application of nature-based solutions, as described earlier.

Furthermore, there is no available systemic analysis of the budget allocation to harmful subsidies and infrastructure developments such as intensive farming, unfavourable land transformations or damaging grey infrastructure development for flood risk mitigation, which undermine conservation efforts and funds.

18.8 Surveillance, Monitoring, Reporting and Applied Research

Launched in 1998, Hungary has a National Biodiversity Monitoring System under the supervision of the ministry responsible for the environment, which also manages the resulting database. The monitoring system was developed under the leadership of the Centre for Ecological Research of the Hungarian Academy of Sciences, with funding from PHARE, resulting in protocols published in a series of 10 thematic books (Horváth *et al.*, 1997). A key development has been the unified National Habitat Classification System (Á-NÉR), revised three times since its compilation in 1997,[59] which became widely used beyond the monitoring programme and includes artificial habitat types as well as natural ones.

Systematic bird monitoring started in the early 1970s, coordinated and implemented mainly by Birdlife Hungary. The Hungarian Bird Ringing Centre is an active member of EURING (European Union for Bird Ringing). The Hungarian bird ringing database holds over 6.6 million records, including all available ringing recoveries since 1950.[60] There are

56 http://alapterkep.termeszetem.hu/

57 European Commission LIFE overview Hungary (updated 11 May 2021) https://cinea.ec.europa.eu/life/life-european-countries_en

58 LIFE17 IPE/BE/000018.

59 https://novenyzetiterkep.hu/node/366

60 http://tringa.mme.hu

nationwide thematic programmes for the Monitoring of Common Birds (MMM), Rare and Colonial Nesting Birds (RTM) and Migrating Waterbirds (VVV), which involve hundreds of volunteer birdwatchers as well as professionals. MMM, which started in 1999, collects data on the presence of common bird species in randomly selected 2.5 × 2.5 km UTM grids at predefined observation points and times. The annual monitoring comprises two field observations in spring and early summer, and one in the winter, the main goal of the programme being to gather quantitative data on population trends of the most common birds in Hungary.[61]

The requirement to identify areas for designation as Natura 2000 sites boosted research into the state of nature in Hungary. Between 2001 and 2002, IBOA, zoological surveys and the CORINE Biotope PHARE projects carried out a major collation and analysis of existing biotic and land cover data. The systematic survey continued owing to projects financed by the National Research and Development Fund. The mapping of flora in Hungary (2002–2005, prolonged to 2014)[62] led by The University of West Hungary, and the Spatial Database of Habitats of Hungary[63] led by the Centre for Ecological Research of the Hungarian Academy of Sciences provided new field data for an advanced national distribution map of habitats and plant species.

Complementing the extensive state assessment surveys, the Wildlife Watch (Vadonleső) online platform is a good example of a successful citizen science project.[64] Started in 2009 by the ministry responsible for the environment, the platform has collected and provided spatial data on the most frequent and easy-to-recognise protected species, such as the Stag Beetle (*Lucanus cervus*), the European Hedgehog (*Erinaceus europaeus*) and the Fire Salamander (*Salamandra salamandra*). The open database currently holds over 11 000 records, and the programme is accompanied by awareness raising and engagement activities, such as campaigns on species of the year and promotions of biodiversity with celebrities.

As in other EU countries, structured, transparent and open biotic data management is a challenge in Hungary. Repeated attempts to set up an efficient Conservation Information System (TIR) have only had partial success due to the complexity of data management solutions, sensitivity of data protection and related rights, and the variety of potential purposes and queries; however, more progress has been achieved recently. The recent version[65] is capable of managing the data from the National Biodiversity Monitoring Programme, and legal designation and land administration data, as well as a platform for the wider public where the most important designations, including Natura 2000 and protected areas, are available. However, the application of the system has remained within the state administration, and several functions (e.g. spatial registration of implemented management measures) are little used.

Another biotic data management initiative, developed through the collaboration of universities and the National Park Directorates, is OpenBioMaps, an open source and free online tool that offers a database structure and GIS data sharing based on voluntary use.[66]

Although Hungary is rather advanced in its biodiversity monitoring and state assessments, information on nature management and its cost-effectiveness – especially on grazing and mowing, innovative forest management and agriculture – is still partial and non-systematic.

18.9 Future Developments

The fact that Hungary has not had an independent ministry for the environment since 2010, and the near-halving of staff in the Environmental Unit of the Ministry for Agriculture in 2018, indicates the limited political support and governmental ownership for conservation and the environment. Future developments and conservation ambitions therefore depend substantially on whether there are changes in political attitudes and approaches. In the current political climate, state conservation focusses on the acceptable operation of the existing nature management, monitoring and education systems. More ambitious aims of phasing out harmful subsidies and infrastructure developments, and preventing the erosion of nature legislation, rely more on continuing challenges from the Green Ombudsman, environmental NGOs, and even the constitutional court.

The maintenance of the NCA, the institutional system of National Park Directorates and the major conservation funding sources is of fundamental importance. However, halting biodiversity loss will need major changes in other sectors and substantial capacity building in environmental governance, management and administration.

If political support for the environment does grow, then a series of development opportunities can be expected to emerge. The wider adoption of nature-based solutions of major importance for climate adaptation and carbon storage, and rapidly growing climate funding, may contribute more to nature conservation. One such example is the potential for reconstructing small- and medium-scale wetlands as a general tool for increasing water retention (discussed in Section 18.6).

[61] www.mme.hu/mindennapi-madaraink-monitoringja-mmm
[62] http://floraatlasz.uni-sopron.hu/
[63] https://novenyzetiterkep.hu/english/node/70
[64] https://xn–vadonles-8sb.hu/
[65] http://web.okir.hu/map/?config=TIR&lang=hu

[66] https://openbiomaps.org/?lang=en

The greatest challenge is the integration of conservation approaches and measures into all relevant sectoral policies and practices. Without a meaningful greening of agriculture, greater and more effective and targeted support from the CAP, stronger implementation of the WFD and a major move towards sustainable forest management, there is little chance of halting the general loss of biodiversity.

18.10 Conclusions

Owing to the greatly transformed landscape of Hungary, a conservation movement was founded relatively early, around 1900, and its development has been strongly connected to sustainable land uses and active nature management. Its results and strengths are numerous. The conservation approach soon became systematic, with consciously designed sectoral development and well-aligned evolution of legislation and institutions. The NCA (1996, in force) was a particular milestone as it determined clear procedures inside protected areas, allocated responsibilities among state and local government bodies and introduced conservation tools for the preservation of non-protected values and sites, besides other important measures. The institutions of nature management changed quite frequently until the late 1990s, when the current system of National Park Directorates was established. It has been operating without radical transformations until today, providing some stability for long-term measures and research, internal capacity and knowledge development. The green NGO sector is also well-structured and still has impact despite decreasing national funding and repeated political attacks. The Green Ombudsman's office and the constitutional court also have a role in awareness raising and the guarding of nature and biodiversity.

The designation of the Natura 2000 network was also a major milestone: it doubled the area of protected habitats to 23%, adequately covering Hungary's natural heritage, and also introduced new funding opportunities (e.g. the LIFE programme) to help fill the financial gaps in the sector. Active habitat management, especially the extensive grazing of the endemic types of Pannonian grassland, coupled conservation with the maintenance of traditional livestock breeds. Therefore the National Park Directorates played an important role in the preservation of the gene pool of such emblematic domestic breeds as the Hungarian Grey Cattle. As recognition of the importance of agricultural management for biodiversity, effective agri-environment type payments emerged prior to EU accession, with the first pilots prepared and launched in the early 2000s.

Although the conceptual basis for conservation is sound, and several successful achievements can be listed in species and habitat protection, and some restoration, the majority of the biodiversity indicators are currently unfavourable. IAS have over-run many habitats, and biodiversity has not benefited sufficiently from high-level policy and land-use decisions, as in many other countries. It is difficult to identify how collaboration with other sectors and decision makers could have been more successful. However, the dominance of restrictive tools against supportive ones (e.g. the lack of a compensation system for species protection in the face of harmful RDP subsidies), insufficient communication and positive relationship management with the public, and limited proactivity in rural development and local collaborations have most likely had negative impacts. The key challenge is clearly the involvement of political decision makers and the transformation to sustainability of all those sectors and businesses that have major impacts on natural resources and landscape-scale environmental conditions, such as water management, agriculture and forestry. Better management and maintenance of the Natura 2000 network and domestic protected areas will continue to be the core tasks for conservation in Hungary, but it is also time to initiate horizontal land-use changes that enable nature to regenerate.

Acknowledgements

This chapter builds on the plentiful publicly available conservation data provided by the Environmental Department of the Ministry for Agriculture – thanks to all those colleagues who contributed to this rich knowledge hub. Thanks also go to Ms Rozália Érdiné Szekeres, to Mr András Schmidt and to Ms Ágnes Zólyomi for their review of the chapter.

References

Botta-Dukát, Z. & Mihály, B. (2004) *Özönnövények 1., A KvVM Természetvédelmi Hivatalának tanulmány kötetei 9.* Budapest: TermészetBÚVÁR Alapítvány Kiadó.

(2006) *Özönnövények 2., A KvVM Természetvédelmi Hivatalának tanulmány kötetei 10.* Budapest: KvVM TvH.

Bölöni, J., Molnár, Zs., Horváth, F. & Illyés, E. (2008) Naturalness-based habitat quality of the Hungarian (semi-) natural habitats. *Acta Botanica Hungarica*, **50** (Suppl.), 149–159.

European Commission (2019) Attitudes of Europeans towards Biodiversity; Special Eurobarometer 481. www.eea.europa.eu/data-and-maps/data/external/special-eurobarometer-481

Horváth, F., Korsós, Z., Kovácsné Láng, E. & Matskási, I. (eds.) (1997) *Nemzeti Biodiverzitás-kutató rendszer I.-X.* Budapest: Magyar Természettudományi Múzeum.

Kovács, E., Kelemen, E. & Pataki, Gy. (2011) Ökoszisztéma szolgáltatások a tudományterületek és a szakpolitikák metszéspontjaiban. *Természetvédelmi Közlemények*, **17**, 1–11.

Mészáros, G. (2001) *Civil szerveződések a környezetvédelemben.* Statisztikai Szemle, 79/3. Budapest: Központi Statisztikai Hivatal.

Ministry for Agriculture (2020) *NATURA 2000 PRIORIZÁLT INTÉZKEDÉSI TERV MAGYARORSZÁG – a 2021–2027-es időszak többéves pénzügyi keretére vonatkozóan.* Budapest: Agrárminisztérium.

Ministry for Rural Development (2013) *A Prioritised Action Framework (PAF) For NATURA 2000. For the EU Multiannual Financing Period 2014-2020 - HUNGARY*. Budapest: Vidékfejlesztési Minisztérium. www.termeszetvedelem.hu/_user/browser/File/Natura2000/PAF/PAF_Hungary_EN_2013-04-18.pdf

Molnár, Zs., Bölöni, J. & Horváth, F. (2008) Threatening factors encountered: actual endangerment of the Hungarian (semi-) natural habitats. *Acta Botanica Hungarica*, **50** (Suppl.), 199–217.

Rakonczay, Z. (2009) *A természetvédelem története Magyarországon, 1872-2002*. Budapest: Mezőgazda Kiadó.

Tóth, P. (2016) *Következtetések (?) és javaslatok a jövő agrár-környezetgazdálkodási kifizetéseire*. Conference presentation: Se veled, se nélküled – az agrár-környezetgazdálkodási kifizetések értékelése természetvédelmi és társadalmi szempontból Budapest, 2016. June 21. www.mme.hu/sites/default/files/images/stories/cikkek/20160623_akg_konferencia/tothp_mme_akg.pdf

Chapter 19

Ireland

Andrew Jackson and Anja Murray

Key Statistics	
Terrestrial area (km²)/% of EU-28*	69 946/1.6%
Marine waters (km²)/% of EU-28*	488 764/10.7%
Population density (p km²)/% of EU-28*	67/58%
GDP per capita (€)/% of EU-28	45 856/171%
Land cover (CORINE land cover classes 2018)	AGR 67%, WET 16%, OSN 8%, FOR 6%, ART 2%
Biogeographical regions	ATL, MATL
Number of HD Annex I habitats/%*	59/26%
Number of HD Annex II, IV and V species/%*	60/5%
Number of BD Annex I bird taxa/%*	33/19%
% HD UCS and decreasing habitats/ % unknown	46%/0%
% HD UCS and decreasing species/ % unknown	15%/13%
% breeding BD Annex I birds decreasing/% uncertain and unknown	18%/14%

* Excluding MAC and MMAC biogeographical regions.
Sources and codes. See Table 1.1.

19.1 The Natural History and Characteristics of Ireland

During the Last Glacial Maximum, c. 27 000 years ago, Ireland's entire terrestrial area and much of its marine area were under ice (Roberts *et al.*, 2020; Ó Cofaigh *et al.*, 2021), and it is only since the melting of the glaciers c. 13 000 years ago that many plants and animals could recolonise (Mitchell, 1995). In consequence of this glaciation and separation by the Irish Sea, Ireland has fewer species than Great Britain and continental Europe.

Almost half of the Republic of Ireland is underlain by carboniferous limestone (Drew, 2008), which, when covered by glacial till, provides the rich soils that sustain Ireland's grass-based livestock farming, the dominant form of agriculture. The limestone, where it lacks overlying soil, forms the 'karst' landscape that is 'The Burren' in the west of Ireland, a botanical jewel of c. 720 km². It is predominantly limestone pavement (8240*),[1] though also plentiful in orchid-rich, dry calcareous grassland (6210*). Seasonal lakes, called turloughs (3180*), fill up in winter months and are important for the many thousands of migratory waterbirds that visit Ireland each winter (Cabot, 1999).

Ireland's climate is determined by the warm waters of the north Atlantic and the Gulf Stream which result in high humidity, high rainfall, mild winters and relatively cool summers. This favours grass and a long growing season, and as a result much of Ireland's agricultural area is grassland. Farming is understood to have begun c. 5 700 years BP, with Neolithic settlers beginning the clearance of the woodlands that dominated the island at that time (Mitchell, 1995). Ireland's geography and political history meant that industrialisation did not occur to the same extent or intensity as in many western European countries (Kirby, 2010). Approximately 67% of Ireland's land area is dedicated to agriculture, of which more than 80% is grassland and rough grazing land (CSO, 2016). A recent predicted distribution suggests that 34% of farmland in the Republic of Ireland is highly or very highly likely to meet High Nature Value (HNV) criteria, with c. 54% of this land coinciding with Natura 2000 sites (Matin *et al.*, 2020).

The oceanic climate has resulted in a high number of rivers, streams and wetlands, and over 12 000 lakes (EPA, 2016). Ireland's varied freshwater habitats host salmonids and other HD species like the Freshwater Crayfish (*Austropotamobius pallipes*), and are home to internationally important numbers of waterbirds, in addition to HD habitats. Ireland's peatlands account for nearly a quarter of the land area (Renou-Wilson, 2018). Because of the waterlogged and acidic conditions, peatlands host highly specialised flora and fauna, including species that are rare globally. Of the nine HD peatland types, eight are present in Ireland. However, only a tiny proportion of peatlands in Ireland remain in near-pristine condition, as a result of impacts from peat extraction, forestry and agriculture.

Ireland has c. 7 000 km of coastline, with its islands, many inlets and estuaries creating a wide range of coastal habitats,

[1] Numbers in brackets are HD Annex I habitat codes (see Appendix for full names). * Indicates a priority habitat.

Table 19.1 Short-term trends (c. 2007–2018) in Habitats Directive Annex I habitats and Annex II, IV and V species with an unfavourable conservation status in Ireland.

	Poor (unfavourable–inadequate)				Bad (unfavourable–bad)			
	Imp	**Sta**	**Dec**	**?**	**Imp**	**Sta**	**Dec**	**?**
HD Habitat assessments								
Number	0	17	10	0	1	5	17	0
%	0%	63%	37%	0%	4%	22%	74%	0%
HD Species assessments								
Number	1	4	4	0	0	4	5	0
%	11%	44%	44%	0%	0%	44%	56%	0%

Key. Imp = improving, Sta = stable, Dec = decreasing, ? = unknown.
Source. Based on Member State HD Article 17 reported data as presented on the EEA national summary dashboard (2021).

including sand dunes, machair coastal grasslands, rocky headlands and sea cliffs, lagoons and estuarine mudflats. The west coast has a high proportion of habitats of international importance for overwintering migratory birds in particular. In addition, Ireland's marine habitats and species are notable internationally, featuring 'whales, dolphins, vast colonies of seabirds, abundant fish and cold-water coral reefs, as well as rich algal and invertebrate communities' (DCHG, 2017a).

Ireland's forest cover fell from c. 80% historically (Mitchell, 1995) to around 1% by 1925 (DAFM, 2020a). Having undergone almost total deforestation, the State's forestry programme, which intensified from the 1950s onwards, was designed to address the dearth of woodlands, to develop an indigenous timber industry and to provide employment in rural areas. Ireland's forestry programmes have relied on planting exotic conifers, predominantly the North American Sitka Spruce (*Picea sitchensis*), including on blanket bogs across upland areas, with this now remaining as a legacy of twentieth century afforestation policy. A 1990 estimate put the area of peatland afforested in Ireland at 200 000 ha (Farrell, 1990). Ireland still has one of the lowest proportions of forest cover in Europe, at only 6.1% according to CLC 2018 data, or 11% according to DAFM (2020a), but the government targets 18% by 2046 (DAFM, 2015a). Afforestation with conifers still predominates: of the current forest cover, about 71% is conifers and 29% broadleaves (DAFM, 2020a); only c. 1.2% is classed as native woodland (Cross, 2012). A new draft Forest Strategy (2022–2030), published in 2022, targets a 'major expansion of native forests and improved diversity of native tree species in new and existing forests' (DAFM, 2022).

19.2 The Status of Nature

19.2.1 Habitat Status and Trends

The past decades have seen dramatic declines in many of Ireland's habitats due to a range of pressures, including agricultural intensification, wetland 'reclamation', arterial and field drainage, ecologically unsuitable grazing levels, burning, unauthorised waste disposal, expansion of conifer plantation and clear-fell forestry, pollution of freshwaters and coastal marine waters, economic growth and associated infrastructure, quarrying and mining, peat extraction, invasive species, increased capacity of the fishing fleet and recreational pressures (Temple Lang, 1970; Fogarty, 2017; NPWS, 2019). While many pressures existed before Ireland's accession to the EEC in 1973 (Temple Lang, 1970), the acceleration of biodiversity loss has coincided with the period since accession. This is reflected in the trends noted below.

The most recent HD Article 17 report,[2] covering 2013–2018, shows that 85% of the 59 assessed habitats had an unfavourable status, with 46% of the assessed habitats demonstrating ongoing declining trends; only 2% showed improving trends, while just 15% had a favourable status (NPWS, 2019). Of particular concern are the 39% that had an unfavourable–bad status, and 74% of these had a declining trend (Table 19.1). Furthermore, six of the declining habitats with an unfavourable–bad status have an EU priority status: lagoons (1150*), fixed dunes (2130*), orchid-rich calcareous grasslands (6210*), active raised bogs (7110*), active blanket bogs (7130*) and alluvial forest (91EO*). These habitats are in need of most urgent action.

19.2.2 Species Status and Trends

For many species and habitats, comprehensive historic data on trends are lacking. The percentage of species in Ireland classified as threatened was in 2010 reported to be relatively low compared with many other OECD countries, though the proportion was higher for freshwater fish, reflecting water pollution, particularly from intensive agriculture (OECD, 2010). Moreover, the relatively lower number of threatened species reported may reflect differences in monitoring; for example, Ireland has just 13 Red Lists in total, with none for many groups (NPWS, 2020d). Based on the taxa for which

[2] See Section 1.3 for background on monitoring under the Nature Directives, status categories and data sources.

there is good information (e.g. birds, bumble bees and some HD species), it can be deduced that there have been significant historical declines in many species, with severe declines in the past 40 years.

Knowledge of the status of HD species has improved as a result of the HD Article 17 reports, the most recent of which reveals that 57% of the 61 species assessed had a favourable status over 2013–2018, including most of the bat species, seals and cetaceans (NPWS, 2019). Of the species assessed, 15% had an unfavourable–inadequate status, another 15% had an unfavourable–bad status, and 13% had an unknown status. As for HD habitats, a particular concern is that a high proportion of species with an unfavourable status had declined over the preceding 12 years (Table 19.1).

A high proportion of freshwater species have decreased, with many having faced severe declines since the 1970s. One indicator of water quality changes is the Freshwater Pearl Mussel (*Margaritifera margaritifera*), a sensitive species dependent on well-oxygenated high-status freshwaters. In 1925, there were known populations in 22 of the 32 counties on the island of Ireland (Jackson, 1925); now this species, protected under HD Annex II, is Critically Endangered in Ireland (NPWS, 2019). Another notable species is the Atlantic Salmon (*Salmo salar*). While Ireland was once famous for salmon fisheries, data reveal a 75% decrease in returns of adult salmon to Irish rivers in the period 1994–2017 (NPWS, 2019).

Declines in a wide range of terrestrial species beyond those listed in HD Annex II are also evident. For example, recent research indicates that more than half of Ireland's bee species have undergone substantial declines since 1980, and 30% of wild bee species are considered threatened with extinction in Ireland (NBDC, 2015). Loss of natural and semi-natural habitats has been a key driver of pollinator declines, as well as general declines in wildflowers (NBDC, 2015).

The latest BD Article 12 reporting data indicate that Annex I bird populations appear to be faring a little better than HD species as about 75% have increasing, stable or fluctuating populations (Table 19.2). Nevertheless, a significant proportion are declining, as well as a number of more common non-Annex I species, of which wading birds are experiencing some of the most severe declines. These species, including Curlew (*Numenius arquata*), Lapwing (*Vanellus vanellus*) and Common Snipe (*Gallinago gallinago*), were once widespread, nesting in damp pastures, hay meadows and peat-bogs. Of the 12 wader species breeding in Ireland, 11 are in serious decline, with the most recent bird atlas showing declines of at least 50% since the 1970s (Balmer *et al.*, 2013). The Curlew, for example, was a common breeder only 30 years ago, but the breeding population is now on the verge of extinction in Ireland (see Section 19.4), and the species is listed as Near Threatened in Europe (BirdLife International, 2021) and globally. Over the 18 years since the annual breeding survey of Ireland's common and widespread birds began, population trend analyses indicate that 37% of species are increasing, 27% are stable, and approximately 26% are in decline (Lewis *et al.*, 2019).

Table 19.2 Short-term trends (c. 2007–2018) in Birds Directive Annex I species in Ireland.

	Breeding				Wintering			
	Inc	Sta/ Flu	Dec	?	Inc	Sta/ Flu	Dec	?
Number	10	5	4	3	3	1	4	8
%	45%	23%	18%	14%	19%	6%	25%	50%

Key. Inc = increasing, Sta = stable, Flu = fluctuating, Dec = decreasing, ? = uncertain and unknown.
Source. Based on Member State Article 12 reported data as presented on the EEA national summary dashboard (2021).

Declines in important winter waterbird populations have been revealed by monitoring carried out since the mid-1990s through the Irish Wetland Bird Survey (I-WeBS) in the Republic of Ireland and WeBS in Northern Ireland. The most recent I-WeBS survey revealed that Ireland has lost around half a million waterbirds, almost 40%, in less than 20 years (Burke *et al.*, 2018). Total numbers have declined by 15% alone since the previous survey period (2006–2007 to 2011–2012). Even Mallard (*Anas platyrhynchos*), a familiar duck, has declined by over 40% in the last 20 years. Habitat loss, climate change and cumulative impacts represent the biggest pressures and threats (Burke *et al.*, 2018).

19.2.3 Drivers and Causes of Change

Intensification of Ireland's agricultural and industrial activities occurred later than in other western European countries and, as such, ecosystems that had 'practically disappeared' elsewhere were still present in Ireland in the 1990s (Schouten, 1994). While damaging impacts of course existed before the 1990s (Temple Lang, 1970), pressures greatly accelerated during that decade, when economic prosperity led to accelerated land-use change. The main threats and pressures reported in relation to EU-protected habitats and species in Ireland over 2013–2018 include agriculture; forestry; extraction and cultivation of 'biological living resources' (including hunting, poisoning, fishing, aquaculture and pollution arising from aquaculture); extraction of resources (including minerals, peat and non-renewable energy resources); energy production processes and related infrastructure; transport systems (including roads and shipping); development and construction; climate change; pollution; and invasive species (NPWS, 2019).

By far the most frequent pressures recorded for HD habitats and BD birds relate to agricultural practices. Unsuitable grazing has been identified as one of the main pressures on terrestrial habitats, with c. 40% of HD habitats impacted by overgrazing and c. 15% (each) by abandonment and undergrazing (NPWS, 2019). Over 70% of HD habitats are impacted by pressures relating to agricultural practices overall, and the pressure is ranked as high importance in more than 50% of habitats (NPWS, 2019), with some of the most damaging impacts resulting from agricultural drainage schemes that have destroyed countless wetlands and river habitats (Cabot, 1999). Additional significant losses have occurred from piecemeal habitat removal and infilling of wetlands (Barron, 2006). Changes in the way grasslands are managed have also impacted biodiversity, as the flatter and most fertile land has been utilised more intensively in recent years. A switch from hay to silage production has driven widespread loss of floristic diversity and the earlier harvesting has also impacted the ability of invertebrates and birds to complete their reproductive cycles during the summer months. For example, the replacement of hay with silage is thought to be a cause of the decline of many of Ireland's bees (FitzPatrick *et al.*, 2007).

In recent years, ambitious growth targets for agricultural output have been set in the *Food Harvest 2020* (DAFF, 2010) and *Food Wise 2025* (DAFM, 2015b) reports, driving increases of up to 50% in dairy outputs and significant increases in the outputs of other agricultural sectors. The expansion of Ireland's dairy herd by c. 500 000 cows (c. 50%) between 2010 and 2019 (CSO, 2012, 2020) and attendant increases in nitrogen inputs have no doubt served as a driver of biodiversity loss, given the negative impacts of long-term nitrogen loading on plant species and insect species richness (Haddad *et al.*, 2000). Impacts on biodiversity are recorded at high frequencies from diffuse pollution to surface and ground waters, and from airborne nitrogen deposition, to which blanket bog, alpine heath and wet heath are particularly vulnerable (NPWS, 2019). The Environmental Pillar – a coalition of 32 national environmental NGOs (eNGOs) – withdrew from Ireland's Agri-Food 2030 Strategy Committee in 2021, having come to the conclusion that 'the draft Strategy [for 2030] is woefully inadequate to meet the social and environmental challenges we face' (Environmental Pillar, 2021). The strategy that ultimately emerged – *Food Vision 2030* – is now the subject of judicial review proceedings brought by Friends of the Irish Environment (FIE), on grounds that include a failure to comply with the Habitats Directive (FIE, 2022).

About half of Ireland's rivers, lakes and estuaries (47%, 49%, 62%) are not achieving 'good or high ecological status' with respect to WFD standards, with agriculture identified as the most significant pressure, affecting 53% of water bodies (EPA, 2019). Pressures on Freshwater Pearl Mussel habitats, which can be taken as indicative of pressures on high-status freshwater sites, come from nutrient enrichment, pollution, riverbank erosion, forestry, road-building, peatland drainage and arterial drainage schemes (Moorkens, 2010).

Afforestation with exotic species followed by clear-felling (see Section 19.1) has contributed to biodiversity loss in Ireland. In recent decades the rate of afforestation in Ireland has been high: over the period 1990–2015, the annual rate of change expressed as a percentage of the total forest area was 2% in Ireland, the highest in the EU (Forest Europe, 2015). Forestry has negatively impacted on biodiversity by drainage and afforestation of peatlands/wetlands, and conversion of marginal species-rich semi-natural grasslands. Bird biodiversity is impacted when exotic conifer plantations replace open habitats, resulting in the loss of specialists and ground-nesting birds, which in an Irish context is typically on marginal farmland in upland areas and along the western seaboard (Kelly, 2019). Forestry is also negatively impacting aquatic biodiversity and water quality (EPA, 2019) due to drainage of sites for planting; nutrient enrichment from run-off of fertilisers; and sedimentation caused by clear-felling and poor forest management. The use of the pesticide cypermethrin is also of concern.

The EPA (2020b) reports that the causes of the bad and deteriorating status in four HD nearshore/coastal habitats (i.e. lagoons, large shallow inlets and bays, halophilous scrub, and fixed grey dunes) are: persistent eutrophication issues (in lagoons, large shallow inlets and bays), loss of species (seagrass and maërl in large shallow inlets and bays), and human-induced impacts and the presence of invasive species (in fixed dunes). In terms of eutrophication, a recent EPA (2021) report highlighted that the loads of total nitrogen and total phosphorus to the marine environment from Ireland's rivers have increased by 24% and 31% respectively since 2012–2014, and that over a fifth (22%) of estuarine and coastal water bodies have unsatisfactory dissolved inorganic nitrogen concentrations. Other pressures identified by the EPA include fishing, aquaculture, climate change, ocean warming, sea level rise, coastal erosion, acidification, noise, dredging and dumping at sea, seaweed harvesting, litter and other forms of pollution.

More widely in the marine environment, bottom contact gear and general fishing pressure have negatively impacted biodiversity and many benthic habitats, including reefs, which are sensitive to disturbance and have long recovery times (DCHG, 2019). Regarding general fishing pressure, 35 out of 74 stocks (47%) are reported to be fished 'sustainably', with the remainder overfished (15%) or unknown (38%) (Marine Institute, 2021). While there has been some improvement in the past decade, the legally binding obligations in the CFP Regulation to end all overfishing by 2020 and to end discarding of unwanted by-catch have not been achieved. In these circumstances, FIE recently argued before the High Court of Ireland that the total allowable catches fixed by the Council of the EU for 2020 in Regulation 2020/123 were in breach of the provisions of the CFP and were therefore illegal. The High Court decided in February 2022 to refer this question of alleged illegality to the CJEU, making this a very significant test case for the future of the CFP across the EU.[3]

19.3 Nature Conservation Governance, Actors, Legislation and Policies

Table 19.3 provides a summary of some of the most important historical nature conservation developments in Ireland, particularly over the last 40 years.

[3] *Friends of the Irish Environment* v *Minister for Agriculture, Food and the Marine, Ireland and the Attorney General* [2022] IEHC 64. Before the CJEU: Case C-330/22 Friends of the Irish Environment.

Table 19.3 Key nature conservation developments in Ireland.

1930	Adoption of the Game Preservation Act 1930 and the Wild Birds Protection Act 1930
1948	An Taisce – the National Trust for Ireland – established as Ireland's first eNGO
1964	Requirement for planning permission introduced nationally for the first time with the coming into force of the Local Government (Planning and Development) Act 1963
1969	Publication of *The Protection of the National Heritage* report by An Foras Forbartha (the National Institute for Physical Planning and Construction Research)
1973	Ireland accedes to the EEC and joins the CAP and CFP
1976	Adoption of the Wildlife Act 1976, which Ireland relied upon as its initial transposition of the Birds Directive
1979	Adoption of the Birds Directive
1981	Publication of national list of Areas of Scientific Interest (ASI)
1992	Adoption of the Habitats Directive
1993	Supreme Court judgment in *MacPharthalain* v *Commissioners of Public Works* [1994] 3 IR 353, meaning in effect that the non-statutory designations of ASIs were unlawful because landowners had not been informed
1992–1994	Nationwide re-survey of former ASIs
1994	Government begins process of publicising proposed Natural Heritage Areas (NHAs), the successor domestic site protection regime to ASIs
1995	Establishment of Dúchas, combining natural and built heritage in a single government agency
1997	Late transposition of the Habitats Directive via the European Communities (Natural Habitats) Regulations 1997
2000	Amendment of the Wildlife Act 1976 to provide legal mechanism for creation of NHA network
2003	Dúchas disbanded and National Parks and Wildlife Service (NPWS) takes over its nature conservation work as part of the Department of the Environment
2007	Major CJEU judgments under the Nature Directives against Ireland in Cases C-418/04 and C-183/05
	Establishment of National Biodiversity Data Centre
2011	Re-transposition of the Nature Directives via the European Communities (Birds and Natural Habitats) Regulations 2011 (1997 Regulations revoked)
2019	First National Biodiversity Conference; declaration by Dáil (lower house of parliament) of a climate and biodiversity emergency
2021	Review of the remit, status and funding of the NPWS
2022	Publication of review report plus *Strategic Action Plan* for the renewal of the NPWS, 2022–2024, committing to a range of actions, including 'up to €55 million' over three budget cycles to realise the plan, plus a commitment to new legislation in relation to Ireland's National Parks and nature conservation more generally

See Table 3.3 for signatory dates of international conventions.

19.3.1 Governance

Criticism of the fragmented governance of nature in Ireland was a consistent refrain from the 1960s to the 1980s (e.g. see An Foras Forbartha, 1969; Temple Lang, 1970; UPTCS, 1987). While a National Parks and Wildlife Service (NPWS) has since been established with primary responsibility for nature conservation, it has never had the resources or political power required. Indeed, the NPWS has been moved regularly from home to home, currently forming part of the Department of Housing, Local Government and Heritage (DHLGH), having moved there in 2020. The NPWS's transition from Dúchas, which combined natural and built heritage in a single agency, to the NPWS of today is worthy of note.

Having been established in 1995, in the early 2000s 'Dúchas [became] hostage,' states Viney (2003), 'to the reckless populism of the IFA [Irish Farmers' Association, a powerful farming lobby] leadership. In a guerrilla war over the future of farming policy, the role of Dúchas in carrying through the EU-driven [SACs and SPAs] made it a handy stick with which to beat the government. In claims that Dúchas was "out of political control", the IFA seized on such measures as those proposed for protection of the scarce hen harrier [*Circus cyaneus*] in the uplands of north Co Kerry and west Co Limerick [. . .]. IFA meetings ended in blatant threats to shoot the harriers and send their corpses to Dúchas. The break-up of [Dúchas], and extinction of its name, was followed by IFA agreement to sign up to a package in the National Agreement – one which included higher environmental payments.'

In addition to having been undermined in this way in the past, the national nature conservation administrative capacity has suffered from a long-term lack of investment (see Section 19.7) and remains highly fragmented. DCHG (2019, at 28–32) and NPWS (2021a) record *inter alia* the following bodies as having a role in nature governance: the NPWS; the National Biodiversity Data Centre (NBDC); the Heritage Council, which provides policy advice to government; Teagasc, the national body providing integrated research, advisory and training services to the agriculture and food industry; Coillte, the commercial state forestry company; Bord na Móna, the semi-state peat company[4]; Inland Fisheries Ireland; the Marine Institute; Bord Iascaigh Mhara, which supports the sea fishing and aquaculture sectors; the Sea Fisheries Protection Authority; Irish Water, Ireland's national water utility; Fáilte Ireland, the state agency responsible for the promotion of tourism and related infrastructure; and the Office of Public Works, whose responsibilities include river, flood and coastal management and the protection of cultural heritage. Most recently, the Maritime Area Planning Act 2021 provides for the establishment of a new Maritime Area Regulatory Authority (MARA), which the government intends to be operational by 2023. MARA will play a key role in the consent process for offshore renewable energy projects.

Ireland's 31 local authorities are responsible for planning and development at county (and in some cases city) level; their development planning and planning permission decisions must of course comply with the Nature Directives and can have considerable biodiversity impacts. Most local authorities have a heritage officer; just four have a dedicated biodiversity officer (DCHG, 2019), though the current government has committed to ensuring that all local authorities have a sufficient number of biodiversity and heritage officers, and increased funding has been provided to this end (Department of the Taoiseach, 2020; DHLGH, 2021c). An Bord Pleanála (Ireland's planning appeals board) is *inter alia* a competent authority with respect to the Nature Directives, hearing first and third party appeals relating to planning applications as well as being the first instance decision maker in respect of strategic infrastructure applications.

Central government departments with biodiversity functions include (details in DCHG, 2019, and NPWS, 2021a): the Department of Housing, Local Government and Heritage (as well as housing the NPWS and natural heritage functions, this department is responsible for spatial planning on land and in Ireland's marine area, for sustainable development, and for the management of water resources 'from source to sea'); the Department of Agriculture, Food and the Marine (DAFM) (responsible for agriculture policy, food, fisheries, aquaculture and forestry, including animal and plant health and animal welfare). The Forest Service, part of the DAFM, manages Ireland's forestry programme. The Department of the Environment, Climate and Communications (DECC) is responsible for sustainability, climate change mitigation and adaptation, and management of the EU LIFE Programme. DECC is the parent department for the Environmental Protection Agency (EPA) and Inland Fisheries Ireland (IFI).

An interdepartmental Biodiversity Working Group drawn from 18 government departments and agencies was reportedly established in 2012. While membership of the group is listed in the *National Biodiversity Action Plan 2017–2021* (DCHG, 2017a), the precise nature of its work is unclear.

Within this fragmented system, the NPWS clearly has primary responsibility for nature conservation. However, its political influence and that of its parent department the DHLGH do not match that of other departments that have functions that have tended to conflict with nature conservation. Moreover, the relatively meagre funding available to the NPWS has meant that it has not been able, historically, to perform certain of its key roles comprehensively or effectively. For example, management planning for Ireland's Natura 2000 sites only recently began in earnest (see Section 19.5), and the NPWS's performance of its consultative role relating to planning applications and forestry appears sometimes hampered by resource constraints. A combination of these factors (i.e. under-resourcing and other departments with overlapping functions) may potentially explain, for example, the need for greater coherence in Ireland between the Nature Directives and marine policies, as identified in the *Nature Scorecard* (2018).

The announcement in May 2022 of significant changes to the NPWS's governance and funding clearly marks an important moment in its history, presenting an opportunity to address difficulties discussed in this chapter. Amongst an array of forthcoming governance changes, the NPWS is to become an executive agency within a government department, it will receive a significant injection of funding over three years (discussed in Section 19.7.3), it is to be restructured, and the government has committed to drafting new legislation to underpin its work (DHLGH, 2022a).

19.3.2 Public Involvement and NGOs

NGOs are often the eyes and ears when it comes to implementation and enforcement of the Nature Directives, which is vital work given the striking reduction in the European Commission's enforcement activities in this area (see Jackson, 2018, Hofmann, 2019). However, compared to some other European countries, such as the UK, the few eNGOs in Ireland that have memberships have tiny memberships, even accounting for differences in population size; no eNGO in Ireland has more than 20 000 paying members.[5] As such, for advocacy and case work on nature conservation, many NGOs rely mainly on core funding provided by the government. However, in 2019, for

[4] Recently rebranded as a 'climate solutions company' in view of its decision to cease peat extraction: www.bordnamona.ie/who-we-are/overview/. However, the company continues to burn its large stockpiles of peat in Edenderry power station.

[5] BirdWatch Ireland (2022), which claims to be the 'largest independent conservation organisation in Ireland', has more than 15 000 members.

example, the 32 NGOs comprising the Irish Environmental Network collectively received just €461 400 in core funding, meaning an average of less than €15 000 per organisation per year. With the Green Party in coalition government since 2020, the total provided by way of core funding (to, now, 34 NGOs) had risen to just over €1 million by 2022, meaning just under €30 000 per organisation per year. Despite this welcome increase, this core funding remains insufficient, meaning that a particularly large proportion of the work is done by volunteers.

Public participation opportunities in environmental decision making in Ireland vary in quality depending on the consent regime. The most effective opportunities for public participation are in the field of planning permission, where planning applications and opportunities for participation are typically well advertised. This is not to say that the public typically feels that their concerns will be reflected in the ultimate decision – often this is not the case – but the planning regime at least provides for transparent and timely opportunities to be heard and to appeal; unusually, Ireland provides for both first party and third party appeals in respect of planning permission decisions. However, in many other consent regimes, public participation has historically functioned less well, such as in relation to forestry and EIA, as discussed in Section 19.6 and online Annex IE.1.[6]

A final point to mention in this area is the ability of the public and NGOs to access justice in nature conservation matters by taking cases to court. Ireland's standard rule on legal costs is that the 'loser pays'.[7] Under this system, an individual or NGO taking a legal action and losing could easily face a bill of more than €100 000,[8] which is clearly prohibitively expensive. In light of obligations under the Aarhus Convention and EU law, Ireland sought to address this by introducing a system of 'costs shifting', which has improved access to justice, albeit proposals have recently been made to change this system (described further in online Annex IE.2).

19.3.3 Legislation and Policies

Modern wildlife law begins in Ireland with the Wildlife Act 1976, which repealed the two major pre-existing pieces of legislation (the Game Preservation Act 1930 and the Wild Birds Protection Act 1930) as well as other earlier legislation (Comerford, 2001). Credit for the Wildlife Act 1976 lies largely with John Temple Lang, an Irish lawyer and ornithologist who would later draft both the Birds and Habitats Directives while working for the European Commission Legal Service in Brussels (Jackson, 2018). As such, Ireland's transposition of the Birds Directive is slightly unusual, in that the Wildlife Act 1976 – which pre-dates the Directive – anticipated at least parts of the Birds Directive.

Ireland first attempted to rely on the Wildlife Act 1976 as its transposition of the Birds Directive, before realising that more would be needed. However, the Irish Regulations made subsequent to the Birds Directive 'do not represent a unified corpus of legislation, as do the regulations made in 1997 to implement the [Habitats Directive]' (Comerford, 2001). The legislature first adopted the European Communities (Conservation of Wild Birds) Regulations 1985, establishing Ireland's first four SPAs, followed by a whole series of amendment regulations in later years adding further SPAs, as well as other statutory instruments giving effect to the Birds Directive in other ways (see Comerford, 2001, pp. 255–265). The result is a complex transposition of the Directive.

Having made very slow progress with the Habitats Directive, Ireland finally secured transposition in February 1997, three years late, via the European Communities (Natural Habitats) Regulations 1997. However, opposition to the Regulations (Laffan and O'Mahony, 2008) and weak political will led to poor implementation (Comerford, 2001). The Regulations were amended several times before being revoked and replaced by the European Communities (Birds and Natural Habitats) Regulations 2011.

The 2011 Regulations, which now represent Ireland's central transposition of the Nature Directives, were introduced explicitly in response to several adverse CJEU judgments regarding poor transposition and implementation of the Nature Directives, including a major Birds Directive case, C-418/04 *Commission* v *Ireland* (DAHG, 2017). This epitomises the tendency to govern reactively in response to domestic and CJEU cases and judgments. At EU level these have included, for example, infringement proceedings (e.g. Cases C-67/99, C-117/00, C-418/04, and C-183/05) and CJEU preliminary rulings following requests from Irish domestic courts (e.g. Cases C-258/11 (Sweetman), C-470/16 (NEPPC), C-164/17 (Grace and Sweetman), C-323/17 (People Over Wind and Sweetman), C-461/17 (Holohan and others), C-254/19 (FIE), and C-721/21 (Eco Advocacy)). Such proceedings have typically resulted in positive changes in Ireland, albeit these have often taken a long time to secure.

Ireland's transposition is complicated by the fact that, in parallel to the 2011 Regulations, it has sought to transpose the Nature Directives via legislative amendments to a range of other consent regimes, including for example planning permission,[9] sea fisheries,[10] forestry,[11] and foreshore development.[12]

6 All annexes are free to access online at CUP: www.cambridge.org/natureconservation

7 Order 99 of SI 15/1986, as amended.

8 For example, see the Klohn case, described by Simons (2012) as a 'common or garden planning judicial review', in which Mr. Klohn fell liable to pay almost €120 000 on losing his case (€32 000 to his own legal team and €86 000 for the respondent's costs): see the Opinion in Case C-167/17 Klohn. Following the CJEU's judgment in that case, the Supreme Court reduced Mr. Klohn's liability for the respondent's costs from €86 000 to €1 250: *Klohn* v *An Bord Pleanála* [2021] IESC 51.

9 Via the Planning and Development Act 2000 (as amended) and Planning and Development Regulations 2001 (as amended).

10 European Union (Birds and Natural Habitats) (Sea-fisheries) Regulations 2013.

11 Forestry Act 2014, Forestry Regulations 2017, and Forestry (Miscellaneous Provisions) Act 2020.

12 Foreshore Act 1933.

The Strategic Action Plan for renewal of the NPWS published in 2022 includes a commitment to 'provide updated and stronger, statutory underpinnings for [Ireland's] National Parks and the work of the NPWS in protecting and conserving threatened and endangered animals, plants and habitats in the State' (DHLGH, 2022a).

Ireland's three National Biodiversity Action Plans (NBAP) – covering 2002–2006, 2011–2016 (after a gap of 5 years), and 2017–2021 – are the major domestic policy interventions, accompanied by some Local Biodiversity Action Plans; a public consultation on a fourth national plan opened in summer 2022. That said, countless other national plans and policies are relevant to biodiversity (e.g. see NPWS, 2014), and reacting to litigation has been at least as important a driver of policy as forward planning. An interim review of the 2017–2021 NBAP, published in 2020, concluded that of the 119 actions, 8 were implemented, 98 were ongoing, and for 13 there had been limited progress, although the review noted that many of the actions were of an open-ended nature (Biodiversity Working Group, 2020). The National Biodiversity Forum in turn audited this 2020 interim review and in February 2021 published a report containing five headline recommendations centred around: improving the governance and stewardship of biodiversity; financing the NBAP and prioritising actions; protecting, recovering and renewing biodiversity and ecosystems; building, connecting and using the biodiversity knowledge base; and engaging communities and the private sector (DHLGH, 2021a).

19.4 Species Protection and Conservation Measures

The Wildlife Act 1976, as amended in 2000, 2010, 2012,[13] 2018 (by the controversial Heritage Act 2018, which enables the Minister to extend the period during which burning and vegetation cutting is permitted) and 2021,[14] remains Ireland's main species protection legislation, alongside the 2011 Regulations discussed earlier. However, in the wake of the 1976 Act there was limited species protection action in practice (Temple Lang, 1988; Comerford, 2001).

Key actions focussed on particular species or groups have included a series of species action and threat response plans, with priority given to species of highest conservation concern (NPWS, 2020a). A Curlew Task Force was also established in Ireland in 2017, resulting in an agreed set of recommendations (Curlew Task Force, 2019).

Government-funded habitat-based action plans have also been developed for groups of birds to address the CJEU's finding in C-418/04 that Ireland had failed to transpose and apply obligations to protect bird habitats outside the SPA network. Implementation has been led by the NGO BirdWatch Ireland, adopting a multi-stakeholder approach.

Programmes to re-introduce three raptor species – Golden Eagle (*Aquila chrysaetos*), White-tailed Sea-eagle (*Haliaeetus albicilla*) and Red Kite (*Milvus milvus*) – have met mixed success (EPA, 2016). The persecution of raptors that led to the local extinction of such species has unfortunately not entirely disappeared (NPWS, 2020b). More generally, wildlife crime remains of serious concern in Ireland.

Other species initiatives have included: an *All-Ireland Pollinator Plan* (in two phases, 2015–2020 and 2021–2025) (NBDC, 2015, 2021); monitoring and conservation efforts by NGOs and the NPWS in relation to Ireland's bat species (Aughney *et al.*, 2018); draft Freshwater Pearl Mussel plans drawn up pursuant to the WFD (EPA, 2009), which have remained in draft since 2009; and EU LIFE-funded projects covering, for example, a number of raptors, Roseate Tern (*Sterna dougallii*), Corncrake (*Crex crex*), Freshwater Pearl Mussel, Atlantic Salmon, Brook Lamprey (*Lampetra planeri*) and Natterjack Toad (*Epidalea calamita*).

Further details of these actions and the species covered are provided in online Annex IE.3.

19.5 Protected Areas and Networks

19.5.1 Protected Area Types and Their Coverage

Prior to 1970, according to Craig (2001), Ireland had just one National Park (Killarney) and no other legally protected areas dedicated to nature conservation (though cf. Temple Lang's (1970) list of nature reserves). Thus, the Nature Directives have clearly transformed nature conservation in Ireland. The Natura 2000 network now covers 13.2% of Ireland's terrestrial area plus about 2.1% of its marine area. The comprehensive reach and detail of the Nature Directives, along with the supporting guidance, case law and enforcement mechanisms, far outweigh national conservation efforts outside the Directives. Ireland's terrestrial network of Natura 2000 sites is now considered almost complete, with the NPWS (2021a) reporting in March 2021 that four remaining SPAs were awaiting formal classification (five were classified later that year), and that two additional active raised bog sites were due to be proposed for designation as SACs.[15] However, according to the NPWS (2021a), it would appear that significant site designation work may remain in the marine area: discussion continues between the Irish Government and the European Commission over additional HD sites for reefs (1170) and Bottle-nosed Dolphins (*Tursiops truncatus*) as well as for SPAs, particularly in the offshore marine area.

[13] The Wildlife (Amendment) Acts 2000, 2010 and 2012.

[14] The Planning and Development, Heritage and Broadcasting (Amendment) Act 2021.

[15] N.B. Many of Ireland's existing proposed SACs, having been adopted by the European Commission as SCIs, still await formal designation at the national level, but they are subject to protection as 'European Sites' under Irish legislation in the meantime – as discussed later.

Table 19.4 The number and extent of international designated sites and protected areas in Ireland.

	IUCN category	Number	Area (km²)
Designated under global and non-EU legislation			
Ramsar Sites	NR	45	670
Biosphere Reserves (exc. buffer and transition zones)	NAs	2	955
OSPAR Marine Protected Areas	NR	19	4 136
Designated under the Nature Directives			
SCIs (terrestrial + marine)	NR	439	7 162 + 9 782
SPAs (terrestrial + marine)	NR	165	4 312 + 1 660
Natura 2000 net terrestrial area	NR	–	9 225
Natura 2000 net marine area	NR	–	10 256
Designated under domestic legislation			
Natural Heritage Areas	IV	148	600
Nature Reserves	NR	80	189
Refuges for Fauna	NR	7	–
Wildfowl Sanctuaries	NR	68	–

Note. NR = Not reported; NAs = not applicable.
Sources. See Section 1.3. Based on data from: Ramsar Convention and UNESCO websites (September 2021); Marine Protected Area Advisory Group (2020); EEA Natura 2000 Barometer, as at end 2020; domestic sites from EPA (2001), DCHG (2016), EEA (2020), NPWS (2020c); IUCN categories from WDPA (March 2022).

Hickie (1997) lists the following protected area designations: SPAs, SACs, Natural Heritage Areas (NHAs, a domestic site-protection system akin to the UK's Sites/Areas of Special Scientific Interest), Ramsar Sites, statutory Nature Reserves, National Parks, Refuges for Fauna, Wildfowl Sanctuaries, Biogenetic Reserves and UNESCO Biosphere Reserves. Of these, SPAs, SACs, NHAs (and proposed NHAs) are the most numerous and prominent in practice (see Table 19.4).

The Wildlife Act 1976 provides for the designation of Nature Reserves on state-owned and privately owned land (where the private owner desires this), as well as Refuges for Fauna, and a relatively small number of sites have been established in this way. The reason for these two separate categories and the differences between them is not entirely clear. The designation 'Refuge for Fauna' is in practice little used (just seven are recorded on Ireland's statute book, all designated in the late 1980s or early 1990s to protect breeding birds on marine cliffs and small islands) and arises where the Minister considers that a particular species of fauna or flora should be specially protected on any land that is, or is contiguous to, a habitat of the species, or that land has features of the landscape that are of major importance for wild flora and fauna including birds, including as corridors or stepping stones. The seven designation orders set out site-specific prohibited activities in some detail. Rather more Nature Reserves have been established, though again the power was used predominantly from 1980 to the early 1990s; the designation orders tend simply to designate the land without specifying in detail the activities that are prohibited. The same general protective obligations apply to public bodies in respect of both Refuges for Fauna and Nature Reserves under s.12 of the Wildlife Act 1976.

In contrast, Ireland's six National Parks are not currently underpinned by dedicated domestic legislation,[16] 'giving rise to issues in relation to management of the parks by the NPWS' (Grant Thornton, 2010). The government has recently committed, in the *Strategic Action Plan* for renewal of the NPWS, to 'provide updated and stronger, statutory underpinnings for [Ireland's] National Parks' (DHLGH, 2022a), with a general scheme of the legislation to be published by June 2023. National Park status has not to date resulted in many of the necessary conservation actions required, especially when it comes to severe problems of managing invasive species (e.g. deer and rhododendron) (DCHG, 2019). Whilst National Parks thus appear, as such, to provide little benefit for biodiversity and are not considered to be protected areas in accordance with IUCN criteria, virtually all the land in Ireland's National Parks overlaps with Natura 2000 designations.

The idea that privately owned land might be impacted involuntarily by nature conservation designations has proven very controversial in Ireland (Laffan and O'Mahony, 2008). As Craig (2001) explains, 'the first attempt at a full inventory of important areas for nature conservation was carried out by An Foras Forbartha in the 1970s, initially on a county-by-county basis. It culminated in a published national list of Areas of

[16] With the exception of the Bourn Vincent Memorial Park Act 1932, which governs part of Killarney National Park.

Scientific Interest (ASIs) [in 1981]. The [non-statutory] ASI designation was set aside as unconstitutional in a 1992 case,[17] primarily because landowners had not been properly notified that their lands were on a list of ASIs. A nationwide resurvey of former ASIs was undertaken from 1992–94 [funded under the LIFE programme] to establish which of the sites were still of significance for conservation and suitable for designation as Natural Heritage Areas (NHAs) and to map the boundaries of these sites. Over 1 100 proposed Natural Heritage Areas were publicly advertised [beginning in 1994], covering just over [9 000 km^2].'

While the identification of proposed NHAs helped to inform Ireland's selection of SPAs under the Birds Directive, and its first list of proposed SCIs under the Habitats Directive, most of the NHAs that did not attain SCI or SPA status remain merely 'proposed' today, apparently as a result of a lack of political will, and as such benefit from minimal legal protection. Indeed, only 75 raised bogs (230 km^2) and 73 blanket bogs (370 km^2) were ever fully designated as NHAs (NPWS, 2020c), and these designations arose only in settlement of infringement proceedings brought by the European Commission (Case C-294/03, brought to enforce C-392/96). Currently 630 sites (650 km^2) remain as proposed NHAs, 25 years after they were first proposed (NPWS, 2020c).[18] Further, in 2014 the government announced plans to 'de-designate' 46 of the raised bog NHAs to allow for peat extraction on these sites, to be 'replaced' by 27 currently undesignated bogs (DAHG, 2014). While the NPWS's (2020c) website states that 'designation [of the pNHAs as NHAs] will proceed on a phased basis over the coming years', there is no sign that work has been or is ongoing on this. The relevant minister commented in early 2022 that 'a review of [the pNHAs] will take place in due course, [but] the current focus of my Department is on fulfilling Ireland's obligations under the EU Habitats and Birds Directives and the designation of Natura 2000 sites' (Noonan, 2022). The discrepancy between the legal protection offered to habitats and species in proposed and designated NHAs (national protection) and those protected under the Natura 2000 network (EU protection) illustrates the weakness of domestic conservation efforts.

Ireland has been slow to formally designate its Natura 2000 sites at the national level: as at 2019, just 61% of Ireland's sites had been formally designated by way of statutory instrument (96% of SPAs but just 48% of SACs; the figure for SACs had risen to 63% by July 2020, and work remains ongoing (NPWS, 2021a)).[19] Pending their formal designation at the national level, such sites nevertheless benefit from protection as 'European Sites' under Irish legislation, a category that includes *inter alia* SACs, SCIs and candidate SCIs.

While the government reported in 2019 that terrestrial protected areas account for around 17% of Ireland's land area (DCHG, 2019), according to analysis by UNEP-WCMC (2022) of the data submitted to the WDPA, just 14.4% (10 123 km^2) of Ireland's terrestrial area is within protected areas.

At sea, its protected coverage is 2.1% based on its Natura 2000 coverage of 10 256 km^2 (Table 19.4) and its area of marine waters (see Key Statistics table). Ireland has also designated 19 OSPAR sites, but these do not appear to add to the area of marine Natura 2000 sites (according to Ireland's Sixth National Report under the CBD, see DCHG (2019, p. 64)). Purely domestic site protection designations are not a feature in Ireland's marine area; for example, some statutory Nature Reserves and Refuges overlap with Natura 2000 sites, and some coastal/marine sites were proposed as NHAs many years ago but have not yet been designated (see Marine Protected Area Advisory Group, 2020, pp. 63–67). While Ireland has not yet established a coherent network of Marine Protected Areas (MPAs) pursuant to the MSFD, a substantial report on expanding Ireland's network was produced for the government in October 2020 by the Marine Protected Area Advisory Group (2020), followed by a public consultation on the report from February to July 2021, with an independent review and analysis to follow. Thereafter, the government has stated that it intends to develop legislation on the identification, designation and management of MPAs (DHLGH, 2022b), and that it aims to expand Ireland's MPA network to cover 30% of Ireland's maritime area by 2030 (DHLGH, 2021d).

In conclusion, Ireland did not achieve the Aichi 2020 targets of 17% coverage on land and 10% at sea. Furthermore, it has a lot more to do when it comes to contributing to the *EU Biodiversity Strategy for 2030* goal of protecting at least 30% of the land and 30% of the sea within the EU.

19.5.2 The Coherence of Protected Areas and Ecological Networks

The Nature Directives impose relatively weak habitat protection obligations outside Natura 2000 (see Section 4.2). Nevertheless, Ireland is the only Member State to have been found in breach: in Case C-418/04, the CJEU found that Ireland had neither transposed nor fully and correctly applied the second sentence of BD Article 4(4), which provides 'Outside these [SPAs], Member States shall also strive to avoid pollution or deterioration of habitats.' Further, Ireland was later found in breach of the EIA Directive for failing to require EIA in respect of land-use changes impacting important corridors and stepping stones, such as hedgerows and wetlands (C-66/06).

While no nationally or globally endorsed single indicator exists to estimate the connectedness of protected areas, DCHG (2019) presents the results of two global indicators and one regional indicator in this regard. For example, the Protected Area Connectedness Index reveals that, compared to 41 other

[17] *MacPharthalain* v *Commissioners of Public Works* [1994] 3 IR 353.

[18] The figure given in DCHG (2019) for proposed NHAs is 1 089; 630 is correct for remaining stand-alone pNHAs.

[19] Legal action is ongoing about this at EU level: see Case C-444/21 *Commission* v *Ireland*.

European states, Ireland has the fifth lowest connectedness index score, and is in the bottom 15% of countries.

Beyond Ireland's contribution to the Natura 2000 network and its domestic protected areas, until recently there had been no serious attempt to develop ecological networks in Ireland. However, Action 1.1.7 of Ireland's *National Biodiversity Action Plan 2017–2021* (DCHG, 2017a) committed to developing 'Green Infrastructure at local, regional and national levels and [promoting] the use of nature based solutions for the delivery of a coherent and integrated network'. Further to this, the National Planning Framework (2018) requires as a national policy objective that integrated planning for green infrastructure and ecosystem services must be incorporated into the preparation of statutory land-use plans. This framework has a horizon of 2040, such that any on-the-ground impacts of this new policy objective should become visible in the coming years.

19.5.3 Approaches to Protection and Levels of Protection

In contrast to the UK, where Natura 2000 sites are underpinned by S/ASSIs, Ireland's Natura 2000 sites are protected almost exclusively via their status as 'European sites'. A search of the Irish Statute Book reveals just over 50 pieces of legislation that use this definition to apply protections to Natura 2000, albeit the central piece of legislation remains the European Communities (Birds and Natural Habitats) Regulations 2011.

In the planning law context, appropriate assessment is required pursuant to the Habitats Directive via Part XAB of the Planning and Development Act 2000. Such assessments are routinely performed by planning authorities informed by *inter alia* a 'Natura Impact Statement' or 'Natura Impact Report' provided by the prospective developer or plan-maker, respectively. In a leading case, the quality of such assessments was scrutinised by the High Court in *Kelly* v *An Bord Pleanála* [2014] IEHC 400, with the Court concluding that Ireland's planning appeals board did not carry out an appropriate assessment that complied with the Habitats Directive.

Another important issue has been planning authorities' apparent general reluctance to find that any plan or project fails the test in HD Article 6(3). Failing this test would of course mean either that permission would need to be refused or that the derogation tests in HD Article 6(4) would need to be satisfied for the plan/project lawfully to proceed. In that regard, it is worthy of note that Ireland has never notified the European Commission of any compensatory measures adopted under HD Article 6(4) (DG Environment, 2020), meaning by implication that Ireland has never justified a damaging project or plan on the basis that it is of overriding public interest in accordance with HD Article 6(4).

Certain developers and planning authorities now seem belatedly to have come round to the view that it may be more efficient to engage early with the potential application of HD Article 6(4) as a means of securing permission. See, for example, Galway Harbour Company's proposal to provide new berths by reclaiming 24 ha of land from marine areas within the Natura 2000 network, in respect of which the developer's proposed HD Article 6(4) compensatory measures are under consideration by the relevant minister (McGrath, 2021). The government also plans to allow peat extraction to continue in certain raised bog SACs by way of Article 6(4) derogations (DCHG, 2017b).

In addition to applying the HD Article 6 appropriate assessment provisions, the 2011 Regulations *inter alia*: impose a general duty on public authorities to exercise their functions in compliance with and so as to secure compliance with the Nature Directives; contain general provisions for the prevention of damage to European sites; and require, in certain circumstances, restoration of land in a European site following damage. Beyond the 2011 Regulations, as noted in Section 19.3, legislation governing a range of different consent regimes in Ireland applies protections to Natura 2000, including for example planning permission, sea fisheries, forestry, and foreshore development.

Legislative protection for Ireland's small number of NHAs is set out in the Wildlife (Amendment) Act 2000. The de facto protection provided for these sites is limited as a result of poor enforcement, evidenced for example by a 2014 review which revealed that just eight of the 75 raised bog NHAs had no active peat extraction plots over the previous seven years (DAHG, 2014). Limited protection has been provided for the larger number of *proposed* NHAs by a recent amendment to the planning regime requiring competent authorities to take account of proposed NHAs in screening for EIA.[20] Also, guidance and a notice have been issued to landowners in the context of infringement proceedings against Ireland relating to agricultural land-use changes without EIA (see C-279/11, para 36), by virtue of which landowners must apply to the Department of Agriculture for a screening decision if proposing to undertake certain activities that may affect a proposed NHA (DAFM, 2018).

19.5.4 Approaches to Managing Protected Areas

Site designation obligations appear to have occupied much of the Irish authorities' time since the adoption of the Nature Directives, meaning that other implementation tasks have yet to be completed, notwithstanding the passage of decades since the Directives' adoption. As at October 2020, just 24% of Ireland's SPAs and 83% of its SACs had site-specific conservation objectives (NPWS, 2021a), and the NPWS records just 43 site-specific management plans in respect of Ireland's 439 SACs (none are recorded for SPAs)(DCHG, 2019),[21] all of which appear to be out of date (covering the period 2005/6 to 2010/11) (NPWS, 2022), plus a collective management plan

[20] The European Union (Planning and Development) (Environmental Impact Assessment) Regulations 2018.

[21] That said, a management plan for 2006–2011 is listed elsewhere for the Owenduff/Nephin Complex SPA and cSAC: see NPWS (2022).

and site-specific restoration plans for the 53 active raised bog SACs (NPWS, 2021a).[22] Similarly, just two of six National Parks have had a management plan, both of which covered 2005–2009, meaning no National Park has a current plan (DCHG, 2019). That said, in its *Strategic Action Plan* for the renewal of the NPWS, the government has recently committed to developing management plans for each of the National Parks (DHLGH, 2022c). No NHAs have management plans, despite the government promising to create site-specific plans for the raised bog NHAs by 2017 (DAHG, 2014). While there is a national conservation objective for active raised bog, there are no site-specific conservation objectives for the NHAs, despite this being promised (DAHG, 2014).

Reliance on generic objectives for sites and the absence of management plans has meant that Ireland's protected area management has tended to be reactive. This has manifested itself, for example, in constrained resources being directed at dealing with infringement actions (NPWS, 2015). Other sectoral measures such as agri-environment schemes, forest management plans and fisheries plans have not in practice comprehensively addressed the gap left by sparse protected area management planning.

19.6 General Nature Conservation Measures

19.6.1 Development Planning and Impact Assessment

Most planning permission decisions in Ireland are taken by local authorities (county councils and city councils) at first instance, with the possibility of an appeal to An Bord Pleanála, Ireland's planning appeals board (the Board). The relevant legislation provides that the applicant for planning permission and any third party who made a submission or observation on the planning application may lodge an appeal with the Board (eNGOs are exempted from this prior participation requirement in certain circumstances). This general third party right of appeal is relatively unusual in an international context.

Compliance with the Nature Directives in the planning context has been heavily litigated in Ireland, resulting in a mature jurisprudence at the national level.[23] While this has undoubtedly led to a learning process on the part of decision makers, planning permission decisions are still regularly overturned in court for failure to comply with the Nature Directives. In terms of SEA, a recent review concluded that the SEA process is now well 'bedded in' and that it is leading to changes in plans/programmes, but that challenges remain, including in respect of the consideration of alternatives and monitoring (EPA, 2020a).

Over the years certain 'politically sensitive' land uses have been taken outside the planning permission regime altogether (e.g. many agricultural land uses, afforestation, and – most recently – large-scale peat extraction). In such cases, dedicated consent regimes have been created that have tended to provide less effective opportunities for public participation (agriculture and forestry examples are provided in online Annex IE.1). Particular problems have arisen with the use of peat, as virtually all of the many companies and individuals extracting peat in Ireland have never obtained planning permission, nor have EIAs been carried out. This is despite the legal requirement to obtain permission for extraction that requires EIA or appropriate assessment on or after 20 September 2012 having been confirmed by the High Court. The government reacted by legislating to take large-scale peat extraction (>30 ha) outside the planning permission regime altogether, to be replaced with a licensing system to be operated by the EPA. Following a legal challenge by FIE, this new legislation was struck down by the High Court in 2019 on the basis that the State had unlawfully granted peat companies an 'enforcement holiday' when it came to EU law, by allowing such companies to continue extracting peat without permission, EIA or appropriate assessment for more than 18 months while they sought a licence under the new system.[24] This has left peat companies in Ireland with significant regulatory hurdles to overcome if they wish to continue peat extraction lawfully. In early 2021, the largest peat company in the country – the semi-state Bord na Móna (2021) – announced a formal end to all peat extraction on its lands.

19.6.2 Habitat Management and Restoration

Most practical habitat conservation measures tend to focus on habitat management and maintenance, rather than proactive restoration (DCHG, 2019), and cover a range of habitats, including notably grasslands and peatlands, as discussed in the following paragraphs. Conservation measures have also led to some improvements in the condition of some coastal marine habitats, four of which are described in Landgrebe-Trinkunaite (2019).

Species-Rich Grasslands and Agri-environment Schemes

Of mounting concern is the ongoing intensification of agriculture in Ireland, in particular in dairy, which has driven significant increases in cattle numbers, increasing chemical fertiliser use and associated losses in grassland biodiversity. Wider impacts are also resulting from pollution of aquatic ecosystems by nitrogen and phosphorus and airborne deposition of ammonia on sensitive habitats.

[22] Legal action is ongoing at EU level regarding Ireland's failure to establish conservation objectives and conservation measures: see Case C-444/21 *Commission* v *Ireland*.

[23] For leading judgments, see *Kelly* v *An Bord Pleanála* [2014] IEHC 400 and *Connelly* v *An Bord Pleanála* [2018] IESC 31.

[24] *Friends of the Irish Environment* v *Minister for Communications, Climate Action and the Environment & Ors.* [2019] IEHC 555 and [2019] IEHC 685.

As such, grassland conservation measures have been ıded in successive iterations of national agri-environment ·mes (AES) under the CAP. A study by Ó hUallacháin *et al.* .6) that compared the various grassland schemes found that absence of specific, measurable objectives made it difficult to clude the extent to which the measures have been effective. ertheless, the authors considered that the schemes' effect- ıess in supporting the maintenance and restoration of bio- ersity in species-rich grasslands had been hampered by: the ence of specific management prescriptions; a lack of geting to sites with the greatest environmental potential; lemanding eligibility criteria; and a failure to differentiate ıncially between management prescriptions that deliver ater benefits for species richness and thus biodiversity value. ıilar criticisms were made by the NPWS (2013), and echo blems in other EU countries (see Section 5.5.2).

As a result, with the exception of some more targeted jects like BurrenLIFE (described later), AranLIFE and the IDE (2020) project, AES in Ireland have generally been atively poor at delivering the objectives of the Nature rectives and wider biodiversity benefits. As noted by the PWS (2013), despite €2.18 billion having been distributed farmers in Ireland under CAP environmental schemes tween 1994 and 2006, the schemes failed to deliver sufficient otection for Ireland's biodiversity and have not ensured that osystems can support a vibrant agricultural sector in the ng term. Further, the NPWS expressed frustration that its st submissions to the Department of Agriculture 'have not rticularly influenced the selection of measures for Natura nds'. A more general criticism of such schemes is that undes- nated land of conservation value is not adequately protected hen a farmer is not in an AES, and 'intensive dairy and arable rms in the south and east have often chosen to remain utside AES' (DCHG, 2019).

A 2019 review of the current general AES scheme ('GLAS'), ı operation since 2014, found that around 35% of farms articipated in the scheme, with participation highest in sheep arms (52%) and lowest in dairy farms (13%) (IGEES, 2019). he scheme has a three-tier hierarchy, with the first tier argeting a number of Priority Environmental Assets – 'pri- narily vulnerable landscapes (including Natura and uplands), pecies at risk (primarily endangered birds), and high-quality watercourses' (ADAS, 2020). Presence of one or more of these ssets on a farm guarantees priority access to the scheme, but environmental assets and actions of lesser importance can also be addressed. GLAS payments are annual and generally up to a maximum of €5 000,[25] with the majority of participants claiming close to the maximum payment (e.g. the average payment was €4 204 in 2017) (IGEES, 2019). Thus, the level of funding at individual farm level is relatively low, with the money being spread widely and relatively evenly amongst scheme participants.

25 'Farmers who undertake particularly challenging actions or who are compelled to take on a high number of compulsory actions, may qualify for an additional payment of up to €2 000 per annum under GLAS+' (ADAS, 2020).

The 2019 review, carried out by the Irish Government Economic and Evaluation Service, stated that it had found evidence that the scheme is effective at targeting high-quality water areas and Natura 2000 sites based on their distribution in regions with high GLAS participation. The review found low participation among dairy farms, which 'have the highest average GHG emissions, highest nitrogen surplus in the soil and the largest farms on average'. In terms of overall effective- ness, the review noted that low additionality and the import- ance of maintaining positive behaviour had been identified as common issues. Regarding additionality, for example, the review noted that some of the actions funded under GLAS may already be in place on participating farms, with addition- ality ranging from 20% to 67%.

A separate evaluation in 2020 suggests that targeting remains an area where improvements could be made (ADAS, 2020). The evaluation cites the idea of targeting 'biodiversity hot-spots' with higher payments as a 'really positive sugges- tion', and recommends reviewing the scope for AES actions to address environmental priorities in water quality through better targeting of the dairy and arable sectors.

Operating alongside the general AES, targeted projects such as BurrenLIFE have delivered benefits for grasslands (DCHG, 2019). A key component of the BurrenLIFE project was the development of a results-based AES approach with payments to participating farmers according to the delivery of measurable biodiversity objectives (e.g. flower indicator species diversity). This has received much positive attention for its success in engaging farmers and other stakeholders in delivering tangible socio-economic and environmental dividends for the region (Dunford, 2016). Indeed, the scheme's success was marked in 2017 with the European Commission recognising BurrenLIFE as the joint best LIFE project in the 25-year history of the programme (European Commission, 2017). The emergence of BurrenLIFE and the way it and its successor programmes have operated are described in detail in Dunford and Parr (2020). The current Burren Programme works with over 300 farmers and forms part of a national programme funded under Ireland's Rural Development Programme.

Following this success, research and pilot schemes have developed and trialled similar results-based approaches in other HNV farming systems, including in Ireland and in Spain (RBAPS, 2019). Some examples are briefly outlined in online Annex IE.4, and further described in O'Rourke and Finn (2020). In early 2021, the government went out to public consultation on a proposed Agri-Environment Results Based Pilot Project to *inter alia* test the suitability and the feasibility of upscaling a results-based model as part of the next national AES to follow on from GLAS.

Peatland Management and Restoration

Ireland contains an estimated 1.7 million ha of peatlands, about 25% of the land area (Renou-Wilson, 2018). Protection of undesignated peatlands for the ecosystem services they provide has until very recently not been pursued in Ireland. Despite the adoption of a *National Peatlands Strategy* in 2015 (DAHG, 2015), there is no decipherable national strategy for peatland protection or restoration outside Natura 2000 and NHA sites, and the actions in the strategy in respect of peatlands outside protected sites remain largely unimplemented. Instead, peat is still being used for electricity generation, for domestic heating, for horticulture, and for use as animal bedding, amongst other things, with around 100 000 ha of peatland utilised for industrial peat extraction in Ireland as at 2018 (Renou-Wilson, 2018). As described earlier, peat extraction remains largely unregulated on undesignated bogs. This said, as mentioned previously, in early 2021 Bord na Móna announced a formal end to all peat extraction on its lands, and in tandem the government and Bord na Móna announced a €126 million project aimed at rehabilitating approximately 40% of Bord na Móna's landholding of 80 000 ha (on top of the approximately 15% that has already been rehabilitated) (Government of Ireland, 2020). Two of Ireland's three peat-fired power stations closed in 2020, and the remaining one (Edenderry) seems likely to stop burning peat when its current permission expires in 2023.

The unregulated situation on undesignated bogs is mirrored in Ireland's SACs and NHAs, many of which remain subject to unlawful extraction ('turf cutting') for domestic home heating purposes. The major peat-extracting companies ceased extracting from these sites when they were designated. What remains is extraction by individuals or (mainly) by hired contractors, mostly with heavy machinery. Ireland's 50 blanket bog SACs remain effectively unregulated when it comes to extraction. In 1999, the government unilaterally granted a 10-year 'derogation' to allow peat extraction to continue in Ireland's then 53 raised bog SACs, without applying the provisions of HD Article 6. In 2009, the then government announced it was putting in place a consultative process to prepare for the phasing out of peat extraction from the raised bog SACs and NHAs. Having made a written submission in response, and dissatisfied with progress, the NGO FIE complained to the European Commission and carried out site visits in 2011 and an aerial survey in 2012, producing reports for the Commission that led to accelerated infringement proceedings. FIE kept the pressure on the European Commission and the government by lodging a complaint with and participating in hearings before the European Parliament's Petitions Committee. This work helped to trigger the establishment by the government of a Peatlands Council in 2011, and led ultimately to the *National Peatlands Strategy* and a *National Raised Bog SAC Management Plan* (DCHG, 2017b). Notwithstanding a compensation and relocation scheme established as part of the government's response, extraction continues unlawfully on Ireland's blanket bog SACs, on 18 of the raised bog SACs, and on six raised bog NHAs (NPWS, 2019; IWT, 2021).

Against this difficult backdrop, several LIFE-funded projects restoring peatlands through tree removal and other actions have been carried out in SACs. In order to address public resistance to the cessation of peat extraction from such areas, community engagement has been an important component of these projects. In addition, in support of Ireland's *Climate Action Plan 2019* (Government of Ireland, 2019), two major LIFE integrated projects with a focus on peatlands commenced in 2020: Wild Atlantic Nature (aiming to restore blanket bogs in 24 Natura 2000 sites), and Peatlands and People.[26]

19.7 Nature Conservation Costs, Benefits and Funding Sources

19.7.1 Nature Conservation Costs

While the National Biodiversity Expenditure Review (NBER) (Morrison and Bullock, 2018) examined expenditure in the period 2010–2015, there has not been a recent comprehensive assessment of the financial needs for conservation in Ireland. However, a 'Financial Needs Assessment' is currently in preparation.[27]

In completing its Prioritised Action Framework (PAF) for Natura 2000 for 2014–2020, Ireland provided cost estimates for its financial needs for Natura 2000 implementation using figures originally sent to the European Commission in 2008, without updating them (NPWS, 2014). Those 2008 figures indicated a total future cost for achieving favourable conservation status in the Natura 2000 network in Ireland of €185 million per annum, the seventh highest estimate of the then 25 Member States (IEEP, 2010). The PAF for 2021–2027 estimates an annual cost of €162 million, which unlike the previous PAF also includes additional costs beyond Natura 2000 for green infrastructure and species-specific measures (NPWS, 2021a).

19.7.2 The Economic Benefits of Nature Conservation

A scoping report in 2008 estimated that Ireland benefits from at least €2.6 billion worth of free goods and services from biodiversity per year (Bullock *et al.*, 2008). Whilst there is no more recent assessment of such economic benefits, Morrison and Bullock (2018) note that the mapping of Ireland's natural capital and ecosystem services, as part of the EU MAES initiative (Section 4.4), could be used to boost arguments for public investment in biodiversity.

19.7.3 Funding Sources for Nature Conservation

Over the six-year period 2010–2015, total national expenditure on biodiversity-related activities is estimated at €1.5 billion, with an annual average expenditure of €250 million, representing an average of just 0.3% of total government expenditure (Morrison and Bullock, 2018). As NPWS (2021b) notes, in this

[26] www.wildatlanticnature.ie and https://peatlandsandpeople.ie
[27] https://biodiversityfinance.ucd.ie

period the NPWS had only 10% of the state funds spent on biodiversity, with 80% being allocated to Department of Agriculture schemes that have proven of limited effectiveness for biodiversity, as noted in the previous section. McGuinness and Bullock (2020) have also commented that 'The prioritisation of terrestrial biodiversity spending over the marine is … starkly apparent.'

EU co-financed funds have provided the vast bulk of financing for nature conservation in Ireland over recent decades, and in particular the European Agricultural Fund for Rural Development (EAFRD), as indicated in Table 19.5 for the 2014–2020 funding period. However, such spending has not generally been effective at biodiversity conservation, and the NBER itself states that agricultural spending generally has been linked to trends towards increased intensification or specialisation (Morrison and Bullock, 2018). While the European Regional Development Fund (ERDF) is a significant potential source of biodiversity funding (with an allocation of €821 million over the 2014–2020 period), the latest PAF does not record any ERDF funding used for nature conservation purposes, despite the fact that Ireland significantly underspent its overall allocation (McGuinness and Bullock, 2020). It is unclear why this was the case. Interreg provided relatively significant funding in the 2014–2020 period, including for projects in which NGOs were partners such as CABB (Cooperation across borders for biodiversity) and CANN (Collaborative action for the Natura network).

Table 19.5 sets out some of the major funding sources for nature conservation in Ireland. Although this does not include funding from NGOs, foundations and business sources, due to insufficient available information, this is unlikely to have significantly reduced the estimated total. More comprehensive accounts of biodiversity funding can be found in Morrison and Bullock (2018) and McGuinness and Bullock (2020).

As discussed earlier, the EU LIFE programme has supported a number of important and effective biodiversity projects. Despite their success, Ireland has had a relatively low number of LIFE nature projects, due principally to a lack of national co-funding for projects. Nevertheless, Ireland has recently secured three large LIFE Integrated Projects, including one that aims to support the implementation of the PAF with respect to blanket bogs, including through their restoration in 24 Natura 2000 sites[28] (€20.6 million with €12.4 million in EU funding). Significant contributions to biodiversity conservation can also be expected from an Integrated Project on peatlands and people,[29] in support of Ireland's *Climate Action Plan* 2019 (€27.8 million with €9.9 million in EU funding), and an Integrated Project aimed at protecting and restoring high ecological status water bodies[30] (€20.4 million, with €9.5 million in EU funding). The government also obtained funding to prepare a Strategic Nature Project (SNaP) funding application, with the full application to be submitted in 2022, to be used 'to integrate and co-ordinate conservation actions in the biodiversity, water, agriculture and marine sectors' (NPWS, 2021a).

The NPWS's budget was severely cut after the onset of the financial crisis: its budget fell from €47 million in 2008 to just €14 million in 2020 (a fall of 71%) (NPWS, 2021b), which compared unfavourably with, for example, government spending on the greyhound industry (€17 million) and horse racing industry (€67 million) (Costello, 2021). Equally, the National Biodiversity Data Centre (NBDC) operates with a small annual budget of c. €500 000 and with considerable precarity (Sargent, 2021). Overall, 'the IUCN has called for all OECD countries to contribute at least 0.3% of their GDP for biodiversity conservation compared to [the] average 0.13% GDP that Ireland currently spends' (Morrison and Bullock, 2018).

Under Ireland's current government, with the Green Party forming part of the coalition, the NPWS's budget for 2022 increased to €47 million (DHLGH, 2021b), bringing it back up to the 2008 level. Further, in May 2022, as part of the Strategic Action Plan for the renewal of the NPWS, the government announced an additional injection of 'up to €55 million' over the following three budgetary cycles to realise the plan (DHLGH, 2022a).

Despite the recent history of relatively low national funding, when the EU sources are taken into account (as given in Table 19.5) it appears that the level of available financing could be potentially sufficient to meet the priority financing needs of €162 million per year estimated in the PAF for 2021–2027 (covering Natura 2000, additional green infrastructure measures beyond Natura 2000, and certain additional species-specific measures). However, the actual adequacy and effectiveness of the funding for meeting Natura 2000 objectives is doubtful, as a substantial proportion (43%) is allocated to payments to areas with natural constraints (M13), which is not included as a relevant measure for Natura 2000 according to the current PAF (NPWS, 2021a). Furthermore, the rest of the funds are primarily allocated to the agri-environment–climate measure, which, as described earlier, has been of variable effectiveness and relevance for Natura 2000. It is also notable that funding for marine Natura 2000 sites is very low. If the adequacy of Natura 2000 funding is doubtful, it seems highly unlikely that funding will be sufficient for broader nature conservation requirements.

19.8 Surveillance, Monitoring, Reporting and Applied Research

According to the EPA (2016), robust baseline biodiversity monitoring systems and comprehensive ecosystem services mapping systems are needed to highlight and protect nature in Ireland, ideally coordinated by a single lead organisation with ring-fenced funding. The closest Ireland comes to this is the regular reporting required under the Nature Directives (plus ongoing work in respect of the EU MAES initiative).

[28] LIFE IP PAF-WILD ATLANTIC NATURE (LIFE18 IPE/IE/000002).

[29] LIFE IP Peatlands and People (LIFE19 IPC/IE/000007).

[30] LIFE-IP Waters of Life (LIFE18 IPE/IE/000003)

Table 19.5 Funding sources for nature conservation in Ireland.

Source and period	Measure/category	Average annual allocation (€ m)		Comments
EU funding instruments		**EU + national**	**EU**	
EAFRD – Rural Development Programme measures – allocations to Priority 4 (restoring, preserving and enhancing agriculture and forest ecosystems) 2014–2020	M13 Areas facing constraints	196.4	104.1	
	M10 Agri-environment-climate	164.8	96.8	On GLAS, Burren Programme and older AES schemes. Inc. for N2k, etc.
	M4 Investments in physical assets	14.4	7.7	
	M11 Organic farming	8.0	4.2	
	M12 Natura and WFD	6.9	4.7	For N2k, etc. Transitional payments: old AES schemes
	M16 Cooperation	6.1	3.4	Locally led projects, e.g. Hen Harrier and Freshwater Pearl Mussel. Some for N2k, etc.
	M1 Knowledge transfer, etc.	4.2	2.2	Training in support of GLAS agri-environment scheme
	M7 Basic services, etc.	0.9	0.5	
	M2 Advisory services, etc.	0.1	0.1	
RDP Priority 4 subtotal		*401.8*	*223.8*	
Interreg 2014–2020		13.2	10.7	For N2k, etc.
LIFE Programme	Nature and biodiversity 1992–2020	2.1	1.3	21 projects
	LIFE Integrated Projects 2020–2028	2.3	1.4	LIFE IP PAF-WILD ATLANTIC NATURE. Two others will also contribute to biodiversity – see text
EMFF 2014–2020	UP1 and UP6	1.3	0.7	For Marine Biodiversity Scheme (2014), expected to be fully used
EU funding instruments subtotal		*420.7*	*237.9*	
National funds within 2014–2020		**Additional national***	**Comments**	
NPWS staffing costs (2014–2020)		7.9	For N2k, etc.	
Bog restoration, relocation, compensation (2014–2020)		7.8	For N2k, etc.	
Forestry (2010–2015)		7.6	See Comment 1 below this table.	
Scientific monitoring and research (2014–2020)		2.7	For N2k, etc.	
NPWS capital expenditure (2014–2020)	National Parks and Nature Reserves	2.5	For N2k, etc.	
Environment Fund (2014–2019)	From levies on plastic shopping bags and the landfill of waste	1.7	For Environmental Awareness. Not all of this spending is on biodiversity	
NPWS Farm Plan Scheme and other NPWS agri-ecology related projects (2014–2020)		1.4	For N2k, etc.	
Species Protection projects in Natura 2000 (2014–2020)		0.2	For N2k, etc.	

Table 19.5 *(cont.)*

Source and period	Measure/category	Average annual allocation (€ m)		Comments
EU funding instruments		**EU + national**	**EU**	
Inland Fisheries Ireland HD programme (2014–2020)		0.3		For N2k, etc.
Additional national total subtotal		*32.3*		
National total (inc. co-financing of EU instruments)		*215.2*		
Total (EU + all national financing)		**>453.1**		Incomplete – see text

Notes. * Excludes co-financing of EU funds unless indicated. 'For N2k, etc.' indicates funding relevant to Natura 2000, related green infrastructure and BHD species (based on PAFs). The subtotals and totals are only approximate indications of the average annual levels of funding as they are based on annualised data from differing periods. *Comments.* 1. Source Morrison and Bullock (2018), where uncertain if this includes any EU co-financing, but assumed no co-financing, as Ireland does not use any RDP forest measures (European Commission Factsheet on 2014–2022 RDP for Ireland, 2022[31]).
Sources. RDP data from European Commission ESIF funds database (23 October 2020); Ireland LIFE factsheet, European Commission (September 2020) and LIFE database; PAFs (NPWS, 2014, 2021a), Morrison and Bullock (2018), DAFM (2020b), McGuinness and Bullock (2020), DECC (2020). See Section 1.3 for further details.

The NPWS, in collaboration with the Northern Ireland Environment Agency, has published 13 Red Lists since 2006 (NPWS, 2020d). There have in addition been four published lists of Birds of Conservation Concern in Ireland (covering the island of Ireland), the most recent in 2021 (Gilbert *et al.*, 2021). Additionally, every 20–25 years a bird atlas is completed, mapping the occurrence of Ireland's and Great Britain's birds during winter and breeding seasons (Balmer *et al.*, 2013). It provides an excellent example of citizen science in action, using records submitted by thousands of volunteer birdwatchers to produce species distribution maps for all species, as well as relative abundance maps for more common species and analyses of change since the previous atlases.

A number of NGOs are involved in additional citizen science work, as for example noted by DCHG (2019) in relation to cetaceans (by the Irish Whale and Dolphin Group), bats (the Vincent Wildlife Trust and Bat Conservation Ireland), and the garden bird survey, countryside bird survey and wetland bird counts (managed by BirdWatch Ireland), as well as the frog survey (by the Irish Peatland Conservation Council).

As well as operating the *All-Ireland Pollinator Plan* (discussed earlier), the NBDC delivers a range of citizen science initiatives, including for example the All-Ireland Bumblebee Monitoring Scheme, the Irish Butterfly Monitoring Scheme, the Butterfly Atlas 2021, Dragonfly Ireland 2019–2024 (also covering damselflies), various specific schemes relating to pollinators,[32] the Spring Flowers Project, and an Irish Hedgehog Survey. Work is advanced on a National Vegetation Database, established by the NBDC in conjunction with the NPWS.

In terms of marine biodiversity, in addition to the Irish Whale and Dolphin Group's work on cetaceans, other citizen science initiatives include the Explore Your Shore project funded by the EPA, which aims to increase knowledge of the distribution of intertidal species, and Seasearch Ireland, a recording scheme for divers and snorkelers. Others provide an overview of the state of the coast (Coastwatch) or compile records of certain species, including: egg cases of sharks, skates and rays; Basking Sharks (*Cetorhinus maximus*); jellyfish; kelp (brown algae in the order Laminariales); and non-native species.

The National Platform for Biodiversity Research provides recommendations for research on agriculture, grasslands, soils, freshwater, the marine, uplands, wetlands and peatlands, and invasive species (BEC Consultants, 2012). More recently, NPWS (2021a) has produced a non-exhaustive list of research required to enable effective management and monitoring of habitats and species listed in the Nature Directives. Research funding is available from a range of sources, summarised in DCHG (2019), which also carries a useful summary of applied research projects.

19.9 Future Developments

Encouragingly, there are indications that public support for nature conservation is increasing in Ireland, with a 2019 EU Barometer finding, for example, that Ireland has seen one of the largest increases (up 6% to 76%) in the proportion of respondents who 'totally agree' with the statement that 'we have a responsibility to look after nature' (European Commission, 2019). The Dáil's (Irish parliament's lower house) declaration of a climate and biodiversity emergency in May 2019 is indicative of this growing public concern, albeit commensurate action has not yet followed in the wake of the declaration.

The current government has committed to a range of actions of relevance to biodiversity, including, for example (see Department of the Taoiseach (2020) for further

31 https://ec.europa.eu/info/food-farming-fisheries/key-policies/common-agricultural-policy/rural-development/country/ireland_en

32 See https://pollinators.ie/record-pollinators/

information): targeting a 51% reduction in the annual level of greenhouse gas emissions in 2030 compared to 2018 levels; developing a new national forest strategy; progressing the establishment of a Citizens' Assembly on Biodiversity Loss (which held its first meeting in May 2022); reviewing the remit, status and funding of the NPWS (which review was published in May 2022 alongside the Strategic Action Plan for the renewal of the NPWS (DHLGH, 2022a) discussed earlier); ensuring that all local authorities have a sufficient number of biodiversity and heritage officers; and reviewing the protection (including enforcement of relevant legislation) of Ireland's natural heritage. Most recently, as noted earlier, the government has stated that it aims to expand Ireland's MPA network to cover 30% (from <3%) of Ireland's maritime area by 2030 (DHLGH, 2021d), albeit the government was criticised after COP15 of the CBD in 2022 for failing to commit to similar expansion in the terrestrial area in Ireland.

In the authors' view the steps announced to date collectively fall well short of the transformative changes required, albeit the establishment of the Citizens' Assembly on Biodiversity Loss and the recent Strategic Action Plan for the NPWS have given cause for hope.[33] As such, despite the considerable legal protection for species and habitats within the Natura 2000 network, unless further significant changes materialise many are likely to continue to deteriorate in the near future, as there is no indication that there will be any reduction in pressures (EPA, 2016) or that the State is significantly increasing conservation actions or allocating the resources required to improve the status of the 50 out of 59 HD habitats that are reported as having an unfavourable conservation status (see Section 19.2). Outside Natura 2000, according to the EPA (2016), 'It is likely that the cumulative impacts of a growing economy will be more evident in the wider countryside rather than the protected areas.' The second national biodiversity conference, held in 2022, presented an opportunity to take stock of these issues and plan for the future.

19.10 Conclusions

EU laws and policies have clearly transformed nature conservation in Ireland since the country joined in 1973. As described earlier, Ireland went from having just one legally protected area for wildlife before EU membership to its network of more than 600 Natura 2000 sites today. While much work is still needed on implementation and enforcement of the Nature Directives, the Directives have undoubtedly positively influenced biodiversity conservation in Ireland, and without them the status of nature would be significantly worse, in the authors' view. This said, other EU policies such as the CAP and the CFP have in parallel played a significant role in driving the serious declines in the conservation status of habitats and species witnessed in Ireland over the past decades. With just 13% of its terrestrial area within Natura 2000, and with marine protected areas in their infancy, Irish nature policy will need to be transformed anew to meet the ambitions of the *EU Biodiversity Strategy for 2030*.

While transformative change is needed across all sectors, agriculture and fishing are particularly worthy of note in the Irish context given the very sizeable areas dedicated to these sectors and the fact that modern agricultural and fishing practices are among the main drivers of biodiversity loss. While Ireland has championed innovative, locally led results-based approaches to AES – exemplified by the ground-breaking Burren Programme – genuinely transformative CAP and CFP reforms, followed by full implementation, are now needed, coupled with a national roll-out of better-targeted results-based AES. This needs to build on (and not undermine) existing successful schemes such as the one in the Burren. On the flipside, as discussed in the conclusions chapter (Section 5.5), the changes in the CAP for the 2023–2027 period are unlikely to improve the situation for biodiversity. This is a seriously discouraging sign. Regarding Ireland's large marine area, full implementation and enforcement of the CFP Regulation, designation and enforcement of MPAs, compliance with the WFD and MSFD, and a phasing out of bottom contact fishing gear are all needed. The outcome of Ireland's CFP Review Group was published in July 2022, but contained disappointingly weak conclusions from an environmental perspective (CFP Review Group, 2022).

Underinvestment in nature in Ireland has contributed over the years to a reactive form of governance, with the State tending to await legal challenges before action is belatedly taken. When action is finally taken, too often the changes made have been simply legislative in nature, with poor subsequent implementation and enforcement. To realise the positive changes needed will require further investment, education and awareness-raising, proactive governance and, in the authors' view, a reorientation away from economic growth maximisation as Ireland's measure of progress, towards a measure that reflects broader goals and respects planetary boundaries. In this regard, the government's commitment to develop new measures of well-being and progress (Department of the Taoiseach, 2020) is welcome and may point to one important path ahead.

Acknowledgements

The authors are grateful to Michael Ewing, formerly coordinator of the Environmental Pillar, and Pádraic Fogarty of the Irish Wildlife Trust for their insightful comments on reviewing this chapter.

References

ADAS (2020) *Evaluation of the Green Low-Carbon Agri-Environment Scheme (GLAS): Synthesis of Evidence*. Report for the Department of Agriculture, Food and the Marine. Dublin: Department of Agriculture, Food and the Marine.

[33] See Fogarty (2017) and Lysaght (2019) for some potential steps in this direction.

An Foras Forbartha (1969) *The Protection of the Natural Heritage.* A report addressed to the Minister for Local Government. Dublin: An Foras Forbartha.

Aughney, T., Roche, N. & Langton, S. (2018) *The Irish Bat Monitoring Programme 2015-2017.* Irish Wildlife Manuals, No. 103. Dublin: National Parks and Wildlife Service.

Balmer, D., Gillings, S., Caffrey, B., Swann, B., Downie, I. & Fuller, R. (2013) *Bird Atlas 2007-2011.* Thetford, UK: British Trust for Ornithology.

Barron, S. (2006) *County Monaghan Wetland Survey.* BEC Consultants. Internal Report, Monaghan County Council. Dublin: Botanical, Environmental & Conservation Consultants.

BEC Consultants (2012) *Combined research priority recommendations.* www.botanicalenvironmental.com/wp-content/uploads/2014/02/NPBR_recomm_combined.pdf

Biodiversity Working Group (2020) *Interim Review of the Implementation of the National Biodiversity Action Plan 2017–2021.* Dublin: Department of Culture, Heritage and the Gaeltacht.

BirdLife International (2021) *European Red List of Birds.* Luxembourg: Publications Office of the European Union.

BirdWatch Ireland (2022) BirdWatch Ireland homepage. https://birdwatchireland.ie/

Bord na Móna (2021) *Bord na Móna announce formal end to all peat harvesting on its lands.* www.bordnamona.ie/bord-na-mona-announce-formal-end-to-all-peat-harvesting-on-its-lands/

BRIDE Project (2020) *Biodiversity Regeneration in a Dairying Environment (BRIDE): Farming with Nature.* www.thebrideproject.ie/

Bullock, C., Kretsche, C. & Candon, E. (2008) *The Economic and Social Aspects of Biodiversity – Benefits and Costs of Biodiversity in Ireland.* Dublin: The Stationery Office.

Burke, B., Lewis, L. J., Fitzgerald, N., Frost, T., Austin, G. & Tierney, T. D. (2018) Estimates of waterbird numbers wintering in Ireland 2011/12 – 2015/16. *Irish Birds,* **11**, 1–12.

Cabot, D. (1999) *Ireland – A Natural History.* London: HarperCollins.

CFP Review Group (2022) Common Fisheries Policy Review Group Report, July 2022. https://assets.gov.ie/236351/33f42e27-6fc1-4983-b6ae-02e2361c8968.pdf

Comerford, H. (2001) *Wildlife Legislation: 1976-2000.* Dublin: Round Hall.

Costello, M. (2021) Government cuts €6 million from racing grant. *The Irish Field,* 12 October 2021.

Craig, A. (2001) The role of the state in protecting natural areas in Ireland: 30 years of progress. *Biology and Environment: Proceedings of the Royal Irish Academy,* **101B** (1/2), 141–149.

Cross, J. (2012) *Ireland's Woodland Heritage: A Guide to Ireland's Native Woodlands.* Dublin: National Parks and Wildlife Service.

CSO (2012) Livestock survey, December 2012. Central Statistics Office. www.cso.ie/en/releasesandpublications/er/lsd/livestocksurveydecember2012/

(2016) Farm structure survey 2016: land utilisation. Central Statistics Office. www.cso.ie/en/releasesandpublications/ep/p-fss/farmstructuresurvey2016/da/lu/

(2020) *Statistical Yearbook of Ireland 2020.* Dublin: Central Statistics Office.

Curlew Task Force (2019) *Action for Curlew in Ireland: Recommendations of the Curlew Task Force.* Dublin: National Parks and Wildlife Service.

DAFF (2010) *Food Harvest 2020: A Vision for Irish Agri-food and Fisheries.* Dublin: Department of Agriculture, Fisheries and Food.

DAFM (2015a) *Forestry Programme 2014-2020: Ireland.* Ref: IRL-DAFM-FS.023. Dublin: Department of Agriculture, Food and the Marine.

(2015b) *Food Wise 2025: A 10-year Vision for the Irish Agri-food Industry.* Dublin: Department of Agriculture, Food and the Marine.

(2018) *Factsheet on Irish Agriculture. August 2018.* Dublin: Department of Agriculture, Food and the Marine. http://edepositireland.ie/bitstream/handle/2262/94910/IrishAgricultureFactsheetFullYear2018010519.pdf?sequence=1&isAllowed=y

(2020a) *Forest Statistics: Ireland 2020.* Dublin: Department of Agriculture, Food and the Marine.

(2020b) *European Maritime and Fisheries Fund 2014-2020.* Dublin: Department of Agriculture, Food and the Marine. www.agriculture.gov.ie/emff/

(2022) *Ireland's Forest Strategy (2022–2030).* Draft for public consultation. Dublin: Department of Agriculture, Food and the Marine.

DAHG (2014) *Review of Raised Bog Natural Heritage Area Network.* Dublin: Department of Arts, Heritage and the Gaeltacht.

(2015) *National Peatlands Strategy.* Dublin: Department of Arts, Heritage and the Gaeltacht.

(2017) *A Programme of measures by Ireland to ensure full compliance with the Judgment of the Court of Justice of the European Union in Case C-418/04* (May 2017 update). Dublin: Department of Arts, Heritage and the Gaeltacht.

DCHG (2016) *Nature reserves – points of interest (dataset).* Dublin: Department of Culture, Heritage and the Gaeltacht, Ireland. https://data.gov.ie/dataset/nature-reserves-points-of-interest

(2017a) *National Biodiversity Action Plan 2017-2021.* Dublin: Department of Culture, Heritage and the Gaeltacht, Ireland.

(2017b) *National Raised Bog Special Areas of Conservation Management Plan 2017-2022.* Dublin: Department of Culture, Heritage and the Gaeltacht, Ireland.

(2019) *Ireland: 6th National Report to the Convention on Biological Diversity.* Dublin: Department of Culture, Heritage and the Gaeltacht. www.cbd.int/doc/nr/nr-06/ie-nr-06-en.pdf

DECC (2020) *Environment Fund Accounts.* Dublin: Department of the Environment, Climate and Communications. www.gov.ie/en/publication/f6b84-environment-fund-accounts/

Department of the Taoiseach (2020) *Programme for Government: Our Shared Future.* www.gov.ie/en/publication/7e05d-programme-for-government-our-shared-future/

DG Environment (2020) Habitats Directive report: notification on compensatory measures Article 6. https://ec.europa.eu/environment/nature/knowledge/rep_habitats/index_en.htm (bottom of page).

DHLGH (2021a) *Minister Noonan welcomes stakeholder review of the National Biodiversity Action Plan.* 16 February 2021. Dublin: Department of Housing, Local Government and Heritage.

(2021b) *Statement from Minister of State for Heritage and Electoral Reform, Malcolm Noonan TD, on Budget 2022 and the future of the National Parks and Wildlife Service, 14 October 2021*. Dublin: Department of Housing, Local Government and Heritage.

(2021c) *Biodiversity funding for local authorities increased to €2.1m*. 6 December 2021. Dublin: Department of Housing, Local Government and Heritage.

(2021d) *Government Takes Next Step Towards Marine Protected Areas forming 30% of Ireland's Maritime Area*. 17 February 2021. Dublin: Department of Housing, Local Government and Heritage.

(2022a) *Ministers announce ambitious new Strategic Action Plan for the National Parks and Wildlife Service*. Last updated 4 May 2022. Dublin: Department of Housing, Local Government and Heritage. www.gov.ie/en/press-release/04792-ministers-announce-ambitious-new-strategic-action-plan-for-the-national-parks-and-wildlife-service/

(2022b) *Marine Protected Areas*. Last updated 25 January 2022. Dublin: Department of Housing, Local Government and Heritage. www.gov.ie/en/publication/e00ec-marine-protected-areas/

(2022c) *Strategic Action Plan for the renewal of the National Parks and Wildlife Service, 2022-2024*. Dublin: Department of Housing, Local Government and Heritage. www.gov.ie/en/publication/fbb81-national-parks-and-wildlife-service-strategic-action-plan-and-review/

Drew, D. P. (2008) Hydrogeology of lowland karst in Ireland. *Quarterly Journal of Engineering Geology and Hydrogeology*, **41** (1), 61–72.

Dunford, B. (2016) *The Burren Life Programme: An Overview*. Dublin: NESC.

Dunford, B. & Parr, S. (2020) Farming for conservation in the Burren. In *Farming for Nature: The Role of Results-Based Payments*, eds. E. O'Rourke & J. Finn, pp. 56–105. Dublin: Teagasc and National Parks and Wildlife Service.

EEA (2020) EUNIS: site search. https://eunis.eea.europa.eu/sites.jsp

Environmental Pillar (2021) The Environmental Pillar withdraws from the problematic 2030 Agri-Food Strategy Committee. 25 February 2021. https://environmentalpillar.ie/the-environmental-pillar-withdraws-from-the-problematic-2030-agri-food-strategy-committee/

EPA (2001) *Biodiversity in Ireland: A Review of Species and Habitats*. EPA, Ireland.

(2009) Freshwater pearl mussel plans 2009-2015. Dublin: Environmental Protection Agency catchments unit. www.catchments.ie/download/freshwater-pearl-mussel-plans-2009-2015/

(2016) *Ireland's Environment: An Assessment*. Wexford: Environmental Protection Agency. https://epawebapp.epa.ie/ebooks/soe2016/files/assets/basic-html/page-1.html#

(2019) *Water Quality in Ireland 2013-2018*. Dublin: Environmental Protection Agency.

(2020a) *Second Review of Strategic Environmental Assessment Effectiveness in Ireland*. Authors: A. González, R. Therivel, A. Gaughran & C. Bullock. Dublin: Environmental Protection Agency.

(2020b) *Ireland's Environment: An Integrated Assessment 2020*. www.epa.ie/our-services/monitoring–assessment/assessment/irelands-environment/state-of-environment-report-/

(2021) *Assessment of the catchments that need reductions in nitrogen concentrations to achieve water quality objectives, June 2021*. Dublin: Environmental Protection Agency.

European Commission (2017) BurrenLIFE wins best-ever EU award. https://web.archive.org/web/20171122105700/https://ec.europa.eu/ireland/news/burren-life-wins-best-ever-eu-award_en

(2019) *Attitudes of Europeans towards biodiversity. Special Eurobarometer 481*. Brussels: Kantar Public Brussels for DG Environment.

Farrell, E. P. (1990) Peatland forestry in the Republic of Ireland. In *Biomass Production and Element Fluxes in Forested Peatland Ecosystems*, Proceedings of a seminar held in Umeå, Sweden, 3–7 September 1990, ed. B. Hånell, pp. 13–18.

FIE (2022) High Court Grants Leave to Challenge the Government's Agriculture Strategy – Food Vision 2030. 1 March 2022. www.friendsoftheirishenvironment.org/press-releases/high-court-grants-leave-to-challenge-the-government-s-agriculture-strategy-food-vison-2030

FitzPatrick, Ú., Murray, T. E., Paxton, R. J. *et al.* (2007) Rarity and decline in bumblebees – a test of causes and correlates in the Irish fauna. *Biological Conservation*, **136**, 185–194.

Fogarty, P. (2017) *Whittled Away*. Dublin: Gill Books.

Forest Europe (2015) *State of Europe's Forests 2015*. Copenhagen: European Environment Agency.

Gilbert, G., Stanbury, A. & Lewis, L. (2021) Birds of Conservation Concern in Ireland 4: 2020–2026. *Irish Birds*, **43**, 1–22.

Government of Ireland (2019) *Climate Action Plan 2019 to Tackle Climate Breakdown*. Dublin: Government of Ireland.

(2020) Cabinet approves €108 million funding for groundbreaking Bord na Móna bog rehabilitation plan, 24 November 2020. www.gov.ie/en/press-release/2aae1-cabinet-approves-108m-funding-for-groundbreaking-bord-na-mona-bog-rehabilitation-plan-minister-ryan-also-announces-that-47-more-projects-in-the-midlands-totalling-278m-are-approved-under-the-just-transition-fund/

Grant Thornton (2010) *Organisational Review of the National Parks and Wildlife Service (NPWS)*. Dublin: Department of the Environment, Heritage and Local Government.

Haddad, N., Haarstad, J. & Tilman, D. (2000) The effects of long-term nitrogen loading on grassland insect communities. *Oecologia*, **124**, 73–84.

Hickie, D. (1997) *Evaluation of Environmental Designations in Ireland*. Dublin: Heritage Council.

Hofmann, A. (2019) Left to interest groups? On the prospects for enforcing environmental law in the European Union. *Environmental Politics*, **28** (2), 342–364.

IEEP (2010) *Costs and Socio-Economic Benefits Associated with the Natura 2000 Network*. London: Institute for European Environmental Policy.

IGEES (2019) *Spending Review 2019: The Green, Low-carbon Agri-environment Scheme*. Dublin: Irish Government Economic and Evaluation Service.

IWT (2021) As the government heads to the COP26 Climate Summit in Glasgow, new information shows that unlawful turf-cutting remains widespread on Special Areas of Conservation. Irish Wildlife Trust, 28 October 2021. https://iwt.ie/press-release-unlawful-turf-cutting-2/

Jackson, A. (2018) *Conserving Europe's Wildlife: Law and Policy of the Natura 2000 Network of Protected Areas.* Abingdon, UK: Earthscan, Routledge.

Jackson, J. W. (1925) The distribution of *Margaritifera margaritifera* in the British Isles. *Journal of Conchology*, **17**, 195–211.

Kelly, F. (2019) *Greening Irish Forestry.* Kilcoole: BirdWatch Ireland.

Kirby, P. (2010) *Celtic Tiger in Collapse: Explaining the Weaknesses of the Irish Model.* Basingstoke, UK: Palgrave Macmillan.

Laffan, B. & O'Mahony, J. (2008) 'Bringing Politics Back In'. Domestic conflict and the negotiated implementation of EU nature conservation legislation in Ireland. *Journal of Environmental Policy & Planning*, **10** (2), 175–197.

Landgrebe-Trinkunaite, R. (2019) Coastal and halophytic habitats: sandbanks which are slightly covered by sea water all the time (1110), estuaries (1130), mudflats and sandflats not covered by seawater at low tide (1140), and large shallow inlets and bays (1160) – Ireland. Case study in *Study on identifying the drivers of successful implementation of the Birds and Habitats Directives*, eds. G. Tucker, T. Stuart, S. Naumann, U. Stein, R. Landgrebe-Trinkunaite & O. Knol. Report to the European Commission, DG Environment on Contract ENV.F.1/FRA/2014/0063, Institute for European Environmental Policy, Brussels. https://circabc.europa.eu/ui/group/173a90fc-40bf-492d-a3a9-df99c4aa8807/library/02f89ff4-2007-4290-9460-e8a66a805e2a

Lewis, L. J., Coombes, D., Burke, B. *et al.* (2019) *Countryside Bird Survey: Status and Trends of Common and Widespread Breeding Birds 1998-2016.* Irish Wildlife Manuals, No. 115. Dublin: National Parks and Wildlife Service.

Lysaght, L. (2019) *The six steps urgently needed to save Irish biodiversity. The Irish Times*, 19 February 2019.

Marine Biodiversity Scheme (2014) *EMFF Operational Programme 2014-2020: Marine Biodiversity Scheme.* Dublin: Department of Agriculture, Food and the Marine.

Marine Institute (2021) *The Stock Book: Annual Review of Fish Stocks in 2021 with Management Advice for 2022.* November 2021. Oranmore: Marine Institute.

Marine Protected Area Advisory Group (2020) *Expanding Ireland's Marine Protected Area Network.* October 2020. A report by the Marine Protected Area Advisory Group for the Department of Housing, Local Government and Heritage.

Matin, S., Sullivan, C. A., Finn, J. A. *et al.* (2020) Assessing the distribution and extent of High Nature Value farmland in the Republic of Ireland. *Ecological Indicators*, **108**, 105700.

McGrath, P. (2021) Ministers consider Galway Harbour extension plan. *RTE*, 5 May 2021. www.rte.ie/news/connacht/2021/0505/1214021-galway-harbour-expansion/

McGuinness, S. & Bullock, C. (2020) *Mobilising Finance for Biodiversity: A policy and institutional review of finance arrangements for biodiversity conservation in Ireland.* University College Dublin.

Mitchell, F. (1995) The role of climate and man in the development of the Irish landscape. In *Science, Green Issues and the Environment: Ireland and the Global Crisis*, eds. D. D. G McMillan, C. O'Rourke, D. J, Fry & N. D. McMillan, pp. 22–26. Kilkenny, Ireland: Tyndall Books.

Moorkens, E. A. (2010) Addressing the conservation and rehabilitation of *Margaritifera margaritifera* populations in the republic of Ireland within the framework of the habitats and species directive. *Journal of Conchology*, **40**, 339–350.

Morrison, R. & Bullock, C. (2018) *A National Biodiversity Expenditure Review for Ireland.* University College Dublin.

National Planning Framework (2018) *Project Ireland 2040: National Planning Framework.* Dublin: Government of Ireland.

Nature Scorecard (2018) *The State of Implementation of the Birds and Habitats Directives in the EU: An analysis by national environmental NGOs in 18 Member States, March 2018.* BirdLife International, EEB, Friends of the Earth Europe, WWF. wwfint.awsassets.panda.org/downloads/Nature_Scorecards_Report_March2018.pdf

NBDC (2015) *All-Ireland Pollinator Plan 2015-2020.* Waterford: National Biodiversity Data Centre.

(2021) *All-Ireland Pollinator Plan 2021-2025.* Waterford: National Biodiversity Data Centre.

Noonan, M. (2022) Written answer of 25 January 2022 by Malcolm Noonan TD in reply to parliamentary question [3087/22] from Holly Cairns TD.

NPWS (2013) Department of Arts, Heritage and the Gaeltacht submission to the Department of Agriculture, Food and the Marine on CAP reform and biodiversity (May 2013).

(2014) *Ireland: a Prioritised Action Framework for Natura 2000 for the EU multiannual financing period 2014-2020.* November 2014.

(2015) Response to the European Commission's 'Evidence Gathering Questionnaire for the Fitness Check of the Nature Directives', March 2015. www.npws.ie/sites/default/files/publications/pdf/Fitness%20Check%2015%204%2015.pdf

(2019) *The Status of EU Protected Habitats and Species in Ireland.* Dublin: National Parks and Wildlife Service.

(2020a) *New Threat Response Plans.* Dublin: National Parks and Wildlife Service. www.npws.ie/publications/species-action-plans

(2020b) *Recording and Addressing Persecution and Threats to Our Raptors: A Review of Incidents.* Irish Wildlife Manuals 126. Dublin: National Parks and Wildlife Service.

(2020c) *Natural Heritage Areas.* Dublin: National Parks and Wildlife Service. www.npws.ie/protected-sites/nha

(2020d) *Red Lists.* Dublin: National Parks and Wildlife Service. www.npws.ie/publications/red-lists

(2021a) *Prioritised Action Framework (PAF) for Natura 2000 in Ireland pursuant to Article 8 of the Habitats Directive for the Multiannual Financial Framework period 2021 – 2027.* March 2021. Dublin: National Parks and Wildlife Service.

(2021b) *Terms of Reference – Review of the NPWS, 2021.* https://npws.ie/sites/default/files/files/Terms-of-Reference-NPWS-Review.pdf

(2022) Available conservation plans. www.npws.ie/protected-sites/conservation-planning/available-plans. Accessed 27 January 2022.

Ó Cofaigh, C. O., Callard, S. L., Roberts, D. H. *et al.* (2021) Timing and pace of ice-sheet withdrawal across the marine–terrestrial transition west of Ireland during the last glaciation. *Journal of Quaternary Science*, **36** (5), 805–832.

Ó hUallacháin, D., Finn, J. A., Keogh, B., Fritch, R. & Sheridan, H. (2016) A comparison of grassland vegetation from three agrienvironment conservation measures. *Irish Journal of Agricultural and Food Research*, **55** (2), 176–191.

OECD (2010) *OECD Environmental Performance Reviews*. Paris: Organisation for Economic Co-operation and Development.

O'Rourke, E. & Finn, J. (eds.) (2020) *Farming for Nature: The Role of Results-Based Payments*. Dublin: Teagasc and National Parks and Wildlife Service. www.npws.ie/farmers-and-landowners/farming-for-nature/book-results-based-payments

RBAPS (2019) RBAPS Project: Developing Results Based Agri-environmental Payment Schemes in Ireland and Spain. https://rbaps.eu/

Renou-Wilson, F. (2018) Peatlands. In *The Soils of Ireland*, eds. R. Creamer & L. O'Sullivan, pp. 141–152. Cham, Switzerland: Springer International.

Roberts, D. H., Ó Cofaigh, C. O., Ballantyne, C. K. *et al.* (2020) The deglaciation of the western sector of the Irish Ice Sheet from the inner continental shelf to its terrestrial margin. *Boreas*, **49** (3), 438–460.

Sargent, N. (2021) Funding failure: Most of Ireland's biodiversity spend goes on schemes that have little impact. *thejournal.ie*, 28 August 2021. www.thejournal.ie/endangered-species-part-2-5520281-Aug2021/

Schouten, M. G. C. (1994) Nature conservation in Ireland: a critical decade. *Biology and Environment: Proceedings of the Royal Irish Academy*, **94B** (1), 91–95.

Simons, G. (2012) *Unresolved issues under the Planning and Development (Amendment) Act 2010*. Paper given at Round Hall conference in Dublin on 10 November 2012. Dublin: Round Hall.

Temple Lang, J. (1970) Conservation of the Environment in Ireland. *Studies: An Irish Quarterly Review*, **59** (235), 279–300.

(1988) Wildlife conservation in the Republic of Ireland. *Studies: An Irish Quarterly Review*, **77** (306), 165–173.

UNEP-WCMC (2022) *Protected Area Profile for Ireland from the World Database of Protected Areas, May 2022*. www.protectedplanet.net

UPTCS (1987) *Our Natural Heritage: A Policy for Nature Conservation in Ireland*. Union of Professional and Technical Civil Servants. Dublin: Brunswick Press.

Viney, M. (2003) *Brooding on the politics of conservation*. *The Irish Times*, 10 May 2003.

Chapter 20

Italy

Elisa Cardarelli, Barbara Calaciura, Daniela Zaghi and Giuseppe Bogliani

Key Statistics

Terrestrial area (km^2)/% of EU-28*	300 578/6.9%
Marine waters (km^2)/% of EU-28*	587 155/12.8%
Population density (p km^2)/% of EU-28*	200/174%
GDP per capita (€)/% of EU-28	26 442/99%
Land cover (CORINE land cover classes 2018)	AGR 52%; FOR 26%; OSN 15%; ART 6%; WET 1%
Biogeographical regions	ALP, CON, MED, MMED
Number of HD Annex I habitats/%*	132/58%
Number of HD Annex II, IV and V species/%*	338/28%
Number of BD Annex I bird taxa/%*	111/63%
% HD UCS and decreasing habitats/ % unknown	38%/4%
% HD UCS and decreasing species/ % unknown	33%/4%
% breeding BD Annex I birds decreasing/% uncertain and unknown	17%/39%

* Excluding MAC and MMAC biogeographical regions.
Sources and codes. See Table 1.1.

20.1 The Natural History and Characteristics of Italy

Italy is a peninsula, bordered and crossed by two important mountain systems (the Alps and Apennines), with two major islands (Sardinia and Sicily) and a collection of minor islands. The variety of its landforms, geology and climatic gradients (temperate in the north and in the Alps and Apennines, and Mediterranean or sub-Mediterranean elsewhere) makes it very rich in terms of habitats and species. Its nature has been historically affected by human activities. Initially, traditional and extensive agriculture increased habitat diversity (e.g. creating wood pastures, heathlands and ponds for livestock or irrigation), but demographic growth since the nineteenth century led to increasing human exploitation and the conversion of large areas of natural and semi-natural habitat, such as wetlands and forests, into agricultural or urban areas (Finlayson *et al.*, 1992). Together with hunting, this had severe impacts on some species groups, such as waterbirds and large carnivores. After World War II, mechanisation and intensification of agricultural practices continued to affect biodiversity in more productive zones, especially on the fertile plains. In contrast, abandonment of mountain and remote areas occurred due to rural depopulation, which led to an increase in forested areas, with benefits for some species such as deer and other ungulates, and the Wolf (*Canis lupus*; see online Annex IT.1[1]), but with loss of suitable areas for open-habitat species (Brambilla *et al.*, 2010).

According to Blasi *et al.* (2017), about 52% of Italy is farmland, most of which is arable land (26.9%), followed by permanent crops such as olive groves, vineyards and fruit trees (7.3%), cropped areas with a significant portion of natural vegetation (6.8%), pastures (1.4%), rice fields (0.9%) and agro-forestry areas (0.6%); the remaining areas (8.4%) are heterogeneous farmland. Forests, shrublands and seminatural grasslands (dry, mesic and wet) occupy 26%, 7.5% and 6% of the surface, respectively. About 3% of the land is covered by damp/humid (hygrophilous) habitats, rocky areas, coastal dunes, sandy shores and marine wetlands. Of the farmland areas, 51% can be classified as High Nature Value (HNV) (Trisorio *et al.*, 2013).

As for marine habitats, Italy has three ecoregions (Adriatic Sea, Ionian and Central Mediterranean Sea, Western Mediterranean Sea) and, thanks to its latitudinal extension, its species range from thermophilous and subtropical elements in the south, to temperate–cold elements in the north (CCN, 2017). There are several marine habitats of conservation interest, such as *Posidonia oceanica* seagrass beds (1120*)[2] and biogenic reefs created by calcareous algae and vermetid molluscs.

Despite the effects of human activities on its nature, Italy is still one of the richest European countries in terms of terrestrial and marine biodiversity (see Section 2.3) and retains high levels of endemism amongst its vascular flora and fauna – 20% and 30%, respectively (CCN, 2017).

[1] All annexes are free to access online at CUP: www.cambridge.org/natureconservation

[2] Numbers in brackets are HD Annex I habitat codes (see Appendix for full names). * Indicates a priority habitat.

Table 20.1 Short-term trends (c. 2007–2018) in Habitats Directive Annex I habitats and Annex II, IV and V species with an unfavourable conservation status in Italy.

	Poor (unfavourable–inadequate)				Bad (unfavourable–bad)			
	Imp	Sta	Dec	?	Imp	Sta	Dec	?
HD Habitat assessments								
Number	11	75	29	9	3	21	72	5
%	9%	60%	23%	7%	3%	21%	71%	6%
HD Species assessments								
Number	16	59	121	10	7	10	67	9
%	8%	29%	59%	5%	7%	11%	72%	10%

Key. Imp = improving, Sta = stable, Dec = decreasing, ? = unknown.
Source. Based on Member State HD Article 17 reported data as presented on the EEA national summary dashboard (2021).

20.2 The Status of Nature

20.2.1 Habitat Status and Trends

The 2013–2018 assessment[3] of the conservation status of HD Annex I habitats (across the four biogeographical regions) showed that only 10% of the assessments reported a favourable status, while 47% and 39% indicated a poor and bad status, respectively. However, it is important to note that there were considerable differences amongst the broad habitat types in terms of their conservation status. No assessments of bog, mire and fen habitats, or wetlands or dunes, were favourable, and only 3% of assessments of grasslands were favourable (i.e. 1 out of 38 assessments). Moreover, 45% of habitats with an unfavourable status showed declines (38% of all HD Annex I habitat types in Italy), with a particularly high proportion amongst those that had a bad status (Table 20.1). These most threatened and declining habitat types include dunes (71%), freshwater habitats (47%) and grasslands (39%).

20.2.2 Species Status and Trends

As regards HD species, the 2013–2018 assessments show that 43% of assessments revealed a favourable conservation status, 36% a poor status and 16% a bad status. Of particular concern is that 59% and 72% of the assessments of species with a poor and bad status, respectively, also showed declines (Table 20.1). The most threatened species groups (i.e. those with a bad status and declining), included fish (39%), amphibians (15%) and vascular plants (12%).

The Article 12 reporting data on birds appear to show a generally less serious situation for Annex I birds compared to HD species. Nevertheless, 17% and 9% of breeding and wintering populations, respectively, declined over 2007–2018 (Table 20.2). It is also of concern that the trends were unknown for a significant proportion of species.

[3] See Section 1.3 for background on monitoring under the Nature Directives, status categories and data sources.

Table 20.2 Short-term trends (c. 2007–2018) in Birds Directive Annex I taxa in Italy.

	Breeding				Wintering			
	Inc	Sta/ Flu	Dec	?	Inc	Sta/ Flu	Dec	?
Number	26	13	15	35	10	5	2	6
%	29%	15%	17%	39%	43%	22%	9%	26%

Key. Inc = increasing, Sta = stable, Flu = fluctuating, Dec = decreasing, ? = uncertain and unknown.
Source. Based on Member State Article 12 reported data as presented on the EEA national summary dashboard (2021).

In the Red List of Italian Vertebrates (which assessed 672 species), there are about twice as many declining species as increasing ones: 27% versus 15% in terrestrial habitats, and 22% versus 12% in marine habitats (Rondinini *et al.*, 2013). However, knowledge of marine taxa is still very incomplete, with 66% of species having unknown trends. The high proportion of Italian vertebrates with a threatened status, with the exception of ungulates and large carnivores like the Wolf, has been confirmed in the latest revision of the Red List (Rondinini *et al.*, 2022). Similarly, data on the demographic trends of plants are scarce, although 65% out of 297 vascular plants (including 151 HD or Bern Convention listed species) and 55% of 99 non-vascular species (including 51 HD or Bern Convention listed species) are considered to be threatened (Rossi *et al.*, 2013). Among Italian endemic vascular plants (1 340 taxa), 6 taxa are extinct and 300 taxa (22%) are threatened with extinction, while 247 taxa (18%) are data deficient (Orsenigo *et al.*, 2018).

The situation seems better for some species groups, such as butterflies, as 6% (18) are considered threatened out of the 289 assessed species (Balletto *et al.*, 2015). However, it is important to note that 55% of the threatened species are endemic (5), subendemic (3) or have a very restricted range in Italy (3) and that changes in land use and agriculture practices threaten the persistence of these populations, especially in wet habitats, rocky habitats and grasslands (Balletto *et al.*, 2015; Bonelli *et al.*, 2018).

There is also evidence that even some common species have experienced rapid declines in abundance. For example, in the 2013 Red List, the Common Toad (*Bufo bufo*) was classified as Vulnerable because of a demographic decline of more than 30% in central and northern populations over the previous decade (Rondinini *et al.*, 2013). In 2020, the farmland bird index (FBI) and the index for species of mountain grasslands (FBI_{pm}) were at 71% and 70%, respectively, of their initial values in the year 2000 (Rete Rurale Nazionale and Lipu, 2021). Eighteen of the 28 species that contribute to the FBI, and six of the 13 species that contribute to the FBI_{pm} have declining populations. Moreover, 70% of FBI and FBI_{pm} species classified as Species of European Conservation Concern (i.e. by BirdLife International, 2017) are declining.

Some of the species that have increased in recent decades are of high conservation importance. These include some ungulates (e.g. Alpine Ibex, *Capra ibex*; Brambilla *et al.*, 2020) and large carnivores (e.g. Wolf; Galaverni *et al.*, 2016), which benefited from the wide-scale abandonment of mountain areas, and, in some cases, targeted conservation projects (Rondinini *et al.*, 2013; see also online Annex IT.1).

20.2.3 Drivers and Causes of Change

Over the last 50 years, major causes of biodiversity decline in Italy have included habitat loss and fragmentation, changes in habitat management, pollution and introduction of invasive alien species (IAS) (CCN, 2017, 2018).

Habitat destruction has occurred as a result of, among other causes, land conversion for urbanisation, affecting 5 000 km^2 over 1990–2014 (CCN, 2017, 2018), with the most severe impacts on lowland habitats, coastal areas and wetlands. Elsewhere, over 1990–2014, about 5 000 km^2 was subject to afforestation due to land abandonment, with the resulting loss of HNV traditional pastures and meadows (CCN, 2017, 2018). Forest habitats have been under pressure from management practices that limit the availability of deadwood for saproxylic species (Rondinini *et al.*, 2014), anthropogenic fires that result in dramatic changes in vegetation and species complexes, and increases in air pollution and hydrogeological risks (CCN, 2018).

Agricultural intensification – including the loss of marginal non-cropped habitats (e.g. hedgerows, isolated trees, and woodlots), changes in cropping timing (e.g. see rice fields in Section 20.6) and high levels of pesticide and fertiliser use (which also increased soil, air and water pollution) – has been the common cause of biodiversity declines in farmland areas (CCN, 2017).

Alien species have caused major problems and continue to be a serious threat, on land and within inland waters and marine ecosystems (Genovesi *et al.*, 2014). In particular, Italian freshwater ecosystems are highly impacted by IAS, with, for example, 51 out of 100 freshwater fish species being exotic (Bianco, 2014). Inland waters are also subject to pressures from sport fishing and pollution (largely from lowland farming), and river engineering (Rondinini *et al.*, 2014; CCN, 2017). Similarly, marine habitats have been affected by unsustainable fishing, pollution, urbanisation of coastal areas and eutrophication, mainly due to nutrient inputs from agriculture, aquaculture and wastewaters (CCN, 2017).

Other drivers of detrimental changes include climate change, which is increasingly causing variations in the distribution of habitats, modifying species' biology and ecology (D'Amen and Bombi, 2009; Rogora *et al.*, 2018) and exacerbating some phenomena such as desertification and fire (CCN, 2018).

20.3 Nature Conservation Governance, Actors, Legislation and Policies

20.3.1 Overview of Nature Developments in Italy

A chronological summary of some of the most important nature conservation policies, laws and other events in the development of nature conservation is provided in Table 20.3.

20.3.2 Governance, Legislation and Policies

In Italy, the State has the power to set standards for nature protection that affect the whole national territory, while the Regions and Autonomous Provinces (R&APs) can further specify these standards based on their specific needs (Cecchetti, 2003). This highly decentralised governance system, coupled with the very horizontal character of nature protection, involves multiple central, regional and local competent institutions with respect to policy, planning, management and implementation. For instance, whilst the ministry with competency for the environment, which is the Ministero dell'Ambiente e della Sicurezza Energetica (MASE; until 1 March 2021 named Ministero dell'Ambiente e della Tutela del Territorio e del Mare – MATTM) is responsible for designation of SACs and SPAs and overseeing the Natura 2000 network at the national level, the R&APs are responsible for the implementation of the Nature Directives (DPR 357/97). They can also delegate some administrative functions, such as the production of Natura 2000 site management plans, to other bodies (e.g. municipalities, or managing authorities of protected areas).

The Permanent Conference for the Relations between the State, the Regions and the Autonomous Provinces of Trento and Bolzano (State-Regions Conference) is the collegiate body for institutional cooperation between the State and the R&APs, providing a forum for coordination and political negotiation between central and regional governments. Moreover, the Nature Protection Directorate of the MASE coordinates national nature conservation policy through thematic Technical Working Groups, comprising representatives of the MASE and other ministries (e.g. agriculture, forestry and

Table 20.3 Key nature conservation developments in Italy.

1922	Royal decrees establish the first two National Parks (Gran Paradiso and Abruzzo, Lazio e Molise)
1923 and 1952	Reorganisation and reform of legislation on forests and mountain areas
1939	Protection of natural heritage
1974	Institution of the first Regional Nature Park (Valle del Ticino Lombardo, Lombardy Region)
1982 and 1992	Provisions for the defence of the sea
1985	First decrees on landscape protection
1986	Establishment of the Ministry of the Environment and regulations on environmental damage
1991	Framework Law on protected areas
1992	Framework Law on the protection of mammals and birds and hunting regulation
2003	Regulation with amendments and additions to Presidential Decree 357/1997 concerning the implementation of the Habitats Directive
2005	Ministerial decree on forestry planning guidelines
2006	Law for the establishment of ecological protection zones beyond the outer limit of the territorial sea
	Environmental regulations (SEA–EIA, water, soil, waste)
2007	Minimum uniform criteria for the definition of conservation measures relating to SACs and SPAs
2010	Publication of the *National Biodiversity Strategy to 2020*
	Implementation of the *Marine Strategy*
2011	Amendment of the penal code by inserting crimes against species and habitats protected under the Nature Directives
2015	Environmental measures for promoting the green economy and limiting the excessive use of natural resources
2017	Transposition of Regulation 1143/2014/EC, laying down provisions aimed at preventing and managing the introduction and spread of invasive alien species
2022	Publication of the *National Biodiversity Strategy to 2030*

See Table 3.3 for signatory dates of international conventions.

fisheries, and economics), ISPRA,[4] R&APs and NGOs. Documents produced by the working groups need the approval of the State-Regions Conference. This mechanism, on one hand, facilitates high standards and the involvement of all relevant institutions. On the other hand, it has often resulted in slow implementation, which has led to the opening of some infringement procedures by the EU Commission (e.g. for transposition and implementation of the Birds Directive; European Commission, 2017). Since the 1970s, the R&APs have played a significant role in nature protection, often anticipating state reforms. This has been particularly evident for protected areas, where northern regions have created systems that were already mature at the time of the approval of the national law in 1991.

The 1986 law[5] establishing the MATTM (now the MASE) introduced key principles (e.g. of prevention and reimbursement of environmental damage), legal procedures (e.g. EIA, environmental damage assessment) and subjective juridic situations. Some of these were innovative for their time, such as citizens' rights to access environmental information and participate in EIA procedures, and environmental associations' legitimacy to appeal (Fonderico, 2007). One of the first legislative acts produced by the new ministry, the 1991 Framework Law on Protected Areas,[6] represents a milestone in nature protection in Italy, because it 'dictates fundamental principles for the establishment and management of protected natural areas, in order to guarantee and promote, in a coordinated form, the conservation and enhancement of the country's natural heritage'.

According to Alonzi *et al.* (2006), nature protection policy developments in Italy have been mainly driven by EU Regulations, Directives and international conventions

[4] The Higher Institute for Environmental Protection and Research (Istituto Superiore per la Protezione e la Ricerca Ambientale – ISPRA) provides technical–scientific support for the MASE.

[5] Law n. 349/1986 'Istituzione del Ministero dell'ambiente e norme in materia di danno ambientale'.

[6] Law n. 394/1991 'Legge-quadro sulle aree protette' and its subsequent partial modification through the Law n. 426/1998 'Nuovi interventi in campo ambientale'.

(Table 20.3). In the last 30 years, they helped change Italian government policies on biodiversity conservation, which is no longer limited to lists of species to be protected, as was the case in the 1980s and 1990s, but also requires the development of management strategies and sustainable development. In 2010, the MASE – together with the R&APs, scientific institutions, NGOs and stakeholders – published the first *National Biodiversity Strategy* (NBS 2011–2020). It included three key themes: biodiversity and ecosystem services, biodiversity and climate change, and biodiversity and economic policies. However, at the end of 2020, only 21% of the objectives were assessed as achieved and 33% of the priorities had actually been implemented, while another 56% were ongoing (MATTM, 2021). The strategy gave impetus to the creation of tools and initiatives both for the enhancement of natural capital, as the base for the development of a green economy, and for setting up environmental accounting, promoting the integration of biodiversity into the programming tools, in the implementation of measures and in territorial planning, as well as specific legislative instruments for the protection of nature, such as the decree to limit the spread of IAS. The second *NBS* (2022–2030; MiTE, 2022) was published in July 2022 and its approval is awaited.

On 8 February 2022, the Italian Parliament approved the amendments to the Constitution and introduced the concept that the Italian Republic protects the environment, biodiversity and ecosystems and specifies a principle of protection for animals.

20.3.3 Stakeholders and Public Involvement

In Italy, the first public concerns over the preservation of cultural and environmental heritage arose after World War II, when rapid economic development and urbanisation started to sharply modify anthropic and natural landscapes (Della Seta, 1999). In the 1960s and 1970s, concerns widened to pollution and wildlife exploitation through hunting, which were then addressed by important environmental NGOs (eNGOs) such as Lega Italiana Protezione Uccelli (LIPU, the BirdLife International partner in Italy), WWF Italia and Legambiente.[7] Public interest and support for nature conservation in Italy have increased, as the national Register of Environmental Organizations at the end of 2022 listed 80 members. This is reflected in eNGO memberships (e.g. in 2022 around 30 000 and 115 000 members and donors for LIPU and Legambiente, respectively) and also in environmental surveillance carried out by specifically trained volunteers (i.e. Guardie Ecologiche Volontarie). There are also many other NGOs covering more specific interests, such as monitoring and research on certain taxa or protection and monitoring at regional/local scales.

In recent decades, the role of NGOs in nature conservation has become more structured; they not only campaign and lobby to conserve nature, but also own or manage reserves and wildlife recovery centres (e.g. WWF Italia manages more than 100 sites totalling 30 000 ha[8]). Moreover, they take legal actions (e.g. Legambiente has more than 200 environmental lawyers[9]) and work in partnership with the statutory sector, private companies and universities to identify sites of conservation concern, develop agri-environment schemes and monitor species and habitats (see Section 20.8). For example, LIPU has produced and is periodically updating the Italian Important Bird Area (IBA) inventory, which has been instrumental in contributing to the designation of SPAs, and is involved in the Farmland Bird Index.

Several hunting associations are active at national or local level, and on some occasions they have participated in conservation programmes on species of hunting interest, in collaboration with eNGOs. Usually, however, the level of collaboration between hunting associations and eNGOs is very low. Hunters and their representatives ask for more concessions, while eNGOs often push in the opposite direction, as in the case of the use of lead ammunition in wetlands and for hunting ungulates, whose carcasses can contaminate eagles and vultures (Andreotti and Borghesi, 2012). Furthermore, there is a recurring opposition also at regional level between hunting and environmental associations regarding the list of huntable species. According to Barca *et al.* (2016), the lack of dialogue and increasingly polarised positions on both sides have made it difficult to assure compliance with EU and national migratory bird hunting laws, and hinder shared solutions.

Hunting activity has been locally associated with the maintenance of valuable habitats, such as coastal wetlands in the northern Adriatic, inland wetlands and forest habitats. Managed private hunting reserves helped prevent the destruction of important biotopes, especially during the economic boom years. Many of these sites are now Natura 2000 sites.

20.4 Species Protection and Conservation Measures

20.4.1 Plants

In Italy, plant species are mainly protected through the implementation of international conventions and EU directives, above all the Bern Convention and Habitats Directive. No plant protection legislation that covers the whole national territory has yet been issued. The matter is actually delegated to the R&APs, almost all of which have developed provisions for conservation of indigenous flora, even if they are characterised by varying approaches (Alonzi *et al.*, 2006). There is a great disparity in the number of protected species and even in the criteria that have been adopted to select them. Furthermore, some measures are rather dated. Contradictory

[7] Legambiente is one of the European Environmental Bureau partners in Italy.

[8] www.wwf.it (date of access 12 April 2022).

[9] www.legambiente.it (date of access 12 April 2022).

local legislation (e.g. among adjacent regions) may result in problems for species conservation at a larger scale.

The Red List of the Italian Flora was updated in 2013, one aim being to overcome regional differences and form the basis for a national strategy for plant conservation activities (Rossi and Gentili, 2008). However, current legal protection remains insufficient to guarantee plant conservation: an assessment of 203 HD vascular plants, bryophytes and lichens revealed that 42% are threatened (Rossi *et al.*, 2016).

20.4.2 Animal Protection

The first systematic measures adopted in Italy for the protection of wild animals living in terrestrial habitats date back to 1939, when the Framework Law on hunting was issued.[10] This law included all wild birds and mammals, with the exception of moles, shrews, dormice, mice and voles. Some species were identified as protected, such as Alpine Ibex, Apennine Chamois (*Rupicapra pyrenaica ornata*), Brown Bear (*Ursus arctos*), Mediterranean Monk Seal (*Monachus monachus*), all species of bats, tits and woodpeckers. All other species could be hunted, including the so-called vermin species, amongst which were Wolf, Otter (*Lutra lutra*), Wild Cat (*Felis silvestris*), several diurnal raptors, herons and cormorants.

Things started to change in 1977, with the approval of Law 968/1977, and then again in 1992, when the new Framework Law 157/1992[11] came into force. The general changes introduced in Law 157 were: (a) the transposition of the Birds Directive; (b) the legal transformation of wildlife from belonging to the person who killed it (*res nullius*) to belonging to the State (*res communis*); (c) all wild mammals and birds becoming protected, with only some species being huntable. Moreover, the list of huntable species took into account whether a species population size can be estimated and the impacts of hunting evaluated. Despite this, the list of huntable species includes several species whose populations are in decline, such as the Skylark (*Alauda arvensis*), the Turtle-dove (*Streptopelia turtur*), which is threatened in Europe and globally, and the Sicilian Rock Partridge (*Alectoris graeca whitakeri*), which is endemic to Italy.

Conservation of wildlife is also complicated by the fact that some administrative functions are delegated to the R&APs, giving them the ability to adapt the national hunting rules. Differences, especially in the Autonomous Provinces of Trento and Bolzano, have emerged, particularly in relation to the Wolf and Brown Bear and the composition of the huntable species list, which includes species protected in the rest of the country, such as the Alpine Marmot (*Marmota marmot*) and the Alpine Ibex. However, these species continue to maintain favourable status in the region. Some R&APs strongly influenced by the hunters' lobby have granted hunting derogations, almost on an annual basis, in breach of BD Article 9, even after several CJEU judgments that declared such derogations illegal. This led to the opening of infringement procedures against the Marche, Lombardy, Veneto and Liguria Regions. The most recent infringement case was closed in 2016 following the European Commission's reasoned opinion to stop derogations granted by regions for trapping birds with nets for their use as live decoys.

As regards other animals, there is no overall regulation at national level and their conservation is mostly dependent on species-specific action plans and regional laws that establish local management measures for fish, amphibians, reptiles, insects and other invertebrates worthy of protection. The Framework Law on fishing dates back to 1931 (Law 1604/1931) and is largely unsuitable for the management and conservation of freshwater fauna. Again, the regions have autonomy on fishing regulation and fish management. This has led to a deterioration in the conservation status of many fish species over recent decades that has been exacerbated by IAS, which are particularly significant in Italian waters (Bianco, 2014).

20.4.3 Hunting and Poaching

The hunting of birds and some mammals is a popular activity in most of Italy, although the number of hunters has declined, from 1.7 million in 1981 to 579 000 in 2015 (Vallini, 2019). However, the impacts of legal hunting on wild species populations are difficult to assess as data on the number of each of the legally huntable species are incomplete and largely unreliable. Even though the regions are obliged to provide the MASE with updated data, very few do so (Sorace and Amadesi, 2016). Despite this, some regional administrations try every year to permit hunting, 'in derogation' to the national law, of species that have sharply declining populations, such as the Turtle-dove, Skylark and Rock Ptarmigan (*Lagopus muta*), without providing adequate data on the size of the populations and the extent of the collection. Every year, the environmental associations oppose this legally and, except in a few cases, win the lawsuits and block the hunting of these species or prevent, in the case of the Turtle-dove, the hunting season starting earlier than the national law allows.

Approximate data, from an expert-based survey, are available on the number of illegally killed birds – an average of 5.6 million individuals (min. 3.4 million to max. 7.8 million), belonging to 43% of known Italian species (Brochet *et al.*, 2016). According to the survey, Italy and Egypt are the countries in the Mediterranean basin in which the greatest number of birds are illegally killed.

In 2017, after EU-Pilot 5283/13/ENV in which the European Commission asked Italy to reduce poaching and to improve the enforcement of the Birds Directive, MASE launched an action plan to address the problem (Conti *et al.*, 2019). The plan aims to strengthen the direct and indirect control of poaching, prevent poaching, monitor poaching

[10] Law n. 1016/1939 'Testo unico delle norme per la protezione della selvaggina e per l'esercizio della caccia'.

[11] Law n. 157/1992 'Norme per la protezione della fauna selvatica omeoterma e per il prelievo venatorio'.

and its results, and provide national coordination of the envisaged actions. In the first half of 2018, 368 wildlife crimes were reported and sanctioned. Poaching has been addressed with new vigour in recent years by the Carabinieri Forestali (the former CFS – National Forestry Corp), in collaboration with eNGOs, focussing above all on seven areas where it was particularly serious. Particularly important successes were achieved on the Strait of Messina, where the killing of migrating birds of prey has been greatly reduced; in southern Sardinia and the Pontine Islands, where the use of snares and traps was widespread; and in the area of the province of Brescia. In the latter area poaching was so widespread that the action of the Carabinieri Forestali only partially reduced the impact on migratory birds. In this, and other areas, poaching is practised to a significant extent by people who have a hunting licence but use illegal hunting methods, such as electronic lures, and kill non-huntable species.

20.4.4 Species Recovery and Action Plans

Actions to conserve species that have an unfavourable conservation status have been initiated in several cases, at least formally, as soon as knowledge was acquired on their population status, threats and required measures. Various action plans on animal species[12] have been prepared, mainly by the MASE through ISPRA and NGOs. Guidelines also have been produced for the translocation of native plant species, the introduction of animal species, the conservation of Italian freshwater fish, the monitoring and conservation of bats, and the control and eradication of IAS.

Practical conservation actions have been carried out using EU (e.g. LIFE), national and local funding. Several re-introductions of locally extinct species or restocking of very small, residual populations have been carried out (e.g. Purple Gallinule (*Porphyrio porphyrio*), Griffon Vulture (*Gyps fulvus*), Bearded Vulture (*Gypaetus barbatus*), Alpine Ibex, Apennine Chamois and Adriatic Sturgeon (*Acipenser naccarii*)). Some of them have not had a favourable outcome, as in the case of the White-headed Duck (*Oxyura leucocephala*), while others have been successful. For example, NGO projects financed by the LIFE programme have led to a remarkable increase in the breeding success and the number of breeding pairs (from 24 in 2011 to 44 in 2019) of the Bonelli's Eagle (*Aquila fasciata*) in Sicily, where egg and chick stealing to supply the falconry black market was the main limiting factor.

Italy has also played a key role in the recovery of the Brown Bear and especially the Wolf, in Europe, as neither species went completely extinct in the country and their populations have since recovered and spread over the last 30 years (see online Annex IT.1 for further details). This has occurred as a result of a mix of land-use changes (including land abandonment and forest expansion), better protection and measures to prevent and compensate for damage such as livestock losses (Paoloni *et al.*, 2014). The Alpine population of the Brown Bear has also benefited from EU LIFE projects that funded translocations of bears from Slovenia.[13]

However, conflicts between humans and these large carnivores are increasing, and this appears to be primarily a cultural issue. The number of incidents and amount of damage are too low to explain the level of conflict that has arisen without including cultural aspects. Therefore, looking to the future, even though conservation biologists have been successful, it is evident that further conservation measures need the additional involvement of a wider range of experts – including communicators, anthropologists and social psychologists – to resolve the increasing conflicts. Moreover, the expansion of predators (wolves especially) close to human settlements is increasing the probability of interactions and incidents, fear, misunderstandings and political exploitation of these very iconic and visible species. The issue of livestock losses will probably decline, but there appears to be an increasing irrational fear due to the proximity of people with these large animals that has been unknown in recent times. This fact is also related to the controversial topic of hybridisation between wolves and dogs. Whilst the scientific community currently does not have a common position on the severity of the problem, it will probably increase, as almost a million stray dogs live close to human settlements in Italy.

20.5 Protected Areas and Networks

20.5.1 Protected Area Types and Their Coverage

Over the last quarter of the twentieth century, Italy developed an extensive network of protected areas – the result of which is summarised in Table 20.4.

The designation and management of protected areas are regulated by Law 394/1991 and following updates, as for Marine Protected Areas (MPAs), Law 979/1982. According to Law 394/1991, protected areas are territories including 'physical, geological, geomorphological and biological elements, or a combination of them, of relevant naturalistic and environmental values'. These areas are subject to a special regime that aims not only to protect their natural and environmental values, but also to sustainably integrate humanity and nature, preserving anthropological, archaeological and historical values, and traditional rural activities. A distinction is then made between:

[12] Marsican Brown Bear (*Ursus arctos marsicanus*), Alpine population of the Brown Bear, Wolf, Eurasian Otter, Italian Roe Deer (*Capreolus capreolus italicus*), Apennine Chamois, Corsican Hare (*Lepus corsicanus*), Lesser Kestrel (*Falco naumanni*), Rock Partridge (*Alectoris graeca*), Grey Partridge, Egyptian Vulture (*Neophron percnopterus*), Eleonora's Falcon (*Falco eleonorae*), Ferruginous Duck (*Aythya nyroca*), Lanner (*Falco biarmicus feldeggii*), Marbled Duck (*Marmaronetta angustrirostris*), Purple Gallinule, Slender-billed Curlew (*Numenius tenuirostris*), Audouin's Gull (*Larus audouinii*).

[13] Ursus/Brenta – URSUS Projects LIFE96 NAT/IT/003152 and LIFE00 NAT/IT/007131.

Table 20.4 The number and extent of international designated sites and protected areas in Italy.

	IUCN category	Number	Area (km^2)
Designated under global and non-EU international legislation			
Ramsar Sites	NR	57	740
Natural and Mixed World Heritage Sites	NA	5	1 647
Biosphere Reserves (exc. buffer and transition zones)	NA	20	1 890
Specially Protected Areas of Mediterranean Importance	IV–V	11	1 362
Designated under the Nature Directives			
SCIs (terrestrial + marine)	NR	2 358	43 959 + 13 713
SPAs (terrestrial + marine)	NR	636	41 273 + 12 819
*Natura 2000 terrestrial area**			*58 447*
*Natura 2000 marine area**			*20 717*
Designated under domestic legislation			
National Parks (Parchi nazionali)	II	25	15 471
Marine Protected Areas (Aree Marine Protette)	IV	29	2 344
State Nature Reserves (Riserve Naturali Statali)	Ia–IV–V	148	1 258
Other National Nature Protected Areas (Altre Aree Naturali Protette Nazionali)	IV	3	25 575
Regional Nature Parks (Parchi Naturali Regionali)	IV–V	143	14 011
Regional Nature Reserves (Riserve Naturali Regionali)	Ia–IV–V	452	2 517
Other Regional Nature Protected Areas (Altre Aree Naturali Protette Regionali)	III–IV–V	140	387

Notes. Nationally designated sites and area relate to both terrestrial and marine domains. World Heritage Sites are natural sites and their area relates to core zones. IUCN types. NA = not applicable; NR = not reported or not assigned. * Taking into account overlaps.
Sources. Based on data from: Ramsar Convention and UNESCO websites (February 2022); SPAMI list updated to 2020, from RAC-SPA[14]; SCIs and SPAs from MASE website (December 2021)[15]; domestic protected areas, SNB – IV Rapporto sull'attuazione della Strategia Nazionale per la Biodiversità (periodo 2017–2018)[16].

1. National Parks that protect terrestrial, inland water and marine areas for which the natural, scientific, cultural and recreational values are of international or national relevance, requiring State action for their protection.
2. Regional Nature Parks that protect areas with natural values of inter-regional or regional importance, and where local culture and traditional activities are part of the system worthy of protection.
3. State and Regional Nature Reserves whose institution aims to protect relevant flora, fauna and ecosystems important for biological diversity or genetic resources.
4. MPAs that protect the marine environment (water, seabed and coastline) in areas with relevant natural, archaeological and cultural values.

Under the law, there is no clear distinction between the Parks and the Reserves, but in practice Reserves are generally smaller than Parks, usually included in wider protected areas, and have stricter protection. The designation, approved with a formal act, of National Parks, State Reserves and MPAs is the responsibility of the MASE, in agreement with the relevant R&APs through the State-Regions Conference. The designation of Regional Parks and Reserves is the duty of the R&APs.

In this domestic legal framework, the IV Report of NBS (2016–2018) reported that Italy had 940 protected areas, with a total terrestrial area of 32 923 km^2 (10.9% of the land) and 28 652 km^2 at sea (4.9% of Italy's marine waters).

Since the 1970s, the criteria used to select nationally designated protected areas have included the presence of rare or threatened habitats and species of conservation interest, often according to assessments by scientific groups. In 1991, the Nature Chart was introduced, under which areas are evaluated in terms of their environmental quality and vulnerability. This has helped with the identification of areas to protect (Cardillo *et al.*, 2017).

There are several other types of protected areas, publicly or privately managed, that contribute to nature conservation. Some of these sites are designated with the specific aim of species or habitat protection (e.g. Nature Oases managed by NGOs) while others integrate environmental goals with

[14] www.rac-spa.org/spami
[15] www.mase.gov.it/pagina/sic-zsc-e-zps-italia
[16] www.mase.gov.it/pagina/documenti-e-atti

maintenance of traditional rural and cultural values (e.g. parks of local interest, such as PLIS in Lombardy and ANPIL in Tuscany). These sites are usually designated with regional, provincial or municipal acts. Moreover, all forests in Italy are protected, in addition to having a significant proportion within protected areas (27% according to INFC (2005)). The management of forested areas is regulated by a specific law (GU 92/2018), ratified and applied by each R&AP.

Following the Nature Directives, the network of domestic protected areas has been complemented, since the mid-1990s, by the Natura 2000 network. The identification of sites for the network was carried out by the MASE with support from two LIFE projects in 1992[17] and 1995.[18] As a result of the projects, Italy was among the first EU Member States to submit to the European Commission, between 1995 and 1998, its initial official lists of proposed Natura 2000 sites (i.e. pSCIs and SPAs).

According to the MASE data presented in Table 20.4, as of December 2021, the combined extent of the SCI and SPA networks, excluding overlapping areas, was 58 447 km^2 on land, which is 19% of the country. Following EU Pilot 8348/16/ENVI, the marine Nature 2000 network has increased significantly, especially in the past few years, from 5 878 km^2 in April 2017 to 20 717 km^2 in December 2021 (CCN, 2021). The marine Natura 2000 network now covers 13% of Italy's Territorial Sea, or 3.5% of its total marine waters. However, according to the European Commission (infringement case 2021/2028) further sites are required, especially to protect reefs and the Mediterranean Monk Seal, Loggerhead Turtle (*Caretta caretta*), Bottlenose Dolphin (*Tursiops truncatus*), Scopoli's Shearwater (*Calonectris diomedea*) and Yelkouan Shearwater (*Puffinus yelkouan*).

Overall, the domestic protected areas, together with the Natura 2000 network (79% of which overlapped with national and regional protected areas in 2016 (MATTM, 2016)), covered 104 000 km^2 in 2020, 20% of the country's terrestrial area, and 11% of its Territorial Sea and Ecological Protection Zone (CCN, 2021). According to more recent data from the WDPA in March 2022, for broadly the same protected area types listed in Table 20.4, the total protected area coverage was 64 856 km^2 on land and 57 094 km^2 at sea (UNEP-WCMC, 2022). This also equates to 21% of the land, whilst marine coverage is estimated to be 9.7% in relation to Italy's total marine waters.

The designation of SCIs as SACs is a multistage process in Italy involving shared agreements between the State and R&APs. Since 2015, after the infringement case 2015/2163 (when only 18% of SCIs were designated as SACs within the six-year deadline required by the Habitats Directive), the MASE and R&APs have engaged in intense cooperation to accelerate the SAC designation process. As a result, by December 2021, 97% of 2 358 SCIs had been designated as SACs.

20.5.2 The Coherence of Protected Areas and Ecological Networks

Even though Natura 2000 increased the area devoted to conservation, some areas with high numbers of species are unprotected and more than 50% of the highly irreplaceable areas are not considered in the system (Maiorano *et al.*, 2007). Only 25% of terrestrial vertebrates are covered by the combination of national protected areas and Natura 2000 (Maiorano *et al.*, 2015). Furthermore, according to Saura *et al.* (2018), in Italy the 'Protected Connected Land' (ProtConn) indicator value, which quantifies the proportion of the land that is within connected protected areas, is between 8% and 12%, and therefore below the CBD Aichi Target 11 of having at least 17% of the land covered by well-connected protected area systems by 2020.

According to the regulation implementing the Habitats Directive, in order to ensure the ecological coherence of the Natura 2000 network, the MASE should define guidelines for the management of areas that have the function of ecologically connecting areas that are of primary importance for wild fauna and flora. This has not yet been accomplished, but many ecological networks have been planned at the provincial and regional level, also supported by the guidelines on the management of functional ecological connection areas drawn up in 2003 by ISPRA (APAT and INU, 2003).

A clear picture of the current state of play of the regional ecological networks is, however, still not available. In order to ensure connectivity beyond regional borders and greater coordination in the development of regional ecological networks, the MASE is planning to define a national strategic vision to be included in the NBS updating, based on the upcoming guidance on a strategic framework for supporting the deployment of EU-level green and blue infrastructure.

20.5.3 Approaches to Protection and Levels of Protection

Hunting is forbidden in all officially listed protected areas,[19] except where necessary for ecological reasons, such as to reduce excessive numbers of Wild Boar (*Sus scrofa*). Spatial development plans and individual projects that may have environmental impacts on protected sites are required to have

17 LIFE92 NAT/IT/013100 – First phase of the implementation of the Habitats Directive in Italy.

18 LIFE94 NAT/IT/001048 – Bioitaly – Extension of the Habitat Italia programme.

19 To be included in the Official List of Protected Areas, a protected area: (1) should be formally designated by a public or private act; (2) should have clear boundaries (with cartography); (3) should include naturalistic values, as listed in Law 394/1991; (4) should adopt conservation measures coherent with Law 394/1991, including amongst others the prohibition of hunting activities; (5) should be managed by a public or private subject; (6) should have a financial report; www.mite.gov.it/pagina/elenco-ufficiale-delle-aree-naturali-protette-0 (12 April 2022).

an SEA and/or EIA. In addition, as a general rule, in Parks and Reserves all activities that could be detrimental for species and ecosystems are forbidden. In practice, the level of protection varies amongst the protected areas, depending on the aims of their designation.

Law 394/1991 prohibits some specific activities in National Parks, including capturing, killing or disturbing animals, collecting plants and damaging vegetation (except for agriculture or forestry), introducing exotic species, opening quarries and waste dumps, modifying water regimes or altering the geophysical or biochemical environment. The same law also prohibits similar activities in MPAs, including fishing – at least in core areas (see below). Any derogations to these prohibitions are included in the regulation of the protected area, which must also be provided with a management plan that not only defines the protection measures needed to achieve its nature conservation aims, but also regulates human activities (e.g. urbanisation, commercial, agricultural, recreational and scientific activities) to ensure they are compatible with the objectives of the site. Indeed, a large proportion of the protected areas are on private property and are important for the country's productivity and economy, particularly for agriculture and tourism. For example, there are 214 535 farms in Natura 2000 areas, which cover almost 47% of the Natura 2000 network (Servadei *et al.*, 2018).

Protected areas are often subject to zoning with differing management strategies based on their biodiversity conservation priorities. For example, part of a site could be a strict reserve, where the integrity of nature is fully maintained and no human activity other than research is allowed; other parts may allow traditional activities (e.g. agriculture or fishing); and, in others, socio-economic activities compatible with the conservation aims of the protected areas could be permitted.

Problems have occurred with respect to Natura 2000 sites as a result of the unsatisfactory implementation of the HD Article 6(3) appropriate assessment procedures. The European Commission found various problems in the implementation of the appropriate assessments (Pilot 6730/14/ENVI), some of which had already been identified in the NBS. The most significant were: poor quality of the studies, inadequate capacity in the competent bodies, fragmentation of competences and difficult interactions, inadequate detailed requirements for the studies and their evaluation, and poor harmonisation of regional rules.

Following the EU Pilot, the Italian authorities, in consultation with the European Commission, have drawn up national guidelines for the implementation of HD Articles 6(3) and 6(4)[20] to improve, clarify, harmonise and streamline appropriate assessment procedures across all regions. Moreover, with the CReIAMO-PA project 'Skills and Networks for Environmental Integration and for the Improvement of Public Administration Organisations', which started in 2018, the MASE also aims to strengthen the administrative capacity for carrying out appropriate assessment procedures (European Commission, 2019).

20.5.4 Approaches to Managing Protected Areas

National, regional and inter-regional parks and reserves are managed through regulations and site management plans, which are obligatory. Plans and regulations are drawn up by the managing body of the protected area, after consultation with the territorially concerned authorities (i.e. regions, municipalities, etc.), and formally approved by the competent authorities (i.e. the MASE and R&APs). Once active, plans of protected areas replace tools of urban and landscape planning of municipalities included in their boundaries.

Since 2007, when only the Belluno Dolomites parks had a plan in force, implementation of plans in National Parks has advanced significantly. As a result, by 2019, 10 National Parks had plans in force, six had adopted plans, seven had drawn up plans or were preparing them, and only two lacked plans (CED PPN, 2016–2017, 2019[21]). As for Regional Nature Parks, in 2019 50% had plans in force, while 25% were preparing them (CED PPN 2016–2017, 2019).

According to the national legislation (DPR 357/97 as amended, and DM 17/10/2007), when Natura 2000 sites, or parts of them, fall within domestic protected areas, conservation measures in force in the protected areas also apply to the Natura 2000 sites, and the Natura 2000 sites' conservation measures are integrated into those of the protected area. Since 2009, significant efforts have been made to carry out this integration, but the coordination of the production of integrated planning and management tools is still ongoing.

Management plans for the Natura 2000 sites are not obligatory, but they can be useful for establishing the necessary conservation measures. In 2002, the MATTM produced the 'Natura 2000 site management guidelines' (DM 3/9/2002), and later a technical manual[22] to assist with their application. A number of plans have been produced by managing bodies in accordance with the guidelines and formally adopted by the R&APs. To prompt the regional process of adopting conservation measures, the ministry has set 'minimum uniform criteria' (DM 17/10/2007) that the R&APs have to follow to identify such measures.

In December 2021, about 97% of the Nature 2000 sites had established conservation objectives and measures, but these are often not clearly defined or quantified. This gap is currently

[20] Linee Guida Nazionali per la Valutazione di Incidenza (VIncA) – Direttiva 92/43/CEE 'HABITAT' articolo 6, paragrafi 3 e 4. Adopted on 28 November 2019 following the agreement between the government and R&APs. https://www.mase.gov.it/sites/default/files/archivio/allegati/rete_natura_2000/linee_guida_nazionali_valutazione_di_incidenza_2019.pdf

[21] http://www.cedppn.polito.it/archivio_parchi_e_aree_protette/italia_2017_2019

[22] MATTM (2005). Manuale delle linee guida per la redazione dei Piani di gestione dei Siti Natura 2000.

subject of the infringement case 2015/2163 (additional nal notice on 24 January 2019). The establishment of more surable objectives and certain mandatory measures are issues that the ministry has prioritised. Furthermore, y plans to update the NBS, reflecting the commitment to w up national strategic conservation objectives that would it achieve favourable conservation status for its habitats species (European Commission, 2019).

Regional innovative approaches in managing Natura 0 sites are taking place. For example, Lombardy and bria have developed an integrated and participatory man-ment strategy that aims to meet the conservation objectives the Natura 2000 network and, at the same time, contribute the objectives of regional development. The strategy has owed Lombardy to obtain the first LIFE Nature Integrated oject in Italy,[23] which includes innovative measures to prove the governance of the Natura 2000 network.

).6 General Nature Conservation Measures

).6.1 Development Planning and Impact ssessment

mong the main threats to species and habitats, the NBS iden-ies land consumption, land-use change and fragmentation. At tional level, land consumption has increased from an esti-ated 2.7% in the 1950s to 7.6% in 2017, including within otected areas, where it is 2.4% (ISPRA, 2018). Moreover, 3% of the national territory is characterised by very high agmentation (ISPRA, 2019).

The NBS recognises that this situation is in part the result f insufficient and inadequate integration of biodiversity pro-ction in the territorial planning instruments. In Italy, regions re responsible for land-use planning, while the State lays own the basic principles. Regions, in turn, regulate the vari-us planning levels (regional, provincial, municipal). The same NBS encourages the MASE to identify as soon as possible the fundamental elements of the territorial arrangement with eference to natural and environmental values' (required by Art. 3 of Law 394/1991) as an indispensable tool to introduce national-level strategic objectives into planning. Although some progress has been made (i.e. the Nature Chart assess-ment of the state of the nature in Italy), these elements have not yet been defined (MATTM, 2020).

One particular problem has been the deficient integration of ecological networks into planning instruments. Whilst there have been numerous designs for ecological networks, trans-forming them from a concept into a real plan component with the strength to tackle fragmentation and to protect natural capital outside protected areas still represents a major challenge (Lombardi and Giunti, 2014). This was also recognised in the 2010 NBS, which included, as a priority action, promoting the use of planning methodologies for large-scale ecological net-works as integral and prescriptive parts of land-use planning.

Although all regions have territorial plans including refer-ences to ecological connectivity/networks, each plan deals with the topic in a different way. Moreover, the networks seem merely to have a theoretical role in the local planning tools, and do not appear to have any practical consequences (ISPRA, 2017). However, some regions, such as Umbria, are giving their ecological network a more binding character by approv-ing it with a legal act.[24] This integrates the ecological network into the land management legislation and requires an SEA, EIA or appropriate assessment before authorising develop-ment plans/projects that may affect the network.

20.6.2 Habitat Management and Restoration

In Italy, human land use has profoundly shaped habitats and continues to affect their diversity and quality. This is true for many habitat types (e.g. lagoons, coasts and forests) but espe-cially for HNV farming systems, where sustainable manage-ment is urgently needed and advocated to conserve their biodiversity. Consequently, measures to support sympathetic practices are included in the CAP Rural Development Programmes (RDPs) of most Italian regions. Within the 2014–2020 CAP, these included the use of the knowledge transfer measure (M1), non-productive investments connected to agri-environment–climate requirements (M4.4), the agri-environment–climate measure (M10), the Natura 2000 and WFD measure (M12) and payments for areas with natural constraints (M13) (see funding allocations in Section 20.7.3).

The most important operations supported through the RDP measures for semi-natural open habitats (e.g. wood pas-tures, heathlands and meadows), traditionally maintained through extensive livestock rearing, have included grazing plans that have regulated livestock density, period of stay and rotation, as well as requiring tree and shrub control and limitations on chemical inputs. The creation of drinking ponds for livestock also benefits amphibians and invertebrates, and the use of sheepdogs and electrified fences helps to reduce livestock losses to large carnivores. However, a recent study conducted in the alpine pastures of Lombardy (Eliante *et al.*,

[23] LIFE14 IPE/IT/000018 – LIFE IP GESTIRE 2020 – Nature Integrated Management to 2020.

[24] DGR 30/11/2005 n. 2003 of the Umbria Region. Approval of the Umbria Region Ecological Network project (RERU), embodied in the PUT (Territorial Urban Planning) LR 22/02/2005 n. 11, amendments to the Regional Law 24/03/2000 n. 27.

DGR 26/11/2008 n. 8/8515 of the Lombardy Region. Approval of the final project on the regional ecological network and of the document 'Regional ecological network and territorial planning of local authorities'.

DGR 30/12/2009 n. 10962 of the Lombardy Region. Regional Ecological Network: approval of the final project, including the Alps and Prealps Sector.

LR 05/02/2013 n.2 of the Marche Region. Rules on the ecological network of the Marche and landscape protection and changes to the Regional Law 15/11/2010 n. 16 '2010 Budget adjustment'.

2017) found that these RDP measures are often too weak to promote the practices that are required to conserve the habitats and biodiversity. Although there are some general operations that can be applied in pasture management (e.g. on grazing plans and livestock surveillance), local conditions (especially land accessibility, the type of livestock and their size, and the presence or absence of large carnivores) create a variety of situations that need to be taken into account in more tailored actions in RDP measures.

Traditionally, flooded ricefields are another biodiverse farmland habitat in Italy as they are a surrogate for natural wetlands and often host vegetation, invertebrates, amphibians and waterbirds of conservation interest (Fasola and Ruiz, 1996; Bambaradeniya *et al.*, 2004). However, the ecological value of the 'modern' paddy fields now appears to be compromised, as shown by declines in species that have been historically common in Italian ricefield areas, including amphibians (Fasola and Cardarelli, 2015) and breeding herons and egrets (Fasola *et al.*, 2018). One of the major threats to the habitat, together with mechanisation and chemical inputs (Cardarelli and Bogliani, 2014; Giuliano *et al.*, 2018), is a change in water management. Traditional cropping using permanent flooding beginning in spring, which creates suitable areas for wetland flora and fauna, is being replaced by techniques that use shallower water for a shorter submersion period (Ranghetti *et al.*, 2018; Ranghetti and Boschetti, 2022). Actions to mitigate the effects of these modern techniques have been developed in recent years (Giuliano *et al.*, 2017) and supported through RDP measures in the regions with large areas of ricefields (e.g. agri-environment–climate measures in Lombardy and Piedmont). Of most importance are measures that create permanently wet water refuges connected to the field and maintain vegetated banks during the cropping season, to which farmers can add accompanying commitments during winter such as flooding the fields, stubble maintenance or maintaining crop cover through the planting of legumes.

Even if RDP measures have helped to maintain or spread environmentally sympathetic practices in agricultural landscapes, in some cases they have been unable to contribute to biodiversity recovery (Assandri *et al.*, 2019). Some agricultural areas, such as ricefields, vineyards and grasslands, may be nowadays so intensively managed that evidence of their benefits to biodiversity in comparison to other arable lands is no longer obvious (Eliante *et al.*, 2017; Assandri *et al.*, 2019). In the future CAP, financial resources need to be better allocated and tailored so that they are appropriate for low natural value areas as well as those with high natural values (e.g. intensive versus extensive meadows).

20.7 Nature Conservation Costs, Benefits and Funding Sources

20.7.1 Nature Conservation Costs

Environmental spending reported in the state budget indicates the attention given to natural capital and ecosystem services at the level of public expenditure. Based on the last Eco-re (MEF, 2019), environmental expenditure amounted to aro €2.5 billion in 2018, 36% of which was related to biodive and landscape protection, and the use and managemen natural resources.

According to Gantioler *et al.* (2010), the annual estim cost for implementing the national Natura 2000 network in I amounted to €182 million (€32 million one-off costs and € million recurrent costs), averaging about €27/ha. More com hensive estimates of the regional costs of the full implementat of Natura 2000 have been carried out by the Lombardy Umbria Regions, through detailed breakdown of actions their costs. The estimated costs of the management of Nat 2000 sites in Lombardy over 2014–2020 were approxima €215 million, equal to about €30 million per year and €92 yr (Gatto *et al.*, 2015). In Umbria, the Natura 2000 costs o 2017–2023 were estimated to be around €78 million, equal €11 million per year and €86 ha/yr (Regione Umbria, 2017).

Estimates of the costs of fully implementing the Nat 2000 network in Italy for the 2014–2020 EU programmi cycle have been reported on the CBD Financial Reporti Framework,[25] based on information from the Prioritised Acti Frameworks (PAFs). For the whole 2014–2020 period, the costs implementing the Natura 2000 network were estimated to ran from €1.8 to 2.5 billion, equalling c. €257–357 million per year

20.7.2 The Economic Benefits of Nature Conservation

Over the last 30 years there have been numerous attempts quantify the economic benefits of nature conservation in Ital in particular in forested areas at a regional scale (CCN, 2018 Financial accounts of stock-flows supplied by ecosystem although extremely difficult to produce, are considered nece sary to demonstrate to public and private administrations th economic, social and environmental costs of biodiversity/hab tat loss. The national system of habitats was valued, for 2017, €13.5 billion (€869/ha), of which €4.9 billion was ascribable t forests, €1.5 billion to pastures and other grassland, and €10 million to wetlands (CCN, 2018). Recreational service (excluding tourism) of green areas (natural, rural and urban were estimated to have a value of €3.3 billion in 2012, wit forests providing the highest contribution (€1.9 billion).

In 2014, the MATTM supported a project that evaluate the status and quantified the financial benefits of the natura capital and environmental flows produced by national MPA (Franzese *et al.*, 2015). During the project, a standardise methodology was developed, based on 'emergy' analysis,[26] tha

[25] Financial Reporting Framework: Reporting on baseline and progress towards 2015: https://chm.cbd.int/pdf/documents/resourceMobilisation/243254/1.

[26] 'Emergy' analysis is an accounting method useful to identify and convert in a single unit of measurement all resources feeding a system.

converted biophysical and trophodynamic characteristics of the marine stocks into monetary units (Vassallo *et al.*, 2017; Buonocore *et al.*, 2019). The economic value of the ecosystem services for each of the 12 MPAs for which the analysis was completed was estimated to vary from €7 to 113 million/yr, giving a total value of €570 million/yr (CCN, 2021).

Economic estimates have also been produced of the stock-flows supplied by some specific habitat/species (ISPRA, 2016; CCN, 2017). For example, land consumption during 2012–2015 caused a loss of agricultural productivity due to reduced pollination services that was estimated to be €20 602/km^2, giving a total of €2.7 million per year at the national scale.

20.7.3 Funding

As shown in Table 20.5, the main EU fund used to support nature conservation is the European Agricultural and Rural Development Fund (EARDF), with some €1.03 billion allocated annually on average over the 2014–2020 funding period in relation to the EAFRD Priority 4 objective for ecosystems.

According to the PAFs, the EAFRD also provides the most significant funding for Natura 2000. The Natura 2000 payments measure (M12) was used in the 2014–2020 RDPs of 11 of the 21 regions (MiPAAF, 2020). More broadly, the RDPs included 359 submeasures/operations related to Natura 2000, 228 of which were specifically focussed on it. Important incentive payments were included in:

- Non-productive investments (M4): creation of agro-ecological infrastructures and ecological corridors (e.g. walls, fences and buffer strips), restoration and management of habitats, mitigation measures for human–large carnivore conflicts.
- Agri-environment–climate measures (M10): management of green infrastructure, protection and improvement of grasslands, wetlands, forests and specific habitats of species, conversion of arable land to grasslands, and integrated farming.
- Improvement of resilience, ecological efficiency, and environmental value of forest ecosystems (M8): diversification of forest structure and composition, preservation of monumental trees (i.e. particularly large or old specimens), creation of ecotones and clearings, establishment of wildlife refuges.
- Cooperation (M16): collective projects to improve conservation status and promote sustainable tourism in Natura 2000 sites, and for the sustainable management of habitats.
- Basic services and village renewal in rural areas (M7): drafting/updating management plans, monitoring and studies of species and habitats, assessment of conservation measures and planning tools, procedures for estimating the economic values of ecosystem services from habitats, recovery of Natura 2000 habitats and priority species protection, restoring and/or creating ecological connection elements, awareness raising activities, recreational and tourist infrastructures.
- Knowledge exchange and training and advice (M1 and M2) on the above issues.

The European Regional Development Fund (ERDF) has not provided substantial funds for nature conservation in Italy over previous programming periods, or over 2014–2020, although some regions have used it to finance the implementation of measures included in the PAFs. The European Maritime and Fisheries Fund (EMFF) and European Social Fund have been used by a few regions in the past, and were expected to have somewhat greater use over 2014–2020.

Although relatively small in terms of total funding, the LIFE programme has been of special importance due to its targeted and tailored nature conservation projects. As of May 2021, Italy had been awarded 318 LIFE Nature and Biodiversity Projects (European Commission, 2021), and currently has two important Integrated Projects underway.

The Italian regulatory framework for the Green Economy (Law 221/2015) promotes payments for ecosystem services (PES), or PES-like schemes, for biodiversity conservation and their integration with other existing tools for public management authorities. In particular, Articles 67 and 70 focus on natural capital accounting systems and PES for the benefit of local communities. These measures aim to resolve the long-standing problem of insufficient financial resources for protected areas. Many recent projects have shown an increasing tendency towards innovative instruments for environmental governance, such as the PES Italian Observatory by the Italian Network of the Ecosystem Service Partnership. However, the application of the payment schemes, which are mainly focussed on securing the provision of ecosystem services through supporting traditional land-use practice or management activity, is still in its early phase (Marino and Pellegrino, 2018).

The EU approach to financing the implementation of the Natura 2000 network is based on existing EU funds rather than developing specific financing instruments. As noted by the European Court of Auditors, Italy has declared that there is a gap between the estimated funding needs for Natura 2000 and the available funds, which reflects the limited integration of Natura 2000 with the EU funding schemes (European Court of Auditors, 2017). Although the total funding potentially available, as identified in Table 20.5, appears to have been more than adequate to meet the requirements for Natura 2000 as identified in the PAFs (c. €257–357 million per year), in practice the resources allocated for 2014–2020 were lower than the regional financial needs for the management of Natura 2000 (MATTM, 2017).

20.8 Surveillance, Monitoring, Reporting and Applied Research

The monitoring of BHD habitats and species is of particular importance, not only to determine the conservation status of

Table 20.5 Funding sources for nature conservation in Italy.

Source and period	Measure/category	Average annual allocation (€ m)		Comments
EU funds		**EU + national**	**EU**	
EAFRD Rural Development Programme measures – allocations to Priority 4 (restoring, preserving and enhancing agriculture and forest ecosystems) 2014–2020	M10 Agri-environment–climate	329.3	163.4	
	M11 Organic farming	291.5	155.2	
	M13 Areas facing constraints	240.9	118.7	
	M8 Investments in forests	79.91	43.71	
	M4 Investments in physical assets	35.4	19.2	
	M16 Cooperation	23.23	10.61	
	M12 Natura and WFD	9.2	5.2	
	M1 Knowledge transfer, etc.	7.1	3.3	
	M15 Forest environment climate	6.3	3.7	
	M2 Advisory services, etc.	5.1	2.4	
	M7 Basic services, etc.	3.8	1.9	
RDP Priority 4 subtotal		1 031.8	527.3	
ERDF (including European Territorial Cooperation) 2014–2020	ERDF	Low use		Implementation of N2k PAFs and management plans, green infrastructure and N2k transboundary cooperation
	Territorial cooperation	Medium use		
European Social Fund 2014–2020		Low use		Training of N2k stakeholders
EMFF 2014–2020		Low use		Management, restoration and monitoring of N2k
LIFE Programme	Nature and Biodiversity Projects (1992–2021)	16.7	9.6	
	Integrated Projects on nature (2016–2027)	4.4	2.6	Two projects: GESTIRE 2020 and IMAGINE UMBRIA
EU funding instrument subtotal		*1 052.9*	*539.6*	
National funds within 2014–2020		**Additional national**	**Comments**	
Various Ministries (Environment, Defence, Agriculture, etc.) 2018		875	Mainly recurring costs of protected areas and of the National Forestry Corp. Uncertain if includes some co-financing of EU instruments above	
Payments for ecosystem services		Sporadic use	Under development	
Bank foundations		Some regions	For habitat restoration, ecological connections, awareness raising	
National total including co-financing of EU instruments		?	Overlaps in funding data	
Total		**c. 1 927.9**	Incomplete and possible overlaps	

Notes. The subtotals and totals are only approximate indications of the average annual levels of funding as they are based on annualised data from differing periods.
Sources. Based on data from: RDP data from European Commission ESIF funds database (23 October 2020); Italy LIFE factsheet (European Commission, May 2021); MiPAAF, 2020; MEF, 2019. See Section 1.3 for further details.

ı component, but also to evaluate the efficacy of conser- on measures applied inside and outside Natura 2000 sites. his framework, during 2015–2016, ISPRA, on behalf of the TTM, produced guidelines to enable the collection of har- nised data at a national level for terrestrial and freshwater cies and habitats, through standardised methodologies for itats (Angelini *et al.*, 2016), plants (Ercole *et al.*, 2016) and mals (Stoch and Genovesi, 2016). The guidelines were pre- ed with the help of the most relevant scientific societies, and sharing the output with the R&APs. The guidance for each ɔitat and species includes descriptive information, data on eats and impacts, and monitoring techniques with recom- ndations for the frequency and number of fieldwork days d number of personnel to be involved.

Over recent years, data from monitoring programmes ried out by scientific societies, institutions and amateurs ve increasingly converged and been organised in georefer- ced and expert-validated databases, at both local and tional scales. An example is a national database on birds,[27] ıose platform also hosts the databases of other animal taxa g. dragonflies, amphibians, reptiles, and marine and terres- al mammals) and specific monitoring programmes on rare d alien species.

Overall, monitoring is currently relatively complete only r some taxa (e.g. birds) and some habitats. Especially for vertebrates, data are incomplete or totally lacking, as moni- ring programmes have only recently started and only include ɔme taxa (e.g. the Butterfly Monitoring Scheme in 2018). For ɔme species with Italian populations of particular inter- ational importance (e.g. Wolf, Brown Bear, Alpine Ibex and pennine Chamois), there are ongoing monitoring projects, ven if the degree of territorial coverage is sometimes incom- lete. The scope of species monitoring therefore needs to be idened, including through public investments that support urther data collection by volunteers.

20.9 Future Developments

The Committee on Natural Capital (CCN, 2018[28]) has recom- mended that, amongst others, the following major issues and priorities are tackled across the whole national territory:

- integration of the physical and economic impacts on habitat and species, and on ecosystem services, in the monitoring and evaluation procedures of investments and public policies;
- evaluation of the positive and negative impacts associated with agronomic management and subsidies to the agricultural sector, in order to inform decisions on the allocation of resources in relation to agricultural policy; and
- implementation of practical policies to control land-use change, (a) strengthening the promotion of actions for environmental restoration, to reduce land consumption and ecosystem fragmentation, and (b) favouring, in the context of territorial planning and project evaluations, those that include environmentally friendly options (nature-based solutions, green infrastructure) compared to traditional infrastructure (grey infrastructure).

As regards the Natura 2000 network, commitments by the MASE include:

- Use of the PAFs as regional strategic programming documents, to better define and quantify conservation objectives and measures for species and habitats, as well as more effective mobilisation of resources to finance priority conservation and restoration measures. Moreover, the MASE has produced the 'Database for the management of the Natura 2000 network', that will make it easier to consult and update the PAFs.
- Definition of an overall national strategy for achieving the favourable conservation status of habitat and species. The MASE intends to provide the regions with a set of objectives at national level as soon as possible.
- Improving governance by providing management bodies with coaching and training to strengthen their capacities and competences, and by supporting the exchange of their management experiences and problems with other countries within the biogeographical processes.

20.10 Conclusions

In Italy, the governance system for nature protection is highly decentralised, with the State having the power to set standards for the whole country, while the R&APs further specify these standards based on their specific needs (Cecchetti, 2003). In this context, the regions have played a significant role in nature protection, sometimes in advance of state actions, for example in terms of the protected areas that were in place in northern regions before the national Framework Law in 1991. It must be said, however, that the position of the 19 regions and the two autonomous provinces has not been homogeneous, and in several cases they have not welcomed the establishment and presence of National Parks or they have insufficiently desig- nated Regional Parks and Reserves where their natural value is high. Actually, nature protection policy developments in Italy have been mainly driven by EU legislation and international conventions (Alonzi *et al.*, 2006). Over the last 30 years, these have helped advance Italian government policies on biodiver- sity conservation, such that they go beyond the protection of listed species to more strategic management and sustainable development.

27 www.ornitho.it

28 In 2015, a Committee for Natural Capital (Comitato per il Capitale Naturale – CCN) was instituted, presided over by the MASE, with the aim to produce an annual report on the status of Natural Capital, including data on its entity (e.g. land-use cover, number of protected areas, . . .) and on its economic value in terms of ecosystem services provided. Moreover, the annual report includes an evaluation of the effects of public policies on Natural Capital and its ecosystem services.

Despite the early proposals for SPAs and SCIs in the 1990s, the identification and establishment of new Natura 2000 sites are still taking place, mostly under pressure from the European Commission, as there are specific BHD habitats and species with inadequate protection in Italy. This is especially the case in the marine environment, where more SCIs and pelagic SPAs are required.

Among the principal achievements of the recent decades has been the publication of the NBS. Progress has been made with its implementation: from the improved protection of habitats and species, to the deepening of knowledge of their conservation status and distribution, to the acceleration in the protection of the marine environment (MATTM, 2021). Efforts have also been made to complete the designation of SACs, integration of domestic protected area and Natura 2000 measures, and integration of the objectives of the Nature Directives into those of other EU directives.

However, given the ongoing habitat degradation and species declines described in Section 20.2, much remains to be done. It seems essential to further improve the actual governance of nature conservation. The dual State–Regions system often results in slow implementation, as well as varying approaches to nature conservation and management at a regional scale. The hierarchical structure, together with the high fragmentation of competences at all levels, makes the effective management of the nature system challenging. It is crucial that central, regional and local administrations cooperate and coordinate more effectively with each other. The identification of national strategic objectives, starting from those set at site level to achieve the favourable conservation status of habitats and species, supported by a clear political commitment, could improve Italy's contribution to the *EU Biodiversity Strategy for 2030*, and at the same time could strengthen national coordination and political guidance at central level. On the other hand, the multilevel governance system has led to regional examples of innovative management, which could be replicated in other Italian and European contexts.

As for the impact of hunting, the number of hunters, nationwide, has more than halved between 1980 and 2020; however, the problem of the lack of congruence between the abundance of the populations of huntable animals and the actual killing remains. Poaching has been addressed with new vigour in recent years, through the activity of the Carabinieri Forestali and environmental NGOs, although in the worst areas it has only partially reduced the impacts on migratory birds. Some important successes have been achieved in reducing the theft of eggs and young from the nests of threatened birds, such as the Bonelli's Eagle, whose breeding population has increased following nest surveillance campaigns in Sicily.

Although substantial funds are potentially available for nature conservation, with the LIFE programme and the EARDF used most, the resources allocated for 2014–2020 were lower than the regional financial needs for the management of Natura 2000 (MATTM, 2017), let alone for broader nature conservation requirements. Greater mobilisation and optimisation of the available financing sources, together w[ith] more effective tracking, therefore remains a major requ[ire]ment to achieve the objectives of the NBS and Na[ture] Directives. It is needed to strengthen coordination at diffe[rent] levels and sectoral competences so that biodiversity and Nat[ura] 2000 are better integrated into the EU funding programm[es].

Overall, greater efforts need to be deployed, especi[ally] considering the new *EU Biodiversity Strategy for 2030*. I[taly] should actively contribute to it, incorporate its targets in [its] national strategy, and actually reach the targets.

Acknowledgements

We are grateful to Dr Graziana Dizonno and Dr Clau[dio] Celada for their precious comments that helped improve t[he] manuscript, and to Dr Mauro Belardi, who supplied inf[or]mation on large carnivores.

References

Alonzi, A., Ercole, S. & Piccini, C. (2006) *La protezione delle speci[e] della flora e della fauna selvatica: quadro di riferimento legislati[vo] regionale*. APAT Rapporti 75/2006. Rome: Agenzia per la protezione dell'ambiente e per i servizi tecnici.

Andreotti, A. & Borghesi, F. (2012) *Il piombo nelle munizioni da caccia: problematiche e possibili soluzioni*. Rapporti ISPRA, 158/2012. Rome: Istituto Superiore per la Protezione e la Ricerca Ambientale.

Angelini, P., Casella, L., Grignetti, A. & Genovesi, P. (eds.) (2016) *Manuali per il monitoraggio di specie e habitat di interesse comunitario (Direttiva 92/43/CEE) in Italia: habitat*. ISPRA, Ser[ie] Manuali e linee guida, 142/2016. Rome: Istituto Superiore per l[a] Protezione e la Ricerca Ambientale.

APAT & INU (2003) *Gestione delle aree di collegamento ecologico funzionale. Indirizzi e modalità operative per l'adeguamento degl[i] strumenti di pianificazione del territorio in funzione della costruzione di reti ecologiche a scala locale*. Manuali e linee guida 26/2003. Rome: Agenzia per la protezione dell'ambiente e per i servizi tecnici e Istituto Nazionale di Urbanistica.

Assandri, G., Bogliani, G., Pedrini, P. & Brambilla, M. (2019) Towar[d] the next Common Agricultural Policy reform: determinants of avian communities in hay meadows reveal current policy's inadequacy for biodiversity conservation in grassland ecosystems. *Journal of Applied Ecology*, **56**, 604–617.

Balletto, E., Bonelli, S., Barbero, F. *et al.* (compilers) (2015) *Lista Rossa IUCN delle Farfalle Italiane – Ropaloceri*. Rome: Comitato Italiano IUCN e Ministero dell'Ambiente e della Tutela del Territorio e del Mare.

Bambaradeniya, C. N. B., Edirisinghe, J. P., De Silva, D. N., Guanatilleke, C. V. S., Ranawana, K. B. & Wijekoon, S. (2004) Biodiversity associated with an irrigated rice agroecosystem in Sri Lanka. *Biodiversity and Conservation*, **13**, 1715–1753.

Barca, B., Lindon, A. & Root-Bernstein, M. (2016) Environmentalism in the crosshairs: perspectives on migratory bird hunting and poaching conflicts in Italy. *Global Ecology and Conservation*, **6**, 189–207.

Bianco, P. G. (2014) An update on the status of native and exotic freshwater fishes of Italy. *Journal of Applied Ichthyology*, **30**, 62–77.

BirdLife International (2017) *European Birds of Conservation Concern: Populations, Trends and National Responsibilities.* Cambridge, UK: BirdLife International.

Blasi, C., Capotorti, G., Alós Ortí, M. M. *et al.* (2017) Ecosystem mapping for the implementation of the European Biodiversity Strategy at the national level: the case of Italy. *Environmental Science and Policy*, **78**, 173–184.

Bonelli, S., Casacci, L. P., Barbero, F. *et al.* (2018) The first red list of Italian butterflies. *Insect Conservation and Diversity*, **11**, 506–521.

Brambilla, A., Hardenberg von, A., Nelli, L. & Bassano, B. (2020) Distribution, status, and recent population dynamics of Alpine ibex *Capra ibex* in Europe. *Mammal Review*, **50** (3), 267–277.

Brambilla, M., Casale, F., Bergero, V. *et al.* (2010) Glorious past, uncertain present, bad future? Assessing effects of land-use changes on habitat suitability for a threatened farmland bird species. *Biological Conservation*, **143**, 2770–2778.

Brochet, A. L., Van Den Bossche, W., Jbour, S. *et al.* (2016) Preliminary assessment of the scope and scale of illegal killing and taking of birds in the Mediterranean. *Bird Conservation International*, **26**, 1–28.

Buonocore, E., Picone, F., Donnarumma, L., Russo, G. F. & Franzese, P. P. (2019). Modelling matter and energy flows in marine ecosystems using emergy and eco-exergy methods to account for natural capital value. *Ecological Modelling*, **392**, 137–146.

Cardarelli, E. & Bogliani, G. (2014) Effects of grass management intensity on ground beetle assemblages in rice field banks. *Agriculture, Ecosystems and Environment*, **195**, 120–126.

Cardillo, A., Augello, R., Bagnaia, R. *et al.* (2017) Carta della Natura: strumento di conoscenza e valutazione del territorio. *Reticula*, **16**, 3–11.

CCN (2017) *Primo Rapporto sullo Stato del Capitale Naturale in Italia.* Rome: Comitato Capitale Naturale.

(2018) *Secondo Rapporto sullo Stato del Capitale Naturale in Italia.* Rome: Comitato Capitale Naturale.

(2021) *Quarto Rapporto sullo Stato del Capitale Naturale in Italia.* Rome: Comitato Capitale Naturale.

Cecchetti, M. (2003) Legislazione statale e legislazione regionale per la tutela dell'ambiente: niente di nuovo dopo la riforma del Titolo V? *Le regioni*, **1**, 318–337.

Conti, M., Testa, S. & Trepuzzano, M. (2019) L'attività di contrasto degli illeciti contro gli uccelli selvatici svolta dalla Sezione Operativa Antibracconaggio e Reati in Danno degli Animali (SOARDA). In *Abstract Book XX CIO – Italian Conference of Ornithology*, eds. R. Balestrieri & G. Bazzi, p. 40. Naples. www.doppiavoce.com/images/Doppiavoce-XX-CIO-web.pdf

D'Amen, M. & Bombi, P. (2009) Global warming and biodiversity: evidence of climate-linked amphibian declines in Italy. *Biological Conservation*, **142**, 3060–3067.

Della Seta, R. (1999) *La Difesa dell'ambiente in Italia. Storia e cultura del movimento ecologista.* Milan: Franco Angeli.

Eliante, FLA & Lipu (2017) *What works in conservation. Condividere le conoscenze e confrontarsi per gestire in modo efficace gli agroambienti.* Report.

Ercole, S., Giacanelli, V., Bacchetta, G., Fenu, G. & Genovesi, P. (eds.) (2016) *Manuali per il monitoraggio di specie e habitat di interesse comunitario (Direttiva 92/43/CEE) in Italia: specie vegetali.* ISPRA, Serie Manuali e linee guida, 140/2016. Rome: Istituto Superiore per la Protezione e la Ricerca Ambientale.

European Commission (2017) *The EU Environmental Implementation Review: Common challenges and how to combine efforts to deliver better results. Italy.* COM (2017) 63 final.

(2019) *The EU Environmental Implementation Review 2019. Country Report – Italy.* SWD, 123 final.

(2021) *Italy. Overview.* https://cinea.ec.europa.eu/system/files/2021-06/Italy_Update_EN_Final_May21_Rev.pdf (downloaded 11 January 2022).

European Court of Auditors (2017) *More efforts needed to implement the Natura 2000 network to its full potential.* Special report 1/2017. Luxembourg: Publications Office of the European Union.

Fasola, M. & Cardarelli, E. (2015) Long-term changes in the food resources of a guild of breeding Ardeinae (Aves) in Italy. *Italian Journal of Zoology*, **82** (2), 238–250.

Fasola, M., Imperio, S., Ranghetti, L., Cardarelli, E. & Boncompagni, E. (2018) Dry rice paddies cause a decline of the breeding herons and egrets in Italy. *The Journal of Heron Biology and Conservation*, **3**, 14.

Fasola, M. & Ruìz, X. (1996) The value of rice fields as substitutes for natural wetlands for waterbirds in the Mediterranean Region. *Colonial Waterbirds*, **19** (Special publication 1), 122–128.

Finlayson, C., Hollis, G. E. & Davis, T. J. (eds.) (1992) *Managing Mediterranean Wetlands and Their Birds.* Slimbridge, UK: International Waterfowl and Wetlands Research Bureau.

Fonderico, F. (2007) *L'Evoluzione della Legislazione ambientale.* Rome: Giuristi Ambientali.

Franzese, P. P., Buonocore, E., Paoli, C. *et al.* (2015) Environmental accounting in Marine Protected Areas: the EAMPA Project. *Journal of Environmental Accounting and Management*, **3** (4), 323–331.

Galaverni, M., Caniglia, R., Fabbri, E., Milanesi, P. & Randi, E. (2016) One, no one, or one hundred thousand: how many wolves are there currently in Italy? *Mammal Research*, **61** (1), 13–24.

Gantioler, S., Rayment, M., Bassi, S. *et al.* (2010) *Costs and Socio-Economic Benefits associated with the Natura 2000 Network.* Report to the European Commission. Brussels: Institute for European Environmental Policy/GHK/Ecologic.

Gatto, P., Maso, D. & Leonardi, A. (2015) *Analisi dei costi per la gestione della Rete Natura 2000 in Regione Lombardia nel periodo 2008-2011 e di stima dei costi nel periodo 2014-2020.* LIFE11NAT/IT/044 GESTIRE. www.naturachevale.it/wp-content/uploads/2016/08/allegato-VIII_C21.pdf

Genovesi, P., Angelini, P., Bianchi, E. *et al.* (2014) *Specie e habitat di interesse comunitario in Italia: distribuzione, stato di conservazione e trend.* ISPRA, Serie Rapporti, 194/2014. Rome: Istituto Superiore per la Protezione e la Ricerca Ambientale.

Giuliano, D., Cardarelli, E. & Bogliani, G. (2018) Grass management intensity affects butterfly and orthopteran diversity on rice field banks. *Agriculture, Ecosystems and Environment*, **267**, 147–155.

Giuliano, D., Rossi, P., Luoni, F., Celada, C. & Bogliani, G. (2017) *Biodiversity Action Plan per le aree risicole dell'Italia Nord-occidentale. Project: Rice for Life: experimentation of good agricultural practices for biodiversity conservation in rice crops.* DOI: 10.13140/RG.2.2.16342.55361

INFC (2005) eds. P. Gasparini & G. Tabacchi. *L'Inventario Nazionale delle Foreste e dei Serbatoi Forestali di Carbonio*. MiPAAF-CFS, CRAMPF. Milan: Edagricole. www.sian.it/inventarioforestale/jsp/home.jsp.

ISPRA (2016) *Consumo di suolo, dinamiche territoriali e servizi ecosistemici, 2016 ed.*, ed. M. Munafò. ISPRA Report 248/2016. Rome: Istituto Superiore per la Protezione e la Ricerca Ambientale.

(2017) *The ecological network in the landscape planning tools, 2017 ed.* Indicator factsheet. http://annuario.isprambiente.gov.it/ (downloaded 20 May 2019).

(2018) *Consumo di suolo, dinamiche territoriali e servizi ecosistemici, 2018 ed.*, ed. M. Munafò. ISPRA Report 288/2018. Rome: Istituto Superiore per la Protezione e la Ricerca Ambientale.

(2019) *Annuario dei dati ambientali, ed. 2018. [Environmental data yearbook]*. Stato dell'ambiente 84/2019. Rome: Istituto Superiore per la Protezione e la Ricerca Ambientale.

Lombardi, L. & Giunti, M. (2014) La traduzione della Rete Ecologica negli strumenti della pianificazione e nelle politiche di settore: dal sistema delle Aree protette al Piano paesaggistico regionale. In *Reti ecologiche e paesaggio per il governo del territorio in Toscana*, eds. E. Falqui & G. Paolinelli, pp. 207–223. ISPRA, ETS Pisa.

Maiorano, L., Amori, G., Montemaggiori, A. *et al.* (2015) On how much biodiversity is covered in Europe by national protected areas and by the Natura 2000 network: insights from terrestrial vertebrates. *Conservation Biology*, **29** (4), 986–995.

Maiorano, L., Falcucci, A., Garton, E. O. & Boitani, L. (2007) Contribution of the Natura 2000 network to biodiversity conservation in Italy. *Conservation Biology*, **21**, 1433–1444.

Marino, D. & Pellegrino, D. (2018) Can payments for ecosystem services improve the management of Natura 2000 Sites? A contribution to explore their role in Italy. *Sustainability*, **10** (3), 665.

MATTM (2016) *Relazione sullo stato di attuazione della legge n. 394 del 1991*. Anno 2016. Rome: Ministero dell'Ambiente e della Tutela del Territorio e del Mare.

(2017) *Strategia nazionale per la biodiversità. III Rapporto (2015-2016)*. Rome: Ministero dell'Ambiente e della Tutela del Territorio e del Mare.

(2020) *Strategia nazionale per la biodiversità. IV Rapporto (2017-2018)*. Rome: Ministero dell'Ambiente e della Tutela del Territorio e del Mare.

(2021) *Strategia nazionale per la biodiversità 2011-2020. Rapporto conclusivo*. Rome: Ministero dell'Ambiente e della Tutela del Territorio e del Mare.

MEF (2019) *L'ecorendiconto dello Stato*. Esercizio finanziario 2018. Rome: Ministero dell'Economia e delle Finanze.

MiPAAF (2020) *FEASR 2014-2020. Report di avanzamento della spesa pubblica dei PSR 2014-2020. Primo trimestre 2020*. Rome: Ministero delle Politiche Agricole, Alimentari e Forestali.

MiTE (2022) *Strategia nazionale per la biodiversità 2030*. Rome: Ministero della Transizione Ecologica.

Orsenigo, S., Montagnani, C., Fenu, G. *et al.* (2018). Red Listing plants under full national responsibility: extinction risks and threats in the vascular flora endemic to Italy. *Biological Conservation*, **224**, 213–222.

Paoloni, D., Ciucci, P., Lapini, L. & Loy, A. (2014) Carnivori. In *Specie e habitat di interesse comunitario in Italia: distribuzione, stato di conservazione e trend*, eds. P. Genovesi, P. Angelini, E. Bianchi, *et al.*, pp. 190–195. ISPRA, Serie Rapporti, 194/2014. Rome: Istituto Superiore per la Protezione e la Ricerca Ambientale.

Ranghetti, L. & Boschetti, M. (2022). Updated trends of water management practice in the Italian rice paddies from remotely sensed imagery. *European Journal of Remote Sensing*, **55**, 1–9.

Ranghetti, L., Cardarelli, E., Boschetti, M., Busetto, L. & Fasola, M. (2018) Assessment of water management changes in the Italian rice paddies from 2000 to 2016 using satellite data: a contribution to agro-ecological studies. *Remote Sensing*, **10**, 416.

Regione Umbria (2017) Piano finanziario. LIFE13 NAT/IT/000371 SUNLIFE. www.life-sun.eu/wp-content/uploads/2019/02/C1_D2_Piano-Finanziario_aggiornatoFebbraio2019.pdf

Rete Rurale Nazionale & Lipu (2021) Common breeding farmland birds in Italy. Update of population trends and Farmland Bird Indicator for National Rural Network 2000-2020. www.reterurale.it/farmlandbirdindex

Rogora, M., Frate, L., Carranza, M. L. *et al.* (2018) Assessment of climate change effects on mountain ecosystems through a cross-site analysis in the Alps and Apennines. *Science of the Total Environment*, **624**, 1429–1442.

Rondinini, C., Battistoni, A., Peronace, V. & Teofili, C. (compilers) (2013) *Lista Rossa IUCN dei Vertebrati Italiani*. Rome: Comitato Italiano IUCN e Ministero dell'Ambiente e della Tutela del Territorio e del Mare.

Rondinini, C., Battistoni, A. & Teofili, C. (2014) *Lo stato della Biodiversità in Italia: l'applicazione dell'approccio Sampled Red List e Red List Index*. www.iucn.it/pdf/Lo_Stato_della_Biodiversita_in_Italia.pdf

(compilers) (2022) *Lista Rossa IUCN dei vertebrati italiani*. Rome: Comitato Italiano IUCN e Ministero dell'Ambiente e della Sicurezza Energetica.

Rossi, G. & Gentili, R. (2008) A partnership project for a new Red List of the Italian Flora. *Plant Biosystems*, **142** (2), 302–304.

Rossi, G., Montagnani, C., Gargano, D. *et al.* (eds.) (2013) *Lista Rossa della Flora Italiana. 1. Policy Species e altre specie minacciate*. Rome: Comitato Italiano IUCN e Ministero dell'Ambiente e della Tutela del Territorio e del Mare.

Rossi, G., Orsenigo, S., Montagnani, C. *et al.* (2016) Is legal protection sufficient to ensure plant conservation? The Italian Red List of policy species as a case study. *Oryx*, **50** (3), 431–436.

Saura, S., Bertzky, B., Bastin, L., Battistella, L., Mandrici, A. & Dubois, G. (2018) Protected area connectivity: shortfalls in global targets and country-level priorities. *Biological Conservation*, **219**, 53–67.

Servadei, L., Ferroni, F., Calvario, E., Martinoya, D. & Vanino, S. (2018) *La Politica di Sviluppo Rurale 2014/2020 per la Biodiversità, Natura 2000 e le Aree protette*. Rapporto Rete Rurale Nazionale. Progetto CREA 23.1 Biodiversità, Natura 2000 e aree protette. Programma LIFE. Rome: Rete Rurale Nazionale. www.reterurale.it/RapportoNatura2000

Sorace, A. & Amadesi, B. (2016) *Analisi dei dati di abbattimento dell'avifauna sottoposta a prelievo venatorio relativi alla stagione 2014-2015 al fine di ottemperare agli obblighi derivanti dalla Direttiva*

2009/147/CE. Report ISPRA. www.isprambiente.gov.it/files/notizie-ispra/notizie-2016/relazione_tesserini_venatori_finale.pdf

Stoch, F. & Genovesi, P. (2016) *Manuali per il monitoraggio di specie e habitat di interesse comunitario (Direttiva 92/43/CEE) in Italia: specie animali*. ISPRA, Serie Manuali e linee guida, 141/2016. Rome: Istituto Superiore per la Protezione e la Ricerca Ambientale.

Trisorio, A., De Natale, F. & Pignatti, G. (2013) Le aree agricole ad alto valore naturale in Italia: una stima a livello regionale. *Agriregionieuropa*, **33**, 70–73.

UNEP-WCMC (2022) *Protected Area Profile for Italy from the World Database of Protected Areas, March 2022*. www.protectedplanet.net

Vallini, M. (2019) Il numero (vero) dei cacciatori. *Armi e Tiro*, **1**, 180–183.

Vassallo, P., Paoli, C., Buonocore, E., Franzese, P. P., Russo, G. F. & Povero, P. (2017) Assessing the value of natural capital in marine protected areas: a biophysical and trophodynamic environmental accounting model. *Ecological Modelling*, **355**, 12–17.

Chapter 21 Latvia

Otars Opermanis, Ilona Mendzina, Ainars Aunins and Inga Racinska

Key Statistics	
Terrestrial area (km^2)/% of EU-28*	64 586/1.5%
Marine waters (km^2)/% of EU-28*	28 348/0.6%
Population density (p km^2)/% of EU-28*	31/27%
GDP per capita (€)/% of EU-28	10 613/40%
Land cover (CORINE land cover classes 2018)	AGR 40%, FOR 36%, OSN 18%, WET 4%, ART 2%
Biogeographical regions	BOR, MBAL
Number of HD Annex I habitats/%*	61/27%
Number of HD Annex II, IV and V species/%*	109/9%
Number of BD Annex I bird taxa/%*	65/37%
% HD UCS and decreasing habitats/ % unknown	21%/8%
% HD UCS and decreasing species/ % unknown	13%/8%
% breeding BD Annex I birds decreasing/% uncertain and unknown	32%/21%

* Excluding MAC and MMAC biogeographical regions.
Sources and codes. See Table 1.1.

21.1 The Natural History and Characteristics of Latvia

Latvia is located on the eastern coast of the Baltic Sea in the Sarmatic mixed forests ecoregion within the biome of mixed and temperate broad-leaved forests of the Palearctic biogeographical realm (Olson *et al.*, 2009). The country is rather flat with a maximum height of 312 m. The climatic gradients are not strong, being most maritime along the Baltic coast and more continental towards the east.

The soils and biota are relatively new, having formed c. 15 000 years ago after the retreat of the ice that had covered the whole country during the last ice age. Latvia's flora and fauna retain elements of all previous post–ice age climatic periods: from the Pre-Boreal period, with tundra and cold steppe conditions, to the conifer dominated Boreal period, to the broad-leaved tree dominated Atlantic period, and then to the Sub-Boreal period when boreal conifers increased their distribution again (Nikodemus *et al.*, 2018).

The uneven terrain and clayey soils of large parts of Latvia, in combination with a climate where precipitation exceeds evaporation, have been favourable for the development of mires (Kalniņa, 2008) and various freshwater ecosystems such as rivers, streams, springs and lakes (Urtāns, 2017). Peat-bogs cover c. 10% of the country and 70% of this area is considered natural (VARAM, 2017). Freshwater occupies 3.6% of the country (Nikodemus *et al.*, 2018); however, a high proportion of rivers have been straightened to speed up drainage and reduce spring floods.

Although modern Latvia has been inhabited by humans for the last 10 000 years, noticeable impacts on the landscape were not observed until c. 5 000 years ago, when mosaic landscapes formed due to the domestication of livestock and small-scale arable farming. Nevertheless, extensive forest cover remained, estimated to be 87% (Nikodemus *et al.*, 2018). The development of agriculture and an increase in the human population were the main drivers of landscape change until World War II. Forest cover reached a minimum in 1920, when it was estimated to be 23% (Saliņš, 1999), and grassland cover peaked in the early twentieth century at an estimated 31% (Boruks, 2004).

The historical and political events of the twentieth century have had a lasting impact on the shape of the Latvian landscape. The Soviet occupation changed the land ownership structure, and the deportation of tens of thousands of rural people resulted in the abandonment of many individual farms. Instead, large agricultural holdings (collective farms) were formed. Tree cover started to develop on the abandoned farmland, leading to continuously increasing forest coverage ever since. Now forest coverage has reached about 52% of the country's territory according to the State Forest Service (VMD, 2018a). According to CLC data for 2018 (see Key Statistics table), forest cover is 36%, but other seminatural habitats comprise 18%, some of which may be regenerating forest. Most of the forest consists of mixed species, with the majority of stands dominated by conifers (c. 52% pine and spruce) followed by birch (31%), with only 1% being hard broad-leaved (Ikauniece, 2017).

The country's return to independence in 1991 once again changed the land ownership structure, dismantling the large collective farms and returning the land to its former private owners. As many of the new landowners did not choose to become farmers, this resulted in further abandonment of farmland. This remained the dominating driving force in land-use change until EU accession in 2004, and had much greater biodiversity impacts than urbanisation and the development of transport infrastructure (Nikodemus *et al.*, 2005). Nowadays, 4% of the country is available for agriculture but abandoned (LAD, 2018), and active farmland has declined to 30% (CSB, 2018a). Of this, 67% is arable and 33% grassland, with 0.6% comprising permanent crops and kitchen gardens. Many meadows and pastures have been lost, such that less than 10% of grassland is now permanent (CSB, 2018a), and most of it is of low natural quality due to drainage and agricultural improvement (Rūsiņa, 2017). However, small-scale farming is still widespread: 77% of all farms are less than 20 ha, and these cover 19% of the farmland (CSB, 2018a).

Heathlands were another group of habitats that reached their historical maximum a century ago, due to deforestation and grazing, when they covered 14% of the country (Skujenieks, 1927). Since the 1950s they have rapidly declined due to land abandonment and vegetation overgrowth (Auniņa and Rove, 2013).

The natural history of the Baltic Sea and Latvia's 500 km of coastline has defined the wide variety of different coastal habitats, from reefs, lagoons and beaches to various types of dunes (Laime, 2017; Nikodemus *et al.*, 2018). As many as 18 EU HD habitat types that occur in Latvia are coastal and they can extend tens of kilometres inland to the ancient shore of the former Baltic Glacial Lake (Rove, 2013). Although widespread in the early nineteenth century, shifting sand dunes no longer occur due to their deliberate stabilisation with pine plantations (Laime, 2017).

For several HD habitats Latvia holds significant proportions of their area in the whole EU Boreal biogeographical region: for example, 87% of the Boreal region area of northern Atlantic wet heaths with *Erica tetralix* (4010),[1] 83% of humid dune slacks (2190), and 75% of wooded dunes (2180).

Latvia has no endemic species but holds core populations of several HD species and BD Annex I birds within their distribution range. Of these, the species with the largest share of their European populations are Lesser-spotted Eagle (*Clanga pomarina*) (23%; Bergmanis *et al.*, 2015) and Corncrake (*Crex crex*) (5%; Keiss, 2005). As a result of range and populations declines elsewhere, Latvia holds important marginal population exclaves isolated from the core populations of the Roller (*Coracias garrulus*) and Freshwater Pearl Mussel (*Margaritifera margaritifera*). Several HD Annex II and BD Annex I species have seemingly gone extinct during the last decades, including the European Mink (*Mustela lutreola*), Siberian Flying Squirrel (*Pteromys volans*) and Willow Grouse (*Lagopus lagopus*).

21.2 The Status of Nature

21.2.1 Habitat Status and Trends

The key historical changes in Latvian landscapes and habitats have not been linear: the declining trend in forest cover in the early twentieth century was reversed after World War I and instead a progressive loss of semi-natural habitats, such as grasslands and heathlands, has been observed. Also, although forest cover has increased, this has not increased HD forest habitats but instead young forests on abandoned farmland and heathlands. In addition to the declines in habitat area, habitat quality (specific structures and functions) has also decreased in the majority of HD habitat types.

These general trends in habitats have been reflected in the most recent Article 17 report[2] by Latvia, which covers 2013–2018. It shows that out of 61 HD habitats, 10% were considered to have a favourable conservation status, 44% a poor status, 38% a bad status and 8% an unknown status. Examination of the status assessments by broad habitat classes indicates that the situation is especially worrying for forests, grasslands, and heath and scrub, as all types of these groups were assessed as unfavourable. Only three of the HD habitats with a poor status showed signs of improving trends, whilst nearly all those with a bad status were declining or had unknown trends (Table 21.1).

21.2.2 Species Status and Trends

The longest-running species trend data in Latvia relate to the large carnivores, other species that are subject to hunting or fishing, and wintering waterbirds.

The limited availability of precise data on population dynamics for most species has led to indirect population trend assessments based on measures such as trends in the species' typical habitats or trends in land use. Such indirectly assessed and expert opinion–based trends are predominant in the Article 17 reports so far. The report for 2013–2018 shows that out of 109 HD species, 43% were considered to have a favourable conservation status, 39% a poor status, 14% a bad and 8% an unknown status. Of particular concern is the high proportion of species that have unfavourable conservation status and are still declining, or have unknown trends, especially those with a bad status (Table 21.1).

Trends in BD Annex I breeding bird populations from Latvia's latest Article 12 report show that most species with known trends have had increasing, stable or fluctuating populations over the short term (Table 21.2). Nevertheless, nearly a third had decreasing trends, whilst about a fifth had unknown

[1] Numbers in brackets are the HD Annex I habitat codes (see Appendix for full names). * Indicates a priority habitat.

[2] See Section 1.3 for background on monitoring under the Nature Directives, status categories and data sources.

Table 21.1 Short-term trends (c. 2007–2018) in Habitats Directive Annex I habitats and Annex II, IV and V species with an unfavourable conservation status in Latvia.

	Poor (unfavourable–inadequate)				Bad (unfavourable–bad)			
	Imp	**Sta**	**Dec**	**?**	**Imp**	**Sta**	**Dec**	**?**
HD Habitat assessments								
Number	3	13	2	9	0	1	11	11
%	11%	48%	7%	33%	0%	4%	48%	48%
HD Species assessments								
Number	4	15	5	18	0	0	9	6
%	10%	36%	12%	43%	0%	0%	60%	40%

Key. Imp = improving, Sta = stable, Dec = decreasing, ? = unknown.
Source. Based on Member State HD Article 17 reported data as presented on the EEA national summary dashboard (2021).

Table 21.2 Short-term trends (c. 2007–2018) in Birds Directive Annex I taxa in Latvia.

	Breeding				Wintering			
	Inc	**Sta/ Flu**	**Dec**	**?**	**Inc**	**Sta/ Flu**	**Dec**	**?**
Number	15	14	20	13	1	0	0	4
%	24%	23%	32%	21%	20%	0%	0%	80%

Key. Inc = increasing, Sta = stable, Flu = fluctuating, Dec = decreasing, ? = uncertain and unknown.
Source. Based on Member State Article 12 reported data as presented on the EEA national summary dashboard (2021).

trends. Trends amongst common species have been revealed by the Latvian breeding bird survey's annual assessments of 106 breeding bird species since 2005. Of these, over 2005–2017, 26% of species increased, 25% were stable, 10% declined, and the trends of 39% species were unclear (Auniņš, 2018). Despite this overall optimistic picture, alarming declines in two-thirds of species with clear trends occurred between 2012 and 2017. Further analyses of these short-term trends show that most affected have been farmland birds and species wintering in sub-Saharan Africa (Auniņš, 2018).

With the development of the common bird monitoring schemes, multi-species wild bird indicators have become available. Both the farmland bird index (1995–2017) and forest bird index (2005–2017) have been stable but with some fluctuations. The farmland bird index declined after Latvia joined the EU and its CAP (Aunins and Priednieks, 2009), increased in the years that followed the economic crisis of 2008, and has considerably dropped again since 2015 (Auniņš, 2018). The forest bird index was increasing until the political decision to increase the limits of allowed forest cutting in state forests in 2008 to cope with the economic crisis. It showed a marked drop over the following two years that matched the increased tree felling volumes (Figure 21.1). Since 2011 the annual tree felling volumes have declined and the forest bird index has increased (Auniņš, 2018).

21.2.3 Drivers and Causes of Change

The observed trends in agricultural habitats and species have primarily resulted from two contrasting agricultural drivers: agricultural abandonment and agricultural intensification. Initially, the return of farmland to the previous owners in the 1990s resulted in rapid abandonment (40% of the farmland in the late 1990s) and other associated impacts. Cattle numbers declined more than two-fold by the mid-1990s, and the application of mineral fertilisers decreased almost 10-fold (Auniņš, 2010; CSB, 2018a). Although the recovery of the agricultural sector in Latvia started from the mid-1990s, the most rapid development was seen after Latvia joined the EU in 2004. Since then, funding allocated to the agricultural sector has increased several times due to EU subsidies and direct payments, which have contributed to the modernisation of farms and acquisition of new lands. Intensification of agriculture also led to increases in the use of agro-chemicals and modern farming machinery (CSB, 2018a, b), as well as a rapid decline in small farms and a corresponding increase in large agricultural holdings (Zdanovskis and Pilvere, 2015; CSB, 2018b) and an increase in agricultural productivity (Nipers *et al.*, 2018).

After the land reform, some forest land was returned to private landowners. Nowadays, the ratio of private to state-owned forests is almost 1:1 (VMD, 2018b). Logging peaked in 1999, declined until 2008, and increased again until 2010, since when it has gradually decreased again (CSB, 2018a) (Figure 21.1). The increased forestry pressure on mature forest stands has been the main reason for the unfavourable conservation status of all HD forest habitat types and many species associated with forests.

In the 1950s the majority (65%) of grasslands were wet (Sabardina, 1957), but this changed as a result of wide-scale intensive drainage, which started in the 1950s and was most intensive between the 1960s and 1980s. Drainage had affected 17% of the meadows and 26% of the pastures by 1967, and 78% of the meadows and 44% of the pastures by 1987 (Boruks, 2004). Currently most farmland and 20% of forests are drained, thus affecting their species communities (Rūsiņa, 2017).

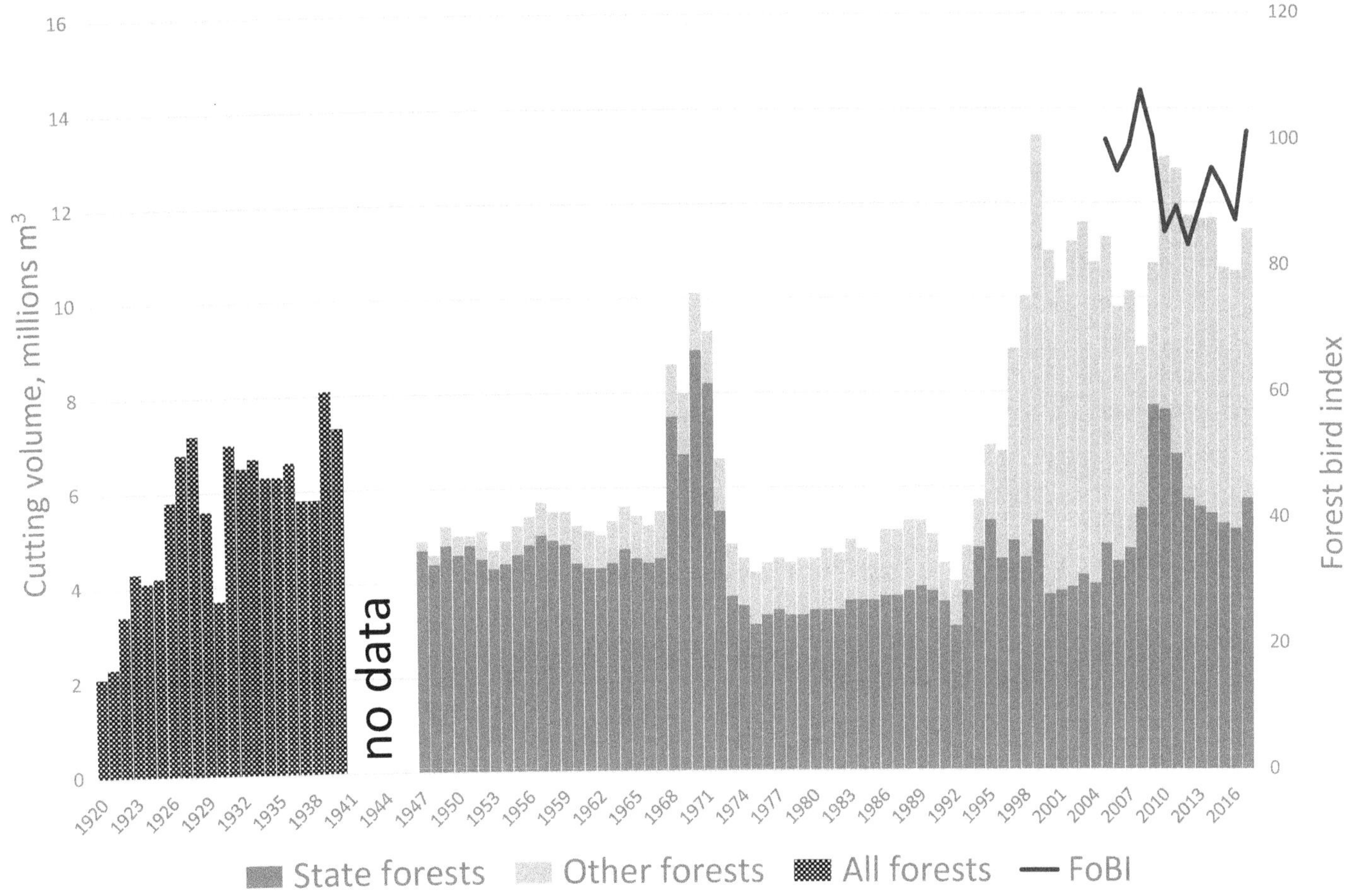

Figure 21.1 Forest exploitation in Latvia in the past century and changes in forest bird index (FoBI) (2005–2017).
Sources. Based on data combined from Saliņš (1999), Strods *et al.* (1999) and CSB (2018a).

The development of three large hydropower stations on the country's largest river, Daugava, in the twentieth century has irreversibly damaged its flow and the natural ecosystems in its valley. The ill-advised state support for the development of small hydropower stations during the 1990s and early 2000s has also damaged many small rivers.

21.3 Nature Conservation Governance, Actors, Legislation and Policies

21.3.1 Governance

Today the main institution in the field of biodiversity conservation is the Ministry of Environmental Protection and Regional Development (MEPRD). The MEPRD was established in 1993 (Table 21.3), elevating the importance of previously neglected environmental issues to the level of 'traditional' sectors (e.g. economy, finance, welfare and agriculture). In the area of environmental protection, nowadays the MEPRD deals with the establishment of requirements and conditions for nature conservation and a clean environment, and ensures that natural resources are used efficiently and sustainably. By means of its subsidiary institutions, the MEPRD controls and implements the policies and environmental legislation mentioned in Section 21.3.3, including legal requirements related to environmental protection, nature conservation, climate change, chemicals, protected areas and the extraction of mineral resources.

While the MEPRD is responsible for the development of nature conservation policy and legislation, their implementation is the responsibility of the Nature Conservation Agency (NCA), under the MEPRD. The NCA: (1) ensures implementation of the unified nature protection policy in Latvia; (2) is responsible for management of all protected areas; (3) supervises and controls international trade in endangered species of wild fauna and flora; and (4) administers certain types of compensation to landowners for restrictions on economic activities in protected areas, as well as for damage caused by migratory species. As in most governmental sectors, nature conservation institutions suffer from understaffing, caused by high demands and underfunding.

Nature conservation is partly mainstreamed in sectoral policies, but inadequately in some. Of particular concern is the Ministry of Agriculture, as it has considerable influence over nature conservation, being in charge of agriculture,

Table 21.3 Key nature conservation developments in Latvia.

1912	First Nature Reserve (Moriscala) established
1973	First National Park (Gauja) established
1977	The Latvian Red Data Book Commission founded
1980	First edition of the Latvian Red Data Book approved by the Academy of Sciences. First Latvian Breeding Bird Atlas project started
1985	Latvian Ornithological Society founded. Latvian Red Data Book published
1990	State Environmental Committee established, the first environmental state institution in Latvia
	State Environmental Inspectorate established, under the State Environmental Committee
1992	First shadow list of biodiversity-rich areas needing protection published by WWF Latvia. This was also the first, in the Latvian language, to stress the need for the management of protected areas
1993	Ministry of Environmental Protection and Regional Development (MEPRD) founded
	Law on the Specially Protected Nature Territories adopted
1995	The National Environmental Policy Plan for Latvia approved by the Government as one of the first strategic planning documents in Latvia
	Accession treaty between the EU and Latvia started the alignment of Latvian environmental legislation to the EU's environmental laws
2000	*National Biodiversity Strategy and Action Plan* (NBSAP) approved by the Government
	Law on Protection of Species and Habitats approved. This law is the main legal act that transposes the provisions of the EU Habitats and Birds Directives. The general provisions of the law are further developed in more than 10 regulations adopted by the Government
2001	First Latvian habitat classification published
2004	EU accession, since when most nature conservation has focussed on the implementation of the EU nature legislation. First Natura 2000 database sent to the European Commission
2005	Latvian Biodiversity Monitoring System developed
2009	Protected area governance changed, with the Nature Conservation Agency becoming in charge of the statewide implementation of nature conservation legislation, including the management of all protected areas and compliance control
2010	*Sustainable Development Strategy of Latvia* 2030 approved
2011	System of Latvian marine Natura 2000 sites established
2014	National Environmental policy concept for 2014–2020 approved
2016	Countrywide systematic mapping of HD Annex I habitats launched

See Table 3.3 for signatory dates of international conventions.

forestry and fisheries. Yet the responsibilities of the MEPRD and the Ministry of Agriculture are strictly divided and do not overlap, reflecting their differing interests and focus. The Ministry of Agriculture strives to support economic activity in rural areas and often sees nature conservation requirements as an impediment to economic development. In contrast, the MEPRD sees nature conservation as an ultimate goal that should be achieved by all means, sometimes neglecting economic considerations. Therefore, whilst it is necessary to balance these interests, it is often not an easy task.

21.3.2 Public Involvement and NGOs

The Latvian Ornithological Society (LOS), which has been the BirdLife International partner since 2001, today is the largest membership-based NGO in Latvia and unites both professional biologists and conservationists with citizens interested in birds and their conservation. It had up to almost 800 active members in the 2000s, but in recent years its membership has averaged about 600. There are a few other thematic national societies (e.g. botanical, entomological and mammalogical), which are mostly formed by professionals and have far fewer members than LOS. Importantly, LOS is equally active in both lobbying for bird conservation and in bringing conservation ideas to the attention of the general public.

After the regaining of independence in 1991, new environmental actors appeared in Latvia. Most important were the Latvian Fund for Nature and WWF Latvia. These NGOs mainly rely on projects for funding and are formed predominantly by professional staff, yet they both cover a wide range of problems and are not specialised in any particular field. LOS and the Latvian Fund for Nature had an important role in Natura 2000 site selection over the pre-accession period, and now they are often involved in the Nature Directive reporting processes, inventories of species and habitats, management plans of protected areas and species action plans.

In 2003, the Environmental Consultative Board at the MEPRD was established. This consists of 20 NGOs that deal with environmental issues. Although the MEPRD provides secretarial support to the Board, it is an independent body providing guidance and advice to the Ministry in the legislative process, including formal negotiation of draft legal acts.

21.3.3 Policy and Legislation

One of the first laws adopted in the field of nature conservation, in the early 1990s, was the Law on Specially Protected Nature Territories.[3] This introduced the concept of site management plans, monitoring of protected areas and outsourcing of management activities to private or non-governmental entities, and provided for compensation for restrictions on economic activities within protected areas. Many of these concepts were innovative at that time and fully in line with the latest developments in global nature conservation policy. In the course of accession to the EU, nature conservation legislation obtained a second 'pillar': the 2000 Law on the Conservation of Species and Biotopes.[4] This transposes the provisions of the Nature Directives, including introducing the concept of favourable conservation status, which at the time of adoption was a novel approach for Latvia. These two laws, and their supplementing regulations, establish quite a comprehensive yet complicated system of legal acts and requirements. A third nature conservation law, which completes the framework, is the Law on Compensation for the Restrictions on Economic Activities in Protected Areas,[5] first introduced in 2005. This established a detailed system for compensation.

The compensation system has been in place since 2006, but in 2013 it was reformed by simplifying the types of compensation and clarifying the sources of funding. Now annual compensation payments have replaced the previous one-off payments. Compensation is paid either from the EU CAP Rural Development Programme (RDP) in Natura 2000 sites (see Section 21.7), or from state or municipal budgets for other areas. In very limited cases, purchase of private land can be carried out as a mean of compensation, but only for certain categories and zones of protected areas.

The long-term framework of national policy in the field of sustainable development is set by the *Sustainable Development Strategy of Latvia* (Anonymous, 2010). In the field of sustainable management of natural values and ecosystem services, the Strategy declared that the objective for Latvia is to become an EU leader in the preservation, increase and sustainable use of natural capital. Under this priority, the following long-term actions are envisaged: management of natural capital, creation of market instruments, capitalisation of natural assets, and promotion of a sustainable lifestyle. However, only a few biodiversity-related indicators are set for this goal: by 2030 protected areas should reach 18% of the country's territory, the index of farmland birds should be greater than 120 (baseline 100 in 1999), organic farming should occur on more than 15% of the total UAA, and forest should cover 55% of the country's territory.

The latest *Environmental Policy Strategy* (EPP) for 2014–2020 is a medium-term policy planning document that covers environmental aspects of sustainable development. The EPP includes environmental protection, nature conservation and climate change policies, and takes into account regional development issues, maritime spatial planning and other aspects of spatial development planning. In the field of biodiversity, the EPP sets only very general goals: to ensure the quality of ecosystems, balancing nature conservation and socio-economic interests as well as further promoting a 'green' country image for Latvia.

There are also other sectoral policies strongly influencing biodiversity and sustainable use of biological resources, for example forestry and rural development policies. Although all policy documents are approved by consensus in the Government, simultaneous achievement of the goals of the forestry and biodiversity conservation policies can be a very challenging task. The RDP is the main policy instrument providing for sustainable management of rural areas, using the support of the European Agricultural Fund for Rural Development (EAFRD). Under this, agri-environment measures and payments for forests in Natura 2000 territories are of particular importance for nature conservation.

In addition, some legislative acts of other fields also contain requirements that relate to nature protection. The most important of these are: the Law on Protection Zones, the Law on Forests, the Fisheries Law, the Law on Environmental Impact Assessment, the Law on Hunting, and the Law on Agriculture and Rural Development. Some of these laws are discussed in relation to specific issues in the following sections.

21.4 Species Protection and Conservation Measures

The key legal act that regulates species protection is the Law on the Conservation of Species and Biotopes. This law aims at: (1) ensuring biodiversity through the conservation of fauna, flora and habitats; (2) regulating the conservation, management and surveillance of species and habitats; (3) promoting the

[3] Likums Par īpaši aizsargājamām dabas teritorijām.

[4] Sugu un biotopu aizsardzības likums.

[5] Likums Par kompensāciju par saimnieciskās darbības ierobežojumiem aizsargājamās teritorijās.

preservation of populations of species in accordance with economic and social preconditions, as well as cultural and historical traditions; and (4) regulating procedures for the listing of protected species and habitats. The law applies to all wild species and protected habitats. In addition, both the Fishery Law and Hunting Law include important conservation aspects relating to harvested fish, birds and mammals, such as lists of species that can be taken, harvest tools, methods and seasons.

The Law on the Conservation of Species and Biotopes, together with associated Cabinet Regulations, transposes the respective articles from the Nature Directives related to species protection (i.e. BD Articles 5–9 and HD Articles 12–16) into Latvian legislation. It also foresees development of national species action plans (for special problem species), which are to be approved by the responsible minister. Up to 2018, 18 species action plans had been approved, including for the Otter (*Lutra lutra*), Eurasian Lynx (*Lynx lynx*), Wolf (*Canis lupus*), European Pond Turtle (*Emys orbicularis*), Hermit Beetle (*Osmoderma barnabita*), Capercaillie (*Tetrao urogallus*), Black Stork (*Ciconia nigra*), Freshwater Pearl Mussel and Black Grouse (*Lyrurus tetrix*), all of which are considered a 'conservation challenge' in Latvia for various reasons.

There has been no attempt to evaluate the efficiency and level of implementation of these action plans. In part this is due to the lack of a common approach in setting favourable reference values at the country level, either in these plans or in other documentation. In Latvia's Article 17 report for 2007–2012, only 6% of favourable reference population values were higher than values at the time (Bijlsma *et al.*, 2017), suggesting that such reference values tended to be automatically accepted without careful consideration.

Latvia has exemptions from the inclusion of the Wolf and Beaver (*Castor fiber*) on HD Annex IV; instead these species are listed under Annex V for Latvia. A similar exemption was also requested for the Eurasian Lynx, but it was not granted. Consequently, the Lynx has been hunted under derogations, with an annual quota based on population monitoring data. Existing Latvian large carnivore management plans have been questioned by the European Commission and NGOs, mostly because they allow a high proportion of the population to be taken (up to 20% of Lynx and up to 50% of Wolves) every year. However, the ongoing monitoring has not reported any significant reduction of either species during the last decade (Ozolins *et al.*, 2014; Bagrade *et al.*, 2016). On the contrary, their numbers have even increased, their range is stable or increasing, and recent genetic studies have not found any genetic separation between different parts of the country. Still, despite this seemingly positive picture, the constantly increasing harvest quota since 2004 for both species looks like an experiment, rather than a long-term plan to maintain the favourable conservation status of the species. Shortly before this chapter was completed, hunting of the lynx was halted in April 2022.

In Latvia, the illegal killing of birds and protected mammal species, including large carnivores, is rare and does not seem to threaten species conservation (BirdLife International, 2017; J. Ozolins, pers. comm.). However, in some locations and years anadromous lampreys and other fish, such as Atlantic Salmon (*Salmo salar*), are illegally harvested during autumn migration despite control efforts by authorities.

21.5 Protected Areas and Networks

21.5.1 Protected Area Types and Coverage

The first documented designation of a protected area to preserve pristine 'wildlife' in Latvia dates back to 1912 with the establishment of the nationally iconic Moricsala Reserve, an island of Lake Usma. Over the following 100 years protected areas gradually but irregularly increased. Until World War II, only a handful of sites had been designated, most of them typically relatively small 'remote wilderness' areas on islands of lakes (Cieceres, Pildas, Istras, Indzera, Kala, Jumurdas, Ezezers and Cirisa) without a specific conservation objective. After the War, with a few exceptions, new terrestrial protected areas were added in packages (Anonymous, 1991, 2019). However, only since 2004 have new protected areas been designated to achieve a functional network aimed at conserving all protected species and habitats such that they have a favourable conservation status. Until then, as in many other countries, each protected area existed 'for itself' without consideration of its role in a broader national or international context. The first electronic database of protected areas and potential new sites in Latvia was developed by the CORINE Biotopes project (1994–1997).

Latvia's current system of protected areas, as summarised in Table 21.4,[6] is very much designed to fulfil obligations under the Nature Directives and to be a part of a wider pan-European network of protected areas, as Latvia also has borders with non-EU countries. Relatively little attention is paid to other international designations. Natura 2000 sites (SACs and SCIs) cover all six existing Ramsar Sites. The only Biosphere Reserve is situated in north-central Latvia and covers 4 744 km^2 (nearly 7% of the country). The whole site includes 27 Natura 2000 sites and is regarded as an experimental area for innovative activities to streamline nature conservation in the wider countryside without formal nature conservation laws.

Almost all protected areas designated under domestic legislation are now included in the Natura 2000 network with identical boundaries. The exceptions are Nature Monuments and also a small number of other sites (five Nature Parks, two

[6] Details on the different protected area types and each site can be found at www.daba.gov.lv/en/protected-areas

Table 21.4 The number and extent of international designated sites and protected areas in Latvia.

	IUCN category	Number	Area (km^2)
Designated under global and non-EU international legislation			
Ramsar Sites	NR	6	1 503
Biosphere Reserves (exc. buffer and transition zones)	NAp	1	?
HELCOM Marine Protected Areas	V/NR	7	4 363
Designated under the Nature Directives			
SCIs (terrestrial + marine)	NR	329	7 421 + 2 664
SPAs (terrestrial + marine)	NR	98	6 606 + 4 280
Natura 2000 terrestrial area			*7 446*
Natura 2000 marine area			*4 388*
Designated under domestic legislation			
Strict Nature Reserve (Dabas rezervāts)	Ib	4	251
Nature Reserve (Dabas liegums)	IV	261	2 320
National Park (Nacionālais parks)	II	4	2 040
Nature Park (Dabas parks)	II (1 IV)	42	1 384
Protected Landscape Area (Aizsargājamo ainavu apvidus)	V	9	1 654
Protected Marine Area (Aizsargājamā jūras teritorija)	IV	7	4 364
Micro-reserve (Mikroliegums)	IV	2 525	441
Nature Monument (Dabas piemineklis)*	III	2 580	

Notes. NR = Not reported; NAp = not applicable; ? = no public data found. * Nature monuments (noble trees, dendrological sites, geological formations and alleys) are counted separately from the Natura 2000 network, although Natura 2000 sites include many of them. The exact area of Nature Monuments is not available because many of them are point-types and linear objects.
Sources. See Section 1.3. Based on data from: Ramsar Convention (December 2021); mpas.helcom.fi (September 2021); EEA Natura 2000 Barometer, as at end 2020; other sites from the nature data management system 'ozols.gov.lv' (10 January 2019); IUCN categories, WDPA (December 2021).

Nature Reserves) that do not host any HD habitat and species or BD Annex I species. In all cases, the boundaries of Natura 2000 site SACs and SPAs coincide. Of all Latvian Natura 2000 sites, only four, or about 1%, are only SPAs; the rest hold at least one SAC trigger species or habitat. Altogether, terrestrial Natura 2000 sites cover 11.5% of the national territory, the second lowest coverage in the EU at the end of 2020 (EEA, Natura Barometer 2021).

Special elements in the Latvian nature protection system are Micro-reserves that can also be attributed to species-specific protection measures. The history of the designation of Micro-reserves dates back to 1973 (Viksne, 1994), starting with the creation of protective buffers around nests of rare bird species. It is relatively easy to designate Micro-reserves, as they are established by an administrative act and not by the government; thus it is possible to react quickly to new discoveries and ensure protection. Micro-reserves can be established for over 200 animal and plant species and a short list of habitats, both on state and privately owned land. They vary from 0.1 to 30 ha in size, depending on their conservation objective. In addition, buffer zones of up to 500 ha for Micro-reserves can be added for sensitive species such as Black Stork, Capercaillie and large eagles. Over recent decades, Micro-reserves have proved to be very effective in maintaining populations of large birds, such as the Lesser-spotted Eagle, White-tailed Sea-eagle (*Haliaetus albicilla*) and Capercaillie. Their breeding pairs/leks are scattered throughout the country and it is difficult to include significant proportions of their national populations within the network of conventional protected areas.

By the end of 2018 there were 2 525 Micro-reserves designated in Latvia with a total area of 441 km^2, or an additional 6% of protected land compared to the area of the existing Natura 2000 network. Only 24 Micro-reserves are included in the Natura 2000 network as independent units.

As a result of the almost complete overlap between domestic protected areas and the Natura 2000 sites, their combined area is virtually the same at 11.6% (excluding the Nature Monuments, which cover a very small total area).

Marine sites comprise seven HELCOM Marine Protected Areas, seven Natura 2000 sites and seven domestic Marine Protected Areas, which are in fact the same sites in name and area. Their area is about 16% of the country's marine waters.

21.5.2 The Coherence of Protected Areas and Ecological Networks

The Latvian Natura 2000 network was formed during 1998–2004 and was based on an analysis of the existing protected areas and subsequent amendments to create a more coherent site network. During this process, the network of protected areas increased from 7.9% to 11.6% of the national terrestrial territory. The scientific approaches to SPA and SCI/SAC selection were similar, however SPA selection was also largely influenced by BirdLife International's inventory of Important Bird Areas (Racinskis, 2004).

The European Commission's assessments by the ETC/BD of the sufficiency of SCIs in the Natura 2000 network for each protected species and habitat consider the network's coherence (number and size of sites, and their connectivity) as one of the criteria in their evaluation (Evans, 2012). Up to now, Latvia has not fulfilled its obligations to designate sufficient SCIs for seven species and habitats, and for five other features it was concluded that additional research should be undertaken. This accounts for 10.2% of the total number of HD Annex I and II features present in Latvia. There is no similar sufficiency assessment process for SPAs, but currently the SPAs and IBAs are a close match, with the exception of two marine sites where further research and consultation are probably necessary.

The network of Natura 2000 sites should be coherent also across country borders (e.g. Ssymank, 2005). However, a study showed that in Latvia there is a relatively low density of Natura 2000 sites along terrestrial political borders, and there is low spatial and functional connectivity of the Natura 2000 site network with the neighbouring countries: Estonia and Lithuania (Opermanis *et al.*, 2012). This is explainable by the fact that in the Baltic States many of the most important wetlands are not situated at, or near, the countries' borders and a relatively low proportion of borders are formed by rivers.

There has been relatively little interest in developing a special country-wide ecological network outside protected areas. This is probably for two main reasons: (1) the Latvian landscape still enables sufficient dispersal of most wild species; and (2) there are already some elements, such as Micro-reserves and protective belts, in place that possibly contribute to the achievement of the aims of ecological networks.

Protective belts are applied to all inland water bodies (i.e. rivers, lakes and artificial reservoirs) and to the Baltic Sea coast. They are multifunctional and aim to reduce the negative effects of pollution, reduce erosion, delimit development in areas with high flood risk and preserve the landscape. The width of protective belts varies from 10 to 500 m depending on the length or area of the water body. Due to their linear character, such belts can serve as vectors for dispersal of wild species between protected areas. Although the functions of protective belts have not been systematically assessed in Latvia, it is obvious that they have been an important instrument for biodiversity conservation (Nikodemus and Brumelis, 2015).

Landscape ecological plans have been developed for two large territories: North Vidzeme Biosphere Reserve (2008; occupying almost 7% of Latvia) and Razna National park (2009; occupying almost 1%). Unfortunately, both plans proved to be more a training exercise for planners, scientists and students; neither plan was ever fully implemented by the authorities, mainly due to their unfortunate timing, as their finalisation coincided with the 2008–2009 economic crisis.

21.5.3 Approaches to Protection and Levels of Protection

In Latvia, in order to establish a Natura 2000 site, the territory must first be designated as one of the domestic protected area categories, or as a Micro-reserve, and included in the relevant legal acts. Depending on the chosen national category, a corresponding level of protection is also established. The protection regime is established by listing the *General Terms* for the site's protection and use, or by *Individual Terms* based on a developed site management plan.

In the legal acts, the General Terms of protection in most cases appear as a list of prohibited activities, assuming that all other possible activities are allowed. Such General Terms apply to protected areas for which Individual Terms for their protection and use have not been drawn up (i.e. no specific site management plan has been developed). Categories of protected areas included in the Natura 2000 network (Table 21.4) differ greatly in their level of protection as defined in the General Terms. Each particular category has been chosen taking into account general site conservation objectives. For example, to protect eutrophic lakes (3150), or generally common species such as White Stork (*Ciconia ciconia*) or Red-backed Shrike (*Lanius collurio*), the Protected Landscape Area category would be chosen. Protected Landscape Areas are typically large territories that can accommodate significant numbers of dispersed breeding pairs, with comparably weak levels of protection. In contrast, designation as a Nature Reserve is a more usual approach for protecting forests, sensitive dune habitats, colonial birds or rare species, such as some raptors.

Individual Terms of protection and use are set for protected areas to safeguard their natural values, taking into account their special characteristics. Individual Terms determine the types of necessary, permitted and prohibited activities and, if necessary, divide protected areas into functional zones. Such zoning can establish stronger protection zones within a generally low-level protected area and vice versa. In most cases, such Individual Terms arise from the adopted site management plan, after undergoing stakeholder consultation.

In comparison, the General Terms are mostly restrictive and more general, and may be better suited to sites where mostly non-intervention management is required. Individual Terms include both restrictive and proactive conservation measures and can be more specific, corresponding to the conservation objectives of certain protected areas.

This is a very sound logical approach, but unfortunately the development of Individual Terms of protection up to now has been only a part of the site management plan development process. In addition, the development of site management plans, and their implementation, is not compulsory for all Natura 2000 sites. Therefore, it is highly likely that there are a number of Natura 2000 sites where proactive conservation measures are required, but are not foreseen because Individual Terms have not been elaborated.

In the second half of 2017, site management plans were developed only for 34% of Natura 2000 sites (NCA, 2017), and the situation has probably changed little since. The Nature Protection Agency has recently analysed this process and prioritised sites that require a review of existing or development of new plans.

21.5.4 Approaches to Managing Protected Areas

Site management plans are developed not only to provide Individual Terms for site protection and possible zoning, but also to define conservation objectives, establish conservation measures (including identification of funding sources and implementing bodies), set the framework for subsequent monitoring and set the time for the next plan review. The process for the preparation and review of site management plans is set in legislation and special guidelines are provided for plan developers and site managers (Racinska, 2002). There are requirements to establish a supervisory committee (including representatives of all stakeholders), organise public hearings, gather and assess public opinions and ideas about the site's management, and enable supervision of the process by the Nature Protection Agency.

Specific knowledge and experience is necessary to choose appropriate management techniques to deal with certain threats and pressures. Over the past two decades a number of seminars have been organised and very useful materials covering the topic of wildlife management techniques have been published in Latvian, sharing the experience from past projects (e.g. Opermanis, 2002; Auniņš, 2008; Cinate *et al.*, 2016) and providing guidelines (e.g. Ikauniece, 2017; Rūsiņa, 2017). However, most information is provided as guidance, because few managers tend to have the time and/or willingness to share their results (Opermanis, 2016). In the past, most published materials on wildlife management focussed on species (e.g. large carnivores, salmon and ducks), whilst the latest publications have mainly been habitat-oriented.

A recurring problem in Natura 2000 site management has been a lack of consistency and country-wide focus when setting site conservation objectives. This has resulted in fragmentation of management efforts when each site is being managed on its own. To rectify this situation, a National Natura 2000 Protection and Conservation Programme has been elaborated (NCA, 2017). This document sets out the priority HD habitats whose active management is most urgent, and the particular Natura 2000 sites where these conservation measures should take place. This document is an integral part of the operational strategy of the Nature Conservation Agency, and a large step towards a unified approach to habitat management in Latvia.

In future, more systematic establishment of measurable site-level conservation objectives is planned, which should provide indicators for the periodic evaluation of management success and adaptive site management planning.

21.6 General Nature Conservation Measures

21.6.1 Development Planning and Impact Assessment

According to the Law on Specially Protected Nature Territories, if a proposed activity would be likely to have a significant effect on a Natura 2000 site, the proponent should undertake an appropriate assessment procedure in accordance with HD Article 6.3. The assessment procedure is largely based on EU guidance (European Commission, 2001). The procedure has the usual stages such as screening, appropriate assessment and review of alternatives and compensatory measures. It also requires public hearings at the appropriate assessment stage. Similar procedures and considerations govern SEAs in regard to potential impacts on Natura 2000 sites.

The coordination of EIAs and SEAs is carried out by the Environment State Bureau (ESB). Initial screening is carried out by the Regional Environmental Boards in cooperation with the Nature Conservation Agency. Appropriate assessment (if the screening results suggest a need for it) is carried out by an external subcontractor. The report is evaluated by the ESB which, based on the consultant's report, makes the final decision and prepares a report to the Cabinet of Ministers and the European Commission, if necessary.

Although the ESB web-page makes reports of individual screenings and appropriate assessments publicly available, it does not offer any statistics on the number of applications and outcomes, for example the number of refused projects and activities. It is important to note, however, that to date no potentially significantly damaging project has gone ahead under the derogation conditions of HD Article 6.4 (i.e. because they are of overriding public interest and there are no alternatives, etc.). This is mainly because the proportions of affected species populations and habitat areas have been very small and therefore their impact was assessed as insignificant. All likely impacts are expected to be effectively managed according to the mitigation hierarchy, with necessary mitigation measures incorporated into the final design of the proposed activity.

There are two concerns about the existing system of impact assessment. Firstly, there should be better understanding of how the proposed activity is evaluated at the screening stage and how it is decided whether it is likely to have significant or non-significant effects on a Natura 2000 site. Regulation

No. 300 of the Government 'Order how to assess impact on the specially protected nature territory of the EU importance' states that the affected species population and habitat area should be weighed against the total resource at the site, national and EU levels. This approach is very sound; however, as no numeric thresholds are provided (e.g. percentage of national population affected when the impact is considered significant), it is hard to understand this decision-making mechanism. Another problem is that there is no public consultation on the draft screening decision and there is no opportunity to appeal against the decision.

Secondly, in the case of mitigation measures, it is not clear if any monitoring of prescribed (mitigation) measures is carried out after permission is granted.

All available spatial information on the distribution of species and habitats of EU importance, and borders of protected areas, is made available and used during development planning in each municipality. Also, the General and Individual Terms of protection of particular protected areas, and any zoning, must be respected.

21.6.2 Habitat Management and Restoration

Habitat management and restoration in Latvia has been implemented since the end of the twentieth century, using national and EU funding mechanisms (see Section 21.7.3). Restoration and management activities have been to a large extent project-based and focussed on solving problems with protected species and habitats in certain locations, without consideration of the importance of such measures in a Latvian or EU context. This has largely been the result of a lack of a comprehensive assessment of site conservation objectives at the site and country levels.

The situation might improve with the elaboration of the National Natura 2000 Protection and Conservation Programme (NCA, 2017), which has assessed all terrestrial habitats in the country and set the priorities for their restoration and management. Detailed guidelines for maintenance of protected habitats have been developed by the LIFE+ project NAT-PROGRAMME. The guidelines include comprehensive descriptions of the characteristics of each habitat, their distribution and possible conservation measures.

Only one country-wide habitat management plan has been elaborated to date, for Fennoscandian wooded meadows (6530*).

A few examples of active habitat management and restoration are presented here in relation to the main broad habitat classes.

Management and Restoration of Coastal and Heath Habitats

Coastal habitat management has been mostly carried out in the open dunes, strengthening the open sand areas through various methods. In several Natura 2000 sites the focus has been on eliminating invasive alien plant species, for example Beach Rose (*Rosa rugosa*) and Low Juneberry (*Amelanchier spicata*). In several coastal areas trees and bushes have been removed to restore grey dunes. Another important element of coastal habitat management is the control of visitor flows, which in several sites has been achieved by constructing nature-friendly infrastructure.

Management of heath habitats has intensified since 2009, focussing on a particular site (Adazi) that hosts the majority of the habitat European dry heaths (4030; Laime, 2017). As this area is also an important military training area, good synergies and balance have been found in the management of the habitat and the military training activities.

Management and Restoration of Rivers and Lakes

The main problems in freshwater habitats are eutrophication and disruption of natural flows and processes. Important experience has been gained from the removal of vegetation and obstructions to flows (including beaver dams) from the rivers. Placement of gravel and stones in streams has also been a popular measure to improve river flows and provide spawning grounds for salmonids and lampreys. Often freshwater restoration activities have been implemented by volunteers and local action groups. Reed removal has been an important measure in several shallow lakes to increase habitat and structural diversity.

Management and Restoration of Grasslands

As discussed earlier, grasslands in Latvia suffer from two opposite processes: intensification and abandonment. To tackle such problems, RDP agri-environment payments for the maintenance of biologically valuable grasslands have been available since 2004. Biologically valuable grasslands is a term used in the RDP context to cover both HD grassland habitats and grasslands that are important as resting, breeding or foraging habitats for birds. However, of 775 km^2 of biologically valuable grasslands identified to date, less than 50% have been applied to the RDP scheme and therefore managed (LAD, 2016).

Biologically valuable grasslands are often in areas that are small, difficult to access and/or manage due to high water tables (e.g. floodplain meadows), or rough terrain and low fertility; thus, they are at high risk of abandonment. On the other hand, more uniform and larger areas with higher soil fertility have a higher risk of cultivation or conversion to arable land; a risk further increased by available CAP payments. To solve such problems, differentiated support payments under the RDP were introduced in 2017 for grassland bird habitats and other biologically valuable grassland types, but it is too early to assess the effects of this change.

Most of the grassland restoration activities have been performed within various projects. Typically, LIFE project grants have supported the initial restoration of grasslands, with ongoing management supported by RDP agri-environment payments. Currently, the GrassLIFE project is aiming to further develop and optimise such restoration and management to improve the conservation status of five declining HD priority grasslands in Latvia.[7] This involves testing innovative methods such as topsoil transfer, deep ploughing and seed transfer, and it will also introduce a mobile grazing approach. The plan is to keep several herds of cattle and sheep, and to deploy them to locations in the country where they are needed to restore grasslands by intensive grazing. The project also seeks to help landowners secure funding for future management of restored grasslands, either by improved means of grassland-related business opportunities or with RDP support.

Management and Restoration of Mire Habitats

Mire habitats are among the most restored habitat types in Latvia, restoration having been underway since the end of the twentieth century with the support of the LIFE programme. The restoration has mostly involved construction of various dams to block ditches that are draining raised bogs (7110*). Having tested various methods of blocking ditches (e.g. wood dams, peat dams and plastic core dams), Latvia has gained considerable experience in the hydrological restoration of mires.

Management and Restoration of Forest Habitats

Various methods of forest restoration and habitat diversification have been tested. This includes the controlled burning of boreal forest habitats, which is an appropriate restoration method for boreal forest and has been proven to be effective in Latvia from a nature protection standpoint (Cinate *et al.*, 2016). However, its wide application is still pending as the method is opposed by the forestry sector and local stakeholders. To support invertebrate populations, there have been actions to make openings in the forest and around large trees. As with mires, another important measure has been restoration of the hydrological regime in drained forests by closing ditches.

21.7 Nature Conservation Costs, Benefits and Funding Sources

21.7.1 Nature Conservation Costs

The assessment of nature conservation costs has evolved along with development of nature conservation in Latvia. In the early 1990s, management of protected territories was largely limited to compliance control, therefore the only costs were those related to the management bodies of National Parks and Strict Nature Reserves and the environmental inspectorate. The establishment of the system of protected areas in the 1990s also coincided with the restitution of private property, therefore private landowners gradually became more vocal about the need for compensation for restrictions on their economic activities in protected areas.

Accession to the EU led to more systematic management planning and implementation of recurring management measures, as well as promoting habitat restoration. It also gave access to EU funding sources, which allowed more systematic compensation for landowners.

According to the Prioritised Action Framework (PAF) for 2021–2027 (NCA, 2021), the total estimated annual costs for Natura 2000 averaged over the period amount to about €97 million (€74 million annual costs, and €23 million one-off costs). These annual costs include €23 million general cross-cutting measures and administrative costs, 56% of which are one-off costs. Averaged annual maintenance and restoration measures for the network are estimated to be €19 million (67% running costs). These maintenance and restoration costs include just €0.78 million for marine and coastal habitats (€1.8/ha averaged over the marine sites) and €19 million for other habitats (€24.9/ha averaged over terrestrial sites). The annual costs of green infrastructure to support the Natura 2000 network are estimated to be €50 million, which are about half the total costs (99% for recurring actions) as they include wide-ranging measures, such as for farmland. The estimated annual costs of additional species-specific nature conservation measures, including compensation for damage, are €4.3 million (64% one-off).

21.7.2 The Economic Benefits of Nature Conservation

There is very little information available on the economic benefits of nature conservation at the national scale in Latvia. Based on EU-level assessments (European Commission, 2013), the overall economic value of the protected areas in Latvia can be expected to be high, but there is a lack of specific national studies to corroborate this assessment. There have been a few efforts to assess some of the benefits, but they are fragmented and project-based.

The LIFE project EcosystemServices has assessed the value of some of the coastal ecosystems and their services (Anonymous, 2017). The LIFE projects REstore and Viva Grass assessed the value of some mire ecosystems and grassland habitats, respectively. Another LIFE Project, Coastlake, produced a socio-economic study that confirmed that the EU and national funding invested in the project was outweighed by the improvement in ecosystem services delivery due to habitat restoration activities in the project areas. The studies show that substantial monetary values can be attached to the protected areas that exceed many times the value of investments into the areas.

A similar conclusion was reached in a study commissioned by the Ministry of Finance of Latvia, which is the responsible authority for the EU Cohesion funds (MoF, 2019). It compared the potential short-term and long-term benefits (both environmental and economic) of investing the same amount of

[7] Xeric sand calcareous grasslands (6120*); semi-natural dry grasslands and scrubland on calcareous substrates, important orchid sites (6210*); species-rich *Nardus* grasslands on siliceous substrates (6230*); Fennoscandian lowland species-rich dry to mesic grasslands (6270*); and Fennoscandian wooded meadows (6530*).

funding in drinking and wastewater treatment facilities or the restoration of habitats. Based on a comparison of different ecosystem services and their economic values, it was concluded that both the short- and long-term net present economic value of the habitat restoration would be many times higher than that of the water treatment facilities.

21.7.3 Funding Sources for Nature Conservation

Table 21.5 provides an overview of the main funding sources used and activities supported. In the context of this chapter, this mainly focusses on the costs related to the implementation of the Nature Directives.

For the period 2008–2018, the main source of funding of biodiversity conservation and sustainable use was the state budget, which funds the biodiversity-related state institutions. Other state budget sources are earmarked funds (e.g. Fishery Fund, Forestry Development Fund, Latvian Environmental Protection Fund), which provide project-based funding for different activities (e.g. elaboration and implementation of management plans, restoration of habitats, supplementing fish stocks). Therefore, national funding provided for nature conservation is scattered over different funds, each of them supporting particular activities. The Latvian Environmental Protection Fund also supports many previously mentioned activities and also capacity enhancement of the state, municipal institutions or NGOs in the field of nature conservation.

Since EU accession, various EU funds and financial instruments have also been used. As land in the protected areas can be both privately and state owned, the range of measures used has depended on the type of land ownership. Therefore, private agriculture and forest lands are mainly supported by the CAP RDP measures (e.g. for the management of biologically valuable grasslands), whereas the building of tourism infrastructure in protected areas has been supported by the European Regional Development Fund and Cohesion Fund. The Cohesion Fund has also supported the development of management plans for Natura 2000 sites and species, and habitat restoration. Although about €6 million per year on average was allocated over 2014–2020 to agri-environment measures for ecosystem conservation (RDP Priority 4), this was a small proportion of the RDP budget allocation under this priority. The vast majority of RDP funding under Priority 4 (84%) was allocated to supporting areas with natural constraints and organic farming (Table 21.5), however the possible benefits for nature conservation from these measures are uncertain or low.

The EU LIFE Programme has been an important source of funding for nature conservation in Latvia. Since 2000, Latvia has participated in the main EU LIFE programme, which has supported 31 traditional projects in the field of nature and biodiversity conservation with €43.5 million total funding (€30 million from the EU).[8] This LIFE funding helped the country prepare for and implement EU nature legislation and has been the main support for the majority of habitat restoration activities. Latvia has been awarded a larger LIFE Integrated Project, LIFE-IP LatViaNature, to support the implementation of the PAF for Natura 2000 for 2021–2027.

The Environmental Policy Concept for 2014–2020 sets objectives for resource mobilisation for nature conservation. By 2020, the funding from different sources (state, EU, private) should have reached €50/ha in the protected areas annually, from a 2014 baseline of €14/ha. This objective is a general total and does not distinguish between different management needs or priorities. It is not known if this objective was achieved.

As described previously, nature conservation measures are mainly supported by public funding, and the vast majority of the direct nature management and restoration activities are dependent on project funding. This reliance on public funds makes nature conservation funding very dependent on political support. Project funding may also be sporadic and lead to uncoordinated funding of nature conservation needs. Such problems have been recognised by the OECD, which concluded that Latvia should develop a comprehensive financing strategy to encourage private sector investment and reduce the reliance on project-specific EU support (OECD, 2019). Although the PAF could be regarded as a financing strategy, it lacks a clear strategy on attracting and/or leveraging the funding necessary to cover at least priority funding needs. The OECD also suggested extending the use of economic instruments for biodiversity management and exploring opportunities to increase payments for ecosystem services. Since one of the working strands of the Integrated Project LIFE-IP LatViaNature is related to voluntary involvement of private landowners in biodiversity management, the eventual conclusions of this project should be seriously taken into account.

21.8 Surveillance, Monitoring, Reporting and Applied Research

The key sources of biodiversity information are species distribution atlases, specific inventories (CORINE Biotopes, Emerald, Natura 2000, country-wide habitat mapping), shadow lists of sites which need protection according to NGOs, the Latvian biodiversity monitoring programme and citizen science data. There have also been a large number of miscellaneous projects, including the development of species and site management plans, LIFE projects and university student theses, many of which have made an important contribution to knowledge of certain species, taxonomic groups, habitats, sites and regions.

Since 2005, the Latvian Biodiversity Monitoring Programme has consisted of three sub-programmes: (1) Natura 2000 monitoring; (2) country-wide monitoring; and (3) special monitoring. The Natura 2000 sub-programme has established surveillance of species and habitats within the Natura

[8] As of 15 March 2021, https://cinea.ec.europa.eu/programmes/life/life-european-countries_en

Table 21.5 Funding sources for nature conservation in Latvia.

Source and period	Measure/category	Average annual allocation (€ m)		Comments
EU funding instruments		**EU + national**	**EU**	
EAFRD Rural Development Programme measures – allocations to Priority 4 (restoring, preserving and enhancing agriculture and forest ecosystems) 2014–2020	M13 Areas facing constraints	37.0	28.4	
	M11 Organic farming	27.8	18.9	
	M10 Agri-environment–climate	12.0	8.1	Inc. for biologically valuable grasslands in N2k and elsewhere
	M12 Natura and WFD	3.4	2.3	Compensation in N2k
	M2 Advisory services, etc.	1.0	0.7	
	M1 Knowledge transfer, etc.	1.0	0.6	
	M8 Investments in forests	0.7	0.5	
RDP Priority 4 subtotal		*82.9*	*59.6*	
Cohesion Fund 2014–2020		3.8	0.7	Various uses – see text
LIFE Programme	Nature and biodiversity 2000–2020	2.1	1.4	
	Integrated Project 2020–2028	2.3	1.4	LIFE-IP LatViaNature
EU subtotal		*91.1*	*63.1*	
National funds within 2014–2020		**Additional national***	**Comments**	
State budget	General expenses of the nature institutions, monitoring, compensatory payments not covered by the EARDF, earmarked activities covered by specific budget funds	4.8		
Additional national subtotal		*4.8*		
National total including co-financing of EU instruments		32.8		
Total		**95.9**		

Notes. * Excludes co-financing of EU funds.
Sources. Based on data from: RDP data from European Commission ESIF funds database (23 October 2020); Latvia LIFE Factsheet, European Commission (March 2021); VARAM (2014). See Section 1.3 for further details.

2000 network. Each site is surveyed at least once every six years, aiming to cover all species and habitats listed in the site's Standard Data Form as having a significant presence. The acquired information is fed into updates of the Natura 2000 database and helps to review site conservation objectives.

Country-wide monitoring aims to provide data on changes in species populations, habitat area and quality, and to be representative of the whole country. It mainly consists of generic monitoring schemes with annual data collection.

The purpose of the special monitoring is to provide information on ecological processes occurring in ecosystems. Indicators are typically umbrella species or species or habitats otherwise important for Latvia. Gathered information includes, for example, reproductive characteristics and the demographic structure of species, or the vegetation structure of a habitat.

The above-mentioned inventories, monitoring, reporting and projects are being implemented by various actors: academic institutions, NGOs, private companies, state agencies and individual experts. Since independence, the state authorities have never had any permanent lead institution dealing with data collection and analysis for nature conservation. Latvia, as a relatively small country, has limited numbers of qualified experts, thus any large campaign or project relies on cooperation amongst several partners or consortiums.

Currently, the most comprehensive databases of species and habitat occurrences are the nature data management system https://ozols.gov.lv maintained by the Nature Conservation Agency, and the nature observation portal https://dabasdati.lv/lv owned and maintained by the Latvian Fund for Nature and Latvian Ornithological Society. Some of the https://ozols.gov.lv data are publicly accessible using a guest account, and additional information is accessible to registered users. Since 2008, the online geoportal for nature observations, https://dabasdati.lv/lv, has become the most important information source on the distribution of wild plant and animal species of various taxa. It is primarily based on citizen science data.

21.9 Future Developments

Similar to the conclusions of the European Commission's Fitness Check of the Nature Directives, the main future focus for nature conservation in Latvia should be on better implementation of the Nature Directives. While the EU legislation is sufficiently transposed, the main challenge remains in successful implementation to ensure the favourable conservation status of species and habitats.

The *Environmental Policy Concept* for 2014–2020 established biodiversity conservation objectives that mainly focus on facilitating implementation of the provisions of the EU nature legislation. Most of the objectives are interlinked and cover all indicated problems. In order to improve the knowledge base, in 2016 Latvia started country-wide systematic mapping of HD Annex I habitats. Results of the mapping will help to minimise the number of features previously reported as unknown, to set country- and site-level conservation objectives, and to solve at least some of the insufficiencies in the Natura 2000 network.

In order to improve management of Natura 2000 sites, Latvia is planning to intensify the development of new management plans for the sites, and the review of existing plans, with the support of the EU's ERDF. Given the sensitivity of the nature conservation provisions and eventual implications for economic activities, Latvia will elaborate internal guiding principles for setting conservation objectives, taking into account all relevant European Commission guidelines and agreed concepts.

As mentioned earlier, the EPP also sets an objective for resource mobilisation for nature conservation needs. In order to better use available funding, including that of the EU funds after 2020, Latvia will elaborate a new PAF. This will also guide the elaboration of the LIFE-IP LatViaNature project, allowing broader and more strategic implementation of the management plans and restoration strategies across the Natura 2000 sites, as well as facilitating the mainstreaming of biodiversity provisions in sectoral policies.

21.10 Conclusions

Over recent history the majority of documented changes in Latvian nature have been linked to socio-economic changes: regaining independence (1991) and subsequent land reform, and joining the EU (2004). At the end of the Soviet Union period, nature was in a relatively good state in Latvia, except in a few locally highly polluted areas. However, this was the result not of well-organised nature conservation policy, but rather the inefficient economy at the time.

Since independence, Latvia has gradually built its nature conservation administration capacities, enforcement structures, scientific support and public participation. With its accession to the EU, Latvia has adopted international nature conservation standards, and funds spent on nature conservation have been higher than ever before. Yet, at the same time, proportionally much more funding has been available for the development of more intensive agricultural and forestry production, and acquiring more land for development. Thus, it is no surprise that, except for a number of really good conservation success stories concerning certain species, habitats or protected areas, the conservation status of the majority of protected species remains unfavourable or is even worsening. In our opinion, in addition to certain concrete tasks such as completion of the Natura 2000 network, establishment of systematic nature conservation planning, and the completion and implementation of site management plans, the greatest challenge is to move nature conservation issues higher up the Latvian political agenda – otherwise economic development and the status of nature will remain unbalanced, if not in direct conflict.

Acknowledgements

The authors thank Jānis Birzaks, Andrejs Briedis, Gatis Eriņš, Viesturs Ķerus, Agnese Priede, Jānis Ozoliņš, Gita Strode and Ģirts Strazdiņš for useful discussions and information during the preparation of this chapter.

References

Anonymous (1991) *Particularly protected nature objects in the territory of the Latvian SSR.* Gandrs. Riga: Environmental Protection Committee of Latvia.

(2010) *Sustainable Development Strategy of Latvia until 2030.* Saeima of the Republic of Latvia.

(2017) *Economic Valuation of Ecosystems and their Services for Pilot Implementation Areas.* Riga: LIFE EcosystemServices.

(2019) *Natura 2000 newsletter.* Number 46. Brussels: European Commission.

Auniņa, L. & Rove, I. (2013) Virsāju biotopi. In *Eiropas Savienības Aizsargājamie Biotopi Latvijā. Noteikšanas rokasgrāmata*, ed. A. Auniņš, pp. 63–103. Riga: Latvian Fund for Nature.

Auniņš, A. (ed.) (2008) *Current Issues in Wildlife Management in Latvia [Aktuālā savvaļas sugu un biotopu apsaimniekošanas problemātika Latvijā].* Riga: Latvian Fund for Nature.

(2010) *Bird Populations in Latvian Farmland: Patterns and Trends.* Riga: University of Latvia.

(2018) Ligzdojošo putnu skaits turpina samazināties: visvairāk cieš Āfrikā ziemojošie un ar lauksaimniecības zemēm saistītie putni. *Putni dabā*, **81**, 10–15.

Aunins, A. & Priednieks, J. (2009) Recent changes in agricultural landscape and bird populations in Latvia: impacts and prospects of EU agricultural policy. *Avocetta*, **33**, 93–98.

Bagrade, G., Ruņģis, D. E., Ornicāns, A. *et al.* (2016) Status assessment of Eurasian lynx in Latvia linking genetics and demography – a growing population or a source-sink process? *Mammal Research*, **61**, 337–352.

Bergmanis, U., Auniņš, A., Petriņš, A. *et al.* (2015) Population size, dynamics and reproduction success of the lesser spotted eagle (*Aquila pomarina*) in Latvia. *Slovak Raptor Journal*, **9**, 45–54.

Bijlsma, R. J., Agrillo, E., Attore, F. *et al.* (2017) *Defining and applying the concept of Favourable Reference Values. Draft technical report.* Wageningen: Wageningen Environmental Research.

BirdLife International (2017) *Review of Illegal Killing and Taking of Birds in Northern and Central Europe and the Caucasus.* Cambridge, UK: BirdLife International.

Boruks, A. (2004) *Dabas apstākļi un to ietekme uz agrovidi Latvijā.* Rīga: Valsts Zemes Dienests.

Cinate, K., Larmanis, V. & Piksena, I. (eds.) (2016) *Important Examples of Wild Species and Habitat Management in Latvia: Forests.* Riga: Nature Conservation Agency.

CSB (2018a) CSB Database [WWW Document]. www.csb.gov.lv/en/statistika/db

(2018b) *Farm Structure in Latvia 2016. Collection of Statistics.* Riga: Central Statistical Bureau of Latvia.

European Commission (2001) *Assessment of plans and projects significantly affecting Natura 2000 sites. A methodological guidance on the provisions of Article 6 (3) and (4) of the Habitats Directive 92/43/EEC.* Luxembourg: Office for Official Publications of the European Communities.

(2013) *The Economic Benefits of the Natura 2000 network – Synthesis Report.* Luxembourg: Publications Office of the European Union.

Evans, D. (2012) *Building the European Union's Natura 2000 network. Nature Conservation*, **1**, 11–26.

Ikauniece, S. (ed.) (2017) *Vadlīnijas aizsargājamo biotopu saglabāšanas vadlīnijas Latvijā. 4. sējums. Meži.* Sigulda: Dabas aizsardzības pārvalde.

Kalniņa, L. (2008) Mire origin and development in Latvia. In *Mire Conservation and Management in Especially Protected Nature Areas in Latvia*, ed. M. Pakalne, pp. 21–25. Riga: Latvian Fund for Nature.

Keiss, O. (2005) Impact of changes in agricultural land use on the Corncrake *Crex crex* population in Latvia. *Acta Universitatis Latviensis*, **691**, 93–109.

LAD (2016) *Bioloģiskās daudzveidības uzturēšana zālājos.* Riga: Lauku atbalsta dienests.

(2018) *LIZ apsekošanas rezultāti novadu griezumā (2017.gads).* Riga: Lauku atbalsta dienests. www.lad.gov.lv/files/zva_17_gat_tabula.pdf

Laime, B. (ed.) (2017) *Vadlīnijas aizsargājamo biotopu saglabāšanas vadlīnijas Latvijā. 1. sējums. Piejūra, smiltāji un virsaji. Dabas aizsardzības pārvalde*, Sigulda, Latvia.

MoF (2019) Biotopu un sugu aizsardzības labvēlīga statusa atjaunošanas pasākumu sociāli ekonomiskais novērtējums. www.esfondi.lv/upload/izvertejumi/biotopu-novertejums-final.pdf

NCA (2017) *National Natura 2000 protection and conservation programme.* [Natura 2000 teritoriju nacionālā aizsardzības un apsaimniekošanas programma]. Sigulda, Latvia: Nature Conservation Agency of Latvia.

(2021) *Prioritised Action Framework (PAF) for NATURA 2000 in Latvia. For the Multiannual Financial Framework Period 2021–2027.* Sigulda: Latvia: Nature Conservation Agency of Latvia.

Nikodemus, O., Bell, S., Grīne, I. & Liepiņš, I. (2005) The impact of economic, social and political factors on the landscape structure of the Vidzeme Uplands in Latvia. *Landscape and Urban Planning*, **70**, 57–67.

Nikodemus, O. & Brumelis, G. (2015) *Dabas aizsardzība [Nature conservation].* Riga: Academic Press of the University of Latvia.

Nikodemus, O., Kļaviņš, M., Krišjāne, Z. & Zelčs, V. (eds.) (2018) *Latvija. Zeme, daba, tauta, valsts.* Riga: Latvijas Universitāte.

Nipers, A., Krievina, A. & Pilvere, I. (2018) Projecting productivity in agriculture in Latvia. *Research for Rural Development*, **2**, 130–137.

OECD (2019) *OECD Environmental Performance Reviews: Latvia 2019.* www.oecd.org/environment/country-reviews/oecd-environmental-performance-reviews-latvia-2019.htm

Olson, V. A., Davies, R. G., Orme, C. D. L. *et al.* (2009) Global biogeography and ecology of body size in birds. *Ecology Letters*, **12**, 249–259.

Opermanis, O. (ed.) (2002) *Important examples of wildlife management in Latvia.* DANCEE. Riga: Ulma.

(2016) Accumulation of knowledge on management of Natura 2000 sites. In *Important Examples of Wild Species and Habitat Management in Latvia: Forests*, eds. K. Cinate, V. Larmanis & I. Piksena, pp. 5–11. Riga: Nature Conservation Agency.

Opermanis, O., MacSharry, B., Aunins, A. & Sipkova, Z. (2012) Connectedness and connectivity of the Natura 2000 network of protected areas across country borders in the European Union. *Biological Conservation*, **153**, 227–238.

Ozolins, J., Mannil, P., Balciauskas, L. & Ornicans, A. (2014) Ecological, social and economic justification of wolf population management in the Baltic region. *Beitrage zur Jagd- und Wildforschung*, **39**, 215–224.

Racinska, I. (2002) *Handbook for management planning in protected areas in Latvia [Rokasgrāmata īpaši aizsargājamo dabas teritoriju dabas aizsardzības plānu izstrādātājiem].* Riga: Ulma.

Racinskis, E. (2004) *Important Bird Areas of European Union Importance in Latvia [Eiropas Savienības nozīmes putniem nozīmīgās vietas Latvijā].* Riga: Latvian Ornithological Society.

Rove, I. (2013) Piejūras un iekšzemes kāpu biotopi. In *Eiropas Savienības AizsargāJamie Biotopi Latvijā. Noteikšanas Rokasgrāmata*, ed. A. Auniņš, pp. 63–103. Riga: Latvijas dabas fonds.

Rūsiņa, S. (ed.) (2017) *Vadlīnijas aizsargājamo biotopu saglabāšanas vadlīnijas Latvijā. 3. sējums. Dabiskās pļavas un ganības.* Sigulda, Latvia: Dabas aizsardzības pārvalde.

Sabardina G. S. (1957) *Lugovaja rastitelnost Latvijskoi SSR.* (In Russian.) Riga: Publishing House of the Academy of Sciences of Latvian SSR.

Saliņš, Z. (1999) *Meža izmantošana Latvijā: vēsture, stāvoklis, perspektīvas.* Jelgava, Latvia: LLU Meža izmantošanas katedra.

Skujenieks, M. (1927) *Latvija. Zeme un iedzīvotāji. A.Gulbja apgādniecība.* Rīga. 752 lpp.

Ssymank, A. (2005) Cross-border implementation and coherence of Natura 2000 in Germany. In *Crossing Borders: Natura 2000 in the Light of EU Enlargement*, eds. M. Leibenath, S. Rientjes, G. Lintz & K. Kolbe-Weber, pp. 25–39. Proceedings of an International Workshop, Dresden. European Centre for Nature Conservation, Leibniz Institute of Ecological and Regional Development.

Strods, H., Zunde, M., Mugurēvičs, E. *et al.* (eds.) (1999) *Latvijas mežu vēsture līdz 1940. gadam.* Riga: WWF - Pasaules Dabas Fonds.

Urtāns, A. (ed.) (2017) *Vadlīnijas aizsargājamo biotopu saglabāšanas vadlīnijas Latvijā. 2. sējums. Upes un ezeri.* Sigulda, Latvia: Dabas aizsardzības pārvalde.

VARAM (2014) Information on the activities supported by the EU Structural funds. www.varam.gov.lv/lat/fondi/kohez/2014_2020/?doc=25728

(2017) Guidelines of sustainable peat use 2018–2050. [Kūdras ilgtspējīgas izmantošanas pamatnostādnes 2018. - 2050. Gadam.] Ministry of Environment Protection and Regional Development.

Viksne, J. (1994) *Important Bird Areas in Latvia.* Riga: Latvian Ornithological Society.

VMD (2018a) *2017.gada publiskais pārskats.* Riga: Valsts meža dienests.

(2018b) Ciršanas apjomi skaitļos un faktos. www.vmd.gov.lv/valsts-meza-dienests/statiskas-lapas/-meza-apsaimniekosana-/valsts-meza-dienests/statiskas-lapas/valsts-meza-dienests/statiskas-lapas/cirsanas-apjomi-skaitlos-un-faktos?id=2638 (Accessed 11 March 2018).

Zdanovskis, K. & Pilvere, I. (2015) Agricultural development in Latvia after joining the European Union. *Research for Rural Development,* **1**, 161–168.

Chapter

22 Lithuania

Rūta Baškytė and Žygimantas Obelevičius

Key Statistics	
Terrestrial area (km^2)/% of EU-28*	64 899/1.5%
Marine waters (km^2)/% of EU-28*	6 437/0.1%
Population density (p km^2)/% of EU-28*	45/39%
GDP per capita (€)/% of EU-28	11 515/43%
Land cover (CORINE land cover classes 2018)	AGR 59%; FOR 29%; OSN 6%; ART 3%; WET 3%
Biogeographical regions	BOR, MBAL
Number of HD Annex I habitats/%*	54/24%
Number of HD Annex II, IV and V species/%*	98/8%
Number of BD Annex I bird taxa/%*	72/41%
% HD UCS and decreasing habitats/ % unknown	7%/0%
% HD UCS and decreasing species/ % unknown	24%/18% (excludes MBAL – as data missing)
% breeding BD Annex I birds decreasing/% uncertain and unknown	29%/0%

* Excluding MAC and MMAC biogeographical regions.
Sources and codes. See Table 1.1.

22.1 The Natural History and Characteristics of Lithuania

Lithuania is one of three Baltic States, located on the eastern coast of the Baltic Sea. It is within the Boreal and Marine Baltic biogeographical regions. The country's climate is cool temperate with moderate warm summers and moderate cold winters.

The land surface of Lithuania was mainly created 10–12 thousand years ago during the last glacial and post-glacial periods, and has since been modified by erosion. Now Lithuania has five main natural landscape types: moraine, sandy plains, coastal plains, hilly moraine highlands, and river valleys. It is mainly a lowland country, but some higher areas (i.e. above 200 m) cover about 21% of the country and rise up to 294 m.

The country is rich in natural and semi-natural habitat types and species as a result of its diverse landforms, geology, soils, climate and landscapes. There are numerous lakes and rivers, and extensive forests and wetlands. As the country lies in the transitional coniferous forest zone, it contains elements of southern taïga and nemoral forest vegetation with the most characteristic forest habitats being Western taïga (9010*),[1] Fennoscandian deciduous swamp woods (9080*) and bog woodland (91D0*). For these reasons its forest habitats are diverse, and their internal structure is complex and often atypical of other areas. Despite the fact that wetlands in Lithuania have faced many threats in the past (described in Section 22.2), they continue to be an integral part of the country's nature. They include important areas of raised bogs (e.g. active raised bogs (7110*)), transition mires (e.g. transition mires and quaking bogs (7140)) and fens (e.g. alkaline fens (7230)). Initially grasslands developed naturally along river floodplains, but then more extensively as a result of agricultural activities.

In recent times, the natural history of the country has been increasingly affected by various human activities, especially as the country's economy has grown. Development in the agriculture and forestry sectors and urbanisation have led to the loss or severe decline of some natural habitats (raised bogs, some forests, small lakes, natural rivers and other wetlands) as a result of wetland drainage, the excavation of raised bogs, felling of small forest patches and creation of vast agricultural fields. Despite all these activities, Lithuania still has around 29% forest coverage (2018 CLC data), although only small areas are natural ancient woodland.

22.2 The Status of Nature

22.2.1 Habitat Status and Trends

According to the latest HD Article 17 report,[2] 22% of habitat assessments in Lithuania had a favourable conservation status over 2013–2018. The country has not therefore come close to achieving its aim of having 48% of habitats in a favourable

[1] Numbers in brackets are the HD Annex I habitat codes (see Appendix for full names). * Indicates a priority habitat.
[2] See Section 1.3 for background on monitoring under the Nature Directives, status categories and data sources.

Table 22.1 Short-term trends (c. 2007–2018) in Habitats Directive Annex I habitats and Annex II, IV and V species with an unfavourable conservation status in Lithuania.

	Poor (unfavourable–inadequate)				Bad (unfavourable–bad)			
	Imp	Sta	Dec	?	Imp	Sta	Dec	?
HD Habitat assessments								
Number	0	3	0	18	0	1	4	16
%	0%	14%	0%	86%	0%	5%	19%	76%
HD Species assessments								
Number	3	7	12	9	0	1	12	0
%	10%	23%	39%	29%	0%	8%	92%	0%

Key. Imp = improving, Sta = stable, Dec = decreasing, ? = unknown.
Source. Based on Member State HD Article 17 reported data as presented on the EEA national summary dashboard (2021).

condition in 2020, as included in its *Landscape and Biodiversity Protection Action Plan for 2015–2020.*[3] Although there has been a slight increase, compared to the 19% over 2007–2012, this may not be a genuine improvement because better information is now available as a result of a national habitats inventory that has since been carried out. Over 2013–2018, 39% of habitat assessments revealed a poor conservation status, and the same proportion of habitats had a bad status, notably alkaline fens (7230), Fennoscandian wooded pastures (9070) and old acidophilous oak woods (9190).

As a result of a lack of funding for surveillance, the overall short-term trends for habitats with an unfavourable status are mainly unknown (Table 22.1). This is mostly because the trends in habitat structure and functions are unknown due to the lack of data prior to the habitats inventory. However, habitats are subject to considerable pressures (as described in Section 22.2.3) and therefore it is likely that many with unknown trends are deteriorating. Given the situation, it is unlikely that the favourable conservation status goal will be reached in the near future unless funding for nature conservation increases significantly (see Section 22.7.3).

22.2.2 Species Status and Trends

In the latest HD Article 17 report, 37% of species assessments revealed a favourable conservation status over 2013–2018, 32% a poor status, 13% a bad status, and 18% an unknown status. Knowledge of the conservation status of most species has improved, although gaps remain for some (e.g. bats), which means that for most species it is not possible to ascertain whether changes in status or trends are genuine or not. The majority of short-term trends of HD species with an unfavourable conservation status are declining, including 92% of those with a bad status (Table 22.1). The situation regarding vascular plants is especially concerning: only 8% had a favourable status, and 80% of those with an unfavourable conservation status had declining short-term trends.

Table 22.2 Short-term trends (c. 2007–2018) in Birds Directive Annex I taxa in Lithuania.

	Breeding				Wintering			
	Inc	Sta/ Flu	Dec	?	Inc	Sta/ Flu	Dec	?
Number	17	30	19	0	1	2	1	0
%	26%	45%	29%	0%	25%	50%	25%	0%

Key. Inc = increasing, Sta = stable, Flu = fluctuating, Dec = decreasing, ? = uncertain and unknown.
Source. Based on Member State Article 12 reported data as presented on the EEA national summary dashboard (2021).

According to the latest BD Article 12 report, most breeding and wintering BD Annex I species have increased or had stable/fluctuating populations over the short term, but a substantial proportion have declined (Table 22.2). Out of 17 common bird species used as farmland bird indicators in the Boreal region, in Lithuania 10 had a declining short-term (2013–2018) trend. The Turtle-dove (*Streptopelia turtur*) decreased most (by 40%), while five species had stable trends and only two increased. As regards the short-term trends of common forest species, out of 19 indicator species in the Boreal region that breed in Lithuania, four had a declining short-term trend with the Lesser Spotted Woodpecker (*Dryobates minor*) decreasing most (by 10%). Thirteen species had stable trends and two increased.

22.2.3 Drivers and Causes of Change

Habitats and species in Lithuania have been influenced by various factors, which are often the result of human activities that have led to habitat loss or change and species declines. According to the *Landscape and Biodiversity Conservation*

[3] *Kraštovaizdžio ir biologinės įvairovės išsaugojimo 2015–2020 metų veiksmų planas, 2015.* www.e-tar.lt/portal/lt/legalAct/4b9e17309cab11e48dcdae4eb2005eaf/cufdjpxPlF

Action Plan 2015–2020[4] and Fifth CBD Report by Lithuania (MoE, 2017), the most important have included: disruption of the natural hydrological regime; loss of, or changes in, traditional farming; and climate change, as well as more habitat-specific threats as described in subsequent paragraphs.

Water bodies have been greatly affected by past pressures. In particular, during Soviet times, 83% of Lithuanian rivers were straightened (Jablonskis *et al.*, 2007) and some areas flooded, which led to the extensive destruction of riverine habitats and losses of associated species. Many shallow and small lakes have also been affected by anthropogenic eutrophication, especially because they are fed by dense networks of water courses (about 1 km/km^2) and therefore susceptible to human activities. Nowadays, water bodies are not undergoing such substantial changes as straightening of rivers, but are threatened by succession and pollution. Currently, the biggest source of pollution that causes eutrophication is diffuse pollution from agricultural activities. While livestock farming is receding, cropland is increasing together with the usage of mineral fertilisers, especially nitrogen-based (Rašomavičius, 2012; Plungė and Gudas, 2018; MoE, 2022).

A significant proportion of bogs, mires and fen habitats in Lithuania have been drained, mainly for peat extraction and land reclamation (Povilaitis *et al.*, 2011). This increased the rate of peat mineralisation with all of its related environmental consequences (Mierauskas, 2009). A more recent important threat for open raised bog habitats (e.g. 7110*) has been colonisation by woody vegetation, as this increased over the last decades of the twentieth century (Taminskas *et al.*, 2008; Edvardsson *et al.*, 2015). Such rapid succession is mainly the result of anthropogenic influences on the raised bogs' hydrology, as well as climate change (Taminskas *et al.*, 2008; Šimanauskienė *et al.*, 2019).

Meadows have been subject to numerous pressures, with the main one being their loss as traditional agricultural systems have disappeared or changed. Where grasslands are abandoned, they become overgrown with trees and shrubs (Rašomavičius, 2012; MoE, 2017, 2022). Losses of meadows and other agricultural changes have led to declines in the Turtle-dove, as mentioned earlier, as well as the Red-backed Shrike (*Lanius collurio*) (MoE, 2022).

Although forest cover is increasing, a large proportion is not in a good condition, one of the main reasons being overexploitation or management that is incompatible with maintaining a favourable conservation status (e.g. clear-cutting, removal of dead or dying trees, and afforestation with more profitable alien tree species). Such intensive forest management, especially clear-cutting, has caused declines in the Black Woodpecker (*Dryocopus martius*) and White-backed Woodpecker (*Dendrocopos leucotos*) (MoE, 2022). In addition, as a result of climate change–induced droughts, Norway Spruce (*Picea abies*) is increasingly being affected by bark beetles. This mainly affects the Fennoscandian herb-rich forest (9050) and its related species. The structure of Fennoscandian natural old broadleaved forests (9020*) and alluvial forests (91E0*) is also being heavily affected by a substantial decline in Ash (*Fraxinus excelsior*) caused by fungal dieback disease. This has strongly contributed to the unfavourable conservation status of the Scarce Fritillary (*Euphydryas maturna*) as the Ash is its main larval host plant (MoE, 2022; Rašomavičius, 2012).

The sandy habitats of the Baltic sea coast are sensitive to natural ecological changes, but are currently most threatened by succession (with embryonic shifting dunes (2110) becoming wooded dunes (2180)), paths over dunes and trampling as a result of intense recreation, especially around towns and resorts (Rašomavičius, 2012; MoE, 2022).

Most of the declines in species are primarily the result of impacts on their habitats. However, additional causes include diseases (e.g. crayfish plague affecting the Noble Crayfish (*Astacus astacus*)), invasive species (also affecting the Noble Crayfish, as well as the plants *Thesium ebracteatum* and *Linaria loeselii*) and decreases in precipitation due to climate change (affecting the moss *Hamatocaulis vernicosus*, plant *Saxifraga hirculus*, and the Fire-bellied Toad (*Bombina bombina*)) (Gudžinskas *et al.*, 2015). Declines in some birds are also considered to be the result of climate change, such as is the case with the Common Tern (*Sterna hirundo*) and Little Tern (*Sternula albifrons*), and predation in the case of the Black Grouse (*Lyrurus tetrix*) (e.g. due to the large population of the Red Fox (*Vulpes vulpes*)) and Little Gull (*Hydrocoloeus minutus*) (e.g. by the invasive American Mink (*Neovison vison*)) (MoE, 2022).

22.3 Nature Conservation Governance, Actors, Legislation and Policies

22.3.1 Development of Nature Conservation in Lithuania

Of particular importance, over the course of Lithuania's long and complicated nature conservation history (summarised in Table 22.3), has been the protection of threatened species and the creation of the protected areas system, the latter being the most effective instrument for nature conservation. The roots of nature protection go back to the very beginnings of the Lithuanian nation, but the first real protected area, Žuvintas Strict Nature Reserve (now a Biosphere Reserve), was established in 1937.

A subsequent very important step was the first Nature Protection Law (1959), which defined some protected area categories (described later in Table 22.5) and established the means for protecting them. By 1960, 89 Nature Reserves had been established and the network developed quite quickly. However, during the Soviet period the reserves were poorly protected. Proactive conservation measures were not implemented, and legal restrictions and requirements were neglected.

[4] Kraštovaizdžio ir biologinės įvairovės išsaugojimo 2015–2020 metų veiksmų planas, 2015. www.e-tar.lt/portal/lt/legalAct/4b9e17309cab11e48dcdae4eb2005eaf/cufdjpxPlF

Table 22.3 Key nature conservation developments in Lithuania.

1937	First protected area – Žuvintas Strict Nature Reserve – established
1955	Nature Protection Commission established under the Soviet Sciences Academy
1957	Nature Protection Committee established under the Soviet of Ministers
1959	The first Nature Protection Law passed
1960	The first Nature Reserves (89) established, the basis for the protected areas network created
1969–1980	Creation of the system of Nature Reserves, enlargement of protected areas
1974	The first National Park established
1980	Long-term scheme for Lithuania's Special Protected Areas prepared
1986	Integrated Nature Protection Scheme approved by the government
1990	Environmental Protection Department at the Parliament established
1991	Four National Parks and one Strict Nature Reserve established
1992	30 Regional Parks and around 128 Nature Reserves established by one decision of the Parliament
1993	The Law on Protected Areas passed
1994	The Ministry of Environmental Protection established
1998	*Lithuanian Biodiversity Conservation Strategy and Action Plan* launched
2002	The State Service for Protected Areas established
2002–2005	Creation of Natura 2000 network started, 26 biosphere polygons for bird protection established
2004	EU Accession. The List of Areas Important for Birds Protection approved by the government
2009	The Informative System of Protected Species created
2011	The inventory of Habitats of EU Importance finished
2014	The *Landscape and Biodiversity Protection Action Plan* for 2015–2020 approved
2015	The *National Environmental Protection Strategy* passed
	The *National Landscape Management Plan* passed
2019	The Law on Special Land Conditions passed

See Table 3.3 for signatory dates of international conventions.

After restoration of **Lithuania as an independent State** (1990), a new period for nature conservation started as environmental and nature protection became a higher priority in Lithuania. The new Constitution (1992) stipulated that the State should concern itself with the protection of the natural environment, its fauna and flora, objects of nature and particularly valuable areas, and should supervise the moderate usage of natural resources as well as their restoration and augmentation.

Replacing the previous Nature Protection Law, the **1992 Law on Environmental Protection**[5] covered the main issues on nature conservation: species protection, protected areas system, Nature Frame (described later), etc. The Environmental Protection Department was also established, later becoming the Ministry of Environmental Protection, and subsequently the Ministry of Environment (MoE). This period was very important in terms of international cooperation, including integration into wider European initiatives, and the ratification of all international conventions for nature protection (see Table 3.3 for dates).

Major steps have since been taken to further develop the system of protected areas over the last 30 years, with especially important decisions taken by the Parliament in the last decade of the twentieth century. This included the passing of the first **Law on Protected Areas**[6] **(1993)** and the designation of many protected areas.

All these important changes in nature conservation were implemented at the same time as radical changes in the socio-

[5] Aplinkos apsaugos įstatymas. www.e-tar.lt/portal/lt/legalAct/TAR.E2780B68DE62/asr

[6] Saugomų teritorijų įstatymas. www.e-tar.lt/portal/lt/legalAct/TAR.FF1083B528B7/asr

economic situation, land reform and the creation of a new legal system, etc.

Another particularly important step in nature conservation was the establishment of the institutional system for protected areas, with administrations for new National Parks and Regional Parks established in 1994–1999, and the State Service for Protected Areas established in 2002 under the MoE.

The next stage in the development of nature conservation was the preparations for, and implementation of, the **EU Nature Directives**, especially the development of the Natura 2000 network. Rapid action was required, because at the start of the negotiations on EU membership the Lithuanian authorities pledged to establish a network of SPAs before the accession date (1 May 2004). Selection of the areas started with the project 'Implementation of [the] Natura 2000 Network on Regional and National Level in Lithuania in 2002–2004', funded by the Danish Cooperation for Environment in Eastern Europe (DANSEE). This was carried out by consulting companies Ornis Consult and NEPCon, with the involvement of Lithuanian scientists and the Lithuanian Ornithological Society (LOD) providing the majority of the bird data.

The first investigations to identify important areas for the protection of HD habitats were made in 2004–2005, but the final steps were carried out over 2011–2015 under an EU-funded special project.[7] Under this, comprehensive data on natural habitats in Lithuania were collected and added to a GIS database.[8]

Requirements for protection of natural habitats and species of EU importance were integrated into the Law on Protected Areas (2001, 2010 and 2016): definitions of habitats, the procedures for establishing the Natura 2000 network, the protection measures, etc.

The main **strategic policy initiatives** were the development of the *Environmental Strategy of Lithuania* (1996), *Biodiversity Protection Strategy of Lithuanian Republic and Action Plan* (MoE, 1998) (BPSAP) and *Strategy for Sustainable Development* (MoE, 2011). These documents identified the main nature/biodiversity conservation challenges (e.g. degradation of natural landscapes, loss of species, negative impacts on bird and fish migration pathways, and the inappropriate use of nature's resources) and raised ambitious objectives and measures in the field of nature conservation. A lot of proposed measures were successfully implemented; some of them are still in progress or have been incorporated in new strategic documents.

Importantly, environmental protection and nature conservation issues became a part of sectoral programmes (e.g. agriculture, forestry, tourism, energy, transport) with proposed specific objectives and measures. Nature conservation objectives were included in key laws, such as the Law of Forests, the Law on Land, and the Law on Territorial Planning. Finally, nature conservation measures have been implemented together with different sectoral development programmes and projects.

Another advance was the integration of nature conservation into the *General Plan of the Lithuanian Republic* (2002) (the main territorial/spatial planning document of the country). A key element of this is dedicated to landscape and biodiversity protection in the context of the Nature Frame policy, which aims to conserve or restore natural structures that are responsible for ecological functions (further described in Section 22.5). The protection of rare and endangered species, establishment of the Natura 2000 network, optimisation of forest structure, avoidance of ecosystem degradation, and prevention of impacts on fish and bird migration pathways became the priorities for biodiversity protection in the General Plan. Other related priorities included optimisation of the protected areas system and protection of wetlands, karst landscapes and natural landscapes in urban areas. Over 2007–2010, general plans for all administrative districts were prepared and approved by the municipalities.

22.3.2 Contemporary (from 2010) Policies, Legislation and Governance

The current nature conservation **legislation** system in Lithuania is rather complicated as it consists of several laws and by-laws. Important provisions regarding the protection of the natural environment come from the Constitution of the Republic of Lithuania – the fundamental law of the State.

The main objectives for protection of biodiversity (coming from international conventions, including the CBD) are integrated into national laws: Law on Protected Fauna, Flora and Fungi,[9] Law on Wild Flora,[10] Law on Wildlife,[11] Law on Protected Areas,[12] Law on Plants National Gene Resources,[13] etc. Each law enables several by-laws (i.e. rule books, acts and regulations). The General Regulations for the Areas Important for Habitats and Birds (for Natura 2000 areas)[14] and individual regulations for protected areas are particularly relevant to nature conservation. Many sectoral laws are also important for biodiversity conservation, such as the umbrella Law on Environment Protection, laws on forests, land, hunting, fisheries, and the Law on Special Land Use Conditions.[15]

Currently the main **strategy** for nature conservation is the *Landscape and Biodiversity Protection Action Plan for*

[7] Inventorisation of Habitats of EU Importance, Setting the Criterions of Favourable Protection Status and Creation of Monitoring System.

[8] www.geoportal.lt

[9] Saugomų gyvūnų, augalų ir grybų įstatymas. This and other legal Acts can be found at www.e-tar.lt/portal/lt/index

[10] Laukinės augalijos įstatymas.

[11] Laukinės gyvūnijos įstatymas.

[12] Saugomų teritorijų įstatymas.

[13] Augalų nacionalinių genetinių išteklių įstatymas.

[14] Bendrieji buveinių ar paukščių apsaugai svarbių teritorijų nuostatai.

[15] Specialiųjų žemės naudojimo sąlygų įstatymas.

2015–2020. Nature protection is also a part of the *National Environmental Protection Strategy* (MoE, 2016), which is the main strategic document covering all sectors of the environment included in the *National Landscape Management Plan*[16] (MoE, 2015), which is very important with respect to the location of national protected areas and Natura 2000 sites. The strategies have been prepared in line with international strategies and conventions (including CBD, EU directives, Ramsar, Bern, CMS and CITES) and have the key goal of stopping the decline of biodiversity and restoring it when possible. The most important and effective strategy for nature conservation is the *Landscape and Biodiversity Protection Action Plan* (LBPAP), with many proposed actions implemented or underway with the support of EU structural funds and state budget resources (see Section 22.7.3); all measures are to be implemented until the end of 2023.

Nature conservation **governance** in Lithuania is mostly the responsibility of governmental institutions at the state level (the MoE) and municipalities at a local level (often working with NGOs) (Baškytė, 2013). The strategic goals of the MoE are to ensure environmental quality, to implement climate change policy, and to protect biodiversity and distinctive landscapes. The MoE's responsibilities cover two key areas: (1) environmental protection (e.g. water, air quality protection, waste and chemical substances management, nature and landscape protection, protected areas, climate change, environmental control) and (2) development of some sectors (e.g. forestry, urban development, construction, architecture). The regulation of EIA, SEA and territorial planning is under the same ministry.

The key unit for nature conservation in the MoE is the Policy Group for Nature Protection. Its activities include policy formation for the conservation of nature and landscape, gene resources, establishment of Natura 2000 areas, creation of the Nature Frame, management of invasive species, ecosystem services, monitoring and GMOs.

The State Service of Protected Areas (SSPA) is the main institution implementing nature conservation policy. It is responsible for management of protected areas, preparation of planning documents for them, running the Protected Areas State Cadastre (i.e. the official state database with GIS data), nature and landscape monitoring, establishment of protected area administrations and supervision of their activities. Currently the SSPA is responsible for the implementation of Species Conservation Plans and the programme for the regulation of invasive species, amongst other things, over the whole country.

Protected area administrations monitor the state of species, habitats and the landscape, and collect information on nature values. They directly implement proposals in planning documents, implement protective measures for habitats and species, nature and cultural values, develop infrastructure for visitors, and implement nature education programmes, etc.

[16] Nacionalinis kraštovaizdžio tvarkymo planas.

Other institutions that contribute to nature protection under the MoE include the Environmental Protection Agency (EPA), Environmental Protection Department, Lithuanian Geology Survey, State Forests Service, and State Forests Enterprise. The EPA covers functions in relation to monitoring and EIA. The management of natural forest habitats in state forests is the responsibility of the State Forests Enterprise. Nine institutions are involved in governance of marine areas, but the EPA is the most important and most relevant to nature conservation.

The other sectors, such as agriculture, energy, transport and tourism development, that have direct or indirect impacts on biodiversity are regulated by the Ministries of Agriculture, Energy, Transport and Communications and the Ministry of Economy and Innovation. Their effective communication and cooperation with institutions responsible for nature protection is therefore essential.

The municipalities are also important actors in nature conservation, including through their participation in territorial (spatial) planning and EIA processes. Municipalities have a statutory duty to consider all biodiversity matters when carrying out all their activities. They can establish and manage municipal nature/landscape reserves, and are involved in management of state protected areas, but in some cases in some municipalities these roles are ignored.

22.3.3 Public Involvement and NGOs

An increasing number of people are joining NGOs involved in nature protection, at regional, national and international levels. All of the NGOs play a particularly important role in promoting the need for biodiversity conservation, and also contribute to practical conservation measures in various ways, including law enforcement, research and monitoring. Some of these NGOs are established on a professional basis (e.g. the Lithuanian Entomologic Society, LOD and the Lithuanian Botanical Society) and seek to conserve flora and fauna by protecting their habitats and working for the sustainable use of natural resources. Some of them are directly involved in important projects for biodiversity, such as wildlife protection, habitat restoration, agri-environment schemes, rural development, nature tourism and environmental education (e.g. Baltic Environmental Forum, Lithuanian Nature Fund, Nature Heritage Fund, and the Association 'Environmental Coalition'). Others (Lithuanian Green Movement, Movement 'For Nature', Community 'Atgaja', Klaipėda Ecological Club 'Žvejonė', Šiauliai Nature and Culture Heritage Protection Club 'Aukuras', and the Nature Protection Association 'Baltic Wolf') organise campaigns against the intensive use of forests and wetlands.

Other stakeholders and the wider public have become more involved in nature conservation over recent years, but often with contrasting opinions. Forests owners are mainly interested in intensive exploitation of forests, and farmers wish to intensify their use of grasslands and other areas. There is also an interest in the excavation of peat, sand and gravel. Such stakeholders have generally opposed the establishment of the Natura 2000 network and other protected areas. In contrast, the wider public is more interested in nature protection, as there is an understanding of the importance of biodiversity protection for the health and wealth of society. Society is aware that this may therefore require some human economic activities to be regulated, restricted or even forbidden. Consequently, there is public opposition to clear-cutting in forests or drainage of wetlands and urban developments in natural areas, and support for increasing protected areas and financial resources for nature protection.

To try and accommodate the various views, all stakeholders and the wider public are consulted during impact assessments (SEA, EIA and HD Article 6(3) appropriate assessments) and can participate in discussions when protected area Planning Schemes and Nature Management Plans are under preparation. Furthermore, a longstanding concept in Lithuania is the joint management of National and Regional Parks. Under this, municipalities, environmental or culture protection organisations, communities and NGOs are involved in their management through the administrations organising various forms of consultations.

22.4 Species Protection and Conservation Measures

22.4.1 Legislation

The Law on Protected Species of Fauna, Flora and Fungi[17] (1997, 2009) sets the main requirements for protecting species and their habitats. This ensures the implementation of the EU legal acts (Nature Directives) by incorporating requirements for the strict protection of species listed in HD Annex IV, and setting the main principles for protection and the responsibilities of institutions. The taking of, and trade in, wild species is controlled, and all rare species and vegetation communities, and their habitats, are protected by law. According to the Law on Protected Species, endangered, vulnerable, rare or endemic species of animal, plant or fungus are recognised and included in the List of Protected Fauna, Flora and Fungi Species of the Republic of Lithuania, as well as those protected under the Nature Directives (including all wild birds) and other international agreements.

Red Data Books have an important formal role in species conservation in Lithuania, as listed threatened species are given legal protection. They also direct attention to them, for example in terms of protection measures, research and monitoring. The first Red List of threatened plant species was approved in 1962. The most recent species Red List came into force in 2019[18] and lists 566 species (14 mammals, 77 birds, 2 reptiles, 5 amphibia, 7 fishes, 120 invertebrates, 224 plants, and 117 fungi and lichen species), and a new Red Data Book (fourth edition) was published in 2021 (Rašomavičius, 2021). In addition to their legal protection, Species Conservation Plans are required for threatened species. These include actions such as breeding and re-introduction programmes, public education, scientific research and monitoring, and the implementation of international agreements. Species Action Plans are also prepared that set out specific measures that are required in particular areas.

The Law on Wild Vegetation[19] (1999) protects and regulates the use of wild vegetation, in order to protect wild plant species and natural communities' diversity, and wild vegetation genetic resources. Under the legislation, the negative impacts of rural development, forestry and other human activities on wild vegetation are to be reduced. This can be achieved by establishing protected areas, implementing measures in territorial planning documents, signing agreements with landowners, and limiting the use of wild vegetation.

The Law on Wild Fauna[20] (1997) determines the protection and use of wild fauna naturally living in the country, or migrating through it, as well as their habitats and breeding sites. All such free-living fauna are owned by the State. Forms of wild fauna use include hunting, fishing, taking from their habitats, keeping in captivity, trade, introduction and re-introduction, and use for science and education. All such uses of fauna are prohibited if they threaten the species.

The Law on Hunting[21] (2002) governs the protection and rational use of hunted fauna. Permission is required to carry out hunting, which must be in accordance with the Hunting Rules that define the game list, hunting limits, periods and methods, etc. Currently 33 species are on the game list in Lithuania; they can be divided into large species, such as Elk (*Alces alces*) and Wild Boar (*Sus Scrofa*), and smaller game, including Red Fox, European Hare (*Lepus europaeus*) and Beaver (*Castor fiber*). The abundance of the Wolf population is regulated according to the Hunting Rules, with the hunting limit approved by the MoE for each hunting season taking into account special conditions and proposals in the species protection plan. If the population falls to 100 or less, hunting is not allowed. Although the populations of the European Hare and Grey Partridge (*Perdix perdix*) are declining, it has not been proven that this is related to hunting. Nevertheless, the species may be removed from the hunting list if the situation does not improve.

According to Lithuanian legislation (the laws on wild fauna and on hunting), landowners may be compensated for damage caused by large carnivores, Bison (*Bison bonasus*), seals and

[17] Saugomų gyvūnų, augalų, grybų rūšių įstatymas. https://e-seimas.lrs.lt/portal/legalAct/lt/TAD/TAIS.46423/asr

[18] www.e-tar.lt/portal/en/legalAct/b3ba7c70b5c811e88f64a5ecc703f89b

[19] Laukinės augalijos įstatymas.

[20] Laukinės gyvūnijos įstatymas.

[21] Medžioklės įstatymas.

waterbirds (but not migrating birds). Compensation may also be paid by hunters if damage is caused by them hunting ungulates. In other cases, compensation is paid from the Environmental Protection Support Programme (EPSP), or from the European Maritime and Fisheries Fund (EMFF) (75%) and Lithuanian State Budget (25%). The EPSP may also fund damage prevention measures.

22.4.2 Species-Specific Conservation Measures

In addition to general conservation measures, a number of targeted actions have been taken for species. These have included a special project – Implementation of Rare Species Protection and Invasive Species Abundance Regulation Measures – which started in 2012. The project, with €8 million from the European Regional Development Fund (ERDF), aimed to improve the general status of biodiversity. Usually each Species Conservation Plan describes and evaluates the species' population and habitat, national and international status, threats, and available research or monitoring data. It then sets out goals, objectives and measures for the species to maintain or improve its favourable conservation status. Over 2012–2017, 41 priority Species Conservation Plans and/or Species Action Plans were prepared and approved by the MoE, including 6 for plants, 7 for invertebrates, 15 for birds, such as Capercaillie (*Tetrao urogallus*) and Black Stork (*Ciconia nigra*), and 6 for mammals (e.g. Bison, Wolf, and Eurasian Lynx (*Lynx lynx*)). The project helped implement 129 Species Action Plans, which improved the state of 23 protected species in 129 areas (not all of them protected areas), prepared three Species Conservation Plans for selected species and controlled the abundance of seven invasive species in 71 areas (Augutis *et al.*, 2015).

Guidance has also been provided to help specialists with the preparation and implementation of Nature Management Plans (for Natura 2000 sites with particularly high numbers of habitats and protected species) and other measures. So far, guidance has been produced for five plants and 42 animal species, consisting of general information on the species and its habitat, criteria for its favourable conservation status, a list of threats and recommended actions.

Practical species-specific actions have included a particularly successful species recovery project[22] that has increased the population of the globally Vulnerable migratory Aquatic Warbler (*Acrocephalus paludicola*) in Žuvintas Biosphere Reserve and in Nemunas Delta Regional Park. Further details of this are provided in online Annex LT.1.[23]

The Bison has been re-introduced into Lithuania (the last was killed in 1854). As a result of the re-introduction, which started in 1969, there were 276 individuals in 2019, with some in captivity and some free-living in herds of 40–50 animals. Now, the goal of the Bison Conservation Plan[24] (2015) is for the population in Lithuania to achieve favourable conservation status. One of its specific objectives is to move part of the population from central Lithuania to western and southern areas, to regulate the free-living and captive breeding herds so the population is healthy, and to foster a positive view on the Bison. A special project is currently underway that is implementing the measures in the Bison Conservation Plan.

22.5 Protected Areas and Networks

22.5.1 Protected Area Types and Their Coverage

Lithuania applies an integrated approach to the conservation of nature and culture values (Baškytė *et al.*, 2019). The system and types of protected areas (Table 22.4) that have evolved over the last 80 years are therefore based on this approach: protecting abiotic and biotic nature, and natural and cultural heritage, as well as typical and unique landscapes (from natural to urban). It is also necessary to note that the understanding of a landscape in Lithuania is much broader than in some other countries. According to the Law on Protected Areas, landscape is a combination of closely related natural components (e.g. landforms, ground waters and water bodies, soil, species and habitats) and/or anthropogenic components (e.g. archaeological remnants, buildings, built infrastructure and land use). In nearly all cases nature conservation is one of the most important functions of protected areas, but sometimes it is difficult to clearly define protected area categories according to the IUCN typology. Consequently, some of the types of protected areas listed in Table 22.5 may not strictly qualify as protected areas according to the IUCN. Nevertheless, with the exception of a few types that have been established primarily for cultural heritage protection, all others contribute to nature conservation.

Domestic national protected areas cover 17.7% of the country's land area, with the main part comprising National and Regional Parks, together with Biosphere Polygons (together 13%). The National Parks represent outstanding values of ethno-cultural regions, and are designated for both landscape and nature conservation. Thus, they have more in common with the National Parks in the UK than those in Finland, Sweden, Norway and many other European countries. Regional Parks are designated for similar values. Nevertheless, National and Regional Parks are very important for nature conservation, and therefore are given a high priority. They are integrated protected areas in which protection of their values is combined with sustainable use of natural and cultural resources, primarily through permitted recreational, agricultural and forestry management activities.

Although they only cover a small proportion of the land, Nature/Landscape Reserves (2.4%) and Strict Nature Reserves

[22] LIFE MagniDucatusAcrola (LIFE NAT/LT/001024).

[23] All annexes are free to access online at CUP: www.cambridge.org/natureconservation

[24] European Bison conservation plan: https://e-seimas.lrs.lt/portal/legalAct/lt/TAD/b3f90932605411e5b316b7e07d98304b/asr

Table 22.4 The number and extent of international designated sites and protected areas in Lithuania.

	IUCN category	Number	Area (km^2)
Designated under global and non-EU international legislation[1]			
Ramsar Sites	I, IV, V, VI	7	656
Biosphere Reserve (exc. buffer and transition zones)	VI	1	92
HELCOM Marine Protected Areas	IV, V, VI	8	1 964
Designated under the Nature Directives			
SCIs (terrestrial + marine)	I, IV, V, VI	498	6 597 + 963
SPAs (terrestrial + marine)	IV, VI	84	5 522 + 1 064
Natura 2000 terrestrial area	I, IV, V, VI	561	8 244
Natura 2000 marine area	IV, VI	8	1 574
Designated under domestic legislation			
State Strict Nature Reserve (Valstybinis gamtinis rezervatas)	Ia	3	191
Nature/Landscape Reserve (Gamtinis/kraštovaizdžio draustinis)	IV, V	514	1 581
Nature Monument (Gamtos paminklas)	III	159	–[2]
National Park (Nacionalinis parkas)	V	5	1 441
Regional Park (Regioninis parkas)	V	30	4 489
Biosphere Reserve (Biosferos rezervatas)	VI	1	186
Biosphere Polygon (Biosferos poligonas)	VI	32	2 364

Notes. 1. All areas designated under international legislation also have domestic protected area status. 2. Nature Monuments (trees, boulders, etc.) occupy a very small area.
Sources. See Section 1.3. Based on data from: UNESCO Biosphere Reserve; State Cadastre on Protected Areas (1 September 2021).

(0.3%) are very important for biodiversity conservation. Some have been established for the protection of particular species or habitats (e.g. raised bogs). In addition, all nature components are afforded protection in Landscape Reserves.

Protected areas in Lithuania are of two levels: state and municipal (local). Municipalities can establish only Nature Reserves and Nature Heritage Objects (a lower level of Nature Monuments).

Special marine protected areas (three biosphere areas in the Baltic Sea and one in the Curonian Lagoon) were established with the aim of conserving BHD species and habitats. Previously only the Curonian Spit National Park, Pajūris Regional Park and Nemunas Delta Regional Park had marine parts in the Baltic Sea and Curonian Lagoon.

International protected areas. Some Lithuanian protected areas have been designated primarily for their international importance, particularly in terms of contributing to the Natura 2000 network. The most valuable areas for the protection of birds were selected as SPAs for the Natura 2000 network in 2002–2004 (Raudonikis, 2004). According to the Law on Protected Areas, SPAs need to be firstly designated under domestic legislation, so this was carried out by the MoE for 38 new protected areas.

The current network of Natura 2000 sites in Lithuania (Table 22.4) covers 12.7% of the land and 24.4% of the marine waters. These overlap with about 75% of the protected area (on land and sea) designated under domestic legislation. However, the designation of the Natura 2000 network is not finished yet, as there is an insufficient area for 21 HD species and 17 HD habitats. Identified HD habitats cover 6.7% of the country. Potential additional Natura 2000 sites are therefore being considered in relation to the latest scientific investigations and inventories.

The 98 km long Curonian Spit, which mainly comprises forests, sand dunes, lagoons and coastal waters, is a Natura 2000 site and of particularly high biodiversity importance. It is also a cultural World Heritage Site.

Based on the most up-to-date compiled data available from September 2021, taking into account overlaps, the **overall coverage of protected areas** was 17.8% on land (11 537 km^2) and 30.5% of the marine waters (1 964 km^2).[25]

22.5.2 The Coherence of Protected Areas and Ecological Networks

Since 1980, and the approval of its *Integrated Scheme of Nature Protection*, Lithuania has developed its Nature Frame strategy

25 Lietuvos Respublikos saugomų teritorijų valstybės kadastras: https://stk.am.lt/portal/ (last accessed 20 December 2022).

Table 22.5 Definitions of the different types of protected area in Lithuania.

Category	Definition and objective	Level of management	Proclamation
Strict Nature Reserve	To protect and conduct research on nature, ensure the unaffected course of natural processes, promote protection of nature. Human activities with the exception of scientific research are prohibited	State	Government
Nature Reserve	Protection of natural sites valuable for scientific research and ecological tourism, nature, landscape and biological diversity, as well as genetic conservation. They can be for a variety of values, e.g. geological, hydrographical, flora, fauna and landscape	State, or municipality	Government, or municipality
Nature Monument	Objects of natural heritage – important landscape elements with special protection. They can be geological, geomorphological, hydrogeological, botanical and/or zoological	State, or municipality	Government, or municipality
National Park	Protection and management of the natural and cultural landscape of national significance representing the natural and cultural values of the country's ethno-cultural regions	State	Government
Regional Park	Protection of the landscape and ecosystems of regional significance from a natural, cultural and recreational point of view	State	Government
Biosphere Polygon	For protection of nature values, for national and regional environmental monitoring in areas of particular geo-ecological or ecological importance	State	MoE

Source. Based on The Law on Protected Areas.

(Kavaliauskas, 1995, 1996). Based on its scientific background, which takes into account the structure of geo-ecosystems, the Nature Frame sets out to conserve and restore the natural structures that are responsible for ecological functions in the environment. The legislative basis for the Nature Frame was first established in the Environmental Protection Law (1992) and was later defined in more detail in the Law on Protected Areas (1993).

The aim of the Nature Frame is to achieve ecological stability in Lithuania by developing a complete interrelated territorial system of zones with ecological compensation functions (e.g. ground water filtration, conservation of biodiversity, recreational resource protection, aesthetic improvements) for maintaining and improving the state of natural ecosystems and minimising human impact on nature.

The Nature Frame is an integrated land management structure that consists of a territorial system of the following geo-ecosystems:

1. Geo-ecological watersheds – core areas – (usually located in higher parts of river basins) areas connecting upper reaches of rivers, surface and ground water watersheds, hilly areas with lakes, raised bogs, littoral zones, karst areas and underground water feeding areas.
2. Areas of inner stabilisation of geo-ecosystems – natural buffers and filters – natural areas important for biodiversity (forests, meadows, wetlands or other natural structures inside of water basins).
3. Migration corridors (e.g. river beds, valleys and ravine systems) that contribute to the intensive migration of different substances, energy flows and species.

All protected areas established for nature and natural landscape conservation (as listed in Table 22.5) and natural recreational areas are connected by the Nature Frame, with the Natura 2000 network forming one of the most important components. Furthermore, it contains all other natural and semi-natural areas and strengthens connections between them, thereby reducing their fragmentation.

Currently the Nature Frame is not a continuous structure, but it does cover about 60% of the country's area and provides a territorial system with centres and axes of ecological stabilisation. Nature conservation, forestry, recreation and regulated farming are priority activities in the Nature Frame, whereas urban and industrial development is prohibited. Implementation is through territorial planning (described later in Section 22.6), where the Nature Frame is taken into account in the preparation of general plans or schemes at national, regional or local levels. It is also considered in SEA and EIA of plans and programmes.

22.5.3 Approaches and Levels of Protection

Until very recently, human activities in protected areas were regulated by the Law on Protected Areas and by-laws, but from 2020 all prohibitions and restrictions for any areas in Lithuania, including protected areas, were transposed to the new Law on Special Land Use Conditions.

Some activities are prohibited in all protected areas (e.g. excavation of minerals and peat, release of GMOs). Other human activities can be stimulated (encouraged), restricted or forbidden depending on their context and the protected area type. These include activities that are: (1) prohibited because they may destroy or damage protected values; (2) based on principles of sustainable development and therefore allowable according to certain conditions (e.g. traditional agricultural management); and (3) restricted, such as construction activities (e.g. the building of houses). The last has been particularly problematical since land reform as many new owners wish to construct houses in attractive locations, but such activities are very damaging and can completely change the nature of an area.

Protected areas with biological values, particularly with vulnerable natural habitats and species, have the strictest regimes. No human activities are allowed in Strict Nature Reserves, with the exception of monitoring, applied scientific research, or biodiversity conservation or restoration actions. A similar protection regime applies to Nature Monuments.

Nature/Landscape Reserves have fewer restrictions; forestry and agriculture activities are allowed within them but subject to restrictions that prohibit particular activities that may destroy or damage their protected values.

In its regulation of protected areas, particularly in National and Regional Parks, Lithuania attempts to find appropriate compromises between the interests of different stakeholders by applying the principles of sustainable development. For example, extensive agriculture in open areas is often encouraged as it is necessary to maintain important semi-natural habitats and their associated species. Furthermore, one of the main goals of National and Regional Parks is to create conditions for eco-tourism and promote traditional ways of living. Although some general restrictions on activities are common to all areas of National and Regional Parks, most are regulated according to functional zones within them. Thus, recreational facilities can be developed in recreational zones, some options exist for extending settlements in residential zones, and more intensive economic activities are foreseen in forestry or agriculture zones. The potential for such appropriate activities is supposed to be carefully evaluated and proposed during the preparation of Planning Schemes (Management Plans) of protected areas. If new construction is not proposed in the plan then it cannot take place.

Under the Law on Protected Areas all citizens have equal access rights to protected areas, and even private owners cannot restrict access. However, the visiting of Strict Nature Reserves is prohibited (only visiting of eco-trails is allowed), and visiting of some Nature Reserves is restricted.

If necessary, land required for conservation needs can be taken under State ownership, in which case the owner has to be properly compensated. If the owner is willing to sell their land, the State has priority to buy it, but this is not usually possible due to limited financial resources. Most opportunities to buy land arise when there are project-related financial resources available and the owners are not interested in carrying out the required conservation activities.

Under the law, **compensation payments** are payable for restrictions on economic activities within protected areas. In general, one-off compensation payments are made for forest owners when the protected area is established, or if the borders are changed, or if new restrictions are set (e.g. on felling). This financial support is paid to private forest owners to cover part of the costs incurred and income forgone as a result of restrictions on the use of forests imposed through implementation of national legislation or the Nature Directives in the respective area (under the CAP Rural Development Natura 2000 measure, see Table 22.6).

22.5.4 Approaches to Managing Protected Areas

As in other aspects of protected areas in Lithuania, an integrated approach is taken to their management. Management measures in protected areas are proposed in strategic or territorial (spatial) planning documents, which according to the Law on Protected Areas can be: (1) Plans of Protected Areas Boundaries and (2) Planning Schemes (Boundaries and Management Plans) for National and Regional Parks, Strict Nature Reserves and Biosphere Reserves. If there is a need to implement urgent measures in Natura 2000 areas, then a strategic planning document may be prepared – a Nature Management Plan or a simpler form of action plan (Management Programme).

All necessary management measures should be foreseen and included in the Nature Management Plans or other planning documents that are prepared, taking into account the conservation objectives of the area, considering economic and cultural aspects, regional and local interests, and the interests of society and landowners.

Special measures for species and habitats (from 2004) are implemented in Natura 2000 areas according to the Nature Management Plans (about 45% of the areas at the time of writing have such plans). Successful results depend not only on the quality of the Nature Management Plans but on the factors affecting their implementation (e.g. financial and human resources).

A high proportion of habitats and species that have an unfavourable status are in privately owned agricultural areas. Some of these areas have been subject to agricultural abandonment (especially in hilly or wet areas), becoming overgrown with shrubs and trees and thereby losing their valuable habitats and species. When active management is urgently needed, the SSPA and protected area administrations have implemented special projects financed by the ERDF. As a result, the condition of around 100 Natura 2000 areas has improved, and 32 projects are continuing until 2022. However, ongoing active management, annual supervision and monitoring will be required to maintain their improved condition over the long term. Some areas, such as meadows, are being effectively managed by farmers after restoration, providing a good example of cooperation between state institutions and communities.

The results of the EU-funded projects have transformed the understanding of the general value of protected areas in Lithuania. As a result of this, nearly all National and

Regional Parks, Strict Nature Reserves and Biosphere Reserve have new visitor centres with attractive thematic exhibitions. There is also a national visitor centre in the capital, which is very popular and acts as an introductory 'gateway' to protected areas.

A management effectiveness evaluation of Lithuanian protected areas in 2006–2007, based on the Rapid Assessment and Prioritisation of Protected Areas Management methodology (Ahokumpu *et al.*, 2007), revealed good results. According to the report, at the time, Lithuania had: created a representative network of protected areas that corresponded to its conservation needs; made significant progress with its administrative structures; been mainly successful in preventing threats and protecting the areas' core values; and created a positive outlook for the future of the whole protected areas system. However, more support and appreciation of protected areas was necessary. After implementation of the evaluation's proposed measures, the management of protected areas has further improved.

22.6 General Nature Conservation Measures

22.6.1 Development Planning and Impact Assessment

Lithuania has very strong traditions and a long experience in territorial planning. Development is based on numerous sectoral strategic and territorial planning documents. The idea is to have one general strategic plan at state level (*National Advancement Programme*, until 2030), which integrates the needs of all sectors (including nature protection), and a *General Plan for the Lithuanian Republic* that sets out proposals for development and protection of specific regions and areas (this is therefore the most important territorial planning document). The main goal of the 2002 *General Plan*[26] was to ensure sustainable development of the country by the best possible use of available resources without harming the natural and cultural environment or the interests of current and future generations, and by coordinating the use of areas and protection at the national level, as well as by retaining and enhancing the identity of the country. The main goals related to nature were the conservation of biodiversity, rational use of biotic resources, and creation/development of the Nature Frame and protected areas system. A new *General Plan* for Lithuania was produced just as this book was being completed. General plans for municipalities have also been prepared, all of which incorporate nature conservation requirements.

Requirements for species and habitats protection in Natura 2000 areas related to implementation of the Nature Directives are transposed into the Law on EIA and to the Special Description of Procedures for Evaluation of SEA (2004). The HD Article 6(3) requirements to assess the impacts of all plans and projects that may negatively impact a Natura 2000 site are part of the Law on Protected Areas and the Law on EIA.

Although the EIA and SEA procedures work quite well, there is a serious problem in that the assessments are not carried out by an independent expert. As is the case in many other EU countries, the assessment of the actual situation in the affected area, and the possible impacts of the activities, is initiated and carried out by the project proponent (e.g. developer). Furthermore, the authorities that make the decisions on the proposals are not able to check the real situation due to insufficient time and financial resources, etc.

22.6.2 Habitat Management and Restoration

According to Lithuanian legislation, all landowners and users have the responsibility to protect nature values, so they need to manage their land appropriately. The Law on Protected Areas allows for individual agreements to be made with landowners in Natura 2000 areas for protection and management of the areas. However, whilst there are legal provisions for the agreements, they require financial resources and are not used widely.

Of key importance to many protected areas, particularly those with grasslands and other open habitats, is their **appropriate management by farmers**. However, currently farmers are insufficiently taking up the agri-environment–climate schemes included in the Rural Development Programme (RDP) due to limited information on the available possibilities. Whilst the protected area administrations try to involve farmers more in nature management, they do not have an administrative mandate for this and are therefore not always able to consult with farmers sufficiently.

The Law of **Forests** (1994) regulates reproduction, protection and usage of forests. It creates legal conditions for forestry in all forests according to sustainable management principles, providing timber, increasing the productivity of forests, and ensuring rational usage of forest resources, biodiversity conservation, landscape stability and environment quality, without damaging ecosystems. The *Lithuanian Forestry Policy and Strategy* (2002) is also based on the principle of sustainable development, and its main goals relate to protecting and increasing forest resources, forest productivity and biodiversity conservation, and informing society. The protection of biodiversity is one of the responsibilities of forest owners.

All forests in Lithuania are divided into four groups: forests of Strict Nature Reserves, forests with special purposes (ecosystem protection and recreational forests), protective forests (e.g. for water bodies or roads), and economic forests. All forests in Lithuania are used according to forest management plans, which take into account requirements of biodiversity protection, including nature protection measures to achieve the favourable status of species and habitats.

According to an inventory over 2001–2005, key forest habitats comprised around 1.2% of forests in Lithuania (Stončius, 2011). These are of very high biodiversity

[26] www.e-tar.lt/portal/lt/legalAct/TAR.737EFE13DAE0

importance as they contain rare and threatened species common to old virgin forests that are not found in intensively managed forests. Despite their importance and identification, their protection has been left up to forest owners; as most are not interested in protecting them, some key forest habitats have been lost. Stončius (2011) found that more than half of key habitats had been damaged by forestry activities, although in all cases they were still valuable for biodiversity. In our view, more recently, consideration of biodiversity in the management of forest habitats by the State Forests Enterprise has decreased due to a greater orientation towards economic priorities.

Of particular importance for **marine nature conservation** was a 2010–2015 LIFE project that prepared an inventory of marine species and habitats to guide the development of the Natura 2000 network in Lithuania's offshore waters.[27] As a result, two marine protected areas were established, as Natura 2000 sites. For each of the areas special plans were developed and restrictions on economic activities were set with the aim of protecting the sites' Baltic Sea reefs and important wintering waterbird populations, including those of the Long-tailed Duck (*Clangula hyemalis*), Common Eider (*Somateria mollissima*) and Velvet Scoter (*Mellanita fusca*), the last two being considered to be threatened in Europe. Additionally, the project elaborated recommendations for marine spatial planning and conservation, and raised public awareness about marine values and conservation needs.

Over 2014–2020, the Lithuanian fishery sector had a special Operational Programme, supported by the EMFF. Most measures related to aquaculture and were not for nature conservation.

A strategic goal of the LBPAP is to decrease or halt the spread of **invasive alien species (IAS)**. To address this, with EU ERDF funding, 50 plans to regulate the abundance of nine IAS are to be prepared and implemented by 2022. However, data on IAS are lacking, and therefore investigations on 100 IAS are to be started in the near future. Priority will be given to implementing the plans in protected areas and the habitats of protected species.

22.7 Nature Conservation Costs, Benefits and Funding Sources

22.7.1 Nature Conservation Costs

Regular nature conservation costs relate to administration, monitoring, preparation of planning documents, direct protection and management activities, and compensation for damage resulting from wolves and other species. Whilst such costs have increased as a result of the additional obligations under the Nature Directives and related legislation, EU accession has provided access to new larger and broader EU-funding opportunities that can contribute to biodiversity conservation, as discussed in Section 22.7.3. Estimates of the overall costs of nature conservation in Lithuania are currently unavailable, but are being included for Natura 2000 in the draft of the first Prioritised Action Framework (PAF) for 2021–2027 (MoE, 2022),[28] which is under preparation.

22.7.2 The Economic Benefits of Nature Conservation

Although they have not been quantified, it is obvious that protected areas present considerable social and economic benefits, not only for the country but for international society as well. More than 1.5 million people visit Lithuanian protected areas every year, not only from within the country but from abroad as well. National and Regional Parks are especially important in this respect for the development of recreation and local activities. The number of visitors to Lithuanian protected areas increased after the establishment of visitor centres and nature schools, and creation of observation towers and eco-tourism trails.

The results of a recent assessment of the socio-economic benefits of the Natura 2000 network (BGI Consulting, 2020)[29] have shown that the network's total benefits (before cost elimination) are equal to €194 million per year, which is 2.2 times the annual socio-economic costs of maintaining the network (€89 million). This gives a net benefit from Natura 2000 sites of €105 million per year. It has also been found that the greatest annual benefits of Natura 2000 sites are generated by the non-consumable values of Natura 2000 sites, such as from walking along nature trials. The greatest value is created by the existence of Natura 2000 sites, regardless of the additional practical benefits they bring.

22.7.3 Funding

Nature conservation and protected areas management is funded by the State and municipalities, EU co-financed projects and some international sources (Table 22.6).

The annual costs of the administration and recurrent management of protected areas are covered by the **state budget** and some special programmes (EPSP and Programme of Funding of General Forestry Needs, which will be part of the state budget from 2022). Financing of nature conservation is a part of the State Investment Programme. Nature conservation needs have grown faster than the available financial resources for daily activities – management of habitats and species and landscape and nature values – and for preparation of planning documents, etc. Although the protected area administrations obtain some revenue for the services they provide visitors, at

[27] DENOFLIT (LIFE09 NAT/LT/000234.)

[28] https://am.lrv.lt/uploads/am/documents/files/PAF%202022%2002%2001.pdf

[29] English summary: https://naturalit.lt/wp-content/uploads/2020/12/BGI_VSTT_Natura-2000_Santrauka_EN_20200916.pdf

Table 22.6 Funding sources for nature conservation in Lithuania.

Source and period	Measure/category	Annual allocation (€ m)*		Comments
EU Funding instruments		**EU + national**	**EU**	
EAFRD Rural Development Programme measures – allocations to Priority 4 (restoring, preserving and enhancing agriculture and forest ecosystems) 2014–2020	M13 Areas facing constraints	48.2	38.0	c. 10% all M13 for N2k etc.
	M11 Organic farming	27.0	20.2	c. 4% all M11 for N2k etc.
	M10 Agri-environment–climate	8.0	6.0	c. 6% all M10 for N2k etc.
	M12 Natura and WFD	2.7	2.0	All M12 for N2k etc.
	M8 Investments in forests	1.8	1.5	c. 20% all M8 for N2k etc.
	M1 Knowledge transfer, etc.	0.3	0.3	Not used for N2k
	M4 Investments in physical assets: 4.4 linked to agri-environment–climate	0.3	0.2	All M4.4 for N2k etc.
	M15 Forest environment climate	0.2	0.1	c.75% all M15 for N2k etc.
	M2 Advisory services, etc.	0.1	0.1	Not used for N2k
	M16 Cooperation	0.1	0.09	Not used for N2k
RDP Priority 4 subtotal		*88.7*	*68.6*	
ERDF, European Social Fund and Cohesion Fund 2014–2020	085 – Biodiversity, nature protection and green infrastructure	6.3	5.8	For N2k etc.
	086 – Protection, restoration and sustainable use of Natura 2000 sites)	1.0	1.0	For N2k etc.
LIFE Programme	Nature and biodiversity (2004–2020)	1.9	0.9	
	Integrated Project NATURALIT (2018–2027)	1.7	1.0	See text
EMFF 2014–2020	1st EU priority: promotion of environmentally sustainable fisheries	1.5	0.5	15% related to N2k etc.
	2nd EU priority: promotion of environmentally sustainable aquaculture	1.2	0.9	25% related to N2k etc.
Interreg 2014–2020	Biodiversity and/or natural heritage (086 – Protection, restoration and sustainable use of N2k sites)	0.09	0.07	All related to N2k etc.
EU subtotal		*102.4*	*78.8*	
Within 2014–2020		**Additional non-EU***	**Comments**	
National funds				
State budget (2019 data)	Administrative costs	6.5	Funding for all PA system	
State budget (2014–2018 data)	Environmental Protection Support Programme	0.4	Inc. compensation for wolves and bison	
	Programme of Funding of General Forestry Needs	0.04		
Others				
Norwegian/EEA funds		1.6	Important for N2k and biodiversity. Only 2015–2018	
Additional national and other subtotal		*8.6*		
National total including co-financing of EU instruments		*32.2*		
Total		**111.0**		

Notes. 'For N2k etc.' indicates allocations relevant to Natura 2000, protection of species and green infrastructure, based on the draft PAF (MoE, 2022). '% for Nk2 etc.' relates to the total use of the RDP measure across all priorities. * Excludes co-financing of EU funds unless indicated. The subtotals and totals are only approximate indications of the average annual levels of funding as they are based on annualised data from differing periods.
Sources. Based on data from: RDP data from European Commission ESIF funds database (23 October 2020); RDP allocations to Natura 2000, ERDF and national funding from 2021–2027 PAF (MoE, 2022), SSPA and MoE data; Lithuania LIFE factsheet, European Commission (February 2021); EMFF – from Statistics of National Paying Agency under the Ministry of Agriculture, Lithuania. See Section 1.3 for further details.

around €1 million per year, this income is not very large compared to overall costs and other sources.

EU financial instruments have been, and still are, the most significant sources of funding, particularly for the implementation of conservation measures over large areas (e.g. the maintenance of required agricultural management). Lithuania started to use **EU Cohesion Policy** funds (from the ERDF) for nature conservation in 2004. In total, over 2004–2020, €133 million were allocated specifically to nature conservation. This major source of funding led to significant improvements in protected areas, through the preparation of planning documents, action plans for species, management of Natura 2000 areas, improved monitoring and management systems, and increased access and facilities for visitors.

The **EAFRD** has been one of the largest sources of financing for nature conservation in Lithuania, particularly for the management of agricultural areas and supporting landowners' involvement in Natura 2000 areas. The total EAFRD allocation for Lithuania for 2014–2020 was around €1.9 billion (MoE, 2022). The most relevant measures for nature conservation have been the agri-environment–climate measure (although uptake by farmers has been low) and the Natura 2000 measure. EARDF support for management measures has been fundamental for agricultural Natura 2000 sites hosting species and habitats that require regular non-intensive management, as well as forested Natura 2000 sites. The Natura 2000 payments have been an especially effective means of convincing private forest owners in Natura 2000 sites to agree to necessary conservation measures that have restricted normal commercial forest management operations.

However, by far the largest proportion of EARFD funding has been allocated to payments to farmers within areas with natural constraints and as support for organic farming. Whilst these payments have not contributed greatly to Natura 2000 management, they have helped maintain environmentally friendly farming systems and promote environmentally friendly technologies, such as conserving organic soils and managing pastures and grasslands in a sustainable way, especially in the karst region of northern Lithuania and the Nemunas delta.

The **LIFE Programme** has been a very important fund for particular Natura 2000 areas and some species and habitats (and there is no similar national financing instrument). Although it potentially enables the involvement of the scientific community and NGOs in nature conservation projects, a major obstacle to this has been obtaining the necessary national co-financing. Nevertheless, since accession in 2004, 17 LIFE nature and biodiversity projects have been awarded to Lithuania. In addition, a large and important 10-year LIFE integrated environment project started in 2018, which is supporting the management of the Natura 2000 network[30] through actions further described in Section 22.9.

[30] NATURALIT LIFE16 (IPE/LT/000016).

Interreg funds have had minor overall impacts, but have been important for protected area administrations that have been carrying out transboundary projects in cooperation with other countries, and involving the public and NGOs. A number of projects are ongoing in the framework of Interreg V-A Latvia–Lithuania–Belarus and Lithuania–Poland.

Around €11 million (during 2015–2018) were provided from the European Economic Area (EEA) and Norwegian Financial Mechanism for measures related to Natura 2000. These included restoration plans for wetland complexes, practical restoration of wetlands and other open habitats, identification of areas with sensitive biodiversity and potential conflicts with wind energy developments, recommendations for conflict management, and capacity building in species monitoring, including the assessment of conservation status.

The availability of EU financial instruments and other external funding sources has been crucial for Natura 2000 and broader biodiversity conservation, because Lithuania is a small country and has insufficient funds for the implementation of the Nature Directives. However, the various funds have not been used to their full potential, and the national allocations have not been sufficient to cover all the measures identified in the Natura 2000 PAF. This shows that a stronger strategic approach is needed, at national and regional levels, to make better use of potential funds for biodiversity conservation, especially within the Natura 2000 network.

22.8 Surveillance, Monitoring, Reporting and Applied Research

The Law on Environment Monitoring[31] (1997) forms the basis for environmental monitoring, which includes monitoring the state of the natural environment (e.g. natural habitats and species) and its changes, as well as the assessment and forecasting of anthropogenic and natural changes, trends and consequences. This is carried out by the government-approved State Environment Monitoring Programme for a six-year period (latest programme for 2018–2023[32]).

Monitoring the state of nature is a significant part of the programme, requiring species and habitats monitoring to be carried out not only within Natura 2000 sites but also outside them. Nature monitoring also includes an assessment of the status of migratory fish, monitoring of game and invasive species, and the status of biological indicators of environmental quality. The MoE is responsible for the overall coordination of the programme, while the MoE and Ministry of Agriculture are responsible for its implementation. Environmental monitoring is coordinated and performed by the Environmental Protection Agency, except in protected areas, where it is coordinated by the SSPA and carried out by protected area staff.

[31] www.e-tar.lt/portal/lt/legalAct/TAR.1A98CE535B1C/asr

[32] www.e-tar.lt/portal/lt/legalAct/d6fdb4b0c89a11e8bf37fd1541d65f38

In order to improve the quality of data collected, in 2016 two manuals on monitoring methods were published by the Environmental Protection Agency: one on BD Annex I birds (Raudonikis *et al.*, 2016), and another on a range of other animals, as well as plants (Balčiauskas *et al.*, 2016). In 2018, the biodiversity monitoring database (BIOMON2) was updated, allowing more efficient data gathering and analysis in relation to BHD species, other rare and protected species and large carnivores (Anuškevičius *et al.*, 2016; MoE, 2022).

Between 2011 and 2015, Lithuania carried out a national inventory of HD Annex I habitats. The inventory included accurate mapping of the habitats throughout the country, and collected data on their structure, functions and typical species. These results allowed the establishment of favourable reference values and relevant national-level conservation objectives for each habitat type. The collected data have also helped to improve the network and management of Natura 2000 areas. Furthermore, they also indicate that the current SCI network might be incomplete for some habitat types and species (MoE, 2017, 2022).

The Nature Research Centre was established in 2009 by merging the Institutes of Botany, Ecology, Geology and Geography with the aim of strengthening scientific nature research. The field of this research includes natural ecosystems, habitats, species, plant communities, structure of populations, genetic diversity, natural resources, including underground resources, etc., research and prognosis.

Effective implementation of the species laws and plans requires reliable data. Therefore, in 2009, the Information System on Protected Species (ISPS) was created with the aim of collecting data on protected species and their localities, and making them available to interested persons. The ISPS data, which are publicly available, are used for preparing protection plans for species, nature management plans, planning documents of protected areas, impact assessments, forestry management plans, and rural development projects, and for taking decisions on planting of forests in open areas, scientific studies and many other uses. However, there is a need to upgrade this database.

22.9 Future Developments

Forthcoming nature conservation actions in Lithuania are likely to be strongly focussed on the completion of the Natura 2000 network (i.e. filling protection gaps for some habitats and species) and, in turn, seeking to achieve and maintain the favourable conservation status of species and habitats. Another target area will need to be to halt, or effectively regulate, the spread of invasive species. In the near future, the ERDF-funded Natura 2000 management projects will continue to be implemented, and planning documents for establishment of new Natura 2000 areas prepared.

Over the next 10 years, it is hoped that more visible results will be achieved through the NATURALIT LIFE Integrated Project, which seeks to improve the Natura 2000 network management system in order to ensure the favourable conservation status of targeted habitats and species. A strong team of specialists, with experience in the protection of BHD habitats and species, will support the state institutions in improving the effectiveness of research, monitoring and, in particular, protection activities.

Steps may also be taken to increase understanding and support for nature conservation among society and landowners, including through further communication between responsible institutions and local communities. Another very important intention is to demonstrate the value of protected areas, including the benefits of Natura 2000 areas.

22.10 Conclusions

The development of nature conservation in Lithuania has reflected its strong tradition in the integrated conservation of nature and landscapes (which are broadly defined), for both intrinsic natural values and cultural reasons, and its related territorial planning. However, major changes in the State have affected conservation issues and given rise to distinct challenges and initiatives. This was particularly the case after the restoration of Lithuania as an independent state, when the socio-economic situation underwent radical changes, land reform started, and a new legal system was developed. At the same time, comprehensive landscape and biodiversity protection was required, resulting in the establishment of a wave of protected areas. Becoming a member of the international community led to new responsibilities and activities as conventions were ratified and international agreements signed. An especially important challenge has been implementing EU legislation.

Building on these traditions, and subsequent actions, major progress has been made, including in relation to the necessary legal acts, strategies and plans, monitoring, the system of protected areas, including the creation of the Natura 2000 network (although it needs further expansion), active nature management and the implementation of Species Conservation Plans. Consequently, no species has gone extinct, and numerous threatened species have increased.

However, unfortunately much more has to be done to improve the situation for many habitats, particularly grasslands, none of which has been assessed as having a favourable conservation status according to the latest HD Article 17 report.

Whilst Lithuania has established the scientific and legal background for nature conservation, and its comprehensive system of protected areas, as in other Central and Eastern European countries it continues to struggle with inadequate administrative and financial capacities compared to other sectors with greater private and political support. EU project funding has therefore been essential in covering financial gaps and enabling large-scale habitat management, species conservation measures and more comprehensive monitoring to take place. However, funding alone cannot address all the problems; better communication amongst institutions and political support are also necessary, as well as setting priorities for nature conservation above private interests. Of particular importance for the future has been the opening up of protected areas for visitors, and other efforts that have led to increasing public support for nature protection and a growth in NGOs. As a result, a new generation of NGO conservation specialists are working with professionals in the protection of habitats and species, as well as practical habitat management.

Acknowledgements

We are grateful to our SSPA colleagues who provided information, particularly Dalia Čebatariūnaitė, Vitalija Valantiejienė and Saulis Skuja.

References

Ahokumpu, A., Brueggeman, J., Gulbinas, Z. & Kotimäki, T. (2007) *Management Effectiveness Evaluation of Lithuanian Protected Areas*. Vilnius: Gilija.

Anuškevičius, D., Klimavičius, A., Augutis, D., Bastytė, D. & Greimas, E. (2016) *Main Requirements in Lithuanian Legislation for Establishment and Management of Sites Protected Under the Habitat and Bird Directives – in relation to Georgian Accession to European Union*. https://greenalt.org/app/uploads/2021/04/Main_requirements_in_LT_legislation.pdf

Augutis, D., Gintaras, A. & Karpuška, R. (2015) *Retųjų rūšių apsaugos geroji praktika Lietuvoje*. Vilnius: Sorre.

Balčiauskas, L., Baranauskas, K., Ferenca, R. *et al.* (2016) *Europos bendrijos svarbos rūšių monitoringo metodikos. Žinduoliai, žuvys, varliagyviai, ropliai, moliuskai, vabzdžiai ir augalai*. Vilnius: Gamtos tyrimų centras, Valstybinė saugomų teritorijų tarnyba prie Aplinkos ministerijos.

Baškytė, R. (2013) Lietuvos gamtinės aplinkos apsauga. In *Lietuvos gamtinė geografija*, ed. M. Eidukevičienė, pp. 226–279. Klaipėda, Lithuania: Klaipėdos universiteto leidykla.

Baškytė, R., Raščius, G., Kavaliauskas, P. & Tukačiauskas, T. (2019) *Protected Areas in Lithuania*. Vilnius: Lututė.

BGI Consulting (2020) *Natura 2000 Tinklo Socio-Ekonominės Naudos Vertinimo Studija. Galutinė Vertinimo Ataskaita*. Vilnius: BGI Consulting.

Edvardsson, J., Šimanauskienė, R., Taminskas, J., Baužienė, I. & Stoffelad, M. (2015) Increased tree establishment in Lithuanian peat bogs – insights from field and remotely sensed approaches. *Science of the Total Environment*, **505**, 113–220.

Gudžinskas, Z., Jukonienė, I. & Žalneravičius, E. (2015) *Mokslinių tyrimų įvertinti ir prognozuoti Europos Bendrijos svarbos augalų rūšių, kurių apsaugai būtina steigti teritorijas, populiacijų ir jų buveinių būklės pokyčius bei natūralių ir antropogeninių veiksnių poveikį jiems. Baigiamoji ataskaita*. https://old.gamta.lt/files/ES_rusiu_bukles_ataskaita_2015%20Augalai%20i%C5%A1%20kito%20projekto.doc

Jablonskis, J., Kovalenkovienė, M. & Tomkevičienė, A. (2007) Lietuvos upių ir upelių vagų tinklas. *Annales Geographicae*, **40** (1), 46–56.

Kavaliauskas, P. (1995) The Nature Frame as a basis for ecological stability in Lithuania. In *EECONET and the Wildlands Project, Proceedings of the European-North American Workshop, Luso-American development Foundation*, ed. G. Bennett, pp. 37–43. Lisbon/Arnhem: Institute for European Environmental Policy.

(1996) Lithuania: the Nature frame. In *Perspectives on Ecological Networks*, eds. P. Novicki, G. Bennett, D. Middleton, S. Rientjes & S. Wolters, pp. 93–99. Tilburg, The Netherlands: European Centre for Nature Conservation.

Mierauskas, P. (2009) *Gamtotvarka. Strateginis ekosistemų valdymas*. Vilnius: Lietuvos gamtos fondas.

MoE (1998) *Biodiversity Conservation Strategy and Action Plan*. Vilnius: Petro ofsetas.

(2011) *Nacionalinė darnaus vystymosi strategija*. Vilnius: Lututė.

(2015) *Nacionalinis kraštovaizdžio tvarkymo planas. Sprendiniai ir rekomendacijos*. Vilnius: Gera idėja.

(2016) *Nacionalinė aplinkos apsaugos strategija*. Kaunas, Lithuania: UAB ARX Reklama.

(2017) *5th National Report to the Convention on Biological Diversity Lithuania*. www.cbd.int/doc/world/lt/lt-nr-05-en.pdf

(2022) "Natura 2000" *Prioritetinių Veiksmų Programa (PVP), Skirta Įgyvendinti Lietuvoje. 2021–2027 m. daugiametės finansinės programos laikotarpiu*. [Unpublished draft.] Vilnius: Lietuvos Respublikos aplinkos ministerija. https://am.lrv.lt/uploads/am/documents/files/PAF%202022%2003%2015-galutinis%20PDF(1).pdf

Plungė, S. & Gudas, M. (2018) *Žemės ūkis ir Lietuvos vandenys. Žemės ūkio veiklos poveikis Lietuvos upių būklei ir taršos apkrovoms į Baltijos jūrą*. https:// old.gamta.lt/files/%C5%BDem%C4%97s_%C5%ABkio_poveikio_ataskaitapdf

Povilaitis, A., Taminskas, J., Gulbinas, Z., Linkevičienė, R. & Pileckas, M. (2011) *Lietuvos šlapynės ir jų vandensauginė reikšmė. Monografija*. Vilnius: Apyaušris.

Rašomavičius, V. (ed.) (2012) *EB svarbos natūralių buveinių inventorizavimo vadovas*. Vilnius: Petro ofsetas.

(ed.) (2021) *Lietuvos raudonoji knyga*. Kaunas, Lithuania: Lututė.

Raudonikis, L. (2004) *Europos Sąjungos reikšmės paukščiams svarbios teritorijos Lietuvoje*. Kaunas, Lithuania: Lututė.

Raudonikis, L., Riauba, G., Brazaitis, G. *et al.* (2016) *Europos Bendrijos svarbos paukščių rūšių monitoringo metodikos*. Vilnius: Lietuvos ornitologų draugija, Valstybinė saugomų teritorijų tarnyba prie Aplinkos ministerijos.

Šimanauskienė, R., Linkevičienė, R., Bartold, M. *et al.* (2019) Peatland degradation: the relationship between raised bog hydrology and normalized difference vegetation index. *Ecohydrology*, **12** (8), e2159.

Stončius, D. (2011) *Kertinės miško buveinės*. Vilnius: Baltijos kopija.

Taminskas, J., Linkevičienė, R., Mažeika, J. & Kibirkštis, G. (2008) Klimato kaitos poveikis Čepkelių raisto hidrometeorologinėms sąlygoms: vertikaliosios vandens apytakos charakteristikoms. *Annales Geographicae*, **40** (2), 50–60.

Chapter 23

The Netherlands

Erik Gerritsen

Key Statistics	
Terrestrial area (km²)/% of EU-28*	37 401/1.9%
Marine waters (km²)/% of EU-28	58 847/1.3%
Population density (p km²)/% EU-28*	426/371%
GDP per capita (€)/% of EU-28	39 548/148%
Land cover (CORINE land cover classes 2018)	AGR 60%, WET 15%, ART 14%, FOR 8%, OSN 3%
Biogeographical regions	ATL, MATL
Number of HD Annex I habitats/%*	52/23%
Number of HD Annex II, IV and V species/%*	81/7%
Number of BD Annex I bird taxa/%*	53/30%
% HD UCS and decreasing habitats/ % unknown	25%/0%
% HD UCS and decreasing species/ % unknown	25%/5%
% breeding BD Annex I birds decreasing/% uncertain and unknown	28%/16%

* Excluding MAC and MMAC biogeographical regions.
Sources and codes. See Table 1.1.

23.1 The Natural History and Characteristics of the Netherlands

The Netherlands borders the North Sea on the western coast of continental Europe, and lies entirely within the deltas of four international rivers: the Rhine, Meuse, Scheldt and Ems. The country has mostly Holocene marine, fluvial and peat soils, with older Pleistocene aeolian soils in areas with higher elevation. The latter are mostly lower sandy hills of up to 110 m in the south, centre and east of the country with relatively impoverished soils due to historic deforestation and overgrazing. The south-eastern Province of Limburg also has a few rocky outcrops of up to 322 m at the northern tip of the Ardennes/Eiffel range. Some of these more elevated areas have siltier aeolian soils with higher fertility. Over half of the Netherlands is just 1 m or less above sea level, originally largely consisting of regularly inundated wetlands and a significant proportion of raised peat-bogs. Since around the sixteenth century, large areas of shallow coastal waters, marshland and bogs have been drained and/or reclaimed from the sea and embanked to form polders. In addition, extensive flood defence works have excluded, or restricted, the influence of tides within most of the estuaries, as well as seasonal flooding along rivers.

The temperate climate, flat ground, and year-round availability of freshwater in the Netherlands provide good conditions for agriculture, which now covers about 60% of the country's land area. Grassland and mixed dairy farming is most widespread and comprises about 60% of the UAA. Some regions have intensive pig and poultry farming, and the country's overall livestock density is the highest in the EU (Eurostat, 2022). Arable agriculture represents 33% of UAA, mostly in the coastal land reclamations. The horticultural sector, while relatively small in terms of land use, is economically significant, in particular through greenhouse agriculture and floriculture.

Most farmland is now of low biodiversity value, but areas remain with the well-known Dutch cultural landscape of large, open and wet permanent lowland meadows and pastures that provide breeding habitats for a wide range of 'meadow bird species'. These areas remain important for species such as Oystercatcher (*Haematopus ostralegus*), Redshank (*Tringa totanus*), Curlew (*Numenius arquata*), Lapwing (*Vanellus vanellus*) and, especially, the nominate race of the Black-tailed Godwit (*Limosa limosa limosa*), with 82% of the EU population,[1] even though agricultural improvements have greatly reduced their populations (discussed in Section 23.6.1). At the same time, the improved grassland provides highly nutritious forage for grazing waterbirds, such as the internationally important populations of wintering geese and swans, leading to increasing conflicts with farmers (discussed in Section 23.4.1).

Despite the historic land reclamations for agriculture, about 15% of the country still consists of some form of wetland habitat, which holds internationally important numbers of

[1] EU breeding population 2013–2018: https://nature-art12.eionet.europa.eu/article12/progress

waterbirds, especially in winter as these areas remain largely ice-free. These include areas of intertidal and shallow-subtidal habitats, such as within the Wadden Sea (c. 270 000 ha) – the largest intertidal area in Europe. It is of international importance for a range of species, but particularly for many breeding, passage, moulting and wintering birds. The Wadden Sea has been estimated to hold 5 to 6 million birds at peak migration, with over 10 to 12 million birds estimated to feed in the area each year, holding in winter almost the entire north-western populations of Common Shelduck (*Tadorna tadorna*) and Pintail (*Anas acuta*) (Kleefstra *et al.*, 2022). Mudflats and shallow subtidal habitat, together with salt marsh, are particularly important for migrating and wintering waders such as Oystercatcher, Dunlin (*Calidris alpina*) and Curlew. The soft sediments hold high densities of molluscs, so the shallow seas in turn support large numbers of wintering Scaup (*Aythya marila*) and Common Eider (*Somateria mollissima*) with winter maxima estimated to be over 100 000 for both species.

The Dutch coast is also important for its large area and variety of dune habitats (such as in the barrier islands of the Wadden Sea), and in particular its fixed coastal dunes with herbaceous vegetation (2130*).[2] Other wetlands include the saline lakes Grevelingen and Veerse Meer, which are important for fish-eating species. Numerous freshwater lakes include the large shallow lakes IJsselmeer (113 341 ha) and Markermeer and IJmeer (68 463 ha), created by their separation from the sea. These hold very large numbers of fish-eating birds, but are particularly important for species that feed on mussels (*Mytilus*), such as Tufted Duck (*Aythya fuligula*) and Scaup.

The Netherlands has one of the highest population densities in Europe, with most people living in cities. Consequently, apart from farmland and wetland, most of the remaining land now consists of urban areas, other settlements, and infrastructure. Forest cover is amongst the lowest in Europe, just 8% of the land (according to 2018 CLC data): 21% mixed, 38% broad-leaved, and 30% coniferous forest (Schelhaas *et al.*, 2014). Forests are primarily managed for nature and recreational objectives by a diversity of public and private owners, and less than a third of forest area still has a timber production function. Other semi-natural habitats, except for dunes, are also now scarce and include heathlands in the areas with poor sandy soils and fragmented remnants of formerly extensive raised bogs, which are now mostly restricted to relatively small protected areas, mainly in the east of the country.

23.2 The Status of Nature

23.2.1 Habitat Status and Trends

As described in Section 23.1, land reclamation, wetland drainage, the clearance of forest and the expansion and development of agriculture and cities have had profound impacts on the Dutch landscape and land use. Changes in the distribution and extent of habitats have been much less significant over the last century, but many habitats have declined in quality as a result of various pressures (discussed in Section 23.2.3). The annual progress report on the implementation of the NNN for 2021 found that, while all types of broad terrestrial ecosystem types had declined in quality since the 1990s (based on qualifying species), in some ecosystem types, like dunes, a higher quality was maintained than in others such as semi-natural grasslands. In addition, while for some broad ecosystem types, such as forests, quality is generally improving, others such as semi-natural grasslands have seen further net declines over the last 15 years (IPO and LNV, 2022).

The HD Article 17 reports[3] reveal that habitats have continued to be impacted by pressures over the last 12 years. Over 2007–2012, only 4% of 52 HD habitat types had a favourable conservation status. Whilst the assessments for 2013–2018 considered that 12% had a favourable status, only two habitats were thought to have shown a genuine improvement. Of those habitats with a favourable status, five were dune habitats (42% of dune habitats) and one was a wetland (3270, rivers with muddy banks). Over 2013–2018, 35% of HD habitats had a poor status, and 54% a bad status. Of particular importance is whether those habitats with an unfavourable status have shown improving or deteriorating trends. Table 23.1 indicates that some were improving and no habitat with a poor status was declining, but of most concern is that 46% of those with a bad status were deteriorating, including estuaries (1130) and large shallow inlets and bays (1160), which the country has a particularly high responsibility for conserving in a European context.

23.2.2 Species Status and Trends

According to long-term monitoring, the Netherlands has experienced a long historic period of biodiversity decline, as shown by the Netherlands Environment Assessment Agency's estimate that Mean Species Abundance decreased from 40% around 1900 to 15% by 2010 (PBL, 2016). However, some authors argue that losses have been significantly higher, as at the turn of the twentieth century ecosystems had already been significantly impoverished (Luiten van Zanden *et al.*, 2021). In recent years, monitored species declines have levelled off on average, with stabilisation of often small remaining populations in protected areas and of the number of threatened species overall.[4] Rural areas and cities have experienced further declines, illustrated for example by the recent addition of once very common and relatively generalist farmland species to the national Red List, such as Common Kestrel

[2] Numbers in brackets are HD Annex I habitat codes (see Appendix for full names). * Indicates a priority habitat.

[3] See Section 1.3 for background on monitoring under the Nature Directives, status categories and data sources.

[4] Rijksoverheid (September 2020) Summary overview of policy-relevant nature indicators for provinces and central government. (In Dutch.) www.clo.nl/indicatoren/nl1617-duiding-provinciale-indicatoren?ond=20877

Table 23.1 Short-term trends (c. 2007–2018) in Habitats Directive Annex I habitats and Annex II, IV and V species with an unfavourable conservation status in the Netherlands.

	Poor (unfavourable– inadequate)				Bad (unfavourable–bad)			
	Imp	Sta	Dec	?	Imp	Sta	Dec	?
HD Habitat assessments								
Number	5	11	0	2	6	8	13	1
%	28%	61%	0%	11%	21%	29%	46%	4%
HD Species assessments								
Number	4	8	6	6	6	6	14	5
%	17%	33%	25%	25%	19%	19%	45%	16%

Key. Imp = improving, Sta = stable, Dec = decreasing, ? = unknown.
Source. Based on Member State HD Article 17 reported data as presented on the EEA national summary dashboard (2021).

(*Falco tinnunculus*), Grey Partridge (*Perdix perdix*) and Skylark (*Alauda arvensis*) (Kleunen *et al.*, 2017).

According to the Living Planet Index for the Netherlands for terrestrial and freshwater species, established in 1990, monitored species populations showed a combined increase of 1.4% between 1990 and 2019, with 161 of 351 recorded species increasing and 126 decreasing in numbers.[5,6] Populations of freshwater and marine species increased most, benefiting from improved water quality, large-scale floodplain and marshland restoration, and better protection from damaging fisheries impacts. For marine species, immigration from growing populations elsewhere also played a significant role. In contrast, species in agricultural areas have declined by 40% since 1990. The data include the population trends used for EU indicators of common farmland birds and grassland butterflies, which both declined by over 30% over the same period.[7,8]

The data from the HD Article 17 species assessments show a less concerning picture than that revealed by the habitat assessments, as regards their overall status. Nevertheless, over the 2013–2018 reporting period, out of 80 HD species only 26% had a favourable conservation status. Of particular concern is that 39% had a bad status, whilst 30% had a poor status, and 5% had an unknown status. Furthermore, as indicated in Table 23.1, a high proportion of species that had an unfavourable conservation status had decreased. The declining species are associated with different habitat types, but are mostly wetland-dependent. For example, of the 13 HD fish species the Netherlands reports on, six had an unfavourable status with decreasing trends.

The most recent BD Article 12 report on bird population trends indicates that most BD Annex I taxa have shown short-term increases (i.e. over 2007–2018), but a substantial proportion have decreased, particularly with respect to breeding populations (Table 23.2). This is especially evident amongst farmland birds, including common species, as mentioned earlier, but long-term declines (since about 1980) and ongoing short-term declines have also been noted for all the typical meadow breeding waders, as mentioned previously. Other farmland birds that have shown long- and short-term declines include Common Kestrel, Grey Partridge and the Corn Bunting (*Emberiza calandra*). The general decline in agricultural biodiversity mentioned earlier is also reflected in the reporting results on HD species.

23.2.3 Drivers and Causes of Change

Most drivers and pressures on biodiversity in terrestrial and freshwater ecosystems in the Netherlands relate to changes in agricultural land management practices, primarily through intensification, and ongoing urbanisation and ecological fragmentation (Wereld Natuur Fonds, 2015, 2020). The three most reported pressures and threats on EU-protected habitats in the Netherlands are agriculture, human-induced changes to water regimes, and natural processes (e.g. vegetation succession) – all impacting over 50% of assessments (EEA, 2021). Pressures and threats on species reported are similar, but with additional high impacts from built infrastructure development. Nitrogen deposition is a key driver of deterioration (see e.g. Van der Burg *et al.*, 2021), mostly driven by livestock farming and fossil fuel combustion by transport and industry.[9] In the marine and coastal environment, the

[5] Rijksoverheid (March 2022) Living Planet Index the Netherlands 1990–2020. (In Dutch.) www.clo.nl/indicatoren/nl1569-living-planet-index

[6] Luiten van Zanden *et al.* (2021) warn that the number of recorded nature observations in the Netherlands has grown significantly since the 1970s, and as a result estimates are likely to be higher. Therefore, evidence of growing populations should be taken with some caution.

[7] Rijksoverheid (March 2022) Trend of farmland birds, 1915–2020. (In Dutch.) www.clo.nl/indicatoren/nl1479-boerenlandvogels

[8] Rijksoverheid (March 2022) Butterflies of grasslands, 1992–2020. (In Dutch.) www.clo.nl/indicatoren/nl1181-dagvlinders-van-graslanden

[9] National Institute for Public Health and the Environment (RIVM) webpage on nitrogen. (In Dutch.) (19 March 2021) www.rivm.nl/stikstof

Table 23.2 Short-term trends (c. 2007–2018) in Birds Directive Annex I taxa in the Netherlands.

	Breeding				Wintering			
	Inc	Sta/ Flu	Dec	?	Inc	Sta/ Flu	Dec	?
Number	20	4	12	7	6	3	3	5
%	47%	9%	28%	16%	35%	18%	18%	29%

Key. Inc = increasing, Sta = stable, Flu = fluctuating, Dec = decreasing, ? = uncertain and unknown.
Source. Based on Member State Article 12 reported data as presented on the EEA national summary dashboard (2021).

most common high-impact pressures are bottom-disturbing fisheries,[10] climate change, mixed-source pollution, human-induced changes in water regimes and invasive alien species (Wereld Natuur Fonds, 2017; I&W and LNV, 2018; EEA, 2021).

Between 1980 and 2019, the Dutch population grew by 22% to over 17 million people[11] and is expected to increase to 20.6 million by 2070,[12] with increased demand for housing and other urban infrastructure. Between 1996 and 2015, the built and semi-built area increased by 15% and is still increasing, mostly at the expense of agricultural land. Mobility is very prominent in the Netherlands, which is among the countries with the highest road density in the EU.[13] In addition to personal transport, the Netherlands has various large seaports, including the largest in Europe (Rotterdam), which generates significant maritime and road transport of goods.

Through government-stimulated innovation, upscaling and mechanisation, food production in the Netherlands grew more than 10-fold between 1950 and 2015.[14] Most farms in the Netherlands have among the highest input intensity per hectare in Europe (European Commission, 2018). The intensification of food production negatively altered habitats in a multitude of ways such as via the direct removal of semi-natural habitats such as grasslands and small landscape elements,[15] intensified land management, modified hydrological management such as through irrigation and drainage, increased nutrient levels from livestock and fertiliser application, and increased use of chemical pesticides.

As a result of the still ongoing development of built infrastructure and intensification of land use, the remaining areas of natural and semi-natural habitat (other than coastal dunes), such as species-rich grasslands, heathlands, bogs and native woodland (other than plantations), are now small and fragmented. Whilst many of these areas are now within protected areas, their habitats and associated species communities remain vulnerable. Their small size and frequent isolation mean that their species populations are at risk of chance events and extinctions unless functional connectivity is sufficient to enable their survival as meta-populations. Habitat ecological processes can also be interrupted by habitat fragmentation and isolation.

Many remaining natural and semi-natural habitats are impacted by external pressures, such as pollution, especially when they are in small sites surrounded by areas under intensive land use – as is often the case in the Netherlands. In this respect, the most significant and widespread pressure is eutrophication from the deposition of airborne nitrogen, the predominant source of which is ammonia from intensive livestock rearing (discussed further in Section 23.6.2). Several decades of acid rain have significantly acidified soils, limiting nutrient availability of especially calcium, magnesium and potassium to plants. The unnaturally high and growing nitrogen levels are further tipping vegetation towards an ever-shrinking number of species. Nitrogen enrichment impacts are further exacerbated by the cessation of grazing and other forms of vegetation management, which would otherwise help to at least slow the rate of nutrient build-up in the vegetation. Dry heaths in particular suffer from a lack of grazing as well as acidification.

The reduction in extensive grazing is a problem for many semi-natural habitats as well as the decline in other traditional forms of use, such as mowing and woodland coppicing. In the absence of such activities, natural succession takes over and can lead to the loss of the particular habitat types and specialist species that the sites are important for as they are replaced by rank vegetation, scrub and eventually poor-quality even-age woodland. Although almost all areas of such habitats are within protected areas, substantial areas are currently not receiving the necessary management[16] and are in poor condition as a result.[17]

[10] Fisheries, especially with dredging techniques, are a grave threat to the seafloor habitats inside and outside Natura 2000 sites. Dredging leads to damage to habitats like shallow water sandbanks (1110) and replacement by invasive alien species like the Atlantic jack-knife clam (*Ensis leei*).

[11] Statistics Netherlands (CBS) (April 2021) Population, households and population development; from 1899. (In Dutch.) www.cbs.nl/nl-nl/cijfers/detail/37556

[12] Statistics Netherlands (CBS) (December 2021) Population prognosis; key figures, 2021–2070. (In Dutch.) www.cbs.nl/nl-nl/cijfers/detail/85089NED

[13] Eurostat (May 2020) The densest motorway networks across EU regions. https://ec.europa.eu/eurostat/web/products-eurostat-news/-/ddn-20200528-1

[14] Statistics Netherlands (CBS) (January 2017) Dutch agricultural production 1950-2015. (In Dutch.) www.cbs.nl/nl-nl/nieuws/2017/05/nederlandse-landbouwproductie-1950-2015

[15] Such as coppice woodland and pollard trees; small waterbodies such as drinking pools, former river branches, and peat pits; hedges (mostly hawthorn) and wooded banks; old dikes or former dikes; duck decoys; or meadow orchards.

[16] New data from Statistics Netherlands and the National Flora Monitoring Network found that while 'open' habitat types (e.g. heaths, semi-natural grassland and open dune) had a vegetation cover of 7% in 1999–2003 this had increased to almost 12% by 2016–2019.

[17] Natural vegetation succession was the second most reported high-impact pressure for combined dune, grassland and heaths and scrub habitats after agriculture in the latest Article 17 reporting (2013–2018).

A common and severe pressure on more natural wetland habitats, such as bogs and fens, is hydrological disruption, especially desiccation. Most of the remaining wetlands in protected areas are surrounded by drained farmland with very low water tables. In such circumstances their permeable peaty or sandy soils are unable to maintain their required damp conditions. In addition, in many farmed peat soil areas drainage has accelerated soil subsidence, which has further aggravated desiccation problems. Other key pressures on wetland habitats are agricultural pollution, in particular with reactive nitrogen, mostly through ammonia (NH_3) released by livestock, and a wide range of invasive alien species, such as numerous plants (e.g. Floating Pennywort [*Hydrocotyle ranunculoides*] and New Zealand Pygmyweed [*Crassula helmsii*], crustacea (e.g. Louisiana Crayfish [*Procambarus clarkia*] and fish species (e.g. Pumpkinseed [*Lepomis gibbosus*].

Important pressures in Dutch coastal and marine habitats have been: both historic and ongoing human changes to water regimes (e.g. through damming of estuaries and sand suppletion); commercial flatfish, shellfish and shrimp fisheries; mixed-source pollution (e.g. from coastal transport); and climate change. Bottom trawling – mainly for Common Sole (*Solea solea*), European Plaice (*Pleuronectes platessa*) and, by lighter vessels, Brown Shrimp (*Crangon crangon*) – has led to significant seabed disturbance with negative knock-on effects on the benthos.[18] This has particularly impacted slowly maturing species, including seven keystone species of native rays and sharks, such as Thornback Ray (*Raja clavata*) and Spiny Dogfish (*Squalus acanthias*). Shellfish fisheries have had significant impacts on migratory bird species such as Red Knot (*Calidris canutus*), Oystercatcher and Common Eider (e.g. Reneerkens *et al.*, 2005). Climate change and changing nutrient composition appear to be causing a decline in gross primary production of phytoplankton with knock-on effects on zooplankton and species dependent on them, including commercial fish species (Capuzzo *et al.*, 2017).

A widespread consequence of the high human population density of the country is the high levels of recreation in the remaining natural areas. This is a particular problem for birds, with Osieck (2000) reporting that about half of Important Birds Areas (IBAs) were threatened by disturbance. Repeated disturbance of large flocks of wintering birds can result in major impacts, as it both increases the birds' energetic requirements and reduces the time that they are able to feed. In summer, the large numbers of visitors to beaches and dunes make large areas of habitat unsuitable for some breeding birds, such as the nationally Red-listed Kentish Plover (*Charadrius alexandrinus*) and Little Tern (*Sternula albifrons*) (Van der Winden *et al.*, 2008).

Last but not least, climate change is a key driver of change in Dutch ecosystems and increasingly impacting on the achievement of nature conservation objectives. For example, distribution changes in many species are believed to be caused by increased temperature and more frequent extreme weather events (see e.g. Besse-Lototsaka *et al.*, 2010; Van Bodegom *et al.*, 2014; Van der Knaap *et al.*, 2018).

23.3 Nature Conservation Governance, Actors, Legislation and Policies

23.3.1 Early Initiatives and Development of Nature Conservation Legislation

As shown in the summary of key developments in Table 23.3, nature conservation in the Netherlands dates back to the late nineteenth century. Its current form mainly originated around 1880, when an increasingly emancipated urban population, inspired by Romanticism and emerging conservation movements in the United States, UK and Germany, started opposing the rapidly accelerating human ecological footprint following the industrial revolution. With increasing organisation, in 1906 this emerging movement successfully prevented the designation of a former lowland peat excavation ('Naardermeer') as a landfill for Amsterdam's household waste. To prevent further threats, the site was acquired for conservation and is considered the country's first 'modern' protected area. To manage the site and acquire additional ones, a new non-governmental Association for the Preservation of Natural Monuments in the Netherlands (hereafter Natuurmonumenten) was established, which exists until this day.[19]

Around the same time, the oldest recorded nature NGO in the Netherlands (1892) raised awareness on the poor state of bird populations and their unsustainable use, including in fashion. The 1912 Bird Act was the first Dutch species protection law, and one of the first of its kind in the world. It was shortly followed in 1914 by the Hunting Act and Useful Species Act, which significantly strengthened protection for various non-bird species. Despite this progressive start on species protection, habitat protection at national level remained mostly a private affair until the adoption of the 1967 Nature Protection Act.[20] It was the first law that made it possible to designate nationally protected areas. Like earlier species protection legislation, the human use value of such areas was the principal policy driver. However, in line with the times, these values did not so much concern primary production but rather recreation and research.

[18] Rijksoverheid (November 2021) Environmental Data Compendium indicator page 'Seabed fauna and fishing North Sea'. (In Dutch.) www.clo.nl/indicatoren/nl1251-bodemfauna-noordzee-en-boomkorvisserij

[19] Today Natuurmonumenten and its now independent regional spin-offs are the largest private nature manager in the Netherlands, jointly managing over 225 000 ha – almost the same area as the State Forest Service.

[20] Wet van 15 november 1967, houdende voorzieningen in het belang van de natuurbescherming.

Table 23.3 Key nature conservation developments in the Netherlands.

1892	Establishment of the Union Combating Horror Fashion, first nature NGO to ultimately become Vogelbescherming Netherlands
1906	First protected area (Naardermeer) and establishment Natuurmonumenten
1912	First Bird Act
1914	Hunting Act and Useful Species Act
1961	Forest Act
1967	First Nature Protection Act
1975	Adoption by Dutch Parliament of a policy memorandum on agricultural nature management (*'Relatienota'*) prompted by major concerns about the rapid decline of nature in agricultural areas
1990	The *National Nature Policy Plan* (NBSAP) for 1990–2000 introduced the National Ecological Network (NEN), which was renamed in 2013 to Netherlands Nature Network (NNN)
1998	Revised Nature Protection Act, revised considerably to transpose the EU Nature Directives and Ramsar Convention and published jointly with the 1998 Flora and Fauna Act
2000	The NBSAP *'Nature for People, People for Nature'*[21] emphasised the societal relevance of nature conservation and the need to mainstream it with other policy areas and economic sectors, and better involve citizens
2013	The *Nature Pact* set out government ambitions in nature conservation up to 2027
	Implementation Agenda Natural Capital addresses CBD commitments and *EU Biodiversity Strategy to 2020*
2015	The Nature Conservation Act combined and replaced the 1961 Forest Act, 1998 Nature Protection Act, 1999 Flora and Fauna Act and formalised the narrower focus on EU nature conservation commitments and decentralisation of implementation to regional level (Provinces)
2016	The new approach to agricultural habitat management with certified collectives entered into force
2019	The ambition statement, *'The Netherlands Nature Positive'*, set out four strategies to accelerate implementation of the *Nature Pact* and tackle other issues
2020	*New Environmental Vision* (NOVI), which will be implemented through a new Environmental Act expected to enter into force on 1 January 2023

See Table 3.3 for signatory dates of international conventions.

After the EU Birds Directive was adopted in 1979, the Netherlands failed to transpose its provisions by the 1982 deadline and was twice taken to court by the European Commission. In the meantime, the 1992 EU Habitats Directive had also been adopted. The two EU Directives ultimately triggered a significant revision of the Nature Protection Act[22] (LNV, 1998a) and the adoption of a new Flora and Fauna Act[23] (LNV, 1998b) in 1998 that merged all (inter)national species protection commitments. Until the new acts entered into force (only in 2005 and 2002, respectively) the EU Nature Directives had direct effect for several years. Afterwards, it became clear that the EU Habitats Directive had not been correctly transposed. In addition, a series of amendments to the Dutch acts had increased their complexity, which further hampered implementation. This resulted in a last revision and merger of both acts into the 2015 Nature Conservation Act,[24] which entered into force on 1 January 2017.

The 2015 Nature Conservation Act is today the principal legal instrument for nature conservation in the Netherlands. Under the Act, which is strongly based on the provisions of the EU Nature Directives, the Netherlands' 12 regional governments (Provinces) are committed to implement the Dutch Nature Network Netherlands (NNN),[25] which includes all Natura 2000 sites but encompasses a larger nationally protected area (described in Table 23.4).

Under the Nature Conservation Act, both the national government and each Province are committed to develop a *Nature Vision* document. The national *Nature Vision* has as a first objective to focus on '… the maintenance and, if necessary, restoration of a favourable conservation status of the species of animals and plants occurring naturally in the Netherlands and the types of natural habitats and habitats of

[21] *Nota natuur, bos en landschap in de 21e eeuw: Natuur voor mensen mensen voor natuur.*

[22] Wet van 25 mei 1998, houdende nieuwe regelen ter bescherming van natuur en landschap (Natuurbeschermingswet 1998).

[23] Wet van 25 mei 1998, houdende regels ter bescherming van in het wild levende planten- en diersoorten (Flora- en faunawet).

[24] Wet van 16 december 2015, houdende regels ter bescherming van de natuur (Wet natuurbescherming).

[25] *Natuurnetwerk Nederland.*

Table 23.4 The number and extent of international designated sites and protected areas in the Netherlands.

	IUCN category	Number	Area (km^2)
Designated under global and non-EU international legislation			
Ramsar Sites	NR	54	914 219
Natural and Mixed World Heritage Sites	NAp	1	?
Biosphere Reserves (exc. buffer and transition zones)	NAp	1	665
OSPAR Marine Protected Areas	NR	5	8 319
Designated under the Nature Directives			
SCIs (terrestrial + marine)	NR	140	3 118 + 12 081
SPAs (terrestrial + marine)	NR	77	4 771 + 8 627
Natura 2000 terrestrial area			*5 495*
Natura 2000 marine area			*15 090*
Designated under domestic legislation			
Nature Network Netherlands excl. N2k (Wet Natuurbescherming and Barro)	IV*	1*	4 000
. . .of which National Parks (Nationaal Park, part of the NNN)	II	21	1 696

Notes. NR = Not reported; NAs = not assigned; NAp = not applicable. Some protected area categories overlap (e.g. all Ramsar Sites in the Netherlands are also protected under the Nature Directives, and the Wadden Sea World Heritage site too). * Under the Nature Conservation Law there is only one NNN with a uniform protection regime; however, only the Netherlands 160 Natura 2000 sites (all IUCN Category IV) with an area of 20 576 km^2 are included in the WDPA.
Sources. See Section 1.3. Based on data from: Ramsar Convention, OSPAR MPA, and UNESCO websites (March 2022); EEA Natura 2000 Barometer, as at end 2020; NNN from Sanders *et al.* (2021); UNEP-WCMC (2021) for other areas and IUCN category.

species occurring in the Netherlands'. Moreover, the Act stipulates that the national *Nature Vision* shall '. . . as far as possible, contain a quantification of the conservation objectives for habitats and species to be protected in Natura 2000 sites . . . and shall provide an understanding of intended changes in conjunction with and in relation to the conservation status of these habitats and species and intended changes to Natura 2000 designation acts . . .'. However, so far no new national nature vision has been published in this new template. The *Nature Vision* published in 2014 before the adoption of the Act (Ministerie van Economische Zaken, 2014) only refers in general terms to nature conservation and the 2013 Nature Pact, an inter-institutional agreement made to devolve competence on the implementation of nature policy from the central government to the 12 Provinces (LNV and IPO, 2013).

All Provinces have by now transposed new provisions under the Nature Conservation Act in their own regulations.[26] The majority have chosen to reuse the pre-existing national policies,[27] which reflect the detailed provisions on Natura 2000 in the Act.[28] The species protection may vary per Province, but the basis is always that all BHD species are protected, with protection for additional species dependent on the specific situation in the Province (e.g. their ecology, flagship species, ambition and finances). Active measures for species conservation are in some Provinces still in preparation as part of the development of the new *Nature Visions* (see also Section 23.4 on species conservation).

23.3.2 Policies and Strategies

Post-war Growth, Agriculture and Nature Policy Response

After northern areas of the Netherlands had suffered from famine in the last winter of World War II, its post-war government strongly focussed on self-sufficiency in food production and the rationalisation of agriculture. Stimulated by this favourable policy framework, investment and knowledge exchange through the Marshall Plan, the country saw rapid increases in food production through mechanisation and economies of scale. The reduction in rural employment led to rapid urbanisation, and an urgent need for housing marked the rise of strong national spatial planning legislation from the 1950s onwards. In the 1960s and 1970s, however, as in many other European countries, awareness grew of the significant negative environmental impacts of growing economies of scale and intensification of farming. In response, in the 1970s, national policies started to better

[26] Webpage on the Nature Conservation Act of the executive agency of the Twelve Provinces: www.bij12.nl/onderwerpen/natuur-en-landschap/natuurwetten-en-regelgeving/provinciale-verordeningen/wet-natuurbescherming/

[27] Assessment by private law firm Stibbe (last updated 3 January 2017): www.stibbeblog.nl/all-blog-posts/environment-and-planning/implementatie-van-de-wet-natuurbescherming-op-provinciaal-niveau/

[28] Own assessment based on sample.

coordinate management of valuable nature and landscapes and agricultural development.

This new ambition was set out in a 1975 policy memorandum '*On the relationship between agriculture and nature and landscape conservation*' (hereafter '*Relatienota*' (CRM, 1975)),[29] which recognised that old cultural landscapes also deserved protection 'because of their ecology, geomorphology, development history, cultural history or amenity value' and that farmers should be financially compensated to do this (L&V, 1975). The memorandum introduced two categories of spatial designations of agricultural land: 'management areas' in which farmers would take relatively small measures, such as maintenance of small landscape elements; and 'agricultural nature reserves', which would be acquired by government and managed by farmers for specific nature objectives, in particular meadow bird protection. The former were agri-environment schemes *avant la lettre* and would inspire the introduction of similar schemes in other European countries and ultimately EU-wide mandatory schemes from 1992 onwards. A target of 100 000 ha was set for each type of management. While this approach was innovative in the European context and resulted in local conservation successes, the identification and designation of the reserves were slow and the voluntary schemes were regarded by farmers as limiting their growth opportunities. As a result, by 1990 less than a quarter of the target had been met (see e.g. De Snoo *et al.*, 2016).

The 1990 Nature Policy Plan and National Ecological Network

Following an increasing concern with the ever-intensifying land and water use, and a growing understanding that nature conservation would not succeed with local reserves and agricultural nature management alone, the 1990 *Nature Policy Plan*[30] set a new and more ambitious direction of travel by introducing national government plans for the acquisition and active restoration of lost nature values beyond traditional protection and sustainable use: so-called 'nature development' (LNV, 1990). The main goal of the plan was sustainable conservation, restoration and development of natural and landscape values by recognising the significance of nature and landscape for human development as well as its intrinsic value.

To achieve this goal, the plan introduced the concept of a National Ecological Network (NEN). Its idea was that by enlarging and better connecting existing nature, and improving the quality of nature, an ecologically coherent network of high-quality natural areas would be established. One of the main triggers for the NEN was the outcome of scientific research demonstrating the impact of habitat fragmentation, based on studies investigating island theory in Dutch landscapes and modelling of meta-populations (e.g. Van Dorp and Opdam, 1987). It was shown that even with good environmental conditions the current protected areas would not sustain the biodiversity levels due to local extinction without recolonisations. The *Nature Policy Plan* set out to augment the then 453 500 ha of existing terrestrial area of natural and semi-natural habitats with a further 275 000 ha through: acquisition, rehabilitation and restoration management on former agricultural land (50%); agricultural nature management (35%); and (other) private nature management (15%). In addition, the NEN included 6.3 million ha of large water bodies (including the Dutch part of the North Sea). While achieving the NEN was not explicitly included in the 1998 Nature Protection Act, it for the first time set out a vision for protected area coherence in the Netherlands, and its inclusion in subsequent spatial planning law meant that it became a long-term policy objective in Dutch nature conservation.

Although the majority of the NEN is owned and primarily managed for nature objectives by the state or private nature management organisations, part of the network is still under active farming. To ensure habitat and species conservation objectives in these areas are met, the Netherlands has developed an elaborate policy for agricultural nature management (see Section 23.6.1).

Responding to CBD Commitments and EU Biodiversity Strategies

The Netherlands is a party to the UN CBD and prepared explicit 10-year national biodiversity strategies and action plans (NBSAPs) in 1990 and 2000. However, no such plan was adopted in 2010. Instead, the country reported progress to the CBD according to the six headline targets of the *EU Biodiversity Strategy to 2020*. After the Dutch parliament asked for a national implementation plan, in June 2013 the government presented an *Implementation Agenda Natural Capital* (Ministerie van Economische Zaken, 2013). Taking the CBD commitments and *EU Biodiversity Strategy* as a starting point, it included 16 actions under four key themes: (1) sustainable production, consumption and supply chains; (2) sustainable fisheries and protection of marine biodiversity; (3) sustainable agriculture and protection of biodiversity; and (4) the valuation of natural capital. From a CBD perspective, it mostly focussed on sustainable use and access and benefit sharing, and much less on nature conservation. From an *EU Biodiversity Strategy* perspective it mostly focussed on targets relating to agriculture and forestry (3), fisheries and seas (4) and halting biodiversity loss globally (6). The document was presented as complementing national nature legislation, the NEN and the *Nature Policy Plan for the Dutch Caribbean* (focussing on 'preservation and strengthening of nature in the Caribbean Netherlands') to cover national and international commitments on conservation and sustainable use.

The 2013 Nature Pact and Decentralisation of Nature Conservation

A deep cut in the nature policy budget – of nearly 70% – by the first government led by Mark Rutte in 2011 triggered a series of new inter-institutional agreements, which were formalised in the 2013 *Nature Pact*[31] (Ministeries BZk *et al.*, 2011; LNV

[29] Full title: *Nota betreffende de relatie landbouw en natuur- en landschapsbehoud.*

[30] *Regeringsbeslissing Natuurbeleidsplan.*

[31] *Natuurpact, ontwikkeling en beheer van natuur in Nederland.*

and IPO, 2013). The two main decisions were a significant reduction in ambition in implementing the NEN, and delegation of responsibilities for nature policy to the 12 Provinces. Under the second Rutte government in 2012, the budget cuts and reduced ambition were partly reversed. However, the decentralisation plans were maintained and in the 2013 *Nature Pact* it was agreed that from 2014 onwards the national government (Ministry of Agriculture, Nature and Food Quality) would only retain responsibility for setting the legal framework and meeting international commitments in nature conservation. The Provinces became responsible for the acquisition, rehabilitation and restoration of 80 000 ha of 'new' nature as well as implementation and enforcement.[32] The NEN was renamed the Nature Network Netherlands (NNN), although it is still sometimes referred to as the NEN. A deadline for completing it was set for 2027, in line with the final implementation deadline of the EU WFD.

Subsequently, in another shared 'ambition statement' presented in 2019 (LNV and IPO, 2019), the Ministry of Agriculture, Nature, and Food Quality and the 12 Provinces outlined a new, more detailed, implementation plan based on four strategies:

1. accelerating implementation of existing agreements on nature conservation in the 2013 *Nature Pact*;
2. improving environmental conditions in and around protected areas (e.g. with regards to nitrogen);
3. broadening nature policy to outside of protected areas (e.g. in farming, cities and water management);
4. better cooperation between different governments (international–local) as well as between key stakeholders.

23.3.3 Governance

The administrative landscape in the Netherlands is, although clearly defined, rather complex with four separate four-yearly elections for the national parliament, 12 provincial parliaments, 380 municipal councils and 23 water boards (responsible for regional water management). The 2015 Nature Conservation Act provided the legal basis for national nature conservation and formalised the new agreements on decentralisation in the 2013 *Nature Pact*. While various responsibilities had already been delegated to the Provinces previously, the Ministry of Agriculture, Nature and Fisheries (LNV) had formally maintained a central coordinating role since 1982. Today, LNV maintains responsibility for international biodiversity policy, including the implementation of the EU Nature Directives.

Another key ministry is that of Infrastructure and Water Management (I&W), responsible for public policy on large-scale infrastructure and the management of water quantity and quality. Importantly, I&W's executive agency, Rijkswaterstaat, is directly responsible for the implementation of nature conservation in all large water bodies, which together make up 69% of the Natura 2000 network in the Netherlands (e.g. Lake IJsselmeer, all major rivers and their deltas, and the Wadden Sea, which are all fully designated under the EU Nature Directives). Together with LNV, I&W is responsible for the reduction of diffuse pollution from agriculture, however LNV carries responsibility for implementation of the Nitrates Directive as well as the Programmatic Approach to Nitrogen (further discussed in Section 23.6.2).

In terms of governance in the field, the site management associations play a key role. Three organisations manage nearly all terrestrial sites: the State Forestry Service (which in 1998 became semi-independent after having been directly reporting to LNV), and the private nature management organisations Natuurmonumenten and De Provinciale Landschappen. In cooperation with regional and local governments, water boards and police services, they coordinate site-level enforcement activities.

23.3.4 Public Involvement, NGOs and Other Stakeholders

The Netherlands has a long history of public involvement in nature conservation. Surveys since 1996 show consistently high support for nature conservation objectives, and in 2017 almost 60% of citizens took part in one or more supporting activities for nature, mostly through small actions in private gardens and clean-ups.[33] Since the 1990s, the membership of nature and environmental NGOs has grown, although after the 2008 financial crisis memberships declined for the largest two membership organisations Natuurmonumenten and Wereld Natuur Fonds (WWF Netherlands). In recent years memberships have been on the rise again. Today nearly 1.9 million citizens are members, and nature organisations can draw on a significant group of volunteers for monitoring, education and nature management.

The large membership organisations, Natuurmonumenten and WWF, and, to a lesser extent, the Provincial Landscape organisations and Vogelbescherming Netherlands (BirdLife Netherlands), are the most powerful public voices for conservation in the Netherlands. Natuurmonumenten combines a large membership base with technical credibility as one of the largest nature managers in the Netherlands. Whilst WWF Netherlands has traditionally been more oriented towards conservation outside of the Netherlands, since the severe 2010–2011 budget cuts its members have demanded a more active domestic stance. This has resulted, among other things, in four *Living Planet Reports* since 2012, which have played an important role in combining evidence from species

[32] The central government only maintains management responsibility for national waters, defence areas and state-owned areas other than those under the jurisdiction of the Ministry of Agriculture, Nature and Food Quality.

[33] Rijksoverheid (April 2019) CLO indicator page 'Active involvement in nature conservation, 2017'. (In Dutch.) www.clo.nl/indicatoren/nl1589-actieve-betrokkenheid-bij-natuurbescherming?ond=20889

conservation organisations and communicating them to a larger audience. Eleven larger conservation NGOs cooperate on public affairs through a common organisation (Groene11). NGOs in the Netherlands play an important role in scrutinising government (in)action, including through initialising legal action. For example, a legal challenge by NGOs in 2017 that the Programmatic Approach to Nitrogen (see Section 23.6.2) conflicted with obligations under the Habitats Directive was upheld in court and led to a major policy shift in the Dutch government's approach to nitrogen pollution (see section 23.6.2).

Other stakeholders are also actively involved in public discourse on nature conservation in the Netherlands. These include a range of Dutch farming unions and organisations (LTO, ZLTO, NJAK, BoerenNatuur), commercial and recreational fisheries organisations (VisNed, Nederlandse Vissersbond, Sportvisserij Nederland), the national hunting association (KNJV), the 10 water companies (which manage large nature areas, especially dunes) and other private sector organisations such as the Rotterdam harbour.

Besides large, organised stakeholder organisations, the Netherlands also benefits from substantial grassroots nature conservation by local citizen groups, for example working on green management in cooperation with local authorities, farmers and landowners. There are also several long-established organisations focussed on nature education and study for adults (e.g. KNNV, IVN, NIVON) and young people (e.g. NJN, JNM), with many local branches. Several of these grassroots organisations, or individuals, have gone to court to stop governmental or private action that is at odds with nature conservation policy objectives. For example, the Programmatic Approach to Nitrogen regulation (described in Section 23.6.2) was annulled after it had been successfully brought to court by a small NGO called Mobilisation for the Environment.

A key guiding principle of the 2013 *Nature Pact* was that nature conservation policy had for too long been too technocratic and should instead be placed back within society – an argument also used to justify the decentralisation decision. One of the central themes of the 2014 *Nature Vision* was to increase the societal significance of nature. A progress assessment of the *Nature Pact* warned that the continued focus on BHD goals and limited implementation of the nature-inclusive intentions risks undermining societal acceptance for nature conservation in the Netherlands (Folkert and Boonstra, 2017). Nonetheless, compared to other Member States, nature conservation in the Netherlands seems very open, and many societal stakeholders are directly or indirectly engaged in implementation.

23.4 Species Protection and Conservation Measures

23.4.1 Basic Measures

The 2015 Nature Conservation Act contains various provisions for basic species protection, including a general duty of care (Article 1.11), various prohibition provisions, and waivers and exemptions. The law focusses on BHD species, but also protects an additional 81 animal and 76 plant species. In total, 500 of the 36 000 animal species occurring in the Netherlands enjoy protection. In addition, the law requires Provinces to take necessary measures for the conservation and restoration of Red Listed species (Article 1.12).

The Nature Conservation Act also provides the legal provisions for hunting, which is only allowed for five species: Pheasant (*Phasianus colchicus*), Mallard (*Anas platyrhynchos*), Wood Pigeon (*Columba palumbus*), European Hare (*Lepus europaeus*) and European Rabbit (*Oryctolagus cuniculus*). Hunting, taking, disturbing, etc., of other species is only allowed in case of damage to agriculture or for nature or safety issues. Licences are required, and derogations are reported. While Provinces are responsible for species protection, enforcement in the field is undertaken by a variety of stakeholders, particularly governmental organisations and/or those owning natural areas such as Rijkswaterstaat, the State Forest Service, and private nature management organisations such as Natuurmonumenten.

An important instrument in streamlining permitting and exemptions by different actors are 'codes of conduct', which set out conditions under which certain recurring activities can be exempt from permitting, such as the taking of certain measures to prevent or limit damage to protected species. For example, in 2020 a new code of conduct for all municipalities was adopted. Under the former Flora and Fauna Act, most municipalities had their own codes, which created a lot of legal uncertainties. Other examples of codes have been developed for drinking water companies, forest owners, extractive industries, and national road and waterway management.

A particularly thorny issue related to species protection has been the conflict between geese conservation and farming interests. As explained in Section 23.1, the Netherlands attracts millions of wintering waterbirds, including over 2 million geese. In the last 50 years, these numbers have grown significantly, especially because of reduced hunting pressures and improved feeding conditions in intensively managed agricultural areas (Buij and Koffijberg, 2019; Hornman *et al.*, 2021). In addition to these wintering visitors, some breeding populations have increased significantly as well, especially Greylag Goose (*Anser anser*), but also Barnacle Goose (*Branta leucopsis*) and the non-native Canada Goose (*Branta canadensis*). With this population growth, goose grazing damage to crops has grown considerably too, mostly to grasslands. The total annual cost of goose damage increased from less than €1 million in 1998/1999 to more than €21 million by 2016/2017, amounting to more than 80% of total wildlife damage between 1998 and 2017 (Buij and Koffijberg, 2019). To prevent damage to crops and airports, the Provinces provide protection exemptions to cull geese, and over 2016–2018 on average at least 329 000 geese were killed per year (Buij and Koffijberg, 2019). Stakeholder positions on culling continue to be miles apart, with Vogelbescherming Netherlands asking for a total ban on culling,[34] and the Dutch farming union (LTO), hunters (KNJV) and federation of private landowners (FPG) asking for a sustainable hunting regime without exemptions.[35]

23.4.2 Action Plans

In the past, national species action plans were prepared in the Netherlands, with the twenty-fifth completed in 2011. In 2008, the approach was replaced by the so-called 'habitat-approach',[36] under which a national list of species is maintained for which Provinces may be required to take active species protection measures depending on their status. The list in 2022 held 435 species, for which 358 active measures are required.[37] Several Provinces have been proactive by choosing flagship species from the list that indicate the habitat quality required for all species of concern under the habitat approach.[38] However, an NGO assessment in 2018 found that the Provinces had insufficiently adopted actions for species with an unfavourable conservation status (BirdLife *et al.*, 2018).

Of particular concern has been the need for action to conserve farmland birds, which have severely declined. Following increasing public and legal pressure, and a request from the Dutch Parliament in 2015, a national action plan to stabilise meadow bird populations was presented by the Ministry of Economic Affairs in 2016.[39] However, Vogelbescherming argued that the Netherlands authorities' principal strategy to protect farmland birds through CAP agri-environment schemes has been ineffective in halting farmland bird declines, which by then had already been clearly demonstrated (e.g. by Kleijn *et al.*, 2001; Kleijn and Lammertsma, 2013), and that the action plan specifically lacked: (1) clear population targets, (2) a dedicated action programme detailing specific measures, (3) adequate financial means, and (4) clear leadership by the national government. A critical legal assessment of the plan (Trouwborst, 2016) led to a formal complaint by Vogelbescherming Netherlands to the European Commission against the Netherlands for breaching its commitments under the Birds Directive[40] (in particular in relation to Articles 2, 3, 4, 5 and 13). Although a new agri-environment scheme was launched in 2016, Vogelbescherming considered that it too was insufficient to sustainably maintain farmland bird populations.[41] This topic is discussed in further detail in Section 23.6.1, in relation to agri-environment scheme management for meadow birds.

Since 2016, five Provinces have developed action plans for farmland birds.[42] In 2020, a group of the six Provinces most important for meadow birds, together with various NGOs, farming organisations and scientists, under the leadership of a former Minister of Environment, proposed a 'battleplan' for the Black-tailed Godwit to reverse the dramatic decline in the breeding population of this national bird of the Netherlands.[43] It is an interesting initiative, as it is evidence-based, widely supported and importantly addresses the critical issue of insufficient economic incentives for farmers to participate in voluntary schemes and for participating farmers to choose 'deep green' measures under the schemes for which economic opportunity costs are too high compared to management subsidies.

23.4.3 Re-introduction Schemes

Re-introduction schemes have been relatively common in the Netherlands, and a recent review found that 18 of 26 known species re-introductions since 1982 were successful (Zekhuis *et al.*, 2021). The most successful re-introduction has been that of the Beaver (*Castor fiber*), the last of which was poached in 1826. Re-introduced in 1988, the species now has a favourable conservation status[44] and has reappeared in all 12 Provinces. In fact, population growth since 2011 has been very rapid, such that the species has become a cause of conflicts again. This triggered the development of a Beaver knowledge portal for water managers in 2021.

Another successful re-introduction has been that of the Otter (*Lutra lutra*) following the adoption of an *Otter Recovery Plan* in 1988, and through cooperation of national and local governments, water boards and nature management organisations. The recovery plan, supported by a significant budget over the years, included measures on habitat restoration and water quality improvement, a breeding programme combined with re-introduction/re-population of the species, connecting habitat and creating safe routes for movement and

---

34 Birdlife Netherlands position on geese and wigeon (date unknown). www.vogelbescherming.nl/over-ons/standpunten/standpunt-ganzen-en-smienten

35 KNJV webpage position on geese (12 June 2019). www.jagersvereniging.nl/nieuws/ganzen-ganzen-ganzen/#jagersvereniging

36 *Leefgebiedenbenadering.*

37 http://minez.nederlandsesoorten.nl/content/soorten-van-de-leefgebiedenbenadering

38 See for example the Province of Utrecht: www.provincie-utrecht.nl/onderwerpen/natuur/natuurbeleid#aandachtsoorten

39 Letter by State Secretary Van Dam (Economic Affairs) of 2 November 2016 to Parliament on the approach taken to conserve meadow birds: https://zoek.officielebekendmakingen.nl/kst-33576-92.html

40 www.vogelbescherming.nl/actueel/bericht/vogelbescherming-dient-europese-klacht-in-tegen-nederland-over-bescherming-van-weidevogels

41 Press release by BirdLife Netherlands of 8 November 2016 '*Vogelbescherming dient Europese klacht in tegen Nederland over bescherming van weidevogels*' : www.vogelbescherming.nl/actueel/bericht/vogelbescherming-dient-europese-klacht-in-tegen-nederland-over-bescherming-van-weidevogels

42 BoerenNatuur web page with Provincial Action Plans: www.boerenlandvogelsnederland.nl/naslagwerken/8/ [Accessed 4 April 2021].

43 Website of the 'battle plan' for the Dutch black-tailed godwit breeding population: https://gruttoaanvalsplan.nl/

44 EEA Habitats Directive Article 17 reporting web tool: www.eionet.europa.eu/article17/

dispersal, and scientific research and educational activities (Zekhuis et al., 2021). Improved water quality resulted mainly from more general international and national policy; however, in areas with Otters, additional measures were taken. The population is still growing and in the winter of 2019/2020 for the first time exceeded the minimum threshold of a viable population in the Netherlands.[45] Nevertheless, traffic mortality and low genetic diversity remain challenges.

Other EU-protected species that have been re-introduced or reinforced with individuals from populations outside of the Netherlands to strengthen remaining populations include Brook Lamprey (*Lampetra planeri*, locally), Atlantic Sturgeon (*Acipenser sturio*), Raven (*Corvus corax*), European Tree Frog (*Hyla arborea*), Noble Crayfish (*Astacus astacus*), Common Spadefoot Toad (*Pelobates fuscus*), White Stork (*Ciconia ciconia*) and Black Grouse (*Lyrurus tetrix*). The Black Grouse is a well-known case of a long-running, challenging and therefore costly re-introduction, which was attempted in two sites (of which one had 11 remaining native birds). Both attempts failed, mainly because of low habitat quality due to nitrification and acidification, and heathlands containing diminishing numbers of large insects, which are important for chick survival in summer (Zekhuis *et al.*, 2021). As a result, new introductions in one site were halted, while they continue in the most promising site in combination with habitat improvement measures.

23.5 Protected Areas and Networks

23.5.1 Protected Area Types and Their Coverage

Table 23.4 presents the number and total area of each type of protected area listed in the WDPA, as of November 2021. The sites under global international protected area legislation (excluding EU legislation) include one natural World Heritage Site (Wadden Sea), one Biosphere Reserve (Maasheggen; the Wadden Sea was withdrawn in 2018) and 54 Ramsar Sites. All of these were designated as Natura 2000 site(s), except the Maasheggen.

The Netherlands has largely completed its Natura 2000 network on land and at sea. In 2019, there were a few minor gaps, mainly concerning species that had only recently been observed in the country and still need to be addressed, plus some potential gaps in the SPA network (European Commission, 2019). At the end of 2020, there were 162 Natura 2000 sites totalling 20 585 km^2, covering 14.7% of the land and inland waters and 25.6% of marine waters.[46]

The Netherlands, in its reporting to the UN CBD, considers the entire Nature Network Netherlands (NNN) as protected, which includes, as well as Natura 2000 sites, the following building blocks:

1. other existing areas designated and managed with a primary nature conservation objective, including all 21 National Parks (which must have at least 1 000 ha contiguous habitats);
2. areas that have been spatially designated by the 12 Provinces for future nature rehabilitation as agreed in the 2013 *Nature Pact*, but which are still under a different land management;[47]
3. areas under the Agricultural Nature and Landscape Management scheme (ANLb),[48] which are foreseen to become part of the NNN;
4. 60 000 km^2 of large water bodies (lakes, rivers, the Dutch part of the North Sea and the Wadden Sea).[49]

The NNN as agreed in 2013 will measure 7 339 km^2 by 2027, of which by 31 December 2020 6 982 km^2 had been restored or rehabilitated, with 357 km^2 remaining still mostly under conventional agriculture (IPO and LNV, 2021). It is debatable whether all of the NNN sites outside the Natura 2000 network qualify as protected areas in accordance with IUCN and CBD criteria. However, as stated by Sanders *et al.* (2021, p. 43), they can be considered to be at least 'other effective area-based conservation measures' (OECMs), and therefore all count towards CBD and EU protected area targets.

In total, the Netherlands has 464 protected areas reported in the WDPA. Taking into account the substantial overlaps, they cover 7 914 km^2 of land and inland waters and 17 247 km^2 of the marine environment (UNEP-WCMC, 2021). Based on the areas given in this chapter's Key Statistics table, these protected areas cover 21.2% of the land and 29.3% of the Netherlands' marine waters. Following the Netherlands' own calculations, which use different definitions for coastlines, protected areas cover 25.9% of the land and 23.8% of coastal and marine areas (Sanders *et al.*, 2021).

23.5.2 The Coherence of Protected Areas and Ecological Networks

As already briefly described, the Netherlands is very densely inhabited with intensive land use for a wide range of functions such as housing, farming, transport and recreation. As a result,

45 Rijksoverheid (November, 2021) Environmental Data Compendium indicator page 'Otter, 2002-2018/2019'. (In Dutch) www.clo.nl/indicatoren/nl1072-otter

46 Since when the new large marine Bruine Bank SPA (1 365 km^2) was announced in 2022.

47 For example where landowners are unwilling to sell their land.

48 *Agrarisch Natuur- en Landschapsbeheer*. The ANLb came into effect on 1 January 2016 and is partly EU co-financed through EARDF.

49 The national framework vision for Infrastructure and Nature (*Structuurvisie Infrastructuur en Ruimte, or SVIR*) formalises the inclusion of the entire Dutch part of the North Sea as NNN, however under a weaker protection regime than the NNN on land, which is why only the Dutch marine Natura 2000 sites are counted in international reporting, see: www.clo.nl/indicatoren/nl1425-begrenzing-van-het-natuurnetwerk-en-natura-2000-gebieden

the country developed a strong legal framework for spatial planning that initially separated protected areas and other land uses. Following poor conservation outcomes due to further ecological fragmentation, plans for a national ecological network (NEN) were first presented in 1990 with the aim of developing a spatially coherent network of qualitatively high nature areas through protection, re-creation and active restoration (LNV, 1990). As mentioned, in 2013, the NEN was renamed the Netherlands Nature Network (NNN).

Informed by improving knowledge of the ecological coherence of the planned network (Reijnen *et al.*, 2000) and the ineffectiveness of implemented measures to increase it (Beentjes and Koopman, 2000), in 2000 the Ministry of Agriculture, Nature and Food Safety proposed seven so-called 'robust connection zones' to better link up existing large nature areas (LNV, 2000). The zones had four different objectives: (1) to help conserve species with networks at a (inter)national scale, (2) to increase the effectiveness of the domestic network, (3) to reduce the risk of large-scale disturbances and natural disasters (in particular those related to climate change) and (4) to better connect breeding areas of Red Deer (*Cervus elaphus*), in particular in the forested but nutrient-poor Veluwe area in the centre of the country to surrounding natural areas in wetlands. Examples were a connection between the Veluwe and the Oostvaardersplassen area (online Annex NL.1)[50] and connections between wetland areas in the north of the Netherlands and Germany. In the end, 12 robust connection zones were included in the 2004 national spatial planning memorandum (Government of the Netherlands 2004).

On the back of the robust connection zones initiative in 2000, a multiannual investment programme for defragmentation was adopted in 2005, which ended in 2021. As part of this programme, the national government, ProRail (responsible for the railway system) and the Provinces worked together to resolve ecological bottlenecks, often in dialogue with municipalities, regional water authorities and nature conservation organisations. In total, more than 500 measures were taken in 176 sites to reconnect natural areas in the Netherlands, such as by creating wildlife passages or tunnels based on a thorough analysis of ecological coherence constraints (MJPO, 2020).

As explained in Section 23.3.2, the first Rutte government (2010–2012) abandoned the robust connection zones as part of a broader downsizing of the ecological network. However, the second Rutte government overturned this decision, and the 2013 *Nature Pact* included a commitment to 'at least a doubling of the task for acquisition, recreation and restoration of natural connection zones'. However, as decision making on implementing the NNN was delegated to the Provinces, some of the initial national plans were kept whilst others were abandoned. For example, the intended connection between the Oostvaardersplassen and Veluwe was abandoned after a successful legal challenge by farmers against the Province of Flevoland.

Whereas the original NEN was intended to measure 7285 km^2, following the 2013 *Nature Pact* the Provinces 'recalibrated' the new designation, which resulted in a reduction of 2% to 6 950 km^2.[51] In the *Nature Pact*, the Provinces committed to develop 800 km^2 of new natural areas in the NNN by 2027, compared to a 2011 baseline. Since then, the Provinces have committed to an additional 695 km^2 in their own plans, so actually an increase of more than 4% over that initially planned. At the end of 2020, the NNN consisted of 6 982 km^2 of nature areas – 95% of the foreseen area.

While significant progress has been made in development of the NNN, its pace has been too slow, so far, to achieve the 2027 target of an additional 800 km^2 of nature. The latest progress report indicates that, by the end of 2021, in total 436 km^2 had been developed, leaving 43% still pending (IPO and LNV, 2022). The latest evaluation of the *Nature Pact* (Folkert *et al.*, 2020) confirms that progress on the NNN has stalled, mainly because mostly agricultural landowners are not sufficiently incentivised by the compensation payments offered. In addition, the assessment estimated that fully completing the NNN would only achieve 65% of conservation objectives under the EU Nature Directives. It advised that additional restoration measures would be required in particular to mitigate the effect of nitrogen emissions and acidification, as well as changes in land management around the network to mitigate the impacts of agriculture on sites (e.g. from drainage and ammonia emissions).

23.5.3 Approaches to Protection and Management

The Netherlands takes a habitats-based approach to conservation, closely related to implementation of the EU Nature Directives, which are transposed in the Nature Conservation Act.

Natura 2000 sites, not as they were intended in the Habitats Directive (and unlike reality in the Netherlands), are presented by the Dutch authorities as a 'network of strictly protected natural areas'. Many of the provisions in the Nature Conservation Act that protect biodiversity in Natura 2000 sites do not apply to the other sites in the NNN (including National Parks), which jointly represent over half of the network. Nevertheless, NNN sites are spatially protected through the national decree on general spatial planning rules,[52] which stipulates that Provinces must ensure that the NNN is spatially designated, including a description of its 'essential characteristics and values', and set rules to 'protect, maintain and develop' it in the preparation of zoning plans (at different levels) and

[50] All annexes are free to access online at CUP: www.cambridge.org/natureconservation

[51] Web-page of the Netherlands Environmental Data Compendium on the delineation of the NEN and Natura 2000 (last update 26 September 2017): www.clo.nl/indicatoren/nl1425-begrenzing-van-het-natuurnetwerk-en-natura-2000-gebieden

[52] Besluit algemene regels ruimtelijke ordening of 22 August 2011 (Barro). Available at: https://wetten.overheid.nl/BWBR0030378/2020-12-01

environmental permitting. These rules can only be deviated from under a similar set of conditions as under the EU Habitats Directive that: (1) there should be a significant overriding public interest; (2) there are no acceptable alternatives; and (3) negative effects on the essential characteristics and values, surface and coherence should be limited and remaining effects equally compensated.

The most significant difference between Natura 2000 and other NNN sites is that the latter have no additional protection against external pressures such as nitrogen pollution, drainage or negative effects of recreation. While nature managers in NNN sites are eligible for nature management subsidies (SNL), as in Natura 2000, another major difference is that there is no obligation for site-specific conservation objectives, management planning, monitoring and reporting beyond what is required under the SNL scheme providing land management.

Currently the only strict nature reserves[53] in the Netherlands are the so-called 'forest reserves' ('*bosreservaat*'), a designation that originated in 1978 but was only formalised in 1987. The main intention of the reserves is to obtain long-term scientific evidence on natural processes in Dutch forests and how they could contribute to more biodiverse and natural forests. Although 60 forest reserves have been designated (the last in 2000) they make up less than 30 km^2, a fraction of the Netherlands' terrestrial surface. At sea, the only strictly protected reserve is a similar 73 km^2 reference area in the Wadden Sea, in which (relatively) undisturbed developments are monitored for scientific purposes.

Conservation objectives for Natura 2000 sites have been set at a national level, and for each site based on the national objectives and the potential of the site. Site objectives are published in designation decisions that are open for public consultation. Objectives as set in the decisions are elaborated in time, size and place in the Natura 2000 management plans (LNV, 2006).

In terms of management planning, the 2010 budget cuts of nearly 70% in combination with the decentralisation process created significant uncertainty and discussion on new responsibilities and resources to implement them. In this same period, the national government was also developing a Programmatic Approach to Nitrogen (PAN) to deal with excess nitrogen deposition in Natura 2000 sites, which was only concluded in 2015 (see Section 23.6.2). As the Netherlands wanted to include measures defined under the PAN in the management plans of relevant Natura 2000 sites, a significant alignment burden was added to the management planning process. Despite these two delaying factors, today management plans are in place for all sites except four marine sites.[54] Progress was aided by a legal deadline in the Dutch Nature Conservation Act that requires management plans to be in place within three years of designation.

The largest share of protected areas is managed by specialised nature management organisations: mostly the State Forest Service (*Staatsbosbeheer*, or SBB),[55] the NGO Natuurmonumenten and their regional counterparts in each of the 12 Provinces.[56] Jointly they manage 4 730 km^2, of which almost 93% is located within the NNN and makes up almost 58% of the terrestrial network. As a result, these three organisations play a critical role in the achievement of nature policy for which the Provinces are responsible. The nature management organisations also manage 250 km^2 of national water bodies that the central government is responsible for.

The main tool supporting and steering nature management in protected areas is the Subsidy system for Nature and Landscape (SNL), through which Provinces set conditions for investment in nature management, agricultural nature management and landscape management for periods of six years. In addition, beneficiaries can apply for one-off investment subsidies for ecological restoration, for example for farmers transitioning from conventional to nature management, which often requires structural changes such as in relation to water management. Subsidies are only available for certified nature management organisations that are managing a set minimum area, certified collectives of (other) private landowners not meeting the set minimum area standard, and certified collectives of farmers for agricultural nature management.

At the core of the system is a national Index for Nature and Landscape that sets uniform standards for 53 different natural and seminatural habitat types, 13 landscape element types, and 18 agricultural nature management types. For example, for each habitat type, the index provides in one place a general introduction; a definition of how it differs from other habitat types; representative areas where the habitat is present in the Netherlands; standard costings for management measures; subsidy conditions and tariffs; and status monitoring indicators. It has helped grow a common terminology used by all parties in the field and facilitate coordination between nature managers and nature authorities.

Following the Index, Provinces develop nature management plans for their territory with maps showing the current

53 Following the definition of strict nature reserves in the IUCN guidelines for applying protected area management categories (Dudley, 2008).

54 Bij12 (2021) Web-page 'Natura 2000 beheerplannen'. www.bij12.nl/onderwerpen/natuur-en-landschap/natura-2000-beheerplannen/

55 Since 1997, SBB has been a 'legal entity with a statutory task' (Dutch: rechtspersoon met wettelijke taak (RWT)), which is a not a government organisation in the strict sense of a public-law entity, however it does perform a government task.

56 These provincial spin-offs were created in the 1920s and 1930s and were called '*Landschap*' (landscape). In 1989, they formally joined forces to form an umbrella foundation (*Stichting Unie van Provinciale Landschappen*), which in 2016 merged with the Dutch network of private regional nature management organisations (*Landschapsbeheer Nederland*) into its current form (*LandschappenNL*). The organisation today owns 13% of nature in the Netherlands and manages 1 141 km^2 including with nearly 80 000 volunteers and supported by 300 000 paying members.

situation, where land managers can apply for subsidies, and what kind of objectives must be achieved in these places. Two important conditions for applicants for subsidies are the obligatory agreement with the Province of a quality assurance plan and an annually updated management plan developed in close cooperation with involved land managers and other stakeholders in the NNN, such as water boards and nature NGOs. Progress on ecological outcomes is evaluated annually between the Province and nature manager, on which basis annual management plans are updated. Unlike previous systems, this allows the Provinces to focus on outcomes, whereas in the past monitoring and evaluation was mostly based on effort/participation. While this is an important step forward, the lack of systematic monitoring of the responses of biodiversity to the management remains an important limitation on the effectiveness of the new approach.

23.6 General Nature Conservation Measures

23.6.1 Habitat Management for Meadow Birds

As described in Section 23.1, the extensive traditionally managed lowland wet pastures and meadows of the Netherlands have been a particularly important terrestrial habitat for biodiversity over the last century or so, especially for breeding 'meadow birds'. These include waders, the most important of which is the Black-tailed Godwit, as well as Oystercatcher, Redshank, Lapwing, Common Snipe (*Gallinago gallinago*) and Ruff (*Calidris pugnax*), ducks such as the Garganey (*Spatula querquedula*), and passerines such as the Yellow Wagtail (*Motacilla flava*).

Since their peak in the 1960s, the populations of these birds have declined considerably, with between 3.3 and 5.7 million breeding pairs of 27 species of (once) common meadow birds lost in Dutch agricultural areas: a decline of 61–73% (RLI, 2013). Despite protection measures, the Black-tailed Godwit population shrank on average 4% per year between 1990 and 2019 to 26 000–33 000 pairs in 2018–2020 (in comparison in 1975 there were still around 120 000 pairs). Some species such as Ruff and Common Snipe have almost disappeared, and the Skylark is now rare in lowland grasslands.

The primary drivers of the declines have been increased drainage and agricultural intensification (e.g. Kleijn and Lammertsma, 2013) including increased use of fertilisers, which has led to faster growing and denser grass swards. This, in turn, has allowed higher grazing intensity and/or earlier and more frequent mowing. These agricultural changes have reduced the extent of breeding habitat. Where the birds do attempt to breed, they are subject to high rates of nest loss and poor habitat for the chicks to feed and hide from predators in, resulting in low chick survival, which is the main cause of the decline in the Black-tailed Godwit (Schekkerman and Müskens, 2000). Additional factors that may have contributed to the declines in meadow birds are: climate change, which has resulted in earlier and faster growing vegetation and earlier mowing (Kleijn *et al.*, 2010); high rates of predation, which are exacerbated by the agricultural changes (Schekkerman *et al.*, 2009) and trees and other structures that can harbour predators (Berg, 1992); and high levels of human disturbance (Holm and Laursen, 2009).

The Dutch government has had a policy since 1975 to protect farmland birds, and has primarily attempted to achieve this through CAP-funded agri-environment schemes. These have followed two main approaches: subsidised nest protection by volunteers, and payments for farmers to delay mowing and other agricultural operations. The former marks nests so that they can be avoided by farmers when carrying out operations such as mowing, or places nest protectors on them on grazed fields to reduce the chances of them being trampled. Whilst these have been found to be effective in reducing clutch mortality rates (Kleijn and Lammertsma, 2013), especially from losses due to mowing or slurry application, the benefits from reduced trampling or losses due to fertiliser applications may be outweighed by losses from predation by both bird and mammal species after nests have been visited (Goedhart *et al.*, 2010).

Restrictions on the first date of agricultural operations, which were included in the very first agri-environment schemes, were initially to reduce the losses of clutches. However, since about 2000, the main objective has been to improve chick survival (as the main factor affecting breeding success) by keeping more tall vegetation in the landscape, as this provides better feeding habitat and cover from predators.

Despite the agri-environment schemes, the population declines have continued as numerous monitoring and research studies have shown that the schemes have been insufficiently effective in relation to most farmland birds, as well as other species such as plants (RLI, 2013; ECORYS Nederland BV *et al.*, 2016). The main problems have been that agri-environment schemes at farm level have been at insufficient scale and often ecologically isolated (between farms with the same measures, as well as between farms and neighbouring natural areas), inadequately addressing the need for reduced habitat fragmentation and landscape-scale impacts. The agri-environment schemes have also focussed too much on delaying cutting, when measures to improve water levels and reduce the amounts of fertiliser used are also necessary (Breeuwer *et al.*, 2009). An important compounding factor has been a lack of continuity, as farmers frequently abandoned nature management due to practical difficulties (e.g. concerning weed control), combining nature management with business activities, poor remuneration, bureaucracy and the sale of their land (Wiertz *et al.*, 2007; PBL, 2009).

Another study demonstrated that agricultural nature management achieved significantly lower conservation results, and at higher costs, than conventional nature management (Kleijn, 2012). Furthermore, as the Black-tailed Godwit populations declined, they have increasingly congregated in favoured areas – usually in and around protected areas targeted at meadow birds (Kleijn and Lammertsma, 2013). Whilst the area of protected areas or nature reserves targeted at meadow birds has been substantial (Van Paassen and Teunissen, 2010) they have been unable to halt the national decline in meadow birds. In this respect, it is important to point out that not all protected areas are under optimal management nor do they all have increasing, or even stable, meadow bird populations. For some this is due to their small size and isolation, which sometimes makes it impossible to raise water levels to meet the requirements of the birds without having significant negative impacts on neighbouring agricultural fields (Kleijn and Lammertsma, 2013). In many cases it is also because potentially suitable protected areas for meadow birds do not have site-specific management objectives set for meadow birds, even in Natura 2000 sites. Therefore the conservation action taken depends entirely on the willingness of the site manager. Since meadow bird conservation is known to be challenging, and therefore costly, site managers often prioritise cheaper management options such as extensive grazing over the maintenance of the necessary wet, flower-rich meadows. More effective nature conservation measures are therefore required both in protected areas and in the wider countryside.

In 2013, the Council for the Environment and Infrastructure (RLI, 2013) presented advice on improving the effectiveness of nature conservation in the Netherlands, which played an important role in changing the direction of policy thinking. Drawing on previous studies, the RLI study made the following recommendations of relevance to meadow birds and other agricultural species:

- Prioritise agricultural nature management in zones alongside protected areas with the same objectives, to foster landscape effects.
- Concentrate agricultural nature management within large, contiguous areas in which the appropriate conditions for meadow and other farmland birds are already in place, or can be re-established.
- Prioritise agricultural nature management to increase the 'permeability' of the agricultural land between nature areas, thus facilitating the interaction between subpopulations of vulnerable species.
- Make significant modifications to the current agricultural nature management activities (such as meadow bird management) in order to enhance their effectiveness.
- Enhance management measures, such that, for example, in important areas for meadow birds they go beyond adjusting mowing times and also address ground water level and the application of fertilisers.
- Withdraw financial support from ineffective agricultural nature management.
- Increase land-users' willingness to take part in agricultural nature management by substantially increasing the minimum duration of the management contracts (provided this will not have the opposite effect).

The Council also suggested working with collectives of land managers on agri-environment schemes in areas with pooled parcels of land and in places with higher conservation potential, and strengthening cooperation and networking opportunities.

To address the weaknesses in the agri-environment schemes, a new scheme was introduced in 2016[57] that focusses on high nature value areas instead of individual farms and aims to stimulate cooperation between farmers in these areas and be more results-driven. Anticipating the new scheme, Dutch farmers established 40 new agri-environment cooperatives, covering the entire countryside. These built on cooperatives that were founded in the 1990s for biodiversity conservation activities, inspired by German and Danish examples.

According to the Ministry, the scheme's core elements are:

- its landscape-scale approach, with conservation measures at habitat level instead of field level;
- the intermediary position for the cooperative as a final beneficiary, allowing increased flexibility at farm and field level with regard to measures and payments;
- the provision of ecological guidance in all relevant phases (i.e. covering management plans, individual contracts and actual activities in the field);
- its tailor-made controls, offering flexibility and the possibility to change the management approach from year to year, and even last-minute changes in the course of one season; and
- increased stakeholder involvement.

However, whilst the new scheme is considered to be an improvement, it is expected to be insufficient to turn around the negative trends in meadow bird populations (Melman *et al.*, 2016). This is because the overall area of land that is expected to be under management and its geographical distribution does not include all core meadow bird breeding areas. As discussed earlier in Section 23.4.2, as a result of this and other weaknesses in the national government's 2016 plan to stabilise meadow bird populations (Trouwborst, 2016), Vogelbescherming submitted a formal complaint to the European Commission against the Netherlands for breaching its commitments under the Birds Directive.

In response to a request from the Dutch Parliament for a more concrete and stronger plan, in March 2017 the government updated the Parliament on its progress, indicating that it had requested the Provinces, Vogelbescherming and BoerenNatuur (the umbrella organisation of the 40 agricultural cooperatives) to agree on a common plan. Following this request, the consortium commissioned Wageningen University and Sovon to model the consequences of implementing four conservation management scenarios for the Black-tailed Godwit (with some additional measures for Lapwing), which would also be expected to benefit other meadow birds (Melman and Sierdsema, 2017). The results indicated that optimisation of current management by reducing

[57] *Agrarisch Natuur- en Landschapsbeheer.*

overgrowth of vegetation and soil impoverishment in protected areas, combined with additional significant but achievable one-off measures (in particular by reducing drainage of both farmland under agricultural management as well as protected areas) could increase the Black-tailed Godwit population in the Netherlands by more than 40%, up to 48 000 pairs.

Based on the preliminary findings of the scenario assessment, the Ministry acknowledged in its letter to Parliament that unforeseen additional measures as well as different management approaches would be required. In particular, management would need to focus on more area-based processes to increase participation in agricultural nature management by farmers, as well as motivated and trained volunteers. Moreover, the Ministry warned that such area-based processes, in particular where measures adjust water levels, are time-consuming because of their impact on individual farmers, agricultural nature management cooperatives, land managers and water boards. The Ministry also indicated that both the (national) government and Provinces are supporting site-based management approaches among others through financing 'improvement plans' for agricultural nature management by cooperatives, the establishment of model field sites to exchange knowledge between cooperatives, research to support predation control, improved monitoring schemes for Black-tailed Godwit and Lapwing, and innovative technologies to support protection such as the use of drones to find and monitor nests.

A recent interim evaluation of the new Dutch agricultural nature and landscape management scheme (Boonstra *et al.*, 2021) concluded that the feasibility and effectiveness of the scheme were improved compared to the previous scheme, and that it slightly improved the ecological conditions for achieving the objectives of the EU Nature and Water Framework Directives. Identified success factors are the area-based and collective approach, and the greater flexibility of the new scheme that allows adaptation to local needs and conditions. As regards potential improvements, the evaluation pointed to the need for better articulation of roles and responsibilities between key participants (collectives, Provinces and NGOs), further simplification to lower the administrative burden to (potential) participants, a greater focus on priority species, and higher standards in management planning among the collectives.

However, even if these improvements are all taken forward, the evaluation also clearly states that much more far-reaching measures are needed to improve environmental conditions in the Dutch countryside. These include large-scale grassland rehabilitation, land swaps and consolidation around core breeding areas, predation control, complementary eco-schemes under the new CAP, and the development of (other) more significant policy and financial instruments to accelerate the transition to more nature-inclusive forms of farming. The findings underline the urgency expressed by the Dutch Court of Auditors in 2021, which severely criticised successive governments over the last 20 years for ignoring progressive evidence on the low effectiveness of increasingly well-funded agri-environment schemes for meadow birds (by maintaining the eligibility of low-impact measures such as nest protection) and the low share of farmland important for meadow birds un the schemes (Algemene Rekenkamer, 2021).

23.6.2 The Programmatic Approach to Dealing w Nitrogen Pollution

As outlined in Section 23.2.3, diffuse nitrogen pollution, mai from livestock farming, has been a significant pressure on e systems in the Netherlands for many decades. Critical loads nitrogen deposition have been exceeded in most natural are hampering the achievement of conservation objectives nitrogen-sensitive habitats in 118 of the 161 Natura 2000 si (Ministerie van Economische Zaken, 2015). While Dutch da farming has significantly reduced nitrogen emissions per litre milk produced, permitting procedures for further econon development became an issue in the early 2000s. In additic the phasing out of the EU milk quota between 2009 and 20 encouraged many dairy farmers to plan increases in productio Alarmed by the potential further conflict between econom production and nature objectives, in 2010 the governme decided to develop a national Programmatic Approach Nitrogen (PAS), which was adopted in 2015.

The overall approach of the PAS was to simultaneous achieve three objectives: (1) provide an opportunity for furth economic growth (and corresponding emissions); (2) tal remediation measures in Natura 2000 sites to remove exce nitrogen from historic pollution; and (3) further reduce emi sion reductions on farms, for example by renovating stock yards, changes to animal feed composition, and changes t grassland fertilisation and grazing/mowing management.

Shortly after the PAS entered into force, various enviror mental NGOs took the Provincial authorities to court for bein in breach of HD Article 6. They argued that, by grantin permits to livestock farms to expand their operations aroun Natura 2000 sites with known exceedance of critical depositio levels, the authorities would knowingly allow further deterior ation before the effect of remediation measures could b proven. After a critical preliminary ruling by the CJEU ir November 2018,[58] the Dutch State Council in May 2019 rule that the permitting under the PAS was indeed incompatibl with requirements under the Habitats Directive.

This led to an instant freeze on permitting of all potentiall nitrogen-polluting projects, not only in farming but also ir other sectors such as building and transport. It caused what is likely the largest political crisis over an environmental issue in the history of the Netherlands to date, triggering emergency measures to unlock economic development.[59] An independent advisory committee (the Remkes Committee) was set up, which

[58] Judgment of the Court (Second Chamber) of 7 November 2018 in Joined Cases C-293/17 and C-294/17: https://curia.europa.eu/juris/document/document.jsf?docid=207424&doclang=EN

[59] See for example the Reuters article of 13 November 2019 '*Netherlands cuts speed limit to reduce nitrogen pollution*'. www.reuters.com/article/us-netherlands-construction-idUSKBN1XN13K

recommended in June 2020 more ambitious and stringent measures to reduce pollution at source. This appears to have been taken to heart, with the entering into force in July 2021 of a dedicated legal act on Nitrogen Reduction and Nature Improvement,[60] a dedicated €25 billion transition fund to 2035 to support it,[61] and a dedicated minister for nature and nitrogen in the fourth Rutte government. However, NGOs and scientists have pointed out that the targets set for emissions reductions are not ambitious enough to achieve favourable conservation status of nitrogen-sensitive HD habitats,[62] and more importantly the approach taken by authorities to achieve them is built on the same basis that resulted in the 2019 PAS annulment.[63] In addition, environmental NGOs backed up by scientific evidence have warned about the negative impacts on nature of too frequent restoration management necessary to counteract current pollution levels, for example on seed banks (e.g. Bobbink, 2021). After nitrogen became an item in the latest national elections, the Coalition Agreement of the fourth Rutte government published in December 2021 raised the ambitions of the law on Nitrogen Reduction and Nature improvement to the level of the recommendation of the Remkes Committee.

23.6.3 Flood Protection in Combination with River Restoration

Several Dutch national programmes and projects combining flood risk reduction and nature restoration have over the last 25 years rehabilitated roughly 12 000 ha of floodplain previously under conventional agricultural use (LNV *et al.*, 2021). The three most significant programmes have been the 1993 NURG programme (*Nadere Uitwerking Rivierengebied*), the 2005 Grensmaas/Zandmaas project (upper Meuse basin) and the 2006 Room for the River decision (*Ruimte voor de Rivier*) (Rhine basin and lower Meuse basin).

A key trigger for the programmes was a 1986 vision document on the Dutch river system (*Plan Ooievaar*) that led to a rapid paradigm shift in river management: giving more space for nature, floods and mineral extraction in floodplains, and helping floodplain farmers to retire early or relocate (De Bruin *et al.*, 1987). In 1991, a new national policy approach was adopted in line with this new vision, which in 1993 resulted in the NURG agreement under which the ministries of Agriculture, Nature Management and Fisheries and of Transport and Water Management agreed to jointly finance over 50 projects restoring 7 000 ha of floodplain by 2015,[64] in total investing over €350 million between 1993 and 2020.

Critical flood events in 1993 and 1995 focussed political minds on scaling up implementation of flood risk reduction measures, ultimately resulting in the €2.3 billion Room for the River decision. It included from the start the dual objective of reducing flood risks and improving environmental quality. Its final evaluation concluded that the quality dimension ensured local support for some quite far-reaching projects, such as dike relocations, long after the immediate interest in flood safety that triggered both projects in 1995 had waned (Folkert and Boonstra, 2017). The similar €1.5 billion Grensmaas/Zandmaas project established in the same period had specific key targets to develop a combined area of at least 1 570 ha of riverine nature. The large-scale rehabilitation and floodplain deepening were found to have prevented floods on the Meuse river during the extreme high waters in July 2021 (Sweco, 2021), while on smaller rivers in the wider Meuse basin over 50 000 people were evacuated and flood damage was estimated to be €350–600 million. The evaluation therefore recommended scaling up the river restoration actions to the Meuse headwaters as a nature-based solution.

A scientific assessment of the biodiversity impacts of the flood risk reduction measures in the Rhine tributaries (Straatsma *et al.*, 2017) found that of all 179 fluvial floodplain sections examined, 137 showed an increase in biodiversity, particularly for fast-spreading species. Pioneer vegetation types on riverbanks, grasslands and meadows, and tall herb fringe communities have benefited most. The largest increases were in bird and mammal species, such as Spoonbill (*Platalea leucorodia*), Corncrake (*Crex crex*), Bluethroat (*Luscinia svecica*) and Beaver, as well as various migratory fish species. Other benefits included increases in insect abundance and diversity, for example in butterflies of grasslands and tall herbs, and dragonflies of flowing water. However, while floodplain forests also increased in extent, many were removed before reaching maturity because they obstructed the flow of water and nullified the restoration projects' intended flood risk reduction benefits. This has hindered the expected return of forest-breeding bird species such as the Black Stork (*Ciconia nigra*), after which *Plan Ooievaar* was named.

23.7 Nature Conservation Costs, Benefits and Funding Sources

23.7.1 Nature Conservation Costs

The gross costs of terrestrial nature conservation measures in the Netherlands between 1999 and 2019 have been on average

[60] Wet van 10 maart 2021 tot wijziging van de Wet natuurbescherming en de Omgevingswet (stikstofreductie en natuurverbetering).

[61] Announced in the government coalition agreement 2021–2025 '*Omzien naar elkaar, vooruitkijken naar de toekomst*'. www.tweedekamer.nl/sites/default/files/atoms/files/coalitieakkoord-2021-2025.pdf

[62] Vlinderstichting and Soorten NL persbericht van 11 April 2021 '*Meer stikstofreductie nodig om natuur te beschermen*', supported by a dedicated study by Van den Burg *et al.* (2021). www.naturetoday.com/intl/nl/nature-reports/message/?msg=27552

[63] Press release Soorten NL of 24 December 2021 '*Het is weer Groundhog Day op het stikstofdossier*'. www.naturetoday.com/intl/nl/nature-reports/message/?msg=28612

[64] The NURG agreement was changed several times and ultimately the target was reduced to 6 700 ha by 2020.

€1 027 million per year.[65] In 2019, these costs were €1 284, of which €616 million or 48% were for nature rehabilitation and management measures, €484 million or 38% were overhead costs (salaries and material), €96 million or 7% were for the acquisition of land for nature rehabilitation, and the remaining 7% for other costs such as mitigation measures, education, and awareness-raising. Of these, the state covered €847 million; €437 million were met by private sector stakeholders. Private investment originated from nature management organisations (74%), farming (19%), water extraction (3%) and other private sectors (4%), leaving a funding gap of €232 million in 2019 or 18%.

23.7.2 The Economic Benefits of Nature Conservation

Since 2010, the Netherlands has increased its understanding of ecosystem services, revealing that nature areas provide both the largest diversity of services and the largest net value in demanded services while representing a smaller area of the country than agricultural areas and cities.[66] No clear trend information since 2010 is publicly available, but trends since 2000 show both improvements (especially in services related to natural areas) and further declines (in services related to agricultural areas).

As part of the development of the Netherlands' national natural capital accounts since 2016, Statistics Netherlands and Wageningen University have attempted to calculate the monetary value of 10 terrestrial ecosystem services in the Netherlands following the international statistical guidelines developed by the UN *et al.* (2014) on the System of Environmental-Economic Accounting: Experimental Ecosystem Accounting (SEEA EEA) (Horlings *et al.*, 2020). One of the most prominent findings was that the cultural values of nature recreation and nature tourism were more than double those of crop and feed production (mostly for dairy) for which the Netherlands is known, even using a conservative estimate of tourism and recreation benefits.

23.7.3 Funding Sources for Nature Conservation

As shown in Table 23.5, the Netherlands largely relies on national funding sources, both for one-off investments and ongoing management subsidies. Nonetheless, co-funding from the European Agricultureal Fund for Rural Development (EAFRD) Rural Development Programme measures has contributed significantly to nature management (LNV, 2021a). In particular, the agri-environment–climate measure (M10) has supported nature management in a broad range of grassland and arable habitats, green infrastructure (e.g. hedges), blue infrastructure (e.g. ponds) and water management (quantitative and qualitative). The non-productive investment measure (M4) has mostly funded hydrological measures in and around Natura 2000 sites, such as rewetting and river/stream restoration. It has also been used for rehabilitation measures for species (e.g. meadow birds) and restoring small landscape elements, in this way adding to and complementing the nature management supported through the agri-environment–climate measure.

The use of other EU funds is more modest compared to the overall portfolio; for example, the Netherlands did not use EU maritime and fisheries (EMFF) or research (Horizon) funding for nature conservation purposes over 2014–2020 (LNV, 2021b). New EU LIFE projects directly benefiting habitats and species were funded almost yearly since 1993, mostly for wetland and dune restoration. However, in general, use of LIFE has only provided a relatively small share of total nature conservation investment. The Netherlands does now have two larger LIFE integrated projects underway, which aim to support the implementation of the Prioritised Action Framework for Natura 2000 in coastal and large inland wetland ecosystems[67] and in terrestrial sites in cooperation with agricultural and other land-users.[68]

Notable is the significant investment in nature by the postal code lottery, which provides annual grants to nature organisations as well as funding individual projects. For example, in 2020, both Natuurmonumenten and their Provincial counterpart LandschappenNL both received €13.5 million in grants, and in 2019, Vogelbescherming (BirdLife Netherlands) and Society for Protection of the Wadden Sea (*Waddenvereniging*) received €15 million for a one-off project to restore an intertidal nature area for breeding birds.

Overall, the Netherlands has invested significant national funds in nature conservation compared to some other European countries, and despite significant funding cuts in recent years these have mostly been reinstated. Following the nitrogen crisis, a new significant budget has been added (see Section 23.6.2), which will hopefully be invested in measures to achieve real reductions in pollution at source (e.g. to buy out the most problematic large livestock farms in nitrogen-sensitive areas). This would reduce the necessary frequency of restoration management and thereby significantly reduce the costs of nature conservation. Although the revised system of agricultural nature management appears to have led to a more cost-effective approach, it is still implemented on an insufficient proportion of farmland to be effective at the national level. To convince a greater proportion of farmers to engage in nature conservation, which is critical considering the country's dire

[65] Statistics Netherlands (CBS) (December 2021) Costs and financing; nature- and landscape management. (In Dutch.) https://opendata.cbs.nl/statline/#/CBS/nl/dataset/70213ned/table?dl=380E

[66] Government of the Netherlands (2021) Environmental data compendium indicator page: www.clo.nl/indicatoren/nl1572-goederen-en-diensten-van-ecosystemen-in-nederland-?ond=20879

[67] DELTA Nature: Integrated approach N2000 Delta Nature to catalyse the implementation of the Netherlands Prioritised Action Framework (LIFE15 IPE/NL/000016).

[68] LIFE IP PAF All4Biodiversity: LIFE IP PAF Biodiversity recovery approach for N2000 sites and surroundings, in cooperation with agricultural and other land-users (LIFE19 IPE/NL/000011).

able 23.5 Funding sources for nature conservation in the Netherlands.

ource and period	Measure/category	Average annual allocation (€ m)		Comments
EU funding instruments		**EU + national**	**EU**	
EAFRD Rural Development Programme measures – allocations to Priority 4 (restoring, preserving and enhancing agriculture and forest ecosystems) 2014–2020	M04 Investments in physical assets	51.2	31.7	See text
	M10 Agri-environment–climate	50.1	38.6	See text
RDP Priority 4 subtotal		*101.4*	*70.3*	
Interreg 2016–2018	Resource efficiency, climate adaptation and environment and resources	>8.5	8.5	Assumed similar funding in other years during 2014–2020
LIFE Programme	Nature and biodiversity projects (1992–2021)	5.9	2.7	47 projects
	Integrated projects – environment (2016–2026)	3.1	1.8	Two projects: DELTA Nature and ALL4Biodiversity
European Regional Development Fund (ERDF) 2014–2020	85 Protection and enhancement of biodiversity, nature protection and green infrastructure	>2.3	2.3	Expenditure for N2k etc. at time of PAF
	86 Protection, restoration and sustainable use of Natura 2000	>1.3	1.3	Expenditure for N2k etc. at time of PAF
EU funding instruments subtotal		*>122.4*	*87.0*	
National funds within 2014–2020		**Additional national***		**Comments**
State budget	Recurring nature management subsidies NNN	350		See 1
	One-off investment subsidies NNN	200		
Postal code lottery		46		
Additional national subtotal		*596*		
National total including co-financing of EU instruments		*631.4*		
Total		**>718.4**		

Notes. * Excludes co-financing of EU funds unless indicated. The subtotals and totals are only approximate indications of the average annual levels of funding as they are based on annualised data from differing periods. *Comments:* 1. Until 2021 these subsidies amounted to 75% of the actual costs for the site manager (so overhead excluded). In other words, 25% is paid by the site managers themselves. Excluding the publicly funded State Forest Service (Staatsbosbeheer) it means the remainder of site management is supported by structural private co-funding by Natuurmonumenten, the 12 Provincial Landscape organisations and other, smaller, private nature managers.

Sources. Based on data from: RDP data from European Commission ESIF funds database (23 October 2020); ERDF, Interreg and national funding from 2021 to 2027 PAF (LNV, 2021b); the Netherlands LIFE factsheet, European Commission (updated April 2021). See Section 1.3 for further details.

state of agro-ecosystems and high ecological fragmentation of the landscape, financial incentives will have to become more competitive than today to encourage farmers to either retire early or adapt to more nature-inclusive approaches.

23.8 Surveillance, Monitoring, Reporting and Applied Research

The Netherlands has one of the most extensive nature monitoring approaches, supported by many professional and volunteer observers in the field. In 1999, the Network Ecological Monitoring was established, in which various relevant public authorities work together to better adapt the collection of nature data to meet the information needs of the government. This has increased the focus of monitoring efforts on specific goals as well as the quality of reporting data, which are made centrally and publicly available through the online Environmental Data Compendium.[69] Regular systemic population monitoring in the Netherlands is undertaken for mammals, breeding birds, reptiles, amphibians, butterflies, dragonflies and freshwater fish.

[69] www.clo.nl/

Since 2014, the Knowledge Network for Nature Restoration and Management (OBN) has provided an independent platform on which policymakers, site managers and scientists cooperate on knowledge development and dissemination for higher-quality management and restoration of natural areas in the Netherlands and in the Atlantic biogeographical region. At the core of the network are eight expert groups: one for wildlife management and seven for the most important landscapes present in the Netherlands. This cross-organisational cooperation has been instrumental in identifying and following up on common practical research needs arising from the field. The national and provincial authorities provide an annual budget of €1.6 million, but the OBN can set its own research priorities. To meet research needs arising from (inter)national legal obligations on nature and the environment, LNV has a renewable contract with Wageningen University and Research of around 30 FTE/year.

23.9 Future Developments

After a period of significant changes, characterised by an increasingly legalistic approach to nature conservation focussed on meeting minimal EU legal requirements and technical fixes to mitigate rather than reduce pressures, the nitrogen 'crisis' appears to have been the event that changed the course of nature conservation policy and practice. The fourth Rutte government's Coalition Agreement has included a commitment to not only finish the NNN, but also extend it by another 400 km^2. In addition, the agreement proposed a new national spatial designation between agriculture and nature. These 'transition zones' should ultimately cover 2 400 km^2 and would be located mostly around Natura 2000 sites to help improve their environmental quality, for instance by having more favourable hydrological conditions (e.g. by preventing seepage) and more extensive forms of agriculture to reduce nutrient pollution. Farmers responsible for local peak load emissions of nitrogen that affect Natura 2000 sites are foreseen to be bought out. Whether the planned extensions of the NNN will be sufficient for conservation remains to be seen, as the Netherlands Environmental Assessment Agency (PBL) estimated that the NNN should be expanded by 1 500 km^2 for BHD habitats and species to reach a favourable conservation status (Van Hinsberg *et al.*, 2020). Such an increase would be in line with the new 30% protected area targets for 2030 proposed in the *EU Biodiversity Strategy to 2030* and the Kunming-Montreal Global Biodiversity Framework for 2030. The Dutch government has promised that the Netherlands will prepare an updated NBSAP in 2023.

Another important question that remains is how this management will be integrated with the achievement of other key societal objectives in the broader countryside. In September 2020, the Ministry of the Interior and Kingdom Relations presented a new *Environmental Vision* (NOVI), which will be implemented through a new Environmental Act expected to come into force on 1 January 2024. The Act will combine into one, but not replace, spatial planning provisions from 26 existing acts, including those from the 2015 Nature Conservation Act. The NOVI takes a 2050 perspective and has four overarching priorities: (1) making space for climate adaptation and the energy transition; (2) a circular and competitive economy; (3) improving multifunctional urban planning, with a focus on better accessibility between cities; and (4) a future-proof rural area in balance with nature and landscape (e.g. through circular agriculture that is better attuned to natural water and soil systems). In particular, the first and fourth priorities, if implemented at a level that matches international commitments (e.g. mitigating greenhouse gas emissions), will have important repercussions for nature conservation in the Netherlands. For instance, reductions in the livestock herd would not only greatly reduce the country's climate, land and water footprints, they would also significantly reduce nitrogen emissions to Dutch nature. Similarly, efforts to rewet peat soils would not only reduce land-based climate emissions but also significantly improve hydrological conditions in lowland peat marshes and improve habitat quality for meadow birds.

Some important factors will make or break the success of such larger integrated sustainability transitions. The first will be the capacity for traditionally opposing interests in natural resource management in the Netherlands to agree on new directions of travel, in particular between nature conservation and farming. Recent surveys (e.g. the *Trouw* newspaper in 2018[70]) showed that farmers are willing, if better supported from the public and market sector, to break current lock-ins. The nitrogen crisis has put this willingness under strain and revived old debates on land sparing and sharing. Some voices are calling for a National Agricultural Network like the NNN in which productive agriculture would be protected against other use functions and allowed to further intensify to reduce land demand (Veerman and Fresco, 2019, cited in van der Boon and Knoop, 2019). Other recent visions call for a stronger zoning of farming too, with much more extensive farming around nature sites, especially on currently drained peat soils and on nitrogen-sensitive soils (Bakker *et al.*, 2021).

A second factor will be one of governance. As the nitrogen crisis showed, the Netherlands have large unsolved national nature conservation and other environmental challenges that impact the root of economic development in the country and pose large planning challenges for the near and more distant future. However, as both nature and agriculture policy are largely delegated to the twelve Provinces, the question is to what extent their common actions will be sufficient. Under the NOVI, a new *National Programme for Rural Areas*[71] has been

[70] Web page with main findings by Geelen Consultancy which executed the 2018 survey 'De Staat van de Boer 1' for newspaper *Trouw*. (In Dutch, 24 April 2021.) www.geelen-consultancy.nl/nieuws/destaatvandeboer/

[71] *Nationaal Programma Landelijk Gebied* (NLPG).

set up in which the national government will set binding targets for the 12 Provinces that are measurable and can be totalled to meet (inter-) national interests. The Provinces then take full responsibility to meet these targets through provincial planning and area-based approaches. The targets set are not negotiable: if a Province will not or cannot deliver on time, the national government will step in and take over. If fully implemented, this new programme would help increase accountability and a fair contribution between Provinces and ensure that cross-border challenges such as nature conservation are not overlooked in regional and local decision making.

23.10 Conclusions

Nature conservation in the Netherlands is a well-established policy and practice, and has been for some decades. The high demands on nature in the densely populated and intensively used delta have meant that conservationists have had to be creative in finding a balance between nature sparing and sharing approaches. This has resulted in some innovative approaches from a European perspective, for example in the adoption of a policy to develop an ecologically coherent national nature network (ecological network), various initiatives to encourage agricultural nature conservation management, and large-scale river restoration in partnership with extractive industries – all of which have inspired similar initiatives in other European countries.

Today, the Netherlands has a protected nature network that spans perhaps as much as 26% of its terrestrial and inland water surface, anchored in a national conservation law and spatial planning policies. Nature conservation is also supported through significant investment, which is mostly from national sources, unlike most other EU countries. This helps fund its substantial professional nature conservation sector and high-quality biodiversity monitoring data that inform evidence-based approaches, with substantial volunteer inputs coordinated by professionals. Jointly, they have helped the recovery of key habitats of international importance such as lowland peatlands, and the return of populations of numerous species including flagship species such as Common Crane (*Grus grus*), White-tailed Sea-eagle (*Haliaeetus albicilla*) and Otter. Biodiversity trends in protected areas now show signs of recovery, whereas outside of these areas biodiversity overall keeps falling – especially on and around farmland – demonstrating what would likely have happened had conservation action not been taken.

Despite these conservation investments, actions and some successes, biodiversity trends still fail to show significant improvements overall, even in the conservation status of protected habitats and species. Some of the main barriers to further successes remain historic degradation (e.g. of soils through acidification and water systems through a wide range of physical modifications); high levels of spatial fragmentation of remaining natural and semi-natural habitats; a range of agricultural pressures and threats both direct (e.g. intensive grassland management) and indirect (e.g. drainage of wetlands); high reactive nitrogen loads from combustion (mostly transport) and livestock; unsustainable fisheries in the marine environment; high levels of disturbance; and increasingly extreme weather events, in particular droughts. Notwithstanding decades of evidence and policy deliberation, the Netherlands still has not found a sustainable balance between nature conservation and its food production and wider economic objectives. This was demonstrated by the successful legal challenge over the Dutch Programmatic Approach to tackling nitrogen pollution. Similar examples have been the country's decades-long struggle to put in place a workable manure policy and meet commitments under the EU Nitrates Directive; the failure to protect farmland birds at a country level; and decision makers' reluctance to transition beyond the highly unsustainable dairy farming on drained lowland peat soils.

While recent years have seen no shortage of attention to biodiversity among the public and stakeholders, not least triggered by the 'nitrogen crisis', it remains to be seen if it will result in a significant change to the status quo on most of the aforementioned issues. Unlike the European Commission, which made climate and biodiversity action two central pillars of its new Green Deal growth strategy, the Dutch government agreement for 2021–2025 (VVD *et al.*, 2021) presents nature under the same heading as agriculture and nitrogen, and emphasises how the country's 'world-renowned' agricultural sector should be given future growth opportunities. Nature conservation is still mostly presented as something undertaken to ensure compliance with EU laws, rather than ensuring the health and well-being of Dutch citizens and their environment. The new Spatial Planning Law is presented as an important tool for more integrated spatial policy including nature objectives, implemented by the first national minister for spatial planning since 2010. In reality, housing development appears to be the overriding national priority. Nature conservation in the terrestrial environment is still largely left to provincial authorities, which, without national support on supraregional goals, investment and new legal instruments, will likely continue to experience challenges to complete the NNN and implement ambitious programmes to increase conservation effectiveness. Without a more focussed, coherent and coordinated strategy at national level, both for protected areas and agricultural nature management, the Netherlands will likely continue to make only slow progress in conserving its natural heritage.

Acknowledgements

We thank the following reviewers for providing helpful suggestions for improvement of the chapter: David Kleijn (Wageningen University & Research), Patrick Nuvelstijn (Vereniging Natuurmonumenten) and Paul Opdam (Emeritus Professor Landscape in Spatial Planning, Wageningen University & Research). In addition, we are grateful for the support provided in identifying relevant contacts and data by Annemiek Adams (Ministry for Agriculture, Nature and Food Quality), Frank Berendse (Professor Emeritus in Nature Conservation and Plant Ecology, Wageningen University & Research) and Wouter Helmer (co-founder and former director at ARK Nature and Rewilding Europe).

References

Algemene Rekenkamer (2021) *Waar is de grutto? Aanpak bescherming weidevogels werkt niet.* The Hague, December 2021.

Bakker, M. M., de Vries, W., Ros, G. *et al.* (2021) *Zoneren biedt landbouw toekomstperspectief.* Tijdschrift Milieu, April 2021, pp. 39–44, Utrecht, the Netherlands: VVM Netwerk van Milieuprofessionals.

Beentjes, R. A. & Koopman, J. C. M. (2000) *Kloppende aders. Een impuls aan de realisatie van Ecologische Verbindingszones in Nederland.* Rapport Projectgroep Ecologische Verbindingszones, Ministerie van Landbouw, Natuurbeheer en Visserij (LNV). The Hague.

Berg, A. (1992) Factors affecting nest-site choice and reproductive success of Curlews *Numenius arquata* on farmland. *Ibis*, **134**, 44–51.

Besse-Lototsaka, A. A., Geertsema, W., Griffioen, A., Veen, M. van der & Verdonschot, P. F. M. (2010) *Natuurdoelen en klimaatverandering.* Wageningen, the Netherlands: Wageningen Alterra, Alterra-rapport 2135.

BirdLife, EEB, FoEE & WWF (2018) *The State of Implementation of the Birds and Habitats Directives in the EU. Nature Scorecard The Netherlands.* Brussels: BirdLife International Europe and Central Asia, European Environmental Bureau (EEB), Friends of the Earth Europe (FoEE), WWF Europe. www.wwf.eu/?326895/The-State-of-Implementation-of-the-Birds-and-Habitats-Directives-in-the-EU

Bobbink, R. (2021) *Effecten van stikstofdepositie nu en in 2030: een analyse.* Nijmegen, the Netherlands: Onderzoekcentrum B-WARE voor Greenpeace Nederland.

Boonstra, F. G., Nieuwenhuizen, W., Visser, T. *et al.* (2021) *Collective approach in progress: Interim evaluation of the agri-environment scheme in the Netherlands.* Wageningen, the Netherlands: Wageningen Environmental Research, Rapport 3066.

Breeuwer, A., Berendse, F., Willems, F. *et al.* (2009) Do meadow birds profit from agri-environment schemes in Dutch agricultural landscapes? *Biological Conservation*, **142**, 2949–2953.

Buij, R. & Koffijberg, K. (2019) *Ganzen en ganzenschade in Nederland; Overzicht van kennis en kennishiaten voor effectief beleid.* Wageningen, the Netherlands: Wageningen Environmental Research, Rapport 2965.

Capuzzo, E., Lynam, C. P., Barry, J. *et al.* (2017) A decline in primary production in the North Sea over 25 years, associated with reductions in zooplankton abundance and fish stock recruitment. *Global Change Biology*, **24**, e352– e364.

CRM (1975) *Nota relatie tussen landbouw, natuuren landschapsbehoud.* The Hague: Tweede Kamer der Staten Generaal 1974-1975 13285 nr. 1-3. Staatsuitgeverij.

De Bruin, D., Hamhuis, D., Van Nieuwenhuijze, L., Overmars, W. Sijmons, D. & Vera, F. (1987) *Ooievaar. De toekomst van het rivierengebied.* Arnhem, the Netherlands: Stichting Gelderse Natuurfederatie.

de Snoo, G. R., Melman, T. C. P., Brouwer, F. M., Van der Weijden, W. J. & Udo de Haes, H. A. (2016) *Agrarisch natuurbeheer in Nederland. Principes, resultaten en perspectieven.* Wageningen, the Netherlands: Wageningen Academic Publishers.

Dudley, N. (ed.) (2008) *Guidelines for Applying Protected Area Management Categories.* Gland, Switzerland: IUCN. WITH Stolton, S., Shadie, P. & Dudley, N. (2013). *IUCN WCPA Best Practice Guidance on Recognising Protected Areas and Assigning Management Categories and Governance Types.* IUCN Best Practice Protected Area Guidelines Series No. 21. Gland, Switzerland: IUCN.

ECORYS Nederland BV, Sweco & Roots Beleidsadvies (2016) *Ex post Evaluatie POP2 2007–2013 Eindrapportage.* Study on behalf of the Ministry of Economic Affairs. Rotterdam, the Netherlands: Ecorys.

EEA (2021) Conservation status and trends of habitats and species, National summary dashboards – Habitats and species, Main pressures and threats. www.eea.europa.eu/themes/biodiversity/state-of-nature-in-the-eu/article-17-national-summary-dashboards/conservation-status-and-trends

European Commission (2018) *CAP Context Indicators 2014–2020. 33. Farming Intensity.* https://agridata.ec.europa.eu/extensions/IndicatorsEnvironmental/FarmingIntensity.html

(2019) *The EU Environmental Implementation Review 2019, Country Report for The Netherlands.* SWD(2019) 133 final.

Eurostat (2022) Livestock density index 2020 (online data code: TAI09). Last update 19 December 2022. https://ec.europa.eu/eurostat/databrowser/view/tai09/default/table?lang=en

Folkert, R. & Boonstra, F. (2017*) Lerende evaluatie van het Natuurpact. Naar nieuwe verbindingen tussen natuur, beleid en samenleving.* The Hague: PBL Planbureau voor de Leefomgeving, PBL-publicatienummer 1769.

Folkert, R., Bouwma, I., Kuindersma, W., van der Hoek, D., Gerritsen, A. & Kunseler, E. (2020) *Lerende evaluatie van het Natuurpact 2020. Gezamenlijk de puzzel leggen voor natuur, economie en maatschappij.* The Hague: PBL Planbureau voor de Leefomgeving, PBL-publicatienummer: 3852.

Goedhart, P. W., Teunissen, W. A. & Schekkerman, H. (2010) *Effect van nestbezoek en onderzoek op weidevogels.* Nijmegen, the Netherlands: SOVON Vogelonderzoek Nederland, Sovon-onderzoeksrapport 2010/01.

Holm, T. E. & Laursen, K. (2009) Experimental disturbance by walkers affects behaviour and territory density of nesting Black-tailed Godwit (*Limosa limosa*). *Ibis*, **151**, 77–87.

Horlings, E., Schenau, S., Hein, L., Lof, M. de Jongh, L. & Polde, M. (2020) *Experimental Monetary Valuation of Ecosystem Services and Assets in the Netherlands.* The Hague: CBS Den Haag/Wageningen, Wageningen University & Research.

Hornman, M., Kavelaars, M., Koffijberg, K. *et al.* (2021) *Watervogels in Nederland in 2018/2019.* Nijmegen, the Netherlands: Sovon Vogelonderzoek Nederland, Sovon rapport 2021/01, RWS-rapport BM 21.08.

I&W & LNV (2018) *Marine Strategy (part 1) Update of current environmental status, good environmental status, environmental targets and indicators 2018–2024.* The Hague: Netherlands national

Ministry of Infrastructure and Water Management (IenW) and Ministry of Agriculture, Nature and Food Quality (LNV).

IPO & LNV (2021) *Zevende Voortgangsrapportage Natuur*. The Hague: Interprovinciaal Overleg (IPO) and Ministerie van Landbouw, Natuur & Voedsellkwaliteit.

(2022) *Achtste voorgangsrapportage natuur*. The Hague: Interprovinciaal overleg (IPO) en het ministerie van Landbouw, Natuur en Voedselkwaliteit (LNV).

Kleefstra, R., Bregnballe, T., Frikke, J. *et al.* (2022) *Trends of Migratory and Wintering Waterbirds in the Wadden Sea 1987/1988 – 2019/2020*. Wadden Sea Ecosystem No. 41. Wilhelmshaven, the Netherlands: Common Wadden Sea Secretariat, Expert Group Migratory Birds.

Kleijn, D. (2012) *De effectiviteit van Agrarisch Natuurbeheer*. Wageningen, the Netherlands: Alterra, Centrum voor Ecosystemen. https://library.wur.nl/WebQuery/wurpubs/430094

Kleijn, D., Berendse, F., Smit, R. & Gilissen, N. (2001) Agri-environment schemes do not effectively protect biodiversity in Dutch agricultural landscapes. *Nature*, **413** (6857), 723–725.

Kleijn, D. & Lammertsma, D. R. (2013) *Conserving the Black-tailed Godwit: Legislation, implementation and effectiveness of conservation in Belgium, Germany and the Netherlands*. Wageningen, the Netherlands: Alterra, Wageningen-UR, Alterra-rapport; No. 2366.

Kleijn, D., Schekkerman, H., Dimmers, W. J. *et al.* (2010) Adverse effects of agricultural intensification and climate change on breeding habitat quality of Black-tailed Godwits *Limosa l. limosa* in the Netherlands. *Ibis*, **152**, 475–486.

Kleunen, A., Foppen, R. & van Turnhout, C. (2017) *Basisrapport voor de Rode Lijst Vogels 2016 volgens Nederlandse en IUCN-criteria*. Nijmegen, the Netherlands: Sovon Vogelonderzoek Nederland, Sovon-rapport 2017/34.

L&V (1975) *Nota betreffende de relatie landbouw en natuur - en landschapsbehoud: Gemeenschappelijke uitgangspunten voor het beleid inzake de uit een oogpunt van natuur- en landschapsbehoud waardevolle agrarische cultuurlandschappen*. The Hague: Ministerie van Landbouw and Visserij.

LNV (1990) *Natuurbeleidsplan. Regeringsbeslissing*. The Hague: Ministerie van Landbouw, Natuurbeheer en Visserij.

(1998a) Wet van 25 mei 1998, houdende nieuwe regelen ter bescherming van natuur en landschap (Natuurbeschermingswet 1998), Staatsblad 1998, 56; Staatsblad 1999, 30.

(1998b) Wet van 25 mei 1998, houdende regels ter bescherming van in het wild levende planten- en diersoorten (Flora- en faunawet). Staatsblad Stb. 1999, 253.

(2000) *Natuur voor Mensen, Mensen voor Natuur. Nota natuur, bos en landschap in de 21eeeuw*. Ministerie van Landbouw. The Hague: Natuurbeheer en Visserij.

(2006) *Natura 2000 targets document – Summary: Setting conservation objectives for the Natura 2000 network in the Netherlands*. The Hague: Ministry of Agriculture, Nature and Food Quality.

(2021a) *Plattelandsontwikkelingsprogramma voor Nederland 2014-2020 (POP3)* Versie 10.1. The Hague: Ministerie van Landbouw, Natuur & Voedselkwaliteit.

(2021b) *Prioritised Action Framework (PAF) for Natura 2000 – The Netherlands*. The Hague: Ministry of Agriculture, Nature & Food Quality. .www.natura2000.nl/sites/default/files/Bibliotheek/Begrippen/PAF%20NL%202021-2027%20FINAL%20(corr).pdf

LNV & IPO (2013) *Natuurpact - ontwikkeling en beheer van natuur in Nederland*. The Hague.

(2019) *Nederland Natuurpositief, Ambitiedocument voor een gezamenlijke aanpak in natuurbeleid*. The Hague: Ministerie van Landbouw, Natuurbeheer en Visserij & Interprovinciaal Overleg, Publicatie-nr. 0919-120.

LNV, SBB & I&W (2021) Waardevolle natuurontwikkeling langs de grote rivieren: 25 jaar kansen pakken en volhouden. Amersfoort, the Netherlands: Staatsbosbeheer. www.staatsbosbeheer.nl/over-staatsbosbeheer/blijf-op-de-hoogte/nieuws/2021/02/25-jaar-nurg

Luiten van Zanden, J., Van Goethem, T., Lenders, R., Schaminée, J. (2021) *De ontdekking van de natuur*, Amsterdam: Prometheus, ISBN:9789044647358.

Melman, Th. C. P., Schotman, A. G. M., Meeuwsen, H. A. M. *et al.* (2016) *Ex-ante-evaluatie ANLb-2016 voor lerend beheer; Een eerste blik op de omvang en ruimtelijke kwaliteit van het beheer in het nieuwe stelsel*. Wageningen, the Netherlands: Wageningen Environmental Research, Rapport 2752.

Melman, D. & Sierdsema, H. (2017) *Weidevogelscenario's: Mogelijkheden voor aanpak van verbetering van de weidevogelstand in Nederland*. Wageningen, the Netherlands: Wageningen Environmental Research, Rapport No. 2769 (Sovon-rapport; No. 2017/22).

Ministerie van Economische Zaken (2013) *Kamerbrief Uitvoeringsagenda Natuurlijk Kapitaal: behoud en duurzaam gebruik van Biodiversiteit*. The Hague: Ministerie van Economische Zaken, 22 June 2013.

(2014) *Rijksnatuurvisie 2014 – Natuurlijk verder*. The Hague: Ministerie van Economische Zaken.

(2015) Besluit van de Staatssecretaris van Economische Zaken en de Minister van Infrastructuur en Milieu van 11 december 2015, nr. DGAN-NB / 15166607, houdende wijziging van het programma aanpak stikstof 2015–2021 (AERIUS, versie 2015), Staatscourant 2015 reference 44608, publicatie 14 December 2015.

Ministeries BZK, MinEZ, MinBiZa & IPO (2011) *Onderhandelingsakkoord Decentralisatie Natuurbeleid*. The Hague.

Ministeries V&W, LNV en VROM (2004) *Meerjarenprogramma Ontsnippering (MJPO), Ministerie van Verkeer en Waterstaat, Ministerie van Landbouw*. The Hague: Natuur & Voedselkwaliteit en Ministerie van Volkshuisvesting, Ruimtelijke Ordening en Milieu.

MJPO (2020) *Eindboek Meerjarenprogramma Ontsnippering 'Natuur verbonden, meer leefruimte voor dieren in Nederland'*. Utrecht, the Netherlands: Meerjarenprogramma Ontsnippering (MJPO).

Osieck, E. (2000) The Netherlands. In *Important Bird Areas in Europe: Priority Sites for Conservation. 1: Northern Europe*, eds. M. F. Heath & M. I. Evans, pp. 465–507. Cambridge, UK: Birdlife International.

PBL (2009) *Natuurbalans 2009*. The Hague: Planbureau voor de Leefomgeving, Publicatienummer 500402017.

(2016) *The GLOBIO model. A technical description of version 3.5*. The Hague: PBL Netherlands Environmental Assessment Agency, PBL publicatie 2369.

Reijnen, R., van der Grift, E., van Veen, M., Pelk, M., Lüchtenburg, A. & Bal, D. (2000) *De weg mét de minste weerstand. Opgave ontsnippering.* Wageningen, The Netherlands: Alterra/ Expertisecentrum LNV.

Reneerkens, J., Piersma, T. & Spaans, B. (2005) *De Waddenzee als kruispunt van vogeltrekwegen.* Den Burg, Texel, the Netherlands: Royal Netherlands Institute for Sea Research (NIOZ), NIOZ rapport 2005 – 4.

RLI (2013) *Nature's imperative: Towards a robust nature policy.* The Hague: Council for the Environment and Infrastructure (RLI), RLI publication no. 2013/02.

Sanders, M. E., Meeuwsen, H. A. M., Roelofsen, H. D. & Henkens, R. J. H. G. (2021) *Progress with the national ecological network: Technical background of the website of the Assessment of the Dutch Human Environment 2020.* Wageningen, the Netherlands: Statutory Research Tasks Unit for Nature & the Environment (WOT Natuur & Milieu), WOt-technical report 202.

Schekkerman, H. & Müskens, G. (2000) Do Black-tailed Godwits *Limosa limosa* produce sufficient young to sustain populations in agricultural grasslands? [In Dutch with English summary.] *Limosa*, **73**, 121–134.

Schekkerman, H., Teunissen, W. & Oosterveld, E. (2009) Mortality of shorebird chicks in lowland wet grasslands: interactions between predation and agricultural practice. *Journal of Ornithology*, **150**, 133–145.

Schelhaas, M. J., Clerkx, A. P. P. M., Daamen, W. P. *et al.* (2014) *Rapport Zesde Nederlandse Bosinventarisatie: methoden en basisresultaten.* Wageningen, The Netherlands: Alterra Wageningen UR (University & Research centre), Alterra-rapport 2545.

Straatsma, M. W., Bloecker, A. M., Lenders, H. J. R., Leuven, R. S. E. W. & Kleinhans, M. G. (2017) Biodiversity recovery following delta-wide measures for flood risk reduction. *Science Advances*, **3**, 10.1126/sciadv.1602762.

Sweco (2021) *Whitepaper Hoogwater Limburg 2021. Wat kunnen we leren?* De Bilt, the Netherlands: Sweco Netherlands.

Trouwborst, A. (2016) *Weidevogels en de Europese en internationale verplichtingen van Nederland: een juridische analyse.* Tilburg, the Netherlands: Tilburg University. Rapport in opdracht van Vogelbescherming Nederland.

UNEP-WCMC (2021) *Protected Area Profile for Netherlands from the World Database of Protected Areas, November 2021.* www .protectedplanet.net

United Nations, European Commission, Food and Agriculture Organisation, International Monetary Fund, Organisation for Economic Co-operation and Development, and World Bank (2014). *System of Environmental-Economic Accounting – Central Framework (SEEA CF).* New York.

Van der Burg, A. B., Berendse, F., Van Dobben, H. F. *et al.* (2021) *Onderzoek naar een ecologisch noodzakelijke reductiedoelstelling van stikstof: Stikstof en natuurherstel.* Wageningen, the Netherlands: WUR (Alterra) for WWF Netherlands, Zeist.

Van der Knaap, Y., Bakker, M., Alam, S. *et al.* (2018) Projected vegetation changes are amplified by the combination of climate change, socio-economic changes and hydrological climate adaptation measures. *Land Use Policy*, **72**, 547–562.

Van der Winden, J., Krijgsveld, K. L., Inberg, H. & Fijn, R. C. (2008) *Beschermingsplan Duin- en kustvogels (Part A and Part B).* Zeist, the Netherlands: Bureau Waardenburg bv/Vogelbescherming Nederland.

Van Bodegom, P. M., Verboom, J., Witte, J. P. M. *et al.* (2014) Synthesis of ecosystem vulnerability to climate change in the Netherlands shows the need to consider environmental fluctuations in adaptation measures. *Regional Environmental Change*, **14** (3), 933–942.

Van Dorp, D. & Opdam, P. F. M. (1987) Effects of patch size, isolation and regional abundance on forest bird communities. *Landscape Ecology*, **1**, 59–73.

Van Hinsberg, A., Egmond, P. van, Pouwels, R., Dirkx, G. H. P. & Breman, B. C. (2020) *Referentiescenario's Natuur: Tussenrapportage Natuurverkenning 2050.* The Hague: Planbureau voor de Leefomgeving (PBL), PBL report No. 3574.

Van Paassen, A. & Teunissen, W. (2010) *Weidevogelbalans 2010.* Landschapsbeheer Nederland, Utrecht/SOVON Vogelonderzoek Nederland, Beek-Ubbergen.

Veerman & Fresco (2019)

VVD, D66, CDA & ChristenUnie (2021) Coalitieakkoord 2021 – 2025 'Omzien naar elkaar, vooruitkijken naar de toekomst'. Den Haag, 15 December 2021.

Wereld Natuur Fonds (2015) *Living Planet Report Natuur in Nederland.* Zeist, the Netherlands: Wereld Natuur Fonds.

(2017) *Living Planet Report. Zoute en zilte natuur in Nederland.* Zeist, the Netherlands: Wereld Natuur Fonds.

(2020) *Living Planet Report Nederland. Natuur en landbouw verbonden.* Zeist, the Netherlands: Wereld Natuur Fonds.

Wiertz, J., Dirkx, G. H. P, Melman, Th. C. P. *et al.* (2007) *Ecologische evaluatie regelingen voor natuurbeheer: programmabeheer en Staatsbosbeheer 2000-2006.* Bilthoven and Wageningen, the Netherlands: Milieu en Natuur Planbureau and Wageningen Universiteit and Researchcentrum, MNP-publicatie 500410002, 500410003.

Zekhuis, M., Van Oort, L. & Hoogenstein, L. (2021) *Gewilde dieren. Herintroducties van dieren in Nederland.* Zeist, the Netherlands: KNNV Uitgeverij.

Chapter 24

Poland

Paweł Pawlaczyk

Key Statistics

Terrestrial area (km^2)/% of EU-28*	311 928/7%
Marine waters (km^2)/% of EU-28*	33 142/0.7%
Population density (p km^2)/% of EU-28*	122/106%
GDP per capita (€)/% of EU-28	10 913/41%
Land cover (CORINE land cover classes 2018)	AGR 59%, FOR 30%, ART 6%, OSN 3%, WET 2%
Biogeographical regions	ALP, CON, MBAL
Number of HD Annex I habitats/%*	81/36%
Number of HD Annex II, IV and V species/%*	175/14%
Number of BD Annex I bird taxa/%*	82/47%
% HD UCS and decreasing habitats/ % unknown	23%/2%
% HD UCS and decreasing species/ % unknown	14%/16%
% breeding BD Annex I birds decreasing/% uncertain and unknown	21%/40%

* Excluding MAC and MMAC biogeographical regions.
Sources and codes. See Table 1.1.

24.1 The Natural History and Characteristics of Poland

Poland is a country in eastern-central Europe that is mostly within the Continental biogeographical region and is dominated by lowland plains formed mainly by glacial deposits (91% of the country). Only a small part of the country, in the Carpathian Mountains in the south, belongs to the Alpine region. Just c. 150 km^2 within the Tatra massif, on the Polish–Slovak border, has a subalpine character and includes Poland's highest point (Rysy Peak at 2 499 m). Poland's marine area is within the Baltic Sea.

Agricultural land currently dominates (59% of the country), but extensive areas of forest remain (30%). However, land cover is not uniform: forests predominate in the mountains as well as the northern and western parts of the country, whereas the centre of the country is mostly agricultural with only small forest patches. Although only about 6% of the country is urbanised, most of the land is affected by farming, forestry and active water management, such that less than 1% can be considered unmanaged wilderness (mainly in non-intervention zones in National Parks).

About 75% of agricultural land is used for arable crops. Meadows and pastures comprise 18% of agricultural land according to official statistics, but they are declining due to abandonment and transformation into arable land. Nevertheless, Poland still has extensive semi-natural and diverse agricultural landscapes, with mosaics of small meadows, pasture, arable fields and orchards, with non-farmed elements such as hedges, shrubs and small ponds.

Natural forests were dominated by deciduous forests in the lowlands (mixed oak–hornbeam forest, oak forest and beech forest – HD 9170, 9160, 9190, 9130, 9110 habitats)[1] and fir–spruce–beech forests (9130) as well as spruce forests (9410) in the mountains. However, as a result of forestry in the nineteenth and twentieth centuries, forests are now dominated by pine (c. 60% of the whole forest area) in the lowlands, and spruce (c. 8%) in the mountains (Milewski, 2018). About 60% of the forest area has stands that are 20–80 years old; stands older than 120 years cover less than 4% (Milewski, 2018). Nevertheless, some forest stands have retained their natural tree composition, and some large-scale forests remain in eastern Poland and the Carpathians with remnants of old-growth stands. These include Białowieża Forest, on the border with Belarus, which is probably the best preserved lowland forest in Europe (Jaroszewicz *et al.*, 2019).

About 4% of the country comprises peatland, over 80% of which has been drained, and most of which is now under agricultural use. Elsewhere, fens predominate (92% of peatlands) and raised bogs comprise about 4.5% of the peatland area; however, active mires that still accumulate peat are extremely rare. Typical peat-forming bog vegetation has been maintained, mainly where drainage was ineffective (Oświęcimska-Piasko

[1] Numbers in brackets are the HD Annex I habitat codes (see Appendix for full names). * Indicates a priority habitat.

Table 24.1 Short-term trends (c. 2007–2018) in Habitats Directive Annex I habitats and Annex II, IV and V species with an unfavourable conservation status in Poland.

	Poor (unfavourable–inadequate)				Bad (unfavourable–bad)			
	Imp	Sta	Dec	?	Imp	Sta	Dec	?
HD Habitat assessments								
Number	3	34	8	5	1	17	19	3
%	6%	68%	16%	10%	3%	43%	48%	8%
HD Species assessments								
Number	6	21	19	42	2	6	17	8
%	7%	24%	22%	48%	6%	18%	52%	24%

Key. Imp = improving, Sta = stable, Dec = decreasing, ? = unknown.
Source. Based on Member State HD Article 17 reported data as presented on the EEA national summary dashboard (2021).

et al., 2006; Joosten *et al.*, 2017). Particularly important remaining mires include the large alkaline fen (7230) in Biebrza valley, and the smaller, but almost natural, fen in Rospuda valley.

There are about 7 000 lakes larger than 1 ha in Poland, most of which are eutrophic lakes (3150), although there are also examples of oligotrophic *Littorelletalia* lakes (3110), calcium-rich stonewort lakes (3140), dystrophic lakes (3160) and brackish coastal lagoons (1150*). Most rivers have been modified for flood control and agriculture intensification on adjacent areas, but some sections of the Vistula and Bug are still in a near-natural state, including their fluvial geomorphological processes, which is rare for such large rivers in Europe. These and some other large river valleys have also retained their semi-natural landscapes dominated by alluvial meadows and pastures, whilst some smaller forest rivers still have almost natural sections.

The Baltic seacoast retains some sections with important natural habitats, including dunes (mosaics of 2110, 2120, 2130*, 2180) and cliffs (1230).

Due to its geographical location, Poland has some habitats and species that also occur in the Boreal, Atlantic, Pannonian and even some Mediterranean biogeographical regions. However, many of them are at their natural range limits, thus many such habitats are represented by atypical, simplified forms, and the populations of many species are marginal.

Only one HD habitat type – 91P0 Holy Cross fir forest (*Abietetum polonicum*) – is endemic to Poland.

Amongst its HD species, the country has all, or at least 90% of the European population of Bison (*Bison bonasus*), Speckled Ground Squirrel (*Spermophilus suslicus*), Swamp Minnow (*Rhynchocypris percnurus*) and Tatra Vole (*Microtus tatricus*), as well as the plants *Carlina onopordifolia*, *Cochlearia polonica*, *Erysimum pieninicum* and *Galium cracoviense*. Poland holds important breeding populations of many bird species, especially of forests and wetlands, including the Common Crane (*Grus grus*), Black Stork (*Ciconia nigra*), Great Snipe (*Gallinago media*), the globally Vulnerable Greater Spotted Eagle (*Aquila clanga*), and 27% of the world population of the globally Vulnerable Aquatic Warbler (*Acrocephalus paludicola*). Polish areas of the Baltic Sea are crucial for wintering waterbirds, including three globally Vulnerable species: Velvet Scoter (*Melanitta fusca*), Long-tailed Duck (*Clangula hyemalis*) and Horned Grebe (*Podiceps auritus*).

24.2 The Status of Nature

24.2.1 Habitat Status and Trends

The HD Article 17 habitat status assessments[2] for 2013–2018 across all the biogeographical regions found that 20% of assessments indicated a favourable status, 43% were poor, 35% were bad and 2% unknown. An additional concern is that a substantial proportion of habitats with an unfavourable conservation status had declining trends over the short term, including in 48% of the assessments that had revealed a bad status (Table 24.1). The declining status concerns all groups of habitats, but especially mires, freshwater habitats, dunes, heathlands and meadows as well as forests. The unfavourable status of most of the habitats was due to their structure and function being assessed as bad.[3]

Unfortunately, at the national scale there are almost no cases where negative habitat trends have been reversed, even where substantial conservation efforts have been made. For example, alkaline fens with typical moss-sedge vegetation (7230) have disappeared over the last decades in many areas, despite conservation efforts and designation of protected areas, including Natura 2000 sites (Wołejko *et al.*, 2019).

Much less up-to-date information is available on non-HD habitats, but Ratyńska *et al.* (2010) considered that 54% of vegetation communities were threatened. The most serious declines affect species-rich meadow and pasture habitats.

[2] See Section 1.3 for background on monitoring under the Nature Directives, status categories and data sources.

[3] https://nature-art17.eionet.europa.eu/article17/

24.2.2 Species Status and Trends

The negative trends of habitats, as described in Section 24.2.1, are reflected also in the status and trends of many species. The HD species status assessments for 2013–2018 revealed that 38% of assessments indicated a favourable status, 34% were poor, and 12% were bad. The species' status was unknown in 16% of assessments. A substantial proportion of species with an unfavourable status showed declines, including in 52% of species assessments that showed a bad status (Table 24.1).

According to the Birds Directive Article 12 report covering 2013–2018, most BD Annex I taxa had increasing, stable or fluctuating short-term trends, but 21% of breeding species declined and the trend of 40% was unknown (Table 24.2). For example, despite its importance, the Aquatic Warbler population only seems to be stable in the Biebrza marshes (Tannenberger and Kubacka, 2018).

Population declines have also been observed in more common bird species. According to the national bird monitoring database,[4] the cumulative Farmland Bird Index has declined, with a decrease of 28% in 2021 compared to the 2000 reference level. The Wetland Bird Index (breeding birds) has also declined by 20%, but over a shorter period, 2007–2018. By contrast, no general decline of common forest birds has occurred (Chylarecki *et al.*, 2018).

The trends of BHD species, and others, are reflected in national Red Lists. As in other European countries, many species are threatened in Poland. For example, an assessment of 765 species of pteridiophytes (vascular) and flowering plants found that 30% are threatened (Kaźmierczakowa, 2016). Similarly, 30% of liverwort species (Klama and Górski, 2018), 33% of moss species (Żarnowiec *et al.*, 2004), 25% of vertebrates, 51% of bivalves, 42% of snails and a similar fraction of insects (Głowaciński, 2002) have been assessed as threatened. As regards birds, 16 species have become extinct and 20% are presently recognised as threatened, including 70% of Charadriiformes and 67% of Anseriformes (Wilk *et al.*, 2020).

There is little information on trends amongst common species other than birds, but some ecologists are of the opinion that insects (in particular wild bees and butterflies), frogs and freshwater fish, as well as oligotrophic plants, are strongly declining in abundance.

Despite the generally declining trends, Poland still hosts strong populations of some charismatic species, including the Bison, the symbol of Polish nature conservation (see online Annex PL.1),[5] and the large carnivores (see Annex PL.2). The populations of some species, such as the Beaver (*Castor fiber*), Wolf (*Canis lupus*), Common Crane (*Grus grus*) and White-tailed Sea-eagle (*Haliaeetus albicilla*) have recovered after strong declines.

Table 24.2 Short-term trends (c. 2007–2018) in Birds Directive Annex I species in Poland.

	Breeding				Wintering			
	Inc	Sta/ Flu	Dec	?	Inc	Sta/ Flu	Dec	?
Number	11	18	15	29	12	5	0	8
%	15%	25%	21%	40%	48%	20%	0%	32%

Key. Inc = increasing, Sta = stable, Flu = fluctuating, Dec = decreasing, ? = uncertain and unknown.
Source. Based on Member State Article 12 reported data as presented on the EEA national summary dashboard (2021).

24.2.3 Drivers and Causes of Change

The main pressures on habitats and species are from the common and uniform intensive use of nearly all agricultural and forested land. Whilst the structure of agricultural land ownership has not changed significantly in recent years, and small farms still dominate, agricultural technologies have become increasingly unified. Former mosaics of various crops have been replaced by rather uniform monocultures, with high proportions of maize and rape. Small semi-natural elements, such as shrubs and small ponds, are often removed by farmers to maximise the areas that are eligible for CAP payments. Under the pressure of farmers, drainage continues, leading to wetland loss. Additionally, the intensification of farming practices, in particular increases in fertiliser and herbicides use, has negative effects. In contrast, many meadows and pastures have been abandoned due to their marginal profitability, resulting in the cessation of their traditional management practices with important detrimental impacts as the affected semi-natural habitats are the most valuable for biodiversity.

Although forest management is assumed sustainable, timber harvests in Polish forests have increased continuously, in particular in state-owned forests, from 1.4% of the stock harvested annually in 1990 to 2.1% in 2018 (PGL LP, 2019). Whilst silvicultural management aids trees' regeneration and growth, it is generally detrimental for old-growth structures and processes, and tree microhabitats and the accumulation of coarse woody debris. The average deadwood volume in Polish forests is estimated at 9.1 m^3/ha (BULiGL, 2021), which is insufficient for many species. As a result, despite good indications of the state of general forest biodiversity, some species with specific ecological requirements (in particular requiring deadwood, old trees, and tree microhabitats) are threatened in forests. Significant changes in forests are also expected due to climate change, which may increase salvage logging because of the increasing frequency and severity of pest outbreaks, such as the bark beetle *Ips typographus* on spruce and *Ips acuminatus* on pine.

Most Polish peatlands are degraded by former drainage and continue to be damaged by further deepening of the ditches. Although Polish peatlands cover 1.2–1.3 million ha (c. 4% of the country area), c. 85% are not accumulating peat due to their drainage (Joosten *et al.*, 2017; Pawlaczyk, 2018; GDOŚ, 2021).

[4] https://monitoringptakow.gios.gov.pl/

[5] All annexes are free to access online at CUP: www.cambridge.org/natureconservation

Instead, the majority are meadows and pastures, or afforested. The greenhouse gases (GHG) emission from degraded peatlands is estimated to be c. 34 million tonnes of CO_2 equivalent (Kotowski, 2021), which means Poland is among the world's 10 worst emitters of GHG from degraded peatlands.

Approximately 99% of river water bodies and 91% of lake water bodies were assessed as having a bad status according to WFD criteria (GIOŚ, 2021). This is partially due to water pollution, but this is diminishing; the main cause is old and recent hydromorphological alterations, conserved by recurrent river maintenance works. There are further substantial potential threats to Polish rivers from governmental investment plans, including to develop inland waterways and to build water reservoirs to prevent droughts and floods.

Pressures from urbanisation and habitat fragmentation by infrastructure are also significant, at least in several regions of Poland. If remnants of natural habitats have survived the other pressures so far, many are now affected by new investments.

The main pressures on habitats and species in the Polish part of the Baltic Sea are related to fisheries. By-catch in nets is a significant threat to the Harbour Porpoise (*Phocoena phocoena*) and for several waterbird species, in particular the important wintering populations of Velvet Scoter and Long-tailed Duck (Marchowski *et al.*, 2020). A more general ecological problem in the Baltic Sea is eutrophication caused by nitrogen and phosphorus originating from the catchment area. In the case of Poland, this pollution comes from almost all of the territory.

Intensive urbanisation has taken place all along the coast, damaging coastal habitats directly and through the increased need for artificial coastal protection measures. In addition, almost all the Baltic coast is visited by tourists in the summer, leaving no space for nesting birds and other wildlife to avoid human presence and disturbance.

In recent years, the pressure of increasing tourism has become important in some other areas, even in National Parks. For example, canoe tourism in Drawa National Park has reduced breeding success of the Goldeneye (*Bucephalla clangula*) (Pawlaczyk, 2014b), hiking in the Tatra mountains increases behavioural stress in the Tatar Chamois (*Rupicapra rupicapra tatrica*) (Pęksa and Ciach, 2015), and tourist trails in Bieszczady reduce the habitat quality of the Brown Bear (*Ursus arctos*) (Selva *et al.*, 2012).

Most of the processes affecting biodiversity are rather complex and contain feedback loops and significant reaction delays, thus identification of the causes and drivers of deterioration is not always easy.

24.3 Nature Conservation Governance, Actors, Legislation and Policies

24.3.1 Overview of Nature Developments in Poland

The history of nature conservation in Poland is complex due to the country's political history. During the nineteenth century and later until 1918, Poland did not exist as an independent state and the territory was part of the Austro-Hungarian Empire, Kingdom of Prussia (part of the German Empire) and the Russian Empire. Until 1945, the present western part of Poland was a part of Germany. Former eastern parts of Poland are presently parts of Ukraine, Belarus and Lithuania. Thus the history of nature conservation of the present Polish territory is not unified. Therefore, although some nature conservation actions occurred before 1945 (e.g. some species protection laws, or the protection of some areas), this account only covers the subsequent main nature conservation developments, as summarised in Table 24.3.

24.3.2 Governance

The governance of nature conservation in Poland reflects the country's general administrative structure that has been in force since 1999 (i.e. a mix of state governance with local governance of communities and regions). Nature is generally the responsibility of the Ministry of Climate and Environment, which is in charge of general policy and supervising National Parks. Since 2008, management of Natura 2000 sites and Nature Reserves, as well as species conservation, has been delegated to a specialised structure that joined the function of environmental authority and executive agency under the General Directorate of Environment Protection (GDOŚ) at the national level with 16 Regional Directorates of Environment Protection (RDOŚ).

Regional government organisations[6] are responsible for landscape protection, such as through Landscape Parks and Protected Landscape Areas (see Section 24.5.1), and general landscape assessment and care. Usually, environmental departments are part of the executive regional offices, but for Landscape Parks there are also specific park or regional Landscape Park directorates with dedicated staff. The RDOŚ supervise landscape protection, reconciling all changes in Protected Landscape Area designations as well as land-use decisions within them and their buffer zones.

Local municipalities are responsible for most aspects of spatial planning and establishing and managing small protected areas: Ecological Areas, Nature-landscape Complexes, and Natural Monuments. Usually, most local communities do not now employ ecologists, and nature conservation is only part of the responsibility of local environmental officers. The RDOŚ also supervise local nature protection, as with regional landscape conservation.

Most forests in Poland are managed by State Forests, a specific independent and self-financing agency. It had been supervised by the Ministry of Environment.

Agriculture, including some key nature conservation aspects (e.g. agri-environment schemes), is the responsibility of the Ministry of Agriculture and Rural Development. An Agency of Restructuring and Modernisation of Agriculture is

[6] 16 *voivodships* – Polish equivalent of provinces.

Table 24.3 Key nature conservation developments in Poland since 1945.

1945	Current territory of Poland established
1946	Decree on plant protection, first list of protected plants
1949	Nature Conservation Act
1952	Start of Bison wild population restitution – first animals released
1956	First list of protected animals
1976	First Landscape Park established (Suwałki Landscape Park)
1977	Declaring first three National Parks as Biosphere Reserves
1990	New Nature Conservation Act
2003	First National Strategy for Biodiversity Conservation and Sustainable Use
2004	Accession to the EU, first part of Natura 2000 sites designated
	New Nature Conservation Act
2006	Beginning of systematic state monitoring of birds as well as HD habitats and species
2007	First extensive and systematic assessment of conservation status of BHD habitats and species
2008	Substantial legislation amendments transposing EU environmental legislation
	General and Regional Directorates of Environment Protection established, as new administration for nature conservation
2009	Rospuda valley case solved – after national court ruling as well as European Commission and CJEU intervention, first substantial change in proposed road project due to Natura 2000 requirements
2017	New Water Act, creation of Polish Water as unified agency managing water bodies and responsible also for achieving environmental objectives
2018	CJEU ruling on Białowieża Forest case – clarification of application of Natura 2000 rules to forestry

See Table 3.3 for signatory dates of international conventions.

responsible for CAP payments and the implementation and monitoring of measures, including monitoring of species and habitats under agri-environment–climate measures.

National Parks have their own specific enforcing authority: the National Parks Guard. In the state forests the Forest Guard prevents offences, but is focussed rather on prevention of timber thefts. The specialised Hunting Guard and Fishing Guard try to prevent offences against game and fish stocks, but are rather focussed on ownership aspects, with only marginal attention to nature conservation. Most of the offences and crimes against nature are under the competence of the police and Prosecutors Office.

Since institutional reform in 2017, another state agency, Polish Waters, has managed all public freshwaters. Freshwater fish populations are managed by various organisations or fishers, leasing rivers or lakes from the Polish Waters agency. A separate maritime administration is responsible for marine areas, including marine Natura 2000 sites. At the national government level, all waters are the responsibility of the Ministry of Infrastructure.

24.3.3 Public Involvement and NGOs

As in many other countries, nature conservation advocacy in Poland is largely provided by NGOs. At least in some cases they have been successful in influencing policy and legislation, and they are also important contributors to collecting funds for nature conservation, implementation of practical conservation measures and nature monitoring.

The beginnings of the Polish naturalists' movement were partly the result of civil scientific, tourism and sightseeing activities at the turn of the nineteenth and twentieth century. The Nature Conservation League, established in 1928, is the oldest of the Polish environmental NGOs. It still exists today and focusses on education and involving school pupils. Most of the present conservation NGOs – including the Naturalists Club, Polish Ecological Club, Society for the Earth, Workshop for All Beings, Society of Nature Friends 'ProNatura', Ecodevelopment Foundation, Society of Nature Conservation 'Salamandra', and the Polish Society for the Protection of Birds –were established in the 1980s and 1990s. Since then, some large international environmental organisations have also been active in Poland: Greenpeace, IUCN and WWF. Despite the huge number of c. 800 organisations declared as environmental, the number of involved citizens and members of NGOs is not large. With the exception of the Nature Conservation League, most nature conservation NGOs have several hundred members and work rather as expert organisations and professional project managers.

In the 2010s, several more or less formal NGO coalitions developed around specific problems, such as the Rospuda valley case (see Annex PL.3), the Białowieża Forest case (Annex PL.4), the 'Let's Live' coalition's campaign to ban bird hunting (and hunting in general), and the 'Save the Rivers Coalition' work against plans for river regulation, canalisation and inappropriate maintenance.

In general, whilst Polish society has shown knowledge of and broad support for nature conservation, this has been weaker for specific cases and practical actions. Awareness of Natura 2000 in Polish society is one of the highest in Europe, with 58% having at least heard of Natura 2000 according to a 2018 Eurobarometer survey, or 68% according to a national survey (Jermaczek, 2018). Nevertheless, as is the case with most European citizens, most Poles do not know exactly what the Natura 2000 network is. In more practical terms, on one hand, 77% of citizens have declared support for existence of wild areas. On the other hand, there is general support for some measures that are harmful for wildlife but implemented under the name of 'regulating nature', such as culling species that are considered to be 'too abundant', cutting down trees that interfere with human intentions, eradication of disease-transmitting animals, and river regulation and 'maintenance' works. Also, local communities are often opposed to nature conservation when it is perceived as a constraint on local economic development.

24.3.4 Policy and Legislation

Although some legislation had been in place previously, shortly after Poland's independence in 1919, protection of several natural features was declared by ministerial decrees. The first modern Nature Protection Act establishing unified legal rules for the whole of Poland entered into force in 1934 (see Annex PL.5 for details).

Following the re-creation of Poland in 1945 as a communist country, with significantly changed borders, eastern territories with 156 nature reserves became part of the Soviet Union. In the north and west, the former German territory with 172 reserves (not all of which were re-established under the Polish law) became part of Poland. In 1946, the decree on plant species protection was issued on the basis of the Act from 1934, declaring 110 species as protected. During 1947–1949 National Parks in Białowieża Forest, Wielkopolska and the Tatra mountains were re-declared. Over the following decades, several new National Parks were established.

In 1949, the new Nature Conservation Act[7] was established, requiring the maintenance, restoration and sustainable use of natural resources. National Parks, Nature Reserves, Natural Monuments and species protection were declared as forms of conservation. The main conservation approach was to establish lists of prohibited activities for each protected area (see Section 24.5.1), although some generic proactive conservation measures were also included.

The political transformation of 1989–1990 was beneficial for nature conservation. The 'round-table agreement' between the communist government and opposition[8] contains a protocol on ecology. Substantial changes were promised, including implementation of EIA procedures, integration of ecological objectives into social, economic and spatial development, introduction of the ecosystem approach into forest management, updating of environmental and nature conservation legislation, protection of the marine environment, a national nature survey, enlargement of the system of protected areas to 30–40% of the national territory, and enlargement of National Parks to at least 1% of the territory. Most of these promises were indeed carried out in the 1990s.

In 1991, the modern Nature Conservation Act[9] was established, replacing the 1949 Act. The Act re-designed nature conservation objectives to include maintaining ecological processes, biodiversity conservation, ecosystem restoration and raising public awareness. The protected area system with differing types of protected areas was established (see Section 24.5.1). The idea of establishing buffer zones around National Parks, Nature Reserves and Landscape Parks was implemented. Management planning of protected areas was also introduced.

The 1990s were the golden decade of nature conservation in Poland. Eight new National Parks, more than 100 Landscape Parks and 300 Nature Reserves were created. Poland ratified the main international conventions on nature conservation (see Table 3.3). It was also the decade when numerous conservation NGOs were established. Nevertheless, the administrative divisions in 1999, and related amendments to the Nature Conservation Act, increased nature conservation administration burdens. In 1999, the rules on the creation of new National Parks were amended, requiring the agreement of all local assemblies, which almost blocked the creation of the new parks for years.

In 2004, Poland joined the EU, and the 2004 Nature Conservation Act[10] was established to implement EU environmental legislation. The main changes were the inclusion and designation of Natura 2000 sites as new types of protected areas, which led to a lot of new areas being under protection. Over the following years the Polish authorities gained experience in implementing the Natura 2000 regulations, whilst in parallel some gaps in the legal transposition of the Nature Directives were identified by the European Commission. Thus, in 2008,

[7] Ustawa o ochronie przyrody z dn. 7 kwietnia 1949.

[8] During the political and economic crisis in the 1980s, 'round-table' negotiations between the communist government and opposition were conducted in the beginning of 1989. The agreement included a commitment to organise a parliamentary election, which was won by the democratic opposition and finally led to Poland's transformation to a democratic country.

[9] Ustawa o ochronie przyrody z dnia 16 października 1991.

[10] Ustawa o ochronie przyrody z dnia 16 kwietnia 2004.

substantial amendments were made to the environmental legislation in order to ensure appropriate impact assessment procedures met the requirements of the Habitats Directive. The 2008 Act on Access to Information and Environmental Impact Assessments was also established. To improve assessment procedures, specialised Directorates for Environmental Protection were established. The case of Rospuda valley (see Annex PL.3), where the building of an Augustów bypass across the fens was finally refused by the Polish courts, demonstrated the power of protection under the Natura 2000 legislation. Landscape protection forms – Landscape Parks and Protected Landscape Areas – became the responsibility of regional government authorities, and the responsibility for Nature Monuments, Ecological Areas and Nature-landscape Complexes was assigned solely to municipalities.

After the election in 2015, the right-wing and Eurosceptic party 'PiS' took over power in Poland, resulting in some decisions that led to important social and environmental conflicts. Contrary to formerly established management plans, the government decided to implement salvage logging after a bark beetle outbreak in part of Białowieża Forest (see Annex PL.4 for details), which was followed by public resistance and legal action by the European Commission and finally by a CJEU ruling[11] in 2018. The restrictions on felling trees in urban and agricultural landscapes were cancelled, which also triggered public protests, as a result of which some restrictions had to be re-introduced. Some new regulations introduced the precedence of sustainable forest management over nature conservation rules, which was the cause of the infringement procedure and application to the CJEU.[12]

To meet CBD requirements, Poland prepared its first ***Biodiversity Strategy*** in 2003. In 2015, an updated strategy, *Programme of Biodiversity Conservation and Sustainable Use*,[13] was established with the aim to contribute more to EU biodiversity targets (including achieving favourable conservation status of at least 10% of habitats and 10% of species). In spite of its rather ambitious goals, implementation of the strategy was slow. The strategy did not contain effective mechanisms for its implementation, and compliance remained voluntary even for nature conservation authorities. Even though the strategy became outdated in 2020, no strategy update has been established; in particular, the targets of the 2021 *EU Biodiversity Strategy for 2030* are yet to be reflected in national strategic planning.

A national *Strategy of wetland conservation* (MS, 2006) was elaborated in 2006, according to the Ramsar Convention's requirements, but has not been implemented successfully. In 2021, a draft of the ambitious new strategy was developed (GDOŚ, 2021).

In general, Poland has no tradition of preparing biodiversity strategies and programmes. Instead they are considered by authorities and wider society as declarative documents that need to be elaborated for others, rather than something that should be actually implemented.

24.4 Species Protection and Conservation Measures

24.4.1 Basic Protection Legislation and Strict Protection of Selected Species

According to the Nature Conservation Act (2004, with further amendments), killing and destroying any animal, plant or fungus, or destroying their habitat, is generally prohibited, with some exceptions such as for management, nature conservation, scientific research, angling, collecting for personal needs, and public health and safety. The Act establishes a system of additional protection for specifically listed species, which prohibits in particular the destruction or damage of the species, or their habitats, in all cases except where specific derogations may apply. The detailed lists of forbidden actions are in general coherent with the requirements of the Nature Directives. The list of protected species comprises 803 animal taxa (mainly species, but sometimes higher systematic groups), 322 taxa of fungi (including lichens) and 716 plant species. The list includes all birds (with the exception of some huntable species) and HD Annex IV species. In specific cases derogations apply.

Strict protection has been given to all large carnivores since 1998. This includes the Wolf, even though the Polish population is excluded from HD Annex IV, and the species has increased and expanded across most of the country. Polish society in general is tolerant of large carnivores, although some stakeholders (foresters and hunters) have become less accepting of their strict protection in recent years (Gosling *et al.*, 2019). Their suggestions to allow Wolf hunting have increased as the species has become more common, but there is an increasingly visible resistance to this in wider society.

Although the Polish population of the Beaver is also excluded from HD Annex IV, it is partially protected. However, the derogation conditions from protection are weak, resulting in annual permits for shooting about 16 000 Beavers. Damage caused by Beavers is compensated by the nature conservation authority, using funds from the state budget. The idea of designating the Beaver as a huntable species is opposed mainly by hunters, because tenants or managers of hunting units are responsible for the damage caused by the huntable species and are obliged to compensate the affected party.

For some species (e.g. birds of prey), so-called 'protective zones' (200–500 m) are created around nests, other breeding sites or wintering spots by the nature conservation authority, with most human activities banned within them. More than 3 500 such zones exists, mainly in state forests.

---

10 Ustawa o ochronie przyrody z dnia 16 kwietnia 2004.
11 Case C-441/17.
12 Case C-432/21.
13 https://monitorpolski.gov.pl/MP/rok/2015/pozycja/1207

With the exception of the special ruling for forest management, Polish legislation seems to be consistent with the species protection requirements of the Nature Directives. Nevertheless, the actual effectiveness of the protection varies. Unintentional, but significant, destruction of protected species' habitats commonly occurs as a result of 'business as usual' activities, such as the maintenance of rivers, ditches and trees. Also, under derogations to control fish-eating animals, several thousand Great Cormorants (*Phalacrocorax carbo*) and Grey Herons (*Ardea cinerea*), and hundreds of Otters (*Lutra lutra*) are culled annually. There have also been incidents of illegal killing of Osprey (*Pandion haliaetus*), Otter and Great Cormorant by owners of fish ponds that have been ineffectively prosecuted.

24.4.2 Regulation of Hunting, Fishing and Other Forms of Exploitation

Hunting is generally regulated by the Hunting Act (1995 with further amendments). There are 22 huntable native species, including deer, Wild Boar (*Sus scrofa*), Red Fox (*Vulpes vulpes*), Golden Jackal (*Canis aureus*), Badger (*Meles meles*), Pine Marten (*Martes martes*), Beech Marten (*Martes foina*), Polecat (*Mustela putorius*), Hazel Grouse (*Tetrastes bonasia*), Mallard (*Anas platyrhynchos*), Common Teal (*Anas crecca*), Common Pochard (*Aythya ferina*), Tufted Duck (*Aythya fuligula*), Coot (*Fulica atra*) and some other typical gamebirds. Additionally, eight alien mammal species are also huntable (e.g. Rabbit (*Oryctolagus cuniculus*).

The country – with the exception of urbanised lands, National Parks and most of the Nature Reserves – is divided into game management units, managed by State Forests or leased to the Polish Hunters Association. Tenants/managers are obliged to carry out sustainable game management (except for the Wild Boar, where there is a policy to reduce numbers as a preventative measure against African swine fever disease) on the basis of annual plans prepared by themselves. The adoption of plans, as well as the preparation of 10-year regional plans of management of larger animals, is the responsibility of State Forests. Game management plans are not subject to HD Article 6(3) appropriate assessments, even in Natura 2000 sites designated for species covered by the plans, which is questioned by some NGOs.

Despite the general sustainability assumption, the populations of Common Teal, Common Pochard, Tufted Duck and Coot seem to be decreasing, and hunting is considered to be a contributory cause (Wylegała and Ławicki, 2019). There is a significant anti-hunting social movement in Poland, especially against bird hunting. According to a social survey in 2019, 94% of Polish citizens are against hunting threatened bird species, and 67% are against bird hunting at all (Gumułka, 2020).

Fish management is regulated by the Inland Fishery Act (1985 with further amendments). All inland public waters, with the exception of National Parks and some Nature Reserves, are divided into fishery units, leased through public tender to interested organisations or persons. Tenants are obliged to implement sustainable fish management on the basis of multiannual plans, which are assessed by designated scientific institutions, including in relation to possible impacts on Natura 2000 sites. The nature conservation authority participates in the assessment.

24.4.3 Species-Specific Conservation Measures

For six species – Curlew (*Numenius arquata*), Great Snipe (*Gallinago media*), Greater Spotted Eagle (*Clanga clanga*), Lesser Spotted Eagle (*Clanga pomarina*), Montagu's Harrier (*Circus pygargus*) and Harbour Porpoise – national action plans have been formally adopted by the General Directorate of Environment Protection. For 11 other endangered species, as well as for some conflict species (e.g. Wolf, Otter, Beaver and Common Crane), informal species action plans/management plans have been prepared. The plans, containing species-related objectives and proposed measures, were mostly prepared as outputs of species conservation projects by relevant project beneficiaries. Nevertheless, even formally adopted plans do not include binding obligations to implement prescribed actions.

The conservation of some species is implemented through larger multiannual projects, many of which are voluntary initiatives of NGOs or State Forests, financed from external sources. For example, State Forests implements conservation projects for the Bison, Osprey and Capercaillie (*Tetrao urogallus*). NGOs have, for example, projects for the European Souslik (*Spermophilus citellus*), Edible Dormouse (*Glis glis*), Common Hamster (*Cricetus cricetus*), Lesser Horseshoe Bat (*Rhinolophus hipposideros*) and some other bats, European Pond Turtle (*Emys orbicularis*), Capercaillie, Black Grouse (*Tetrao tetrix*) and the re-introduction of the Eurasian Lynx (*Lynx lynx*) in western Poland.

24.5 Protected Areas and Networks

24.5.1 Protected Area Types and Their Coverage

Poland has developed over the last 100 years an extensive and multi-role network of protected areas under the responsibility of various actors (as described in Section 24.3.2). The current components, listed in Table 24.4 and further described in the following paragraphs, include seven types of protected areas developed under national legislation, Natura 2000 sites and other sites designated under international legislation. In many cases these components overlap, creating a combined legal regime. WDPA data (not including Biosphere Reserves, Natural Monuments, Ecological Areas and Nature Landscape complexes) from April 2022 indicate that, taking into account overlaps, the terrestrial network covers 123 355 km^2, which is 39.5% of the land (UNEP-WCMC, 2022). This comprises domestic site coverage of 32.2%[14] and partially overlapping

[14] According to the EEA EU Biodiversity Strategy dashboard, based on CDDA data (which are the European source for the WDPA) from 2021.

Natura 2000 coverage of 19.6%. Marine protected areas primarily comprise Natura 2000 sites, which cover 7 237 km^2 – 21.8% of Poland's marine waters. Although this is relatively high coverage, on land it is predominantly from Landscape Parks and Protected Landscape Areas (IUCN Category V), which are unlikely to provide effective protection for habitats and species.

Protected Areas Designated under Domestic Legislation

National Parks are rather large areas (>1 000 ha) designated for conservation of all elements of nature and the landscape. Nature conservation is their main goal, in most cases overriding all other activities. Designation requires a decision by parliament, followed by a governmental decree to establish the park's borders.

Each National Park has its own directorate and specialised staff, and generally manages land owned by the state within its boundaries. Overall, 85% of the National Parks' territory belongs to the state, although privately owned land is more substantial in some. Most human activities are prohibited, except for active nature conservation measures prescribed in the Protection Plan, and on private land. Nevertheless, in some cases conservation involves activities similar to forestry or farming, and some game animals are hunted for population control. Fishing on some lakes is still accepted as a 'traditional use' and angling is allowed in indicated waters. Visitor access is limited to indicated places and trails. The 23 National Parks in Poland cover about 1% of the country and vary in size from 2 146 ha (Ojców) to 59 223 ha (Biebrza). Woliński and Słowiński National Parks also include marine areas.

Nature Reserves are smaller areas designated exclusively for nature or landscape conservation, and have individual conservation objectives. The land still belongs to various owners but is strictly supervised by the regional conservation authority (RDOŚ). The legal regime is similar to National Parks. Nature Reserves cover about 0.5% of Poland, with their individual area ranging from 0.1 ha to 8 604 ha (average 113 ha). Although legally possible, no marine areas have been included yet.

Landscape Parks are areas designated by the regional government (Voivodship Assembly) for landscape protection with some elements of biodiversity conservation, integrated with sustainable land use. Land uses such as agriculture and forestry are still continued, and the protection is achieved mainly by land-use planning. Each park has a specific management body (directorate), which may supervise compliance, implement its own conservation projects, organise stakeholder collaboration for landscape care, and implement education measures. Landscape Parks cover 8% of the national territory.

Protected Landscape Areas are large areas designated for general landscape protection. The protection regime is only through some general restrictions on land-use planning, which lack clear biodiversity-related objectives, measures and monitoring. Only some of them have generic 'biodiversity conservation recommendations' declared. The contribution of this kind of protected area to biodiversity conservation is limited; nevertheless they are formally counted as part of the protected areas system and reported to WDPA. The areas cover 22% of the land.

Natural Monuments are protected natural features of outstanding importance (e.g. monumental trees, erratic boulders, rocks, caves, waterfalls or springs) that are designated by the local community. Most of them are trees (94%).

Ecological Areas are smaller fragments of usually unmanaged land important for biodiversity (e.g. swamps, fens, natural ponds, dunes and oxbows), designated by the local community. Designated Ecological Areas cover 0.2% of the country.

Nature-landscape Complexes have unique landscapes, natural and/or cultural values, and are designated by the local community. Protection is achieved mainly by land-use planning, usually with no limitations on forestry, farming and hunting. They vary from small areas (several hectares) to a few thousand hectares, and overall cover 0.5% of the country.

Natural Monuments, Ecological Areas and Nature-landscape Complexes are usually small protected areas that are designated and managed by local communities. They are not reported to the WDPA, but locally may be significant in protecting local biodiversity.

Natura 2000 Sites

Natura 2000 sites in Poland are a separate type of protected area, which may totally or partially overlap other protected areas. Based on the data in Table 24.4, at the end of 2020, terrestrial SPAs covered 15.5% of the country, and SCIs 11.0%, giving an overall Natura 2000 coverage of 19.6% of the land. Marine Natura 2000 sites covered 22.6% of Poland's marine waters, with overlapping 13.5% SCI coverage and 22.6% SPA coverage.

There is no indicative list of forbidden activities. If necessary, the authority has the competence to stop each action undertaken without appropriate assessment in case of likely significant negative impact on the site.

Sites Designated under Global and Non-EU International Legislation

Nine Polish marine Natura 2000 sites that contain the majority of the marine area in the network are also designated as HELCOM Marine Protected Areas,[15] covering 8 075 km^2 in total and c. 7 200 km^2 of the sea (53.3% of the Polish territorial and internal sea waters but only 7.4% of the EEZ). They also include the marine parts of Woliński and Słowinski National Parks.

The areas designated under global legislation as Ramsar Sites, World Heritage Sites and Biosphere Reserves have no specific status in Polish national legislation.

[15] http://mpas.helcom.fi/

Table 24.4 The number and extent of international designated sites and protected areas in Poland.

	IUCN category	Number	Area (km²)
Designated under global and non-EU international legislation			
Ramsar Sites	Ib, II, IV, NR	19	1 529
Natural and Mixed World Heritage Sites	NAp	2	630
Biosphere Reserves (exc. buffer and transition zones)	NAp	11	925
HELCOM Marine Protected Areas	II, NR	9	8 075
Designated under the Nature Directives			
SCIs (terrestrial + marine)	NR	864	34 266 + 4 339
SPAs (terrestrial + marine)	NR	145	48 428 + 7 224
Natura 2000 net terrestrial area		998	61 220
Natura 2000 net marine area		18	7 237
Designated under domestic legislation			
National Parks (park narodowy)	II, V, NAs	23	3 151
Nature Reserves (rezerwat przyrody)	IV, NAs	1502	1 702
Landscape Parks (park krajobrazowy)	V, NAs	126	25 204
Protected Landscape Areas (obszar chronionego krajobrazu)	V, NAs	387	69 265
Natural Monuments (pomnik przyrody)	III	34 898	Not measured
Ecological Areas (użytek ekologiczny)	NAs	8291	552
Nature-landscape Complexes (zespół przyrodniczo-krajobrazowy)	NAs	327	1 180

Note: NR = Not reported; NAp = not applicable; NAs = not assigned.
Sources. See Section 1.3. Based on data from: Ramsar Convention, UNESCO and HELCOM websites, WDPA (IUCN Categories), all accessed November 2021; EEA Natura 2000 Barometer, as at end 2020; others, Statistics Poland[16] 2021.

The sites included on the Ramsar list are usually already protected as one of the national forms. The 19 Polish Ramsar Sites cover c. 2.7% of the estimated area of wetlands.

The UNESCO natural World Heritage Sites are Białowieża Forest and part of the multinational Ancient and Primeval Beech Forests of the Carpathians and Other Regions of Europe. The Białowieża Forest World Heritage Site is the transboundary Polish–Belarusian site covering a total area of 141 885 ha (42% in Poland). The status proved to be useful in the conflict over the logging of the forest (described earlier and in Annex PL.3). The Polish part of the Ancient and Primeval Beech Forests of the Carpathians and Other Regions of Europe World Heritage Site is composed of four forests in Bieszczady National Park (3 471 ha in total).

Biosphere Reserves in Poland do not have an official protected areas status, but are rather areas of stakeholder collaboration, usually organised by nature conservation bodies. The core zones are always formally protected with a non-intervention regime, as Nature Reserves or strict protection zones of National Parks, whereas buffer zones and transition areas have varied status (active conservation zones of national parks, landscape parks or unprotected). The collaboration and integrated sustainable development stimulated under the Biosphere Reserve flag is potentially a useful mechanism, but so far its effectiveness outside core zones is still in question due to the lack of legal obligations and enforcing instruments.

24.5.2 The Coherence of Protected Areas and Ecological Networks

Despite its very extensive coverage, the Polish network of protected areas is commonly assessed as still being insufficient. The national biodiversity strategy (*Programme of Biodiversity Conservation and Sustainable Use for 2015–2020*),[17] indicated an urgent need to establish at least three additional National Parks and enlarge several others. There are also suggestions from others for 13–15 additional National Parks in various parts of Poland (Holeksa *et al.*, 2017, and literature cited therein). Nevertheless, these recommendations have not been implemented.

Similarly, analyses of the Nature Reserve network suggest that hundreds of additional Nature Reserves are still necessary (Jermaczek, 2016; Holeksa *et al.*, 2017, and literature cited

[16] www.stat.gov.pl

[17] https://monitorpolski.gov.pl/MP/rok/2015/pozycja/1207

therein). In 2018, the Polish NGO Naturalists Club started a project to help develop the Nature Reserves network, by preparing and publishing regional proposals.[18] In 2021, a coalition of Polish NGOs was established to support an appropriate Polish contribution to the 10% strictly protected areas target of the *EU Biodiversity Strategy for 2030*.

Although updated several times in response to the European Commission's demands over 2004–2020, the Natura 2000 network also remains insufficient. Consequently, an official infringement file has been opened by the Commission for insufficient identification of SCIs for 54 HD habitats and 57 HD species. Whilst SPAs cover all Important Bird Areas (IBAs) known in 2004, they do not include 30 new sites identified in a 2010 IBA update (Wilk *et al.*, 2010).

The concept of a comprehensive ecological network has been permanently present over the course of Poland's development of its protected areas system. In the 1970s, the idea of integrating protected areas into a unified system as a basis for general land-use planning was proposed (Różycka, 1977). In the 1990s, this idea was extended through an ECONET-POLAND network project (Liro, 1998). The ideas were included in some studies of national and regional development and land-use planning. Presently, the older concepts have been substituted by the Natura 2000 network with the additional ecological corridors connecting sites (Jędrzejewski *et al.*, 2011). Nevertheless, the effectiveness of the protection of ecological corridors is limited as it is not based on a binding legal requirement. Only HD Article 6(3) appropriate assessment procedures concerned with Natura 2000 sites enforce in some cases the need to analyse project impacts on ecological connectivity.

24.5.3 Approaches to Protection and Levels of Protection

Domestic Protected Areas

The diversity of types of protected areas reflects a gradation from a land-sharing to a land-sparing approach, as well as various levels of responsibility for the site. The conventional domestic forms of protection, developed before EU accession, are based on fixed lists of activities that are banned within the protected area.

The National Parks and Nature Reserves, in particular those designated on public land, are the areas intended almost exclusively for nature conservation. Their general legal protection approach is to ban all activities, so that non-intervention is the default, with exceptions only for activities that are necessary for conservation. Conservation measures are planned and implemented only if the need is convincing and clear. Three zones may be established if necessary: strict protection (non intervention zone), active protection (with either non-intervention or active conservation measures applied), and a so-called 'landscape protection zone' (private land with continued management by landowners). National Parks and Nature Reserves are commonly considered to be the most effective domestic protected areas, although they have some limitations, as described below.

Firstly, restrictions on activities in domestic protected areas do not apply to private land, even in National Parks and Nature Reserves, which sometimes means the protection of such land is illusory. Whilst National Parks are large enough to protect whole landscapes and ecosystem complexes (unlike Nature Reserves), they are not wild areas and are often fragmented by roads and other infrastructure. In such cases the traffic and, in particular, road renovation, may be detrimental to the Park's fauna (e.g. Biebrza, Kampinos and Magura National Parks). National Parks and some Nature Reserves are also commonly affected by tourism and recreation, which is usually allowed but limited to selected trails and places. However, the restrictions do not in practice limit visitor numbers, which has caused problems in some sites (as discussed in Section 24.2.3). Furthermore, in the Tatra and Karkonosze National Parks there has been notable pressure to develop ski infrastructure. Although forestry is normally prohibited (unless required for management or restoration), some damaging salvage logging has been allowed in response to natural disturbances, such as forest fires, wind-throws and insect outbreaks.

Landscape Parks, Landscape Protected Areas and Nature-landscape Complexes are examples of an 'integrated' (land-sharing) approach to nature conservation within protected areas, integrating some conservation measures with sustainable land use. The general protection system is based on a list of prohibited activities, but with general derogations for sustainable land uses, including forestry, farming, hunting and fisheries. Projects considered to have significant effects on the environment are generally forbidden, with the exception of projects of public interest. In practice, use of these derogations is widespread. Additionally, in many cases the small and scattered developments or improvements (e.g. dispersed habitation, small water abstractions, water course maintenance and small water reservoirs) that are not prohibited by the protected area rules can still generate significant cumulative impacts.

In general, the effectiveness of these types of protected area seems to be weak, as forest management and farming practices within them are the same as outside. Furthermore, the Landscape Parks' management plans have no legal power to enforce any modifications. Nevertheless, due to better recognition of their natural values and existence of management teams supervising the areas (see also below) the protection of Landscape Parks is better than that of Landscape Conservation Areas. At least some parks' teams are active conservation managers, implementing specific projects on biodiversity and landscape conservation and/or raising public awareness on biodiversity issues.

Natural Monuments and Ecological Areas are rather effectively protected, generally just by 'labelling' the subject of protection and increasing awareness of it, to prevent its destruction. Implementation of proactive conservation measures is much more problematic, but not necessary in most cases.

[18] www.kp.org.pl/pl/rezerwaty-przyrody-czas-na-comeback/

Natura 2000 Sites

Natura 2000 sites have specific protection through the transcribed provisions in the Nature Directives (e.g. HD Articles 6 (2) and 6(3)). There is no indicative list of forbidden activities. Instead protection is primarily through the appropriate assessment system and the general ban on plans and projects with potentially significant detrimental impacts. If necessary, the nature authority has the powers to stop potentially damaging actions without an appropriate assessment. In some famous cases, the legislation in the Habitats Directive has prevented some damaging impacts, such as the plan to build a road across the Rospuda valley (see Annex PL.3). Due to these rules and experiences, on the one hand Natura 2000 sites are perceived by society as rather effectively protected; on the other hand, some consider these protected areas as an obstacle to economic development, especially because Natura 2000 has doubled the area of effective protected areas, which is a source of conflict (Grodzińska-Jurczak and Cent, 2011).

Nevertheless there are numerous cases where the appropriate assessment procedure has not worked so well. In particular, the quality of early screening of projects, especially by local authorities, has in some cases been doubtful, thus some small projects affecting Natura 2000 sites have avoided assessments. Special legislation has also allowed implementation of some specific projects before the appropriate assessment has been finalised. Shortcomings also result from political pressure for investments to go ahead. Presently (2022) some controversial projects affecting Natura 2000 sites are strongly supported by the government, for instance construction of dams on the Vistula river and the Wisłoka river, a navigable channel crossing the Vistula Spit, the S-16 express road crossing Masurian Lakeland and Biebrza Valley, new coal mines in Polesie and the barrier along the Polish–Belarusian border.

Projects that have damaged Natura 2000 sites have only been allowed to go ahead due to overriding public interests and a lack of alternative options (i.e. under HD Article 6(4) derogations) a few times: to construct three roads (A2 highway, S3 express road, regional road in Grądy w Dolinie Odry site), for one water management project (Wroclaw Water Node), and to extend a military base in Powidz (Gniezno lakeland). The compensatory measures that were carried out included designation of some new Natura 2000 sites, and restoration of a river section and some alluvial forests.

24.5.4 Approaches to Managing Protected Areas

Traditionally, management of protected areas in Poland has mainly relied on prohibiting destructive activities within them. Whilst this remains the case for some, a more contemporary approach based on management planning, establishing conservation objectives, implementation of relevant measures and monitoring of results has been more widely implemented since the 1980s for National Parks and Nature Reserves. The need to implement the Habitats Directive's site conservation objective approach for Natura 2000 sites was a further motivation to increase the clarity of management planning schemes.

The preparation of 10–20-year management plans is now obligatory for National Parks, Nature Reserves, Landscape Parks and Natura 2000 sites. All plans must follow the structure and content set out in regulations, and are eventually established as binding legal acts. They set out the conservation objectives, identify threats to the protected features, and list conservation measures. The plans are developed by relevant conservation authorities (in practice by contracted external experts), usually based on a field survey, and then consultations with relevant stakeholders. The site conservation objectives for Natura 2000 sites are based on common national criteria for favourable conservation status of species and habitats, with some flexibility for local conditions.

Despite the legal obligation, the plans have not yet been completed for all relevant protected areas. As of 2021, almost all National Parks had at least draft plans, but less than 50% of Nature Reserves, less than 30% of Landscape Parks and c. 55% of Natura 2000 sites were covered. There have also been problems with the quality of some plans, for example being too generic and setting unverifiable objectives. For domestic protected areas there is no common monitoring standard. In Natura 2000 sites, monitoring is based on specific, mostly measurable, sets of indicators (see Mróz, 2013). However, even if the sites' objectives are not reached, adaptive management is not always implemented, due to the substantial burden associated with changing a plan.

Only National Parks and some Landscape Parks have dedicated site managers (i.e. management teams working in, and exclusively for, the site). Most Landscape Parks, all Nature Reserves and Natura 2000 sites are managed by regional conservation authorities, with personnel that remotely manage many sites.

Last but not least, the standard public budget available for nature conservation is usually only sufficient to cover basic staff costs, and is inadequate to complete management plans, let alone implement the required measures. Even obligatory measures are not implemented if there is a lack of funds. Therefore, despite having a common system of protected area management, the effective practical management of many sites is dependent on externally funded projects, in particular using EU funds (as further discussed in Section 24.7.3). This is particularly the case on private land, where effective conservation is very difficult without incentive payments. Currently the CAP Rural Development Programme (RDP) agri-environment–climate measure funded schemes are the only such incentive mechanisms that are widely used (see Section 24.6.2).

24.6 General Nature Conservation Measures

24.6.1 Development Planning and Impact Assessment

Land-use planning in Poland has a hierarchical structure, encompassing national, regional (voivodship) and local levels.

Nevertheless, only detailed local land-use plans are legally binding and generally prevent investments that are not compliant with the plans. All plans are subject to an SEA, and those that overlap protected areas, or protective zones around them, require an opinion from the nature conservation authority.

At the national level, the *National Spatial Development Concept*[19] has been developed and adopted by the government (last updated 2011). One of its objectives is the development of a spatial structure that supports the maintenance of high-quality natural environment and landscape values. Development of the protected area network is assumed, including establishment of three new National Parks and enlargement of some others, establishment of new Landscape Parks, and legal protection of ecological corridors. Nevertheless, at the same time, the concept declares transport network and general economic development intentions, which may conflict with biodiversity needs, but are not addressed at this planning level.

Regional spatial development plans are prepared by the regional authorities. The quality and topicality of the plans varies, but in most cases the present system of protected areas, as well as ecological corridors, are properly reflected. In some plans the needs of new protected areas are also expressed quite well. However, at the same time, the plans include project ideas that may damage the environment (e.g. new water reservoirs).

At the local level, two spatial planning instruments exist. The whole territory of each community is covered by so-called Studies of Conditions and Directions of Spatial Development.[20] These documents should describe all natural and social conditions of the community and develop spatial development policy, although they are not legally binding. Some elements of 'green infrastructure' may be considered, although the term is not used in the legislation and not commonly used in the practice of planning. If necessary, detailed and legally binding Land Use Plans[21] are developed, determining details of possible land use and investments.

Despite this rather well-designed spatial development system, some problems appear in practice. Firstly, if the area is not covered by a Land Use Plan, the individual local authority can make its decision on a project without taking into account higher-level plans. Thus gaps in Land Use Plan coverage over most of Poland are the primary threat undermining the system. Secondly, the legal regime allows exceptions for some kinds of public investments (e.g. roads and flood prevention infrastructure), which may be built even when contrary to the plans. Thirdly, despite SEA procedures, the quality of biodiversity information in the assessment reports is not always sufficient.

The authorisation procedure for individual development projects generally reflects the requirements of the EU EIA and Habitats Directives. However, there are weaknesses, the main one being the insufficient quality of EIA reports, especially for smaller projects of local significance. For larger investments, a problem has been created by special legislation (several acts for specific developments, e.g. roads, flood defence and the Vistula Spit navigation channel) that has allowed implementation of some specific projects before finalisation of the EIA procedure. Even the eventual cancellation of environmental consents by a final court ruling has not prevented the continuation of an investment project. In 2021, the decision to build the barrier along the Polish–Belarusian border was made without any environmental impact assessment.

24.6.2 Habitat Management and Restoration

Tackling Agricultural Issues

Given the high proportion of agricultural land in Poland, and the impacts that agricultural practices are having on biodiversity, implementation of the CAP has been under high scrutiny since EU accession in 2004. Of particular relevance have been the scale and effectiveness of RDP agri-environment schemes. The Polish 2014–2020 RDP included a number of measures that contributed to biodiversity conservation, of which the agri-environment–climate measure funded schemes for maintaining natural habitats and bird habitats were most important, particularly for Natura 2000 sites (see Section 24.7.3). Payments under the RDP measure 'Investments in farms in Natura 2000 sites' also supported the schemes by helping farmers purchase equipment for meadow management and infrastructure for livestock. Schemes have also contributed to maintaining natural habitats outside Natura 2000 sites.

Several variants of the agri-environment schemes have been available. One of them was just extensive grazing or mowing in SPAs. Others have been more tailored, and were designed by certified experts who confirmed the presence of target habitats or birds (e.g. various meadows and mires, waders, Aquatic Warbler, and Corncrake (*Crex crex*)) and then prescribed the optimal management measures for them. Average annual payments were from €150/ha for extensive meadows to about €300/ha for more demanding habitat management (MRiRW, 2014). In fact, no scheme has addressed climate objectives, despite the urgent need to prevent CO_2 emissions from degraded peatlands in agricultural use.

In 2018, the agri-environment–climate measure funded schemes were implemented on approximately 1 million ha, including about 295 000 ha in Natura 2000 sites (about 50% of the assumed Natura 2000 site requirements, GDEP (2021)). Although detailed monitoring results have not been published, the schemes seem to have been partially effective. However, this was reduced by low compliance with the detailed rules of the particular schemes and experts' recommendations. The main problems and potential future development of agri-environment schemes, including necessary improvements, have been identified and summarised by Stalenga *et al.* (2016).

---

[19] pl: Koncepcja Zagospodarowania Przestrzennego Kraju.

[20] pl: Studium uwarunkowań i kierunków zagospodarowania przestrzennego.

[21] pl: Miejscowy plan zagospodarowania przestrzennego.

Forest Habitats and Forestry

Forest management is based on the Forest Act, which includes biodiversity conservation as an objective and assumes sustainable use of forest resources. For state-owned forests, which comprise 81% of the forest area, management plans are prepared by the forest manager for each 10-year period and accepted by the Ministry of Environment. For other forests, the plan is prepared and accepted by the county-level administration. Since 2009, due to a Polish Administrative Court rule, forest management plans have been generally subject to SEA and Natura 2000 impact assessments. However, according to a CJEU ruling,[22] the assessment methodology does not fully meet the requirements of the Habitats Directive. Additionally, the Directive's rules are sometimes disregarded by foresters and the Ministry, with some cases of forest management plans having been authorised despite significant negative impacts on a Natura 2000 site according to the regional conservation authority (Pawlaczyk, 2017).

Forest management standards (Silviculture Rules[23], Stands Protection Rules[24]) are established by State Forests, the specialised public agency, and followed in state-owned forests and usually in others. Small clear-cuts (up to 6 ha) are commonly used in pine forests. In mountain forests clear-cuts are not normally carried out (with the exception of salvage logging after disturbances). In deciduous forests a shelterwood system prevails. Despite a commonly declared preference by Polish foresters for natural regeneration, more than 90% of stands are regenerated by tree planting. Although State Forests declare that they care for forest biodiversity elements (e.g. protected species, deadwood and tree microhabitats), in practice managed forests have a significantly simplified structure and reduced biodiversity (Pawlaczyk *et al.*, 2016).

State Forests manages 77% of the forest area. It is a self-financing agency, using timber income for management measures and investments in managed forests, paying a low tax (c. €7/ha of forest annually) and a relatively small contribution to the state budget (2% of its income from timber sales). With its c. 7.6 million ha of forest under management, 26 400 employees and an annual budget of c. €2.3 billion, State Forests is the largest forest owner in the EU and one of the biggest companies in the Polish economy – giving it significant political influence. In practice, State Forests' opinion has often led to exemptions from the general rules applicable to forest management. At the local level, State Forests' opinion has often significantly influenced the development of natural protected areas and approval of nature conservation management plans, often blocking the creation of new National Parks and Nature Reserves (Pawlaczyk *et al.*, 2016).

Conflicts over forest management in state-owned forests have arisen in a number of places. The most well-known has concerned Bialowieża forest and the use of salvage logging and sustainable forest management rather than the protection of natural processes (Annex PL.4). Other conflicts have arisen in intensively harvested forests in Bieszczady and the Carpathian foothills. In some places the forest management has been contested not only by conservationists, but also by the general public interested in natural forest wildlife and landscape values.[25]

On the other hand, State Forests are involved in some nature conservation projects. The agency has carried out national conservation projects for the Bison, Osprey, Capercaillie and Black Grouse, as well as some local wetland restoration projects. For decades it has cooperated with an NGO (the Eagle Conservation Committee) to establish protective zones around the nests of several species of birds of prey, the Black Stork and some other rare birds.

Water Management, Freshwaters and Other Wetlands

Hydrological modifications, through former river regulation and other alterations, and ongoing typical river and ditch maintenance practices, are common pressures on biodiversity, including in protected areas. Thus, water management seems to be of special importance.

Since 2017, all inland waters in Poland have been managed by the specialist agency Polish Waters. The general water management model follows WFD requirements and is based on a six-year River Basin Management Plan (RBMP) that establishes the environmental objectives to be achieved for each water body. The first RBMPs in Poland did not contain specific objectives for protected areas, and were therefore incompatible with WFD requirements. The second RBMPs, for 2015–2021 (delayed until December 2016), do comprehensively identify the water management needs of Natura 2000 sites, National Parks, Landscape Parks, Protected Landscape Areas and Nature Reserves. Due to inadequate knowledge, their objectives are not always clear, but they are likely to be improved as site management plans are further developed.

Many river management practices, in particular river maintenance measures, are problematic from the point of view of biodiversity. Common practices that damage river ecosystems include mud, sand and gravel extraction, embankment strengthening, removal of woody debris and the removal of trees from river banks. However, in 2018, new guidelines for good river maintenance practices were prepared (Biedroń *et al.*, 2018; Prus *et al.*, 2018), which would probably improve the situation – if followed by river managers.

[22] CJEU judgement on case C-441/17 on the Białowieża Forest but applicable also to all forest management plans in Poland.

[23] pl: Zasady Hodowli Lasu.

[24] pl: Instrukcja Ochrony Lasu.

[25] http://lasyiobywatele.pl/

24.6.3 Pollution Challenges

One of the most important environmental problems in Poland is pollution, especially nitrate and phosphate pollution from the point of view of natural habitats (Szalińska *et al.*, 2018).

Poland has had substantial problems with the implementation of relevant EU legislation to control water pollution. The Nitrates Directive was transposed to Polish legislation by the Water Act, by the Fertilisers Act and by the Decree on Details of Fertilisers' Use. After several attempts to designate and enlarge the Nitrate Vulnerable Zones (where nitrogen inputs from agricultural sources should be reduced), they were eventually found to be insufficient by the CJEU in 2014.[26] Poland therefore decided to establish and apply an action programme (under Article 5 of the Nitrates Directive) throughout the whole national territory. The Programme of Reduction Nitrogen Pollution of Waters was established by a government decree in 2018. The programme lacks active measures but is rather a set of requirements for farmers, establishing limitations on fertiliser use in buffer zones around large water bodies and channels, limiting fertiliser use periods and conditions for fertiliser use on slopes, and establishing requirements for fertiliser storage and for farm nitrogen management planning.

Despite the national programme, the problem does not seem to have been fully resolved. In particular, the lack of buffer zones along small ditches may be causing rapid nitrogen flows along the ditches to water bodies. Also, stronger limitations are necessary for sensitive habitats, such as oligotrophic waters (3110), oligo-mesotrophic waters with *Chara* species (3140) and rivers with *Ranunculion fluitantis* and *Callitricho-Batrachion* vegetation (3260). However, this is not possible according to the present legislation, even in particular Natura 2000 sites. The programme also lacks proactive measures. Nevertheless, as mentioned earlier, some of the farmers' investments needed to meet the requirements may be financially supported by RDP measures.

The artificially high levels of phosphates in waters are even more detrimental to aquatic ecosystems due to changes in the N:P ratios, which leads to eutrophication and promotes cyanobacterial blooms in lakes, reservoirs and the Baltic. Whilst some of the measures to reduce nitrogen pollution described above will reduce phosphate levels, more needs to be done to tackle this problem. In particular, restoration of wetland buffer zones along water courses is necessary (Biedroń, 2020; Jabłońska *et al.*, 2020).

24.6.4 Invasive Alien Species

Like other European countries, Poland is affected by IAS, with about 1 000 alien plant species recorded in Poland (27% of the present flora). Of these, 84 have been assessed as invasive or potentially invasive (Tokarska-Guzik *et al.*, 2012). Amongst animals, 305 alien species have been recorded, 119 of which are invasive or significant for other reasons (Głowaciński *et al.*, 2012). Some types of wetlands are highly affected by IAS (Dajdok and Pawlaczyk, 2009).

A database of alien species present in Poland has been completed[27] and national risk assessments conducted for 118 of the most important species. Several externally financed projects have tackled IAS problems, eradicating American Mink (*Neovison vison*) in some National Parks, *Spiraea tomentosa* in Drawa forest, *Rosa rugosa* in Hel penisula dunes, and Caucasian hogweeds (*Heracleum mantegazzianum* and *Heracleum sosnowskyi*) in several localities.

In 2022, a specific Act on Alien Species entered into force, implementing the EU regulation on Invasive Alien Species. Under this, the introduction of alien species is generally prohibited, but for most of the species there are extensive derogation provisions related to gardening, the management of parks and other green areas, agriculture, aquaculture and forest management. Lists of IAS of European and of national importance have to be established by the Ministry of the Environment (although this has not yet been done). Such species cannot be kept, bred or sold, and citizens are obliged to inform the relevant authority of the species' presence. Conservation authorities, State Forests, water authorities, local communities and landowners are required to implement relevant preventive measures to control or eradicate such IAS.

Despite the established legislation, there is still no comprehensive national strategy to tackle the IAS problem.

24.6.5 Habitat Restoration and Rewilding

Although Poland has no comprehensive restoration strategy, individual restoration projects have been implemented. Restoration of peatlands, including the restoration of their hydrological conditions, has been part of nature conservation projects implemented by NGOs, several National Parks and State Forests. However, in 2017, the restored peatland was only c. 7 500 ha (Peters and Unger, 2017), compared to c. 1.3 million ha of degraded and drained peatlands in Poland (Kotowski, 2021). An ambitious vision of extensive peatland rewetting and restoration, to reduce GHG emissions, was recently presented in the draft of the new *Wetlands Conservation Strategy* (GDOŚ, 2021).

Only a few river restoration projects have been carried out: re-creating meanders, supplementing gravel sediments, supplementing coarse debris or restoring connectivity. Building fish passes is more common. In 2020, *The National Programme of Waters Restoration* was developed (Biedroń, 2020), supplemented by restoration guidance (Pawlaczyk, 2020). It was estimated that, in order to meet the WFD's targets of good ecological status or potential, c. 90% of Polish rivers need hydromorphological restoration. At the very least, harmful maintenance works need to be halted to enable some natural recovery.

[26] Case C-356/13.

[27] www.iop.krakow.pl/ias

The idea of rewilding is not very popular in Poland. Two areas have been declared as 'rewilding' areas by some stakeholders: Oder delta in north-western Poland and Bieszczady in the south-east. In practice, only the ecotourism components based on some wilderness features have been implemented. Overall, the protection of the remaining wilderness, road-less areas, and increasing non-intervention approaches in protected areas, is promoted by only a few experts in Poland (Jermaczek, 2014; Pawlaczyk, 2014a; Jermaczek and Kwaśny, 2019).

24.7 Nature Conservation Costs and Funding

24.7.1 Nature Conservation Costs

There has been no comprehensive analysis of the costs of biodiversity conservation in Poland. Only the budget of the National Parks is clearly discernible in the official state budget. In 2019, the total expenditure of the 23 National Parks was €68 million (€215/ha), varying from €50/ha to €1 200/ha of protected area. About 35% of their costs were covered directly from the state budget, with 33% from each park's income – mainly from entrance tickets and the sale of timber harvested for conservation purposes. The parks were also donated c. €10 million by State Forests. The remaining budget for specific conservation projects was funded by national or EU financial mechanisms (see Section 24.7.3).

The financial needs for implementing Natura 2000 have been fairly reliably estimated for the Prioritised Action Framework (PAF) for 2021–2027 (GDEP, 2021) by adding the costs of all prescribed measures in the site management plans and extrapolating the costs to the sites without plans. This has given a total annual cost estimate of c. €1 062 million. This includes the general cross-cutting measures for the network, which are estimated to cost c. €246 million per year averaged over the period, of which 98% are one-off costs (mainly research, etc.). Maintenance and restoration measures for the network are estimated to be c. €752 million per year on average (94% running costs). The costs of additional species-specific measures are estimated to average c. €64 million per year (69% recurrent costs). The PAF cost estimates do not include the cost of green infrastructure to support the Natura 2000 network, although this is requested as part of the standardised PAF format.

24.7.2 The Benefits of Nature Conservation

Quantitative assessments of the value of ecosystem services provided by protected areas and biodiversity in general are not well-developed or common in Poland, and there is no tradition of using such estimations in decision making. There are only a few examples of such valuations of natural protected areas, contrasting with more common valuations of other environmental components (Żylicz, 2017). Buszko-Briggs *et al.* (2004) estimated the economic value of recreation services of Białowieża Forest, finding that the value was about 27 times larger than the profits from the timber sold therein in the early 2000s.

Interesting attempts to introduce ecosystem service valuations have been proposed for water management, in guidance on river maintenance good practices (Biedroń *et al.*, 2018, and references therein). More recently, a cost-benefit analysis has demonstrated substantial benefits from restoring wetland buffer zones along rivers through their reduction of nitrogen and phosphorus pollution impacts on freshwaters and the Baltic Sea (Jabłońska *et al.*, 2020). A similar analysis was carried out of the national wetlands restoration programme proposed in the new *Wetlands Conservation Strategy* (GDOŚ 2021): in 10 years the potential benefits would be an estimated c. €16.5 billion, three times more than the strategy's implementation costs.

24.7.3 Funding Sources for Nature Conservation

Since Poland's accession in 2004, the main funding sources for nature conservation have been EU funds (with national-co-financing), as indicated in Table 24.5.

By far the largest source of EU funding has been from the European Agricultural Fund for Rural Development (EAFRD) for RDP measures. Most of the EAFRD budget (46%) has been allocated to payments to farmers within areas facing constraints, which support biodiversity only indirectly, reducing land abandonment. As described earlier, the agri-environment–climate measure has been especially important for funding habitat management and some species conservation, especially within Natura 2000 sites.

Poland decided not to use the RDP forest-environment measures, claiming that private forests are not significant for biodiversity at a national scale. Nevertheless, the absence of this funding instrument has hindered Natura 2000 management planning in some sites in central and eastern Poland. Also the Natura 2000 payments measure is not used in Poland, due to a lack of binding obligations for farmers in Natura 2000 sites that need to be compensated.

Over 2014–2020, €622.9 million, supplemented by €123.9 million from national sources, were allocated to Poland for nature and biodiversity from the ERDF and Cohesion Fund (through the Operational Programme Infrastructure and Environment, and regional operational programmes). Although not quantifiable, much of this was used for financing Natura 2000 network needs, such as preparing more than 400 management plans and implementation of some conservation measures.

Over 2003–2020, the LIFE financial mechanism supported 49 LIFE nature projects that have benefited Poland, with a total value of c. €143 million (€79 million from the EU). Of these projects, about 60% targeted habitats (with an emphasis on grasslands, rivers and mires) and the rest targeted species (especially birds) except for one project on invasive species.

The European Maritime and Fisheries Fund (EMFF) has been little used for nature conservation purposes. This is

Table 24.5 Funding sources for nature conservation in Poland.

Source and period	Measure/category	Average annual allocation (€ m)		Comments
EU funding instruments		**EU + national**	**EU**	
EAFRD Rural Development Programme measures – allocations to Priority 4 (restoring, preserving and enhancing agriculture and forest ecosystems) 2014–2020	M13 Areas facing constraints	283.3	180.3	20% overlapping N2k, probably indirectly supporting agricultural biodiversity
	M10 Agri-environment–climate	195.2	124.2	Very important for N2k and meadows outside N2k. 35% for N2k
	M11 Organic farming	100.0	63.6	Indirectly supporting agricultural biodiversity
	M4 Investments in physical assets	29.9	24.2	Biodiversity benefits unclear
	M8 Investments in forests	4.3	2.7	Biodiversity benefits unclear
	M1 Knowledge transfer, etc.	1.7	1.1	Biodiversity benefits unclear
RDP Priority 4 subtotal		*614.4*	*396.1*	
ERDF/Cohesion fund 2014–2020 – national operational programme	85 Protection and enhancement of biodiversity, nature protection and green infrastructure	138.3	117.2	For N2k etc. Allocation not fully spent
	86 Protection, restoration and sustainable use of Natura 2000	26.6	22.6	Allocation not fully spent
ERDF 2014–2020	16 Regional Operational Programmes	27.3	21.8	For N2k etc. Allocation not fully spent; biodiversity benefits not always clear
Interreg V	Protection and enhancement of biodiversity, green infrastructure and N2k	13.3	11.8	For N2k etc. Incl. Swiss and Norwegian funds
EMFF 2014–2020	2.5 Aquaculture providing environmental services	12.3	9.2	For N2k etc. Important for fish pond biodiversity, both in N2k and outsideAllocation finally not used (see text)
	4.2 Local development strategies	12.0	10.2	For N2k etc., but relevance unknown, allocation not fully spent
	1.4 Conservation and restoration of biological diversity	1.6	1.6	For N2k etc. Allocation not fully spent
	Others of some relevance to N2k etc. (1.1, 1.2, 1.3, 6.2 and 6.3)	2.7	0.9	Some biodiversity benefits unclear
LIFE Programme	Nature and biodiversity projects (2003–2020)	7.9	4.4	
EU funding instruments subtotal		*856.5*	*595.7*	
National funds within 2014–2020		**Additional national***	**Comments**	
State budget		42.5	Covers staff and other basic requirements	
State Forests	Donation to National Parks	6.3		
National Fund for Environmental Protection and Water Management		3.2		
Regional Funds for Environmental Protection and Water Management		2.2		
Additional national subtotal		*54.2*		
National total including co-financing of EU instruments		*315.0*		
Total		**910.7**		

Notes. 'For N2k etc.' indicates funding relevant to Natura 2000, related green infrastructure and BHD species (based on PAF). * Excludes co-financing of EU funds unless indicated. The subtotals and totals are only approximate indications of the average annual levels of funding as they are based on annualised data from differing periods.

Sources. Based on data from: RDP data from European Commission ESIF funds database (23 October 2020); RDP allocations to Natura 2000, ERDF, EMFF and national funding from 2021–2027 PAF (GDEP, 2021); Poland LIFE factsheet, European Commission (updated May 2021). See Section 1.3 for further details.

despite traditional carp fish ponds being important habitats for some species in the Polish Natura 2000 network. In recognition of this, support for environment-friendly inland aquaculture (fish ponds) was foreseen in the Operational Programme 'Fisheries and the Sea' under the EMFF. However, the allocation of €92 million was cancelled in 2017 by the government and transferred to marine fisheries for other purposes. This decision was criticised by the nature conservation sector and NGOs because it leaves Natura 2000 fish ponds without any financial support.

The main national funding source for nature conservation is the national state budget. This only covers basic needs, such as maintaining the national protected area administrations' and relevant nature conservation authorities' offices, staff and logistics. The implementation of practical conservation measures usually relies on external funding. Additionally, the regional and local communities' budgets cover some basic needs of Landscape Protected Areas and local protected areas.

The other main domestic source for conservation is the National Fund for Environmental Protection and Water Management (NFOŚiGW), which is a state legal body established in 1989 that uses revenue from environmental penalties and fees. The annual budget of the fund is about €750 million, but only a minor proportion is dedicated to biodiversity conservation and protected areas. Interested bodies or legal persons can apply to the fund for project financing (usually up to 75–85% of the project costs) in response to calls and priority programmes. At the regional level, there are similar Voivodships Funds (WFOŚiGW).

There has been no significant private funding of Natura 2000 or broader nature conservation in Poland. Some conservation NGOs have used their own funds (usually earned from providing expert opinions, or preparing management plans, etc.) for conserving nature on their own land, and as co-financing for externally funded projects. Several commercial companies have provided small grants for nature conservation (e.g. to enhance their business image), but the scale of this funding has been very small.

In general, financing is very much reliant on the EU. Although the availability of such EU funds substantially improved the situation for Poland after EU accession, funding is still far below the estimated nature conservation needs for Natura 2000 alone (GDEP, 2021). The problems concern not only the overall amount of funding available, but also its uptake and effectiveness. Some funds marked as 'promoting nature' in fact only indirectly and slightly benefit biodiversity, or are neutral (e.g. compensating natural constraints for farmers, supporting organic farming, infrastructure investments in Natura 2000 and local development), or are even harmful (e.g. improved water management and some tourism developments). Moreover, the limited capacity of the nature conservation authorities and conservation NGOs reduces the actual funding of nature conservation more than the amounts of money potentially available.

24.8 Research and Monitoring

There has been no systematic assessment of nature covering the whole of Poland, and therefore information on species and habitats is based on scattered publications and records, and partial surveys. Existing syntheses and atlases, etc., mostly present the sum of such scattered information rather than the results of systematic field surveys. Some regional biodiversity surveys of local communities (*gmina*) have been carried out, mainly in the 1990s. In 2007, a survey of the Polish State Forest was conducted, mainly by State Forests' employees, but the results are not fully credible. Usually, surveys of natural features in protected areas are done as part of the development of management plans. In 2009, a standard for spatial nature conservation data was elaborated, and most data collected in protected areas are now coherent with this standard. Although a comprehensive system for the public presentation of available data on species and habitats does not exist yet, one is in development (except for some sensitive species).

Before accession to the EU, biodiversity monitoring in Poland was limited. After 2004, wide monitoring programmes for the BHD habitats and species were developed to meet the requirements of the Nature Directives. This monitoring responsibility falls under the General Inspectorate for Environmental Protection, although the practical work is carried out by contracted scientific institutes.

The most advanced biodiversity monitoring is on birds; it covers most species and is based on 2 500 randomly located plots to produce a representative assessment of populations.[28] For non-bird species, more than 800 plots are monitored for plants, and about 7 000 for animals, although these focus mainly on Natura 2000 sites. All HD species are covered, but some of them only as part of wider groups (*Cladina* spp., *Lycopodium* spp., *Sphagnum* spp.). HD habitat monitoring is based on more than 5 000 monitoring plots, located mostly in Natura 2000 sites.[29] Monitoring methodological guidelines have been prepared for each HD Annex I type of habitat, with some examples translated into English (Mróz, 2013). The monitoring plots for habitats and non-bird species are not randomised, thus they do not provide representative assessments (in particular, information from outside Natura 2000 sites is under-represented), but they enable changes in the most interesting locations to be examined.

24.9 Future Developments

The future of nature conservation in Poland is not fully predictable. Since 2015, substantial political tensions have appeared in Polish society. The ruling party, 'PiS', seem to be right-wing populists, promoting citizens' right to use nature, although taking into account the principle of sustainable development. As a result, agriculture, forestry, hunting and

28 https://monitoringptakow.gios.gov.pl/

29 http://siedliska.gios.gov.pl/

infrastructure development is preferred over the development of protected areas and nature conservation. The government, whilst considering the EU alliance to be important for Poland's development, criticises regulatory initiatives and opposes the main developments in EU environmental policy. The involvement of Poland in the future implementation of EU environmental strategies and policy may be through declared intentions, but little action. In particular, the Polish government opposes the Green Deal and implementation of the *EU Biodiversity Strategy for 2030*, especially the 10% strict protection target. On the other hand, some ideas of nature conservation development, such as creating new National Parks, development of the protected areas network including strict protection of some areas, transforming water management towards an ecosystem approach, and establishing the office of nature ombudsman, are still promoted by conservation NGOs and left-wing parties.

Almost all political forces in Poland seem to agree on increasing economic development, developing infrastructure, and river regulation, which results in increasing threats to nature. Due to the lack of a tradition of strategic thinking in Poland, even officially formulated and adopted strategies – such as the biodiversity strategy, wetland conservation strategy, transport development strategy and others – are not regarded seriously.

Nevertheless, the key elements of the nature conservation system in Poland remain intact, seemingly stabilised by EU membership and requirements under EU environmental law. In particular, Natura 2000 works as a protective umbrella also for some overlapping nationally designated sites.

Climatic changes provide an additional instability factor. In recent years, some changes (e.g. a lack of snow in winter, summer droughts and recurrent floods) have become visible to most people, rapidly increasing public awareness. This has been followed by increasing support for nature conservation, but also for harmful investments such as river regulation and building reservoirs. Due to climate change, the validity of some assumptions concerning the management of protected areas and species is now doubtful. As a result of the above-mentioned issues, most Polish nature conservationists feel that the future is very unpredictable.

24.10 Conclusions

The development of the nature conservation system in Poland seems to have been not very spectacular, but robust. Despite the declarations of some Polish politicians, the nation's biodiversity is not the best in Europe, but is still impressive. Even if landscapes in general are strongly transformed, some natural pearls such as Białowieża Forest, forests in the Carpathians, Biebrza and Rospuda fens, and Słowiński National Park dunes are still well preserved. Furthermore, the density of the smaller valuable natural sites still seems to be above the European average, and some species that are rare in most of Europe remain quite common in Poland.

The restoration of the Bison population, which is now 90% of the EU's, has been an unquestionable triumph of Polish nature conservation. Other important achievements have included the strict protection and recovery of all large carnivore species (despite some opposing voices in recent years) and the reversal of declines in many other species over the last 70 years, such as Beavers, Ravens (*Corvus corax*), various birds of prey and the Common Crane. Despite these successes, the effectiveness of nature conservation over recent decades in halting and reversing declines in habitats and species has been limited. Most of the BHD species and habitats remain in an unfavourable conservation status, which is apparent from the monitoring of the habitats and species of EU importance and the increasing lengths of consecutive editions of national Red Lists.

Due to some European Commission enforcement interventions, the Natura 2000 network has achieved at least average EU-level coverage (see Table 5.4) and relatively effective protection. Management plans are being successively established, and will soon cover all sites. Polish National Parks are rather well managed. Nevertheless, in other areas Poland has lagged behind, particularly in the national designation of protected areas with stronger conservation regimes. In Poland, these comprise National Parks and Nature Reserves, which only have 1.5% coverage on land – well below most other European countries. Further development of the system is urgently required. Chaotic spatial development and the weakness of regulating tools seem to be an important threat. Although some marine protected areas are established, their management is still insufficient and the general status of marine biodiversity remains poor.

A feature of Poland is the state ownership of most of the forests, managed by the State Forests agency, the largest forest owner in Europe. Forest management is sustainable in the sense of forest stands and timber resources, but in practice it sometimes damages biodiversity. This damage has been difficult to address, as State Forests is the most influential actor on the national political scene, thus forestry interests in many cases override nature conservation needs.

The EU CAP has substantially changed Polish agriculture, aiding intensification, whilst at the same time agri-environment schemes have been particularly important in reversing the increasing abandonment of pastures and meadows with high natural values. Nevertheless, agriculture intensification continues, along with even more intense river alteration in the agricultural landscape.

In general, despite some efforts, the declining trends of most threatened habitats and species are not being reversed. The need for more restoration, especially of rivers and peatlands, is increasingly recognised, as actual restoration remains the exception.

Acknowledgements

Andrzej Kepel, Wiktor Kotowski and Przemysław Nawrocki reviewed the first version of this chapter and provided substantial comments. Their input is greatly appreciated.

References

Biedroń, I. (ed.) (2020) *Krajowy Program Renaturyzacji Wód Powierzchniowych*. Warsaw: Wody Polskie.

Biedroń, I., Dubel, A., Grygoruk, M., Pawlaczyk, P., Prus, P. & Wybraniec, K. (2018) *Katalog dobrych praktyk w zakresie robót hydrotechnicznych i prac utrzymaniowych wraz z ustaleniem zasad ich wdrażania*. Warsaw: Ministry of Environment.

BULiGL (2021) *Wielkoobszarowa inwentaryzacja stanu lasów w Polsce, wyniki za okres 2016-2020*. Sękocin Stary: Biuro Urządzania Lasu i Geodezji Leśnej.

Buszko-Briggs, M., Giergiczny, M., Zięzio, J. & Żylicz, T. (2004) *Wartość ekonomiczna Puszczy Białowieskiej*. Warsaw: WWF Polska.

Chylarecki, P., Chodkiewicz, T., Neubauer, G. *et al.* (2018) *Trendy liczebności ptaków w Polsce*. Warsaw: GIOŚ.

Dajdok, Z. & Pawlaczyk, P. (ed.) (2009) *Inwazyjne gatunki roślin ekosystemów mokradłowych Polski*. Świebodzin, Poland: Naturalists Club.

GDEP (2021) *Priority Framework Actions (PAF) for Natura 2000 in Poland. Article 8 of Council Directive 92/43/EEC on the conservation of natural habitats and of wild fauna and flora (Habitats Directive) for the Multiannual Financial Framework 2021-2027*. Warsaw: General Directorate for Environmental Protection.

GDOŚ (2021) *Projekt strategii ochrony mokradeł w Polsce na lata 2022-2032*. Warsaw: Generalna Dyrekcja Ochrony Środowiska.

GIOŚ (2021) *Informacja na temat funkcjonowania systemu ochrony wód w Polsce*. Warsaw: Główny Inspektor Ochrony Środowiska.

Głowaciński, Z. (ed.) (2002) *Czerwona lista zwierząt ginących i zagrożonych w Polsce*. Kraków, Poland: Institute of Nature Conservation.

Głowaciński, Z., Okarma, H., Pawłowski, J. & Solarz, W. (2012) *Gatunki obce w faunie Polski*. Kraków, Poland: Institute of Nature Conservation.

Gosling, E., Bojarska, K., Gula, R. & Kuehn, R. (2019) Recent arrivals or established tenants? History of wolf presence influences attitudes towards the carnivore. *Wildlife Society Bulletin*, **3** (4), 639–650.

Grodzińska-Jurczak, M. & Cent, J. (2011) Expansion of nature conservation areas: problems with Natura 2000 implementation in Poland? *Environmental Management*, **47**, 11–27.

Gumułka, P. (2020) *Wniosek o moratorium na polowania na dzikie ptaki w Polsce*. Bystra, Poland: Fundacja Niech Żyją.

Holeksa, J., Wojterska, M. & Kasprowicz, M. (2017) Rozmieszczenie parków narodowych a możliwości ochrony pełnego zróżnicowania siedlisk leśnych w Polsce. *Roczniki Bieszczadzkie*, **25**, 279–290.

Jabłońska, E., Wiśniewska, M., Marcinkowski, P. *et al.* (2020) Catchment-scale analysis reveals high cost-effectiveness of wetland buffer zones as a remedy to non-point nutrient pollution in north-eastern Poland. *Water*, **21** (629), 1–24.

Jaroszewicz, B., Cholewińska, O., Gutowski, J. M., Samojlik, T., Zimny, M. & Latałowa, M. (2019) Białowieża Forest - relic of the high naturalness of European forests. *Forests*, **10**, 849.

Jędrzejewski, W., Nowak, S., Stachura, K. *et al.* (2011) *Projekt korytarzy ekologicznych łączących Europejską Sieć Natura 2000 w Polsce*. Białowieża, Poland: Mammals Research Institute.

Jermaczek, A. (2014) Obszary dzikości – warunek skutecznej ochrony antropofobnej fauny. *Przegląd Przyrodniczy*, **25** (4), 104–129.

(2016) Ochrona rezerwatowa w Polsce – czy dokądś zmierzamy? *Przegląd Przyrodniczy*, **27** (4), 3–17.

(2018) Co Polki i Polacy wiedzą o ochronie przyrody? *Przegląd Przyrodniczy*, **29** (4), 28–41.

Jermaczek, A. & Kwaśny, Ł. (2019) Czy sieć Natura 2000 w Polce pokrywa się z obszarami bez infrastruktury – potencjalnymi obszarami dzikimi? *Przegląd Przyrodniczy*, **30** (4), 108–131.

Joosten, H., Tannenberger, F. & Moen, A. (ed.) (2017) *Mires and Peatlands of Europe: Status, Distribution, and Nature Conservation*. Stuttgart, Germany: Schweizerbart Science Publishers.

Kaźmierczakowa, R. (ed.) (2016) *Polska czerwona lista paprotników i roślin kwiatowych*. Kraków, Poland: Institute of Nature Conservation.

Klama, H. & Górski, P. (2018) Red list of liverworts and hornworts of Poland (4th edition). *Cryptogamie, Biologie*, **39** (4), 415–441.

Kotowski, W. (2021) *Oszacowanie emisji gazów cieplarnianych z użytkowania gleb organicznych w Polsce oraz potencjału ich redukcji*. Warsaw: WWF Poland.

Liro, A. (1998) *Strategia wdrażania krajowej sieci ekologicznej ECONET-POLSKA*. Warsaw: IUCN Poland.

Marchowski, D., Jankowiak, Ł., Ławicki, Ł., Wysocki, D. & Chylarecki, P. (2020) Fishery bycatch is among the most important threats to the European population of Greater Scaup *Aythya marila*. *Bird Conservation International*, **30** (2), 176–193.

Milewski, W. (ed.) (2018) *Forests in Poland 2018*. Warsaw: The State Forest Information Centre.

MRiRW (2014). *Program Rozwoju Obszarów Wiejskich na lata 2014-2020*. Warsaw: Ministerstwo Rolnictwa i Rozwoju Wsi.

Mróz, W. (ed.) (2013) *Monitoring of Natural Habitats. Methodological Guide*. Warsaw: General Inspectorate for Environment Protection.

MS (2006) *Strategia ochrony obszarów wodno-błotnych wraz z planem działań (na lata 2006-2013)*. Warsaw: Ministerstwo Środowiska.

Oświęcimska-Piasko, Z., Piórkowski, H., Dembek, W. *et al.* (2006) *System Informacji Przestrzennej o Mokradłach Polski*. Instytut Melioracji i Użytków Zielonych w Falentach, Zakład Ochrony Przyrody Obszarów Wiejskich. www.gis-mokradla.info

Pawlaczyk, P. (2014a) Ochrona dzikości (wilderness) – nowy paradygmat europejskiej ochrony przyrody? *Przegląd Przyrodniczy*, **25** (4), 17–35.

(2014b) Czy ochrona naturalnych procesów w przekształconym krajobrazie ma sens? Doświadczenia z planowania i realizacji ochrony Drawieńskiego Parku Narodowego. *Przegląd Przyrodniczy*, **25** (4), 42–77.

(2017) *Dokumentacje strategicznej oceny na środowisko planów urządzenia lasu w latach 2012–2017*. Rzeszów, Poland: Fundacja Dziedzictwo Przyrodnicze.

(2018) *Legal Regulatory Framework of Peatland Exploitation, Draining and Restoration in Poland*. Świebodzin, Poland: Naturalists Club Poland.

(2020) *Podręcznik dobrych praktyk renaturyzacji wód powierzchniowych*. Warsaw: Wody Polskie.

Pawlaczyk, P., Bohdan, A. & Grzegorz, A. (2016) *Próba oceny zarządzania najcenniejszymi lasami w Polsce*. Białystok, Poland: Stowarzyszenie Pracownia na rzecz Wszystkich Istot, Oddział Podlaski.

Pęksa, Ł, & Ciach, M. (2015) Negative effects of mass tourism on high mountain fauna: the case of the Tatra chamois *Rupicapra rupicapra tatrica*. *Oryx*, **49** (3), 500–505.

Peters, J. & Unger, M. (2017) Peatlands in the EU Regulatory Environment. Survey with case studies on Poland and Estonia. *BfN Skrpiten*, **454**, 1–103.

PGL LP (2019) *Sprawozdanie finansowo-gospodarcze za 2018 r.* Warsaw: Państwowe Gospodarstwo Leśne Lasy Państwowe.

Prus P., Popek Z. & Pawlaczyk P. (2018) *Dobre praktyki utrzymywania rzek*. 2nd ed. Warsaw: WWF Poland.

Ratyńska, H., Wojterska, M., Brzeg, A. & Kołacz, M. (2010) *Multimedialna encyklopedia zbiorowisk roślinnych Polski*. Warsaw: NFOSiGW.

Różycka, W. (1977) Propozycja formowania Ekologicznego Systemu Obszarów Chronionych w planach zagospodarowania przestrzennego. *Człowiek i Środowisko*, **1** (4), 53–72.

Selva, M., Zwijacz-Kozica, T., Sergiel, A., Olszańska, A. & Zięba, F. (2012) *Program ochrony niedźwiedzia brunatnego Ursus arctos w Polsce, projekt*. Warsaw: SGGW.

Stalenga, J., Brzezińska, K. & Jobda, M. (2016) *Rekomendacje zmian w programie rolnośrodowiskowym*. Puławy, Poland: IUNG.

Szalińska, E., Orlińska-Woźniak, P. & Wilk, P. (2018) Nitrate vulnerable zones revision in Poland – assessment of environmental impact and land use conflicts. *Sustainability*, **10**, 3297.

Tanneberger, F. & Kubacka, J. (eds.) (2018) *The Aquatic Warbler Conservation Handbook*. Potsdam, Germany: Brandenburg State Office for Environment (LfU).

Tokarska-Guzik, B., Dajdok, Z., Zając, M. *et al.* (2012) *Rośliny obcego pochodzenia w Polsce ze szczególnym uwzględnieniem gatunków inwazyjnych*. Warsaw: General Directorate of Environment Protection.

UNEP-WCMC (2022) *Protected Area Profile for Poland from the World Database of Protected Areas, April 2022*. www.protectedplanet.net

Wilk, T., Chodkiewicz, T., Sikora, A., Chylarecki, P. & Kuczyński, L. (2020) *Czerwona Lista Ptaków Polski*. Marki, Poland: Ogólnopolskie Towarzystwo Ochrony Ptaków.

Wilk, T., Jujka, M., Krogulec, J. & Chylarecki, P. (2010) *Ostoje ptaków o znaczeniu międzynarodowym w Polsce*. Marki, Poland: Ogólnopolskie Towarzystwo Ochrony Ptaków.

Wołejko, L., Pawlaczyk, P. & Stańko, R. (2019) *Alkaline Fens in Poland – Diversity, Resources, Conservation*. Świebodzin, Poland: Naturalists Club.

Wylegała, P. & Ławicki, Ł. (2019) *Głowienka, czernica, cyraneczka, łyska – stan populacji w Polsce i wpływ gospodarki łowieckiej*. Poznań, Poland: IUCN Polish Committee.

Żarnowiec, J., Stebel, A. & Ochyra, R. (2004) Threatened moss species in the Polish Carpathians in the light of a new Red-list of mosses in Poland. In *Bryological Studies in the Western Carpathians*, eds. A. Stebel & R. Ochyra, pp. 9–28. Poznań, Poland: Sorus.

Żylicz, T. (2017) *Wartość ekonomiczna przyrody. Zarządzanie Publiczne*, **39** (1), 114–124.

Chapter 25

Portugal

Ana Guimarães Ferreira and Domingos Leitão

Key Statistics	
Terrestrial area (km^2)/% of EU-28*	89 073/2%
Marine waters (km^2)/% of EU-28*	315 220/7%
Population density (p km^2)/% of EU-28*	117/101%
GDP per capita (€)/% of EU-28	17 023/64%
Land cover (CORINE land cover classes 2018)*	AGR 48%, OSN 29%, FOR 18%, ART 4.0%, WET 1%
Biogeographical regions (excluding Macaronesian)	ATL, MATL, MED
Number of HD Annex I habitats/%*	88/39%
Number of HD Annex II, IV and V species/%*	228/19%
Number of BD Annex I bird taxa/%*	81/46%
% HD UCS and decreasing habitats/ % unknown	64%/2%
% HD UCS and decreasing species/ % unknown	15%/32%
% breeding BD Annex I birds decreasing/% uncertain and unknown	17%/65%

* Excluding MAC and MMAC biogeographical regions.
Sources and codes. See Table 1.1.

25.1 The Natural History and Characteristics of Portugal

Portugal consists of a mainland area (a.k.a. continental Portugal) and two Atlantic regions: the archipelagos of Madeira and the Azores. However, as Madeira and the Azores are distinct autonomous regions that occur within the Macaronesian biogeographical region, which is not covered in this book (see Chapter 1), they do not generally feature in this chapter.

Mainland Portugal is located in the extreme south-west of Europe, in the transition between two biogeographical regions: the Atlantic and the Mediterranean. Its considerable biophysical diversity results from the differing climate of these two regions and the country's extensive latitudinal range. It also has diverse soils and a varied topography, although a large proportion is relatively mountainous (i.e. above 300 m), especially in the north and centre of the country, the highest point reaching 1 993 m. It is crossed by four major river basins: the Mondego, Douro, Tagus and Guadiana. Portugal's mainland climate is temperate, with annual average temperatures ranging from 7 °C in the central interior highlands to 18 °C along the south coast. Average rainfall ranges from 3 100 mm in the mountainous north-west regions to 450 mm on the south coast.

By European standards, Portugal has a rich and diverse flora and fauna, to which the island territories, located in the Macaronesian biogeographical region, make a significant contribution. The islands hold habitats that are absent or very rare in mainland Portugal, such as laurel forests, lava fields, volcanic caves and peatlands. They are also very rich in terrestrial endemic species (particularly flora – 81 species in Azores and 155 in Madeira – and invertebrates), and in marine life, particularly seabirds and sea mammals (Jardim and Francisco, 2000; Fagundes and Vaz, 2017; Saavedra *et al.*, 2018). The wildlife in these two Atlantic archipelagos is different from that of mainland Portugal, and subject to different influences, and hence the description of nature conservation in these areas is beyond the scope of this book.

With its long coastline (942 km), Portugal hosts important stretches of threatened coastal habitats, including estuaries, coastal lagoons, mudflats and sandflats, reefs, vegetated sea cliffs with endemic flora and several dune types. Important inland habitats are mainly related to native forests, particularly oak forests and *montados* (sclerophyllous grazed forests with evergreen *Quercus* spp.), with Cork Oak (*Quercus suber*), Pyrenean Oak (*Quercus pyrenaica*) and Holm Oak (*Quercus rotundifolia*). Although scarce or declining, important areas of highly threatened Mediterranean habitats still occur, such as Mediterranean temporary ponds, pseudo-steppes, and some specific scrub habitats.

Some Portuguese habitats are species-rich and/or important for supporting species populations of national and global significance. These include marine and freshwater/wetland habitats, as they support 40% of the HD species that occur in Portugal. Estuaries and coastal lagoons are also key staging

Table 25.1 Short-term trends (c. 2007–2018) in Habitats Directive Annex I habitats and Annex II, IV and V species with an unfavourable conservation status in Portugal (excluding assessments in MAC and MMAC regions).

	Poor (unfavourable– inadequate)				Bad (unfavourable–bad)			
	Imp	Sta	Dec	?	Imp	Sta	Dec	?
HD Habitat assessments								
Number	1	13	41	0	0	0	38	0
%	2%	24%	75%	0%	0%	0%	100%	0%
HD Species assessments								
Number	1	26	25	45	1	2	20	6
%	1%	27%	26%	46%	3%	7%	69%	21%

Key. Imp = improving, Sta = stable, Dec = decreasing, ? = unknown.
Source. Based on Member State HD Article 17 reported data as presented on the EEA national summary dashboard (2021).

posts for large populations of long-distance migratory birds, mainly waders, but also gulls, terns and duck species. Scrub habitats, *montados* and other oak forests are particularly important for top predators, such as large eagles, vultures, Iberian Wolf (*Canis lupus signatus*) and the globally Endangered Iberian Lynx (*Lynx pardinus*). Some semi-natural habitats, created by High Nature Value (HNV) farming systems, such as pseudo-steppe (i.e. low-intensity dry grasslands and cereal cropland), support important populations of some BD Annex I species, including Little Bustard (*Tetrax tetrax*), Lesser Kestrel (*Falco naumanni*) and the globally Vulnerable Great Bustard (*Otis tarda*).

25.2 The Status of Nature

25.2.1 Habitat Status and Trends

According to the report submitted by Portugal on the conservation status of HD habitats and species over 2013–2018,[1] only 23% of the assessed habitats had a favourable status in the mainland territory. The majority of habitats had an unfavourable status, with 44% considered poor and 31% considered bad. The status of the remaining 2% was unknown. Most of the coastal habitats had an unfavourable conservation status (82%), including those that are extremely important for wildlife and ecosystem services (e.g. estuaries, coastal lagoons, several types of salt marshes and other intertidal habitats). Of particular concern is the situation with bogs and mires, sand dunes and forests, as all habitat types in these groups had an unfavourable status. Other habitat groups that had a high number of types with an unfavourable status were coastal habitats (13 of 16 types), freshwater habitats (8 of 11 types), particularly oligotrophic waters, natural dystrophic lakes and temporary Mediterranean ponds, grasslands (6 of 10 types), scrub and heath (3 of 5 types) and sclerophyllous scrubs (6 of 8 types). A considerable proportion of these habitats had a declining trend (Table 25.1).

25.2.2 Species Status and Trends

As further discussed in Section 25.8 on monitoring and research, Portugal has few biodiversity long-term monitoring programmes, other than for birds, and available data have not been adequately analysed and reported on. The status of species populations is therefore mainly now known through the reports required under the EU Birds and Habitats directives, although they are not based on systematic monitoring. According to the 2013–2018 report submitted by Portugal, only about a quarter of HD species had a favourable conservation status in the mainland territory (27%), whilst 41% had an unfavourable conservation status. The status of many species was unknown (32% overall), particularly amongst invertebrates and non-vascular plants (both groups with 100% unknown status) and amphibians (67% of species with an unknown status). Fish stand out negatively, as 94% of the species had an unfavourable status, and to a lesser extent molluscs and reptiles, as for each group over half had an unfavourable conservation status. Out of the species that had an unfavourable status, a considerable proportion also had a declining trend, whilst the trend of many was unknown (Table 25.1).

The 2013–2018 monitoring results for all birds in mainland Portugal, as reported in accordance with Article 12 of the Birds Directive, indicate that 23% of breeding species and 17.5% of key wintering species had declined over the short term. As regards BD Annex I taxa, 17% of breeding populations decreased over the short term, whilst only two species had declining wintering populations (Table 25.2). It is also of concern that the population trends remain uncertain or unknown for a large proportion of all birds: 47% for breeding birds and 50% for wintering populations. The proportion with unknown trends is even higher for breeding Annex I species, even though they should be a focus of attention.

[1] See Section 1.3 for background on monitoring under the Nature Directives, status categories and data sources.

Table 25.2 Short-term trends (c. 2007–2018) in Birds Directive Annex I taxa in Portugal.

	Breeding				Wintering			
	Inc	Sta/ Flu	Dec	?	Inc	Sta/ Flu	Dec	?
Number	9	2	11	41	8	3	2	10
%	14%	3%	17%	65%	35%	13%	9%	43%

Key. Inc = increasing, Sta = stable, Flu = fluctuating, Dec = decreasing, ? = uncertain and unknown.
Source. Based on Member State Article 12 reported data as presented on the EEA national summary dashboard (2021).

According to the common bird index, on mainland Portugal over 2004–2020, common bird populations were generally stable when assessed across all bird species and for farmland and forest birds subsets. However, some farmland birds have shown a moderate decrease (less than 5% average decrease per year). This also reflects the situation with BD Annex I species, as those showing declines included the Little Bustard, Great Bustard and Black-bellied Sandgrouse (*Pterocles orientalis*).

25.2.3 Drivers and Causes of Change

From a country originally covered by deciduous and evergreen oaks, most of Portugal has become a mix of planted pine and eucalyptus forest, heath and some remaining oak cover, mainly due to forestry and agriculture over the twentieth century (AFN, 2010). In the first half of that century, widespread clearing of native forests occurred over large parts of the country, mainly for cereal farming and pine plantations, which was followed by large-scale planting of eucalyptus over most of the centre and large parts of the south (Schmidt, 2008; Tavares, 2013). From the first plantations in the 1930s, the area of eucalyptus has increased to more than 700 000 ha (AFN, 2010), and together with Maritime Pine (*Pinus pinaster*) plantations, comprises roughly 50% of the forest area of mainland Portugal.

More recently, the CAP initiated a new burst of agricultural intensification, such as a threefold increase in irrigated crops since 1986 (DGADR, 2019). Irrigation has destroyed biodiversity-rich HNV farming systems, including low-intensity cereal pseudo-steppe and traditional olive groves, leading to observed strong declines of associated species, such as Montagu's Harrier (*Circus pygargus*), Little Bustard, Turtle-dove (*Streptopelia turtur*) and Woodchat Shrike (*Lanius senator*). Breeding populations of farmland birds have also exhibited a recent negative trend, including in several common species such as the House Sparrow (*Passer domesticus*), Goldfinch (*Carduelis carduelis*), Barn Swallow (*Hirundo rustica*) and Little Owl (*Athene noctua*) (Meirinho et al., 2013; Silva, 2018; Alonso *et al.*, 2019, 2021).

A major driver of river biodiversity losses has been the country's electrification programme, which started in the 1940s and has consisted mainly of the construction of large hydroelectric plants, mostly in northern and central Portugal. In 2018, there were about 260 large dams in the country (i.e. over 15 m high) or reservoirs over 1 million m^3 (APA, 2021). The resulting habitat changes, disruption of water flow and associated disturbance of fish migration have led to strong population declines in many native fishes and other species associated with river systems. This situation has been aggravated by the introduction of invasive alien species (IAS), such as Wels Catfish (*Siluris glanis*), Pumpkinseed fish (*Lepomis gibbosus*), Common Bleak (*Alburnus alburnus*), Signal Crayfish (*Pacifastacus leniusculus*) and Red Swamp Crayfish (*Procambarus clarkii*), which are leading to losses of native wildlife in Portuguese rivers (Bernardo *et al.*, 2011; Anastácio *et al.*, 2018; Martelo *et al.*, 2021).

Land claim for housing and industry has been the main cause of habitat destruction in coastal areas during most of the last century, when there was a strong exodus from inland to the coast. This trend was particularly intense in the large towns of the west coast, and in the southern region of the Algarve (especially in the 1970s and 1980s). Lagoons, estuaries and rivers have been particularly affected, with some large tourist developments targeting natural sites on the south coast of the Algarve, including some protected areas, such as the Ria Formosa Natural Park, Vilamoura reedbeds, Salgados lagoon and Alvor lagoon. Whilst some developments were stopped by court rulings brought through civil actions, most were carried out.

Pressures affecting marine ecosystems and species have been more intense since the 1970s. They are mostly related to overfishing, and by-catch mortality that affects many species of fish, seabirds and other marine life. For example, by-catch in fishing nets was apparently the main cause of the extinction of the large population of Common Murre (*Uria aalge*) on Berlenga Island (Catry *et al.*, 2010), and this, and other fishing industry-related impacts, remain a threat to the Razorbill (*Alca torda*) and the globally Critically Endangered Balearic Shearwater (*Puffinus mauretanicus*) (Catry *et al.*, 2010; DBIO, 2017). Oil drilling, deep-sea mining and offshore wind power are recent pressures on marine habitats, which need further evaluation.

Most species that are threatened at a European level and occur in Portugal are found in pasturelands and scrubland, native forests and wetlands – often requiring mixes of these habitats in remote areas. Most are semi-natural ecosystems and landscapes that require renaturalisation or appropriate low-intensity management, to ensure that they remain suitable for their sensitive threatened species, including iconic top predators and scavengers, such as Iberian Wolf, Iberian Lynx, eagles and vultures. Most of the threatened plant species have rather different and very specific requirements, as they depend more on natural habitats in mountain ranges and coastal habitats. All these habitats, and many of their threatened species, are under pressure from current human activities, particularly agricultural intensification, forestry, fisheries and energy production and transportation.

Climate change effects will further significantly increase pressures on species and habitats, particularly in river basins, coastal areas, forests, and farmland and grassland areas (Santos *et al.*, 2002). Freshwater stress, sea level rise, increasing forest fires and decreasing grassland productivity are climatic change consequences that are expected in Portugal. These have the capacity to affect species, displacing some and increasing biodiversity loss, including in protected areas (Council of Europe, 2009; Alagador *et al.*, 2014).

The environmental changes described here are reflected in the Habitats Directive reporting by Portugal for 2013–2018, which indicates that the pressures and threats having high impacts on HD habitats and species were mainly related to IAS, agriculture and development, construction and use of residential, commercial, industrial and recreational infrastructure and areas. Regarding birds, many of the same pressures affect BD Annex I species, particularly agriculture, forestry and infrastructure construction. Additionally, energy development, human-induced changes in water regimes, and direct persecution (including with illegal methods) exact a toll on some bird species.

25.3 Nature Conservation Governance, Actors, Legislation and Policies

25.3.1 Overview of Nature Developments in Portugal

The history of nature conservation in Portugal has been studied and documented in some detail in a number of texts, and it is only possible to briefly outline the key elements and decisive events here (some of which are summarised in Table 25.3). For more detailed overviews see Silva (2000) on protected areas, Queirós (2002) on the development of the government's environmental policy as reflected in the financial support given to environmental projects since the 1980s, Borrego (2010) on environmental policies after the 1992 Rio de Janeiro Conference, and Tavares (2013) for a thorough historical review of environmental and nature conservation policies from the 1960s until the present.

25.3.2 Governance

The first governmental department specifically dedicated to environmental and nature conservation issues was created in 1974. This was the Sub-secretariat of Environment, within the State Secretariat of Housing and Urbanism, under the Ministry

Table 25.3 Key nature conservation developments in mainland Portugal.

1948	Creation of the first major Portuguese environmental NGO, LPN
1967	Deadly flash floods affect Lisbon and initiate discussion of disorderly urban development and environmental pressures
1970	First environmental law (Law 9/70)
1971	Designation of the first national protected area
	Definition of Public Water Domain
1974	Revolution replaces authoritarian regime with a democratic one
	Creation of a sub-secretariat of Environment
1975	Creation of a National Service of Parks, Reserves and Landscape Heritage
1983	Creation of the National Ecological Reserve (REN)
1986	Accession to the European Community
	First criminal cases against nature taken to court
1987	New general environmental law (Law 11/87) later updated by Law 19/2014
1990	Adoption of legislation on water and air quality and EIA
1992	First LIFE Nature projects kick-started the establishment of Natura 2000 in the country
1991, 1997	Transposition into Portuguese law of the Birds and Habitats Directives (revised in 1999)
2008	Adoption of Sectorial Plan for the Natura 2000 Network
2012	Fusion of former nature conservation and biodiversity agency with forestry agency

See Table 3.3 for signatory dates of international conventions.

of Social Equipment and Environment (later upgraded to Environment State Secretariat, in 1975). In the same year, the National Service of Parks, Reserves and Landscape Heritage was created (renamed in 1993 as the Institute for Nature Conservation, ICN, and in 2007 as the Institute for Nature Conservation and Biodiversity, ICNB), with responsibility for the definition and management of the network of protected areas.

The national authority currently directly responsible for nature conservation and forests in mainland Portugal is the Institute for Nature Conservation and Forests (ICNF), which resulted from the fusion of the former nature conservation and biodiversity agency, ICNB, with the forestry agency in 2012. The agency is under two different Ministries, the Ministry of the Environment and Energetic Transition and the Ministry of Agriculture, Forests and Rural Development. It has administrative and financial autonomy and is responsible for proposing, monitoring and enforcing the implementation of nature conservation and forest policies in mainland Portugal and the adjacent marine area; the regulation and licensing of hunting and fishing in inland waters; and the management of the national network of protected areas and the Natura 2000 network. In the case of marine areas, these tasks are undertaken in cooperation with the Directorate General for Natural Resources and the Portuguese Institute of the Sea and the Atmosphere, under the Ministry of the Sea, which is responsible for defining and implementing the *National Marine Strategy*. The Portuguese Environmental Agency (APA) is the public institute that proposes, develops and monitors environmental and sustainable development policies on air, water, waste, climate change and pollution, as well as being responsible for environmental assessments and citizen participation.

A major governance challenge still faced both at regional and national level is the need to improve information sharing and coordination of actions between public entities that are responsible for inspection and monitoring in relation to agriculture, the sea and the environment. A national IMPEL Platform (European Network for the Implementation and Enforcement of Environmental Law) has been in place since 2016, but further work is still needed. An example is the management of coastal areas, which depends on several entities (local municipalities, APA, ICNF, port authorities and the maritime authority) but lacks effective coordination (Libreiro, 2018).

25.3.3 Public Involvement and NGOs

Public involvement in environmental issues is relatively recent in Portugal and first emerged in response to concerns over suburban growth from the 1960s. On an autumn night in 1967 one of the deadliest flash floods in history ravaged the city of Lisbon, destroying houses, cars and private businesses, and claiming around 700 lives and displacing more than 1 000 people. It was the result of unusually heavy rain, exacerbated by unplanned development, ground sealing, and destruction of vegetation within infiltration areas. The disaster resulted in an unprecedented mobilisation of civil society and a broad discussion about urban development and its associated environmental pressures that has since been seen as a turning point in environmental awareness in the country (Soromenho-Marques, 1998; Schmidt, 2008).

Most of the Portuguese environmental NGOs are young, having been created after the 1974 revolution. One of the oldest, the Nature Protection League (LPN), resulted from some leading scientists lobbying the government over the cutting of a privately owned native forest for charcoal production. This led to the forest being purchased by the state and the creation in 1948 of LPN by the scientists. Over its first few decades, LPN focussed almost exclusively on scientific work (Schmidt, 2008) and was responsible for the first inventories of natural heritage, which would later be used as a basis for the creation of several national protected areas. Until 1970, it only had around 150 members, but this increased after the 1970 European Conservation Year raised the profile of the organisation and its work. From this time onwards, it also became gradually more interventionist.

Given its size and troubled recent history, Portugal has a considerable number of environmental NGOs, operating at local, regional and national levels. As of 29 September 2021, the Portuguese environmental NGO database included 106 legally registered organisations. Although the trend has been increasing over the last 20 years, membership of the Portuguese environmental NGOs is still rather small compared with other European NGOs, as even the larger organisations have fewer than 5 000 members.

Most of the larger Portuguese environmental NGOs operate nationwide in mainland Portugal. The Portuguese Society for the Study of Birds (SPEA) is the only one operating in all Portuguese territories. The six largest – LPN, SPEA, Quercus, Spatial Planning and Environment Study Group (GEOTA), Nature Association Portugal (ANP) and the Protection Fund for Wildlife (FAPAS) – are united under a platform, called 'Conservation 6', for campaigning on high-profile issues. Most of these organisations are also affiliated to international confederations, including the European Environmental Bureau (LPN, Quercus, GEOTA and ZERO), BirdLife International (SPEA) and WWF (ANP).

Site protection is a very important long-term result of the work of Portuguese NGOs. The only National Park in Portugal, Peneda-Gerês, was created due to the work of LPN. They made an inventory of natural values and proposed the creation of a protected area to the government, arguing that there was a need for the country to keep up with growing recognition in Europe of the importance of environmental protection. Other important Natural Parks and Natural Reserves were created after the work of LPN and other organisations. After Portugal's accession to the EU, the Portuguese NGOs also played an important role in the designation of Natura 2000 sites, especially SPAs for birds. In particular, SPEA developed in Portugal the Important Bird Areas (IBA) programme (see Section 25.5.1).

Low membership has been one of the weaknesses of Portuguese NGOs, but other ways to involve citizens have been explored. SPEA has been very successful in involving people in citizen science projects. Most of the regular monitoring, such as the common bird census, atlas and other bird counts, is done with the help of hundreds of volunteer birdwatchers and ornithologists. Other NGOs, like Quercus, ATN and LPN, have managed to buy land with appeals and donations.

The larger Portuguese NGOs originally focussed mainly on nature protection, with most of the early advocacy and campaigns on protecting sites, habitats and species. One iconic campaign in the late 1970s led by LPN to save the Lynx of Malcata resulted in the creation of a Natural Reserve in the area. Until the early 2000s only a few NGOs worked on other issues, like waste and residue management, energy and transport, or land planning. More recently, mainly with the help of umbrella organisations (e.g. BirdLife International, EEB and WWF), the Portuguese NGOs have engaged with other national policy themes, and with EU policy. However, as in many other European countries, there has been little success in changing significantly harmful policies. In particular, Portuguese NGOs have failed to mobilise public opinion to put pressure on politicians for real changes in the CAP over the last 20 years. The situation is similar for fisheries, forestry, hunting and energy. For some reason, Portuguese citizens are willing to exert some pressure to save particular species or sites, but are less interested in getting involved in discussions on ways of achieving more fundamental changes. It seems clear that Portuguese NGOs need to invest more in working together, and in new ways of involving people, if they want to increase their influence on decision makers and change policies.

25.3.4 Policy and Legislation

Environmental legislation has been in force in Portugal dating back to the nineteenth century, but most of it related to forests, the hunting and protection of game, and the protection of water resources – the last being a focus between 1933 and 1974.

External influences have been a key driver of change in environmental legislative provisions (Soromenho-Marques, 2005). In particular, the 1970 European Conservation Year and the Stockholm Conference on the Human Environment led to the first true Portuguese environmental law in the same year. Law 9/70 protects nature and its resources through National Parks and other types of protected area. Major progress in environmental policy was achieved after the 1974 revolution, with numerous legislative acts related to nature conservation, including the creation of a National Network of Protected Areas (see Section 25.5.1). However, around this time many of the pressures on biodiversity increased, such as from industry, tourism, urbanisation and the planting of eucalyptus (as described earlier), which resulted in an unprecedented decrease in the populations of game and other wildlife species (Mansinho and Schmidt, 1997).

In the 1980s, after many years of insufficient integration of environmental concerns into development policies, a significant effort was made to solve accumulated environmental problems. Legislation was passed to regulate and control land use, including municipal and regional master plans and the creation of National Ecological Reserves (REN), which integrate areas of ecological value, sensitivity and susceptibility to natural hazards.

Portugal's accession to the European Community in 1986 was a landmark in the development of the country's environmental policies, which were given greater consistency and strength, and later consolidated in the 1990s (Queirós, 2002). In particular, nature conservation only gained high importance after the adoption of the Birds and Habitats Directives, which were adhered to by the Portuguese administration from their outset and formally transposed to the Portuguese law in 1991 and 1997, respectively, later revised in 1999[2] and re-adjusted in 2005.[3] They gave the public a new and effective weapon to fight against destructive development projects and policies, backed up by the European Commission's infringement procedures. Several cases were raised against the Portuguese State, some of which led to unprecedented stringent control and compensatory measures and CJEU rulings against Portugal that resulted in clear nature conservation gains. For example, the case of the Vasco da Gama bridge over the Tejo estuary (Melo, 2000) and the ruling concerning the A2 motorway[4] resulted in more effective compensation measures for damaging projects (in relation to HD Article 6(4)). Other cases led to an increase in Natura 2000 sites for several species (e.g. some dry-grassland birds and birds of prey).

In some respects Portugal has been ahead of the EU, in recognising the need for all sectors of society to contribute to biodiversity conservation. Thus, Law 11/87,[5] which defines the basis of the country's environmental policy, requires the government to draft a *National Strategy for the Conservation of Nature and Biodiversity* that integrates nature conservation with other relevant sectoral policies. This was some two decades before the *EU Biodiversity Strategy for 2010*, which set out similar policy integration concepts. However, it took until 2001 for the Portuguese Strategy to be adopted by the Council of Ministers, and until 2018 for it to be revised,[6] since when it has never been fully implemented.

[2] Decreto-Lei n.º 140/99, Diário da República n.º 96/1999, Série I-A de 1999-04-24.

[3] Decreto-Lei n.º 49/2005, Diário da República n.º 39/2005, Série I-A de 2005-02-24.

[4] Case C-239/04 Commission v Portugal, paragraphs 16–25.

[5] Lei n.º 11/87, Diário da República n.º 81/1987, Série I de 1987-04-07.

[6] Resolução do Conselho de Ministros nº 55/2018, Diário da República n.º 87 Série I, de 7 de maio.

In 2008, new laws were passed on nature conservation and biodiversity that clarified previous legislation. In particular, a *Sectoral Plan for the Natura 2000 Network* (PSRN2000)[7] defined the strategy for the management of the mainland protected areas, taking into account their natural value and long-term maintenance requirements (EEA, 2015). The plan provides guidelines for the management of habitats and species within Natura 2000 sites and requires their inclusion in existing municipal and special land-use plans.[8] Another new law redefined the juridical regime of the REN,[9] strengthening its importance and creating a financial regime that favours areas integrated in the REN, allowing more equitable compensation in case of expropriation. However, this was amended in 2012 by a new decree,[10] which was much criticised by the main environmental NGOs because it gave more power to municipal authorities to declassify areas from the REN, and reduced the participation of the national REN steering committee (on which environmental NGOs and other independent organisations are represented).

Despite all the progress in terms of legislation, there have been implementation problems. The PSRN2000 and other sectoral and special plans (e.g. the rural development plans and the municipal spatial development plans) are not sufficiently comprehensive and precise to achieve adequate levels of protection. For instance, although the PSRN2000 includes a large amount of information, it has information gaps and uses a very large scale, which often results in the prescribed management guidelines being too generic. Another problem is that development decisions have often not been in accordance with approved plans, which are instead updated to accommodate ongoing development. Such is the case with national interest projects ('PIN' projects), which have been subject to a special regime since 2005 (OECD, 2011). HD Article 6(4) 'overriding public interest' criteria have not always been well justified, and generally any dam, motorway, railway or airport has been considered of overriding public interest, which hampers the assessment of their real importance (BirdLife International *et al.*, 2018). Examples include the Baixo Sabor dam and the tourism resorts Herdade da Comporta, Costa Terra and Herdade do Pinheirinho, all within Natura 2000 sites (Bastos *et al.*, 2012). Finally, as discussed in the previous section, general environmental awareness and involvement remains low, which, together with an inadequately effective justice system, creates a sense of impunity (Schmidt, 2008). Human and equipment resources for implementing environmental laws are also scarce, such as is the case with the staffing level of ICNF, which is clearly insufficient for the management and surveillance of the whole protected area network.

25.4 Species Protection and Conservation Measures

25.4.1 Basic Protection Legislation and Strict Protection of Selected Species

The basic law that addresses nature conservation and the protection of biodiversity is the national decree 142/2008,[11] which defines environmental offences relating to the disturbance or destruction of habitats and nests, egg collection, taking in the wild, exploitation, killing, transportation and trade of the species listed in the National Registry of Natural Classified Values. The list includes all species considered by the national nature conservation authority to be threatened according to international criteria. However, for unknown reasons, ICNF has not yet finalised or approved this list, which means that in practice the actual implementation of protection measures is not adequately regulated.

A few plant species are covered by specific protection laws: Holly (*Ilex aquifolium*), Common Yew (*Taxus baccata*), Holm Oak and Cork Oak. The Iberian Wolf is the only animal species that is covered by a specific protection law in Portugal.

Despite the generally good protection legislation in Portugal, there is no systematic approach to setting favourable reference values, objectives or surveillance for the conservation of species covered by the Nature Directives. The current approach to surveillance and law enforcement takes into account the complaints registered at the Department of Inspection of Agriculture, the Sea, Environment and Land Planning (IGAMAOT), public and media reports, and the requests of the Ministry or other public entities. It is therefore a reactive, rather than a proactive, protection strategy (Holzgraefe *et al.*, 2017).

25.4.2 Regulation of Hunting and Fishing, and Approaches to Game Management

Hunting is regulated by a national law decree.[12] The hunting days, hunting season and the list of species that can be taken, when and by which means, are defined for three-year periods. Measures to ensure sustainable taking of hunted species are defined by the hunting exploitation plans of each hunting area (*planos de ordenamento e exploração cinegética*), hunting management plans (*planos de gestão*) and annual exploitation plans (*planos anuais de exploração*), which are submitted for the approval of ICNF. Surveillance and law enforcement are under the responsibility of the National Guard (Guarda Nacional

[7] Resolução do Conselho de Ministros n.º 115-A/2008, Diário da República n.º 139/2008, 1º Suplemento, Série I de 2008-07-21.

[8] www2.icnf.pt/portal/pn/biodiversidade/rn2000/p-set/q-e

[9] Decreto-Lei n.º 166/2008, Diário da República n.º 162/2008, Série I de 2008-08-22.

[10] Decreto-Lei n.º 239/2012, Diário da República n.º 212/2012, Série I de 2012-11-02.

[11] Decreto-Lei n.º 142/2008, Diário da República n.º 142/2008, Série I de 2008-07-24.

[12] Decreto-Lei n.º 167/2015, Diário da República n.º 163/2015, Série I de 2015-08-21.

Republicana, GNR), nature wardens, forest wardens, forest police, maritime police and municipal police. According to the legislation, monitoring of the impacts of hunting can be done through the examination of the reports that each hunting area submits to the competent authority. However, it is not clear if these data are in reality used for monitoring purposes.

Fishing in inland waters is also regulated by a national law decree.[13] The angling season, the species that can be taken, when and by what means, are defined annually. Each recreational and professional angling area has a management and exploitation plan where measures to ensure sustainable taking of fish species are defined. Surveillance and law enforcement are the responsibility of GNR.

Whilst hunters and fishers could have an active role in conserving species, in reality this only happens through specific projects, many funded by the LIFE programme, for limited periods and areas. A good example is the creation of a network for sustainable hunting by the transnational Spanish/Portuguese LIFE project Innovation Against Poison. More often, hunters and fishers have had negative impacts, such as by successive restocking of hunting and fishing species, which contribute to the spread of disease and pressure on native species (Guerrero-Casado *et al.*, 2013; Carro *et al.*, 2019).

Predator control in hunting areas also contributes to the accidental mortality of protected species, as there are doubts about the selectivity of the trapping methods normally used legally to control predators,[14] and this activity is inadequately supervised. These predator control activities, together with illegal hunting (which affects species like the Iberian Wolf, Iberian Lynx and birds of prey), are the main species conservation problems related to hunting in Portugal.

25.4.3 Species Action/Recovery Plans

A number of species action plans have been developed and approved by ICNF, but they have not always promoted key practical conservation actions, and some have only been partly implemented, due to a lack of regular and adequate funding. Other plans have been prepared but not endorsed to date.

For example, an action plan for the conservation of the Iberian Wolf in Portugal was published and endorsed in 2017.[15] Habitat fragmentation and loss of connectivity between wolf population nuclei were among the main threats identified, together with destruction, degradation and disturbance of breeding sites, and lack of coordination between legal instruments for land management and planning – the guidelines for the conservation of the Wolf are not adequately integrated in the municipal territorial plans. Whilst the plan diagnoses the threats and limiting factors, and sets adequate objectives, most proposed activities are too generic, many indicators measure effort rather than actual results, and there is no estimation of the required human or financial resources. Updated information is also needed, as there have only been two national Wolf population censuses, in 1994/1996 (ICN, 1997) and 2002/2003 (Pimenta *et al.*, 2005). This information is needed in view of infrastructure developments (dams, roads, highways and windfarms) that have meanwhile further reduced connectivity, which is of particular concern for the vulnerable subpopulation south of the river Douro and the apparently decreasing number of Wolf packs in the Alvão-Padrela nucleus.

The action plan for the conservation of the Iberian Lynx in Portugal[16] was published and endorsed in 2015. Recommended activities relate mainly to very generic communication and awareness actions, rather than aiming to reduce the main causes of mortality (hunting and road traffic) or addressing the problem of unsupervised predator controls in hunting areas.

A specific programme for the conservation of the globally Vulnerable Spanish Imperial Eagle (*Aquila adalberti*) has been underway since 2008.[17] Ongoing conservation measures, identified by a working group in collaboration with an equivalent group in Spain, include modification of dangerous powerlines, cooperation with hunting estate owners and managers, controlling access to nest areas and population monitoring. A wider programme to reduce bird collisions and electrocutions has been supported by a major electricity producer and distributor in Portugal, EDP (Energias de Portugal), with the involvement of NGOs and the nature conservation authorities. A recent LIFE project (LIFE Imperial) includes among its objectives the preparation of a national action plan for the species.

Action plans have also been developed and approved for the Spanish Minnowcarp (*Anaecypris hispanica*) (Cardoso *et al.*, 2014), cave bats (Palmeirim and Rodrigues, 1992) and the Bottlenose Dolphin (*Tursiops truncatus*) population in the Sado estuary.[18]

25.4.4 Dealing with Human–Fauna Conflicts: the Case of the Wolf

There has been some success in reducing conflicts between the Wolf in north-east Portugal and shepherds and herders through the establishment of a Mastiff Dog Distribution Programme to defend the livestock. This was firstly carried out in the framework of a cooperation protocol between the Transmontano Mastiff Breeders' Association and the Montesinho Natural

13 Decreto-Lei n.º 112/2017, Diário da República n.º 172/2017, Série I de 2017-09-06.

14 http://datazone.birdlife.org/userfiles/file/hunting/HuntingRegulations_Portugal.pdf

15 Despacho nº 9727/2017, Diário da República n.º 215/2017, Série II de 2017-11-08.

16 Despacho nº 8726/2015, Diário da República n.º 153/2015, Série II de 2015-08-07.

17 www2.icnf.pt/portal/pn/biodiversidade/patrinatur/especies/aves/ag-imp

18 Despacho nº 21997/2009, Diário da República n.º 192/2009, Série II de 2009-10-02.

Park. A comparison between the localities where dogs were delivered and where Wolf attacks have occurred seems to indicate that the increase in the number of dogs has significantly contributed to reductions in livestock losses to Wolves (European Commission, 2017). Subsequently, the working protocol used in Montesinho Natural Park has been adopted in other areas of the country where the Wolf is present. The cooperation protocol helped to boost the rural economy and rural cultural traditions by allowing the maintenance of a traditional activity and encouraging related activities, such as dog breeding by shepherds and other professionals.

The protocol is no longer in place, and since 2018 there has been a new payment scheme that compensates farmers for their losses.[19] Nevertheless, some activities started under the protocol have continued, such as an online database of available mastiff puppies, an annual mastiff dog show in the district's capital and several smaller shows in rural fairs throughout the region, which are now part of the cultural events calendar.

25.5 Protected Areas and Networks

25.5.1 Protected Area Types and Their Coverage

Protected areas were defined for the first time in Portugal in 1970, through a law[20] that attributed the Portuguese State the responsibility to protect nature and natural resources over the whole national territory, particularly through the designation of National Parks and other types of protected areas. The first national protected area to be designated was the Peneda-Gerês National Park, in 1971 (as mentioned in Section 25.3.3).

In 1976, the types of protected areas were redefined; in 1978, legislation was published defining the way protected areas should be governed. For the first time, it included the need to have specific legally binding spatial plans for the areas (*Planos de Ordenamento*), which establish protection regimes and zones where particular uses and activities are prohibited, conditional or promoted.

The following six different types of protected areas are designated under domestic legislation (with numbers and areas, and equivalent IUCN types indicated in Table 25.4):

- A **National Park** is a large area containing mostly representative samples of typical natural regions, natural and human landscapes, biodiversity and geological elements, with scientific, ecological or educational value. Its main aim is the protection of its natural values and the integrity of ecosystems. The only Portuguese National Park in mainland Portugal is Peneda-Gerês.
- A **Natural Park** includes mainly natural or semi-natural ecosystems, where the long-term conservation of biodiversity might depend on human activities.
- A **Natural Reserve** is an area that is not permanently or significantly inhabited and which contains ecological, physical and biological characteristics or others with scientific, ecological or educational value.
- A **Protected Landscape** contains landscapes resulting from a harmonious blend between human action and nature, resulting in a relevant aesthetic, ecological or cultural value.
- A **Natural Monument** is a natural feature of outstanding or unique value because of its inherent rarity, representative of aesthetic qualities or ecological, scientific or cultural significance.
- A **Private Protected Area** is an area of privately owned land, not included in any other protected area type, which hosts natural values that because of their rarity, scientific, ecological, social or scenic value require specific conservation and management activities. Faia Brava is the only Private Protected Area in the country.

Protected area types can be designated at a national level, with proposals made or assessed by ICNF (National and Natural Parks, Natural Reserves, Protected Landscapes and Natural Monuments), at a regional level when involving more than one municipality (Regional Protected Landscapes and Regional Natural Parks) or a local level (Local Natural Reserves and Local Protected Landscapes). ICNF later decides whether to include the areas in the National Network of Protected Areas.

Currently, the terrestrial Natura 2000 network in mainland Portugal covers 18 913 km^2, which is 21% of the land area according to the proposed Prioritised Action Framework (PAF) (ICNF, 2020). Despite this high coverage, the network is not yet complete as the PAF states that there is still a need to identify and propose additional sites (mostly marine) and update the boundaries of existing sites. Most of the terrestrial and marine protected areas in the mainland were designated under domestic legislation before the Natura 2000 network was completed, and followed different criteria. Nevertheless, more than 90% of the protected areas in the National Network of Protected Areas fall within Natura 2000 sites. However, as the national network covers less than 8% of the land area, most Natura 2000 sites are outside it.

The IBA programme on land, and associated projects, were the basis for SPA designations and enlargements in the whole country. Similarly, SPEA's marine IBA programme,[21] carried out in collaboration with partners, particularly university and government bodies, was pioneering and played an important role in defining the marine SPAs. Portugal has 17 marine IBAs covering 14 551 km^2 of marine area (Ramírez *et al.*, 2008), but the total marine area protected by national legislation and the overlap between marine IBAs and SPAs is still low. Work remains in progress to increase the number of marine areas in the Natura 2000 network and to designate the existing marine IBAs. Currently, Portugal's marine Natura 2000 network within the Atlantic biogeographical marine region covers 33 450 km^2, which is 10.2% of its marine area within the region (ICNF, 2020).

[19] https://files.dre.pt/1s/2017/11/21300/0588705888.pdf

[20] Lei nº 9/70, Diário do Governo n.º 141/1970, Série I de 1970-06-19.

[21] http://lifeibasmarinhas.spea.pt/en/

Table 25.4 The number and extent of international designated sites and protected areas in mainland Portugal.

	IUCN category	Number	Area (km^2)
Designated under global and non-EU international legislation			
Ramsar Sites	*	18	1 174
Biosphere Reserves (exc. buffer and transition zones)	*	6	1 549
OSPAR	*	5	535
Designated under the Nature Directives			
SCIs (terrestrial + marine)	*	63	15 526 + 29 741
SPAs (terrestrial + marine)	*	42	9 206 + 6 269
Natura 2000 net (terrestrial)	*	82	18 913
Natura 2000 net (marine)	*	17	33 450
Designated under domestic legislation			
National Parks (Parque Nacional)	II	1	696
Natural Parks (Parque Natural)	VI	13	5 965
Natural Reserves (Reserva Natural)	Ib	9	741
Protected Landscapes (Paisagem Protegida)	V	2	20
Natural Monuments (Monumento Natural)	III	7	12
Regional/local Protected Landscapes	V	11	251
Regional Natural Parks	VI	1	248
Local Natural Reserves	Ib	2	1
Private Protected Area (Área Protegida Privada)	*	1	5

Note. * Variable: only the sites that overlap with sites designated under domestic legislation have IUCN categories clearly indicated in national authority data.
Sources. See Section 1.3. Based on data from: Ramsar and Biosphere, Guerra *et al.* (2019) (data refer to 2017 and 2018); OSPAR MPA database (December 2021); SCIs and SPAs, PAF 2021–2027 (ICNF, 2020); domestic sites, ICNF.[22]

Environmental NGOs in Portugal also have their own nature reserves, which are mostly included within Natura 2000 or the National Network of Protected Areas. LPN owns 20 km^2 of land for bustards and other dry-grassland birds, Quercus owns 6 km^2 of Cork Oak and Holm Oak forest, plus several micro-reserves, and ATN (Association Transhumance and Nature) owns more than 4 km^2 of land for vultures and eagles. Several other NGOs have smaller nature reserves elsewhere in the country. Besides wildlife protection, these privately owned reserves have other important roles in education and awareness, nature tourism and, in particular, testing nature-friendly ways of farming, grazing and exploiting forest products.

The total terrestrial protected area in mainland Portugal, summing up all protected area types and levels of protection and accounting for overlaps, currently covers 20 171 km^2, which is 22% of the terrestrial area of the country.[23] While it was not possible to calculate the total marine protected area within the marine area, excluding areas in the marine Macaronesian biogeographical region, we estimate this area to be largely coincident with the net marine Natura 2000 area, which is 10.2% of the marine area.

25.5.2 The Coherence of Protected Areas and Ecological Networks

The protected areas designated under the National Network of Protected Areas are considered to be core areas (*áreas nucleares*), and the Natura 2000 sites, National Agricultural Reserve areas (RAN), National Ecological Reserve areas (REN) and public water domain areas (where use restrictions apply) are connection areas (*áreas de continuidade*). The connection areas have been of vital importance to guarantee the conservation of sensitive areas such as riparian areas and steep slopes susceptible to erosion that often host important habitats. Together, the core areas and connection areas make up the Fundamental Network of Nature Conservation (*Rede Fundamental de Conservação da Natureza*), which covers a significant part of the national territory (Mota, 2016).

22 https://geocatalogo.icnf.pt/catalogo.html (accessed December 2021).
23 https://geocatalogo.icnf.pt/geovisualizador/areas_classificadas.html (December 2021)

Despite the recognition of the Fundamental Network, landscape connectivity issues are not well planned or implemented in practice. They are not integrated into the management plans of domestic protected areas or Natura 2000 sites. Instead, the REN is meant to support the connection between the core areas, such as through the incorporation of green and blue infrastructure elements (e.g. sustainable use areas and natural connectivity features), which must then be indicated in regional and local plans. Special committees, involving local authorities, central and regional public agencies, manage the application of the REN regulation and address conflicts (IEEP *et al.*, 2011). However, the REN only weakly protects connectivity because its constraints can be altered through a simple procedure in response to economic, social and cultural requirements[24] – an option that has often been used (Birdlife *et al.*, 2018).

25.5.3 Approaches to Protection and Levels of Protection

In mainland Portugal, most protected areas are on private land, and therefore the main approach to protecting nature within them is through the use of spatial plans (*Planos de Ordenamento de Áreas Protegidas, POAP*) that set out zones with varying restrictions on land use depending on the type of site and its specific context and requirements. The zones include strictly protected core areas (roughly matching the Strict Nature Reserve IUCN category), sustainable use areas (where only low-impact activities are allowed), specific intervention areas (e.g. those requiring particular rehabilitation measures) and buffer zones. The zoning plans are prepared either by in-house ICNF staff or by external experts, and overseen by an ICNF-led committee including representatives of the municipalities involved and other governmental sectors (land planning, agriculture, rural development, tourism and infrastructure). Official endorsement takes place after a public consultation process.

Out of the 47 domestic protected areas in mainland Portugal, 25 have a zoning plan (all the National Parks, Natural Parks, Natural Reserves and Protected Landscapes). No management plans have yet been adopted for any Natura 2000 areas in mainland Portugal. The remaining protected areas (Natural Monuments and regional and local protected areas) do not require a zoning plan, as they are subject to the municipal planning regime. Legislation from 2017[25] requires that the relevant content of the protected areas' spatial plans must be transposed into the municipal master plans by July 2020. This work is currently ongoing. The effectiveness of the protected area zoning system depends on the balance of power between the environmental sector and other policy sectors, the level of awareness of the local populations and the strength of the NGOs operating in each area. Advocacy and policy work by NGOs has been successful in preventing some damaging projects or, in the case of Natura 2000 sites, securing improved impact mitigation and compensation measures. In these cases, the mobilisation of public opinion has also been very important, together with the European Commission's infringement procedures and, more recently, legal actions in the Portuguese courts of justice. No systematic analysis has been carried out of the extent to which protected areas have been destroyed or damaged, but some damaging projects have been allowed to go ahead. A recent case is the dredging of the Setúbal harbour, which has been approved by the competent authority (APA) for socio-economic reasons despite recognition of important environmental impacts on an adjacent Natural Reserve and a Natura 2000 site.[26,27]

25.5.4 Approaches to Managing Protected Areas

The management of Portuguese protected areas is based on the spatial plans described in Section 25.5.3 (*planos de ordenamento*), which include general conservation objectives and management guidelines for each protected area as well as its zoning. The ongoing integration of these plans within municipal or intermunicipal plans has been criticised by NGOs, who fear that it is ineffective in protecting nature due to a large proportion of protected areas falling outside municipal jurisdiction.[28] Where Natura 2000 sites overlap with domestic protected areas, the corresponding zoning or management plans apply. However, as previously mentioned, only a small proportion of Natura 2000 sites are included in the National Network of Protected Areas and therefore most lack such plans. Instead, the areas are subject to municipal master plans (*planos diretores municipais*), which are currently undergoing a revision process. Presently, less than 40% of Natura 2000 sites have management plans, and most of them are not being fully implemented.

The designation of SACs in mainland Portugal was not completed within the Habitats Directive's six-year deadline. In response, in September 2019, the CJEU ruled that Portugal had failed to designate 61 SCIs as SACs and adopt the necessary conservation measures for them. Furthermore, the Court also stated that the conservation measures adopted by Portugal, in particular the Natura 2000 Sectoral Plan (as well as others referred to by the Portuguese authorities), did not meet the specific ecological requirements of the BHD habitat types and species concerned and could not therefore be considered as 'necessary conservation measures'.[29]

24 https://dre.pt/dre/detalhe/decreto-lei/239-2012-191514

25 Lei nº 74/2017, Diário da República n.º 157/2017, Série I de 2017-08-16.

26 https://zero.ong/dragagens-do-estuario-do-sado-sedimentos-contaminados-tem-destino-nao-contemplado-na-declaracao-de-impacte-ambienta/

27 www.europarl.europa.eu/doceo/document/E-9-2020-000044_EN.html

28 https://www.wilder.pt/historias/organizacoes-ambientais-contra-mudancas-nos-planos-de-ordenamento-das-areas-protegidas/

29 Judgment of the Court (9th Chamber) of 5 September 2019, European Commission v Portuguese Republic, C-290/18.

On 16 March 2020, a regulatory decree[30] classified the SCIs in mainland Portugal as SACs, and set requirements for management plans to be prepared for them within two years at the latest. At the time of completion of this chapter (January 2022), eight management plans had been prepared and are currently subject to public consultation.

25.6 General Nature Conservation Measures

25.6.1 Development Planning and Impact Assessment

The first EIA law in Portugal was published in 1990, in order to implement the EIA Directive and its rather complex requirements. Although the SEA law was introduced in 2007, to implement the SEA Directive, few large projects and plans have been subject to this environmental evaluation; unlike the EIA, very few citizens are aware of the procedure and its public consultation requirements.

Over recent years, Portugal has streamlined its EIA process, in particular through the introduction of a Single Environment Permit scheme in 2015.[31] The scheme aims to integrate, harmonise and simplify processes and procedures in order to facilitate their interpretation and application. It also aims to strengthen cooperation between the various public competent bodies and services involved in environmental matters. However, environmental NGOs consider that its main aim is to speed up the licensing of the so-called 'projects of national interest' as it sets very short deadlines for the assessments, including the possibility of tacit approval after a given period has elapsed.

Even without the streamlining, the EIA process in Portugal has been weak in many ways. Firstly, as is typical in many countries, the EIA work and report is performed by the promoter of the project, which creates an incentive to underestimate potential impacts. Also, in many cases, it is seen as just a formal step that is necessary in order to proceed with a decision already taken at political level, rather than a process that genuinely aims to support decision making and reduce impacts. Furthermore, EIAs rarely assess the no-project alternative, or sufficient project alternatives, or properly apply the precautionary principle. Another common flaw in Portugal has been a failure to evaluate cumulative impacts. As a result, many large projects (e.g. motorways, railways and irrigation schemes) promoted by the government have been divided up into sub-projects that by themselves have lower impacts, but all together would have been more difficult to approve. Public consultation procedures are also undermined by poor information and advertisement, lack of debate and short public consultation periods that are often scheduled during the summer vacation.

As regards appropriate assessments and Natura 2000 sites, some damaging projects have been approved (in accordance with HD Article 6(4) derogation provisions) because they have been claimed to be of overriding public interest. However, whilst the public interest has often been stated in governmental declarations, proper justification and supporting data are lacking. A well-known example was the construction of the Sabor hydropower dam, which went ahead despite its severe impacts on the *Rio Sabor e Maçãs* SPA and its claimed national importance for energy production never being confirmed (SPEA, 2016). On top of this, mitigation and compensation measures are often confused (Aragão and van Rijswick, 2014) and impact minimisation and particularly compensation measures have usually been weak and ill-designed. It has not been unusual for compensation measures to be unrelated to the impact of the project, or in areas where the government already has a legal obligation to implement similar measures. For example, no alternative locations were assessed for the railway connection between Évora Norte and Elvas/Caia (overlapping SPA Torre da Bolsa and IBA Planície de Évora), despite significant negative impacts on priority bird species, and the compensation measures focus on different areas[32] (SPEA, 2017). Furthermore, in a few cases compensation measures have not been fully implemented or have been substantially delayed, as was the case with the Sabor dam (BirdLife International *et al.*, 2018). Lastly, post-project monitoring typically has no consequences, even if it reveals that greater negative impacts have occurred than foreseen.

Public intervention in EIAs and appropriate assessments is mostly through environmental NGOs, with the larger organisations usually participating in consultations when projects may impact Natura 2000 sites and national protected areas. NGOs use the process to call for impacts to be minimised and then properly compensated, and, in the most important cases affecting Natura 2000 sites, encourage further public intervention and raise complaints to the European Commission. As a result of such actions, and the Commission's intervention, important improvements in compensation have been obtained, such as through SPA enlargement (e.g. the second crossing of the Tejo estuary, and the A2 motorway to Algarve) or the purchase of land for conservation (e.g. the second crossing of the Tejo estuary and the Alqueva dam). However, this important role of the European Commission was not applied to the Sabor dam, as in 2008 the Barroso Commission closed the infringement process despite the project's limited contribution to energy production and the important damage to Natura 2000 sites. Since then, the Commission has been less active in initiating infringement proceedings with EU Member States (as discussed in Chapter 5). Portuguese NGOs have therefore turned to national courts to halt potentially damaging projects, but the

30 Decreto Regulamentar n.º 1/2020.

31 Decreto-Lei nº 75/2015, Diário da República n.º 90/2015, Série I de 2015-05-11.

32 http://siaia.apambiente.pt/AIADOC/AIA2966/aprecia%C3%A7%C3%A3o%20da%20pron%C3%BAncia2018313173434.pdf

judicial processes are very slow and judges are generally not sufficiently well-informed to evaluate environmental damage procedures.

In conclusion, whilst Portugal has a comprehensive legal architecture for impact assessments in accordance with the EIA, SEA and Habitats Directive, there is still much to be done to improve the effectiveness of the procedures in fully preventing, minimising and compensating for damage to the environment and biodiversity. In spite of NGO and wider public participation, and the intervention of the European Commission in some cases, improvements have been small and the situation has remained roughly the same as when the EIA and SEA Directives began being implemented in the country.

25.6.2 Habitat Management and Restoration

Habitat management and restoration only started significantly in the 1990s and was mainly for game and fishing interests. The resulting increases in small game (partridges and rabbits, etc.) benefited some threatened birds of prey, such as the Eagle-owl (*Bubo bubo*), Bonelli's Eagle (*Aquila fasciata*) and Spanish Imperial Eagle. However, at the same time, some have been negatively impacted by game restocking activities (as described earlier in Section 25.4.2). Coastal sand dunes, particularly in more popular beach areas, have also benefited from protection and restoration management over the last couple of decades.

Agriculture management for nature conservation is also rather new in the country. Pioneer work was done, starting in the 1990s, by environmental NGOs and farmers in Alentejo, which resulted in good models for the conservation of dry-grassland birds, particularly for bustards and the Lesser Kestrel (SPEA, 2007; Castro *et al.*, 2012). However, such nature-friendly schemes have not been implemented more widely due to inadequate support from CAP RDP funds. This is despite the important role that better RDP measures could play in conserving wildlife. For example, the 2014–2020 Portuguese RDP was criticised by NGOs for providing the majority of financial investments for new dams, irrigation systems and river catchment management (even though existing irrigation systems are underused), as well as high investment in new afforestation schemes that include eucalyptus on farmland. In contrast, although Natura 2000 covers 18% of the country's agricultural land, it is only allocated 2.6% of RDP support, and not all of this would contribute to the required habitat management. Furthermore, it included agri-environment–climate measures for basic integrated production and efficient water use with no added value compared to existing practices (SPEA, 2014).

With the flexible approach being taken in the post-2020 CAP, it seems that stronger political backing is required for the agriculture authorities to provide adequate agri-environment schemes for farmers to contribute more to biodiversity conservation.

Forest management for wildlife also only started this century, with the introduction of Forest Stewardship Council (FSC) and other forest certification schemes. These were driven by wood and paper markets and introduced management practices for biodiversity in all types of exploited forest, from extensive Cork Oak *montados* to highly intensive eucalyptus stands. Although forest management certification impacts are yet to be fully assessed, the results of an analysis of audit reports in 2011 indicated that the process had improved over time, with fewer non-compliances detected, and that certification had contributed to enhancing forest management and environmental practices (Marques, 2011).

The most urgent requirement for habitat management and restoration concerns rivers, as they all suffer from dams, over-extraction of water for agriculture (e.g. contributing to low flows), riverine vegetation destruction, river bank soil erosion or water pollution from agriculture, industry and other sources (Oliveira *et al.*, 2007). As a result, according to the National Water Plan,[33] a significant number of water bodies (47%), including the main rivers, do not have a 'good' status, which puts at risk the fulfilment of the objectives of the Plan (Guerra *et al.*, 2019). This is due to problems with the issuing of permits for damaging economic activities, inadequate implementation of restrictions and conditions imposed by the permits, and deficient inspections by the competent authorities. Although measures to address the improvement of the status of the water bodies have been included in the River Basin Management Plans, as required under the WFD, some of these have not been implemented, in particular due to the recent difficult economic situation in the country (Guerra *et al.*, 2019).

Few restoration projects have been carried out for threatened habitats and sites in mainland Portugal, but some important projects, most of them led by NGOs and several funded by LIFE, have included the restoration of wet heaths in the central and northern mountains (LIFE Higro), the protection and restoration of temporary ponds on the south-west coast (LIFE Charcos), and habitat restoration of Berlenga island, off the west coast (LIFE Berlengas). There is still much to be done to restore and properly manage natural habitats elsewhere, especially sand dunes, lagoons and small estuaries in the most densely populated coastal areas.

25.6.3 Invasive Alien Species

In mainland Portugal, over the last two centuries, the number of exotic species has increased significantly, amounting in 2018 to about 772 plant taxa (species, subspecies, varieties and hybrids), which is 21% of the total native flora (de Almeida, 2018), as well as many vertebrates and invertebrates. Of particular concern are freshwater fish communities, as 13

[33] Decreto-Lei nº 76/2016, Diário da República n.º 215/2016, Série I de 2016-11-09.

(25%) are alien species. Some of these exotic species are considered invasive, posing a serious threat to ecosystems.

In 1999, Portugal recognised the problem and issued a law, recently revised in 2019,[34] that regulates the introduction of non-native (exotic) species, lists the exotic species introduced into the country, indicates which are considered invasive and prohibits the introduction of new species (with some exceptions). Furthermore, the legislation prohibits the possession, cultivation, growing and trade of species that are considered invasive or of ecological risk.

However, measures to supervise and control the trade of non-native species and address the threat caused to native habitats and species are rarely planned and implemented. Law enforcement is very weak and it is not unusual to find invasive plant species being sold in garden centres[35] or read in the news about exotic birds or reptiles being released into the wild (Barros, 2008). A national platform to collect and disseminate information on IAS, with the cooperation of citizens, has been recently developed to address the lack of awareness of the problems caused by non-native species.[36] A transnational project (LIFE Invasaqua) is being carried out in Portugal and Spain to improve early detection and rapid responses in the management of IAS in freshwater and estuarine habitats, through awareness and training of key groups, namely surveillance agents, river and estuary users and educators.

There are no large-scale IAS eradication programmes in mainland Portugal, but some species have been removed from small areas, particularly on islands and islets (e.g. eradication of Black Rat, *Rattus rattus*, from the Berlenga islands). Although few, these successful IAS eradications have provided well-known benefits for the habitats and threatened species in the areas concerned. However, some eradication projects have failed, such as of the Red-eared Slider (*Trachemys scripta elegans*) turtle species in the Algarve.

The main approach to addressing IAS problems in Portugal is through control. This is carried out intensively in small areas, or more generally over large parts of the territory, such as for the Water Hyacinth (*Eichhornia crassipes*) in the lower Tejo basin, and the Asian Hornet (*Vespa velutina nigrithorax*).

25.7 Nature Conservation Costs, Benefits and Funding Sources

25.7.1 Nature Conservation Costs

The financial needs for the implementation of Natura 2000 in the mainland territory and adjoining marine areas of Portugal were estimated in 2008 to be €136 million per year (ICNF, 2013), based on the questionnaires sent to Member States by the European Commission. The estimate was updated in the draft PAF (ICNF, 2020) for 2021–2027 to €176 million per year. Most of the costs (88%) are allocated to recurrent habitat management and monitoring. The estimated one-off costs include management (scientific studies, administration, consultation and preparing management plans), land purchase and compensation payments, and infrastructure needed for the improvement/restoration of habitat or species as well as public access and interpretation, etc.

It is important to note that the PAF estimates only cover the costs of conserving BHD habitats and species, mainly within Natura 2000 sites, therefore leaving out a significant number of areas protected under domestic legislation, as well as biodiversity in the wider countryside. Although the PAF includes some cost estimates for green infrastructure outside Natura 2000, these are low (less than half a million euro per year) and only cover two ecosystem categories (woodlands/forest and freshwater habitats). As they are insufficiently described, it is difficult to assess how adequate they are.

25.7.2 The Economic Benefits of Nature Conservation

Few studies have been carried out of the economic benefits of nature conservation in Portugal. Nevertheless, there is some evidence that nature conservation can contribute to the local economy. The project LIFE Priolo, dedicated to the recovery of the habitat of the Azores Bullfinch (*Pyrrhula murina*), invested €2.9 million in the eastern part of São Miguel island, which led to new jobs, new businesses and protected ecosystem services with an annual value of about €4 million (Cruz *et al.*, 2008). Marine protected areas on the west coast of mainland Portugal, with a management cost of less than €2 million per year, have been found to generate at least €7 million per year from tourism and leisure activities alone (Gomes, 2015).

25.7.3 Funding Sources for Nature Conservation

Information on the use of financial resources allocated to nature conservation in the country is scarce, and the summary provided in Table 25.5 is therefore largely based on the information provided in the 2021–2027 Portuguese PAF, and European Commission databases.

EAFRD RDP measures are the most important primary means of supporting the active management of Natura 2000 sites in mainland Portugal, given that around 80% of the terrestrial sites have an agricultural or forestry use, and the majority depend on specific agricultural or forestry management to achieve their conservation objectives. In particular, as indicated in Table 25.5, the vast majority of funding comes from the agri-environment–climate measure. However, a large proportion of RDP funding has been allocated to measures that have little benefit for biodiversity and habitat management, especially within Natura 2000 sites, such as organic

[34] Decreto-Lei nº 92/2019, Diário da República n.º 130/2019, Série I de 2019-07-10.

[35] www.dn.pt/lusa/plantas-invasoras-poem-biodiversidade-em-risco-tema-em-debate-durante-uma-semana-8744199.html

[36] https://invasoras.pt/

Table 25.5 Funding sources for nature conservation in mainland Portugal.

Source and period	Measure/category	Average annual allocation (€ m)		Comments
EU funding instruments		**EU + national**	**EU**	
EAFRD Rural Development Programme measures – allocations to Priority 4 (restoring, preserving and enhancing agriculture and forest ecosystems) 2014–2020 mainland only	M10 Agri-environment – climate	119.2	94.3	
	M8 Investments in forests	36.8	28.2	
	M11 Organic farming	23.6	18.8	
	M13 Areas facing constraints	23.0	19.1	
	M4 Investments in physical assets	16.2	12.7	
	M12 Natura and WFD	8.0	6.2	
	M16 Cooperation	1.2	1.0	
	M2 Advisory services, etc.	0.8	0.6	
	M1 Knowledge transfer, etc.	0.3	0.2	
	M15 Forest environment climate	0.08	0.06	
RDP Priority 4 subtotal		*229.4*	*181.3*	
European Regional Development Fund (ERDF) 2014–2020	85 Protection and enhancement of biodiversity, nature protection and green infrastructure	3.7	3.2	For N2k etc.
	86 Protection, restoration and sustainable use of Natura 2000	3.0	2.5	
LIFE Programme 1992–2020	Nature and biodiversity projects – mainland only	>2.1	>1.5	51 projects in mainland PT led by PT
EMFF 2014–2020	Protection and restoration of biodiversity	1.7	1.3	For N2k etc.
	Organic Aquaculture and Environmental Services	0.2	0.1	
EU funding instruments subtotal		*>240.1*	*>184.2*	
Within 2014–2020		**Additional non-EU***	**Comments**	
National				
Environment Fund 2014–2020		0.2	Plus 0.5 for co-financing of LIFE projects. No funding in 2019	
Other national public funding 2014–2018	Administration, management compensation etc. (see note 1)	0.2	For N2k etc.	
Public/Private Partnership financing		Minor		
Others				
European Economic Area Grants 2014–2020	Sustainable development plans for national Biosphere Reserves	0.3	40% for N2k etc.	
	Ceira River Basin management to adapt to climate change	0.4	20% for N2k etc.	
Additional national and others subtotal		*c. 1.1*		
National total including co-financing of EU instruments		*>56.3*		
Total		**>241.2**		

Notes. 'For N2k etc.' indicates funding relevant to Natura 2000, related green infrastructure and BHD species (based on the PAF). * Excludes co-financing of EU funds unless indicated. The subtotals and totals are only approximate indications of the average annual levels of funding as they are based on annualised data from differing periods. 1. Includes administrative costs (including salaries); site/species/habitat monitoring; information, education and training; management and protection measures not co-financed by the EU (may include national co-financing from the Environment Fund); compensation measures for damage caused by protected species (Iberian Wolf).

Sources. Based on data from: RDP data from European Commission ESIF funds database (23 October 2020); LIFE Programme public database; other funds from PAF 2021–2027 (ICNF, 2020). See Section 1.3 for further details.

ning and protecting water resources. An average of only 2.3 million per year is dedicated exclusively to Natura 2000 es, but most of it (i.e. €8.3 million) is for area payments that k any management conditions.[37] Only €4 million per year s been recently invested in specific agri-environment iemes that contribute to the management of Natura 00 farmland habitats. This is only 6% of the management eds estimated in the PAF. Despite the high level of EAFRD nding under Objective 4 for agricultural and forest ecosys- ns, Portugal has not invested enough RDP money in the anagement of Natura 2000.

The 2014–2020 RDP for mainland Portugal (PDR, 2020) ng-fenced €111 million for Natura 2000. However, this was sufficient given that the PAF prepared by the national nature nservation authority estimated that €102 million per year ould be necessary to cover just the needs for adequate agri- ilture and forestry management within Natura 2000. Despite is, the RDP budget was not reinforced for Natura 2000. In ddition, most of these funds were allocated for area payments, nd only one-third were invested in the direct management of irmland and forest habitats. This was a missed opportunity to se EAFRD funds for habitat management in Natura 2000 ites. It was also the result of political decisions that were ontrary to the views of organisations representing civil society nd farmers, as well as against the European Court of Auditors' recommendations (ECA, 2011, 2018).

The EU European Regional Development Fund (ERDF) ias provided some biodiversity financing, with an averaged allocation of €6.7 million per year over the 2014–2020 programming period (Table 25.5). The funding has been mainly for regional operational programmes linked to Natura 2000 infrastructure and information and communication, developing species and habitats indicators and monitoring systems, and managing threatened habitats and species (e.g. the Iberian Lynx and marine birds) and marine protected areas (MPAs). An example of a successful ERDF-funded project was the transnational Future of the Atlantic Marine Environment (FAME) project[38] with partners from five countries: UK, Ireland, France, Spain and Portugal. Between 2010 and 2012, it monitored seabirds, produced comprehensive maps to inform the designation of MPAs, communicated with stakeholders to minimise the impact of human activities on important areas for marine biodiversity and developed recommendations for the management of MPAs. ERDF Interreg funding was also used between 2010 and 2013 to develop a coordinated network of MPAs in the Atlantic arc under the transnational project, Marine protected areas in the Atlantic Arc (MAIA).[39]

The ERDF and the Cohesion Fund are also relevant for investments related to the assessment and restoration of biodiversity and ecosystem services, as well as for one-off conservation investments. However, the use of these funds for nature conservation is still minimal, and there is no effective governmental strategic plan to address this. Instead, environmental NGOs and scientific societies are starting to use such structural funds for nature conservation investments. Whilst this might prove successful, it will take time for these small organisations to adapt to the highly complex and demanding regulations originally designed for public bodies.

Finally, despite its relatively low funding levels, the LIFE programme has probably been the most effective EU financial instrument for nature conservation inside Natura 2000 in Portugal. It has been widely and successfully used, mainly by environmental NGOs in partnership with public bodies (e.g. environmental authorities, universities and municipalities), to identify priority sites and restore habitats and species. Despite this, two major problems have constrained the use of the LIFE fund, and await solutions. Firstly, whilst NGOs have a high capacity to implement LIFE projects, they have problems obtaining the necessary co-financing and have to rely on governments and donors. However, the Portuguese government on the mainland, and the regional government in Madeira, are normally unable to guarantee national co-financing. In contrast, the regional government of the Azores has a completely different approach and has been able to guarantee co-financing and to establish long-lasting partnerships with private organisations to implement LIFE projects. The second major challenge has been sustaining the long-term benefits of LIFE projects once they have closed. Many times, important after-project commitments from the environmental authorities have been greatly delayed or insufficiently implemented. Again, better strategic planning of actions and investments is required to sustain the results of LIFE projects.

National funds for nature conservation that are additional to the co-financing of EU-funding instruments are very low (Table 25.5). Although a few foundations and corporations recently invested in nature conservation, funding nearly stopped in the economic crisis in 2011. More recently, a new economic and financial regime for nature conservation and biodiversity and a related fund, known as the Environmental Fund (supported by industry taxation and carbon credits), was created in 2016.[40] This made it possible for NGOs and municipalities to submit nature conservation proposals for state-funded grants. In 2017 (the first year), the fund invested €3.9 million in nature conservation (Fundo Ambiental, 2018). In 2018, the total budget allocated to nature conservation in the Environmental Fund was €700 000 (with €500 000 allocated for co-financing of LIFE Nature and Biodiversity projects, and €200 000 for other

37 Programa de Desenvolvimento Rural 2014–2020: www.pdr-2020.pt

38 https://spea.pt/projetos/fame/

39 www.maia-network.org/homepage/the_project/goals_and_activities

40 Decreto-Lei nº 42-A/2016, Diário da República n.º 155/2016, 1º Suplemento, Série I de 2016-08-12.

projects). However, in 2019, no money was made available in the Environmental Fund to co-finance LIFE projects, resulting in several projects already approved by the Commission having severe financial constraints.

In conclusion, recent available funding for Natura 2000, and wider nature conservation needs, has been insufficient. This has been due to very low national funding, with inadequate co-financing constraining the use of LIFE programme funds, etc., and the use of most of the EU RDP budget for general measures.

25.8 Surveillance, Monitoring, Reporting and Applied Research

Although HD Article 11 requires Member States to undertake surveillance of the conservation status of BHD habitats and species, there is currently no systematic monitoring system in place in Portugal. The main biodiversity index being produced is the Common Bird Index (CBI), based on the Common Bird Census data. As a proxy measure of the state of biodiversity, the CBI is used as a baseline indicator for the *National Strategy for Sustainable Development* and the RDP. The Index had been published by the National Statistics Institute (INE – Instituto Nacional de Estatística) until 2009, and then stopped due to budget cuts. More recently, SPEA published the State of Birds in Portugal (Alonso *et al.*, 2019).

The mid-winter waterbird count, which started in 1976, is probably the oldest wildlife monitoring scheme in the country. The second oldest monitoring programme is the cave-dwelling bat monitoring scheme, which has undertaken annual bat surveys since 1987. Other programmes ran for a few years and were later abandoned, like the Index for Fresh Water Fish; others have performed over longer periods, such as some atlas studies (breeding bird atlas 1989 and 2008, amphibian and reptile atlas 2008). Besides these schemes, long-term monitoring has been carried out only for a few threatened or iconic species, such as the Wolf, Iberian Lynx, White Stork (*Ciconia ciconia*), Black Stork (*Ciconia nigra*), Golden Eagle (*Aquila chrysaetos*), Spanish Imperial Eagle and Great Bustard.

More recently, other monitoring schemes have started, mainly through citizen science programmes promoted by NGOs. SPEA coordinates the Christmas and New Year Bird Count (2002) and the Common Bird Census (2004), Arenaria covers coastal waders (2009), RAM covers seabirds (2009) and Noctua covers owls and nightjars (2010). Together with ICNF, the APAA (Portuguese Bird Ringers Association) coordinates the PEEC – Constant Effort Ringing Stations (2002), a monitoring programme based on bird ringing. The Portuguese Botanical Society is currently coordinating the first national atlas of flora. There are also three major public access wildlife databases on the internet: PortugalAves/eBird (exclusively for birds), Biodiversity4All (covering different taxonomic groups) and Flora-on (covering vascular plants).

25.9 Future Developments

In the near future, the protected areas zoning plans will be integrated within municipal plans through the new special protected area programmes, but the way this will be carried out is currently unknown. There is some concern amongst the main Portuguese NGOs that the restructuring risks weakening the safeguarding of the sites' natural values.

In the longer term, there is some hope for nature conservation in Portugal, as annual reports on the state of the environment (Fernandes *et al.*, 2017, 2018; Guerra *et al.*, 2019) indicate that there seems to be a growing trend in the population's interest in the conservation and sustainable use of biodiversity. This is reflected in consistent increases in the total number of visitors to protected areas, which amounted to 420 905 in 2018. Furthermore, the ecosystem services provided by the natural habitats and protected areas are now being recognised, studied and evaluated in Portugal, together with their educational and recreational value. The most obvious case is the Cork Oak *montado*, which provides a unique forest product as well as economic, health and social welfare benefits, whilst at the same time maintaining a valuable forest with important wildlife and important carbon stores. Estuaries and coastal lagoons have roles in terms of water quality, flood prevention and as nurseries for fish and other seafood, all of them with economic value for human communities. Whilst the increasing recreational interest in nature can increase pressures on fragile habitats and species, it will also increase awareness of its benefits that may encourage more government investment in nature conservation.

25.10 Conclusions

Nature conservation in Portugal, although it has a very long history, has assumed public visibility and relevance only recently. Progress with nature conservation has also been rather modest, in part due to an absence of a culture of public participation, including in environmental matters. The effects of this are still visible today, with weak public involvement in issues such as planning decisions and EIA consultations, as well as difficulties with the integration of environmental and other policies.

Accession to the EU was clearly the most important event in recent nature conservation history in Portugal. It introduced stronger nature conservation laws and resulted in an increased level of funding that otherwise would never have been invested in nature conservation. So far, the enforcement of the EU Nature Directives and the investments from the LIFE programme and other EU funds have mainly contributed to the identification and designation of additional protected areas, improved the capacity of NGOs to prevent destructive developments, and helped the recovery of a few highly threatened species and habitats, such as marine birds in Berlengas and steppe birds in Alentejo. Efforts to disseminate nature and environmental information, and involve the public in practical activities, seem now to be starting to pay off, with signs of growing interest and involvement of the population in nature conservation issues. But there is still much to be done.

As agricultural changes, especially farmland intensification, continue to be the main driver of habitat destruction and species declines, a major change in agriculture policy is the most urgent requirement to protect wildlife, as well as soil and water and other natural resources. This urgency is accentuated by the high vulnerability of the country to the effects of climate change and the risk of further afforestation encouraged by land abandonment and climate sequestration goals, at the expense of HNV farmland and other open semi-natural habitats. IAS problems also require stronger laws and improved awareness and law enforcement to prevent further introductions and deliberate spreading of problematic species. Many IAS will need comprehensive, costly and time-consuming programmes of control or eradication in order to decrease their negative impacts on ecosystems. Marine habitat and species protection is still at a crossroads in Portugal. The network of MPAs needs to be consolidated and several important areas are still awaiting designation. More general actions are also required to increase public awareness and involvement in nature conservation (despite the encouraging signals mentioned above), the appreciation of ecosystem services and the value of nature, and the protection and monitoring of species and habitats.

To meet these requirements, nature conservation in Portugal needs a real strategy with clear and measurable goals and an operational plan that prioritises actions and identifies the most important actors and financial sources. With more funding from EARDF and LIFE, better planning and greater involvement of civil society, the status of habitats, species and ecosystem services could certainly be improved over the coming decades. Based on these findings and lessons, the priority actions include completing the Natura 2000 site designation process, putting in place clearly defined conservation objectives and the necessary conservation measures for the sites, further developing and promoting smart and streamlined implementation approaches (especially as regards appropriate assessment procedures and species permitting procedures) and strengthening the integration of biodiversity concerns into other policies.

Acknowledgements

We would like to thank Helder Costa, Luís Costa and Jorge Palmeirim for their invaluable comments on the draft of this chapter, and Paulo Cabral for supplying information on area statistics of the protected area network.

References

AFN (2010). *5.º Inventário Florestal Nacional*. Lisbon: Autoridade Florestal Nacional.

Alagador, D., Cerdeira, J. O. & Araújo, M. B. (2014) Shifting protected areas: scheduling spatial priorities under climate change. *Journal of Applied Ecology*, **51**, 703–713.

Alonso, H., Coelho, R., Gouveia, C., Rethoré, G., Leitão, D., & Teodósio, J. (2021) *Relatório do Censo de Aves Comuns 2004-2020*. Lisbon: Sociedade Portuguesa para o Estudo das Aves.

Alonso, H., Teodósio, J., Andrade, J. & Leitão, D. (2019) *O estado das aves em Portugal, 2019*. Lisbon: Sociedade Portuguesa para o Estudo das Aves.

Anastácio, P. M., Ribeiro, F., Capinha, C. *et al.* (2018) Non-native freshwater fauna in Portugal: a review. *Science of The Total Environment*, **650** (2), 1923–1934.

APA (2021) *As barragens em Portugal*. https://apambiente.pt [Accessed 2 December 2021].

Aragão, A. & van Rijswick, F. M. W. (2014) Compensation in the European Union: Natura 2000 and Water Law. *Utrecht Law Review*, **10** (2), 155–160.

Barros, M. C. (2008) *As perigosas invasoras verdes*. Downloaded from www.dn.pt on 28 November 2021.

Bastos, R., Dias, J., Baptista, M. & Batista, C. (2012) Ocupação do Litoral do Alentejo, Portugal: passado e presente. *Revista da Gestão Costeira Integrada/Journal of Integrated Coastal Zone Management*, **12**, 99–116.

Bernardo, J. M., Costa, A. M., Bruxelas, S. & Teixeira, A. (2011) Dispersal and coexistence of two non-native crayfish species (*Pacifastacus leniusculus* and *Procambarus clarkii*) in NE Portugal over a 10-year period. *Knowledge and Management of Aquatic Ecosystems*, **401** (28), 1–13.

Birdlife International, EEB, Friends of the Earth Europe & WWF (2018) *The State of Implementation of the Birds and Habitats Directives in the EU*. An analysis by national environmental NGOs in 18 Member States. www.wwf.eu/wwf_news/media_centre/?uNewsID=326895

Borrego, C. (2010) *A política ambiental de Portugal no espaço europeu: atitudes e desafios*. Parede, Principia, Europa: Novas Fronteiras Portugal: 25 anos de integração europeia Centro de Informação Europeia Jacques Delors, **26/27**, 177–182.

Cardoso, A. C., Carrapato, C., Farinha, J. C., Rito, P. & Silva, E. (2014) *Plano de Ação do Saramugo (Anaecypris hispanica) / 2012-2016*. Mértola, Portugal: ICNF & Somincor.

Carro, F., Ortega, M. & Soriguer, R. C. (2019) Is restocking a useful tool for increasing rabbit densities? *Global Ecology and Conservation*, **17**, 1–8.

Castro, P., Mouro, C. & Jacinto, S. (2012) *Conservação da Abetarda, Sisão e Peneireiro-das-torres nas estepes cerealíferas do Baixo Alentejo*. Projecto LIFE07/NAT/P/654. Relatório Técnico da Acção D1 FASE II. Lisbon: Centro de Investigação e Intervenção Social / CIS-IUL.

Catry, P., Costa, H., Elias, G. & Matias, R. (2010) *Aves de Portugal. Ornitologia do território continental*. Lisbon: Assírio & Alvim.

Council of Europe (2009) *Protected Areas and Climate Change*. T-PVS/Inf (2009) 10 rev.

Cruz, A., Benedicto, J. & Gil, A. (2008) *O Projecto LIFE Priolo – Avaliação socioeconómica de um sítio Natura 2000*. Lisbon: Sociedade Portuguesa para o Estudo das Aves.

DBIO (2017) *LIFE+Marpro*. Layman's Report. www.marprolife.org (28 November 2021).

de Almeida, J. (2018) New additions to the exotic vascular flora of continental Portugal. *Flora Mediterranea*, **28**, 259–278.

DGADR (2019) *Sistema de Informação sobre o regadio.* Lisbon: Direção-Geral de Agricultura e Desenvolvimento Rural.

ECA (2011) *Is agri-environment support well designed and managed?* European Court of Auditors, Special Report: 7. Luxembourg: European Court of Auditors.

(2018) *Opinion No 7/2018, concerning Commission proposals for regulations relating to the Common Agricultural Policy for the post-2020 period.* COM(2018) 392, 393 and 394 final.

EEA (2015) *State of Nature in the EU. EEA Technical Report No 2/ 2015.* Luxembourg: Publications Office of the European Union.

European Commission (2006) *Halting the loss of biodiversity by 2010 and beyond. Sustaining ecosystem services for human well-being.* COM/2006/0216 final.

(2017) *Linking Natura 2000 and cultural heritage. Case studies.* Luxembourg: Publications Office of the European Union.

Fagundes, A. I. & Vaz, P. (2017) *Contagens RAM nos Açores e Madeira entre 2014 e 2016.* Lisbon: Sociedade Portuguesa para o Estudo das Aves.

Fernandes, A. C., Guerra, M. D., Ribeiro, R. & Rodrigues, S. (2017) *Relatório do estado do ambiente 2017.* Lisbon: Agência Portuguesa do Ambiente.

(2018) *Relatório do estado do ambiente 2018.* Lisbon: Agência Portuguesa do Ambiente.

Fundo Ambiental (2018) *Relatório de Atividades e Gestão 2017.* Lisbon: Fundo Ambiental, República Portuguesa. www.fundoambiental.pt (28 November 2021).

Gomes, R. C. (2015) *Avaliação dos Serviços dos Ecossitemas em Áreas Marinhas Protegidas: dois casos de Estudo na Costa Centro de Portugal Continental.* Lisbon: Sociedade Portuguesa para o Estudo das Aves.

Guerra, M. D., Ribeiro, R. & Rodrigues, S. (2019) *Relatório do estado do ambiente 2019.* Lisbon: Agência Portuguesa do Ambiente.

Guerrero-Casado, J., Letty, J. & Tortosa, F. S. (2013) European rabbit restocking: a critical review in accordance with IUCN (1998) guidelines for re-introduction. *Animal Biodiversity and Conservation,* **36** (2), 177–185.

Holzgraefe, G., Quintela, A. V., Bayley, K. *et al.* (2017) *Nature protection in permitting and inspection of industrial installations – Implementation of Art. 6(3) of the Habitats Directive (phase 4). Roadmap for a Planning Tool for Inspection of Natura 2000 sites.* Brussels: European Union Network for the Implementation and Enforcement of Environmental Law.

ICN (1997) *Conservação do lobo em Portugal. Instituto da Conservação da Natureza. Projeto realizado ao abrigo do programa Life. Relatório final.* Lisbon: Instituto da Conservação da Natureza.

ICNF (2013) *Portugal – Mainland Only (including contiguous marine area). Format for a Prioritised Action Framework (PAF) for Natura 2000 for the EU Multiannual Financing Period 2014–2020.* Lisbon: Instituto da Conservação da Natureza e das Florestas.

(2020) *Proposta de: QUADRO DE AÇÃO PRIORITÁRIA (QAP) PARA A REDE NATURA 2000 em Portugal continental e espaço marítimo adjacente em conformidade com o artigo 8.º da Diretiva 92/43/CEE do Conselho, relativa à preservação dos habitats naturais e da fauna e da flora selvagens (Diretiva Habitats) no âmbito do Quadro Financeiro Plurianual para o período 2021–2027.* Lisbon: Instituto da Conservação da Natureza e das Florestas.

IEEP, Ecologic Institute, GHK, Syzygy, TAU Consultora Ambiental, University of Antwerp & VITO (2011) List of Identified Green Infrastructure Initiatives by Member State. In *Green Infrastructure implementation and efficiency, Final Report to the European Commission under Service Contract ENV.B.2/SER/2010/0059, Annex I.I.*

Jardim, R. & Francisco, D. (2000) *Flora Endémica da Madeira.* Setúbal, Portugal: Múchia Publicações.

Libreiro, A. S. C. (2018) Que modelo de gestão para as zonas costeiras – o caso da frente Atlântica do concelho de Almada – Costa da Caparica. [*The case of the Atlantic front municipality of Almada – Costa da Caparica] In Proceedings of the 25th APDR Congress, pp. 20–27.* Lisbon: APDR.

Mansinho, I. & Schmidt, L. (1997) Portugal: réinventer le rural par l'environnement. In *Vers un Rural Postindustriel. Rural et Environnement dans Huit Pays Européens,* ed. M. Jovillet, pp. 261–308. Paris: L'Harmattan.

Marques, J. A. G. (2011) *Forest Certification as a Promotion Tool for Sustainable Forest Management in Portugal. Mestrado em Ecologia e Gestão Ambiental.* Lisbon: Universidade de Lisboa.

Martelo, J., Costa, L. M., Ribeiro, D. *et al.* (2021) Evaluating the range expansion of recreational non-native fishes in Portuguese freshwaters using scientific and citizen science data. *BioInvasions Records,* **10** (2), 378–389.

Meirinho, A., Leal, A., Marques, A. T. *et al.* (2013) *O estado das aves comuns em Portugal 2011 – Relatório do projecto Censo de Aves Comuns.* Lisbon: Sociedade Portuguesa para o Estudo das Aves.

Melo, J. J. (2000) The Vasco da Gama bridge on the Tagus estuary: a paradigm of bad decision-making, but good post-evaluation. *World Transport Policy and Practice,* **6** (2), 20–31.

Mota, P. M. F. (2016) *Avaliação do desempenho das políticas de Ambiente em Portugal. Dissertação para obtenção do Grau de Mestre em Engenharia do Ambiente.* Lisbon: Faculdade de Ciências e Tecnologia da Universidade Nova de Lisboa.

OECD (2011) *OECD Environmental Performance Reviews: Portugal 2011.* Paris: OECD Publishing.

Oliveira, J. M. (coord.), Santos, J. M., Teixeira, A. *et al.* (2007) *Projecto AQUARIPORT: Programa Nacional de Monitorização de Recursos Piscícolas e de Avaliação da Qualidade Ecológica de Rios.* Lisbon: Direcção-Geral dos Recursos Florestais.

Palmeirim, J. M. & Rodrigues, L. (1992) *Plano Nacional de Conservação dos Morcegos Cavernícolas. Estudos de Biologia e Conservação da Natureza, 8.* Lisbon: Serviço Nacional de Parques, Reservas e Conservação da Natureza.

PDR (2020) *Programa de Desenvolvimento Rural do Continente para 2014-2020.* www.pdr-2020.pt (2 December 2021).

Pimenta, V., Barroso, I., Álvares, F. *et al.* (2005) *Situação populacional do lobo em Portugal: resultados do censo nacional 2002/2003. Relatório Técnico.* Lisbon: Instituto de Conservação da Natureza/ Grupo Lobo.

Queirós, M. (2002) O Ambiente nas políticas públicas em Portugal. *Finisterra,* **37** (73), 33–59.

Ramírez I., Geraldes P., Meirinho, A. *et al.* (2008) *Áreas Marinhas Importantes para as Aves em Portugal.* Projecto LIFE04NAT/

PT/000213. Lisbon: Sociedade Portuguesa Para o Estudo das Aves.

Saavedra, C., Santos, M. B., Valcarce, P. *et al.* (2018) *Macaronesian Roof Report 2018. Annual MISTIC SEAS II Report 2018.* Madrid: MITECO/IEO.

Santos, F. D., Forbes, K. & Moita, R. (2002) *Climate Change in Portugal. Scenarios, Impacts and Adaptation Measures. SIAM Project.* Lisbon: Gradiva.

Schmidt, L. (2008) *Políticas Ambientais em Portugal – processos e insucessos entre o 'global' e o 'nacional'.* VI Congresso Português de Sociologia. Lisbon: Associação Portuguesa de Sociologia.

Silva, C. P. (2000) Áreas Protegidas em Portugal: Que papel? Conservação versus desenvolvimento. *GeoINova*, **2**, 27–44.

Silva, J. P. (2018) EU protected area network did not prevent a country wide population decline in a threatened grassland bird. *PeerJ*, **6**, 1–13.

Soromenho-Marques, V. (1998) *O futuro frágil – Os desafios da crise global de ambiente.* Lisbon: Publicações Europa-América.

(2005) *Raízes do Ambientalismo em Portugal, Metamorfoses. Entre o Colapso e o Desenvolvimento Sustentável.* Mem Martins: Publicações Europa-América.

SPEA (2007) *Conservação do sisão no Aentejo. Projecto nº LIFE02/NAT/P/8476 Relatório Técnico Final.* Lisbon: SPEA.

(2014) *Parecer SPEA sobre a proposta de Programa de Desenvolvimento Rural 2014-2020 para Portugal Continental. Parecer.* Lisbon: SPEA.

(2016) Responses to the evidence gathering questionnaire for the Evaluation Study to support the Fitness Check of the Birds and Habitats Directives. Available at https://ec.europa.eu/environment/nature/legislation/fitness_check/index_en.htm

(2017) *Consulta Pública – Processo de AIA 2966, Nova ligação ferroviária entre Évora Norte e Elvas/Caia – Corredor Internacional Sul. Parecer.* Lisbon: SPEA.

Tavares, B. R. (2013) *O ambiente e as políticas ambientais em Portugal: contributos para uma abordagem histórica. Dissertação de Mestrado em Cidadania Ambiental e Participação.* Lisbon: Universidade Aberta.

Chapter 26

Romania

Erika Stanciu, Ioan-Cristian Ioja, Mariana Tintarean and Mihai Pop

Key Statistics	
Terrestrial area (km²)/% of EU-28*	238 369/5.5%
Marine waters (km²)/% of EU-28	29 560/0.6%
Population density (p km²)/% of EU-28*	84/73%
GDP per capita (€)/% of EU-28	7 391/28%
Land cover (CORINE land cover classes 2018)	AGR 57%, FOR 30%, ART 6%, OSN 5%, WET 3%
Biogeographical regions	ALP, BLS, CON, PAN, STE, MBLS
Number of HD Annex I habitats/%*	87/38%
Number of HD Annex II, IV and V species/%*	245/20%
Number of BD Annex I bird taxa/%*	104/59%
% HD UCS and decreasing habitats/ % unknown	7%/2%
% HD UCS and decreasing species/ % unknown	15%/9%
% breeding BD Annex I birds decreasing/% uncertain and unknown	?/?

* Excluding MAC and MMAC biogeographical regions.
Sources and codes. See Table 1.1.

26.1 The Natural History and Characteristics of Romania

Romania is located in central eastern Europe and has an ecological identity defined by the Carpathian Mountains (28% of the country), the Danube River (38% of its length is in Romania, including most of the Danube Delta), the Black Sea (5% of its total shoreline) and the Steppic biogeographical region (17% of the country). Romania is the only country in the EU with part of its territory in the Steppic biogeographical region. It has a varied topography, with about 48% comprising mountains, and 30% lowland plains, including the Danube Delta. Its highest point is 2 544 m in the Făgăraş Mountains. The climate varies regionally, but is generally temperate continental.

Romania is the only EU country with territory in five European terrestrial biogeographical regions: Continental (53% of the country), Alpine (23%), Steppic (17%), Pannonian (6%) and Black Sea (1%) (Primack *et al.*, 2008). These have distinctive assemblages of species and habitats, which contribute to Romania's varied biodiversity. In terms of EU biodiversity, the Steppic and Black Sea bioregions are especially important, as they hold more unique species and habitats when compared to the Continental and Alpine bioregions, which are more widespread in Romania and Europe overall. The Steppic biogeographical region is located in south-east Romania and characterised by grasslands (with *Stipa* and other short grasses) and locally *Artemisia* steppe. The Black Sea bioregion includes the Danube Delta, the second largest delta in Europe, with huge expanses of reed (*Phragmites*), as well as a diverse range of other habitats (including 29 HD habitats) (Gâştescu and Ştiucă, 2008).

The country's natural conditions and habitats have been significantly changed by human interventions over millennia. Until relatively recently (i.e. the beginning of the nineteenth century), much of the natural forest was intact. The rest of the country had widespread distinctive traditional cultural landscapes and High Nature Value (HNV) farming systems (Hartel *et al.*, 2020). These include extensive semi-natural pastures (HNV type 1), mainly in the uplands, and small-scale mosaics of pastures, meadows, rotational arable crops, scrub, pastoral woodlands and orchards (HNV 2) (Page *et al.*, 2012). The original natural forests were more widely felled for timber and for agriculture, such that by the beginning of the twentieth century, forest cover was reduced to less than 25% of the country, and less than 10% in lower-lying regions (Primack *et al.*, 2008). From the early twentieth century, wetlands, such as within the Danube floodplain, underwent widespread conversion to agriculture (Habersack *et al.*, 2016; Tănăsescu and Constantinescu, 2020). During the communist era (1947–1989), the most important pressures on biodiversity were reforestation using native and non-native species (including coniferous species), agricultural expansion and intensification (nearly all in state or collective farms, except the mountains), conversion of the Danube floodplain into ble land (80% of wetlands were lost), construction of large-le infrastructure projects (such as the Iron Gates I and II ms on the Danube), industrialisation and urbanisation. cause of its fertile soil, the steppe has mostly been converted arable agriculture. Since 1990, the country's conversion to a arket economy has led to other detrimental pressures on odiversity, including uncontrolled urbanisation, illegal gging, poaching, and further widespread agricultural intensi-ation, with abandonment of traditional agricultural practices ojă *et al.*, 2010, 2011). As further described later, habitats and ecies have also been impacted by the spread of invasive alien ecies (e.g. *Amorpha fruticosa, Ailanthus altissima, Ambrosia temisiifolia, Lepomis gibbosus, Rapana venosa, Mnemiopsis idyi*), hunting and fishing, persecution of some predators, d other pressures.

Despite the detrimental human impacts, Romania remains ne of the most biodiverse countries in Europe as natural and mi-natural habitats still cover about 47% of the country, ostly comprising forests and pastures (especially in the puseni and Carpathian Mountains) and the Danube Delta Popovici *et al.*, 2016). Of particular importance are Romania's orests, about 78% of which have a high degree of naturalness, ith 840 km^2 (1.5%) of these pristine or quasi-pristine forests National Forest Inventory of Romania 2013–2018).[1] It has een estimated that up to two-thirds of the virgin forests of urope are located in Romania, mostly beech–fir–spruce and ountainous spruce forests in the Carpathians (Veen *et al.*, 2010). Such relatively large continuous areas of untouched rimary forests are very rare in Europe. Furthermore, the ountry has a large proportion of the total area of some HD orest habitat types (e.g. 91V0 Dacian beech forests, 91X0* Dobrogean beech forests, and 91Y0 Dacian oak and hornbeam orests).[2]

The country also retains some of Europe's largest areas of raditionally managed grasslands, including hay meadows, and other forms of HNV farming (Page *et al.*, 2012; Dahlstrom *et al.*, 2013). The EEA estimated in 2012 that 36% (52 000 km^2) of Romania's agricultural area still consisted of HNV farm-land. However, this was based on 2006 CORINE land-use data, so the current area is likely to be significantly lower than this. The main types of remaining HNV farmland include upland semi-natural grasslands (HNV 1), such as semi-natural dry grasslands and scrubland facies on calcareous substrates with important orchid sites (6210*), mountain pastures with *Nardus* spp. (6230*) and mountain hay meadows (6520) (e.g. Brinkmann *et al.*, 2009). HNV type 2 mixed farming systems primarily remain in the Transylvanian basin and north-east, with scattered areas in the predominantly arable farmland elsewhere. They are important for their lowland hay meadows (6510), but also their relatively high species richness and land-scape diversity. In the south-east, along the Black Sea coast, the arable farmland supports large numbers of migrating and wintering waterbirds (HNV type 3), including the Red-breasted Goose (*Branta ruficollis*).

Romania also has almost 170 km^2 of wooded pastures, which is the largest national area of this biodiverse habitat in central eastern Europe, and the third largest in Europe (Plieninger *et al.*, 2015). These wooded pastures are important elements of the ecological and cultural identity of Romania and can contain high densities of large old trees (Hartel *et al.*, 2018).

These extensive areas of habitat are in turn able to support large populations of some species that are rare elsewhere in Europe. Most notably, Romania has the highest populations of large carnivores in Europe: Brown Bear (*Ursus arctos*), Wolf (*Canis lupus*) and Eurasian Lynx (*Lynx lynx*) (Popescu *et al.*, 2019). The huge and diverse habitats of the Danube Delta are especially important for their threatened fish species, including sturgeons (*Acipenser* spp.) and the globally Endangered Danube Salmon (*Hucho hucho*), and breeding, passage and wintering waterbirds, particularly wintering ducks and geese (Muntean, 2000). For example, the Delta holds two-thirds of the European breeding population of the Ferruginous Duck (*Aythya nyroca*) and about 10% of the Dalmatian Pelican (*Pelecanus crispus*) (BirdLife International, 2017), both of which are globally Near Threatened. The Delta and Black Sea coast are also important for other migrating birds, such as raptors, as it lies on one of the most important migration corridors for birds in Europe (the Via Pontica).

In the Steppic biogeographical region, the natural steppe vegetation (including HD habitat 62C0 – Ponto Sarmatic steppe) only remains in fragmented patches, mainly in protected areas such as the Macin Mountains National Park in the Dobrogea region (Sărbu *et al.*, 2009). These areas are of high importance for a number of steppe species, such as the Common Tortoise (*Testudo graeca*) and Marbled Polecat (*Vormela peregusna*), both of which have a globally Vulnerable threat status. The Romanian Hamster (*Mesocricetus newtoni*), which is Near Threatened globally, can be also found there, as well as on the arable lands of Dobrogea. Romania is a stronghold for the globally Endangered European Souslik (*Spermophilus citellus*), which inhabits most of the Steppic region and parts of the Continental and Pannonic regions.

26.2 The Status of Nature

Although scientific information on various species and habitats has been collected since the early nineteenth century, this has not been compiled and there are few systematic reports on the status of nature in Romania prior to EU accession in 2007. Accession led to more active biodiversity surveying and monitoring, including to identify sites for the Natura 2000 network and to prepare national reports on the status of BHD habitats and species as required under Article 12 of the Birds Directive

1 http://roifn.ro/site/rezultate-ifn-2/

2 Numbers in brackets are HD Annex I habitat codes (see Appendix for full names). * Indicates a priority habitat.

Table 26.1 Short-term trends (c. 2007–2018) in Habitats Directive Annex I habitats and Annex II, IV and V species with an unfavourable conservation statu Romania.

	Poor (unfavourable–inadequate)				Bad (unfavourable–bad)			
	Imp	Sta	Dec	?	Imp	Sta	Dec	?
HD Habitat assessments								
Number	2	32	10	2	0	3	2	1
%	4%	70%	22%	4%	0%	50%	33%	17%
HD Species assessments								
Number	13	127	54	33	0	6	38	9
%	6%	56%	24%	15%	0%	11%	72%	17%

Key. Imp = improving, Sta = stable, Dec = decreasing, ? = unknown.
Source. Based on Member State HD Article 17 reported data as presented on the EEA national summary dashboard (2021).

and Article 17 of the Habitats Directive.[3] This chapter is therefore mainly based on the information gathered for these reports. However, Romania's first reports were not required until 2014 (for 2007–2013), and trends are only based on short-term assessments. Furthermore, the authors of this chapter consider that these 'official results' should be treated with some caution as many of the habitats and species have not yet been adequately assessed in practice.

26.2.1 Habitat Status and Trends

The most recent report submitted by Romania (2013–2018) based on the 173 assessments carried out across the biogeographical regions shows that 68% of the habitat assessments showed a favourable conservation status, 27% a poor status, and 3% a bad status. For three assessments the habitats' status was reported as unknown. There was relatively little variation in the status of the habitats across the five terrestrial biogeographical regions, but there was marked variation in the proportion of broad habitats with a favourable status. The situation is especially concerning for HD bogs, mires and fens, as 83% of the assessments revealed an unfavourable status. A significant proportion of assessments were also unfavourable for heath and scrub habitats (56%), forests (44%), dunes (40%) and marine coastal habitats (37%).

Of particular importance is whether those habitats with an unfavourable status have improving or deteriorating trends. As indicated in Table 26.1, many of the habitat assessments that reported an unfavourable status also reported a decreasing trend, including 33% of those with a bad status.

26.2.2 Species Status and Trends

The most recent HD Article 17 report submitted by Romania (2013–2018) indicates that, over 621 species assessments, 46% were considered to have a favourable conservation status, 37%

Table 26.2 Short-term trends (c. 2007–2018) in Birds Directive Annex I ta in Romania.

	Breeding				Wintering			
	Inc	Sta/ Flu	Dec	?	Inc	Sta/ Flu	Dec	?
Number	10	8	5	60	1	0	4	15
%	12%	10%	6%	72%	5%	0%	20%	75%

Key. Inc = increasing, Sta = stable, Flu = fluctuating, Dec = decreasing, ? = uncertain and unknown.
Source. Based on Member State Article 12 reported data as presented on th EEA national summary dashboard (2021).

a poor status and 9% a bad status. Another 9% had an unknown status. Most alarmingly, 72% of those with a bad status were also decreasing, as were 24% of those with a poor status (Table 26.1).

Of the HD species groups, it is clear that the situation for fish is particularly concerning, as only 11% out of 111 assessments were favourable. Moreover, 24% had a bad status, the most important ones including five sturgeon species and the endemic and Critically Endangered Romanian Darter (*Romanichthys valsanicola*).

The BD Article 12 report from Romania for 2013–2018 indicates that the short-term population trends of most birds, including Annex I listed taxa, are uncertain or unknown (Table 26.2). The declining breeding Annex I species include two species of raptor that have relatively important European populations in Romania: Lesser Spotted Eagle (*Clanga pomarina*) and Red-footed Falcon (*Falco vespertinus*).

26.2.3 Drivers and Causes of Change

Over the last few decades, pressures on biodiversity in Romania have rapidly increased, largely as a result of economic development and changes in land ownership. As further described in the following paragraphs, the most significant have been widespread agricultural intensification, abandonment of grazing in

[3] See Section 1.3 for background on monitoring under the Nature Directives, status categories and data sources.

some regions, inappropriate forest management, increasing urbanisation and infrastructure, inappropriate renewable energy projects, unsustainable hunting and fishing, and increasing uncontrolled tourism in natural areas. Alongside these pressures, Romania faces the common threats of invasive alien species (IAS), climate change and pollution (especially in aquatic ecosystems).

Economic drivers, as well as social trends, have especially affected land use and land management practices, causing the expansion and intensification of arable agriculture in many areas. For example, agriculture has become increasingly large-scale and intensive in western Transylvania and southern Romania, with land consolidation as one of the main drivers (Szocs and Miklos, 2015). Despite EU CAP measures to limit grassland losses, they are still happening and affecting natural and semi-natural habitats. In remaining grassland areas, agricultural practices that were highly beneficial for biodiversity, such as traditional haymaking, are now being replaced by less 'nature-friendly' methods. This is threatening the special biodiverse grasslands of Romania (Feurdean *et al.*, 2018) and their species, including ground-nesting birds such as the Corncrake (*Crex crex*) (Bellebaum *et al.*, 2018). A further problem has been that in many cases CAP payment agencies have misinterpreted payment conditions and required farmers to remove shrubs and trees from their land for them to receive their direct income support payments (Beaufoy *et al.*, 2015). At the same time, inappropriate use of state funding for animal husbandry, combined with a lack of efficient governance of community grasslands, has led to a significant increase in sheep farming, replacing traditional cattle herding in many areas, with significant detrimental impacts on semi-natural grassland habitats (Manolache *et al.*, 2020).

In remote and less productive farming areas, land primarily functions as a means of subsistence, mainly due to the extremely small land holdings (averaging 2.15 ha per household) and correspondingly low numbers of livestock (2–4 cattle and 10–12 sheep) (Membretti and Iancu, 2017). In addition, many young people are migrating from rural to urban areas, reducing the workforce available to maintain subsistence farming (Maanen *et al.*, 2006). The marginal economic value of these systems, along with social drivers, is resulting in widespread agricultural abandonment, which leads to scrub encroachment and a decline in the extent and quality of biodiverse semi-natural habitats (Angelstam *et al.*, 2013; Loos *et al.*, 2016). This is apparent in hilly and mountain areas, and is affecting the valuable cultural landscapes of Transylvania and Bucovina. Grasslands in northern and central parts of Romania have some of the highest risks of abandonment in the EU (Hatna and Bakker, 2011).

Low productivity land and steppic habitats, as well as some other lowland grasslands, are being affected by afforestation, encouraged by national and local authorities, especially when privately funded (Feurdean *et al.*, 2018).

Another important driver has been the major change in land ownership following the land restitution process initiated in 1991. Most of the agricultural land is now in the private ownership of individuals and various legal entities (74%), whilst most of the forest land remains owned by the state (48%) and public authorities (16%) (Ministry of Waters, Forests and Environmental Protection, 2019). Significant impacts have occurred where forests have changed from large-scale management based on centralised state interests to smaller private forest management units (ranging from less than one to several thousand hectares per owner). Such impacts include the loss of mature forests to legal and illegal logging. Evidence of the scale of illegal logging comes from National Forest Inventory data, which show that the average timber harvest has been around 38 million m^3 per year, whilst the annual allowance has varied between 18 and 20 million m^3 (Romanian Forest Research and Management Institute, 2018). The additional harvesting may have occurred both in forest habitats and wooded pastures.

Many sensitive forest habitats are inappropriately managed, such as those that fringe alpine pastures. These used to be considered of high importance for soil protection and were strictly protected until about 2005; since then some under private management have been lost, affecting species that depend upon them, such as the Capercaillie (*Tetrao urogallus*). At the same time, the high fragmentation of land ownership has led to more human disturbance in most of the forests, with fewer and fewer undisturbed areas, affecting many wild species and generating human–wildlife conflicts.

Wetlands and rivers have been subject to serious pressures from hydropower production, fisheries and fish farming, or direct water use. The main causes of declining fish populations have been habitat degradation, because of infrastructure development on rivers, and unsustainable exploitation over recent decades. The drainage and conversion of wet meadows to agriculturally improved grassland or arable land was a national policy in the twentieth century, and is still considered an important strategy to increase the extent of productive land. National policies on flood prevention have also promoted the commercial extraction of sand and gravel from riverbeds, significantly damaging river ecosystems, ironically sometimes with the support of the National Environmental Fund – even in Natura 2000 sites (e.g. Lower Siret).

As regards future threats, pressures are increasing to allow more economic activities in protected areas. Conservation management objectives are considered to be secondary to socio-economic objectives associated with mining activities, such as for limestone (e.g. Buila-Vânturarița) and granite (e.g. Măcin Mountains). National strategies are being promoted for mass tourism investments without prior assessments of their potential impact on high biodiversity areas, especially in the remaining wilderness areas (e.g. the Făgăraş Mountains). Even though more emphasis is being given to the implementation of environmental legislation, there is still weak enforcement of these requirements, largely due to poor-quality impact assessment studies (Nistorescu, 2012).

Political changes also have to be considered as major drivers of biodiversity change in Romania, with some

Table 26.3 Key nature conservation developments in Romania.

1904	First Nature Reserve – Codrul Secular Slătioara
1930	First nature protection law – Law no. 213/1930 on the natural monuments protection from Romania
1973	Environmental Protection Law
1996	First *National Strategy for Biodiversity Conservation*
	Establishment of the first protected area administration, for the Danube Delta Biosphere Reserve
1998	Biodiversity Departments established within each Environmental Protection Agency at county level
2000	Second *National Strategy for Biodiversity Conservation*
	List of national protected areas in the *National Master Plan*
	First Protected Area Law and legal documents with the list of protected areas to date[4]
	First administration teams for the management of National and Natural Parks established within the National Forest Administration
2001	The first contracts for delegated protected area management signed by the ministry responsible for the environment with NGOs and other entities
2004	Expansion of the national protected area network, first detailed description of the limits and zoning of National Parks and Natural Parks
2005	Protected area management (administration) teams established for most of the National Parks and Natural Parks
2007	EU accession – required adaptation of the environmental and protected areas legislation to the Nature Directives and WFD, as well as SCI and SPA designation
2013	Third *National Strategy for Biodiversity Conservation 2013–2020*
2016	Total ban on hunting strictly protected large carnivore species, but without comprehensive human–wildlife conflict mitigation measures
	Legal framework approved for the Establishment of the National Agency of Natural Protected Areas (ANANP)
2018	Actual establishment of ANANP
	Changes in the Protected Area Law led to ANANP taking over protected area management
2020	Total ban on hunting for all migratory bird species – as a result of a court ruling initiated by NGOs

See Table 3.3 for signatory dates of international conventions.

decisions having high impacts on biodiversity conservation and protected areas management. For example, the prioritisation of economic interests with inadequate consideration of biodiversity conservation requirements in many domains (e.g. hunting, forestry, agriculture, urbanisation and mining), delayed decisions related to protected areas (e.g. management plan approvals) and the ongoing refusal to allocate state funds to protected area management are all strongly influenced by the political situation (Hossu *et al.*, 2017, 2018).

26.3 Nature Conservation Governance, Actors, Legislation and Policies

26.3.1 Overview of Nature Developments in Romania

Table 26.3 provides a summary of some of the most important nature conservation developments in Romania, which are further described in this section.

26.3.2 Policies and Legislation

Nature conservation can be said to have started at the very beginning of the twentieth century, with the designation of the first Nature Reserve in 1904. This was followed in 1930 by the first nature protection law, issued at the initiative of pioneering naturalists (e.g. Emil Racoviță, Alexandru Borza and Andrei Popovici-Bâznoșanu), and the establishment of the Commission for the Protection of Natural Monuments, within the Romanian Academy, at the same time. Almost 15 000 ha were protected in 36 protected areas (including Retezat, the first National Park), and 15 plants and 16 animals were declared strictly protected as 'Natural Monuments'. Efforts to extend the protected area network apparently continued in the communist period, with the number and extent of registered protected areas showing a significant increase (Toniuc *et al.*, 1992). However, these initiatives had little practical value during these times, as

[4] Law no.5/2000 on the approval of the National Spatial Planning Plan – Section III – protected areas.

most, if not all, the protected areas were 'paper protected areas' in that they had no management authorities or defined objectives, and no active management measures were taken within them. Some general references to protected areas and biodiversity conservation were included in the Environmental Protection Law (1973), but with limited efforts for implementation during the communist times.

The situation started to change during the mid-1990s, when the first *National Strategy for Biodiversity Conservation* was adopted in 1996, reflecting CBD objectives (Ministry of Environmental Protection, 1996). Based on it, the first projects were funded by the World Bank through the Global Environmental Facility (GEF). As a result, the Danube Delta Biosphere Reserve Administration was established in 1996, and the first three National and Natural Park administrations were established in 1999/2000 and served as models for 19 new management bodies for National and Natural Parks established in 2004/2005 by the National Forest Administration, Romsilva.

The first *Biodiversity Strategy* also triggered the development of the first Romanian law for protected areas and biodiversity conservation[5] and the official recognition of the existing protected areas through a dedicated section of the *National Territorial Plan*, approved in 2000. Official recognition was thus provided for 17 areas designated as Biosphere Reserves, National and Natural Parks, as well as 827 Nature Reserves and Natural Monuments, covering 4.1% of Romania in 2004.

In 2000, the second *National Strategy for Biodiversity Conservation* was developed, having as one of its key objectives the harmonisation of the national legislation with the Nature Directives and the WFD (Ministry of Environmental Protection, 2000).

In 2013, a *National Strategy and Action Plan for Biodiversity Conservation for 2014–2020* was developed, with support from a UNDP/GEF project (Ministry of Environment and Climate Change, 2014). Ten strategic objectives were considered, of which the most relevant to nature conservation discussed here were: (1) developing the legal and institutional framework and ensuring financial resources; (2) ensuring consistent effective management of protected areas; (3) ensuring a favourable conservation status of protected wildlife species; (4) control of invasive species; (5) supporting and promoting knowledge, innovations and traditional practices; (6) communication, education and public awareness. Even though less than 40% of the actions planned through the Strategy were actually implemented, they led to considerable progress in biodiversity conservation legislation in Romania.

Romania ratified all the main global agreements on biodiversity conservation (see Table 3.3). Romania has also taken part in some regional initiatives for biodiversity conservation, such as the *EU Strategy for the Danube Region*, the *Strategic Action Plan for the Black Sea Region*, the Convention on the Protection and Sustainable Development of the Carpathians and the Alpine Convention.

[5] Emergency Ordinance no. 236/2000 on the regime of protected natural areas, conservation of natural habitats, wild flora and fauna.

26.3.3 Public Institutions Involved in Biodiversity Conservation and Governance Arrangements

Institutions

The existing institutional framework for biodiversity conservation is relatively new. During the communist era, the central authority for the environment delegated its responsibilities to county-level Environmental Protection Agencies (EPAs). In 2004, a National Environmental Protection Agency (NEPA) was established under the ministry responsible for the environment, to coordinate the activity of these agencies as preparation for the implementation of EU legislation (Hossu *et al.*, 2017).

The central public authority for environmental protection, which now is called the Ministry of Environment, Waters and Forests, is the body responsible for implementing international environmental legislation and developing national policies for biodiversity conservation, including the management of protected areas. The EPAs have responsibilities for collecting and analysing biodiversity data, compiling biodiversity reports for various national and international bodies, and analysing potential impacts of programmes and projects on biodiversity (EIAs and/or HD Article 6(3) appropriate assessments). Until 2018, county EPAs were also responsible for protected areas that did not have designated management bodies.

Currently, and for most of the last 30 years, the central authority for environmental protection has been combined institutionally with both the forest management authority Romsilva and that responsible for waters, as part of the same ministry. To some extent, this has helped with the harmonisation of environmental, forest and water related policies and programmes. Even so, this has been a challenge: many gaps remain and a more integrated legal framework is needed. For example, there is a clear legal requirement for harmonising projects and plans (including natural resource management plans) with the protected area management plans. Whilst the required adaptation of forest management plans has started in some protected areas, the process is very slow and far from being a regular practice. There is significant reluctance in the forest sector to accept that harvesting operations require clearance from protected area managers and environmental authorities to prevent negative impacts on the environment, including on biodiversity. Indeed, changes were made in the environmental legislation to 'clear the way' for forest harvesting operations, which are no longer subject to approval from protected area managers and that of environmental agencies.

Some sector-related biodiversity conservation functions also reside with the Ministry of Agriculture and Rural Development, which coordinates agricultural policy and manages the CAP payment schemes. This ministry has a key role in planning and implementing Natura 2000 payments, but the process of developing the framework is very slow, with only a few active pilot payments (for further information, see Section 26.7).

In 2001, the Environmental Guard was established, a governmental agency responsible for environmental law enforcement, with an important role in regulating and preventing illegal activities with impacts on biodiversity. The Romanian Police have a specialised Forest and Water Resource Protection Service, whose main objective is to 'protect the natural ecosystems' of the country, but with most activities focused on illegal timber harvesting.

The regional Forest Guards – governmental agencies established in nine regions in 2004 and responsible for forest law enforcement – also have an important role in biodiversity conservation by controlling forest management measures established through protected area management plans. In addition, these agencies have played an important role in the process of including pristine and quasi-pristine forest areas in the National Catalogue of Pristine and Quasi-pristine Forests[6] for strict protection.

In 2016, a new governmental agency was created – the National Agency for Natural Protected Areas (ANANP).[7] Its original main responsibilities were to develop and coordinate the implementation of national strategies and management approaches to achieve an effective and efficient protected area network, and to supervise those protected areas that were not subcontracted to third parties. As further described later, its functions changed in 2018.

The only dedicated protected area management authority established and funded by the state is the Danube Delta Biosphere Reserve Authority, which was established in 1996 with GEF World Bank support. For all other protected areas, the national environmental authority has delegated their management to various entities through a system established in 2001 (see online Annex RO.1 for details).[8] Until 2018, there were two forms of delegating management responsibilities: (1) administration teams with associated stakeholder platforms in a participatory governance system for large Natura 2000 sites, National Parks and Natural Parks (i.e. IUCN categories II and V); and (2) custodianships for the other protected areas (i.e. mainly IUCN categories I, III, IV and some V). EPAs were responsible for overseeing protected areas without dedicated management teams or custodians.

From 2004 to 2018, Romsilva, the state company managing all state forests across the country, had the responsibility, delegated by the environmental public authority, for the management of 12 National Parks, 10 Natural Parks, and 63 Natura 2000 sites. Romsilva was thus the main administrator of protected areas in Romania during that period. It has since retained responsibility for the National and Natural Par One National Park is managed by a county council.

The delegated management system had not been fu functional, partly because the delegated entities lacked st financial support for their protected area responsibilit (Popescu *et al.*, 2014; Rozylowicz *et al.*, 2017, 2019), a because there was no functional coordinating entity in pla nor a functional system for evaluating management effectiv ness. The relative inefficiency of the delegated manageme system led between 2014 and 2017 to 134 cancelled manag ment contracts. This reflected both a weak selection system f the delegated entities and a lack of monitoring and assessme of their activities and performance.

Although delegating protected area management to vario private entities had been a trend within other former com munist European countries (Stanciu and Ioniță, 2014 Romania made significant changes in its legislation i 2018 and changed the approach to a predominantly centralise protected area management system. The newly establishe ANANP became the main management entity for protecte areas – a major change by the government without any cor sultation and transparency that has imposed a centralise protected areas governance system in which NGOs and othe private entities have no influence over economic interests. Th change also resulted in more than 1 000 people engaged in th delegated protected area management system being replace with fewer than 400 newly hired ANANP staff, most of whom had no experience in protected area and conservation manage ment. Despite the problems that occurred with some custo dians, as described above, most were carrying out substantia management activities in the field and secured some project based or voluntary support. In contrast, ANANP has very limited resources and was not set up to manage individua protected areas.

A challenge related to all public institutions in Romania has been their very limited capacity generally, as well as their limited ability and willingness to collaborate with other relevant stakeholders for the sustainable management of natural resources (Hossu *et al.*, 2017) and to solve environmental conflicts in protected areas (Iojă, 2017; Hossu *et al.*, 2018).

26.3.4 Stakeholder and Public Involvement

At the beginning of the twentieth century, nature conservation was primarily led by scientists, especially those associated with the Romanian Academy. Later, in communist times, wider interest grew amongst the research institutes and universities through their research. Civil society, represented by NGOs, became active after 1990, with a growing contribution to biodiversity conservation across the country.

There are currently 1 233 NGOs that work in environmental protection in Romania, ranging from conservation of biodiversity to environmental awareness, waste management, monitoring of public institutions, and public involvement in the decision-making process and in conservation biology.

[6] www.mmediu.ro/categorie/paduri-virgine/207

[7] Agenția Națională pentru Arii Naturale Protejate.

[8] All annexes are free to access online at CUP: www.cambridge.org/natureconservation

Although most are focussed on projects at the local/regional level, several are active at the national level and have made significant contributions to both policy making and implementation of conservation projects. Some of the most prominent NGOs actively involved in nature conservation are: the Romanian Ornithological Society, ADEPT Foundation (HNV farmed landscapes of Transylvania), WWF Romania (protection of pristine forests, freshwater wetlands, ecological corridors, Natura 2000 network and protected area management), Fauna & Flora International (landscape-level conservation, and human–wildlife conflicts), Greenpeace (forest management and pristine forest conservation), Foundation Conservation Carpathia – FCC (e.g. forest conservation, land acquisition and wild fauna management), ProPark Foundation for Protected Areas (capacity building and participatory management planning), MILVUS Group Bird and Nature Protection Association, Association for the Conservation of Biological Diversity (large carnivore management, research and conservation activities), Romanian Bat Association, Romanian Herpetological Society, Society of Oceanographic Explorations and Protection of the Marine Environment-Oceanic Club and Mare Nostrum (marine environment conservation activities), EKE Satu Mare (protected area management and watchdog role), Planting Good Deeds (reforestation activities), and the Romanian Ecotourism Association. A special role is played by organisations that are raising and managing funds for conservation projects (i.e. the Romanian Environmental Partnership Foundation and the Civil Society Development Foundation).

Conservation projects have been developed and implemented by a number of research institutions, some of which are described later.

In 2003, 55 biodiversity conservation NGOs joined efforts and established the Natura 2000 NGO Coalition. The coalition contributed significantly to the designation of the Natura 2000 sites and had an instrumental role in increasing the area of the network proposed by the ministry. Subsequently, members of the coalition contributed to management planning and management of the protected areas. The coalition also has a watchdog role, with some successful court cases on record in Natura 2000 sites, as a result of which some potentially harmful projects were stopped.

26.4 Species Protection and Conservation Measures

26.4.1 Hunting Regulation

The 45 years of the communist era adopted restrictive measures that provided a high level of protection for species. In particular, legislation aimed at poaching and trafficking of trophy species, such as Brown Bear, Northern Chamois (*Rupicapra rupicapra*), Red Deer (*Cervus elaphus*) and Capercaillie was highly efficient. There were also measures that protected the habitats of game species, such as specific measures for forests important for Capercaillie, or non-disturbance zones in forests for certain game species. In the first years after 1989 there was a relaxation in hunting legislation, with quotas increasing for most species.

Since 1997, when Romania became a party to the Bern Convention, the management of game species has improved and quarry species have been grouped into (1) game species with general management provisions and (2) game species that require special management (i.e. species with hunting restrictions). With the first Protected Areas, Habitat and Wild Species of Flora and Fauna Law issued in 2000, Romania transposed the species conservation requirements from the Nature Directives into national legislation, as well as giving protection to some species of national interest.[9]

Wildlife management and related hunting activities are planned and implemented at the level of the game management unit (GMU) (Salvatori *et al.*, 2002, 2020), defined as the basic administrative unit for game management according to the national hunting legislation. There are over 2 100 GMUs in Romania (National Parks and the Danube Delta are not included), covering approximately 92% of the country's surface (Cazacu *et al.*, 2014). Hunting quotas for each game species are established annually by the responsible national authorities, based on population abundance reports made by game managers (Salvatori *et al.*, 2020). For strictly protected species, derogations are established by the central authorities, mainly with the declared goal of preventing human–wildlife conflicts, but this system also allows commercial hunting.

The most well-known and debated derogations are those given by the central authorities for hunting Brown Bears and some birds of conservation interest. Between 1955 and 1989 about 85 bears were hunted per year, with an increase to about 212 individuals in the following years until 1996, when Romania joined the Bern Convention, but with no rules implemented for conservation purposes. From 1997 until EU accession (2007), an average of 163 bears were shot per year, with special licences for reducing human–wildlife conflicts. After 2007, even though commercial hunting of this strictly protected species was officially stopped, 'intervention quotas' were approved for about 277 bears per year. Although the officially declared purpose of the quota was only to address conflict situations, trophy hunting took place. No 'intervention quotas' have been approved for this species since 2016.

Hunting quotas for bird species were challenged by environmental NGOs (Asociatia Alianța pentru Combaterea Abuzurilor, Milvus Group and OTUS) because these were not based on scientific data and there is no information regarding the potential impact on the populations. In 2019, hunting was banned for four bird species: Skylark (*Alauda arvensis*), Redwing (*Turdus iliacus*), Scaup (*Aythya marila*)

[9] Ordonanța de Urgență nr. 236 din 24 noiembrie 2000 privind regimul ariilor naturale protejate, conservarea habitatelor naturale, a florei și faunei salbatice.

and Goldeneye (*Bucephala clangula*).[10,11] In 2020, the hunting quotas for 36 migratory bird species were invalidated by a court ruling (Brasov County)[12] based on the lack of monitoring results for these populations, hence it was not possible to demonstrate that hunting is not damaging the populations. A subsequent ruling from another court (Tg. Mures county) also invalidated the quotas for Greater White-fronted Goose (*Anser albifrons*) and Greylag Goose (*Anser anser*). Although these two species were already included in the first case, the ruling is important as it shows that two independent courts came to the same conclusion. Consequently, now only two resident bird species can be hunted: Pheasant (*Phasianus colchicus*) and Grey Partridge (*Perdix perdix*).

26.4.2 Species Conservation Initiatives

Today, the main conservation management measures for species remain protection of nesting areas for birds, limiting illegal killing, limiting supplementary feeding of mammals (e.g. Wild Boar (*Sus scrofa*) and Red Deer) and compensation payments for wildlife damage from protected species. Supplementary feeding has increased game abundance in some regions, but also led to habituation of animals to humans, leading to more human–wildlife conflicts in the context of increased disturbance and habitat fragmentation (Pop *et al.*, 2018; Popescu *et al.*, 2019). A further problem is that the compensation payment scheme is linked to a very bureaucratic procedure that results in delayed payments and to lower acceptance of wildlife species, especially predators, by local communities.

Interest in large carnivore conservation grew following EU accession, and all the biogeographical populations of the Wolf, Brown Bear and Eurasian Lynx were reported as having a favourable conservation status over 2013–2018. Nevertheless, they are threatened by habitat degradation and fragmentation, livestock grazing, feral dogs, poaching, loss of food resources and conflicts with humans (Iojă, 2017). Furthermore, the present management framework for large carnivores (as well as other mammals and birds) is a result of an unhealthy combination of inadequate ecological information (due to limited resources and biased data) and policymakers taking decisions in response to demands with inadequate consideration of the species' ecology. Decision makers and the other stakeholder groups, including the general public, need appropriate ecological, social and economic studies to integrate management measures, and the necessary resources for efficient implementation (Rozylowicz *et al.*, 2017).

As a result of concerns amongst the general public and professional groups, there is pressure on the authorities to implement proper management solutions, and new objectives and conservation measures for the Brown Bear and Wolf have been discussed (Popescu *et al.*, 2019). In 2018, Romania approved action plans for the Brown Bear and Wolf, developed under EU LIFE projects.[13] In September 2021, the Ministry of Waters, Forests and Environment in partnership with the Kronstadt Local Public Forest Administration, the National Forest Administration and the National Forest Research and Development Institute Marin Dracea started a project[14] on the genetic estimation of the Brown Bear population, as required in the action plan. Some efforts are being made to improve reactions related to habituated bears,[15] but no effective solutions have been found yet. Nevertheless, as a result of the activity of Romanian conservationists, international conventions and the Nature Directives, steps are being made to base large carnivore management policies on scientific data.

A few other species now have management plans, including the Lesser Spotted Eagle, Pygmy Cormorant (*Microcarbo pygmeus*), Ferruginous Duck, Dalmatian Pelican and some bat species (*Rhinolophus ferrumequinum, Rhinolophus hipposideros, Myotis myotis, Myotis oxygnathus, Myotis bechsteinii, Barbastella barbastellus* and *Miniopterus schreibersii*). However, there are no coordinated efforts and no resources are allocated for their implementation. CAP-funded Rural Development Programme (RDP) agri-environment schemes have implemented conservation measures for *Maculinea* butterflies, Corncrake, Lesser Grey Shrike (*Lanius minor*), Red-footed Falcon, Red-breasted Goose, Lesser Spotted Eagle, Great Bustard (*Otis tarda*) and common birds of arable farmland, but the funds offered through these schemes are generally inadequately taken up by farmers and therefore are not contributing as much to the conservation of the species as intended.

26.5 Protected Areas and Networks

26.5.1 Protected Area Types and Their Coverage

As described earlier, Romanian biodiversity conservation essentially began in the 1930s through the establishment of protected areas. By 2000, the surface of terrestrial protected areas was about 4% of the country's area and increased to almost 8% by 2007 (Iojă *et al.*, 2010). In 2007, the legislation for protected areas and biodiversity conservation was revised to ensure full harmonisation with the Nature Directives.

[10] https://milvus.ro/ciocarlia-de-camp-sturzul-viilor-rata-cu-capul-negru-si-rata-sunatoare-au-scapat-de-vanatoare-incepand-de-azi/

[11] Case 502/64/2019 and https://portal.just.ro/64/SitePages/Dosar.aspx?id_dosar=6400000000017586&id_inst=64

[12] Cases 326/64/2020, 452/64/2020 and 565/64/2021.

[13] LIFE FOR BEAR (LIFE13NAT/RO/001154) and WOLFLIFE (LIFE13NAT/RO/000205).

[14] www.mmediu.ro/app/webroot/uploads/files/2021-09-03_comunicat%20proiect%20estimare%20populatie%20urs.pdf

[15] Governmental Emergency Ordinance 81/2021 regarding immediate intervention methods for preventing and stopping Brown Bear attacks on people and goods in inhabited areas – responsibilities for solving conflictual situations are transferred to the local authorities.

Table 26.4 The number and extent of protected areas in Romania.

	IUCN category	Number	Area (km²)
Designated under global and non-EU international legislation			
Ramsar Sites	–	20	17 777
Natural World Heritage Sites	–	2	4 009
Biosphere Reserves (exc. buffer and transition zones)	–	1	7 322
Designated under the Nature Directives			
SCIs (terrestrial + marine)	–	435	40 310 + 6 188
SPAs (terrestrial + marine)	–	171	37 118 + 1 630
Natura 2000 terrestrial area			54 214
Natura 2000 marine area			6 362
Designated under domestic legislation			
Scientific Reserve (Rezervatie stiintifica)	I	32	166
National Parks (Parc national)	II	13	3 174
Natural Monuments (Monument al naturii) and Nature Reserves (Rezervatie naturala)	III, IV	916	2 921
Natural Parks (Parc natural)	V	14	7 694

Sources. See Section 1.3. Based on data from: Ramsar Convention and UNESCO websites (December 2021); EEA Natura 2000 Barometer, Natura 2000 as at end 2021; domestic protected areas, ANANP (2021).[18]

The current protected area system in Romania is summarised in Table 26.4. In addition to the Natura 2000 sites, and other international sites, the following types of protected area of national interest are designated according to the Protected Area Law[16]:

- **Scientific Reserve** (IUCN category I), provides strict protection of natural ecosystems.
- **National Park** (IUCN category II), aims to conserve natural ecosystems and maintain natural processes, with limited human interventions in core areas, allowing only for sustainable tourism, education and traditional use of grazing areas for local communities. Core area coverage is less than 75% in most National Parks.
- **Natural Monument** (IUCN category III), designated to conserve exceptional natural features (spectacular rocks, gorges, lakes, fossils sites, caves, waterfalls, etc.).
- **Nature Reserve** (IUCN category IV), established for the conservation of natural and semi-natural habitats and valuable species that need active conservation management measures. This type includes marine reserves.
- **Natural Park** (IUCN category V), established for the conservation of natural and cultural landscapes and to promote responsible use of natural resources for the sustainable development of local communities.

In combination, the national protected areas cover 8% of the national terrestrial area, with Natural Parks covering 3.2%, National Parks 1.6%, Natural Monuments and Nature Reserves together 1.2%, and Scientific Reserves 0.07%.[17] Less than 1% of Romanian sea areas are under national protection.

Romania has designated a large number of Natura 2000 sites, which at the end of 2020 totalled 60 577 km², covering 22.7% of its land area and about 21.5% of its marine waters. Many of the terrestrial protected areas of national interest overlap with the Natura 2000 sites. Before the Natura 2000 site designations, only small parts of the marine area were included in one relatively small Strict Reserve and partly in the Danube Delta Biosphere Reserve.

Taking into account the overlaps between sites, the total protected area[19] was estimated to amount to 58 225 km² on land, and 6 866 km² in the marine environment according to the WDPA in December 2021. This equates to 24.4% coverage of the land and 23.2% of Romania's marine waters.

Despite the relatively large Natura 2000 area, SCI coverage has minor–moderate insufficiencies for many HD habitats and species. The coverage of SPAs is also considered insufficient in relation to Important Bird Areas (IBAs). To address this, specialist organisations and institutions (SOR, Milvus and the National Danube Delta Institute for Research and Development) prepared proposals in 2012 for new SPAs, but these have not been designated so far.

[16] Emergency Government Ordinance No. 57/2007 on the regime of protected natural areas, conservation of natural habitats, wild flora and fauna, with further amendments.

[17] http://ananp.gov.ro/ariile-naturale-protejate-ale-romaniei/

[18] http://ananp.gov.ro/ariile-naturale-protejate-ale-romaniei/

[19] Although this excludes the Biosphere Reserve, there is a high level of overlap with other sites.

It is important to note that, whilst the protected area network currently exceeds the CBD's Aichi terrestrial and marine coverage targets, there are problems with the effectiveness of its protection and management in practice, as further discussed in Section 26.6.4.

26.5.2 The Coherence of Protected Areas and Ecological Networks

The coherence of the Natura 2000 network has been evaluated by Rozylowicz *et al.* (2019), who identified network hubs (protected areas that are critical for the coherence of the entire network), module hubs (protected areas that are important for coherence at the bioregion level), connector sites and peripheral sites. The authors concluded that there is an obvious need for prioritisation of conservation measures in key SCIs/SPAs that constitute the backbone of the network. However, there is insufficient information about the coherence of the wider protected area and ecological network, and an integrated assessment is needed.

There have been several conservation initiatives that have aimed to improve the coherence of the broader range of protected areas in some parts of Romania and the wider region, including initiatives for large carnivores and the Lower Danube Green Corridor. These and some supporting studies are described in online Annex RO.2. Despite these initiatives, and a legal requirement in the protected area law for the identification of ecological corridors in Romania, no corridors have been officially designated so far, nor management measures approved.

26.5.3 Approaches to Protection and Levels of Protection

According to the Protected Area Law, the management of a protected area should be based on a management plan (described further in Section 26.6.4) that is developed by the manager and approved by the ministry, or based on conservation measures approved by the ministry for target species and habitats. For areas that have multiple and/or overlapping designations, integrated management plans have to be developed, defining harmonised management measures. Once the protected area management plan is approved, landowners and other stakeholders have to harmonise management activities that might impact biodiversity values with the management plan.

Since 2001, the law has provided for, and defined, a zoning system for National and Natural Parks. Zoning is an important management tool for promoting the rules established for the protected areas and for preventing and controlling pressures and threats. Strictly protected zones are defined for areas where natural processes are to prevail, within which no activities other than research, education and visitors are allowed. Zones with 'integral protection' are defined for areas where no commercial harvesting and use is allowed for natural resources, but subsistence activities (collecting wild berries and fruits) and grazing by livestock owned by locals allowed. Buffer zones are defined where the sustainable use natural resources is allowed, and sustainable developm zones for areas included in settlements or in tourism reso or where more intensive land management practices allowed. The internal zoning of the existing National a Natural Parks was legally approved in 2004. Zoning is bei re-analysed in the management planning process and has to revised with stakeholder participation.

Before 2004, the ministry responsible for the environme and local EPAs were responsible for the management of pr tected areas. Capacity limitations led to inadequate manag ment, sometimes with damaging development proje implemented even in the strictly protected core zones. Som of the most notorious examples include an illegally built mo astery in the core area of Ceahlau National Park (2006) ar illegal infrastructure development in the core area of th Rodna National Park, affecting an alpine lake (2005).

The situation improved when the first Protected Areas La was approved and with the establishment of the first thr National and Natural Park management teams by the Nation Forest Administration (Romsilva), followed by the developme of management plans for three under the 1999–2004 GE Biodiversity Conservation Management Project. Thus, a part decentralised (delegated) protected area management syste was established for the National and Natural Parks. This syste was gradually extended to some of the Nature Reserves an Natura 2000 sites and various entities were delegated to manag them (as described in online Annex RO.1).

After EU accession in 2007, the requirement for HD Article (3) appropriate assessment procedures was included in th Protected Area Law. Hence, given that the majority of the pro tected areas of national interest are included in the Natura 200 sites, a stronger level of legal protection was secured for the site due to the Nature Directives. Any projects or plans that migh have a negative impact on a Natura 2000 site have to undergo ar appropriate assessment and be approved by ANANP, based or the official agreement of the protected area administrator. As a result, the impetus for economic development in the core zones of domestic protected areas, which overlap greatly with Natura 2000 sites, has declined, especially in National Parks.

However, there are still unresolved issues related to the Natura 2000 network, mainly because of the limited capacity of ANANP as well as the need to designate all SCIs as SACs, based on the approval of the management plans and the updated Natura 2000 standard forms. Also, many of the impact assessments of proposed economic developments have been either superficial or their recommendations have not been followed. In some cases, active interventions by NGOs were needed to provide proper assessments, for example in the case of some proposed micro-hydropower plants on rivers with HD species (Iojă, 2017; Hossu *et al.*, 2018). Therefore, it is necessary to improve the transparency and quality of impact assessment studies, to limit and mitigate conflicts between stakeholders and to build trust in the EIA procedures (Nita *et al.*, 2015).

26.5.4 Approaches to Managing Protected Areas

The framework for the management of protected areas is established through the management plans, which have to be developed in a participatory process, involving the Scientific and Consultative councils, for the large and complex protected areas (IUCN category II and V). Local environmental agencies and relevant ministries are involved before the ministry responsible for environment approves the management plan. Public consultation is mandatory in the process and is carried out by the environment agencies as part of the SEA procedure (required for all strategic planning documents, including protected area management plans) but with limited influence on the management plan content. The level of stakeholder participation (e.g. public institutions, NGOs and landowners) in the planning process varies, with only a few examples of good practice. According to Hossu *et al.* (2017), problems in the consultation process have included: (1) poorly organised consultations by managers and/or public authorities; (2) limited stakeholder interest or understanding of the real impact of conservation measures; and (3) limited conflict management capacity.

The main approaches to implementing management measures in protected areas can be grouped as follows: (1) *active* management, when the managing entity has the capacity and the resources (mainly from conservation projects, e.g. LIFE projects, or those funded by the Norwegian or Swiss governments) to implement management actions; (2) *participatory* management, when a transparent approach and well-defined procedures enable the involvement of various stakeholders in the protected area's management, and help designated managers carry out active management; (3) *passive* management, where pressures and threats remain unaddressed to avoid conflicts and the protected area staff focus on educational or other, non-conflicting, activities.

As of 2021, 307 management plans have been approved, mostly developed with EU funds, for 286 Natura 2000 sites (205 SCIs, i.e. 47%, and 81 SPAs, i.e. 47%)[20] and overlapping protected areas of national interest (i.e. National and Natural Parks and Nature Reserves). In many cases, links between the biodiversity values and pressures and threats are not clearly indicated in the management plan, hence the proposed management measures are rather general. In 40% of the management plans, measures for species and habitats of conservative interest are established, but only in a few cases are these sufficiently detailed to include practical measures to maintain or improve their conservation status. Management plans have also been developed for five of the 12 marine protected areas, but knowledge on marine species, habitats, pressures and threats is even more limited, and there is a critical need for further data collection and monitoring (Teampău, 2020; Văidianu *et al.*, 2020). International management of Black Sea natural resources is an important challenge that needs substantial involvement from state authorities.

According to the national legislation, landowners and natural resource managers have the responsibility to implement conservation measures as established through management plans or the approved list of conservation measures. However, there are no supporting funds available from the state budget for management measures, other than compensation payments to private forest owners (see Section 26.6.2). Therefore, implementation of conservation measures is very challenging, especially on private land. Some EU funding is available through the CAP RDP measures, and the voluntary agri-environment schemes have helped to maintain some species and habitats of conservation interest (see Section 26.7), but the RDP Natura 2000 payment measure has not been used to compensate for conservation management measures imposed on private landowners. The delay in developing and implementing the Natura 2000 payments is seriously hindering conservation management measures. For landowners whose forests are included in the core zones of national protected areas, some compensation payments from the state budget are available for non-intervention management or if forest activities are limited by nature conservation objectives or other objectives.

Similar problems occur for managers of public land, such as the National Forest Administration, Romsilva. Romsilva has obligations to manage state forest in a profitable way and to allocate financial resources for the management of protected areas in its jurisdiction. It does not receive any financial support for the management of protected areas from the government, but it can access EU funds that can contribute to Natura 2000 management.

The Protected Area Law requires all relevant sectoral plans to be harmonised with the protected area management plans (i.e. for them to reflect conservation management measures) and implemented in their daily activities. However, the only sectoral legislation that is harmonised with the Protected Area Law is for forestry, which specifically requires harmonisation of forest management plans with protected area management plans.

Since 2018, when ANANP ostensibly replaced the custodians (described in Section 26.3.3), implementation of management plans has stagnated and EU funds allocated for the management of Natura 2000 sites have been used only to a small extent. A notable exception is that of the National and Natural Parks and two Natura 2000 sites that are still delegated to Romsilva and some other entities, which are continuing to implement their management plans.

26.6 General Nature Conservation Measures

26.6.1 Development Planning and Impact Assessment

Romania's national planning policy was influenced by the communist legacy, EU accession and the recent global economic crisis. During the communist period, Romania had a top-down planning system under the total control of the state authorities. For a few years after the fall of the communist

[20] http://ananp.gov.ro/wp-content/uploads/Anexa-8-Planuri-de-management_specii-aprobate-1.pdf

regime the void in the legal framework slowed down the transfer to a decentralised approach. Then the EU accession process, which started in 2000, triggered several institutional and legislative reforms. A first law on territorial planning was adopted in 2001, clarifying the responsibilities of national and local public authorities. Furthermore, stricter environmental legislation was adopted, influencing territorial planning activities. However, the governance system remained somewhat centralised (Grădinaru *et al.*, 2020).

Under the EU's influence, administrative decentralisation gave additional land-use decision-making powers to local authorities, but the local authorities lacked the knowledge and human resources to carry out planning and implementation. Concomitantly, strong global economic growth, coupled with capital inflows due to the increase in Romania's attractiveness for foreign investments, amplified the expansion of built-up areas, resource extraction and urbanisation. It soon became clear that the planning system could not keep up with the growing demand for new housing and service areas and the pressure of private interests. Furthermore, the territorial planning was not integrating transport, environment or social aspects. Uncontrolled development, including in protected areas, started to become an issue (Iojă *et al.*, 2010).

After EU accession in 2007, the EU exerted a strong influence over domestic policies and practices, particularly at the national level. Over recent decades, the development planning system in Romania has been adapted to fulfil EU obligations. Thus the importance of EIA in the administrative process of development planning increased, especially in protected areas where environmental policy limited development plans (Grădinaru *et al.*, 2020). At a local level, spatial planning increasingly integrates relevant provisions from the protected area management plans. The main issue is the lack of resources for renewing territorial plans for all of the administrative units within the new legal framework, hence in some protected areas there are still conflicting provisions regarding development projects.

26.6.2 Habitat Management and Restoration

Habitat management is one of the main challenges in Romania for a number of reasons, some of which have been discussed previously, including the governance system and legislation that lead to sectoral and centralised approaches to land management, as well as the complex issues related to land ownership. The same factors also constrain habitat restoration, even when promoted by national authorities and when funds have been secured through, for example, LIFE projects. Illustrative examples of particular importance in Romania are given in the following paragraphs from a range of habitats.

Coastal Habitats and the Danube Delta

Nature conservation in the Danube Delta, and other areas of the Romanian coast, is both particularly important and challenging, because of the sensitivity of their distinct habitats and numerous rare species, and the high social and economic importance of the area (Tănăsescu and Constantinesc 2020). The main challenge is to find a balance between touris infrastructure development and the need to conserve existi natural habitats. Because the area is increasingly targeted infrastructure developers, property values have become ove inflated, reaching levels that cannot be matched by potenti compensation mechanisms that would aim to promote conse vation measures. Besides anthropogenic pressures, a furth challenge arises from the natural dynamics of the delta ar coastal system, which constantly change the habitats throug erosion, accretion and changes in water salinity (Góme Baggethun *et al.*, 2019).

Many of the conservation measures for coastal habita have focussed on the Danube Delta due to its unique biodive sity and landscape values (Teampău, 2020). It has also been place of conflict between stakeholders looking for profitab investments in commercial fishing, tourism or agriculture, an local communities whose livelihoods have been threatened as result. Management efforts have focussed on developing pressure and threat monitoring system, awareness-raisin actions and tourism infrastructure development.

The Danube Delta Biosphere Reserve is one of the area where the need to integrate biodiversity conservation measure and support and benefit-sharing with local communities i urgent. Stand-alone, short-term conservation projects, even i well-planned and funded, do not have a long-term impact i social and economic issues are not properly addressed Currently, the Danube Delta is the largest Biosphere Reserve in Romania, a key Ramsar Site and a UNESCO World Heritage Site; however, all management efforts undertaken in the Danube Delta are still far from being able to demonstrate effective biodiversity conservation. The development of new housing and tourism facilities with negative impacts on the environment (e.g. destruction and fragmentation of habitats, waste and wastewater management), the increased attraction as a tourist destination, the widespread impact of invasive species, along with bird poaching and overfishing, continue to contribute to the decline of species and habitats in this region (Teampău, 2020).

Grassland and High Nature Value Farmland

As discussed in Section 26.1, Romania has some of the most extensive and best-preserved HNV farmland in Europe, including semi-natural pastures and meadows, and diverse mosaics of traditional croplands, orchards and pastoral woodlands. As well as being of exceptional biodiversity importance (Page *et al.*, 2012), they are also high cultural value farmlands (Rolo *et al.*, 2020), because their management practices entail local knowledge, values and cultural ingredients that are inextricably linked to the farming landscapes.

The ADEPT Foundation is a leading Romanian NGO that promotes the HNV farmland concept and works closely with authorities and local communities to help conserve and increase the social, economic and environmental values of Europe's most important HNV farmed landscapes by encouraging the sustainable use of natural resources (ADEPT Transilvania *et al.*, 2016). Finding ways to maintain traditional farming systems is critical not only for securing the survival of vibrant rural communities and high-quality food production systems, but also for conserving species, habitats and ecosystems services.

Despite the efforts of some organisations, during the last programming period of the CAP many grasslands were converted into arable land or were abandoned and are now gradually turning into scrub and forests.[21] There are no official detailed data on grassland transformation to arable land, or how many hectares have been abandoned, but a recently started project[22] aims to collect such information. At the same time, some of the arable land has been abandoned and is starting to go through natural succession.

Forests

Forests cover 65 920 km^2, 28% of Romania's territory, with 48% being state forests, 18% owned by local authorities and 34% private forests (e.g. individuals, communities, schools and churches).[23] Whilst there are valuable forest habitats in various parts of the country, the Romanian Carpathians have retained the largest tracts of natural broadleaf and mixed forests, including relatively large areas of old growth and virgin/quasi-virgin forest. The Romanian forest management system is based on ecologically sound forestry practices, which has resulted in the relatively good conservation status of forest habitats, especially in the Alpine bioregion (Strat and Mihăilescu, 2017).

The national forest management system, applicable to both state and private forests, incorporates measures to secure the continuous delivery of forest ecosystem services, including measures for biodiversity conservation, mainly by separating forests managed for timber from 'protective forests', that is, forests that are managed mainly to deliver ecosystem services and for biodiversity conservation (Environmental Ministry Order 3397/2012). Protective forests comprise 66% of forest land.[24]

Habitats and species targeted in protected area management plans have also to be considered in the forest management plans (required for all state forests and for private forests over 100 ha), and conservation measures must be implemented by forest managers in all forests – not only in the protective forests, but also in the 'production forests' (i.e. those managed mainly for timber). Conservation measures aim to reduce the frequency and intensity of logging activities and include recommendations for the conservation of forest-dependent species (e.g. rules/management measures for raptor nesting sites, or for retaining minimum amounts of deadwood).

In the early 2000s, the Forest Stewardship Council (FSC) voluntary certification scheme was introduced and promoted in Romania and was adopted by the National Forest Administration and private forest managers. The FSC scheme contributed to forest ecosystem conservation, especially through identification of high conservation value forests (HCVFs) and by providing management recommendations for maintaining/restoring them – recommendations that are included in the forest management plans. HCVFs include forests that have a high biodiversity and/or social value, for example forest habitats that are rare or endangered or have a high importance for rare species, or have an important role for protecting unique water sources for local communities. Recently the Programme for the Endorsement of Forest Certification (PEFC) system has been promoted, with the expectation that it will support responsible forest management by owners of small areas of forest.

The same management system[25] is applied to both public and private forests, but there are intense debates about adapting the management system to increase possibilities for private owners with small areas of forest to use the forest resources. One important measure contributing to the conservation of valuable forest habitats and species is the compensation payment scheme from the state budget. This applied for a brief period in the late 1990s and has been reactivated since 2016 for strictly protected forests and for those where only very limited interventions are allowed. The restrictive measures are applicable to private forests that are critical for maintaining biodiversity and/or environmental services. The scheme is available for about 3% (223 200 ha) of the total area of managed forests, owned by private persons or private owner associations. Additionally, a voluntary Forest Environment and Climate Scheme is being implemented with funds allocated from the European Regional Development Fund (ERDF), encouraging designation of non-disturbance areas and using non-intrusive collection methods for the harvested timber. However, this scheme has had limited impact, mainly because of low interest amongst forest owners and/or because it has not been promoted properly.

Despite the generally nature-friendly management approaches required by the forest legislation, Romania has faced a major challenge in recent decades from high levels of uncontrolled and illegal logging (as discussed in Section 26.2.3). To address this issue, efforts are being made to improve the wood tracking system and reduce illegal logging. Starting in 2014, the government activated a timber transport survey tool for citizens, with a mobile app named Forest Radar

---

21 About 46 000 ha, according to Eurostat, as calculated within the 'Working Together for a Green Europe' project, based on information provided by Eurostat and FAOSTAT.

22 Working Together for a Green Europe – implemented by Milvus Group in partnership with the Transylvanian Carpathia Society Satu Mare with funds from the Active Citizens Fund Romania, a programme funded by Iceland, Liechtenstein and Norway through the EEA Grants 2014–2021.

23 www.mmediu.ro/categorie/starea-padurilor/209

24 www.mmediu.ro/categorie/starea-padurilor/209

25 The Forest Law – Codul Silvic, Law no 46/2008.

(linked to the Emergency Call Service 112); this was upgraded in 2016 to a more transparent mobile app and a new satellite alert system signalling forest disturbances (Forest Inspector). This tool allows public involvement in identifying and reporting illegal harvesting and timber transport, with relatively good impacts. For example, a Greenpeace report on illegal logging in Romanian forests[26] found that 42% of the officially registered incidents of illegal timber transport were reported by citizens. Although it seems that there is a large community of citizens and organisations involved in this process (the Facebook Forest Radar group has more than 10 000 members), these tools only tackle a small proportion of illegal logging.

Only an integrated approach will solve the illegal logging problem, with solutions derived from a thorough analysis of the drivers and integrated into strategic planning in the relevant sectors (e.g. forest management, energy and construction). For example, one of the main drivers of illegal logging is the lack of firewood on the market: only about 50–60% of the need is currently met. This has to be addressed both through forest management–related measures, and by considering energy sources and energy efficiency–related aspects.

It is also important to have as soon as possible a formal recognition of the importance of large old trees (in and outside forests) as well as better integration of forestry and isolated large tree protection and CAP measures. For example, the CAP payment condition GAEC 7 (Section 4.3.8) recognises the biodiversity and wider environmental value of scattered trees on farmlands. However, as implemented in Romania, GAEC 7 allows the removal of scattered trees with the permission of the forestry authorities. Foresters favour the removal of large old trees that are often perceived as looking 'unhealthy' with their holes and dead branches.

26.7 Nature Conservation Costs, Benefits and Funding Sources

26.7.1 Nature Conservation Costs

The overall costs of nature conservation in Romania have not been estimated. However, the costs of implementing the Natura 2000 network have been estimated through the development of Prioritised Action Frameworks (PAFs). According to the PAF for 2014–2020, the required budget was €301 million per year (MoE, 2013). The total financial needs for implementing Natura 2000 for 2021–2027 are estimated to be €544 million per year averaged over the period, of which 28% are for one-off costs and the remainder are running costs (MoE, 2019). These annual costs cover general cross-cutting measures for the network (26%), habitat maintenance and restoration measures (57%), green infrastructure to support the Natura 2000 network (14%) and additional species-specific measures (3%).

26.7.2 The Economic Benefits of Nature Conservation

There are no comprehensive evaluations of the economic benefits of nature conservation in Romania. The *NATURE4DECISION-MAKING (Demonstrating and promoting natural values, to support the decision-making process in Romania' – N4D)* project, implemented by NEPA in 2015–2016, provided resources for mapping and biophysical evaluation at a national level of ecosystems and their services, as part of the EU Mapping and Assessment of Ecosystems and their Services (MAES) initiative. However, the results have not been used so far to determine the economic benefits of nature conservation.

The National Centre for Sustainable Development developed a methodology for rapid assessment for ecosystem services in protected areas in the *EcoServNat* project (Adamescu *et al.*, 2016). This methodology was initially applied in three Natura 2000 sites and then extended to another 10, but the project had limited impacts on other protected areas. Recommendations were developed for ecosystem service payment mechanisms for several sectors: agriculture, forestry, energy and tourism.[27]

26.7.3 Funding Sources for Nature Conservation

Table 26.5 provides an overview of the main sources of funding (primarily during the 2014–2020 funding period) that have been used for nature conservation in Romania, as described in the following paragraphs.

Until recently, Romania did not have a dedicated state budget for biodiversity conservation, other than funding for salaries of staff with specific responsibilities in public institutions and agencies, and an annual budget for the Danube Delta Biosphere Reserve Administration. Since 2018, some funding has been allocated for the newly established ANANP, but only for salaries and basic administrative needs. The National Forest Administration covers the costs for the administration of the 12 National Parks and 10 Nature Parks under its management, and one county council provides some budget for one National Park. These funds do not cover practical conservation management needs. For example, the variable annual budget, averaging c. €3.2 million over 2011–2015 for the management of the Danube Delta Biosphere Reserve has been insufficient to carry out practical management, monitoring and enforcement activities for this large (c. 500 000 ha) and complex area (Rozylowicz *et al.*, 2019). Entities that were responsible for protected areas based on custodianship contracts from about 2004 to 2018 had to secure financial resources for management planning and for

[26] www.greenpeace.org/static/planet4-romania-stateless/2019/07/4d0a0d0c-4d0a0d0c-raport-taieri-ilegale-2016.pdf

[27] http://valueecoserv.cndd.ro/rezultate/

management activities, mainly from EU project funds and private donors.

External funding for conservation has been available from the 1990s through the GEF support for key conservation actions up until 2016 (totalling €630.6 million – not included in Table 26.5) (MoE, 2013). The actions included the development of the protected area legislation, and the first protected area management teams, management plans and management activities. Since then, EU funds have helped to develop national databases for species and habitats, elaborate management plans for protected areas, develop action plans for species, manage and restore habitats, increase public awareness, and develop support for biodiversity conservation. The ERDF has been particularly important for Natura 2000 over the two funding periods since Romania became a Member State, with €215 million allocated for 2007–2013 and €344 million for 2014–2020 (see online Annex RO.3 for further details).

Although much lower, the funding for LIFE projects has also been very important and effective, with an investment of €66 million (€33 million from the EU) supporting 46 nature and biodiversity projects in Romania since the first in 1999 (further described in Annex RO.3). There have also been multinational LIFE nature projects that have involved Romania. The European Social Fund was used, through a SIPOCA project[28] over 2007–2013 (€3.4 million), to support capacity development in the ministry responsible for the environment, to improve the implementation of biodiversity policies.

Funding for active habitat conservation management, inside and outside Natura 2000 sites, has mainly been provided by the European Agricultural Fund for Rural Development (EAFRD). As discussed in relation to species conservation in Section 26.4.2, the agri-environment–climate measure has been used to fund practical conservation measures for several threatened species, as well as HNV farmlands. However, these measures are voluntary for farmers and have not been widely taken up. The schemes also only cover the ecological needs of a few BHD habitats and species. A relatively small proportion of RDP funding has been allocated to forest–environment–climate measures, and take-up has also been low. A further problem has been that Romania has not yet used the RDP Natura 2000 payments measure to compensate landowners for necessary restrictions on the use of their land, such as those identified in management plans, which has been a major constraint on their implementation. By far the largest proportion of RDP funding has been allocated to areas with natural constraints, which was extended during the 2014–2020 programming period, and used half of the budget. Although this probably helps to reduce abandonment of HNV farmland and covers a large proportion of Natura 2000, its nature conservation benefits are uncertain.

26.8 Surveillance, Monitoring, Reporting and Applied Research

The most comprehensive national-level inventory and monitoring efforts have been developed as a result of the surveillance and reporting obligations under the EU Nature Directives. These have been supported by a number of past and ongoing EU ERDF projects, coordinated by the ministry responsible for the environment. These projects have developed monitoring guidelines, carried out field surveys and prepared the necessary reports. The monitoring and reporting (2013 and 2020 reports) on HD habitats and species were carried out by a consortium of public institutions led by the Romanian Academy's Institute of Biology, with the fieldwork carried out by individual experts, research institutes and specialised NGOs subcontracted by the project beneficiaries. A Monitoring Information System for Species and Habitats of Community Interest (SIMSHAB) was also developed through the same project, but it is not open for public access.

The monitoring and reporting on the conservation status of birds were managed by the ministry responsible for the environment in partnership with the National Centre for Sustainable Development Foundation, with the fieldwork carried out by specialised NGOs (Romanian Ornithological Society, and Milvus Group Bird and Nature Protection Association).

Besides the national monitoring activities, site-level monitoring plans have been developed and are being implemented through EU co-funded projects, assessing mostly the conservation status of BHD species and habitats within the individual Natura 2000 sites.

These monitoring actions, carried out due to EU obligations, do not fully cover the biodiversity monitoring needs of Romania. There is no integrated surveillance and monitoring system for biodiversity. Some Natura 2000 sites and domestic protected areas have developed local surveillance and monitoring systems, with some data collected in national or regional monitoring projects and included in an Integrated Biodiversity Information System (IBIS) database. However, most of the existing information is only available on request from the national authority. There are also NGOs (e.g. Romanian Ornithological Society, Milvus Group Bird and Nature Protection Association, Romanian Society of Herpetology, Association for Biological Diversity Conservation) that are carrying out local and regional monitoring of certain taxa, some allowing restricted access to their databases.[29] The Romanian Ornithological Society and Milvus coordinate a few national monitoring programmes that have been running

[28] Ministry of Environment (2016–2020), Developing the administrative capacity of the Ministry of Environment to implement biodiversity policy (SIPOCA 22), Proiectul"Dezvoltarea capacității administrative a Ministerului Mediului de a implementa politica în domeniul biodiversității" | Ministerul Mediului.

[29] The Milvus database is open to the public: https://openbirdmaps.ro

Table 26.5 Funding sources for nature conservation in Romania.

Source and period	Measure/category	Average annual allocation (RON m)		Average annual allocation (€ m)		Comments
EU funding instruments		**EU + national**	**EU**	**EU + national**	**EU**	
EAFRD Rural Development Programme measures – allocations to Priority 4 (restoring, preserving and enhancing agriculture and forest ecosystems) 2014–2020	M13 Areas facing constraints	1 114.6	929.6	226.4	188.8	
	M10 Agri-environment–climate	589.4	491.7	119.7	99.9	
	M11 Organic farming	177.1	147.6	36.0	30.0	
	M15 Forest environment climate	63.4	52.8	12.9	10.7	
	M1 Knowledge transfer, etc.	0.4	0.3	0.1	0.1	
RDP Priority 4 subtotal		*1 944.8*	*1 622.1*	*395.1*	*329.5*	
ERDF/Cohesion Fund 2014–2020	Cat. 4.1 Enhancing biodiversity conservation	221.7	188.5	45.0	38.3	For N2k, etc.
EMFF 2014–2020	M11.10 Aquaculture providing environmental services	35.0	17.5	7.1	3.6	Compensation in N2k
Interreg	Relevant to biodiversity conservation	16.9	14.3	3.4	2.9	14 projects
LIFE Programme 1999–2020	Nature and biodiversity projects	14.8	7.4	3.0	1.5	46 projects only involving Romania
European Social Fund 2014–2020	Measure 119. Investments in institutional capacity	2.3	1.9	0.5	0.4	For Ministry of Environment
EU funding instruments sub-total		*2 235.5*	*1 851.7*	*454.1*	*376.2*	
National public funds within 2014–2020		**Additional non-EU* (RON m)**		**Additional non-EU* (€ m)**		
National Forestry Administration 2014–2020	Administration of 22 protected areas	15.8		3.2		
ANANP 2017–2020	Staff and administration	10.3		2.1		Range: €0.6 in 2017 – €4.3 in 2020
Danube Delta Biosphere Reserve 2011–2015	Staff and administration	c. 15.8		c. 3.2		Range: €2.1 in 2012 – €4.5 in 2015
Others						
European Economic Area grants 2009–2014	RO02 Services for biodiversity and ecosystems	12.5		2.5		
Swiss Romanian Cooperation Programme 2009–2014	Grants for environmental NGOs	0.3		0.06		
Additional national and others subtotal		*54.6*		*11.1*		
National total including co-financing of EU instruments		*425.7*		*86.5*		
Total		**2 290.1**		**465.2**		

Notes. Currency conversion rate (29 May 2021): €1 = RON 4.93. 'For N2k etc.' indicates funding relevant to Natura 2000, related green infrastructure and BHD species (based on PAF). * Excludes co-financing of EU funds unless indicated. The subtotals and totals are only approximate indications of the average annual levels of funding as they are based on annualised data from differing periods.

Sources. Based on data from: RDP data from European Commission ESIF funds database (23 October 2020); RDP allocations to Natura 2000, ERDF, EMFF and national funding from 2021 to 2027 PAF (MoE, 2019); Romania LIFE factsheet, European Commission (updated February 2021). See Section 1.3 for further details.

for more than 10 years and now collect information annually.[30] Although these schemes are based on the joint efforts of a small number of volunteers, they provide valuable information about many bird species.

Independently from the monitoring of species and habitats of EU and national nature conservation interest, some monitoring is carried out by natural resource managers and public authorities to set quotas. This includes annual monitoring of game species by game managers, and fish stocks in the Black Sea and Danube by research institutions. Most of these monitoring results are not available to the public, and most NGOs consider the methodologies and the results to be unreliable.

Databases compiled through the various monitoring activities need consistent development and standardisation to be considered as part of a national surveillance system of biodiversity (Popescu *et al.*, 2016).

26.9 Future Developments

Despite the progress made in the last two decades, there are still many challenges for nature conservation in Romania. One of the key challenges is related to the legal framework for biodiversity conservation (i.e. harmonisation of sectoral policies and effective law enforcement). SACs need to be designated by the ministry responsible for the environment, although there is still no government strategy in this direction. A new Biodiversity Strategy should also be developed within the framework established by the *EU Biodiversity Strategy for 2030*.

Strengthening and further developing a transparent, delegated, participatory governance and management system for biodiversity conservation inside and outside protected areas is key for achieving or maintaining the favourable status of BHD habitats and species, as highlighted in the *Vision for Protected Areas in Romania* (Coalitia Natura 2000, 2020). Building the capacity of entities actively involved in, or influencing, biodiversity conservation is also key for effective stakeholder involvement and conflict resolution. ANANP should become the key player in the development and implementation of a coherent vision and strategy for the protected area system, in close cooperation with various entities at local and national levels. Allocating the resources needed by this agency and to support the development of substantial institutional and human resource capacity is of critical importance.

Steps also must be taken to increase the network's resilience to climate change and other threats and to harmonise economic development plans with biodiversity conservation objectives.

The state budget allocation has to be secured, and additional funding sources identified, most likely from EU funds allocated for the sustainable management of protected areas. Funds are needed not only to implement management measures defined through the management plans (including restoration work), but also to compensate private landowners whose economic interests are affected by conservation measures. Natura 2000 payments should be activated as soon as possible.

The surface of strictly protected areas and strictly protected zones within the core areas of existing protected areas has to be increased to meet the 10% EU target. Speeding up the strict protection of the remaining pristine forests is critical and will contribute to achieving this 10% target. Another important step is to develop and implement short- and medium-term strategies to increase the core areas in National Parks to 75% of their area, as recommended by IUCN (Dudley, 2008).

Efforts to designate and properly manage ecological corridors have to be intensified, starting with the clarification of the legal framework needed for effective ecological corridors.

A national plan for ecological restoration should be developed to prioritise this type of conservation work, as was required under Target 2 of the *EU Biodiversity Strategy to 2020*. Restoration should target the most endangered types of habitats, including lowland grasslands (mostly steppic and pannonic) and riverine forests. There is also a need to establish a few protected areas of national interest in the lowlands and in the hilly areas, because these areas are generally very biodiverse and less represented in the protected area network. Restoration of degraded arable land and artificial forests should also be included in the plan.

An integrated scientific approach for managing and monitoring biodiversity conservation should be initiated by the ministry responsible for the environment through long-term collaboration with research institutes, universities, museums and specialised NGOs. These efforts would ideally result in the development of a national biodiversity database, including the information needed for the EU reports. In addition, establishing and implementing a well-defined monitoring programme, able to detect changes and trends in the status of populations and habitats, will enable authorities to act in a timely manner to prevent further species declines or habitat degradation.

Biodiversity conservation must be integrated with social and economic development plans and must become one of the key concerns for natural resource managers. EU support tools (e.g. agro-environment packages) are key for this process and continuous efforts are required to improve and further develop such tools using the management plans.

26.10 Conclusions

Romania has an outstanding biodiversity, with a great diversity of natural and cultural landscapes, large tracts of wilderness, pristine forests and biodiversity hotspots such as the Danube Delta. This has been subject to increasing pressures, especially after 1989.

The legal framework for nature conservation significantly improved through its harmonisation with EU legislation. Progress is being made also in terms of concrete conservation actions, especially since Romania is an EU Member State.

30 https://milvus.ro/en/birds/monitoring/

However, further steps are needed to harmonise and strengthen the implementation of sectoral legislation with the Protected Area Law and other relevant legal provisions, which the European Commission could play an important role in driving and supporting. Whilst the institutional framework changed to some extent with the establishment of ANANP, the management of individual protected areas has worsened significantly since this agency was appointed to manage an important part of the protected area network. The highly decentralised protected area management system, although underfunded, worked reasonably well until 2018. Returning to delegated management of protected areas would allow a highly participatory system, with important professional and financial contributions from various entities such as NGOs, universities, research institutes, local authorities or natural resource managers and landowners.

The obligation to designate the Natura 2000 network and the decision of the Romanian government to consider them as protected areas led to a significant extension of the protected area system, with good coverage of BHD habitats and species. Future improvements in protected area coherence and representativity, as recommended in the *EU Biodiversity Strategy to 2030*, should consider carefully also species and habitats of national interest and ecological corridors.

Through the management planning processes initiated mainly due to the European Commission requirements to prepare SAC designation, important progress was made in habitat and species inventories and monitoring, as well as in establishing management measures to maintain and improve their conservation status, but most remain unimplemented due to a lack of funding.

Over the last few years, progress has been made in protecting large carnivore species, both through legal protection and by identifying management measures through pilot projects funded mainly by the EU. This EU support is still very much needed for the implementation of the recently developed national action plans for the protection of Brown Bear and Wolf, especially identifying solutions for the increasing human–wildlife conflicts.

Although the Romanian research institutes and universities are making efforts to initiate and implement applied research projects, a coherent strategy for prioritising research needs and coordinating activities with those of the conservation management organisations would contribute significantly to further nature conservation work.

The EU's financial support has provided, and continues to provide, the most important sources of funding for conservation actions, enabling species and habitat inventories and monitoring, management planning, some practical habitat management and species measures, awareness raising, education and community outreach, with some financial contribution from Romania's state budget. However, Romania has not taken full advantage of the available funding. In particular, it needs to use more RDP funds for nature conservation, including by developing an efficient Natura 2000 payment system. There are also opportunities to use other funding lines relevant for land use and natural resource management or for encouraging sustainable development of communities whose livelihoods are still very much dependent on natural resources.

During the past decades Romania has concentrated much of its biodiversity conservation efforts on establishing the protected area network. Whilst it is vital to maintain and improve this network, authorities/decision makers and civil society have to recognise as soon as possible that biodiversity conservation is not only about protected areas. Biodiversity conservation has to become a national priority, that is, it has to be considered important by all relevant sectors to have a country that offers conditions for a high quality of life. To achieve this aim, it is of paramount importance to improve the legal framework and law enforcement, increase funding, monitor and report in a transparent way the status of biodiversity, harmonise sectoral policies and economic development with nature conservation measures, and secure conditions for highly professional, experienced public servants in the biodiversity sector, as well as in key sectors managing natural resources.

Acknowledgements

Thanks are due to numerous individuals who have contributed to the compilation of this chapter and who have provided feedback, comments and suggestions during preparation. Many people from various entities involved in biodiversity conservation have offered information, for example from the Association for the Conservation of Biological Diversity (ACDB), Milvus Group, WWF Romania, EKE Satu Mare, Romanian Ornithological Society, and the Ministry of Environment, Water and Forests. The authors thank the reviewers, Viorel Popescu and Tibor Hartel, for their contribution. We are also very grateful to Owen Mountford, who provided very valuable support in developing and refining the chapter.

References

Adamescu, M., Cazacu, C., Arhre, G. *et al.* (2016) *Ghid metodologic pentru evaluarea rapidă a serviciilor ecosistemice în ariile protejate din România.* Bucharest: Fundația Centrul Național pentru Dezvoltare Durabilă (CNDD).

ADEPT Transilvania, WWF Romania & ProPark Foundation (2016) *Rural Development and High Nature Value Farmland in Romania.* Romania: Fundația ADEPT Transilvania.

Angelstam, P., Elbakidze, M., Axelsson, R. *et al.* (2013) Maintaining cultural and natural biodiversity in the Carpathian Mountain Ecoregion: need for an integrated landscape approach. In *The Carpathians: Integrating Nature and Society towards Sustainability*, eds. J. Kozak, K. Ostapowicz, A. Bytnerowicz & B. Wyżga, pp. 393–424. Berlin: Springer Press.

Beaufoy, G., Blom, S., Hartel, T. *et al.* (2015) *Europe's wood pastures: condemned to a slow death by the CAP? A test case for EU agriculture and biodiversity policy.* Romania: European Forum on

Nature Conservation and Pastoralism, and Pogány-havas Association.

Bellebaum, J. & Koffijberg, K. (2018) Present agri-environment measures in Europe are not sufficient for the conservation of a highly sensitive bird species, the Corncrake *Crex crex*. *Agriculture, Ecosystems & Environment*, **257**, 30–37.

BirdLife International (2017) *European Birds of Conservation Concern: Populations, Trends and National Responsibilities.* Cambridge, UK: Birdlife International.

Brinkmann, K., Păcurar, F., Rotar, I., Ruşdea, E., Auch, E. & Reif, A. (2009) The grasslands of the Apuseni Mountains, Romania. In *Grasslands in Europe of High Nature Value*, eds. P. Veen, R. Jefferson, J. de Smidt & J. van der Straaten, pp. 226–237. Leiden: Brill.

Cazacu, C., Adamescu, M., Ionescu, C. *et al.* (2014) Mapping trends of large and medium size carnivores of conservation interest in Romania. *Annals of Forest Research*, **57** (1), 97–107.

Coaliţia Natura 2000 (2020) *Viziunea Federaţiei Coaliţia Natura 2000 pentru sistemul de management al reţelei de arii protejate din România 2019-2028.* Romania: Fundaţia Coaliţia Natura 2000.

Dahlstrom, A., Iuga, A. M. & Tommy, L. (2013) Managing biodiversity rich hay meadows in the EU: a comparison of Swedish and Romanian grasslands. *Environmental Conservation*, **40**, 194–205.

Dudley, N. (ed.) (2008) *Guidelines for Applying Protected Area Management Categories.* Gland, Switzerland: IUCN.

Feurdean, A., Ruprecht, E., Molnár, Z., Hutchinson, S. M. & Hickler, T. (2018) Biodiversity-rich European grasslands: ancient, forgotten ecosystems. *Biological Conservation*, **228**, 224–232.

Gâştescu, P. & Ştiucă, R. (2008) *Delta Dunării. Rezervaţie a biosferei.* Bucharest: CD Press.

Gómez-Baggethun, E., Tudor, M., Doroftei, M. *et al.* (2019) Changes in ecosystem services from wetland loss and restoration: an ecosystem assessment of the Danube Delta (1960–2010). *Ecosystem Services*, **39**, 100965.

Grădinaru, S., Peilei Fan, R., Iojă, I. C., Niţă, M. R., Suditu, B. & Hersperger, A. M. (2020) Impact of national policies on patterns of built-up development: an assessment over three decades. *Land Use Policy*, **94**, 104510.

Habersack, H., Hein, T., Stanica, A. *et al.* (2016) Challenges of river basin management: current status of, and prospects for, the River Danube from a river engineering perspective. *Science of the Total Environment*, **543**, 828–845.

Hartel, T., Hanspach, J., Moga, C. I. *et al.* (2018) Abundance of large old trees in wood-pastures of Transylvania (Romania). *Science of the Total Environment*, **613**, 263–270.

Hartel, T., Nita, A. & Rozylowicz, L. (2020) Understanding human-nature connections through value networks: the case of ancient wood-pastures of Central Romania. *Sustainability Science*, **15**, 1357–1367.

Hatna, E. & Bakker, M. (2011) Abandonment and expansion of arable land in Europe. *Ecosystems*, **14**, 720–731.

Hossu, C. A., Iojă, I. C., Niţă, M. R., Hartel, T., Badiu, D. L. & Hersperger, A. M. (2017) Need for a cross-sector approach in protected area management. *Land Use Policy*, **69**, 586–597.

Hossu, C. A., Iojă, I. C., Susskind, L. E., Badiu, D. L. & Hersperger, A. M. (2018) Factors driving collaboration in natural resource conflict management: evidence from Romania. *Ambio*, **47**, 816–830.

Iojă, I. C. (2017) *Methodology for the management and resolution of environmental conflict in Natura 2000 network.* Bucharest: University of Bucharest, Centre for Environmental Research and Impact Studies.

Iojă, I. C., Pătroescu, M., Rozylowicz, L. *et al.* (2010) The efficacy of Romania's protected areas network in conserving biodiversity. *Biological Conservation*, **143**, 2468–2476.

Iojă, I. C., Rozylowicz, L., Pătroescu, M., Nita, M. & Onose, D. (2011) Agriculture and conservation in the Natura 2000 network: a sustainable development approach of the European Union. In *Agricultural and Environmental Informatics, Governance and Management: Emerging Research Applications*, eds. Z. Andreopoulou, B. Manos, N. Polman & D. Viaggi, pp. 339–358. Hershey, USA: IGI Global.

Loos, J., Abson, D. J., Dorresteijn, I. *et al.* (2016) *Sustainable Landscapes in Central Romania: A Social-ecological Study on the Future of Southern Transylvania.* Lueneburg, Germany: Pensoft.

Maanen, E. van, Predoiu, G., Klaver, R. *et al.* (2006) *Safeguarding the Romanian Carpathian Ecological Network. A vision for large carnivores and biodiversity in Eastern Europe.* Braşov: ICAS Wildlife Unit, A&W ecological consultants.

Manolache, S., Nita, A., Hartel, T., Miu, I. V., Ciocanea, C. M. & Rozylowicz, L. (2020) Governance networks around grasslands with contrasting management history. *Journal of Environmental Management*, **273**, 11115

Membretti, A. & Iancu, B. (2017) From peasant workers to amenity migrants. Socialist heritage and the future of mountain rurality in Romania. *Journal of Alpine Research*, **105-1**, 1–13.

Ministry of Environmental Protection (1996) *Strategia Nationala si Planul de Actiune pentru conservarea Biodiversitatii si utilizarea durabila a componentelor sale.*

(2000) *Strategia Naţională de Conservare a Biodiversităţii.* www.mmediu.ro/img/attachment/32/biodiversitate-54784ffea5918.pdf

Ministry of Environment and Climate Change (2014) *Strategia naţională şi Planul de acţiune pentru conservarea biodiversităţii 2014 - 2020.* http://mmediu.ro/new/wp-content/uploads/2014/05/NBSAP.pdf

Ministry of Waters, Forests and Environmental Protection (2019) *Raport privind starea pădurilor Romaniei 2017.* www.mmediu.ro/categorie/starea-padurilor/209

MoE (2013) *PRIORITISED ACTION FRAMEWORK (PAF) FOR NATURA 2000 in ROMANIA pursuant to Article 8 of Council Directive 92/43/EEC on the conservation of natural habitats and of wild fauna and flora (the Habitats Directive) for the Multiannual Financial Framework period 2014 - 2020.* Bucharest: Ministry of Environment.

(2019) *Prioritised Action Framework (PAF) for Natura 2000 in Romania for the multiannual financial framework period 2021–2027.* Bucharest: Ministry of Environment.

Muntean, D. (2000) Romania. In *Important BIRD AREAS in Europe: Priority Sites for Conservation. Volume 2: Southern Europe*, eds. M. F. Heath, M. I. Evans, D. G. Hoccom, A. J. Payne & N. B. Peet, pp. 481–502. Cambridge, UK: ICBP – BirdLife International.

Nistorescu, M. C. (2012) *Effectiveness of environmental impact assessment in Romania and simple means to improve it.* Report in UNDP-GEF Project MP5-PIMS 3069, 'Strengthening the Capacity to Integrate Environment and Natural Resources Management for Global Environmental Benefits (CB2)'.

Nita, A., Buttler, A., Rozylowicz, L. & Patru-Stupariu, I. (2015) Perception and use of landscape concepts in the procedure of Environmental Impact Assessment: Case study-Switzerland and Romania. *Land Use Policy*, **44**, 145–152.

Page, N., Bălan, A., Huband, S., Popa, R., Rakosy, L. & Sutcliffe, L. (2012) Romania. In *High Nature Farming in Europe: 35 European Countries – Experiences and Perspectives*, eds. R. Oppermann, G. Beaufoy, G. Jones, pp. 346–357. Ubstadt-Weiher, Germany: RegionalKultur Verlag.

Plieninger, T., Hartel, T., Martin-Lopez, B. *et al.* (2015) Wood-pastures of Europe: geographic coverage, social-ecological values, conservation management, and policy implications. *Biological Conservation*, **190**, 70–79.

Pop, I. M., Bereczky, L., Chiriac, S. *et al.* (2018) Movement ecology of brown bears (Ursus arctos) in the Romanian Eastern Carpathians. *Nature Conservation*, **26**, 15–31.

Popescu, V., Pop, M., Chiriac, S. & Rozylowicz, L. (2019) Romanian carnivores at a crossroads. *Science*, **364**, 1041.

Popescu, V. D., Artelle, K., Pop, M., Manolache, S. & Rozylowicz, L. (2016) Assessing biological realism of wildlife population estimates in data-poor systems. *Journal of Applied Ecology*, **53**(4), 1248–1259.

Popescu, V. D., Rozylowicz, L., Niculae, I. M., Cucu, A. L. & Hartel, T. (2014) Species, habitats, society: an evaluation of research supporting EU's Natura 2000 Network. *PLoS ONE*, **9** (11), e113648.

Popovici, E. A., Bălteanu, D. & Kucsicsa, G. (2016) Utilizarea terenurilor şi dezvoltarea actuală a agriculturii. In *Romania. Natură şi Societate*, ed. D. Bălteanu, pp. 329–376. Bucharest: Romanian Academy Press.

Primack, R., Pătroescu, M., Rozylowicz, L. & Iojă, I. C. (2008) *Essential in Biological Conservation/Fundamentele conservarea diversității biologice.* Bucharest: AGIR Press.

Rolo, V., Hartel, T., Aviron, S. *et al.* (2020) Challenges and innovations for improving the sustainability of European agroforestry systems of high nature and cultural value: stakeholder perspectives. *Sustainability Science*, **15**, 1301–1315.

Romanian Forest Research and Management Institute (2018) National Forest inventory of Romania – second cycle. http://roifn.ro/site/ifn-ciclul-ii/

Rozylowicz, L., Nita, A., Manolache, S., Ciocanea, C. M. & Popescu, V. D. (2017) Recipe for success: a network perspective of partnership in nature conservation. *Journal for Nature Conservation*, **38**, 21–29.

Rozylowicz, L., Nita, A., Manolache, S., Popescu, V. D. & Hartel, T. (2019) Navigating protected areas networks for improving diffusion of conservation practices. *Journal of Environmental Management*, **230**, 413–421.

Salvatori, V., Balian, E., Blanco, J. C. *et al.* (2020) Applying participatory processes to address conflicts over the conservation of large carnivores: understanding conditions for successful management. *Frontiers in Ecology and Evolution*, **8**, 182.

Salvatori, V., Okarma, H., Ionescu, O., Dovhanych, Y., Find'o, S. & Boitani, L. (2002) Hunting legislation in the Carpathian Mountains: implications for the conservation and management of large carnivores. *Wildlife Biology*, **8** (1), 3–10.

Sârbu, A., Negrean, G. & Sârbu, I. (2009) The grasslands of the Dobrogea, Romania. In *Grasslands in Europe of high nature value*, eds. P. Veen, R. Jefferson, J. de Smidt & J. van der Straaten, pp. 218–225. Zeist, the Netherlands: KNNV Publishing.

Stanciu, E. & Ioniță, A. (2014) *Governance of Protected Areas in Eastern Europe. Overview on Different Governance Types, Case Studies and Lessons Learned.* Study commissioned to ProPark, Romania, by the German Federal Agency for Nature Conservation (BfN). BfN-Skripten.

Strat, D. & Mihăilescu, S. (2017) Conservation status of habitat types of European Community Interest in the Alpine Biogeographic Region of Romania. *Forum Geografic*, **16** (1), 59–69.

Szocs, B. & Miklos, A. (2015) *Land Grabbing in Romania.* Romania: Eco Ruralis.

Tănăsescu, M. & Constantinescu, S. (2020) The human ecology of the Danube Delta: a historical and cartographic perspective. *Journal of Environmental Management*, **262**, 110324.

Teampău, P. (2020) Trouble in paradise: competing discourses and complex governance in the Romanian Danube Delta. *Marine Policy*, **112**, 103522.

Toniuc, N., Oltean, M., Romanca, G. & Zamfir, M. (1992) List of protected areas in Romania. *Ocrotirea mediului înconjurător*, **36**, 23–33.

Văidianu, N., Tătui, F., Ristea, M. & Stănică, A. (2020) Managing coastal protection through multi-scale governance structures in Romania. *Marine Policy*, **112**, 103567.

Veen, P., Fanta, J., Raev, I., Biriş, I. A., de Smidt, J. & Maes, B. (2010) Virgin forests in Romania and Bulgaria: results of two national inventory projects and their implications for protection. *Biodiversity and Conservation*, **19**, 1805–1819.

Chapter 27

Slovakia

Viera Šefferová Stanová and Rastislav Rybanič

Key Statistics

Terrestrial area (km^2)/% of EU-28*	49 026/1.1%
Marine waters (km^2)/% of EU-28*	None
Population density (p km^2)/% of EU-28*	111/96%
GDP per capita (€)/% of EU-28	14 083/53%
Land cover (CORINE land cover classes 2018)	AGR 47%, FOR 41%, ART 6%, OSN 5%, WET 1%
Biogeographical regions	ALP, PAN
Number of HD Annex I habitats/%*	66/29%
Number of HD Annex II, IV and V species/%*	195/16%
Number of BD Annex I bird taxa/%*	73/41%
% HD UCS and decreasing habitats/ % unknown	10%/2%
% HD UCS and decreasing species/ % unknown	25%/1%
% breeding BD Annex I birds decreasing/% uncertain and unknown	35%/2%

* Excluding MAC and MMAC biogeographical regions.
Sources and codes. See Table 1.1.

27.1 The Natural History and Characteristics of Slovakia

Slovakia, like some other central European countries, has a rich biological diversity for a small territory. This is a result of its varied relief, altitudinal scale and geology (being in the Carpathian Arc of mountains and the Pannonian lowlands) and historically diverse land use. The Carpathian Mountains, which stretch from the west to the east across the north and take up more than two-thirds of the country, hold a large proportion of the remaining areas of relatively undisturbed nature and have high species endemism. The highest range within the Carpathians is known as the Tatra Mountains in Slovakia, and rises to an altitude of 2 655 m (Gerlachovský štít Peak). Two biogeographical regions occur in Slovakia: the Alpine biogeographical region in the Carpathian Mountains (71% of the country) and the Pannonian biogeographical region (29%), which is in the south of the country and an important agricultural area.

Several altitudinal vegetation zones and many diverse habitats occur in Slovakia, ranging from forest steppes, floodplain forests, sand dunes, salt marshes and wetlands in the lowlands, through oak and European natural beech forests in the lower mountains, to coniferous forests and unique alpine and sub-alpine ecosystems in the highest mountains. Slovakia is one of the most forested countries of Europe, with more than 41% forest cover that has increased over the last decades. The forests are characterised by their high level of species and habitat diversity, with abundant coniferous (36%) and broad-leaved species (64%) (MoARD, 2021b). Primeval forests cover 106 km^2, which represents 0.47% of the total forest area (Jasík and Polák, 2011).

Agriculture covers the largest proportion of the land: 47% based on CLC 2018 data, or 39% according to the MoARD (2018), who estimate that it is around 71% arable land and 28% permanent grasslands, the latter having decreased slightly in area. Whilst most grasslands have been agriculturally improved, Slovakia retains a relatively large area of semi-natural grasslands. Furthermore, some semi-natural calcareous grasslands are amongst the richest in the world in terms of their plant species diversity, setting a world record for fine-scale species richness: 52 and 63 species in 0.25 m^2 and 0.5 m^2, respectively, in the Slovenský raj National Park (Chytrý *et al*., 2015).

High Nature Value (HNV) grasslands have been identified through a National Grassland Inventory, which was carried out over 1998–2006 (by DAPHNE – Institute of Applied Ecology). Based on an analysis of satellite images, about 3 060 km^2 of semi-natural grasslands were mapped, which is 57% of the total permanent grassland area in Slovakia, as registered in LPIS. Since 2007, for the needs of the Rural Development Programme (RDP), and due to equivalent or very similar ecological and management requirements, grassland habitats were classified into seven ecological groups of HNV grasslands, with a total area of 1 664 km^2 (MoARD, 2014).

Additionally, some traditional agricultural landscapes (TAL) have been retained in some regions of Slovakia and these sites still hold a rich diversity of farmland flora and

Table 27.1 Short-term trends (c. 2007–2018) in Habitats Directive Annex I habitats and Annex II, IV and V species with an unfavourable conservation status in Slovakia.

	Poor (unfavourable–inadequate)				Bad (unfavourable–bad)			
	Imp	Sta	Dec	?	Imp	Sta	Dec	?
HD Habitat assessments								
Number	0	29	4	18	0	1	6	3
%	0%	57%	8%	35%	0%	10%	60%	30%
HD Species assessments								
Number	14	95	42	21	1	12	39	17
%	8%	55%	24%	12%	1%	17%	57%	25%

Key. Imp = improving, Sta = stable, Dec = decreasing, ? = unknown.
Source. Based on Member State HD Article 17 reported data as presented on the EEA national summary dashboard (2021).

fauna. TAL in Slovakia cover an area of 444 km^2, which constitutes 1.8% of Slovak agricultural land (Špulerová *et al.*, 2017).

Slovakia is one of a few European countries with abundant natural habitats inhabited by relatively stable populations of large carnivores, such as Brown Bear (*Ursus arctos*), Wolf (*Canis lupus*) and Eurasian Lynx (*Lynx lynx*). Due to its rich habitat diversity, 226 bird species breed in Slovakia – amongst the highest number in Europe (BirdLife International, 2015).

27.2 The Status of Nature

The most recent national report (2013–2018) provided by Slovakia as required under the Habitat and Birds Directives[1] was mostly based on the data from the Complex Information and Monitoring System of Habitats and Species of the Community Interest (KIMS),[2] as further described in Section 27.8. However, the previous national reports were mostly based on expert evaluations. Therefore, as the most recent assessment can be expected to have been more reliable, comparisons with previous results should be undertaken with care. In particular, the apparent overall reported deterioration is mainly due to improved knowledge through the establishment of the monitoring system, refinement of expert estimates and identification of relevant data for species and habitats that were previously unknown (Černecký *et al.*, 2020).

27.2.1 Habitat Status and Trends

According to the EEA dashboard summary of the 2013–2018 assessments of the conservation status of habitats overall across both biogeographical regions, 38% of the 101 habitat assessments revealed a favourable status. Of the remaining assessments, 50% showed a poor status and 10% a bad status, whilst 2% had an unknown status. It is notable that a very high proportion of assessments showed a favourable status amongst the rocky habitats (93%) and heath and scrub (86%). The proportion of assessments that showed a favourable status was much lower for other habitat groups: forests (34%), freshwater (27%), grasslands (23%) and zero for all others.

27.2.2 Species Status and Trends

Overall, across all 195 HD species assessments over 2013–2018, 23% revealed a favourable status, 54% a poor status and 22% a bad status; 1% had an unknown status. The species groups with the highest proportions of assessments that showed a favourable status were the mammals (56%) and non-vascular and vascular plants (33% and 32% respectively). The most threatened groups (i.e. species with a bad status) were fish (57% of assessments).

Closer examination of the trends of species with an unfavourable status (Table 27.1) reveals that nearly a quarter of species with a poor status showed a short-term decline. As was the situation with the HD habitats, of particular concern is that a high proportion (57%) of assessments of species with a bad status also showed declines.

The short-term trends (2007–2018) in BD Annex I bird populations (Table 27.2) show that, whilst most species increased or were stable or fluctuating, a substantial proportion decreased (35%). Examination of the species-specific data on the EEA dashboard indicates that species that have increased include some that have benefited from nature conservation actions, such as the Eastern Imperial Eagle (*Aquila heliaca*) and White-tailed Sea-eagle (*Haliaeetus albicilla*). Many of the Annex I species that declined are farmland birds, such as the Corncrake (*Crex crex*), White Stork (*Ciconia ciconia*), Saker Falcon (*Falco cherrug*), Lesser Grey Shrike (*Lanius minor*) and Red-backed Shrike (*Lanius collurio*). This also reflects decreases in more common farmland birds, such as Lapwing (*Vanellus vanellus*), Grey Partridge (*Perdix perdix*), Turtle-dove (*Streptopelia turtur*) and Corn Bunting (*Emberiza calandra*).

[1] See Section 1.3 for background on monitoring under the Nature Directives, status categories and data sources.
[2] www.biomonitoring.sk

Table 27.2 Short-term trends (c. 2007–2018) in Birds Directive Annex I taxa in Slovakia.

	Breeding				Wintering			
	Inc	Sta/ Flu	Dec	?	Inc	Sta/ Flu	Dec	?
Number	8	31	22	1	3	5	0	1
%	13%	50%	35%	2%	33%	56%	0%	11%

Key. Inc = increasing, Sta = stable, Flu = fluctuating, Dec = decreasing, ? = uncertain and unknown.
Source. Based on Member State Article 12 reported data as presented on the EEA national summary dashboard (2021).

In forest habitats, declines have been observed in the Capercaillie (*Tetrao urogallus*), White-backed Woodpecker (*Dendrocopos leucotos*) and Three-toed Woodpecker (*Picoides tridactylus*), which are mostly due to intensive forest management.

The short-term trends in wintering Annex I bird taxa are less concerning (Table 27.2). However, it is difficult to draw any conclusions for wintering bird species as their presence in Slovakia depends largely on weather conditions, which are certainly influenced by the changing climate.

27.2.3 Drivers and Causes of Change

Agriculture and forestry are the most extensive human activities in the country and have a major influence on biodiversity. Other human actions that have had substantial impacts on biodiversity include a range of activities affecting wetlands, and the construction of infrastructure (e.g. roads and small hydropower plants) that has increased habitat fragmentation.

Intensification has been a dominant feature of Slovak agriculture since the 1950s, when small strips of fields around villages were amalgamated into large blocks of intensive arable land. Similarly, large areas of intensive hay meadows were established in the valley basins, where most of the agricultural production was concentrated. Remote patches of semi-natural grasslands have also been abandoned due to their difficult access and because agriculture has primarily focussed on intensification and productivity increases elsewhere, through the increased use of monoculture crops, fertilisers and pesticides, re-sowing of grasslands with hybrid and non-native seed mixtures, concentrated grazing, and converting grasslands to arable land (Orsillo, 2008). Political changes in 1989 also led to a decrease in permanent grasslands due to a reduction in livestock numbers. For grasslands the main recent pressures identified are natural processes related to abandonment, followed by shrub encroachment, lack of proper management by grazing or mowing, intensive agriculture, alien and problematic species and the construction of infrastructure (MoE, 2021).

Forest habitats are adversely affected by an increasing frequency of extreme climate conditions and weather events (e.g. storm damage and long periods of drought), which can lead to abiotic disturbances, such as windthrows, that then trigger harmful biotic impacts, such as pest outbreaks. This is contributing to bark beetle outbreaks, which are causing spruce stands to die out, including in natural mountain spruce climax forests. This is now considered the main forestry problem, having escalated since 2006 (Gubka *et al.*, 2013). As a result of such problems, incidental felling represents more than 47% of total annual felling (MoARD, 2021b). It is expected that the most serious problem affecting forests in the future will be their state of health.

Forest biodiversity is also being affected by inappropriate forest management and felling practices. At present, 75% of Slovak forests are managed by shelterwood management (MoARD, 2021b). Problems persist in the way that logging is carried out, such as the use of inappropriate logging techniques and inappropriate route tracing causing excessive erosion, soil contamination and damage to the forest ecosystem. Even some remote sites with the remains of old-growth natural forests are being affected by logging and increasing access through the creation of forest roads. Forest habitats in the lowlands and uplands are also locally strongly negatively influenced by invasive alien plant species.

Wetlands have been detrimentally affected by sediment excavation, land-use change, urbanisation and technical water management measures, such as the construction of dams and small hydropower plants that have not taken environmental issues into account. For example, the Gabčíkovo Water Project was a major construction programme that dammed and canalised the upper part of the middle Danube. When completed in 1992, it destroyed most of the 230 km^2 of wetlands, mainly floodplain forests and river tributaries. Two-thirds of the wetlands became dry, discontinuous or severely changed, which ultimately destroyed the original floodplain ecosystem and led to a decline in biodiversity (Balon and Holčík, 1999). The most significant pressure on surface waters has been diffuse agricultural pollution, which has also affected groundwater. In terms of impacts on surface waters, the most significant were altered habitats due to morphological changes and organic, nutrient and chemical pollution. For groundwater the most significant impacts were from chemical pollution and abstraction exceeding the available groundwater resource (European Commission, 2020).

27.3 Nature Conservation Governance, Actors, Legislation and Policies

27.3.1 Development of Nature Conservation in Slovakia

The first protected areas and forest legislation came into existence under the Austro-Hungarian Empire (Table 27.3). When Slovakia became part of the First Czechoslovak Republic in 1918, state nature conservation was intrinsically linked to heritage preservation (codified for the first time in a ministerial regulation in 1919). This resulted in a number of

Table 27.3 Key nature conservation developments in Slovakia.

1876	The first protected area, Kvetnica in the Vysoké Tatry Mountains, was created by the Hungarian Carpathian Association
1879	The first Forest Act in the Austro-Hungarian Empire was passed, which with partial adjustment remained valid in Slovakia until 1960
1919	Establishment of the governmental Nature Protection Bureau, which was tied to heritage preservation
1932	The first European transboundary nature park was created in the Pieniny Mountains of Poland and Slovakia
1933	The first inventory of protected areas was published, containing the 18 oldest protected areas
1948	The first National Park, Tatranský národný park (Tatra National Park), was announced
1955	The legal framework for nature protection was created in Act No. 1/1955 Coll. SNR on State Nature Protection
1981	Foundation of the independent State Nature Conservation Centre
1992	Establishment of the Ministry of the Environment of the Slovak Republic
1995	Inscription of bilateral (with Hungary) World Natural Heritage Site Caves of Aggtelek Karst and Slovak Karst
1997	The first *National Biodiversity Strategy*
1998	Slovakia hosted CBD COP 4 in Bratislava
2000	The State Nature Conservancy of the Slovak Republic was established
2002	Act No. 543/2002 on Nature and Landscape Protection transposed the EU Nature Directives
2004	Slovakia joined the EU
	The national lists of SPAs and pSCIs were adopted
2007	Inscription of transnational World Natural Heritage site (with Ukraine) Primeval Beech Forests of the Carpathians
2011	First addition to national list of pSCIs to complete Natura 2000 network
2012	Completion of SPA network in Slovakia
2014	Updated *Biodiversity Strategy* to 2020 (and the *Action Plan*) adopted
2017	Second addition to national list of pSCIs to complete Natura 2000 network
2020	Amendment of the Forest Act introduced criteria for extensive close-to-nature forest management
2021	Establishment of the Old Growth Forest of Slovakia nature reserve
2022	Amendment of the Nature and Landscape Act established the National Park Administrations as independent organisations and transferred state forest land to the National Parks
	Third addition to national list of pSCIs to complete the Natura 2000 network

See Table 3.3 for signatory dates of international conventions.

organisational arrangements in which nature conservation developed in parallel with the conservation of heritage. One of the most important steps was its statutory provision under the first nature protection legislation (Act No. 1/1955). Whilst professional organisations had earlier dealt with nature conservation, an important development was the creation of the independent State Nature Conservation Centre in 1981. This was followed by a rapid growth in the protected areas system, with some limited international cooperation (Burkovský and Stockmann, 2006; Stockmann, 2013).

After political changes in 1989, many changes in nature conservation took place, including its transfer from the cultural and heritage to the environmental sector, and a reorientation towards landscape ecology principles. The Ministry of the Environment was established in 1992. In 1993, Czechoslovakia split up into the Czech Republic and the Slovak Republic.

The 1955 Act was replaced in 1994 by the new act on Nature and Landscape Protection (Act No. 287/1994), which established the principles of state administration and protected area management in nature protection in Slovakia. Subsequently, in order to harmonise the legislation with the EU *acquis* and considering other international obligations, the new Nature and Landscape Protection Act No. 543/2002 was adopted. As further described in Section 27.3.2, this enabled the later creation of the Natura 2000 network in Slovakia.

Following ratification of the CBD in 1994, the first National Report and the first *National Biodiversity Strategy* (Straka and Guziová, 1998) were prepared and adopted by the Slovakian Parliament in 1997. The first *Action Plan* for 1998–2010 was adopted in 1998 and was a turning point in the approach to nature conservation, widening the focus from protected areas and species to biodiversity as a whole.

A further significant boost to the biodiversity agenda in Slovakia was the successful organisation of the Fourth CBD COP in Bratislava in May 1998.[3]

EU accession negotiations with Slovakia commenced in October 1999. A new professional nature conservation organisation with a national scope, the State Nature Conservancy of the Slovak Republic, was therefore established in 2000 to help with the intensive preparations necessary to conform to the EU's nature conservation legislation and establish the Natura 2000 network. This included significant inventory work, through engagement with experts in NGOs and the wider scientific community, which resulted in the scientific proposals for the Natura 2000 network (further discussed in Section 27.5.1). Slovakia joined the EU in May 2004.

27.3.2 Current Nature Conservation Legislation and Policies

Nature conservation legislation in Slovakia is based on species and territorial protection, both of which are regulated by **Act No. 543/2002 on Nature and Landscape Protection,**[4] hereafter referred to as **'the Nature and Landscape Act'**, as amended by later regulations. It covers general nature protection obligations, species protection and protected areas including Natura 2000 sites, and sets conditions for the management and protection of these areas. The Act regulates the competences of the government authorities and municipalities, and the rights and obligations of citizens in nature and landscape protection, with the aim being to support the preservation of biodiversity, to create conditions for sustainability, restoration and rational use of natural resources, preservation of natural heritage and characteristic landscape features, and to reach and maintain ecological stability.

Besides the above-mentioned Act there is a wide range of environmental and sectoral legislation that is highly relevant for nature protection, including the Forest Act, Act on the Prevention and Management of the Introduction and Spread of Invasive Alien Species (adopted in 2019), Act on the Conservation of Species of Wild Animals and Plants by Regulating Trade Therein, Act on environmental impact assessment (covering EIA and SEA procedures), Act on Hunting, Act on Fishing, Act on the Protection and Use of Agricultural Land, Water Act, Mining Act, Act on Spatial Planning and Building Order, and the Act on the Prevention and Remedying of Environmental Damage.

The ***Updated National Strategy for the Protection of Biodiversity to 2020*** (MoE, 2014) was directly based on the CBD's *Strategic Plan for Biodiversity 2011–2020* and the *EU Biodiversity Strategy to 2020.* It was prepared for the purposes of defining national targets for the protection of biodiversity, taking into consideration its state and trends in Slovakia, within the EU, and also at the global level, and reflecting national and international commitments. It was implemented by an *Action Plan* approved by Government Decree No. 442 on 24 August 2014.[5] The update of the biodiversity strategy is expected after adoption of the Global post 2020 biodiversity framework.

The basic planning document in nature conservation in Slovakia is the *Framework for Nature and Landscape Protection*[6] (the Nature and Landscape Act, §54). The most recent version was adopted in 2006 and it should have been updated after 10 years. However, the Ministry of Environment (MoE) has struggled to prepare a widely accepted document and the work has been delayed. In 2019, the MoE started negotiating on a document that should guide development and implementation of the nature conservation policy until 2030 (MoE and SNC SK, 2019).

The main strategic document on environmental protection is the ***Strategy of the Environmental Policy of the Slovak Republic until 2030*** (so-called *'Envirostrategy 2030'*, MoE, 2019). It defines a vision until 2030, which takes into account possible, probable, and desired future development, identifies the fundamental systemic challenges, sets the objectives until 2030 and proposes a framework for measures to improve the current situation. It also contains basic result indicators that will enable verification of achieved results. The biggest environmental challenges in Slovakia, and hence topics that will be prioritised within the environmental policy by 2030, are issues such as waste management, air quality and habitat and species conservation, especially in forest, meadow and wetland ecosystems.

The main government-approved framework document setting out water management policy is the ***Proposed Orientation, Principles and Priorities of the Slovak Republic Water Management Policy by 2027*** (MoE, 2015). Measures to achieve the policy goals are subject to further planning and conceptual documents, mainly River Basin Management Plans or the Water Plan of Slovakia.[7]

27.3.3 Nature Conservation Governance

The state organisation of nature protection in Slovakia is directly or indirectly managed by the **MoE** and is divided into state administration (decision-making) bodies, an enforcement body (the **Slovak Inspectorate of the Environment**) and management bodies, of which the **State Nature Conservancy** (SNC) and National Parks fall under the MoE. The MoE is responsible for protecting the environment and landscape, as well as ecological aspects of spatial planning, water management, flood prevention, protection of water quality and quantity, rational use of waters and fishing.

---

3 www.cbd.int/meetings/COP-04

4 Zákon č. 543/2002 Z. z. o ochrane prírody a krajiny.

5 www.minzp.sk/files/oblasti/ochrana-prirody-a-krajiny/biodiverzita/1_vlastny_ap-biod_aug_2014.pdf

6 Koncepcia ochrany prírody a krajiny.

7 https://ec.europa.eu/environment/water/participation/map_mc/countries/slovakia_en.htm

The SNC is an expert management agency that focusses on nature conservation and landscape protection actions, including territorial, species and habitat protection, management planning, environmental education, training and promotion, monitoring, information collection and database maintenance. It supports the decision making by the environment authorities and municipalities by providing opinions based on expert analysis and data, although they are not binding (Vozár, 2019). As of 2022, National Park Administrations were established for each park as independent organisations managing all state-owned land in zoned National Parks in cooperation with the SNC.

The Slovak Inspectorate of Environment was established by the MoE as an independent control and enforcement body for the environment. In the field of nature conservation, the inspectorate has the responsibility to enforce compliance with the law and administrative decisions. If violation of a law is detected and proved, it may impose fines. More serious nature protection law offences are addressed by the Environmental Police Force and prosecutors according to the Criminal code.

The state's administration of nature conservation is carried out at four levels: municipal, district, regional and national. National-level state administration in nature conservation is implemented by the MoE. Municipal-level responsibilities for nature conservation are very limited: approval for tree felling outside managed forests and designation of municipal protected areas. Since 2013, district and regional levels of state administration have been under the **Ministry of Interior,** which is now responsible for 8 regional and 79 district offices and their environmental departments, which carry out the state administration of the environment, including on nature conservation issues. Also of relevance to nature conservation are the land, forestry and agricultural departments that are under the supervision of the Ministry of Agriculture and Rural Development (MoARD) and have administrative responsibility for hunting, forestry, agriculture and protection of agricultural land.

From the outset, systematic intersectoral cooperation has been carried out in relation to measures under the CAP and the preparation of Rural Development Programmes. However, the development of the Prioritised Action Framework (PAF) in 2014 for financing Natura 2000 for the first time brought together all the relevant ministries to focus on Natura 2000. Since then intersectoral cooperation has also been established for several other issues, such as a working group dealing with bird crime and another on the preparation and implementation of management plans for large carnivores. However, there are many issues for which further cooperation is required, mainly integration of biodiversity considerations into other policies and their associated funds (notably agriculture, forestry and infrastructure planning).

27.3.4 Public Involvement and NGOs

There is a relatively well-established environmental NGO sector in Slovakia. The oldest is the Slovak Union of Nature and Landscape Protectors (SZOPK), which was established in 1969, but most NGOs originated after the political changes in 1990. Some NGOs are research- and science-oriented, and provide important contributions to national biodiversity inventories and monitoring programmes (e.g. the Slovak Ornithological Society/BirdLife Slovakia, Raptors Protection of Slovakia, Slovak Bat Conservation Society, Prales o.z. and DAPHNE – Institute of Applied Ecology). The management of protected areas and other aspects of nature conservation are the focus of some other NGOs, such as the Regional Association for Nature Conservation and Sustainable Development (BROZ), For Nature (Pre prírodu), OZ Tatry, the Carpathian Altruists Conservation Association (KOZA) and Združenie Slatinka. There are also NGOs that have a strong involvement in protected area policies, for example Greenpeace Slovakia, WWF Slovakia, Friends of the Earth-CEPA, Centre for Sustainable Alternatives (CEPTA), the Institute for the Protection of Nature and VLK (WOLF) Forest Protection Movement. VIA IURIS promotes effective public participation in decision and policy making and has provided legal assistance in cases concerning access to information and construction of waste incinerators, small hydropower plants and activities in protected natural areas.

In Slovakia, stakeholders and the wider public routinely participate in environmental impact assessments and decision-making procedures, such as those relating to tree felling, protected areas and other activities with potentially significant environmental impacts. Participatory approaches have also been used for the establishment of the Natura 2000 network, and are a legal obligation for the preparation of management plans for all protected areas including SPAs and SCIs/SACs (as further described in Section 27.5.4). This can also lead to some problems and conflicts, such as during the Natura 2000 designation process. Although it involved wide-reaching public consultations with local citizens, a common criticism was that the process relied too much on technical information from experts, while the opinions of local citizens were ignored, such that the state was not prepared to modify its original plans. Nevertheless, the consultations raised awareness within the protected area administrations of the existing local stakeholders and demonstrated the capacity of private actors to articulate their interests (Yakusheva, 2019).

27.4 Species Protection and Conservation Measures

27.4.1 Legislation

General species protection is established under the Nature and Landscape Act, which prohibits unnecessary and unauthorised catching, killing or damage of most groups of plants and animals and their habitats where they occur naturally. If cases of killing and damage are recorded, the Nature Conservation Authority is required to restrict the activity or stop it. The general protection requirements in the legislation also include provisions for energy distributors, telecommunication operators, transport and waterways developers and managers to plan new or upgrade existing infrastructure carefully and monitor its impact. If they discover that it unintentionally kills migratory birds or animals, they are obliged to take the necessary technical measures to minimise these adverse effects. These legal requirements led to cooperation between the SNC, NGOs and infrastructure operators (with support from the LIFE ENERGY project). This reduced bird electrocution risks nationally over 30% of electricity pylons, and led to other measures to prevent bird collisions with overhead power lines (Gális *et al.*, 2019).

The national legislation for the protection of species is fully aligned with the requirements in the Nature Directives. These species protection provisions are strictly applied, and derogations are issued in accordance with the relevant conditions, with details provided in regular derogation reports.[8] Derogations have been mainly issued for research purposes, deterring birds causing damage (e.g. Great Cormorant (*Phalacrocorax carbo*)), building insulation and renovation, and shooting dangerous individual Brown Bears. No objections have been raised by the European Commission with regard to species protection and derogations in Slovakia, except in relation to the Wolf.

As a result of the EU accession negotiations, the Wolf population in Slovakia is listed in HD Annex V, rather than Annex IV, and was therefore not as strictly protected as in most other Member States. As a result, legal hunting continued, based on a yearly quota issued by the MoARD. However, questions were raised on the application of the protection regime for the Wolf based on HD Annex IV as the Wolf was hunted in Natura 2000 sites designated for the species. A European Commission infringement process was launched against Slovakia in 2013 for breaching the Habitats Directive with respect to the Wolf. While this resulted in a Wolf hunting ban in Natura 2000 sites, the species remained inadequately protected elsewhere. The case was solved by the implementation of a stricter protection regime and an increase in non-hunting zones for the Wolf, reducing the hunting quota and further protection conditions applied to hunting (Guimarães *et al.*, 2016). Following public and NGO demands, as of 2021, a new by-law under the Nature and Landscape Act[9] has given the Wolf year-round protection and prohibited its hunting.

27.4.2 Species-Specific Conservation Measures

Recovery programmes for endangered plant and animal species are developed in compliance with the Nature and Landscape Act. They are for species especially in need of rapid conservation action and are developed by the SNC, in cooperation with experts and stakeholders, and approved by the MoE and published on its website.[10] They usually cover a five-year period and aim to identify the main pressures and reasons for the species' decline, eliminate negative factors and determine the measures necessary to improve their status and to ensure that their population remains stable or increases.

Recently, recovery programmes for several animal species have been developed by the SNC and adopted by the MoE for: Danube Clouded Yellow (*Colias myrmidone*), Apollo (*Parnassius apollo*), European Pond Turtle (*Emys orbicularis*), Capercaillie, Red-footed Falcon (*Falco vespertinus*), Black Grouse (*Lyrurus tetrix*), Eurasian Bittern (*Botaurus stellaris*) and Ferruginous Duck (*Aythya nyroca*). Updated recovery programmes include analysis of implementation in the previous implementation period; however, to date, no systematic evaluation of these documents has taken place in Slovakia.

Management plans for species were introduced in 2014 as a new initiative for species that are not immediately endangered in Slovakia but require conservation management. Management plans for the large carnivores Brown Bear, Wolf and Eurasian Lynx[11] have been prepared with stakeholders under the lead of the SNC and approved by the MoE. Special cross-sectoral working groups were created to prepare the plans for each species, including all the relevant stakeholders and the recognised experts, who discussed and agreed population density estimates and hunting quotas. The management plans have helped to solve some controversial issues regarding the setting of hunting quotas for the Wolf and its protection (see Section 27.4.1). The plans also improved the measures used to reduce the damage to farmers' livelihoods caused by large carnivores and the compensation payments for killed animals.

A successful re-introduction of the European Bison (*Bison bonasus*) was carried out in 2004. Since then the Slovak population has become self-sustaining and mixed with wild Polish populations (Perzanowski *et al.*, 2006).

27.5 Protected Areas and Networks

27.5.1 Protected Area Types and Their Coverage

More than 90% of today's protected areas were established between 1970 and 1990. This happened in a top-down fashion, with no or only very poor discussion with stakeholders as all land was considered as state-owned. Nature conservation organisations were established only to provide professional supervision and define technical requirements. Effective tools for managing protected areas were lacking (Meessen *et al.*, 2015).

After the political transformation, private land ownership rights on protected areas were fully restituted, resulting in their current diversified ownership structure (Kluvánková-Oravská

[8] https://cdr.eionet.europa.eu/sk/eu/habides/

[9] Vyhláška MŽP SR č. 170/2021, ktorou sa vykonáva zákon č. 543/2002 Z. z. o ochrane prírody a krajiny v znení neskorších predpisov.

[10] www.minzp.sk/ochrana-prirody/druhova-ochrana/programy-zachrany/

[11] www.minzp.sk/ochrana-prirody/druhova-ochrana/programy-starostlivosti/

et al., 2009). The Nature and Landscape Act introduced after 2002 the possibility to establish private and municipal protected areas. In a private protected area, 100% of the land is private and the owner supports (also financially) the designation and management of the area. Despite the simplicity of the idea, the actual implementation has been fairly complex and bureaucratic, and it takes several years to declare a private protected area (Otto and Chobotova, 2013).

Currently, as summarised in Table 27.4, there are a variety of domestic protected areas in Slovakia that differ in their size, purpose and degree of protection. There are five levels of increasing protection according to the Nature and Landscape Act, with the first level of protection being valid over the whole territory of Slovakia, and the fifth (strictest) level of protection only applying to Protected Sites, Nature Reserves, Natural Monuments and Protected Landscape Elements. The highest level of protection applies where protected areas with different levels of protection overlap.[12] The Nature and Landscape Act defines those activities that are prohibited in relation to each level of protection. For example, in the first level of protection, permission is needed to significantly affect wetlands or habitats of European and national importance, but most activities are allowed. In contrast, the strictest fifth level of protection includes prohibition of building, logging, hunting, angling and even public access except on designated trails. To reduce direct impacts on the sites, buffer zones are established around National Parks and Nature Reserves with their protection one level lower than their respective protected area.

A classification of large-scale protected areas (National Parks and Protected Landscape Areas) according to the IUCN protected area management categories was carried out by SNC over 2014–2016. As a result of this, only two of the nine National Parks were classified in IUCN category II, while the others are in category IV or V (Kadlečík, 2017; UNEP-WCMC, 2022). A new category of protected area, 'Natural Park', was introduced in a recent amendment of the Nature and Landscape Act applicable from 2020. As of December 2022, no proposals for Natural Parks had been presented.

In contrast to SPAs, Slovak authorities have not considered it necessary to distinguish SCIs/SACs as a separate category of domestic protected area; instead, most sites are designated as an appropriate type of domestic protected area. Furthermore, Natura 2000 is considered to be integrated into the domestic system of protected areas as there is a high degree of overlap in both area and approach to conservation in Natura 2000 sites and domestic protected areas. These overlapping systems complicate management.

Table 27.5 lists the types of protected areas in Slovakia and their coverage. The proportion of large protected areas (i.e. National Parks and Protected Landscape Areas) represents 22.5% of the total area of Slovakia. The combined total area of all domestic protected areas with higher levels of protection (i.e. second to fifth levels), taking into account overlaps, amounted to 11 471 km^2 in 2020 (23.4% of the Slovak territory).[13] There have been some more recent additions to the protected area network, for example, in 2021, the Old Growth Forest of Slovakia nature reserve, which preserves the last remnants of primeval and old growth forests. It includes 76 state-owned forest localities in various parts of Slovakia with a total area of 6 462 ha.

At the end of 2020, the Natura 2000 network in Slovakia covered 14 633 km^2, which is 29.8% of the land area, with Birds Directive SPAs covering 26.3% and Habitats Directive SCIs/SACs covering 12.5%. Agricultural and forest areas within Natura 2000 sites accounted for 16% of the country's utilised agricultural area (including natural grassland) and 46% of the total forest area (European Commission, 2020). While the designation of the SPAs is complete, more SCIs will be identified and designated as their sufficiency is 93% (MoE, 2021). Slovakia is committed to completing the network; after field surveys were carried out to fill in the missing data, new SCIs have been proposed and negotiations with owners started in 2020.

Twenty sites have been designated under international protected area legislation, overlapping considerably with the Natura 2000 network and domestic designations. Two Biosphere Reserves are transboundary: the Tatry Biosphere Reserve (Slovakia and Poland), and the first trilateral site in the world, the East Carpathians Biosphere Reserve (Slovakia, Poland and Ukraine). There are two World Heritage Sites that are shared with other countries: the Caves of Aggtelek Karst and Slovak Karst (with Hungary), and the Ancient and Primeval Beech Forests of the Carpathians and Other Regions of Europe (with 18 other European countries).

According to a spatial analysis of the WDPA data by UNEP-WCMC (2022), the protected area network in Slovakia in February 2022 amounted to 18 396 km^2, which is 37.5% of the country.[14] Whilst the WDPA includes the protected area types in Table 27.5 (except Biosphere Reserves and the seven Municipal Protected Areas) it also includes buffer zones, which would not be regarded as protected areas in many countries. In the case of Slovakia, as described previously, the buffer zones do receive some level of protection, so their inclusion in the protected area coverage statistic is justified.

27.5.2 Protected Area Designation Process and Criteria

The main criteria for the designation of protected areas are nature and biodiversity values, such as the presence of threatened species and habitats or other values (e.g. landscape, aesthetic features, geology, caves or natural waterfalls). Each protected area category, as described in Table 27.4, has its predefined scope and criteria for identification in the Nature

[12] www.sopsr.sk/web/?cl=114

[13] www.enviroportal.sk/indicator/detail?id=121&print=yes

[14] www.enviroportal.sk/indicator/detail?id=121

Table 27.4 Definitions of the different types of protected area in Slovakia.

Category (and IUCN type)	Definition	Designation	Level of protection (see text)
Protected Landscape Area (IUCN V)	A large area, usually over 1 000 ha, with ecosystems important for the conservation of biological diversity and ecological stability, with characteristic landscape features or specific forms of historical settlements	Regulation of the Government of the Slovak Republic	Second level
Natural Park (IUCN III)	A larger area, usually over 500 ha, with ecosystems influenced by human activity important for the conservation of habitats and species	Regulation of the Government of the Slovak Republic	Second and third level, depending on what the Government declares in accordance with the regulation
National Park (IUCN II, V, VI)	A large area, usually over 1 000 ha, primarily with ecosystems that are not significantly changed by human activities, or with a unique natural landscape and structure, which form supra-regional bio-centres and hold the most important natural heritage, where nature conservation is a priority in relation to other activities	Regulation of the Government of the Slovak Republic	Third level
Protected Site (IUCN III, IV)	A location, usually up to 1 000 ha, where HD habitats or habitats of national interest are present, or where the habitats of HD species, bird species or species of national interest are present and where the favourable conservation status of these habitats depends on human management	Regulation of the Government of the Slovak Republic	Second, third, fourth or fifth level, depending on what the Government declares in accordance with the regulation
Nature Reserve (IUCN Ia, Ib or V)	A site, usually up to 1 000 ha, which contains natural habitats, or habitats only a little altered by human activities, or HD habitats, habitats of national interest, habitats of HD species, birds or species of national interest	Regulation of the Government of the Slovak Republic	Fourth or fifth level, depending on what the Government declares in accordance with the regulation. IUCN category I
Natural Monument (IUCN III)	A small and/or linear area, usually up to 50 ha, which is of scientific, cultural, environmental, aesthetic or landscape importance. Includes all natural waterfalls and caves	Regulation of the Government of the Slovak Republic	Fourth or fifth level, depending on what the Government declares in accordance with the regulation
Protected Landscape Element (IUCN III, V)	An important landscape element that fulfils the function of a bio-centre or bio-corridor or interaction element, in particular of local or regional importance	Regulation of the Government of the Slovak Republic	Second, third, fourth or fifth level, depending on what the Government declares
Special Protection Area (SPA)	Habitats of BD Annex I birds and other SPA trigger species such as migratory birds, in accordance with BD Article 4	Declared at a national level by a regulation of the Government in accordance with the Birds Directive	First level, but the Government with its regulation specifies prohibited activities that may affect the birds and their habitats
Private Protected Area (IUCN II, IV or VI)	An owner of land that meets the conditions laid down in the Nature and Landscape Act for one of the categories of protected areas (with the exception of a National Park or Protected Landscape Area) can develop a conservation proposal in order to designate a Private Protected Site, a Private Nature Reserve or a Private Nature Monument and submit it to the Regional Environment Authority	The Regional Environmental Authority must declare the private protected area with its Generally Binding Decree establishing also a level of protection and details of its territorial protection	The level of protection corresponds to the selected category of protected area and ranges from second to fifth level
Municipal Protected Area (JUCN III)	A locally important site, usually with an area of up to 100 ha, with a cultural, scientific, ecological, aesthetic or landscape-forming significance	Municipal protected areas are declared by the municipality by a generally binding regulation	First level of protection. The municipality specifies within its regulation the required management measures

Source. Based on Vozár and Alberton (2013), updated by authors based on legislation changes after 2014.

Table 27.5 The number and extent of protected areas in Slovakia.

	IUCN category	Number	Area (km^2)
Designated under global and non-EU legislation			
Ramsar Sites	NR	14	407
Natural and Mixed World Heritage Sites	NA	2	105
Biosphere Reserves (exc. buffer and transition zones)	NA	4	546
Designated under the Nature Directives			
SCIs	NR	642	6 151
SPAs	NR	41	13 105
Natura 2000 area			14 633
Designated under domestic legislation			
Nature Reserve (prírodná rezervácia)	I	593	950
National Park (národný park)	II	9	3 216
Protected Site (chránený areál)	IV	172	110
Natural Monument (prírodná pamiatka)	III	331	16
Protected Landscape Area (chránená krajinná oblasť)	V	14	5 225
Protected Landscape Element (chránený krajinný prvok)	III	1	0.03
Private Protected Area (súkromné chránené územie)	V	3	0.69
Municipal Protected Area (obecné chránené územie)		7	1.75
Natural Park (prírodný park) – none designated yet		0	0

Note. NR = Not reported; NA = not applicable.
Sources. See Section 1.3. Based on data from: Ramsar Convention and UNESCO websites (November 2021); EEA Natura 2000 Barometer, as at end 2020; State Nature Conservancy of the Slovak Republic;[15,16] UNEP-WCMC (2022), e.g. for IUCN category.

and Landscape Act. Before 2014, the competences for protected area designation were distributed among the municipalities, regional environment administrative offices, MoE and the national government. Since then, the designation of most types of protected areas has been the sole responsibility of the government, and MoE is responsible for the process. The exception is for municipal and private protected areas, where the municipality and the regional environmental authority, respectively, have the designation power.

A conservation proposal (or 'conservation project' as directly translated from '*projekt ochrany*') is delivered to the competent nature conservation authority (MoE) which leads to the designation process in consultation with owners (and/or administrators or tenants) of the land, the affected municipality and the concerned governmental authorities. The owners and other stakeholders, affected municipalities and the concerned governmental authorities have the right to submit comments. The state authority has discretion over how the comments are taken into account in the subsequent process of designation (Vozár and Alberton, 2013). As a result of changes in legislation, since 2014 private owners have the right to veto the designation of protected areas with the strictest level of protection (i.e. fourth and fifth). However, at the same time, new options have been created for compensating private landowners for income forgone due to the designation and protected area restrictions. The following options are available: sell the land to the state, rent it, sign a stewardship contract and manage the land in line with nature protection requirements, or receive financial compensation.

The process for the identification and designation of SPAs and SCIs for inclusion in the Natura 2000 network required a major data collection exercise and analysis, involving many experts and institutions. As this was rather complex and involved several steps, it is described in detail in online Annex SK.1.[17]

27.5.3 The Coherence of Protected Areas and Ecological Networks

Ecological networks are represented by the Territorial System of Ecological Stability (TSES), which is a system of landscape and nature protection that was developed in the late 1970s in the

[15] www.biomonitoring.sk/Home/ProtectedSites
[16] http://worldprotectedareas.sopsr.sk/en/unescos-mab-programme/biosphere-reserves/

[17] All annexes are free to access online at CUP: www.cambridge.org/natureconservation

former Czechoslovakia (Míchal, 1994). It is one of the approaches to building green infrastructure. TSES is basically defined in the Nature and Landscape Act, and also in all sector-specific planning laws, as a spatial structure of ecosystems related to each other that ensure diversity in the landscape. The main components are bio-centres (clusters of biodiversity-rich areas), bio-corridors (corridors of various types connecting bio-centres), interactive elements (landscape elements acting as stepping stones for biodiversity movements), and the ecostabilising measures (measures proposed for re-naturalisation of landscapes with low ecological stability) of national, regional or local interest. The main aim is to create ecological networks in intensively utilised landscapes through the revitalisation of existing natural habitats and the creation of new ones, along with proposals for their protection and management.

TSES is a planning tool for the regulation of developments in the landscape used by various sectors – particularly nature conservation, spatial development, agriculture and forestry – especially for land consolidation and re-parcelling purposes. As the maintenance and enhancement of TSES is in the public interest, developers and others responsible for activities that affect ecosystems are obliged, at their own expense, to prevent and limit their damage and destruction. Implementation of TSES is mostly through land consolidation projects such as restoring arable land to grassland on steep slopes, planting trees in the landscape, the creation of bio-corridors, etc.

New guidelines for the development of documentation on Regional TSES were prepared by the Slovak Environment Agency in 2014 (SAŽP, 2014) and adopted by the MoE. Consequently, a number of Regional TSES have been adopted and/or implemented.[18] However, the practical creation of TSES faces various problems ranging from insufficient resources for field implementation to deficient legislation and limited cooperation among involved sectors (Izakovičová, 2007; Izakovičová *et al.*, 2019).

27.5.4 Approaches to Managing Protected Areas

In Slovakia, management plans are required for all protected areas. For each SCI they are developed together with the conservation proposal, the aim of which is to establish a national category of protected area for each site. It is a time-consuming and demanding process requiring expert, administrative and financial resources. Therefore, there have been delays with the preparation and adoption of the management plans. By May 2021, only a small proportion of the required plans had been prepared and approved by the government: 2 out of 9 National Parks, 20 out of 41 SPAs, and 88 out of 642 SCIs (MoE, 2021).

As required under the Nature and Landscape Act, management plans of protected areas contain measurable conservation objectives and measures with responsibilities, as well as their timeframe and required financial resources. Ecological requirements and threats are generally described for all conservation objectives. Whilst the lifetime of the management plans is set at 30 years, they can be updated in response to monitoring data, and re-evaluated if necessary. Monitoring requirements are defined for each management plan based on the planned measures. Reporting on management planning and the evaluation of the plan's implementation is carried out by SNC annually for the MoE, but the results are not publicly available.

Specific targets for BHD habitats and species, for which the sites have been designated, have been set within each SCI/SAC and SPA. By combining information from each site's standard data form (SDF) and the national reports under BD Article 12 and HD Article 17, a framework for setting the priorities has been established. This considers the need to improve or maintain the status of individual habitats/species at the local level and compares it with wider biogeographical requirements. In this way, broad goals are identified for habitats based on their representativeness, relative area and conservation status, and for species related to their population, isolation and conservation status. Lists of such conservation goals and their prioritisation (low, medium, high, unknown) form the basis for the creation of management plans in each individual site. It is recommended that the listed goals set in this way are then modified when setting site-level objectives if necessary, depending on local conditions (MoE, 2021).

Management plans are prepared by state environmental authorities, mainly the SNC. The MoE supervises the preparation of management plans for National Parks, Protected Landscape Areas and Natura 2000 sites, which require approval by the government. Plans are very often prepared in close cooperation with environmental NGOs as part of LIFE programme and other EU-funded projects. Stakeholder participation is legally required in the development of management plans for all protected areas including SPAs and SCIs/SACs. Stakeholders are consulted on the management plans when they have been developed, through passive consultation only and not through a collaborative effort in their development. Nevertheless, it is a complicated and time-consuming procedure, in part due to the complex and varying patterns of land ownership and land tenure in the country.

Forest management plans (FMPs) are a particularly important instrument as they are obligatory for all forests, and some 49% of Slovak forest land overlaps with Natura 2000 sites. FMPs may contribute a lot to protected area management, provided they respect the requirements for the conservation and improvement of the habitats and species that are the subject of conservation within the Natura 2000 site concerned. They are procured by the forest land-user and authorised by the Regional Forest Offices, and they cannot be adopted without a binding statement of the nature conservation administration concerning their compatibility with the Nature and Landscape Act. For example, if a non-intervention regime is appropriate for the habitats and species that are the focus of protection within a given area, this can be applied with adequate compensation.

[18] For details see: www.sazp.sk/zivotne-prostredie/starostlivost-o-krajinu/zelena-infrastruktura/dokumenty-uses-v-sr.html

In practice, it has often been difficult to reach agreement on forest management in relation to nature protection requirements, the interests of owners and the opinion of the forestry departments. Excessive logging in some Natura 2000 sites has occurred due to inappropriate FMPs. The biodiversity impacts of such forestry practices, and the inadequate preparation and implementation of protected area management plans, are illustrated by the decline over recent decades in the Capercaillie, an umbrella species of natural mountain spruce climax forests. The population of thousands recorded in the 1970s has declined by about 70% to just a few hundred (Mikoláš *et al.*, 2015). The main cause has been the increase in the extent and intensity of logging in natural mountain forests, which has created large-scale forest clearings and transformed them into artificial forests with low biodiversity.

Consequently, the contradictory approaches to forest management in protected areas have been identified as one of Slovakia's three biggest environmental challenges (European Commission, 2019). Whilst some FMPs include sufficient measures, many will need to be substantially modified in accordance with the principles of nature-friendly forest management (MoE, 2021). As the life cycle of FMPs is 10 years, all nature protection requirements can be incorporated during their renewal process, namely in the description of the stands and the plan of economic measures.

The practical management of protected areas is carried out by various bodies, with only a small proportion directly by SNC. Management of state-owned forest land is by public forest enterprises established by the MoARD, whose primary mission is profit-making exploitation of all forest lands, including protected areas. A similar situation exists in other state-owned areas, including agricultural land and water bodies. As several administrations with varying interests may be involved in activities within protected areas, this can lead to conflicting objectives and strained relationships (Vozár and Alberton, 2013).

A partial solution to this situation would be to transfer the management of state-owned land in protected areas solely to the SNC, and this was included as one of the long-term goals in the *Envirostrategy* (MoE, 2019). The transfer of state forest land to National Parks under the administration of the nature protection bodies was approved by the Parliament in one of the latest amendments of the Nature and Landscape Act, in December 2021. In April 2022, all National Park Administrations were established as legally independent nature conservation bodies. In Tatranský National Park and Slovenský Raj National Park – the two National Parks with approved management plans and zonation (zone A is strictly protected and zone B allows limited human interventions) – all state-owned land was transferred from state forest to National Park management. In the remaining seven National Parks, the transfers only covered areas with the strictest fourth and fifth levels of protection. Meanwhile, the management planning and zonation within the remaining seven National Parks is in preparation and once completed all state-owned land will be transferred from state forest land to the National Park administrations. The amendment has significantly improved nature conservation of National Parks in Slovakia.

27.6 General Nature Conservation Measures

27.6.1 Development Planning and Impact Assessment

Spatial planning in relation to ecological networks and green infrastructure in Slovakia is carried out through the preparation of TSES documents (as described in Section 27.5.3), at national, regional and local level. However, the system is not financed properly, so the development of regional and local documents is very slow and implementation is hampered by complicated land ownership issues. Izakovičová *et al.* (2019) identified several other barriers to the system's integrated management of landscapes. There is a lack of experience with the processing of landscape ecological studies, low promotion and poor explanation of the reasons for landscape ecological planning, ineffective quality control of documentation processing and of the integration of results into the next planning stages, and insufficient or under-use of landscape ecological plans in territorial planning and land consolidation projects.

The potential impacts of other plans and projects on nature and biodiversity are scrutinised in SEA and EIA procedures. These consider impacts on species and habitats of EU and national importance, protected areas, wetlands, ecological networks and migration corridors. Where negative impacts are identified they should be mitigated or compensated (in accordance with the EIA and SEA Directives).

Like other EU Member States, Slovakia initially struggled with proper and full implementation of appropriate assessments under HD Article 6(3). The appropriate assessment process in Slovakia is implemented within EIA and SEA procedures for all plans and projects that fall under the EIA and SEA Directives (Sundseth and Roth, 2013). For specific activities that are not projects in the sense of the EIA Directive, the need for an appropriate assessment may be identified through steps according to Section 28 of the Nature and Landscape Act. In such cases, appropriate assessments are carried out through special EIA procedures that apply to these cases. The influence of developers, lack of data and inadequate time for dedicated surveys were identified as the main weaknesses in the EIA procedures in Natura 2000 sites in Slovakia by Michalcová and Baus (2014).

Some improvements in appropriate assessment practices have been achieved with the preparation of a dedicated guidance document (SNC, 2016) and stricter requirements from the authorities. The latest development in this area has been the establishment of special authorisation requirements for consultants preparing appropriate assessment documents in accordance with the latest amendment of the Nature and Landscape Act, which came into force in January 2020 and has been applied since September 2021.

27.6.2 Habitat Management and Restoration

The Use of Rural Development Measures

RDP measures have been important for species, habitat and landscape conservation, and have been the main funding source for the management of the Natura 2000 network, as well as wider biodiversity objectives (see Section 27.7.3). Agri-environment schemes have been designed partly to address management needs in protected areas and other important sites, and have become the main tool for conserving biodiversity within farmland and, to a lesser extent, forests.

The most important RDP measures for nature conservation over 2014–2020 were the area-based schemes, as summarised below (based on Šefferová Stanová and Šeffer, 2019; see Table 27.6 for funding levels):

- **Conservation of habitats of natural and semi-natural grasslands** (M10.1): a very successful key measure for achieving the favourable status of HD grassland habitats (covered 17 types), as well as grassland habitats of national importance. Tailored to seven ecological types of grasslands. Coverage objective 150 000 ha, 2019 uptake of 135 234 ha.
- **Ground Squirrel [European Souslik] (*Spermophilus citellus*) habitat conservation** (M10.2): a new scheme, targeted to areas where this globally Endangered species occurs. Coverage objective 3 000 ha, but only 750 ha uptake in 2019, as farmers considered the scheme to be quite complicated.
- **Protection of habitat of the Great Bustard on arable land** (M10.3). Coverage objective up to 2 000 ha, uptake was 1 788 ha in 2019. The scheme focussed on two SPAs with the species, which is on the edge of extinction in Slovakia.
- **Natura 2000 and Water Framework Directive compensation payments** (M12) are two submeasures for compensation payments for Natura 2000 agricultural and forest areas that aimed to support about 31 100 ha (uptake was 1 843 ha in 2019) of land within Natura 2000 sites, with the fourth and fifth level of protection for agricultural areas, and the fifth level of protection on forest land. This covered a small fraction of Natura 2000 sites as there is 312 528 ha on farmland and 1 271 286 ha on forest land, located across all five levels of protection. Although the environmental authorities wanted to also include Natura 2000 sites with the second level of protection, this was not possible due to lack of funds.
- **Forest-environmental and climate services and forest protection** (M15): covered two schemes for forest protection in Natura 2000 sites, located in the third and fourth levels of protection. It supported activities that go beyond mandatory forest management requirements (e.g. natural rejuvenation of trees to at least 60% of the restored area, excluding forestry activities in nesting sites of threatened bird species). Again, the scheme was small in scale, covering only 24 000 ha, with uptake of 22 175 ha.

In addition to the above key measures, support for organic farming (M11) has some benefits for biodiversity and helps reduce water pollution. Slovakia currently has 200 000 ha under organic farming systems, which is 9.9% of agricultural land (MoARD, 2021a), with RDP support over 157 000 ha. A substantial proportion of farmers in Slovakia (65%) receive payments for being within areas with natural and other constraints (M13). The allocations for this have used up the largest share of RDP environmental funding, amounting to 52% over 2014–2020 (Table 27.6).

A constraint on the use of RDP agri-environment measures for semi-natural grasslands has been the ineligibility of some small-scale mountain grasslands in remote areas because they are not included in LPIS. As a result, although some have been managed using project-related funds, or the SNC's limited financial resources, the majority are abandoned. Financial support for low-intensity farming of hardly accessible semi-natural grasslands and mosaic structures with natural and structural elements is insufficient (Šatalová *et al.*, 2021).

Wetland Restoration

The revitalisation of wetlands and water courses, including the re-opening of dead river branches and restoration of functional floodplains, are amongst the top priorities of *Envirostrategy 2030*. Wetland restoration projects have been mainly funded by the LIFE Programme and focussed on the lowland areas of Slovakia. Detailed published information on the scale of restoration and its impact, as well as the national restoration programme and its financing, is lacking.

In order to reverse wetland losses and degradation, and declines in wetland management, the updated *Slovak Wetland Management Programme 2015–2024* identified four main strategic goals based on the proposed *Ramsar Strategic Plan*:[19]

1. Addressing the drivers of wetland loss and degradation.
2. Effective conservation and management of the Ramsar Site network.
3. The wise/sustainable use of all wetlands.
4. Enhancing implementation.

Many measures have aimed in particular to remove migration barriers on water courses belonging to, or related to, the Natura 2000 network, these being integrated into the *Water Plan of the Slovak Republic* to support the restoration of river continuity. The measures are being implemented by the Slovak Water Management Enterprise. So far, initiatives on lateral connectivity and flooding of selected dead arms have been launched to a lesser extent. In 2020 and 2021, a controlled flood of the inland Danube delta took place (MoE, 2021).

19 www.minzp.sk/files/sekcia-ochranyprirodyakrajiny/dohovory/ramsar/program-starostlivosti-mokrade-2024.pdf; www.minzp.sk/oblasti/ochrana-prirody-krajiny/medzinarodne-dohovory/ramsarsky-dohovor/dokumenty/program-starostlivosti-mokrade-slovenska-ramsarsky-strategicky-plan/

27.7 Nature Conservation Costs, Benefits and Funding Sources

27.7.1 Nature Conservation Costs

There has been no systematic or regular assessment of all the costs of nature conservation in Slovakia, and only fragmented information is available on the topic. In 2017, the Environmental Policy Institute at the MoE assessed national expenditure on environmental protection, including nature protection (Haluš *et al.*, 2017). Based on these results, nature conservation spending by state statutory agencies between 2010 and 2016 was approximately €16.5 million per year on average (the highest spending was in 2015). Slightly over half of this (51%) was covered by the Slovak state budget and the rest by EU and other foreign funds, and was spent through statutory state nature and environmental agencies, namely the SNC, Slovak Museum of Nature Protection and Speleology (SMOPAJ), Slovak Environment Agency and Zoo Bojnice. These organisations used state budget funds between 2010 and 2016 mainly to cover labour and insurance costs (17% and 6%, respectively) and purchase of goods and services (8%).

The PAF for 2021–2027 (MoE, 2021) provides a more recent estimate of the costs of nature conservation in relation to the Natura 2000 network and supporting actions, which is not limited to the costs of the environmental agencies (as above). The general cross-cutting measures and administrative costs for the network are estimated to be €14.5 million per year averaged over the period; 81% of this is for annual running costs, with the remainder being one-off costs. Maintenance and restoration measures for the network are estimated to be €110.4 million per year on average (83% running costs, 17% one-off costs). Thus the total average estimated annual costs of the network itself amount to €124.9 million per year. Although costs per unit area will vary considerably, this gives an indicative average annual cost of €85 per ha.

The PAF for 2021–2027 also includes the costs of green infrastructure to support the Natura 2000 network (e.g. to increase coherence and cross-border connections), which it estimates to be €63.9 million per year (93% recurring actions, 7% one-off costs). The cost of additional species-specific nature conservation measures, and compensation for damage by species, is estimated to average €53.6 million per year (68% recurrent costs, 32% one-off costs). The average estimated annual total costs for all the PAF measures for the Natura 2000 network amount to €242.4 million.

27.7.2 The Economic Benefits of Nature Conservation

Several studies have evaluated the economic benefits from nature conservation, primarily in individual protected areas (Považan *et al.*, 2014, 2015; Solár *et al.*, 2016). Whilst no comprehensive nationwide assessment of these economic benefits exists, it is clear that the large ecological network, with numerous protected areas, is an important source of green jobs and revenue from visitors and recreation activities, and other nature-based businesses, especially in National Parks and Landscape Protected Areas.

Mapping and assessment of ecosystem services slowly developed between 2012 and 2019 in response to Target 2 of the *EU Biodiversity Strategy to 2020*. This included an assessment of ecosystems in Slovakia (Černecký *et al.*, 2019), an assessment of values from the water sector (Bujnovský, 2018) and a catalogue of assessed ecosystem services in Slovakia (Mederly *et al.*, 2019). An estimation of the economic benefits of ecosystem services is under preparation.

27.7.3 Funding Sources for Nature Conservation

Table 27.6 provides an overview of the main funds used for nature conservation in Slovakia, with estimates of the funding allocations, mainly over the 2014–2018 EU programming period.

Since EU accession in 2004, EU funds have been a major source of financing for nature conservation in Slovakia, particularly for the Natura 2000 network and other protected areas. EU funds that were widely known as the **Structural Funds** (i.e. European Regional Development Fund (ERDF), European Social Fund and Cohesion Fund) have been the main source of funding for projects and capital investments. The volume of funds drawn from the EU Structural Funds over 2007–2015 and the Environmental Fund (national project based funding) financing over 2013–2016 amounted to €57 million (Haluš *et al.*, 2017). Almost 40% was used for centres for the individual protected area administrations, the construction of educational trails, improvements to and extension of visitor infrastructure in protected areas and caves, and the digitisation and improvement of the collections of the Slovak Museum of Nature Protection and Speleology. Monitoring and research projects represented the second largest group (26%). About 30% of total expenditure went to practical nature conservation (i.e. management of sites and species), of which €15 million was used to prepare and implement protected area management plans. The smallest part (€2 million) was allocated to measures for improving the status of selected species of plants and animals. Projects that focussed on education and awareness raising (€2.4 million) had a relatively small share (Haluš *et al.*, 2017).

During the 2014–2020 financing period, Slovakia used the **EU Cohesion Fund** Operational Programme Quality of Environment (OPQE) for its priority activities in the Natura 2000 network and for other biodiversity conservation actions. The OPQE supported the preparation and implementation of management plans for Natura sites, preparation and implementation of action plans for priority species and habitats, enhancement of the monitoring and reporting framework, building green infrastructure, and management and eradication of alien invasive plants.

Table 27.6 Funding sources for nature conservation in Slovakia.

Source and period	Measure/category	Average annual allocation (€ m)		Comments
EU funding instruments		**EU + national**	**EU**	
EAFRD Rural Development Programme measures – allocations to Priority 4 (restoring, preserving and enhancing agriculture and forest ecosystems) 2014–2020	M13 Areas facing constraints	65.6	49.0	
	M10 Agri-environment–climate	20.3	15.0	
	M8 Investments in forests	15.1	11.1	
	M11 Organic farming	14.7	10.8	
	M4 Investments in physical assets	6.4	4.8	
	M12 Natura and WFD	1.3	0.9	
	M15 Forest environment–climate	0.7	0.6	
	M1 Knowledge transfer, etc.	0.4	0.3	
	M16 Cooperation	0.3	0.2	
	M2 Advisory services, etc.	0.2	0.1	
RDP Priority 4 subtotal		*125.0*	*92.8*	
ERDF 2014–2020	OP Quality of Environment (Codes 85 and 86)	22.3	19.0	N2k etc.
	Integrated Regional OP (Code 85)	5.4	4.7	N2k etc.
	Cross-border cooperation (Codes 85, 86, 87 and 88)	6.0	5.1	N2k etc.
	Interreg V-A – nature and biodiversity measures in cities	5.9	5.1	N2k etc.
LIFE Programme	Nature and biodiversity projects (2004–2020)	4.3	2.7	27 projects
	Integrated project (2020–2030)	1.7	1.0	LIFE19 IPE/SK/000003
Danube Strategy Implementation	Regional projects with Slovak involvement	0.3	0.3	6 projects
European Maritime and Fisheries Fund (EMFF)	Productive investments in aquaculture	0.5	0.4	N2k etc.
EU funding instruments subtotal		*171.4*	*131.1*	
National funds within 2014–2020		**Additional national***		**Comments**
State Budget 2014–2020	Compensation to owners and farmers and land purchase	4.9		
	Budgeted for SNC and SMOPAJ	3.9		
Environmental Fund 2014–2020	Project-based funding	0.7		
Green Education Fund 2017–2019	Project-based funding	0.1		
Additional national subtotal		*9.6*		
National total including co-financing of EU instruments		*49.9*		
Total		**181.0**		

Notes. 'For N2k etc.' indicates funding relevant to Natura 2000, related green infrastructure and BHD species (based on the PAF). * Excludes co-financing of EU funds unless indicated. The subtotals and totals are only approximate indications of the average annual levels of funding as they are based on annualised data from differing periods.

Sources. Based on data from: RDP data from European Commission ESIF funds database (23 October 2020); RDP allocations to Natura 2000, ERDF, EMFF and national funding from 2021 to 2027 PAF (MoE, 2021); Slovakia LIFE factsheet, European Commission (updated February 2021); MoE, 2020, www.envirofond.sk. See Section 1.3 for further details.

Various instruments under the **ERDF** have included allocations for nature conservation and green infrastructure activities such as the Integrated Regional Operational Programme and cross-border cooperation. Interreg V-A supported green infrastructure development in cities and projects implemented in regional cooperation under the Danube Strategy. Six projects implemented under the Danube Strategy have focussed on fostering the restoration and management of ecological corridors and supporting environmentally friendly and safe transport systems and balanced accessibility of urban and rural areas (e.g. projects ConectGreen[20] and TRANSGREEN[21]).

The European Agricultural Fund for Rural Development (EAFRD) through RDP measures has been the main source of support for nature conservation and restoration by farmers and foresters. As described earlier, the agri-environment-climate measure has been especially important and effective through its tailored schemes, such as for the management of HNV grasslands and habitats of threatened species. However, just over half of the EAFRD funding allocated to ecosystems under Priority 4 during 2014–2020 was allocated to payments to farmers in areas with natural and other constraints, which provides uncertain environmental benefits.

Although relatively small compared to some other EU funds, the **LIFE Programme** has provided tailored and targeted financing for nature conservation in Slovakia. Up to February 2021, the country had been awarded 27 nature and biodiversity projects with a total value of €73 million (€46 million from the EU). The projects have traditionally focussed on conservation interventions in Natura 2000 sites, or priority birds, or other species or habitats.

As described earlier, state funding is available for the SNC and some other state authorities for nature conservation purposes. Another source of public financing is the Environmental Fund,[22] the purpose of which is to support a variety of activities related to environmental protection, including in protected areas (Vozár and Alberton, 2013). The Environmental Fund has provided nature conservation financing of approximately €1–2 million annually, with the amount varying according to project applications and fund availability. The SNC also gains some income from entrance fees (mainly from caves open to the public), which amounts to around €3 million per year according to the SNC annual reports.

The various national and EU budget sources for nature conservation, including funding for Natura 2000, seem to be inadequate for the needs of Slovak nature conservation. Furthermore, the capacity to absorb and spend funds efficiently needs to be increased as well. The allocation and availability of funds should be increased in future with adequate increases to meet the expected needs of implementing the *EU Biodiversity Strategy for 2030*. The main challenges will be connected with the increased area of land managed directly for nature conservation, further structural changes in protected areas administration and expected restoration commitments. However, two main areas need to be urgently improved, namely implementation of conservation measures for the benefit of species and habitats (especially restoration of habitats, species-specific measures and removal of threats requiring management of sites), and overall coordination and prioritisation of the use of the funds.

27.8 Surveillance, Monitoring, Reporting and Applied Research

The systematic monitoring of species and habitats of European interest in Slovakia started with a 2009–2015 project[23] financed by the EU ERDF. Monitoring methodologies for individual species and habitats were elaborated (Saxa *et al.*, 2015), permanent monitoring sites (TML) were identified and subsequently verified, and monitoring of habitats and species of European importance was carried out. The Complex Information and Monitoring System of Habitats and Species of the Community Interest (KIMS) database system was also created. The monitoring subjects are 65 HD habitat types, 146 HD animal species and 49 HD plant species within more than 11 000 TML inside and outside of protected areas (MoE, 2021). Part of KIMS also includes data from the grassland and peatland information system (managed by DAPHNE – Institute of Applied Ecology) and database of vegetation records (managed by the Institute of Botany of the Slovak Academy of Sciences). The original monitoring project did not include systematic monitoring of the birds needed for reporting under BD Article 12, and the proper management of bird species. However, this has now been revised and the bird monitoring component in KIMS has been under development since 2021.

With the repeated monitoring of BHD habitats and species it is not only possible to detect and assess changes in them, but also broader biodiversity trends. This information from protected areas, as well as the rest of the country, is an invaluable asset for nature conservation. It provides an important baseline for organisations dealing with nature protection (especially SNC), acting in accordance with the Nature and Landscape Act. It also creates a framework for informing other stakeholders. In this respect, SNC and MoE created in 2012 two consultation bodies for the key players like NGOs, academic institutions and hunters involved in data collection under the Nature Directives. Also, all the information is publicly available through a dedicated website[24] of SNC, which

[20] Project ConnectGREEN: www.interreg-danube.eu/approved-projects/connectgreen

[21] Projet TRANSGREEN: www.interreg-danube.eu/approved-projects/transgreen

[22] Established by Act No. 587/2004 Coll. on the Environmental Fund.

[23] Preparation and Establishment of Monitoring of Species and Habitats and Improvement of Accessibility of the Result to the Public.

[24] www.biomonitoring.sk

provides a strong incentive for enhancing the conservation status of BHD species and habitats.

SOS/BirdLife Slovakia collects bird data and organises censuses of birds (Common Bird Census, Waterbird Census, voluntary bird monitoring of Important Bird Areas, monitoring of SPAs and another schemes). Raptor Protection Slovakia (RPS) also contributes to monitoring of birds of prey and owls through dedicated censuses and LIFE project–related monitoring.

In addition, other important habitat monitoring programmes include the National Inventory and Monitoring of Forests (based on FAO survey methods) and database of primary forests.[25]

27.9 Future Developments

Whilst there are challenges ahead for nature conservation in Slovakia, there are signals in some important strategic documents that give nature conservation an increasingly important role in the future.

The biggest environmental challenges in Slovakia, and hence areas that will be prioritised within environmental policy to 2030, include waste management, air quality, and habitats and species conservation, especially in forests, meadows and wetlands (MoE, 2019). In the framework of the *Envirostrategy 2030*, it is expected that the protection of biodiversity will be improved and measures to prevent the deterioration of species and habitats will be taken. As previously described, one of the latest amendments to the Nature and Landscape Act has established independent administrations for each of the National Parks, and responsibility for the management of state-owned land once they have been zoned. The system of protected areas and degrees of protection will be also reviewed and simplified by 2024.

In the National Parks, the biggest challenge ahead is their zoning to align them with IUCN criteria, which is necessary for most of them. The core zones will be without human intervention and should represent at least 50% of the total area of each National Park by 2025 and 75% by 2030. Logging will be prohibited in core zones and environmentally friendly land management will be encouraged in surrounding areas with active management of habitats. Logging will also be under increasing public and institutional control. Another aim is to ensure that the total value of forest ecosystem services does not decrease in all forested areas. In agricultural areas, the aim is to protect and restore traditional landscape features.

New impetus for further nature and biodiversity measures is also anticipated as a result of the EU *Biodiversity Strategy for 2030* and the global post-2020 biodiversity framework. This will require an updating of Slovakia's biodiversity strategies. It is clear that biodiversity-related polices will need to be further integrated into other sectors, and implementation more closely bound to climate change adaptation and mitigation goals and measures.

As regards water policy, green measures – such as the revitalisation and re-naturalisation of water courses and adjacent riverine landscapes, along with the necessary technical infrastructure – will be included in the flood prevention system. Drought and water scarcity impacts will be reduced by means of water retention, better landscape planning and more responsible water management.

The national *Climate Change Adaptation Strategy* (MoE, 2018) counts on effective nature protection to achieve several goals, including promoting synergies between adaptation and mitigation measures. Ecosystem-based adaptation approaches will also be used, such as integrated water retention in landscapes, green infrastructure development, measures to increase ecological connectivity and integrated monitoring systems for adaptation planning based on ecosystem services assessments.

In accordance with the proposed *Framework of Nature and Landscape Protection* (MoE and SNC SK, 2019), further improvements in the nature and landscape conservation policy are expected, namely increasing the effectiveness of nature protection in achieving its goals, especially in protected areas and in species and habitats protection. Furthermore, there are plans to improve legislation and implementation of landscape protection and its integrated management.

27.10 Conclusions

Nature conservation in Slovakia has undergone a long and exciting journey. The majority of today's domestic protected areas were established between 1970 and 1990. This happened in a top-down fashion, with no, or only very poor, consultation processes among relevant stakeholders as all land was considered as state-owned (Meessen *et al.*, 2015). After political changes in 1989, many changes in nature conservation took place. In particular, Slovakia's accession to the EU in 2004 brought great support for the establishment of the Natura 2000 network, as a result of which, with nearly 30% coverage, it is amongst the EU countries with the highest proportion of its land area within the network (see Table 5.4). Slovakia has made fairly good use of the increase in EU funding after its accession into the EU, such as for protected area support and management planning, the mapping of HNV grasslands and development of agri-environment schemes. Another important development was that of the monitoring system.

However, there are also challenges and unsolved problems that need further attention from decision makers and institutions, especially to increase the efficiency of administrative and management planning processes and focus more on consistent real habitat and site management and restoration work, as well as species recovery activities. It is important for Slovakia to re-evaluate the national system of targeted protected areas (including the current zoning strategy and levels of protection), further develop conservation research and monitoring of the implementation of conservation measures (especially of restoration success and progress towards site-specific

25 www.pralesy.sk

conservation objectives), and improve cooperation with landowners and users (Švajda and Fenichel, 2011). There is also a need to improve the communication and participation of stakeholders in all decision-making processes, including the compensation of landowners in protected areas (Otto and Chobotová, 2013). Additional goals will be to adapt nature conservation policy to the changing climate and other developments, using innovative and more integrated approaches. In this context, a key challenge is to increase and secure long-term sustainable financing of nature conservation and implementation of innovative funding schemes.

Improving cross-sectoral cooperation, to achieve more ambitious joint objectives, also remains a challenge for nature conservation in Slovakia, especially in relation to water management, agriculture, forestry, spatial planning, transportation and tourism. Good inspiration for such mainstreaming can be found in the CBD's agenda. The aim is to continue to implement this basic principle, as natural ecosystems need to be protected and at the same time used in a sustainable way, whilst conforming with current domestic and international legal obligations and policy.

Acknowledgements

We would like to thank Zuzana Guziová (Slovak Raptor Protection) and Ján Kadlečík (SNC) for reviewing the text and providing valuable comments, as well as Ján Šeffer (DAPHNE – Institute of Applied Ecology) for overall contributions to the chapter. Furthermore, we would like to thank also Jan Černecký (SNC), Jana Durkošová and Peter Jány (both MoE) for providing additional information.

References

Balon, E. K. & Holčík, J. (1999) Gabčíkovo River Barrage System: the ecological disaster and economic calamity for the Inland Delta of the Middle Danube. *Environmental Biology of Fishes*, **54**, 1–17.

BirdLife International (2015) *The European Red List of Birds.* Luxembourg: Office for Official Publications of the European Communities.

Bujnovský, R. (2018) Estimation of benefits from the actual use of inland water ecosystem services in the Slovak Republic. *Ekológia* (Bratislava), **37**, 201–218.

Burkovský, J. & Stockmann, V. (2006) Dve významné výročia štátnej ochrany prírody. *Enviromagazín*, **3**, 3–6.

Černecký, J., Čuláková, J., Ďuricová, V. *et al.* (2020) *Správa o stave biotopov a druhov európskeho významu za obdobie rokov 2013 – 2018 v Slovenskej republike.* Banská Bystrica, Slovakia: ŠOP SR, 109 pp.

Černecký, J., Gajdoš, P., Špulerová, J. *et al.* (2019) Ecosystems in Slovakia. *Journal of Maps*, **16** (2), 28–35.

Chytrý, M., Dražil, T., Hájek, M. *et al.* (2015) The most species-rich plant communities in the Czech Republic and Slovakia (with new world records). *Preslia*, **87**, 217–278.

European Commission (2019) *The EU Environmental Implementation Review 2019 Country Report – SLOVAKIA.* SWD(2019) 116 final.

(2020) *Commission recommendations for Slovakia's CAP strategic plan.* SWD(2020) 392 final.

Gális, M., Naďo, L., Hapl, E., Šmídt, J., Deutschová, L. & Chavko, J. (2019) Comprehensive analysis of bird mortality along power distribution lines in Slovakia. *Raptor Journal*, **13**, 1–25.

Gubka, A., Nikolov, Ch., Gubka, K. *et al.* (2013) History, present and expected future of forests in Slovakia. *American Journal of Plant Sciences*, **4**, 711–716.

Guimarães, N., Álvarez, F., Finďo, S., & Urban, P. (2016) Ecology and management of the Grey wolf (*Canis lupus*) in Slovakia. *Folia Oecologica*, **45**(1). ISSN: 1336–5266.

Haluš, M., Bendíková, M., Gális, M. *et al.* (2017) *Revízia výdavkov na životné prostredie. Záverečná správa.* Bratislava: Inštitút environmentálnej politiky (IEP) MŽP SR, Útvar hodnoty za peniaze (ÚHP) MF SR.

Izakovičová, Z. (2007) Landscape ecology contribution to the concept of territorial system of ecological stability. *Životné Prostredie*, **41** (3), 131–134.

Izakovičová, Z., Miklós, L., Miklósová, V. & Petrovič, F. (2019) The integrated approach to landscape management – experience from Slovakia. *Sustainability*, **11**(17), 4554.

Jasík, M. & Polák P. (eds.) (2011) *Pralesy Slovenska.* Banská Bystrica, Slovakia: FSC Slovensko.

Kadlečík, J. (2017) Prvé hodnotenie chránených území Slovenska podľa manažmentových kategórií IUCN. *Chránené územia Slovenska*, **89**, 25–30.

Kluvánková-Oravská, T., Chobotová, V., Banaszak, I., Slávikova, L. & Trifunovova, S. (2009) From government to governance for biodiversity: the perspective of Central and Eastern European Transition Countries. *Environmental Policy and Governance*, **19**, 186–196.

Mederly, P., Černecký, J., Špulerová, J. *et al.* (2019) *Katalóg ekosystémových služieb Slovenska.* Banská Bystrica, Slovakia: ŠOP SR, UKF v Nitre, ÚKE SAV.

Meessen, H., Švajda, J., Kohler, T. *et al.* (2015) Protected areas in the Slovak Carpathians as a contested resource between metropolitan and mountain stakeholders. *Journal of Alpine Research*, **103** (3), 1–17.

Míchal, I. (1994) *Ekologická stabilita 2.* Veronica, Brno.

Michalcová, Z. & Baus, P. (2014) The determination of main obstacles for the developing of high-quality documentation during EIA Process on Natura 2000 Sites. *GeoScape*, **8** (1), 1–8.

Mikoláš, M., Svitok, M., Tejkal, M. *et al.* (2015) Evaluating forest management intensity on an umbrella species: Capercaillie persistence in central Europe. *Forest Ecology and Management*, **354**, 26–34.

MoARD (2014) *Stanovenie území s vysokou prírodnou hodnotou v poľnohospodárstve a v lesnom hospodárstve Slovenskej republiky.* Bratislava: Ministry of Agriculture and Rural Development.

(2018) *Report on agriculture and food sector in the Slovak Republic 2017.* Bratislava: Ministry of Agriculture and Rural Development.

(2021a) *CAP STRATEGIC PLAN 2023-2027 (Draft).* Bratislava: Ministry of Agriculture and Rural Development.

(2021b) *Report on the Forest Sector of the Slovak Republic 2020 – GREEN REPORT.* Bratislava: Ministry of Agriculture and Rural Development of the Slovak Republic and National Forest Centre.

MoE (2014) *Aktualizovaná národná stratégia ochrany biodiverzity do roku 2020.* Bratislava: Ministry of Environment of the Slovak Republic.

(2015) *Orientation, Principles and Priorities of the Slovak Republic Water Management Policy by 2027.* Bratislava: Ministry of Environment of the Slovak Republic, Published by Slovak Environment Agency.

(2018) *Stratégie adaptácie Slovenskej republiky na zmenu klímy.* Bratislava: Ministry of Environment of the Slovak Republic.

(2019) *Envirostrategy 2030. Strategy of the Environmental Policy of the Slovak Republic until 2030.* Bratislava: Ministry of Environment of the Slovak Republic.

(2021) *Prioritný Akčný Rámec (PAF) pre sústavu Natura 2000 v Slovenskej Republike.* Bratislava: Ministry of Environment of the Slovak Republic.

MoE & SNC SK (2019) *Koncepcia ochrany prirody a krajiny do roku 2030 - Návrh.* Bratislava: Ministry of Environment of the Slovak Republic.

Orsillo, N. P. (2008) *Agricultural intensification in communist Czechoslovakia and its impact on the environment.* Master's thesis. Masaryk University, Brno, Czech Republic.

Otto, I. M. & Chobotova, V. (2013) Opportunities and constraints of adopting market governance in protected areas in Central and Eastern Europe. *International Journal of the Commons,* **7** (1), 34–57.

Perzanowski, K., Adamec, M. & Pčola, Š. (2006) Stav a reštitúcie cezhraničnej poľsko-slovenskej populácie zubra hrivnatého. [The state and restitutions of the trans-boundary Polish-Slovak population of the European Bison.] In *Výskum a ochrana cicavcov na Slovensku VII.* Zborník referátov z konferencie (Zvolen 14.-15. 10. 2005), eds. M. Adamec & P. Urban, pp. 159–164. Banská Bystrica, Slovakia: Štátna ochrana prírody SR.

Považan, R., Getzner, M. & Švajda, J. (2014) Value of ecosystem services in mountain national parks. Case study of Veľká Fatra National Park (Slovakia). *Polish Journal of Environmental Studies,* **23** (5), 1699–1710.

(2015) On the valuation of ecosystem services in Muránska Planina National Park (Slovakia). *Journal on Protected Mountain Areas Research and Management,* **7**, 61–69.

Šatalová, B., Špulerová, J., Štefunková, D., Dobrovodská, M., Vlachovičová, M. & Kozelová, I. (2021) Monitoring and evaluating the contribution of the rural development program to high nature value farmland dominated by traditional mosaic landscape in Slovakia. *Ecological Indicators,* **126**, 107661.

Šefferová Stanová, V. & Šeffer, J. (2019) *Evaluation of the impact of the CAP on habitats, landscapes, biodiversity. AGRI-2018-0492. Case study Slovak Republic.* Brussels: Unpublished report, IEEP.

Saxa, A., Černecký, J., Galvánková, J., Mútňanová, M., Balážová, A. & Gubková Mihaliková, M. (eds.) (2015) *Príručka metód monitoringu biotopov a druhov európskeho významu.* Banská Bystrica, Slovakia: Štátna ochrana prírody Slovenskej republiky.

SAŽP (2014) *Metodické pokyny pre vypracovanie dokumentov RÚSES.* Banská Bystrica, Slovakia: Slovenská agentúra životného prostredia.

SNC (2016). *Metodika hodnotenia významnosti vplyvov plánov a projektov na územia sústavy Natura 2000 v Slovenskej republike.* Banská Bystrica, Slovakia: Štátna ochrana prírody Slovenskej republiky.

Solár, J., Janiga, M. & Markuljaková, K. (2016) The socioeconomic and environmental effects of sustainable development in the Eastern Carpathians and protecting its environment. *Polish Journal of Environmental Studies,* **25**, 291–300.

Špulerová, J., Bezák, P., Dobrovodská, M., Lieskovský, J. & Štefunková, D. (2017) Traditional agricultural landscapes in Slovakia: why should we preserve them? *Landscape Research,* **42**, 891–903.

Stockmann, V. (2013) *Dejiny ochrany prírody na Slovensku. Štátna ochrana prírody na Slovensku.* Banská Bystrica, Slovakia: Štátna ochrana prírody Slovenskej republiky.

Straka, P. & Guziová, Z. (1998) *National Biodiversity Strategy of Slovakia.* Bratislava: Ministry of Environment of the Slovak Republic.

Sundseth, K., and Roth, P. (2013) *Study on Evaluating and Improving the Article 6.3 Permit Procedure for Natura 2000 Sites.* Main report and case study compilation. Brussels: Ecosystems Ltd.

Švajda, J. & Fenichel, E. P. (2011) Evaluation of integrated protected area management in Slovak National Parks. *Polish Journal of Environmental Studies,* **20**, 1053–1060.

UNEP-WCMC (2022) *Protected Area Profile for Slovakia from the World Database of Protected Areas, February 2022.* www .protectedplanet.net

Vozár, I. (2019) Súčasné nastavenie organizácie štátnej správy starostlivosti o životné prostredie - vybrané otázky. In *Výkon štátnej správy starostlivosti o životné prostredie - vybrané otázky,* ed. I. Vozár. Banská Bystrica, Slovakia: Via Iuris.

Vozár, I. & Alberton, M. (2013) *Analysis of National Institutional Frameworks and Legislations affecting biodiversity and ecological connectivity in the Carpathian countries. National report Slovakia.* BioREGIO project report.

Yakusheva, N. (2019) Managing protected areas in Central Eastern Europe: between path-dependence and Europeanisation. *Land Use Policy,* **87**, 104036.

Chapter

28 Slovenia

Peter Skoberne

Key Statistics	
Terrestrial area (km^2)/% of EU-28*	20 267/0.5%
Marine waters (km^2)/% of EU-28*	214/0.005%
Population density (p km^2)/% of EU-28*	102/89%
GDP per capita (€)/% of EU-28	18 465/69%
Land cover (CORINE land cover classes 2018)	FOR 56%; AGR 34%; OSN 6%; ART 3%; WET 0.6%
Biogeographical regions	ALP, CON, MMED
Number of HD Annex I habitats/%*	60/27%
Number of HD Annex II, IV and V species/%*	200/16%
Number of BD Annex I bird taxa/%*	71/40%
% HD UCS and decreasing habitats/ % unknown	31%/0%
% HD UCS and decreasing species/ % unknown	24%/18%
% breeding BD Annex I birds decreasing/% uncertain and unknown	18%/46%

* Excluding MAC and MMAC biogeographical regions.
Sources and codes: See Table 1.1.

28.1 The Natural History and Characteristics of Slovenia

The Republic of Slovenia is situated in the south-eastern part of the Alps and on the northernmost part of the Balkan Peninsula. One of its main characteristics is its high biological and landscape diversity within a relatively small area, mainly due to the transitions between the Mediterranean, Alps, Dinaric region and Pannonian plain. This is associated with its wide-ranging altitudes (from 0 to 2 864 m) and differing climates: Mediterranean in the west, alpine in the north-west and north, and continental in the central part and in the east. It also has a varied and influential geology, the most remarkable being the carbonate rocks (44%), which give rise to well-developed karstic phenomena such as caves, sinkholes, dolines, poljes[1] and intermittent rivers. More than 13 000 caves are registered, the longest cave system being M-16 (c. 42 km), while the deepest is Čehi 2 (1 505 m).[2] An additional contribution to the country's species, habitat and landscape diversity comes from its small and shallow area of the Adriatic Sea (south-east of the Gulf of Trieste).

Slovenia is a water-rich country, as most of the country is situated at the headwaters of rivers (e.g. the Soča, Sava, Savinja, Krka and Reka). Most of the natural lakes are of glacial origin in the Alps, but are small and scarce due to the limestone character of the mountains. Of special importance is Cerkniško jezero, an intermittent karstic lake within a polje, which can cover up to 30 km^2. Fens and swamps are small and scattered, but important contributors to the richness. Of special interest are some larger raised bogs in the upland areas of Pokljuka (Šijec, Blejsko barje), Jelovica (Za Blatom) and Pohorje (Lovrenška jezera), as they are at the south-eastern border of their distribution in the Alps.

Forest remains the predominant land cover (58% in 2021[3]), with 24% of forests under public ownership (by the state or communes).[4] Most forests (70%) are beech, beech–fir or beech–oak forests, which have an important production potential.[5] The most representative are Illyrian *Fagus sylvatica* forests (*Aremonio-Fagion*) (91K0).[6] There are some 70 tree species indigenous to the Slovenian territory. During recent times, forest cover has increased, from 47% in 1961 to 53% by 1990.[7] While forest expansion in the 1960s was the result of forestry practices, its increase since then has mainly been due

[1] Dolines are closed depressions in karst areas formed by solution of the surficial rock or other processes, with poljes being the largest such depressions covering up to several hundred square kilometres (Kranjc, 2013).

[2] www.katasterjam.si/

[3] http://kazalci.arso.gov.si/sl/content/povrsina-gozda-7

[4] www.zgs.si/eng/slovenian_forests/forests_in_slovenia/forest_ownership/index.html (last accessed 4 March 2021).

[5] www.zgs.si/gozdovi_slovenije/o_gozdovih_slovenije/gozdnatost_in_pestrost/index.html

[6] Numbers in brackets are HD Annex I habitat codes (see Appendix for full names). * Indicates a priority habitat.

[7] www.uradni-list.si/glasilo-uradni-list-rs/vsebina/1996-01-0632/program-razvoja-gozdov-v-sloveniji-nprg

agricultural abandonment and natural regeneration of tive tree species. Traditional sustainable forest management actices and low production demand (0.7% of GDP in 2016[8]) ve also helped to maintain the relatively high quality of the rest ecosystems and associated species, including large carni-res: Brown Bear (*Ursus arctos*), Wolf (*Canis lupus*) and rasian Lynx (*Lynx lynx*).

About one-third of Slovenia is agricultural land: 36% arable nd, 6% permanent plantations (orchards) and 58% meadows d pastures (Travnikar *et al.*, 2019). Natural grasslands are w only found above the forest line and in wetlands. All other odiverse grasslands are dependent on traditional extensive anagement through haymaking and grazing. About 70% of ricultural land is within upland and mountain farms, hich cover 20–30% of the national territory. According to timates based on CORINE land cover data and agricultural nd use, 60–80% of the UAA in Slovenia was High ature Value (HNV) farmland in 2008.[9] This proportion is ow likely to be lower as some HNV farmland has been griculturally improved, or abandoned (contributing to the orest expansion).

A typical characteristic of Slovenian farmland is the small ize of farms (averaging 7.0 ha in 2020)[10] and its mosaic andscape. This is the result of the country's natural character nd long history of human colonisation together with various and uses and practices that are adapted to the natural condi-ions. The pattern of the main bio-regions reflects the diversity of Slovenian landscapes, yet the transition between the Alps nd the other regions is particularly evident as 'the pre-alpine andscapes' (Marušič *et al.*, 1998). Within these regions micro-factors have further built up the mosaic of small characteristics and diverse landscape structures. In recent years, there have been changes in the farmland landscape, with the consolida-tion of farms, amalgamation of fields and removal of features such as hedges and walls, which have reduced its diverse patchwork character.

Slovenia has a modest level of species endemism, in part because it was a refuge for Balkan and Alpine species during glaciation/interglaciation periods. About 60 plant and about 850 animal species are endemic (Majić Skrbinšek, 2009). Most of the endemic taxa are troglobiotic (i.e. restricted to caves), many of them being highly specialised and with a very narrow distribution range. The aquatic troglobiotic fauna, with its 200 species, is by far the richest in the world (Sket *et al.*, 2004), while the terrestrial troglobiotic fauna, with 150 species, is second only to the more southern areas of the Dinaric karst. Five Slovenian cave systems have been listed among the 20 richest cave systems for troglobiotic fauna in the world. The Postojna-Planina cave system, with its 50 aquatic and 35 terrestrial species, is by far the most species-rich. The best-known and iconic troglobiotic species is the Olm (*Proteus anguinus*). This amphibian is the largest troglobiotic species in Europe, and has a black pigmented subspecies with developed eyes, *Proteus anguinus parkelj* (Bizjak Mali and Sket, 2019). The Olm is globally Vulnerable, threatened mostly by water pollution and potentially by various pathogens that are of global concern to amphibians. The very limited distribution of the black subspecies makes it particularly vulnerable.

28.2 The Status of Nature

28.2.1 Habitat Status and Trends

According to the results of the HD Article 17 national report for 2013–2018,[11] out of 89 assessments of the conservation status of 45 HD habitats (across the three biogeographical regions) 38% showed a favourable status, 31% a poor status and 30% a bad status. Of particular concern is that a high proportion of the habitats with an unfavourable status also had decreasing trends, including 67% of the assessments with a bad status (Table 28.1).

The most threatened HD habitats (i.e. those with a bad status and declining trends) were amongst the grasslands (44% of assessments), followed by freshwater habitats (31%) and then bogs, mires and fens (18%). As forests comprise about 60% of the Natura 2000 area, their conservation status is especially important, yet only 23% of assessments showed a favourable status, and 12% had a bad status and declining trend. The worst situation is with riparian forests, in particular alluvial forests with *Alnus glutinosa* and *Fraxinus excelsior* (91E0*), especially in the Continental region, and riparian mixed forests of *Quercus robur*, etc. (91F0), as they are closely linked to river dynamics and all human activities that influence water flow.

These HD Article 17 assessments have two major draw-backs. Firstly, only the specific HD habitat types are included, so the information is incomplete. Secondly, as quantitative data are lacking, they are mostly based on the opinion of the experts at the Institute for Nature Conservation. Therefore, whilst the results fit with the general view that wetlands and grasslands are under the most pressure, they are not sufficient to adapt policies to improve the situation.

28.2.2 Species Status and Trends

According to the HD Article 17 national report for 2013–2018, 29% of HD species assessments revealed a favourable status, 39% a poor status and 14% a bad status. The status of 18% of species was unknown. As with HD habitats, a high proportion of the species with an unfavourable status also had decreasing trends, including 79% of those with a bad status (Table 28.1).

[8] www.stat.si/StatWeb/en/News/Index/6996 (last accessed 10 June 2020)

[9] kazalci.arso.gov.si/en/content/high-nature-value-farmland-areas (last accessed 4 January 2021).

[10] www.stat.si/StatWeb/news/Index/9459 (last accessed 19 December 2022).

[11] See Section 1.3 for background on monitoring under the Nature Directives, status categories and data sources.

Table 28.1 Short-term trends (c. 2007–2018) in Habitats Directive Annex I habitats and Annex II, IV and V species with an unfavourable conservation statu Slovenia.

	Poor (unfavourable–inadequate)				Bad (unfavourable–bad)			
	Imp	Sta	Dec	?	Imp	Sta	Dec	?
HD Habitat assessments								
Number	1	14	10	3	0	6	18	3
%	4%	50%	36%	11%	0%	22%	67%	11%
HD Species assessments								
Number	0	10	43	76	0	0	37	10
%	0%	8%	33%	59%	0%	0%	79%	21%

Key. Imp = improving, Sta = stable, Dec = decreasing, ? = unknown.
Source. Based on Member State HD Article 17 reported data as presented on the EEA national summary dashboard (2021).

The most threatened species groups were reported to be amphibians and grassland arthropods (butterflies), which corresponds with their habitat trends. As a result of monitoring of selected butterflies, which started in 2012, in 2019 the conservation status of the False Ringlet (*Coenonympha oedippus*), Scarce Fritillary (*Euphydryas maturna*), Large Blue (*Maculinea/Phengaris arion*), Scarce Large Blue (*Maculinea/Phengaris teleius*) and Dusky Large Blue (*Maculinea/Phengaris nausithous*) were all assessed as unfavourable (Zakšek *et al.*, 2019). As national monitoring of non-bird species is limited to HD Annex II species, the trends of most other species are unknown or based mostly on expert opinions.

Although more monitoring data are available for birds (see Section 28.8.1), the BD Article 12 national report for 2013–2018 indicated that the short-term trends for the majority of BD Annex I species were uncertain or unknown (Table 28.2). Amongst those with known trends, a substantial proportion of the breeding and wintering populations of BD Annex I taxa had declined. Most alarming is evidence of steep declines in certain BD Annex I species breeding numbers, mostly related to grasslands, such as Corncrake (*Crex crex*) and Ortolan Bunting (*Emberiza hortulana*), as well as forests, including Hazel Grouse (*Bonasa bonasia*), White-backed Woodpecker (*Dendrocopos leucotos*), Three-toed Woodpecker (*Picoides tridactylus*) and Capercaillie (*Tetrao urogallus*) (Denac *et al.*, 2015).

Data from farmland bird monitoring have also shown declines in more common species.[12] Over the 14 years from 2008 to 2021, the farmland bird index declined moderately from its starting level of 100 to 76.8, whilst the index for the subset of grassland birds declined to 56.3. The 14-year trend of farmland birds shows a moderate decline, despite a stable trend over the last 8 years.

Whilst a substantial proportion of species of conservation concern have declined over recent decades, some have shown positive population trends. This is the case for all large

Table 28.2 Short-term trends (c. 2007–2018) in Birds Directive Annex I ta in Slovenia.

	Breeding				Wintering			
	Inc	Sta/Flu	Dec	?	Inc	Sta/Flu	Dec	?
Number	7	11	9	23	3	2	2	8
%	14%	22%	18%	46%	20%	13%	13%	53%

Key. Inc = increasing, Sta = stable, Flu = fluctuating, Dec = decreasing, ? = uncertain and unknown.
Source. Based on Member State Article 12 reported data as presented on th EEA national summary dashboard (2021).

carnivores in Slovenia. The Wolf and Brown Bear have recovered and are spreading. Active management measures are in place to prevent and compensate for damage, and regulate the population where necessary. As a result, both species have a favourable conservation status in both terrestrial biogeographical regions.

The Eurasian Lynx became extinct at the beginning of the twentieth century and was re-introduced in 1973. Since then, the population has spread over Slovenia to Croatia and to Bosnia. However, the population has recently declined because of genetic inbreeding (Sindičić *et al.*, 2013). A LIFE Lynx re-introduction project[13] is therefore being carried out to address this, the 2021 results from which are promising.

The Beaver (*Castor fiber*) was extirpated in the eighteenth century (Kryštufek, 2003). After re-introduction in Posavina (Croatia) in 1996 (Popp, 2011), the species spread to Slovenia and colonised areas along most of the rivers in the central and eastern part of the country.

28.2.3 Drivers and Causes of Change

Globalisation and market economics have been the main drivers of change, and remain so now. These have led to

[12] http://kazalci.arso.gov.si/en/content/clone-farmland-bird-index-1 (last accessed 19 December 2022).

[13] www.lifelynx.eu/

pressures on nature, as have the socio-economic situation and attitudes of people, and the increasing consequences of climate change.

The resulting pressures on forests are still low compared to other habitats, but there is rising interest in increasing exploitation, which might weaken the 'green' forestry legislation. In recent years, the main negative changes have been due to highly damaging events, such as those caused by snow and wind, and resulting bark beetle outbreaks. The contravention of forest plans has also been increasing.

Changes in agricultural policy and technology, and the influence of the market economy, have driven significant changes in farming systems and practices and the farmland landscape. As traditional extensive livestock systems are not economically competitive compared with modern intensive farming, some farmers have abandoned traditionally managed marginal agricultural land. In contrast, areas of potentially more productive land have been intensified, with increased use of fertilisers and machines (e.g. hay baling and mowing at night) and the removal of small structures (e.g. hedges and dry stone walls). These processes have caused rural depopulation of the countryside and affected the appearance of karst and alpine pastures as well. Needless to say, biodiversity at all levels has reflected these changes.

In the past, wetlands and floodplains were under considerable pressure from the expansion of agriculture, such as in the Vipavska dolina and Pesnica valley at the end of 1980s as a result of large-scale drainage. In recent years, interest in conversion of large areas has declined, but smaller areas remain threatened. As a result of climate change there is a new pressure, from increasing interest in irrigating farmland, which may affect the water regime of rivers and streams.

Generally, Slovenian surface waters are relatively unpolluted, with 49% of water bodies considered to have had a good ecological state over 2015–2021. Compared to the previous reporting period (2009–2015), good ecological status was achieved on 9% fewer water bodies.[14] Hydro morphological changes and general degradation have been the main causes where surface waters have had a moderate or low ecological state. However, as mentioned for the Olm, water pollution from sewage and agriculture has affected caves and their characteristic sensitive species. Whilst EU legislation is promising regarding sewage water treatment, even from dispersed settlements, it is not sufficient to prevent eutrophication of underground waters.

Some of the most fundamental pressures on rivers have come from hydropower plants. As a result, only a few semi-natural large river ecosystems remain in Slovenia. Currently, to meet ambitious renewable energy targets, the energy sector is focussed on using the remaining hydro potential of the larger Slovenian rivers (e.g. Sava and Mura). Consequently there is a collision of interests between nature conservation and the renewable energy sector and climate change objectives.

An important pressure, with wider impacts, comes from increasing infrastructure and urbanisation. As a transport corridor, Slovenia has a high demand for international rail and road links. In addition, as the population is rather dispersed, it is developing its internal road network as well. These built developments have negative impacts through their fragmentation of ecosystems and increasing disturbance. Over the last decade, the detrimental effects of light pollution have also become more apparent on species such as bats and nocturnal insects.[15] Although this has been partly addressed by the Decree on limit values of light pollution of the environment, unnecessary illumination in rural areas needs to be minimised.

Whilst hydropower, other energy projects, and urban and infrastructure developments have had undoubted impacts, they can be mitigated or even avoided by a good planning system.

Tourism and outdoor activities can contribute to improving attitudes to nature; however, they also present another important pressure. To some extent, this has been longstanding in the mountains, as mountaineering is a traditional Slovenian recreational sport. A network of mountain huts and shelters, as well as hiking routes, is maintained by clubs associated with the Alpine Association of Slovenia (Planinska zveza Slovenije). Over the last decade, visits to the mountains have increased and many other outdoor activities have become more common, such as running, trekking, cycling, mountain-biking, paragliding, parachuting, winter sports and off-road driving. The increased number of visitors is stimulating investments in facilities, such as car parks and resorts, and further increasing disturbance, pollution and fire risks. Tourist pressures are greatest at peak times, and at the most attractive and well-known places. However, some tourists are changing their visits to very early or very late in the day, and to other times of the year – further increasing the periods of disturbance. Visitor management in natural areas therefore appears to be a growing challenge for the future.

28.3 Nature Conservation Governance, Actors, Legislation and Policies

28.3.1 Development of Nature Conservation in Slovenia

Table 28.3 lists the most important events that have affected nature conservation in Slovenia, some of which are further described in the following paragraphs. Online Annex SL.1 provides a more detailed description of the early developments in nature conservation (up to 1990).

[14] http://kazalci.arso.gov.si/en/content/chemical-and-ecological-status-surface-water-bodies-0

[15] Improving the conservation status of nocturnal animals (moths and bats) by reducing the effect of artificial lighting at cultural heritage sites. LIFE project NAT/SI/000378 (2014).

Table 28.3 Key nature conservation developments in Slovenia.

1892	Karl Auersperg's forester, Leopold Hufnagel, defined in the forest plan forest reserves (intentional no-use)
1896	Protection of Edelweiss (*Leontopodium alpinum*) in Princely County of Gorizia and Gradisca (Austro-Hungarian Empire)
1907	Albin Belar published the abstract of the first inventory of natural monuments in Carniola, prepared in 1906
1920	*Memorandum of the Section for Nature and Natural Monument Conservation* produced by the Slovene Museum Society
1922	Protection of plant and animal species, protection of caves
1924	Protection of Dolina Triglavskih jezer by a 20-year contract
1946	Establishment of the authority for nature conservation
1960	First Nature Conservation Act
1961	Renewal of the protection of the Dolina Triglavskih jezer as Triglav National Park
1976	Inventory of important natural heritage. Nature conservation introduced into the spatial planning process
1981	Natural and Cultural Heritage Act
	Extension of the Triglav National Park and establishment of a management authority, beginning of actual management of protected areas; about 4% of national territory was included
1991	Slovenia declared independence and separated from the Socialistic Republic of Yugoslavia
1994	Responsibilities for nature conservation moved from the cultural to the environmental authorities
1999	Nature Conservation Act (NCA) adopted, under which the *National Nature Protection Programme* (NNPP) was adopted within the framework of the *National Environment Protection Action Programme* (NEEP)
2001	*National Biodiversity Strategy* adopted by the government
2002	The Institute for Nature Conservation, established in 1999, started to operate
2004	EU accession: adaptation of the legislation (Nature Conservation Act amended), first Natura 2000 proposal
2005	Resolution on the *National Environmental Action Plan* 2005–2012
2007	Operational Programme – Natura 2000 management programme for 2007–2013 adopted
2010	Amendment of the Triglav National Park law
2013	First amendment of the Natura 2000 network after the biogeographical seminars
2015	Natura 2000 Management Programme (2015–2020) adopted by the government
	Second amendment of the Natura 2000 network after the bilateral biogeographical seminar
2020	The NEEP with measures for 2030 was adopted by the Parliament, containing an updated NNPP

See Table 3.3 for signatory dates of international conventions.

As in the rest of Europe, nature conservation ideas first appeared at the end of the nineteenth century. The beginning of a systematic approach was then disturbed by World War I. After the break-up of the Austro-Hungarian monarchy, in 1920 a group of naturalists at the Slovene Museum Society (an NGO) prepared a concise national nature conservation memorandum (Beuk, 1920). This was the first national strategy in the field of nature conservation, and was the foundation for the first protected area in the Dolina Triglavskih jezer (1924), as well as protection of species and caves (1922).

After World War II, an office for nature conservation was established within the National History Museum and gradually developed into an expert body. It was initially an independent consultative body within the government structure, and then under the Ministry of Culture after 1981.

The next milestone was the independence of Slovenia from Yugoslavia in June 1991, although no major changes in nature conservation administration initially occurred, as it had been a state rather than a federal-level responsibility. At the end of 1994, the responsibility for nature conservation moved from the Ministry of Culture to the Ministry of the Environment. Consequently, nature conservation departments within the regional cultural heritage offices were merged in 2002 to form a newly established Institute for Nature Conservation (Zavod RS za varstvo narave). Gradually, the present structure of competences was built (see Section 28.3.2).

After Slovenia became independent, international cooperation increased. At the beginning, IUCN with the Central and Eastern European Programme played an important role, as did BirdLife International. Contacts with international governmental organisations followed, such as the Council of Europe, UNESCO and UNEP. International treaties ratified by Yugoslavia were notified (see Table 3.3).

Independence also led to a long process of land restitution. Over a few decades, state ownership changed in favour of private ownership, which has had significant consequences for protected area establishment and management.

The accession process to the EU started in 1996, with accession negotiations underway from 1998. In 1999, intensive work on the legal transposition of the Nature Directives started, as did preparations for the identification of Natura 2000 sites. For SPAs under the Birds Directive, DOPPS (the Slovenian BirdLife partner) drafted a proposal based on the Important Bird Areas (IBA) inventory (Božič, 2003). A complex project was launched to collect data to identify proposed SCIs in accordance with the Habitats Directive, which was completed in 2001 (Kryštufek, 2001). In 2003, the Ministry of the Environment and Energy started the Natura 2000 project to prepare the proposal for Natura 2000 sites and to establish a broad consultation with stakeholders and the public. Most relevant specialists and organisations (scientific and NGO) were involved in the process, and all results were published on the Internet.[16] Whilst this created transparency, there were nevertheless some tense debates during the process.

The transposition of the Nature Directives required significant amendments to the Slovenian environmental and nature conservation legislation. This resulted in the adoption of an amended Nature Conservation Act[17] (NCA) in 2004, as well as a proposal for the Natura 2000 network. Slovenia joined the EU on 1 May 2004.

28.3.2 Organisation of Nature Conservation in Slovenia (Governance)

Governmental Bodies

The competent body for nature conservation in the Parliament is the Committee on Infrastructure, Environment and Spatial Planning.

At the executive level, nature conservation is organised as a department within the Directorate for the Environment at the Ministry of the Environment and Spatial Planning (MESP). The Ministry is responsible for guiding policy, implementing legislation, proposing legal instruments and administering the state budget for the environment. The Slovenian Environment Agency (ARSO) operates under the auspices of the Ministry. It carries out administrative tasks (legal procedures) and some reporting in the field of nature conservation. In 2021, administrative tasks were moved from ARSO to MESP. Another body under the MESP is the Inspectorate for the Environment and Spatial Planning, which is responsible for law enforcement.

The next level of governance is public institutions, with the Institute for Nature Conservation being the main national professional organisation in the field of nature conservation since 2002. It gives special attention to the most valuable natural aspects of nature, and the most threatened areas and species. It educates and raises awareness about the importance of nature conservation and the sustainable co-existence of man and nature, and carries out various national and international projects enhancing nature. It also prepares the reports required under the Nature Directives.

The Institute for Nature Conservation has seven regional units that provide high-quality coverage over the entire country. Their work is coordinated by a central unit, which ensures uniform methods and functioning of the Institute externally and internally. The Institute employs about 100 people, most of them university trained, predominantly from the spheres of biology, forestry, geography, geology and landscape architecture.

The Slovenia Forest Service (Zavod za gozdove Slovenije) is the competent authority for special forestry issues. Additionally, it monitors large carnivores and plays a leading role in their management, including evaluations of damage to livestock and advising on damage control and prevention measures.

The Fisheries Research Institute of Slovenia (Zavod za ribištvo Slovenije) performs public service activities in the fields of freshwater and marine fisheries. It also engages in commercial activities in the areas of sport and recreational fishing and fish farming, and various research and expert tasks in the field of fisheries biology.

With the use of CAP Rural Development Programme (RDP) agri-environment–climate measures, the importance and engagement of the Slovenian Agricultural Chamber (Kmetijsko gozdarska zbornica Slovenije) have increased. It has a network of agricultural advisers giving advice to individual farms. However, there is still a lot of work to be done to increase its nature conservation benefits in practice.

There are 11 public management authorities for larger protected areas (see Section 28.5.3). They carry out administrative and practical management, visitor management and inspections. They are financed mainly by the state and partly from their own sources, as well as projects. There is no coordinating body for these management authorities; however, the MESP closely follows the implementation of working programmes of state-established authorities. Each authority is guided by a council comprising delegates of the founder (government or municipality) and key stakeholders.

Non-governmental Nature Conservation Organisations

After independence was declared, some private enterprises started to work in different areas of nature conservation. Some specialised in environmental impact assessment and appropriate assessment as consultancies (e.g. Oikos, Limnos and Aquarius), whereas others were non-profit and dedicated to educational and public relations activities, public

---

[16] https://natura2000.gov.si/natura-2000/razvoj-nature-2000/strokovne-podlage (last accessed 12 April 2020).

[17] Zakon o ohranjanju narave: www.pisrs.si/Pis.web/pregledPredpisa?id=ZAKO1600

participation, or research and management in nature (e.g. Symbiosis, the Centre for Cartography of Fauna and Flora, and Lutra).

At the beginning of the nineteenth century, there were few active NGOs other than some mountaineering societies (Slovenian and German) and others supported by cavers, fishers and hunters. In 1934, the Natural History Society (Prirodoslovno društvo Slovenije) was founded and became the leading NGO in nature conservation for decades. Since 1979, there has also been a very active ornithological society, DOPPS (Društvo za opazovanje in proučevanje ptic Slovenije), which is the BirdLife International partner in Slovenia. It has over 1 000 members and supporters, as well as a small professional team.

A lot of new and specialised biological NGOs appeared after Slovenia became independent. These are relatively small (mostly 20–60 members) interest groups covering plants, bats, butterflies, amphibians and reptiles, and large carnivores. They have dedicated, technically qualified and experienced members, and focus their work on nature.

From 1964, a broad public discussion took place about a proposal for a dam on the upper Soča river near Bovec. In 1966, the decision was made against building the dam. Since then, the general public has responded to numerous cases where development projects might jeopardise natural assets (e.g. Cerkniško and Planinsko polje, Mura river and Volovja reber). Civil society has also become more engaged in nature conservation over recent decades. Umanotera (the Foundation for Sustainable Development) plays an important role in channelling the views of civil society. There is no regular coordination of NGOs in this field, but they may organise themselves in relation to specific issues (e.g. Mura river, Pohorje, Sava river and large carnivores).

Some NGOs have played an important role in triggering infringement procedures by the European Commission in relation to the Nature Directives. In line with the Aarhus Convention, NGOs are now more frequently taking advantage of the option to request information, and want to participate in national decision-making and court cases. NGOs can also ask for the special status of an organisation acting in the public interest with respect to nature, which allows them to take part in relevant decision-making procedures and provides advantages when applying for nature conservation projects. However, the government defines the conditions under which the status is granted. In 2021, more restrictive conditions were adopted and the number of NGOs with this status dropped from 48 in 2020 to 30 in 2021.

28.3.3 Legislation

The general obligation to take care of nature is stipulated in Article 73 of the Constitution: 'Everyone is obliged in accordance with the law to protect natural sites of special interest, rarities, and cultural monuments. The state and local communities shall promote the preservation of the natural and cultural heritage.'

The basic legal instrument for nature conservation in Slovenia is the 1999 NCA, which has since been amended several times.[18] It regulates two key areas: conservation of biodiversity, and valuable natural features (e.g. geological phenomena, caves, gorges, waterfalls, lakes and exceptional trees).

Biodiversity conservation measures refer to the protection of wildlife (plants and animals), including their genetic material, habitats and ecosystems, facilitation of the sustainable use of biodiversity components and the maintenance of natural balance. The NCA regulates the protection of species (see Section 28.4) and their habitats, and establishes the ecological network and system of protected areas (see Section 28.5). The Act also lays down the responsibilities of the state and local communities. It is important to stress that landowners must be able to demonstrate that required ecological and social functions (such as in forest plans) are being met on their property, including by allowing free public access.

The component of the NCA dedicated to valuable natural features addresses the procedure for their legal evaluation and designation, competences (state or local importance), conservation measures (contractual protection, protection, temporary protection and restoration), as well as their protection and development guidelines. There is a general obligation that biodiversity conservation measures and the system for the protection of valuable natural features are integrated into spatial planning and the use and exploitation of natural assets.

The majority of the EU Nature Directives' obligations were transposed into the NCA through amendments in 2004. This required changes to the legal system (e.g. for appropriate assessments and the Natura 2000 network), which would probably not have been adopted by the Parliament if they were not part of the accession process. The NCA is related to, and harmonised with, other areas that have an influence on nature: forestry, water management, planning and housing, infrastructure, etc. Therefore, it is important to stress that other sectoral legislation has an important role in reaching nature conservation goals.

Although the NCA has a very solid legal base, the implementation of such a powerful instrument requires substantial human and technical capacity, as well as solid knowledge about biodiversity. Unfortunately, this is lacking. Furthermore, not all foreseen by-laws are adopted, and thus some parts of the NCA are inoperative. Law enforcement is also relatively weak as the capacity of the inspectorates and nature protection wardens is generally low. However, there have been some good examples of targeted investigations, such as in relation to illegal trapping of cave invertebrates. Although those responsible were not identified, the action had an important deterrent effect.

[18] Zakon o ohranjanju narave (Uradni list RS, št. 96/04 – officially consolidated text, 61/06 – ZDru-1, 8/10 – ZSKZ-B, 46/14, 21/18 – ZNOrg, 31/18, 82/20 and 3/22 - ZDeb): www.pisrs.si/Pis.web/pregledPredpisa?id=ZAKO1600; English translation: www.pisrs.si/Pis.web/npbDocPdf?idPredpisa=ZAKO6877&idPredpisaChng=ZAKO1600&type=doc&lang=EN (last accessed 28 March 2022).

28.3.4 Strategic Documents

The first strategic document for nature conservation in Slovenia was the *Memorandum* (Spomenica) by Beuk (1920). Since then, the basic goals have not changed substantially but were broadened in line with international developments (e.g. with more emphasis on biodiversity, habitat conservation and ecological networks).

The NCA requires the *National Nature Protection Programme* (NNPP), covering the conservation of biodiversity and the protection of valuable natural features, which is adopted by the National Assembly of Slovenia. It is a 10-year programme integrated with the *National Environmental Protection Programme* (NEPP), which determines, based on a nature conservation status assessment, the goals and guidelines for:

- biodiversity conservation by means of the programme of measures for the protection of plant and animal species and their habitats and ecosystems;
- the protection of valuable natural features by means of the programme for the establishment of protected areas and the restoration of valuable natural features;
- the manner of fulfilling international obligations;
- education in the field of nature conservation;
- public awareness of the importance of nature conservation; and
- the provision of financial resources for nature protection.

The first NNPP was adopted within the framework of the *National Environment Protection Programme* in 1999,[19] followed by the Resolution on the *National Environmental Action Plan 2005–2012* in 2005.[20] The Resolution incorporated content of the *National Biodiversity Strategy* (MESP, 2002) and was based on a review document (Hlad and Skoberne, 2001).

Following a gap between plans (when the previous plan applied), the current NEPP with programmes of measures for 2030 was adopted by the Parliament in 2020,[21] following discussions involving civil society. This includes a chapter with a revised NNPP, which follows global (e.g. Aichi targets) and EU strategic documents (e.g. the *EU Biodiversity Strategy for 2030*). The NEPP also includes an annex with a *Strategic Plan of Biodiversity Conservation in Slovenia,* [22] which is an action plan for the implementation of the NNPP.

A strength of the NEPP is that it links national programmes for the environment, nature conservation and water management, but its implementation has been mostly weak because of a lack of coordination.

[19] www.pisrs.si/Pis.web/pregledPredpisa?id=NACP5

[20] http://pisrs.si/Pis.web/pregledPredpisa?id=RESO41

[21] www.pisrs.si/Pis.web/pregledPredpisa?id=ODLO1985

[22] www.uradni-list.si/_pdf/2020/Ur/u2020031.pdf (pp. 1840–1849).

28.4 Species Protection and Conservation Measures

In terms of wildlife, there is a general protection regime in the NCA for all native species, such that it is prohibited to destroy a plant or animal species, to reduce their populations or habitats, or to otherwise worsen their living conditions to such an extent that they become threatened. In addition, plants and animals and their habitats cannot be killed or harmed intentionally without a justified reason. Although violation of these general rules is regarded as a serious offence, they are not applied in practice.

Under the NCA, the minister responsible for nature conservation may intervene and prescribe the least disturbing manner of, and conditions for, an activity affecting nature. Sustainable use of plants and animals is possible under certain conditions. The introduction of plants or animals of non-indigenous species is prohibited.

Threatened plant or animal species are legally defined by ministerial regulation (based on the Red List) with a list of measures (e.g. physical protection, providing suitable places for breeding, feeding and wintering, re-population, and re-introduction). One of the possible measures is legal protection of the threatened plant or animal species beyond that conferred by the general NCA provisions. There are two governmental decrees: one on protected animal species[23] and one on plants.[24] They transpose all requirements for strict protection according to the Nature Directives and other international treaties (e.g. the Bern Convention). Besides strict protection of the species, there is an obligation to maintain their habitat. The Decree on protected wild animal species also gives protection to all beetles and butterflies living above the forest line, as well as all troglobiotic species.

The state is liable for damage caused by protected animals when it cannot be prevented by obligatory preventive measures. Procedures for damage compensation are set by by-laws.

The MESP is the competent authority for protected species, whereas the Ministry for Agriculture, Forestry and Food is responsible for hunting and fishing and domestic livestock.

Over the last few decades, the populations of Brown Bear and Wolf in Slovenia have spread in line with their strict protection. The estimated population of the bear in spring 2020 was 993 individuals (Jerina *et al.*, 2020), with very high densities in some areas (>0.5 individuals/km^2). In the 2020–2021 season, the total Wolf population was estimated at around 138 individuals in 12 Wolf packs, 2 of which are transboundary (Bartol *et al.*, 2021).

[23] Decree on protected wild animal species (Official Gazette, 46/04, 109/04, 84/05, 115/07, 32/08): www.pisrs.si/Pis.web/pregledPredpisa?id=URED2386

[24] Decree on protected wild plant species (Official Gazette, 46/04, 110/04, 115/07, 36/09 in 15/14): www.pisrs.si/Pis.web/pregledPredpisa?id=URED3192

As Slovenia has no large wilderness areas, these large carnivores' habitat is shared with people and their co-existence depends on their social acceptance. This is a very dynamic, fragile process and requires a sophisticated management system with trust-building communication. Management of large carnivores includes various measures to prevent and minimise damage for stock breeders, compensation systems, and intervention groups, as well as the possibility of limited and strictly controlled population reductions in accordance with the HD Article 16 derogations. With support from LIFE projects, the management and monitoring of the Brown Bear and Wolf have improved greatly over the last decade. Despite this and the very good data and positive trends in the populations of both species, discussions on their population regulation remain controversial, rather emotional, and politically sensitive. Decisions on using derogations are regularly legally challenged in the national court.

28.5 Protected Areas and Networks

28.5.1 Protected Area Types and Their Coverage

Legally Recognised Nature Areas

According to the NCA, there are various possibilities for legally recognising important nature at specific sites. Legal recognition is a condition to use legal measures (e.g. protection). Outstanding natural phenomena – such as old trees, solitary rocks, gorges and waterfalls – are legally recognised as **Valuable Natural Features**. In addition, as a matter of law, all caves are Valuable Natural Features of national importance under the Underground Cave Protection Act.[25]

Areas of a habitat type, or parts of a larger ecosystem unit that significantly contribute to biodiversity conservation, can be designated as an **Ecologically Important Area**. These can be designated if they are: areas of habitat types that are exceptionally biodiverse or well-preserved, where there are habitats of threatened or endemic plant or animal species and habitats that are internationally important (according to ratified international treaties), or which in any other manner contribute to biodiversity conservation; areas of a habitat type or large ecosystem unit that significantly contribute to natural balance by being evenly biogeographically distributed with regard to other ecologically important areas, and contribute to the ecological network; habitats of protected animal or plant species; animal migration routes; or areas that significantly contribute to the gene flow among populations of plant or animal species. These areas form a national ecological network, which covers 52% of the national territory.[26]

The legal consequence of recognition as an Ecologically Important Area is advisory, mostly in relation to spatial and sectoral planning. However, as the areas are primarily identified according to expert opinion, rather than data, their actual influence in legal procedures is rather low. Nevertheless, the recognition of the areas could be very useful for nature conservation if used consistently. As they have no defined legal regime, they are not considered in the IUCN category system.

Designation as a protected area provides the most important measure for the protection of some Valuable Natural Features and Ecologically Important Areas. Their main purpose is to maintain biodiversity and Valuable Natural Features through a legal regime. The NCA establishes the types of domestic protected areas, and related responsibilities, procedures, conservation measures, guidelines for different use and management requirements. On that basis, a protection act is adopted for each protected area. Ecologically Important Areas that are needed to maintain HD species and habitats are additionally designated as Natura 2000 sites, and therefore benefit from their strong legal protection.

Legally Protected Areas

Table 28.4 lists the current protected area types that generally meet IUCN and CBD criteria. Domestic categories of protected areas generally correspond to the IUCN management categories; however, this is not always the case. For a reliable comparison each protected area should be assessed individually in relation to its management goals.

Protected areas form two groups: small-scale protected areas, which comprise Natural Monuments, Strict Nature Reserves and Nature Reserves; and large-scale protected areas, which are National Parks, Regional Parks and Landscape Parks. Small-scale protected areas can be nested within large-scale protected areas.

A **Strict Nature Reserve** is an area of naturally preserved geotopes, habitats of threatened, rare or representative plant or animal species, or an area important for biodiversity conservation where natural processes occur without human influence. All activities, including access, are prohibited. Currently there is only one very small Strict Nature Reserve in Slovenia.

Nature Reserves are areas of geotopes, habitats of threatened, rare or representative plant or animal species, or an area important for biodiversity conservation that is maintained through sustainable human activity. Only activities that maintain the character of the reserve are allowed.

Natural Monuments are areas containing one or more Valuable Natural Features that have an outstanding form, size, content or location, or that are a rare example of a Valuable Natural Feature. Activities that might adversely affect the state of a Valuable Natural Feature, or reduce its aesthetic value, are prohibited. All other activities, including visits, are allowed.

A **National Park** is a large area possessing numerous Valuable Natural Features and high biodiversity. Nature in its near-original state, with intact ecosystems and natural processes, is present in the majority of the park (i.e. wilderness areas). Smaller areas with human influence may occur, but in harmony with nature. At least two protection zones are defined, including a core zone with no economic activities,

[25] Underground Cave Protection Act (Official Gazette, 2/04, 61/06 – ZDru-1, 46/14 – ZON-C and 21/18 – ZNOrg): www.pisrs.si/Pis.web/pregledPredpisa?id=ZAKO2068

[26] Urša Mežan, Slovenian Environmental Agency (ARSO), personal communication (17 September 2020).

Table 28.4 The number and extent of international designated sites and protected areas in Slovenia.

	IUCN category	Number	Area (km^2)
Designated under global and non-EU international legislation			
Ramsar Sites	IV, V	3	82
Natural and Mixed World Heritage Sites	III, Ia	2	12
Biosphere Reserves (exc. buffer and transition zones)	V	4	975
Designated under the Nature Directives			
SCIs (terrestrial + marine)	IV	324	6 637 + 1
SPAs (terrestrial + marine)	IV	31	5 069 + 5
Natura 2000 terrestrial area			7 676
Natura 2000 marine area			6
Designated under domestic legislation			
Strict Nature Reserve (Strogi Naravni Rezervat)	Ia	1	0.2
Nature Reserve (Naravni Rezervat)	Ia (IV)	56	5 639
Natural Monument (Naravni Spomenik)	III	1 161	?
National Park (Narodni Park)	II/V	1	840
Regional Park (Regijski Park)	V	3	430
Landscape Park (Krajinski Park)	V	46	1 251

Note. ? = Area unknown as only point data available.
Sources. See Section 1.3. Based on data from: Ramsar Convention and UNESCO websites (January 2022); Kirchmeir and Kovarovics (2016); SCIs and SCAs, M. Petkovšek, ZRSVN, personal communication (May 2022); National designations, Ministry of the Environment and Spatial Planning, ARSO, personal communication (January 2021).

although recreation and visitors are allowed. There is currently only one in Slovenia (Triglavski narodni park). Its protection dates back to 1924, when the Alpine Conservation Park was founded, and to 1981, when the law on Triglav National Park defined the park's present borders. Currently, there are no proposals for another National Park.

Regional Parks are extensive areas of ecosystems and landscapes characteristic of the region, with large areas in a natural state and areas of Valuable Natural Features interwoven with parts of nature where human influence is relatively substantial but in harmony with nature.

Landscape Parks are areas with an emphasis on the high-quality and long-term mutual interaction of people and nature and with high ecological, biotic and landscape value (cultural landscape).

Domestic terrestrial protected areas according to the NCA covered 2 701 km^2 in June 2021, about 13% of the Slovenian land area.[27] In combination with 7 676 km^2 of Natura 2000 sites (Table 28.4), the total of protected areas in line with the IUCN definition was about 8 048 km^2 in March 2022, giving 40% coverage.[28]

As the total Slovenian coastline is about 40 km, the country's marine area is not very significant. Within its marine waters there are very small domestic protected marine areas (most of which are an integral part of an adjoining terrestrial protected area), which comprise about 3 km^2, 3.1% of the marine waters.[29] Complementing this, the whole sea under Slovenian jurisdiction is designated as an Ecologically Important Area.

Slovenia has designated substantial areas as SCIs and SPAs, which cover 33% and 25% of the land area, respectively. The terrestrial and marine network is just about complete, as there are only some minor insufficiencies due to some gaps in knowledge (sufficiency index 2017: 97%; Petkovšek, 2017). Together the Natura 2000 sites cover 38% of the land area, which is currently a greater proportion than any other EU Member State (Table 5.4). Most of the Natura 2000 network is covered by forest (about 70% in 2016), followed by grasslands (13%). The marine component of the Natura 2000 network covers about 6 km^2, which is 2.8% of Slovenia's marine waters.

About 18% of the Natura 2000 network in Slovenia overlaps with domestic protected areas (Petkovšek, 2017). This means that the Natura 2000 sites have an important role in

[27] http://kazalci.arso.gov.si/en/content/protected-areas-9 (last accessed 15 April 2022).

[28] http://kazalci.arso.gov.si/en/content/nature-areas-under-protection-1 (last accessed 15 April 2022).

[29] B. Vidmar, ZRSVN, personal communication (May 2022).

ensuring the management of those areas (see Section 28.5.3). However, as the majority of the Natura 2000 network is outside the domestic network, the need to implement the Nature Directives has led to a considerable expansion in the overall protected area network.

Additionally, there are designated areas of international importance (Table 28.4). Of particular importance are the two World Heritage Sites: Škocjanske jame (designated for its caves, huge underground canyons, other karst phenomena, and archaeology), and old beech forests (Snežnik-Ždrocle, Krokar). These international sites overlap considerably with other protected area categories.

Lastly, forest reserves under the forestry legislation should be mentioned, even though they are not part of the protected areas system. They are designated under the Decree on protective forests and forests with a special purpose[30] and form a network of forest plots excluded from management and dedicated to research.

28.5.2 The Coherence of Protected Areas and Ecological Networks

Protected areas were established rather spontaneously, following grassroots initiatives, proposals by experts within nature conservation authorities and, in some cases even, political interests. Despite this rather unsystematic approach, most of the best-preserved and important areas for nature conservation are included in the network. Whilst some places need protection, a substantial increase in new protected areas is not expected.

The processes of defining Ecologically Important Areas and consecutively the Natura 2000 network were different, as the latter was based on analysis of data on BHD habitats and species. In this case, their coherence was considered and addressed through the biogeographical seminar process (see Section 4.2.3). The approach in Slovenia was to designate large areas for highly mobile species (e.g. birds and large carnivores) and for dynamic ecosystems to accommodate their natural requirements.

The structure of the Slovenian landscape still supports ecological connectivity between sites for most plants and animals, so the main nature conservation challenge is to maintain it. However, special emphasis, for instance, is given to fish corridors in the case of planning dams, and large carnivores with respect to major infrastructure projects. These issues can be dealt with in the planning of projects, but it is much more difficult to halt the loss of connectivity elements in the agricultural landscape as the agricultural policy measures, such as cross-compliance conditions on CAP payments, have been ineffective.

[30] www.pisrs.si/Pis.web/pregledPredpisa?id=URED6654

28.5.3 Approaches to Protection and Manageme

The NCA defines different levels of protection, from general very specific, that are set by by-laws that deal specifically w biodiversity (species and habitats), Valuable Natural Featu and protected areas. The general protection measures appli to each type of protected area are as described earlier.

The idea is that other fields of legislation (environme water management, hunting, fishing, agriculture, forest housing and planning, etc.) complement nature conservati measures. In practice, implementation is not ideal, but t principle of integration of nature conservation requiremen into other sectoral policies has a long-term perspective.

Approaches to Managing Domestic Protected Areas

The management type of each protected area is individual defined in its protection act. In compliance with the protectic act, the authority that established the protected area ma directly manage the area through a unit or through a publ institution established for this purpose, entrust the manag ment to an appropriate existing public institution, or gran a concession.

For most larger protected areas, a public managemen authority has been established by the government or loca community (11 in 2020). There is no central body for coordin ating the management of protected areas. However, all man agement institutions are part of a Community of the Park (Skupnost parkov), a voluntary association for coordination o views and cooperation.

Each protected area is supervised by a council on which there are representatives of the designation authority and ke stakeholders. The composition and terms of reference of th council are defined by the protection act.

Management authorities have competence exclusively fo implementation of the protection act. All other responsibilities, such as for forest management, are implemented by other authorities (e.g. Slovenia Forest Service). However, they have to include protected area restrictions in their plans (e.g. forestry management plans) and the relevant protected area authority takes part in the process. The Institute for Nature Conservation gives consent to the draft plan. The real challenge is implementation of the plan in the field.

Management plans are obligatory for National Parks and Regional Parks and recommended for Landscape Parks. The protected area authorities prepare a management plan for a 10-year period, which is adopted by the council and government or local community. Public participation is obligatory, and the format of the management plan and adoption procedure is set out in the NCA. Most of the protected areas with management authorities have adopted their management plans.

Unless there is a particular need, most other domestic protected areas have no active management, but all must be considered in planning and permitting procedures.

In 2009, a report on the effectiveness of protected areas (Kus Veenvliet and Sovinc, 2009) was prepared, which used the WWF RAPPAM methodology (Ervin, 2003) and assessed nine large-scale protected areas (covering 78% of the surface of protected areas). The report emphasised implementation weaknesses, mostly due to insufficient personal and financial capacities (low political profile), and laid down a set of recommendations to enhance management effectiveness.

Approaches to Managing Natura 2000

As more than one-third of Slovenia is included in the Natura 2000 network, it is not practical to manage these sites in the same way as domestic protected areas. Instead of a management plan for each site, an integral (master) Management Programme for Natura 2000 is prepared and adopted by the government for a period of seven years, corresponding to the EU's Multiannual Financial Framework (MFF) period. The first programme was adopted in 2007 (for 2007–2013), the second in 2015 (for 2014–2020),[31] and the third one is under preparation.

The essence of the programme is detailed elaboration of conservation objectives for each species and/or habitat type with a significant presence in each Natura 2000 site. Each is linked to specific measures with identified responsible actors. There are several types of management approach, depending on the nature of the actors. Authorities concerned with forestry, fishing, hunting and water management must include Natura 2000 measures in their plans and implement them regardless of the ownership. Agricultural management is dealt with by using the Management Programme to tailor RDP agri-environment–climate schemes to BHD species and habitat types. Nature conservation bodies, especially protected area management authorities, must include their allocated specific measures in their annual working programmes and report on them.

The Management Programme also lists project needs, as well as plans for monitoring and research. In this way the Programme is used to identify suitable financial mechanisms (e.g. see EU funds in Section 28.7.3), giving a basis for arguments for their need and possible prioritisation, as well as the drafting of the Prioritised Action Framework (PAF) for Natura 2000.

The integrated approach to management is in line with the 'Rio' spirit: every subject is responsible for biodiversity conservation. The burden of implementation of measures is distributed between actors. However, the process is sophisticated and orientated towards the long term. There are currently gaps in implementation, and differences between sectors. The best experience is in forestry as the forestry planning and management process is traditional and well-developed. The situation is improving regarding water management, as the WFD is already streamlining activities in the same direction. The most challenging situation is in agriculture, because the implementation of measures identified in the Management Programme is dependent on the take-up of suitable agri-environment–climate schemes, which are voluntary for farmers. As discussed further in Section 28.6.2, their uptake of the schemes has been too low to achieve significant results.

Another challenge of the Management Programme is its complexity. There are many actors that need to be involved in implementing the Programme, yet many of them are unaware of their obligations. Therefore, intense and constant communication (trust building) is needed to implement such a partnership approach.

For communication with stakeholders, it is of utmost importance that nature conservation reasoning is simple, concrete and understandable. The same is true for measures that stakeholders must accept and implement. The Programme attempts to take this approach by describing goals and measures precisely, but it is very demanding to do this in a simple, measurable, realistic and reasonable way. Thus, the quality of the Programme and the level of successful implementation in the field directly depend on the quality of the regular monitoring and knowledge of experts involved. Nevertheless, the integrated approach seems to be the best long-term option, while management of every Natura 2000 site through some administrative bodies is not feasible in Slovenia.

28.6 General Nature Conservation Measures

28.6.1 Development Planning and Impact Assessment

The basic strategic document for planning is the *Spatial Planning Strategy of Slovenia*, adopted by the government in 2004.[32] Since 2015, a new spatial development strategy until 2050 has been in preparation. As the process is participatory and complex, its adoption is foreseen in 2022.

In the framework of the principles of the *Spatial Planning Strategy*, large projects of state interest are included in a particular plan (state spatial plan) and local communities must adopt their own spatial plans in line with the Strategy. Each of these plans must also be assessed regarding natural assets (natural values, domestic protected areas, Natura 2000 and ecologically important areas). The effectiveness of protection depends on the quality of a particular plan and its implementation.

Impact assessments are integrated at all levels of assessments: planning level through SEA, project level through an EIA, and for activities requiring the consent of nature authorities or a permit. In general, the assessment procedures are carried out in accordance with the provisions of the Environmental Protection Act. Specific details regarding nature are also set out in the NCA and under a special by-

31 www.natura2000.si/natura-2000/life-upravljanje/program-upravljanja/

32 Ordinance on Spatial Planning Strategy of Slovenia (Official Gazette, 76/04, 33/07 – ZPNačrt in 61/17 – ZUreP-2): www.pisrs.si/Pis.web/pregledPredpisa?id=STRA12

law.[33] In this way, appropriate assessments according to HD Article 6(3) are regulated. Besides this, possible negative influences on domestic protected areas' goals and protected species are assessed.

All types of impact assessment on nature are carried out by specialised expert groups (mostly private companies). Although no special licence is needed, a university biological background and five years of experience in environmental assessments on nature are required by law. Nature impact assessments are verified in the planning or permit adoption process by the Institute for Nature Conservation, and the final decision is given by a competent authority.

In general, the legal framework for impact assessments is appropriate, however fine-tuning based on experience is needed. There is scope to speed up the process in cases where negative impacts are unlikely, as well as strengthening the quality control of assessments and requirements for assessment providers. Another weakness concerns the quality of data and their interpretation. Whilst the available data on habitats and species may have been adequate for designating an area, they may not be sufficient to assess potential impacts. This can lead to incorrect assessments or the excessive risk-averse use of the precautionary principle. Additionally, most of the available data are for BHD species, and therefore species that are only nationally protected tend to be neglected.

In some earlier cases (e.g. for the Škofljica motorway), positive HD Article 6(3) appropriate assessments for projects affecting Natura 2000 sites were given because they included extensive compensation areas for potential damage, which were treated as mitigation measures. This avoided the need to show that potentially damaging projects met all the HD Article 6(4) requirements to go ahead, including being of overriding public interest and the absence of alternatives. However, this changed following court decisions (notably in the Netherlands[34]) that clarified the distinction between mitigation and compensation measures, and that the latter should not be considered as part of the appropriate assessment.

As the HD appropriate assessment procedure is one of the most important tools to conserve Natura 2000, whilst reducing business risks for investors, more effort is needed to strengthen it, including improving the quality and quantity of monitoring, public databases on species and habitats, and the quality of assessment providers.

28.6.2 Habitat Management and Restoration

Forest Management

Most of the forest area is planned for sustainable production (91%), with just 8% protected to prevent erosion and 0.8% in forest reserves dedicated for research.[35] In the last two categories there is no regular human interference. The Slovenian forestry tradition is based on the principles of close-to-nature management, using selective cutting and limiting exploitation to such a degree and in such a way as to ensure the preservation of all natural forest constituents. The multifunctionality of forests (ecological, social and economic) is also legally recognised, clearcutting has been forbidden since 1948, and plantations of non-native species are not allowed. Unlike many countries, the management rules are the same for private and state-owned forests, framed by the Forest Law.[36]

The basic strategic document for forest management is the *National Forest Programme* (NFP), which follows a tradition from the nineteenth century and was most recently adopted by the Parliament in 2007.[37] Based on this, the competent ministry prepares five-year operational programmes and reports to the Parliament on their implementation. For each of 14 areas, the Slovenia Forest Service prepares a draft 10-year plan, which must consider nature conservation. These are adopted by the government after a thorough consultation phase, and require consent from the Institute for Nature Conservation. The obligations of the Forest Act for all owners include maintaining and improving the conservation status of Natura 2000 sites. Some measures, such as allowing natural succession and retaining trees as habitats for certain species, are financed by the National Forest Fund (see Section 28.7.3).

A major challenge to forest management in Slovenia is that 80% of the forest is under private ownership, with 413 000 forest owners on 286 000 forest estates (2020).[38] Therefore, forest management involves a huge number of people, except in the 20% managed by a state company – the Slovenian State Forests (Slovenski državni gozdovi). Furthermore, as private forest estates are small, averaging only 3 ha (and often fragmented), it is difficult to achieve their conservation goals as well as having profitable timber production.

Agri-environment Schemes

As mentioned earlier in relation to the Management Programme for Natura 2000 sites, the management of agricultural land has been highly dependent on RDP agri-environment schemes, which are financed through the RDP agri-environment–climate measure (see Section 28.7.3 for funding levels).

During the 2014–2019 programming period, the agri-environment–climate measure was used to support four schemes dedicated to the maintenance of BHD habitats and species in Natura 2000 sites: special grassland habitats,

33 Rules on the assessment of acceptability of impacts caused by the execution of plans and activities affecting nature in protected areas (Official gazette, 130/04, 53/06, 38/10 and 3/11): www.pisrs.si/Pis.web/pregledPredpisa?id=PRAV5539

34 Case C-521/12.

35 www.gozd-les.com/slovenski-gozdovi/statistika-gozdov/povrsina-gozdov (last accessed 20 April 2020).

36 www.pisrs.si/Pis.web/pregledPredpisa?id=ZAKO270

37 www.uradni-list.si/glasilo-uradni-list-rs/vsebina?urlurid=20075510

38 www.zgs.si/eng/slovenian_forests/forests_in_slovenia/forest_ownership/index.html (5 April 2021).

grassland habitats of butterflies, habitats of birds of humid extensive meadows, and litter meadows. Other schemes that contributed to the conservation and enhancement of biodiversity in agricultural landscape were for: steep meadow habitats, hummocky meadows, livestock rearing in areas with large carnivores, mountain pasture, old extensive meadow orchards with trees with tall trunks, and the preservation of hedges.

Analysis of the outcomes in relation to the schemes' objectives showed that, since 2019, the objectives were achieved in only five areas, giving only a 5% success rate (MKGP, 2019). The reasons for this were, mainly: poor enrolment of farmers in the schemes offered; inadequate financial evaluation of the targeted nature conservation operations of the schemes; lack of compatibility of species' ecological needs and the farmers' expectations regarding land management; and lack of technologies and knowledge about sustainable agricultural use. Despite efforts to transfer knowledge and provide advice, farmers were also put off by the additional administration and controls, insufficient knowledge of the content and importance of individual operations, length of contractual obligations, complex instructions and the system of penalties for compliance infringements (MKGP, 2019).

The economic disparities and inconsistency in agricultural policy have been the main reasons for the rapid disappearance of species-rich meadows in some Natura 2000 sites (e.g. Ljubljansko barje, Goričko and Šentjernejsko polje) despite CAP greening requirements. This is primarily the result of ploughing grasslands and intensifying their use. In 2019, 21 784 ha (28% of grasslands in Natura 2000) were defined as Environmentally Sensitive Permanent Grasslands (MKGP, 2019). However, many of these grasslands are small and scattered and were not covered by the CAP measures. There was an exception for organic farmers and small farmers regarding ploughing grasslands.

28.7 Nature Conservation Costs, Benefits and Funding Sources

28.7.1 Nature Conservation Costs

Estimations of all the costs related to nature conservation, including communication and education needs and covering all objectives, are extremely difficult to obtain and not yet available. However, estimates of the costs of implementing the Natura 2000 network, and related green infrastructure, have been prepared within the European Commission's PAF. The Slovenian PAF for 2021–2027 (MESP, 2019) estimates the total Natura 2000-related costs for the period to be €59–88 million per year on average. However, it should be borne in mind that whilst the Natura 2000 network will help many species other than those targeted by the Nature Directives, not all nationally important species are covered; therefore, their protection will incur additional costs.

28.7.2 The Economic Benefits of Nature Conservation

As elsewhere, the economic value of nature conservation was underestimated in Slovenia in the past. Especially since the launch of the study on *The Economics of Ecosystems and Biodiversity* (TEEB), more systematic attention has been given to this topic, particularly to the indirect benefits. In Slovenia, some projects have assessed ecosystem services in protected areas, wetlands and forests, but there is no comprehensive state-level review covering all ecosystems and services.

In general, there remains a lack of awareness amongst decision makers and the business sector regarding the benefits of nature. They are taken for granted, with no consideration that decisions can cause the deterioration of ecosystem services and biodiversity loss. This is despite the fact that nature is extremely valuable for people's quality of life in Slovenia, where many enjoy mountaineering, trekking, gathering plants and mushrooms, cycling, running, fishing, photography, etc. Furthermore, the green image of Slovenia, and its protected areas, is promoted by the tourist industry, but there is no system for using revenues to conserve nature and maintain visitor infrastructure in protected areas. Further strengthening of the economic perspective of nature is therefore needed, as economic arguments are more understandable by the general public.

28.7.3 Funding Sources for Nature Conservation

Table 28.5 provides a summary of the main sources of funding used for nature conservation in Slovenia.

The state budget for nature conservation is approximately €8 million per year and is administered by the MESP. It includes core financing of the Institute for Nature Conservation, public management bodies of protected areas, and co-funding of Natura 2000-related activities of other public bodies (e.g. the Fisheries Research Institute of Slovenia, and Slovenia Forest Service). It also includes about €1 million for annual salaries of employees at the MESP and bodies within the ministry working on Natura 2000 tasks.

The other main source of public national financing for nature is the Forest Fund, which was established in 2016 under the Management of State Forests Act.[39] The fund is mostly financed from the sale of state forests and payments made by Slovenian State Forests, the company that manages state forests. The company pays the state an annual fee of 20% of the revenues from the sale of timber from the state forests. According to the law, the funds are used for measures in private forests in Natura 2000 sites in accordance with the Natura 2000 Management Programme and the Forest Investment Programme.

Only a small proportion of the state budget financing can be used for practical conservation measures, regular

[39] www.pisrs.si/Pis.web/pregledPredpisa?id=ZAKO7109

monitoring or applied research. Such activities and investments are therefore mostly dependent on project financing, primarily from EU-funding instruments (Table 28.5). The largest source of funding has been the European Regional Development Fund (ERDF), which has for example been used for restoration projects, building of corridors (e.g. green bridges for large carnivores) and educational infrastructure in Natura 2000 sites. The LIFE Programme has also been particularly important for implementation of the Nature Directives and especially Natura 2000. Since the first projects in 2000, up to March 2021 there have been 26 traditional projects awarded to Slovenia in the priority area Nature and Biodiversity, with a total investment of €62 million (60% from the EU). There is also currently one LIFE integrated environment project over 2014–2026 enhancing the implementation of the Natura 2000 management programme, with a total budget of €17 million (€10.2 million EU contribution).

Over the 2014–2020 programme period, Slovenia also took part in four cross-border Operational Programmes with all neighbouring countries (Croatia, Italy, Austria and Hungary) and also in five transnational cooperation areas (Alpine space, Adriatic Ionian, Central Europe, Danube Area and Mediterranean area), as well as in Operational Programme Interreg Europe. Altogether, Slovenia participated in 59 international projects, which contributed to the maintenance and restoration of BHD habitats and species, the integrity of Natura 2000 sites or the coherence of the network. Total funding from the projects that were focussed on the implementation of EU nature policy and associated green infrastructure (Priority 6d) amounted to c. €12 million (85% EU contribution).

In addition to funding projects, EU funds are also by far the most important source of financing for habitat management and ongoing restoration, primarily through European Agricultural Fund for Rural Development (EARFD) RDP measures. Funds allocated to measures under Priority 4 (i.e. restoring, preserving and enhancing ecosystems) are indicated in Table 28.5, the largest proportion of which go to payments to farmers in areas facing natural and other constraints. Whilst these payments have no specific environmental requirements, they help to maintain some important HNV farming systems (MESP, 2019). The most important measure for maintaining biodiversity is the agri-environment–climate measure, which had an allocation of nearly €30 million per year. As mentioned earlier, this supports a range of key habitat management and species conservation schemes, which unfortunately have not been sufficiently taken up. The forestry measure (M8) has been used (e.g. for forest birds), but with only €1.1 million allocated per year under the RDP Priority 4. According to the PAF, a further €5.5 million per year under other RDP priorities is benefiting Natura 2000. Despite their availability, both the forest environment climate measure (M15) and the Natura 2000 compensation measure (M12) were not used. Thus it appears that Slovenia has not made full use of all the RDP funding opportunities.

In broad terms, the total annual amount of funding allocated over the last funding period for measures that could contribute to nature conservation (c. €104 million) appears to be adequate for Natura 2000 (€59–88 million) based on the recent PAF (MESP, 2019). However, most of the available funds were allocated for areas facing constraints, with uncertain biodiversity benefits, and for agri-environment–climate measures, with low take-up. Furthermore, whilst a lot was achieved through project funding, such funding remains unstable and uncertain and may not correspond to priorities or be sustainable in the long term. The reliance on project funding means that the priorities are highly influenced by project opportunities and not necessarily by nature conservation needs. As a result, it is difficult to manage the general policy and secure long-term objectives.

Irrespective of the total funding available, the core state budget for nature is far too slim to properly implement legislation, access all potentially available external funding, and ensure the long-term viability of monitoring, data support and public services. Existing EU financial mechanisms do not support such functions and cannot make up for the particularly costly obligations of countries such as Slovenia with high proportions of the EU's Natura 2000 network.

The available funding also appears to be insufficient for wider nature conservation objectives (as well as geology and geomorphology) as financial priorities are focussed towards BHD species and habitats.

28.8 Surveillance, Monitoring, Reporting and Applied Research

28.8.1 Monitoring and Reporting

There is a well-established tradition of classical biological sciences although this has been very fragmented between various research institutions, NGOs and individual experts. More systematic collection of data, needed for the conservation of species and habitats, started in 2001 to help prepare for EU accession and the requirements of the Nature Directives. The ornithological community was best able to respond to this need as an ornithological atlas had been completed (Geister, 1995), since updated (Mihelič, 2019).[40] Some bird monitoring schemes were also already in place, including the International Waterbird Census (since 1988), and covering farmland birds and selected BD Annex I species (Denac *et al.*, 2015).

To fill knowledge gaps, regular population monitoring is also now organised for HD Annex II species, following the first surveys that provided baseline information. Monitoring protocols have been developed for each of 132 species (e.g. bats, beetles, butterflies, dolphins and large carnivores), which are adapted to the available resources and specifically define the monitoring areas and frequencies. For example, transect monitoring data are available for some butterflies (Butterfly Society:

[40] https://atlas.ptice.si/atlas/index.php?r=site/page&view=about

Table 28.5 Funding sources for nature conservation in Slovenia.

Source and period	Measure/category	Average annual allocation (€ m)		Comments
EU funding instruments		**EU + national**	**EU**	
EAFRD Rural Development Programme measures – allocations to Priority 4 (restoring, preserving and enhancing agriculture and forest ecosystems) 2014–2020	M13 Areas facing constraints	36.1	27.1	See text
	M10 Agri-environment–climate	29.6	22.2	Low uptake
	M11 Organic farming	9.4	7.1	
	M8 Investments in forests	1.1	0.8	See text
	M16 Cooperation	0.8	0.6	
	M2 Advisory services, etc.	0.7	0.6	
	M1 Knowledge transfer, etc.	0.4	0.3	
RDP Priority 4 subtotal		*78.1*	*58.6*	
ERDF 2014–2020	85 Protection and enhancement of biodiversity, etc.	4.3	3.4	For N2k etc.
	86 Protection, restoration and sustainable use of Natura 2000	3.6	2.9	For N2k etc.
LIFE Programme	Nature and biodiversity projects (2000–2020)	3.0	1.8	Award period
	Integrated projects – 1 (2014–2026)	1.9	1.1	LIFE-IP NATURA.SI
EU Interreg	Cross-border and transnational Priority 6d projects for EU nature policy and associated green infrastructures	1.7	1.5	For N2k etc.
EMFF 2014–2020	Art. 80.1.b protection of marine environment	0.04	0.03	For N2k etc.
EU funding instrument subtotal		*92.5*	*69.3*	
National public funds within 2014–2020		**Additional non-EU***	**Comments**	
State budget (2019)	Administered by the MESP for nature conservation	8.0	Mostly salaries. c. 62% on N2k etc.	
National Climate Fund (2018–2019)	Climate mitigation/adaptation measures, related to N2k	1.7	€3.35 million from 2018, annual amount uncertain	
National Forest Fund (2019)	Allocation for measures in N2k forests.	1.8		
Others				
European Economic Area and Norwegian grants (2014–2021)	Restoration of the Natura 2000 network ecosystems	0.3	For N2k.	
Additional national and others subtotal		*11.7*		
National total including co-financing of EU instruments		*34.7*		
Total		**104.3**		

Notes. * Excludes co-financing of EU funds. 'For N2k etc.' indicates funding relevant to Natura 2000, related green infrastructure and BHD species (based on the PAF). The subtotals and totals are only approximate indications of the average annual levels of funding as they are based on annualised data from differing periods.

Sources. Based on data from: RDP data from European Commission ESIF funds database (23 October 2020); ERDF and EMFF allocations to Natura 2000 and national funding from 2021 to 2026 PAF (MESP, 2019); LIFE Slovenia factsheet, European Commission (March 2021). See Section 1.3 for further details.

DPOMS[41]). All monitoring reports are publicly available.[42] Although it is necessary to expand the monitoring schemes to more species (e.g. plants, dragonflies and HD Annex IV and national protected species), this is limited mostly by finances, but also by the availability of experts. It is extremely important to secure continuity and comparability of monitoring, which is always a challenge in an unpredictable political environment.

Nature conservation planning and decision making are also dependent on a good overview of ecosystems and their distribution. Mapping for nature conservation purposes started in 1999 (Ljubljansko barje), mainly financed by the MESP, while some mapping was initiated by other users (e.g. local communes and investors in large projects). The field mapping uses an agreed typology (Jogan *et al.*, 2004) and method with digital orthophotos.[43] All habitat types are mapped, except for forest areas as they are already covered by a forestry sector mapping system, although it is not entirely compatible with the nature conservation method. Up until 2020, about 3 800 km^2 of non-forest areas had been mapped and all data made available online.[44] A process is underway to update the method using remote sensing in addition to field mapping.

The main international reporting on nature conservation is related to the Nature Directives and is carried out by the Institute for Nature Conservation. The Slovenian Environmental Agency (ARSO) is the Eionet focal point of the EEA, regularly reporting data on protected areas. At a national level there is an online platform providing information on environmental indicators in Slovenia,[45] including nature.

28.8.2 Data Management

In addition to essential support data (e.g. base maps), nature conservation is crucially dependent on the availability and quality of two main groups of data: basic data on nature, and data on nature conservation actions and their impacts.

Basic data on nature include geological, hydrological, pedological data, habitat type maps, species catalogues, distribution data and population data. There is a considerable amount of historical and recent data on these components, but their availability is hindered by their dispersal amongst different institutions and NGOs. There is no central point to obtain an overview of existing available data. The largest, but still incomplete, database is held by the Centre for Cartography of Fauna and Flora (CKFF), which in March 2022 had more than 1.9 million localised distribution records for about 27 000 plants and more than 64 000 animals[46] and corresponding species catalogues as well as a list of sources (over 8 200 items).

Nature conservation data include information on the conservation status of species (e.g. level of threat, protected species and international status), areas with a nature conservation status (e.g. status of Valuable Natural Features, Ecologically Important Areas, Natura 2000 sites, domestic protected areas and international designations), permit procedures and inspection interventions. Most of these data are available to the public through an online viewer (Atlas okolja[47]) managed by ARSO, with downloadable GIS files.[48] However, a more user-friendly public interface for the current standard information is lacking, and many other nature conservation data are not available.

The lack of data coordination and availability hinders the efficient implementation of nature conservation legislation and decision making, such as in relation to impact assessments and planning. It can even be a risk to investors, as late in the impact assessment and planning procedures new relevant data can appear that can reveal possible negative impacts of a development that were not known of. This can lead to conflicts and potentially biodiversity losses that could have been avoided.

In 2020, a LIFE project (LIFE Narcis) was approved with the aim of organising a common state-level information platform for nature conservation.

28.9 Future Developments

Although the legal system for nature conservation is well-established, some fine tuning is expected as a response to actual situations and to improve some recognised inefficient processes. Adoption of the Natura 2000 management programme (2021–2027) is foreseen in 2022 together with supporting activities to enhance its efficient implementation. Strengthening of law enforcement is planned as well.

Strategic documents for invasive alien species and management of the Brown Bear are planned to be adopted in the near future.

28.10 Conclusions

Slovenia has a long history of nature conservation, starting with sporadic ideas in the nineteenth century that were incorporated into a programme (the *Memorandum*) in 1920, which was the foundation for the first protected area and protection of species and caves. During the Kingdom of Yugoslavia and after World War II, institutional, legal and civil bodies for nature conservation gradually developed.

Independence in 1991 and political changes introduced social and economic changes: privatisation and a market

41 Društvo za proučevanje in ohranjanje metuljev Slovenije.

42 www.natura2000.si/natura-2000/natura-2000-v-sloveniji/monitoring/

43 That is, aerial photos or satellite images with a uniform scale.

44 http://zrsvn-varstvonarave.si/informacije-za-uporabnike/katalog-informacij-javnega-znacaja/habitatni-tipi/izvedena-kartiranja-habitatnih-tipov/ (last accessed 20 September 2020).

45 http://kazalci.arso.gov.si/en

46 www.ckff.si/zbirka.php (last accessed 30 March 2022).

47 http://gis.arso.gov.si/atlasokolja/profile.aspx?id=Atlas_Okolja_AXL@Arso (last accessed 5 September 2020).

48 http://gis.arso.gov.si/geoportal/catalog/main/home.page (last accessed 5 September 2020).

economy that gradually influenced nature and the cultural landscape, with the abandonment of extensive agriculture and more rapid urbanisation. This process was accelerated by accession to the EU in 2004. In nature conservation, the EU legislation emphasised the need to focus on habitats and species of EU concern, whilst the CBD drives a broader biodiversity agenda.

Another important milestone was in 1994, when the responsibility for nature conservation moved from the cultural to the environmental sector and the Nature Conservation Act (NCA) was adopted. The organisational structure has persisted and has its strengths, but there are insufficient staff to fully implement the NCA. Whilst civil society is well-engaged in nature conservation and active, its influence is dispersed and uncoordinated.

Slovenia has established a large terrestrial protected area network (40% coverage), comprising the system of domestic protected areas (13% coverage) and the partially overlapping Natura 2000 network (38%), the highest proportion in the EU. These sites also form part of the wider national ecological network of legally recognised Ecologically Important Areas, which covers 53% of national territory. Whilst not legally protected, their designation is taken into account in planning, albeit with relatively weak effects. Only a few percent of Slovenia's small marine area is protected, in small areas – most of which are an integral part of the coastal protected areas.

The management of Natura 2000 sites is based on a comprehensive application of the integration principle – in part due to the size of the network. All relevant sectors are directly responsible for the implementation of conservation measures identified as being required for each site in the Management Programme for Natura 2000, adopted by the government for each MMF period. This approach works well for forestry (affecting 70% of Natura 2000), as a good planning and management system has been in place for a long time and is compulsory for all forest owners. However, in practice implementation might differ from the planned measures, partly due to weak inspection. For agriculture there is no such system, so implementation depends mostly on CAP-funded agri-environment schemes. As these are voluntary and not competitive with other CAP measures, uptake has been very limited and their actual nature conservation impacts are modest. Integration with other sectors (e.g. water management and infrastructure) needs a lot more trust building, which is expected to gradually occur with positive results.

Perhaps as a result of its longstanding legislation, large protected area network and management traditions and current system, a relatively high proportion of Slovenia's extensive forests and other semi-natural habitats have a favourable conservation status (EEA, 2020). However, most could be improved, including some forest types (e.g. riparian forests), and others are threatened, especially amongst freshwater habitats and semi-natural grasslands. Furthermore, relatively high proportions of those HD habitats and species that do have an unfavourable status also have a declining trend.

The main challenge for Slovenia is to improve the implementation of its existing nature conservation systems, particularly strengthening the inspection and enforcement regime for environmental legislation. This requires not only adequate professional staff and means, but also the necessary consolidated scientific information in a central system that is publicly available. Better communication between all actors is also an essential element of long-term trust building. As a rule, the political profile of nature conservation is low and focussed on the short term, thus better cooperation between all actors and greater public support are of utmost importance.

Acknowledgements

Responsibility for conserving nature goes beyond selected governmental bodies, dedicated NGOs and enthusiastic individuals. Everyone is taking part in this story with their own personal decisions. My first tribute is to all those who have participated in these movements, whenever it was. Special thanks go to my colleagues and friends with whom I worked with over 40 years. Special thanks are also due for the contributions and advice from Jana Kus, Andrej Sovinc, Tomaž Mihelič, Marijan Govedič, Mladen Kotarac and Andrej Bibič.

References

Bartol, M., Černe, R., Hanc, Ž. *et al.* (2021) *Stanje ohranjenosti volkov v Sloveniji v sezoni 2020/2021*, Končno poročilo. Ljubljana: Zavod za gozdove Slovenije.

Beuk, S. (1920) Spomenica Odseka za varstvo prirode in prirodnih spomenikov. *Glasnik Muzejskega društva za Slovenijo*, **1** (1–4), 69–75.

Bizjak Mali, L. & Sket, B. (2019) History and biology of the 'black proteus' (*Proteus anguinus parkelj* Sket & Arntzen 1994; Amphibia: Proteidae): a review. *Folia Biologica et Geologica*, **60** (1), 5–37.

Božič, L. (2003) *Mednarodno pomembna območja za ptice v Sloveniji 2. Predlogi Posebnih zaščitenih območij (SPA) v Sloveniji.* Ljubljana: DOPPS, Monografija DOPPS 2.

Denac, K., Mihelič, T., Kmecl, P. *et al.* (2015) *Monitoring populacij izbranih vrst ptic -popisi gnezdilk 2015. Poročilo.* Ljubljana: Ministrstvo za kmetijstvo, gozdarstvo in prehrano.

EEA (2020) *State of nature in the EU. Results from reporting under the nature directives 2013–2018.* Copenhagen: European Environment Agency.

Ervin, J. (2003) WWF: *Rapid Assessment and Prioritization of Protected Area Management (RAPPAM) Methodology.* Gland, Switzerland: WWF.

Geister, I. (1995) *Ornitološki atlas Slovenije.* Ljubljana: Državna založba Slovenije.

Hlad, B. & Skoberne, P. (eds.) (2001) *Biological and Landscape Diversity in Slovenia.* Ljubljana: Ministrstvo za okolje in prostor Republike Slovenije, Agencija RS za okolje.

Jerina, K., Majić-Skrbinšek, A., Stergar, M. *et al.* (2020) *Strokovna izhodišča za upravljanje rjavega medveda (Ursus arctos) v Sloveniji (obdobje 2020–2023). Ekspertiza.* Ljubljana: Biotehniška fakulteta

Univerze v Ljubljani, Gozdarski inštitut Slovenije, Zavod za gozdove Slovenije.

Jogan, N., Kaligarič, M., Leskovar, I., Seliškar, A. & Dobravec, J. (2004) *Habitatni tipi Slovenije – tipologija.* Ljubljana: Ministrstvo za okolje, prostor in energijo, Agencija RS za okolje.

Kirchmeir, H. & Kovarovics, A. (eds.) (2016) *Nomination Dossier 'Primeval Beech Forests of the Carpathians and Other Regions of Europe' as extension to the existing Natural World Heritage Site 'Primeval Beech Forests of the Carpathians and the Ancient Beech Forests of Germany' (1133bis).* Klagenfurt, Austria: E.C.O. and Environment Agency Austria.

Kranjc, A. (2013) 6.10 Classification of Closed Depressions in Carbonate Karst. *Treatise on Geomorphology*, **6**, 104–111.

Kryštufek, B. (ed.) (2001) *Raziskava razširjenosti evropsko pomembnih vrst v Sloveniji.* Ljubljana: Prirodoslovni muzej Slovenije.

(2003) *Strokovno izhodišče za vzpostavljanje omrežia NATURA 2000 – Bober (Castor fiber).* Ljubljana: Prirodoslovni muzej Slovenije.

Kus Veenvliet, J. & Sovinc, A. (2009) *Protected area management effectiveness in Slovenia, Final report of the RAPPAM analysis.* Gland, Switzerland: WWF.

Majić Skrbinšek, A. (2009) *Report on the implementation of the Bern Convention in Slovenia.* Strasbourg: Council of Europe. Document T-PVS/Inf (2009)12.

Marušič, I., Ogrin, D. & Jančič, M. (1998) *Regionalna razdelitev krajinskih tipov Slovenije.* Ljubljana: Ministrstvo za okolje in prostor, Urad RS za prostorsko planiranje; Univerza v Ljubljani, Biotehniška fakulteta, Oddelek za krajinsko arhitekturo.

MESP (2002) *Biodiversity Conservation Strategy of Slovenia.* Ljubljana: Ministrstvo za okolje, prostor in energijo.

(2019) *Prioritised Action Framework (PAF) for Natura 2000 in Slovenia for the Multiannual Financial Framework period 2021 – 2027.* Ljubljana: Ministry of the Environment and Spatial Planning. www.natura2000.si/fileadmin/user_upload/Dokumenti/LIFE_IP_NATURA_SI/Rezultati/C.3.4_IP_PAF_Slovenia_2021-2027_final.pdf

Mihelič, T. (ed.) (2019) *Atlas ptic Slovenije: popis gnezdilk 2002–2017.* Ljubljana: Društvo za opazovanje in proučevanje ptic Slovenije.

MKGP (2019) *Analiza ciljev in ukrepov Programa upravljanja območij Natura 2000 2015–2020 za obdobje 2015–2019, Sektor kmetijstvo.* LIFE integrirani projekt za okrepljeno upravljanje Nature 2000 v Sloveniji. LIFE17 IPE/SI/000011. Ljubljana: Ministrstvo za kmetijstvo, gozdarstvo in prehrano.

Petkovšek, M. (2017) Slovensko omrežje Natura 2000 v številkah. *Varstvo narave*, **30**, 99–126.

Popp, L. (2011) *Dabar (Castor fiber, Linné, 1758) u Hrvatskoj: stanje i perspektiva petnaest godina nakon reintrodukcije.* Undergraduate thesis. University of Zagreb, Faculty of Science.

Sindičić, M., Polanc, P., Gomerčić, T. *et al.* (2013) Genetic data confirm critical status of the reintroduced Dinaric population of Eurasian lynx. *Conservation Genetics*, **14** (5), 1009–1018.

Sket, B., Paragamian, K. & Trontelj, P. (2004) A census of the obligate subterranean fauna of the Balkan Peninsula. In *Balkan Biodiversity: Pattern and Process in the European Hotspot*, eds. H. I. Griffiths, B. Kryštufek & J. M. Reed, pp. 309–322. Dordrecht, the Netherlands: Springer.

Travnikar, T., Bedrač, M., Brečko, J. *et al.* (2019) *Slovensko kmetijstvo v številkah – 2019.* Ljubljana: Kmetijski inštitut Slovenije.

Zakšek, B., Verovnik, R., Zakšek, V., Kogovšek, N. & Govedič, M. (2019) *Monitoring izbranih ciljnih vrst metuljev v letu 2019.* Miklavž na Dravskem polju: Center za kartografijo favne in flore.

Chapter 29 Spain

Concha Olmeda and Juan Carlos Blanco

Key Statistics	
Terrestrial area (km^2)/% of EU-28*	499 008/11.4%
Marine waters (km^2)*/% of EU-28*	583 416/12.7% (total inc. MMAC 1 068 464)
Population density (p km^2)/% of EU-28*	89/77%
GDP per capita (€)/% of EU-28	23 332/87%
Land cover (CORINE land cover classes 2018)*	AGR 48%, OSN 27%, FOR 21%, ART 3%, WET 0.9%
Biogeographical regions (except MAC & MMAC)	ALP, ATL, MATL, MED, MMED
Number of HD Annex I habitats / %*	111/49%
Number of HD Annex II, IV and V species / %*	325/27%
Number of BD Annex I bird taxa / %*	110/63%
% HD UCS and decreasing habitats/ % unknown	17%/18%
% HD UCS and decreasing species/ % unknown	27%/14%
% breeding BD Annex I birds decreasing/% uncertain and unknown	28%/6%

* Excluding MAC and MMAC biogeographical regions.
Sources and codes. See Table 1.1.

29.1 The Natural History and Characteristics of Spain

Spain is the second largest country in the EU, with most of its European territory within the Iberian peninsula of south-west Europe, and a smaller area in the nearby Balearic Islands in the Mediterranean Sea. This European territory mostly falls within the Mediterranean biogeographical region (80%), as well as the Atlantic and Alpine regions, and adjacent marine Atlantic and Mediterranean regions. The Canary Islands archipelago, in the Atlantic and Macaronesian biogeographical region, is also part of Spain, but is not geographically part of Europe so is not covered in this book and chapter (where available data allow).

The country is located in one of the 25 biodiversity hotspots in the world, thanks to its high rate of endemism (IUCN, 2020) and very high level of biodiversity due to its geographical position (e.g. only 14 km away from Africa), geological diversity, variable climate, topography and soils, extensive coastline and existence of islands. Oceanographic and biogeographical factors also give rise to a great diversity of marine habitats and species across its marine area, which is almost double its land area. Its 8 000 km of coastline also has a wide variety of ecosystems, which include intertidal habitats, beaches, cliffs, dunes and salt marshes. Spain also has a diverse range of inland wetlands, with more than 75 000 km of rivers and at least 1 500 wetlands. Much of the land is dominated by high plateaus and mountain ranges (with a highest point in the Sierra Nevada of 3 478 m), and large areas have a very low human population density, since a large proportion of the population is concentrated in the main cities and along the coast.

Human activities have moulded the landscape, which is now predominantly a mosaic of woodland, scrub and grasslands, together with cultivated land and permanent crops. Large areas of High Nature Value (HNV) farmland include mountain pastures and meadows, pseudo-steppe grasslands and low-intensity dry cereals, wooded pastures (*dehesas*), and some olive groves and other permanent crops (Beaufoy *et al.*, 2012). However, these habitats have declined in recent decades as a result of agricultural intensification, including the expansion of irrigated agriculture. The current extent of HNV farmland is uncertain, but has been estimated to be between 145 000 and 250 000 km^2 (Keenleyside *et al.*, 2014), which is 29–50% of the country's mainland area.

According to the CORINE land cover data for 2018, forest covers around 21% of the territory (107 000 km^2). About 30% of forests are coniferous, 28% are broadleaved and 21% are mixed, with 17% of forests being plantations, about a third of which are of non-native introduced species, especially *Eucalyptus* spp. and *Pinus radiata* (FAO, 2015). The latest National Forest Inventory (IFN, 2020) detected a notable increase in woodland area at the cost of a decrease in cultivated area and grasslands, with an annual increase in forest area of 1.2% over 2004–2018. Native broadleaf trees (oaks, chestnut

Table 29.1 Short-term trends (c. 2007–2018) in Habitats Directive Annex I habitats and Annex II, IV and V species with an unfavourable conservation status in mainland Spain.

	Poor (unfavourable–inadequate)				Bad (unfavourable–bad)			
	Imp	Sta	Dec	?	Imp	Sta	Dec	?
HD Habitat assessments								
Number	4	57	23	46	0	9	14	17
%	3%	44%	18%	35%	0%	23%	35%	43%
HD Species assessments								
Number	16	106	66	50	3	38	83	23
%	7%	44%	28%	21%	2%	26%	56%	16%

Key. Imp = improving, Sta = stable, Dec = decreasing, ? = unknown.
Source. Based on Member State HD Article 17 reported data as presented on the EEA national summary dashboard (2021).

and beech, etc.) have increased considerably in both surface area and biomass since the previous inventory. In the northern provinces, however, the expansion of eucalyptus plantations has been very high (currently covering over 5 000 km^2).

Spain holds a large proportion of HD habitat types, including some habitats more typical of north Africa, such as the *Tetraclinis articulata* forests (9570*)[1] that only occur in Spain and Malta in the EU, or the arborescent matorral with *Zyziphus* (5220*), which is only found in south-east Spain, Sicily and Cyprus in the EU. Some semi-natural habitats are very characteristic of the Iberian Peninsula, including the grazed *dehesa* forests (6310), the Galicio-Portuguese oak woods with *Quercus robur* and *Quercus pyrenaica* (9230) and the Cork Oak (*Quercus suber*) forests (9330).

There is a high diversity of plant and animal species and a high proportion are endemic, thus giving Spain a particularly high responsibility for the conservation of the biodiversity in western Europe. According to the *National Inventory of Biodiversity* (2017), the number of vascular plants in Spain is over 9 000 species (85% of those known in the EU), of which about 1 500 are endemics. Among vertebrates, 64% of amphibian species are endemic to Spain and Portugal, while 58% of fish and 56% of reptiles are endemic species in Spain. There are 11 species of endemic mammals in the Iberian Peninsula, including the globally Endangered Iberian Lynx (*Lynx pardinus*), Endangered Pyrenean Desman (*Galemys pyrenaicus*) and Near Threatened Cabrera's Vole (*Microtus cabrerae*).

Spain's avifauna is also of outstanding importance. The majority of the world population of the globally Vulnerable Spanish Imperial Eagle (*Aquila adalberti*) occurs in Spain, as well as the largest European breeding populations of the Near Threatened Cinereous Vulture (*Aegypius monachus*) and 90% of Europe's Griffon Vultures (*Gyps fulvus*) (Del Moral, 2017; Del Moral and Molina, 2018).

The country is the stronghold for most steppe bird species in western Europe (Bota *et al.*, 2005). For example, it harbours a large proportion of the total European breeding population of the globally Vulnerable Great Bustard (*Otis tarda*), Vulnerable Dupont's Lark (*Chersophilus duponti*) and Near Threatened Little Bustard (*Tetrax tetrax*). It is therefore of great concern that most of the Spanish steppe bird populations have declined sharply since the 1990s.

Other globally threatened bird species include the Critically Endangered Balearic Shearwater (*Puffinus mauretanicus*), which breeds exclusively in the Balearic Islands. Spain also has important but precarious breeding populations of the Vulnerable Marbled Teal (*Marmaronetta angustirostris*) and Endangered White-headed Duck (*Oxyura leucocephala*), although the latter species is currently increasing.

Overall, despite recent changes, Spain still has relatively large areas where there is a favourable balance between nature and traditional human activities. It therefore still harbours a rich diversity of animals, plants and cultural landscapes, which makes it an important destination for nature tourism.

29.2 The Status of Nature

29.2.1 Habitat Status and Trends

Knowledge about the status of HD habitat types has improved over the last decade. According to the latest Article 17 report[2] submitted by Spain on the conservation status of HD habitats (2013–2018), out of 224 assessments across the biogeographical regions (excluding the Macaronesian terrestrial and marine regions) only 6% revealed a favourable status, 76% showed an unfavourable status (58% poor and 18% bad) and 18% were unknown. Out of the habitats that had an unfavourable status, a considerable proportion had a declining or an unknown trend, including 35% with a bad status (Table 29.1). The situation is particularly serious amongst bog habitats,

[1] Numbers in brackets are HD Annex I habitat codes (see Appendix for full names). * Indicates a priority habitat.

[2] See Section 1.3 for background on monitoring under the Nature Directives, status categories and data sources.

dunes and forests as more than 80% of the assessments for these habitats revealed an unfavourable status. The status of coastal habitats is also of concern as no assessment showed a favourable status.

29.2.2 Species Status and Trends

The latest report on the status of biodiversity in Spain (MAPAMA, 2017) reveals that biodiversity is declining, as elsewhere in Europe and globally. Approximately 12% of vascular plants are nationally threatened and 11% of terrestrial vertebrates overall, with higher levels amongst amphibians (14%), reptiles (16%) and mammals (17%). Furthermore, seven species are currently designated as Critically Endangered in Spain: European Mink (*Mustela lutreola*), Lesser Grey Shrike (*Lanius minor*), Marbled Teal, Cantabrian Capercaillie (*Tetrao urogallus cantabricus*), Fan Mussel (*Pinna nobilis*), Spengler's Freshwater Mussel (*Margaritifera auricularia*) and *Cistus heterophyllus* subsp. *carthaginensis*.

As regards the HD species that occur in European Spain, of over 558 assessments across the biogeographical regions (excluding the Macaronesian terrestrial and marine regions) during 2013–2018, only 17% of the assessments showed a favourable conservation status, 69% had an unfavourable status (43% poor and 26% bad) and 14% had an unknown status. Furthermore, a significant proportion of the species with an unfavourable status also were declining (Table 29.1). It is notable that the situation is particularly serious for freshwater fish species: all assessments indicated an unfavourable status, and 55% with a bad status also had a declining trend.

Despite the generally bad situation, it should be acknowledged that some conservation efforts have had very positive results. For example, some species that had a critical status have recovered over recent decades, such as the Iberian Lynx and Brown Bear (*Ursus arctos*) (see Section 29.4.3).

The latest available results on the status of birds, as reported in accordance with Article 12 of the Birds Directive, indicate that across all bird species (excluding populations in the Canary islands) 24% increased over the short term (i.e. 2007–2018), 39% had stable or fluctuating trends, 28% decreased, and 9% had uncertain or unknown trends. Longer-term trends over about the last 30 years have been similar, with the same proportion having decreased, but there were more species with increasing trends (37%), balanced by fewer species with stable trends (23%).

Of most importance is the situation regarding the trends in BD Annex I taxa (Table 29.2). Despite these species being the focus of nature conservation measures, over a quarter have shown short-term decreases in their breeding populations. Nearly a quarter of wintering Annex I taxa have also shown short-term declines.

As is the case all over Europe, a marked decline of farmland birds has been noticeable in Spain. Agricultural habitats have the largest proportion of declining species and species with the highest population declines, such as the Skylark (*Alauda arvensis*), Turtle-dove (*Streptopelia turtur*), Calandra Lark (*Melanocorypha calandra*), Little Bustard and Corn Bunting (*Emberiza calandra*) (SEO/BirdLife, 2019). Of major concern is that the population of the globally Near Threatened Little Bustard has declined by 50% over the last decade, which triggered its inclusion in CMS Appendices I and II in 2020.

In contrast, some bird species have experienced increasing trends since the 1980s, including the Spanish Imperial Eagle, Bonelli's Eagle (*Aquila fasciata*) now with 63% of the European population in Spain in 2018, and the globally Endangered Egyptian Vulture (*Neophron percnopterus*) (SEO/BirdLife, 2019). Section 29.4.3 provides more information on the recent recovery of some threatened species.

Table 29.2 Short-term trends (c. 2007–2018) in Birds Directive Annex I taxa in mainland Spain.

	Breeding				Wintering			
	Inc	Sta/ Flu	Dec	?	Inc	Sta/ Flu	Dec	?
Number	25	37	26	6	19	6	9	4
%	27%	39%	28%	6%	50%	16%	24%	11%

Key. Inc = increasing, Sta = stable, Flu = fluctuating, Dec = decreasing, ? = uncertain and unknown.
Source. Based on Member State Article 12 reported data as presented on the EEA national summary dashboard (2021).

29.2.3 Drivers and Causes of Change

The direct drivers of biodiversity change in Spain include changes in land use, pollution, climate change, the overexploitation of natural resources and the introduction of invasive alien species (IAS) (MMARM, 2011). Changes in land use, which result in habitat loss, fragmentation and degradation, are the greatest pressures on terrestrial biodiversity in Spain.

Forest ecosystems suffer multiple pressures and threats, including fragmentation, environmental pollution, climate change and extreme weather events, overexploitation of resources, forest fires and competition with introduced species and varieties (MAGRAMA, 2014a). They also lack sustainable forest management that properly considers the ecological requirements of forest species.

As in other parts of Europe, agricultural improvements, specialisation and intensification (e.g. ploughing and reseeding of grasslands, conversion to crops, irrigation, increased use of fertilisers and pesticides, and reduced use of rotations and fallow and grazing of crop stubbles) are the major causes of farmland bird and invertebrate species declines. A recent study found a significant positive relationship between the Spanish farmland and cereal crop bird indices (e.g. the fallow specialist Little Bustard) and the amount of fallow in the landscape, highlighting the need to maintain fallow land in European agro-ecosystems to conserve farmland bird populations (Traba and Morales, 2019). The use of very harmful pesticides is another issue of great concern, in particular the widespread use of neonicotinoid insecticides, which have a devastating effect on bees and other pollinator species.

In contrast, many semi-natural grasslands are becoming scrubland communities as a result of abandonment of traditional grazing due to the lack of economic profitability of this activity in some areas. Some semi-natural grasslands and scrubland have also been converted to forest plantations, especially in northern Spain.

The deployment of renewable energy infrastructure (e.g. solar panels) in grassland areas is increasing rapidly, with direct consequences on the habitat of steppe birds.

Nutrient enrichment through the deposition of atmospheric nitrogen is a significant threat to some terrestrial ecosystems in Europe, and is thought to be important in Spain, although information is lacking. A recent study tried to assess the effects of critical nitrogen loads on protected habitats in National Parks and pointed out that biodiversity in high-altitude National Parks could be endangered by eutrophication. The habitats considered to be most threatened are natural grasslands, whilst other particularly threatened habitats are some medium- and high-altitude scrublands and heathlands (García-Gómez *et al.*, 2017). Spain has recently established its commitments to reduce atmospheric emissions pollutants, including nitrogen oxides and ammonia, and set out measures to achieve this in accordance with EU legislation.

The rise of IAS is affecting areas of high ecological value in many regions of Spain. Historically, there have been both intentional and accidental introductions of exotic species, which include, across all of Spain: 25 species of fish (30% of fish species), 15 species of amphibians, 32 species of reptiles, 13 species of birds, 15% of mammals, and 15% of plants.

Inland aquatic ecosystems are the most vulnerable in mainland Spain, being affected by the synergistic effects of all the main direct drivers of change. The main impacts are from pollution, overexploitation of water resources, and the destruction and degradation of habitats by infrastructure (e.g. dams and river channelisation). Faced with these threats, however, recent water policies have made significant progress in reducing pressures and improving water quality. The biodiversity of inland water ecosystems is also significantly affected by IAS (as mentioned above).

Coastal and marine habitats and species are subject to significant pressures and impacts from various human activities, especially coastal developments related to tourism, fishing practices, maritime traffic and transport and recreational navigation, exploration and exploitation of the seabed, development of infrastructure for power generation, and marine litter. Pollution from agriculture and sewage are also still important pressures on coastal and marine ecosystems. One of the most catastrophic examples, due to diffuse pollution from agriculture together with untreated sewage, has been the recent collapse due to eutrophication of the coastal lagoon ecosystem of Mar Menor on the Mediterranean coast, which has been suffering from this pressure for decades (Giménez Casalduero and Martínez Fernández, 2020). The Spanish government has recently approved an ambitious programme with more than 70 actions to try to recover this ecosystem.

Whilst many drivers of biodiversity loss have increased over the last 50 years, appreciation of nature has noticeably increased in Spain, as in other western countries. As a result of changed attitudes many carnivores and raptors, which were considered as vermin and harshly persecuted, have become symbols of nature and the subject of successful conservation and recovery plans. As a result of improved legal protection, law enforcement and the conservation of their habitats, species such as the Wolf (*Canis lupus*), Brown Bear (in the Cantabrian Mountains), Otter (*Lutra lutra*), birds of prey like the Bearded Vulture (*Gypaetus barbatus*), Cinereous Vulture, Griffon Vulture and Spanish Imperial Eagle have partially recovered and hold relatively high populations in Spain in the European context (as further described in Section 29.4.3).

In contrast, other less charismatic species have started, or continue, to decline due to various causes, including those described earlier. In some cases, the causes are not fully understood, as is the case with the Cantabrian Capercaillie, which is heading towards extinction probably as a result of climate change and edge of distribution range effects (Ludwig *et al.*, 2006). As further described in Section 29.6.4, climate change is expected to be an increasingly serious cause of ecosystem degradation and biodiversity loss in future.

29.3 Nature Conservation Governance, Actors, Legislation and Policies

29.3.1 Nature Conservation Developments, Legislation and Policy

Table 29.3 provides an overview of the most important nature conservation developments in Spain. The earliest of these date back to the nineteenth century and mainly concern the protection of game species and the preservation of forests and water resources. The Forest Act (*Ley de Montes*) adopted in 1863 established a legal regime where forest and scrub areas of public interest (*Montes de Utilidad Pública*) were classified. These have since been managed by the forest administration and have played a fundamental role in nature conservation, with many areas now being protected due to their good ecological condition.

After the declaration of Yellowstone National Park (USA) in 1872, the conservation of natural areas received more attention in Spain and led to the enactment of the National Parks Act in 1916 and the establishment of the first National Parks in 1918.

Other initiatives at the beginning of the twentieth century focussed on game species such as the Southern Chamois (*Rupicapra pyrenaica*) and the Spanish Ibex (*Capra pyrenaica*) of Sierra de Gredos, for which reserves were established.

In the late 1960s and in the 1970s, nature conservation advanced considerably in Spain thanks to the naturalist and broadcaster Félix Rodríguez de la Fuente, who started to

Table 29.3 Key nature conservation developments in Spain.

1863	Forest Act (*Ley de Montes*) established forest and scrub areas of public utility
1905	First hunting reserve established to protect the Spanish Mountain Goat
1916	National Parks Act
1918	Establishment of the first National Parks
1950	Spanish Mountain Goat conservation programme established, with the creation of numerous reserves
1965	Creation of the Doñana Biological Station, a wildlife research centre
1966	A Ministerial Order prohibited hunting the Iberian Lynx, several waterfowl and raptors
1969	Establishment of Doñana National Park
1970	Hunting Act abolished the concept of vermin species
1986	Accession to the European Community
1989	Act on Conservation of Natural Areas, Wild Flora and Fauna
2007	Natural Heritage and Biodiversity Act
2010	Marine Environment Protection Act
2011	*Natural Heritage and Biodiversity Strategic Plan*

See Table 3.3 for signatory dates of international conventions.

educate the Spanish public on the need to preserve nature. At that time, the government established the Doñana Biological Station (1965), a wildlife research centre based in Seville, which initiated the scientific conservation of vertebrates in Spain. All raptors were protected in 1966 and the Doñana National Park was established in 1969. The Hunting Act of 1970 abolished the concept of vermin under which the extermination of certain wild species that were considered harmful to agriculture and hunting was promoted and paid for. The Act also included the Wolf as a game species, which gave it a certain degree of protection and allowed its recovery.

The accession of Spain to the European Community in 1986 led to further major developments in environmental legislation, particularly in nature conservation through the adaptation of the provisions of the Birds Directive. In 1989, a new Act on Conservation of Natural Areas and Wild Flora and Fauna[3] was adopted, which replaced the previous law of 1975 on Natural Protected Areas. It extended the legal protective regime for natural resources beyond mere protected natural areas and addressed the necessary articulation of nature conservation policy within the new distribution of powers between the State and the Autonomous Communities. It also established the main categories of protected areas at the time, the regime for the protection of wild species and some tools to ensure their conservation, such as the compilation of recovery plans.

The recent basic legislation on nature conservation, the Natural Heritage and Biodiversity Act[4] (NHBA) adopted in 2007 and modified by Act 7/2018, has further developed and established the protection regime for species, habitats and natural areas. According to the Act, the Autonomous Communities (regions) are responsible for the designation of protected areas, including Natura 2000 sites, and for establishing the necessary conservation measures, except for marine offshore sites that are under the competences of the State administration. National Parks are established by law of the national parliament and managed by the regions. The NHBA also establishes the general rules for the conservation of threatened species. Within the framework of this law, conservation strategies have been developed for specific threatened species, and to combat some key general threats (see Section 29.4.3).

A *Natural Heritage and Biodiversity Strategic Plan*[5] (MMARM, 2011) was implemented from 2011 to 2017. Its objectives and actions comply with Spain's international and EU commitments on biodiversity, in particular those derived from the CBD's *Strategic Plan for 2011–2020* and the *EU Biodiversity Strategy to 2020*. It also gave rise to new objectives and actions to improve knowledge and conservation of species and habitats, to strengthen management of protected areas and to streamline biodiversity into other policies and sectors. An evaluation of the Strategic Plan (MAGRAMA, 2017) showed partial achievement of the objectives and targets set in the plan. At the time of writing (December 2022), a *Strategic Plan on Natural Heritage and Biodiversity to 2030* was approved, which is in line with the EU and global biodiversity targets.

29.3.2 Governance

Since the 1980s, Spain's public administration has been highly decentralised, including in the field of the environment. According to the Spanish Constitution (1978), the State (national parliament and government) has competency over basic legislation for environmental protection while the Autonomous Communities (i.e. regions) have the powers to establish additional protection measures and to further develop the basic legislation. Currently, the Ministry for Ecological Transition and Demographic Challenge (MITECO) is the national body responsible for nature conservation.

The national administration is responsible for coastal and marine protection, including marine protected areas if they do not have ecological continuity with adjacent terrestrial sites

[3] Ley 4/1989, de 27 de marzo, de Conservación de los Espacios Naturales y de la Flora y Fauna Silvestres.

[4] Ley 42/2007, de 13 de diciembre, del Patrimonio Natural y de la Biodiversidad.

[5] Plan Estratégico del Patrimonio Natural y la Biodiversidad. www.miteco.gob.es/es/biodiversidad/temas/conservacion-de-la-biodiversidad/valoracion-y-aspectos-economicos-de-la-biodiversidad/cb_vae_plan_estrategico_patrimonio_nat_bio.aspx

that are protected by regional administrations. It also has important powers relating to water management on rivers that flow through different regions and are managed at the level of broader hydrographic basins.

Most nature protection policy and the implementation of nature conservation measures, however, is mainly a regional responsibility. The regions exercise their nature protection competences independently, and coordination among them, and with the national administration, is often a challenge. There are, nevertheless, institutional and cooperation mechanisms to facilitate coordination and joint implementation of conservation measures. The State Commission for Natural Heritage and Biodiversity is a consultation and cooperation body between the State and the regions to ensure efficient decision-making processes in nature conservation issues. It is supported by specialised committees on natural protected areas, wetlands, wild flora and fauna, forestry, conservation of genetic and forest resources, forest fires, access to and utilisation of genetic resources, and biodiversity inventory.

A Scientific Committee exists to provide advice on the listing of taxa under the Spanish Catalogue of Threatened Species and the List of Wild Species under Special Protection. This Committee is also the advisory body for the Alert Network on IAS.

In addition to the above-mentioned committees, a large number of working groups are actively addressing specific conservation issues, which include IAS, green infrastructure, habitat monitoring, plant conservation, critically endangered species, the Wolf, Brown Bear, Bearded Vulture, bats, species distribution, habitat fragmentation, inventory of biodiversity, indicators and forest mapping. All these working groups include representatives from the national and the regional level, as well as research institutions and scientific societies, with a view to ensuring a good exchange of experiences and cooperation.

Lastly, a commission has been created for ensuring coordination among different ministries in the elaboration, implementation and monitoring of the Marine Strategies[6] elaborated under the Marine Environment Protection Act[7] (which includes transposition of the MSFD).

For water issues, the main consultation and participation body is the National Water Council, created in 1985, which meets regularly and tries to ensure wide engagement of competent authorities and stakeholders, the professional and scientific community and NGOs in water-related issues.

Cooperation and coordination on nature conservation among all the competent administrations is an important challenge taking into account the almost federal character of the Spanish political system. Therefore, efforts need to be maintained and reinforced to ensure effective and common approaches, for instance to establish national- and biogeographical-level conservation objectives and strategies for achieving the favourable conservation status of BHD habitats and species, species protection and Natura 2000 management and connectivity.

29.3.3 Public Involvement and NGOs

Public involvement in environmental issues is relatively new in Spain. Over recent decades, participatory approaches have been incorporated in decision-making processes as a way to strengthen the bonds between diverse areas of knowledge and social actors in natural resources management and environmental governance.

Access to information, public participation and access to justice in environmental matters is regulated by national Law 27/2006, which guarantees the public participation in the field of the environment.

The State Council for Natural Heritage and Biodiversity[8] ensures the participation of stakeholders, professional and scientific experts, companies and NGOs.

Recently, the designation of marine SACs by the MITECO has included an ad hoc participatory process, which involved the organisation of workshops with relevant stakeholders during the elaboration of each site management plan.

The regions have their own rules and procedures on participatory approaches and promote public participation in relevant decisions for nature conservation, in particular for new legislation or regulations. For the declaration of natural protected areas, as in general for the approval of site management plans, a procedure for public consultation must be established.

NGOs dedicated to nature conservation issues have strengthened their capacities in the last decades. There are, nowadays, many environmental NGOs operating at different scales: local, regional and national. The largest Spanish NGOs are involved in the protection and monitoring of species and natural areas as well as in the development of awareness campaigns. Several Spanish NGOs dedicated to nature conservation are united under a platform called Ecologistas en Acción (Ecologists in Action) and work together, mainly campaigning on high-profile issues.

The Spanish Ornithological Society (SEO)/BirdLife, is one of the oldest NGOs (founded in 1954) and the largest in terms of membership (almost 15 000 members). It coordinates national surveys on bird species all over the country and runs a programme focussing on common birds. It is also active in the conservation and management of natural areas, including for instance through the preparation of agri-environment schemes that help conserve waterbirds and steppe birds. SEO/BirdLife has created a label that promotes nature-friendly products in SPAs, such as the rice grown in the Ebro Delta and

6 www.boe.es/boe/dias/2012/05/11/pdfs/BOE-A-2012-6263.pdf

7 Ley 41/2010, de 29 de diciembre, de protección del medio marino.

8 Consejo estatal para el patrimonio natural y la biodiversidad. www.miteco.gob.es/es/biodiversidad/temas/conservacion-de-la-biodiversidad/conservacion-de-la-biodiversidad-en-espana/cb_esp_cons_est_patrimon_natur_bio.aspx

pasta produced from wheat grown in steppe areas. It also develops intensive campaigns and works with fishers to reduce the by-catch of seabirds. One of its most relevant recent projects was the identification of the network of marine Important Bird Areas (IBAs), which has been the basis of the designated network of Spanish marine SPAs.

WWF Spain also has a long track record in nature conservation projects and campaigns in Spain, addressing various work areas from species conservation to forests, agriculture, water, oceans, climate and energy. It actively supports the conservation of threatened species, such as the Iberian Lynx, working in cooperation with the national and regional administrations, stakeholders and landowners. It has recently developed a proposal for a network of ecological corridors connecting Natura 2000 sites in Spain (WWF, 2018).

Some NGOs are more specialised and have worked successfully on the conservation of particular species over recent decades, achieving very good results. This is the case for Fundación Oso Pardo (Brown Bear Foundation), which has been very active since the 1990s, working in cooperation with all the regional administrations to combat the main threats to the Brown Bear. Fundación para la Conservación del Quebrantahuesos (Foundation for the Conservation of the Bearded Vulture) has been working since 1995 in the Pyrenees and more recently in Picos de Europa National Park (described further in Section 29.4.3).

Other NGO and scientific societies are very active in the study, monitoring and conservation of specific groups of wildlife, including the Spanish Mammal Society (SECEM), the Society for Bat Conservation (SECEMU), the Society for Cetacean Conservation (SEC) and the Herpetological Society (AHE).

Several NGOs are also implementing joint actions to fight specific drivers of biodiversity loss. This is the case of Programa Antidoto, which focusses on the issue of wildlife poisoning, in which eight of the main Spanish NGOs participate.

Despite significant efforts, both by the public administrations and NGOs, there remains a need to make participatory processes more effective and increase awareness about nature among relevant stakeholders and the public. This is evident for instance from the results of a survey carried out in a LIFE Project (Activa Red Natura 2000[9]), which found that only 21% of the population knew about the Natura 2000 network in 2017.

29.4 Species Protection and Conservation Measures

29.4.1 Basic Protection Legislation, and Strict Protection of Selected Species

In accordance with the NHBA, the State and the Autonomous Communities must ensure the conservation of all native wild species, addressing the preservation of their habitats and establishing specific protection measures when required. The destruction or deterioration of breeding sites, resting places or wintering areas is prohibited for all wild species under the Special Protection Regime. This includes species, subspecies and populations of scientific, ecological or cultural value, due to their uniqueness, rarity or degree of threat, as well as those that are included in the EU Nature Directives (BD Annex I, HD Annexes II and IV) and international conventions ratified by Spain.

Sanctions applied in the case of non-compliance with species protection legislation include fines of up to €2 million for causing the death of threatened species. The penal code also includes considerable penalties for such crimes, including prison sentences of up to two years.

29.4.2 Regulation of Hunting, Fishing and Other Forms of Exploitation

The national law includes provisions that concern the protection of species in relation to hunting and fishing. Hunting and fishing is regulated by the regions, including relevant areas, dates and the species for which these activities are permitted, which cannot be any of the species included in the Special Protection Regime List or the ones forbidden by the EU.

As regards predator control, which is mainly practised on the Red Fox (*Vulpes vulpes*), the methods allowed are standardised and require an individual accreditation. Finally, the use of lead ammunition has been forbidden since 2001 in Ramsar Sites, the Natura 2000 network and domestic protected areas.

In the marine environment, novel protection measures for cetaceans were established in 2007 through a Royal Decree.[10] These include a protection area around these animals at sea and the regulation of whale watching to minimise disturbance. This prohibits any boat from approaching cetaceans within 60 m and establishes other measures to avoid disturbance.

29.4.3 Conservation Strategies and Species Recovery Plans

Recovery Plans have been adopted and implemented since the 1990s for endangered animal species in Spain. Conservation strategies are also developed under the coordination of the national administration for species that occur in several regions. Ad hoc working groups have also been set up to promote their implementation.

The NHBA establishes the List of Wild Species under Special Protection and the Spanish Catalogue of Threatened Species. Recovery Plans must be adopted for nationally Endangered and Critically Endangered species, whilst Vulnerable species require Conservation Plans. The MITECO provides resources for the conservation of Endangered species.

[9] LIFE11INF/ES/665: http://activarednatura.es/

[10] Real Decreto 1727/2007, de 21 de diciembre, por el que se establecen medidas de protección de los cetáceos.

For Critically Endangered species, actions and projects targeting their recovery are declared of general interest, thus receiving national support.

Recovery Plans for species address habitat protection and, where appropriate, the designation of Critical Areas, such as breeding sites and resting places, which are especially protected from disturbance and deterioration. For example, the strategy for the conservation of the Brown Bear in the Cantabrian Mountains (January 2019) considers as Critical Areas: (1) hibernation areas, areas with regular presence of females with cubs in their first weeks outside the den, as well as refuge areas and resting sites used by bears in different seasons; and (2) crucial areas of ecological corridors.

The regions have also established their own catalogues of threatened species and compile and approve conservation and recovery plans for them.

Strategies and plans for flora and fauna species that are subject to exploitation or collection must include measures to ensure that those activities are compatible with their favourable status.

Conservation Strategies have been adopted for threatened animals or species of particular relevance in Spain (e.g. Cantabrian subspecies of the Capercaillie, Marbled Teal, Spengler's Freshwater Mussel and European Mink, Spanish Imperial Eagle, Pyrenean Desman, Iberian Lynx, Wolf, Brown Bear, Balearic Shearwater, Bearded Vulture and White-headed Duck). It is notable, however, that some taxa are under-represented, such as small mammals, fishes, amphibians and invertebrates. Recently, a number of strategies have been adopted for the conservation of groups of protected plant species in various environments (coastal, aquatic, rocky areas and high mountains). Other national strategies are in place to tackle important threats to biodiversity, including several IAS (see Section 29.6.3) and the use of poisoned baits.

For marine species, Conservation and Recovery Plans and Conservation Strategies are also adopted by the national administration with the protection instruments included in the Marine Environment Protection Act.[11] The Ministry of Ecological Transition is also responsible for the implementation of species plans in the marine environment. The Conservation Plan for the Killer Whale (*Orcinus orca*) in the Gibraltar Strait and the Gulf of Cadiz, approved in 2017,[12] is the first conservation plan adopted for a marine species. It includes, among other measures, the regulation of recreational whale watching, and supports the recovery of the Atlantic Bluefin Tuna (*Thunnus thynnus*), the Killer Whale's main food resource in the region, through the allocation of the fishing quota of Bluefin Tuna for the Strait fleet between July 1 and August 31 (within the International Commission for the Conservation of Atlantic Tuna, ICCAT), and updating fishery management plans to allow sufficient food resources for the Killer Whale in the area.

Some Examples of Species Recovery

Spain has a long tradition and proven success with recovery programmes for some threatened species for which the country has a significant responsibility, such as the Iberian Lynx, Brown Bear, Spanish Imperial Eagle and Bearded Vulture. These recoveries have been possible due to a change in attitude towards predators since the 1990s, and sustained proactive conservation efforts for more than two decades (largely financed by the EU LIFE programme), mainly to combat remaining illegal persecution and accidental deaths of the target species. Further details of these successful species recovery programmes are provided in the following paragraphs.

The Iberian Lynx, which was considered by IUCN to be the most threatened cat in the world, is an endemic species of the Iberian Peninsula. Illegal hunting, habitat destruction and road accidents reduced its populations during the twentieth century. Then the collapse of its staple prey, the European Rabbit (*Oryctolagus cuniculus*), caused by several diseases, led the lynx to the very verge of extinction in the 1980s. In 2002, there were less than 100 individuals, living in two isolated populations in Doñana National Park and in the Sierra Morena (both in Andalusia) in an area of 500 km^2 (Guzmán *et al.*, 2004).

Since 1994, several projects have been carried out in different regions, confirming the extinction of the populations of the species except in Andalusia, where conservation efforts have been focussed since 2002. The LIFE *Lince Andalucía* project (2002–2006) launched a captive breeding programme and implemented urgent actions to avoid the extinction of the lynx. Currently, there are breeding facilities in Andalusia (2), Extremadura (1) and Portugal (1). The success of captive breeding allowed for an ambitious re-introduction programme. The LIFE *Reintroducción Lince Andalucía* (2006–2011) consolidated the two existing populations of Doñana and Sierra Morena, mainly combating human-caused mortality and focussing on habitat improvement and reinforcement of rabbit populations. The LIFE *Iberlince* project (2011–2018) carried out a large number of re-introduction campaigns and recovered several areas where the lynx had become extinct, both in Spain and in Portugal. Recently, the region of Andalusia, along with several partners, started a new project (*Life Lynxconnect* 2020–2023) to connect areas currently occupied by the species.

Thanks to these actions, the Iberian Lynx has shown a dramatic increase, both in numbers and in range, from less than 100 individuals in 2002 to 1 111 in 2020[13] over a range of more than 3 000 km^2 in three Autonomous Regions of Spain (Andalusia, Castilla-La Mancha and Extremadura) and in one area of Portugal. As a result of this recovery, the IUCN

[11] Ley 41/2010, de 29 de diciembre, de protección del medio marino.

[12] Orden APM/427/2017, de 4 de mayo, por la que se aprueban las medidas de protección y el Plan de Conservación de las orcas del Estrecho y Golfo de Cádiz.

[13] MITECO. *Último censo de lince ibérico en España y Portugal.* Madrid: Ministerio para la Transición ecológica y el Reto Demográfico.

downgraded the Iberian Lynx's threat status from Critically Endangered to Endangered (Rodríguez and Calzada, 2015). Some further discussion of the re-introduction elements of these conservation programmes is provided in Section 29.4.5.

Another example of a successful recovery of a threatened species concerns the **Spanish Imperial Eagle**, another endemic species of the Iberian Peninsula. At the beginning of the 1990s, there were only about 150 breeding pairs and the species suffered serious threats due to the alteration of its habitat, shortage of its main prey (the rabbit), electrocution on powerlines and high mortality caused by poisoning. Since 1992, coordinated efforts have been carried out among the national and regional administrations across the entire species range. As a result, the eagle population increased to 530–540 breeding pairs in 2020, equating to more than 1 000 mature individuals. The population has experienced a long-term (1980–2018) increase of c. 500% (BirdLife International, 2022). The species recolonised Portugal in 2003, where it had not bred for more than 20 years, and has slowly increased since, with 17 breeding pairs recorded in 2020 (BirdLife International, 2022). These positive trends are largely attributed to the actions carried out to reduce mortality associated with powerlines and the use of poison, and the efforts made to recover the species' feeding habitat, driven decisively by a set of LIFE projects over the last 25 years (with a total cost of €16.2 million).

The recovery of the **Brown Bear** in the Cantabrian Mountains is another very positive experience. This population, which was considered Critically Endangered until recently, is divided into two subpopulations separated by an area of about 80 km, where breeding females are absent. The main historical threats have been illegal persecution and habitat fragmentation, which kept the two subpopulations genetically unconnected. Since the 1990s, many conservation projects and measures have been implemented in this area by the regional administrations and by the NGO Fundación Oso Pardo. In total, 20 LIFE projects were carried out between 1992 and 2020, with a total cost of €19 million. These efforts have reduced poaching and conflicts between bears and human traditional activities. Females with cubs of the year (monitored every year since 1989) increased from 10 in the biennium 1993–1994 to 79 in 2017–2018 (González *et al.*, 2016; Fundación Oso Pardo web-page). Both subpopulations are now reconnected due to the recovery of forest areas in the main corridors. As a result of this and the population increase, males (but not females) nowadays disperse regularly between both subpopulations, which is improving their genetic condition, especially in the much smaller eastern nucleus. The main cause of the recovery is the dramatic decrease in human-caused mortality, which has allowed an average annual rate of increase of 10% (González *et al.*, 2016). As a result, IUCN recently downgraded the threat status of the Cantabrian population to Endangered (Huber, 2018).

The nationally Critically Endangered **Bearded Vulture** increased in the Spanish Pyrenees from 17 breeding pairs in 1979 to 128 in 2014 (FCQ, 2018) due to the reduction in human-caused mortality achieved by conservation campaigns, and a successful supplementary feeding programme. The vultures were illegally shot and frequently accidental victims of poisons used illegally to control predators that allegedly preyed on game species. Since Bearded Vultures have a long lifespan (up to 40 years) and low reproductive rates, the death of mature individuals has a particularly strong effect on their population dynamics. Re-introduction projects have also been carried out for this species in Spain, which are described in more detail in Section 29.4.5.

29.4.4 Dealing with Conflicts

Some wildlife species interact with particular human activities and cause conflicts that need to be addressed with appropriate measures. When considering solutions to these interactions, the general species protection provisions need to be taken into account. In accordance with the Habitats Directive and national law, exceptional authorisations can only be granted if it is demonstrated that there are no other satisfactory solutions, and the selected solution will not prevent the population concerned achieving a favourable conservation status.

The **Wolf** is typically the species that causes most conflicts with humans throughout its range in Spain. Unlike other countries in western and central Europe, wolves were not exterminated in Spain, but they were very much reduced during the twentieth century. In 1970, there were a few hundred individuals in the north-west of the country and a few remnant and isolated populations in the south. As persecution decreased over the last five decades, the Wolf population in the north-west increased and its range expanded southward. The northern population crossed the river Duero around the year 2000, and in 2019 five breeding packs were detected in the autonomous community of Madrid, not far from the city of 5 million people. At the same time, the very small populations that survived in the 1970s in the south of Spain became extinct, including the Sierra Morena population. Nowadays, there are at least 300 Wolf packs in Spain (and 60 more contiguous packs in the north of Portugal).

As in other European countries, wolves kill livestock, causing conflicts with farmers. The resulting damage is compensated through a monetary payment (except in some areas of Castilla y León), and some measures to prevent attacks are used by many farmers. The management of livestock can play an important role in the prevention of attacks by wolves; where livestock are properly guarded, the attacks are lower. However, it is difficult to implement protective measures for extensive livestock systems. Some projects implemented in Spain, such as the LIFE project COEX,[14] included the lease of electric and conventional fences for extensive livestock raising practices (including high-value fighting bulls), the lease of livestock-guarding dogs, the development of a management plan for

[14] LIFE04 NAT/IT/000144.

stray dogs and the revision of compensation procedures. In spite of these efforts, Wolf depredation on livestock is still a crucial challenge.

According to the Habitats Directive, the Wolf is included in Annex V north of the river Duero, and in Annexes II and IV (strictly protected) south of the river Duero. Since Annex V allows some flexibility, some regions north of the Duero (e.g. Galicia, Cantabria, La Rioja and Castilla y León) have managed the Wolf as a game species until recently, with around 200 individuals killed per year according to quotas established by the authorities. South of the Duero, just a few individuals were culled every year through derogations (HD Article 16) in the region of Castilla y León, in order to reduce the increasing depredation of cattle, mainly in Ávila province. Nevertheless, in September 2021, all Spanish Wolves were included by the national government in the List of Wild Species under Special Protection, as a result of which hunting is forbidden in the entire country and culling for damage control is heavily restricted. To reduce conflicts caused by wolves, the European Commission launched a Wolf management platform for stakeholders in Ávila province during 2019–2021. This kind of social mediation adds to the other well-established measures (e.g. damage prevention and compensation, and selective culling) to alleviate ongoing Wolf conflicts.

29.4.5 Species Re-introductions

Several re-introduction or reinforcement projects have managed to restore some populations of extinct or threatened species in Spain. Some ongoing examples are described in an online annex (ES.1),[15] concerning the Iberian Lynx, extinct populations of Bearded Vulture, and a project to restore the Brown Bear population in the Pyrenees. Scientifically planned re-introductions and reinforcements have proven to be a good method to restore extinct or remnant populations that are unable to recover by themselves. The first steps of the re-introduction/reinforcement programmes are usually fraught with problems, but after a difficult learning period they usually give good results if they have been planned with a medium- or long-term perspective. In the re-introductions/reinforcements of species that cause conflicts with people, like large carnivores, addressing social issues is at least as important as solving the biological and ecological problems.

29.5 Protected Areas and Networks

29.5.1 Protected Area Types and Their Coverage

Table 29.4 lists the main types of protected areas that occur in Spain, as well as their number and coverage (including the Canary Islands, as it is not straightforward to focus on European Spain with the available data). Protected areas are defined and regulated on a national basis by the NHBA, which classifies them into three main types: Natural Protected Areas (NPAs), Natura 2000 sites, and protected areas designated under international instruments. NPAs are classified into five basic categories by the national law: Parks (including Natural Parks and National Parks), Nature Reserves, Marine Protected Areas (MPAs), Natural Monuments and Protected Landscapes.

Among all the NPAs, the best known are the **National Parks** and **Natural Parks**, not only for their conservation management history but also for their public use and the socio-economic development to which they contribute. National Parks are governed by specific legislation of the national parliament.[16] In principle they are less transformed than Natural Parks, where more uses are allowed, although the use of natural resources can be limited, prohibiting in any case those incompatible with the purposes that justified their creation. Hunting in National Parks is not allowed but it can be done in Natural Parks. Social enjoyment, as in National Parks, is part of the objectives of Natural Parks, facilitating the entry of visitors with some limitations to guarantee their conservation. All the autonomous communities have declared Natural Parks and have responsibilities for their planning and management, although some are managed by the provinces.

Nature Reserves are designated for the conservation of ecosystems, communities or biological elements that, due to their rarity, fragility, importance or uniqueness, deserve special protection.

MPAs are designated for ecosystems, communities or biological or geological elements of the marine environment, including intertidal and subtidal areas, which, due to their rarity, fragility, importance or singularity, deserve special protection.

Natural Monuments are designated to protect areas or important natural elements due to their singularity, rarity or beauty, such as single especially notable trees, geological formations, palaeontological and mineralogical deposits, and other elements of special interest due to their scientific, cultural or landscape values. These categories can overlap; for example, Nature Reserves and Natural Monuments can be included in other protected area types, such as Natural or National Parks.

Protected Landscapes are designated to protect natural, aesthetic and cultural values, in accordance with the Landscape Convention of the Council of Europe, to preserve the harmonious interaction between nature and culture in a given area and to maintain traditional practices that contribute to the preservation of their values and natural resources.

According to the latest update of the National Inventory of Protected Areas (MITECO, 2022), in 2020, there were 1 824 NPAs in Spain (including the Canary Islands), covering 74 551 km^2 on land (14.7% of the area), and 52 572 km^2 at sea,

[15] All annexes are free to access online at CUP: www.cambridge.org/natureconservation

[16] The National Parks Act. Ley 30/2014, de 3 de diciembre, de Parques Nacionales.

Table 29.4 The number and extent of protected areas in all Spain (2020).

	IUCN category	Number	Area (km²)
Designated under global and non-EU international legislation			
Ramsar Sites	NR, IV	76	3 130
Natural World Heritage Sites	III	4	768
Biosphere Reserves (core and buffer zones)	VI	52	72 147
OSPAR Marine Protected Areas	NR	15	28 242
SPAMI	NAs	9	1 484
Designated under the Nature Directives			
SCIs (terrestrial + marine)	NR	1 468	118 636 + 54 751
SPAs (terrestrial + marine)	NR	658	102 508 + 51 986
Natura 2000 terrestrial area			138 460
Natura 2000 marine area			84 322
Designated under domestic legislation			
Natural Protected Areas (NPAs)			
Natural Parks (Parque natural)	Ib, II, IV, V, VI, NAs	133	36 389
National Parks (Parque nacional)	II	15	4 628
Nature Reserves (Reserva natural)	Ia, Ib, III, IV, V, VI, NAs	230	1 498
Natural Monuments (Monumento natural)	III, IV, V, NAs	264	867
Protected Landscapes (Paisaje protegido)	III, V, NAs	62	1 797
Marine Protected Areas (Área marina protegida)	VI	2	48 963
Others			
30 types listed in the WDPA		829	<c. 22 000*

Notes. NR = Not reported; NAs = not assigned; NAp = not applicable. * Based on sum of areas in WDPA; actual combined area will be lower due to overlaps. Nature Reserves include those listed as Integral, Management, Partial, Special and Wildlife in the WDPA.
Sources. Based on data from: Ramsar Convention website (May 2022); OSPAR website (database update February 2022); World Heritage Sites, Biosphere Reserves, SPAMI, SCIs and SPAs, from National Inventory of Protected Areas (update 2020); others and IUCN categories from WDPA (April 2022).

which is 4.9% of all Spanish marine waters. Natura 2000 sites, of which there are 1 857, cover 27.3% of the national terrestrial area and 7.9% of the marine area. As a result, Spain provides about 20% of the whole EU terrestrial Natura 2000 network, which is the largest contribution of any Member State.

Virtually all NPAs declared under the national or regional legislation are part of the Natura 2000 network. Furthermore, Natura 2000 areas are considered NPAs by the national legislation.

Since most of the regions have developed their own legislation on protected areas, in addition to these basic categories established by national law, there are currently more than 40 different types of protected areas designated in European Spain. For instance, some regional administrations have established other categories of area-based conservation measures such as micro-reserves, protected biotopes, protected wetlands, ecological corridors, singular natural areas, sites of scientific interest, sites of national interest and private reserves. However, some of these areas possibly do not meet IUCN protected area criteria, for instance the ecological corridors.

In addition to the protected areas established under nature legislation, there are other designations that provide some biodiversity protection. These include Marine Reserves established under fisheries law (described below) and Fluvial Reserves declared as water policy measures that aim to preserve sections of rivers with little or no human intervention and in very good ecological condition (most of which are in NPAs or Natura 2000 sites). They are the responsibility of the relevant river basin management body.

Up-to-date national public data on the combined area of all protected areas do not appear to be available. According to WDPA data, taking into account overlaps, the total coverage over all Spain of protected areas (except Biosphere Reserves) is 142 503 km^2, which is 28.1% of the land (UNEP-WCMC, 2022). The reliability of this estimate is a little uncertain, as the WDPA includes 30 types of protected area in addition to

NPAs, some of which may not meet IUCN/CBD protected area criteria. Nevertheless, as the Natura 2000 coverage alone is 27.3%, the vast majority of the area can be reliably considered to meet protected area standards.

Spain has made a big effort to protect its marine biodiversity, including through its Natura 2000 marine sites and the declaration of a large area in the Mediterranean, a Cetacean Migration Corridor, declared in 2018 as an MPA and Specially Protected Areas of Mediterranean Importance (SPAMI) under the Barcelona Convention. It is a continuous strip of marine water under national jurisdiction with a total area of 46 385 km^2 and an average width of about 85 km that runs between the Catalan and Valencia coast and the Balearic archipelago. With this new MPA and its Natura 2000 sites, Spain has protected over 12% of its total marine waters (MITECO, 2022).

Marine Reserves are established under fisheries legislation, with protection measures that include the elimination of industrial fishing and the most damaging fishing gear (e.g. bottom trawling and purse seining), and regulation of artisanal fishing and other activities such as diving. According to the WDPA, there are currently 12 reserves covering 1 125 km^2, of which about 100 km^2 have higher protection levels where fishing and diving are prohibited. Marine Reserves occur within other protected areas, such as Natural Parks and Natura 2000 sites, and are also part of the MPA network.[17]

Overall, taking into account overlaps, the total coverage of the marine protected areas in all Spain is 131 789 km^2 (MITECO, 2022), which is 12.3% of its total marine waters.

Ministerial statements have indicated that Spain supports and is fully committed to reaching the target of at least 30% coverage of the land and 30% of the sea area within protected areas, in line with the *EU Biodiversity Strategy to 2030* and proposals under the CBD for the post-2020 global biodiversity framework.

29.5.2 The Coherence of Protected Areas and Ecological Networks

With over 27% of the European land area included in the Natura 2000 network, one might expect that the network of terrestrial protected areas in Spain would be coherent. At the national scale, coherence would mean that the full range of variation in specific habitats and species is represented; appropriate dispersal, gene flow and migration between sites is supported; and threatened species and habitats are protected. However, there has been no analysis of whether this is the case. Furthermore, protected areas in Spain are also designated by the regional administrations, which apply their own criteria, making the assessment of coherence more complex.

To ensure ecological connectivity, a *National Strategy for Green Infrastructure, Ecological Connectivity and Restoration* was adopted in 2020 (MITECO, 2020b). It provides criteria and guidelines for: identification and conservation of green infrastructure in the terrestrial and marine area; promotion of climate change mitigation and adaptation; defragmentation of strategic areas for connectivity; restoration of degraded ecosystems; and prioritisation of habitat restoration.

Some regional initiatives regarding ecological connectivity have also been carried out, such as the Green Infrastructure Programme in Catalonia (Generalitat de Catalunya, 2017), the Plan for Improvement of Ecological Connectivity in Andalusia (Junta de Andalucía, 2018), the Network of Ecological Corridors of the Murcia Region (ATECMA, 2007), and the Network of ecological corridors in the Basque Country (Gurrutxaga, 2005). Recently, WWF (2018) also identified a network of ecological corridors for forest species, especially mammals.

Spain strongly supports a specific target on ecological connectivity under the post-2020 global biodiversity framework (MITECO, 2020b).

29.5.3 Approaches to Protection and Managing Protected Areas

The NHBA defines the national legal framework for protected areas, while the regional governments have also developed their own legislation regimes, which results in some different approaches and procedures for their protection and management.

In any case, all protected areas must be declared through legislative acts by the competent administration, which must also develop adequate planning and management to achieve their conservation objectives. The national administration, currently the MITECO, is responsible for wholly marine protected areas under national sovereignty or jurisdiction, whilst the regions are responsible for terrestrial protected areas in their territory, and for terrestrial protected areas that extend over marine waters that have ecological continuity with the former.

The NHBA requires management plans to be adopted for National and Natural Parks, Nature Reserves and MPAs. The use of natural resources may be limited in these areas, prohibiting those activities that are not compatible with the objectives that justified their designation. Given their special function, Nature Reserves have stricter protection, and the exploitation of resources is usually prohibited in these reserves. Management plans are also adopted in Spain for all Natura 2000 sites in accordance with national law.

Management plans for protected areas usually include a zoning system, regulation of activities and the necessary conservation measures. According to the national legislation, the management plans for National and Natural Parks prevail over urban planning.

Peripheral protection zones may be established around protected areas to buffer impacts from outside areas. In order to promote local development, areas of socio-economic influence may be established, which include a special economic

[17] In accordance with 33/2014 Act, which amends 3/2001 Act on the protection of the marine environment.

regime (e.g. subsidies for certain activities) and appropriate compensation for limitations on economic activities in the protected areas (e.g. certain agriculture and forestry practices).

Management plans are prepared and adopted through some legislative act by the regional or national administration, and the regulations included in the plans are mandatory. There is, however, a wide variation of approaches and procedures for the development and adoption of management plans by the relevant authorities. These may involve only basic consultations, or have a wide participatory approach, depending on the case.

A good participatory approach has been recently used in the preparation of management plans for marine Natura 2000 sites by the national administration. This involved an intensive participatory process including workshop discussions with all relevant stakeholders (e.g. fishing, tourism and energy sectors, local authorities, ports, scientific institutions, NGOs and civil society organisations). The aim of this process, which took around two years in some cases, was to reach an agreement with the main actors in order to ensure effective management of the conservation areas. This has proved very useful to define the necessary conservation measures and possible regulation of activities, ensuring also better governance of these marine sites.

National Parks were formerly managed by the national administration and had a very high level of protection until the Constitutional Court ruled in 2004 that they should be managed by the regions. Since then, the regional governments manage their budget, approve their management plans, appoint the director and decide on the possible restrictions (e.g. the regulation of vehicle access, livestock and camping). In some cases, conservation in National Parks has been degraded, as political pressures have more weight at the regional and local levels of administration than at the national level, and the interests of biodiversity conservation are subordinated to other interests, such as tourism, development and the economy. Nowadays, National Parks are often managed as Natural Parks or regional parks.

Although the central administration still has a coordinating role for National Parks and transfers annually the funds for their management to the Autonomous Regions, in practice its intervention capacity is very low. This is clear in the National Parks that span several Autonomous Regions, such as the Picos de Europa National Park, the first one created in Spain and one of the most visited. It has currently three directors, appointed by the three regional administrations, who actually manage the National Park as if it were three independent natural areas.

29.6 General Nature Conservation Measures

29.6.1 Development Planning and Impact Assessment

Spatial planning is carried out in Spain at the national level (e.g. the general plan for roads, renewable energy plan and hydrological plan) or at the regional level (e.g. land-use plans and port development plans). These plans are subject to SEA, which must consider their potential impacts on protected areas and other important natural elements, such as those that contribute to ecological connectivity.

The national legislation on environmental assessment (21/2013 Act) promotes streamlining all environmental assessment procedures, including the HD Article 6(3) appropriate assessment of plans' and projects' effects on Natura 2000 sites. There are guidelines to integrate these assessments in the EIA procedures (MAPAMA, 2018). Criteria and methodologies for the assessment of adverse effects on the integrity of Natura 2000 sites were also published in 2019 (MITECO, 2019).

The assessment of the *National Strategic Infrastructure Plan* (2005–2020) included an evaluation of habitat fragmentation for various species for which ecological connectivity is an important issue (e.g. Iberian Lynx, Brown Bear and some other mammals). Roads were classified considering their potential negative effect on these species and relevant recommendations were provided to mitigate impacts. A national Group on Habitat Fragmentation and Transport Infrastructure also plays an important role in coordinating and producing technical recommendations and guidelines on preventing the impacts of transport infrastructure on landscape connectivity.[18]

The previously mentioned *National Strategy for Green Infrastructure, Ecological Connectivity and Restoration* provides guidelines to ensure that sectoral planning contributes to maintaining ecological connectivity and the functionality of ecosystems, to promote defragmentation of strategic areas for connectivity and restoration of degraded ecosystems.

In 2020, the MITECO launched the SAFE Plan, a citizens' science initiative to evaluate fauna mortality in the transport network and identify hotspots for road kills. The project is coordinated by the Doñana Biological Station (EBD) and implemented by scientific societies such as SEO/BirdLife, the Society for Study and Conservation of Mammals (SECEM) and the Spanish Herpetological Society (AHE).

29.6.2 Habitat Management and Restoration

Historically, habitat management and restoration were mainly focussed on forests, and to a lesser extent on dunes, wetlands and seagrass meadows.

Afforestation and reforestation have been used as a measure to counteract land degradation and desertification, and as mitigation against climate change and biodiversity loss. This has been coordinated by the government through a national restoration programme for forests. The total reforested area over 1983–2013 amounted to 1 762 km^2, which represents 6% of the area currently classified as forest (MITECO, 2020a). However, ecological restoration criteria have not always been

[18] www.miteco.gob.es/es/biodiversidad/temas/ecosistemas-y-conectividad/conectividad-fragmentacion-de-habitats-y-restauracion/fragm-documentos-grupo-trabajo.aspx

properly considered in the design of reforestation schemes. This can have negative impacts on biodiversity, mainly due to the inappropriate choice of species (sometimes non-native) and lack of sustainable forest management. Moreover, reforestation carried out on agricultural areas, wasteland, abandoned agricultural areas or pastures has in some cases led to the modification of valuable habitats, such as dry grasslands and steppe areas. Recently the situation has improved and some programmes (e.g. those receiving EU funds in some regions) include particular requirements on the selection of areas and choice of species, including an appropriate percentage of shrub species.

Coastal dune restoration interventions date back to the beginning of the twentieth century, but as their main objective was fixing mobile dunes, they did not always follow principles of ecological restoration. The actions have aimed to, and continue to, restore the coastline affected by storms, adapt to climate change and restore dune habitats for nature conservation. Since 2007, a *Coastal Dune Restoration Manual* (MMA, 2007) has provided restoration techniques for these ecosystems.

In the last three decades, **wetland restoration** has been promoted throughout the country and a *Strategic Plan for the Conservation and Rational Use of Wetlands* (MMA, 1999) was launched in 1999. Wetland restoration projects have generally followed basic principles of ecological restoration, such as the use of native species, identification of local reference ecosystems, recovery of the hydrological dynamics and reinforcement of populations of threatened species. However, there have been very few cases of landscape-scale restoration to address pressures on wetlands, such as from agricultural pollution.

The vast majority of wetland restoration projects have been financed with EU funds (mainly LIFE). However, despite the number of restoration projects carried out, the restored area is still very small (slightly more than 60 km^2), compared to an estimated 7 000 km^2 of wetlands that have been lost over the whole country (Mediterranean Wetlands Observatory, 2012).

River restoration has been especially promoted in the last two decades. Initially it focussed on the restoration of riverbanks and later gradually progressed towards restoration of entire river stretches, where flow, sediment and obstacle removal have played a primary role. The *National River Restoration Strategy* (MMARM, 2010) has promoted the removal of more than 200 dams and weirs, the recovery of old river beds and riparian galleries, eliminating channels and reconnecting meanders, ensuring minimum flows and floods for ecological purposes. Some projects have also contributed to the recovery of river habitats for protected species, such as the European Mink. In addition, scientific and technical support for river restoration is provided by the Iberian Center for River Restoration, founded in 2009, which is part of the European Center for River Restoration.

Marine ecosystem restoration experiences have been mainly focussed on seagrasses. Measures in place for the protection and conservation of *Posidonia* meadows (1120*) include the prohibition of bottom trawl fishing and the regulation of boat mooring in marine protected areas.

The installation of artificial reefs in the sea has also been used to protect certain habitats from bottom trawling or to promote recolonisation by marine species.

Restoring ecosystems across land and sea is one of Spain's priorities for the future, under the framework of the EU's intended Nature Restoration Plan (MITECO, 2020a).

29.6.3 Invasive Alien Species

Spain has included 185 taxa in its Catalogue of Invasive Alien Species,[19] which also requires the elaboration of strategies against IAS, giving priority to the ones affecting protected areas. Six national strategies have been developed and adopted[20] addressing Water Hyacinth (*Eichhornia crassipes*), Pampas Grass (*Cortaderia selloana*), invasive snakes in islands, Asian Hornet (*Vespa velutina*), American Mink (*Neovison vison*), Zebra mussel (*Dreissena polymorpha*) and invasive plants in dune systems. Actions for endangered species threatened by IAS (e.g. the European Mink by the American Mink) are also being implemented. An alert system to detect IAS at an early stage has been established.

Despite the strategies, and some actions, a lack of management and research programmes is making it difficult to deal with the serious problems posed by IAS. Further financial and human resources are therefore required.

In the application of the EU IAS Regulation, the risk analysis model suggested by the European Commission is widely used to assess proposals for new species to be included in the List of IAS of Union concern. Work is also ongoing to prepare a prevention strategy on IAS entry pathways.

29.6.4 Climate Change Adaptation

Climate change is expected to have a significant impact on biodiversity in Spain. In particular, direct impacts on vegetation are expected due to increases in temperature and reduced rainfall, and declines in water resources. Two main trends are envisaged: the 'mediterranisation' of northern regions and desertification in the south (MITECO, 2020a). Many inland aquatic ecosystems will change from permanent to seasonal, and some others will disappear. The most vulnerable habitats are temporary wetlands, high-mountain lakes, lagoons, rivers and streams, and coastal wetlands depending on ground water. In terrestrial ecosystems, major impacts are expected in the ecosystems with ecological or geographical restrictions, such as in high mountains or some arid areas. The most vulnerable habitats are predicted to be high-mountain vegetation, deciduous forests and shrubland sensitive to drought; sclerophyllous

[19] Real Decreto 630/2013, de 2 de agosto, por el que se regula el Catálogo español de especies exóticas invasoras.

[20] www.miteco.gob.es/es/biodiversidad/temas/conservacion-de-especies/especies-exoticas-invasoras/ce-eei-estrategia-planes.aspx

and laurel forests in the south and south-east; and coastal vegetation. Phenology will be altered, as will interactions among species; changes in distribution of terrestrial and aquatic species will occur, as well as the extinction of local species. The most vulnerable animal populations are expected to be those of highly threatened habitats, particularly in mountain areas (MMA, 2006).

A first *National Plan for Adaptation to Climate Change* was adopted in 2006 (MMA, 2006) and has been developed through three successive work programmes up to 2020. The third programme (MAGRAMA, 2014b) included an evaluation of the vulnerability of biodiversity with regard to IAS and the development of guidelines to integrate adaptation into ecological restoration and ecosystem connectivity efforts and into the planning and management of protected areas. Specific adaptation projects have been developed since 2015 through ad hoc programmes (PIMA Adapta) on biodiversity, aquatic ecosystems, National Parks and coastal areas.

Since 2016, a *National Strategy for Adaptation of the Coast to Climate Change* (MAGRAMA, 2016) has been in place, which defines the specific objectives, general guidelines and measures to contribute to increasing the resilience of coastal and marine ecosystems, with special attention to endemic, threatened and protected species.

A new adaptation plan has been prepared for 2021–2030 (MITECO, 2020a), which proposes 81 action lines in various areas, including human health, water, biodiversity and protected areas, forest protection, combating desertification, agriculture, livestock, fishing and food. It also proposes new governance to improve the coherence of climate action in the public and private sector.

Spain will need to carry out major efforts to try to limit the effects of climate change, which are expected to be particularly severe in the Mediterranean region, considered one of the global 'hotspots' in terms of climate change. Recognising this, the Spanish Council of Ministers approved in January 2020 a *Declaration on Climate and Environmental Emergency* in Spain, which includes 30 priority action lines to combat climate change with cross-cutting policies. Among these priorities, the national government will strengthen synergies with rural development policies, the protection of biodiversity and the recognition of ecosystem services, and will prepare a new Strategy to Combat Desertification, a National Forest Strategy and a Strategy against the Demographic Challenge.

29.7 Nature Conservation Costs, Benefits and Funding Sources

29.7.1 Nature Conservation Costs

The total costs of conservation measures for Natura 2000 and other areas beyond the network have been calculated as part of the development of Prioritised Action Frameworks (PAFs) for each autonomous community or city. According to the combined PAF for the whole of Spain for 2022–2027 (MITECO, 2020c), the full costs of implementing Natura 2000 amount to €1.4 billion per year on average (81% for recurrent costs). Of these costs, 24% are foreseen for horizontal measures and administration, 49% are for habitat maintenance or restoration in Natura 2000 sites, 21% are for supporting green infrastructure, and 6% are allocated to species-specific measures. The costs of habitat measures within Natura 2000 sites comprise €14 million for marine and coastal habitats (€1.7/ha over all marine sites) and €663 million for other habitats (€48/ha over all terrestrial sites).

29.7.2 The Economic Benefits of Nature Conservation

A study on the economic benefits of terrestrial Natura 2000 sites over all Spain estimated that the benefits derived from their designation and management have an annual value of around €1.9 billion (García, 2019), which is higher than their estimated management costs. The study also points out that the Natura 2000 network in Spain may employ around 600 000 people.

29.7.3 Funding Sources for Nature Conservation

According to national statistics, the total expenditure on biodiversity and landscape protection in Spain in 2020 amounted to €861 million,[21] which indicates a significant deficit considering the financial need estimated for Natura 2000 alone, as indicated in Section 29.7.1.

Funding for nature conservation in Spain comes mainly from public sources, both EU and national funds (Table 29.5). Due to the recent economic crisis, national funding for biodiversity has declined over the last decade. Although potential EU funding of biodiversity conservation has been substantial, it has been insufficient in practice due to its limited integration into the funding instruments. Over the 2014–2020 period, an evaluation of the degree of inclusion of priorities and measures identified in the PAF for Natura 2000 in the Rural Development Programmes (RDP) concluded that the funds dedicated to biodiversity had not met the needs of the network (Bi+De, 2015). It is also notable that a significant part of RDP funding is allocated to investments in forests, organic farming and areas facing constraints, which do not actually contribute significantly to nature conservation, in general and in Natura 2000.

A similar conclusion was reached that funding for marine biodiversity measures from the 2014–2020 Operational Programme of the European Maritime and Fisheries Fund (EMFF) was inadequate (ATECMA, 2014). Although the European Regional Development Fund (ERDF) has contributed to financing some actions for Natura 2000, particularly actions related to development of infrastructure and

[21] www.ine.es/dyngs/INEbase/es/operacion.htm?c=estadistica_C&cid=1254736177048&menu=ultiDatos&idp=1254735976603

Table 29.5 Funding sources for nature conservation in all Spain.

Source and period	Measure/category	Average annual allocation (€ m)		Comments
EU funding instruments		**EU + national**	**EU**	**Data for all of Spain**
EAFRD Rural Development Programme measures – allocations to Priority 4 (restoring, preserving and enhancing agriculture and forest ecosystems) 2014–2020	M10 Agri-environment–climate	186.0	122.4	
	M8 Investments in forests	173.5	114.8	
	M11 Organic farming	130.1	90.7	
	M13 Areas facing constraints	116.0	79.6	
	M4 Investments in physical assets	15.2	9.9	
	M7 Basic services, etc.	14.7	9.9	
	M12 Natura and WFD	7.5	5.6	
	M16 Cooperation	3.7	2.3	
	M1 Knowledge transfer, etc.	2.6	2.1	
	M15 Forest environment climate	2.2	1.6	
	M2 Advisory services, etc.	0.8	0.4	
RDP Priority 4 subtotal		*652.3*	*439.4*	
ERDF 2014–2020	85 Protection and enhancement of biodiversity, nature protection and green infrastructure	46.3	31.2	For Nk2 etc.
	86 Protection, restoration & sustainable use of N2k	0.1	0.1	
LIFE Programme	Nature and biodiversity projects (1992–2020)	16.7	10.3	
	One Integrated project (2017–2024)	3.4	1.7	LIFE-IP INTEMARES
ERDF-Interreg 2014–2020		6.5	5.2	For Nk2 etc.
EMFF 2014–2020	E.g. Innovation in fisheries and aquaculture, protection of marine biodiversity and ecosystems, waste removal, advisory services and partnerships	3.5	2.6	For Nk2 etc. EMFF Arts. 27, 28, 39, 40a, 40.1(b–g, i), 47, 49, 50 and 54
EU funding instruments subtotal		*728.8*	*490.4*	
National public funds within 2014–2018		**Additional national***	**Comments**	
National funds for the General State Administration (2014–2017)	Budgets dedicated to the protection and improvement of the natural environment assigned to the General Directorate of Biodiversity, Forests and Desertification, DG of the Coast and the Sea, DG of Water and National Parks	32.5		
National total including co-financing of EU instruments		270.9		
TOTAL		**761.3**		

Notes. * Excludes co-financing of EU funds unless indicated. 'For Nk2 etc.' indicates funding relevant to Natura 2000, related green infrastructure and BHD species (based on the PAF). The subtotals and totals are only approximate indications of the average annual levels of funding as they are based on annualised data from differing periods.

Sources. Based on data from: RDP data from European Commission ESIF funds database (23 October 2020); ERDF, EMFF and national funding from 2021–2027 PAF (MITECO, 2020c); Spain LIFE factsheet, European Commission (May 2021). See Section 1.3 for further details.

restoration activities, its contribution to nature conservation has been relatively low overall. Minor use has been made of the European Social Fund (ESF) for nature conservation, which has only funded a few actions related to environmental training and for environmental businesses in some natural areas.

The contribution of the LIFE Programme to financing nature conservation in Spain has been very important. Since the creation of this fund in 1992, LIFE has co-financed 282 nature and biodiversity projects over Spain (including the Canaries), with a total value of over €483 million (€297 million EU funding) and with the participation of numerous partners, including both governmental and non-governmental organisations. LIFE projects have contributed to the recovery of threatened species, as discussed earlier in this chapter. It is also worth mentioning that a large LIFE Integrated Project (INTEMARES)[22] is underway to support the conservation of marine habitats and species, and the consolidation of the marine Natura 2000 network.

In recent decades, there have been some limited efforts to obtain private funding for nature. For example, the Biodiversity Foundation of the MITECO has supported the Spanish Business and Biodiversity Initiative. The Initiative facilitates cooperation among private companies, NGOs, the scientific community and the administration to join efforts to improve and maintain Spain's natural heritage. Over 2016–2018 the Initiative supported more than 2 000 projects with more than €480 million.

29.8 Surveillance, Monitoring, Reporting and Applied Research

Traditionally, species monitoring and surveillance in Spain has mainly targeted threatened species that have conservation or recovery plans. This involves specialist NGOs, experts from relevant regions and neighbouring countries, and coordination meetings run by the national authorities. As an example, each autonomous region funds the monitoring of Wolf pack numbers as a proxy of the total population, but the frequency of the monitoring depends on each region.

Some NGOs also monitor non-threatened species, using volunteers led by scientists, in programmes that are sometimes funded by administrations, private companies and other irregular funding sources. For example, SEO/BirdLife maintains 13 monitoring programmes for bird species and their habitats, based on citizen science, which involve several thousand collaborators and generate millions of records annually. In the same way, since 2009, the Spanish Herpetological Society (AHE) has monitored Spanish amphibians and reptiles (project SARE), applying a standardised methodology, which is open to the participation of professionals and volunteers. This programme is partially funded by the MITECO. In recent years, some NGOs have used funds for applied research that have come from national taxes and been channelled through the MITECO.

Monitoring and reporting on BHD habitats and species in accordance with the Nature Directives was mainly through expert judgement and the collation of available data from existing programmes, and therefore lacked a proper systematic approach. Moreover, many species and habitats were reported as having an unknown status in the previous reporting periods. However, the situation has improved slightly over the last decade. Guidelines for monitoring and assessment of the conservation status of threatened and protected species were adopted in 2012, and further updated in 2018. These aim to coordinate and aid the exchange of information and surveillance results among regional and national authorities. More recently, more robust methodologies for monitoring the conservation status of habitat types in Spain have been developed,[23] and standardised protocols are being established and agreed with all the regions to provide more reliable assessments from now on. In this biodiversity monitoring work, the important role played by scientific societies in Spain needs to be underlined.

Despite these recent efforts, much remains to be done to improve knowledge and monitoring of natural habitats and species, especially in the marine environment, and also to assess the effectiveness of conservation measures and management plans implemented in protected areas. At the time of writing (January 2022), an ambitious system of biodiversity monitoring is being set up including remote sensing and satellite earth observation data from the EU's Earth Observation Programme – Copernicus.

29.9 Future Developments

A new *Strategic Plan on Natural Heritage and Biodiversity to 2030* was adopted in Spain in December 2022, which incorporates the relevant actions to achieve the objectives and targets set in the *EU Biodiversity Strategy for 2030* and the *Farm to Fork Strategy*.

Science will play a pivotal role in shaping environmental policy in Spain, and a Strategy on Science, Innovation and Biodiversity is currently being prepared by the MITECO. Other priorities include the new System for Biodiversity Monitoring and a strong emphasis on ecosystem restoration under the UN Decade on Ecosystem Restoration. Another intention is to increase Spain's presence at, and contribution to, international conventions and processes.

There is also a commitment to reach the goal of 30% protection of marine waters by 2030 through the planning, declaration and management of new MPAs.

22 LIFE15 IPE/ES/000012.

23 www.miteco.gob.es/es/biodiversidad/temas/ecosistemas-y-conectividad/Seguimiento_habitats_metodologia.aspx

29.10 Conclusions

Spain is one of the most biodiverse countries in the EU and Europe. It has a diverse range of habitats and species, and it is of major importance for some endemic and threatened species, such as the Iberian Lynx, birds of prey and steppe birds. Some of these iconic species have been the focus of conservation and recovery programmes, which have been successful due to changes in attitude towards predators and concerted actions in key areas. However, less attention has been paid to the conservation of commoner and less charismatic species.

Recognising its natural heritage, significant progress has been made in the designation and management of protected areas. In Spain, over 28% of the terrestrial area and 12.3% of marine waters are protected. However, the protected areas need to be managed more effectively, and human and financial resources increased to achieve this. Despite this need for strategic investment, national funding has declined over the last decade due to the recent economic crisis.

Efforts aimed at the conservation of HD habitats began later, and their development has been hampered in many cases by a lack of knowledge about their management needs. A particularly high proportion of grassland habitats have an unfavourable status and require decisive wide-scale actions to conserve and restore them and allow the recovery of their associated species. Bird species associated with traditional HNV farmland are also suffering significant declines, caused by agricultural intensification or the lack of appropriate management of their habitats; this has been particularly serious in the case of the Little Bustard and other steppe birds. The national and regional governments need to strengthen the use of available agricultural measures to promote a recovery of biodiversity, and the new CAP has a crucial role in this.

In addition to conservation actions, restoration efforts are needed to improve the status of many habitat types and species in Spain and to make them more resilient to climate change. Some existing restoration initiatives need to be strengthened and new ones promoted, in line with the new EU agenda on restoration. Ecological connectivity also plays a crucial role in the restoration of biodiversity and ecosystems functioning. It will therefore be important for the *National Strategy on Green Infrastructure and Ecological Connectivity and Restoration* to boost relevant efforts in the coming years.

Significant progress has been made in Spain in recent years with the creation of the marine protected area network. Efforts are continuing to further boost marine conservation and to identify and designate new MPAs to achieve the EU and global goals to protect 30% of the sea area under national jurisdiction. Management plans are being developed and updated for marine protected areas, but further efforts in terms of staff, management bodies and stakeholder cooperation are required to ensure their effectiveness.

Acknowledgements

We are grateful to Borja Heredia and Yolanda Cortés for their review and valuable suggestions on this chapter.

References

ATECMA (2007) *Identificación y diagnóstico de la red de corredores ecológicos de la Región de Murcia. Documento de Síntesis.* Región de Murcia.

(2014) *Integración de las medidas de conservación necesarias para la Red Natura 2000 en el Programa Operativo del Fondo Europeo Marítimo y de la Pesca (FEMP).* Madrid: Fundación Biodiversidad.

Beaufoy, G., Caballero, J. & Oñate, J. J. (2012) Spain. In *High Nature Value Farming in Europe*, eds. G. Oppermann, G. Beaufoy & G. Jones, pp. 382–405. Ubstadt-Wiher, Germany: Verlag.

Bi+De (2015) *Evaluación del grado de inclusión de las prioridades del MAP para la Red Natura 2000 en los Programas de desarrollo rural 2014–2020 de España.* Madrid: Fundación Biodiversidad.

BirdLife International (2022) Data Zone. Spanish Imperial Eagle Aquila adalberti. http://datazone.birdlife.org/species/factsheet/spanish-imperial-eagle-aquila-adalberti/text (last accessed 11 January 2022).

Bota, G., Morales, M. B., Mañosa, S. & Camprodon, J. (eds.) (2005) *Ecology and Conservation of Steppe-land Birds.* Barcelona: Lynx Edicions & Centre Technològic Forestal de Catalonia.

Del Moral, J. C. (ed.) (2017) *El buitre negro en España, población reproductora en 2017 y método de censo.* Madrid: SEO/BirdLife.

Del Moral, J. C. & Molina, B. (eds.) (2018) *El buitre leonado en España, población reproductora en 2018 y método de censo.* Madrid: SEO/BirdLife.

FAO (2015) *Global Forest Resources Assessments. Country reports 2015 – Spain.* Food and Agriculture Organization. www.fao.org/forest-resources-assessment/past-assessments/fra-2015/country-reports/en/

FCQ (2018) *Conservación del quebrantahuesos como recurso para el mantenimiento de los servicios ecosistémicos.* Fundación para la Conservación del Quebrantahuesos, Final Report Life 12/NAT/ES/000322.

García, S. (2019) *Beneficios económicos de la Red Natura 2000 en España.* Madrid: Ministerio para la Transición Ecológica.

García-Gómez, H., González-Fernández, I., Vivanco, M. G. *et al.* (2017) Riesgo de efectos en los hábitats terrestres de la Red de Parques Nacionales. *Ecosistemas*, **26** (1), 55–65.

Generalitat de Catalunya (2017) *Programa d'Infraestructura verda de Catalunya* 2017–2021.

Giménez Casalduero, F. & Martínez Fernández, J. (2020) El colapso del Mar Menor. Historia de una joya ecológica maltratada. *Mètode*, **106** (3), 22–29.

González, E. G., Blanco, J. C., Ballesteros, F., Alcaraz, L., Palomero, G. & Doadrio, I. (2016) Genetic and demographic recovery of an isolated population of brown bear *Ursus arctos* L., 1758. *PeerJ*, **4**, e1928.

Gurrutxaga, M. (2005) *Red de Corredores Ecológicos de la Comunidad Autónoma de Euskadi.* Gobierno Vasco.

zmán, J. N., García, F. J., Garrote, G., Pérez de Ayala, R. & Iglesias, M. C. (2004) *El Lince ibérico (Lynx pardinus) en España y Portugal. Censo–diagnóstico de sus poblaciones*. Madrid: DGCN, Ministerio de Medio Ambiente.

ıber, D. (2018) *Ursus arctos. The IUCN Red List of Threatened Species* 2018: e.T41688A133236401. https://dx.doi.org/10.2305/IUCN.UK.2018-2.RLTS.T41688A133236401.en (last accessed 9 January 2020).

N (2020) *Inventario Forestal Nacional*. Madrid: Ministerio para la Transición Ecológica y el Reto Demográfico.

JCN (2020) Country focus. *Spain*. www.iucn.org/regions/europe/resources/country-focus/spain

ınta de Andalucía (2018) *Plan director para la mejora de la conectividad ecológica en Andalucía. Una estrategia de infraestructura verde. Áreas estratégicas para la mejora de la conectividad ecológica*. May 2018.

eenleyside, C., Beaufoy, G., Tucker, G. & Jones, G. (2014) *High Nature Value farming throughout EU-27 and its financial support under the CAP*. London: Institute for European Environmental Policy report for DG Environment.

udwig, G. X., Alatalo, R. V., Helle, P., Lindén, H., Lindström, J. & Siitari, H. (2006) Short- and long-term population dynamical consequences of asymmetric climate change in black grouse. *Proceedings of the Royal Society of London B: Biological Sciences*, **273** (1597), 2009–2016.

MAGRAMA (2014a) *Quinto informe nacional sobre la diversidad biológica*. Madrid: Ministerio de Agricultura, Alimentación y Medio Ambiente.

(2014b) *Plan Nacional de Adaptación al Cambio Climático. Tercer programa de trabajo 2014–2020* Madrid: Ministerio de Agricultura, Alimentación y Medio Ambiente.

(2016) *Estrategia de adaptación al cambio climático de la costa española*. Madrid: Ministerio de Agricultura, Alimentación y Medio Ambiente.

(2017) *Informe de seguimiento del Plan Estratégico del Patrimonio Natural y de la Biodiversidad*. Madrid: Ministerio de Agricultura, Alimentación y Medio Ambiente.

MAPAMA (2017) *Informe anual 2017 sobre el estado del Patrimonio Natural y de la Biodiversidad en España*. Madrid: Ministerio de Agricultura, Pesca, Alimentación y Medio Ambiente.

(2018) *Recomendaciones sobre la información necesaria para incluir una evaluación adecuada de repercusiones de proyectos sobre Red Natura 2000 en los documentos de evaluación de impacto ambiental de la A.G.E. Medio Ambiente. Guía destinada a promotores de proyectos/consultores*. Madrid: Ministerio de Agricultura, Pesca, Alimentación y Medio Ambiente.

Mediterranean Wetlands Observatory (2012) *Mediterranean Wetlands Outlook – 2012. Synthesis for Decision-makers & Technical Report*. Tour du Valat, France.

MITECO (2019) *Guía metodológica de evaluación de impacto ambiental en Red Natura 2000. Criterios utilizados por la Subdirección General de Biodiversidad y Medio Natural para la determinación del perjuicio a la integridad de Espacios de la Red Natura 2000 por afección a Hábitats de interés comunitario*. Madrid: Ministerio para la Transición Ecológica.

(2020a) *Plan nacional de adaptación al cambio climático 2021–2030*. Madrid: Ministerio para la Transición Ecológica y el Reto Demográfico.

(2020b) *Estrategia nacional de Infraestructura verde y de la conectividad y restauración ecológicas*. Madrid: Ministerio para la Transición Ecológica y el Reto Demográfico.

(2020c) *Marco de Acción Prioritaria (MAP) para la financiación de la Red Natura 2000 en España para el marco financiero plurianual para el período 2021–2027*. Madrid: Ministerio Para la Transición Ecológica y el Reto Demográfico.

(2022) Inventario de espacios naturales protegidos, Red Natura 2000 y áreas protegidas por instrumentos internacionales. www.miteco.gob.es/es/biodiversidad/temas/inventarios-nacionales/inventario-espanol-patrimonio-natural-biodiv/sistema-indicadores/05c-IE-espacios-protegidos.aspx (last accessed 11 May 2022).

MMA (1999) *Plan Estratégico Español para la Conservación y el Uso Racional de los Humedales, en el marco de los ecosistemas acuáticos de que dependen*. Madrid: Ministerio de Medio Ambiente.

(2006) *Plan Nacional de Adaptación al Cambio Climático*. Madrid: Ministerio de Medio Ambiente.

(2007) *Manual de restauración de dunas costeras*. Madrid: Ministerio de Medio Ambiente.

MMARM (2010) *River restoration. Basis of the National Strategy for River Restoration*. Ministerio de Medio Ambiente, y Medio Rural y Marino.

(2011) *Plan estratégico del patrimonio natural y de la biodiversidad* 2011–2017. Ministerio de Medio Ambiente y Medio Rural y Marino.

Rodríguez, A. & Calzada, J. (2015) *Lynx pardinus. The IUCN Red List of Threatened Species* 2015: e.T12520A50655794. https://dx.doi.org/10.2305/IUCN.UK.2015-2.RLTS.T12520A50655794.en (last accessed 9 January 2020).

SEO/BirdLife (2019) *Programas de seguimiento y grupos de trabajo de SEO/BirdLife 2018*. Madrid: SEO/BirdLife. https://doi.org/10.31170/0073

Traba, J. & Morales, M. B. (2019) The decline of farmland birds in Spain is strongly associated to the loss of fallow land. *Scientific Reports*, **9**, 9473. https://doi.org/10.1038/s41598-019-45854-0

UNEP-WCMC (2022) *Protected Area Profile for Spain from the World Database of Protected Areas, March 2022*. www.protectedplanet.net

WWF (2018) *Autopistas salvajes. Propuesta de WWF España para una Red Estratégica de Corredores Ecológicos entre espacios Red Natura 2000*. http://awsassets.wwf.es/downloads/AutopistasSalvajesInforme.pdf?_ga=2.28409596.1131200152.1641936812-1092613832.1641936812

Chapter 30

Sweden

Mats O. G. Eriksson and Mia Pantzar

Key Statistics	
Terrestrial area (km^2)/% of EU-28*	449 718/10.3%
Marine waters (km^2)/% of EU-28*	155 625/3.4%
Population density (p km^2)/% of EU-28*	22/19%
GDP per capita (€)/% of EU-28	41 981/157%
Land cover (CORINE land cover classes 2018)	FOR 56%, OSN 18%, WET 15%, AGR 9%, ART 1%
Biogeographical regions	ALP, BOR, CON, MATL, MBAL
Number of HD Annex I habitats/%*	89/39%
Number of HD Annex II, IV and V species/%*	165/13%
Number of BD Annex I bird taxa/%*	66/38%
% HD UCS and decreasing habitats/ % unknown	35%/0%
% HD UCS and decreasing species/ % unknown	25%/1%
% breeding BD Annex I birds decreasing/% uncertain and unknown	22%/6%

* Excluding MAC and MMAC biogeographical regions.
Sources and codes. See Table 1.1.

30.1 The Natural History and Characteristics of Sweden

Sweden is the third largest country in Europe (excluding Russia), and one of the most sparsely populated. Much of the population is concentrated in the regions around the two largest cities, Stockholm and Göteborg, and Skåne County, where the density reaches similar levels to western and central Europe.

The range and distribution of habitats and species in Sweden reflect its climate and the legacy of past glaciation and ongoing coastal uplift. Along the coast of the Gulf of Bothnia this uplift has resulted in a unique habitat, 'Natural forest of primary succession stages of land-upheaval coast' (9030*),[1] found only in Sweden and Finland in the EU. The long distance (1 572 km) between north and south results in large differences in climate: south and central Sweden are characterised by a temperate climate, while the mountain region in the northernmost parts has an arctic climate.

The Swedish landscape is dominated by forests, which cover 63%[2] of the land surface; 85% of forest is subjected to commercial forestry (Statistics Sweden, 2019). Three-quarters are coniferous forests with roughly equal distribution between Norway Spruce (*Picea abies*) and Scots Pine (*Pinus sylvestris*). Open mires are another dominant feature of much of the landscape, covering 7% of the surface. Heathland, bare rocks and similar habitats cover 12%, mainly concentrated in mountainous areas and coastal archipelagos. Agricultural land covers 9% (according to CORINE 2018 data), or 6% according to Statistics Sweden (2019), with an 85% dominance of arable land and the remaining 15% being grazed pastures. Freshwater lakes and streams cover 9% of Sweden, making Sweden together with Finland the two countries richest in freshwater habitats in the EU. About 50% of the surface area, or almost the entire northern half of Sweden, is included in the Reindeer Herding District, where a customary law entitles the Sami people to carry out reindeer husbandry.[3]

The **Boreal** biogeographical region covers some 77% of the land surface, with a dominance of coniferous forests. Although most of the forested land is subject to commercial forestry, nuclei of high nature conservation values remain, including Western taïga (9010*) that covers almost 4% of the surface area. Mires and wetlands cover about 11%, with a dominance of bog woodland (91D0*), transition mires and quaking bogs (7140) and active raised bogs (7110*), plus Aapa mires (7310*) in the northernmost parts. Another characteristic element is freshwater lakes, primarily of oligotrophic character (3110, 3130), as well as many small dystrophic lakes (3160) in

[1] Numbers in brackets are HD Annex I habitat codes (see Appendix for full names). * Indicates a priority habitat.

[2] 56% forest cover according to the CORINE data in the Key Statistics table, but some young forest may be included in the 18% 'other semi-natural habitats' category.

[3] The Sami Parliament (Sametinget) website provides definitions of key terms: www.sametinget.se/ordforklaringar

connection to mires. The freshwater lakes in this region host substantial proportions of the EU populations of Black-throated Diver (*Gavia arctica*) and Osprey (*Pandion haliaetus*).

An important feature of the Boreal region is the large natural rivers that have not been exploited for hydropower production, with unregulated water flow from their sources to their outlets (3210). The rivers Piteälven, Vindelälven, Kalixälven and Torneälven are the most important ones in relation to their length and catchment area, benefiting from a specific legal status as 'National Rivers'.[4]

This Boreal forest-dominated region is interspersed with small- and medium-scale agricultural land where semi-natural habitats characterised by hay-cutting or grazing (or a combination of both) have evolved. Today, the remaining fractions represent some of the most species-rich terrestrial habitats in Sweden, including Fennoscandian wooded pastures (9070) and wooded meadows (6530*).

The archipelagic areas along the coasts of the Skagerrak, the Baltic Sea and the Gulf of Bothnia – scattered with islets of bare rocks – are other characteristic features, hosting habitats such as large shallow inlets and bays (1160), sea cliffs (1230) and Baltic islets and small islands (1620).

The **Continental** biogeographical region comprises only about 2% of Sweden, including a narrow strip along the coast of Kattegat, most of Skåne and Blekinge counties and the southern half of Öland island. This region was dominated historically by deciduous forests of primarily European Beech (*Fagus sylvatica*), Pedunculate Oak (*Quercus robur*) and Sessile Oak (*Quercus petraea*). Today, large parts of the woodlands have been converted into coniferous plantations. Much of the region is characterised by urban areas and arable land. Remnants of deciduous forests still exist and are locally abundant, including beech forests (9110, 9130) and oak forests (9160, 9190).

Part of Öland island is characterised by Nordic alvar (6280*), including open areas of calcareous flatrocks with a very thin layer of soil; 45% of the area of this habitat in the EU is found here.

The fjeld or mountain ridge along the border with Norway is in the **Alpine** biogeographical region, comprising about 21% of the land surface. In contrast to mountainous areas in continental Europe, the treeline is dominated by forests of Mountain Birch (*Betula pubescens* subsp. *czerepanovii*, 9040). Alpine heaths (4060) dominate above the treeline, interspersed with siliceous and calcareous grasslands (6150, 6170) and *Salix* scrub (4080) in wet depressions and along water courses. Palsa mires (7320*) are rare but typical for the region, characterised by peat mounds of permafrost and confined to areas with a mean annual temperature below −1 °C. Although the Swedish Alpine region is often considered to be in pristine condition, more recent understanding reveals a long period of human impact, with reindeer herding being the predominant land use.

The Swedish marine areas are divided between the **Marine Atlantic** (including parts of the Skagerrak, Kattegat and Öresund) and the **Marine Baltic**, which covers the Baltic Sea including the Gulf of Bothnia. Due to differing inflows of freshwater, these are characterised by a declining salinity gradient: from 25–30‰ in the northern parts of the Skagerrak to 2–3‰ in the northernmost parts of the Gulf of Bothnia. The difference in salinity is the main reason for the varying biodiversity between the Marine Atlantic and the Marine Baltic biogeographical regions (see Table 2.2).

Some open sea areas are of high importance as wintering sites for several species of waterfowl, in particular the Long-tailed Duck (*Clangula hyemalis*), as well as important areas for the European population of Harbour Porpoise (*Phocoena phocoena*). The Baltic Sea also hosts an endemic population of the arctic Ringed Seal (*Phoca hispida botnica*).

30.2 The Status of Nature

30.2.1 Habitat Status and Trends

Article 17 reporting for 2013–2018 covered 89 HD habitats,[5] with 186 assessments carried out over the five biogeographical regions.[6] The approaches for setting favourable reference values and assessing conservation status are described by Westling *et al.* (2020). Overall, 23% of the assessments revealed a favourable conservation status, 23% had a poor status and 54% a bad status. Almost half of the habitat assessments also revealed a declining trend (Table 30.1).

The assessment results for forests should be considered from the perspective that 20% of Sweden's land area qualifies as HD forest habitats, and that it hosts a substantial part of the EU's total forest area. It is therefore particularly concerning that all the 15 HD forest habitats reported for the Boreal or Continental biogeographical regions had an unfavourable status, with no indications of improvement for most of them.

For bogs, mires and wetlands in the Boreal and Continental regions, seven out of eight HD habitats, including active raised bogs (7110*), had an unfavourable status, with an ongoing declining trend. Also, in these regions, most of the eight lake and water course habitats had an unfavourable status.

Of the 32 assessments of the 14 grassland habitats in Sweden, 26 (81%) reported a bad status. As in most of the EU, pressures are linked to a lack of recurring management in terms of grazing or mowing, in combination with their fragmented and small areas compared to their historical distribution.

In general, the conservation status of habitats in the Alpine biogeographical region was markedly more favourable than

[4] Miljöbalk (Environmental Code), chapter 4, 6§.

[5] See Section 1.3 for background on monitoring under the Nature Directives, status categories and data sources.

[6] This may therefore over-represent habitats occurring in more than one biogeographical region.

Table 30.1 Short-term trends (c. 2007–2018) in Habitats Directive Annex I habitats and Annex II, IV and V species with an unfavourable conservation status in Sweden.

	Poor (unfavourable–inadequate)				Bad (unfavourable–bad)			
	Imp	Sta	Dec	?	Imp	Sta	Dec	?
HD Habitat assessments								
Number	0	20	17	6	3	21	48	29
%	0%	47%	40%	14%	3%	21%	48%	29%
HD Species assessments								
Number	11	14	12	8	10	19	57	10
%	24%	31%	27%	18%	10%	20%	59%	10%

Key. Imp = improving, Sta = stable, Dec = decreasing, ? = unknown.
Source. Based on Member State HD Article 17 reported data as presented on the EEA national summary dashboard (2021).

the other regions, especially for wetlands and freshwater habitats, as well as two grasslands primarily found in the Alpine region (6150, 6170). Most of the habitats in fjeld areas of the Alpine region also had a favourable status, with two notable exceptions: Palsa mires (7320*), that occur only in Sweden and Finland in the EU, and permanent glaciers (8340), both of which suffer from impacts of climate change.

The ecologically important large Fennoscandian natural rivers (3210) in the north have an unfavourable status with an ongoing negative trend. This is problematic since these rivers, in their natural state, provide a mosaic of different habitats with excellent conditions for high biodiversity.

All the nine marine habitats in Swedish waters had an unfavourable status, with an ongoing negative trend for, *inter alia*, biogenic reefs (subtype under 1170). For sandbanks (1110) characterised by eelgrass (*Zostera* spp.), the situation remains problematic due to previous and still ongoing declines in their area.

30.2.2 Species Status and Trends

Species Listed in HD Annexes

In total, 165 HD species were covered by the Article 17 reporting for 2013–2018. Out of the 274 assessments carried out, 47% revealed a favourable status, whilst 16% had a poor status and 35% had a bad status. The status was unknown in 1% of the assessments. Furthermore, a large proportion of assessments showed that species with an unfavourable status had negative trends, especially amongst those with a bad status (Table 30.1).

The overall poor status of forest habitats spills over to the situation for various species depending on natural forests. Among 11 beetles linked to forest habitats, 10 were considered to have an unfavourable status, with 3 having a declining trend.

Similarly, the precarious situation for grasslands is reflected in the status of species linked to these habitats. For example, out of seven butterfly species that were assessed and are connected to semi-natural habitats, all had a bad conservation status. For at least two vascular plant species, *Arnica montana* and *Botrychium simplex*, their bad status and declining trend are due to habitat deterioration and lack of management.

Among the 11 fish species assessed, all, except 2, had an unfavourable status in one or several of the biogeographical regions, with a particularly difficult situation for the Sea Lamprey (*Petromyzon marinus*), of which only small fragments of the previous population remain. For the Freshwater Pearl Mussel (*Margaritifera margaritifera*) – an important indicator species – and Thick-shelled River Mussel (*Unio crassus*) it is still too early to conclude if recent restoration measures (both inside and outside protected areas) will help them recover from their currently continuing bad conservation status; the species' very long life cycles make assessments of the impacts of measures challenging.[7]

Birds

The BD Article 12 assessment for 2013–2018 covered 259 breeding and 31 wintering species, including 67 and 7 species listed in BD Annex I, respectively, that were assessed with reference to short-term trends (c. 2007–2018). Across all species, whilst nearly two-thirds had increasing, stable or fluctuating breeding populations, 29% had declined. The situation was slightly better for Annex I species, with 22% of taxa having declined (Table 30.2). Amongst key wintering species, 10% showed short-term declines. No declines were observed amongst the seven wintering Annex I taxa.

For several bird species in the agricultural landscape, indicators based on the same data as for the BD Article 12 assessments show both short-term and long-term declines (Green *et al.*, 2019). In forested landscapes, the situation is more complex: a previous and long-term positive trend for generalist species has levelled out, and species with more specific habitat

[7] Both species have been subject to LIFE Nature projects: LIFE04NAT/SE/000231 (2004–2009) and LIFE10NAT/SE/00046 (2012–2016).

Table 30.2 Short-term trends (c.2007–2018) in Birds Directive Annex I taxa in Sweden.

	Breeding				Wintering			
	Inc	Sta/ Flu	Dec	?	Inc	Sta/ Flu	Dec	?
Number	15	33	15	4	4	2	0	1
%	22%	49%	22%	6%	57%	29%	0%	14%

Key. Inc = increasing, Sta = stable, Flu = fluctuating, Dec = decreasing, ? = uncertain and unknown.
Source. Based on Member State Article 12 reported data as presented on the EEA national summary dashboard (2021).

requirements are declining. Those with declining trends include quite widespread species, such as Black Woodpecker (*Dryocopus martius*) and Three-toed Woodpecker (*Picoides tridactylus*), which indicate the overall impaired condition of forest habitats.

For wintering species, although it is not a BD Annex I species, the continuously declining trend of wintering Long-tailed Duck raises concern, especially given the international importance of its wintering grounds in the Baltic Sea.

30.2.3 Drivers and Causes of Change

The following overview is primarily based on the national HD Article 17 report for 2013–2018 (Westling *et al.*, 2020).

The 2013–2018 report confirmed previous assessments that had shown the precarious status of several forest habitats that have declined in area and are of low quality, primarily due to commercial forestry and clearcutting practices. This development was further highlighted by a number of species linked to forest habitats being added to the recently updated national Red List, including vascular plants, bryophytes and fungi (Eide *et al.*, 2020). Other pressures include a lack of natural disturbances, such as fire and flooding. Restoration measures, such as controlled burning, are therefore carried out to re-create or mimic various disturbance processes and initiate their recovery (Westling *et al.*, 2020).[8]

The agricultural landscape in Sweden has seen a process of change that is very similar to elsewhere in Europe. A key driver of biodiversity declines has been a long-term decrease in semi-natural meadows and pasture, from just over 15 000 km^2 in the 1890s to only a fragment today (Statistics Sweden, 2019). This is also reflected in the high number of grassland species of vascular plants on the national Red List (Eide *et al.*, 2020).

For bogs and mires, the currently poor status is primarily related to historic impacts, such as drainage, but also to more recent and ongoing pressures, including peat extraction and eutrophication from long-distance airborne nitrogen emissions from, *inter alia*, road traffic and industry. The overall result is quite large-scale deterioration, for example in terms of overgrowth by vegetation. Restoration measures, such as blocking of ditches and clearing overgrowth, are required to improve the situation.[9]

The unfavourable status of several freshwater habitats and species can be related to ongoing pressures, such as water level regulation (mostly for hydropower production) and external impacts including eutrophication caused by run-off from agriculture and forestry, but also to remaining impacts from acid precipitation during the twentieth century (see Section 30.6.2). As elsewhere, small-scale modifications of water courses for hydropower, such as for mills and sawmills, have a long tradition. The exploitation of water courses for electricity generation since the early twentieth century has led to large-scale impacts, including impoundments, which affect water levels and flows, and act as migration barriers for anadromous fish.[10] Historical pressures that are important to recognise include physical modifications along floatways for timber transport. Most water courses in Sweden have at some point been cleared of stones and blocks, embanked and straightened for transporting timber. Clearing floatways ended in the second half of the twentieth century, yet a lack of natural flow dynamics remains a significant issue for, for instance, Fennoscandian natural rivers (3210) and their associated aquatic and riparian species.

The poor status of marine and coastal habitats and species is largely due to commercial fisheries (including the effects of bottom trawling, by-catch and incidental killing), high nutrient levels (primarily phosphorus and nitrogen from agriculture and urban wastewater, as confirmed by the European Court of Auditors, 2016) and physical exploitation (such as dredging). These pressures result in significant changes in benthic species and habitats as well as cascade effects. For instance, large parts of the Swedish Baltic are suffering from anoxic conditions and algal blooms during the summer months, as a result of eutrophication caused by nutrient run-off. Absence of large predatory fish, caused by intensive fishing and other pressures, is reinforcing eutrophication.

Only more recently has attention been paid to the challenges related to invasive alien species (IAS). Particular concern has been expressed for marine habitats and species, not least with reference to the Pacific Oyster (*Crassostrea gigas*). For freshwaters, the long-term decline in the population size and range of the Noble Crayfish (*Astacus astacus*) is primarily due to fungal infection by *Aphanomyces astaci*, carried by the introduced Signal Crayfish (*Pacifastacus leniusculus*).

Among terrestrial habitats, IAS may pose a particular problem for Nordic alvar (6280*). It was considered to have a favourable status in the Continental region in previous

[8] Such measures have also been supported through LIFE Nature funding, e.g. LIFE13NAT/SE/000065, 2015–2019.

[9] Experiences gained in some LIFE Nature projects, e.g. LIFE08NAT/SE/000268 (2010–2015).

[10] Sweden's second River Basin Management Plan provides further details about the magnitude of these pressures: https://ec.europa.eu/environment/water/participation/map_mc/countries/sweden_en.htm

assessments, but now has a poor status due to invasive *Cotoneaster* species. For dune habitats, as well as sand beaches and *Salicornia* mudflats (1310), continuously declining trends can, amongst other things, be linked to IAS, such as *Rosa rugosa* and *Cotula coronopifolia*.

It is apparent from the above summary that most pressures and drivers of change in the conservation status of Swedish habitats and species relate to intensive and/or changing land and sea use or IAS. However, as regards the Wolf (*Canis lupus*) and other carnivore species, poaching is another serious problem that has to be adequately handled if their long-term status is to be maintained (Liberg *et al.*, 2011, 2020).

30.3 Nature Conservation Governance, Actors, Legislation and Policies

30.3.1 The Development of Nature Conservation in Sweden

As in most countries in Europe, nature protection and conservation in a more modern sense became an issue of concern in Sweden at the transition between the nineteenth and twentieth centuries. Following a proposal in 1904, the Parliament adopted regulations introducing legal protection of certain Natural Monuments and the establishment of National Parks in 1909, with 10 sites designated the same year. The focus was on *protection* of individual objects and sites, rather than *conservation* (see Table 30.3 for an overview of key milestones).

After the initiatives of 1909, further development was relatively slow. Funds for implementation were lacking and the legislation was found to be largely insufficient in scope. Also, the revised Nature Protection Act of 1952 was soon found to be unfit for its purpose, although separate legislation adopted in parallel for the protection of shoreline zones along lakes, rivers and the sea is still in force.

The 1952 legislation was replaced by a new Nature Conservation Act in 1964, with a stronger focus on conservation and management. Its framework for site protection and management is still the cornerstone of the current approach (Section 30.5.1). A State Committee for Nature Conservation was established in 1963, when also the first staff with specialist competence in nature conservation were recruited by the regional County Administrative Boards (CABs). Four years later, in 1967, the Swedish Environmental Protection Agency (SEPA) replaced the state committee as one of the first national agencies with a primary focus on the environment. SEPA has a much broader scope than nature conservation, also covering water and air quality, waste treatment, etc. SEPA was given an overall governing, planning and monitoring role, and the regional CABs the main responsibility for implementation.

The new legal framework introduced both political and financial incentives for a more ambitious approach. The number of legally protected sites (primarily Nature Reserves) has increased steadily from the 1970s until today (Section 30.5.1). In parallel, the need to consider nature values in forestry and agricultural land use gained increased acceptance. Over time, it became the responsibility of the sectoral boards and agencies for forestry, agriculture and fisheries to administer these aspects.

The next primary change came with the adaptation of the legal framework to transpose the Nature Directives after Sweden joined the EU in 1995, including establishing a national network of Natura 2000 sites.

In 1999, Swedish environmental policy underwent significant restructuring. The Parliament adopted the so-called '*Generational Goal*' that outlines the overarching objective of all environmental policy in Sweden. In the same year, the Government adopted a set of 15 (later 16) more precisely defined environmental quality objectives to help deliver the generational goal. These were later revised, more precisely defined, and split into interim targets (Ministry of the Environment, 2012), including specific ones for biodiversity and ecosystems services set for 2020.

It was also in 1999 that a new framework for environmental policy entered into force – the Swedish Environmental Code.[11] The Code harmonised previous rules and principles from civil law with relevance to the environment, but without major changes in content or structure.

In 2014, a *National Strategy for Biodiversity and Ecosystem Services* was adopted, with undertakings in line with the CBD Aichi Targets for 2011–2020 and the *EU Biodiversity Strategy to 2020*.[12]

Sweden has, together with the other EU Member States, given its support to the new *EU Biodiversity Strategy for 2030*.[13] The main national initiative taken so far is a memorandum wherein the Government declared its position to various commitments in the strategy.[14]

30.3.2 Today's Nature Conservation Governance

Four governmental agencies are involved in the implementation of national nature conservation policy. SEPA has the mandate to develop policy instruments and to support, coordinate and oversee the implementation of policy to ensure compliance with the Environmental Code and with national environmental objectives. SEPA is also the formal national contact point with the European Commission and various international conventions.

SEPA works in close collaboration with the 21 CABs for the implementation of nature conservation policy. They are responsible for the protection, management and enforcement of protected areas, including the Natura 2000 network, as well as for a range of environmental licensing, compliance monitoring and enforcement at the regional level.

[11] Miljöbalk (Environmental Code) 1998:808. Nature conservation legislation is primarily in chapters 7 and 8.
[12] Regeringens proposition (Government Bill) 2013/14:141.
[13] Through the Council of the European Union conclusions, 16 October 2020.
[14] Regeringskansliet Faktapromemoria (Government Offices Memorandum) 2019/20:FPM43.

Table 30.3 Key nature conservation developments in Sweden.

1909	First nature protection legislation. First National Parks established
	Swedish Society for Nature Conservation (Svenska Naturskyddsföreningen) established
1945	Swedish Ornithological Society (nowadays BirdLife Sverige) established
1952	Revised nature protection legislation
	Legislation about protecting a zone of 100–300 m along the shoreline of lakes, rivers and the sea
1963	A State Committee for Nature Conservation established. First nature conservation staff employed by the regional County Administrative Boards (CABs)
1964	New Nature Conservation Act
1967	The Swedish Environmental Protection Agency (SEPA, Naturvårdsverket) established
1978	Governmental decision about protection of the four large 'national' rivers: Kalixälven, Torneälven, Piteälven and Vindelälven
1990	The SLU Swedish Species Information Centre established
1993	Revised Forestry Act, where objectives related to forestry and environmental aspects are given equal value
1994	National Mire Protection Plan adopted
1995	Sweden joined the EU. The two Nature Directives apply in Sweden
1999	A system of environmental quality objectives decided by the Parliament
	Legislation about nature conservation incorporated in the newly established Environmental Code
	Species Protection Ordinance in force (revised in 2007)
2004	The WFD in force in Sweden
2009	The first marine National Park, Kosterhavet, inaugurated
2010	The MSFD transposed into Swedish law through the Marine Environment Ordinance[15] (Havsmiljöförordningen)
2011	The Swedish Agency for Marine and Water Management (SwAM, Havs- och Vattenmyndigheten) established
2014	*National Strategy for Biodiversity and Ecosystem Services* adopted by the Government
2020	Adoption of a national plan for ensuring hydropower plant licences' compliance with environmental requirements and EU directives

See Table 3.3 for signatory dates of international conventions.

The Swedish Forest Agency (SFA) has the responsibility to guide landowners on matters related to the conservation of species and habitats, and to establish Habitat Protection Areas (under the Environmental Code, Sections 30.4.1 and 30.5) on forest land.

The Swedish Board of Agriculture (SJV) is responsible for regulations regarding consideration of natural and cultural values in agricultural activities. It plays the main role in the national implementation of the CAP Rural Development Programme (RDP), and thus the allocation of agri-environment funds for the management of semi-natural habitats.

The Swedish Agency for Marine and Water Management (SwAM) has the main responsibility for management of habitats and species in marine and freshwater environments, including issues related to the implementation of the two Nature Directives. The agency is also responsible for, for instance, Maritime Spatial Planning (MSP) in Swedish waters.

The fulfilment of targets linked to the previously mentioned 16 environmental quality objectives is assessed annually within a system of national environmental monitoring coordinated by SEPA and the regional CABs, involving about 30 public authorities (further explained in Section 30.8). The environmental monitoring also supports the assessment of progress with national undertakings under the EU biodiversity strategies. The Swedish Species Information Centre[16] (a scientific centre of the Swedish University of Agricultural Sciences, SLU) has the main responsibility for compiling information for reporting under HD Article 17 and BD Article 12.

30.3.3 Stakeholders and Public Involvement

Sweden has a strong tradition of citizens forming and joining associations, not least those related to nature and outdoor life. The Swedish Society for Nature Conservation (SSNC) is the

[15] Havsmiljöfördordning (2010/1341).

[16] www.artdatabanken.se/en/

largest environmental association and NGO, with about 230 000 members. From having a strong academic profile when established in 1909, the Society has widened its scope in terms of membership and coverage. Today, the SSNC covers environmental and related socio-economic issues in a broader sense than purely nature conservation. Since it was established, it has played an important role in public debate and in lobbying.

Besides the SSNC, BirdLife Sweden, established in 1945 and with about 16 000 members, plays an increasingly important role in disseminating aspects related to the protection and management of bird populations, as well as in public debate and in lobbying.

Other NGOs focus on, for instance, plants, fungi and insects. From initially having had a quite narrow and scientific profile, these NGOs have expanded their membership base. Most encourage their members to take active roles in various citizen science activities, including schemes that provide data for environmental monitoring. Similarly, Swedish hunters and recreational fishers provide important information for wildlife inventories and monitoring.

Since being established in 1971, the Swedish section of WWF has initiated several national projects, often with the ambition of starting an activity that is later taken on by other stakeholders (including public bodies, when relevant).

The Swedish Association for Hunting and Wildlife Management, which numbers approximately 155 000 members, was established as early as 1830. It has a high profile in lobbying, sometimes in consensus with the other environmental NGOs, sometimes in opposition, not least when it comes to the management of the populations of large carnivores.

As regards landowners, the Federation of Swedish Farmers (LRF) and three regional forest owners' associations organise around 250 000 individuals. They are important stakeholder organisations when it comes to the protection and management of forests and semi-natural habitats, and strong lobbyists, defending the right of ownership and involvement in local and regional nature conservation planning, as well as fair rules and levels of compensation.

The Swedish Anglers Association, with more than 65 000 members and several affiliated local clubs, plays an important role in opinion building and participation in various activities related to monitoring and management of freshwater and marine habitats. Its primary focus is on the status and survival of fish stocks, but it has adopted an increasingly broader perspective over time and with reference to the management and conservation of aquatic habitats.

Despite the relatively high number and membership of civil society organisations and NGOs related to nature conservation, communication with and involvement of the public and other stakeholders remain a challenge. The draft Swedish Prioritised Action Framework (PAF) for 2021–2027 identifies a need to improve and increase communication with key nature conservation stakeholders, including farmers, forestry owners, fishers and the public (SEPA, 2021a). This conclusion reflects the fact that the primary pressures on Swedish habitats and species, as mentioned previously, relate to different types of land and sea use.

30.3.4 Policy and Legislation

The system of area protection, introduced in the legislation of 1964, has been amended over time, most importantly by transposing the Nature Directives after Sweden's accession to the EU in 1995. This includes a framework for establishing a national Natura 2000 network and conducting appropriate assessment procedures in line with HD Article 6(3), incorporated into the Environmental Code in 1999 (further details in Section 30.5.3). Species protection is also primarily regulated in the Environmental Code and the Species Protection Ordinance (2007:845), although separate legislation for hunting and fishing has precedence when it comes to taking or killing birds, mammals and fish species (further explained in Section 30.4.1).

The Swedish Forestry Act from 1993 is relevant to the protection and management of forests. It introduced a principle of equal weight between production and environmental objectives, where 'environmental' includes aspects related to biodiversity, good water quality, cultural heritage and social values. Although the principle has been widely accepted, difficulties remain when it comes to balancing these two objectives in practice. This is indicated not least by the fact that all the 15 HD forest habitats reported for the Boreal or Continental biogeographical regions had an unfavourable status in the 2019 reporting, with no indication of improvement for most of them. In 2020, a government-appointed commission of inquiry proposed several changes to the Forestry Act (and other pieces of legislation) to strengthen landowners' rights and enable more flexible protection measures and nature conservation in forests (Ministry of the Environment, 2020). The inquiry and its proposals are discussed in more detail in Section 30.6.2.

30.3.5 Thematic Strategies and Plans

Within the legal framework explained in Section 30.3.4, thematic plans and strategies have been launched for various groups of habitats.

Based on a nationwide wetland inventory, started in 1981 and covering a total of 35 000 sites and about 43 000 km^2 (Gunnarsson and Löfroth, 2009), a national protection plan for mires was adopted in 1994 and revised in 2006 (SEPA, 2007a), and further adjusted after that. Although the implementation of the plan was supported by a LIFE Nature project during 1995–1999,[17] the ambition to fully achieve the plan's targets was still not met by 2020 (SEPA, 2021b).

[17] Mire Protection Plan (LIFE95NAT/S/000507).

A nationwide survey of natural forests was carried out during 1978–1981, identifying 314 sites with a total area of c. 450 km^2 (Bråkenhielm, 1982). Following intensive debates, funding for initiating the protection of these sites could finally be raised during the 1990s. In parallel, following the acquisition of better knowledge about declining trends in the status of forest habitats and forest-dependent species, the target adopted based on the 1978–1981 survey was found to be much too low. A new ambition of setting aside 9 000 km^2 from commercial forestry by 2010 was consequently included in the environmental quality objectives adopted in 1999. To meet the target, a more coherent national strategy for the protection of forests was jointly launched by SEPA and SFA in 2005, with a target to protect 4 000 km^2 of productive forest land (SEPA and SFA, 2005). The strategy was revised in 2017 (SEPA, 2017a), and a milestone of protecting 1 500 km^2 during 2012–2020 was set up. Nature values were still a primary criterion for selection of sites, but with greater emphasis on, *inter alia*, proximity to urban areas, recreational values and cultural values as additional criteria. The 2017 strategy to a large extent relies on voluntary protection efforts by landowners. Targets for protection were still not fully met in 2020 (SEPA, 2021b).

A national strategy for the protection of freshwater habitats was launched in 2007 (SEPA, 2007b). It identified 573 sites of particular importance and expressed an ambition that 50% should be adequately protected by 2010. However, a 2017 review revealed that the target was far from being met and that the current level of ambition was largely insufficient (SwAM, 2017). The review also found that formally protected sites sometimes had insufficient or inadequate regulations.

As in most coastal EU Member States, the protection of marine sites has been slow. SwAM adopted a national action plan in 2016 with the aim to protect 10% of Swedish waters (including the EEZ) by 2020 (SwAM, 2016). By the end of 2020, 14.8% of the marine areas were covered by some kind of protection (see also Section 30.5.1). Whilst the 10% coverage target was met in spatial terms, there is ongoing debate on whether the protection is sufficient and adequate.

SwAM is currently developing a national strategy to reduce physical damage of Swedish coastal waters and promote their biological restoration. The initiative, partly linked to Sweden's nature restoration commitments at EU level, is based on comprehensive knowledge recently developed and compiled on the impacts of physical modification of the coastline. The exercise revealed, among other things, that the rate of destruction of important coastal habitats far exceeds that of their protection or restoration (Törnqvist *et al.*, 2020).

To summarise: the high ambition of most of the thematic plans listed here is currently far from being fulfilled. As discussed later, some of the main reasons relate to a lack of funding, inadequate management of sites, insufficient mapping of areas of high nature value outside formal protection, and a lack of follow-up and enforcement of rules in protected areas. This is also reflected in a 2021 review that found that hardly any of the environmental quality objectives will be met, including those related to wetlands, forests, the agricultural landscape and the status of species. Neither was the interim target related to the protection of nature by 2020 fully met (SEPA, 2021b).

30.4 Species Protection and Conservation Measures

30.4.1 Legal Framework

The legal framework for species protection is chapter 8 of the Environmental Code, with more detailed regulations in the Species Protection Ordinance from 2007 (with amendments after that).[18] It covers relevant articles of the Habitats Directive on species protection, including derogations, and the handling and trade of species listed in HD Annexes IV and V. Experience has shown that its interpretation leaves room for ambiguity, and the CJEU has judged parts of the Swedish interpretation of the species protection provisions of the EU Nature Directives to be incorrect (see the CJEU ruling in the joined cases C-473/19 and C-474/19). A governmental proposal to revise the national regulation was circulated for consultation during 2021 (Ministry of the Environment, 2021), but the Parliament has not yet decided on a revised legislation.

As regards the killing of birds, mammals and fish, the relevant hunting and fishing legislation has precedence, in accordance with regulations in the Hunting Act from 1987[19] and the Fisheries Act from 1993,[20] both with later amendments and with more detailed regulations provided in the Hunting Ordinance[21] and the Fisheries Ordinance.[22]

The basic principle in the hunting regulation is that all mammals and bird species are legally protected, with exceptions outlined in the Hunting Ordinance, including rules for derogations in line with the Nature Directives, and open seasons. The Species Protection Ordinance still applies to aspects not covered by the regulation for hunting. Following a large number of amendments over the years, the current regulation is regarded as complex and difficult to apply in practice. Thus, SEPA has published a strategy on how to address this complexity (SEPA, 2018a).

In addition to fish species, the Fisheries Act also regulates the taking of crustacean and mollusc species of potential economic importance. The Government or competent authority can ban or limit fishing of certain species in particular areas and with the use of specific tools, etc. In marine protected areas, fishing can be regulated with support of either the Fisheries Act or the Environmental Code. In accordance with Article 11 of the Common Fisheries Policy (CFP), Sweden has

[18] Artskyddsförordning (2007:845), with further guidelines published by SEPA (2009a).
[19] Jaktlag (1987:259).
[20] Fiskelag (1993:787).
[21] Jaktförordning (1987:905).
[22] Fiskeförordning (1994:1716).

proposed fisheries measures in several marine protected areas in which other Member States have fishing interests, including for Natura 2000 sites, such as Hoburgs bank och Midsjöbankarna. These areas are of international importance to migrating seabirds as well as an important breeding ground for the critically endangered Baltic population of Harbour Porpoise. In 2020, the European Commission opened a formal infringement procedure against Sweden in relation to insufficient protection of this HD Annex II and IV species.

As mentioned previously, the Forestry Act includes regulations with reference to nature values,[23] and SEPA and SFA (2016) have elaborated common guidelines about their practical application. In principle, the proposed considerations shall not exceed a level where the landowner may qualify for compensation payments. In such cases, alternative measures, such as time-limited agreements or some kind of site protection (further explained in Section 30.5.1), may apply.

30.4.2 Species Management

For rare or endangered species for which site protection or management measures in forestry, fishing or agriculture are insufficient, specific Action Programmes (*åtgärdsprogram*) have been prepared, under the supervision of SEPA and SwAM and in collaboration with the regional CABs and the SLU Swedish Species Information Centre, and formally adopted by SEPA or SwAM. There are programmes for about 300 species, and for a few habitat types.[24] Over the years, EU LIFE programme support has been provided for initiating or implementing Action Programmes for some species.[25]

For a few species, for which management involves particularly delicate balances between various stakeholder groups (and therefore may be politically controversial), SEPA has elaborated very detailed Management Plans (*förvaltningsplaner*), to be implemented in collaboration with the regional CABs and relevant stakeholders.[26] The plans include objectives related to favourable reference values, distribution and range, proposed measures, and responsibilities for the implementation of the plan during a pre-determined time period. In 2013, the Parliament decided on population size reference values for the large carnivore species[27] as part of a decision about a sustainable carnivore policy – a disputed approach for handling such a scientific/technical issue. By January 2022, Management Plans had been adopted for four species of carnivores – Brown Bear (*Ursus arctos*), Wolf, Eurasian Lynx (*Lynx lynx*) and Wolverine (*Gulo gulo*) – and the Golden Eagle (*Aquila chrysaetos*). The management of the Wolf population is coordinated with competent authorities in Norway. Some of the Management Plan's objectives are controversial and under constant debate. For example, the Wolf is practically excluded from the Reindeer Herding Area (SEPA, 2016a), which covers almost the entire northern half of Sweden. This may make the necessary genetic input from emigrating wolves from Finland and Russia difficult.

Plans have also been adopted for two more common and widespread species deemed to cause particularly serious damage: Great Cormorant (*Phalacrocorax carbo*) and Wild Boar (*Sus scrofa*).

The most problematic types of damage include crop damage (primarily by geese, Common Crane (*Grus grus*) and Wild Boar), killing of reindeer and livestock as well as huntable wildlife (by the above-mentioned large carnivores) and damage to fishing gear or stocks (primarily by seals and Great Cormorant). The primary approach to minimising such damage is through various kinds of preventive measures, in combination with compensation payments when losses are not prevented. In addition, derogations are given for killing animals in order to prevent damage – an issue under constant debate when it comes to the large carnivore species. In addition to case-by-case derogations in order to prevent serious damage (primarily under HD Article 16(1)(b)), derogations in line with HD Article 16(1)(e) have been given for the Wolf since 2010, within an annual quota decided by SEPA (except for 2019 and 2020) and with reference to 'strictly supervised conditions, on a selective basis and to a limited extent'. This practice has been criticised (Darpö and Epstein, 2015; Hansson, 2016) and is subject to an ongoing complaint case at the European Commission.

Re-introduction programmes have a century-old tradition in Sweden. These included a programme for the Beaver (*Castor fiber*), which was initiated as early as the 1920s, after the species had gone extinct during the 1870s. Since the 1960s, re-introduction or restocking programmes have been implemented, including for the Peregrine Falcon (*Falco peregrinus*), Eagle-owl (*Bubo bubo*), Lesser White-fronted Goose (*Anser erythropus*) and White Stork (*Ciconia ciconia*). A common feature of these programmes is that they have been highly dependent on volunteers for most of the work. At smaller scales and more locally, other species, including Freshwater Pearl Mussel, have been subject to re-introductions in rivers where the species had gone extinct in the past.

30.5 Protected Areas and Networks

30.5.1 Protected Area Types and Their Coverage

The system of site protection established in the legislation of 1964 is still the cornerstone. Chapter 7 of the Environmental Code defines the types of protected areas that can be formally

[23] Skogsvårdslag (1979:429, 1993:553), 30§.

[24] www.naturvardsverket.se/om-oss/publikationer/atgardsprogram/ and www.havochvatten.se/hav/fiske–fritid/arter/arter-och-artskydd/hotade-arter-och-naturtyper-med-atgardsprogram.html

[25] For example, Arctic Fox (*Vulpes lagopus*, LIFE98NAT/SE/005371 and LIFE03NAT/SE/00073), Freshwater Pearl Mussel (LIFE04NAT/SE/000231), Harbour Porpoise (LIFE08NAT/SE/000261) and Thick-shelled River Mussel (LIFE10NAT/SE/00046).

[26] www.naturvardsverket.se/amnesomraden/jakt-och-vilt/forvaltningsplaner-for-vilt

[27] www.riksdagen.se/sv/dokument-lagar/arende/betankande/en-hallbar-rovdjurspolitik_H101MJU7

designated, and the Ordinance about Spatial Protection[28] regulates who can designate nature conservation measures and what restrictions can be imposed within them, etc. By the end of 2020, 65 222 km^2, or 14.6% of Sweden's terrestrial territory, benefited from some kind of legal protection in accordance with chapter 7 of the Environmental Code. At the regional level, the percentage coverage is significantly higher in the alpine region (46%) than in the rest of the country (5–6%) (SEPA, 2021b). At sea, 22 995 km^2 is legally protected (Hedeklint and Olsson, 2021), which is 14.8% of Sweden's marine waters (as given in the Key Statistics table).

Sweden has the following types of protected areas (the Swedish names, numbers and areas of which are indicated in Table 30.4):[29]

- **National Parks** are mostly large areas of state-owned land, with the primary focus on high nature conservation values in a close to natural state. Currently, they cover 1.6% of Sweden's total surface.
- **Nature Reserves** are by far the most widespread protected areas, covering 9.5% of the country's surface. They focus primarily on areas of high nature values, but social and recreational aspects are also important designation criteria, mainly in urban regions. Decisions are mostly taken by the CABs, but several municipalities have a delegation that gives them competency over decisions and management. Land can be either in private or public (state or municipality) ownership. On private land, the landowner has the right to compensation payments.
- **Nature Conservation Areas** were identified under an old instrument with weaker restrictions than the Nature Reserves. Today, they cover 0.3% of the land surface. There is a slow but ongoing process to transform these areas to Nature Reserves.
- **Natural Monuments** were introduced in the first nature protection legislation of 1909 and consist of many small sites, mostly less than 1 ha.
- **Sanctuaries for Wildlife or Plants** are areas with restrictions in place for the protection of species that are rare or sensitive to disturbance. Most have been established to avoid disturbing breeding birds, and most have restrictions that only apply to part of the year (e.g. the reproduction period). Some Sanctuaries overlap with or are encapsulated in areas benefiting from other types of legal protection.
- **Habitat Protection Areas** are small areas (often <5 ha) of importance for rare or threatened plant or animal species. A general protection applies, in principle wherever they occur, to seven easily identifiable habitats,[30] and the CABs, SFA and the municipalities can decide to protect another 35 specific biotopes.[31] Currently, 99% are on forest land. Although these areas only cover 0.1% of the total land surface, the network includes some key sites for the species concerned. Most areas are in private ownership, with compensation payments for land-use restrictions.

Initially, the designation of sites was guided by general knowledge about nature values and rare or threatened species, often in combination with short-term risks for damage or exploitation. With the elaboration of thematic surveys from the 1990s onwards, the focus shifted to proactive measures, with habitats and species listed in relevant BHD annexes as an important criterion for priority.

A general protection from construction of new buildings, certain alterations of existing buildings and activities that may cause significant change to living conditions for animals or plants applies to a zone of 100 m along lakes, water courses and coasts, with local extensions of up to 300 m. This shoreline protection was introduced in 1952, and although the primary objective is to ensure public access, these restrictions provide important corridors of green infrastructure for habitats and species. The system has recently been under review, but a governmental decision about revised legislation is still pending.

In addition, general protection against the construction of hydropower plants applies in 42 Swedish rivers.

Complementary to the measures above, SEPA or SFA can enter time-limited nature conservation agreements with private landowners. These civil law agreements (*naturvårdsavtal*) cover just over 1 700 km^2, with compensation payments for temporary restrictions on land use. The maximum length is 50 years, but most agreements apply for shorter periods. They can be useful for the protection of a site during specific succession stages when it can be expected to temporarily host threatened or rare species.

Based on the data in Table 30.4, the national Natura 2000 network covers about 12.3% of the land surface. Of this area, 63% also benefits from legal protection within domestic protected areas (Hedeklint and Olsson, 2021). For most of the Natura 2000 sites without any additional domestic legal protection, restrictions are judged to be sufficient under other planning instruments or regulations. Marine components of the Natura 2000 network cover 13.0% of Sweden's total marine waters. However, the coverage varies greatly between the marine regions: 11% in the Baltic region under Swedish jurisdiction and 30% in the Atlantic region (SEPA, 2021b). A large majority of the marine sites are located relatively close to shore, with the exception of a few very large offshore sites.

The first proposal for sites to be included in the Swedish Natura 2000 network, presented in 1995, almost exclusively

28 Förordning (1998:1252) om områdesskydd.

29 See skyddadnatur.naturvardsverket.se/ for a map viewer showing all protected sites under the various legal instruments.

30 Tree-lined alleys, stone walls/stone fences, cultivation cairns, small groves, small ponds, and small wetlands and springs.

31 See www.naturvardsverket.se/vagledning-och-stod/skyddad-natur/biotopskyddsomraden/, for, *inter alia*, detailed descriptions of the biotopes that are covered.

Table 30.4 The number and extent of international designated sites and protected areas in Sweden.

	IUCN category	Number	Area (km²)
Designated under global and non-EU international legislation			
Ramsar Sites	IV	68	66 459
Natural and Mixed World Heritage Sites	III	2	12 703
Biosphere Reserves (exc. buffer and transition zones)	VI	4	4 451
OSPAR MPA		10	2 422
HELCOM MPA		28	8 386
Designated under the Nature Directives			
SCIs (terrestrial + marine)	Ia, Ib, III, IV, V	4 002	55 023 + 20 175
SPAs (terrestrial + marine)	Ia, Ib, III, IV, V	550	26 445 + 14 468
Natura 2000 terrestrial area			*55 534*
Natura 2000 marine area			*20 263*
Designated under domestic legislation			
National Parks (Nationalparker)	II	30	7 435
Nature Reserves (Naturreservat)	Ia, Ib, III, IV, V	5 241	50 104
Nature Conservation Areas (Naturvårdsområden)	IV, V	89	1 496
Natural Monuments (Naturminnen)	III, IV	1 358	Small sites
Habitat Protection Areas (Biotopskyddsområden)	IV	8 502	316
Sanctuaries for Wildlife and Plants (Djur- och växtskyddsområden)	IV	1 565	

Note. No areas given for Natural Monuments, as these are mostly very small sites, and Sanctuaries, as they often overlap with other protected areas, primarily Nature Reserves.
Sources. See Section 1.3. Based on data from: Ramsar Sites, SEPA;[32] World Heritage Sites, Biosphere Reserves, OSPAR MPAs and HELCOM MPAS, UNEP-WCMC (2022); EEA Natura 2000 Barometer, Natura 2000 as at end 2020; sites under national protection are by end of 2020 (Hedeklint and Olsson, 2021). All web-links downloaded 8 January 2022.

listed sites already benefiting from legal protection. Thus, the principle that sites should be proposed solely on scientific criteria was ignored. As a consequence, several deficiencies were identified during the biogeographical seminars, resulting in substantial 'homework' to propose additional sites, with a focus on under-represented BHD habitats and species. By 2019, the EU Environmental Implementation Review considered the terrestrial Swedish Natura 2000 network to be complete, whilst the marine network was still under assessment (European Commission, 2019). In 2020, the European Commission questioned whether the Swedish network of marine SPAs had been designated based on the areas most important for birds (in accordance with BD Articles 4.1 and 4.2).[33] As a result, SEPA is carrying out an assessment of 20 marine Important Bird and Biodiversity Areas (IBAs)[34] to evaluate whether these areas are eligible for SPA designation.

The Swedish network of Ramsar Sites covers c. 15% of the country's surface. A large proportion of their area (72%) overlaps with Natura 2000 sites, and 63% benefits from domestic legal protection, primarily as Nature Reserves or as parts of National Parks.

Although not a protection type per se, the system of so-called 'key habitats', launched and administered by SFA in the early 1990s, plays an important role in the protection of forest sites with high nature values.[35] The system is used in forest inventories to identify sites with important nature values. It includes more than 50 nature types for which consultation is required before carrying out any forestry practices. Any resulting restrictions should ideally be under the level that qualifies for compensation payment. Certified forest companies (i.e. all larger Swedish forest companies) are often bound by their certification standard not to cut or sell timber from identified key habitats. After some debate, including

[32] www.naturvardsverket.se/amnesomraden/skyddad-natur/olika-former-av-naturskydd/vatmarker-skyddade-enligt-ramsarkonventionen/

[33] Letter of formal notice from the European Commission to Sweden, 14 May 2020. C(2020) 2503 final.

[34] http://datazone.birdlife.org/site/results?cty=202

[35] www.skogsstyrelsen.se/miljo-och-klimat/biologisk-mangfald/nyckelbiotoper/ and www.skogsstyrelsen.se/sjalvservice/karttjanster/skogens-parlor/ for maps.

issues about their legal status and landowners' right to compensation payments, more strict legal procedures have been proposed, as previously mentioned, within the framework of a government appointed commission of inquiry into the future direction of the national forest policy (Ministry of the Environment, 2020).

For a few habitats, partly overlapping with HD habitats, specific action programmes have been elaborated by SEPA or SwAM in collaboration with the regional CABs. These are similar to the Action Programmes described earlier for species.

30.5.2 The Coherence of Protected Areas and Ecological Networks

As mentioned previously, in large parts of the country, protected areas are often small and relatively far apart. The thematic plans and programmes for various nature types (Section 30.3.5) have primarily focussed on individual sites rather than any explicit ambition to include aspects of ecological coherence. Although the need to consider ecological coherence that enables dispersal and genetic exchange within species has been recognised for some decades, it has only recently been taken into account in Swedish nature conservation planning.

Following a government initiative in 2014 and under the coordination of SEPA, the CABs presented regional action plans for green infrastructure in 2019. The ambition was to facilitate sustainable use of all areas in Sweden. The plans identify nuclei of high value for various habitats and landscape elements, and propose measures to maintain or create dispersal corridors and connectivity, and to maintain ecosystems services. Much of this information has recently been made available in an online map tool, showing regions with concentrations of high nature values in forests, grasslands, freshwaters, coastal habitats and marine areas.[36] The draft PAF from 2021 (SEPA, 2021a) points out that there is little knowledge about the location and status of, for instance, BHD habitats and species in the 'green infrastructure' outside protected areas.

Also, the National Board of Housing, Building and Planning (Boverket) has developed guidance on ecosystem services and green infrastructure in planning and building,[37] and some municipalities have taken initiatives to apply the proposed approach in spatial planning.

30.5.3 Approaches to Protection and Levels of Protection

Site-specific regulations are elaborated for each protected area. They vary substantially between sites, depending on what is deemed necessary to address damage or deterioration to habitats and/or species in relation to potential threats.

For most sites, the focus is on restricting land use, which requires negotiations and agreements with landowners and other stakeholders. Depending on the level of restrictions, protected sites can either remain in private ownership (with compensation payments) or the land is bought by the State or (more rarely) by the municipality. Several Nature Reserves and most Habitat Protection Areas have stayed in private ownership, while the National Parks are always located on State-owned land. Inevitably, the negotiations have involved conflicts with landowners and other stakeholders, and sometimes have led to compromises in both delineation of protected sites and adjustments of restrictions and regulations.

The principle of public access to both private and public land, '*allemansrätten*', is a basic premise for land use in Sweden. It allows anyone to hike in most parts of the country as well as camp for single nights. However, the principle also imposes a responsibility not to damage trees or arable land, or to disturb livestock or reindeer. The right is a customary law with a long historical tradition, but it was protected in the Constitution only in 1994. Consequently, restricting public access in protected areas is rare in Sweden. Rather, the public is actively encouraged to visit most of the site, although certain activities may be restricted. If strict regulation of public access is deemed necessary, for instance to minimise disturbance of wildlife, this is regulated by establishing Sanctuaries for Wildlife or Plants.

Procedures for appropriate assessments in accordance with HD Article 6(3) and derogations in line with HD Article 6(4) are transposed in the Environmental Code.[38] However, it took many years for adequate routines to be established, as manifested in thoroughly revised guidelines in 2017 (SEPA, 2017b). Over time, several cases have raised public debate and resulted in national court cases. A few complaints have also been submitted to the European Commission, but none has been brought before the CJEU. Detailed information about 'key cases' that contributed to national precedent is included in the guidelines of 2017.

30.5.4 Approaches to Managing Protected Areas

About 80% of Swedish protected areas are managed by the State, via the regional CABs. Another 12% are managed by municipalities, and the remaining sites by various foundations or state agencies other than the CABs (Hedeklint and Olsson, 2021). Conservation Plans and Management Plans are the main guiding documents for the implementation of conservation measures:[39]

36 https://ext-geoportal.lansstyrelsen.se/arcgis/apps/MapSeries/index.html?appid=24cf35e88cd7423f9cd46d5a8b9e5620

37 The guidance is provided as a web-based information tool on Boverkets website www.boverket.se, 'Ekosystemtjänster i den byggda miljön – vägledning and metod'.

38 Miljöbalk (Environmental Code), chapter 7, 27–29§§.

39 In English texts, including European Commission documents, 'management plan' can refer to both kinds of plans. In this text and in order avoid ambiguity, 'Conservation Plan' and 'Management Plan' are consistently used as explained here.

- **Conservation Plans** (*bevarandeplaner*) should be elaborated for each Natura 2000 site and include information on their designated BHD species and habitats, degree of conservation and site-specific objectives. They are guidance documents that do not benefit from any legal status. For the majority of Natura 2000 sites that also benefit from additional legal protection, the Conservation Plans constitute a framework for the more detailed Management Plans.
- **Management Plans** (*skötselplaner*) are more detailed plans that are prepared for each National Park and Nature Reserve. They are legally binding documents, and they include a description of the site, detailed conservation objectives for its habitats and species, management measures (including an approximate time-schedule) and monitoring measures. Often, the plans include maps showing zones of different restrictions or specific management needs. The plan also identifies responsibilities for funding.

Both kinds of plans are prepared and approved by the regional CABs, after consultation with landowners and other stakeholders, including the municipalities. As a complement to the formal communication of draft plans, consultations may include personal contacts and public meetings. The CABs also have the main responsibility for implementing and updating the plans. Work is underway to gradually transfer site Conservation and Management Plans from what are now commonly individual PDF documents into a digital format that will make it easier to compare and compile information on multiple sites, to update the plans, etc.

The management of protected areas is primarily financed with State funds. As farmers and landowners are the primary managers of semi-natural grasslands in Sweden, recurring biotope management, such as grazing or mowing of semi-natural habitats, is carried out through the RDP agri-environment scheme in most cases. There are no additional or 'earmarked' financial mechanisms for Natura 2000 sites.

For semi-natural habitats in the Natura 2000 network, a 2018 review found that about 80 000 ha were in need of management, but only 60% were taken up in the agri-environment scheme (SEPA, 2018b). The low take-up was most striking in remote areas, and the primary reasons were, *inter alia*, a lack of interest among farmers and insufficient levels of remuneration for landowners. The system was also found to suffer from administrative deficiencies, such as inadequate definitions of meadow and pasture habitats eligible for compensation.

Other challenges, such as a lack of animals for grazing, major historical losses of habitat and land abandonment, are too great and complex to be solved through CAP measures alone. Swedish habitat types and species associated with the agricultural landscape generally have a poor and deteriorating status. Major efforts, both in restoration and management, are needed in the whole landscape to improve the status of Swedish semi-natural grasslands (SEPA, 2021a). In fact, in 2017, SLU found, in a study funded by SJV, that semi-natural grasslands with agri-environment payments only had marginally better quality than those without. Between the periods 2006–2010 and 2011–2015, the correlation between receiving payments and indicators such as species richness of butterflies, bumble bees and vascular plants was not statistically significant (Glimskär *et al.*, 2017).

30.6 General Nature Conservation Measures

30.6.1 Development Planning and Impact Assessment

Protected areas are routinely taken into account in planning processes at regional/CAB and local/municipal levels, as well as at the national level. Findings of relevance for protected areas must also be considered in strategic assessments and EIAs for individual plans and projects.

The governmental agency with the overall responsibility for spatial planning and environmental impact analyses is the National Board of Housing, Building and Planning. The Planning and Building Act provides the legal framework, in its most recent version from 2010 (with later amendments).[40] The Act defines concepts of public and individual interests and how these should be weighted – an important aspect in nature conservation. It also defines a framework for spatial planning at the municipal level, where the Municipal Comprehensive Plans play a key role.

The Municipal Comprehensive Plans also identify areas of national interest for, for instance, nature conservation, as well as various kinds of land use and exploitation and how these should be balanced, in accordance with the Environmental Code.[41] Recent amendments also prescribe that the plans need to include adaptation measures to future climate change. Although these plans are guiding documents subject to political decisions, and have no binding legal status, experience has shown that they can play a role in public debate about controversial plans and projects.

The SEA and EIA Directives are transposed into national legislation, within the framework of the Environmental Code,[42] with details further elaborated in the EIA Ordinance.[43] The National Board of Housing, Building and Planning provides a website with links to further details and guidance.[44] SEPA's handbook on assessments and supervision of Natura 2000 sites includes more detailed guidance on how to apply the EIA rules when a Natura 2000 site may be affected by a plan or project (SEPA, 2017b).

40 Plan- och bygglag (2010:900).
41 Miljöbalk (Environmental Code), chapters 3 and 4.
42 Miljöbalk (Environmental Code), chapter 6.
43 Miljöbedömningsförordning (2017:966).
44 www.boverket.se/sv/PBL-kunskapsbanken/planering/detaljplan/miljobedomningar/6-kap-MB/

Several chapters in the Environmental Code include rules for offsetting and compensation measures. Damages to values of public interest, such as areas important for, *inter alia*, biodiversity, green infrastructure and ecosystem services, shall be compensated for by creating new or protecting equivalent values, as explained in SEPA's guidelines on ecological compensation (SEPA, 2016b). The guidelines include a specific chapter on how to apply Article 6(4) when Natura 2000 sites may be affected by a plan or project.

According to a review of the period 2015–2018, ecological compensation measures were required for only 3% of permits or exemptions given to projects and plans in protected areas in accordance with the Environmental Code,[45] and 4% of exemptions from the Species Protection Ordinance, but as high as 73% for Habitat Protection Areas (Sjöholm, 2019). Proportions of approvals with demands for compensation were similar during the period 2011–2014 (SEPA, 2015), although this review also indicated that intrusions in areas without formal protection were only rarely compensated for, regardless of existing high nature values.

Recent research projects financed by SEPA (Öckinger *et al.*, 2021; Widenfalk *et al.*, 2021), reviewing both national and international studies, found little evidence of positive impacts on biodiversity values and ecosystems services from ecological compensation. At the national level, the implementation of compensation measures has been hampered by legal and bureaucratic constraints, as well as ambiguous standards. Recommended improvements include clearer standards and demands for systematic and long-term monitoring.

30.6.2 Habitat Management and Restoration

This section discusses three examples of habitat management and restoration that are particularly relevant for Sweden.

Management of Natural Forest Habitats

Large-scale exploitation of forests can be traced back at least to the seventeenth century when large volumes of timber were extracted for various purposes, not least charcoaling for iron smelting. The political and economic power of forestry has been under more or less continuous debate ever since. It was not until the introduction of large-scale clearcutting during the twentieth century, however, that a rapid loss of natural forests began.

To conserve biodiversity in productive forests, a three-level approach was established during the 1980s: protected forest areas at a landscape scale, reserves set aside within the matrix of productive forests, and key habitats at the smallest scale (e.g. Simonsson, 2016). However, a 2019 review indicated that the approach had failed over much of the area and proposed adjustments for the future, including better integration between the conservation approach and production, and enhanced diversification of forest management alternatives (Felton *et al.*, 2020).

The approach to date has resulted in an overall unfavourable conservation status of most forest habitats, as concluded from the recent HD Article 17 reporting (Section 30.2.1), with coniferous forest habitats such as western taïga (9010*) and Fennoscandian herb-rich spruce forests (9050) in particular focus. 'Passive management' with little or no intervention has been a key management principle, as illustrated by SEPA's guidelines for the management of forests and woodlands in protected areas (SEPA, 2013). Over time, this concept has been debated. Experience has shown that intervention that allows for the maintenance or restoration of key habitat elements, such as dead and decaying wood, is a prerequisite for achieving a favourable conservation status for several HD forest habitats (Westling *et al.*, 2020).

In 2020, the Government appointed a commission of inquiry to propose a future direction of the national forestry policy. The commission's concluding report proposed several changes to the Forestry Act that would strengthen landowners' rights and enable more flexible protection measures and nature conservation in forests (Ministry of the Environment, 2020). Another mandate was to propose changes that would promote a growing circular forest-based bio-economy. A particularly controversial proposal is to protect remaining natural sub-montane forests (about 5 000 km^2) in the Alpine region bordering Norway, while making voluntary set-asides the primary protection tool in all other Swedish forests and ending inventories of sites with important nature values in forests (so-called 'key habitats', see Section 30.5.1). Protecting these remaining sub-montane forests could help the long-term conservation of related HD habitats and species currently suffering from an unfavourable status, as well as the maintenance of important ecosystem services in this region. However, questions have been raised, for instance, about whether relying on voluntary set-asides would ensure protection of the most ecologically valuable and 'best' located sites.[46] The Government is still considering the inquiry's proposals and the subsequent consultation responses. It remains to be seen what, if any, changes are made to Swedish forest policy. So far, however, the proposals made have not helped to dampen the polarised debate.

Liming of Lakes Affected by Acid Precipitation

During the twentieth century, freshwater lakes and streams in Sweden were impacted by acid precipitation related to anthropogenic emissions. This had serious impacts on biodiversity and ecological processes – declining fish stocks being striking (e.g. Muniz, 1990). An estimated 50 000–100 000 lakes have been affected in Europe, primarily in Sweden, Norway, Finland and Scotland. In Sweden, c. 18 000 lakes and several water courses have suffered serious impacts on biodiversity (SEPA, 2011).

[45] Miljöbalk (Environmental Code), chapter 7, 7§ and 29§ and chapter 16, 9§.

[46] See, for instance, SEPA's consultation response on the Government's website (in Swedish): https://www.regeringen.se/remisser/2021/02/remiss-av-skogsutredningens-betankande-sou-202073/

As general measures to reduce acidifying airborne pollutants were considered insufficient and too slow to initiate a recovery process, a large-scale lime treatment programme was initiated during the 1970s. Research showed that the measures had some success, including in relation to the recovery of fish stocks. The liming programme has also helped improve the conditions for the Freshwater Pearl Mussel, which is particularly sensitive to low pH (SEPA, 2021b). However, it is unlikely that the areas once impacted will ever fully recover (see several chapters in Henrikson and Brodin, 1995).

Today, acidification of freshwaters is almost a non-issue in the national debate, although liming is still a long-term need in order to maintain the positive achievements (Fölster *et al.*, 2020). A large number of lakes and rivers are still limed annually at a cost of about €15.7 million per year (SEPA, 2021a) – an illustration that it may take several decades to repair past neglect and that full restoration might not be possible.

The acid rain issue necessitated international cooperation and coordination on a level never seen before, with new and pioneering interactions between science and politics. Concepts well-known today, such as 'critical load', were established. The issue of acid rain in a 50-year perspective of developing and implementing environmental concepts and policies has been outlined by Grennfelt *et al.* (2020).

Modern Permits for Hydropower Production

Domestic hydropower has been the basis of the Swedish electricity mix for many decades. About 75% of the water courses in Sweden are exploited for hydropower production (Jonsson, 2015). There are some 2 100 hydropower production plants in Sweden, of which the 200 largest ones in the north contribute 95% of the production capacity. The great majority of plants are very small. Most plants were built around the turn of the last century and some decades thereafter (Lindblom and Holmgren, 2016). While the electricity they produce is positive from a climate change mitigation point of view, these plants have serious impacts on biodiversity and ecological functions. Most striking are impoundments and barrier effects linked to them, and artificial regulation of the water flow, with impacts on anadromous fish species as well as benthic and littoral flora and fauna. Also, interactions between river and adjacent terrestrial habitats, such as regular flooding, are disturbed (e.g. Näslund *et al.*, 2013).

In 2021, the Swedish electricity consumption was 166 TWh, provided largely by hydropower and nuclear power (together 73%).[47] In 2016, the Swedish Parliament agreed on a long-term energy policy with a target to achieve an energy production based entirely on renewable sources by 2040.[48] Hydropower production is foreseen to play an increasingly important role, not least due to the challenges of continued nuclear power production. Thus, concerns for biodiversity must be balanced against the importance of continuing to decarbonise Swedish electricity production.

As an attempt to address this dilemma, and as part of the parliamentary agreement of 2016, the Government adopted new legislation[49] in 2017 aiming to align hydropower production to modern environmental requirements, including those of the Environmental Code, the WFD and the Nature Directives. To ensure a holistic approach to this major review, the Government tasked SwAM, the Swedish Energy Agency and Svenska Kraftnät to draft a national plan (SwAM *et al.*, 2019). The final plan was adopted by the Government in June 2020 and the revision of the first set of licences will begin in 2022. Reviewing all plants is expected to take 20 years. In parallel, 450 Conservation Plans of Natura 2000 sites currently impacted by hydropower production will be revised in order to, if needed, elaborate and clarify the sites' ecological needs and measures necessary to achieve the sites' conservation objectives. In March 2021, SwAM published guidance to support the CABs in their revision of the Conservation Plans in question.

The entire process of revising the hydropower licences will be fully financed by the hydropower industry, including a fund through which larger operators will support the costs borne by small-scale operators.

30.7 Nature Conservation Costs, Benefits and Funding Sources

30.7.1 Nature Conservation Costs

Nature conservation incurs costs across a range of sectors and at various governance levels. The draft PAF for 2021–2027 identifies priority financing needs of SEK 8 352 million (€812 million) per year (SEPA, 2021a). Horizontal measures and administrative costs related to Natura 2000 are estimated to account for 71% of the total annual costs, Natura 2000 site-related maintenance and restoration measures for species and habitats for another 16%, additional green infrastructure measures beyond Natura 2000 for 10%, and additional species-specific measures not related to specific ecosystems or habitats for 3%.

SEPA stresses that its cost estimates in the most recent PAF draft are limited by the assumption that the 2021 level of national financing for measures will remain the same for each year of the rest of the period, and that funding within the CAP framework will have the same limits in Sweden every year as it did in 2021. This is contrary to the intended use of the PAF, which is to specify the financing needed to implement the Natura 2000 network and to achieve the objectives of the HD.

[47] www.ekonomifakta.se/fakta/energi/energibalans-i-sverige/elproduktion/, downloaded 18 March 2022.

[48] www.regeringen.se/artiklar/2016/06/overenskommelse-om-den-svenska-energipolitiken/

[49] Governmental proposal 2017/18:243, data.riksdagen.se/dokument/H503243, downloaded 29 January 2022.

The costs of managing Swedish protected areas are primarily incurred at their implementation level (the CABs) and include physical management, measures to ensure access to sites by visitors, and public communication. CABs report their use of state funds to SEPA on an annual basis. In recent years (2016–2018), habitat maintenance and management has been by far the largest cost, using 40–45% of funds allocated to managing protected areas, with grazing and restoration measures being the two largest habitat-related costs (SEPA, 2019). Most CABs have engaged external contractors through public procurement to carry out this work.

30.7.2 The Economic Benefits of Nature Conservation

There has been no overall estimation to date of the economic benefits of nature conservation in Sweden, neither is there any systematic, recurring or long-term follow-up of such impacts (SEPA, 2021c). SEPA has, however, elaborated a possible method for assessing the economic impacts on rural development, employment and nature tourism of state investments in protecting and supporting valuable nature (SEPA, 2021c). Whether this, or any other approach, is adopted and deployed remains to be seen.

Individual studies have explored and estimated the economic value of different nature management measures and ecosystem services, and how these can be better reflected in decision making.[50] These may give indirect indications of the economic benefits of nature conservation in Sweden. Furthermore, a SwAM study has shown that maritime spatial planning in the Baltic Sea 'most likely' has long-term benefits from a socio-economic perspective (Hasselström *et al.*, 2019).

30.7.3 Funding Sources for Nature Conservation

The Swedish government allocates state funds relevant to nature conservation via different statutory agencies, as part of a budget expenditure item 'General environmental and nature conservation'. SEPA administers two funds under this allocation, for the protection and management of valuable nature, respectively. For 2021, almost 89% of the fund for protection was allocated to redeeming land when creating new protected areas and included purchasing land and compensating landowners, as well as contributing to municipalities, CABs and foundations' expenses for acquisition of land and compensation payments.[51]

In addition to the two funds administered by SEPA, other relevant funding streams include allocations on measures for marine and freshwater environments, for which SwAM is responsible. Nature conservation is only one part of these funds, however.

For the 2014–2020 period, the state spending allocation to nature conservation measures or projects not benefiting from any EU co-funding was an estimated SEK 3 457 million (€342 million) per year[52] (SEPA, 2021a).

Part of the state funds are allocated to implementing actions at regional and local levels and are made available to other actors through various initiatives. For example, Swedish municipalities can apply for funding through the Local Nature Conservation Programme (LONA) and its marine/freshwater equivalent (LOVA). Since the start of the initiative in 2004, LONA funds have been used for a range of initiatives, such as management and green infrastructure plans, awareness raising campaigns and habitat restoration measures. Municipalities are required to contribute 50–90% match funding using their own finances.

The EU funds most frequently used in Sweden for nature conservation-related purposes in the 2014–2020 funding period have been the European Agricultural Fund for Rural Development (EAFRD), the LIFE programme fund and the European Agricultural Guarantee Fund (EAGF), as shown in Table 30.5. The largest funding allocations were for the RDP measures, and in particular agri-environment-climate measures, which, as described earlier, have been important for managing semi-natural habitats, although a substantial proportion of scheme funding has not been taken up and the effectiveness of some schemes needs to be improved.

According to the draft PAF for 2021–2027 by SEPA, the European Agricultural Guarantee Fund (EAGF) has supported the maintenance of pastures and meadows through direct income payments to eligible farmers[53] of about SEK 963 million (€95.3 million) per year during the 2014–2020 funding period. However, such payments are not generally considered to be very beneficial for biodiversity, and are not normally included in PAFs. Therefore, as SEPA does not indicate to what extent the EAGF payments have benefited Natura 2000 areas per se, they are not included in the funding table here (as is also the case in other country chapters).

As indicated several times already, the LIFE programme has been very important for the Swedish Natura 2000 network by enabling large-scale restorations and maintenance that have improved the status of valuable habitats and endangered species (SEPA, 2019, 2021a).

Interestingly, the European Maritime and Fisheries Fund (EMFF) (previously the European Fisheries Fund (EFF)) has not been used much in Sweden, despite both the 2013 and the draft 2021 PAFs stressing the need for additional measures in marine Natura 2000 sites. This has been partly due to the difficulty of combining national co-finance and EU finances

50 See the SEPA-funded initiative 'Värdet av ekosystemtjänster': www.ecosystemservices.se

51 Information from SEPA, www.naturvardsverket.se/om-miljoarbetet/uppfoljningar/resultat-av-anslagen-for-vardefull-natur/anslag-for-vardefull-natur/, downloaded 16 January 2021.

52 This figure includes the relevant staff costs of CABs. Exchange rate, €1 = SEK 10.11, on 29 May 2021.

53 In accordance with Swedish Regulation (2014:1101) on EU direct payments for farmers.

Table 30.5 Current funding sources for nature conservation in Sweden.

Source and period	Measure/category	Averaged annual allocation (SEK m)		Averaged annual allocation (€ m)		Comments
EU funding instrument		**EU + national**	**EU**	**EU + national**	**EU**	
EAFRD – Rural Development Programme – allocations to Priority 4 (restoring, preserving and enhancing agriculture and forest ecosystems) 2014–2020	M13 Areas facing constraints	1 030.9	570.2	102.0	56.4	
	M10 Agri-environment–climate	1 030.3	581.6	101.9	57.5	See 1
	M11 Organic farming	543.8	302.3	53.8	29.9	
	M1 Knowledge transfer, etc.	88.9	44.4	8.8	4.4	
	M2 Advisory services, etc.	41.8	20.9	4.1	2.1	
	M7 Basic services, etc.	33.0	16.5	3.3	1.6	
	M8 Investments in forests	14.7	7.3	1.5	0.7	See 2
	M4 Investments in physical assets	8.9	4.4	0.9	0.4	See 3
	M16 Cooperation	8.0	4.0	0.8	0.4	
RDP Priority 4 subtotal		*2 800.2*	*1 551.7*	*277.0*	*153.5*	
LIFE Programme	Nature and biodiversity projects 1995–2019	105.3	53.4	10.4	5.3	92 projects
	Integrated projects – environment 2016–2025	47.2	9.8	4.7	1.0	Two projects
EMFF 2014–2020	Art. 40. 1b-g. Measure I.18B	2.9	1.8	0.3	0.2	For N2k etc. See text
ERDF Interreg V 2014–2020	85 Protection and enhancement of biodiversity, nature protection and green infrastructure	21.5	15.5	2.1	1.5	For N2k etc. See 4
	86 Protection, restoration and sustainable use of N2k	2.0	1.0	0.2	0.1	
	Other categories (87+88)	7.8	6.8	0.8	0.7	
EU funding instrument subtotal		*2 986.9*	*1 639.9*	*295.4*	*162.2*	
National public funds		**Additional national* (SEK m)**		**Additional national* (€ m)**		**Comments**
National public funds 2014–2020	Government-funded measures	4 165.3		412		See 5
National total inc. co-financing of EU instruments		*5 512.3*		*545.2*		
TOTAL		**7 152.3**		**707.4**		

Notes. Currency conversion rate (29 May 2021): €1 = SEK 10.11. 'For N2k etc.' indicates funding relevant to Natura 2000, related green infrastructure and BHD species (based on the PAF). * Excludes co-financing of EU funds unless indicated. *Comments*: 1. Less than 50% of M10 is spent on maintenance of semi-natural habitats. 2. Primarily spent on prescribed burning, restoration of sites and activities to increase the proportion of deciduous forests. 3. Around 95% spent on fencing to protect livestock from large carnivores. 4. Figures from SEPA (2021a); only including funds for agreed projects at the time of writing the PAF and with managing authorities located in Sweden. 5. An estimate of the total financing allocated to implement EU nature policy and associated green infrastructure, including staff costs for CABs.

Sources. Based on data from: RDP data from European Commission ESIF funds database (23 October 2020); LIFE Sweden factsheet (March 2021), and LIFE Public database; SEPA (2021a). See Section 1.3 for further details.

for these types of projects. There is an ambition to try to make better use of the funds available through the EMFF in the future.

Lastly, it is important to emphasise (as SEPA also does in the 2021–2027 PAF), that members of the public, farmers, landowners, private companies, organisations and municipalities spend significant amounts of money and time on nature conservation in different ways. These sources of funding are not monitored and are therefore not reflected in these funding data.

Whilst a direct comparison between the funding sources identified in Table 30.5 and the cost estimates in Section 30.7.1 is not straightforward, the available funding seems to be insufficient even to meet the annual requirements for Natura 2000 identified in the PAF (i.e. SEK 3 457 million, €342 million), let alone wider nature conservation needs. Furthermore, a large proportion of the EU funding is allocated to supporting farmers in areas with natural constraints, and organic farming, which may not be contributing significantly to current biodiversity goals (SEPA, 2021a). Such conclusions are also in line with the annual reviews of the national environmental quality objectives, which identified a lack of funds for adequate management as a main reason for failures to meet the objectives set for forests, wetlands and grasslands, for example (SEPA, 2021b). Similarly, a lack of resources is also preventing sufficient monitoring of compliance with, and enforcement of, existing rules in many protected areas, in particular related to physical construction, recreational activities and fisheries in the marine environment (SEPA, 2021a).

30.8 Surveillance, Monitoring, Reporting and Applied Research

Environmental monitoring in Sweden is coordinated by SEPA, involving about 30 other governmental agencies and a range of different stakeholders. The primary aim is to provide data for the periodical assessments of the 16 national environmental quality objectives (Section 30.3.2), and the reporting requirements of EU policies and international conventions. The monitoring is organised into 10 themes, covering a wide range of aspects related to biodiversity and other environmental issues. SwAM has the coordinating role for themes that are linked to marine and freshwater environments.

Several institutions, including SLU, SwAM, the Geological Survey of Sweden, SFA and the Swedish Environmental Research Institute (IVL), are engaged in the storing of biodiversity-related data and making them publicly available.[54]

Key examples of the many activities of relevance for the monitoring and storing of Swedish biodiversity data are listed below.

- The **National Inventory of Landscapes in Sweden** (NILS) is a nationwide programme monitoring the condition of, and changes in, the terrestrial landscape.[55] It is based on aerial and field surveys and provides various indicators of relevance for status assessments of habitats and species.
- The **Swedish Bird Monitoring Programme** carries out surveillance of long-term trends. Species-specific indices from 1975 and onwards on the majority of bird species breeding in Sweden are openly available.[56] The programme involves the national waterfowl surveys, established in 1967. Both schemes are primarily based on volunteers carrying out the fieldwork.
- The **Swedish Butterfly Monitoring Scheme** was initiated in 2010, with the objective to follow trends in the national populations. It provides data to the EU Grassland Butterflies Population Index and is also based on volunteers' participation.[57]

Other and equally important sources of information include the baseline inventory of Natura 2000 and protected areas conducted during 2004–2008 (SEPA, 2009b), the national Action Programmes for threatened species (Section 30.4.2), the National Wetland Survey (Gunnarsson and Löfroth, 2009), the National Freshwater Lake Register,[58] the Swedish National Forest Inventory[59] and the national TUVA database for semi-natural habitats.[60] The last is based on the results of a national survey carried out in 2002–2004 by SJV (Persson, 2005), which provides the baseline for the protection and management of semi-natural habitats.

The assessments under HD Article 17 and BD Article 12, as well as the regular updates of the national Red List at five-year intervals, are coordinated by the SLU Swedish Species Information Centre. The Centre plays a key role in monitoring the status of species and habitats, in collaboration with SEPA, several scientific institutions and NGOs. It relies to a large extent on data from the above-mentioned information sources, although the contribution by volunteer-based activities related to rare and threatened plant and animal species, as well as information in the Species Observations System (*Artportalen*[61]), is crucial. Overall, most of the Centre's assessments are based on data collected across the country from different types of landscapes, rather than restricted to protected sites or the Natura 2000 network. In the marine

54 SLU provides a list of information sources, including web-links: www.slu.se/miljoanalys/statistik-och-miljodata/miljodata/

55 www.slu.se/en/Collaborative-Centres-and-Projects/nils/

56 www.fageltaxering.lu.se/

57 www.dagfjarilar.lu.se/english

58 Hosted by the Swedish Meteorological and Hydrological Institute: www.smhi.se/data/hydrologi/vattenwebb/damm-och-sjoregister-1.116177

59 Hosted by SLU: www.slu.se/en/Collaborative-Centres-and-Projects/the-swedish-national-forest-inventory/

60 Hosted by SJV: https://jordbruksverket.se/e-tjanster-databaser-och-appar/e-tjanster-och-databaser-stod/tuva

61 www.artportalen.se/

environment, many assessments are still based on expert judgements due to lack of field data.

The Swedish system of surveillance and applied research on the status and trends of species and habitats is coherent overall, but there is room for improvement. Despite numerous sources of information, several knowledge gaps remain when it comes to the distribution and the status of habitats and species. This is particularly true for marine habitats, forest habitats outside protected areas, certain grassland habitats, and freshwater habitats, as identified in the PAF for 2021–2017 (SEPA, 2021a).

30.9 Future Developments

30.9.1 The Short-Term Perspective

As for most policy areas, successful nature conservation is dependent on political and public support, primarily to ensure the necessary funding. A quick analysis may give an erroneous impression of consensus across the Swedish political spectrum that biodiversity issues are highly prioritised, and that no abrupt changes in general direction are to be expected in the foreseeable future. However, there are considerable differences between the political parties, as indicated by drastic changes in funding and short-term priorities, with changed political constellations behind the national budget during the last decades. Constraints have been more serious during periods when Sweden has been governed based on a budget influenced by right-wing parties. Further, the parliamentary instability of recent years has caused unpredictability in both funding and strategic direction, which continues to hamper cost-effective nature conservation and restoration.

As indicated throughout this chapter, several regulatory and other changes are expected in the coming years that will influence Swedish nature conservation. The review of over 2 000 hydropower plant licences, a revised Forestry Act, changes to the 70-year-old shoreline protection regulation and potentially revised species protection legislation are key examples.

Other short-term developments that may have implications for Swedish nature conservation relate to modernisation of mapping and monitoring. As mentioned previously, many of the field data collected for surveillance and monitoring are still dependent on the participation of volunteers, not least the knowledge of retirees who have taken part in the data collection over several decades and whose expertise is not being replaced at the same rate as it is disappearing. However, the development and deployment of modern surveillance and monitoring tools are gaining pace, which will hopefully be cost-effective and support adaptive management of protected areas. In the marine and coastal environment, where Swedish nature conservation struggles with some of its most significant knowledge gaps, both site-based and biogeographical monitoring are under revision. New methods are being developed, partly satellite-based, to produce high-resolution maps of larger areas, based on in-field data.[62]

EU-level collaboration on nature conservation continues to have a huge impact on Swedish efforts and strategies. Although no new national biodiversity strategy has yet been published, to reflect the commitments made in the *EU Biodiversity Strategy for 2030*, work has begun to prepare for the coming changes at EU level. One important example is in anticipation of the EU nature restoration law and its various targets. The 2019 CJEU ruling on species protection and ongoing infringement cases against Sweden may also result in short- to medium-term changes.

30.9.2 The Long-Term Perspective and Impacts of Climate Change

Inevitably, efforts to conserve nature and stop the loss of biodiversity must consider the impacts of climate change. Most scenarios foresee increased precipitation affecting large parts of Sweden, with more frequent episodes of intense rainfall. South-east Sweden may be an exception, where a warmer climate and increased evaporation may lead to an overall drier climate than today.[63] With reference to biodiversity and ecological processes, a vegetation season prolonged by one to two months, or more in the southernmost parts of the country, may have large impacts on the status and range of already vulnerable species and habitats. Many species may face difficulties in adapting at the same rate at which climate zones move northwards.

The terrestrial habitats in the Alpine biogeographical region are expected to be some of the most seriously affected by climate change in Europe. Increases in temperature will likely lead to shorter periods of snow cover and an extended vegetation period in this region (European Commission, 2013; Nila *et al.*, 2019). For freshwater lakes and streams, higher water temperatures and darkening of water colour or 'browning' by elevated concentrations of dissolved organic carbon may lead to serious impacts on lake ecosystems (e.g. Weyhenmeyer *et al.*, 2016). All these changing conditions need to be taken into account in the planning and management of protected areas in Sweden and the overall network design, as well as species protection measures.

Conclusions about the implications at the national level from recent IPCC and IPBES reports underline the importance of realising potential synergies between measures for climate mitigation and adaptation, and those for the preservation of biodiversity (Bergström *et al.*, 2020). In Sweden, the carbon

---

62 See more information on SwAM's website, including 'Nationell marin kartering' and environmental monitoring of benthic habitats: https://www.havochvatten.se/overvakning-och-uppfoljning/miljoovervakning/nationell-marin-kartering.html.

63 A general overview of future climate impact in Sweden has been published by SEPA (Bernes, 2016).

storage potential of boreal forests is the subject of particular attention. As previously mentioned, there is an ongoing debate about the future direction of Swedish forestry policy and how best to achieve objectives related to both combating climate change and the protection of nature. There is a strong push both at national and EU level to increase bio-energy production, as one component in meeting climate targets. Especially in the case of forest-based energy, this must be carefully managed to avoid the further degradation of ecosystem functions and loss of biodiversity. Even from a climate change mitigation perspective, there is debate about the most effective use of forests. One recently published model indicates that even moderate harvest reductions would result in substantial climate mitigation benefits, while increased harvesting would be counter-productive (Skytt *et al.*, 2021).

Balancing the climate benefits of energy and materials based on raw forest biomass, but also wind- and hydropower production, with their impacts on biodiversity, will continue to be a major challenge. More attention to the wider climate change mitigation and adaptation benefits of protecting nature and developing and deploying systematic methods of measuring socio-economic benefits of nature conservation might help to inform this delicate balancing act. Concepts such as ecosystem services and green infrastructure are gradually being incorporated into Swedish planning and decision making, but, as elsewhere, this is a slow process.

30.10 Conclusions

With a 50-year perspective, Sweden has developed a comprehensive system for nature conservation administration and the management of species and habitats, and has come reasonably far in monitoring their status and trends.

There has without doubt been major progress, not least after the accession to the EU and the implementation of the Nature Directives. At the same time, few of the national environmental quality objectives have been achieved, and the more specific needs identified in the thematic plans for various groups of habitats have hardly been fulfilled. There are still insufficiencies in protection, such as for forests, marine areas and freshwaters with high values, as well as inadequate regulations and insufficient enforcement of rules in some protected sites.

The remaining insufficiencies have been clearly illustrated by the recent assessments reported under the Nature Directives, which show that many habitats and species still have an unfavourable status, and some of them continue to decline. Whilst important successes over recent decades must not be ignored, obviously these efforts have not been enough to enhance the conservation status of many habitats and species. The need for continued and additional measures is very great: at different scales (from local measures to international cooperation) and both inside and outside formally protected areas.

Swedish nature conservation is arguably at a crossroads, with important but difficult trade-offs between political agendas, as well as changes to important pieces of legislation anticipated in the coming years. Knowledge gaps remain and there is an obvious need for further research, but it is not an excuse for refraining from action. Although issues related to political priorities and fluctuating levels of public funding raise concern, recorded positive trends verify that restoration and management measures can be effective.

Naturally, large-scale trends and processes, including land abandonment, eutrophication and climate change, continue to pose challenges for conservation efforts. To address these and ensure long-lasting impacts, traditional site- and species-based protection measures and habitat management as well as restoration must be combined with ambitious policies to improve the overall state of the Swedish environment.

Acknowledgements

Three reviewers have assisted with remarks on previous drafts: Ulf Gärdenfors (Professor Emeritus in Conservation Biology and former deputy Director of the SLU Swedish Species Information Centre), Klas Hjelm (former Senior Adviser and Head of the Nature Conservation Department at SSNC), and the late Lennart Henrikson (self-employed consultant and former expert in aquatic conservation and management at WWF-Sweden).

References

Bergström, L. Borgström, P., Smith, H. G. *et al.* (2020) *Climate change and biodiversity – conclusions from the IPCC and IPBES in a Swedish perspective.* Stockholm: SMHI and Naturvårdsverket Klimatologi 56.

Bernes, C. (2016) *En varmare värld. Växthuseffekten och klimatets förändringar. Monitor 23.* Stockholm: Naturvådsverket.

Bråkenhielm, S. (ed.) (1982) *Virgin forests. An inventory of old natural forests in Sweden.* (In Swedish, summary in English.) Stockholm: Naturvårdsverket Report 1507.

Darpö, J. & Epstein, Y. (2015) Thrown to the wolves – Sweden once again flouts EU standards on species protection and access to justice. *Nordic Environmental Law Journal*, **2015** (1), 7–20.

Eide, W., Ahrné, K., Bjelke, U. *et al.* (2020) *Status and Trends for Species and Their Habitats – Red-Listed Species in Sweden 2020.* (In Swedish, summary in English.) Uppsala, Sweden: SLU Artdatabanken SLU.

European Commission (2013) *Guidelines on Climate Change and Natura 2000. Dealing with the impact of climate change. On the management of the Natura 2000 network of areas of high biodiversity value.*

(2019) *The EU Environmental Implementation Review 2019 Country Report – SWEDEN.* SWD(2019) 117 final.

European Court of Auditors (2016) *Combating eutrophication in the Baltic Sea: further and more effective action needed.* Luxembourg: European Court of Auditors, Special Report 03, 2016.

Felton, A., Löfroth, T., Angelstam, P. *et al.* (2020) Keeping pace with forestry: multi-scale conservation in changing production matrix. *Ambio*, **49**, 1050–1064.

Fölster, J., Wallman, K. & Moldan, F. (2020) *Målvattendragsundersökningen 2010–2016. Bedömning av surhet och försurning i kalkade vattendrag*. Göteborg, Sweden: Havs- och Vattenmyndigheten Report 2020:29.

Glimskär, A., Berg, Å., Żmihorski, M., Cronvall, E. & Eriksson, Å. (2017) *Quality changes in seminatural meadow and pastures with and without agri-environmental payments*. (In Swedish, summary in English.) Jönköping, Sweden: Jordbruksverket UTV17.4.

Green, M., Haas, F. & Lindström, Å. (2019) *Övervakning av fåglarnas populationsutveckling. Årsrapport för 2018*. Lund University, Sweden: Department of Biology.

Grennfelt, P., Engleryd, A., Forsius, M., Hov, Ø., Rodhe, H. & Cowling, E. (2020) Acid rain and air pollution: 50 years of progress in environmental science and policy. *Ambio*, **49**, 849–864.

Gunnarsson, U. & Löfroth, M. (2009) *The national wetland survey – results from 25 years of surveys. National final report for the wetland surveys (VMI) in Sweden*. (In Swedish, summary in English.) Stockholm: Naturvårdsverket Report 5925.

Hansson, K. (2016) *Assessing total legitimacy in Swedish large carnivore policy*. Thesis for the Master's seminar in Political Science at Umeå University, Sweden. Fall semester 2016.

Hasselström, L., Cole, S. & Löfmarck, A. (2019) *Samhällsekonomisk konsekvensanalys av förslag till havsplan Östersjön*. Göteborg, Sweden: Havs- och Vattenmyndigheten Report 2019:8.

Hedeklint, K. & Olsson, B. (2021) *Protected Nature 2020*. (In Swedish, summary in English.) Stockholm: Sveriges officiella statistiska meddelanden MI 41 2020A01.

Henrikson, L. & Brodin, Y. (eds.) (1995) *Liming of Acidified Surface Waters: A Swedish Synthesis*. Berlin: Springer-Verlag.

Jonsson, M. (2015) *Rikedomar runt rinnande vatten. De ekonomiska värdena av en miljöanpassad vattenkraft*. Stockholm: Sportfiskarna, WWF, Älvräddarna and Naturskyddsföreningen.

Liberg, O., Chapron, C., Wabakken, P. *et al.* (2011) Shot, shovel and shut up: cryptic poaching slows restoration of a large carnivore in Europe. *Proceedings of the Royal Society B*, **279**, 910–915.

Liberg, O., Suutarinen, J., Åkesson, M. *et al.* (2020) Poaching-related disappearance rate of wolves in Sweden was positively related to population size and negatively to legal culling. *Biological Conservation*, **243**, 108456.

Lindblom, E. & Holmgren, K. (2016) *Den småskaliga vattenkraftens miljöpåverkan och samhällsnytta. En syntesstudie*. Stockholm: IVL Svenska Miljöinstitutet Report B 2258.

Ministry of the Environment (2012) *Svenska miljömål – preciseringar av miljökvalitetsmålen och en första uppsättning etappmål*. Stockholm: Regeringskansliet Ds 2012:23.

(2020) *Stärkt äganderätt, flexibla skyddsformer och naturvård i skogen*. Stockholm: Regeringskansliet SOU 2020:73.

(2021) *Skydd av arter – vårt gemensamma ansvar*. Stockholm: Regeringskansliet SOU 2021:51.

Muniz, I. P. (1990) Freshwater acidification: its effects on species and communities of freshwater microbes, plants and animals. *Proceedings of the Royal Society of Edinburgh, Section B: Biological Sciences*, **97**, 227–254.

Näslund, I., Kling, J. & Bergengren, J. (2013) *Vattenkraftens påverkan på akvatiska ekosystem – en litteratursammanställning*. Göteborg, Sweden: Havs- och vattenmyndigheten Report 2013:10.

Nila, M. U. S., Beierkuhnlein, C., Jaeschke, A., Hoffmann, S. & Hossain, M. L. (2019) Predicting the effectiveness of protected areas under Natura 2000 under climate change. *Ecological Processes*, **8**, 13.

Öckinger, E., Josefsson, J., Widenfalk, L. *et al.* (2021) *When can ecological compensation contribute to the conservation of biodiversity and ecosystem services?* (In Swedish, summary in English.) Stockholm: Naturvårdsverket Report 6996.

Persson, K. (2005) *Meadows and pastures survey, 2002–2004*. (In Swedish, summary in English.) Jönköping, Sweden: Jordbruksverket Report 2005:1.

SEPA (2007a) *Mire Protection Plan for Sweden. Main report from revision 2006*. (In Swedish, summary in English.) Stockholm: Naturvårdsverket Report 5667.

(2007b) *National Strategy for the Protection of water-connected environments*. (In Swedish, summary in English.) Stockholm: Naturvårdsverket Report 5666.

(2009a) *Handbok för artkyddsförordningen. Del 1. fridlysning och dispenser*. Stockholm: Naturvårdsverket Handbook 2009:2, edition 1.

(2009b) *Base inventory of Natura 2000 and protected areas 2004–2008. Description of a completed project*. (In Swedish, summary in English.) Stockholm: Naturvårdsverket Report 5990.

(2011) *National plan for liming 2011–2015*. (In Swedish, summary in English.) Stockholm: Naturvårdsverket Report 6449.

(2013) *Management of forests and other wooded land in protected areas*. (In Swedish, summary in English.) Stockholm: Naturvårdsverket Report 6561.

(2015) *Tillämpning av miljöbalkens bestämmelser om ekologisk kompensation*. Stockholm: Naturvårdsverket Report 6667.

(2016a) *Nationell förvaltningsplan för varg. Förvaltningsperioden 2014–2019. Justerad version januari 2016*. Stockholm: Naturvårdsverket.

(2016b) *Ecological compensation. A guidance about compensation at loss of nature values*. (In Swedish, summary in English.) Stockholm: Naturvårdsverket Handbook 2016:1, edition 1.

(2017a) *National strategy for formal protection of forests. Revised version 2017*. (In Swedish, summary in English.) Stockholm: Naturvårdsverket Report 6762.

(2017b) *Förutsättningar för prövningar och tillsyn i Natura 2000-områden*. Stockholm: Naturvårdsverket Handbook 2017:1, edition 1.

(2018a) *Strategi för svensk viltförvaltning*. Stockholm: Naturvårdsverket.

(2018b) *Agri-environmental support and valuable grasslands – How do they function for the work with favourable conservation status?* (In Swedish, summary in English.) Stockholm: Naturvårdsverket Report 6822.

(2019) *Återrapportering av skydd och åtgärder för värdefull natur 2016–2018*. Stockholm: Naturvårdsverket Report 6876.

(2021a) *Prioritised Action Framework (PAF) for Natura 2000 in Sweden pursuant to Article 8 of Council Directive 92/43/EEC on the conservation of natural habitats and of wild fauna and flora (the Habitats Directive) for the Multiannual Financial Framework period 2021–2027 (Draft version 2021-09-23)*. Stockholm: Naturvårdsverket.

(2021b) *Miljömålen – Årlig uppföljning av Sveriges nationella miljömål 2021 – Med fokus på statliga insatser.* Stockholm: Naturvårdsverket Report 6968.

(2021c) *Landsbygdsutveckling och naturturism. Redovisning av ett regeringsuppdrag.* Stockholm: Naturvårdsverket.

SEPA & SFA (2005) *Nationell strategi for formellt skydd av skog.* Stockholm: Naturvårdsverket and Skogsstyrelsen.

(2016) *Gemensamma riktlinjer för handläggningen av artskyddsärenden i skogsbruket. Version 1.1 2016-07-05.* Stockholm: Naturvårdsverket and Skogsstyrelsen.

Simonsson, P. (2016) *Conservation measures in Swedish forests. The debate, implementation and outcomes.* Doctoral thesis, Swedish University of Agricultural Sciences, Umeå University, Sweden. *Acta Universitatis Agriculturae Sueciae* 2016:103.

Sjöholm, L. (2019) *Ecological compensation and the application of regulations for compensation in the Environmental Code, 2015–2018.* (In Swedish, summary in English.) Stockholm: Naturvårdsverket Report 6898.

Skytt, T., Englund, G. & Jonsson, B.-G. (2021) Climate mitigation forestry – temporal trade-offs. *Environmental Research Letters*, **16**, 114037.

Statistics Sweden (2019) *Land Use in Sweden.* 7th ed. (In Swedish, summary in English.) Örebro, Sweden: Statistiska Centralbyrån.

SwAM, Swedish Energy Agency & Svenska Kraftnät (2016) *Handlingsplan för marint områdesskydd.* Göteborg, Sweden: Havs- och vattenmyndigheten Report 2016.

(2017) *Skydd av värdefulla sjöar och vattendrag. Redovisning av genomförandet av ett uppdrag i Havs- och vattenmyndighetens regleringsbrev för 2017.* Communication 2017-11-28. Göteborg, Sweden: Hav- och vattenmyndigheten.

(2019) *Förslag till nationell plan för omprövning av vattenkraft. Med beskrivning av vattenmiljö och effektiv tillgång till vattenkraftsel samt identifierade behov av fortsatt arbete.* Göteborg, Sweden: Havs- och Vattenmyndigheten, Energimyndigheten and Svenska Kraftnät.

Törnqvist, O., Klein, J., Vidisson, B. *et al.* (2020) *Fysisk störning i grunda havsområden. Kartläggning och analys av potentiell påverkanszon samt regional och nationell statistik angående störda områden.* Göteborg, Sweden: Havs- och Vattenmyndigheten Report 2020:12.

UNEP-WCMC (2022) *Protected Area Profile for Sweden from the World Database of Protected Areas, January 2022.* www.protectedplanet.net

Westling, A., Toräng, P., Jacobson, A. *et al.* (2020) *Species and habitats of Sweden in the EU Habitats Directive. Result from reporting to EU in 2019 of conservation status, 2013–2018.* (In Swedish, summary in English.) Stockholm: Naturvårdsverket.

Weyhenmeyer, G. A., Müller, R. A., Norman, M. & Tranvik, L. J. (2016) Sensitivity of freshwaters to browning in response to future climate change. *Climate Change*, **134**, 225–239.

Widenfalk, L., Josefsson, J., Blicharska, M. *et al.* (2021) *Ecological knowledge for ecological compensation.* (In Swedish, summary in English.) Stockholm: Naturvårdsverket Report 6995.

Chapter 31

United Kingdom

Graham Tucker, Wyn Jones, Susan Davies and Joan Edwards

Key Statistics	
Terrestrial area (km²)/% of EU-28*	244 573/5.6%
Marine waters (km²)/% of EU-28*	885 430/19.0%
Population density (p km²)/% of EU-28*	261/228%
GDP per capita (€)/% of EU-28	31 375/117%
Land cover (CORINE land cover classes 2018)	AGR 55%, OSN 16%, WET 12%, ART 9%, FOR 8%
Biogeographical regions	ATL, MATL
Number of HD Annex I habitats/%*	83/37%
Number of HD Annex II, IV and V species/%*	94/8%
Number of BD Annex I bird taxa/%*	55/31%
% HD UCS and decreasing habitats/ % unknown	23%/1%
% HD UCS and decreasing species/ % unknown	17%/22%
% breeding BD Annex I birds decreasing/% uncertain and unknown	27%/2%

* Excluding MAC and MMAC biogeographical regions.
Sources and codes. See Table 1.1.

31.1 The Natural History and Characteristics of the United Kingdom

The United Kingdom (UK) consists of Great Britain (i.e. England, Scotland and Wales) and Northern Ireland. The nearby Isle of Man, Bailiwick of Guernsey and Bailiwick of Jersey are not part of the UK (or the EU), but are Crown dependencies, and for brevity are not covered in this chapter. Gibraltar, which is a UK Overseas Territory, and the UK Sovereign bases on Cyprus are not covered in this chapter, but some information on the Sovereign bases is included in Chapter 10 on Cyprus.

For its size, the UK has a rich variety of habitats (see Proctor, 2013, and Lake *et al.*, 2015, for descriptions). This primarily results from its diverse landforms, topography, geology and soils, and climatic gradients (principally from cool and wet oceanic in the north-west to warmer and drier in the south-east). Humans have influenced these natural habitats since Neolithic times (c. 6 000 years ago), creating new semi-natural habitats, such as wood pasture, heathland and meadows. Much of the original natural broad-leaved forest was cleared for agriculture, so that by 1 250 woods covered only a few percent of England (Rackham, 1986). Many lowland wetlands were drained, including a large proportion of the extensive fens in East Anglia during the seventeenth century. Further habitat change occurred following the industrial revolution, in particular the expansion and intensification of agriculture and plantation forestry of mainly non-native conifers after the World War II. As has been widely documented, this led to the destruction of large areas of the remaining semi-natural habitat, including 80% of the UK's lowland heathland since 1800 (Webb, 1986), and 97% of lowland meadow in England and Wales between the 1930s and 1984 (Fuller, 1987).

Now the UK has a relatively high but uneven population density, and most lowlands, especially in England, are dominated by farmland habitats (mainly intensively managed arable agriculture in the east and grasslands in the west, with little remaining mixed farming). About 9% of the UK is built-up urban areas and villages, and only about 8% is woodland[1] (of which 42% is non-native coniferous woodland). Lowland semi-natural habitats such as heathlands, calcareous and acid grasslands, and fens and mires are now much reduced and fragmented. Nevertheless, extensive areas of semi-natural habitat remain, mainly along the wilder parts of the coast and in the uplands (particularly in Scotland and Wales). Consequently, the UK hosts significant proportions of the European and biogeographical ranges of habitats such as estuaries, sea cliffs, heaths, blanket mires and machair. The majority of the world's chalk streams, a globally rare habitat, also occur in England.

Whilst the UK has a high diversity of habitats, terrestrial habitats tend to have fewer species than their equivalents in

[1] 2018 CORINE landcover data. Land used for forestry, which includes open space and felled areas awaiting restocking, was about 13% in 2021. www.forestresearch.gov.uk/tools-and-resources/statistics/forestry-statistics/

continental Europe as a result of the Ice Age and the UK's later separation from the continent around 5 000 BCE (Simmons, 2001). At their greatest extent, ice sheets covered most of the country, and the remaining open areas in the south were too cold for most of the UK's current flora and fauna. With the retreat of the ice, the land bridge to the continent was lost before many species were able to recolonise.

Historic habitat changes and other pressures have led to further species impoverishment. In particular, many forest species have disappeared or become restricted to small areas. For example, all the larger carnivores and many other mammals and large birds became extinct as a result of forest loss and/or hunting. A significant number of species that are present in the UK now are deliberate or accidental introductions, including most vascular plants (Stace, 2019). Most non-native plants do not appear to have harmed the native flora of common habitats (Maskell *et al.*, 2006), but some have, perhaps most notably the Common Rhododendron (*Rhododendron ponticum*). The Rabbit (*Oryctolagus cuniculus*) has also had a major impact on some habitats and associated species. The accidently introduced Dutch elm disease (i.e. the fungus *Ophiostoma novo-ulmi*), has profoundly changed the UK landscape since the 1970s, killing off most elm trees (*Ulmus*) in waves of disease outbreaks.

Despite the natural constraints, habitat losses and other human impacts, the UK has particular importance for numerous species. For example, the wet climate makes Britain rich in bryophytes, lichens and some fungi, and the extensive blanket bogs and upland heathlands are very important for breeding waders. Estuaries and other tidal coastal habitats provide extensive, ice-free and food-rich habitats in winter, and therefore support large populations of migratory waterbirds that are relatively short distances from their Arctic breeding grounds.

The UK's marine waters[2] are more than three and half times its land area. These include transition zones between north-eastern, cold-water communities and south-western, temperate-water communities, and waters that range from shallow intertidal areas to thousands of metres in depth beyond the continental shelf. Tidal ranges are high: typically 1–5 m, but up to 15 m in some locations. As a consequence, marine habitats are highly productive and diverse, ranging from sheltered soft sedimentary shores and dynamic benthic habitats to higher energy environments and hard-rock subtidal and coastal habitats. Despite ongoing pressure from destructive fishing practices and other disturbing activities, these habitats support diverse characteristic marine communities and abundant populations of many species. For example, cliffs and predator-free small islands provide suitable nesting sites for seabirds within the range of productive waters, and as a result the UK holds large proportions of the European and global populations of several species, such as 80% of the world's Manx Shearwaters (*Puffinus puffinus*). The UK also has an estimated 40% of the global population of the Grey Seal (*Halichoerus grypus*) and a high diversity of cetaceans.

31.2 The Status of Nature

31.2.1 Habitat Status and Trends

As mentioned in Section 31.1, there have been profound historical changes in the UK's habitats, with the main trend being a progressive loss of natural and semi-natural habitats. Over the last 40 years or so, the rate of habitat conversion has declined, in part as a result of changes in agricultural and forestry policy and the effects of nature conservation measures. However, some habitats have continued to decline in extent, such as lowland grasslands (Stroh *et al.*, 2019) and coastal salt marshes, and most habitats have shown widespread declines in quality (Marren, 2002; Burns *et al.*, 2013; Hayhow *et al.*, 2016, 2019). Evidence of this comes from the assessments of the conservation status of HD habitats. The UK HD Article 17 report[3] for 2013–2018 reveals that out of 77 HD habitats (excluding those in Gibraltar), 80% had a bad status, 10% had a poor status, and only 8% were considered to have a favourable conservation status. One habitat type had an unknown status. The situation is especially concerning for HD bogs, mires and fens, and grasslands – all types of these habitats had a bad status, as did nearly all heathland and scrubland, forest and freshwater habitats.

Of particular importance is whether those habitats with an unfavourable status had improving or deteriorating trends. Table 31.1 indicates that whilst most habitats with an unfavourable status had stable or improving trends, 29% of those with a bad status had declined (including all eight dune habitats with a bad status).

31.2.2 Species Status and Trends

Changes in the distribution and abundance of most species groups have been reviewed by respective experts in Hawksworth (1974, 2001) and MacLean (2010). Not surprisingly, these reviews revealed major changes across most taxa, especially over more recent decades as surveys have been repeated and monitoring schemes established (see Section 31.8). Whilst declines appear to have been predominant, it has been difficult to draw reliable and generally representative conclusions on the overall state of biodiversity in the UK from such accounts. To address this, since 2013 most of the relevant conservation and research organisations have collaborated and pooled their monitoring data to produce UK *State of Nature* reports every three years (Burns *et al.*, 2013; Hayhow *et al.*, 2016, 2019).

The key findings from the 2019 *State of Nature* report are summarised below (the 2013 and 2019 results being similar). Overall, the indicator of average abundance for 697 terrestrial

[2] Including the seabed beyond the EEZ to the edge of the continental shelf.

[3] See Section 1.3 for background on monitoring under the Nature Directives, status categories and data sources.

Table 31.1 Short-term trends (c. 2007–2018) in Habitats Directive Annex I habitats and Annex II, IV and V species with an unfavourable conservation status in the United Kingdom.

	Poor (unfavourable–inadequate)				Bad (unfavourable–bad)			
	Imp	**Sta**	**Dec**	**?**	**Imp**	**Sta**	**Dec**	**?**
HD Habitat assessments								
Number	2	5	0	1	13	26	18	5
%	25%	63%	0%	13%	21%	42%	29%	8%
HD Species assessments								
Number	1	7	9	7	3	3	7	3
%	4%	29%	38%	29%	19%	19%	44%	19%

Key. Imp = improving, Sta = stable, Dec = decreasing, ? = unknown.
Source. Based on Member State HD Article 17 reported data as presented on the EEA national summary dashboard (2021).

and freshwater species for which long-term abundance data were available (primarily birds, mammals, butterflies and some moths) fell by 13% between 1970 and 2016. Over this period, 58% of species declined (of which 41% were moderate or strong declines), 42% of species increased (with 26% showing strong or moderate increases) and 33% showed little change. Comparison of trends across the monitored taxa groups reveals that declines have been greatest in butterflies and moths (see also Fox *et al.*, 2013, 2015). There has also been a tendency across taxa for the greatest declines to occur in habitat specialists and lowland species.

More recent data on birds reveal ongoing declines in a wide range of species, notably including upland birds, seabirds, wintering waterbirds, Afro-Palearctic long-distance migrants and some woodland specialists (Stanbury *et al.*, 2021). Some of the most substantial declines have occurred in farmland species, as shown by the UK farmland bird indicator,[4] which declined by 55% over 1970–2019. Although the rate has slowed since about 1985, the decline continues. As a result, 14 of 26 farmland breeding bird species are Red-listed in the UK (Stanbury *et al.*, 2021).

The 2019 *State of Nature* report, and other assessments of the condition of the UK's marine ecosystems (UKMMAS, 2010; DEFRA, 2019a), indicate that marine taxa have shown mixed population trends with complex regional variations. Long-term data from the Continuous Plankton Recorder programme have revealed major changes in plankton communities (Beaugrand *et al.*, 2002), resulting in an ecosystem regime shift in the North Sea (Beaugrand, 2004). Several seabird species have also been in decline, leading to a 28% decrease in the UK Breeding Seabird Indicator between 1986 and 2018 (DEFRA, 2019b). Similarly, some fish species are declining, most notably large-bodied predators at the top of food webs (e.g. sharks and rays), as well as many pelagic species. Most seal and cetacean species appear to be stable, but the situation for some populations is more varied and uncertain.

Despite the data gaps, the terrestrial and marine data show that biodiversity in the UK is in an alarming state of flux, with a high proportion of species showing changes in abundance and range. Of particular concern are the trends in the relative abundance of species identified as priorities by each UK country, as revealed by the official UK biodiversity indicator (based on 224 out of 2 890 priority species).[5] By 2019, the abundance indicator had declined by 61% compared to its 1970 baseline, although there was no further significant change over 2014–2019. The 2019 *State of Nature* report also includes a review of all the species that have been assessed using current regional IUCN Red List criteria in terms of their risk of extinction in Great Britain. Of the 8 431 species assessed using these criteria, 15% were considered to be threatened.

These declines are reflected in the UK's assessments of the conservation status of its HD species. Over the 2013–2018 reporting period, out of 93 HD species (excluding those in Gibraltar), 35% had a favourable conservation status, but 26% had a poor status and 17% had a bad status (the remaining 21% had an unknown status). Of further concern is that a high proportion of species that had an unfavourable conservation status were declining (Table 31.1). The UK's most recent BD Article 12 report on bird population trends also indicates that a significant proportion of BD Annex I taxa had shown short-term declines, in particular with respect to wintering populations (Table 31.2).

31.2.3 Drivers and Causes of Change

Numerous studies have investigated and identified the causes of habitat loss and degradation, and changes in the abundance and range of various species groups. Despite such a range of evidence, it has been difficult to draw representative conclusions on what have been the most important overall drivers and causes of biodiversity change. Research carried out to support the 2016 *State of Nature* report therefore attempted to address this issue by identifying the causes of long-term

[4] https://jncc.gov.uk/our-work/ukbi-c5-birds-of-the-countryside/

[5] https://jncc.gov.uk/our-work/ukbi-c4a-species-abundance/

Table 31.2 Short-term trends (c. 2007–2018) in Birds Directive Annex I taxa in the UK.

	Breeding				Wintering			
	Inc	Sta/ Flu	Dec	?	Inc	Sta/ Flu	Dec	?
Number	29	5	13	1	6	3	8	4
%	60%	10%	27%	2%	29%	14%	38%	19%

Key. Inc = increasing, Sta = stable, Flu = fluctuating, Dec = decreasing, ? = uncertain and unknown.
Source. Based on Member State HD Article 12 reported data as presented on the EEA national summary dashboard (2021).

population changes (c. 1970–2012) amongst a sample of 400 terrestrial and freshwater species from a variety of taxonomic groups (Burns *et al.*, 2016). Based on a literature review and expert evidence, the research found that changes in habitat management, and in turn its quality for the species concerned, have had greater impacts than changes in habitat extent over the last 40 years or so. By far the most detrimental changes have been the result of agricultural intensification (e.g. the loss of mixed farming systems, changes in cropping timing, intensification of grazing regimes, increased levels of pesticide and fertiliser use, and declines in extent of marginal non-farmed habitats). Many such impacts are the legacy of agricultural changes that took place several decades ago, but more recent developments have also been damaging. For example, the widespread use of systemic neonicotinoid pesticides on oilseed rape since 2002 has contributed to increased population extinction rates in wild bee species in the UK (Woodcock *et al.*, 2016). Other important land management–related causes of change have been reductions in extensive grazing systems (but countered in some locations by agri-environment schemes); drainage of wetlands (especially in the uplands); over-abstraction of water; and cessation of traditional habitat management practices (e.g. coppicing of woodland).

Where biodiversity impacts have been the result of habitat loss, this has often been the result of urbanisation leading to the loss of lowland habitats such as heathlands. Development of post-industrial sites, such as spoil heaps, can also lead to the loss of habitat for some rare specialist species. Although not mentioned in the Burns *et al.* (2016) study, another effect of urbanisation is light pollution, which is detrimental for a wide range of invertebrates (Bruce-White and Shardlow, 2011) and has been found to be a cause of moth declines in the UK (Macgregor *et al.*, 2017; Wilson *et al.*, 2018). Some population declines have also occurred as a result of afforestation, particularly of lowland heaths and upland habitats. In contrast, some species have benefited from habitat creation schemes (most notably reedbeds and eutrophic lakes, often following gravel extraction), but these have been at a limited scale for most other habitat types.

The Burns *et al.* (2016) research revealed that climate change has also been a major driver of change in UK species over the study period, but its impacts have been mixed so far. As a result of the UK's geographical position, more species have expanded their range into the UK from the south than have yet been lost in northern and upland locations. However, as climate change progresses, the effects are expected to become increasingly negative for more species, such as upland birds and seabirds (Pearce-Higgins, 2021) (see climate adaptation in Section 31.6.3).

The 2019 *State of Nature* report did not update the Burns *et al.* (2016) study, but provided evidence of a number of other pressures whose importance was perhaps underestimated by the 2016 study or not covered (i.e. marine ecosystems and species). In this respect, perhaps one of the most overlooked causes of ecosystem degradation is air pollution, and particularly the impacts of eutrophication, acidification and toxicity on vegetation resulting from the deposition of reactive nitrogen. Yet its effects are widespread and impact many semi-natural habitats and associated species (e.g. Field *et al.*, 2014; Pescott *et al.*, 2015). Efforts to reduce ammonia emissions from agriculture, now the main source of reactive nitrogen, have had little effect so far.

The 2019 report also highlights the environmental damage caused by pathogens and invasive non-native species (more widely referred to in Europe as alien invasive species – IAS). The number of IAS has continued to increase (averaging 10–12 new species per year), leading to an estimated 1 849 established alien species in Great Britain in 2012, approximately 12% of which are considered to have ecological impacts (Roy *et al.*, 2012). A variety of habitats have been seriously affected by highly competitive alien plants, such as freshwaters by the New Zealand Pygmyweed (*Crassula helmsii*), whilst a number of species of high conservation importance have declined as a result of predation or competition by alien animal species or their transmitted pathogens.

The population levels of some birds (notably some seabirds, gamebirds and waders) may also be limited by some native predators that occur in particularly high densities in the UK, most notably the Red Fox (*Vulpes vulpes*) and Carrion Crow (*Corvus corone*) (Roos *et al.*, 2018). The reasons for the high densities of these species are not certain, but are likely to include the absence of higher predators and the presence of suitable landscapes and human-associated food sources, such as waste and huge numbers of released gamebirds, about 57 million annually according to Aebischer (2019).

In the marine environment, observed increases in most commercially fished demersal species are believed to be the result of more sustainable fishery management. Fishing also continues to have a variety of impacts on non-target species and marine benthic habitats, some of which have led to fundamental long-term ecosystem changes (Kaiser *et al.*, 2000; Sewell and Hiscock, 2005; Robinson and Frid, 2008), with continuing high levels of disturbance from trawling and scallop dredging. Levels of by-catch are also of concern for some cetacean species (ICES, 2019) and seabirds (Bradbury *et al.*, 2017). Non-fishing–related pressures include dredging, sand

and gravel extraction, dumping, pollution, and the effects of wind turbines on benthic habitats, seabirds and cetaceans (Hiscock, 2014; OSPAR, 2017), although there may be possible benefits from the absence of fishing around wind farms. Pollution pressures from coastal sewage releases have declined, but eutrophication continues from agricultural run-off. Other current pollutants include plastics, and a range of toxic chemicals, such as polychlorinated biphenyls, that have been observed to occur in high levels in cetaceans (Jepson *et al.*, 2016). Climate change is probably the main cause of the observed regime change and geographic shifts in planktonic populations, resulting in knock-on food-chain impacts on some fish (Perry *et al.*, 2005; Beaugrand *et al.*, 2008; Capuzzo *et al.*, 2018) and, in turn, on several British seabirds (MacDonald *et al.*, 2015).

31.3 Nature Conservation Governance, Actors, Legislation and Policies

31.3.1 Overview of Nature Conservation in the UK

Nature conservation in the UK has a long and complex history as much of the currently relevant legislation dates to the 1940s and 1980s, and there are separate legal systems for England and Wales; Northern Ireland; and Scotland (Table 31.3). Further complications have arisen since devolution in the 1990s with greater divergence in conservation policy and legislation. Most recently, the UK's departure from the EU is leading to further profound changes in some aspects of land, sea and environmental policy. In view of this particularly complex and detailed story, this chapter is longer than others. However, it is not possible to extend it sufficiently to provide a detailed account for each of the UK countries. Instead, it provides an overview of the main developments and examples of their effects and most important lessons. For more detailed overviews, see Stamp (1974), Moore (1987), Evans (1992), Sheail (1998) and Marren (2002) and, for an account focussed on marine nature conservation, Earll (2018).

31.3.2 Governance

Unlike in most European countries, nature conservation in Great Britain has largely been the responsibility of semi-autonomous government-funded agencies. The first of these was the Nature Conservancy, which was established in 1949 by Royal Charter to provide advice and undertake research on nature conservation matters. Later in the same year, under the National Parks and Access to the Countryside Act, it was given powers to establish and manage nature reserves for the study or preservation of flora, fauna and geological or physiographical features of special interest – now known as National Nature Reserves (NNRs). It was also given the duty to notify the relevant local planning authority of similar areas of special interest that for the time being were not nature reserves. These areas came to be known as Sites of Special Scientific Interest (SSSIs) in Great Britain. Local government authorities were also given powers to establish, manage and protect Local Nature Reserves (LNRs) with the agreement of the Nature Conservancy.

In 1974, the Nature Conservancy was split, forming the Nature Conservancy Council (NCC) and the Institute of Terrestrial Ecology (ITE). The NCC retained the Nature Conservancy's advisory and executive functions, and research needed to support those functions, whilst the ITE took on its other research. According to Marren (2002), the NCC initially had a rather low key primarily advisory role, but its powers and resources were increased greatly by the 1981 Wildlife and Countryside Act (further discussed in Section 31.3.4).

A fundamental change in the organisation of nature conservation occurred as a result of the 1990 Environment Protection Act, which split the NCC and led to the creation of English Nature, the Countryside Council for Wales and, following the Natural Heritage (Scotland) Act 1991, Scottish Natural Heritage. Although these have remained semi-autonomous as non-departmental government bodies, a relevant government department provides each with its funds and appoints its board members. The culture within the agencies has therefore reflected government views to some extent.

The division of the NCC was highly controversial, being justified by the UK Government as a way of decentralising conservation governance. However, this pre-empted the governmental devolution that occurred in the late 1990s, and many conservationists felt that the real aim was to weaken nature conservation and defuse objections to afforestation of the internationally important areas of blanket bog habitat in the north of Scotland (Mackay, 1995; Stroud *et al.*, 2015).

The Environment Protection Act also created the Joint Nature Conservation Committee (JNCC) to coordinate some activities in Great Britain (e.g. on monitoring and reporting) and to provide advice on international issues. It was formally given a UK coordinating role in 2006. The JNCC's responsibilities were also extended to offshore Natura 2000 issues (i.e. beyond the 12 nautical mile limit) in 2008 and to wider offshore marine issues in 2009 (except in Scotland, where all offshore nature conservation issues have since been the responsibility of the Scottish Government).

A number of changes have subsequently occurred to the agencies that were initially established, leading to gradual divergences between them in their powers and interactions with other environmental authorities. In England, Natural England is now responsible for nature conservation matters – as well as landscape, access and recreation issues – and provides advice on the design and delivery of land management schemes. Although its nature conservation powers cover all areas and habitats, it is currently complemented by other non-departmental public bodies, namely the Environment Agency (e.g. in relation to air and water pollution, and the management of water resources and rivers), the Forestry Commission (which manages the publicly owned forest estate, as well as

Table 31.3 Key nature conservation developments in the UK.

1904	Royal Society for the Protection of Birds received royal charter
1912	Society for the Promotion of Nature Reserves formed (later to become the Wildlife Trusts)
1915	List of 'Areas Worthy of Protection' in the British Isles created by Charles Rothschild
1949	Nature Conservancy created by Royal Charter; National Parks and Access to the Countryside Act
1967	Countryside (Scotland Act)
1968	Town and Country Planning Act; Countryside Act
1973	UK joined the European Common Market; the Nature Conservancy was split into the Nature Conservancy Council (NCC) and the Institute for Terrestrial Ecology
1977	Publication by the NCC of *Nature Conservation Review* – a catalogue of 'key sites'
1979	EC Birds Directive; Wildlife Link was established
1981	Wildlife and Countryside Act
1984	NCC published *Nature Conservation in Great Britain*
1985	The Agriculture Act introduced Environmentally Sensitive Areas
1988	Marine Nature Conservation Review published by NCC
1990	The first 'environment' White Paper: *This Common Inheritance*; Environmental Protection Act 1990 split NCC
1991	Creation of English Nature, the Countryside Council for Wales and the Joint Nature Conservation Committee (JNCC)
1992	Habitats Directive; UK signed CBD at Earth Summit; Scottish Natural Heritage (SNH) established
1994	*Biodiversity: the UK Action Plan* launched, and white papers published in response to the Earth Summit; Habitats Directive transposed by the Conservation of Natural Habitats (& c.) Regulations (and in Northern Ireland in 1995)
1995	The Environment Act led to the creation of the Environment Agency and the National Parks Authorities (England and Wales) and the Scottish Environmental Protection Agency (in 1996)
1997	*Wildlife Charter* published by Wildlife Link
2000	The Countryside and Rights of Way Act (CROW) (England and Wales)
2001	Department of Environment, Food and Rural Affairs (DEFRA) was formed
2002	The Environment (Northern Ireland) Order with similar provisions to CROW Act; *Northern Ireland Biodiversity Strategy; Working with the grain of nature. A biodiversity strategy for England*
2004	Nature Conservation (Scotland) Act with similar provisions to CROW Act; *Scotland's Biodiversity: It's in Your Hands* published
2006	Natural Environment and Rural Communities (NERC) Act (England Wales) gave public bodies biodiversity duties
2007	Offshore Marine Conservation (Natural Habitats &c.) Regulations
2008	*Biodiversity – the UK Approach* published by JNCC and DEFRA
2009	Marine and Coastal Access Act, implementing EU legislation (i.e. the MSFD) as well as OSPAR
2010	*Making Space for Nature: A Review of England's Wildlife Sites and Ecological Network* (Lawton *et al.*); significant overhaul of the 1994 Habitats Regulations (England and Wales)
2011	*UK National Ecosystem Assessment*; DEFRA White Paper *The Natural Choice;* DEFRA publishes *Biodiversity 2020: A strategy for England wildlife and ecosystem services*
2012	*UK Post-2010 Biodiversity Framework* produced; the *National Planning Policy Framework* published by the UK's Department of Communities and Local Government
2013	Natural Resources Wales established; *2020 Challenge for Scotland's Biodiversity* published
2014	Planning (Wales) Act 2014
2015	Well Being of Future Generations Act in Wales; *Nature Recovery Plan for Wales; Valuing Nature: A Biodiversity Strategy for Northern Ireland*
2016	In a national referendum on 23 June, the UK voted to leave the EU

	Environment (Wales) Act
	Natural England's *Conservation Strategy for the 21st Century*
2017	*A Green Future: Our 25 Year Plan to Improve the Environment* published by DEFRA
	Natural Resources Policy published by Welsh Government
2019	Habitats Regulations for England and Wales, Northern Ireland and Scotland updated to make them operable post-Brexit.
2020	On 31 January, the UK left the EU and entered a transition period, which ended on 31 December
	Agriculture Act (2020) and Fisheries Act (2020)
	The Environment Strategy for Scotland: vision and outcomes published by the Scottish Government
2021	Environment Act (2021), and Office for Environmental Protection (OEP) created (for England and Northern Ireland).
	Draft *Environment Strategy for Northern Ireland* published by DAERA
	Environmental Standards Scotland created
	UK Prime Minister signed Leaders' Pledge for Nature, the '30 by 30' protected area commitment, and G7 agreement to 'halt and reverse biodiversity loss' (i.e. become Nature Positive) by 2030
2022	*Nature Recovery Green Paper* and *Environmental Targets* for England published by DEFRA
	Biodiversity Strategy to 2045 published by the Scottish Government

See Table 3.3 for signatory dates of international conventions.

regulating private forestry and carrying out research) and the Marine Management Organisation (MMO), which covers marine planning, permitting, fisheries and marine protected areas (MPAs).

In Scotland, there is a similar division of powers and responsibilities amongst NatureScot (rebranded from Scottish Natural Heritage (SNH) on 1 May 2020), the Scottish Environment Protection Agency, and Forestry and Land Scotland, Scottish Forestry, Crown Estate Scotland and Marine Scotland.

Wales has taken a more integrated approach, abolishing the Countryside Council for Wales and the Welsh functions of the Environment Agency and Forestry Commission, and replacing them with Natural Resources Wales (NRW) in 2013, which also took over marine licensing functions from the Welsh Government.

Although they are supposed to be 'arms-length' bodies, there is wide agreement in the conservation community that government influence over all the country conservation agencies has increased over the past 10 years. This has resulted in more of a focus on maintaining the economic and social benefits of biodiversity, and a more accommodating approach to the wishes of landowners and businesses. The latter is particularly apparent in *Natural England's Conservation Strategy for the 21st Century* (Natural England, 2016). The functions and ability of all the agencies have also been reduced substantially by severe budget cuts (see Section 31.7.3). In the case of Natural England, its budget almost halved after 2010 (Marren, 2020), although it has increased again since 2019. Natural England's role in policy making has also been cut and DEFRA has taken over its external communications. Consequently, a Select Committee of the House of Lords (2018) agreed that Natural England 'no longer has a distinctive voice' and had 'concerns regarding its ongoing ability to perform core regulatory functions'.

Unlike in Great Britain, in Northern Ireland nature conservation, landscape, wider environment and forestry matters have always been dealt with directly by a government department. From 1974 until 2016, this was the Department of the Environment. Since then, nature conservation and most related issues have been the responsibility of the Department of Agriculture, Environment and Rural Affairs (DAERA). The Northern Ireland Environment Agency (NIEA), an Executive Agency of DAERA, provides support through site management, monitoring, awareness raising and scientific and technical advice. DAREA also receives advice from the Council for Nature Conservation and the Countryside.

The country conservation agencies in England, Wales and Scotland, together with the relevant government department in Northern Ireland, are often known as the UK Statutory Nature Conservation Bodies, and therefore are referred to as SNCBs hereafter.

Local authorities throughout the UK also have an important role to play in nature conservation, particularly regarding development planning and control (Section 31.6.1). Related to this, they work with the Wildlife Trusts, Local Environmental Record Centres (see Section 31.8) and other partners to identify Local Wildlife Sites, which are not statutory protected areas but receive some protection through the planning system (see Section 31.6.1). They also have responsibility for the designation and management of LNRs, although in practice these play a relatively minor role in the protected area network in the UK. According to the legislation described in Section 31.3.4, they have a statutory duty to give consideration to biodiversity matters when carrying out their activities. However, the impact of austerity on local government in undertaking their biodiversity obligations has been considerable: ranger services and ecological advisers have been cut, so much so that in 2011–2012 only 35% of local planning authorities had ecologists on their staff (ALGE, 2013), and numbers have fallen further since then.

As a result of the UK's withdrawal from the EU, compliance with environmental legislation is no longer scrutinised by the European Commission, or subject to infringement procedures and rulings of the CJEU, leaving a potential 'governance gap' (Nesbit *et al.*, 2017). To address this, new bodies have been created to fulfil a similar role: Environmental Standards Scotland (October 2021) and the Office for Environmental Protection (November 2021), which has a remit covering England and Northern Ireland. In Wales, a permanent environmental governance oversight body is being developed, with an Environment Commissioner appointed in March 2021 as a temporary measure. These bodies are discussed further in relation to the post-Brexit legislative framework.

31.3.3 Public Involvement and NGOs

The UK is renowned for being a nation of animal lovers, and this, together with wider related interests in natural heritage, landscapes and recreation, and inspiring natural history TV, has led to considerable public support for nature conservation. This has resulted in the development of numerous strong environmental NGOs, who have played a crucial role in further raising public awareness and promoting the need for nature conservation. As is evident in numerous places in this chapter, they have also been successful in influencing policy and legislation, as well as contributing to enforcement, practical conservation measures, research and monitoring, and funding.

NGO influence dates back to the early twentieth century, when campaigning by the Royal Society for the Protection of Birds (RSPB) led to some important early bird protection legislation. Despite its focus on birds, the RSPB is increasingly adopting a broader 'whole nature' remit, and has had considerable influence on overall nature conservation issues. This is primarily due to its huge membership, which grew particularly rapidly in the 1970s and 1980s, reaching 100 000 in 1972 and 1 million in 1997. At the end of March 2021, the RSPB had 1.13 million members and an income for the financial year of £142 million[6] (€163 million)[7] (RSPB, 2021). As well as engaging in campaigning, the RSPB has occasionally mounted legal challenges, some of which had substantial EU-wide implications. Perhaps most importantly, the Lappel Bank case[8] led to the CJEU ruling that Member States must not take social or economic considerations into account when designating an SPA. Subsequent cases established a similar principle in relation to the designation of SCIs/SACs.

The Wildlife Trusts, of which there are now 46, are part of the federation of Wildlife Trusts and have also played important roles, both at regional levels and nationally (see Sands, 2012, for a detailed account). This started in 1912, when Charles Rothschild put forward the idea that nature reserves should be set up to protect important areas for habitats and rare species from development (Rothschild and Marren, 1997). To take this forward he founded the 'Society for the Promotion of Nature Reserves', which carried out the first ever national survey of wildlife sites. In later years, after a number of developments and name changes, this Society gave rise to the current Wildlife Trusts.

In more recent years the Wildlife Trusts' campaigning, in coalition with a wide range of environmental bodies, has led to important developments in domestic legislation, especially in the marine environment. The Trusts also tend to be the nature NGOs most active in local conservation issues, such as scrutinising and commenting on development proposals. The Wildlife Trust setup in Scotland has diverged and is formed by a single body (Scottish Wildlife Trust) that focusses more on influencing national policy, managing its reserves and landscape scale restoration.

Other influential campaigning NGOs covering broad environmental subjects include Friends of the Earth, Greenpeace and WWF-UK. There are now also many other NGOs covering a very wide variety of more specialist interests, such as certain habitats (e.g. Wildfowl and Wetlands Trust, and the Woodland Trust), taxa (e.g. Amphibian and Reptile Conservation, Bat Conservation Trust, Buglife, Butterfly Conservation, and Plantlife) and surveys, monitoring and research (e.g. the British Trust for Ornithology, and the Botanical Society of Britain and Ireland). Marine NGOs include the Marine Conservation Society and the Whale & Dolphin Conservation Society. Specialist NGOs ClientEarth and Wild Justice focus on legal activity.

Some hunting organisations have contributed to nature conservation, such as the Game and Wildlife Conservation Trust (in particular through farmland bird conservation research and advocacy) and the British Association for Shooting and Conservation (who facilitated the development of a rational wildfowling regime on the coasts). However, as discussed later, there have been increasing conflicts between nature conservation interests and some in the hunting community over the impacts of commercial shooting.

To help facilitate collaboration between NGOs on tackling government policy, Wildlife Link was formed around 1979 to represent the views of its member organisations. It now consists of four collaborating sister bodies covering each 'country'. In combination, their members have the support of over 8 million people in the UK.

As well as campaigning, many NGOs work in partnership with the statutory sector, such as in the development of agri-

[6] Slightly less than previous years due to COVID-19 restrictions on some income-raising activities.

[7] 29 May 2021 exchange rate: €1 = £1.151.

[8] *C44/95 Regina* v. *Secretary of State for the Environment.*

environment schemes, research, survey and monitoring. They also own and/or manage substantial areas as nature reserves. For example, the National Trust is the third largest landowner in the UK, with more than 250 000 ha, much of which is in coastal areas. The RSPB and the Wildlife Trusts also have sizeable nature reserves, and manage 158 300 ha and 103 576 ha, respectively. Together these areas cover about 2.1% of the UK's land area.

31.3.4 Policy and Legislation

National and EU Policy and Legislation to 2010

Whilst wildlife and animal welfare legislation dates back to the nineteenth century, much of it was related to hunting, protection of game and protection of birds. The National Parks and Access to the Countryside Act 1949 was the first major piece of nature conservation legislation, which is still of relevance today. It enabled the designation and protection of NNRs and LNRs, and also required the Nature Conservancy, and its successors, to notify the local planning authorities of SSSIs in the area so that they could be considered in the planning process. However, as further described in online Annex UK.1,[9] the establishment of the series of NNRs and LNRs was slow, and the protection of SSSIs proved to be weak (Moore, 1987). This was especially the case in relation to agricultural and forestry activities, as most were exempt from the planning system.

Significant steps were required to address the weak protection afforded to SSSIs, and to prepare for implementation of the 1979 Bern Convention (which the UK ratified in 1982) and EU obligations under the 1979 Birds Directive. This resulted in the 1981 Wildlife and Countryside Act (WCA), which applies to Great Britain. Following devolution, the WCA has been modified in Scotland through the Nature Conservation (Scotland) Act 2004 and the Wildlife Natural Environment Act (Scotland) 2011.

A similar important legislative development in Northern Ireland was the passing of the Wildlife (Northern Ireland) Order 1985, which addressed species protection, and the Nature Conservation and Amenity Lands (Northern Ireland) Order 1985. The latter provided the SNCB, in consultation with the Council for Nature Conservation, with new powers to declare, protect and manage Areas of Special Scientific Interest (ASSIs) in broadly equivalent ways to SSSIs under the WCA.

The WCA and related laws in other parts of the UK are widely regarded as the most important developments in UK nature legislation. Although subject to numerous amendments, they remain the key laws governing the protection and conservation of species and their habitats. Existing species protection laws were consolidated, measures to protect sites were improved, and new powers to establish Marine Nature Reserves were introduced. Under the legislation, owners and occupiers of ASSI/SSSIs must be notified of the extent of the site, its scientific interest, and those activities that may damage those interests. In turn, owners and occupiers must inform the relevant SNCB where they propose to undertake any of the activities on land within the site. This was referred to as a 'reciprocal' notification.

In the case of ASSIs/SSSIs, the provisions that aimed to ensure their protection and appropriate management were not robust, but the WCA is considered to have at least slowed the rate of their loss and damage. This is because, in practice, it did not rely on a strict compulsory approach, but instead emphasised the development of voluntary actions by the landowner through engagement and cooperation with the conservation agencies (Cranbrook, 2001; Marren, 2002). Such a pragmatic approach was necessary given that much of the land designated as an ASSI/SSSI in the UK is under private ownership and used for socio-economic purposes (e.g. for grazing livestock, forestry, water storage, hunting, fishing and other forms of recreation). These landowners and economic sectors have powerful representatives in unions, parliament and government.

In 1992, the Habitats Directive was approved and its legal transposition completed in the Conservation of Natural Habitats (& c.) Regulations in 1994 in Great Britain, and analogous legislation for Northern Ireland in 1995 (hereafter these are referred to as the Habitats Regulations).[10] The regulations concerning the protection and management of sites and the protection of species largely reflected the provisions in the amended WCA, adding the tests set out in the Directive to apply the precautionary approach to decisions on development affecting SACs, SPAs and protected species populations. This was logical given that the WCA was the means by which the UK met its obligations under the Bern Convention and Birds Directive. However, it was apparent that the protection was inadequate as many SSSIs continued to suffer damage (Rowell, 1991). In response, in 1997, the NGOs working collectively under the auspices of Wildlife Link published a *Wildlife Charter* that contained proposals to improve SSSI legislation.

As a result of the evidence of ongoing damage to SSSIs, NGO campaigns and the risks of infraction measures from the European Commission, the UK Government brought in the Countryside and Rights of Way Act 2000 (CROW Act), which only applies in England and Wales. Although the new Act continued to emphasise that the management of SSSIs should

[9] All annexes are free to access online at CUP: www.cambridge.org/natureconservation

[10] Most recently the Conservation of Habitats and Species Regulations 2017 in England and Wales (including the adjacent territorial sea) and to a limited extent in Scotland (reserved matters) and Northern Ireland (excepted matters). In Scotland, the Habitats Directive was transposed through a combination of the Habitats Regulations 2010 (in relation to reserved matters) and the Conservation (Natural Habitats &c.) Regulations 1994. Transposition of the Habitats and Birds Directives in offshore waters was primarily through the Conservation of Offshore Marine Habitats and Species Regulations 2017 and, for certain activities, the Offshore Petroleum Activities (Conservation of Habitats) Regulations 2001. Subsequently, these have been amended by 2019 regulations.

normally be achieved through voluntary agreements, it introduced robust provisions to prevent damage by owners and occupiers, damage by third parties, and damage through neglect – all reinforced by comprehensive enforcement provisions. Similar provisions are contained in the Nature Conservation (Scotland) Act 2004 and in the Environment (Northern Ireland) Order 2002.

Biodiversity Strategies and Plans, and the NERC Act

Alongside developments in legislation, a complementary driver of nature conservation policy and actions, mainly from the 1990s until 2010, was the development of the UK biodiversity action plan (BAP) framework.[11] Taking on board some NGO proposals (Wynne *et al.*, 1993), the government produced a *UK Biodiversity Action Plan* (UKBAP) (Anonymous, 1994), which was the first such plan to be produced by any country in reaction to obligations under the CBD. The overall goal of the plan was 'To conserve and enhance biodiversity within the UK and to contribute to the conservation of global biodiversity through all appropriate mechanisms.' To achieve this, it included a list of 59 broad conservation and enhancement targets for the UK Government and its nature conservation agencies to achieve in partnership with NGOs and other stakeholders.

The UKBAP led to a considerable amount of activity, including the establishment of various UK and country-level groups, involving a wide range of stakeholders and resulting in the development, over 1995–1999, of 45 Habitat Action Plans[12] and 391 Species Action Plans[13] for identified priority habitats and species, as well as new initiatives regarding public access to data and information. A key element of the process was to encourage local stakeholder-driven target-setting actions, which led to the creation of at least 170 local biodiversity partnerships and local BAPs, as well as national and local records centres. This created a national–local shared sense of purpose that, together with substantial government funding, triggered a huge amount of effort and broadened the range of species subject to direct conservation actions. For example, the UKBAP is considered to have had 'enormous benefits for moth conservation' (Fox *et al.*, 2013). However, more generally, practical implementation of the plans was mixed and their outcomes were patchily quantified (despite the development of a sophisticated web-based Biodiversity Action and Reporting System). Over time, government enthusiasm for the plans waned. As a result, the UKBAP website and plans are now archived[14] and are no longer widely driving concrete actions, although some local authorities are still using outputs from the process.

Devolution also led to changes, resulting in the UKBAP being gradually replaced with tailored national biodiversity strategies for England (DEFRA, 2002a), Northern Ireland (DOENI, 2002) and Scotland (Scottish Executive, 2004), and a broader *Environment Strategy for Wales* (Welsh Assembly Government, 2006). To complement the national actions, and to help meet shared challenges, *Biodiversity – the UK Approach* was produced (Biodiversity Indicators Partnership, 2007) and set out overall objectives and agreed requirements for work at a UK level.

To give legal weight to the BAP process, the Nature Conservation (Scotland) Act 2004 places a duty on public bodies to further the conservation of biodiversity, taking into account the Scottish Biodiversity Strategy. Through the Wildlife and Natural Environment (Scotland) Act 2011, every public body in Scotland is required to produce a publicly available report, once every three years, on compliance with the biodiversity duty. Although an evaluation of the biodiversity reports due in 2015 was undertaken for the Scottish Government (Daly *et al.*, 2016), this has not been repeated so it is difficult to assess the extent of public bodies' contributions to the overall goals. In England, the Natural Environment and Rural Communities (NERC) 2006 Act gives public authorities a similar duty, but this is more ambiguously worded as a requirement to *have regard* to the purpose of conserving biodiversity. Both Acts also require lists of species of principal importance to be identified, which are updated lists of priority species from the BAP process. In theory, the biodiversity duties apply to all biodiversity components, but in practice the focus is on the habitats and species of principal importance and protected areas. In England at least, these provisions have not been very effective because of the limited awareness and understanding among public bodies of the duty, its weak wording, and the lack of clear reporting requirements and enforcement measures (House of Lords, 2018). In recognition of this, the Environment Act 2021 (described in Section 31.3.4) has introduced measures to strengthen the biodiversity duty and require reporting.

The 2006 NERC Act originally applied equally to both England and Wales, but was amended in Wales by the Environment (Wales) Act 2016, which created an 'enhanced' biodiversity duty, requiring all public authorities in Wales to *seek to* maintain and enhance biodiversity (consistent with the exercise of their functions), and also making an explicit link between conserving biodiversity and promoting the resilience of ecosystems. The 2016 Act also requires each public authority in Wales to publish and keep under review a plan on how it intends to comply with the biodiversity duty, and to report every three years on what steps it has taken. There are now similar requirements in England as a result of the Environment Act 2021.

The 2010–2020 Agenda

As a post-2010 biodiversity agenda emerged, a substantial change occurred across the UK in the rationale for nature

[11] A detailed list of key developments up to 2012 is provided on the JNCC website: http://jncc.defra.gov.uk/page-5155

[12] https://webarchive.nationalarchives.gov.uk/ukgwa/20120306225937/http://jncc.defra.gov.uk/page-5718

[13] https://webarchive.nationalarchives.gov.uk/ukgwa/20120306225139/http://jncc.defra.gov.uk/page-5717

[14] https://webarchive.nationalarchives.gov.uk/ukgwa/20120315152705/http://jncc.defra.gov.uk/default.aspx?page=5155

conservation and its broad aims and approach. This initially arose from a UK Government White Paper, *The Natural Choice* (HMG, 2011a), which aimed to develop a new vision for nature conservation, taking into account the CBD's 2010 Strategic Plan and Aichi targets, and the EU's 2011 *Biodiversity Strategy to 2020*. It also took on board the concepts of The Economics of Ecosystems and Biodiversity initiative (e.g. TEEB, 2011) and reflected the influential findings of the *UK National Ecosystem Assessment* (UK NEA, 2011) – the first of its type in the world and following from the global *Millennium Ecosystem Assessment*.[15] This national assessment clearly demonstrated the wide-ranging social and economic importance of ecosystem services in the UK and the factors that are leading to their deterioration.

Reflecting its influences, the White Paper made a strong case for nature conservation and ecosystem restoration and appeared to move nature conservation towards the political mainstream. This emboldened DEFRA to set out an ambitious 2020 mission 'to halt overall biodiversity loss, support healthy well-functioning ecosystems and establish coherent ecological networks, with more and better places for nature for the benefit of wildlife and people'. Later in 2011, *Biodiversity 2020: A strategy for England wildlife and ecosystem services* was published (DEFRA, 2011a), which aimed to achieve the biodiversity objectives set out in the *Natural Environment White Paper*. In Scotland, the *25 Year Biodiversity Strategy* was supplemented by initiatives to contribute to the CBD Aichi targets (Scottish Government, 2013, 2015) and then replaced by the *Biodiversity Strategy to 2045: Tackling the Nature Emergency* (Scottish Government, 2022). Building on its first biodiversity strategy in 2002, *Valuing Nature: A Biodiversity Strategy for Northern Ireland to 2020* (DOENI, 2015) responded to the CBD and EU biodiversity targets.

As further described in online Annex UK.2, these strategies had many similarities, including emphasis on providing ecosystem services and maintaining natural capital, inspiring and involving the wider public, and integrating nature conservation across sectors. To advance the last of these, an additional element in Scotland has been the production of a *Land Use Strategy for 2016–2021* (Scottish Government, 2016) with a strategic vision on how to realise the full potential of Scotland's land for multiple benefits for the economy, environment and communities.

Wales has focussed principally on sustainability, economic and social well-being, and an integrated 'whole environment' approach to the use, management and valuation of natural resources. This has included a package of legislation implementing the sustainability principles set out in the CBD, including the Environment Wales Act 2016 referred to earlier, the Planning (Wales) Act 2014, the Well Being of Future Generations Act 2015 (promoted by a Commissioner) and the Welsh Government's *Natural Resources Policy* in 2017.[16] Within this broader policy framework, a *Nature Recovery Plan for Wales*[17] was published by the Welsh Government in 2015.

In 2012, a *UK Post-2010 Biodiversity Framework* was produced (JNCC and DEFRA, 2012). It identified 23 areas of work that the SNCBs agreed to contribute to with a UK focus. Apart from these agreed work areas, the framework showed a further shift away from a coordinated and overarching UKBAP towards country-level priorities and approaches that vary from place to place.

The increasingly ecosystem services–focussed and rather utilitarian rationale for nature conservation boosted political support and, together with NGO pressure, led to some ambitious strategies and targets. However, it was not matched by correspondingly bold measures and investments or effective implementation. For example, Scotland's Land Use Strategy was considered pioneering at the time, but lost momentum due to limited adoption by key parts of government. Of more concern has been the backward step resulting from the major reduction in public funding across the UK (see Section 31.7.3). Overall there is little evidence that the strategies for 2020 added significant value to existing measures (e.g. protected areas and other environmental legislation, landowner and NGO actions), and consequently their headline 2020 targets and most of their other objectives were not achieved.

Marine Nature Conservation

Marine nature conservation has lagged behind that undertaken on the land and freshwater. Initially this was because it was not on the government's policy agenda: impacts on nature were largely out of sight and therefore the seas were ignored and even treated as a dumping ground. Whilst the WCA 1981 introduced provisions to designate statutory Marine Nature Reserves, these were made almost unworkable by a policy of making designation dependent on the complete agreement of local stakeholders (Mitchell, 2018). This resulted in the establishment of just three small reserves (Lundy, Skomer and Strangford Lough). More fundamentally, the governance systems, legislation and policies at the time were more suited to nature conservation on land (Cole-King, 1995; Earll, 2018).

The first large-scale and effective marine nature conservation measures were introduced in the UK to comply with obligations under the Birds Directive and, especially, the Habitats Directive later. These led to major improvements in the legal protection of a wide range of species and habitats, and the first step towards the establishment of a meaningful network of MPAs in the UK. However, the limited inclusion of marine habitats in HD Annex I, and the way they were classified, caused difficulties in the UK and led to false starts, wasted money and unnecessary measures (Hiscock, 2018). The

[15] www.millenniumassessment.org/en/Reports.html#

[16] https://gov.wales/sites/default/files/publications/2019-06/natural-resources-policy.pdf

[17] https://gov.wales/nature-recovery-action-plan-2015

influence of the Directive was also initially reduced by the UK limiting its application to its Territorial Sea (12 nautical miles). This was challenged by Greenpeace through a judicial review[18] in 2000 that recognised the need to extend conservation activities out to the 200 nautical mile limit (which includes some important deep-sea HD habitats). Subsequently, the decision was confirmed by the CJEU,[19] such that the requirement applies to all EU waters (De Santo and Jones, 2007).

In the late 1990s, it was clear that wider-ranging and more ambitious measures were required for marine nature conservation that complemented those in the Nature Directives and replaced domestic legislation unsuited to marine nature conservation requirements (Edwards, 2018). A working group, including the relevant authorities, marine users and NGOs, was set up to review marine nature conservation in the UK, from which emerged *A Strategy for the Conservation and Sustainable Development of our Marine Environment* (DEFRA, 2002b). The strategy addressed domestic and international needs (including EU obligations and the commitments made at the Fifth North Sea Conference in March 2002), and put forward an ambitious vision of achieving 'clean, healthy, safe, productive and biologically diverse oceans and seas' through an ecosystem-based management approach. It also included commitments to identify and designate relevant areas in UK waters as MPAs to form a network of well-managed sites by 2010, as well as to improve the integrated management of the wider marine environment.

DEFRA announced in 2004 that it would bring forward a Marine Bill to help implement the strategy. This progressed slowly, but eventually led to the Marine and Coastal Access Act 2009 (England and Wales) as a result of focussed and coordinated advocacy by NGOs through Wildlife and Countryside Link, cross-party support in the UK Parliament, effective leadership from DEFRA officials working with the Welsh Government, and buy-in from industry (Edwards, 2018). Importantly, the legislation reformed the marine licensing regime, provided a statutory basis for marine planning (see Section 31.6.1), created the MMO and enabled the designation of Marine Conservation Zones (MCZs) within UK Secretary of State Waters,[20] with the aim of creating a coherent network of sites covering a wider range of marine habitats and species than under the Nature Directives.

As marine conservation matters are devolved, subsequently similar legislation was put in place through the Marine (Scotland) Act 2010 with a general duty to 'further the achievement of sustainable development, including the protection and, where appropriate, enhancement of the health of that area'. The Act covers marine planning, licensing and protection, including the creation of Nature Conservation Marine Protected Areas (NCMPAs) alongside Demonstration and Research MPAs and Historic MPAs, as well as conservation of seals and sea fisheries responsibilities. The Marine Act (Northern Ireland) 2013 allows for the designation of MCZs similar to those in Secretary of State Waters.

Brexit and Post-2020 Plans: Including the 25 Year Environment Plan and Environment Act 2021

The referendum decision on 24 June 2016 to leave the EU, and the eventual exit on 31 January 2020, means that the Nature Directives and the rest of the EU legislation have no longer applied since the end of the transition period on 31 December 2020. To prepare for this, and avoid initial problems, amendments were made in 2019[21] to the relevant components of the WCA, Habitats Regulations and Offshore Regulations to ensure they would be operable and effective after Brexit and the end of the transition period. Most of the changes involved transferring functions from the European Commission to the appropriate national authorities. Procedures that are mostly very similar to those under the Nature Directives were established for, amongst others, setting site- and network-level objectives, designating SACs (SPAs are already classified nationally), obtaining an opinion on whether a plan or project affecting priority habitats or species is justifiable for 'imperative reasons of overriding public interest' (IROPI), reporting and arrangements for amending the schedules to the Regulations (i.e. prohibited means of capture and killing) and the annexes of habitats and species to the Nature Directives that apply to the UK. Any such changes will require an affirmative statutory instrument and be debated in the relevant parliament.

Importantly, whilst there will no longer be a Natura 2000 network, existing and potentially new SPAs and SACs will be included in a new National Site Network (also known as a UK Site Network) and retain the same protections. Furthermore, the aim of ensuring that the network is coherent will be retained, in order to meet such international obligations as the Bern Convention (i.e. being part of the Emerald Network), OSPAR, CMS and Ramsar Convention.

Over the longer term, the decision to leave the EU has given the UK considerable scope to change its approaches to nature conservation. In response to this (and the 2015 and 2017 Conservative Party manifesto intentions 'to be the first generation to leave the environment in a better state than we inherited it'), the UK Government published its *25 Year Plan to Improve the Environment* (HMG, 2018). Notably, this received the very highest level of political support with a foreword from the then Prime Minister, Theresa May.

The plan, which primarily relates to England, adopts a 'natural capital' agenda (following recommendations from

[18] *R* v. *Secretary of State for Trade and Industry, ex p Greenpeace Ltd* (No 2) (2000) 2 CMLR 94 (QBD).

[19] Case C-6/04 *Commission* v. *UK*.

[20] That is, within the territorial and offshore waters of England, and the offshore waters of Northern Ireland.

[21] Through the Conservation of Habitats and Species (Amendment) (EU Exit) Regulations 2019, and related amendments in Scotland and Northern Ireland.

the UK Government–established Natural Capital Committee) and sets out a rational evidence-based case for action on many important environmental challenges. It includes some ambitious goals, although most do not need to be achieved for 25 years. These include reversing the loss of marine biodiversity, restoring 75% of protected areas to favourable condition, and creating or restoring a Nature Recovery Network, including by creating an additional 500 000 ha of wildlife-rich habitat outside protected areas. Policies and actions to achieve these goals are proposed, including strengthening the development planning and control framework, such that built developments result in net gain in biodiversity where this is feasible (see Section 31.6.1), publishing a Strategy for Nature (building on the existing BAP) and designing and delivering a new Environmental Land Management scheme. The last is probably the most important initiative as its aim is to eventually redirect much of what would have been the post-2020 CAP direct payments to payments that provide environmental public goods, including biodiversity. However, the plan lacks clear indications of how its objectives will be achieved and shorter-term targets to drive forward and monitor achievements (Tucker and Baldock, 2018). Its implementation and eventual impacts are therefore highly dependent on further supporting policy and new legislation, target setting and substantial funding (discussed in Section 31.9 on future developments).

To give the *25 Year Environment Plan* a statutory footing (encouraged by NGOs) and address a number of Brexit-related governance issues,[22] the UK Government announced in May 2018 its intention to introduce an Environment Bill. However, although a draft of the governance and principles part of the Bill was first published in 2018, due to a General Election and disruption from COVID-19, it took until 9 November 2021 for the full Environment Act to be passed.

The first part of the Act includes a new domestic framework for environmental governance, which mainly applies to England but has provisions that apply to Wales and Northern Ireland, and some relating to Great Britain and the UK as a whole. The Act addresses the need to consider environmental principles in the development of future legislation and policy, in particular those that are included in the EU treaties (see Section 4.1) but not included in the legislation taken across in the EU Withdrawal Act. Taking into account a consultation on the Bill, the Act obliges ministers to have due regard in policy making to a government statutory Policy Statement on Environmental Principles that will set out how policy making should interpret and apply five key principles: the integration of environmental protection into policy making, taking preventative action to avert environmental damage, the precautionary principle, rectifying environmental damage at source, and the polluter pays principle.

[22] Including commitments in the European Union (Withdrawal) Act 2018 and subsequent EU–UK Trade and Cooperation Agreement.

The actual environmental implications of the legislation on the principles are uncertain, as the legal duty is to give 'due regard' to the Environmental Principles Policy Statement, which is a government policy document.[23] In the EU, the Treaty obligations relate to the legal principles themselves, as informed by legal jurisprudence, rather than government policy.

Another fundamental governance requirement that the Act addresses is the need for a body to oversee compliance with environmental legislation. This has resulted in the establishment of the Office for Environmental Protection (OEP), a new public body for England and Northern Ireland, which provides advice on environmental policies and monitors progress in achieving environmental improvement plans and targets. It also has a 'watchdog' role, investigating complaints, and can bring cases of its own volition to ensure that the government and public authorities comply with environmental law. Cases may be considered in a novel legal mechanism, Environmental Review, which will resemble judicial review, but have certain differences.

Whilst the arrangements have improved considerably from initial proposals, the OEP does not have the same level of enforcement power as the European Commission and CJEU. There are also concerns that it will not be sufficiently independent, as the Secretary of State will determine its budget and its board. Furthermore, the government has powers to 'guide' its enforcement work. Unlike Environmental Standards Scotland, which has a similar role, the UK Government has not safeguarded the OEP's independence through law.

Part 1 of the Environment Act 2021 also introduces a new statutory cycle of monitoring, planning and reporting to ensure continuing improvement to the environment. This includes obligations to set legally binding long-term environmental targets (expected in 2022 for at least 15 years) in four priority areas (air quality, water, biodiversity, and resource efficiency and waste reduction) and to produce statutory Environmental Improvement Plans (EIPs) in England and Wales. EIPs are also required in Northern Ireland under Part 2 of the Act. The plans are supposed to include interim targets (which are not legally binding) and set out the steps to be taken to improve the natural environment, achieve the targets and monitor and report on them. The plans will be updated every five years. The first EIP is the 2018 *25 Year Environment Plan*, but it lacks interim targets. In response to this, NGO lobbying and an HM Treasury–commissioned review of the economics of biodiversity (Dasgupta, 2021; HM Treasury, 2021), the Act now includes a new specific duty to set and meet a target for 2030 relating to the abundance of species (to be set by 31 October 2022).

The remaining part of the Act makes provision for specific improvements in the environment, including for nature

[23] Whilst a consultation draft was published in 2021, it was not finalised until January 2023. www.gov.uk/government/publications/environmental-principles-policy-statement/environmental-principles-policy-statement.

conservation, which link closely to the *25 Year Environment Plan* and aim to supplement existing legislation. The following provisions are of most relevance to nature conservation in the England:

- Amendments to section 40 of the 2006 NERC Act Biodiversity Duty that strengthen the biodiversity duty on public bodies so that it goes beyond conservation to enhancement. Also, the Act shifts the focus of the duty to an active requirement to seek the further conservation and enhancement of nature. There is also a requirement for public bodies to give regard to Local Nature Recovery Strategies, Species Conservation Strategies and Protected Site Strategies (see later), and report on how they have sought to conserve and enhance biodiversity.
- A mandatory planning permission requirement for developments, including major infrastructure developments, to achieve a net gain in biodiversity (i.e. a 10% increase in biodiversity after development, compared to the level of biodiversity prior to the development taking place, as measured by a DEFRA metric), although the legal requirement to retain the gain is only for 30 years.
- Requirements for local authorities (which in practice will normally be County Councils) to develop statutory Local Nature Recovery Strategies, which must include a statement of biodiversity priorities for the plan area (e.g. describing and prioritising existing biodiversity and opportunities for recovery and enhancement), with the aim of providing a comprehensive vision of the current and future potential biodiversity value of the area and some proposed practical steps that different bodies may choose to support to help make improvements. The strategies will also support the delivery of a Nature Recovery Network by acting as a key tool to help local partners better direct investment and actions.
- Species Conservation Strategies and Protected Site Strategies, which aim to improve the protection of the most vulnerable species and habitats from developments, whilst also reducing delays.
- Powers for the Secretary of State to amend Regulation 9 and Part 6 of the Habitats Regulations 2017 (as amended by the 2019 Regulations) to re-focus the Regulations to support delivery of domestic biodiversity priorities. Regulation 9 sets out duties for relevant public authorities to exercise their nature conservation functions in compliance with the Nature Directives. Part 6 sets out the requirements for the assessment of plans and projects on European protected sites (i.e. SPAs and SACs).
- Conservation Covenants, which are defined as voluntary, legally binding private agreements between landowners and responsible bodies, which conserve the natural or heritage features of the land, enabling long-term conservation.

Overall, it is clear that the Environment Act 2021 includes a range of measures that have the potential to make a major step towards achieving the aims of the *25 Year Environment Plan*. However, their actual impacts will depend on further planning, funding and practical implication issues, as discussed later in relation to some of the specific measures and in Section 31.9 on future developments. There is also concern that the Secretary of State's powers may be used in future to weaken some aspects of the Habitat and Species Regulations.

Discussions and initiatives have also taken place in the other UK countries on the implications of Brexit and the adjustments to legal and governance arrangements that flow from this. Following a stakeholder engagement and consultation period, *The Environment Strategy for Scotland: vision and outcomes* was published by the Scottish Government in 2020,[24] which includes 'restoring nature' as part of its 2045 ambition. It also contains a commitment to embed EU environmental principles and governance into Scots Law. The Scottish Government has also established an independent public body to scrutinise compliance with environmental legislation – Environmental Standards Scotland. In 2021, an initial set of indicators for monitoring progress against the Environment Strategy was published with a commitment to reflect the latest position on a continuous basis (Scottish Government, 2021). Plans in Northern Ireland and Wales are less advanced and are therefore discussed further in Section 31.9 on future developments.

31.4 Species Protection and Conservation Measures

31.4.1 Basic Protection Legislation and Strict Protection of Selected Species

Protection measures for birds, other animals and plants are provided for in Part 1 of the WCA (which still applies to Great Britain, with modifying legislation in Scotland), and in similar legislation enacted in Northern Ireland (see online Annex UK.3 for details). With some exceptions, these provide basic protection for all wild birds, such as from killing and destruction of their nests. Selected animals and plants (including HD Annex IV listed species), and some bird species, are also given higher levels of protection, such as from disturbance and loss of their site of shelter. In general, once instigated, these species protection measures have been largely effective. However, there have been problems updating the lists of protected species, and as a result some threatened species are not covered. Large numbers of some common perceived 'pest' bird species have, and continue to be, killed under quite loosely defined so-called 'General Licences', which have been subject to legal challenges. So far, these have resulted in some changes to the licences in England and in Wales.

Over recent decades, few species have been significantly impacted by illegal hunting, persecution or other illegal activities. This is probably mainly due to social attitudes, public 'policing' of wildlife protection and the work of influential

[24] www.gov.scot/publications/environment-strategy-scotland-vision-outcomes/pages/2/

NGOs. However, there are exceptions, the most important of which has been the illegal persecution of raptors. In contrast, there have also been problems with the overly inflexible and inappropriate protection of some HD Annex IV species. These issues are therefore further described below.

The Persecution of Raptors

Despite their legally protected status, birds of prey continue to be persecuted in Britain, mainly in the uplands in relation to protecting the interests of commercial shooting of Red Grouse (*Lagopus lagopus scotica*) (Avery, 2015; RSPB, 2018; Murgatroyd *et al.*, 2019). This continues to be responsible for the absence of the Hen Harrier (*Circus cyaneus*) from large areas of suitable habitat, especially in England (Potts, 1998; Redpath *et al.*, 2010; Fielding *et al.*, 2011). Furthermore, a review of evidence concluded that illegal killing is also reducing the population and breeding success of the Golden Eagle (*Aquila chrysaetos*), Peregrine Falcon (*Falco peregrinus*) and Goshawk (*Accipiter gentilis*) on some grouse moors (GMMRG, 2019). Intensive grouse moor management also leads to other detrimental environmental impacts (Thompson *et al.*, 2016), including inappropriate burning of some habitats (see uplands discussion in Section 31.6.2).

The high rates of illegal persecution are considered to be largely due to inadequate enforcement combined with low fines for those who are caught, particularly in England. In Scotland the situation is marginally better, due to stronger enforcement and perhaps the incorporation of vicarious liability into the law, which means that estate owners can be held responsible for the illegal actions of their employees (Tingay, 2015). Although there have been few prosecutions, vicarious liability may have a deterrent effect. In addition, the use of some General Licences in Scotland can be withheld if NatureScot has reason to believe that certain people, or certain areas of land, have been implicated in wildlife crime; this is also considered to be an important additional deterrent. Yet, despite such measures, raptor persecution in Scotland persists, as for example indicated by a study that found that a very high proportion of satellite-tagged Golden Eagles disappeared in suspicious circumstances over grouse moors in Scotland (Whitfield and Fielding, 2017).

To investigate this and other issues, in 2017 the Scottish Government established an independent Grouse Moor Management Group to review the environmental impact of grouse moor management practices and to offer advice on the option of licensing grouse shooting businesses. The Group's report (GMMRG, 2019) stopped short of recommending the immediate introduction of a licensing system, but sent a strong signal that licensing should be introduced if certain raptor species fail to reach favourable conservation status on or near to grouse shooting estates over the next five years. The Scottish Government responded to the report in 2020 and took a stronger line, committing to introduce a licensing system within five years due to continuing raptor persecution in Scotland (Scottish Government, 2020a). This, and measures for hare shooting, muirburn management and the use of traps for land management, are being addressed through a Wildlife Management (Grouse Moor) Bill.

No recent steps have been taken to further investigate or regulate grouse shooting and its related impacts on species and habitats in England. Instead, a Hen Harrier Action Plan has been developed by DEFRA and other stakeholders (DEFRA, 2016a). Although it includes some widely supported actions (i.e. monitoring, diversionary feeding and actions to improve protection), it has been highly controversial as it also intends to reintroduce the species to southern England (which has a questionable rationale) and to carry out brood management. Brood management involves removing eggs or chicks from grouse moors when breeding numbers reach a density that is expected to significantly reduce grouse numbers, and then releasing the reared birds later on in unoccupied suitable habitat. This is considered to be inappropriate by many conservationists, including the RSPB,[25] and resulted in two legal challenges, which were dismissed,[26] as were their subsequent appeals.

Habitats Directive Strictly Protected Species: The Case of the Great Crested Newt

In contrast to the situation with raptors, there have been some problems in the UK with an overly strict and inflexible protection of some HD Annex IV species. In accordance with HD Article 16 (discussed in Section 4.2), if it is not possible for a development or other activity to avoid harming these species, a derogation can then be obtained for certain reasons, provided that it will not be detrimental to the maintenance of the species' favourable conservation status. However, in the UK, inadequate knowledge of the distribution and status of some HD Annex IV species, and an absence of defined favourable conservation status requirements, has led to derogation licensing that has sought to maintain the status quo by protecting every individual, rather than trying to conserve the population concerned.

This has been a particular problem with the Great Crested Newt (*Triturus cristatus*), as this is locally common in parts of England and north-east Wales. As a result, developers have been required to capture and/or exclude all Great Crested Newts from the impact zone, and to relocate them to suitable habitat as nearby as possible, creating new habitat if necessary. However, evidence suggests that such measures are often ineffective (Lewis *et al.*, 2017) and have contributed little to the long-term conservation of the species' populations (Watson, 2008; Beebee, 2015). Furthermore, the approach has been costly and caused delays, resulting in adverse publicity and hostile attitudes amongst developers and some politicians[27]

[25] https://community.rspb.org.uk/ourwork/b/martinharper/posts/the-rspb-s-position-on-the-hen-harrier-action-plan

[26] *Royal Society for the Protection of Birds, R (On the Application Of) v. Natural England* [2019] EWHC 585 (Admin) (15 March 2019).

[27] For example, in a speech on 30 June 2020, the then Prime Minister Boris Johnson referred to 'newt-counting delays' as an example of red tape that is constraining the productivity and the prosperity of this country.

to newts, protected species more generally, and the Nature Directives as a whole.

To address this issue in England, a more flexible policy is now allowing an approach that focusses on conserving local newt populations through area surveys of their distribution and then avoiding developments in key areas and providing strategic advance compensatory habitat where necessary (Cameron *et al.*, 2019). The approach is being implemented through District Level Licensing schemes, according to defined standards (Natural England, 2019), an example of which is described by Tew *et al.* (2018).

31.4.2 Species Action/Recovery Plans and Re-introductions

A wide range of species re-introductions have taken place in the UK, several of which started in Scotland due to the lower potential for human–wildlife conflicts as well as the large expanses of suitable habitat for the target species. These have included the White-tailed Sea-eagle (*Haliaeetus albicilla*) and Beaver (*Castor fiber*), as described in Gaywood *et al.* (2016) and online Annex UK.4. Recently, such projects in Scotland have been informed by the National Species Reintroduction Forum, which represents a range of stakeholders from the land-use, conservation and science sectors. The Forum has produced the Scottish Code for Conservation Translocations and accompanying best practice guidance (NSRF, 2014). In addition to re-introductions, species recovery projects have also bolstered small populations and helped species recolonise their former range, such as for Red Kite (*Milvus milvus*) in England, Scotland and Northern Ireland (see online Annex UK.4), Common Crane (*Grus grus*) in England, and Sand Lizard (*Lacerta agilis*) in England and Wales. Currently, the Back from the Brink Project in England aims to save 20 species from extinction, as well as benefiting many more.

31.5 Protected Areas and Networks

31.5.1 Protected Area Types and Coverage

Terrestrial Protected Areas

The UK has developed over the last 60 years an extensive and multi-role network of protected areas that provides a key mechanism for protecting and managing important areas for biodiversity (and geodiversity) on publicly owned and private land. These and other designated areas were reviewed[28] in a project coordinated by the IUCN UK National Committee that provided an independent assessment of which areas qualify as protected according to the IUCN definition and criteria (Dudley, 2008) and ascribed management categories and governance types to each, where possible (Crofts *et al.*, 2014). This was in line with IUCN advice that ideally there should be an independent and informed process to verify and confirm decisions regarding protected areas at national level. Those protected areas that were considered to meet the definition and criteria are included in Table 31.4. The total areas of each type of domestic protected area are sourced mainly from the WDPA[29] because no UK-level compilation of such summary information is publicly available. It is possible that there may be some discrepancies with more up-to-date information held by the SNCBs. Nevertheless, it is assumed that the WDPA data provide a reasonable picture of protected areas in the UK.

The terrestrial protected areas designated under domestic legislation that are of primary importance for nature conservation in Great Britain are NNRs and SSSIs. In Northern Ireland, the equivalent protected areas are ASSIs and statutory Nature Reserves, which comprise Nature Reserves and NNRs.[30] All of these areas are designated for their special interest (i.e. conservation importance) in terms of their flora, fauna, or geological or geomorphological features. The vast majority of sites are designated for their flora and fauna and can be considered to be biodiversity-focussed protected areas.

NNRs are supposed to have the best examples of natural and semi-natural habitats, and accordingly be given the highest level of protection (Marren, 2018). Sites for designation as NNRs were identified through a survey that resulted in *A Nature Conservation Review* (Ratcliffe, 1977). This milestone in UK conservation identified 750 of the best sites throughout Great Britain, but in practice it proved to be far too difficult and expensive to designate them all as NNRs. However, all were subsequently notified as SSSIs. NNRs have since been designated by the SNCBs when they are considered to be the best sites and/or have a particular role in terms of research or providing opportunities for education or other forms of public enjoyment of nature.

Another layer of protected area designation that is available for terrestrial biodiversity across the UK under domestic legislation is the establishment of LNRs by local authorities, with the advice and support of the SNCBs. Although their primary purpose is the protection of nature and they are generally considered to give high protection (Lawton *et al.*, 2010), the IUCN UK Committee project noted that this designation type does not guarantee long-term protection and management, and recommended that inclusion in the WPDA should be on a site-by-site basis (Crofts *et al.*, 2014).

[28] With reference to Statements of Compliance prepared for some types of protected area, available at https://iucn-nc.uk/projects/putting-nature-on-the-map/

[29] www.protectedplanet.net/

[30] The first statutory Nature Reserves in Northern Ireland were declared in 1970 under the provisions of the Amenity Lands Act (Northern Ireland) 1965. This was superseded by the Nature Conservation and Amenity Lands (Northern Ireland) Order 1985, which introduced the concept of NNRs for sites of 'national importance' while retaining the Nature Reserve designation. In practice they have similar levels of protection.

Table 31.4 The number and extent of international designated sites and protected areas in the UK.

	IUCN category	Number	Area (km^2)
Designated under global and non-EU international legislation			
Ramsar Sites	NR[a]/IV[b]	150	7 938
Natural and Mixed World Heritage Sites	NAp[a]	3	269
Biosphere Reserves (exc. buffer and transition zones)	NA[a]	7	454
OSPAR Marine Protected Areas	NR/Ia/IV[a]	382	320 088
Designated under the Nature Directives			
SCIs (terrestrial + marine)	?[a]/IV[b]	660	13 176 + 121 745
SPAs (terrestrial + marine)	?[a]/IV[b]	276	16 091 + 21 582
SCI + SPA (former Natura 2000) terrestrial area			21 008
SCI + SPA (former Natura 2000) marine area			132 130
Designated under domestic legislation			
Areas of Special Scientific Interest/Sites of Special Scientific Interest	III/IV	7 016	24 842
National Nature Reserves	?[a]/II/III/IV	361	3 007
Local Nature Reserves	?[a]/IV	1 809	593
National Parks	V	15	23 151
Areas of Outstanding Natural Beauty (in England and Wales only)	V	39	20 563
Marine Conservation Zones	?[a]/NR	97	32 743
Marine Protected Area (West Coast of Scotland MPA)	?[a]/NR	1	107 718
Nature Conservation Marine Protected Areas	?[a]/IV NR	35	68 015
Demonstration and Research Marine Protected Area	?[a]/NR	1	157
Other			
Nature Reserves managed by NGOs	Ia/II/III/IV/V/VI	947	2 885

Note. NR = Not reported; NAs = not assigned; NAp = not applicable. a. Sites lack categories ascribed by Crofts *et al.* (2014), as well as in the WDPA. b. IV as all sites are SSSIs/ASSIs.

Sources. See Section 1.3. Based on data from: JNCC Ramsar Site list (January 2022 update);[32] UNESCO website (January 2022); OSPAR website (February 2022 update); EEA Natura 2000 Barometer, as at end 2019; MCZs and NCMPAs data from UK SNCB websites (January 2022); other sites from WDPA (UNEP-WCMC, January 2022); IUCN categories based on Crofts *et al.* (2014) or WDPA dataset.

SPAs have been identified and classified since the mid-1980s based on guidelines developed by JNCC,[31,32] on behalf of the SNCBs. Three national and systematic reviews of the network have been undertaken (Stroud *et al.*, 1990, 2001, 2016), progressively refining and extending the scope of the network in response to the needs of 151 BD Annex I and/or migratory species. Yet, whilst the second review (2001) made substantive recommendations for classification for multiple species and sites, limited progress has been made to implement these, especially in England and Wales, and the network remains substantially incomplete. Further recommendations were made in the third review (2016), which identified insufficiencies to be addressed by the four countries. A separate marine SPA review will also provide additional recommendations. There is substantial overlap between the UK SPA and Important Bird Area (IBA) networks, primarily due to very similar identification criteria (JNCC, 1999).

SCIs have been identified and adopted since the mid-1990s (and subsequently designated as SACs), with site identification and submission to the European Commission being coordinated at a UK level by JNCC. The UK's SCI network was largely completed on land, but was considered to be incomplete at sea (European Commission, 2019). Given the UK's departure from the EU, it is uncertain if further sites will be added to the current national network.

According to the European Commission's data, at the end of 2019, the combined area of Natura 2000 sites (i.e. SCIs and SPAs) in the UK covered some 21 000 km^2 on land (8.6%) and 132 130 km^2 at sea (14.9% of marine waters). Some Ramsar Sites could be added to this area, as under government policy they are afforded the same levels of protection and management as Natura

[31] https://jncc.gov.uk/our-work/special-protection-areas-overview/#spa-classification-selection-guidelines-for-spas

[32] https://jncc.gov.uk/our-work/ramsar-sites/

2000 sites. However, most Ramsar Sites are in any case designated as Natura 2000 sites. Compared to other countries, the UK has a very low level of Natura 2000 coverage, although this is partly because protected areas' boundaries were more tightly drawn around the areas of high nature conservation interest.

Based on the currently most up-to-date public JNCC data,[33] as of March 2021 the terrestrial coverage of biodiversity-focussed protected areas (i.e. ASSIs/SSSIs, NNRs, SPAs and SACs) was 25 950 km^2, 10.6% of the UK land area (England 6.5%, Northern Ireland 9.8%, Wales 10.6% and Scotland 17.6%).

The IUCN UK Committee project also concluded that National Parks (in England, Scotland and Wales) and Areas of Outstanding Natural Beauty (AONBs) in England and Wales (but not Northern Ireland) also qualify as protected areas (Crofts *et al.*, 2014). National Parks are primarily designated to conserve and enhance their natural beauty, wildlife and cultural heritage, as well as promoting opportunities for the public understanding and enjoyment of the areas (hence IUCN category V). Thus, they have broader objectives than National Parks in most other European countries, which tend to be more often IUCN category II areas.

AONBs are designated to conserve and enhance the natural beauty of the landscape (also IUCN category V). Although National Parks and AONBs have some statutory responsibilities to contribute to nature conservation, given their scope and objectives, their added protection of biodiversity is weak. The inclusion of AONBs in protected area statistics is especially debatable as the IUCN UK project considered that they 'only just met the requirements' and their recognition was in part based on a 'commitment to nature conservation which will be reflected in future management plans from each AONB manager' (Crofts *et al.*, 2014). More recently, a government-commissioned review of National Parks and AONBs concluded that both were 'not currently delivering on their duty in relation to nature' (Glover, 2019). As a result, the review's first proposal is for National Parks and AONBs to 'have a renewed mission to recover and enhance nature. . .'. Until this happens there seems to be no justification for including AONBs as protected areas, and the justification for including National Parks is questionable.

Despite this, since 2013 National Parks and AONBs have been included in the UK's protected area coverage indicator[34] (which is a UK Government official statistic), used, for example, to report in relation to CBD Aichi Target 11. As they cover large proportions of the national territory, when combined with ASSI/SSSIs and NNRs they increase substantially the protected area coverage. The indicator also includes National Scenic Areas (NSAs) in Scotland and AONBs in Northern Ireland, even though they did not qualify as protected areas according to IUCN UK. On the other hand, the JNCC indicator does not include LNRs or private areas that are managed by nature NGOs, contrary to the review's recommendations. In theory, the protected area indicator and WDPA should include information on the UK's terrestrial protected areas that is consistent with the project's recommendations, because a verification procedure for the inclusion of protected areas was established (Crofts *et al.*, 2014). This does not appear to have been acted on by the governments of the UK and other relevant parties, as the indicator and WDPA still include some sites that the IUCN UK review ruled out (e.g. NSAs, Regional Parks and Heritage Coasts).

According to the UK protected area indicator, as of 31 March 2021, the total protected area coverage was 67 930 km, which equates to 27.8% of the UK's land area. However, given the above, this overestimates the area that meets the IUCN's protected area definition. This is despite the fact that NGO areas and LNRs are not included, because the former overlap greatly with other areas, and LNRs are very small.

A review of protected area coverage and effectiveness in the UK came to a similar conclusion that the UK protected area indicator provides a misleading assessment of the degree to which the UK met Aichi Target 11 (Starnes *et al.*, 2021). The study estimated that, as of December 2020, protected area coverage was 61 147 km^2, or 25.0% of the land area, when areas not meeting IUCN protected area criteria were excluded. Furthermore, only 11.4% of land was within areas designated primarily for nature conservation (i.e. IUCN categories 1a–IV). When the condition of these areas is taken into account, only 4.9% of the UK land area is in effectively managed protected areas, as indicated by the proportion within biodiversity-focussed protected areas that have a favourable condition (i.e. 43–51%).

Starnes *et al.* also queried whether AONBs in England and Wales and National Parks meet the IUCN protected area criteria, noting that the 1995 Environment Act removed the Sandford Principle (Sandford, 1974) that nature conservation in National Parks should be given priority if there are conflicts with other management objectives. In any case, they strongly suggest that governments in general should not use IUCN protected area categories V and VI to artificially inflate protected area statistics – as appears to be the case in the UK since 2013. As documented in many other chapters in this book, this is common practice in many countries.

A more recent policy paper on protected areas, by the British Ecological Society, also concluded that significant reforms are required to National Parks and AONBs before they can count towards the UK Government's target to protect 30% of land and sea in all four countries (Bailey *et al.*, 2022). Moreover, the authors consider that the main effort to meet the target 'is less about designating new areas, but more about transforming existing areas so they can deliver for nature', and they provide recommendations to do so.

The limited contribution that National Parks and AONBs are currently making to nature conservation has been recognised by the UK Government in its full response to the 2019 Glover review (DEFRA, 2022a). Consequently, it is accepted that currently they should not count towards protected area targets, and therefore amendments will be made to their statutory purpose so that they do drive nature recovery.

[33] https://jncc.gov.uk/our-work/ukbi-c1-protected-areas/
[34] https://jncc.gov.uk/our-work/ukbi-c1-protected-areas/

Despite this, and some other positive responses to the review, more will need to be done to achieve this goal. According to Hampson (2022), this should focus on increasing resources, improving leadership and expertise in the areas' boards, and strengthening management planning, target setting, monitoring and reporting, as well as the duty of public bodies to contribute to nature conservation.

Marine Protected Areas

Until the identification of marine SACs, the UK had designated an extremely small proportion of its marine environment as protected areas. This was partly because some government legal advisors considered that ASSIs/SSSIs could not legally extend below mean low water mark. To fill this gap, the WCA 1981 included powers to establish Marine Nature Reserves. Despite this, and the publication of a directory of potential reserves (Gubbay, 1986) and an official Marine Nature Conservation Review (Hiscock, 1998), only three were established (for the reasons discussed earlier). Consequently, the Nature Directives' requirements to designate SPAs and SACs resulted in the establishment of the first suite of MPAs, which at the end of 2019 covered about 14.9% of the UK's marine waters. However, the SPA component for birds at sea is still incomplete due to slow action by government and its SNCBs.

The 2009 and subsequent marine legislation has enabled the designation of Marine Conservation Zones (MCZs) and, in Scotland, Nature Conservation MPAs (NCMPAs). Their broad aim is to build on Natura 2000 marine sites and provide an ecologically coherent network of well-managed MPAs (Chaniotis *et al.*, 2018), based on principles derived from OSPAR guidance (OSPAR, 2006). In this respect, important design requirements of the MCZ/NCMPA process included protecting a wider range of marine habitats and species than were included in the Habitats Directive, and ensuring their adequate representation and replication. Although such requirements were incorporated into the guidance for the designation of MCZs (Natural England and JNCC, 2010), according to Hiscock (2018) the lists of broad habitats and species of conservation importance requiring the designation of MCZs were poorly thought through (as were the HD lists) and omitted many that are worthy of protection (e.g. rare or declining). In 2014, Scottish Ministers adopted a list of 81 priority marine features (many of which are features characteristic of the Scottish marine environment) that need to be taken into consideration through the Scottish Government's *National Marine Plan*[35] and the *Strategy for Marine Nature Conservation in Scotland's Seas.*[36]

Over recent years, a substantial number of MCZs have been designated in England and Northern Ireland, and similar NCMPAs in Scotland (Table 31.4). These include the three Marine Nature Reserves that have been converted to MCZs. One of these is in Wales (Skomer), but efforts to establish other MCZs in Wales are ongoing.

Based on JNCC data, as of 28 February 2022, there were 374 UK MPAs (i.e. MCZs, NCMPAs, and SACs and SPAs with marine components), covering 338 545 km^2, which is 38% of UK waters.[37] Coverage of the UK Territorial Sea was 47%. There will be further work to complete the network, but the priority now is to implement effective robust management in the designated areas.

31.5.2 The Coherence of Protected Areas and Ecological Networks

The concept of ecological networks has not been widely integrated into nature conservation policies. This is in part a result of scepticism towards simplistic models for ecological networks and limited evidence of their efficacy, in particular in relation to ecological corridors. Whilst there is growing international evidence that corridors can increase movements of some plants and animals between habitat patches (Gilbert-Norton *et al.*, 2010; Resasco, 2019), it remains uncertain whether they increase the viability of otherwise isolated populations (Haddad and Tewksbury, 2006). It is also unclear whether typical habitat corridors can increase the movements of habitat specialists, which tend to be the species of most conservation concern. Given such uncertainties, and limited funds, nature conservation actions have tended to focus on maintaining core areas of high-quality habitat and reducing pressures on them (e.g. as proposed by Hodgson *et al.*, 2009), primarily through the designation and management of protected areas.

Despite the UK's reliance on its protected area network for nature conservation, no assessment of its adequacy has been undertaken. To address this in England, DEFRA commissioned a review by Sir John Lawton and a panel of experts to ascertain whether or not the wider network of wildlife sites comprises a coherent and resilient ecological network (Lawton *et al.*, 2010). Their review recognised that England's terrestrial wildlife site network (which is similar across the UK) comprises three tiers:

1. Biodiversity-focussed statutory protected areas (i.e. SSSIs, NNRs, LNRs, Ramsar Sites and Natura 2000 sites) and land managed for nature conservation by NGOs.
2. Local Wildlife Sites[38] designated using selection criteria by local partnerships for their high biodiversity value (4 322 in England, >5% land area (The Wildlife Trusts, 2018)) and listed Ancient Woodlands, both of which are not legally

[35] www.gov.scot/publications/scotlands-national-marine-plan/

[36] www.webarchive.org.uk/wayback/archive/20160107013417mp_/http://www.gov.scot/Resource/Doc/295194/0115590.pdf

[37] https://jncc.gov.uk/our-work/uk-marine-protected-area-network-statistics/

[38] Also often known as County Wildlife Sites, Sites of Nature Conservation Interest or similar names, and also occurring in Northern Ireland, Scotland and Wales: www.wildlifetrusts.org/local-wildlife-sites

protected areas, but receive some protection through the planning system.

3. National Parks and AONBs, which give some protection to important areas of habitat, as well as buffering or connecting Tier 1 and 2 sites.

It was also recognised that to be resilient, especially in the face of climate change, the sites need to be part of a coherent ecological network, which should support the full range of biodiversity, be of adequate size, and have sufficient ecological connections between sites to enable species movement. To maintain such a network, the sites need long-term protection and appropriate management, and to be valued by, and be accessible to, people.

Lawton *et al.* concluded that the SSSI series in England, important though it is, did not comprise a coherent and resilient ecological network, nor did the wider combined network of Tier 1, 2 and 3 sites. In summary, the network needed 'more, bigger, better and joined' sites.

Although more than 12 years have passed since the Lawton review, few substantive practical actions have been taken to implement its recommendations on the expansion of the site network, even though the UK Government's 2011 White Paper included actions to address this. In particular, in 2012 the UK Government established 12 Nature Improvement Areas that were supposed to provide strategic landscape-scale improvements in the network. Whilst the initiative led to the creation of new partnerships and stimulated considerable volunteer activity, the £7 million (€8 million) provided for the areas was clearly inadequate. When the funding stopped after just three years, only 4 265 ha of priority habitat had been restored or created (CEP, 2015). Despite some other initiatives, there has been no significant increase in the area of terrestrial biodiversity-focussed protected areas in England or elsewhere in the UK.

Several NGOs have made ambitious calls to increase and enhance the network aspects of protected areas and habitats in the wider environment. For example, the Wildlife Trusts championed the Living Landscape initiative for many years and now campaign for Nature Recovery Networks, which aim to create naturally functioning landscapes by conserving core areas of high-quality wildlife habitat, enhancing connections between core areas and increasing permeability across the whole landscape (The Wildlife Trusts, 2020). Recently, Natural England has provided guidance to support the *25 Year Environment Plan's* ambition to create a Nature Recovery Network (Crick *et al.*, 2020), which is now relevant to the Local Nature Recovery Strategy required under the Environment Act 2021. The guidance has confirmed the general principles and priorities identified by the Lawton review, but has split the need for 'joined' sites into two, because providing 'stepping stones' and improving the 'permeability' of the landscape matrix (i.e. the ability for wildlife to move through it) is usually more important than providing physical corridors. Thus, according to the guidance, the current general hierarchy for designing coherent nature networks is 'Better site quality > Bigger sites > More sites > Stepping stones and more permeable matrix > Corridors'.

So far, limited funding, high land prices and other constraints have constrained the scale of ecological network and nature recovery initiatives to marginal increases in habitat area and connectivity. This even happened when the Scottish Government adopted an ambitious proposal from a Scottish Wildlife Trust conference in 2013 and included a commitment in its third *National Planning Framework* to create a national ecological network (see online Annex UK.5). Despite this policy backing, progress with the development of the network at a national scale has been lamentably slow (Davies *et al.*, 2018). Whilst some bottom-up initiatives, such as Cairngorms Connect and Buglife Scotland's B-Line project, are leading to network enhancements, these are not sufficient to achieve the desired holistic national objectives. The consultation process on a fourth *National Planning Framework* has just commenced and acknowledges the need for any emerging spatial strategies to promote nature-based solutions to climate change and to biodiversity (Scottish Government, 2020b).

31.5.3 Approaches to Protection and Levels of Protection

Terrestrial Protected Areas

Almost all UK terrestrial protected areas, including SPAs, SACs, Ramsar Sites and others of international importance, are given statutory protection through their underpinning designation as ASSI/SSSIs. These sites are protected through a number of means. Firstly, in the development planning process (see Section 31.6.1) they are taken into account in the preparation of plans and in decisions on individual projects, which are also informed by the findings of SEA, EIA and appropriate assessment reports where applicable. Secondly, they are protected from potentially damaging activities (listed as Operations Requiring Consent – see Section 31.5.4) that are undertaken by owners and occupiers without written consent from the SNCBs. Thirdly, they are given protection from damage by the actions and regulatory decisions of public and statutory bodies who, as well as being under a general duty to conserve ASSI/SSSIs, are also required to notify the relevant SNCB before carrying out, or authorising, any operations likely to affect a site (including development or land management works). Although they are not ultimately prevented from proceeding against statutory nature conservation advice, they must give it considerable weight and justify any departure from it. Fourthly, ASSIs/SSSIs are given protection from damage by the actions of third parties through the provisions in the WCA and later supplementary legal provisions in individual countries, which make it a criminal offence to intentionally or recklessly destroy or damage the features for which an ASSI/SSSI is designated.

SPAs and SACs have additional more stringent protection through the Habitats Regulations, which transposed the Nature Directives but have been retained with amendments since Brexit. Of particular relevance are Habitats Regulations Assessments, which involve an initial screening stage followed by an appropriate assessment if proposals are likely to have a significant adverse impact on a listed BHD habitat or species within a Natura 2000 site (in accordance with HD Article 6 (3)). Once the Regulations began to be tested in the 2000s through the planning process, they had profound implications for decision making, both at the project level and then later for strategic plans (see online Annex UK.6 for details).

HD Articles 6(3) and (4) presented considerable legal and administrative challenges in the UK, and their implementation evolved substantially in response to learning, inquiry decisions and domestic and European case law (Jones, 2012). For example, an innovation has been the development of strategic mitigation measures. This was first used, in the form of Strategic Alternative Natural Greenspace, to mitigate the potential impacts of increased recreational disturbance from multiple housing developments near SPAs (online Annex UK.6; Ricketts and Bischoff, 2012). Such experiences have shown that correct implementation requires both a comprehensive understanding of the ecology of the site and considerable knowledge of the precedent decisions, which has been a particular challenge for the local planning authorities due to their limited resources. Support has therefore been provided through the development of best practice guidance, including an in-depth online handbook and journal,[39] as well as guidance from relevant environmental authorities and professional ecological consultants. As a result of the accumulation of experience, expertise and guidance, and despite perceptions amongst some politicians that the Nature Directives were putting ridiculous costs on business, a government assessment of the regulations concluded that 'the Directives are working well, allowing both development of key infrastructure and ensuring that a high level of environmental protection is maintained' (HMG, 2012). On the other hand, based on a review of CJEU and Commission opinions and guidance, Clutten and Tafur (2012) found that an increasingly wide interpretation of the meaning of 'imperative reasons of overriding public interest' (the IROPI test) was taken in the UK.

Marine Protected Areas and Networks

In the marine environment, problems arose from the wording in HD Article 6(3) that refers to impacts from *any plan or project*. For many years, fishing was not deemed to be a project or a plan by the UK authorities, and therefore its impacts on Natura 2000 sites were not regulated under these provisions. Consequently, for example, the benthic habitats of several SACs were damaged by scallop dredging. However, following the CJEU ruling on the Waddenzee case (C-127/02, described in Section 4.2.3) and numerous complaints to the European Commission from The Wildlife Trusts, in 2004 DEFRA concluded it had to consider fishing as a project or plan in Natura 2000 sites. Then, after a campaign begun in 2008 by the Marine Conservation Society (MCS) and ClientEarth, in 2012 DEFRA announced what is called the 'revised approach to fishing', which aims to ensure all commercial fishing operations are managed in line with HD Article 6 (Clark *et al.*, 2016). This approach is further described in online Annex UK.6.

MCZs and SACs are managed by the Inshore Fisheries and Conservation Authorities (IFCAs) and MMO. In general, in inshore waters IFCAs apply by-laws (measures necessary to meet the objectives of each site) for fishing. The MMO licenses all other activities, as well as fishing in offshore waters. The authorities then instigate and enforce by-laws as necessary to meet the objectives.

The approach being followed for NCMPAs in Scotland is gaining support and is being implemented successfully (Gubbay, 2018). Following extensive and diverse stakeholder consultations by way of a Marine Strategy Forum, Marine Scotland produced a handbook outlining the aim of producing a management plan for each MPA. This consultative approach has been underpinned by a range of regulatory measures set out in the Marine (Scotland) Act 2010, which require Scottish ministers and public authorities to exercise their functions in a way that achieves sustainable development and, if appropriate, enhances the health of the Scottish marine area. Public bodies have duties to further the achievement of the conservation objectives and can only authorise activities that do not present a risk to achieving the conservation objectives. Scottish ministers can also make Marine Conservation Orders, which may involve prohibiting, restricting or regulating any activity (such as a damaging form of fishing) in any way that is necessary to achieve a site's objectives.

Despite the UK's extensive network of MPAs, only a tiny area within them is designated as no-take zones, and many are subject to other multiple and damaging uses. To address this, the UK Government contracted an independent review into the possible introduction of 'Highly Protected Marine Areas' (HPMAs) in Secretary of State Waters. These aim to allow marine ecosystems to recover to a mature state by taking a whole-site approach (i.e. protecting all habitats and species within them) and prohibiting all damaging activities. The review by Benyon *et al.* (2020) concluded that such sites should be designated, and made recommendations on their selection, protection and management, such as through early engagement with stakeholders. In 2021, the UK Government response[40] indicated that it accepted most of the recommendations and would begin introducing HPMAs by identifying a number of locations within English waters to pilot the

[39] www.dtapublications.co.uk/

[40] www.gov.uk/government/publications/government-response-to-the-highly-protected-marine-areas-hpmas-review/government-response-to-the-highly-protected-marine-areas-hpmas-review

approach, using powers under the Marine and Coastal Access Act (2009). In 2022, Scotland agreed to make 10% of its sea bed HPMA.

31.5.4 Approaches to Managing Protected Areas

The setting of conservation objectives and identification of required management measures for ASSI/SSSIs, SPAs and SACs are the responsibility of the SNCBs. Although the procedures and forms of presentation vary amongst the countries, the key elements are much the same. Normally, the objectives are set through a streamlined process that does not require the development of management plans for ASSI/SSSIs. Of key importance is the citation for each site, which includes a description of the site and its features (habitats, species or geological features) for which it is legally notified, and which forms the basis for the site's conservation objectives. Owners are provided with a list of the operations that may damage the features of interest, and therefore would require prior consent from the SNCB for them to take place. Objectives for each feature and the corresponding management requirements are then set out in some form of a site management statement. As many sites are privately owned, the objectives and management measures tend to be very broad and qualitative, and aspirational rather than prescriptive. This is particularly the case in England and Northern Ireland, where the documents set out the agencies' 'Views about management', which indicate general desirable management principles.

A more robust management approach is taken for SPAs and SACs; although management plans are not a legal requirement, they are being increasingly used, both for single sites and at the habitat level (e.g. Forestry Commission management plans). However, the plans are relatively narrowly focussed, such as the Site Improvement Plans produced for all Natura 2000 sites in England as part of an EU LIFE project (IPENS). These describe the main issues affecting the condition of the site and identify the priority actions and their costs and possible funding sources. NRW carried out similar work for all Natura 2000 sites in Wales. NatureScot has produced conservation advice packages and conservation and management advice documents. These advise on the management needed to achieve the conservation objectives and are scheduled for review every 10 years.

In the interests of transparency, the location, features, operations requiring consent, conservation objectives, management measures and condition of features for ASSI/SSSIs, SPAs and SACs are made public via the Internet (e.g. via SiteLink[41] in Scotland).

A high proportion of protected areas are grasslands, heaths and other habitats that require active management, usually to maintain the vegetation of interest and prevent its replacement by natural succession. In the past, many such semi-natural habitats were maintained by agricultural use, predominantly low-intensity grazing, sometimes in combination with hay cutting. Unfortunately, such agricultural systems are now rarely financially viable, and so additional financial support is required to maintain them (and also to reduce pressures to plough or plant the site). Whilst the SNCBs have some funds for supporting management, most funding is obtained through targeted CAP Rural Development Programme (RDP) measures (see Section 4.3.8). Where an owner or occupier is unwilling to carry out management, the SNCB can require it to be done. However, the use of statutory powers to enforce positive management practices has supposedly been rarely required (National Audit Office, 2008) and is certainly rarely used. More proactive efforts tend to be made for NNRs, whose management is typically the responsibility of the SNCB. These are managed in one of three ways: directly by the SNCB, or through an agreement between the SNCB and the owners (many sites being privately owned), or by an approved body such as the RSPB and National Trust (Marren, 2020).

A particularly important development was the Labour Government's introduction of **Public Service Agreement (PSA) targets** in England in 1998 (with a similar programme for government measures in Scotland). These aimed to ensure that departmental spending was tied to achieving specific targets. Crucially, each PSA defined how the targets were to be achieved, who was responsible for them, and how performance against these targets was to be measured.

In 2000, a PSA target was set to bring at least 95% of England's SSSI area into 'favourable' or 'unfavourable recovering' condition by 2010. The target's scope and level remain a bit of a mystery because it covered all SSSIs and so went beyond the requirements relating to Natura 2000 sites. Furthermore, it was very ambitious given that only 57% of the SSSI area was in such condition at the time, and there were many problems to overcome (CJC Consulting *et al.*, 2004). Nevertheless, PSAs were treated very seriously by government, and DEFRA and English Nature (now Natural England) had no alternative but to embark on a programme of strategic planning, targeting of measures and very intensive action. The practical actions primarily focussed on ensuring that each SSSI had the required management in place. Improvements on private land were often made through voluntary agreements supported by carefully selected and tailored agri-environment schemes (see Section 31.6.2). The CROW Act duty for public bodies to conserve SSSIs was also brought into play, and an important role was played by a Major Landowners Group (e.g. Forestry Commission, the Ministry of Defence and NGOs).

Natural England claimed that it had met its target, since, by December 2010, 95.8% of the SSSI area in England was said to be in favourable or unfavourable recovering condition (Natural England, 2011). However, this figure was not all it seemed, since changes were made to the definition of 'unfavourable recovering condition' from that set out in the Common Standards Monitoring framework (see Section 31.8). Instead of being dependent on monitoring evidence showing that an observable recovery had at least started to take place, it

[41] https://sitelink.nature.scot/

meant only that conservation measures, usually in the form of an agri-environment agreement, were in place that would be expected to achieve the desired condition. That is clearly not a realistic assumption as agri-environment schemes and other measures do not always achieve their expected outcomes. Therefore, the true proportion of SSSIs that are actually recovering in England is lower than that quoted. Nevertheless, it is clear that the PSA target in England (and similar targets elsewhere) triggered a well-directed surge in effort to address the poor condition of many SSSIs, and that undoubtedly has had substantial benefits (discussed further in Kirby *et al.*, 2010). In fact, it was probably the most important nature conservation initiative since the WCA.

PSAs were abolished in June 2010 by the newly elected Conservative and Liberal Democrat coalition. However, England's *Biodiversity Strategy* included a target of 'at least 50% of SSSIs in favourable condition, while maintaining at least 95% in favourable or recovering condition by 2020'. Neither component of the target was met by March 2021, as only 38.4% were in favourable condition and the area in unfavourable recovering condition declined after 2016 to 91.4% (DEFRA, 2021). Elsewhere, the condition of ASSIs/SSSIs remains an internal target of all the SNCBs and is a country-level indicator in Northern Ireland and Scotland. In Northern Ireland, in March 2020, 58% of ASSI biological features were reported as being in favourable condition or unfavourable recovering condition, with 54% in favourable condition (DAERA, 2021a). The NatureScot website[42] states that the proportion of natural features in 'favourable condition' on protected sites as of 31 March 2022 was 77.9%, with virtually no change since 2009. However, unlike elsewhere in the UK, this statistic actually includes the features that have an unfavourable recovering condition. Subtracting them gives 65.0% in favourable condition.

31.6 General Nature Conservation Measures

31.6.1 Development Planning, Control and Mitigation

The Treatment of Biodiversity in the Planning System

The planning system provides an important instrument for protecting biodiversity, as well as potentially enhancing and restoring it. It is a complex system that has been changed by governments many times over the years and has diverged in detail amongst the UK countries, especially since devolution. Nevertheless, across the country, a **plan-led approach** has been followed since the 1947 Town and Country Planning Act. Under this, the planning authority prepares the development plan, in accordance with national planning policy and guidance, and identifies priority areas for future development. Plans are prepared by Local Planning Authorities, or National Park authorities (or county planning authorities in Wales). Scotland also has an intermediate level of regional planning. Planning decisions are then made by the planning authority (through elected councillors) according to the plan, unless there are justifiable reasons to do otherwise. Public consultation takes place on the preparation of the plan and on actual proposals. National Significant Infrastructure Projects are treated differently in England and Wales, through a more streamlined process and national Planning Inspectorates.

The main contribution that the planning system makes to nature conservation is through the consideration of the need to protect areas of high biodiversity value when preparing the plans. As discussed in Section 31.3.2, such a requirement dates back to the legislation that created SSSIs in 1949, which included the need to notify planning authorities about them. This did not initially give them strong protection, and many damaging developments were given approval. Local Wildlife Sites, whilst not a statutory protected area, are also supposed to be protected to some extent through the planning system, but such protection has been weak, as illustrated by the loss of 130 Local Wildlife Sites and the reduction in size of 62 others in Derbyshire alone between 1984 and 2008 (Wildlife Trust evidence, cited in Lawton *et al.*, 2010).

Over time, planning policies and decisions have given greater weight to nature conservation issues. This is in part due to the increasing awareness of the social benefits of nature, as well as improved ecological impact assessments as a result of the SEA and EIA Directives, and better information (e.g. on habitat and species distributions from record centres, see Section 31.8). Impact assessments have also benefited from the development of best practices, through training and accreditation of professional ecologists (by the Chartered Institute for Ecology and Ecological Management), and the preparation of detailed practical guidance (CIEEM, 2018) and a British Standard for Biodiversity (BSI, 2013).

Planning policy has also increasingly recognised the importance of the wider wildlife site network (i.e. beyond statutory protected areas), as well as ecological network/green infrastructure components. In particular, the importance of green infrastructure in urban areas has been widely accepted for some time and incorporated into urban spatial planning guidance and support initiatives (Natural England, 2009). Some local plans have sought to protect and restore habitats that act as wildlife corridors and stepping stones (Piper *et al.*, 2006), but in practice the full potential of such opportunities has often not been realised (Lee, 2010). In response to NGO initiatives, such as the Living Landscapes programme of the Wildlife Trusts and the Lawton review, recent policy developments and plans have given greater attention to ecological network requirements and larger-scale actions. However, practical actions on the ground and biodiversity benefits have mostly been very limited. In recognition of this, the Environment Act 2021 in England has introduced a statutory requirement for local authorities, through partnerships, to

[42] https://www.nature.scot/doc/proportion-scotlands-protected-sites-favourable-condition-2022

produce Local Nature Recovery Strategies (as outlined in Section 31.3.4). As a result of the above-mentioned planning policy and legislative changes, and improved knowledge and practices, areas identified for strategic development in local plans now usually avoid protected areas, ancient woodlands and other areas of significant biodiversity importance. Proposals for developments outside the development areas that would have adverse effects on biodiversity are normally rejected or modified. However, this is not always the case. Exceptions can apply to the nature protection planning policies, and some requirements are not clearly defined, leaving local authorities considerable leeway in their interpretation. Even ASSI/SSSIs can still be threatened. Local Wildlife Sites and other ecological network elements that lack statutory protection remain particularly vulnerable to development, especially from small-scale developments or changes in land use insufficient to be considered as causing significant harm to biodiversity.

The current UK Government has indicated its desire to liberalise development planning and simplify procedures in England. A recent Environment Secretary also indicated in a 2020 speech on nature recovery that the government wishes to simplify the assessment of environmental impacts.[43] However, a radical simplification of the planning system proposed in the 2021 Planning Bill was withdrawn due to widespread public concern, including in relation to its potential environmental impacts and implications for local democracy. Despite this decision, some streamlining of the planning system seems likely in the future, at least in England.

Development Mitigation and No Net Loss/Net Gain of Biodiversity

Although biodiversity impacts as a result of housing, commerce and infrastructure developments have been longstanding and cumulatively substantial, especially outside protected areas, the UK has not had effective policies to offset residual losses (i.e. achieve at least no net loss after avoidance and mitigation). In fact, studies have shown that even intended biodiversity mitigation measures have often been poor and ineffective (Treweek and Thompson, 1997; Drayson and Thompson, 2013). Local authorities' powers to require offsetting (most notably through the 2000 CROW Act and 2006 NERC Act) have been rather weak and ambiguous. As a result, offsetting has been ad hoc and rare (except as regards HD Annex IV species). To encourage and support local authorities in using their powers to require offsets for significant harm to biodiversity, DEFRA launched a number of initiatives that are summarised in online Annex UK.7 and described more fully in Baker *et al.* (2018). These included a study of the appropriate design of an offsetting scheme (Treweek, 2009), the development of a habitat-based metric for measuring biodiversity losses and expected gains from offsets (DEFRA and Natural England, 2012), and a two-year pilot biodiversity offsetting initiative in 2012, in six areas, to test approaches and gather evidence (Baker *et al.*, 2014).

Despite the various initiatives, the scale of offsetting did not increase under this voluntary regime, which provided a strong indication that a clearer mandatory requirement was required for it to take place at scale (Ecosystem Markets Task Force, 2013; Baker *et al.*, 2014). Recognising this, and that envisaged housing and infrastructure developments will inevitably lead to further substantial unavoidable impacts, the UK Government announced in its *25 Year Environment Plan* its intention to strengthen planning policy and consult on introducing a mandatory requirement in England for developments to result in a net gain in biodiversity.

Following a consultation, provisions for achieving net gain have been included in the Environment Act (2021), with a requirement to achieve at least a 10% increase in biodiversity as measured by a revised version of the DEFRA biodiversity metric. The requirement will apply to all developments on land that require planning permission, including those approved through the Nationally Significant Infrastructure Project (NSIP) regime (e.g. major road and rail schemes, airports and ports). The legal requirement comes into force in November 2023 for all projects except NSIPs, where the requirement will be introduced no later than 2025. However, the legal requirement will only be to provide a net gain for 30 years. This is not consistent with the former offsetting design principles (DEFRA, 2011b); it severely limits the potential for creating many habitat types (unless created and banked in advance) and could be said to merely delay a development's impact. Instead, it reflects the practical need for landowners to offer sites, as few were willing to offer their land for longer-term offsetting during the pilots (Baker *et al.*, 2014), and the costs that developers are willing to meet. At the time of writing an approach was being prepared for net gain for developments within the marine environment.

Despite its limitations, the net gain policy should at least be an additional disincentive to significantly damage biodiversity, as offsets can be costly and in practice some gains will probably be retained for longer than 30 years. It is also intended that the policy will contribute to the proposed Nature Recovery Network as it has the potential to provide a substantial new funding stream, conservatively estimated to be £90–470 million per year (€103–541 million) (Duke *et al.*, 2013), although most of this will be spent on offsetting losses rather than the gain component. The location of biodiversity gains should also be informed by the development of the forthcoming Local Nature Recovery Strategies, which will help direct net gain investment so that it has the greatest benefit for local wildlife and people. Time will tell if the legislation and policy achieves its potential, as effective offsetting depends on careful regulation, high procedural standards, institutional support and enforcement (BBOP, 2012; Rayment *et al.*, 2014; Wende *et al.*, 2018). Given the current Conservative Government's preference for light touch regulation, this is

[43] www.gov.uk/government/speeches/george-eustice-speech-on-environmental-recovery-20-july-2020

not guaranteed. A further concern is that the required level of support and oversight may not be available, especially from the environmental departments of local authorities, which have a considerable role to play in delivering net gain, but have been subject to the most significant budget cuts. Whilst the government has committed to fully fund the new responsibilities placed upon authorities arising from the net gain requirement, it is uncertain what this means in practice and whether it will address the historic shortfall in capacity of ecological and planning expertise.

Marine Spatial Planning

In contrast to the UK's decentralised local-scale planning in the terrestrial environment, large-scale strategic spatial planning is applied to the marine environment. Marine planning is currently carried out under the UK Marine and Coastal Access Act 2009 (which applies in England, Wales and UK offshore waters), the Marine (Scotland) Act 2010 and the Marine Act (Northern Ireland) 2013 to support the sustainable development of marine resources through an ecosystem-based approach. To facilitate policy coherence, it is guided by a *UK Marine Policy Statement* (HMG, 2011b). Plans are being developed at the national level in Wales (by the Welsh Government) and Northern Ireland (by the Department of the Environment) and are under review in Scotland (by Marine Scotland). England has no national plan and is developing 11 regional plans under the coordination of the MMO.

The marine plans aim to guide decision making to support the achievement of environmental objectives. Each plan, which is subject to an SEA and 'plan level' appropriate assessment, requires the characterisation of the marine plan area, including trends in activities and the status of the environment, and sets the direction for the licensing and consenting of developments and regulated activities. An important element of the process is significant stakeholder engagement (compared to UK terrestrial planning) in order to help shape each plan's objectives and policies.

Unfortunately, at a time of increased demand for sea space from competing industries (GOS, 2018), the first iteration of marine plans across the UK had insufficient environmental information and consideration of the precautionary principle, and lacked the spatial element critical to ensuring industry needs could co-exist and progress sustainably in the future (WWF, 2017). Achieving the sustainable use of UK seas requires the development of a new regional marine spatial planning programme that considers MPAs, resource areas and fishing – underpinned by a clear hierarchy of policies that deliver good environmental status – which the current plans do not. Arguably, however, the biggest ongoing and future issue for marine spatial planning, especially in the North Sea, is the major and rapid escalation of offshore renewable energy development, especially wind farms, and its potential impacts on seabird populations (Green *et al.*, 2016), other marine wildlife and fisheries.[44] The challenge is how and where to deploy renewable technologies to achieve carbon reduction targets while avoiding significant risks to biodiversity of conservation concern (Gove *et al.*, 2016).

31.6.2 Habitat Management and Restoration

It is not possible to describe in sufficient detail all the various habitat management and restoration actions that have been undertaken. Instead, some actions of particular importance are described, where there are significant lessons to learn. Other important measures that could have been covered include Catchment Sensitive Farming[45] to reduce water pollution (Environment Agency, 2019), sustainable forest management through certification and the UK Standard (Forestry Commission, 2017), and IAS control and eradication (Anonymous, 2015).[46] Neither is it possible to mention all the measures that are being taken in each of the UK countries. Instead, the few examples provided here illustrate some of the most important initiatives.

Habitat Conservation and Restoration in the Uplands and Peatlands

A high proportion of the UK's most important and extensive areas for terrestrial biodiversity are in the uplands (defined as unenclosed uncultivated land usually over 240 m above sea level), and in other areas of peatlands, such as the Flow Country of the far north of Scotland. These areas include extensive areas of blanket bog, lakes, grassland, dwarf shrub heath (much of which is *Calluna*-dominated 'heather moorland'), rocky habitats, and montane habitats with Arctic tundra vegetation in the highest areas. Many of these habitats, and their characteristic species, are of international importance (Ratcliffe and Thompson, 1988; Thompson *et al.*, 1995). Despite this, they have been subject to high levels of environmental damage, including drainage, liming and conversion of marginal moorland to agricultural land; afforestation (mainly large-scale plantations of non-native coniferous species); inappropriate burning; overgrazing (mainly by sheep, but also deer in Scotland); and acidification and eutrophication from airborne deposition of sulphur and nitrogen (e.g. Brown *et al.*, 2001; Bain *et al.*, 2011; UK NEA, 2011; RoTAP, 2012; Glaves *et al.*, 2013; Martin *et al.*, 2013; Werritty *et al.*, 2015). In Scotland, between around 1947 and 1988, heather moorland declined by 23%, due mainly to afforestation and conversion to rough grassland, while blanket mire decreased by around 21%, mainly due to afforestation and drainage (SNH, 2009). The combined effects of increased sheep grazing (and less by cattle) and eutrophication have probably been the main cause of an increasing dominance of grass species over dwarf shrub species and mosses in some areas. These changes have, in turn, had

[44] www.nsrac.org/wp-content/uploads/2020/05/03-1718-Engagement-with-TenneT-Proposal-for-NSea-Wind-Power-Hub.pdf

[45] https://catchmentbasedapproach.org/

[46] www.nonnativespecies.org/home/index.cfm

detrimental impacts on many species, such as reducing the abundance and diversity of upland birds (Newton, 2020).

These activities were incentivised by CAP subsidies that were coupled to the number of livestock present, as well as grants for moorland drainage and conversion from national and EU public funds. Following the EU's switch from coupled CAP payments to mainly area-based single farm payments (in 2005), sheep numbers declined (Condliffe, 2009) by 22% over 1998–2009 in Scotland (Holland *et al.*, 2011). Afforestation was also driven to some extent from the 1950s to 1980s by financial incentives by way of tax breaks, as well as post-war forestry policy, and led to major habitat losses, most famously in the Flow Country. Over the UK, at least 9% of deep peatland was drained for forestry (Hargreaves *et al.*, 2003), whilst in Scotland it could have been as much as 17% (Vanguelova *et al.*, 2018).

Since then, agricultural and forestry policies have changed, and a large proportion of the most important upland and peatland areas have much more effective protection, especially the Natura 2000 sites. Some large-scale habitat restoration programmes have taken place, involving ditch-blocking, vegetation re-establishment (in severely eroded peatlands) and even the removal of planted trees in the most important sites. Most of this has been funded by EU LIFE projects and RDP measures, and by some national sources (e.g. UK Heritage Lottery Fund). Some restoration has been driven and funded by water companies,[47] as overgrazing and inappropriate burning management led to erosion and discoloured water quality, which is expensive to treat. However, to date, all these schemes have addressed only a relatively small proportion of the degraded habitats, and much larger scale measures are required, as set out in IUCN's 2018–2040 *UK Peatland Strategy*.[48]

There are now indications that major government-funded restoration may occur as a result of growing recognition of the natural capital and ecosystem services that uplands and peatlands provide – especially in relation to flood regulation, carbon storage, recreation and tourism (UK NEA, 2011) – the low economic value of current livestock production, and the cost-effectiveness of restoration (Bain *et al.*, 2011; NCC, 2015). Scotland's first *National Peatland Plan*[49] in 2015 sets out a long-term vision (2050) for managing and restoring peatlands through a joint approach involving landowners and managers, supported by the necessary policies and guidance. In 2020, the Scottish Government committed £250 million (€287 million) over the next 10 years, with a target of restoring 250 000 ha of degraded peatland by 2030. On 11 March 2020, the UK Chancellor of the Exchequer announced that £640 million (€736 million) will be invested in a new Nature for Climate Fund, which includes £50 million (€57 million) for the restoration of at least 35 000 ha of peatland.

Agri-environment Schemes

The UK has been a leader in the development and use of agri-environment schemes. Like many other EU countries, it has used them as the principal means of supporting conservation and restoration measures on farmland and most other semi-natural habitats, because of the high level of CAP funding that is available (see Section 31.7.3).

Following some pilots, the first schemes were introduced in the UK in 1987 and supported environmentally beneficial management within targeted Environmentally Sensitive Areas. Although these concentrated interventions helped to create landscape-scale benefits, they have been replaced with schemes that are not restricted to specific areas. This was partly to meet the political need for CAP agri-environment funding to be available to all farmers – an issue that is still driving policy today, as discussed under future developments in Section 31.9. 'Higher-level' schemes have been used to target specific habitats and species through tailored approaches and competitive applications (e.g. favouring protected areas), but their coverage has been limited as they are expensive and unpopular with farmers. To address this, 'entry-level' schemes that are open to all farmers and have simpler and less demanding requirements have also been introduced. These various schemes have led to a variety of outcomes and lessons; these cannot be fully described here, but more information is provided in online Annex UK.8, and detailed accounts are provided by Vickery *et al.* (2004), Boatman *et al.* (2008), Wilson *et al.* (2009) and Newton (2017). Dicks *et al.* (2013) summarise the effects of specific agri-environment interventions, based on studies in the UK and elsewhere in northern Europe.

Monitoring of the schemes has revealed that some have substantially helped to maintain semi-natural habitats and the populations of declining species in the UK. Most notably, the higher-level schemes have probably prevented the extinction of Stone Curlew/Eurasian Thick-knee (*Burhinus oedicnemus*), Corncrake (*Crex crex*) and Cirl Bunting (*Emberiza cirlus*) (Wilson *et al.*, 2009). In all three cases, Evans and Green (2007) consider that their recovery was dependent on three key factors: (1) a thorough diagnosis of the reasons for their population decline; (2) design and testing of the intended management measure; and (3) targeted deployment of the tested measures in the right locations and at the right scale. These and other successes were also achieved through partnerships between government agencies, NGOs and landowners, and the design of agri-environment schemes that 'fit' in well with existing farm practices (Evans *et al.*, 2002). Good-quality technical advice is also needed to ensure that farmers achieve the desired objectives (Boatman *et al.*, 2014). In more recent years, some successful landscape-scale habitat improvement measures for threatened butterflies have been implemented through higher-level agri-environment schemes (Fox *et al.*, 2015).

47 For example, United Utilities' Sustainable Catchment Management Programme (SCaMP): www.unitedutilities.com/corporate/responsibility/stakeholders/catchment-systems-thinking/catchment-management/

48 www.iucn-uk-peatlandprogramme.org/uk-strategy

49 https//bit.ly/3hRvFWg

Unfortunately, the effects of the schemes on other more widely dispersed farmland species have been less beneficial, and whilst local increases have been observed in some bird species, population-level increases have not occurred (Baker *et al.*, 2012; Bright *et al.*, 2015; Colhoun *et al.*, 2017; Walker *et al.*, 2018). These overall results may have been partly due to the entry-level scheme options being less effective than targeted and tailored measures (Still and Byfield, 2010; Pywell *et al.*, 2012). A further problem with the entry-level schemes has been that participants have tended to select the easiest options, which also happen to have the least biodiversity benefits, especially for declining farmland specialist birds that require in-field habitat improvements (Davey *et al.*, 2010).

The other main reason for the limited overall impacts of the schemes on most species has been the limited scale of coverage of the most beneficial measures (Newton, 2017). Sharps *et al.* (2023) estimated that, to increase farmland bird populations by 10% over 10 years, higher-tier agri-environment scheme coverage of 17% is required in pastoral landscapes and 34% in arable landscapes, if focussed on areas with high abundances of target species. Without targeting, the required coverage increases to 26% and 47% respectively.

In conclusion, to halt the loss of farmland biodiversity, well targeted and designed agri-environment schemes are needed over much larger proportions of the countryside – and this requires much more funding. This could be possible as a result of the UK's departure from the EU and the freedom it allows to use what were CAP direct payments for agri-environment–type schemes instead. However, as discussed in Section 31.9, it is uncertain if the funding will be sufficient.

Forest Expansion

As mentioned in Section 31.1, UK woodland cover is only about 8%, and forest land about 13%, which is amongst the lowest in Europe. Furthermore, only about half is native woodland, and 20% is semi-natural (Forestry Commission, 2017). In addition to protecting and sustainably managing existing semi-natural woodland (e.g. through protecting designated Ancient Woodland areas, the planning system and the UK Forest Standard), a key nature conservation objective is to increase the area of native woodland. Woodland expansion has high public support, and tree-planting targets have been included in recent political party manifestos, including that of the current Conservative UK Government (an additional c. 30 000 ha per year). In 2018, the Welsh Government published its *Woodlands for Wales* strategy[50] with a target of increasing woodland cover in Wales by 2 000 ha per year until at least 2030. Various public grants have been available for some time for woodland creation, and in 2011 a Woodland Carbon Code was launched to facilitate the sale of carbon credits in the UK. Despite the support and incentives, the rate of planting has remained well below that needed to reach the UK targets.

Whilst the potential contribution of forest expansion to the UK's climate targets has been long recognised, it has gained more prominence recently, such as in the UK Government's *Net Zero Strategy: Build Back Greener* (HMG, 2021). To advance this, the *England Trees Action Plan* (UK Government, 2021) committed to increasing tree-planting rates from 13 410 ha across the UK in 2020–2021, to 30 000 ha each year by the end of the Parliament. In support, more than £500 million (€575 million) of the £640 million (€736 million) Nature for Climate Fund will be dedicated to trees, with the aim of trebling woodland creation rates by the end of the Parliament. The *Net Zero Strategy* has more recently indicated that the Nature for Climate Fund will be increased to £750 million (€863 million). Whilst tree-planting certainly has an important role to play, the broader potential for nature based solutions seems to have been under-played, such as more ambitious restoration of peatland (see earlier) and other habitats (Stafford *et al.*, 2021). Furthermore, there are indications that the various grants, tax breaks and other tree-planting incentives are now leading to a rapid expansion of plantations for the commercial carbon offsetting market.

The nature conservation benefits of the planned woodland expansion are questionable, because much is targeted towards recreational, aesthetic and climate-related objectives, as reflected in the areas prioritised by the Woodland Carbon Fund in England. Planting for carbon offsetting also tends to be for fast-growing trees, including non-native species. Where native species are being planted, most, especially in England, are on former highly fertilised farmland, which constrains the ecological quality of the woodland, including its understorey plants. There is also concern over the predominance of planting rather than targeted regeneration in areas where the latter approach would be suitable (Farjon and Hill, 2019). Currently, except for a couple of schemes in Scotland, relatively little native woodland expansion is planned for upland areas, where judiciously targeted afforestation with the appropriate species could contribute more to biodiversity conservation priorities and multiple ecosystem services (e.g. water quality and reduced run-off, as well as increased carbon sequestration).

Wetland and Coastal Habitat Creation and Compensation

Current predictions of climate change and associated sea level rise, along with the growing realisation that the maintenance of flood and coastal defences is not necessarily an economically viable prospect, have meant that the traditional 'hold the line' policy is no longer officially regarded as a realistic long-term option for coastal management in some areas (Ledoux *et al.*, 2005). One response to this has been to allow controlled flooding of lowland areas of low economic value that are subject to high rates of coastal erosion, a process known as 'managed realignment'. This has provided opportunities for the restoration or creation of intertidal habitats, such as salt marsh and mudflats. To help make the most of these opportunities, including compensating for habitat losses from sea

[50] https://gov.wales/sites/default/files/publications/2018-06/woodlands-for-wales-strategy_0.pdf

level rise in Natura 2000 sites, strategic planning in England has been undertaken through a series of Coastal Habitat Management Plans,[51] which have informed Shoreline Management Plans that provide the long-term management framework for coastal flooding and erosion. These, and similar initiatives elsewhere, have stimulated and guided some large-scale and successful coastal habitat creation schemes in the UK (summarised in Miles and Richardson, 2018), such as in Wallasea Island (Ausden *et al.*, 2015b).

However, managed realignment has unavoidably led to the loss of some areas of freshwater wetland habitat and associated species, some of which are protected under the Nature Directives. This has particularly affected the Eurasian Bittern (*Botaurus stellaris*) as a high proportion of its small UK population was in a small number of affected coastal reedbeds. In response, a LIFE-funded strategic programme of reedbed improvement and creation was carried out to provide habitat for the displaced coastal population, as well as to create new 'stepping-stone' sites to link up small and isolated bittern populations elsewhere (Wotton *et al.*, 2009). More recently, some wetland creation schemes have resulted from innovative mineral planning policies and the creation of the Nature After Minerals partnership[52] between NGOs and the mineral extraction industry.

Rewilding

Over recent years there has been growing interest in some quarters in rewilding (i.e. restoring natural processes and allowing them to dominate, rather than undertaking intensive conservation management to achieve specific biodiversity targets). Whilst it has not been favoured by the SNCBs, it has been promoted most enthusiastically by a number of nature conservation commentators (e.g. Taylor, 2005; Monbiot, 2014; MacDonald, 2019) and the Rewilding Britain NGO, which formed in 2015. Rewilding ideas have mainly focussed on the idea of halting or radically reducing the intensity of farming, and in some situations replacing domestic livestock with semi-wild equivalents. Rewilding may also re-introduce 'ecosystem engineers' such as the Wild Boar (*Sus scrofa*) and Beaver, and some are promoting the re-introduction of high-level predators – most feasibly the Eurasian Lynx (*Lynx lynx*).

So far, most initiatives and proposals have focussed on the uplands because impacts on farming and food production are relatively low (being unprofitable in many areas without subsidies) and there are potentially significant gains in terms of associated ecosystem services. There have been a number of relatively small-scale pilots, such as Carrifran Wildwood in the Southern Uplands of Scotland (Ashmole and Ashmole, 2009) and Ennerdale in the English Lake District. The new Cairngorms Connect project, mentioned earlier in relation to ecological networks, also contains some rewilding elements. In mid-Wales, a larger-scale project was launched to restore some 10 000 ha of land from the Pumlumon massif moorland, through wooded valleys to the Dyfi Estuary, and out into Cardigan Bay. However, the original project (Summit to Sea[53]) received some serious objections from local farmers, who claimed that their views had not been listened to and that their livelihoods were at risk. Such problems are indicative of the controversial nature of rewilding and the claims by some that rewilding projects do not attempt to consult with or enrol local stakeholders sufficiently. To rebuild trust and support in the Pumlumon area, the Summit to the Sea project was replaced in 2022 with a new partnership and project co-designed with local people, named Tir Canol.[54]

Rewilding in the lowlands is severely hampered by the very high costs of land (and opportunity costs in most situations) and also ecological problems, such as the high nutrient status of farmed soils, hydrological limitations and habitat fragmentation. Despite such challenges, there has been one well-publicised large-scale successful case, at the 1 400 ha Knepp Estate in Sussex in southern England (Tree, 2018). The private owners decided to give up their intensive arable farming (partly due to the very heavy clay soils) and allow natural regeneration and mixed grazing by hardy traditional breeds of sheep, pigs and cattle. This has resulted in the creation of a mosaic of grassland, scrubland and open woodland habitat and a spectacular colonisation of the area by a number of threatened species, most notably the Purple Emperor (*Apatura iris*), Turtle-dove (*Streptopelia turtur*) and Common Nightingale (*Luscinia megarhynchos*). However, as there is currently no clear policy for rewilding agricultural land, it seems unlikely that such initiatives will be repeated sufficiently to make a significant UK-scale contribution to nature conservation.

31.6.3 Climate Change Adaptation

The detrimental impacts of climate change on species are expected to become more widespread and severe as ecosystem and food web disruption increases (Ockendon *et al.*, 2014) and range shifts are limited by a lack of suitable habitat, as found in butterflies in Britain (Mair *et al.*, 2014). Such constraints were included in a model-based analysis of climate impacts on 402 species, which indicated that whilst 42% may expand their range, 35% are likely to contract their range, and three-quarters of upland species are at risk from climate change (Pearce-Higgins *et al.*, 2017).

The serious long-term implications of climate change have been recognised for some time (e.g. reviewed in IACCF, 2010) and have led to initiatives to help biodiversity adapt to its direct and indirect effects. Adaptation principles and approaches have been broadly agreed (e.g. Hopkins *et al.*, 2007; Mitchell *et al.*, 2007; Smithers *et al.*, 2008). Key

[51] Supported by the Living with the Sea LIFE project.
[52] https://afterminerals.com/
[53] https://www.endangeredlandscapes.org/project/summit-to-sea/
[54] https://tircanol.cymru/

recommendations are to identify the habitats and species that are most vulnerable to climate change, and then to increase their resilience (primarily by reducing pressures and improving habitat quality) and their ability to accommodate change by helping species move to, and colonise, new areas with suitable climatic conditions (see online Annex UK.9). Overall, it is clear that the best way to assist with climate change adaptation, in terms of both resilience and accommodation, is to provide networks of good-quality habitat, with protected areas playing a fundamental role (Hopkins *et al.*, 2007; Smithers *et al.*, 2008; Lawton *et al.*, 2010; Hodgson *et al.*, 2011; Johnston *et al.*, 2013; Ausden *et al.*, 2015a). Thomas *et al.* (2012) have provided empirical evidence that protected areas facilitate species range expansions. Whilst protected area conservation objectives may need to be more flexible and dynamic, this is certainly possible within the UK's policy and legislative framework (Carey *et al.*, 2016).

In practice, there is little evidence of substantial biodiversity adaptation measures having been taken. Most notably, as discussed earlier, the recommendations of Lawton *et al.* (2010) have been little implemented. Nor has action been taken on more specific climate change adaptation recommendations that were made in a follow-up study for DEFRA (Carey *et al.*, 2016), which included suggestions for strengthening the protection of Local Wildlife Sites as a means of creating a more resilient protected area network. The SPA network also remains incomplete, despite studies that reaffirmed its importance in conserving UK bird populations in the face of climate change (Pearce-Higgins *et al.*, 2011; Johnston *et al.*, 2013). Proposed actions in the *25 Year Environment Plan* promote general restoration and connectivity through corridors, rather than prioritising making sites bigger and better, as recommended by Lawton *et al.* (2010). In conclusion, in order to significantly reduce climate change impacts on biodiversity in the UK, it appears that climate adaptation actions need to be more focussed on delivering key recommended evidence-based actions, and a step change made in their implementation.

31.7 Nature Conservation Costs and Funding

31.7.1 Nature Conservation Costs

A number of estimates have been made of the costs of achieving certain nature conservation targets in the UK. These include estimates by the SNCBs in the Prioritised Action Frameworks (PAFs) for 2015–2020[55] of annual costs of implementing the Natura 2000 network in England of £230 million (€264 million), in Scotland of £132 million (€152 million), in Wales of £136 million (€156 million) and in offshore marine sites of £0.57 million (€0.66 million), totalling £499 million (€574 million). Estimates for Northern Ireland were not included in the 2015–2020 PAF.

[55] Available at https://hub.jncc.gov.uk/assets/1b21d5b4-e87a-42db-a3af-894f10d40e4e

A study of the costs of land management requirements to achieve a broader range of environmental objectives estimated that the costs of achieving the targets would amount to £2.19 billion (€2.52 billion) annually (Rayment, 2017). Of this total, £2.10 billion (€2.42 billion) was for actions that would significantly contribute to biodiversity conservation, with the main costs being for the maintenance, restoration and expansion of UKBAP priority habitats. This is close to a previous estimate of £2 billion (€2.3 billion) per year for similar objectives (Cao *et al.*, 2009).

31.7.2 The Benefits of Nature Conservation

Although the estimated costs of nature conservation in the UK are substantial, several studies have shown that such investments provide considerable benefits that outweigh their costs. For example, a willingness to pay–based study estimated the benefits of SSSIs in England and Wales as being £956 million (€1 100 million) per annum (GHK *et al.*, 2011), compared to the estimated annual public expenditure on them at the time of £111 million (€128 million). Furthermore, the study estimated that the annual benefits would increase by £769 million (€884 million) if SSSIs were all restored to favourable condition.

A study by Christie *et al.* (2011) looked at the broader benefits of nature conservation across the country by estimating the value of ecosystem services that would be delivered by implementing the UKBAP. The annual benefits of expenditure on the UKBAP at the time were estimated at £1.36 billion (€1.56 billion), which would increase to £2.1 billion (€2.42 billion) if investment increased sufficiently to fully achieve the UKBAP targets. This is 2.5 times the annual £837 million (€963 million) costs of fully implementing the UKBAP as estimated by GHK (2010).

31.7.3 Funding Sources for Nature Conservation

The main source of funds for practical conservation measures in the UK over the last few decades was the CAP European Agricultural Fund for Rural Development (EAFRD), together with UK co-funding, as illustrated by the 2014–2020 EU budget allocations in Table 31.5. As discussed earlier, the agri-environment–climate measure was the principal means of managing semi-natural habitats, and it therefore received the largest RDP allocation over 2014–2020. It made up about 72% of the total funding allocated to Priority 4 on ecosystems, although not all of it provides significant biodiversity benefits. The agri-environment–climate measure was sometimes supported by other measures, such as M4 investments (e.g. for scrub clearance and tree-planting). The other RDP measures of potential importance for nature conservation were the forest measures (M8 and M15), which funded forest creation and the sustainable management of existing woodlands. However, their allocation was relatively small (6%) and their coverage was therefore low, except in Scotland where they aimed to cover 40% of forests.

Table 31.5 The principal funding sources for nature conservation in the UK (whilst an EU Member State).

Source and period	Measure/category	Average annual allocation (£ m)		Average annual allocation (€ m)		Comments
EU funding instruments		**EU + national**	**EU**	**EU + national**	**EU**	
EAFRD Rural Development Programme measures – allocations to Priority 4 (restoring, preserving and enhancing agriculture and forest ecosystems) 2014–2020	M10 Agri-environment–climate	397.1	309.7	456.9	356.4	
	M4 Investments in physical assets	52.2	49.3	60.1	56.7	1.
	M13 Areas facing constraints	35.0	29.3	40.2	33.7	2.
	M8 Investments in forests	26.8	22.6	30.8	26.0	3.
	M11 Organic farming	9.3	7.6	10.8	8.8	
	M15 Forest environment climate	7.5	7.0	8.6	8.0	4.
	M16 Cooperation	5.8	4.3	6.7	4.9	
	M7 Basic services, etc.	4.7	4.6	5.4	5.3	
	M2 Advisory services, etc.	4.4	3.9	5.1	4.5	Supports M10
	M1 Knowledge transfer, etc.	0.8	0.5	1.0	0.5	Supports M10
RDP Priority 4 subtotal		*543.7*	*438.7*	*625.6*	*504.8*	
LIFE Programme	Nature and biodiversity projects 1992–2020	9.0	5.3	10.3	6.1	
EU Funding instruments subtotal		*552.7*	*444.0*	*635.9*	*510.9*	
National public funds (>£1 million p.a.)		**Additional national annual***				**Comments**
		(£ m)		**(€ m)**		
Forestry Commission and NI Forest Service 2010/11		84.8		97.5		5.
SNCBs 2010/11		49.0		56.4		
Heritage Lottery Fund 1994–2017		50.0		57.5		6.
Environment Agency E & W, SEPA 2010/11		24.3		28.0		5.
Natural Environment Research Council 2010/11		7.8		9.0		5.
Landfill Community Fund 1996–2020		5.2		5.9		7.
Big Lottery Fund 2010/11		2.1		2.4		8.
Others (>£1 million p.a.)						
NGOs 2019/20		266.0		306.1		9.
Additional national subtotal		*489.1*		*562.8*		
National total including co-financing of EU instruments		*597.8*		*687.9*		
Total		**1 041.8**		**1 198.7**		

Notes. Currency conversion rate (29 May 2021) €1 = £1.151. * Excludes co-financing of EU funds unless indicated. The subtotals and totals are only approximate indications of the average annual levels of funding as they are based on annualised data from differing periods. *Comments:* 1. E.g. used to reduce pollution and support M10. 2. Uncertain contribution to nature conservation. 3. Includes forest creation. 4. E.g. measures to deal with alien species and deer control. 5. Projected BAP spending. 6. Awarded for nature and landscapes. 7. Only England and Wales data available. Some funding may be included under NGO expenditure. 8. Projected UK BAP spending. Some funding may be included under NGO expenditure. 9. Net of government funding, but some overlap with other funds.
Sources. Based on data from: RDP data from European Commission ESIF funds database (23 October 2020); ERDF and EMFF from PAFs for 2014–2020 for England (2016 version), Northern Ireland (2013), Scotland (2016) and Wales (2015) (JNCC); UK LIFE factsheet, European Commission (updated March 2021); GHK (2010) for SNCBs, Environment Agency, Forestry Commission and SEPA; Heritage Lottery Fund;[56] ENTRUST;[57] NGO = UK Biodiversity Indicator E2 (2021). See Section 1.3 for further details.

56 www.heritagefund.org.uk/our-work/landscapes-parks-nature (last accessed 5 January 2022).

57 www.entrust.org.uk/landfill-community-fund/statistics/ (last accessed 5 January 2022).

Although other EU funds, in particular the European Regional Development Fund, Social Fund and European Maritime and Fisheries Fund, could have been used to fund nature conservation–related projects, such opportunities were infrequently taken up in most of the UK. In contrast, the UK made considerable use of the EU LIFE Programmes, albeit in an uncoordinated way, with a very high share of successful applications (ICF, 2019). From its start in 1992, until the end of 2020, 82 LIFE nature and biodiversity-themed projects had been awarded in the UK, with a total co-funded value of £260 million (€299 million) and 59% EU contribution. Over the years, a large proportion of these projects focussed on habitat restoration or the removal of IAS. Although the total value of LIFE projects was relatively small compared to that of RDP funding, their targeting of habitats, species and sites of high importance that require conservation measures and their bespoke design led to a high level of cost-effectiveness. Thus, although their overall impact has not been assessed, it is clear that they had a disproportionate benefit for nature conservation, especially where they developed, trialled and refined measures that were then scaled up and secured longer-term funding (such as through RDP measures).

The second most important source of funding, after co-financed EU funds, has been from the nature NGOs. According to the UK biodiversity indicator on expenditure,[58] NGO spending on nature conservation actions (excluding administration and training) in 2019–2020 amounted to £266 million (€306 million), and had increased by 12% over the previous five years. This was about 26% of total UK nature funding. Much of this expenditure is through the use of the NGOs' own revenues, such as from membership and site entry fees, legacies, project funding appeals and sales of merchandise. Another major source of funding for NGOs, particularly for projects, is from foundations, which route most of the spending through the larger NGOs (this is therefore not double counted in Table 31.5). Detailed breakdowns of the sources and allocation of UK foundation funding are regularly provided by the Environmental Funders Network.[59]

The largest public source of funding for nature conservation in the UK is the national lottery, and in particular the National Heritage Lottery Fund, which since 1994 has invested £1.8 billion (€2.1 billion) in landscapes and nature (some of which may be included under NGO expenditure). Such projects are considered to have made a significant contribution to the implementation of the UKBAP and other strategic nature conservation priorities (Harding and Spencer, 2017).

In addition to co-funding EU funds, the SNCBs, and some other public bodies, have made some of their own funds available for nature conservation activities, such as for the management of NNRs and other protected areas where agri-environment schemes and other sources cannot be used. However, such funds are small compared to the RDP and other measures, and have declined because of austerity-related cuts. One-off projects have sometimes been funded as a result of particular initiatives, such as £7 million (€8 million) in 2012 for the Nature Improvement Areas in England. Under the Green Recovery Challenge Fund, DEFRA recently provided £80 million (€92 million) over two rounds for short-term projects to support nature and nature-based solutions and jobs. In practice, most public funding for habitat restoration or creation has come from broader environmental actions, such as coastal flood defence realignment schemes (as described earlier).

Another significant and innovative public source of funding has been the landfill tax, which was introduced in 1996 to discourage the disposal of waste to landfill. Landfill operators are able to use a proportion of their landfill tax liability for community and environmental projects in the vicinity of a landfill site. This is no longer a UK scheme, and overall UK biodiversity spending figures are not available. Spending on projects with a biodiversity objective in England and Wales has so far amounted to £134 million (€154 million), however tightening of Treasury rules on fund administration has made it much harder for smaller NGOs to access the funds, and several landfill distributors have decided to no longer accept biodiversity-focussed funding applications. With a similar rationale to the Landfill Tax, a UK Aggregates Levy was introduced in 2002. Up to 10% of the revenue generated was then allocated to an Aggregates Levy Sustainability Fund, which was used for community and biodiversity projects. For uncertain reasons, the fund was abolished in 2011.

To help bolster the limited funds available for nature conservation, some attempts have been made to develop market-based funding instruments (Comerford *et al.*, 2010; Helm, 2019). These include Payments for Ecosystem Services (PES) schemes, which aim to create revenue streams that encourage the provision of valuable but otherwise 'free' services, such as carbon sequestration and clean water. A successful example of such a scheme has been the Sustainable Catchment Management Programme (SCaMP) of the United Utilities water company, which funded moorland restoration.[60] PES schemes have been supported by DEFRA through the provision of best practice guidance and case studies (DEFRA, 2013a,b; Smith *et al.*, 2015) and a series of PES pilots (DEFRA, 2016b). Despite these efforts, it has proved difficult to apply PES schemes at a scale that will make a major contribution to biodiversity funding. However, this may change in the National Parks as a financing facility (Revere[61]) has been set up that aims to provide £240 million (€276 million) of private-sector funding by 2030 for ecosystem restoration and resulting ecosystem services.

Substantial funding for mandatory biodiversity net gain requirements for built developments in England will also

[58] https://jncc.gov.uk/our-work/ukbi-e2-biodiversity-expenditure/ (last accessed 5 January 2022).

[59] www.greenfunders.org/resources/

[60] www.unitedutilities.com/corporate/responsibility/stakeholders/catchment-systems-thinking/catchment-management/

[61] https://revere.eco/

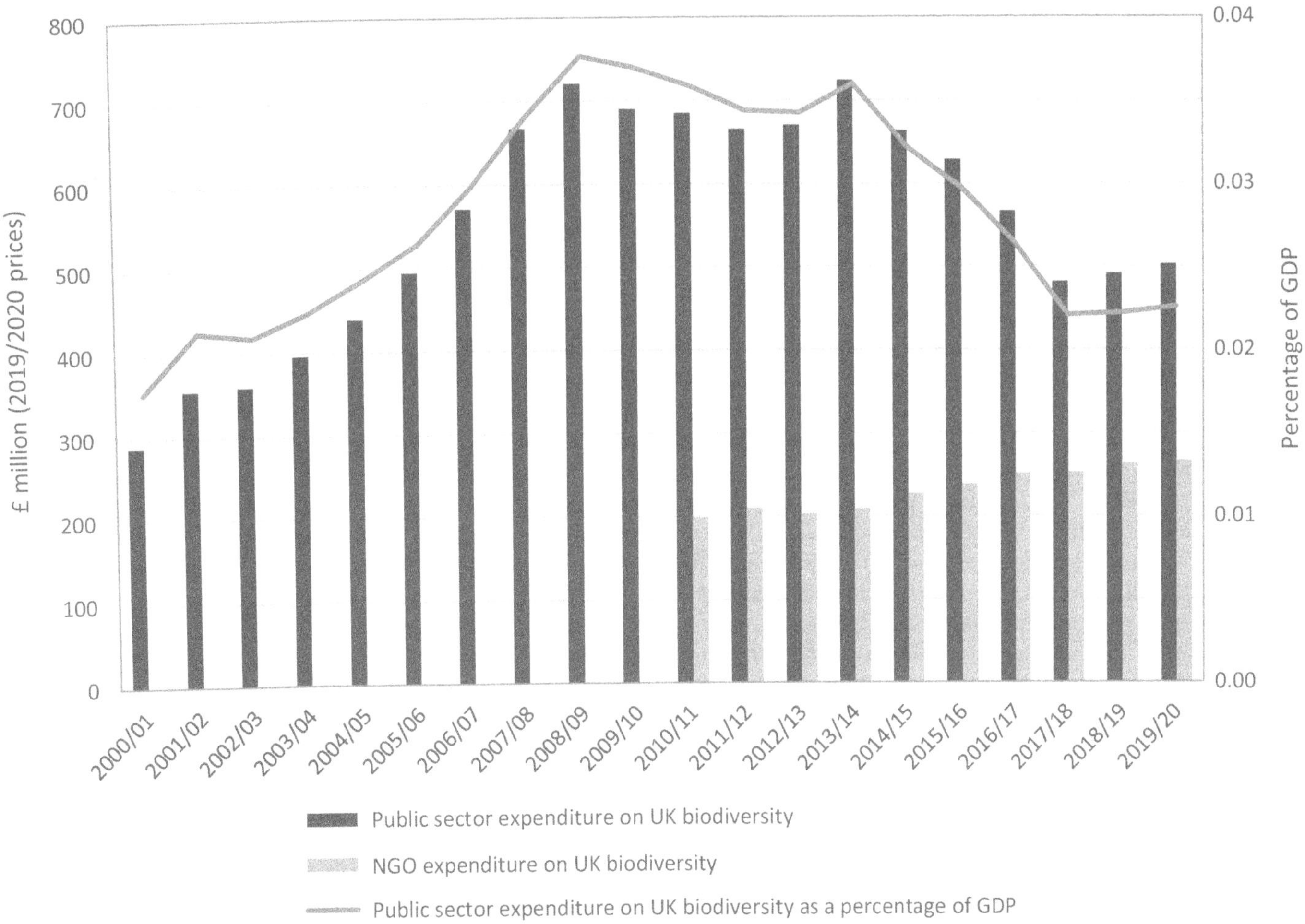

Figure 31.1 Total financial year expenditure on UK biodiversity, and public expenditure as a percentage of GDP.
Notes. Expenditure deflated using UK GDP deflator. Where possible, NGO expenditure is net of government funding.
Source. Based on DEFRA and HM Treasury data as presented in the JNCC E2 expenditure indicator https://jncc.gov.uk/our-work/ukbi-e2-biodiversity-expenditure (January 2020).

become available from November 2023, but most of this will be offsetting losses, and the actual gain requirement is only 10% and only for 30 years.

Over recent decades there have been substantial changes in the availability of public sector funding for nature conservation in the UK, as revealed by the UK biodiversity expenditure indicator (Figure 31.1). From the start of the period covered by the indicator in 2000, there were annual increases to a peak of expenditure in the financial year of 2008/2009, equivalent to £721 million (€829 million) adjusted to current prices, and 0.038% of GDP. Since then, spending has fallen almost every year, in terms of amount and percentage of GDP, to a low in 2017/2018, one-third lower than its 2008/2009 level. This was primarily the result of the spending cuts that were a response to the 2008 banking crisis. Whilst the cuts affected all government departments, they were particularly severe for DEFRA and Natural England (Marren, 2020). As a consequence, concerns were raised that Natural England may have been unable to perform its core regulatory functions (House of Lords, 2018). Similar declines in funding for SNH and other agencies in Scotland have been documented by Scottish Environment Link.[62] Perhaps as a result of such concerns, Natural England's funding more than doubled over 2019/2020–2021/2022.[63] However, part of this increase seems to be to deal with anticipated increases in housing and other development proposals. Furthermore, as indicated in Figure 31.1, the overall increase in public funding for biodiversity over the last two years over the UK has been small.

Even before the cuts in public spending, it has been widely acknowledged that there has been a significant nature conservation funding gap despite the UK's relatively high use of CAP funds and substantial funding from the NGOs and foundations. In relation to the previous 2007–2013 EU programming

[62] www.scotlink.org/funding-the-nature-and-climate-emergency-reversing-a-decade-of-austerity-for-the-environment/

[63] www.gov.uk/government/publications/natural-england-action-plan-2021-to-2022/natural-england-action-plan-2021-to-2022

period, the funding gap was estimated to be £273 million per year (€314 million), but only with respect to UKBAP target costs (GHK, 2010). According to the more recent estimate by Rayment (2017) of the annual costs of broader biodiversity objectives of £2.10 billion (€2.42 billion), and the approximate estimate of £1.04 billion (€1.20 billion) available funding per year (Table 31.5), it appears that the funding gap is at least £1 billion per year. This is almost certainly a significant underestimate, because the main source of funding is the RDP measures, and not all of their spending will significantly contribute to nature conservation objectives.

As a result of Brexit, fundamental changes are taking place in the financing of nature conservation the implications of which are currently uncertain. It has been expected that funding for nature conservation through land management schemes would increase as the equivalent of previous CAP direct payments is increasingly used to provide public goods, including biodiversity. However, as discussed later in Section 31.9, there are indications that much of the funding will be used for rather general environmental measures, leaving less for more focussed and ambitious nature conservation actions. It is also unknown whether more biodiversity-focussed projects such as those under the LIFE Programme will continue, and whether they will receive similar levels of funding. There would therefore be a major funding gap if such projects were not replaced, irrespective of the outcome of the envisaged land management schemes (ICF, 2019).

31.8 Research and Monitoring

One of the undoubted strengths of UK nature conservation is the amount of information that is available on ecosystems, habitats and species, in terms of their location, status and trends and ecological relationships (see Preston *et al.* (2012) for a review of 50 years of biological recording). Whilst habitats have not been mapped as thoroughly as in some other countries, the UK has perhaps the best information available on species, largely as a result of the high number of keen amateur naturalists. Some ad hoc species sightings are reported directly to Local Environmental Record Centres, but most information is obtained through surveys and monitoring schemes coordinated by natural history societies and other NGOs (sometimes with government support), some with the assistance of the Biological Records Centre, set up in 1964. According to Preston *et al.*, in 2012 there were 80 recording schemes and societies. Huge amounts of unpaid time are put into these surveys by the volunteer observers that undertake them (which are not included in the estimates of nature conservation costs described earlier).

In recent times, the wider public have been increasingly involved in data gathering through citizen science surveys, such as of birds and butterflies and the spread of IAS (Dickinson *et al.*, 2012). To coordinate and link the activities of the various record centres, schemes and societies, the National Biodiversity Network was established in 2000. This compiles and integrates most biodiversity data and makes them available through an online searchable database (with 204 million occurrence records in May 2022).[64]

As a result of such efforts, national distribution atlases (usually based on 10 × 10 km grids) have been published for all vertebrates and vascular plants, many invertebrates, and bryophytes, lichens, charophytes and seaweeds. Furthermore, some of these have been repeated, allowing quantitative assessments to be made of changes in distribution range. In the case of birds, three national atlases have been produced, with the last two providing spatial abundance estimates, as well as occurrence data (Gibbons *et al.*, 1993; Balmer *et al.*, 2013). Many regional and county atlases, and county floras, have also been produced, often using finer-scale 2 × 2 km squares. In addition, systematic sample-based monitoring (i.e. surveillance) schemes have also been established for some time, including for birds, butterflies, moths and mammals. A National Plant Monitoring Scheme has also been running since 2015.

The available biodiversity information has formed the basis for identifying conservation priorities (e.g. Red Lists and BAP priority habitats and species), detecting changes and their causes, setting targets, and assessing performance (including through formal indicators[65]). Spatial data on species occurrence also play a key role informing SEAs and EIAs, and targeting agri-environment schemes. However, the use of biodiversity in planning is hampered by a lack of suitable and up-to-date habitat maps for most of the UK, including protected areas. Some local authorities have therefore carried out their own habitat mapping (e.g. in Warwickshire) and used the information to identify important areas for biodiversity, green infrastructure and the potential for restoration through offsetting.

The monitoring of SSSIs/ASSIs, Natura 2000 sites and Ramsar Sites is primarily carried out by SNCB staff, using a cost-effective standardised system that assesses the condition of features in each site (or subunit) according to a Common Standards Monitoring framework (JNCC, 1998, 2022; online Annex UK.10). Each site is supposed to be assessed at least every six years, but due to funding cuts the frequency of site assessments has declined. For example, in England nearly half (47%) of SSSIs have not been assessed in the last six years.[66]

The UK also has a long history of carrying out ecological and applied research of high relevance to nature conservation through research institutes, such as the Centre for Ecology and Hydrology, universities and some NGOs (e.g. the British Trust for Ornithology and the RSPB). Knowledge-sharing initiatives

64 https://nbnatlas.org/

65 For example, see https://jncc.gov.uk/our-work/uk-biodiversity-indicators/

66 https://unearthed.greenpeace.org/2018/09/07/half-england-sssi-sites-not-monitored/

and networks have been established to identify research priorities (e.g. Sutherland *et al.*, 2006), to develop collaborative work and to disseminate results beyond academic journals. In this respect, a particularly innovative initiative has been the establishment of conservationevidence.com. This provides online summaries of the observed effects of specific conservation interventions from scientific studies, as well as synopses of the evidence relating to topical subjects (e.g. farmland conservation and peatland conservation).

31.9 Future Developments

31.9.1 Plans, Targets and Funding

Despite the recent Environment Act (2021), at the time of finalisation of this chapter in 2022 there remains considerable uncertainty over the future of nature conservation in the UK – largely due to contrasting high ambitions and weak proposed measures to achieve them. On the one hand, a step change in nature conservation and restoration aims appears to have been signalled in the *25 Year Environment Plan*, the 2030 biodiversity target, the signing of the Leaders' Pledge for Nature, the '30 by 30' protected area commitment, the G7 agreement to 'halt and reverse biodiversity loss' (i.e. become Nature Positive) by 2030, and various other UK and country-level announcements on ecosystem restoration. On the other hand, as concluded by the EAC (2021), whilst the policies set out in the *25 Year Environment Plan* and Environment Bill (now Act) are a welcome start, they 'do not represent the transformative change required to bend the curve of biodiversity loss'. Instead, 'Action needs to be stepped up in scale, ambition, pace and detail.' The OEP (2022a) also concluded that implementation of the plan has been slow. Similarly, whilst some measures to achieve the UK's biodiversity targets have been identified by the SNCBs (Brotherton *et al.*, 2021), they conclude that 'Becoming Nature Positive by 2030 requires significantly greater action and investment in nature now.'

The achievement of the UK and country targets will therefore be dependent on the development of further policy measures and delivery mechanisms. In England this will include the 2023 EIP update, which will build on the recent publication of, and consultations on, the *Nature Recovery Green Paper* (DEFRA, 2022b) and finalised legally binding environmental targets (DEFRA, 2022c), as required under the Environment Act. The Green Paper proposals are not described in detail here, because they are not well developed and may be changed considerably. Indeed, some rethinking seems necessary as the proposals are widely considered to be insufficient to achieve the ambitions described above and instead focus more on simplifying regulations and measures (e.g. in relation to protected area types, species protection, the Habitats Regulations Assessment and governance). The merger, or other reforms, of Natural England and the other statutory environmental bodies is also put forward as an idea for consideration. Not surprisingly, conservation NGOs are concerned by this focus, stating that 'there is no time to waste with reform' and that it is more important to build on what exists (WCL, 2022a). Whilst the OEP (2022b) recognises that there is some logic to replacing the current mix of domestic and EU derived legislation, it also considers that it could be counter-productive. Consequently, in addition to detailed recommendations, it advises the government that 'reforms should only be pursued that will clearly improve on current protections and support nature's recovery at pace'.

The environmental targets are more positive, in accordance with the claim by the then Secretary of State, George Eustice, that they will aim to 'halt the decline in nature', and therefore be a 'world leading measure'.[67] The main biodiversity targets for 2042 are to ensure that species abundance is higher than in 2022, and at least 10% higher than in 2030;[68] improve the Red List Index for species extinction risk compared to 2022;[69] create at least 500 000 ha of wildlife-rich habitat outside protected areas; and have 70% of designated features (c. 150 types of habitat and species) in the MPA network in favourable condition. However, most of the targets are widely considered to lack ambition, including by the OEP (2022c), and there are clear weaknesses and gaps. Most obviously, if species abundance continues to decline until 2030 (as expected), species abundance in 2042 could be little more than current levels. Furthermore, the habitat restoration target is not as ambitious as in the *25 Year Environment Plan* as it is no longer focussed on priority habitats, and some 'wildlife-rich habitats' as defined for the target can be of low biodiversity value.

The most critical factor that will probably affect the implementation of the *25 Year Environment Plan* (as updated in 2023), and other country initiatives, will be the level of funding that is provided. In this respect there are mixed signals. Encouraged by the HM Treasury review of the economics of biodiversity (Dasgupta, 2021; HM Treasury, 2021) and increasing recognition of the role that nature-based solutions can contribute to meeting the UK's *Net Zero Strategy* for greenhouse gas emissions, there have been some significant nature conservation and broader ecosystems restoration funding announcements. These include the large increase in the budget for Natural England (as mentioned) and an additional £150 million (€173 million) to the £650 million (€748

[67] www.gov.uk/government/speeches/environment-secretary-speech-at-delamere-forest-on-restoring-nature-and-building-back-greener

[68] Based on a set of over 1 100 species across all habitat types, including birds, bats, butterflies, moths, freshwater invertebrates and plants, excluding marine species except marine birds (due to inadequate reliable data).

[69] Compared to 2022, based on Red List assessments of 6 500 species, including birds, mammals, reptiles, amphibians, some invertebrates, vascular plants, bryophytes, lichens and some fungi.

million) announced for the Nature for Climate Fund. Of concern is that government ministers have announced to the media that they wish to cut civil service numbers back to 2016 levels, the lowest level since World War II, and equivalent to losing one in six jobs.[70] Based on previous cuts to public services, those related to the environment are at risk of being disproportionately targeted.

There are also concerns over the lack of a replacement for LIFE nature-type projects, and in particular whether the anticipated redirection of CAP direct payments will actually be complete and provide the equivalent level of funding for environmental land management over the long term, which is needed to meet the stated biodiversity ambitions. All the UK countries are currently going through agricultural policy transition. This is most advanced in England, with the Agriculture Act (2020) and the current development of the Environmental Land Management scheme that will make payments to farmers for public goods (including biodiversity) as direct payments are gradually phased out by 2008. Whilst the government has committed to spend £2.4 billion (€2.8 billion) each year on the agricultural schemes during this Parliament, the future level of funding is unknown and the definition of public goods is broad. Furthermore, there is no regulatory baseline equivalent to CAP cross-compliance conditions on payments. This could undermine the 'polluter pays' principle and result in a high proportion of future funding being spent on achieving environmental standards that operated as regulatory requirements under the CAP (Baldock and Hart, 2020).

At the time of writing, the development of the three Environmental Land Management schemes was incomplete, with the Sustainable Farming Initiative, which focusses on soil health, being most advanced. This is open to all farmers and will provide payments to those who choose to comply with one of three levels of soil standards. Reaction to it from nature conservationists has been that it shows a 'shocking lack of ambition'.[71] The House of Commons Environment, Food and Rural Affairs Committee (2021) also had concerns that simplicity and high uptake seem to have been prioritised at the risk of providing limited benefits, as was the case with entry-level agri-environment schemes. Announcements by DEFRA on the plans for the other two schemes, Local Nature Recovery and Landscape Recovery, were more encouraging as they provide funding for measures that have the potential to significantly improve and restore habitats (including through large-scale and long-term projects) and tackle declining species. With sufficient take-up, DEFRA estimates that by 2042 the schemes have the potential to create or restore up to 300 000 ha of habitat and bring over half of SSSIs into favourable condition. However, the proposals currently lack detailed objectives and measures, and it is uncertain whether the schemes will actually have the budget, payment rates and uptake to achieve their objectives over the required scale and timetable.

The achievement of environmental objectives for the marine environment is dependent on the UK's Fisheries Act 2020 and the related negotiations with the EU over fishing rights. The Fisheries Act is underpinned by a commitment to sustainability with legally binding objectives for fishing to be environmentally sustainable and to take an ecosystem approach in order to minimise, and where possible reverse, negative effects on the environment. There is also a specific by-catch objective to tackle the incidental capture of undersized fish and marine wildlife. To help achieve these objectives, the MMO has been given by-law–making powers to prohibit or restrict fishing in offshore English waters, both inside and outside MPAs (with equivalent powers for Scottish ministers). As part of the Fisheries Act, the UK fisheries policy authorities are required to develop a Joint Fisheries Statement to set out the policies for achieving the fisheries objectives, which will also include a framework for working with international partners for the sustainable management of shared fishing resources. However, the Act lacks the type of legal sanctions included in the CFP that are required to ensure authorities meet their overall objectives.

31.9.2 Longer-Term Divergence from the EU and within the UK

So far, changes to nature conservation legislation have been rather limited since Brexit, with the key elements of the Nature Directives and other EU legislation retained in the amended Habitats Regulations. Over the longer term, further changes may be likely as the UK Government stated in the *25 Year Environment Plan* that Brexit 'presents a unique opportunity to set in motion the behavioural and institutional changes necessary to build sustainable, enduring growth as well as improvement in our well-being'. More recent UK Government announcements have given further indications of likely deregulation and divergence. As a result, there is particular alarm amongst environmental groups over the proposed Retained EU Law (Revocation and Reform) Bill, as it has a clear deregulatory purpose and proposes to give ministers considerable powers to scrap or change EU laws with little parliamentary scrutiny (WCL, 2022).

There is already concern that the Environment Act includes provisions that allow the Secretary of State to amend the Habitats Regulations, which have resulted in the Green Paper proposals to simplify Habitats Regulations Assessments. There is also the potential to alter the lists of habitats and species that receive protection, which were based on the annexes of the Nature Directives. Whilst further legislative changes that greatly weaken nature conservation might not be considered politically acceptable, the same effect might be achievable more subtly.

Without the common need to comply with EU targets, there may also be further divergence in the policies and

[70] www.bbc.co.uk/news/uk-politics-61432498

[71] The Wildlife Trusts, the National Trust and the RSPB www.wildlifetrusts.org/news/brexit-farming-promises-broken

legislation of the Devolved Administrations. The Welsh Government's environmental agenda was already taking a different direction by the time of the EU referendum in 2016, with policy and new underpinning legislation increasingly focussed on broader sustainability concepts. At the time of writing, the Welsh Government is still considering arrangements for environmental governance in Wales and whether to create something similar to the OEP in England. New legislation covering agriculture and fisheries in Wales is also being developed.

In Northern Ireland, consultation has taken place on the *Draft Environment Strategy* (DAERA, 2021b), the first EIP as required under the UK Environment Act. The Strategy proposes a wide range of interrelated actions and targets, including a new Biodiversity Strategy for 2030 linking to the CBD, nature recovery plans and actions plans, increasing protected areas to 30% on land and at sea by 2030, increasing favourable conservation status within protected sites and restoring habitat in the wider environment. However, the document lacks detail, and NGOs consider that it is unambitious, has insufficient specific, time-bound and binding targets and actions, and lacks a clear indication that it will be adequately resourced and monitored.[72]

Following a successful 'Fight for Scotland's Nature' campaign, led by ScotLink, the Scottish Government has also committed to produce a Natural Environment Bill in 2023/2024 that, if passed by the Scottish Parliament, will set legally binding nature recovery targets. In the meantime, Scottish ministers have expressed their commitment to protecting Scotland's position as a climate change leader and to continuing to contribute to EU-wide environmental policies and standards. The First Minister has established an Environmental Council to advise Scottish ministers on international best practice, with a view to ensuring that Scotland does lead in response to the climate and nature crises.

31.10 Conclusions

The UK has been a leader in many aspects of nature conservation in Europe, especially in terms of the involvement of nature conservation NGOs, which have driven policy and legislation, assisted enforcement and managed sites for nature conservation. They have also played a fundamental role in establishing detailed and wide-ranging scientific information on species – possibly the best in the world. The difficult issue of conserving biodiversity on private land has also been tackled to some extent, through the progressive development of protected area legislation, combined with softer nature conservation agency-led voluntary approaches and persuasion, backed up with financial incentives (most notably agri-environment schemes). Following the 'Earth Summit' in 1992, the UK was one of the first signatories to the CBD to produce a detailed and ambitious biodiversity action planning framework. In recent years it has championed the concepts of ecosystem services, carried out a detailed national ecosystem assessment, and established the first natural capital committee in the world.[73]

However, although the UK has prepared many useful studies, strategies and plans, it has been weak in their implementation. This is particularly the case with the designation of biodiversity-focussed protected areas, as the proportional coverage of the country is amongst the lowest in Europe. Progress in increasing their coverage, protection and management in recent decades has largely been driven by EU legislation, especially in the marine environment (with additional pressure from NGO campaigns). Inadequate public funding has been a significant constraint on nature conservation, despite the UK's relative wealth. This problem worsened after 2010 due to government cuts, which disproportionately affected and undermined the effectiveness of the environmental authorities. Moreover, the public funding of biodiversity strategies and initiatives, such as the Nature Improvement Areas and the ecological network in Scotland, has been nowhere near sufficient to achieve their objectives. Whilst NGO funding and volunteer actions for nature conservation have been substantial and have increased, they have not made up for the inadequate public funding. In the last couple of years there has been an increase in public funding for the environment, especially for tree-planting, but there is little indication that it will grow further to fill the existing funding gap, let alone be sufficient to meet the UK's new ambitions.

Other obstacles to progress have included government antipathy to regulation, especially over the last decade, and an over-reliance on voluntary approaches, even though evidence indicates that they are often ineffective (McCarthy and Morling, 2015). The independence and powers of the SNCBs have also declined, as have their policy influence and enforcement roles and capacity. NGOs have been reluctant to campaign forcefully on such issues.

The net effect of nature conservation in practice has, over the last 40 or so years, led to successes in halting and reversing declines in some habitats, especially wetlands and their associated species, and recoveries of some other species (e.g. several raptors and a few scarce farmland birds). However, these successes are limited, and many increases in species are largely the result of NGO management in reserves or targeted actions elsewhere for restricted range species. Broader policy measures for the wider environment have been less successful, and consequently most semi-natural habitats and species of EU and national conservation importance show ongoing and

[72] www.rspb.org.uk/about-the-rspb/about-us/media-centre/press-releases/rspb-ni-open-letter-to-first-minister-and-deputy-first-minister-calls-out-unambitious-environment-strategy/

[73] Which ran from 2012 to December 2020. Many of its functions have now been transferred to the OEP.

significant declines, as do many more common species (e.g. many pollinating insects, farmland and forest birds and seabirds).

Since the UK decided to leave the EU, some aspects of the environmental agenda have been more positive, particularly in relation to climate objectives, but also biodiversity. In England this mainly revolves around DEFRA's updated *25 Year Plan for the Environment* and the Environment Act (2021), which is key to implementing the Plan as well as setting the legally binding targets discussed previously. Similar policies are being developed in Northern Ireland, Scotland and Wales, but their ambition and approaches are less certain at the moment. However, there are concerns that the legislation and other measures, institutions and funding may not be sufficient to meet the ambitions. The UK may once again fall down on its implementation. Implementation may also be weakened by responses to the COVID-19 pandemic, including the Government's emphasis on building and deregulation as a route to economic recovery.[74] More recently, political attention has turned to energy and food security and prices – heightened by the war in Ukraine. This may also undermine commitment to environmental objectives and actions, and encourage further weakening of environmental regulation, especially without the influence of the EU's environmental legislation and oversight. The future will therefore provide an interesting comparative experiment to see whether nature conservation in the UK benefits or suffers from the direct and indirect effects of Brexit, and what will transpire from the diverging approaches in different parts of the UK.

Acknowledgements

This chapter has benefited greatly from the thoughtful and constructive comments that we received from additional contributors and numerous reviewers. We therefore thank them all for this, especially as it coincided with the challenging time of the COVID-19 outbreak. We are particularly grateful for the contributions that we received from Andrew Baker on the UK Habitats Regulations Assessments, Nicola French on managed realignment, Donal Griffin, Bob Earl and Euan Dunn on some marine issues, and Kate Jennings on protected areas. Valuable comments were received on drafts of the chapter from Mark Avery, David Baldock, Richard Benwell, Peter Brotherton, Pete Chaniotis, Adam Cole-King, Stephen Grady, Martin Harper, Kaley Hart, David Hill, Ian Jardine, Clunie Keenleyside, Peter Marren, Will Peach, Matt Rayment, Matt Shardlow, David Stroud and James Williams.

References

Aebischer, N. J. (2019) Fifty-year trends in UK hunting bags of birds and mammals, and calibrated estimation of national bag size, using GWCT's National Gamebag Census. *European Journal of Wildlife Research*, **65** (4), 1–13.

ALGE (2013) *Implications of the Comprehensive Spending Review on biodiversity work within local government. Main Findings. Financial Year 2011 to 2012.* London, Association of Local Government Ecologists.

Anonymous (1994) *Biodiversity: the UK action plan.* London: HMSO.

(2015) *The Great Britain Invasive Non-native Species Strategy.* York, UK: The Great Britain Non-native Species Secretariat.

Ashmole, M. & Ashmole, P. (2009) *The Carrifran Wildwood Project. Ecological Restoration from the Grassroots.* Jedburgh, UK: Borders Forest Trust.

Ausden, M., Bradbury, R., Brown, A., Eaton, M., Lock, L. & Pearce-Higgins, J. (2015a) Climate change and Britain's birdlife: what might we expect? *British Wildlife*, **26**, 161–174.

Ausden, M., Dixon, M., Fancy, R. *et al.* (2015b) Wallasea: a wetland designed for the future. *British Wildlife*, **26**, 382–389.

Avery, M. (2015) *Inglorious. Conflict in the Uplands.* London: Bloomsbury.

Bailey, J. J., Cunningham, C. A., Griffin, D. C. *et al.* (2022) *Protected Areas and Nature Recovery. Achieving the Goal to Protect 30% of UK Land and Seas for Nature by 2030.* London, UK: British Ecological Society.

Bain, C. G., Bonn, A., Stoneman, R. *et al.* (2011) *IUCN UK Commission of Inquiry on Peatlands.* Edinburgh: Peatland Programme.

Baker, D. J., Freeman, S. N., Grice, P. V. & Siriwardena, G. M. (2012). Landscape-scale responses of birds to agri-environment management: a test of the English Environmental Stewardship scheme. *Journal of Applied Ecology*, **49**, 871–882.

Baker, J., Papadopoulou, L. & Sheate, W. (2018) United Kingdom. In *Biodiversity Offsets – European Perspectives on No Net Loss of Biodiversity and Ecosystem Services*, eds. W. Wende, G. Tucker, F. Quétier, M. Rayment & M. Darbi, pp. 211–239. Cham, Switzerland: Springer.

Baker, J., Sheate, W. R., Bennett, T. *et al.* (2014) *Evaluation of the Biodiversity Offsetting Pilot Programme.* London: Collingwood Environmental Planning & IEEP Report to DEFRA.

Baldock, D. & Hart, K. (2020) *Risks and opportunities of a post-EU environmental regulatory regime for agriculture in England.* London and Brussels: Institute for European Environmental Policy.

Balmer, D. E., Gillings, S., Caffrey, B. J., Swann, R. L., Downie, I. S. & Fuller, R. J. (2013) *Bird Atlas 2007–11: The Breeding and Wintering Birds of Britain and Ireland.* Thetford, UK: BTO Books.

BBOP (2012) *Standard on Biodiversity Offsets.* Washington, DC: Business and Biodiversity Offsets Programme (BBOP).

Beaugrand, G. (2004) The North Sea regime shift: evidence, causes, mechanisms and consequences. *Progress in Oceanography*, **60**, 245–262.

Beaugrand, G., Edwards, M., Brander, K., Luczak, C. & Ibanez, F. (2008) Causes and projections of abrupt climate-driven ecosystem shifts in the North Atlantic. *Ecology Letters*, **11**, 1157–1168.

Beaugrand, G., Reid, P. C., Ibañez, F., Lindley, J. A. & Edwards, M. (2002) Reorganization of North Atlantic marine copepod biodiversity and climate. *Science*, **296**, 1692–1694.

Beebee, T. (2015) The Great Crested Newt: an ongoing conservation dilemma. *British Wildlife*, **26**, 230–236.

[74] See, for example, Boris Johnson www.gov.uk/government/speeches/pm-economy-speech-30-june-2020

Benyon, R., Barham, P., Edwards, J. *et al.* (2020) *Benyon Review into Highly Protected Marine Areas – Final Report.* DEFRA.

Biodiversity Indicators Partnership (2007) *Biodiversity – The UK Approach.* London: Department for Environment, Food and Rural Affairs.

Boatman, N., Ramwell, C., Parry, H. *et al.* (2008) *A review of environmental benefits supplied by agri-environment schemes.* York/Cheltenham, UK: Central Science Laboratory/Countryside and Community Research Institute.

Boatman, N., Short, C., Elliott, J. *et al.* (2014) *Agreement scale monitoring of Environmental Stewardship 2013–14 – assessing the impact of advice and support on the environmental outcomes of HLS agreements.* Natural England contract LM0432.

Bradbury, G., Shackshaft, M., Scott-Hayward, L. *et al.* (2017) *Risk Assessment of Seabird Bycatch in UK Waters.* Slimbridge, UK: Wildfowl & Wetlands Trust. Report to DEFRA.

Bright, J. A., Morris, A. J., Field, R. H. *et al.* (2015) Higher-tier agri-environment scheme enhances breeding densities of some priority farmland birds in England. *Agriculture, Ecosystems & Environment*, **203**, 69–79.

Brotherton, P., Anderson, H., Galbraith, C. *et al.* (2021) *Nature Positive 2030 – Evidence Report.* Peterborough, UK: Joint Nature Conservation Committee.

Brown, A., Hall, R. & Townshend, D. (eds.) (2001) *State of Nature: The Upland Challenge.* Peterborough, UK: English Nature.

Bruce-White, C. & Shardlow M. (2011) *A Review of the Impact of Artificial Light on Invertebrates.* Peterborough, UK: Buglife-The Invertebrate Conservation Trust.

BSI (2013) BS 42020:2013. *Biodiversity. Code of practice for planning and development.* British Standards Institution.

Burns, F., Eaton, M. A., Barlow, K. E. *et al.* (2016) Agricultural management and climatic change are the major drivers of biodiversity change in the UK. *PLOS ONE*, **11**, e0151595.

Burns, F., Eaton, M. A., Gregory, R. D. *et al.* (2013) *State of Nature report.* The State of Nature partnership.

Cameron, R., Burgess, K. & Payne, B. (2019) The rationale for Great Crested Newt district level licensing. *In Practice: Bulletin of the Chartered Institute of Ecology and Environmental Management*, **103**, 22–27.

Cao, Y., Elliott, J., McCracken, D., Rowe, K., Whitehead, J. & Wilson L. (2009) *Estimating the Scale of Future Environmental Land Management Requirements for the UK.* ADAS & SAC Report for the Land Use Policy Group.

Capuzzo, E., Lynam, C. P., Barry, J. *et al.* (2018). A decline in primary production in the North Sea over 25 years, associated with reductions in zooplankton abundance and fish stock recruitment. *Global Change Biology*, **24**, e352–e364.

Carey, P. D., Griffiths, G. H., Vogiatzakis, I. N. *et al.* (2016) *Priority Habitats, Protected Sites and Climate Change: Three Investigations to Inform Policy and Management for Adaptation and Mitigation.* London: DEFRA. DEFRA CR0439.

CEP (2015) *Monitoring and Evaluation of Nature Improvement Areas: Final Report (2012–15).* London: Collingwood Environmental Planning. Defra Research Project WC1061.

Chaniotis, P., Cioffi, B., Farmer, R., Cornthwaite, A., Flavell, B. & Carr, H. (2018) Developing an ecologically-coherent and well-managed Marine Protected Area network in the United Kingdom: 10 years of reflection from the Joint Nature Conservation Committee. *Biodiversity*, **19** (1–2), 140–147.

Christie, M., Hyde, T., Cooper, R. *et al.* (2011) *Economic Valuation of the Benefits of Ecosystem Services Delivered by the UK Biodiversity Action Plan. DEFRA Project: NE0112.* Aberystwyth, UK: Aberystwyth University Report to Defra.

CIEEM (2018) *Guidelines for Ecological Impact Assessment in the UK and Ireland: Terrestrial, Freshwater, Coastal and Marine Version 1.1.* Winchester, UK: Chartered Institute of Ecology and Environmental Management.

CJC Consulting, Tucker, G., Willis, K. & Doyle, P. (2004) *Cost Effectiveness Study of Approaches for Delivery of PSA Target Relating to SSSIs.* Oxford, UK: CJC Consulting.

Clark, R., Humphreys, J., Solandt, J.-L. & Weller, C. (2016) Dialectics of nature: the emergence of policy on the management of fisheries in English European Marine Sites. *Marine Policy*, **78**, 11–17.

Clutten, R. & Tafur, I. (2012) Are imperative reasons imperilling the Habitats Directive? An assessment of Article 6(4) and the IROPI exception. In *The Habitats Directive – A Developer's Obstacle Course?*, ed. G. Jones, pp. 167–182. Oxford, UK: Hart Publishing.

Cole-King, A. (1995) Marine protected areas in Britain: a conceptual problem? *Ocean & Coastal Management*, **27**, 109–127.

Colhoun, K., Mawhinney, K., McLaughlin, M. *et al.* (2017) Agri-environment scheme enhances breeding populations of some priority farmland birds in Northern Ireland. *Bird Study*, **46**, 545–556.

Comerford, E., Molloy, D. & Morling, P. (2010) *Financing Nature in an Age of Austerity.* Sandy, UK: RSPB.

Condliffe, I. (2009) Policy change in the uplands. In *Drivers of Environmental Change in Uplands*, eds. A. Bonn, T. Allott, K. Hubacek & J. Stewart, pp. 59–89. London: Routledge.

Cranbrook, Earl of. (2001) Fifty years of statutory nature conservation in Great Britain. In *The Changing Wildlife of Great Britain and Ireland*, ed. D. L. Hawksworth, pp. 1–22. London and New York: Taylor & Francis.

Crick, H. Q. P., Crosher, I. E., Mainstone, C. P. *et al.* (2020) *Nature Networks Evidence Handbook.* York, UK: Natural England, Research Report NERR081.

Crofts, R., Dudley, N., Mahon, C. *et al.* (2014) *Putting Nature on the Map: A Report and Recommendations on the Use of the IUCN System of Protected Area Categorisation in the UK.* IUCN National Committee UK.

DAERA (2021a) *Northern Ireland Environmental Statistics Report. May 2021.* Belfast: Department of Agriculture, Environment and Rural Affairs. And data tables at: www.daera-ni.gov.uk/publications/northern-ireland-environmental-statistics-report-2021

(2021b) *Draft Environment Strategy for Northern Ireland.* Belfast: The Department of Agriculture, Environment & Rural Affairs.

Daly, E., Fenn, T. & Miller, J. (2016) *Evaluation of the Compliance and Quality of Biodiversity Duty Reports 2015. Final Report.* Edinburgh: The Scottish Government.

Dasgupta, P. (2021) *The Economics of Biodiversity: The Dasgupta Review.* London: HM Treasury.

Davey, C. M., Vickery, J. A., Boatman, N. D., Chamberlain, D. E. Parry, H. R. & Siriwardena, G. M. (2010) Assessing the impact of Entry Level Stewardship on lowland birds in England. *Ibis*, **152**, 459–474.

Davies, S., Hughes, J. & Wilson, B. (2018) Creating a National Ecological Network in Scotland. *British Wildlife*, **29**, 411–417.

De Santo, E. M. & Jones, P. J. S. (2007) Offshore marine conservation policies in the North East Atlantic: emerging tensions and opportunities. *Marine Policy*, **31**, 336–347.

DEFRA & Natural England (2012) *Biodiversity Offsetting Pilots. Technical Paper: The metric for the biodiversity offsetting pilot in England.* London: DEFRA.

DEFRA (2002a) *Working with the Grain of Nature: A Biodiversity Strategy for England.* London: Department for Environment, Food and Rural Affairs (DEFRA).

(2002b) *Safeguarding Our Seas – A Strategy for the Conservation and Sustainable Development of Our Marine Environment.* London: DEFRA.

(2011a) *Biodiversity 2020: a strategy for England's wildlife and ecosystem services.* London: DEFRA.

(2011b) *Biodiversity offsetting: guiding principles for biodiversity offsetting.* London: DEFRA.

(2013a) *Payments for Ecosystem Services: A Best Practice Guide.* London: DEFRA.

(2013b) *Payments for Ecosystem Services: A Best Practice Guide. Annex Case Studies.* London: DEFRA.

(2016a) *Joint action plan to increase the English hen harrier population.* Bristol, UK: DEFRA, Uplands Stakeholder Forum, hen harrier sub-group.

(2016b) *Defra's Payments for Ecosystem Services Pilot Projects 2012–15: review of key findings.* London: DEFRA.

(2019a) *Marine strategy part one: UK updated assessment and Good Environmental Status – consultation document.* London: DEFRA.

(2019b) *Wild Bird Populations in the UK, 1970–2018.* London: DEFRA.

(2021) ENV09 – England biodiversity indicators. London: DEFRA. www.gov.uk/government/statistical-data-sets/env09-england-biodiversity-indicators

(2022a) *Landscapes review (National Parks and AONBs): government response.* London: DEFRA.

(2022b) *Nature Recovery Green Paper: Protected Sites and Species.* London: DEFRA.

(2022c) *Environmental targets consultation summary of responses and government response.* London: DEFRA.

Dickinson, J. L., Shirk, J., Bonter, D. *et al.* (2012) The current state of citizen science as a tool for ecological research and public engagement. *Frontiers in Ecology and the Environment*, **10**, 291–297.

Dicks, L. V., Ashpole, J. E., Dänhardt, J. *et al.* (2013) *Farmland Conservation Synopsis: Evidence for the Effects of Interventions in Northern Europe.* Synopses of Conservation Evidence, Volume 3. Exeter, UK: Pelagic Publishing.

DOENI (2002) *Northern Ireland Biodiversity Strategy.* Belfast: Department of the Environment for Northern Ireland.

(2015) *Valuing Nature: A Biodiversity Strategy for Northern Ireland to 2020.* Belfast: Department of the Environment Northern Ireland.

Drayson, K. & Thompson, S. (2013) Ecological mitigation measures in English Environmental Impact Assessment. *Journal of Environmental Management*, **119**, 103–110.

Dudley, N. (ed.) (2008). *Guidelines for Applying Protected Area Management Categories.* Gland, Switzerland: IUCN. WITH Stolton, S., Shadie, P. & Dudley, N. (2013) *IUCN WCPA Best Practice Guidance on Recognising Protected Areas and Assigning Management Categories and Governance Types.* Best Practice Protected Area Guidelines Series No. 21. Gland, Switzerland: IUCN.

Duke, G., Conway, M., Dickie, I. *et al.* (2013) *EMTF Second Phase Research: Opportunities for UK Business that Protect and/or Value Nature. Final Report.* London: ICF GHK.

EAC (2021) *Biodiversity in the UK: bloom or bust?* House of Commons Environmental Audit Committee. First Report of Session 2021–22.

Earll, B. (2018) *Marine Conservation: People, Ideas and Action.* Exeter, UK: Pelagic publishing.

Ecosystem Markets Task Force (2013) *Realising Nature's Value. The Final Report of the Ecosystem Markets Task Force.*

Edwards, J. (2018) Joan Edwards. In *Marine Conservation: People, Ideas and Action*, ed. B. Earll, pp. 84–94. Exeter, UK: Pelagic Publishing.

Environment Agency (2019) *Catchment Sensitive Farming Evaluation Report – Water Quality, Phases 1 to 4 (2006–2018).* York, UK: Natural England.

European Commission (2019) *The EU Environmental Implementation Review 2019 Country Report – United Kingdom.* SWD(2019) 118 final.

Evans, A. D. & Green, R. E. (2007) An example of a two-tiered agri-environment scheme designed to deliver effectively the ecological requirements of both localised and widespread bird species in England. *Journal of Ornithology*, **148**, Supplement 2, 279–286.

Evans, A. D., Armstrong-Brown, S. & Grice, P. V. (2002) The role of research and development in the evolution of a 'smart' agri-environment scheme. *Aspects of Applied Biology*, **67**, 253–264.

Evans, D. (1992) *A History of Nature Conservation in Britain.* 2nd ed. London and New York: Routledge.

Farjon, A. & Hill, L. (2019) Natural woodland generation as an alternative to tree-planting. *British Wildlife*, **30**, 177–185.

Field, C., Dise, N., Payne, R. *et al.* (2014) The role of nitrogen deposition in widespread plant community change across semi-natural habitats. *Ecosystems*, **17**, 864–877.

Fielding, A., Haworth, P., Whitfield, P., McLeod, D. & Riley, H. (2011) *A Conservation Framework for Hen Harriers in the United Kingdom.* Peterborough, UK: Joint Nature Conservation Committee, Report 441.

Forestry Commission (2017) *The UK Forestry Standard. The Governments' Approach to Sustainable Forestry.* Edinburgh: Forestry Commission.

Fox, R., Brereton, T. M., Asher, J. *et al.* (2015). *The State of the UK's Butterflies 2015.* Wareham, UK: Butterfly Conservation and the Centre for Ecology & Hydrology.

Fox, R., Parsons, M. S., Chapman, J. W., Woiwod, I. P., Warren, M. S. & Brooks, D. R. (2013) *The State of Britain's Larger Moths 2013*. Wareham, UK: Butterfly Conservation and Rothamsted Research.

Fuller, R. M. (1987) The changing extent and conservation interest of lowland grasslands in England and Wales: a review of grassland surveys 1930–1984. *Biological Conservation*, **40**, 281–300.

Gaywood, M. J., Boon, P. J., Thompson, D. B. A., Strachan, I. M. (eds.) (2016) *The Species Action Framework Handbook*. Perth, UK: Scottish Natural Heritage.

GHK (2010) *Costs of the UK Biodiversity Action Plan – Update*. London: GHK Consulting Ltd.

GHK, Christie, M., ADAS, IEEP, Minter, R. & the Research Box (2011) *Benefits of Sites of Special Scientific Interest*. London: GHK Consulting Ltd.

Gibbons, D. W., Reid, J. B. & Chapman, R. A. (1993) *The New Atlas of Breeding Birds in Britain and Ireland: 1988–1991*. London: T & A D Poyser Ltd.

Gilbert-Norton, L., Wilson, R., Stevens, J. R. & Beard, K. H. (2010). A meta-analytic review of corridor effectiveness. *Conservation Biology*, **24**, 660–668.

Glaves, D., Morecroft, M., Fitzgibbon, C., Owen, M., Phillips, S. & Leppitt, P. (2013) *The effects of managed burning on upland peatland biodiversity, carbon and water*. Peterborough, UK: Natural England Upland Evidence Review NEER004.

Glover, J. (2019) *Landscapes Review. Final Report*. London: DEFRA.

GMMRG (2019) *Grouse Moor Management Review Group. Report to the Scottish Government*. Edinburgh: Scottish Government.

GOS (2018) *Foresight Future of the Sea: a report from the Government Chief Scientific Adviser*. London: Government Office for Science.

Gove, B., Williams, L. J., Beresford, A. E. *et al.* (2016) Reconciling biodiversity conservation and widespread development of renewable energy technologies in the UK. *PLoS ONE*, **11** (5), e0150956.

Green, R. H., Langston, R. H. W., McCluskie, A., Sutherland, R. & Wilson, J. D. (2016) Lack of sound science in assessing wind farm impacts on seabirds. *Journal of Applied Ecology*, **54**, 1635–1641.

Gubbay, S. (1986) *A Coastal Directory for Marine Nature Conservation*. Ross-on-Wye, UK: Marine Conservation Society.

(2018) Sue Gubbay. In *Marine Conservation: People, Ideas and Action*, ed. B. Earll, pp. 72–83. Exeter, UK: Pelagic Publishing.

Haddad, N. M., & Tewksbury, J. J. (2006) Impacts of corridors on populations and communities. In *Connectivity Conservation*, eds. K. R. Crooks & M. A. Sanjayan, pp. 390–415. Cambridge, UK: Cambridge University Press.

Hampson, D. (2022) The Landscapes Review – an opportunity for change? *British Wildlife*, **33** (4), 265–269.

Harding, M. J. & Spencer, K. S. (2017) *What Has the Heritage Lottery Fund Done for Nature?* London: Report by Hummingbird/OHES for the Heritage Lottery Fund.

Hargreaves, K. J., Milne, R. & Cannell, M. G. R. (2003) Carbon balance of afforested peatland in Scotland. *Forestry*, **76**, 299–317.

Hawksworth, D. L. (ed.) (1974) *The Changing Flora and Fauna of Britain*. The Systematics Association Special Volume 6. London: Academic Press.

(ed.) (2001) *The Changing Wildlife of Great Britain and Ireland*. London and New York: Taylor & Francis.

Hayhow, D. B., Burns, F., Eaton, M. A. *et al.* (2016) *State of Nature 2016*. The State of Nature partnership.

Hayhow, D. B., Eaton, M. A., Stanbury, A. J. *et al.* (2019) *The State of Nature 2019*. The State of Nature partnership.

Helm, D. (2019) *Green and Prosperous Land: A Blueprint for Rescuing the British Countryside*. London: William Collins.

Hiscock, K. (2014) *Marine Biodiversity Conservation: A Practical Approach*. Abingdon, UK: Routledge.

(2018) Keith Hiscock. In *Marine Conservation: People, Ideas and Action*, ed. B. Earll, pp. 61–71. Exeter, UK: Pelagic Publishing.

(ed.) (1998) *Marine Nature Conservation Review: Benthic Marine Ecosystems of Great Britain and the North-East Atlantic*. Peterborough, UK: Joint Nature Conservation Committee.

HM Treasury (2021) *The Economics of Biodiversity: The Dasgupta Review – Government response*.

HMG (2011a) *The Natural Choice: securing the value of nature*. London: HM Government, the Stationery Office.

(2011b) *UK Marine Policy Statement*. HM Government. https://assets.publishing.service.gov.uk/government/uploads/system/uploads/attachment_data/file/69322/pb3654-marine-policy-statement-110316.pdf

(2012) *Report of the Habitats and Wild Birds Directives Implementation Review*. London: HM Government, Department for Environment, Food and Rural Affairs.

(2018) *A Green Future: Our 25 Year Plan to Improve the Environment*. London: HM Government.

(2021) *Net Zero Strategy: Build Back Greener*. London: HM Government.

Hodgson, J. A., Moilanen, A., Wintle, B. A. & Thomas, C. D. (2011) Habitat area, quality and connectivity: striking the balance for efficient conservation. *Journal of Applied Ecology*, **48**, 148–152.

Hodgson, J. A., Thomas, C. D., Wintle, B. A. & Moilanen, A. (2009) Climate change, connectivity and conservation decision making: back to basics. *Journal of Applied Ecology*, **46**, 964–969.

Holland, J. P., Morgan-Davies, C., Waterhouse, T., Thomson, S., Midgley, A. & Barnes, A. (2011) *An Analysis of the Impact on the Natural Heritage of the Decline in Hill Farming in Scotland*. Inverness, UK: Scottish Natural Heritage.

Hopkins, J. J., Allison, H. M., Walmsley, C. A., Gaywood, M. & Thurgate, G. (2007) *Conserving biodiversity in a changing climate: guidance on building capacity to adapt*. London: DEFRA.

House of Commons Environment, Food and Rural Affairs Committee (2021) *Environmental Land Management and the agricultural transition*. Second Report of Session 2021–22.

House of Lords (2018) *The Countryside at a Crossroads: Is the Natural Environment and Rural Communities Act 2006 Still Fit for Purpose?* London: The Select Committee on the Natural Environment and Rural Communities Act 2006, House of Lords.

IACCF (2010) *Biodiversity and Climate Change in the UK*, eds. D. A. Procter, J. M. Baxter, H. P. Q. Crick, D. Mortimer, F. Mulholland & C. A. Walmsley. Peterborough, UK: Inter-agency Climate Change Forum/JNCC.

ICES (2019) *Working Group on Bycatch of Protected Species (WGBYC)*. Copenhagen: International Council for the Exploration of the Sea. *ICES Scientific Reports*, **1** (51).

ICF (2019) *Evaluation of EU LIFE Fund in the UK. Natural England Commissioned Report 266*. ICF Consulting Services Limited report for Natural England.

Jepson, P. D., Deaville, R., Barber, J. L. *et al.* (2016) PCB pollution continues to impact populations of orcas and other dolphins in European waters. *Scientific Reports*, **6**, 18573.

JNCC (1998) *A statement on common standards monitoring*. Peterborough, UK: Joint Nature Conservation Committee.

(1999) *The Birds Directive. Selection guidelines for Special Protection Areas*. Peterborough, UK: Joint Nature Conservation Committee.

JNCC (on behalf of the Common Standards Monitoring Inter-agency Working Group1) (2022) *A Statement on Common Standards for Monitoring Protected Sites (2022)*. Peterborough, UK: Joint Nature Conservation Committee. https://hub.jncc.gov.uk/assets/0450edfd-a56b-4f65-aff6-3ef66187dc81

JNCC & DEFRA (on behalf of the Four Countries' Biodiversity Group) (2012) *UK Post-2010 Biodiversity Framework*. Peterborough, UK: Joint Nature Conservation Committee.

Johnston, A., Ausden, M., Dodd, A. M. *et al.* (2013) Observed and predicted effects of climate change on species abundance in protected areas. *Nature Climate Change*, **3**, 1055–1061.

Jones, G. (ed.) (2012) *The Habitats Directive – A Developer's Obstacle Course?* Oxford, UK: Hart Publishing.

Kaiser, M. J., Ramsay, K., Richardson, C. A., Spence, F. E. & Brand, A. R. (2000) Chronic fishing disturbance has changed shelf sea benthic community structure. *Journal of Animal Ecology*, **69**, 494–503.

Kirby, K., Jefferson, R., Larwood, J., Russell, D., Le Bas, B. & Wright, R. (2010) What has the SSSI improvement programme achieved for nature conservation in England? *British Wildlife*, **22**, 16–25.

Lake, S., Liley, D., Still, R. & Swash, A. (2015) *Britain's Habitats – A Guide to the Wildlife Habitats of Britain and Ireland*. Woodstock, UK: Princeton University Press.

Lawton, J. H., Brotherton, P. N. M., Brown, V. K. *et al.* (2010) *Making Space for Nature: A Review of England's Wildlife Sites and Ecological Network*. London: DEFRA.

Ledoux, L., Cornell, S., O'Riordan, T. *et al.* (2005) Towards sustainable flood and coastal management: identifying drivers of, and obstacles to, managed realignment. *Land Use Policy*, **2**, 129–144.

Lee, R. (2010) *Delivering landscape-scale habitat restoration and creation through spatial planning*. Sandy, UK: Royal Society for the Protection of Birds.

Lewis, B., Griffiths, R. A. & Wilkinson, J. W. (2017) Population status of great crested newts (*Triturus cristatus*) at sites subjected to development mitigation. *Herpetological Journal*, **27**, 133–142.

MacDonald, A., Heath, M. & Edwards, M. (2015) Climate driven trophic cascades affecting seabirds around the British Isles. *Oceanography and Marine Biology – An Annual Review*, **53**, 55–80.

MacDonald, B. (2019) *Rebirding: Rewilding Britain and Its Birds*. Exeter, UK: Pelagic Publishing.

Macgregor, C. J., Evans, D. M., Fox, R. & Pocock, M. J. O. (2017) The dark side of street lighting: impacts on moths and evidence for the disruption of nocturnal pollen transport. *Global Change Biology*, **23**, 697–707.

Mackay, D. (1995) *Scotland's Rural Land Use Agencies*. Aberdeen, UK: Scottish Cultural Press.

Maclean, N. (ed.) (2010) *Silent Summer. The State of Wildlife in Britain and Ireland*. Cambridge, UK: Cambridge University Press.

Mair, L., Hill, J., Fox, R., Botham, M., Brereton, T. & Thomas, C. D. (2014) Abundance changes and habitat availability drive species' responses to climate change. *Nature Climate Change*, **4**, 127–131.

Marren, P. (2002) *Nature Conservation: A Review of the Conservation of Wildlife in Britain 1950–2001*. London: The New Naturalist, HarperCollins.

(2018) National Nature Reserves in crisis? *British Wildlife*, **29**, 314–319.

(2020) Will Natural England survive? *British Wildlife*, **31**, 173–178.

Martin, D., Fraser, M. D., Pakeman, R. J. & Moffat, A. M. (2013) *Impact of moorland grazing and stocking rates*. York, UK: Natural England, Evidence Review, Number 006.

Maskell, L. C., Firbank, L. G., Thompson, K., Bullock, J. M. & Smart, S. M. (2006) Interactions between non-native plant species and the floristic composition of common habitats. *Journal of Ecology*, **94**, 1052–1060.

McCarthy, D. & Morling, P. (2015) *Using Regulation as a Last Resort: Assessing the Performance of Voluntary Approaches*. Sandy, UK: Royal Society for the Protection of Birds.

Miles, R. & Richardson, N. (2018) *Sustainable Shores*. Sandy, UK: RSPB. Technical Report.

Mitchell, R. (2018) Roger Mitchell. In *Marine Conservation: People, Ideas and Action*, ed. B. Earll, pp. 47–60. Exeter, UK: Pelagic Publishing.

Mitchell, R. J., Morecroft, M. D. & Acreman, M. (2007) *England Biodiversity Strategy – towards adaptation to climate change*. London: DEFRA.

Monbiot, G. (2014) *Feral: Rewilding the Land, the Sea and Human Life*. Chicago: University of Chicago Press.

Moore, N. W. (1987) *The Bird of Time: The Science and Politics of Nature Conservation*. Cambridge, UK: Cambridge University Press.

Murgatroyd, M., Redpath, S. M., Murphy, S. G., Douglas, D. J. T., Saunders, R. & Amar, A. (2019) Patterns of satellite tagged hen harrier disappearances suggest widespread illegal killing on British grouse moors. *Nature Communications*, **10**, 1094.

National Audit Office (2008) *Natural England's Role in Improving Sites of Special Scientific Interest*. London: The Stationery Office.

Natural England (2009) *Green Infrastructure Guidance*. York, UK: Natural England.

(2011) *Protecting England's Natural Treasures – Sites of Special Scientific Interest*. York, UK: Natural England.

(2016) *Conservation 21. Natural England's Conservation Strategy for the 21st Century*. York, UK: Natural England.

(2019) *Natural England Technical Information Note TIN176. A Framework for District Licensing of Development Affecting Great Crested Newts.* York, UK: Natural England.

Natural England & JNCC (2010) *Marine Conservation Zone Project Ecological Network Guidance.* Peterborough, UK: JNCC and Natural England.

NCC (2015) *The State of Natural Capital Protecting and Improving Natural Capital for Prosperity and Wellbeing.* UK: Natural Capital Committee. Third report to the Economic Affairs Committee.

Nesbit, M., Ankersmit, L., Friel, A. & Colsa, A. (2017) *Ensuring compliance with environmental obligations through a future UK-EU relationship.* London: Institute for European Environmental Policy.

Newton, I. (2017) *Farming and Birds.* London: The New Naturalist, Harper Collins.

(2020) *Uplands and Birds.* London: Harper Collins (New Naturalist).

NSRF (2014) *The Scottish Code for Conservation Translocations. Best Practice Guidelines for Conservation Translocations in Scotland Version 1.1.* Edinburgh: Scottish Natural Heritage on behalf of the National Species Reintroduction Forum.

Ockendon, N., Baker, D. J., Carr, J. A. *et al.* (2014) Mechanisms underpinning climatic impacts on natural populations: altered species interactions are more important than direct effects. *Global Change Biology*, **20**, 2221–2229.

OEP (2022a) *Taking stock: protecting, restoring and improving the environment in England.* Worcester, UK: The Office for Environmental Protection.

(2022b) *Advice on the Nature Recovery Green Paper.* Worcester, UK: The Office for Environmental Protection.

(2022c) *Advice on environmental targets.* Worcester, UK: The Office for Environmental Protection.

OSPAR (2006) *Guidance on Developing an Ecologically Coherent Network of OSPAR Marine Protected Areas.* (Reference number 2006-3). London: OSPAR Commission.

(2017) *Intermediate Assessment 2017 (IA 2017).* London: OSPAR Commission.

Pearce-Higgins, J. W. (2021) *Climate Change and the UK's Birds.* Thetford, UK: British Trust for Ornithology.

Pearce-Higgins, J. W., Beale, C. M., Oliver, T. H. *et al.* (2017) A national-scale assessment of climate change impacts on species: assessing the balance of risks and opportunities for multiple taxa. *Biological Conservation*, **213**, 124–134.

Pearce-Higgins, J. W., Johnston, A., Ausden, M. *et al.* (2011) *Final Report to the Climate Change Impacts on Avian Interests of Protected Area Networks (CHAINSPAN) Steering Group.* Defra Ref: WC0750/CR0440.

Perry, A. L., Low, P. J., Ellis, J. R. & Reynolds, J. D. (2005) Climate change and distribution shifts in marine fishes. *Science*, **308**, 1912–1915.

Pescott, O. L., Simkin, J. M., August, T. A., Randle, Z., Dore, A. J. & Botham, M. S. (2015) Air pollution and its effects on lichens, bryophytes, and lichen-feeding Lepidoptera: review and evidence from biological records. *Biological Journal of the Linnean Society*, **115**, 611–635.

Piper, J. M., Wilson, E. B., Weston, J., Thompson, S. & Glasson, J. (2006) *Spatial planning for biodiversity in our changing climate.* Peterborough, UK: English Nature, Research Report R677.

Potts, G. R. (1998) Global dispersion of nesting Hen Harriers *Circus cyaneus*; implications for grouse moors in the U.K. *Ibis*, **140**, 76–88.

Preston, C. D., Roy, D. B. & Roy, H. E. (2012) What have we learnt from 50 years of biological recording? *British Wildlife*, **24**, 97–106.

Proctor, M. (2013) *Vegetation of Britain and Ireland.* London: The New Naturalist Library, Harper Collins.

Pywell, R. F., Heard, M. S., Bradbury, R. B. *et al.* (2012). Wildlife-friendly farming benefits rare birds, bees and plants. *Biology Letters*, **8**, 772–775.

Rackham, O. (1986) *The History of the Countryside: The Classic History of Britain's Landscape, Flora and Fauna.* London: Orion Books Ltd.

Ratcliffe, D. A. (ed.) (1977) *A Nature Conservation Review.* Cambridge, UK: Cambridge University Press.

Ratcliffe, D. A. & Thompson, D. B. A. (1988) The British uplands: their ecological character and international significance. In *Ecological Change in the Uplands*, eds. M. B. Usher & D. B. A. Thompson, pp. 9–36. Oxford, UK: Blackwell Scientific Publications.

Rayment, M. (2017) *Assessing the costs of Environmental Land Management in the UK. Final Report.* A report for the RSPB, the National Trust and The Wildlife Trusts.

Rayment, M., Haines, R., McNeil, D., Conway, M., Tucker, G. M. & Underwood, E. (2014) *Study on specific design elements of biodiversity offsets: biodiversity metrics and mechanisms for securing long term conservation benefits.* London: ICF & IEEP Report to the European Commission.

Redpath, S., Amar, A., Smith, A., Thompson, D. B. A. & Thirgood, S. (2010) People and nature in conflict: can we reconcile hen harrier conservation and game management? In *Species Management: Challenges and Solutions for the 21st Century*, eds. J. M. Baxter & C. A. Galbraith, pp. 335–350. Edinburgh: The Stationery Office.

Resasco, J. (2019) Meta-analysis on a decade of testing corridor efficacy: what new have we learned? *Current Landscape Ecology Reports*, **4**, 61–69.

Ricketts, S. & Bischoff, S. (2012) SANGs: The Thames Basin Case Study. In *The Habitats Directive – A Developer's Obstacle Course?*, ed. G. Jones, pp. 119–138. Oxford, UK: Hart Publishing.

Robinson, L. A. & Frid, C. L. J. (2008) Historical marine ecology: examining the role of fisheries in changes to North Sea benthos. *Ambio*, **37**, 362–371.

Roos, S., Smart, J., Gibbons, D. W. & Wilson, J. D. (2018) A review of predation as a limiting factor for bird populations in mesopredator-rich landscapes: a case study of the UK. *Biological Reviews*, **93**, 1915–1937.

RoTAP (2012) *Review of Transboundary Air Pollution: Acidification, Eutrophication, Ground Level Ozone and Heavy Metals in the UK.* Edinburgh: Centre for Ecology and Hydrology.

Rothschild, M. & Marren, P. (1997) *Rothschild's Reserves. Time and Fragile Nature.* Colchester, UK: Balaban Publishers & Harley Books.

Rowell, T. A. (1991) *SSSIs: a health check*. London: Wildlife Link.

Roy, H. E., Bacon, J., Beckmann, B. *et al.* (2012). *Non-native Species in Great Britain: establishment, detection and reporting to inform effective decision making*. Wallingford, UK: NERC Centre for Ecology & Hydrology.

RSPB (2018) *The Illegal Killing of Birds of Prey in Scotland 2015–17*. Sandy, UK: Royal Society for the Protection of Birds.

(2021) *Annual Report 2020–2021*. www.rspb.org.uk/globalassets/downloads/annual-report-2021/rspb-annual-report-2020-2021_digital.pdf?sourcecode=GENWEB0026

Sandford, J. C. E. (1974) *Report of the National Park Policies Review Committee*. London: HM Stationery Office.

Sands, T. (2012) *Wildlife in Trust: 100 Years of Nature Conservation*. London: Elliott & Thompson.

Scottish Executive (2004) *Scotland's biodiversity: it's in your hands. A strategy for the conservation and enhancement of biodiversity in Scotland*. Edinburgh: The Scottish Executive.

Scottish Government (2013) *Challenge for Scotland's biodiversity: a strategy for the conservation and enhancement of biodiversity in Scotland*. Edinburgh: The Scottish Government.

(2015) *Scotland's biodiversity: a route map to 2020*. Edinburgh: Scottish Government.

(2016) *Getting the best from our land: A Land Use Strategy for Scotland 2016–2021*. Edinburgh: The Scottish Government.

(2020a) *Scottish Government response to the report from the Grouse Moor Management Group*. Edinburgh: The Scottish Government.

(2020b) *Scotland's Fourth National Planning Framework Position Statement*. Edinburgh: The Scottish Government.

(2021) *The Environment Strategy for Scotland: Initial Monitoring Framework*. Edinburgh: Scottish Government.

(2022) *Biodiversity Strategy to 2045: Tackling the Nature Emergency*. Edinburgh: The Scottish Government.

Sewell, J. & Hiscock, K. (2005) *Effects of Fishing within UK European Marine Sites: Guidance for Nature Conservation Agencies*. Plymouth, UK: Marine Biological Association.

Sharps, E., Hawkes, R. W., Bladon, A. J. et al. (2023) Reversing declines in farmland birds: How much agri-environment provision is needed at farm and landscape scales?. *Journal of Applied Ecology*, **00**, 1–13. https://doi.org/10.1111/1365-2664.14338

Sheail, J. (1998) *Nature Conservation in Britain: the formative years*. London: The Stationery Office.

Simmons, I. G. (2001) *An Environmental History of Great Britain: From 10,000 Years Ago to the Present*. Edinburgh: Edinburgh University Press.

Smith, S., Rowcroft, P., Everard, M. *et al.* (2015) *Payments for Ecosystem Services: A Best Practice Guide*. Updated edition. London: DEFRA.

Smithers, R. J., Cowan, C., Harley, M., Hopkins, J. J., Pontier, H. & Watts, O. (2008) *England Biodiversity Strategy Climate Change Adaptation Principles*. Bristol, UK: DEFRA.

SNH (2009) *Trend Note: a summary of land cover change in Scotland from 1947–1988*. Edinburgh: Scottish Natural Heritage.

Stace, C. A. (2019) *New Flora of the British Isles*. 4th ed. Iver, UK: C&M Floristics – Pemberley Books.

Stafford, R., Chamberlain, B., Clavey, L. *et al.* (eds.) (2021) *Nature-Based Solutions for Climate Change in the UK: A Report by the British Ecological Society*. London: British Ecological Society.

Stamp, D. (1974) *Nature Conservation in Britain*. 2nd ed. London: The New Naturalist, Collins.

Stanbury, A., Eaton, M., Aebischer, N. *et al.* (2021) The status of our bird populations: the fifth Birds of Conservation Concern in the United Kingdom, Channel Islands and Isle of Man and second IUCN Red List assessment of extinction risk for Great Britain. *British Birds*, **114**, 723–747.

Starnes, T., Beresford, A. E., Buchanan, G. M., Lewis, M., Hughes, A. & Gregory, R. D. (2021) The extent and effectiveness of protected areas in the UK. *Global Ecology and Conservation*, **30**, e01745.

Still, K. & Byfield, A. J. (2010) Is Environmental Stewardship working for rare and threatened plants? *Aspects of Applied Biology*, **100**, 279–286.

Stroh, P., Walker, K., Smith, S., Jefferson, R., Pinches, C. & Blackstock, T. (2019) *Grassland Plants of the British and Irish Lowlands: Ecology, Threats and Management*. Durham, UK: Botanical Society of Britain & Ireland.

Stroud, D. A., Bainbridge, I. P., Maddock, A. *et al.* (eds.) (2016) *The Status of UK SPAs in the 2000s: The Third Network Review*. Peterborough, UK: Joint Nature Conservation Committee.

Stroud, D. A., Chambers, D., Cook, S. *et al.* (eds.) (2001) *The UK SPA Network: Its Scope and Content*. Peterborough, UK: Joint Nature Conservation Committee.

Stroud, D. A., Pienkowski, M. W. & Mudge, G. P. (1990) *Protecting internationally important bird sites: a review of the network of EC Special Protection Areas in Great Britain*. Peterborough, UK: Nature Conservancy Council.

Stroud, D. A., Reed, T. M., Pienkowski, M. W. & Lindsay, R. A. (2015) The Flow Country: battles fought, war won, organisation lost. In *Nature's Conscience – The Life and Legacy of Derek Ratcliffe*, eds. D. Thompson, J. Birks & H. Birks, pp. 401–439. Norfolk, UK: Langford Press.

Sutherland, W. J., Armstrong-Brown, S., Armsworth, P. R. *et al.* (2006) The identification of one hundred ecological questions of high policy relevance in the UK. *Journal of Applied Ecology*, **43**, 617–627.

Taylor, P. (2005) *Beyond Conservation: A Wildland Strategy*. London: Earthscan.

TEEB (2011) *The Economics of Ecosystems and Biodiversity in National and International Policy Making*. London and Washington: Earthscan.

Tew, T., Biggs, J. & Gent, T. (2018) District licensing for great crested newts – delivering a big idea. *In Practice: Bulletin of the Chartered Institute of Ecology and Environmental Management*, **100**, 33–37.

Thomas, C. D., Gillingham, P. K. & Bradbury, R. B. *et al.* (2012) Protected areas facilitate species' range expansions. *Proceedings of the National Academy of Sciences*, **109**, 14063–14068.

Thompson, D. B. A., MacDonald, A. J., Marsden, J. H. & Galbraith, C. A. (1995) Upland heather moorland in Great Britain: a review of international importance, vegetation change and some objectives for nature conservation. *Biological Conservation*, **71**, 163–178.

Thompson, P. S., Douglas, D. J. T., Hoccom, D. G., Knott, J., Roos, S. & Wilson, J. D. (2016) Environmental impacts of high-output driven shooting of Red Grouse *Lagopus lagopus scotica*. *Ibis*, **158**, 446–452.

Tingay, R. E. (2015) *Natural injustice – Paper One: a review of the enforcement of wildlife protection legislation in Scotland*. Perth, Scotland: Scottish Environment LINK.

Tree, I. (2018) *Wilding: The Return of Nature to a British Farm*. London: Picador.

Treweek, J. R. (2009) *Scoping study for the design and use of biodiversity offsets in an English context*. Final Report to Defra (Contract NE 0801). London: Defra.

Treweek, J. R. & Thompson S. (1997) A review of ecological mitigation measures in UK Environmental Statements with a view to sustainable development. *International Journal of Sustainable Development and World Ecology*, **4**, 40–50.

Tucker, G. M. & Baldock, D. (2018) The 25-Year Environment Plan – what promises for nature? *British Wildlife*, **29**, 245–250.

UK Government (2021) *The England Trees Action Plan 2021–2024*. https://assets.publishing.service.gov.uk/government/uploads/system/uploads/attachment_data/file/987432/england-trees-action-plan.pdf

UK NEA (2011) *The UK National Ecosystem Assessment: Technical Report*. Cambridge, UK: UNEP-WCMC.

UKMMAS (2010) *Charting Progress 2 – the state of UK seas*. Report by the UK Marine Monitoring and Assessment Strategy community. London: DEFRA.

Vanguelova, E., Chapman, S. J., Perks, M. *et al.* (2018) *Afforestation and Restocking on Peaty Soils – New Evidence Assessment*. Edinburgh: ClimateXChange.

Vickery, J. A., Bradbury, R. B., Henderson, I. G., Eaton, M. A. & Grice, P. V. (2004) The role of agri-environment schemes and farm management practices in reversing the decline of farmland birds in England. *Biological Conservation*, **119**, 19–39.

Walker, L. K., Morris, A. J., Cristinacce, A., Dadam, D., Grice, P. V. & Peach, W. J. (2018) Effects of higher-tier agri-environment scheme on the abundance of priority farmland birds. *Animal Conservation*, **21**, 183–192.

Watson, H. (2008) Great Crested Newts and their protection: are we getting it all wrong? *In Practice: Bulletin of the Institute of Ecology and Environmental Management*, **60**, 25–28.

WCL (2022a) *Response to Consultation on Nature Recovery Green Paper*. Wildlife and Countryside Link, May 2022. www.wcl.org.uk/docs/WCL_NatureRecoveryGP_Response_May2022.pdf

WCL (2022b) *What price deregulation? The environmental and social costs of the Retained EU Law Bill*. Wildlife and Countryside Link. https://www.wcl.org.uk/docs/What_price_deregulation_Link_briefing_12_12_22.pdf

Webb, N. (1986) *Heathlands: A Natural History of Britain's Lowland Heaths*. London: Collins.

Welsh Assembly Government (2006) *Environment Strategy for Wales*. Cardiff: Welsh Assembly Government.

Wende, W., Tucker, G., Quétier, F., Rayment, M. & Darbi, M. (2018) *Biodiversity Offsets – European Perspectives on No Net Loss of Biodiversity and Ecosystem Services*. Cham, Switzerland: Springer.

Werritty, A., Pakeman, R. J., Shedden, C., Smith, A. & Wilson, J. D. (2015). *A Review of Sustainable Moorland Management*. Battleby, UK: Scottish Natural Heritage.

Whitfield, D. P. & Fielding, A. H. (2017). *Analyses of the fates of satellite tracked golden eagles in Scotland*. Scottish Natural Heritage Commissioned Report.

The Wildlife Trusts (2018) *The status of England's Local Wildlife Sites 2018*. Newark, UK: The Wildlife Trusts.

(2020) *Nature_Recovery_Network_Handbook*. www.wildlifetrusts.org/sites/default/files/2020-11/Nature_Recovery_Network_Handbook_FINAL_Sept%202020.pdf

Wilson, J. D., Evans, M. G. & Grice, P. V. (2009) *Bird Conservation and Agriculture. Ecology, Biodiversity and Conservation*. Cambridge, UK: Cambridge University Press.

Wilson, J. F., Baker, D., Cheney, J. *et al.* (2018) A role for artificial night-time lighting in long-term changes in populations of 100 widespread macro-moths in UK and Ireland: a citizen-science study. *Journal of Insect Conservation*, **22**, 189–196.

Woodcock, B., Isaac, N., Bullock, J. *et al.* (2016) Impacts of neonicotinoid use on long-term population changes in wild bees in England. *Nature Communications*, **7**, 12459.

Wotton, S., Brown, A., Burn, A. *et al.* (2009) Boom or bust – a sustainable future for reedbeds and Bitterns? *British Wildlife*, **20**, 305–315.

WWF (2017) *Delivering ecosystem-based marine spatial planning in practice: an assessment of the integration of the ecosystem approach into UK and Ireland Marine Spatial Plans*. Woking, UK: WWF.

Wynne, G., Avery, M., Campbell, L. et al. (1993) *Biodiversity Challenge: An Agenda for Conservation Action in the UK*. 1st ed. Sandy, UK: RSPB.

Appendix Habitats Directive Annex I Habitat Types Referred to in This Book

Natural habitat types of Community interest whose conservation requires the designation of Special Areas of Conservation (SACs). For a full list see https://eur-lex.europa.eu/legal-content/EN/TXT/?uri=CELEX:01992L0043-20130701

The code corresponds to the Natura 2000 code. '*' Indicates a priority habitat type.

1. COASTAL AND HALOPHYTIC HABITATS

11. Open sea and tidal areas

1110 Sandbanks which are slightly covered by seawater all the time

1120 * *Posidonia* beds (*Posidonia oceanica*)

1130 Estuaries

1140 Mudflats and sandflats not covered by seawater at low tide

1150 * Coastal lagoons

1160 Large shallow inlets and bays

1170 Reefs

12. Sea cliffs and shingle or stony beaches

1230 Vegetated sea cliffs of the Atlantic and Baltic Coasts

13. Atlantic and continental salt marshes and salt meadows

1310 *Salicornia* and other annuals colonising mud and sand

1320 *Spartina* swards (*Spartinion maritimae*)

1330 Atlantic salt meadows (*Glauco-Puccinellietalia maritimae*)

15. Salt and gypsum inland steppes

1520 * Iberian gypsum vegetation (*Gypsophiletalia*)

1530 * Pannonic salt steppes and salt marshes

16. Boreal Baltic archipelago, coastal and landupheaval areas

1620 Boreal Baltic islets and small islands

1630 * Boreal Baltic coastal meadows

2. COASTAL SAND DUNES AND INLAND DUNES

21. Sea dunes of the Atlantic, North Sea and Baltic coasts

2110 Embryonic shifting dunes

2120 Shifting dunes along the shoreline with *Ammophila arenaria* ('white dunes')

2130 * Fixed coastal dunes with herbaceous vegetation ('grey dunes')

2140 * Decalcified fixed dunes with *Empetrum nigrum*

2150 * Atlantic decalcified fixed dunes (*Calluno-Ulicetea*)

2160 Dunes with *Hippophaë rhamnoides*

2170 Dunes with *Salix repens* ssp. *argentea* (*Salicion arenariae*)

2180 Wooded dunes of the Atlantic, Continental and Boreal region

2190 Humid dune slacks

21A0 Machairs (* in Ireland)

22. Sea dunes of the Mediterranean coast

2250 * Coastal dunes with *Juniperus* spp.

23. Inland dunes, old and decalcified

2320 Dry sand heaths with *Calluna* and *Empetrum nigrum*

2340 * Pannonic inland dunes

3. FRESHWATER HABITATS

31. Standing water

3110 Oligotrophic waters containing very few minerals of sandy plains (*Littorelletalia uniflorae*)

3130 Oligotrophic to mesotrophic standing waters with vegetation of the *Littorelletea uniflorae* and/or of the *Isoëto-Nanojuncetea*

3140 Hard oligo-mesotrophic waters with benthic vegetation of *Chara* spp.

3150 Natural eutrophic lakes with *Magnopotamion* or *Hydrocharition*–type vegetation

3160 Natural dystrophic lakes and ponds

3180 * Turloughs

32. Running water – sections of water courses with natural or semi-natural dynamics (minor, average and major beds) where the water quality shows no significant deterioration

3210 Fennoscandian natural rivers

3230 Alpine rivers and their ligneous vegetation with *Myricaria germanica*

3260 Water courses of plain to montane levels with the *Ranunculion fluitantis* and *Callitricho-Batrachion* vegetation

3270 Rivers with muddy banks with *Chenopodion rubri* p.p. and *Bidention* p.p. vegetation

32A0 Tufa cascades of karstic rivers in the Dinaric Alps

4. TEMPERATE HEATH AND SCRUB

4010 Northern Atlantic wet heaths with *Erica tetralix*

4030 European dry heaths

4060 Alpine and Boreal heaths

4080 Sub-Arctic *Salix* spp. scrub

5. SCLEROPHYLLOUS SCRUB (MATORRAL)

52. Mediterranean arborescent matorral

5220 * Arborescent matorral with *Zyziphus*

6. NATURAL AND SEMI-NATURAL GRASSLAND FORMATIONS

61. Natural grasslands

6120 * Xeric sand calcareous grasslands

6150 Siliceous alpine and boreal grasslands

6170 Alpine and subalpine calcareous grasslands

62. Semi-natural dry grasslands and scrubland facies

6210 Semi-natural dry grasslands and scrubland facies on calcareous substrates (*Festuco-Brometalia*) (* important orchid sites)

6220 * Pseudo-steppe with grasses and annuals of the *Thero-Brachypodietea*

6230 * Species-rich *Nardus* grasslands, on siliceous substrates in mountain areas (and submountain areas in Continental Europe)

6240 * Sub-Pannonic steppic grasslands

6250 * Pannonic loess steppic grasslands

6260 * Pannonic sand steppes

6270 * Fennoscandian lowland species-rich dry to mesic grasslands

6280 * Nordic alvar and precambrian calcareous flatrocks

62A0 Eastern sub-Mediterranean dry grasslands (*Scorzoneratalia villosae*)

62B0 * Serpentinophilous grassland of Cyprus

62C0 * Ponto-Sarmatic steppes

63. Sclerophyllous grazed forests (dehesas)

6310 *Dehesas* with evergreen *Quercus* spp.

64. Semi-natural tall-herb humid meadows

6410 *Molinia* meadows on calcareous, peaty or clayey-silt-laden soils (*Molinion caeruleae*)

6430 Hydrophilous tall herb fringe communities of plains and of the montane to alpine levels

6440 Alluvial meadows of river valleys of the *Cnidion dubii*

6450 Northern boreal alluvial meadows

6460 Peat grasslands of Troodos

65. Mesophile grasslands

6510 Lowland hay meadows (*Alopecurus pratensis, Sanguisorba officinalis*)

6520 Mountain hay meadows

6530 * Fennoscandian wooded meadows

6540 Sub-Mediterranean grasslands of the *Molinio-Hordeion secalini*

7. RAISED BOGS AND MIRES AND FENS

71. Sphagnum acid bogs

7110 * Active raised bogs

7130 Blanket bogs (* if active bog)

7140 Transition mires and quaking bogs

72. Calcareous fens

7210 * Calcareous fens with *Cladium mariscus* and species of the *Caricion davallianae*

7230 Alkaline fens

7240 * Alpine pioneer formations of the *Caricion bicoloris-atrofuscae*

73. Boreal mires

7310 * Aapa mires

7320 * Palsa mires

8. ROCKY HABITATS AND CAVES

82. Rocky slopes with chasmophytic vegetation

8240 * Limestone pavements

83. Other rocky habitats

8310 Caves not open to the public

8330 Submerged or partially submerged sea caves

8340 Permanent glaciers

9. FORESTS

90. Forests of Boreal Europe

9010 * Western Taïga

9020 * Fennoscandian hemiboreal natural old broad-leaved deciduous forests (*Quercus, Tilia, Acer, Fraxinus* or *Ulmus*) rich in epiphytes

9030 * Natural forests of primary succession stages of landupheaval coast

9040 Nordic subalpine/subarctic forests with *Betula pubescens* ssp. *czerepanovii*

9050 Fennoscandian herb-rich forests with *Picea abies*

9070 Fennoscandian wooded pastures

9080 * Fennoscandian deciduous swamp woods

91. Forests of Temperate Europe

9110 *Luzulo-Fagetum* beech forests

9120 Atlantic acidophilous beech forests with *Ilex* and sometimes also *Taxus* in the shrublayer (*Quercion robori-petraeae or Ilici-Fagenion*)

9130 *Asperulo-Fagetum* beech forests

9150 Medio-European limestone beech forests of the *Cephalanthero-Fagion*

9160 Sub-Atlantic and medio-European oak or oak–hornbeam forests of the *Carpinion betuli*

9170 *Galio-Carpinetum* oak–hornbeam forests

9180 * *Tilio-Acerion* forests of slopes, screes and ravines

9190 Old acidophilous oak woods with *Quercus robur* on sandy plains

91A0 Old sessile oak woods with *Ilex* and *Blechnum* in the British Isles

91C0 * Caledonian forest

91D0 * Bog woodland

91E0 * Alluvial forests with *Alnus glutinosa* and *Fraxinus excelsior* (*Alno-Padion, Alnion incanae, Salicion albae*)

91F0 Riparian mixed forests of *Quercus robur, Ulmus laevis* and *Ulmus minor, Fraxinus excelsior* or *Fraxinus angustifolia*, along the great rivers (*Ulmenion minoris*)

91H0 * Pannonian woods with *Quercus pubescens*

91I0 * Euro-Siberian steppic woods with *Quercus* spp.

91J0 * *Taxus baccata* woods of the British Isles

91K0 Illyrian *Fagus sylvatica* forests (*Aremonio-Fagion*)

91N0 * Pannonic inland sand dune thicket (*Junipero-Populetum albae*)

91P0 Holy Cross fir forest (*Abietetum polonicum*)

91S0 * Western Pontic beech forests

91V0 Dacian Beech forests (*Symphyto-Fagion*)

91X0 * Dobrogean beech forests

91Y0 Dacian oak and hornbeam forests

92. Mediterranean deciduous forests

9230 Galicio-Portuguese oak woods with *Quercus robur* and *Quercus pyrenaica*

93. Mediterranean sclerophyllous forests

9330 *Quercus suber* forests

9370 * Palm groves of *Phoenix*

9390 * Scrub and low forest vegetation with *Quercus alnifolia*

93A0 Woodlands with *Quercus infectoria* (*Anagyro foetidae-Quercetum infectoriae*)

94. Temperate mountainous coniferous forests

9410 Acidophilous *Picea* forests of the montane to alpine levels (*Vaccinio-Piceetea*)

9430 Subalpine and montane *Pinus uncinata* forests (* if on gypsum or limestone)

95. Mediterranean and Macaronesian mountainous coniferous forests

9530 * (Sub-) Mediterranean pine forests with endemic black pines

9560 * Endemic forests with *Juniperus* spp.

9570 * *Tetraclinis articulata* forests

9590 * *Cedrus brevifolia* forests (*Cedrosetum brevifoliae*)

Index

Acacia saligna, 222, 231
Accipiter gentilis. *See* Goshawk
Acer negundo. *See* Box Elder
Acipenser naccarii. *See* Adriatic Sturgeon
Acipenser ruthenus. *See* Sterlet
Acipenser sturio. *See* Atlantic Sturgeon
Acrocephalus paludicola. *See* Aquatic Warbler
Action Plan for nature, people and the economy, 84, 116
adaptive management, 53, 82, 176, 317, 334, 443, 504, 630
Adriatic basin, 200–201
Adriatic Sea, 17, 33, 201, 214, 415, 574
Adriatic Sturgeon, 421
Aegean, 353
Aegean Sea, 33
Aegean_Levantine Sea, 17
Aegypius monachus. *See* Cinereous Vulture
Aesculapian Ratsnake, 243
Africa, 50, 436, 594
agricultural land
  habitat management and restoration
    Austria, 151–152
    Belgium, 173
    Croatia, 212–213
    Germany, 343
    Hungary, 385–386
    Italy, 425–426
    Poland, 505
    Portugal, 526
    Slovakia, 567
    Slovenia, 586–587
    UK, 659–660
agriculture, biodiversity pressures. *See also* pollution
  abandonment, 20, 23, 34, 96, 110–111, 145, 151, 184, 195, 201–202, 221, 238, 249, 274, 311, 313, 355, 376, 395, 415, 417, 434–436, 444, 461, 493, 535–537, 557, 575, 591, 596
  agricultural expansion, 19–21, 111, 219, 254, 469, 534, 537, 577, 634
  agricultural improvements/intensification, 19, 22–23, 96–97, 110, 129, 145, 151, 160–161, 163, 184, 212, 221, 257, 276, 302, 313, 331, 334, 355, 394–395, 404, 407, 415, 420, 436, 470–471, 482, 495, 516–517, 531, 534, 536, 557, 593, 595, 610, 634, 637
  bioenergy, 20, 22, 239
  conversion of pastures/grassland to cropland, 22, 110, 151, 184, 255, 595
  conversion of pastures and meadows to silage, 19, 22, 110, 163, 302, 396
  conversion of semi-natural habitats to intensive agricultural land, 163, 239, 374, 493, 516, 658
  decline in rotations and fallow, 19, 97, 110, 130, 202, 334, 595
  drainage, 19, 95, 257, 313–314, 334, 394, 396, 435–436, 471, 482, 495, 537, 577, 659
  fertiliser use, 19, 144, 163, 260, 313, 334, 404, 417, 453, 471, 482, 495, 557, 577, 595, 637
  forage harvesting replacing grazing, 22, 334
  industrialisation, 19–20, 23, 110, 237
  irrigation, 19, 95, 376, 471, 516, 526, 595
  loss of landscape features (hedges, etc.), 19, 237, 302, 312–313, 334, 417, 495, 575, 577
  monocultures/low crop diversity, 22, 334, 376, 495
  over-grazing, 95–96, 184, 376, 394–395, 468, 658–659
  pesticides, 19, 22, 144, 163, 239, 313, 323, 417, 471, 495, 557, 595, 637
  ploughing and reseeding of semi-natural grassland, 22, 95, 110, 313–314, 587, 595
  specialisation, 19, 110, 129, 160, 163, 257, 407, 595
Aichi targets. *See* Convention on Biological Diversity targets
Ailanthus altissima, 535, *See also* Tree of Heaven
Alauda arvensis. *See* Skylark
Albania, 1, 26, 30–31, 41, 49, 125, 357, 364
Alburnus alburnus. *See* Common Bleak
Alca torda. *See* Razorbill
Alcedo atthis. *See* Common Kingfisher
Alces alces. *See* Elk
Alectoris chukar. *See* Chukar
Alectoris graeca whitakeri. *See* Sicilian Rock Partridge
Alnus glutinosa, 160, 291, 575, 680
Alnus incana, 291
Alpine biogeographical region, 13, 15, 200
  Austria, 142–143, 148
  Croatia, 200, 209–210
  Finland, 292–293
  France, 311–312
  Germany, 332–333
  Poland, 493
  Romania, 534, 547
  Slovakia, 555
  Spain, 593
  Sweden, 613, 621, 630
Alpine Ibex, 417, 420, 429
Alpine Longhorn Beetle, 142
Alpine Marmot, 420
Alps, 13, 15, 17, 20, 29, 32–33, 142–143, 149, 151, 311, 415, 574–575
Alytes obstetricans. *See* Common Midwife Toad
Ambrosia artemisifolia, 535
Amelanchier spicata. *See* Low Juneberry
American Bullfrog, 163
American Mink, 275–276, 288, 304, 453, 507, 606
Amorpha fruticosa, 535, *See also* False Indigo-bush
Anas acuta. *See* Pintail
Anas crecca. *See* Common Teal
Anas platyrhynchos. *See* Mallard
Andorra, 30–31, 125
Angelica palustris, 243
Anser albifrons albifrons. *See* Greater White-fronted Goose
Anser anser. *See* Greylag Goose
Anser brachyrhynchus. *See* Pink-footed Goose
Anser erythropus. *See* Lesser White-fronted Goose
Apennine Chamois, 420–421, 429
Apennines, 13, 32, 415
Apollo, 360, 561
appropriate assessment (HD Article 6(3)). *See* Natura 2000
Apus apus. *See* Common Swift
aquaculture, biodiversity pressures, 395–396, 537
aquaculture, environmentally friendly, 185, 510
Aquatic Warbler, 50, 458, 494–495, 505, 513
Aquila adalberti. *See* Spanish Imperial Eagle
Aquila chrysaetos. *See* Golden Eagle
Aquila fasciata. *See* Bonelli's Eagle
Aquila heliaca. *See* Eastern Imperial Eagle

Arctic biogeographical region, 15
Ardea alba. See Great White Egret
Ardea cinerea. See Grey Heron
Arenaria interpres. See Turnstone
Arhus Convention. *See* Convention on Access to Information, Public Participation in Decision Making and Access to Justice in Environmental Matters
Arnica montana, 614
Artemisia laciniata, 148
Asclepias syrica. See Common Milkweed
Ash, 453
Asia, 17, 50
Asian Hornet, 527, 606
Asp, 284
Aspen, 291
Aspius aspius. See Asp
Astacus astacus. See Noble Crayfish
Athene noctua. See Little Owl
Atlantic, North-east (MSFD region), 13, 17, 90
Atlantic biogeographical region, 15, 33
  Belgium, 161
  Denmark, 254–255
  France, 311–313
  Germany, 332
  Portugal, 514
  Spain, 593
Atlantic marine region, 13, 18, 522
  Denmark, 256
  Germany, 332
  Spain, 593
  Sweden, 613, 621
Atlantic Ocean, 24–25, 28–29, 50, 115
Atlantic Salmon, 243, 266, 395, 400, 440
Atlantic Sturgeon, 479
Atriplex semibaccata, 222
Aurochs, 14, 19, 243
Austria, 26, 30–32, 34, 43, 49, 108–109, 116, 125, 130, 142–158, 588
Austropotamobius pallipes. See Freshwater Crayfish
Avocet, 312, 332
Aythya ferina. See Common Pochard
Aythya fuligula. See Tufted Duck
Aythya marila. See Scaup
Aythya nyroca. See Ferruginous Duck
Azores, 1, 107, 125, 514, 529
Azores Bullfinch, 527

Badger, 500
Balaenoptera musculus. See Blue Whale
Balearic Shearwater, 50, 516, 594, 600
Balkan, 29, 32–33
Balkan Chamois, 360
Balkan states, 41, 197
Balkans (mountains and peninsula), 13, 181, 574
Ballast Water Management Convention (BWM Convention). *See* International Convention for the Control and Management of Ships' Ballast Water and Sediments
Baltic Dunlin, 257, 275, 288
Baltic Sea, 13, 17, 21, 24–25, 45, 50, 90, 115
  Denmark, 256, 258, 263
  Estonia, 273, 275
  Finland, 293–294, 300, 304, 307, 309
  Germany, 332, 334
  Lithuania, 451, 459, 463
  Poland, 493–494, 496, 507–508
  Sweden, 613, 615, 620–621, 627
Baltic Sea Action Plan, 45
Baltic states, 13, 29, 33, 41, 107, 300, 442
Barbastella barbastellus, 542
Barbel, 162
Barbus barbus. See Barbel
Barcelona Convention. *See* Convention for the Protection of the Mediterranean Sea against Pollution
bark beetle outbreaks, 144, 153, 237, 239, 249, 252, 453, 495, 499, 557, 577
Barn Owl, 234
Barn Swallow, 516
Barnacle Goose, 280, 332, 477
Basking Shark, 409
bats, conservation of, 50, 79
  Belgium, 168
  Estonia, 280
  France, 318
  Spain, 521
Bay of Biscay, 13, 17
Bearded Vulture, 360, 421, 596, 598, 600–602
Beaver, 14, 79, 114, 148, 168, 205–206, 243, 261, 274–275, 323, 333, 457, 478, 485, 495, 499–500, 511, 576, 620, 649, 661
Beech Marten, 500
bees, conservation of
  Belgium, 168
  Ireland, 400
Belarus, 1, 29–33, 49, 125, 502
Belgium, 19, 26, 30–31, 34, 42, 49, 108–109, 116, 125–126, 128–129, 160–179
  Brussels Capital Region, 160
  Flanders, 116, 160–161
  Wallonia, 160–161
Bern/Berne Convention. *See* Convention on the Conservation of European Wildlife and Natural Habitats
Beroe ovata, 25
Betula pendula, 291
Betula pubescens, 291
Betula pubescens subsp. *czerepanovii. See* Mountain Birch
Białowieża forest, 493, 498–499, 502, 506, 508, 511
biodiversity indicators, 8, 54, 90, 101, 106–107, 110, 118, 132, 251, 269, 275, 307, 327, 447, 465, 590, 624, 629, 635–636
  birds, 23–24, 110, 118, 144, 157, 238, 274, 313, 417, 436, 452, 470, 495, 516, 636
  butterflies, 110, 347, 470, 629
  SEBI indicators (Streamlining European Biodiversity Indicators), 83, 178
biodiversity offsets, 117–118
  Finland, 302, 308
  France, 322, 326
  Germany, 342–343
  UK, 657–658
Biodiversity Strategies
  Belgium, 167
  Bulgaria, 186
  Croatia, 204
  Cyprus, 225
  Denmark, 260
  Estonia, 277–278
  Finland, 297
  Germany, 338
  Greece, 358, 367
  Hungary, 379
  Ireland, 400
  Italy, 419
  Lithuania, 455–456
  the Netherlands, 475
  Poland, 499
  Portugal, 519–520
  Slovakia, 559
  Slovenia, 581
  Spain, 597
  Sweden, 618–619
  UK, 643–644
biogeographical regions, 6, 13
Biosphere Reserves. *See* UNESCO Man and Biosphere Programme
bird species, number in each European country, 31
BirdLife International, 66, 108, 123
Birds Directive (BD), 3, 59, 135
  aims (Article 2), 3, 63–64, 75
  Annex I birds, 3, 7–8, 27, 29, 34, 77, 106–107, 109–110, 113–114, 116–117
  Annex II birds that may be hunted, 75–78, 113–114, 122–123
  Annex III birds that may be traded, 75
  Articles, summarised versions, 62
  drivers and development, 4, 51, 60, 70
  funding, 80–82, 112, 114, 134–137, *See also* nature conservation funding
  habitat measures for birds (Article 3), 69
  impacts on bird populations, 83–84, 109
  influence on the development of the Habitats Directive, 65
  monitoring and reporting, 82–83, 134
    conservation status, 5–8, 77, 106–109, 116
    conservation status categories, 5
    implementation reports (Article 12), 83
  objectives and key measures, 3, 64
  Special Protection Areas (SPAs)
    coherence requirements (Article 4(3)), 66–67, 69
    identification and classification (Article 4), 66–69
    inclusion in the Natura 2000 network, 66
    management and protection (Articles 3(3) and 4(4)), 69–75, 402
    overall coverage, 126, *See also* protected area types and coverage
  species protection measures, 75–78
    derogation provisions and conditions (Article 9), 76–77, 121, 420
    general protection (Article 5), 75
    hunting regulation (Article 7), 75–78, 114, 121–123
    trade restrictions (Article 6), 75
  strengths and weaknesses, 111–114
  supporting measures, 79–80
    research topics (Annex V), 79

Birds Directive (BD) (cont.)
 transposition and role of case law, 65–66, 72, 127
Bison, 14, 243, 457–458, 494–495, 500, 506, 511, 561
Bison bonasus. *See* Bison
Black Cherry, 386
Black Francolin, 226
Black Grouse, 163, 238, 291, 440, 453, 479, 500, 506, 561
Black Locust, 374, 386
Black Rat, 527
Black Sea biogeographical region, 16
 Bulgaria, 534
Black Sea coast, 14, 29, 182, 184, 534
 Bulgaria, 535
Black Sea marine region, 13, 17–18, 24–25, 46, 50, 90, 92, 115
 Romania, 545
Black Stork, 162, 200, 274, 279, 288, 333, 440–441, 458, 485, 494, 506, 530
Black Woodpecker, 29, 453, 615
Black-bellied Sandgrouse, 516
Blackcap, 226–227
Black-tailed Godwit, 275, 288, 317, 468, 478, 482–484
Black-throated Diver, 613
blue infrastructure, 320, 322–323, 423, 486, 524
Blue Whale, 28
Bluefin Tuna, 600
Bluethroat, 485
Bombina bombina. *See* Fire-bellied Toad
Bonasa bonasia. *See* Hazel Grouse
Bonelli's Eagle, 421, 430, 526, 595
Bonn Convention. *See* Convention on Migratory Species (CMS)
Boreal biogeographical region, 15, 132, 292, 618
 Estonia, 273
 Finland, 291, 293
 Latvia, 435
 Lithuania, 451
 Sweden, 612–613
Bos primigenius. *See* Aurochs
Bosnia and Herzegovina, 1, 26, 30–31, 41, 49, 125, 576
Botaurus stellaris. *See* Eurasian Bittern
Botrychium simplex, 614
Bottlenose Dolphin, 201, 423, 521
Box Elder, 386
Branta bernicla. *See* Brent Goose
Branta canadensis. *See* Canada Goose
Branta leucopsis. *See* Barnacle Goose
Branta ruficollis. *See* Red-breasted Goose
Brent Goose, 332
Brexit, 5, 666, 668
 UK post-2020 plans and 2021 Environment Act, 645–647
Britain, 634–677, *See also* United Kingdom
Brook Lamprey, 256, 400, 479
Brown Bear, 19, 27, 111, 142, 148, 155, 189, 200, 206, 274, 280, 291, 298, 318, 333, 360, 420–421, 429, 496, 535, 541–542, 556, 561, 575–576, 581, 590, 595–596, 598–602, 605, 620
Brown Shrimp, 472
Bubo bubo. *See* Eagle-owl
Bucephalla clangula. *See* Goldeneye
Bucharest Convention. *See* Convention on the Protection of the Black Sea against Pollution
Bufo bufo. *See* Common Toad
Bufotes viridis. *See* Green Toad
built/infrastructure developments, 85, 110, 128, 136, 144, 163, 182, 202, 222, 243, 270, 279, 288, 293, 322, 324, 342, 355, 374, 376, 389, 470–471, 511, 521, 537, 544, 546, 577, 646, 657, 664
Bulgaria, 16, 26, 29–30, 32–34, 43, 49, 107–108, 123, 125, 127, 129, 134–135, 155, 181–199
Burhinus oedicnemus. *See* Eurasian Thick-knee/ Stone Curlew
butterflies, conservation of, 34, 113
 Croatia, 213
 Romania, 542
 Slovenia, 581, 587
 UK, 659
Buxbaumia viridis, 142

Cabrera's Vole, 594
Calandra Lark, 595
Calidris alpina. *See* Dunlin
Calidris alpina schinzii. *See* Baltic Dunlin
Calidris canutus. *See* Red Knot
Calonectris diomedea. *See* Scopoli's Shearwater
Canada Goose, 477
Canary Islands, 1, 107, 602
Canis aureus. *See* Golden Jackal
Canis lupus. *See* Wolf
Cantabrian Capercaillie, 595–596, 600
CAP (Common Agricultural Policy), 19, 61, 82, 88, 95–100, 410
 Agricultural Fund for Rural Development (EAFRD), 43
 Areas facing Natural Constraints (ANC) (M13)/Less Favoured Areas (LFA), 99–100, 130, 152–153, 195, 197, 214, 249, 313, 345, 369, 407, 425, 446, 465, 549, 567, 570, 607, 629
 CAP 2014-2020
  Priority 4, 97, 99
 CAP 2023-2027, 96–97, 130, 136
  CAP Strategic Plans (CSPs), 96, 131
  eco-schemes, 96, 131, 484
 environmental conditionality (cross-compliance), 96–97, 130, 584, 668
 European Agricultural and Guarantee Fund (EAGF), 95, 627
 European Agricultural Fund for Rural Development (EAFRD), 10, 80–81, 96, 99, 131, 151, 153, 175, 214, 232, 267, 284, 286, 288, 305, 326, 345, 369, 389, 407, 427, 439, 465, 486, 508, 527, 529, 549, 570, 588, 627, 662
 objectives and reforms, 95–96, 131
 organic farming measure (M11), 153, 567, 607
 payments linked to production, 96
 Pillar 1 greening measures (green direct payments), 96–97, 130, 152, 302, 587
  Ecological Focus Areas (EFAs), 97, 130, 152
  environmentally sensitive permanent grassland (ESPG), 97, 130, 213, 587
 RDP advisory services measure (M2), 427
 RDP agri-environment/agri-environment–climate (M10), 11, 42, 71, 77–99, 101, 103, 121, 129–131, 137, 139
  Austria, 152–153, 157
  Belgium, 161, 166–167, 171, 173
  Bulgaria, 184, 189, 195
  Croatia, 205, 213
  Cyprus, 232
  Czech Republic, 249
  Denmark, 267, 274
  Estonia, 276, 283, 286
  Finland, 292, 302, 305, 307
  France, 316–317, 320, 326
  Germany, 343, 345
  Greece, 367, 369
  Hungary, 384–386, 389, 391
  Ireland, 404–405, 407
  Italy, 419, 425–427
  Latvia, 439, 444, 446
  Lithuania, 456, 462, 465
  the Netherlands, 475, 478, 482–484, 486
  Poland, 496, 504–505, 508, 511
  Portugal, 526–527
  Romania, 542, 545, 549
  Slovakia, 567, 570–571
  Slovenia, 579, 585–588, 591
  Spain, 598
  Sweden, 617, 624, 627
  UK, 637, 642, 655, 659–660, 662, 666
 RDP basic services measure (M7), 232, 427
 RDP forest-environment-climate measure (M15), 100, 132, 153, 195, 389, 567, 662
 RDP investments in forests measure (M8), 100, 132, 249, 427, 588, 662
 RDP knowledge transfer measure (M1), 425, 427
 RDP Nature 2000 measure (M12), 71, 99, 132, 195, 213, 224, 234, 249, 286, 427, 461, 465, 508, 545, 549, 551, 567, 588
 RDP non-productive investments measure (M4.4), 99, 207, 213, 425, 427, 486, 662
 RDP organic farming measure (M11), 99, 130, 197, 214, 249, 267, 302, 313, 324, 345, 369, 446, 465, 529, 629
 Rural Development Programme (RDP) measures, 10, 71, 97–100, 118, 129–130, 132–133, 137, 175, 195, 213, 234, 249, 286, 288, 305, 369, 426, 446, 486, 508, 526–527, 545, 567, 570, 588, 627, 655, 659, 664
 Rural Development Programmes (RDPs), 11, 96, 173, 207, 224, 232,

248, 318, 326, 367, 405, 425, 439, 526, 607
Rural Development Regulation, 96, 100, 130
set-aside, 96, 234, 313, 334
Capercaillie, 75, 202, 238, 274, 291, 317, 440–441, 458, 500, 506, 537, 541, 557, 561, 566, 576
Capra ibex. *See* Alpine Ibex
Capra pyrenaica. *See* Spanish Ibex
Capreolus capreolus. *See* Roe Deer
Carassius auratus. *See* Goldfish
Carduelis carduelis. *See* Goldfinch
Caretta caretta. *See* Loggerhead Turtle
Carlina onopordifolia, 494
Carpathians, 13, 15, 17, 29, 32, 143, 151, 493, 511, 535, 547, 555
Carrion Crow, 637
Castor fiber. *See* Beaver
CBD (Convention on Biological Diversity), 1, 46, 52–54, 111, 116
Aichi targets, 54–55, 119
11 – protected areas, 9, 54, 68, 115, 123–124, 128, 423, 651
15 – ecosystem restoration, 116, 304
9 – IAS, 94
date since countries have been a Party, 49
in situ conservation (Article 8), 53
Other effective area-based conservation measures (OECM), definition of, 9
post-2020 global biodiversity framework, 52–56, 101, 119
protected area, definition of, 8–9, 124
Strategic Plan and targets for 2010, 54, 100
Strategic Plan and targets for 2020 (Aichi targets), 9, 55
sustainable utilisation, definition of, 75
CDDA (European Nationally Designated Areas Inventory), 8, 10, 123–124
Cedar of Lebanon, 232, *See* *Cedrus brevifolia*
Cedrus brevifolia, 220, 230, 232, *See also* Cedar of Lebanon
Celtic Seas, 13, 17
Central Mediterranean Sea, 17, 415
Cervus elaphus. *See* Red Deer
cetaceans (whales and dolphins), conservation of, 50, 56
Ireland, 400, 409
Italy, 423
Portugal, 521
Spain, 599, 604
Cetorhinus maximus. *See* Basking Shark
Chalara fraxinea, 203
Charadrius alexandrinus. *See* Kentish Plover
Charadrius hiaticula. *See* Common Ringed Plover
Chelonia mydas. *See* Green Turtle
Chersophilus duponti. *See* Dupont's Lark
Chukar, 226
Ciconia ciconia. *See* White Stork
Ciconia nigra. *See* Black Stork
Cinereous Vulture, 189, 360, 594, 596
Circus aeruginosus. *See* Marsh Harrier
Circus cyaneus. *See* Hen Harrier
Circus pygargus. *See* Montagu's Harrier
Cistus heterophyllus subsp. *Carthaginensis*, 595
CITES. *See* Convention on International Trade in Endangered Species of Wild Fauna and Flora
CJEU (Court of Justice of the European Union) cases, *66*
Basses Corbiéres (C-374/98), 70
C-10/96, 77
C-103/00, 360
C-157/89, 76
C-157/96, 73
C-182/02, 77
C-182/10, 73
C-247/85, 77
C-252/85, 77
C-279/11, 403
C-294/03, 402
C-3/96, 68
C-323/17, 73
C-334/04, 363
C-344/03, 77
C-371/98, 68
C-38/99, 76
C-392/96, 402
C-435/92, 76
C 473/19, 619
C-474/19, 619
C-504/14, 360
C-518/04, 360
C-66/06, 402
C-674/17, 298
C-79/03, 77
C-849/19, 363
Commission v. *France* (C-241/08), 71, 317
Commission v. *Ireland* (C418/04), 66, 399–400, 402
Commission v. *UK* (C-6/04), 65, 68
Eco Advocacy (C-721/21), 399
FIE (C-254/19), 399
Germany C-244/0571
Grace and Sweetman (C-164/17), 399
Holohan and others (C-461/17), 399
Lappel Bank (C-44/95), 68, 641
Leybucht Dykes (C-57/89), 70
NEPPC (C-470/16), 399
People Over Wind and Sweetman (C-323/17), 399
Santoña Marshes (C-355/90), 68
Sweetman (C-258/11), 399
Waddenzee (C-127/02), 72–73
Clanga clanga. *See* Greater Spotted Eagle
Clanga pomarina. *See* Lesser Spotted Eagle
Clangula hyemalis. *See* Long-tailed Duck
climate change
adaptation, 35, 46, 51, 67, 99, 101, 113, 128, 222, 298, 344, 364, 511, 571, 653
Finland, 304
France, 323
Hungary, 386–388
Spain, 607
UK, 661–662
biodiversity impacts, 25, 110, 144–145, 161, 164, 376, 453, 472, 482, 606, 614, 630, 637–638
biodiversity pressures, 24, 109, 111, 115, 144, 174, 184, 203, 221–222, 293–294, 314, 355, 395–396, 417, 471–472, 495, 517, 537, 577, 595, 630
ecosystem-based mitigation and adaptation, 101, 304, 308, 571, 604–605, 631, 653, *See also* nature-based solutions
carbon sequestration and storage, 116, 136, 175, 285, 303–305, 308, 630, 659, 664
effects, 24–25, 144–145, 222, 303, 308, 366, 375, 377, 472, 517, 630
mitigation, 99, 304, 626, 631
coastal habitats
habitat management and restoration
Latvia, 444
Romania, 546
Spain, 606
UK, 660–661
Cochlearia polonica, 494
Coenonympha oedippus. *See* False Ringlet
Colias myrmidone. *See* Danube Clouded Yellow
collectivisation, 19, 23, 237
Coluber cypriensis, 220
Columba palumbus. *See* Woodpigeon
Common Bleak, 516
Common Crane, 302, 489, 494–495, 500, 511, 620, 649
Common Eider, 255, 275, 298, 463, 469, 472
Common Hamster, 173, 500
Common Kestrel, 469–470
Common Kingfisher, 66
Common Milkweed, 386
Common Murre, 516
Common Pochard, 122, 261, 500
Common Rhododendron, 635
Common Ringed Plover, 312
Common Scoter, 255
Common Shelduck, 469
Common Snipe, 395, 482
Common Sole, 472
Common Spadefoot Toad, 162, 479
Common Swift, 163
Common Tern, 453
Common Toad, 239, 417
Common Tortoise, 535
Common Whitethroat, 313
Common Yew, 520
Conference on the Human Environment in Stockholm, 48, 59, 259, 519
conservation status. *See* Birds Directive and Habitats Directive, monitoring and reporting
Continental, 15, 332
Austria, 142–144, 157
Belgium, 160, 163
Croatia, 200–201, 209–210
Czech Republic, 237
Denmark, 254–255
France, 311–312
Poland, 493
Romania, 534
Slovenia, 575
Sweden, 613, 618
Convention for the Prevention of Pollution from Ships (MARPOL 73/78), 44

Convention for the Protection of the Marine Environment of the North East Atlantic (OSPAR Convention), 47, 56, *See also* protected areas, types and their coverage
Convention for the Protection of the Mediterranean Sea against Pollution (Barcelona Convention)/ Convention for the Protection of Marine Environment and the Coastal Region of the Mediterranean, 45, 56
 Specially Protected Areas of Mediterranean Importance (SPAMI), 8, 45, *See also* protected areas, types and their coverage
Convention on Access to Information, Public Participation in Decision Making and Access to Justice in Environmental Matters (Aarhus Convention), 48, 186, 337, 380, 399, 580
Convention on International Trade in Endangered Species of Wild Fauna and Flora (CITES), 44, 85
Convention on Long-Range Transboundary Air Pollution (CLRTAP), 46, 94
Convention on Migratory Species (CMS), 45, 49, 51, 112, 122, 134, 595
 action on lead gunshot, 51
 Agreements
  ACAP, 49–50
  ACCOBAMS, 49–50
  AEWA, 49–51, 77, 123
  ASCOBANS, 49–50
  EUROBATS, 49–50
  Wadden Sea Seal, 49–50
 COP resolutions, 50
 MoUs and Action Plans, 50
  African-Eurasian Landbirds, 50
  migratory raptors, 50–51
 protection for globally threatened species, 50
Convention on the Conservation of European Wildlife and Natural Habitats (Bern Convention), 41, 45, 49, 51–52, 77–78, 111, 121–122, 224, 226, 333, 419, 541, 581, 642
 Emerald Network, 52, 59, 112, 645
 enforcement/case files, 51, 185, 227, 359–360, 364
 influence on the Birds and Habitats Directives, 51
 protection of species listed in appendices, 51, 113, 204, 279
 Tunis Action Plan 2013–2020, 78
Convention on the Protection of the Black Sea against Pollution (Bucharest Convention), 46
Convention on Wetlands (Ramsar Convention), 8, 41–49, 66, 75
 Ramsar Sites, 41, *See also* protected areas, types and their coverage
Coracias garrulus. *See* Roller
Coregonus oxyrinchus. *See* Houting
CORINE, 5–6, 21–22, 26
CORINE Biotopes, 66, 187, 190, 390, 440, 446
Cork Oak, 514, 520, 526, 530, 594
Corn Bunting, 145, 257, 470, 556, 595
Corncrake, 202, 213, 312, 400, 435, 485, 505, 537, 542, 556, 576, 659
Cortaderia selloana. *See* Pampas Grass
Corvus corax. *See* Raven
Corvus corone. *See* Carrion Crow
Cotula coronopifolia, 616
Coturnix coturnix. *See* Quail
Council of Europe, 1, 13, 51–52, 100, 112
Council of Europe Landscape Convention, 602
Council of the European Union, 91, 93
COVID-19 pandemic, 11, 55, 61, 119, 135–136
Crangon crangon. *See* Brown Shrimp
Crassula helmsii. *See* New Zealand Pygmyweed
Crex crex. *See* Corncrake
Cricetus cricetus. *See* Common Hamster
Croatia, 26, 30–31, 34, 41, 43, 49, 108–109, 125–126, 134, 200–218, 576, 588
Curlew, 150, 317, 395, 468–469, 500
Cypriot Mouse, 220
Cypripedium calceolus. *See* Lady's Slipper Orchid
Cyprus, 26, 30–31, 33–34, 43, 49, 107–108, 122, 125, 127, 129, 219–236, 634
Cyprus Grass Snake, 221, 223
Cyprus Mouflon, 220, 223–224, 234
Cyprus Scops-owl, 220
Cyprus Warbler, 220–221, 234
Cyprus Wheatear, 220
Czech Republic, 21, 23, 26, 30–31, 33–34, 43, 49, 69, 108–109, 116, 119, 125, 128, 148, 237–254

Dalmatian Pelican, 193, 535, 542
Dama dama. *See* Fallow Dear
Damon Blue, 381
dams, 24, 35, 114, 202, 221, 276, 284, 314, 516, 521, 526, 557, 596
 removal of, 288, 314, 323, 606
Danube Clouded Yellow, 561
Danube Delta, 16, 28, 143, 534–535, 541, 543, 546, 548, 551
Danube Salmon, 535
data sources, 5–11
deer, effects of high densities, 24, 163, 658
Dendrocopos leucotos. *See* White-backed Woodpecker
Denmark, 9, 26, 30–31, 34, 41–42, 49–50, 108–109, 116–117, 123, 125, 136, 254–272, 332
Desmoulin's Whorl Snail, 256
development planning and impact assessment
 Austria, 151
 Belgium, 173
 Bulgaria, 193
 Croatia, 212
 Cyprus, 230–231
 Czech Republic, 247
 Denmark, 265
 Estonia, 283
 Finland, 301–302
 France, 322
 Germany, 342
 Greece, 365–366
 Hungary, 384–385
 Ireland, 404
 Italy, 425
 Latvia, 443–444
 Lithuania, 462
 Poland, 504–505
 Portugal, 525
 Romania, 545–546
 Slovakia, 566–567
 Slovenia, 585–586
 Spain, 605
 Sweden, 624–625
 UK, 656–658
Dianthus arenarius ssp. *bohemicus*, 243
Dicranum viride, 142
Dinaric Alps and region, 13, 200, 574
Directive on Environmental Crime, 87
Directive on the conservation of natural habitats and of wild fauna and flora. *See* Habitats Directive
Directive on the conservation of wild birds. *See* Birds Directive
disease/pathogens, 20, 24, 203, 453, 521, 575, 600, 635, 637
disturbance, biodiversity pressures, 20, 22, 111, 129, 185, 230–231, 238, 257, 260, 262, 277, 298, 334, 472, 482, 489, 496, 537, 577, 599
Douglas Fir, 314
drainage, 374
dredging, 276, 284, 293, 396, 524, 637
Dreissena polymorpha. *See* Zebra Mussel
drivers and causes of change
 Austria, 144–145
 Belgium, 163–164
 Bulgaria, 184–186
 Croatia, 203
 Cyprus, 221–222
 Czech Republic, 239
 Denmark, 257–258
 Estonia, 276–277
 Finland, 293–294
 France, 313–314
 Germany, 334–335
 Greece, 355–356
 Hungary, 375–377
 Ireland, 395–396
 Italy, 417
 Latvia, 436–437
 Lithuania, 452–453
 the Netherlands, 470–472
 Poland, 495–496
 Portugal, 516–517
 Romania, 536–538
 Slovakia, 557
 Slovenia, 576–577
 Spain, 595–596
 Sweden, 615–616
 UK, 636–638
Dryobates minor. *See* Lesser Spotted Woodpecker
Dryocopus martius. *See* Black Woodpecker
dumping at sea, 56, 396, 638
Dunlin, 312, 332, 469
Dupont's Lark, 594
Dusky Large Blue, 576
Dwarf Mountain Pine, 257

Eagle-owl, 526, 620
Eastern Bloc, 19, 23, 119, 121
Eastern Europe, 1, 22–23

Eastern Imperial Eagle, 143, 197, 381, 556
ecological corridors and networks, 66, 69, 113
  Austria, 150
  Belgium, 164, 169, 171
  Bulgaria, 187, 189–190
  Croatia, 204
  Czech Republic, 243, 246
  Denmark, 260, 264–265
  Estonia, 282–283
  Finland, 302
  France, 113, 316, 320
  Germany, 340
  Hungary, 383–384
  Ireland, 403
  Italy, 423, 425
  Latvia, 442
  Lithuania, 113, 460
  the Netherlands, 113, 475, 479–480
  Poland, 503, 505
  Portugal, 523
  Romania, 544, 551
  Slovakia, 564–566, 570
  Slovenia, 580, 582, 591
  Spain, 599–601, 603–604
  Sweden, 621, 623
  UK, 652–653, 656–657, 662, 669
ecosystem services, 35, 101, 110, 118, 121, 307
ecosystem-based approach, 56, 87, 91–93, 115, 129, 658
Edible Dormouse, 500
Egypt, 122, 417
Egyptian Fruit bat, 221
Egyptian Vulture, 360, 369, 595
EIA Directive, 72, 85–86, 129
  implementation issues, 129, 151, 185, 192, 322, 366, 385, 402–404, 498, 505, 525–526, 546, 566, 656, 666
Eichhornia crassipes. See Water Hyacinth
Eleanora's Falcon, 354
Elk, 261, 291, 457
Emberiza calandra. See Corn Bunting
Emberiza cirlus. See Cirl Bunting
Emberiza hortulana. See Ortolan Bunting
Empetrum hermaphroditum, 142
Emys orbicularis. See European Pond Turtle
England, 118, 131, 136, 634–677
English Channel, 13, 17
Environment Commissioner, Virginijus Sinkevičius, 135
Environment for Europe Ministerial Conferences, 42, 100
Environmental Action Programme (EAP), 42, 59, 81, 100–101
environmental crime/wildlife crime, 87, 122, 227, 381, 400, 421, 599, 648
Environmental Liability Directive (ELD), 86, 89
environmental principles, 60, 646
Epidalea calamita. See Natterjack Toad
Equus ferus ferus. See Tarpan
Erica tetralix, 435
Erinaceus europaeus. See European Hedgehog
Erithacus rubecula. See Robin
Erysimum pieninicum, 494
Estonia, 26–27, 30–31, 33–34, 43, 49, 108–109, 116, 125–126, 130, 273–290
EU air pollution policy and legislation, 93–94
  Air Quality Directive, 87, 93
  Large Combustion Plants Directive, 94
  National Emissions Ceilings Directive (NECD), 87, 94
EU biodiversity strategies and targets, 60
  1998 European Community Biodiversity Strategy, 100
  2010 EU target and Biodiversity Strategy, 100–101
  EU Biodiversity Strategy to 2020,
    achievements, 116–119
    Target 1 improve habitats and species, 102, 112, 116
    Target 2 Action 6 green infrastructure, 116
    Target 2 Action 7a biodiversity proofing, 117
    Target 2 Action 7b no-net-loss, 117
    Target 2 ecosystem restoration, 102, 116–119, 133, 137, 174, 568
    Target 2 MAES, 118
    Target 3 agriculture and forestry, 102, 118, 132
    Target 4 fisheries, 102, 118–119
    Target 5 IAS, 94, 102, 118
  EU Biodiversity Strategy for 2030, 119–121, 124, 132–133, 135, 370, 488, 503, 511, 571, 581, 597, 604, 609, 616, 630
    EU Nature Restoration Plan – key commitments by 2030, 120
    Nature protection – key commitments by 2030, 120
    new governance framework, 120
EU Biodiversity Strategy Dashboard, 9
EU Circular Economy Action Plan, 61
EU Common Agricultural Policy (CAP). *See* Common Agricultural Policy
EU environmental legislation, 134–135, 137
  infringement cases, 61, 65, 68, 79, 122, 127–128, 135
    Austria, 150, 156
    Belgium, 166
    Bulgaria, 185, 187, 192, 198
    Cyprus, 225, 227, 229
    Estonia, 284
    Germany, 332, 337, 339, 342
    Greece, 358, 360, 363, 365, 367, 370
    Ireland, 399, 402–404, 406
    Italy, 418, 420, 423, 425
    Malta, 227
    Poland, 499, 503
    Portugal, 519, 524–525
    Slovakia, 561
    Slovenia, 580
    Sweden, 620, 630
  instruments that influence biodiversity, 86
  nature conservation legislation, 3, 61–84, 111–114
  pilot cases
    Cyprus, 225
    Italy, 420, 423–424
  pilot mechanism and infringement procedures, 65, 135
EU environmental policy, its development, 59–61
EU Farm to Fork Strategy, 61, 119, 121, 131, 609
EU forest policy, 85, 99, 132, 136
  EU Forest Strategy, 85, 99, 132
EU maritime policy and legislation,
  Common Fisheries Policy (CFP), 56, 87, 92–93, 396, 410
    European Fisheries Fund (EFF), 80–81, 627
    European Maritime and Fisheries Fund (EMFF), 80–81, 92–93, 116, 133, 176, 195, 214, 232, 305, 326, 369, 427, 458, 463, 486, 508, 607, 627, 664
    European Maritime, Fisheries and Aquaculture Fund (EMFAF), 93
    measures, 128, 410, 463, 619
  Integrated Coastal Zone Management (ICZM), 91, 366
  Marine Strategy Framework Directive (MSFD), 13, 17, 87, 90–91, 93, 112, 118, 129, 167, 172, 263, 282, 402, 410
    good environmental status, 87, 90–92, 102, 116–118, 129, 167, 172, 658
    marine protected areas (MPAs). *See* protected areas, types and their coverage
    strengths and weaknesses, 115–116
  Maritime Spatial Planning Directive (MSPD), 87, 91–92, 116, 129
EU Nature Restoration Law (proposal), 61, 119, 131, 137, 630
EU Platform on Coexistence between People and Large Carnivores, 79
EU Soil Strategy, 60
EU Sustainable Development Strategy, 100
EU Treaties
  Amsterdam (1997), 43, 60–61, 100
  Lisbon (2007), 43
  Maastricht (1992), 42, 60
  Rome (1957), 42, 59–60, 95
  Single European Act (1986), 42, 60
  TEU (Treaty on the European Union), 43, 70, 92
  TFEU (Treaty on the Functioning of the European Union), 43, 65, 80
EU water policy and legislation,
  Environmental Quality Standards Directive, 115
  Floods Directive, 86, 115
  Groundwater Directive, 115
  Nitrates Directive, 86, 90, 115, 323, 489, 507
  Urban Waste Water Treatment Directive (UWWTD), 86, 90, 115
  Water Framework Directive (WFD), 60–91, 98, 112, 153, 222, 265–266, 284, 334, 366, 389, 391, 400, 410, 476, 496, 506–507, 526, 539, 585, 626
    ecological status, 24, 89, 91, 114, 157, 396, 407, 577
    river basin management plan (RBMP), 89, 98, 115, 222, 506, 526, 559

EU water policy and legislation (cont.)
strengths and weaknesses, 114–115
water status, 89, 114–115
water policy Fitness Check, 114–115
Euphydryas aurinia. *See* Marsh Fritillary
Euphydryas maturna. *See* Scarce Fritillary
Eurasian Bittern, 561, 661
Eurasian Lynx, 27, 142, 148, 150, 200, 206, 217, 238, 243, 274, 280, 291, 297–298, 304, 333, 440, 458, 500, 535, 542, 556, 561, 575–576, 620, 661
Eurasian Thick-knee/Stone Curlew, 148, 659
Eurasian Wild Horse. *See* Tarpan
European Beech, 160, 332, 613
European Breeding Bird Atlas (2nd), 29
European Commission, support and enforcement role, 65, 68, 74, 77–79, 121, 123, 127, 135, 525
European Community, 1, 42, 51, 60–61, 94, 100
European Council, 101
European Environment Agency (EEA), 1, 5, 8, 21, 60, 67
European Green Deal, 61, 101, 119, 131, 356, 370, 511
European Hare, 223, 226, 360, 457, 477
European Hedgehog, 390
European Mink, 275–276, 280, 288, 435, 595, 600, 606
European Mouflon, 376, 386
European Parliament, 60, 93, 119, 406
European Plaice, 472
European Pond Turtle, 440, 500, 561
European Rabbit, 477, 500, 526, 600–601, 635
European Rewilding Network, 173
European Souslik, 148, 243, 500, 535, 567
European Topic Centre on Biological Diversity (ETC/BD), 7–8, 64, 67
European Tree Frog, 261, 479
European Union, 49
extinction debt, 35, 276, 284

Fagus sylvatica, 574, *See also* European Beech
Falco cherrug. *See* Saker Falcon
Falco eleonorae. *See* Eleanora's Falcon
Falco naumanni. *See* Lesser Kestrel
Falco peregrinus. *See* Peregrine Falcon
Falco tinnunculus. *See* Common Kestrel
Falco vespertinus. *See* Red-footed Falcon
Fallow Deer, 376
False Indigo-bush, 185, 386
False Ringlet, 576
Fan Mussel, 595
farmland birds, conservation of, 130
the Netherlands, 478, 482–484
UK, 659–660
Federation of Associations for Hunting and Conservation of the EU (FACE), 77, 122–123
Felis silvestris. *See* Wildcat
Fen Orchid, 381
Fenton's Wood White, 142
Ferruginous Duck, 185, 200, 202, 535, 542, 561
Finland, 26, 29–31, 34, 43, 49, 107–109, 124–125, 132, 291–310, 612, 614, 625
Finnish Forest Reindeer, 291
Fire Salamander, 390
Fire-bellied Toad, 261, 267, 453
fires, 24–25, 185, 231–232, 303, 313, 366, 503, 595
fish, conservation of, 79, 118
Austria, 153
Denmark, 266
Estonia, 284
France, 323
Greece, 360
Poland, 507, 510
Slovenia, 584
Sweden, 626
fisheries, biodiversity pressures, 28, 115–116, 164, 270, 396, 410, 596
bottom trawling/dredging, 21, 115, 164, 258, 263, 472, 615, 637, 654
by-catch, 21, 24, 110, 115, 121, 164, 277, 496, 516, 615, 637
overfishing/unsustainable use, 21, 24–25, 92–93, 110, 115, 118, 277, 355, 417, 489, 516, 537, 546
fisheries, measures to address pressures, 56, 78, 92–93, 118, 214, 228, 230, 256, 298, 320, 366, 370, 396, 599, 604, 606, 637, 654, 668
Fitness Check of the Birds and Habitats Directives, 33, 61, 83–84, 110, 112, 116, 126, 135, 448
lessons, 114
Floating Pennywort, 163, 472
flood prevention/protection/defence, 20, 257, 334, 345, 485, 537, 660–661
FOREST EUROPE process, 132
Forest Stewardship Council (FSC), 174, 188, 193, 526, 547
forestry
forest certification, 171, 174, 188, 193, 526, 547, 622, 658
management plans, 131–132, 169, 173–174, 462, 506, 539, 545, 547, 565–566
sustainable forest management, 24, 99, 132, 169, 173–174, 188, 212, 499, 506, 575, 658
forestry, biodiversity pressures, 121
abandonment of traditional management (e.g. coppicing), 20, 24, 248, 471, 637
afforestation, 21, 95, 144, 276, 374, 376, 394, 396, 421, 453, 526, 537, 574, 594, 606, 637–638, 658
bioenergy, 132, 145, 308, 314, 631
clear-cutting, 20, 110, 132, 145, 239, 303, 376, 394, 396, 453, 615, 625
conversion of semi-natural forest to intensive forestry, 23–24, 27, 132
deadwood removal, 20, 143–144, 152, 293, 303, 417, 495
decline in old growth/old trees, 276, 293, 303, 495, 548, 557
drainage, 130, 257, 275, 291, 293, 303, 396
fertiliser use, 396
forest roads, 303, 557
high harvest rates, 132, 274, 276, 288, 303, 314, 436, 453, 495, 506, 566, 595
illegal logging, 185, 230, 535, 537, 547–548
intensification, 20, 110, 132, 274, 276, 293, 303, 314, 376, 453, 557
non-native trees, 20, 132, 182, 314, 374, 386, 394
eucalyptus, 20, 516, 519, 526, 593
Pinus, 20, 516, 593
pesticides, 396
plantations, 20, 23, 182, 237, 239, 386, 394, 396, 435, 516, 593, 596, 613, 634, 658, 660
forests
biodiversity changes, 302–303
habitat management and restoration, 131–132
Austria, 152
Belgium, 173–174
Bulgaria, 193
Croatia, 212
Estonia, 283–284
Finland, 303–304
Germany, 343–344
Greece, 366
Latvia, 445
Poland, 506
Portugal, 526
Romania, 548
Slovakia, 567
Slovenia, 586
Sweden, 625
UK, 660
pest outbreaks, 185, 495, 503, 557
primeval/primary/virgin, 27, 132, 149, 239, 463, 535, 547, 555, 571
protection
Bulgaria, 193
Finland, 303–304
Greece, 366
Lithuania, 462–463
Sweden, 625
France, 26, 29–30, 32–33, 42, 49, 69, 108–109, 117–118, 122, 125, 127–128, 311–330, 529
Francolinus francolinus. *See* Black Francolin
Fraxinus angustifolia. *See* Narrow-leaved Ash
Fraxinus excelsior, 160, 575, 680, *See also* Ash
Freshwater Crayfish, 393
Freshwater Pearl Mussel, 243, 275, 395–396, 400, 408, 435, 440, 614, 620, 626
Fulica atra. *See* Coot

Galemys pyrenaicus. *See* Pyrenean Desman
Galium cracoviense, 494
Gallinago gallinago. *See* Common Snipe
Gallinago media. *See* Great Snipe
Garganey, 482
Gavia arctica. *See* Black-throated Diver

Gelochelidon nilotica. *See* Gull-billed Tern
Gentianella bohemica, 148, 243
Germany, 13, 19, 21, 26, 28, 30–34, 41–42, 49–50, 70–71, 107–109, 118, 125, 127, 136, 148, 155, 206, 261, 270, 331–352, 472
Geronticus eremita. *See* Bald Ibis
Glis glis. *See* Edible Dormouse
Golden Eagle, 275, 279–280, 297, 400, 530, 620, 648
Golden Jackal, 189, 500
Golden Plover, 75
Goldeneye, 496, 542
Goldfinch, 144, 516
Goldfish, 386
Goosander, 275
Goshawk, 648
governance, legislation and policy, 134–137
  Austria, 147
  Belgium, 164
  Bulgaria, 186–188, 192–193
  Croatia, 203–205
  Cyprus, 222–225
  Czech Republic, 239–247
  Denmark, 258–260
  Estonia, 277–279
  Finland, 294–297
  France, 314–317
  Germany, 335–338
  Greece, 356–359
  Hungary, 377–380
  Ireland, 396–400
  Italy, 417
  Latvia, 437, 439
  Lithuania, 453–456
  the Netherlands, 472–477
  Poland, 496, 498–499
  Portugal, 517, 519–520
  Romania, 538–540
  Slovakia, 557–560
  Slovenia, 577–581
  Spain, 596–598
  Sweden, 616–619
  UK, 638–647
grassland
  habitat management and restoration
    Estonia, 284
    Ireland, 404–405
    Latvia, 444–445
    the Netherlands, 482–484
grassland and HNV farmland
  Romania, 546–547
Great Britain, 13, 17, 29, 33, 393, 634–677
Great Bustard, 27, 50, 143, 155, 381, 385–386, 515–516, 530, 542, 567, 594
Great Cormorant, 168, 280, 381, 500, 561, 620
Great Crested Newt, 79, 121, 242, 648
Great Snipe, 494, 500
Great White Egret, 375, 381
Great White Pelican, 182
Greater Spotted Eagle, 279–280, 288, 297, 494, 500
Greater White-fronted Goose, 332, 542
Greece, 14, 26, 29–30, 32–33, 42, 49, 108–109, 119, 125–126, 129, 133, 353–373
green infrastructure, 116–117, 129
  Belgium, 164, 167, 171, 174–175
  Bulgaria, 193–194, 197
  Croatia, 213
  Cyprus, 232
  Denmark, 260
  Estonia, 285
  Finland, 302, 304, 307
  France, 316, 320, 322–323
  Germany, 340
  Hungary, 379
  Ireland, 403, 406
  Italy, 427, 429
  Latvia, 445
  the Netherlands, 486
  Portugal, 524, 527
  Romania, 548
  Slovakia, 565–566, 568, 571
  Spain, 604–605, 607, 610
  Sweden, 623, 626–627
  UK, 656
Green Toad, 275
Green Turtle, 220–221
Greenpeace judgement (UK High Court), 65, 645
Grey Heron, 168, 500
Grey Partridge, 167–168, 312, 333, 375, 457, 469–470, 542, 556
Grey Seal, 28, 256, 275, 280, 288, 332, 635
Grey-headed Woodpecker, 163
Greylag Goose, 317, 477, 542
Griffon Vulture, 189, 221, 323, 360, 421, 594, 596
Grus grus. *See* Common Crane
Gull-billed Tern, 332
Gulo gulo. *See* Wolverine
Gypaetus barbatus. *See* Bearded Vulture
Gyps fulvus. *See* Griffon Vulture

habitat/ecosystem degradation, 163, 184–185, 203, 334, 355, 365, 374, 376, 455, 537, 542, 577, 595–596, 636
habitat fragmentation, 22, 35, 110, 129, 144, 161, 163, 184, 203, 231, 238, 258, 314, 374, 417, 425, 470, 475, 480, 489, 496, 521, 542, 546, 557, 577, 595, 601
habitat loss/destruction/conversion, 110–111, 129, 131, 144, 163, 184–185, 314, 322, 374, 395, 415, 417, 452, 500, 516, 531, 595, 600, 619, 635–637, 659
habitat management and restoration, 135
  Austria, 151–153
  Belgium, 173–175
  Bulgaria, 193
  Croatia, 212–213
  Cyprus, 231–232
  Czech Republic, 248
  Denmark, 265–266
  Estonia, 283–285
  Finland, 302–304
  France, 322–323
  Germany, 343–344
  Greece, 366–367
  Hungary, 385–386
  Ireland, 404–406
  Italy, 425–426
  Lithuania, 462
  the Netherlands, 482–484
  Poland, 505, 507–508
  Portugal, 505
  Romania, 546–548
  Slovakia, 567–568
  Slovenia, 586–587
  Spain, 605–606
  Sweden, 625–626
  UK, 658–662
habitat status and trends, 110, 118
  Austria, 143–144
  Belgium, 161
  Bulgaria, 182–183
  Croatia, 201
  Cyprus, 220
  Czech Republic, 238
  Denmark, 255
  Estonia, 274
  Finland, 292–293
  France, 312
  Germany, 332–333
  Greece, 354
  Hungary, 375
  Ireland, 394
  Italy, 416
  Latvia, 435
  Lithuania, 451
  the Netherlands, 469
  Poland, 494
  Portugal, 515
  Romania, 536
  Slovakia, 556
  Slovenia, 575–576
  Spain, 594–595
  Sweden, 613–614
  UK, 635
habitat types (HD Annex I)
  1110 Sandbanks, 164, 413, 614
  1120* *Posidonia* beds, 18, 201, 220, 354, 415, 606
  1130 Estuaries, 413, 469
  1140 Mudflats, 161, 164, 413
  1150* Coastal lagoons, 201, 222, 394, 494
  1160 Large shallow inlets, 413, 469, 613
  1170 Reefs, 201, 400, 614
  1230 Vegetated sea cliffs, 494, 613
  1310 *Salicornia*, 616
  1320 Spartina swards, 255
  1330 Atlantic salt meadows, 255
  1520* Iberian gypsum vegetation, 230
  1530* Pannonic salt steppes, 16, 143, 201
  1620 Boreal Baltic islets, 613
  1630* Boreal Baltic coastal meadows, 15, 273, 292
  2110 Embryonic shifting dunes, 201, 255, 453, 494
  2120 Shifting dunes, 161, 494
  2130* Fixed coastal dunes, 394, 469, 494
  2140* Decalcified fixed dunes, 255
  2160 Dunes with *Hippophaë*, 161
  2180 Wooded dunes, 435, 453, 494
  21A0* Machairs, 15
  2250* Coastal dunes with *Juniperus*, 360
  2320 Dry sand heaths, 255
  2340* Pannonic inland dunes, 16, 143, 201
  3110 Oligotrophic waters, 494, 507, 612
  3130 Oligotrophic to mesotrophic waters, 612
  3140 Hard oligo-mesotrophic waters, 494, 507
  3150 Natural eutrophic lakes, 161, 442, 494
  3160 Natural dystrophic lakes, 494, 612
  3180* Turloughs, 393
  3210 Fennoscandian natural rivers, 613–615
  3230 Alpine rivers with *Myricaria*, 201
  3260 Water courses – plain to montane, 201, 507
  3270 Rivers with muddy banks, 201, 469

habitat types (HD Annex I) (cont.)
32A0 Tufa cascades of karstic rivers, 201
4010 North Atlantic wet heaths, 332, 435
4030 European dry heaths, 201, 332, 444
4060 Alpine and Boreal heaths, 142, 332, 613
4080 Sub-Arctic *Salix* scrub, 613
5220* Mattoral with *Zyziphus*, 230, 594
6120* Xeric sand calcareous grasslands, 161
6150 Siliceous alpine and boreal grasslands, 311, 613–614
6170 Alpine and subalpine calcareous grasslands, 201, 332, 613–614
6210* Semi-natural dry grasslands, 273–274, 332, 393–394, 535
6220* Pseudo-steppe, 16, 201
6230* Species-rich *Nardus* grassland, 201, 535
6240* Sub-Pannonic steppic grasslands, 201
6250* Pannonic loess steppic grasslands, 16, 201, 375
6260* Pannonic sand steppes, 16, 201, 375
6270* Fennoscandian lowland grasslands, 274
6280* Nordic Alvar, 15, 273–274, 615
62A0 E. sub-Mediterranean dry grasslands, 201
62B0* Serpentinophilous Cyprus grassland, 220
62C0* Ponto-Sarmatic steppes, 535
6310 Dehesas, 16, 594
6410 *Molinia* meadows, 161, 201, 312, 375
6430 Hydrophilous tall herbs, 312, 332
6440 Alluvial meadows, 375
6450 N. boreal alluvial meadows, 15, 273–274
6460 Peat grasslands of Troodos, 220, 222
6510 Lowland hay meadows, 161, 312, 332, 375, 535
6520 Mountain hay meadows, 161, 312, 332, 376, 535
6530* Fennoscandian wooded meadows, 274, 444, 613
6540 Sub-Mediterranean grasslands, 201
7110* Active raised bogs, 273, 394, 445, 451, 453, 612–613
7130* Blanket bogs, 15, 394
7140 Transition mires, 273, 451, 612
7210* Calcareous fens, 273
7230 Alkaline fens, 273–274, 451–452, 493–494
7240* Alpine pioneer fens, 311
7310* Aapa mires, 15, 612
7320* Palsa mires, 613–614
8240* Limestone pavements, 393
8310 Caves not open to the public, 161, 201
8330 Sea caves, 201
8340 Permanent glaciers, 614
9010* Western taïga, 15, 274, 451, 612, 625
9020* Fennoscandian hemiboreal broad-leaved forest, 453
9030* Natural forests of upheaval coast, 15, 612
9040 Nordic subalpine/ subarctic forests, 613
9050 Fennoscandian herb-rich forests with *Picea*, 303, 453, 625
9070 Fennoscandian wooded pastures, 15, 274
9080* Fennoscandian deciduous swamp woods, 15, 274, 303, 451
9110 *Luzulo-Fagetum* beech forests, 15, 160, 493, 613
9120 Atlantic acidophilous beech forests, 311
9130 *Asperulo-Fagetum* beech woods, 160, 493, 613
9150 Medio-European limestone beech forests, 257
9160 Sub-Atlantic and medio-European oak and oak–hornbeam forests, 201, 493, 613
9170 *Galio-Carpinetum* oak–hornbeam forests, 15, 257, 493
9180* *Tilio-Acerion* forests of slopes, 201, 311
9190 Old acidophilous oak woods, 160, 452, 493, 613
91A0 Old sessile oak woods, 15
91C0* Caledonian forest, 15
91D0* Bog woodland, 303, 451, 612
91E0* Alluvial forests, 160–161, 200, 453, 575
91F0 Riparian mixed forests, 161, 200–202, 575
91H0* Pannonian woods, 16
91I0 Euro-Siberian steppic woods, 375
91J0* *Taxus baccata* woods, 15
91K0 Illyrian *Fagus sylvatica* forests, 200, 574
91N0* Pannonic sand dune thicket, 375
91P0 Holy Cross fir forest, 494
91S0* Western Pontic forest, 16
91V0 Dacian beech forest, 535
91X0*Dobrogean beech forest, 535
91Y0 Dacian oak and hornbeam forests, 535
9230 Galicio-Portuguese oak woods, 311, 594
9330 *Quercus suber* forests, 594
9370* Palm groves of Phoenix, 354
9390* Scrub with *Quercus alnifolia*, 220
93A0 Woodlands with *Quercus infectoria*, 220
9410 Acidophilous montane to alpine *Picea* forests, 201, 493
9430* Subalpine and montane *Pinus uncinata* forests, 142
9530* (Sub-) Mediterranean pine forests with black pine, 222
9560* Endemic forests with *Juniperus*, 232
9570* Tetraclinus articulata forests, 594
9590* Cedrus brevifolia forests, 220
Habitats Directive (HD), 3, 135
aims (Article 2(2)), 3, 63
Annex II species, 3, 5, 67
Annex IV species, 3, 5, 63, 78–79, 121
Annex V species, 3, 5
Articles, summarised versions, 62
drivers and development, 51, 60, 65, 70
favourable conservation status, 63, 67, 70, 79–80, 82, 112, 134
application to the Environmental Liability Directive, 86, 89
definition and assessment, 7, 64, 107, 134
favourable reference values, 7, 64, 230, 440, 466, 520, 613, 620
funding, 80–82, 112, 114, 132–137, *See also* nature conservation funding
Prioritised Action Frameworks (PAFs), 82, 133
provisions for SCIs (Article 8), 80
HD habitats (Annex I habitat types), 3, 5–6, 11, 13, 27, 66, 97, 106–107, 113, 116–117
HD species (Annex II and/or Annex IV and V), 5–6, 29, 34, 106–107, 113–114, 116–117
monitoring and reporting, 82–83, 134
conservation status, 5–8, 106–109, 116, 118
conservation status categories, 5
surveillance and reports (Article 17), 83–84
Natura 2000, coherence (Article 3), 67, 112–113
Natura 2000, coherence measures, 67
landscape features (Articles 3(3) and 10), 67, 136
Natura 2000, establishment of the network, 67–69, *See also* Natura 2000 coverage
Natura 2000, identification and designation of sites (Article 4), 66–67
biogeographical seminars, 67
designation of SCIs as SACs (Article 4(4)), 67
SCIs (Article 4(2)), 67
Standard Data Form, 67, 69, 447
Natura 2000, site management and protection, 69–75, *See also* Natura 2000
assessment of plans and projects (Article 6(3)), 70–73, 88, 127
conservation objectives, 69–71, 73–74
conservation priorities (Article 4(4)), 69
derogation conditions and compensatory measures (HD Article 6(4)), 73–75, 85, 127
deterioration and disturbance, prevention of (Article 6(2)), 70–71, 75, 127, 317
necessary conservation measures (Article 6(1)), 70
objectives and key measures, 3, 64
species protection measures, 78–79
derogation provisions and conditions (Article 16), 79, 207, 582, 602, 620, 648
management measures for Annex V species (Article 14), 79
strict protection of Annex IV species (Articles 12 and 13), 78, 121, 360

strengths and weaknesses, 111–114
supporting measures, 79–80
transposition and role of case law, 65–66, 72, 127
Haematopus ostralegus. *See* Oystercatcher
Haliaeetus albicilla. *See* White-tailed Sea-eagle
Halichoerus grypus. *See* Grey Seal
Hamatocaulis vernicosus, 453, *See also* Slender Green Feather Moss
Harbour Porpoise, 164, 256, 332, 334, 496, 500, 613, 620
Harlequin, 386
harmful subsidies, 93, 114, 129–130, 328, 386, 389–390
Harmonia axyridis. *See* Harlequin
Hazel Dormouse, 162, 256
Hazel Grouse, 163, 280, 291, 500, 576
HD habitats (Annex I habitat types). *See also* habitats (HD Annex I)
heathland
habitat management and restoration
Latvia, 444
Helsinki Convention on the Protection of the Marine Environment of the Baltic Sea Area, 45, 56
hen harrier, 397
Hen Harrier, 122, 312, 408, 648
Heracleum mantegazzianum, 507
Heracleum sosnowskyi, 507
High Nature Value (HNV) farming, 27–28, 34, 96–100, 110, 118, 129–131, 142, 152, 161, 173, 273, 276, 284, 302, 312, 313, 331, 332, 343, 347, 393, 415, 425, 515, 531, 534, 535, 546, 549, 555, 575, 588, 593
loss, 110, 129, 161, 276, 313, 331, 343, 417, 516, 575
support for, 34, 96–97, 99–100, 130, 152–153, 161, 175, 213, 326, 343, 546, 549, 570, 588
High Nature Value (HNV) forest, 28, 111
Hirundo rustica. *See* Barn Swallow
Holly, 520
Holm Oak, 514, 520
Horned Grebe, 494
House Sparrow, 516
housing, 21, 144, 220–221, 231, 312, 471, 489, 516, 654, 657
Houting, 256, 267
Hucho hucho. *See* Danube Salmon
Hungarian Meadow Viper, 381
Hungary, 26, 30–32, 34, 43, 49, 108–109, 116, 125, 130, 204, 214, 374–392, 588
hunting, biodiversity pressures, 28, 121, 224, 239, 260, 274, 277, 361, 395, 415, 420, 430, 477, 521, 535, 537, 542, 647, *See also* under species issues
birds, 110, 122, 226–227, 500, 648
mammals, 206, 275, 600
hunting, regulation, 75–79, *See also* species issues
Hydrocoloeus minutus. *See* Little Gull
Hydrocotyle ranunculoides. *See* Floating Pennywort
hydropower, 24, 111, 144–145, 185, 237, 298, 437, 525, 537, 557, 577, 613, 615, 621, 626, 631
Hyla arborea. *See* European Tree Frog

Iberia, 13, 22, 29, 32–33, 593–594, 600–601
Iberian coast, 13, 17
Ice Age, The/glaciation effects, 13–18, 29, 181, 254, 291, 311, 353, 393, 434, 575, 612, 635
Iceland, 1, 26, 28–30, 33, 49, 125, 197
Ilex aquifolium. *See* Holly
illegal killing/persecution, biodiversity pressures, 110, 121–122, 217, 226–227, 260, 298, 361, 400, 420, 440, 500, 521, 601, 648, *See also* poaching, *See also* species issues
Impatiens glandulifera. *See* Indian Balsam
Important Bird Areas (IBAs), as a reference for the identification of SPAs, 66, 68, 126
Indian Balsam, 386
Industrial Emissions Directive, 86, 89
Integrated Pollution Prevention and Control Directive, 86
International Convention for the Control and Management of Ships' Ballast Water and Sediments (Ballast Water Management Convention or BWM Convention), 48
International Council for Bird Preservation, 68
International Panel on Biodiversity and Ecosystem Services (IPBES), 21, 54–55
International Waterfowl and Wetlands Research Bureau, 68
invasive alien species (IAS), biodiversity pressures, 20, 25, 110, 144, 161, 163, 184, 203, 221–222, 247, 276, 314, 367, 374–375, 386, 417, 471–472, 507, 516, 526, 535, 537, 595, 615, 637
invasive alien species (IAS), measures to address, 87, 94–95, 118
Cyprus, 231
Czech Republic, 247–248
Finland, 292–293
Greece, 367
Hungary, 386
Lithuania, 463
Poland, 507
Portugal, 526–527
Spain, 606
Invasive Alien Species (IAS) Regulation, 48, 87, 94–95, 118
Ionian Islands, 362
Ionian Sea, 17, 353, 415
Ips acuminatus, 495
Ips typographus, 495
Ireland (Republic of), 13, 17, 26, 29–30, 32, 34, 42, 49, 66, 71, 93, 108–109, 116, 125, 127–128, 130, 135, 393–414, 529
Irish Sea, 50
Italy, 14, 19, 26, 30–33, 42, 49, 108–109, 113, 122, 125, 135, 148, 214, 415–433, 588
IUCN (The International Union for Conservation of Nature), 8–9, 11, 52, 113, 119, 123–124, 126

Juncker, Jean-Claude, 84

Karpathos Frog, 360
Kattegat, 13, 17
Kazakhstan, 1
Kentish Plover, 472
key developments in nature conservation, summary of, 397, 418, 438, 578
Austria, 145
Belgium, 165
Bulgaria, 186
Croatia, 203
Cyprus, 225
Czech Republic, 240
Denmark, 259
Estonia, 278
Finland, 295
France, 315
Germany, 336
Greece, 357
Hungary, 378
Lithuania, 454
the Netherlands, 473
Poland, 497
Portugal, 517
Romania, 538
Slovakia, 558
Spain, 597
Sweden, 616–617
UK, 639
key events in nature conservation policy and legislation in Europe, 42
Key statistics, 5–7
BD Annex I birds, 5
biogeographical regions, 5, 13
conservation status standard indicator of progress, 5–8
gross domestic product (GDP) per capita, 5
HD habitats (Annex I), 5
HD species (Annexes II and/or Annex IV and V), 5
land cover (CORINE), 5
marine waters area, 5, 13
national table
Austria, 142
Belgium, 160
Bulgaria, 181
Croatia, 200
Cyprus, 219
Czech Republic, 237
Denmark, 254
Estonia, 273
Finland, 291
France, 311
Germany, 331
Greece, 353
Hungary, 374
Ireland, 393
Italy, 415
Latvia, 434
Lithuania, 451
the Netherlands, 468
Poland, 493
Portugal, 514
Romania, 534
Slovakia, 555
Slovenia, 574
Spain, 593
Sweden, 612
UK, 634
population density, 5
table explanation, 6
terrestrial area, 5
key terms, definitions, 5–11
Killer Whale, 25, 600
Kosovo, 1, 26, 30–31
Kyiv Resolution, 43, 47, 100

Lacerta agilis. *See* Sand Lizard
Ladigesocypris ghigii, 360
Lady's Slipper Orchid, 280, 381
Lagocephalus sceleratus. *See* Silver-cheeked Toad-fish
Lagopus lagopus. *See* Willow Grouse
Lagopus lagopus scotica. *See* Red Grouse
Lagopus muta. *See* Rock Ptarmigan
lakes
  habitat management and restoration
    Greece, 367
    Latvia, 444
    Sweden, 625–626
Lampetra fluviatilis. *See* River Lamprey
Lampetra planeri. *See* Brook Lamprey
land use change/land cover change/land conversion, 19, 21–22, 29, 110–111, 152, 161, 163, 173, 221, 334, 344, 355, 366, 395, 402, 416, 435, 557, 595
landscape features, 69, 79, 97, 101, 184, 559, 571
Lanius collurio. *See* Red-backed Shrike
Lanius minor. *See* Lesser Grey Shrike
Lanius senator. *See* Woodchat Shrike
Lapland, 294, 303
Lapwing, 122, 257, 333, 395, 468, 482–484, 556
Larch, 242
Large Blue, 275, 280, 576
Larix decidua. *See* Larch
Latvia, 9, 26, 30–31, 34, 41, 43, 49, 108–109, 116, 123, 125, 132, 434–451
Lawton, John, 127, 652–653, 656, 662
Lebanon, 122
legislation. *See also* governance, legislation and policy
Lepomis gibbosus, 535, *See also* Pumpkinseed
Leptidea morsei. *See* Fenton's Wood White
Lepus europaeus. *See* European Hare
Lepus timidus. *See* Mountain Hare
Lesser Grey Shrike, 542, 556, 595
Lesser Horseshoe Bat, 142, 500
Lesser Kestrel, 189, 360, 515, 526
Lesser Spotted Eagle, 279, 297, 435, 500, 536, 542
Lesser Spotted Woodpecker, 452
Lesser White-fronted Goose, 360, 620
Levantine Viper, 225
Liechtenstein, 1, 26, 30–31, 49, 125, 197
light pollution, 163, 577, 637
Limosa limosa. *See* Black-tailed Godwit
Linaria loeselii, 453
Lionfish, 231
Liparis loeselii. *See* Fen Orchid
Lissotriton vulgaris. *See* Smooth Newt
Lithobates catesbeianus. *See* American Bullfrog
Lithuania, 26, 30–31, 33–34, 41, 43, 49, 69, 108–109, 125, 127, 132–133, 451–467
Little Bustard, 515–516, 594–595, 610
Little Gull, 453
Little Tern, 202, 257, 453, 472
Loggerhead Turtle, 201, 220, 354, 357, 360, 369, 423
Loiseleuria procumbens, 142
Lolium perenne, 19
Long-tailed Duck, 261, 463, 494, 496, 613, 615
Louisiana crayfish, 472
Low Juneberry, 444
Lower Danube Green Corridor, 188, 193–194, 544
Lucanus cervus. *See* Stag Beetle
Luronium natans. *See* Floating Water-plantain
Luscinia svecica. *See* Bluethroat
Lutra lutra. *See* Otter
Luxembourg, 1, 26, 30–31, 34, 42, 49, 108–109, 125–126
Lynx lynx. *See* Eurasian Lynx
Lyrurus tetrix. *See* Black Grouse

Macaronesian biogeographical region, 1, 5, 514, 593
Macaronesian islands, 29
Macaronesian marine region, 1, 5, 13, 29, 90, 523
Macrovipera lebetina. *See* Levantine Viper
Maculinea arion. *See* Large Blue
Maculinea/Phengaris nausithous. *See* Dusky Large Blue
Maculinea/Phengaris teleius. *See* Scarce Large Blue
Madeira, 1, 107, 125, 514, 529
Magallana gigas. *See* Pacific Osyter
Mallard, 395, 477, 500
Malta, 1, 26, 30–31, 34, 43, 49, 65, 107–108, 122, 125–126, 135, 227
Manx Shearwater, 28
Marbled Polecat, 535
Marbled Teal, 594
Margaritifera auricularia. *See* Spengler's Freshwater Mussel
Margaritifera margaritifera. *See* Freshwater Pearl Mussel
marine ecosystems, 21, 24–25, 34, 90–91, 110, 115, 134, 201, 353, 516, 596, 606–607, 636–637
  habitat management and restoration
    Spain, 606
marine nature conservation. *See also* EU Maritime Policy and Legislation
  Cyprus, 227
  Lithuania, 463
  Spain, 599–600, 606
  Sweden, 619
  UK, 644–645, 652, 654–655, 658
marine protected areas (MPAs). *See* Protected Areas, types and coverage
marine Regional Seas Conventions and North Sea Ministerial Conferences, 56–57
marine regions, 13, 18
Maritime Pine, 516
Marmaronetta angustirostris. *See* Marbled Teal
Marmota marmota. *See* Alpine Marmot
Marsh Fritillary, 267
Marsh Harrier, 167
Martes foina. *See* Beech Marten
Martes martes. *See* Pine Marten
Mauremys caspica. *See* Striped-neck Terrapin
Mauremys rivulata. *See* Striped-neck Terrapin
Mediterranean biodiversity hotspot, 18, 219, 353, 593
Mediterranean biogeographical region, 13, 16–17, 25, 29, 32–33, 122
  Croatia, 200–201, 209–210
  Cyprus, 219
  France, 311–313
  Greece, 353
  Italy, 420
  Portugal, 514
  Slovenia, 574
  Spain, 593, 607
Mediterranean islands, 14, 311
Mediterranean Monk Seal, 28, 50, 220–223, 225–227, 234–235, 354, 360, 369, 420, 423
Mediterranean Sea marine region, 13, 17–18, 24–25, 28–29, 33, 50, 90, 92, 115
  Cyprus, 234
  Greece, 354, 360
  Italy, 415
  Spain, 593, 604, 607
Melanitta fusca. *See* Velvet Scoter
Melanitta nigra. *See* Common Scoter
Melanocorypha calandra. *See* Calandra Lark
Meles meles. *See* Badger
Mergus merganser. *See* Goosander
Mesocricetus newtoni. *See* Romanian Hamster
meta-populations, 35, 70, 471
Microcarbo pygmaeus. *See* Pygmy Cormorant
Microtus cabrerae. *See* Cabrera's Vole
Microtus tatricus. *See* Tatra Vole
Milvus milvus. *See* Red KIte
mineral extraction (gravel, etc.), 185, 255, 258, 395, 506, 537, 637–638, 661
mining, 128, 136, 182, 376, 383, 394, 516, 537
Miniopterus schreibersii, 542
mires
  habitat management and restoration, 507
    Estonia, 285
    Ireland, 405–406
    Latvia, 445
    UK (peatlands and uplands), 658–659
mitigation hierarchy, 74, 117–118, 322, 342, 443
Mnemiopsis leidyi, 25, 535
Moldova, 1, 30–31, 41, 49, 125, 193
Monachus monachus. *See* Mediterranean Monk Seal
Montagu's Harrier, 167, 312, 500, 516
Montenegro, 1, 26, 30–31, 49, 125
Motacilla flava. *See* Yellow Wagtail
Mountain Birch, 291, 613
Mountain Hare, 291
multilateral environmental agreements (MEAs), 41, 44, 121
Mus cypriacus. *See* Cypriot Mouse
Muscardinus avellanarius. *See* Hazel Dormouse
Mustela lutreola. *See* European Mink
Mustela putorius. *See* Polecat
Myotis bechsteinii, 542
Myotis myotis, 542
Myotis oxygnathus, 542

Najas flexilis. *See* Slender Naiad
Narrow-leaved Ash, 200, 203
National Biodiversity Strategies and Action Plans

(NBSAP). *See* Biodiversity Strategies
Natrix natrix cypriaca. *See* Cyprus Grass Snake
Natterjack Toad, 168, 275, 280, 288, 400
Natura 2000, 112, 116, *See also* Habitats Directive
appropriate assessments (HD Article 6(3)), implementation issues, 129
Croatia, 212
Cyprus, 224, 231
Czech Republic, 247
France, 317, 320
Greece, 365–366
Ireland, 403–404
Italy, 424
Lithuania, 443
Poland, 504
Portugal, 525–526
Romania, 544
Slovakia, 566, 586
Sweden, 623
UK, 654
coherence, 67, 69–70, 73–74, 79, 112–113, 126, 128
coherence measures, 67
compensatory measures (HD Article 6(4)), implementation issues, 127, 172, 211, 247, 320, 343, 403, 504, 519–520, 525, 586, 625, 654
costs, 132–133
establishment of the network, 67–69
funding, 80, 82, 132–133
identification and designation of sites (HD Article 4), 66–67
integrated funding model, 80, 82, 133
overall coverage
marine, 69, 124
terrestrial, 68, 119, 124
Prioritised Action Frameworks (PAFs), 10–11, 82, 131, 133
Austria, 153
Belgium, 175–176
Bulgaria, 194, 197
Croatia, 213
Cyprus, 232
Czech Republic, 248
Denmark, 266–267
Estonia, 285–286
Finland, 304, 307
Germany, 344–345
Greece, 367, 369
Hungary, 388–389
Ireland, 406–407
Italy, 426–427, 429
Latvia, 445–446
Lithuania, 463
Poland, 508
Portugal, 527, 529
Romania, 548
Slovakia, 568
Slovenia, 587–588
Spain, 607
Sweden, 618, 626
UK, 662
site management and protection, 69–75, 127–128, 130
conservation objectives, 128
Natura 2000 barometer, 8, 68
natural history, 13
Austria, 142–143
Belgium, 160–161
Bulgaria, 181–182
Croatia, 200–201
Cyprus, 219–220
Czech Republic, 237–238
Denmark, 254–255
Estonia, 273–274
Finland, 291–292
France, 311–312
Germany, 331–332
Greece, 353–354
Hungary, 374–375
Ireland, 393–394
Italy, 415–416
Latvia, 434–435
Lithuania, 451
the Netherlands, 468–469
Poland, 493–494
Portugal, 514–515
Romania, 534–535
Slovakia, 555–556
Slovenia, 574–575
Spain, 593–594
Sweden, 612–613
UK, 634–635
natural history, European overview, 13–25
nature conservation benefits, 133, 136
Belgium, 175
Bulgaria, 194
Croatia, 213
Cyprus, 232
Czech Republic, 248–249
Denmark, 266–267
Estonia, 285–286
Finland, 305
France, 324
Germany, 344–345
Greece, 367
Hungary, 388–389
Ireland, 406
Italy, 426–427
Latvia, 445–446
Lithuania, 463
the Netherlands, 486
Poland, 508
Portugal, 527
Romania, 548
Slovakia, 568
Slovenia, 587
Spain, 607
Sweden, 627
UK, 662
nature conservation costs
Austria, 153
Belgium, 175
Bulgaria, 193–194
Croatia, 213
Cyprus, 232
Czech Republic, 248
Denmark, 266
Estonia, 285
Finland, 304–305
France, 323–324
Germany, 344
Greece, 367
Hungary, 388
Ireland, 406
Italy, 426
Latvia, 445
Lithuania, 463
the Netherlands, 485–486
Poland, 508
Portugal, 527
Romania, 548
Slovakia, 568
Slovenia, 587
Spain, 607
Sweden, 626–627
UK, 662
nature conservation funding, 80–82, 119, 121, 131–137
Austria, 153–156
Belgium, 175–176
Bulgaria, 194–197
Croatia, 214
Cyprus, 232–234
Czech Republic, 249–251
Denmark, 267–269
Estonia, 286
Finland, 305–307
France, 324–326
Germany, 345–347
Greece, 367–370
Hungary, 389
Ireland, 406–407
Italy, 427
Latvia, 446
Lithuania, 463–465
the Netherlands, 486–487
Poland, 508–510
Portugal, 527–530
Romania, 548–549
Slovakia, 568–570
Slovenia, 587–588
Spain, 607–609
Sweden, 627–629
UK, 662–666
funding sources
Cohesion Fund, 80–82, 133, 195, 249, 283–285, 288, 389, 446, 508, 529, 568
EU LIFE Programme, 11, 80–82, 121, 128, 133, 174, 189, 195, 214, 232, 249, 267, 288, 326, 345, 389, 407, 427, 446, 465, 508, 529–530, 549, 570, 588, 609, 627, 664, 666
EU Multiannual Financial Framework (MFF), 80, 117, 133, 136
European Agricultural Fund for Rural Development (EAFRD), 80–81, *See also* Common Agricultural Policy (CAP)
European Fisheries Fund (EFF). *See* Common Fisheries Policy (CFP)
European Maritime and Fisheries Fund (EMFF). *See* Common Fisheries Policy (CFP)
European Regional Development Fund (ERDF), 80–82, 128, 134, 155, 175, 195, 232, 249, 286, 305, 326, 369, 389, 407, 427, 461, 465, 508, 529, 549, 568, 570, 588, 607, 664
European Social Fund (ESF), 81, 133, 427, 609, 664
Horizon 2020 Framework Programme for Research and Innovation, 81, 446
Interreg, 155, 157, 166, 174–175, 214, 249, 367, 369, 407, 465, 529, 570, 588
payments for ecosystem services (PES), 176, 324, 327, 427, 446, 664
Nature Directives, 3, 135, *See also* Birds Directive and Habitats Directive
benefits, 114
BHD habitats and species, 5
costs, 114
Fitness Check, 43, 114, *See* Fitness Check of the Birds and Habitats Directives
funding, 80–82, 112, 114, 132–137, *See also* nature conservation funding
provisions for SCIs (Article 8), 113
monitoring and reporting, 82–83, 134
conservation status, 5–8, 106–109
conservation status categories, 5
objectives and key measures, 3, 64

Nature Directives (cont.)
  standard indicator of progress, 8, 107, 109
  strengths and weaknesses, 111–114
  supporting measures, 79–80
  transposition and role of case law, 65–66, 72, 127
nature-based solutions (NBS), 136
  flood protection
    the Netherlands, 485
  Hungary, 386–388
Neogobius melanostomus. *See* Round Goby
Neolithic, 18, 254, 393, 634
Neophron percnopterus. *See* Egyptian Vulture
Neovison vison. *See* American Mink
net gain (of biodivesity), 118, 136, 646, 657–658
Netherlands, 19, 26, 30–31, 34, 41–42, 49–50, 69, 108–109, 113, 116, 125, 135, 332, 468–492
New Zealand Pygmyweed, 472, 637
NGOs, stakeholders and the wider public, 134–135, 137
  Austria, 147
  Bulgaria, 188
  Croatia, 205
  Cyprus, 223–224
  Czech Republic, 241–242
  Denmark, 258–259
  Estonia, 279
  Finland, 294–296
  France, 315–316
  Germany, 335
  Greece, 359
  Hungary, 380–381
  Ireland, 398–399
  Italy, 419
  Latvia, 438
  Lithuania, 456–457
  the Netherlands, 476–477
  Poland, 497
  Portugal, 518
  Romania, 540–541
  Slovakia, 560
  Slovenia, 579–580
  Spain, 598–599
  Sweden, 617–618
  UK, 641–642
no net loss (of biodiversity), 61, 89, 117, 136, 322, 657
Noble Crayfish, 453, 479, 615
noise, biodiversity pressures, 115, 163, 396
Nordic countries, 295, 300, 305
North America, 17
North Macedonia, 1, 26, 30–31, 41, 49, 125, 357, 364
North Sea, 13, 21, 50, 56, 115, 164, 167, 172, 258, 332, 636, 658
North Sea Ministerial Conferences, 56
Northern Chamois, 541
Northern Ireland, 395, 634–677
Norway, 1, 13, 26, 28, 30–31, 41, 49, 122, 125, 197, 251, 620, 625
Norway Spruce, 291, 332, 344, 453, 612
Numenius arquata. *See* Curlew
Numenius tenuirostris. *See* Slender-billed Curlew
Nyctereutes procyonoides. *See* Racoon Dog

Odocoileus virginianus. *See* White-tailed Deer
Odum, H. T., 246
OECM (Other effective area-based conservation measures). *See* protected areas, definitions and criteria
Oenanthe cypriaca. *See* Cyprus Wheatear
Oenanthe fluviatilis. *See* River Water-dropwort
Olm, 575, 577
Oostvaardersplassen, 17, 480
Ophiostoma novo-ulmi (Dutch elm disease), 635
Orcinus orca. *See* Killer Whale
organic farming, 131, 343, 376, *See also* Common Agricultural Policy (CAP)
  biodiversity impacts, 131, 153, 249, 302, 510
  EU target, 101
Oriental Plane, 220
ORNIS Committee, 76–77
Ortolan Bunting, 313, 576
Oryctolagus cuniculus. *See* European Rabbit
Osmoderma eremita, 280
OSPAR Convention. *See* Convention for the Protection of the Marine Environment of the North East Atlantic
Osprey, 279, 297, 500, 506, 613
Otis tarda. *See* Great Bustard
Otter, 79, 115, 148, 156, 168, 243, 266, 275, 297, 318, 381, 420, 440, 478, 489, 500, 596
Otus cyprius. *See* Cyprus Scops-owl
Ovis aries musimon. *See* European Mouflon
Ovis gmelini ophion. *See* Cyprus mouflon
Ovis orientalis ophion. *See* Cyprus Mouflon
Oxyura leucocephala. *See* White-headed Duck
Oystercatcher, 468–469, 472, 482

Pacifastacus leniusculus. *See* Signal Crayfish
Pacific Oyster, 25, 615
Padus serotina. *See* Black Cherry
Pampas Grass, 606
Pandion haliaetus. *See* Osprey
Pan-European 2020 Strategy for Biodiversity, 47
Pan-European Biological and Landscape Diversity Strategy (PEBLDS), 42, 100, 133
Pannonian biogeographical region, 16
  Czech Republic, 237
  Hungary, 374
  Romania, 534
  Slovakia, 555
Pannonian plain, 574
Paris European Summit, 59
Parnassius apollo. *See* Apollo
Passer domesticus. *See* House Sparrow
Passer montanus. *See* Tree Sparrow
peat extraction, 273–274, 276, 393–395, 403–404, 406, 453, 615
Pelecanus crispus. *See* Dalmatian Pelican
Pelecanus onocrotalus. *See* Great White Pelican
Pelobates fuscus. *See* Common Spadefoot Toad
Pelophylax cerigensis. *See* Karpathos Frog
Perdix perdix. *See* Grey Partridge
Peregrine Falcon, 185, 239, 333, 620, 648
Petromyzon marinus. *See* Sea Lamprey
Phalacrocorax carbo. *See* Great Cormorant
Phasianus colchicus. *See* Pheasant
Pheasant, 75, 477, 542
Phocoena phocoena. *See* Harbour Porpoise
Phoenicolacerta troodica, 220
Phoenix theophrasti, 354
Picea abies, 303, *See also* Norway Spruce
Picea sitchensis. *See* Sitka Spruce
Picoides tridactylus. *See* Three-toed Woodpecker
Picus canus. *See* Grey-headed Woodpecker
Pine Marten, 500
Pink-footed Goose, 167
Pinna nobilis. *See* Fan Mussel
Pintail, 469
Pinus mugo. *See* Dwarf Mountain Pine
Pinus pinaster. *See* Maritime Pine
Pinus radiata, 593
Pinus sylvestris. *See* Scots Pine
Plant Protection Products Regulation, 86
plants, conservation of, 34, 78–79, 113
  Bulgaria, 189, 194
  Cyprus, 229
  Denmark, 266
  Italy, 419
  Sweden, 621
Platalea leucorodia. *See* Spoonbill
Platanus orientalis. *See* Oriental Plane
Pleuronectes platessa. *See* European Plaice
Pluvialis apricaria. *See* Golden Plover
Pluvialis squatarola. *See* Grey Plover
poaching, 185, 223, 226, 420–421, 430, 535, 541–542, 546, 601, 616, *See also* hunting and illegal killing
Podiceps auritus. *See* Horned Grebe
poisoning, 185, 360, 381, 395, 601
Poland, 13, 15, 21, 23, 26, 29–34, 43, 49, 108–109, 119, 123, 125, 127, 129, 132–133, 135, 204, 465, 493–512, 558, 562
Polecat, 500
policy. *See also* governance, legislation and policy
pollinators, conservation of, 101
  Belgium, 168
  Ireland, 400
polluter pays principle, 60, 136, 646, 668
pollution, 28, 114, 355, 417, 595
  acid rain and acidification, 20–21, 93–94, 471, 479–480, 489, 615, 625–626, 637, 658
  air, 20, 24, 46, 93–94, 110, 136, 239, 327, 417, 637, *See*
  from agriculture, 20–21, 144, 275, 417, 470, 472, 484, 489, 557, 577, 596, 606, 638
  from aquaculture, 395, 417
  ammonia, 21, 24, 93–94, 163, 265, 404, 471–472, 480, 637

chemicals, 110, 115, 284, 334, 557
critical load, 24, 94, 111, 121, 163, 334, 484, 626
DDT (Dichlorodiphen yltrichloroethane), 19, 239, 275, 288, 294, 333
eutrophication, 21, 93, 293, 471
freshwater habitats, 20, 163, 256, 294, 334, 444, 453, 507, 577, 615
marine habitats, 21, 115, 163, 276, 294, 334, 396, 417, 496, 507, 596, 615
terrestrial habitats, 21, 163, 257, 293, 341, 596, 615, 637, 658
heavy metals, 21, 25
industrial discharges/effluent, 20, 24
lead poisoning from gunshot, 24, 123, 419
marine litter/ocean waste, 25, 115, 334
nitrates, 86, 90, 506
nitrogen dioxide/nitrogen oxides, 21, 94, 239
nutrients (nitrogen and phosphorus), 20, 25, 110, 115, 163, 257–258, 265, 270, 294, 334, 396, 404, 496, 506–507, 615
oil, 276
ozone, 93–94
polychlorinated biphenyls (PCBs), 21, 25, 275, 288, 294, 334, 638
sewage, 20–21, 24, 577, 596, 638
sulphur, 21, 24, 93–94, 239, 294, 658
toxic contaminants, 25, 110, 115, 333, 638
from transport, 470, 489
from waste, 56, 90, 115, 144, 184–185, 202, 294, 394, 472, 546, 571, 615, 637
water, 24, 182, 185, 284, 394, 417, 496, 507, 526, 537, 567, 575
pollution, measures to address, 86–87, 114, 260, 596
Denmark, nutrient pollution, 265
lead gunshot, prohibition, 51, 123, 599
the Netherlands, nitrogen pollution, 484–485
Poland, water pollution, 506
UK, water pollution, 658
Polyommatus damon. See Damon Blue
Populus tremula. See Aspen
Porphyrio porphyrio. See Purple Gallinule
Portugal, 16, 19, 26, 30–31, 33–34, 42, 49, 107–109, 113, 125, 133, 514–534, 594, 600
Posidonia oceanica, 18, 415
powerlines, biodiversity pressures (collisions and electrocution), 186, 243, 314, 601
precautionary principle, 56, 60, 73, 127, 384, 525, 642, 646, 658
predators
effects of high densities, 238, 482, 637
persecution of, 110, 185, 304, 521, 535, 600–601, 648
pressures on habitats and species, conclusions, 110–111
pressure indicators, past trends and outlook, 110
Prioritised Action Frameworks (PAFs). *See* Natura 2000, funding
Procambarus clarkia. See Louisiana crayfish
Procambarus clarkii. See Red Swamp Crayfish
Programme for the Endorsement of Forest Certification (PEFC), 174, 547
protected areas, approaches to management, 128
Austria, 151
Belgium, 172–173
Bulgaria, 192
Croatia, 211–212
Cyprus, 230
Czech Republic, 246–247
Denmark, 264–265
Estonia, 283
Finland, 301
France, 321–322
Germany, 341–342
Greece, 364–365
Hungary, 384
Ireland, 403–404
Italy, 424–425
Latvia, 443
Lithuania, 461–462
the Netherlands, 480–482
Poland, 504
Portugal, 524–525
Romania, 545
Slovakia, 565–566
Slovenia, 584–585
Spain, 604–605
Sweden, 623–624
UK, 655–656
protected areas, approaches to protection, 127–128
Austria, 150–151
Belgium, 172
Bulgaria, 191–192
Croatia, 210–211
Cyprus, 229–230
Czech Republic, 246
Denmark, 264
Estonia, 282–283
Finland, 300–301
France, 320–321
Germany, 340–341
Greece, 364–365
Hungary, 383–384
Ireland, 403
Italy, 423–424
Latvia, 442–443
Lithuania, 460–461
the Netherlands, 480–482
Poland, 503–504
Portugal, 524
Romania, 544–545
Slovakia, 562–563
Slovenia, 582–584
Spain, 604–605
Sweden, 623
UK, 653–655
protected areas, coherence
Austria, 150
Belgium, 171–172
Croatia, 210
Cyprus, 229
Czech Republic, 246
Denmark, 264
Estonia, 281–282
Finland, 300
France, 320
Germany, 340
Greece, 364
Hungary, 383
Ireland, 402–403
Italy, 423
Latvia, 442
Lithuania, 459–460
the Netherlands, 479–480
Poland, 500–502
Portugal, 523–524
Romania, 544
Slovakia, 564–565
Slovenia, 584
Spain, 604
Sweden, 623
UK, 652–653
protected areas, definitions and criteria, 8–9, 123–124
protected areas, IUCN categories, 9–10, 127
protected areas, targets, 8–9, 119, 124, 128
protected areas, types and their coverage, 8–10, 115, 123–124, 126–128
Austria, 149–150
Belgium, 169–171
Bulgaria, 189
Croatia, 207–210
Cyprus, 227–229
Czech Republic, 243–246
Denmark, 261–264
Estonia, 280–281
Finland, 298–300
France, 318
Germany, 338–340
Greece, 362–364
Hungary, 381–383
Ireland, 400–402
Italy, 421–423
Latvia, 440–442
Lithuania, 458–459
the Netherlands, 479
Poland, 500–502
Portugal, 522–523
Romania, 542–544
Slovakia, 561–564
Slovenia, 582–584
Spain, 602–604
Sweden, 620–623
UK, 649–652
Protected Planet website, 8, 124, 128
Proteus anguinus. See Olm
Pseudotsuga menziesii. See Douglas Fir
Pterocles orientalis. See Black-bellied Sandgrouse
Pterois miles. See Lionfish
Pteromys volans. See Siberian Flying Squirrel
Puffinus mauretanicus. See Balearic Shearwater
Puffinus puffinus. See Manx Shearwater
Puffinus yelkouan. See Yelkouan Shearwater
Pumpkinseed, 472, 516
Purple Gallinule, 421
Pusa hispida. See Ringed Seal
Pusa hispida saimensis. See Saimaa Seal
Pygmy Cormorant, 182, 542
Pyrenean Desman, 594, 600
Pyrenean Oak, 514
Pyrenees, 13, 15, 17, 29, 32, 311, 318, 599, 601–602
Pyrrhula murina. See Azores Bullfinch

Quail, 221
Quaternary period, 13, 181, 254
Quercus alnifolia, 220
Quercus infectoria, 220
Quercus petraea. See Sessile Oak
Quercus pyrenaica, 594, *See also* Pyrenean Oak
Quercus robur, 160, 200, 332, 575, 594, 613
Quercus rotundifolia. See Holm Oak
Quercus suber. See Cork Oak

Raccoon Dog, 276, 304
Raja clavata. See Thornback Ray
Ramsar Convention/Sites. *See* Convention on Wetlands

Rangifer tarandus fennicus. See Finnish Forest Reindeer
Ranunculion fluitantis, 507
Rapana venosa, 535
raptors/birds of prey, conservation of, 50, 111
  Bulgaria, 189
  Estonia, 275
  Hungary, 381
  Ireland, 400
  Italy, 421
  Poland, 499
  Romania, 547
  Spain, 596–597
  UK, 648
Rattus rattus. See Black Rat
Raven, 479, 511
Razorbill, 516
recreation, 257, 262, 267, 270, 302, 305, 334, 339–341, 355, 394, 426, 453, 460–463, 469, 472, 480, 503, 508, 530, 568, 596, 629, 654, 659
Recurvirostra avosetta. See Avocet
Red Deer, 261, 376, 480, 541–542
Red Fox, 226–227, 360, 453, 457, 500, 599, 637
Red Grouse, 648
Red Kite, 338, 400, 649
Red Knot, 332, 472
Red List (Red Data Book) assessments, 7, 11, 35, 63
  Belgium, 161
  Bulgaria, 182–184
  Croatia, 202–203
  Cyprus, 221
  Czech Republic, 238, 242
  Denmark, 257
  European, 29, 106
  Finland, 292–294, 308
  France, 312, 317, 327
  Germany, 333
  Greece, 355, 370
  Ireland, 394, 409
  Italy, 416, 420
  Lithuania, 457
  Poland, 495, 511
  Sweden, 615, 629
  UK, 636
Red Squirrel, 291
Red Swamp Crayfish, 516
Red-backed Shrike, 442, 453, 556
Red-breasted Goose, 28, 182, 189, 192, 197, 535, 542
Red-eared Slider, 527
Red-footed Falcon, 536, 542, 561
Redshank, 468, 482
Redwing, 541
Regional Seas Conventions, 45–47, 56, 90
renewable energy, 25, 51, 185, 314, 365, 537, 577, 596, 658
reservoirs, 20, 314, 496, 511, 516
rewilding, 17, 34, 135
  Czech Republic, 243, 248
  Denmark, 262
  France, 327
  Hungary, 383
  the Netherlands, 480
  Poland, 507–508
  UK, 661
Rhinolophus ferrumequinum, 542
Rhinolophus hipposideros, 142, 500, 542, *See also* Lesser Horseshoe Bat
Rhododendron ponticum. See Common Rhododendron
Rhodopes (Rhodope mountains), 13, 181
Rhynchocypris percnurus. See Swamp Minnow
ricefields
  habitat management and restoration
    Italy, 425–426
Ringed Seal, 275, 613
Riparia riparia. See Sand Martin
river cannelisation/channel re-engineering, 20, 24, 237, 266, 314, 498, 557, 596
River Lamprey, 256
River Water-dropwort, 266
rivers
  habitat management and restoration, 507
    Austria, 152–153
    Denmark, 266
    Estonia, 284–285
    Latvia, 444
    the Netherlands, 485
    Poland, 506
    Spain, 606
  hydropower licensing
    Sweden, 626
Robin, 226
Robinia pseudoacacia. See Black Locust
Rock Ptarmigan, 420
Roe Deer, 280, 304, 376
Roller, 221, 435
Romania, 16, 26–27, 29–30, 34, 43, 49, 107–108, 123, 125, 128, 133–134, 155, 193, 534–554
Romanian Darter, 536
Romanian Hamster, 535
Romanichthys valsanicola. See Romanian Darter
Rosa rugosa, 257, 444, 507, 616
Rosalina alpine. See Alpine Longhorn Beetle
Roseate Tern, 400
Rospuda valley/fens, 135, 494, 498–499, 504, 511
Round Goby, 277
Rousettus aegyptiacus. See Egyptian Fruit Bat
Rowan, 291
Rupicapra pyrenaica. See Southern Chamois
Rupicapra pyrenaica ornata. See Apennine Chamois
Rupicapra rupicapra. See Northern Chamois
Rupicapra rupicapra balcanica. See Balkan Chamois
Rupicapra rupicapra tatrica. See Tatar Chamois
Russia, 1, 13, 50, 300, 305

Saimaa Seal, 292, 298
Saker Falcon, 143, 189, 197, 556
Salamandra salamandra. See Fire Salamander
Salmo farioides, 360
Salmo lourosensis, 360
Salmo macedonicus, 360
Salmo pelagonicus, 360
Salmo peristericus, 360
Salmo salar. See Atlantic Salmon
Sámi people, 297–298, 303, 612
Sand Lizard, 275, 280, 649
Sand Martin, 168
Sandwich Tern, 257, 332
Sardinian Warbler, 234
Saxicola rubetra. See Whinchat
Saxifraga hirculus, 453
Scandes, 15
Scandinavia, 13, 15, 25, 32, 121
Scarce Fritillary, 142, 243, 453, 576
Scarce Large Blue, 576
Scaup, 469, 541
Sciurus vulgaris. See Red Squirrel
Scopoli's Shearwater, 423
Scotland, 15, 123, 625, 634–677
Scots Pine, 291, 332, 612
SEA Directive, 72, 85–89, 129
  implementation issues, 129, 283–284, 404, 462, 505, 525–526, 545, 656, 666
  REFIT evaluation, 88, 129
Sea Lamprey, 256, 614
seabirds, conservation of, 56, 92, 118
  Croatia, 214
  UK, 92
seals, conservation of, 50
  Cyprus, 227
  Finland, 298
Serbia, 1, 26, 29–30, 41, 49, 125, 155, 214
Sessile Oak, 332, 613
Siberian Flying Squirrel, 274–275, 279–280, 291, 435
Sicilian Rock Partridge, 420
Signal Crayfish, 516, 615
Silene holzmannii, 360
Siluris glanis. See Wels Catfish
Silver-cheeked Toad-fish, 231
Sitka Spruce, 394
Skylark, 226, 257, 313, 333, 420, 470, 482, 541, 595
Slender Green Feather Moss, 256
Slender Naiad, 256
Slender-billed Curlew, 50
Slovakia, 21, 23, 26, 30–31, 33–34, 43, 49, 69, 107–109, 113, 125, 128, 130, 132–133, 151, 555–573
Slovenia, 26, 30–34, 43, 50, 108–109, 126, 132, 206, 214, 574–592
Smooth Newt, 239
solar energy, 222, 535, 596
Solea solea. See Common Sole
Solidago canadensis, 386
Solidago gigantea, 386
Somateria mollissima. See Common Eider
Sorbus aucuparia. See Rowan
Southern Chamois, 596
Spain, 14, 16, 19, 26, 29, 32–34, 42, 50, 68, 80, 108–109, 116–117, 122, 126, 130, 529, 593–611
Spalax leucodon. See Lesser Mole-rat
SPAMI (Specially Protected Areas of Mediterranean Importance. *See* Convention for the Protection of the Mediterranean Sea against Pollution (Barcelona Convention)/ Convention for the Protection of Marine Environment and the Coastal Region of the Mediterranean
Spanish Ibex, 596
Spanish Imperial Eagle, 521, 526, 530, 594–596, 600–601
spatial planning, 128–129
  Austria, 151
  Belgium (maritime/marine), 167
  Bulgaria, 193
  Croatia, 205, 212
  Cyprus, 222
  Czech Republic, 246–247
  Denmark, 265
  Estonia, 283, 288
  Finland (land-use planning), 301
  France, 315, 320, 322
  Germany (landscape planning), 337, 342–343
  Hungary, 383
  Ireland, 398
  Lithuania, 455
  maritime/marine, 87, 91–92, 115–116, 129, 621

the Netherlands, 474–475, 480, 488–489
Poland, 496, 503–505
Romania, 546
Slovakia, 559, 562, 572
Slovenia, 580, 585
Spain, 605
Sweden, 621, 624
UK (maritime/marine), 658
Spatula querquedula. *See* Garganey
species, areas with high richness in species, endemics and threatened species, 32
species-area relationship, 33
species issues
dealing with conflicts, 79
Austria, large carnivores, 157
Belgium, fish-eating species, Beaver and Wolf, 168–169
Croatia, large carnivores, 207
Hungary, e.g. Wolf, Otter and Great Cormorant, 381, 389
Italy, large carnivores, 421
the Netherlands, geese on farmland, 477–478
Portugal, Wolf, 521
Romania, large carnivores, 541, 552
Spain, large carnivores, 601–602
Sweden, crop damage, fisheries and large carnivores, 620
UK, Great Crested Newt, 121, 648–649
hunting, 4, 75–79, 114, 121–123
Belgium, 168
Croatia, 206
Cyprus, 223–224, 226, 228
Denmark, 260, 262
Estonia, 280, 288
France, 316–317
Greece, 360–362
Italy, 419–421, 430
Lithuania, 457
Poland, 500
Portugal, 521
Romania, 541–542
Spain, 599
Sweden, 618–619
UK, 641, 648
illegal killing/persecution, 51, 78, 87, 114, 122
Cyprus, songbirds, 226–227
Ireland, raptors, 400
Latvia, fish, 440
Poland, fish-eating species, 500
Portugal, predators, 521
Spain, Brown Bear, 601
UK, raptors, 648
large carnivores, management
Croatia, 206–207
Denmark, 261
Finland, 298
France, 318
Romania, 542, 552
Slovenia, 581–582
poisoning, addressing, 51
Greece, 360
Portugal, 521
Spain, 599–600
powerlines, reducing impacts
Czech Republic, 239, 243
Portugal, 521
Slovakia, 561
Spain, 601
Species of European Conservation Concern (SPEC), 28
species protection legislation (including hunting and fishing regulations), 148
Belgium, 167–172
Bulgaria, 188–189
Croatia, 205–206
Czech Republic, 242–243
Denmark, 260–261
Estonia, 279–280
Finland, 297–298
France, 317–318
Germany, 338
Greece, 359–360
Hungary, 381
Ireland, 400
Italy, 419–421
Latvia, 439–440
Lithuania, 457–458
the Netherlands, 477
Poland, 499
Portugal, 520
Romania, 541–542
Slovakia, 560–561
Slovenia, 581–582
Spain, 599
Sweden, 619–620
UK, 647–648
species recovery and re-introduction programmes, 121
Austria, 148–149
Belgium, 167–168
Bulgaria, 189
Croatia, 206
Denmark, 261
Estonia, 280
France, 317–318
Hungary, 381
Ireland, 400
Italy, 421
Latvia, 440
Lithuania, 458
the Netherlands, 478–479
Poland, 500, 506
Portugal, 521
Romania, 542
Slovakia, 561
Spain, 599–602
Sweden, 620
UK, 649
species status and trends, 110
Austria, 144
Belgium, 161–163
Bulgaria, 183–184
Croatia, 202
Cyprus, 220–221
Czech Republic, 238–239
Denmark, 256–257
Estonia, 274–276
Finland, 293
France, 312–313
Germany, 333–334
Greece, 354–355
Hungary, 375
Ireland, 394–395
Italy, 416–417
Latvia, 437
Lithuania, 452
the Netherlands, 469–470
Poland, 495
Portugal, 515–516
Romania, 536
Slovakia, 556–557
Slovenia, 575–576
Spain, 595
Sweden, 614–615
UK, 635–636
Speckled Ground Squirrel, 494
Spengler's Freshwater Mussel, 595, 600
Spermophilus citellus. *See* European Souslik
Spermophilus suslicus. *See* Speckled Ground Squirrel
Spiny Dogfish, 472
Spiraea tomentosa, 507
Spoonbill, 200, 332, 485
Squalus acanthias. *See* Spiny Dogfish
Stag Beetle, 162, 261, 390
Steppic biogeographical region, 16, 534–535
Romania, 534
Sterlet, 155
Sterna dougallii. *See* Roseate Tern
Sterna hirundo. *See* Common Tern
Sternula albifrons. *See* Little Tern
Strategic Action Plan for the Black Sea Region, 539
Streptopelia turtur. *See* Turtle-dove
Striped-neck Terrapin, 221, 223
Strix aluco. *See* Tawny Owl
surveillance, monitoring, reporting and research, 134–135, 407–409
Austria, 156
Belgium, 176–178
Bulgaria, 197–198
Croatia, 214–216
Cyprus, 234
Czech Republic, 251–252
Denmark, 269–270
Estonia, 286–288
Finland, 307–308
France, 326–327
Germany, 347
Greece, 370
Hungary, 389–390
Italy, 427–429
Latvia, 446–448
Lithuania, 465–466
the Netherlands, 487–488
Poland, 510
Portugal, 530
Romania, 549–551
Slovakia, 570–571
Slovenia, 588–590
Spain, 609
Sweden, 629–630
UK, 666–667
Sus scrofa. *See* Wild Boar
Sustainable Hunting Initiative, 77, 121
Swamp Minnow, 494
Sweden, 26, 29, 31, 34, 41, 43, 50, 94, 108–109, 126–127, 132, 612–631
Switzerland, 1, 26, 31–32, 50, 122, 126
Sylvia atricapilla. *See* Blackcap
Sylvia borin. *See* Garden warbler
Sylvia communis. *See* Common Whitethroat
Sylvia melanocephala. *See* Sardinian Warbler
Sylvia melanothorax. *See* Cyprus Warbler
Syria, 122

Tadorna tadorna. *See* Common Shelduck
Tarpan, 14
Tatar Chamois, 496
Tawny Owl, 29
Taxus baccata. *See* Common Yew
Testudo graeca. *See* Common Tortoise
Tetraclinis articulata, 594
Tetrao urogallus. *See* Capercaillie
Tetrao urogallus cantabricus. *See* Cantabrian Capercaillie
Tetrax tetrax. *See* Little Bustard
Thalasseus sandvicensis. *See* Sandwich Tern
Thesium ebracteatum, 453
Thornback Ray, 472
Three-toed Woodpecker, 557, 576, 615

Thunnus thynnus. See Bluefin Tuna
tourism, 153, 161, 163, 201–202, 219–221, 305, 355, 365, 367, 381, 424, 446, 496–497, 503, 510, 519–520, 527, 537, 546, 572, 577, 596, 605, 659
  ecotourism/nature tourism, 232, 285, 312, 326, 456, 461, 463, 486, 508, 523, 594, 627
Trachemys scripta. See Yellow-bellied Slider Turtle
Trachemys scripta elegans, 527
transport (roads, etc.), biodiversity pressures, 20, 22, 35, 110, 164, 206, 221–222, 314, 395, 435, 504, 517, 521, 557, 577, 596, 605, *See also* pollution
transport (roads, etc.), measures to address, 171, 340, 363
Tree of Heaven, 185, 386
Tree Sparrow, 144
Tringa totanus. See Redshank
Triturus cristatus. See Great Crested Newt
Tufted Duck, 469, 500
Turdus iliacus. See Redwing
Turkey, 1, 21
Turnstone, 312
Tursiops truncatus. See Bottlenose Dolphin
Turtle-dove, 78, 122, 221, 226, 317, 361, 420, 452–453, 516, 556, 595, 661
turtles, conservation of, 50, 92
  Croatia, 214
  Cyprus, 224, 227, 230, 234
  Greece, 360
Tyto alba. See Barn Owl

Ukraine, 1, 23, 29, 33, 50, 126, 155, 193
UNEP, 52
UNEP World Conservation Monitoring Centre (UNEP-WCMC), 8–9, 123
UNESCO Man and Biosphere Programme (MAB), 44
  Biosphere Reserves, 8, 10, 44, *See also* protected areas, types and their coverage
United Kingdom (UK), 1, 9–10, 14, 26, 29, 31–34, 41–43, 50, 61, 65, 68, 70, 79, 92, 108–109, 116, 121–123, 126–127, 130–131, 133, 135, 634–677
United Nations, 52
United Nations Convention on the Law of the Sea (UNCLOS), 46, 56
United Nations Economic Commission for Europe (UNECE), 94
United Nations Framework Convention on Climate Change (UNFCC), 46
unsustainable harvesting, 185
urban sprawl, 22, 314, 385
urbanisation, 23, 110, 163, 182, 201, 221, 308, 312, 322, 417, 419, 435, 451, 470, 496, 519, 535, 537, 557, 577, 591, 637
Uria aalge. See Common Murre
Ursus arctos. See Brown Bear

Vanellus vanellus. See Lapwing
Velvet Scoter, 261, 275, 463, 494, 496
Vera, Frans, 14
Vertigo moulinsiana. See Desmoulin's Whorl Snail
Vespa velutina nigrithorax. See Asian Hornet
Vipera schweizeri, 360
Vipera ursinii rakosiensis. See Hungarian Meadow Viper
Volga Delta, 28
von der Leyen, Ursula, 61
Vormela peregusna. See Marbled Polecat
Vulpes vulpes. See Red Fox

Wadden Sea/Waddenzee, 28, 50, 72–73, 255, 263, 332, 334, 469, 476, 479, 481, 654
Wales, 634–677
water abstraction, 20, 24, 114, 293, 526, 557, 637
Water Hyacinth, 527, 606
water management, 366
  Poland, 506
waterbirds/waterfowl, conservation of, 41, 50, 111, 367
  Bulgaria, 193
  the Netherlands, 482–484
  Spain, 598
Wels Catfish, 516
Western Europe, 19, 22–23, 121, 135
Western Mediterranean Sea, 17, 415
wetlands, 110
  drainage, 19, 111, 254, 257, 273, 279, 293, 303, 355, 396, 451, 457, 495, 577, 615, 637, 658
  habitat management and restoration
    Bulgaria, 193
    Croatia, 213
    Denmark, 266
    France, 322
    Romania, 546
    Slovakia, 567–568
    Spain, 606
    UK, 660–661
Whinchat, 312
White Stork, 243, 442, 479, 530, 556, 620
White-backed Woodpecker, 453, 557, 576
White-headed Duck, 182, 421, 594, 600
White-tailed Deer, 304
White-tailed Sea-eagle, 148–149, 156, 200, 259, 270, 275, 279, 288, 297, 333, 400, 441, 489, 495, 556, 649
Wild Boar, 14, 168, 238, 261, 274, 360, 376, 423, 457, 500, 542, 620, 661
Wildcat, 205–206
wildlife crossing/green bridge, 206, 322, 340, 588
Willow Grouse, 435
wind energy/turbines, 145, 151, 176, 185, 203, 222, 334, 338, 342, 516, 521, 631, 638, 658
wise use, 48, 75, *See also* sustainable use
Wolf, 19, 27, 111, 147–148, 157, 168, 189, 200, 206, 217, 238, 256, 259, 261, 274, 280, 291, 298, 318, 333, 351, 381, 415, 417, 420–421, 429, 440, 457–458, 495, 499–500, 512, 515–516, 520–521, 528, 530, 535, 542, 556, 561, 575–576, 581, 596–598, 600–602, 609, 616, 620
Wolverine, 27, 291, 298, 620
Woodchat Shrike, 516
Woodpigeon, 226, 477
World Commission on Environment and Development, 52
World Database on Protected Areas (WDPA), 8–10, 123–124
World Heritage Convention, 44
  World Heritage Sites, 8, 44, *See also* protected areas, types and their coverage
World Summit on Sustainable Development ('Rio + 10' summit), 54, 100
WWF, 52

Yelkouan Shearwater, 423
Yellow Wagtail, 482
Yellow-bellied Slider Turtle, 231

Zamenis longissimus. See Aesculapian Ratsnake
Zebra mussel, 606

For EU product safety concerns, contact us at Calle de José Abascal, 56–1°,
28003 Madrid, Spain or eugpsr@cambridge.org.

www.ingramcontent.com/pod-product-compliance
Lightning Source LLC
LaVergne TN
LVHW062133200926
848484LV00007B/1807